Tadpoles

THE BIOLOGY OF ANURAN LARVAE

Edited by

Roy W. McDiarmid and Ronald Altig

The University of Chicago Press
Chicago and London

Roy W. McDiarmid is research zoologist, project leader, and curator of amphibians and reptiles for the U.S. Geological Survey, Patuxent Wildlife Research Center, at the National Museum of Natural History. He is also past president of the American Society of Ichthyologists and Herpetologists.

Ronald Altig is professor of biological sciences at Mississippi State University.

The University of Chicago Press, Chicago 60637
The University of Chicago Press, Ltd., London

Printed in the United States of America

08 07 06 05 04 03 02 01 00 99   1 2 3 4 5

ISBN: 0-226-55762-6 (cloth)

Library of Congress Cataloging-in-Publication Data

Tadpoles : the biology of anuran larvae / edited by Roy W. McDiarmid and Ronald Altig.
p. cm.
Includes bibliographical references (p.) and index.
ISBN 0-226-55762-6 (cloth : alk. paper)
1. Tadpoles. I. McDiarmid, Roy W. II. Altig, Ronald.
QL668.E2T15 1999
597.813′9—dc21   99-17655
CIP

⊗ The paper used in this publication meets the minimum requirements of the American National Standard for Information Sciences—Permanence of Paper for Printed Library Materials, ANSI Z39.48-1992.

We dedicate this volume to our mothers, Orpha and Eileen, for their encouragement and support from our formative years onward.

CONTENTS

PREFACE

If we had been clairvoyant about ten years ago, we might have had the wisdom to side-step a scene that occurred on the steps of the Ruthven Building during the annual meeting of the American Society of Icthyologists and Herpetologists at the University of Michigan in June 1988. As well as we can remember, it was during a conversation with several friends that Norm Scott suggested we quit complaining and issued the following challenge: "Why don't you two guys just write a book on tadpole biology?" Sometime later that evening, one of us asked the other: "Are we going to write a book on tadpole biology?" The other one answered: "I guess so." And so it began!

We had talked about such a project, but this scenario seems to be the way the idea was actually conceived. As soon as we returned home, we divided the field of tadpole biology into large categories, and after adjusting a few topics, we found authors willing and able to write chapters on those subjects. Eventually, 14 authors prepared the 12 chapters that follow. We plunged ahead naively and before long a prospectus presented to the University of Chicago Press was reviewed and accepted. Now we were committed; authors and editors were dependent on us and we on them. Regrettably, we did not meet all of our self-imposed deadlines and still recall the several doses of anxiety and frustration that resulted. Early on, friends pointed out that our original plans were conceived through heavily tinted glasses. We reluctantly agree now and receive little satisfaction from the adage that hindsight is always clearer! Too much time has passed since we started, and the problems we encountered along the way are becoming fuzzy even in our minds. It matters little now as the book is finished, and we are happy that we decided to prepare it. We also are optimistic and excited that this multiauthored volume that we conceived and helped produce summarizes current knowledge about the organisms that have kept us mentally intrigued for many years. During the preparation of this book, we learned many things about ourselves, about others, and about tadpoles. We became acutely aware of how large and diverse the study of anuran tadpoles really is, how fast the field is growing, and how large and where the numerous holes are in our knowledge base. Even though the bibliography is large, the reader needs to be aware that all available references have not been cited. Given the rate at which new studies on tadpole biology are being published, it is disappointing to realize how soon the book will be out of date. On the other hand, if this compilation contributes measurably to an increase in the knowledge of tadpole biology, then becoming obsolete quickly, in fact, is an outcome that we welcome.

We never intended nor do we now want to imply now that all things are known about tadpoles or that the authors know all there is to know. For the most part, we were interested in subjects pertaining exclusively to tadpoles. Topics (e.g., classical embryology) that apply to tadpoles and all vertebrates were excluded. A thorough review of each topic coupled with a reasonable review of the available literature were primary goals, and we urged all the authors to maintain an evolutionary viewpoint and cover their topics broadly. We did not demand that authors support all the ideas presented in the book nor did we totally agree between ourselves. Many subjects (e.g., the primitive vs. derived state of various larval traits and their value in estimating anuran phylogeny, morphology of the ancestral tadpole, the role that physiological and ecological factors have played in the evolution and maintenance of the tadpole stage in anuran life history, to name but a few) are too uncertain to take a single view, and we urged each author to recognize and generate new ideas. We suggest that certain comments that appear to be contradictory between chapters are actually alternative opinions, perspectives, or suggestions by the authors, and we have tried to cross-reference such topics for easy access.

Some chapters stand alone reasonably well (e.g., chap. 6; discussion of the nervous system) because other areas of study have not yet incorporated such information, and others are tightly intertwined (e.g., behavior + ecology + physiology). Throughout the book, we have used current taxonomic names (D. R. Frost 1985 and subsequent updates) when citing studies and publications in the text. Our staging notations most commonly are based on Gosner (1960) unless noted otherwise near the start of each chapter. Through it all, we strived to produce a volume with a uniform format and style of presentation. We often have lamented about the burden of not having a single reference for basic information on tadpoles when we started our careers. We trust that our efforts in producing a reference volume devoted to the entire spectrum of studies involving tadpoles will further stimulate this growing field of research.

Roy W. McDiarmid and Ronald Altig

ACKNOWLEDGMENTS

The preparation of a reference volume certainly cannot be done alone, and we graciously acknowledge those persons who were instrumental in the preparation of this book in three separate groups—those who contributed to the entire project, those who assisted Roy W. McDiarmid in Washington, D.C., and those who helped Ronald Altig in Mississippi.

V. Bennett from the public relations office at Southampton College of Long Island University in Southampton, New York, excitedly provided us with a transcript of the commencement address that Kermit the Frog made at the College in 1996. The quotation from Kermit's speech that begins the first chapter effectively introduces the theme of the entire book—addressing the scholars of tadpole biology.

R. F. Inger and R. J. Wassersug discussed our preliminary ideas regarding organization and content of the book and provided useful advice about potential contributors. M. H. Wake reviewed an early prospectus of the book and an early draft of chapter 5. She and Inger were available for consultation along the way, and we greatly appreciated their continued interest and guidance. Unpublished information was supplied by R. I. Crombie, L. Haugen, and H. R. De Silva; specimens were provided by R. F. Inger (Field Museum of Natural History) and loaned by J. P. Rosado (Museum of Comparative Zoology). The provision of current electronic versions of *Amphibian Species of the World,* by D. R. Frost, greatly expedited much of our work on the entire book. Persons who supplied photographs for our use are acknowledged in the appropriate chapters.

Perhaps most important, two women at the University of Chicago Press deserve special mention. We have no understanding of how either Susan Abrams or Christie Henry acquired their unusual mix (each from a unique recipe) of expertise, patience, and personality. They calmly and commonly transformed what seemed like insurmountable barriers into stepping stones, excised several troubling concerns with scarcely a notice, and guided our progress with something that resembled nonchalance. After each experience, we stood in admiration and would have applauded loudly had our hands not been busy scratching our heads in wonderment. We have not met two more capable editors, and we really appreciated their support. Others who also worked patiently with our many foibles, Theresa Biancheri, Leslie Keros, Karen Boyd, and probably many others we did not even know also deserve our thanks.

In Washington, D.C., Catherine "Kate" Spencer not only designed and executed many excellent illustrations herself but assisted McDiarmid with such tasks as setting standards, creating electronic files, and insuring common labeling for all illustrations, micrographs, and other graphics. She tracked many permissions for use of illustrations from publishers and read figure captions and associated text for clarity. Her contributions are noticeable throughout the text. T. Kahn also provided illustrations in chapters 2 and 3 and helped with others. P. C. Ustach checked the accuracy of most references in an early draft of the Literature Cited and provided the cartoon in chapter 1. T. Touré pursued some especially difficult references for us. L. K. Overstreet provided access to certain rare books in the Special Collections, and other librarians in the Natural History Branch, Smithsonian Library, facilitated our work immeasurably. C. Phillips provided secretarial support, and the staff of the Information Resources Management group at Patuxent Wildlife Research Center, particularly K. Boone, aided in the final preparation of illustrations.

In Mississippi, E. C. Akers, M. E. Audo, J. P. Maxwell, and C. L. Taylor helped with the horrendous amount of library work that was required to verify accuracy of literature citations. S. Hall of the Mississippi State University library staff worked several minor library miracles, and T. Tuberville (Savannah River Ecology Laboratory) also aided with final literature clarifications. G. Blankenship, J. Cotton, and T. Blake helped in many secretarial ways. W. J. Diehl offered many clarifications and opinions. R. B. Thomas read much of the manuscript as yet another check for the infinite number of errors and nonconformities that could exist. The graduate students S. Forbes, T. Schauwecker, and L. Williams calmly assisted with many impromptu demands.

The University of Chicago Press solicited three reviews for the entire manuscript, and the editor told us those names after the reviews were returned. Although differing in emphasis and style, all the reviews were strongly positive and very helpful. We and all other authors took great pride in the commendations that were given the manuscript. One of the reviewers even taught us how to tell when a period is upside down! To M. L. Crump, M. A. Donnelly, and R. F. Inger, we succinctly state, "Thank you very much."

CONTRIBUTORS

Ross A. Alford
Department of Zoology and Tropical Ecology
James Cook University
Townsville, Queensland 4811
Australia

Ronald Altig
Department of Biological Sciences
Mississippi State University
Mississippi State, MS 39762

Andrew R. Blaustein
Department of Zoology
Oregon State University
Corvallis, OR 97331

David F. Bradford
U.S. Environmental Protection Agency
National Exposure Research Laboratory
P.O. Box 93478
Las Vegas, NV 89193

David Cannatella
Texas Memorial Museum
University of Texas
Austin, TX 78712

Joseph Freda
Environmental Resources Analysts
601 County Road 57
Notasulga, AL 36866

Reid N. Harris
Department of Biology
James Madison University
Harrisonburg, VA 22807

Karin vS. Hoff
Department of Biological Sciences
University of Nevada–Las Vegas
Las Vegas, NV 89154

Michael J. Lannoo
The Muncie Center for Medical Education
Indiana University School of Medicine
Ball State University
Muncie, IN 47302

Roy W. McDiarmid
U.S. Geological Survey
Patuxent Wildlife Research Center
National Museum of Natural History
10th and Constitution Avenue, NW
Washington, D.C. 20560

Susanne Richter
Department of Anatomy and Morphology
Institute of Zoology
University of Vienna
Althanstrasse 14
A-1090 Wien, Austria

Giselle Thibaudeau
Department of Biological Sciences
Mississippi State University
Mississippi State, MS 39762

Gordon R. Ultsch
Department of Zoology
University of Alabama
Tuscaloosa, AL 35487

Bruno Viertel
Institut für Zoologie
Fachbereich 21, Biologie
Johannes Gutenberg–Universität Mainz
Saarstrasse 21
D-55099 Mainz, Germany

Tadpoles

1

INTRODUCTION
The Tadpole Arena

Roy W. McDiarmid and Ronald Altig

When I was a tadpole growing up back in the swamps, I never imagined that I would one day address such an outstanding group of scholars.

Commencement address by Kermit the Frog at Southampton College, New York, 1996

Introduction

Kermit was excited about being invited to address a group of outstanding scholars at a commencement address. With a similar feeling, we use a statement from the famous Muppet to introduce the theme of this compilation to another group of outstanding scholars—tadpole biologists. Kermit remarked that, like the rest of his kindred, he had started out as a tadpole—an odd organism with a composite head and body, a muscular tail without vertebrae, and dorsal and ventral fins that lack bony supports. Tadpoles always have eyes and usually external nares. Even though only the rear legs are visible during the free-swimming period, the front legs are developing beneath the operculum and follow a somewhat similar developmental trajectory through tadpole ontogeny. A spiracle(s), which may occur in assorted positions on different tadpoles, provides an exit for water that is pumped through the respiratory and food-trapping structures. Neuromasts occur in specific patterns over a skin rich in mucous glands. A characteristically lengthy intestine usually arranged in a double, conical spiral and a large liver are major and often visible components of the viscera. A structurally variable and evolutionarily unique oral apparatus typically comprised of soft and keratinized parts facilitates the harvesting of a myriad of food sources for a rapidly growing larva. A considerable assortment of morphological variations is superimposed on this rather simple body plan and reflects phylogeny and adaptations to either fast-flowing, slow-moving, or sometimes stagnant, ephemeral freshwater habitats. That is the legacy of Kermit, and that is the theme of this book.

Victor C. Twitty was trained as a classical amphibian embryologist, but he made a transition into field biology late in his career. In his popular book, *Of Scientists and Salamanders,* Twitty (1966) offered a precaution to laboratory biologists not to dabble in field research unless they were willing to risk being lured away from the lab bench forever. Tadpoles present a challenging lure for the innate curiosity of many biologists, and we succumbed to such enticement. Both of us were entangled early in our careers. McDiarmid's first published paper included a description of an unknown tadpole of a Mexican hylid frog, but his first taste of the elixir we know as tadpole field biology came while wading in fast-flowing streams in tropical forests of Costa Rica; he was hooked securely by the time he began mucking around with the high tadpole diversity in rain forest ponds in Ecuador and Peru in the 1980s. Altig's initial symptoms appeared as he marveled at the tadpoles of *Bufo canorus* in the alpine meadows at 3000 m near Tioga Pass, California, in 1965 and became highly inflamed while waist deep in the swamps and backwaters of lowland South Carolina in 1967.

A few of our colleagues have admitted nurturing an interest derived from an early childhood fascination with watching a polliwog change into a frog in a fishbowl right before their eyes; others have conceded, usually quietly, that they too were enthralled early on by the magic of metamorphosis. Unfortunately, most of these associates lost interest in this most intriguing topic at about the same rate that a tadpole changes into a frog. We devote this book to these latter souls who had the opportunity to reach the stars but through their own misguided ventures became mired in the gloom of other disciplines. For those who missed the satisfaction of even briefly pursuing a field of such biological intrigue, we hope that the following pages will be the first step in redirecting your idle wanderings.

Tadpoles, the ephemeral, feeding, nonreproductive larvae in the life cycle of a frog, are known by many conventional names (e.g., *têtard* in France, *kaulquappen* in Germany, *girino* in Italy, *renacuajo* in Mexico, and *ketou* in Mandarin China) and some less formal but nonetheless interesting local names (e.g., *guzarapo* in parts of Argentina and *bunbulun* in Honduras). The terms "tadpole"—the toad that seems all head—and "polliwog"—the head that wiggles—are familiar descriptors that came down to us from Middle English. It is regrettable that being called a "tadpole" or "taddy" is an insult in some countries. Even with this global language recognition, the extraordinary public appeal that frogs entertain, and an apparent surge in interest in all aspects of the biology of these organisms, tadpoles remain rather poorly known. Field biologists are reluctant to collect what they have little chance of identifying, and this single factor probably has resulted in tadpoles being studied much less than adult frogs. Tadpoles are simply not as vocal or visible as their parents, and those of many species of frogs have not been described or their identifications are otherwise difficult. The result is that the adult stages in the life cycle of a frog are more familiar to most people than are tadpoles.

Because early researchers were from Europe and North America, much of the descriptive terminology and early appreciation of tadpoles was from "typical" forms (e.g., species of *Bufo* and *Rana*). Many of these are monotonously similar, a feature that makes distinguishing among species an even more difficult and tedious task. Few of the northern temperate species show even a hint of the morphological diversity that characterizes certain tropical forms. As a result, tadpoles were ignored by most herpetologists, and little attention has been paid to their distinguishing features. In many areas of the world even today, tadpoles are infrequently or only secondarily sampled during faunal inventories, and some presumably knowledgeable field biologists discourage their collection entirely. When collections are made, they frequently languish on laboratory shelves because of a lack of interest or adequate keys for identification. As a consequence, few workers ever experience the exhilaration that comes from finding bizarre new tadpoles while examining collections from isolated mountains in Venezuela, remote forests in west Africa, or rocky streams in Madagascar.

Things are changing! Although the literature on tadpoles is widely scattered and variable in coverage, research on these organisms has increased greatly in the last two decades. Of the approximately 2,500 citations in this synthesis of our knowledge about tadpoles, about 72% were published since 1980 (i.e., last 18 years) and 25% since 1990. Research on many aspects of tadpole biology is increasing, and we anticipate that this book will build on that interest, point out obvious gaps in our biological comprehension of these marvelous creatures, and provide additional momentum to this research effort.

Why are tadpoles good subjects for investigation? Of about 4,400 (D. R. Frost, personal communication) different kinds of frogs known in the world today, about three-quarters of these species have a tadpole for some period during their development, and tadpole descriptions are available for about a third of those. From this perspective one might assume that we have a basic understanding of tadpole morphology, but new variations and unusual forms are being described regularly. Most tadpole descriptions are based on aspects of coloration; shapes of the body; fins and tail musculature; and positions of the spiracle, eyes, and vent. We are learning that the complex oral apparatus and buccopharyngeal structures of tadpoles are diverse on both micro- and macromorphological scales, but interspecific, ontogenetic, and geographic patterns of variation have not been described adequately (see Gollman and Gollman 1991, 1995, 1996). A diligent worker has a reasonable chance of identifying tadpoles from some regions of the world, but reference collections and keys to the tadpoles from most tropical latitude countries are woefully inadequate. The complex life cycle of most amphibians, particularly frogs, allows the larval and adult stages to occupy entirely different ecological settings, and the evolutionary interaction between these two stages provides a situation unique among tetrapods. Small size, general abundance, low trophic position, relatively short life span, absence of sexual behavior, and ease of culture make tadpoles particularly good subjects to investigate the nature of many complex biological processes and interactions. The value of understanding ontogeny has become more apparent in recent years, especially in studies dealing with developmental and functional morphology and evolution, and tadpoles often have been primary research subjects in these investigations. In many ways, tadpoles are the only vertebrate analogs to the larvae of holometabolous insects. Their considerable developmental plasticity and complex life cycle provide an opportunity to investigate intriguing questions about environmental and populational influences on growth, larval duration, morphology, and metamorphosis but also present complications to an already intriguing evolutionary compromise. Studies of anuran larvae have contributed immensely to many areas of biological investigation, and tadpoles have been prime subjects in defining general principles and the nature of interactions operating at the individual, population, and community levels of ecology. Research projects range from the analysis of DNA and the role of tadpoles in energy dynamics in wetland environments; through a diverse array of anatomical, physiological,

Fig. 1.1. Fossil tadpole from Utah (Middle Eocene, Evacuation Creek Member, Green River Formation, near Watson, Uintah Co.).

and ecological studies; to the uncompleted tasks of description and identification. Studies of tadpole morphology and the evolution of the tadpole as a life-history stage of a frog have been obstructed by the paucity of paleontological materials (e.g., Estes et al. 1978; Paicheler et al. 1978; Ŝpinar 1972, 1980; Wassersug and Wake 1995; see our fig. 1.1).

Understanding morphology, especially that of the oral disc, is crucial to comprehending the feeding ecology of tadpoles and therefore likely is the foremost factor in interpreting most, perhaps all, aspects of their biology. Variations in the timing of morphological change in tadpoles have strongly influenced the evolutionary history of frogs and certainly are central to species definitions and diagnoses. Therefore, describing the basic morphological patterns of tadpoles diversity and interpreting them in the light cast by their physical and biological environments are central themes in this book. The large morphological diversity across taxa coupled with the morphological conservatism found in some groups (e.g., species of *Bufo*) add to the conundrum. Ecorphological guilds that assist our understanding of the diversity of tadpoles have been proposed, but a functional and evolutionary understanding of most of the features used to define these categories is still needed.

The lack of a standardized terminology has been a continual hindrance to tadpole studies and species identifications. Neither of the attempts to formulate a standardized terminology (Altig 1970 and Van Dijk 1966) are adequate because only tadpoles from a restricted geographic region were described in any detail. Some terms have been adopted from other groups (e.g., operculum and spiracle) and do not refer to homologous structures. In our opinion these terms should be retained because they have become entrenched in the literature without causing as much confusion as a change would produce. Some other common descriptive terms are misused, redundant, and inaccurate, and we recommend that they be modified or discarded. It is clear that any new descriptive terminology should be formulated accurately and accompanied by adequate descriptions and illustrations.

Considering all of the above, the benefits of a comprehensive review of the biology of this group of organisms seem obvious. A synthesis of the published information on the biology of tadpoles at a time when the research on these organisms seems to be in an early phase of exponential growth is imperative. We and our coauthors hope to prepare the foundation for the focus of future work. Our overall objectives in writing this book are (1) to provide a detailed description of the basic tadpole body plan, including most aspects of external and internal morphology from hatching through metamorphosis; (2) to summarize what is known about the development, behavior, ecology, and environmental physiology of tadpoles with special emphasis on the evolutionary consequences of the tadpole stage; (3) to synthesize available information on tadpole biodiversity; and (4) to present a standardized terminology and exhaustive literature review of tadpole biology.

A Chronology of Tadpole Research

This short review of the study of tadpoles serves as a prelude to that which is expanded upon in the following chapters. Our intent is to focus on the progression of ideas relating primarily to tadpole morphology and identification and not to develop a formal history of all topics. We do not attempt to reference all of the important papers that have appeared on specific aspects of tadpole biology; these will be included in the appropriate chapter. Much of the early material outlined below was extracted from the summaries written by Héron-Royer and Van Bambeke (1889) and Nichols (1937).

In the classic *Historiae Animalium* (1551–1604), Gesner illustrated (1586) an adult and metamorph of a frog (see Goin et al. 1978). Rösel von Rosenhof (1753–1758) published a historically important work on the life history of European frogs and provided the first description of developmental series of several European species from egg to adult. Even though these drawings clearly showed that zoologists knew that tadpoles were the young of frogs, some confusion on this point remained until much later. For example, in 1796 Hutchinson discussed the generally accepted view that the frog fish of Surinam, *Rana* [= *Pseudis*] *paradoxa* (fig. 1.2), somehow changed from a frog to a fish and then after a few years back to a frog. While Hutchinson clearly understood that this animal was nothing more than a peculiar tadpole that gradually changed into a frog, and seemingly so illustrated it, one has to wonder if Hutchinson purposely positioned the spread toes of the left hind leg (animal facing to the right on the page) above the back of the tadpole to make it look like a fish's dorsal fin. In addition, we hasten to point out that some of the illustrations depict a tadpole with almost no tail fins, toothlike structures on the fishlike lower jaw, front leg that looks very much like a pectoral fin, and what appear to be scales on the tail. Also, the sequence of eight illustrations shows a recently transformed metamorph progressing into a fishlike tadpole! Our review suggests that progress on understanding the nature of tadpoles was being made but slowly and reluctantly, if this paper is any indication. The fact that the tadpole was drawn in right lateral view, thus precluding the illustration of the spiracle on the left side, also is of note. Because of this common structural asymmetry, it has become conventional to illustrate tadpoles in left lateral view.

The earliest figure and description of a tadpole's mouthparts were those by Swammerdam (1737–1738, pl. 40, fig. 1). Nearly a century elapsed before Saint-Ange (1831) described tadpole mouthparts and Dugès (1834) described and partially illustrated larval mouthparts in his comparative study of the osteology and myology of amphibians. These classical works were followed by the first histological study of the development of labial teeth and jaw sheaths (Vogt 1842). Van Bambeke (1863) pioneered the use of larval traits, including mouthparts, as taxonomic characters and was the first to note interspecific differences among European species. There followed several studies on tadpoles of European frogs. Among the more noteworthy are papers by Leydig (1876) on structural variation in mouthparts and Lataste (1879) on structural differences in individual labial teeth of *Discoglossus pictus;* Lataste also generated familial names based on characteristics of the spiracles (e.g., Mediogyrinidae [midventral spiracle] and Laevogyrinidae [sinistral spiracle]). Keiffer (1888) discussed the development and distribution of labial teeth on the oral disc of *Alytes obstetricans,* and Schulze (1888) did the same for *Pelobates fuscus.* Gutzeit (1889) expanded on these comparative works and studied the development of jaw sheaths. Papers by Héron-Royer and Van Bambeke (1881, 1889) described and compared the formation and structure of the mouthparts of tadpoles of 22

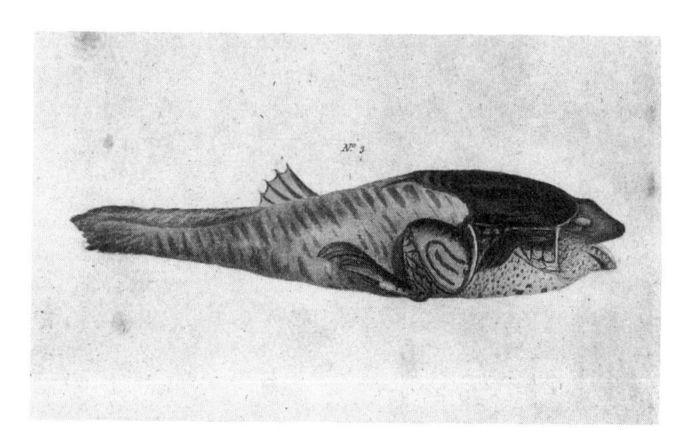

Fig. 1.2. An early rendition of the tadpole of *Pseudis paradoxa* from Hutchinson 1797, *The Natural History of the Frog Fish of Surinam,* pl. 3.

kinds of frogs (15 currently recognized European and 2 African species) and built on the earlier work of Van Bambeke by recognizing that certain morphological traits were useful in distinguishing among tadpoles of different species. They also pointed out those larval characteristics useful at the generic and familial levels. Boulenger (1892) recognized the value of being able to identify tadpoles and presented a synopsis of useful larval characters for distinguishing among species. He clarified and expanded the results of previous work, affirmed the value of the position of the spiracle and vent tube in distinguishing among groups, argued for including patterns of pigmentation in species descriptions, was the first to propose a labial tooth row formula, discussed the relative value of the distribution of neuromasts in species characterizations, and wrote a key to the tadpoles of 19 species of European anurans. This paper had a significant influence on future workers. The excellent illustrations of the tadpoles and mouthparts, some of the best ever published, were reissued in his monograph on the tailless batrachians of Europe (Boulenger 1897–1898). Boulenger described only what he called "mature" tadpoles, and only recently have we started looking at all stages of development in our deliberations on tadpole morphology.

Interest in the tadpoles of North American frogs also was developing in the later part of the nineteenth century. At about the same time that Héron-Royer and Van Bambeke (1881) published their important paper on European tadpoles, Mary Hinckley independently had begun to study the larvae of anurans in Massachusetts (Hinckley 1880). Her subsequent paper describing the mouthparts of tadpoles of seven species was the first comparative work on North American anuran larvae (Hinckley 1881) and marked the advent of comparative developmental studies of tadpoles in the New World. She quickly followed with developmental studies of two other species (Hinckley 1882, 1884). We have observed that older literature, even though informative, often is slighted and that curious observations reported therein too often are missed. For example, Hinckley (1882:95) reported that *Rana sylvatica* occasionally has dual spiracles, a condition not known in ranid tadpoles. In his classic studies on North American Anura, A. H. Wright (1914, 1932) published detailed descriptions and photographs of tadpoles and drawings of mouthparts of species occurring near Ithaca, New York, and from the Okefenokee region of southern Georgia. In a separate synopsis, A. H. Wright (1929) described, illustrated, and presented a key to tadpoles of 38 species mostly from the eastern and southern United States. Some information on tadpoles appeared in the first (1933) and second (1942) editions of the *Handbook of Frogs and Toads* by A. A. Wright and Wright, but it wasn't until the third edition (1949; reprinted in 1995) by A. H. Wright and Wright that a key and photographs of tadpoles and their mouthparts appeared. Unfortunately, A. H. Wright continued to use only "mature" larvae because "half-grown larvae . . . are often quite abnormal in the usual characters used in larval descriptions." While the early works by A. H. Wright were truly useful, his preoccupation with variation and measurements and complex ways of expressing relative positions and lengths of tooth rows were quite confusing. After a close examination of the synopsis by A. H. Wright (1929), Nichols (1937:16) lamented that ". . . not less than fifty different expressions will be encountered, each being used from one to twenty times. Of these fifty or more ways of expressing the relative measurements used, not less than fifteen are either mathematical impossibilities or of very uncertain meaning." Nichols continued, ". . . the number of different terms applied to what is presumably the chord length of a given row or of the upper beak is apparently infinite . . ." Nichols examined how tooth rows change as tadpoles grow and evaluated the relative taxonomic contribution of tooth row ratios.

As variation in certain traits became better understood and more species were studied, knowledge of the North American tadpole fauna increased. In 1952, Orton summarized traits and wrote a key to the genera of tadpoles known from the United States and Canada, and in 1971, Altig and Brandon did a similar paper for Mexico. Altig (1970) attempted to standardize terminology and presented a key to tadpoles of 72 species of frogs from the United States and Canada; he followed that in 1987 with one for about 100 species from Mexico. Today, most regional compilations and field guides include descriptions and illustrations of some tadpoles.

As tadpoles became better understood, workers began to include information derived from their study in broader taxonomic and morphological schemes. Noble (1926b, 1927, 1931) argued that larval characteristics were frequently useful in clarifying intrageneric relationships and discussed the link between adaptive tadpole types and various aquatic habitats; Noble (1927) concluded that certain larval characteristics (e.g., the position of the vent, increase in the number of larval tooth rows, etc.) were not very useful to discerning relationships. Two important papers by Orton appeared in the 1950s that inspired many biologists and were instrumental in setting in motion the current interest in tadpole biology. On the basis of internal and external features of their anatomy, Orton (1953) allocated tadpoles of about 600 species of frogs to four basic larval types and explored their patterns of adaptive radiation. She followed up on some of the earlier ideas by Noble and commented on the appearance of convergent forms in different lineages in similar environments. Later she expanded on these ideas and argued for the consideration of larval traits in reconstructing anuran phylogeny (Orton 1957). Even though certain aspects of Orton's larval categories have been criticized (I. Griffiths and De Carvalho 1965; Kluge and Farris 1969), Starrett (1973) formalized Orton's groups into taxonomic categories: (Type 1, Xenoanura; Type 2, Scoptanura; Type 3, Lemmanura; Type 4, Acosmanura) and presented an evolutionary scenario of their suspected relationships (Starrett and others have used roman numerals to indicate these types, but we follow Orton's original usage). O. M. Sokol (1975) argued that Types 3 and 4 were more primitive than 1 and 2, that they were closest to the ancestral anuran larval type, and that Types 1 and 2 either were derived both from 4 or independently 1 from 3 and 2 from 4. Sokol also proposed what he called phyletically neutral terms for groups proposed by Orton and Starrett: pipoid (for Type 1, Xenoanura), microhyloid (for Type 2, Scoptanura), discoglossoid (for Type 3, Lemmanura), and ranoid (for Type 4, Acosmanura). Altig and Johnston (1986) surveyed the morphological traits of tadpoles of the world and three years later presented a scheme of ecomorphological guilds that effectively summarized the morphological diversity of tadpoles from an adaptive perspective (Altig and Johnston 1989).

Regional studies in other parts of the world have contributed significantly to our general appreciation of tadpole biodiversity and reported faunas with tadpoles that are often quite different from those in the Holarctic. Some studies from other geographic regions are mentioned here, and others are cited in chapter 12, which treats tadpole diversity. Among the more important contributions to descriptions of Asian tadpoles are the monographs by Okada (1931) and Maeda and Matsui (1989) and the descriptions in the monumental work by Bourret (1942) on the frogs of Indochina. C.-C. Liu (1950) described and illustrated tadpoles of many frogs of western China, and Inger (1985) reviewed the tadpole fauna of Borneo. In an excellent work that included color photographs of each species, Chou and Lin (1997) reviewed the tadpoles of Taiwan.

Starting about 1960, works by several researchers on the African tadpole fauna have augmented our knowledge considerably. Representative among these are contributions by Amiet (1970), Channing (1986), Lamotte et al. (1959a), Perret (1966), Van Dijk (1966), and Wager (1986) to mention but a few. Research on tadpoles of the Australopapuan region includes papers by A. A. Martin and Littlejohn (1966), Méhëly (1901), Tyler et al. (1983) and others. Similar patterns of investigation have been repeated in Central and South America. E. H. Taylor (1942), Starrett (1960), and Duellman (1970) provided descriptions and other data on tadpoles of many Middle America species, and their examples have been followed by many researchers continuing to study the amphibian fauna of that region. Detailed documentation of the larvae of South American amphibians was provided early in the works of Kati Fernández and her associates (Fernández 1926; Fernández and Fernández 1921; Scott-Birabén and Fernández-Marcinowski 1921), who described the morphology, character variation, and ecology of many Argentine tadpoles. Work of this sort has continued with contributions by Bokermann (1963), Cei (1962, 1980), Duellman (1978), Echeverría et al. (1987), Formas and Pugín (1978), Heyer et al. (1990), Izecksohn et al. (1979), Lavilla (1984a), Lescure (1981), Peixoto (1982), and others.

As knowledge of and interest in tadpoles increased, other workers began to focus on different aspects of their biology. Starting in the early 1970s, two major research efforts began that have added considerably to our understanding of tadpole biology. Richard Wassersug, with his students and colleagues, published studies on the functional and evolutionary aspects of tadpole morphology, including studies on their buccal structures (1976a, 1980), swimming (Wassersug and Hoff 1985), and feeding ecology (Lannoo et al.

1987). The other major thrust began with the publication of two major papers, one on the structure and dynamics of a larval amphibian community (Wilbur 1972) and the other on the ecological aspects of amphibian metamorphosis (Wilbur and Collins 1973). From these exemplary efforts, many papers, most notably by Henry Wilbur, his colleagues, and students, have defined the dynamics of tadpole populations and the nature of the interactions between environmental and biological factors that determine community structure. These research directions are but two of many that have continued to expand our knowledge of these fascinating organisms. Much of what is currently known about the biology of anuran larvae was summarized by Duellman and Trueb (1986). These and many other topics are reviewed in detail in the chapters that follow.

One additional subject concerns anuran phylogeny and the impact that studies of larvae have had (e.g., Inger 1967 and O. M. Sokol 1975) and will have on resolutions of anuran phylogeny. Although advancing slowly, our understanding of the relationships among major lineages of anurans (e.g., L. S. Ford and Cannatella 1993) is progressing toward a unified hypothesis, but once the systematics of familial categories stabilizes, understanding relationships within those groups will demand an immense amount of additional study. Although views of tadpole biology expressed throughout this book are couched in phylogenetic terms, only the information on the skeletomuscular systems (chap. 4) is treated in this context. The profound and sometimes well documented differences between larval and adult anurans have contributed to a disturbing perspective that somehow tadpoles and frogs need to be treated as separate entities. If we are to make any more progress in understanding anuran phylogeny, this has got to change!

Summary

It seems obvious to us that the study of tadpole biology is progressing rapidly on many fronts, and we predict that we are not yet near the asymptote of this "growth curve." Anuran eggs and larvae are used in many areas of study not covered in this book (e.g., early embryology and developmental endocrinology). Molecular techniques are being used to evaluate hypotheses of frog relationships with adult specimens, but little work has been done that uses the more common and easily collected larvae in these analyses. Certain of these techniques would be helpful in determining which larvae represent which adults when time and facilities do not encourage rearing tadpoles. The use of monoclonal antibodies has increased resolution and ontogenetic understanding of anatomical data but have not been used much for unraveling the intricacies of developmental change that are manifest as the individual grows from a zygote to a metamorph. Tadpoles will continue to be used in studies of community and population ecology and may hold the key to understanding the perceived decline of amphibians in some areas of the world. It is this exciting juncture in understanding the biology of tadpoles that stimulated the preparation of this reference volume.

In 1946, Grace Orton published a note entitled "The Unknown Tadpole." Above the text, a cartoon of a tearful tadpole states, "Wish somebody'd pay some attention to me!" We hope that this book is the attention that she thought was warranted, and we gratefully acknowledge the academic boost that her studies gave to the study of tadpoles. It is not surprising that two of the quotations that introduce the themes of the chapters are attributed to Orton (1953, 1957); two others were taken from the book on the ecology and life history of the common frog by R. M. Savage (1962). In this spirit, we recognize the absorbing writing of Clifford Pope and his germane summation of our fascination with tadpoles, even if in a cleverly underhanded way: "In spite of relative shapelessness, softness due to lack of bony skeleton, and hopeless immaturity, these potential frogs and toads have their points of interest" (Pope 1940:150). The following cartoon extends Pope's and Orton's views with our own.

ACKNOWLEDGMENTS

K. Adler, Department of Neurobiology and Behavior, Cornell University, Ithaca, New York, arranged for a photograph of plate 3 from Thomas Hutchinson's treatise of *The Natural History of the Frog Fish of Surinam* from his original copy. J. Gardner, University of Edmonton, Alberta, Canada, kindly provided the photograph of the fossil tadpole. Paul Ustach, Department of Biology, Utah State University, Logan, Utah, effectively captured our attitudes about tadpoles in the final figure.

2

RESEARCH
Materials and Techniques

Roy W. McDiarmid and Ronald Altig

Hard to find, hard to preserve, hard to identify, they are the most troublesome form of life.

Pope 1940:417

Introduction

Clifford Pope's comment about tadpoles holds true almost 60 years later. A properly preserved tadpole is the first requisite for many kinds of studies, and yet producing such specimens remains an illusive task. In contrast to adult frogs, the small amount of skeletomuscular tissue, large lymph sinuses, and massive gut of tadpoles frequently make it difficult to obtain well-preserved, undistorted specimens that will persist for a long period. The care and use of preserved tadpoles differ in many ways from techniques successfully used for adults, and specific fixatives sometimes must be used to process properly a specimen for study. The further manipulation and examination of appropriately prepared specimens provide other challenges. Most tissues of tadpoles, let alone the flimsy, yolky mass of embryos, usually require specific treatments and techniques for study. In this chapter we review some of the basic methods and approaches to doing tadpole research and offer practical guidelines that we have found particularly useful.

In the first section we briefly review staging systems and ecomorphological classifications of tadpoles that are useful reference points for comparative studies of their morphology, ecology, and development. Next we comment on topics relative to the proper fixation, preservation, storage, and maintenance of tadpoles as subjects of research. Examination of tadpoles requires some basic equipment, techniques, and a fair dose of "what works for you." In the third section, we describe some of the approaches that we and others have used to examine tadpole structure and allow the reader to work out the remainder. Any mentions of specific brands are simple suggestions without implication of endorsement. Information on eggs and embryos is included because of their increasing importance to the study of tadpole biology.

Inadequate tadpole descriptions are common, and even in well-studied areas of the world, tadpoles of many species have not been described adequately or at all. In the next section, we make certain recommendations on how to describe and illustrate a tadpole. Because most researchers do not have access to extensive preserved collections, they must rely on published descriptions for identifications. Consequently, we urge authors preparing new descriptions to include a suite of basic features in a standard format and call for an overall increase in the amount of detail in such descriptions. We believe that particulars of geographic and ontogenetic variation are especially important to include and encourage mention of other features as well. Further details of measurements and external morphology (e.g., nares, spiracle, and vent tube arrangements) can be found in chapter 3, and we refer the reader to added comments and perspectives by Van Dijk (1966). A synthesis of the morphological diversity of tadpoles is in chapter 12, together with an introduction to literature pertinent to their identification.

In the last two sections we review certain facets of collecting, sampling, and rearing tadpoles for research. Tadpoles, especially if they can be identified, probably provide a more expeditious means of sampling the occurrence of species at a given site than do adults. To understand many facets of tadpole biology, sampling should be repetitive and as far as possible quantitative. This section provides specific comments and reference in this regard. Rearing of tadpoles often is required to obtain specimens of specific stages for future examination and to provide specimens for laboratory study. The significance of having ontogenetic series is becoming more appreciated, and we discuss methods that hopefully will provide typical specimens.

Staging and Other Classifications

Staging

Staging is the recognition of certain morphological landmarks that appear useful in comparing the sequence of events in a developmental continuum. With the use of a staging system, tadpoles that are widely disparate in size and developmental period and of different species that have completed similar developmental steps can be compared. It is not possible to devise a system that applies exactly to all taxa under all conditions because the appearance of morphological structures at different temperatures varies temporally within taxa, while their sequence and even presence may vary among species. Of the more than 45 staging tables that have been produced (e.g., Duellman and Trueb 1986:128), four (Gosner 1960, general; Nieuwkoop and Faber 1956, 1967, *Xenopus;* Shumway 1940, *Rana;* and A. C. Taylor and Kollros 1946, *Rana*) are cited most commonly. We hold that the system proposed by Gosner (1960) works reasonably well and regard it as meeting a prime objective of allowing interspecific comparisons (see Witschi 1962). Unless one wants to document the appearance of morphological features that temporally appear very close together and within a short developmental sequence, increased developmental resolution, as exemplified by the Nieuwkoop and Faber system (1967) for *Xenopus laevis,* usually is not required. We remain convinced of the value of a staging system that can be applied across all taxa (i.e., universal applicability). We recommend that the Gosner system be adopted or at least be used as an adjunct with any other system. We provide approximations of Gosner's stages to those in the other three commonly cited staging systems (see table 2.1) but recommend that investigators apply the Gosner system to their organisms with specimens in hand, rather than translate to Gosner from another system with this table. Even if accuracy declines as a specimen deviates further from a "typical tadpole" morphotype, having the author's estimation of equivalent stages is informative. All staging notations in this chapter refer to those of Gosner (1960), whose system is illustrated in figure 2.1.

Table 2.1 Suggested approximations of Gosner (1960) stages to those of Nieuwkoop and Faber (1967, *Xenopus laevis,* 66 stages, fertilization to metamorphic completion), Shumway (1940, *Rana sylvatica,* 25 stages, fertilization to opercular closure), and Taylor and Kollros (1946, *Rana pipiens,* 25 stages, hind limb bud appearance to metamorphic completion). See table II in Nieuwkoop and Faber (1967) for other comparisons among other staging systems.

Gosner	N/F	Shumway	T/K
1	1	—	
2	1^{+}	2	—
3	2	3	—
4	3	4	—
5	4	5	—
6	5	6	—
7	6	7	—
8	7	8	—
9	9	9	—
10	10	10	—
11	11	11	—
12	12	12	—
13	13	13	—
14	14	14	—
15	17	15	—
16	19	16	—
17	23	17	—
18	26	18	—
19	30	19	—
20	32	20	—
21	35	21	—
22	41	22	—
23	43	23	—
24	44	24	—
25	46	25	—
26	48	—	I
27	49	—	II
28	50	—	III
29	51	—	IV
30	52	—	V
31	53	—	VI
32	53^{+}	—	VII
33	53^{++}	—	VIII
34	54	—	IX
35	54^{+}	—	X
36	54^{++}	—	XI
37	55	—	XII
38	56^{+}	—	XIII
39	56	—	XIV
40	56^{+}	—	XV–XVII
41	58	—	XVIII–XIX
42	59	—	XX
43	61	—	XXI
44	62	—	XXII
45	63	—	XXIII–XXIV
46	66	—	XXV

Developmental Categories

Because tadpoles are ephemeral, larval, and nonreproductive (but see Bokermann 1974), common terms that imply sexual maturity (i.e., adult, mature) are inappropriate. Likewise, written estimates of size (i.e., large, small) are not very informative even if the reader knows the normal range of sizes of the species in question. A notation of stage, accompanied by some measurement (e.g., body length; see chap. 3), is the primary means of recording what the current morphology is along the developmental continuum. In certain circumstances, qualitative categories that denote longer periods of development and are defined by major morphological change also are useful when defined accurately.

For use in developmental and ecological studies, we prefer the following: embryo (Gosner 1 to about 20), hatchling (about 21–24), tadpole (25–41), and metamorph (42–46). The stage at which hatching occurs is dependent on species and environmental conditions. Hatching, not stage, therefore becomes the absolute criterion. Size at metamorphosis is assumed to refer to Gosner 46 unless otherwise noted. "Larva(e)" or "larval" are generic terms that most often apply to the tadpole but may encompass hatchling stages. Larval caecilians and salamanders are not tadpoles, and their development is treated with other staging tables (e.g., see Duellman and Trueb 1986:127–133). Even though the hatchling period is short, we have found it useful to recognize this period as distinct from those of the embryo (intracapsular) and tadpole (mobile and feeding). In many ecologically studies, the terms "hatchling" and "hatchling tadpole" most often refer to Gosner 24–26. Direct-developing endotrophs are embryos until they hatch as froglets, and other qualitative categorizes have not been proposed. We suggest that metamorph not be used as a synonym of froglet for direct-developing endotrophs.

Another useful set of terms based on developmental periods is premetamorphic, prometamorphic, and metamorphic. Regrettably, developmental biologists define these terms relative to patterns of thyroxine production and morphologists use them relative to structural landmarks. The different meanings delimit categories with a different sequence of stages and necessitate clarification at each usage. The morphological terminology as used by Etkin (1968) is defined by stages of Taylor and Kollros 1946 (Gosner stages in parentheses; see table 2.1) as follows: premetamorphosis, about stages V–XI (25–35), growth of trunk and tail; prometamorphosis, about stages XI–XX (36–41), rapid growth of limbs; and metamorphosis, about stages XX–XXV (42–46), eruption of front limbs to completion of metamorphosis (metamorphic climax of some authors). Because of interspecific variation in developmental patterns and plasticity of growth, these terms should at qualified by some measure of the size of the tadpole.

Ecomorphological Categories

Only a few attempts have been made to understand the primitive/derived nature of larval characters (e.g., Donnelly et al. 1990a; Duellman and Trueb 1983; I. Griffiths 1963; I. Griffiths and De Carvalho 1965; Inger 1967; Kluge 1988; Kluge and Farris 1969). Many similarly appearing structures of tadpoles seemingly are the result of convergent evolution, but the homologous nature of many such traits will depend on phylogenetic analyses of frogs and the incorporation therein of more larval-based characters (also see discussion in chap. 4). Some workers have attempted to interpret morphology in the context of tadpole ecology with mixed results. The rather weak resolution of these categories will increase as our understanding of functional morphology improves and phylogenetic hypotheses are presented.

Orton (1953) formalized the concept of ecomorphological types for tadpoles by depicting arboreal, carnivorous, mountain-stream, nektonic, and surface-feeding tadpoles; she also recognized direct development as a category for frogs. In her view, the tadpole types were derived from a generalized form (i.e., a typical lentic-benthic tadpole like that of *Rana catesbeiana*). She also recognized morphological convergence (umbelliform and suctorial) in several families and formulated phylogenetic types based on four major patterns of mouthpart and spiracular morphology in tadpoles.

Starrett (1973) expanded the morphological base of Orton's types and proposed formal taxonomic names. The tadpoles of Xenoanura (Orton's Type 1; pipids and rhinophrynids) have separate branchial chambers posterior to the heart with separate spiracular openings to the outside, front legs that develop posterior to the branchial chambers, and a simple slitlike mouth. Type 2 tadpoles or Scoptanura (microhylids) have separate branchial chambers that open posteromedially into a common opercular tube that exits via a single midventral spiracle (see *Otophryne* below), front legs that develop posterior to the branchial chambers, and variable fleshy oral structures that usually are much less complex than the following types. The Lemmanura (Type 3; ascaphids and discoglossids) have separate branchial chambers that are joined by two short opercular tubes that open immediately as a single ventral spiracle near the posterior extent of the branchial chamber, the limbs develop close to the peribranchial wall and possibly break through this barrier during metamorphosis, and the oral apparatus is complex with jaw sheaths and labial teeth, some rows of which are bi- or multiserial. Last, the Acosmanura (Type 4; all remaining families) have the branchial chambers joined internally by a common ventral chamber below the heart that opens via a single spiracle on the left side; the forelimbs develop within the branchial chamber, and the oral apparatus resembles that of Type 3 tadpoles except all tooth rows are uniserial. O. M. Sokol (1977b) discussed the anatomy and relationships among these groups and substituted nontaxonomic names for those of Starrett (1973). Wassersug (1989b) debated the morphology of Orton's Type 2 category and with Pyburn (1987) proposed a fifth type for the unusual tadpole of *Otophryne pyburni* (see description in chap. 12).

Other ecomorphological classifications are based on various ecological, geographic, and taxonomic assemblages (e.g., Altig and Johnston 1989; Lamotte and Lescure 1989; C.-C. Liu 1950; Van Dijk 1972). McEdward and Janies (1993) discussed the terminology associated with the general concept of "larva." Viertel (1982, 1987) presented generic diagnoses and correlated morphology of the branchial filtering system of European tadpoles with Orton's four types. Inger et al. (1986a) recognized five feeding types in four microhabitats within a stream community of 29 species of Bornean tadpoles. Lavilla (1985, 1988) diagnosed genera of tadpoles based on a suite of morphological features. Lamotte and Lescure (1989) recognized ecomorphological categories of lotic tadpoles, although several of their assemblages seem imprecise based on our field studies and correlations between tadpole morphology and ecology.

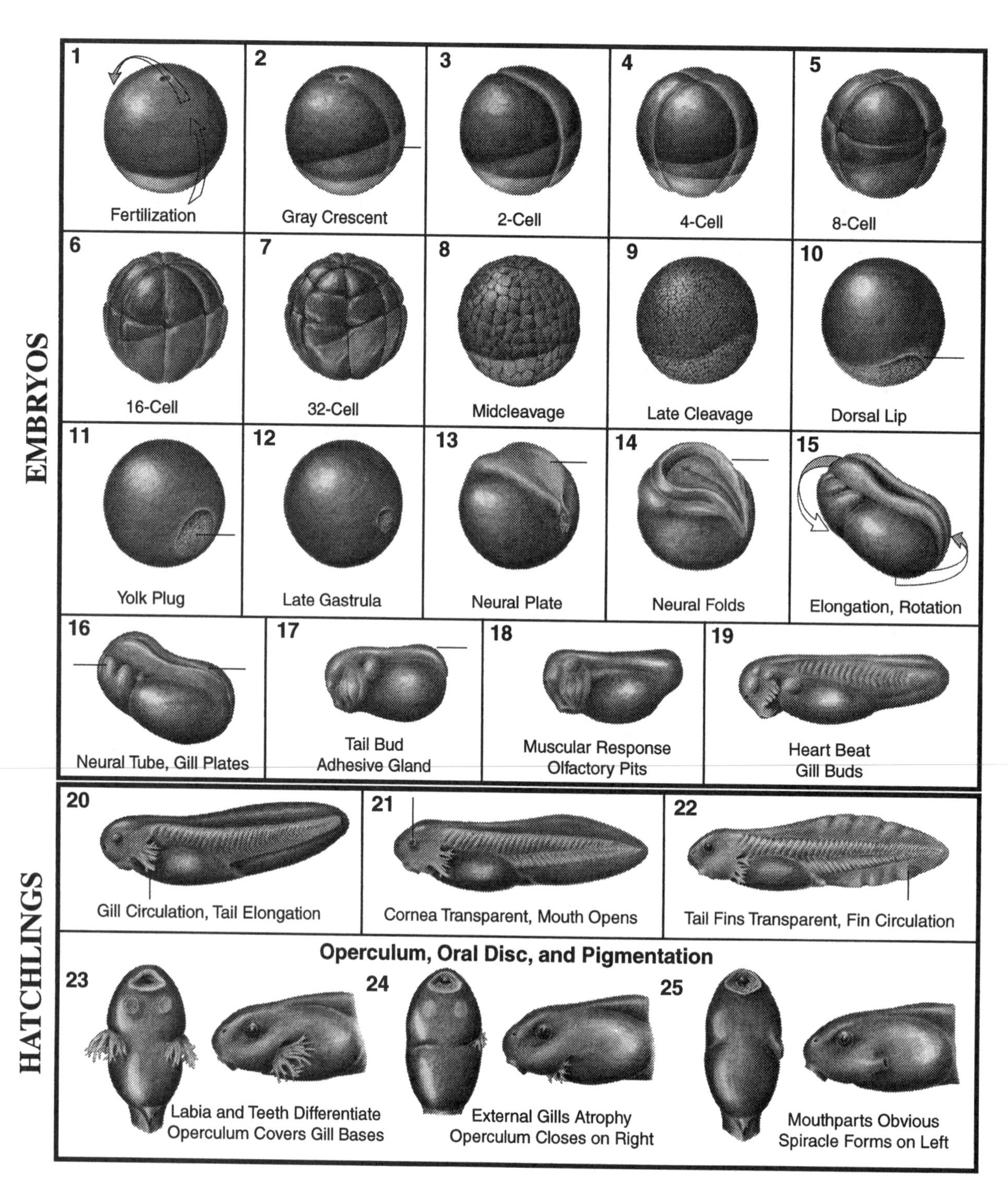

Fig. 2.1. The Gosner (1960) staging system recommended for use with exotrophic tadpoles; developmental stages are based on *Bufo valliceps* raised at 25° C. Translations to other staging systems are shown in table 2.1. Embryos: stages 1–19, 41.5 h (6.2%) developmental time. After fertilization (stage 1) and release of the second polar body (2), the zygote undergoes cleavage to stage 9 without an increase in size. Germ layers begin to form during gastrulation (10–12), which is followed by neural tube formation (13–15). Sensory structures appear during stages 16–19 and somitogenesis occurs in stages 18–19). Hatching may occur as early as stage 16. Hatchlings: stages 20–25, 73.5 h (11.0%) developmental time. This period represents the transition from a relatively immobile embryo to an active, feeding tadpole; external gills atrophy and the spiracle forms. Structures associated with feeding and swimming appear and pigments begin to form larval color patterns. Tadpoles: stages 26–41, 384 h (57.6%) developmental time. This is the longest part of the larval period and is marked by growth and limb development. Metamorphs: stages 41–46, 168 h (25.2%) developmental time. During this crucial period the tadpole loses its larval characteristics and takes on adult structures; the tail begins to atrophy (43), larval feeding structures are replaced by adult jaws and tongue (41–43), and forelimbs and hind limbs become functional. This period typically is marked by a passage from the aquatic to the terrestrial environment. Metamorphosis is complete (46) when the tadpole has become a froglet. Original drawings of stages 1–25 were provided by Linda Trueb (Duellman and Trueb 1986); depictions of stages 26–46 were redrawn by Kate Spencer; additional drawings were prepared by Ty Thierry.

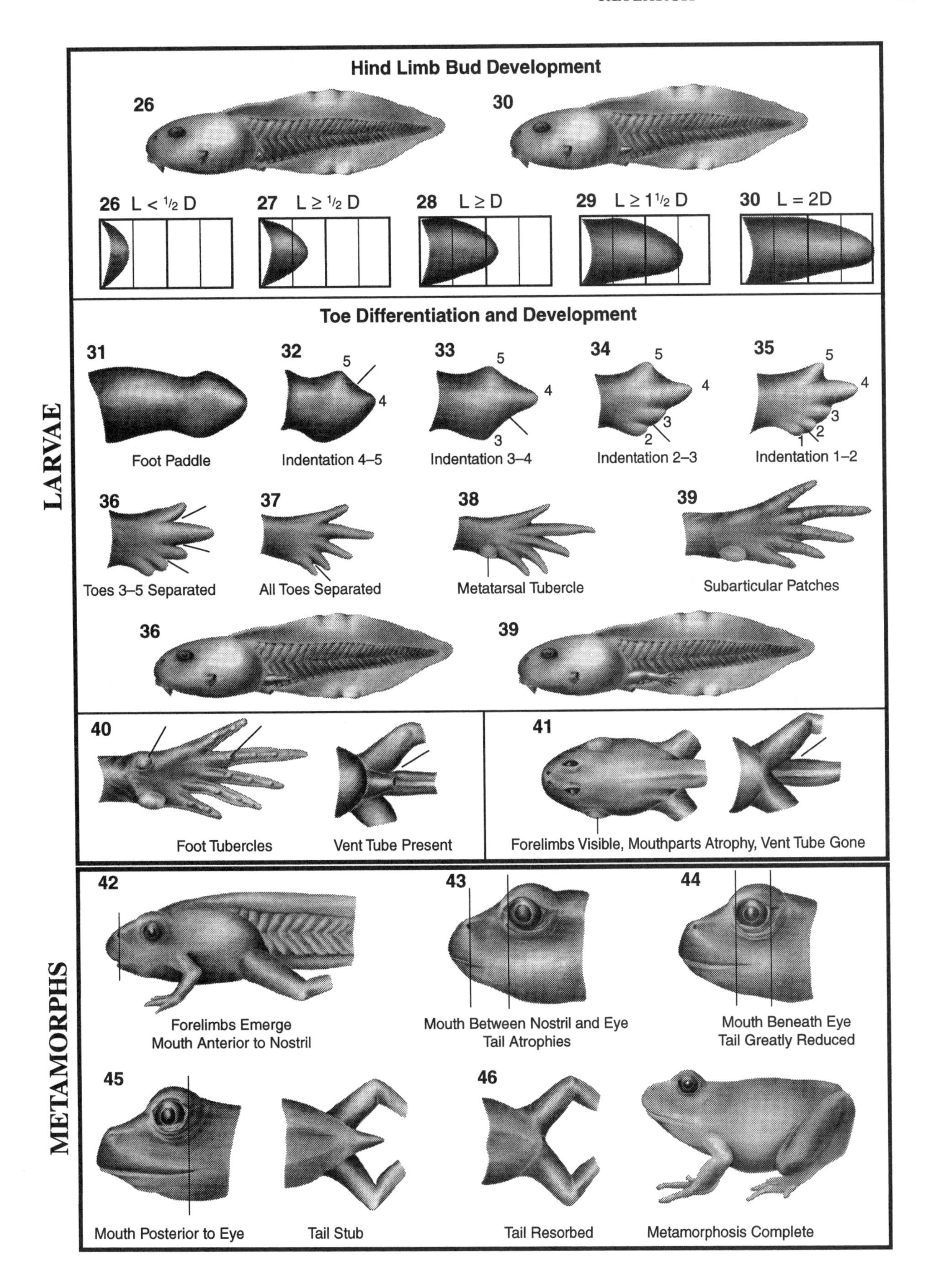
Hind Limb Bud Development
26
30
26 L < 1/2 D
27 L ≥ 1/2 D
28 L ≥ D
29 L ≥ 1 1/2 D
30 L = 2D
Toe Differentiation and Development
31
Foot Paddle
32
Indentation 4–5
33
Indentation 3–4
34
Indentation 2–3
35
Indentation 1–2
36
Toes 3–5 Separated
37
All Toes Separated
38
Metatarsal Tubercle
39
Subarticular Patches
36
39
40
Foot Tubercles
Vent Tube Present
41
Forelimbs Visible, Mouthparts Atrophy, Vent Tube Gone
LARVAE
42
Forelimbs Emerge
Mouth Anterior to Nostril
43
Mouth Between Nostril and Eye
Tail Atrophies
44
Mouth Beneath Eye
Tail Greatly Reduced
45
Mouth Posterior to Eye
Tail Stub
46
Tail Resorbed
Metamorphosis Complete
METAMORPHS

Altig and Johnston (1989) considered the source of developmental energy as the ultimate discriminator of anuran developmental modes. They distinguished endotrophic (embryos that derive all developmental nutrition from parental sources) from exotrophic (tadpoles that obtain energy by oral consumption of many types of food that rarely is associated in any way with either parent) groups. Six developmental guilds of endotrophs (chap. 7) were based on the degree of loss of the tadpole morphotype, activities of hatchlings and early larvae, site of development, and the sex of the parent and its association with the embryo. A revised scheme of ecomorphological guilds (table 2.2) recognizes 15 guilds of exotrophic tadpoles; three are common to lentic and lotic environments, five are found in lentic habitats, and seven are in lotic environments. These guilds are based on the general morphology (body and tail, eye position, oral apparatus) and known or presumed feeding behavior and ecology of the tadpoles.

Most of these ecomorphological classifications are based on similarity of appearance. Their phylogenetic bases have not been studied but histories undoubtedly have played a major role in determining the evolution of certain morphologies (e.g., mouthparts in ranoids). Some guilds (e.g., lentic suspension raspers, lotic fossorial forms) seem to reflect a common history; others (e.g., neustonic forms) clearly do not. Unfortunately, few workers have included larval traits in their phylogenetic constructs. We suggest that extensive research is needed to demonstrate the functional basis for certain morphological structures and to elucidate the developmental patterns common to such morphologies. Such lines of investigation should increase the resolution of ecomorphological classifications and enhance our understanding of all aspects of tadpole biology.

Fixation, Preservation, and Storage

Tadpoles

Determination of which fixative and preservative invariably will produce a nondistorted tadpole that will persist for a long period without clearing has not been made, and tests using some of the newer fixatives need to be conducted on tadpoles. An evaluation of preservation effects on larvae, comparable to those done on adults (e.g., J. C. Lee 1982), has not been performed (but see Gotte and Reynolds 1998). We have seen specimens of tadpoles that were fixed either in formalin (with and without buffer) or in alcohol that appear fine after many years in a collection and other specimens preserved in either fixative that soon after collection are nearly worthless for morphological study and almost certainly will not endure long-term storage. Generally speaking, specimens that are stored in alcohol that is strong enough to provide a properly hardened specimen at fixation often are too rigid for easy examination (e.g., discs are difficult to open and teeth break free easily from their ridges). With long-term storage in alcohol, tadpoles seem to become extremely distorted through dehydration and often appear as "raisins" in the bottom of a jar. Given the gap in our knowledge, we continue to rely on a known solution and advocate neutral buffered 10% formalin for both fixation and preservation (storage) of tadpoles. A commonly used recipe to make 18 liters (about 5 gallons) of 10% buffered formalin is: 1800 ml formalin (37% = full strength commercial), 16,200 ml distilled water, 72 g sodium phosphate monobasic, and 117 g sodium phosphate dibasic (anhydrous). Buffering of the fluid counteracts the formic acid formed when formalin oxidizes. The pH of the solution should be checked when the solution is made and once or twice a year while in use. Tadpoles are very watery by nature and will dilute the initial concentration of formalin in the fixation process. This dilution effect should be countered either by increasing the ratio of fixative to tissue volume or the initial concentration of the fixative. We frequently fix large series or many, large specimens in the field with formalin that is stronger than 10% (sometimes as high as 30%–50%) and then transfer specimens to 10% after about an hour. The goal is to fix a tadpole so that the thinnest fins and tail-tips are stiff (i.e., will not fold or bend when the specimen is removed from fluid). A firm specimen is always preferable to a soft one, both for immediate examination and long-term storage. In our experience, some tadpoles respond differently to fixation with the same concentration of buffered formalin than others; for example, microhylid tadpoles almost never fix as firmly as do ranids of the same size and at the same concentration of formalin. In any case, the process of fixation should not result in a distorted specimen. The person interested in searching for the "perfect" tadpole fixative may find the following citations useful: Bragg (1949), F. J. Gordon (1934), Khan (1982b), and Tyler (1962).

With the present concern over the use of formalin and the problems associated with carrying corrosive fixatives on aircraft, several nonformalin and nonaldehyde fixatives have been developed and commonly are used with class dissection materials and histological preparations. These fluids must be used without dilution, so the bulk in transport may be a problem. On the short term, the fins and bodies of larvae (*Pseudacris* and *Ambystoma*) fixed in STF (Streck Laboratories, Omaha, Nebraska) had good conformation but were noticeably softer than specimens preserved in formalin. This would be expected in a fixative that does not cross-link proteins. Pigment changes were similar or slightly less drastic than those seen in specimens preserved in formalin. Although we cannot predict how specimens will respond over time, we are skeptical that these fixatives will work satisfactorily for long-term storage and are not promoting them as a preservative. On the other hand, normal histological work, polymerase chain reactions, and antigenic examinations can be performed on tissues preserved in STF.

Through the process of saponification (= "clearing" in curatorial parlance) tadpoles sometimes become nearly transparent and take on a soft, soapy consistency. In our opinion, this problem is severe in some collections. Because most concepts of "what is correct" for preservation are derived from short-term, untested protocols (see DiStefano et al. 1994), we are unable to recommend specific solutions to rectify this problem. W. R. Taylor (1977) provided some information based on tests on fishes, and it is clear that a number of factors (e.g., pH, the buffering agent) relative to

Table 2.2 Ecomorphological diversity of the anuran tadpoles as revised from Altig and Johnston (1989). The list of example taxa in each group is seldom exhaustive.

I. **ENDOTROPH:** developmental energy to produce a free-living juvenile (= froglet) derived entirely from maternal sources of energy, usually vitellogenic yolk
  1. **Viviparous:** froglet birthed from oviducts; additional nutrition provided by maternal oviducal materials after yolk exhaustion, placenta not present; development highly modified, "fetus" appropriate for advanced unborns: Bufonidae (part), *Nimbaphrynoides occidentalis*
  2. **Ovoviviparous:** froglet birthed from oviducts; no extra-ova nutrition provided by mother; development highly modified: Bufonidae (part), *Nectophrynoides tornieri, N. viviparus;* and Leptodactylidae (part), *Eleutherodactylus jasperi*
  3. **Paraviviparous:** froglet hatched at various sites in or on the mother's body, usually the site of egg deposition: in stomach (e.g., Myobatrachidae: *Rheobatrachus* spp.), in dorsal skin (e.g., Pipidae: *Pipa* spp.), on dorsal skin (e.g., Hylidae: *Hemiphractus* spp.), or within a complete (e.g., Hylidae: *Gastrotheca* spp.) or incomplete dorsal pouch (e.g., Hylidae: *Flectonotus* spp.); larva may exhibit partial tadpole morphotype during development
  4. **Exoviviparous:** froglet "birthed" from some site other than the site of egg deposition and embryo (1) moves of own accord into dorsal pouches with inguinal ports of the father (e.g., Myobatrachidae: *Assa darlingtoni)* or onto parent's dorsum (e.g., Leiopelmatidae [part]: *Leiopelma archeyi, L. hamiltoni,* Sooglossidae [part]: *Sooglossus sechellensis)* or (2) is swallowed into the vocal sac of the father (e.g., Rhinodermatidae: *Rhinoderma* spp.)
  5. **Direct developer:** froglet hatched from egg capsules of terrestrially deposited eggs; parental care present but eggs not intimately associated with parent's body; embryo develops immediately toward a frog morphotype: Arthroleptidae (part), *Arthroleptis* spp.; Microhylidae (part): Asterophryninae, Brevicipitinae, Genyophryninae, and Melanobatrachinae; Leptodactylidae (part), *Eleutherodactylus* (most species); Myobatrachidae (part): *Bryobatrachus nimbus* and *Myobatrachus gouldii;* Ranidae (part): *Platymantis* spp.; and Sooglossidae (part): *Sooglossus gardneri*
  6. **Nidicolous:** froglet metamorphoses from non-feeding larva usually confined to a nest site where eggs deposited; parental care present but eggs or larva not intimately associated with parent's body; morphology of larva varies widely: Bufonidae (part): *Altiphrynoides malcolmi* and *Pelophryne;* Leptodactylidae (part): *Eupsophus, Vanzolinius,* and *Zachaenus;* Microhylidae (part): all Cophylinae and Microhylinae (part): *Synapturanus;* Myobatrachidae (part): *Philoria;* Ranidae (part): *Rana* (part) and *Phrynodon*

II. **EXOTROPH:** developmental energy derived from ingested food as a free-living larva (= tadpole) after yolk supplies are exhausted during embryological and hatchling stages
  A. **Lentic and lotic habitats:** three guilds, including the very large benthic assemblage involving many taxa from several families, occur in both lentic or lotic water with no morphological differentiation related to either habitat
    1. **Benthic:** rasp food from submerged surfaces; keratinized mouthparts present; mostly at or near the bottom, pools and backwaters in lotic sites; body depressed, eyes dorsal, fins low with rounded or slightly pointed tip; dorsal fin originates at or near the dorsal tail-body junction: Bufonidae (part): *Bufo* spp. (part); Dendrobatidae (part): *Dendrobates* spp.; Leptodactylidae (part): *Leptodactylus* (part); Myobatrachidae (part): *Pseudophryne* and *Uperoleia;* Pelobatidae (part): *Pelobates* spp., *Scaphiopus* spp., and *Spea* spp. (part); Ranidae (part): *Rana* (part)
    2. **Nektonic:** rasp food from submerged surfaces; keratinized mouthparts present; live somewhere within the water column, quiet backwaters in lotic sites; body compressed, eyes lateral, fins high with pointed tip with or without a flagellum; dorsal fin often originates well anterior to dorsal tail-body junction: Hylidae (part): *Hyla, Phrynohyas* (part), *Pseudacris,* and *Scinax* (part); and Hyperoliidae (part): *Kassina*
    3. **Neustonic:** filter particles in or near the surface film with up-turned (= umbelliform) oral disc with or without keratinized mouthparts; body depressed, eyes usually lateral, tail long with low fins: Arthroleptidae (part): *Leptodactylodon* sp.; Dendrobatidae (part): *Colostethus nubicola;* Hylidae (part): *Phasmahyla guttata* and *P. jandaia;* Megophryidae (part): *Megophrys* spp.; Microhylidae (part): *Microhyla heymonsi;* and Rhacophoridae (part): *Mantidactylus opiparus* group
  B. **Lentic only:** inhabits various microhabitats in nonflowing systems
    4. **Arboreal:** inhabits isolated pockets of water (i.e., phytotelmata), elevated or not (e.g., bromeliad cisterns, tree holes, holes in logs and stumps, bamboo stumps, leaf axils, fallen leaves, nut husks, etc.), sometimes spend time out of water
      a. **Type 1:** body elongate, tail long with low fins; keratinized mouthparts present, reduced, or absent (Microhylidae); sometimes spend time out of water; various modifications for macrophagy: Hylidae (part): *Hyla bromeliacia* and *Osteopilus brunneus;* Hyperoliidae (part): *Acanthixalus spinosus;* Microhylidae (part): *Hoplophryne* spp.; and Rhacophoridae (part): *Nyctixalus* sp., *Mantidactylus* (part) spp., *Philautus* spp., and *Theloderma* sp.
      b. **Type 2:** Body robust, tail short with moderate fins; keratinized mouthparts present with LTRF $\leq 2/2$, jaw sheaths sometimes modified for carnivory: Hylidae (part): *Anotheca spinosa* and *Hyla zeteki*
      c. **Type 3:** small, lightly pigmented, nonelongate tadpoles that inhabit pockets of water in large snail shells, fallen leaves, and tree holes; fleshy crown encircles eyes and nares: Bufonidae (part): *Mertensophryne* spp. and *Stephopaedes anotis*
      d. **Type 4:** inhabit humid microhabitats on and under bark of trees; keratinized mouthparts present; tail long and attenuate with reduced fins; hind legs develop precociously like Semiterrestrial tadpoles: Ranidae (part): *Indirana gundia*
      e. **Type 5:** tadpoles closely resemble lentic forms in the included taxa, although have complete marginal papillae, but breed in small isolated pockets of water: Hylidae (part): *Agalychnis calcarifer* and *Phrynomedusa fimbriata*
    5. **Carnivorous:** feeds on macroinvertebrates and conspecific and heterospecific tadpoles, either rasp the prey apart or engulf intact (does not include opportunistic scavenging of dead organisms); keratinized mouthparts present and usually modified (see table 10.2 and fig. 3.9): Hylidae (part): *Hyla marmorata* group; Leptodactylidae (part): *Ceratophrys* spp.; Myobatrachidae (part): *Lechriodus fletcheri;* Pelobatidae (part): *Spea* spp. (part); Pipidae (part): *Hymenochirus boettgeri;* and Ranidae (part): *Hoplobatrachus tigerinus*
    6. **Macrophagous:** presumably feed by taking larger bites (compared with smaller particles generated by rasping tadpoles) of attached materials on submerged substrates, sometimes at least facultatively oophagous; oral disc nearly terminal, jaw sheaths present, LTRF 0/0–0/1, marginal papillae greatly reduced to absent: Hylidae (part): *Hyla leucophyllata, H. microcephala,* and *H. parviceps* groups; and Ranidae (part): *Occidozyga* sp.
    7. **Suspension feeder:** feeds almost entirely by sitting quiescently in the water column and pumping water through buccopharyngeal structures to entrap small suspended particles; keratinized mouthparts lacking; body depressed; eyes lateral
      a. **Type 1:** oral disc derivatives present as hemispherical labial flaps pendant over the mouth; body strongly depressed, circular in dorsal view; eyes lateral, nares absent for most of ontogeny; fins low with pointed tip; dorsal fin originates at or near dorsal

Table 2.2 *continued*

tail-body junction; spiracle midventral near vent: Microhylidae (part): all Dyscophinae, most Microhylinae, all Phrynomerinae, and some Scaphiophryninae: *Paradoxophyla palmata*

b. **Type 2:** oral disc absent; spiracles dual and lateral; social schooling common; active day and night: Pipidae (part): *Pipa* (part), *Silurana,* and *Xenopus;* and Rhinophrynidae: *Rhinophrynus*

8. **Suspension-rasper:** feed partially by filtering suspended particles from within the water column and rasping submerged surfaces; jaw sheaths and labial teeth (usually 2/3) present, marginal papillae with anterior gap; eyes lateral; tail flagellum common; ventral fin often higher than dorsal fin: Hylidae (part): *Agalychnis* spp. (part), *Pachymedusa dacnicolor,* and *Phyllomedusa* spp. (part)

B. **Lotic only:** inhabits various microhabitats in flowing water systems

9. **Clasping:** marginal papillae with anterior gap; LTRs commonly 5 but as numerous as 8/8 and usually with anterior rows more numerous than lower rows (e.g., 9/3); inhabit medium to slow currents, position maintenance via the oral disc minor, body globular to slightly depressed: Ranidae (part): *Rana palmipes* and *R. pustulosus* groups; and Rhacophoridae (part): *Boophis* spp. (part)

10. **Adherent:** marginal papillae small and complete; LTRF commonly 2/3; inhabit faster water than Clasping tadpoles, position maintenance via oral disc common to continuous, body often depressed: Hylidae (part): *Duellmanohyla* spp., *Hyla* spp. (part), *Plectrohyla* spp., and *Ptychohyla* spp.; Myobatrachidae (part): *Taudactylus* spp.; and Rhacophoridae (part): *Mantidactylus* spp. (part)

11. **Suctorial:** marginal papillae small and complete; LTRF > 2/3 to a maximum of 17/21; inhabit faster water than Clasping or Adherent tadpoles, position maintenance via oral disc continuous, body depressed: Arthroleptidae (part): *Trichobatrachus robustus;* Ascaphidae: *Ascaphus truei;* Heleophrynidae: *Heleophryne* spp.; Hylidae (part): *Hyla* spp. (part); Megophryidae (part): *Scutiger* spp.; and Ranidae (part): *Conraua* spp. (part), and *Petropedetes* (part)

12. **Fossorial:** marginal papillae with anterior gap; LTRF 2/3 or less; inhabit leaf mats in slow water areas; position maintenance via oral disc absent, body vermiform: Centrolenidae: *Centrolene* spp., *Cochranella* spp., and *Hyalinobatrachium* spp.

13. **Gastromyzophorous:** marginal papillae complete or with anterior gap, sometimes weakly developed posteriorly; LTRF ≥ 2/3; jaw sheaths often modified; inhabit fast and often turbulent water via adhesion with the oral disc and an abdominal sucker; position via attachment organs continuous: Bufonidae (part): *Ansonia* sp. (part), *Atelophryniscus chrysophorus, Atelopus* spp., and *Bufo* sp. (part); Ranidae (part): *Amolops* spp., *Huia* spp., *Meristogenys* spp., and *Rana* sp. (part)

14. **Psammonic:** marginal papillae mostly lateral; LTRF 0/0; jaw sheath derivatives present as hypertrophied serrations without a basal sheath; bury in sand in shallow, slow flowing areas of small creeks: Microhylidae (part): *Otophryne pyburni*

15. **Semiterrestrial:** marginal papillae with anterior gap; LTRF usually 2/3; jaw sheaths usually with a high and narrow arch; inhabit rock faces, leaves, and the forest floor that provide damp or wet surfaces with little free water: Leptodactylidae (part): *Cycloramphus* spp. (part) and *Thoropa* spp. and Ranidae (part): *Nannophrys ceylonensis*

the initial fixative and solution (preservative) used for storage likely are involved. If specimens in your collection are not clearing, the best advice we can offer is to maintain the status quo. There is no way to recover cleared specimens.

Several techniques are available for rehydrating totally dried specimens. We have had mixed results in limited tests and none of the rehydrated specimens we have examined is totally satisfactory. However, if the dried, previously preserved specimen is valuable, rehydration is worth trying. We have had moderate success with potassium hydroxide at about 1% solution. After the dried tadpole has returned to a satisfactory condition, one must stop the action of the hydroxide by immersing the specimen in about a 3% solution of acetic acid (i.e., similar to household vinegar). Nonpreserved, dried specimens seem to return to their former shapes in potassium hydroxide better than preserved ones, and after preservation, they can be used for detailed study.

As noted above, an absolute protocol for fixation, preservation, and storage of tadpoles is not available. In our experience, specimens preserved in alcohol and transferred to 10% formalin do not develop problems over the first few years, and usually these specimens have better body contours than they did once the alcohol dehydration is restored by rehydration in formalin. We have never tried the reverse because we maintain that fixation and storage is best in formalin. Unfortunately, we seldom know if the alcohol stored specimens that we successfully have transferred to formalin were fixed originally in alcohol or formalin. We also have noted that specimens preserved in alcohol seem to have quite different color than those kept in formalin. We do not know how the color of alcohol-fixed and -stored tadpoles compares to their color in life but suspect that coloration is more lifelike in alcohol-fixed specimens than formalin-fixed and -stored specimens. Van Dijk (1966) transferred tadpoles fixed in Bouin's to 70% alcohol for long-term storage. Clearly there is much to be learned about the best solution for fixation and storage of tadpoles.

Efficient curation and maintenance of tadpole collections require some modification of normal curatorial practices. Because tadpoles are stored in formalin and usually in smaller jars than are adults, for efficiency and safety we suggest that they be physically separated from the rest of the collection. We have found cardboard bin boxes of various sizes to be very useful in this regard. Small containers of tadpoles can be stored by species in bin boxes and handled as a unit. Physical separation of containers of formalin-stored from alcohol-stored specimens reduces the problem of mistakenly mixing fluids during routine collection maintenance. Because the pH of formalin should be checked routinely, the entire box of tadpoles can be moved to a properly ventilated hood for examination. The use of bin boxes also prevents small bottles from inadvertently being knocked off of shelves during specimen retrieval and cuts down significantly on breakage and formalin spills. Normal curatorial practices and computer cross-referencing can effectively compensate for any inconvenience of separate storage areas.

Larval samples frequently are stored in small containers of fluid, and evaporation from improperly fitting lids can have a devastating effect on fragile specimens in a relatively short period of time. The risk of loss is high with most screw-top vials with which we are familiar. In lieu of increased curatorial vigilance and to avoid losses from evaporation, we recommend that small jars rather than vials be used for tadpole storage. If multiple samples comprise a collection (i.e., several samples of the same taxon from the same site) and especially if developmental series of embryos and small tadpoles are part of the collection, we advocate storage of each lot in a separate vial within a larger jar. We do not recommend that tags be tied to individual tadpoles. Rather, we urge that tadpole samples be treated as lots (i.e., all specimens of the same species that were collected from the same specific site/habitat at the same time is a lot), and that lots rather than single specimens (unless the sample is a single specimen) be cataloged into museum collections. A single, individually numbered tag that is associated with the lot and its accompanying data should be placed into an appropriate size vial with the tadpole(s), filled with fluid, and cotton plugged. The individual vials that make up the series can then be placed into an appropriate size jar which is itself filled with formalin. Although this procedure makes retrieval of an individual sample more difficult, it avoids a common problem of specimen loss through evaporation, which occurs too often in small containers. A label in each larger jar can include the explicit data for the sample (e.g., species, locality, date[s] of collection, number of specimens, developmental stage, etc.) plus any ancillary information about the sample (e.g., whether the tadpoles were field-collected or -reared, their habitat, behavioral notes, specific impressions of coloration, and anything else that seems remotely important). Information is lost if specimens of a single species sampled at different times from the same site (i.e., developmental series) are subsequently mixed. A high rag-content bond paper can be used for the labels and tags (e.g., Paxar Corporation, Vandalia, Ohio) and a permanent, carbon-based ink used to record associated data.

If specimens are to be shipped from the field or sent on loan to an investigator, embryos and tadpoles must be shipped in fluid. Fragile specimens wrapped in cotton or cheesecloth often are damaged or even destroyed in shipment. In preparation for transport, the lids of field or shipping vials should be taped because vibrations from transport often will shake them loose. It is best to leave a small air space in each vial, as filled vials may burst with changes in pressure. Check the lids on vials closely just before shipping. The small, screw-capped vials commonly used for small specimens have a bad habit of breaking off right below the cap. Clear flint glass vials, small plastic bottles with screw-capped lids, or frosted plastic scintillation vials appear to be good alternatives; Whirl-pak bags also are a viable option, especially for large samples or specimens.

Eggs

No satisfactory study of procedures for preserving eggs (i.e., short- and long-term effects of different fixatives or preservatives on eggs and jellies) exists. Sometimes the jelly envelopes of eggs that presumably were preserved in the same way persist, and other times they completely disintegrate. These disparate results suggest possible interspecific differences in the chemical structure of the jelly or perhaps the age (stage of development) of the egg. Possibly they reflect the consequences of minor differences in fixation. We have preserved amphibian eggs in 10% buffered formalin with satisfactory results initially, but the jelly layers often disintegrate within a few months. Care needs to be exercised if a large number of eggs, or eggs developing in thick jelly or dense foam, are fixed in a relatively small volume of formalin because the formalin will be diluted by the large amount of water contained in the jelly. As with large tadpoles, we sometimes use slightly stronger formalin to overcome the dilution effect. Khan (1982b) preserved eggs in 5% formalin, but his recommendation of removing the jelly coats to obtain adequate fixation eliminates those interesting and poorly studied structures. Because of the tedious nature of this task, the likelihood of damaging the developing embryo, and the loss of information about the jelly layers, we recommend increasing the strength of the fixative rather than removing the jelly coat of the egg. Khan (1982b) suggested preservation in Bouin's fixative (75 ml saturated aqueous picric acid, 25 ml concentrated formalin, 5 ml glacial acetic acid) for 5 min and then storage in 1:4 solution of Bouin's and water. With this approach the specimens will be colored yellow, and the danger of working with picric acid should always be recognized. Although our tests have run for only three years, we have the impression thus far that good and apparently stable specimens are obtained with a full-strength, nonaldehyde fixative (such as STF; also see Ligname 1964).

Methods of Observation

Tadpoles

A good microscope with variable magnification and strong, preferably fiber-optic lights, also of variable intensity, are prime requisites. Varying the direction, type (i.e., substage, dark-field, and polarized), and intensity of light often allows one to see morphological details more effectively; we find ring-lights ineffective because their light is too weak and diffuse. Altig prefers to elevate the light box so that one fiber-optic arm rests between the oculars of the microscope. This arrangement shines the light at about the same angle of observation and keeps the arm of the light out of the way of specimen manipulation. Ocular reticles of various kinds (e.g., grid, line scale, protractor; Edmund's Scientific) are handy or required depending on what is being done. Remember that reticle units must be transformed to millimeters and that measurements are accurate only if the plane of the object being measured is parallel with the plane of the objective.

How one observes a specimen depends on personal preference, equipment, and handedness; we present our ideas only as a guide. The tadpole is grasped by the base of the tail with forceps held in the right hand. The most functional forceps should not require excessive pressure for closure, and it takes practice to learn the appropriate "feel." The tips of the forceps should be rough and nearly parallel so as to restrain, and care must be taken not to pinch the tail off of small or weakly preserved specimens. Because the mouthparts provide much of the needed information, we examine all facets of the oral disc first. A pin or fine probe is usually

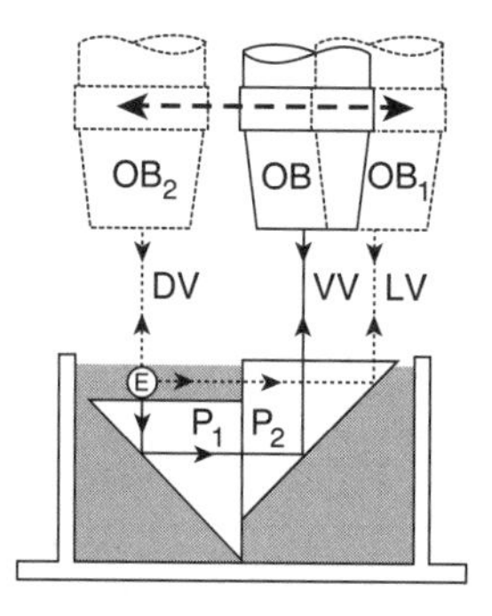

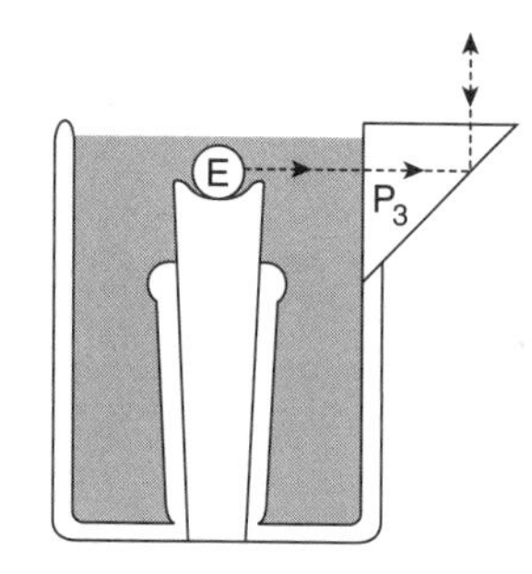

Fig. 2.2. Schematic diagrams of two easily constructed devices for examining developing eggs and embryos. In the devise on the left (after Schechtman 1934), different views of the same embryo are possible by moving the microscope or the container. In the devise on the right (after Daniel and Burch 1933), a developing embryo can be viewed by rotating the pedestal on which it sits. The container is filled with water during observations. Abbreviations: E = egg/embryo; OB = objective in position to obtain ventral view (VV) of embryo; OB_1 = objective in a position to obtain lateral view (LV) of embryo; OB_2 = objective in a position to obtain dorsal view (DV) of embryo; P_1, P_2, and P_3 = right-angle prisms.

required to open and manipulate the oral structures for viewing (McDiarmid prefers to examine oral structures with the rounded end of an insect pin). After looking at the oral disc, we typically examine the vent tube and then roll the specimen slightly to see the spiracle (typically sinistral) and fin configurations. The tadpole can then be repositioned for dorsal viewing, with the nares, eyes, and general body shape examined in sequence.

It is difficult to position fragile embryos for observation. An apparatus proposed by Schechtman (1934; fig. 2.2) allows views from three perspectives without moving the specimen. Two 90° prisms are mounted as illustrated in a small tank constructed from pieces of glass. It is desirable to keep the tank and prisms clean, so fluids should not be left to evaporate in the tank, and frequent flushing with strong alcohol is advised. Gluing the components together with minimal amounts of silicon cement allows reasonably easy disassembly for cleaning. A specimen placed on the submerged, upper surface of the left-hand prism can be viewed directly, from the side and from the bottom by moving the tank progressively to the right. For some detailed observations, one may wish to view the specimen directly, and pieces of a glass slide can be used to stabilize the embryo.

Daniel and Burch (1933) described another apparatus that would be useful if one commonly observes small embryos; in its entirety, the apparatus allows for top and side views because the embryo can be rotated. In a simpler form (fig. 2.2), the apparatus allows top (directly) and one side view (via the prism) unless the embryo is moved manually. A support with an indented top holds the embryo in a small container. A 90° prism that is silvered on the oblique side is secured in a notch cut into the side of the chamber. Live specimens can be rendered temporarily immobile with 0.2 mg of MS-222/10 ml (Sigma A5040, St. Louis, Missouri) of water (Bantle et al. 1990).

Large yolky eggs and embryos of endotrophic anurans are notoriously difficult to prepare for scanning electron microscopy (SEM; explanations of SEM procedures are outside the realm of this chapter; see Postek et al. 1980). Moury and Hanken (1995; and see Wollweber et al. 1981) developed a protocol based on *Eleutherodactylus coqui* that produced excellent preparations, and this procedure should work equally well on all embryos. Jelly layers of live embryos were removed prior to fixation; the outer layer was removed with a dejellying solution (0.63 g cysteine HCl, 0.12 g NaCl, adjusted to pH 7.9–8.1 with 5 N NaOH). The tough, elastic, middle layer that resisted chemical removal was torn off manually with sharpened watchmaker's forceps. Most of the viscous inner layer was removed by a second treatment with the dejellying solution. Because these embryos are so large and fragile, they were partially fixed before attempting to take off the vitelline membrane with forceps. This step made the solidified yolk less likely to rupture. Older embryos (Townsend/Stewart 5 and older; 1985) are less delicate, and the vitelline membrane could be removed prior to fixation.

Embryos were fixed in half-strength Karnovsky's fixative (Karnovsky 1965) in 0.1 M cacodylate buffer (pH 7.4) for 1.0–1.5 h. They were postfixed for 1 hr in 1% OsO_4 in cacodylate buffer and dehydrated rapidly through an ethanol series. Specimens were critical-point dried with liquid CO_2 as the exchange medium, mounted on stubs, and sputter-coated with gold. Overlying epidermis and subsequent layers can be removed with tape or rubber cement after critical-point drying (Tosney 1978) or the specimens can be freeze-fractured. Addressing the procedures for handling specimens preserved in formalin, R. J. Wassersug (personal communication) suggested that as many as 10 cycles in the critical-point dryer were needed to avoid distortion of many types of watery samples from tadpoles. Increasing postfixation in OsO_4 and the number of cycles in the critical-point drier are options that can be tried. Bantle et al. (1990) stated that lyophilization of *Xenopus* embryos (fixed in gluteraldehyde and OsO_4 dehydrated via an alcohol series) produced less shriveling than did critical point drying (also A. Kemp 1977). Use of chemical dehydration (e.g., Nation 1983) and the scanning electron microscopes that can handle wet samples are other options that need investigation; we have obtained both good and bad results with these two techniques but suggest that wet SEM does not work well at higher magnifications. The embryo illustrated in figure 7.2A was photographed with a wet SEM. Here and elsewhere one should note that special fixatives are required if the maximum usage of some sorts of specimens is to be realized (see Dent's fixative below).

Birefringence in polarized light makes skeletal muscle and collagen fibers appear bright white against a dark background (e.g., Carr and Altig 1991, 1992). Muscle cells, even at the magnification of dissecting microscopes, can usually be distinguished from collagen fibers by the cross striations. Other tissues appear highly transparent, but one should avoid using any kind of paper toweling because the extreme birefringence of the cellulose fibers is annoying. Dissecting and compound microscopes can be rigged easily and inexpensively for this type of lighting if they have a removable head. One plastic polarizing filter (e.g., Edmund's Scientific, Barrington, New Jersey) should be cut to lie over both lenses in the head of the scope; the other filter is placed over a substage light. Be sure to remove the plastic backing on the filters, and protect the surface of the filters from scratches. A variable-intensity, strong light, preferably fiber optics, is needed. While looking through the microscope with the light adjusted to reasonable brightness, the lower filter is rotated slowly until the point of extinction is reached (i.e., the polarization planes of the two filters are at 90° to each other). Depending on the quality of the filters, the field will appear dark blue to black. Specimens must be observed while in glass containers because plastic depolarizes the light and ruins the effect. Tissue is birefringent relative to the plane of polarization, so all fibers may not be visible when viewed in one position; vary the light intensity and rotate the specimen container to guarantee seeing all aspects of the specimen correctly. Only specimens thin or transparent enough to transmit light can be observed in this fashion (see BABB below), and clearing in glycerin greatly enhances the process. Clearing with KOH and especially enzymes ruins the birefringence.

Klymkowsky and Hanken (1991) provided a summary of techniques for examining a multitude of features of amphibian larvae. Perhaps the use of antibody stains for cartilage and various muscle groups in embryos is one of the most innovative (e.g., Hanken et al. 1992). These stains seemingly work best if the specimen is prepared using Dent's fixative (1 part dimethyl sulfoxide, 4 parts 100% methanol). Also, a clearing agent called Murray's Clear or BABB renders even large, darkly pigmented tadpoles largely transparent without destroying muscle birefringence in polarized light, and the specimens remain soft enough to dissect. The recipe is 1:2 mixture of benzyl alcohol ($n_0 = 1.54035$) and benzyl benzoate ($n_0 = 1.5681$). If you prepare this clearing agent, make sure to mix the chemicals together well and only use the fluid under a hood and avoid skin contact. The protocol for formalin-fixed and preserved specimens is: eviscerate carefully and wash in water to remove excess formalin; dehydrate in methanol series, finishing with two 15-min washes in 100% methanol; soak in 7:3, 1:1, and 3:7 methanol-BABB. The amount of time a specimen is immersed in the latter three baths varies with specimen size. Make sure all solutions are thoroughly mixed, and leave the specimen in a given fluid until it sinks. In our experience sizable tadpoles can be moved safely from a water-based fixative directly into 100% methanol.

Data about cartilaginous and osseus elements form a major component of systematic and developmental studies of amphibians. Clearing the tissue and then staining for cartilage (e.g., Alcian Blue 8GX) and bone (e.g., Alizarin Red S) or both kinds of structures simultaneously is a common technique (Klymkowsky and Hanken 1991). Scadding (1993) presented a simple appearing method for staining for cartilage with Victoria Blue B, and Song and Parenti (1995) described a method to demonstrate bone, cartilage, and nerves simultaneously.

Although the oral discs of tadpoles contain many important and interesting structures, they often are difficult to study because preserved tadpoles typically die with their mouths closed and oral discs collapsed or folded. A closed disc sometimes cannot be opened without damage to the oral apparatus, and in cases where jaw muscles are particularly large, the jaw sheaths (e.g., centrolenid tadpoles) are notoriously difficult to pry apart without causing damage. Although freeze-drying removes the specimens from being used for most other purposes, it (lyophilizing; Hower 1967) overcomes some of these problems if one has access to live specimens. Altig (1975) placed anesthetized tadpoles in a lyophilizer in enough water to cover them and froze and degassed them at the same time. In the process, tadpoles were "fixed" with mouths open and oral discs extended. Upon drying, the disc and other structures are opaque white and contrast sharply with the dark keratinized structures. The oral disc then was carefully sliced from the specimen and mounted dry on a slide beneath an elevated coverslip. Other delicate structures (e.g., spiracle, vent tube, etc.) on the tadpole can be treated similarly. If kept dry, these preparations are stable and can be photographed easily.

Van Dijk (1966) made permanent mounts for examination of the oral disc and other soft parts as follows: remove parts, stain with Light Green saturated in 96% alcohol, dehydrate, clear in methyl benzoate or xylene, and mount on a slide with glycerin (also see Saruwatari et al. 1997; D. W. Slater and Dornfield 1939). Because glycerin evaporates so slowly, such preparations survive for long periods with no further treatment. Mounting such preparations, or any small delicate structures, with Mowiol (CALBIOCHEM 475904, San Diego, California) is a handy option because dehydration is not required. Mowiol is a clear, permanent, water/glycerin soluble medium prepared as follows: mix 2.4 g of Mowiol in 6 ml glycerin, add 6 ml water and leave at room temperature for 2 hr, add 12 ml of water and stir, or put in an oven at about 100°C until dissolved. After a specimen is soaked for 1 hr in water to remove excess formalin, it can be placed on a slide, covered with Mowiol and a coverslip, and allowed to dry flat for 24 hr. Especially if the specimen has not experienced glycerin clearing beforehand, additional medium may have to be added the next day to fill gaps produced by water evaporation. The coverslip should rest on pieces of microscope slide if the specimens are particularly thick. The mixture will harden in the container after several days, so it should be mixed in the proportions above as needed. We have used this medium to good advantage for preparing muscles for viewing with polarized light. We have specimens prepared five years ago that show no signs of deterioration or discoloring.

Jaw sheaths can be removed intact from the underlying cartilages by soaking a dead tadpole (not preserved) in water for a day or so, or preserved ones for up to a week or more. Examinations of labial teeth of intact specimens by SEM allow only part of the tooth to be examined, but it is easier to see the front and back surfaces of the head of a tooth in these preparations than in preparations of extracted teeth that lie on their sides. Teeth can be extracted and examined primarily in lateral view as follows: set an automatic pipetter to about 20 μl and use the finest tip possible. If needed, a piece of capillary tubing can be glued onto the tip to serve as a fine extension. Knock teeth loose by rubbing across the tooth row with a fine pin, pick up the teeth with the pipetter, place them within small inked circles on a slide, air dry, and mount. For SEM, place the teeth on a small circular coverslip and sputter-coat them after drying. As above, inked circles on the coverslip also facilitate locating teeth. A faster preparation but one that usually does not produce material suitable for SEM is as follows: snip out a small section of tooth row with iridectomy scissors, place this piece on a slide with a coverslip and press hard enough to dissociate the tissue and force the teeth apart and into one plane. These preparations can be mounted in glycerin or Mowiol (see above). All keratinized structures can be cleared with KOH or dilute sodium hypochlorite (= household bleach). We recommend some preliminary experimentation with unimportant specimens because structures will dissolve if the bleach is too strong or they are left in the solution for too long a time.

The small size, rather delicate construction, and short larval period prohibits the use of certain field research methods (e.g., a number of marking techniques and radiotelemetry) in tadpoles that have proven useful in studying adults. Adequate marking techniques (i.e., those that are inexpensive, fast, feasible in the field, and provide individual recognition, long retention, and low mortality) either are not available or pose certain problems if large numbers of uniquely marked individuals are desired. The use of fluorescent elastomers (Anholt et al. 1998b; Binckley et al. 1998) is an exciting new option. As a result of the difficulty of marking, ecological traits as basic as population size and mortality are seldom quantified. The following references and citations in these papers may provide some ideas worth exploring (Anholt et al. 1998a; Bagenal 1967; Castell and Mann 1994; Giles and Attas 1993; Rice et al. 1998; Schlaepfer 1998). Partially as a result of these technical difficulties, behavioral information (e.g., interspecific, ontogenetic, and habitat variations in activity and foraging) is scarce and our understanding of basic ecological concepts (e.g., dispersal, microhabitat selection and use, species interactions) inadequate. The advent of small, relatively cheap video equipment should allow many of these kinds of studies to be made in the laboratory or field, often without the need for marked individuals. Even commercially available cameras work well in dim light, and infrared light sources are readily available. Perusal of the exciting and readable book edited by Wratten (1994) will stimulate ideas for numerous studies.

Field or laboratory examination, marking individuals, and other procedures that require handling of live tadpoles are greatly enhanced if the tadpole is narcotized, and the following references address this idea (Abraham and Rafols 1995; Anholt et al. 1998; Raines et al. 1993; Rice et al. 1998). We have found that MS-222 works well, but we have no ideas of any side effects.

Eggs

Orienting eggs and embryos in the jelly and seeing the various jelly layers are the two primary difficulties encountered when working with eggs. Removing the jelly usually is the least frustrating task and provides a better view of the embryo; without the jelly, the embryo will sink to the bottom of the observation chamber. Two insect pins pushed through the jelly at opposite angles and scissored apart is the easiest method of jelly removal. Because of the rigidity of the vitelline membrane (which usually does not impede viewing), one seldom will stick a pin into the embryo. Even so, removal of jelly from large, yolky eggs of endotrophs can be tedious. The transparent jelly is difficult to see, and a thick zone near the embryo is quite glutinous. If embryos are fixed well, a specimen can be rolled onto a dry area of a paper towel and the jelly will distort because of its stickiness (Guyer 1907). Pins can then be stuck through the jelly. A recipe for removing jelly layers is given above. Although based on zebra fish, the manual by Westerfield (1994) offers a number of techniques useful in the observations of eggs and small embryos.

Other Data

Lateral line organs vary in visibility depending on their size and how they contrast with pigment patterns of the tadpole. Dark-field illumination of skin samples placed on slides (Lannoo 1985) makes these organs very apparent, and Stone (1933) used histological stains. We also have found lateral line organs to be especially visible on albino tadpoles. Photographs taken of translucent specimens fixed in formalin and stained according to Bantle et al. (1990) appear quite good: rinse off formalin, add 3 drops of Hucker's Crystal Violet (equal volumes of Solutions A [2 g Crystal Violet in 20 ml 95% ETOH] and B [0.8 g ammonium oxalate in 80 ml distilled water]) for 30 sec, rinse with water until solution is clear, cover specimens with dilute acid alcohol (acid alcohol = 3% v/v HCL in 95% ETOH + 95 ml water; dilute acid alcohol = 5 ml acid alcohol + 95 ml water) for 2–10 sec until desired decolorization occurs, rinse thoroughly, and return to formalin. Various detailed differences in shapes of chromatophores and their positions within the skin can be seen by comparing white, phase-contrast, and dark-field illumination. Filipski and Wilson (1985) provided a technique for staining nerves in cleared specimens.

Illustrations

As is the case with most small organisms, a clear illustration or photograph is an important component of a useful description and this is especially true of tadpoles. Because of the typical asymmetry of the spiracle, all tadpoles should be drawn, photographed, and discussed as in left lateral view (see below). We hope that a model for a standard method of describing a tadpole will appear from our efforts to standardize terminology (see chap. 3 and Glossary) and our attempts to present descriptions in a standard format (e.g., Cadle and Altig 1991; McDiarmid and Altig 1990). Such a presentation enhances the use of the information by providing the same information in the same sequence with the same terminology. Obviously, anything not specifically included in this format should be added if it is of interest in any way. When warranted, inclusion of more details about external nares, spiracle(s), and vent tubes (e.g., G. F. Johnston and Altig 1986) is encouraged.

We recommend that at least two illustrations accompany each tadpole description: a lateral view of the left side of the tadpole and a face view of the oral apparatus. The lack of such figures is frustrating because authors sometimes fail to describe key morphological elements in the written description or present the information in such a way as to be difficult to find or dangerous to interpret by the reader. Some authors also include a dorsal, a ventral, or both views of the tadpole and, depending on body shape, this can be helpful. The lateral view should clearly show the position and orientation of the spiracle and can be used to depict the position of the vent tube. We exhort authors and illustrators to represent the tadpole as it appeared in life or shortly after proper fixation and to ignore the artifacts of preservation that too often are shown in published illustrations in the literature. The intent is to provide a useful and representative illustration to accompany the species description, not to illustrate one preserved specimen with its individual and artifactual traits. This often requires some latitude in illustrating a tadpole from preserved specimens and close coordination between the author and illustrator. Given the paucity of material of certain species in collections and the often imperfect condition of each specimen, we often have been forced to publish illustrations that combine information from two or three individuals. We argue that including artifactual folds, bends, and wrinkles, showing the desiccating results (e.g., dehydration of the caudal muscle bundles, shrunken bellies) of too strong a fixative or preservative, or depicting the results of attempted predation (e.g., nipped tail tips and torn fins) in illustrations can be misleading and often make it difficult to impossible to use the illustrations as originally intended (i.e., to aid in identification or to illustrate morphological characters not easily included in the description). Missing parts on the fins or tail can be depicted with dotted lines to show the normal extent of the structure.

Illustrations of the oral apparatus should depict the disc with the labia open and the jaw sheaths separated enough to see their shape and extent. Folded margins or a partially open disc often conceal important structures. Composite illustrations of specimens of the same size and stage sometimes are necessary to show the number, morphology, and extent of labial tooth rows, which often are damaged or developmentally impaired in laboratory-reared specimens and field-collected samples of certain taxa from some habitats and areas. We also have found that some field-collected tadpoles retained in the laboratory will lose their keratinized mouthparts (see Annandale and Rao 1918). As with lateral body illustrations, we think that an informative depiction of the oral disc of the species is preferable to one that shows the disc as it appears in a preserved individual, unless that is representative.

Recently, we have begun photographing live tadpoles with good results. Photographs capture details of color and pattern caused by nonmelanic pigments in the skin and eyes that often are lost in preserved specimens. Patterns too subtle to see with the unaided eye become visible and can be used to distinguish among similar appearing tadpoles (e.g., the uniformly black appearing tadpoles of many species of *Bufo* are not so similar when viewed in life). Also the transparent bellies of tadpoles often noted in descriptions are not always so in life. In the following paragraph, we present a brief description of the apparatus that we have used to obtain consistently good, illustrative photographs of lateral views of tadpoles. The system works well with color or black and white film, in the lab or the field, and with live or preserved specimens. For field photography one should carry a source of clean water and should photograph in deep shade to avoid reflections.

A new aquarium (e.g., 36 × 21 × 26 cm) can be modified easily and used exclusively for photography, or a smaller narrower tank that holds less water can be constructed from pieces of glass. A glass partition down the middle of the aquarium will produce two usable photo chambers if that is desirable. Cut a piece of glass that fits into the tank and stands parallel with the front; place the base of that glass next to the base of the front glass and run a bead of clear silicon glue along its back base; prop the glass insert at about a 45° angle until the glue dries into a flexible hinge. This piece of hinged glass can be moved to form a V-shaped space between it and the front of the tank. By changing the position of the hinged glass relative to the front of the tank, appropriate size space can be created to accommodate different size specimens. An adjustable bracket to hold the glass in the desired position is handy and can be constructed of Plexiglas or comparable materials. We have found that a sheet of clear acetate (e.g., document protector available in office supply stores) that is bent (not folded) in half and stapled along the free edges can be inserted between the glass sheets to form a platform on which the tadpole rests. We have glued a thin, uniform layer of pale-colored soil on the center of the acetate sheet to appear as substrate in the photographs. The finely granulate surface also tends to reflect diffuse light back toward the venter of the animal. We also use a piece of Plexiglas that has been sprayed with flat black paint to insert into the tank as background; this allows one to see clear fins better. Different lighting effects also can be obtained by changing the slant of the Plexiglas background. To use the apparatus, clean all surfaces and fill the tank nearly full with clean water; place the stapled edge of the acetate sheet at the bottom of the crevice between the front sheets of glass; swing the hinged glass forward and lock in position with a bracket; place a lightly anesthetized (MS-222, Sigma) tadpole, left side facing out, in the space between the two pieces of glass; adjust the angle of the moveable glass to accommodate the size of the tadpole and raise the specimen to an appropriate position in the tank; position the specimen with a spatula or similar tool making sure the ventral fin is on the ridge or at least not behind it. You are now ready to photograph the tadpole. We have had success with a single-lens reflex camera with extension tubes, close-up lens, or a macro lens and a dedicated electronic flash that is independent (not mounted) of the camera. A through-the-lens (TTL) metered flash makes things immensely easier. Position the flash at a steep angle near the top of the aquarium and shoot at large f/stops (high numbers) to increase depth of field. Reflections can be reduced by elevating the back of the tank about 1 cm and using dark clothing and black tape over shiny camera parts. Always shoot a test roll and look at results before attempting important photographs. Many tadpoles have bright iridophore pigments in several layers that reflect strongly and much of their tail fins are quite transparent. Trial and error, accompanied by detailed notes on exposure, background, and position of the light source, usually gives quality results.

Drawings of eggs and their jelly layers are always going to be somewhat diagrammatic because the structures appear as a series of concentric circles; subtleties of the membranes and layers are difficult to view let alone depict. With strong substage lighting and contrast filters, we have had good luck photographing amphibian eggs. Because characteristics of the jellies change with age and embryonic development, these data should be provided with the illustrations.

Finally, we recommend that illustrated specimens be measured, staged, and stored separately from the remainder of a lot of preserved tadpoles. Putting a label that indicates the tadpole is the voucher for an illustration or photograph in the vial with the specimen and citing the reference to the publication in the primary museum database also are good practices.

Collecting and Sampling

Tadpoles

In this section we address specific collecting techniques and some additional details of sampling methodology to be used in conjunction with that recommended by H. B. Shaffer et al. (1994). With regard to quantitative data, perhaps the most important thing to keep in mind is the importance of using a standard technique with a detailed explanation of methods so that data from many workers and sites can be compared. Olson et al. (1997) summarized sampling methods for amphibians occurring in lentic habitats of the Pacific Northwest, and readers interested in sampling protocols will find much useful information in that volume. Tadpoles often occur in the most unlikely places, and one often does not know the biology well enough to predict where tadpoles occur. As a result we recommend searching for tadpoles in every conceivable microhabitat. When collecting in streams, be sure to work through leaf packs, gravel beds, and around and under larger stones as thoroughly as possible. R. F. Inger (personal communication) successfully used electroshockers in tropical streams in Borneo, and this technique works in some streams in the North America. Respiratory toxins like rotenone work well but often are unwieldy (transport), difficult to contain in many aquatic habitats, and kill many other organisms.

Nets needed for collecting vary with the purpose and habitat. Small aquarium nets are handy for collecting small tadpoles, particularly when they are off the bottom. Metal-frame delta nets are required for working through heavy vegetation or in deep sediments. We have found that a large metal "tea strainer" or colander is the most useful tool. Frequently they are available in markets of many countries. At times the handles need to be reinforced, but typically they are strong enough to move through the water fast enough to catch tadpoles, have small enough mesh to capture all but embryos, and stand open for easy examination of the catch. Seines are useful for larger tadpoles in large bodies of water but frequently are damaging to habitats in smaller ponds (see Wassersug 1997). Wire mesh or net containers of decaying leaves may be useful sampling devises for some tropical tadpoles (e.g., centrolenids and mantelline rhachophorids). Adams et al. (1997) provide information on the use of funnel traps for surveying and monitoring amphibian larvae.

C. Gascon (personal communication) used bottom nets to sample tadpoles quantitatively in Manaus, Brazil. Wooden frames 1.5 × 2.5–3.0 m with a fine nylon mesh bag attached were laid on the bottom of ponds before they filled with water. At each visit by two persons, the frame would be grabbed and quickly lifted out of the water to capture the tadpoles in the entire water column. Gascon recommended calibration by repeated sampling of small ponds at different periods with the bottom net and then exhaustively dipnetting and seining the pond until all tadpoles were collected. A comparison of the bottom-net sample to the total catch (actual tadpole density) will allow evaluation of the accuracy of the sampling device for that pond. If there is good correspondence between the sample and the total catch, then a bottom net can be used to follow tadpole populations quantitatively at one site through time, or to compare abundances among sites. The method should be calibrated for each pond and probably for different species because of the different behaviors of tadpoles and the modification of behaviors in the presence of other species (see chaps. 9 and 10).

Eggs

Surprisingly few anuran eggs have been described, and much of the available information seems to be not very accurate. For example, it often is not clear whether the ovum, the jelly coats, the membranes, or the ovum plus its jelly layers are being discussed. Data on the morphological diversity and ecology of eggs and embryos are meager, and the relationships between egg morphology and reproductive tract function are inadequately appreciated. Recall that the first membrane external to the ovum is the vitelline membrane and of ovarian origin, and material external to the vitelline membrane is of oviducal origin. Comparative research on the oviduct and its relation to jelly formation in anuran eggs, similar to studies by Sever (1994), would be most welcome.

The physical properties of egg jelly change with time. To be useful, an egg description needs to contain an estimate of the time since oviposition. The exact stage of the enclosed embryo also should be given. The number of jelly layers usually is counted as the number of visible membranes that enclose each envelope, but rather simple manipulations show that the actual number of layers may be much greater (e.g., Steinke and Benson 1970). One may not always be able to discern the jelly structure from intact eggs. Staining of intact eggs, let alone sections, allows one to evaluate jelly layers and their consistency better than any other method (e.g., Kramer and Windrum 1955; Lillie 1965; Luna 1968; Pearse 1968). Sarikas (1977) used staining techniques in the analysis of egg jellies of two species of *Ambystoma*. It is unfortunate that membrane homologies have not been studied, and much more detailed research is needed before an explicit terminology can be recommended. At a minimum, we urge the recognition of the difference between membranes and layers. Membranes are in fact membranous, while layers are the jelly zone(s) between membranes. There may be visible differences in density of the jelly so that there is more than one layer between actual membranes. Dissections should be made to allow qualitative (e.g., watery vs. gel) evaluations of consistency at least of the major layers. Similar qualifications can be made for the membranes.

Most anuran eggs can be picked up or dipped carefully, but surface film eggs are most easily collected by allowing them to flow into a partially submerged cup or bucket. The stickiness of the jellies often makes netting such eggs a problem. Our unpublished data show that the method by which surface eggs float at the surface differs among taxa, and careful notes should be taken. A large-bore pipette (= "turkey baster") is handy for rapid manipulation of nonsticky eggs and particularly embryos. Khan (1982b) collected floating eggs by putting a plate beneath the floating eggs and gently lifting to expel excess water. Eggs attached to submerged vegetation are best collected by cutting the plants and placing the leaves or stems into water-filled plastic bags. Single eggs or small clumps that sink to the bottom often are difficult to find; when eggs are located, a wide-mouthed pipette is the best collecting device. Strings of eggs from species of *Bufo* can be cut into suitable lengths and put into water-filled bags for transport. Many species attach eggs to leaves over water. These can be collected by detaching the leaf or cutting out the section containing the eggs. Eggs of nidicolous and direct-developing species can be carefully collected with a spoon or similar device and carried to the laboratory in any container that protects them from damage.

The small size of eggs, stickiness of the jellies, and frequent mixing of clutches make it difficult to determine clutch size in the field. When egg number is high, probably the easiest estimate of clutch size is to calculate the mean of several weights of known numbers of eggs and extrapolate the number in the weight of the total clutch.

Rearing

Culturing tadpoles or using them in experimental situations in the laboratory is not difficult, but doing these things in the field in such a way as to produce typical specimens is not easy. Even so, we frequently collect eggs of known parents and rear tadpoles to get individuals of various stages (developmental series) or raise unknown tadpoles through metamorphosis in the field. These samples often are essential for associating an unknown larval morphology with the proper adult. Samples preserved at different stages also are needed to understand the development of some structures and to follow changes in pattern and coloration. Although based on the axolotl, information contained in chapters 19–25 in Armstrong and Malacinski (1989) on breeding, rearing, and diseases of amphibians may be useful (also see Flores-Nava et al. 1994 and Read 1994).

As is usually the case, efforts to duplicate natural conditions often are successful for maintaining tadpoles but frequently are too difficult or costly to be practical. Population density (e.g., Adolph 1931; Beebee 1991; Hourdry and Guyetant 1979; Miranda and Pisanó 1993), temperature (e.g., Harkey and Semlitsch 1988), light (e.g., Edwards and Pivorun 1991), pH (Andrén et al. 1988; Cummins 1989), dissolved oxygen (Bradford and Seymour 1988b; Crowder et al. 1998), and diet (e.g., Leibovitz et al. 1982; Marshall et al. 1980; Syuzyamova and Iranova 1988; Wimberger 1992, 1993) have major effects on development, internal, and external larval morphology (Hisaoka and List 1957; Martinez et al. 1992; Martinez et al. 1990; Underhill 1966, 1967), intestinal commensals (Hegner 1922), and the eventual adult anatomy (Smirnov 1992). Other papers by Smirnov (1990, 1992, 1994, 1995, 1997) raise interesting ontogenetic questions relative to the evolution of various osteological features.

Environmental factors apparently influence the formation and retention of keratinized mouthparts, but with the exception of temperature, data on the causes of these aberrations are lacking (Bresler and Bragg 1954). Muto (1971) saw deformities of the digits and spiracles of tadpoles reared at high temperatures. Cultured tadpoles often have mouthparts with the keratinized structures missing or deformed in various ways, and occasional field collections are found that include a large percentage of deformed mouthparts; this situation often occurs in situations where a contaminant (see C. L. Rowe et al. 1996, 1998a, b) is not known to occur. Field-collected tadpoles kept in the laboratory often lose keratinized mouthparts (ranids seem most affected), and there are occasional field collections (K. R. Lips, personal communication) in which every tadpole either lacks the keratinized structures or the structures lack pigmentation.

To avoid some of these complications, we recommend raising tadpoles in the water from which they were collected, providing them natural foods (e.g., algal-covered rocks, aquatic plants, etc.) to eat, and changing the water often. Nidicolous larvae from phytotelmata (e.g., most cophyline microhylids) often are difficult to rear (see Krügel 1993 for data on the biological conditions in phylotelmata), and using very shallow water taken from the natal site and keeping the tadpoles in dimly lighted situations seem to improve the chance of success. We have successfully reared tadpoles on a variety of commercial and artificial foods: rabbit pellets, trout chow, Frog Brittle (NASCO, Fort Atkinson, Wisconsin), and various fish foods, such as ground goldfish food, TetraMin (Blacksburg, Virginia), and Spirulina (Argent Chemical Laboratories, Redmond, Washington). Many authors and several general references suggest feeding tadpoles boiled lettuce, but in our experience, boiled lettuce tends to foul water quickly, is inconvenient, and may not provide all the needed nutrients to get through metamorphosis. Briggs and Davidson (1942) reported that boiled spinach produced the most rapid growth and development of *Rana pipiens*. In contrast, Berns (1965) reported that a spinach diet adversely affected the viability of metamorphs from calcium oxalate kidney stones they developed, and Borland (1943) found that feeding cabbage produced goiters. Harkey and Semlitsch (1988) used a finely ground mixture of one part Gerber High Protein (Fremont, Michigan) baby cereal, one part TetraMin tropical fish flakes, and two parts Hartz (Secaucus, New Jersey) guinea pig pellets to rear *Pseudacris ornata*. We have been very successful at rearing the filter-feeding tadpoles of *Gastrophryne carolinensis* by sprinkling finely pulverized rabbit pellets on the surface of the water. Powdered leaves of a couple of species of *Urtica* (nettle powder) is commonly used by European amphibian culturalists (also Martinez et al. 1994).

Even if one has adequate facilities, the correct materials, and follows directions well, one may meet with considerable difficulties if the experiment requires cultured algae for the tadpole. What happens spontaneously where it is not wanted often defies explanation, and what may seem like an ideal but uncontrolled change is not easy to replicate intentionally. We have had relatively little experience in this field and much of that has been somewhat frustrating, but we include references to get you started. Millamena et al. (1990) addressed the growth and preparations of algae as food sources. Stein (1975) is a valuable source book on the subject of culturing algae, and growth media and culturing methods for freshwater algae are reviewed in the chapter by H. W. Nichols (1975). Trainor et al. (1991) evaluated the carrying capacity of a common culture medium (Bristol's or Bold's basal medium) for specific species of algae with that of Miracle-Gro plant food. The productivity of each medium varied with the algal species, but the ease of obtaining the medium may be a benefit in certain cases. Live algal stocks can be obtained from several biological supply companies, but for more assurance of correct identifications and single-species cultures, we suggest contacting the culture collections at the University of Texas (Starr and Zeikus 1993). This reference also includes recipes for various solutions and media and a large bibliography.

In other parts of this volume authors take the traditional track of describing the feeding of tadpole as detritivorous, herbivorous, or omnivorous. We present an alternative idea on this subject and note the paucity of substantial data on the natural diet of tadpoles (see also chaps. 9 and 10). In part, the lack of data arises from the fact that few herpetologists are competent in phycology and associated fields, but the common practice of examining gut samples or fecal material may actually be uninformative or misleading. In a gross sense, most tadpoles have been classified as herbivores or detritivores, but we question if data on materials that can be counted and measured visually have much to do with actual digestion and assimilation by tadpoles. Even cultured algae and artificial foods, let alone natural dietary items, quickly accumulate enormous populations of soft-bodied aufwuchs, bacteria, fungi, protozoans (e.g., Nathan and James 1972), and even molecules. Tadpoles may consume the larger items that can be seen merely as carriers of digestible material; microcarnivorous or "moleculovorous" may be better descriptors once adequate data are available. Obviously there is much work to be done in this area. Epiphytic (= aufwuchs) communities that so commonly serve as food for tadpoles can be prepared for SEM examinations by the method of Veltkamp et al. (1994).

Rearing tadpoles in the laboratory is somewhat easier after various of the above factors are considered. Overcrowding that will produce a stunted population (e.g., Adolph 1931) should be avoided unless that is the intent. Because tadpoles are ammoniotelic, one has to be cautious about waste build-up relative to tadpole density, container volume, and the amount of uneaten food (e.g., Flores-Nava et al. 1994); if bubbles or a bacterial scum start to accumulate on the surface, the water definitely needs to be changed. Most sources of water must be aged, filtered, or chemically treated to remove all ionic and chemical toxicants; iron deposits from particularly hard water running through old plumbing systems can be a problem. Letting water sit overnight is sufficient to allow chlorine to dissipate. Other factors that potentially are detrimental and need to be considered include certain brands of paper towels, some detergents, and even some rubber gloves (Sobotka and Rahwan 1994). An ionically balanced solution (e.g., Holtfreter's or Ringer's) should be used for culturing embryos that have been surgically manipulated, and Del Pino et al. (1994) noted that urea was necessary for the culture of *Gastrotheca* embryos.

Eggs deposited on leaves (e.g., centrolenids, some hylids) can be transported to a field laboratory or camp and reared. We have been successful at fastening small, egg-laden leaves with clothes pins to the edges of plastic cups or other containers with the leaf tip in the water. We also have placed larger leaves in bottles of water with the eggs suspended over another container. Eggs on leaves should be kept out of the sun and moistened frequently; a spray bottle that produces a fine mist works well. If the incubation period is lengthy, eggs may need to be removed and placed in shallow water because the tadpoles often will die if the leaf starts to dry up and turn brown. Some unpigmented eggs of both exotrophic and endotrophic species are highly sensitive to light even for short periods (Matz 1975; R. Saunders, personal communication). Avoid touching the surfaces of aerial and direct developing eggs as they are quite prone to fungal infestations.

Summary

The study of anuran tadpoles oftentimes requires methods and techniques different from those used with adult frogs. Those used to working with small organisms probably will already be familiar with many of our suggestions, but for others we offer some approaches that we have found useful in the study of tadpoles. Our hope is that they will facilitate additional collecting and research. In spite of the impression gained from looking at this book, large gaps remain in our basic knowledge of tadpole biology. These gaps result in part from the general lack of adequate collections, an unfounded impression that tadpoles are nearly impossible to identify and therefore not worth collecting, a reluctance on the part of some museum curators to invest much time in larval collections, and a general emphasis on the larger and later stages in tadpole development. Hopefully, some of the information in this and subsequent chapters will put these notions to rest. We encourage field biologists to collect and preserve tadpoles from throughout the world. Ontogenetic series of even the most common species are infrequently available, and basic information on the eggs and hatchlings of most species is lacking. Our understanding of ontogenetic patterns of variation is inadequate and descriptions of fin shape, color, and pattern in life, including the color of the iris, are needed on a geographic scale as well. Basic information about the ecological conditions at the time and place of collection (e.g., water depth and clarity, pond shading, aquatic microhabitats, other organisms in the habitat, etc.) is essential to understanding the relationships between ontogenetic, geographic, and specific patterns of variation. Morphology tells us something about systematics and larval ecology and both of these topics are aided by the study of well-preserved samples, including developmental series, from a spectrum of habitats and localities. We hope that this primer will encourage some to read this book and open sufficient windows for others to begin to answer the interesting questions about anuran tadpoles that are posed along the way.

ACKNOWLEDGMENTS

We thank the many friends and associates who have spent time in the field with us looking for tadpoles; many of those were memorable occasions. We also thank those colleagues who have gone out of their way to collect and preserve tadpoles from remote areas of the world and send them to us for study. R. P. Reynolds and S. W. Gotte of the United States National Museum graciously supplied literature and other information that is handy for a museum curator. M. L. Crump and K. R. Lips supplied other ideas and information gleaned from their experiences in the field. Finally, we would like to thank Linda Trueb for making available part of her drawings on the Gosner stages of development used in this chapter, and Kate Spencer and Ty Theirry for recreating and expanding the second part of that presentation.

3

BODY PLAN

Development and Morphology

Ronald Altig and Roy W. McDiarmid

Larval adaptive radiation and convergent adaptation have produced an abundance of modifications . . . and other specializations of less obvious ecological significance.

Orton 1957:81

Introduction

The morphological diversity of tadpoles is immense, especially when one considers that this ephemeral stage of the life history is rather simple in overall construction. Body shapes and fin configurations seem to vary across habitats, and the morphological diversity in all components of the oral apparatus is extraordinary. Even so, this diversity often goes unrecognized because most tadpoles in the northern hemisphere, where the study of these organisms began, are morphologically conservative. Tadpoles that inhabit lentic waters and feed typically by rasping or scraping material from underwater surfaces are the rule. The low diversity in northern latitudes of tadpoles in streams in general and rheophilous forms specifically contrasts sharply with that of faunas from more tropical regions. Suspension feeders also are uncommon in most temperate regions.

In this chapter we review the diversity and development of external features of exotrophic tadpoles and argue for a standardized terminology (also see Glossary and synonymous terminology included therein). In chapter 12 we review current knowledge of tadpole diversity at the familial and generic levels and provide some information on morphological variation. Aspects of buccopharyngeal papillae and skin structure are included in chapter 5, and neuromasts, sometimes mistaken as glands (Pillai 1978), are discussed in chapter 6.

The details of development of the body and tail are not reviewed because they have been so well illustrated in numerous studies of classical embryology (B. I. Balinsky 1981). In addition to the information available in most basic embryology texts and the details presented in various staging tables, the following selected papers pertain to development of the body and various of its parts: Keller (1991), general embryology; Brock (1929) and J. A. Hall et al. (1997), gill cavities and operculum; R. S. Cooper (1943), olfactory organ; De Jongh (1967) and Twitty (1955), eye; Kessel et al. (1974), epidermal cilia; Lieberkund (1937) and Thiele (1888), adhesive gland; Paterson (1939b), olfaction and barbel of *Xenopus;* Smithberg (1954), tail; Watanabe et al. (1984), oropharyngeal membrane; Zwilling (1940), olfactory organ and (1941) otic vesicles. We also omit reviewing the massive literature on developmental embryology (Davidson 1986), endocrinology, and immunology (e.g., Manning and Al Johari 1985), even though developmental patterns may be one of the strongest signals to amphibian origins and relationships (e.g., Hanken 1986).

A general developmental scenario is as follows. No analyses of shape prior to stage 25 when the tadpole morphotype is attained or after stages 41–42 when the frog morphotype is emerging have been published. As a result, our concepts of how body shape changes during these periods (< 25 and > 42) are based on qualitative descriptors and measurements, usually of length. Embryonic elongation starts with neurulation and continues as the tail begins to form about stage 19. By stage 20 the fins are apparent; the eyes are fully formed by stage 22 and the operculum closes by stage 25 (Brock 1929). Developmental patterns are quite uniform among most types of tadpoles except rheophilous forms. Nodzenski and Inger (1990) showed that the oral apparatus forms earlier in embryology and remains much later into metamorphic ontogeny in rheophilous forms than in more typical tadpoles. Comparable alterations in the timing of developmental formation and metamorphic atrophy of various structures occur in the odd tadpole of *Lepidobatrachus laevis.* The eruption of the front legs through the opercular wall is the first obvious major metamorphic change in body shape, and atrophy of the oral apparatus and the accompanying changes in head shape follow immediately in most tadpoles. Either the right (Braus 1906; Echeverría 1994) or usually the left leg erupts first in most tadpoles. In tadpoles of *Ascaphus* both legs emerge through the midventral spiracle on the chest. In *Xenopus,* the legs emerge through newly formed openings well anterior of the dual spiracles. Later in development the eyes elevate, and the tail starts to atrophy. Borkhvardt and Malashichev (1997) found no support for the idea that the order of arm eruption affected the epicoracoid arrangement in *Bombina.*

Smithberg (1954) addressed the origin and development of the tail, and Yoshizato (1986) reviewed the large literature on the endocrinological, biochemical, and morphological aspects of tail regression (also A. Nishikawa et al. 1989). The progressive loss of the tail and initial appearance of a frog body reciprocally influences swimming and hopping abilities (Wassersug and Sperry 1977; compare with swimming in chap. 9) and thereby impacts predator avoidance. Changes to the adult configuration of the rectus abdominis (figs. 3.4A–C; Carr and Altig 1992; K. Lynch 1984) and the appearance of several associated muscles occur mostly during metamorphosis.

The development and use of standard measurements and descriptive terminology that are applicable across all taxa are crucial to fostering good communication. Likewise, a measurement without discretely and consistently definable landmarks is misleading at best. If an alternate measurement is more illustrative of a given morphology, then its landmarks also must be defined clearly. We perceive tadpole morphology to be divided logically into two major units, body and oral apparatus, and we present pertinent information including measurements and terminology useful in their description separately.

Morphology and Ontogeny: Body, Tail, and Limbs

Terminology

Several components of the body are defined by measurements and require clarification. Three points have been used to denote the posterior end of the body, and each has its use when properly defined. Because the posterior body wall of a tadpole usually follows an anterodorsal curve beginning near the vent, various descriptors (e.g., body length and points of fin origin) become confusing without discrete definitions and landmarks. A decision as to what is body and what is tail is needed, and the multiple ways of doing this have produced noncomparable measurements. Also, we do not like the use of snout-vent length for two reasons. First, if a measurement appropriate for a tadpole has the same name as one fitting for a frog, a correlation between the two measurements is suggested. This is certainly not true for all body-based measurements and at all stages in a tadpole. Second, the landmarks of “snout” and “vent” often are difficult to define in certain taxa, and any resulting measurements likely will not be comparable across taxa. The oral disc is terminal or nearly so in some tadpoles and thereby interferes with selection of the anterior landmark. If attention is paid to determining where the snout actually is and to ensuring that the measurement does not include any anterior fleshy projections of the disc, then we believe that the snout can serve as a definable anterior landmark.

A larger problem arises if the vent tube is used as the posterior landmark. The vent tube of some tadpoles exits the abdomen well to the right of the midline, and the tube often is far removed from the plane of the ventral fin (e.g., some phyllomedusine hylids). Using this sort of vent tube as a landmark would produce a very erroneous measurement of body length compared with other tadpoles. Even comparing measurements of a tadpole having a dextral vent tube with one having a medial vent tube could cause discrepancies. If the tip of the tube is used, the apparent body lengths of two tadpoles of the same actual size would vary as a function of vent tube length. Also, depending on how the posterior body wall slopes anteriorly from near the vent, either measurement using the vent tube as the landmark would exaggerate body length to various degrees.

At the same time, we do not suggest measuring to the dorsal terminus of the body. This point is not always discretely definable because of how the body blends into the tail, and use of this landmark would underestimate body length. Finally, measuring from the tip of the snout to the terminus of the edge of the intestinal mass is not consistently definable, if at all, and could not be used on the many tadpoles with opaque bellies.

We suggest that the most consistent measurement of body length among all body shapes and with discrete landmarks is taken from the tip of the snout, giving consideration to placement of the oral disc, to the junction of the posterior body wall with the axis of the tail myotomes. This axis is defined by an imaginary line connecting the apices of

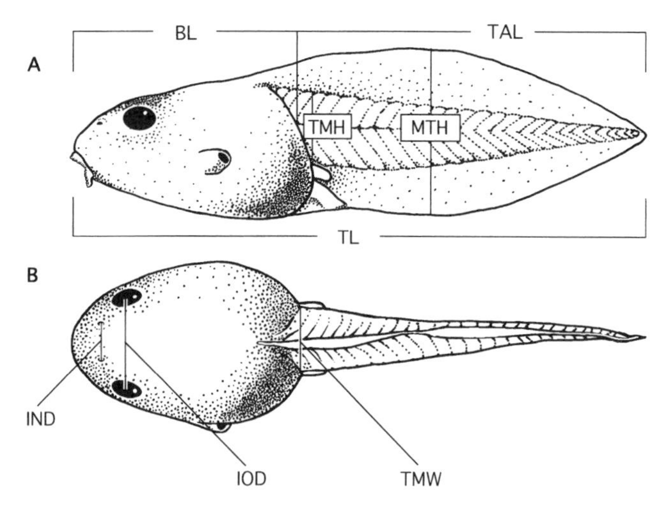

Fig. 3.1. Primary landmarks and measurements of a tadpole body. (A) Lateral view. (B) Dorsal view. Abbreviations: BL = body length, IND = internarial distance (measured between centers of narial apertures), IOD = interorbital distance (measured between centers of pupils), MTH = maximum tail height, TAL = tail length, TL = total length, TMH = tail muscle height, and TMW = tail muscle width.

the tail myotomes (fig. 3.1A). The junction of the posterior body wall with the tail axis is termed the body terminus. Also, this measurement of body length of tadpoles at Gosner 35 or greater correlates strongly with metamorphic snout-vent length. Because the body changes shape at metamorphosis, this measurement most closely reflects the size of the frog that is hidden within a tadpole once the tadpole is about as big as it is going to get.

By using this convention for body length, we define tail length as that distance from the body terminus to the absolute tail tip (fig. 3.1A). The appropriate relative qualifiers for tail length are short and long. For example, tadpoles of *Atelopus* and *Bufo* have short tails and those of *Cardioglossa* and *Cochranella* have long tails; a review of the relative tail lengths of tadpoles illustrated in figure 12.1 will give the reader an idea of variability. When evaluating the maximum, vertical extent of the tail fins, the appropriate terms are low (not short, which more commonly implies length) versus high (= tall). Thin versus thick or narrow versus broad, both pairs of which indicate transverse measurements, are inappropriate, as are shallow versus deep, terms which connote depth. Tail shape can be described by measuring the greatest height of the fin(s) above or below the tail musculature at a specified distance from the body terminus or tail tip. The dorsal fin origin is defined relative to its closest landmark (e.g., the plane of the eyes, the plane of the spiracular aperture, the plane of the tail-body junction, or distally on the tail muscle), and the origin of the ventral fin is judged relative to the junction of the ventral body wall and the tail muscle. Tail muscle height (fig. 3.1A) is measured vertically from the junction of the body wall with the ventral margin of the tail muscle, and tail muscle width (fig. 3.1B) is measured transversely at the same plane.

As noted above, vertical dimensions (e.g., body and fins in typical lateral view) are described as heights (i.e., high or tall vs. low, not wide or deep) and longitudinal measurements are referred to as lengths (see oral apparatus below for one exception). Body shape can be described by indicating the greatest height in lateral view and greatest width in dorsal view and the location of those points from the tip of the snout (e.g., globular, round, and oval). A body is depressed if wider than high (not dorsoventrally compressed, depressed, or flattened) and compressed if higher than wide (not laterally compressed, depressed, or flattened). When body height equals body width, the body is equidimensional. A cylindrical body is approached only in a few nektonic forms (e.g., *Hyla leucophyllata* group); intervening contours usually disrupt an actual cylindrical shape.

General body shape and specifically snout shape are two important but subjective concepts that are commonly used without specifically defining the modifying terms; studies of the shapes of other objects (e.g., elliptical, oval, oblong, prolate, and round; see Maritz and Douglas 1994; Preston 1968; Tatum 1975) should aid in generating well-defined terms.

Likewise, alternative ways of measuring eye diameter (= orbit diameter, not corneal diameter) and interorbital distance (i.e., between medial margins of the orbits, between centers of the pupils, or between medial margins of the cornea) produce quite different results. Eye diameter should include only the orbit viewed while looking perpendicular to the pupillary axis. Comparable measurements of the eye between darkly and lightly pigmented species, of those with a small eye and large cornea (or vice versa), and of those with eyes of different orientations and positions, often are difficult to obtain. To facilitate comparisons, we suggest that interorbital distance is most consistent when measured from the centers of the pupils (fig. 3.1B). For the same reasons, internarial distance (i.e., variation in narial diameter, position, and orientation) and measurements using the location of the spiracle (i.e., inability to define consistently other landmarks of the spiracle) should be measured from the centers of the nares (fig. 3.1B) and spiracular aperture(s).

Body Morphology

Most growth of a tadpole occurs during the exponential phase of a sigmoid curve and is quite isometric (Strauss and Altig 1992; see Alford and Jackson 1993). This period of maximum growth with minimal development is preceded (i.e., embryogenesis through hatchling) and followed (i.e., metamorphosis) by periods of significant development and little growth. Once a tadpole reaches a certain threshold, its metamorphic potential is modulated by many ecological factors (see chaps. 9, 10, and 11).

In lateral view, the general shape (curve) of the snout from about the plane of the eyes to the base of the upper labium and how that curve changes near the oral disc varies considerably among taxa. Extremes include some benthic and nektonic forms in which snout shape is a rather sharp and uniform curve and some suctorial forms in which the snout slopes gradually for most of its length and then turns abruptly downward at the oral disc. In dorsal view, the shape of the snout varies from nearly square to uniformly rounded with different degrees of curvature; some are quite pointed.

The posterior margin of a tadpole's body generally curves anterodorsally from near the vent to a point where it blends with the tail musculature. The shape of the posterior part of the body varies among taxa as a function of this curve. Variations in body shape apparently correlate with phylogeny and ecology. In general, nektonic forms are compressed and have lateral eyes (fig. 3.2A) and high fins, whereas others that spend considerable time in midwater, such as some suspension feeders (e.g., microhylids, rhinophrynids, and pipids) are depressed and have lateral eyes and low fins. Benthic forms are depressed with dorsal eyes and low fins; extremes in this group are lotic-suctorial forms (e.g., *Ascaphus* and *Heleophryne*).

Other than well-fed tadpoles being more plump than others, the case of cannibal versus normal morphotypes of *Spea* tadpoles is the most notable example of intraspecific variation in body shape (Bragg 1956; Pfennig 1990a, b; also see Heusser 1971). In the cannibal morphs, the widest point of the body is near the plane of the eyes instead of near the center of the abdomen, as in omnivorous tadpoles of the same species or in most tadpoles in general. This shape change is caused by immense hypertrophy of jaw musculature; labial teeth are few to absent (highly variable but typically a number of tooth rows in typical tadpoles), and the jaw sheaths are robust and incised (less robust and uniform in omnivores).

Some tadpoles have a constriction of the body wall at the plane of the spiracle. Others that appear to have this constriction do not; much of the pattern of tadpoles is derived from chromatophores in layers deeper than the integument, and some researchers mistake the outline of the buccopharyngeal apparatus seen through the skin as a body constriction. Altig and Johnston (1989) evaluated body shape in their formulations of ecomorphological guilds. Even though strong patterns (e.g., Lannoo et al. 1987; Wassersug et al. 1981a; Wassersug and Heyer 1983) linking shape and ecology have been reported, no tests of these interactions exist. H. Liu et al. (1996, 1997) were the first to examine the hydrodynamics of a generalized, three-dimensional tadpole shape.

Tadpoles have some striking morphological modifications of the body wall. Tadpoles of *Amolops* sensu lato, some *Ansonia, Atelopus,* and *Atelophryniscus* are gastromyzophorous—the belly is modified as a sucker that assists the tadpole in maintaining position in fast-flowing water (fig. 3.3I); the oral disc is also large in these tadpoles. The tadpoles of a few other species (e.g., *Bufo veraguensis;* Cadle and Altig 1991) have a morphology that might be a precursor to gastromyzophory. The report by I. Griffiths (1963) of an abdominal sucker formed only of lateral body walls in *Nyctimystes cheesmanae* appears erroneous. The tadpole of *Ansonia* "sucker" (Inger 1992a) has a peculiar, fingerlike projection at the anterolateral corners of the belly sucker. These structures are similar to folds that commonly occur in the margin of the oral disc of suctorial tadpoles and may enhance adherence of the sucker to rocky substrates. Based on studies of *Amolops* (Annandale and Hora 1922; Noble 1929), muscles and tendons allow the roof of the sucker to be raised to create a suction. Gastromyzophorus tadpoles use the oral disc to move like suctorial tadpoles but the timing and mechanics of oral disc and sucker operations have not been examined. Numerous, small blunt papillae are concentrated posteriorly in the roof of the sucker, and some *Amolops* have keratinized areas on the roof as well. The rim of the sucker is hollow and presumably is associated in some way with the lymphatic system.

In some semiterrestrial tadpoles the margins of the belly along the posterior (e.g., *Thoropa petropolitana*) or lateral and posterior (e.g., *Cycloramphus valae*) areas extend into a flap (fig. 3.3E). Lavilla (1988) categorized these tadpoles as gastromyzophorous, but we question if this structure can act as a sucker because it does not appear that the musculature can elevate the center of the belly. These flaps may only increase contact area for cohesion to wet surfaces and perhaps reduce turbulent flow. A short passage from Sampson (1900:693) provides an interesting historical perspective (the tadpole was incorrectly identified; *Hyla* (*Ololygon*) *abbreviata* is an apparent synonym of *Thoropa miliaris;* D. R. Frost, personal communication): "The abdomen of the tadpole is flat and serves as a sucker, so that if the perpendicular rock is slightly moist, the tadpole can move rapidly over it without legs. The tail is round, with a fin only at the end; on the ventral side, the fin is anteriorly converted into a sole, which probably aids the tadpole in adhering to the rock." The lack of information on how the tadpole moved (e.g., with or without tail undulations) is frustrating, and the curious reference to the fin being converted to a sole is unusual.

Astylosternine ranids have "lateral sacs" (Amiet 1971) of various proportions along the length of the body. In *Astylosternus corrugatus* the sacs are distinctly delimited from the entire length of the body, while in *Leptodactylodon* and *Trichobatrachus* the sacs are less distinct. When viewed with transmitted light, these sacs appear empty and could not provide any means of adhesion because of their shape and lateral position. No function has been suggested. Tadpoles of *A. corrugatus* are depressed and eel-shaped. If the sacs are connected to the large lateral lymph sinuses (chap. 5), they may serve to wedge the tadpole more effectively among small rocks in the streams where it lives.

The tadpole of *Hoplophryne rogersi* has a unique structure on their ventrolateral surface near the plane of the wall of the spiracular chamber (fig. 3.3H). The bilateral, fingerlike projection is attached to the subhyoideus muscle and covered with scalelike cells. Noble (1929) suggested that the structures act as friction pads used in squirming locomotion upon damp surfaces in confined spaces.

Four tadpoles have dorsal body modifications. Faivovich and Carrizo (1992) described a fleshy, roughly triangular flap projecting vertically from near the tip of the snout in the tadpole of *Chacophrys pierotti,* a ceratophryine leptodactylid. The black, *Bufo*-like, schooling tadpoles of *Schismaderma carens* have a semicircular flap that begins near the back of the eye and extends in an arc across the top of the head to the other eye (fig. 3.3G). The tadpoles of species of *Merten-*

sophryne and *Stephopaedes,* small African bufonids that commonly breed in phytotelmata, have a circular, fleshy "crown" that encircles the eyes and nares (Channing 1978, 1993; Grandison 1980). We suggest that the function of this structure remains conjectural because none of the suggestions (e.g., attachment, floatation, and respiration) seem feasible.

Even though the dorsal and lateral profiles of the tadpole snout vary considerably and their shapes appear to be phylogenetically and ecological correlated, no descriptive terms have been proposed. Viewed from above, the snout varies from rather acute to broadly rounded. In lateral profile (eyes to base of upper labium) the snout varies from uniformly rounded to sloping, the trajectory of which turns abruptly near the oral disc.

Tail Morphology

Some aspects of tail size and shape have been discussed previously. Here we explore the tail structure in more detail. The tail muscle is composed of bilateral myotomic muscle masses divided by V-shaped myosepta, the points of which face anteriorly; the epaxial portion is smaller than the hypaxial portion. The unsupported fins are composed of loose connective tissue covered by epidermis (Lebedinskaya et al. 1989; Rhodin and Lametschwandtner 1993; Yoshizato 1986). Fins extend from various points on the body and tail posteriorally to slightly beyond the tip of the notochord and tail muscle. Dorsal fins originate at or near the tail-body junction (most common), at various points anterior to the dorsal tail-body junction as far forward as the plane of the eyes (e.g., *Kassina senegalensis* and *Scinax rubra* group), or at various points along the tail muscle (see fig. 12.1 and sketches of examples of tadpoles in various genera within the taxonomic accounts). The margins of low tail fins usually are nearly parallel with the margins of the tail muscle. Extremes in this case are represented by suctorial species, *Poyntonia paludicola,* and various semiterrestrial forms (e.g., *Arthroleptides, Cycloramphus, Nannophrys,* and *Thoropa*). The anterior part of the dorsal fin of the tadpoles of *Occidozyga* (sensu stricto) rises abruptly (M. A. Smith 1916b) and then slopes evenly to the tip. The same descriptors apply to the ventral fin, which most commonly originate at the ventral terminus of the body but may originate anywhere on the tail or anterior to the vent tube at various points on the abdomen (e.g., phyllomedusine hylids, pipids, some *Rhacophorus,* and *Xenopus* in which a distinct lobe may be present).

The tail tip takes many shapes: broadly or narrowly rounded to various degrees or extended into a flagellum. A flagellum comprises the terminal part of the tail muscle and reduced fins; it is more or less distinctly separated from the anterior portion of the tail by a prominent reduction in fin height. The flagellum often is pigmented less and may undulate independently of the remainder of the tail. A flagellum may be present in tadpoles of several ecological types with low (e.g., *Hyla leucophyllata* group, phyllomedusine hylids, and *Xenopus*) or high (e.g., some *Hyla, Kassina,* and *Scinax*) fins. Duellman (1978) used xiphicercal to describe a tail that narrows abruptly to a distinct flagellum. Complimentary terminology for other forms of fins is not available, and such usages may be inappropriate. If tadpole tails were classified by the same terminology as fish tails, all tadpoles would be diphycercal, although presumably not homologous (in diphycercal fishes, all medial fins fused into one unit). Terms suggested by Van Dijk (1966) apply only to curvature of the tail axis. Fins of lotic tadpoles are lower, thicker, and stiffer than the taller, thinner more flexible ones of lentic forms. The amount and distribution of connective tissue in the tails of lotic versus lentic forms have not been studied. Kinematics comparing swimming between the extremes of tail morphology surely would be worth investigating, but only pond forms have been studied (chap. 9).

Van Dijk (1966) and Lambiris (1988a, b) noted that the basal portion of the tail muscle and adjacent fins of *Hemisus* tadpoles has a sheath of thick connective tissue visible in live and preserved specimens (see figures in chap. 12). This tail sheath is more common than previously suspected but when present, is visible only under specific, but undetermined, circumstances of preservation. We know of no explanation of what the structure is, even if it is connective tissue, or of how it functions, but it does occur in other taxa (e.g., O. L. Peixoto, personal communication, *Hyla marmorata* group; I. Griffiths and De Carvalho 1965, *Stereocyclops;* personal observation, *Gastrophryne carolinensis* and *Rana sphenocephala*).

Anyone who works with tadpoles from different habitats soon questions the influence of environmental factors on morphology. Unfortunately, few definitive data are available. Tadpole color and coloration obviously vary in different environments, and Van Dijk (1966) questioned if tadpoles raised in shallow water have lower fins than normal. Tadpoles of *Rana chiricahuensis* captured in flowing water differed in the degree of development of abdominal musculature, pigmentation, and shape compared with conspecifics from still water (R. D. Jennings and Scott 1993). Martinez et al. (1994) reported skeletal deformities in reared tadpoles that were likely nutritional.

Limbs

The hind limbs of tadpoles are exposed during much of tadpole development, and their pattern of development (see Alberch and Gale 1983; B. I. Balinsky 1972) is of primary importance in comparing growth rate among tadpoles of different sizes and taxa under differing conditions (e.g., Gosner 1960; see fig. 2.1). Even though the front limbs are covered by the operculum, they follow a similar developmental pattern. Based on our unpublished data, development of the front limbs lags behind that of the hind limbs by about one stage (i.e., length-diameter changes in limb buds in Gosner stages 26–30 and toe differentiation in stages 33–37) in spite of the fact that only four digits occur on the front limb. The apparent differential in limb development increases greatly in semiterrestrial tadpoles, but the adjustments in timing and pattern of the growth trajectories required to reach proper metamorphic conditions are not known. Drewes et al. (1989) found a difference of about five stages between comparable development of hind and front limbs of *Arthroleptides,* and Lawson (1993) saw tadpoles of *Petropedetes* with only fully developed hind legs hopping about on the forest floor. Definitive data are lacking, but it appears that finger and toe webbing of metamorphs is equal to or less than that of adults. Inhibitory versus stimulatory modulation of limb growth by hormones of the pituitary and adrenal hormones is stage dependent (M. L. Wright et al. 1994; also Elinson 1994). L. Muntz (1975) recognized four behavioral stages in leg mobility and muscle and nerve development during leg myogenesis (also Dunlap 1967; Hughes and Prestige 1967; L. Muntz et al. 1989). Comparisons of her data on *Xenopus* to those from species of different metamorphic patterns and ecology would be informative. Also, the ontogeny of pigment patterns during limb development, although largely unstudied, may be useful in species identifications.

Eyes and Nasolacrimal Duct

Eye position and eye orientation are different concepts (fig. 3.2A). Position indicates where the structures are located on the head and orientation describes their manner of placement or alignment (e.g., facing direction). Eye position is generally described as dorsal or lateral, although a continuum exists. Dorsal eyes may face dorsally (upward), dorsolaterally, or laterally, while lateral eyes always face laterally. The eyeball of tadpoles is covered by a two-layered cornea that is continuous with the body epidermis. Dorsal eyes have no part of the eye or cornea in the dorsal silhouette, and lateral eyes have some part of the eye or cornea included in the dorsal silhouette (fig. 3.2). Dorsal eyes exemplify benthic tadpoles in both lentic and lotic systems, while lateral eyes are most common in lentic forms that spend considerable time in the water column (e.g., many hylids and microhylids). Relative positions of dorsal or lateral eyes can be qualified verbally or by measurements. Lateral eyes are typically larger and have more cornea curvature and lenticular protrusion than dorsal eyes. Eye size ranges from minute (e.g., centrolenids) to quite large (e.g., some nektonic hylids). The exceptionally small eyes of centrolenids are located quite close to the midline, and because of their orientation and an apparent late closure of the choroid fissure, they often appear C-shaped in dorsal view. The extent that the eye protrudes from the surface of the head, the curvature of the cornea, and the lenticular protrusion also vary. We detect some correlations with ecology, but definitive data to support these observations are lacking.

Van Dijk (1966) discussed an umbraculum and two types

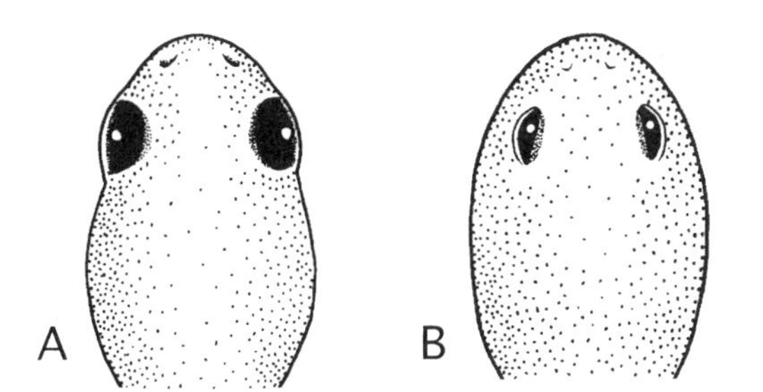

Fig. 3.2. Dorsal views of stylized tadpole bodies showing eye positions. (A) Lateral. (B) Dorsal.

of elygia, all of which are assumed to shade the eye from intense light. The umbraculum (e.g., *Rana vertebralis*) is a projection of the dorsal pupillary margin. Elygia may occur as a hemispherical area of melanophores extending laterally from the middorsal margin of the iris (ocular) or in the dorsal cornea (epidermal). The corneas of an unidentified hylid tadpole (personal observation) are mostly pigmented; if not a case of teratology, this may be an extreme form of epidermal elygium. Wassersug et al. (1981a) noted similar pigmented corneas in *Theloderma carinensis*. Recently we have observed an ocular elygium in a number of other taxa (e.g., some *Bufo, Hyla, Mantidactylus, Pseudacris,* and *Rana*), although often it is visible only with bright illumination and at magnification of live specimens. The eyes of *Staurois natator* are covered with "thick skin . . . but in stages 36 onward there is a small, clear window over the eyeball" (Inger and Tan 1990:4).

The tadpole iris commonly has bright, metallic pigments arranged in complicated patterns that seem to be influenced by the topography of underlying blood vessels. Interspecific patterns in the arrangement and distribution of color probably exist, but these features (e.g., Duellman 1978; Gallardo 1961; Scott and Jennings 1985) have seldom been examined. Although the entire eye can be moved, the round pupil has no ability to adjust its size relative to light intensity.

During mid- to late metamorphosis, either dark pigmentation appears or the surrounding pigmentation decreases in intensity to accent a shallow groove that extends in an arc from the anterior corner of the eye toward the naris. This nasolacrimal duct develops in association with the Harderian gland (Schmalhausen 1968; see Baccari et al. 1990) that lubricates the eye.

Integument and Chromatophores

Numerous unicellular glands throughout the epidermis of tadpoles produce mucus; chapter 5 has more details of integumentary histology. Small serous glands often are scattered generally over the body surface (e.g., *Ascaphus* and *Poyntonia paludicola*) or concentrated along the dorsal fin, within the ventral fin, along the dorsum of tail muscle, and dorsolaterally and ventrolaterally on the body (e.g., Inger 1966; Liem 1961; S. J. Richards 1992; some *Amolops, Huia,* and *Meristogenys, Hyla, Hylarana, Litoria, Rana, Phyllomedusa,* and *Physalaemus*). Tadpoles of some *Amolops* have keratinized spinules on the dorsum that may abate turbulence as water flows over the body; they also have various patterns of small, densely arranged papillae and light keratinization in the roof of the belly sucker. Short papillae occur on the snout of the tadpoles of *Hoplophryne rogersi* (Noble 1929).

Close examination of certain tadpoles reveals a circular area of slightly contrasting color anterolateral to the base of the vent tube. This slightly raised area appears glandular, but it may be associated with the lateral line system. It is most obvious in darkly pigmented, neotropical, lotic hylids.

Bright coloration and distinctive patterns are important sexual and species recognition features for many adult vertebrates but apparently are of little importance to larval or immature ones. Dorsal coloration in tadpoles is mostly of muted tones of black, brown, gray, and green arranged in various patterns; ventrally most tadpoles are coppery, silver, or white. Altig and Channing (1993) described the major components of tadpole color pattern; compared to adults the coloration of tadpoles is somber and presumably functions in some aspect of crypsis (i.e., camouflage). Altig and Channing (1993) also identified common patterns that they associated with background matching, disruption, and countershading. Other colors (e.g., gold, orange, and red), when present, usually are comparatively bright or contrasty. After considering the coloration and the behavior of the tadpole of *Xenohyla*, Izecksohn (1997) considered it to be a leaf mimic, and the coloration is also surprisingly convergent with young tadpoles of *Pseudis paradoxa*.

Specific pigments sequestered in chromatophores provide color and pattern (Bagnara et al. 1978). Melanophores contain black or brown melanins derived from tyrosine; xanthophores and erythrophores contain synthesized yellow and red pteridines plus dietary carotenoids; and iridophores contain reflective purine platelets. Chromatophore type can be identified in TEM sections by the structure of the pigment-containing organelles (S. K. Frost and Robinson 1984). Filiform, foliose, punctate, rodular, and stellate are terms that are useful for describing the shapes of chromatophores, especially of melanophores, that remain in formalin-preserved material. The presence and apparent shape and size of chromatophores differ when viewed with incident, transmitted, dark-field, or polarized illumination. The type of pigment and its distribution within the cell and the density, location, orientation, shape, and size of the chromatophores modify the perceived color and pattern. Much of the color and pattern of tadpoles commonly is derived from chromatophores positioned in subintegumentary areas (e.g., walls of the gill chambers, peritoneum, blood vessels, nerves, and gut). Published figures often show these differences without mention in the text.

Other components of coloration probably serve to delay (e.g., yellow ocelli at the base of the tail in *Rana alticola*) or misdirect (e.g., black tail tips of some *Acris* tadpoles) a predator attack. Caldwell (1982) published one of the few quantitative studies on the distribution and occurrence of color pattern in tadpoles. She argued that a polymorphic pattern involving black tail-tips in *Acris* is maintained by disruptive selection and functions as a deflection mechanism to misdirect predatory attacks by certain dragonfly naiads from the head of the tadpole to the tail. Integumentary glands that presumably secrete noxious substances occur in some tadpoles (e.g., Liem 1961, *Rana chalconota;* personal observation, *Physalaemus petersi*) and their location on the body or tail often is highlighted by bright contrasting coloration (see photographs of *Rana alticola* and an Indian ranid in figs. 1H and I in Altig and Channing 1993). Such patterns of aposematic coloration are well known among other organisms and their scattered occurrence among tadpoles is not surprising. Brodie and Formanowicz (1987) argued that the intense black coloration and cutaneous toxicity of many bufonid tadpoles and their tendency to aggregate in slow moving schools are traits indicative of aposematic coloration (see Wassersug 1971, 1973; see chap. 9). The bright red coloration of many centrolenid tadpoles is attributable to blood capillaries and not pigment in the skin (Villa and Vallerio 1982; personal observation). There also is some evidence that the bright golden band on small tadpoles of *Rana heckscheri* may serve as a visual cue to promote spatial orientation of schooling individuals. Altig and Channing (1993) reviewed generalizations about the associations of color and pattern with habitat, behavior, and ontogeny.

Metachrosis (i.e., change in color via changes in the distribution of pigments within chromatophores) is limited, although considerable blanching and darkening relative to diel light cycles occur. The body of many tadpoles becomes paler at night, while the tail, particularly the distal third, often gets darker. The tail of the tadpoles of *Hyla gratiosa* has punctate melanophores near the base and stellate ones distally; it is the stellate population of melanophores that dilates at night that produces the intensely black tail of this species. Detailed studies of the distribution and morphology of melanophores are needed to serve as an aid to identification and to help explain certain aspects of behavioral ecology. For example, a chainlike pattern of melanophores visible only at magnification occurs in discoglossid tadpoles (Bytinski-Salz and Elias 1938). Also, the tadpoles of some species of *Leptodactylus* have melanophores arranged in rodular bundles that are oriented at various angles to each other. As tadpoles approach metamorphosis, adult pigments begin to appear.

Tadpole coloration is known to vary ecologically and geographically but detailed studies are lacking. Conspecifics inhabiting turbid water often appear unpigmented whereas those in clear or tannin-stained water even a few meters away often are quite differently and frequently prominently colored. The reddish or yellowish colors that occur in the fins of tadpoles of certain species (e.g., *Hyla chrysoscelis, H. cinerea, H. femoralis, H. versicolor,* and *Kassina senegalensis*) develop most intensely in individuals that develop in tannin-stained waters and seldom in those from turbid water. Experiments by McCollum (1993) and McCollum and Van Buskirk (1996) seemed to indicate that the reddish or yellowish pigments developed only in the presence of predatory odonate naiads.

Understanding details of how color pattern develops during hind limb ontogeny would be useful in distinguishing among species and probably provide interesting data on pattern development in earlier stages as well. For example, hatchlings of a number of North American hylids have dorsally banded tail muscles. This pattern subsequently is lost in most species but usually retained by tadpoles of species of *Acris, Hyla avivoca,* and *Pseudacris crucifer.* A similar case occurs in young salamanders of various species of *Ambystoma* with only *A. talpoideum* retaining the bands throughout larval ontogeny. Unfortunately, such studies are uncommon (e.g., Altig 1972). Ontogenetic changes in tadpole coloration usually are not profound after about stage 25. Tadpoles of *Hyla gratiosa* between stages 25–28 have a single black saddle on the tail muscle that disappears later in ontogeny when tail metachrosis appears (see above). Small tadpoles of *Rana vaillianti* have bands across the body and tail, and small tadpoles of *Pseudis* are prominently banded in contrast to the unicolored, larger ones. The black rim on the tail fins of

Rana heckscheri tadpoles begins to develop at about the same stage (28) that the prominent golden band of smaller tadpoles begins to become less prominent; whether there is an ontogenetic change in the orientation of individuals in schools is unknown.

External Nares

The olfactory sacs and both external and internal nares usually are present during ontogeny, but the external nares of most microhylid tadpoles do not open until metamorphosis; *Metaphrynella* (Berry 1972), *Ramanella* (Kirtisinghe 1958), and *Uperodon* (Mohanty-Hejmadi et al. 1979) are reported to have open nares during tadpole ontogeny. Berry (1972) also stated that the nares of *Rana glandulosa* do not open until just before metamorphosis. Embryos of *Gastrophryne carolinensis* and *Microhyla heymonsi* have external narial openings (personal observation), but the canal is blocked; this external opening subsequently closes at the surface for most of larval ontogeny, and, for a while, one can detect where the opening was on SEM photomicrographs. The narial openings presumably form via induction from the olfactory ectoderm (Zwilling 1940), but the mechanism for closure of the external opening and subsequent reopening late in ontogeny is not known. The external nares of the nidicolous cophyline microhylids are open throughout ontogeny, although it is not known if they are blocked within the tube during at least early stages.

External nares are situated at various points on the snout. Tadpoles with cylindrical bodies and long snouts (e.g., *Hyla leucophyllata* group) have the nares much closer to the tip of the snout and more lateral than seen in most tadpoles; the nares of *Xenopus* are distinctly parasagittal and near the mouth (see illustrations in chap. 12). Most often the narial aperture is rounded, but it may be oval, roughly triangular (e.g., *Bufo*), or transversely elliptical (e.g., *Xenopus*). Interspecific differences in snout shape and surface topography cause the aperture to face anterolaterally, dorsally, dorsolaterally, or laterally. The nares of *Bufo* are quite large for the size of the tadpole, and comparisons of relative size among species often can be used for identification (e.g., *Rana pipiens* vs. *R. catesbeiana* groups). Other details of narial structure have not been examined in detail (e.g., Channing et al. 1988; G. F. Johnston and Altig 1986). For example, the narial opening may be flush with the surrounding surface, countersunk, or marked with a marginal rim that varies interspecifically in prominence and shape (fig. 3.3F). Other types of narial ornamentation range from a single, medial papilla to various arrangements of several papillae. Some descriptions suggest that the authors have looked through the transparent skin and mistaken pigmentation around the nasal canal for a tubular external naris. The narial rims of *Ascaphus* and certain other suctorial forms extend as distinct tubes (e.g., fig. 3.3F).

Operculum and Spiracle

"Operculum" is the gill covering of tadpoles and a term that has different meanings in other groups; it is not homologous with the gill covering of fishes or the ear element in frogs (chap. 4). "Spiracle," the exit for respiratory water in tadpoles, forms as a result of the growth and fusion pattern of the operculum with the body wall; it is another borrowed term and not homologous with the hyoid pouch derivative in elasmobranchs. We retain both terms because they have been used for a long time for tadpoles and changing the names likely would produce more confusion than that caused by the multiple meanings that exist in different groups today.

The number, location, and morphological configuration of the spiracle(s) depend on where and how the opercular fold fuses with the body wall (Starrett 1973). Starrett observed a frequent association of spiracular position with feeding mode and fin shape, especially of midwater and surface feeding tadpoles. She commented that an excurrent spiracular flow from a single sinistral spiracle would tend to rotate and thereby decrease the efficiency of a tadpole feeding in the water column and observed that most midwater filter-feeding tadpoles have either dual lateral or a single midventral spiracle. The position and number of the spiracle(s) generally are consistent enough among related species and within certain higher taxa to be useful for species identification and as phylogenetic traits. Exceptions include a midventral spiracle in *Flectonotus fissilis,* rather than the sinistral condition typical of hylid tadpoles and a spiracle that opens on the left side of the tail muscle into the same tube as the vent in *Stereocyclops,* rather than midventrally as in most microhylid tadpoles (I. Griffiths and De Carvalho 1965).

By far the most common spiracle morphology is a single midlateral opening on the left side (= sinistral or laevogyrinid) but at various positions along the body. At least phyllomedusine hylids, pelobatids, and some myobatrachid and leptodactylid tadpoles have a single spiracle situated far below the midlateral area in a parasagittal position. G. F. Johnston and Altig (1986) coined the term paragyrinid for this condition. A single spiracle situated anywhere from the chest (e.g., *Ascaphus*) to near the vent (most microhylids) is termed midventral, medioventral, or mediogyrinid. Dual, lateral spiracles (= amphigyrinid) occur in the tadpoles of rhinophrynids, pipids, and species of *Lepidobatrachus* (Leptodactylidae), although the mode of formation and therefore homology in the latter case (Ruibal and Thomas 1988) is different from the first two. Lavilla and Langone (1991) described the ontogenetic changes in the spiracle of *Elachistocleis ovalis,* and Inger and Frogner (1979) cautioned workers about using spiracular characteristics of small microhylid larvae because of ontogenetic changes. For example, the edge of the ventral wall of the spiracular tube of *Paradoxophyla palmata* is entire, but older individuals have four streamers extending posteriorly (personal observation). The unique spiracle of *Otophryne pyburni* (Pyburn 1980) includes a free tube that is the longest spiracular tube known (i.e., extends well past the body) and a seemingly sinistral position unique within microhylids. Wassersug and Pyburn (1987) showed that the spiracle in *Otophryne* starts out in the expected medioventral position and ontogenetically moves to the left side. These authors suggested that the tadpole feeds passively while embedded in the bottom of sandy streams. Water flowing over the orifice of the long spiracle trailing upwardly in the water acts as an aspirator to pull water into the mouth. Other tadpoles with rather long spiracular tubes

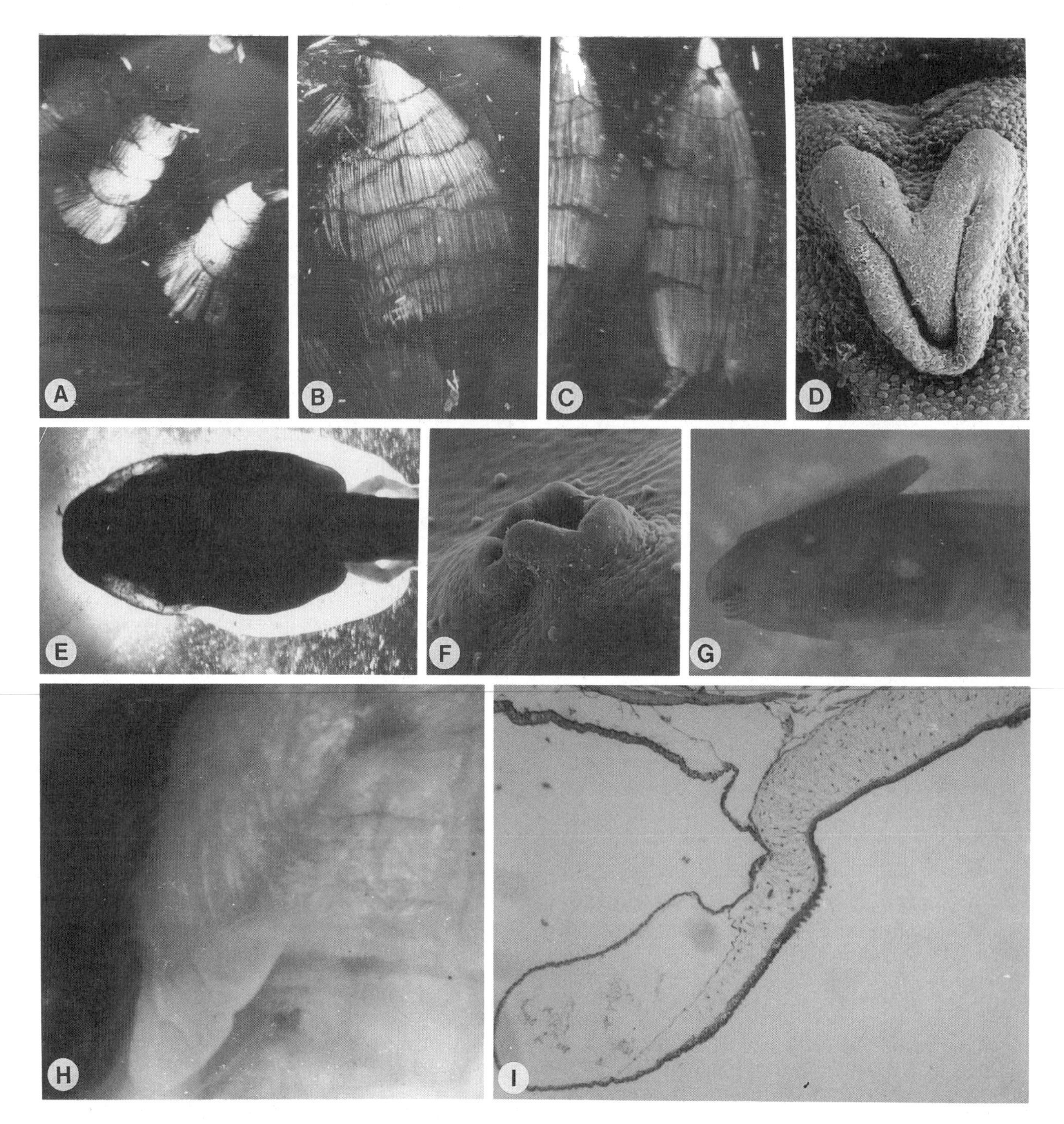

Fig. 3.3. Representative anatomical and morphological features of select tadpoles. Typical arrangement of the rectus abdominis muscle (origin uppermost in all frames, viewed with polarized light) in (A) *Gastrophryne carolinensis* (Microhylidae), (B) *Agalychnis callidryas* (Hylidae), and (C) *Lepidobatrachus laevis* (Leptodactylidae). (D) Adhesive gland of *Scaphiopus holbrookii* (Pelobatidae). (E) Abdominal flap (dorsal view, transmitted light) of *Thoropa petropolitana* (Leptodactylidae). (F) Narial aperture of *Heleophryne purcelli* (Heleophrynidae). (G) Head flap of *Schismaderma carens* (Bufonidae). (H) Belly "finger" (right side, ventral view, head at top) of *Hoplophryne rogersi* (Microhylidae). (I) Belly sucker (anterior to right, longitudinal section) of *Atelopus ignescens* (Bufonidae).

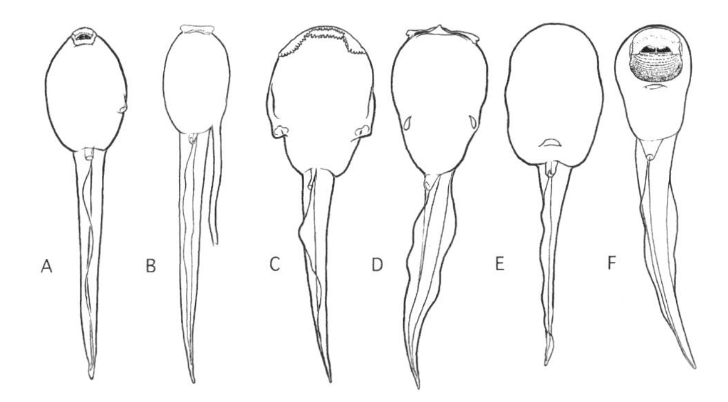

Fig. 3.4. Basic patterns of spiracular tube arrangement in anuran larvae. (A) Single, sinistral, *Dendrobates tinctorius* (Dendrobatidae). (B) Single, sinistral with long spiracular tube, *Otophryne pyburni* (Microhylidae). (C) Dual, lateral, *Lepidobatrachus llanensis* (Leptodactylidae). (D) Dual, lateroventral, *Rhinophrynus dorsalis* (Rhinophrynidae). (E) Single, posterior ventral, *Kaloula pulchra* (Microhylidae). (F) Single, midventral (on chest), *Ascaphus truei* (Ascaphidae).

include *Acris* and *Heleophryne*. *Scaphiophryne* (Scaphiophryninae) has many odd features (e.g., jaw sheaths) for a microhylid tadpole, including a spiracle that is between midventral and sinistral. Representative spiracle arrangements are shown in figure 3.4.

Because of differences in the pattern of fusion between the opercular fold and body wall, the spiracular aperture may face various directions and have different shapes and tube configurations. The different developmental patterns that produce dual, lateral spiracles, a single medial spiracle at various sites, or a sinistral spiracle are understood in general, but details of which tissues are involved and the patterns of formation among taxa are unclear (Inger 1967; Nieuwkoop and Faber 1967; O. M. Sokol 1975; Starrett 1973). The inner or centripetal wall of a sinistral, paragyrinid, or midventral spiracle may be absent (e.g., *Ascaphus;* phyllomedusine hylids), present only as a partial or complete ridge, or present as the wall of a tube of various lengths that posteriorly or terminally is free from the body wall with the posterior tip free from the body wall (fig. 3.5A–D). The lateral wall may end anterior to, at the same plane as, or posterior to the insertion of the medial wall. The spatial limits of the spiracular walls determine the shape and orientation of the aperture. In certain cases the aperture faces laterally and is commonly smaller than the bore of the spiracular tube.

Functional explanations to account for the different spiracular arrangements are few. Starrett (1973) suggested that midventral (e.g., microhylid) or dual lateral (e.g., pipid) spiracles in suspension-feeding, nektonic species may allow the tadpole to avoid disturbance of planktonic or particulate food items. At least some of these forms can maneuver solely by the force of the spiracular jet (Starrett 1973; personal observation, several taxa). Wassersug and Viertel (1993) suggested that folds in the inner walls of the long spiracle of *Otophryne* may provide elasticity to modulate water flow. Based on tadpoles raised from a single clutch of *Stereocyclops* eggs, I. Griffiths and De Carvalho (1965) suggested that a genetic link exists between the variations in spiracular tube length and middorsal stripe.

Vent Tube

An anus is the posterior opening of the mammalian digestive tract. Amphibians have a common collecting chamber—a cloaca—into which the digestive, urinary, and reproductive tracts empty. The aperture through which feces exit to the outside is not the anus in the adult and probably not in the tadpole either. Without knowing more details of the development and origin, the use of proctodeum in association with the vent is also debatable. For tadpoles, we suggest the terms vent and vent tube in deference to the same notation (i.e., snout-vent length) used for frogs. See previous discussions of why body and tail length measurements should not involve the vent or vent tube as a landmark.

The digestive tract of a typical tadpole exits the abdominal cavity through a tube, the aperture of which usually is located in the sagittal plane and commonly associated with the ventral fin (fig. 3.5). The location of the aperture usually is consistent enough within taxa to be useful for identification. The primary character states, recognized more than

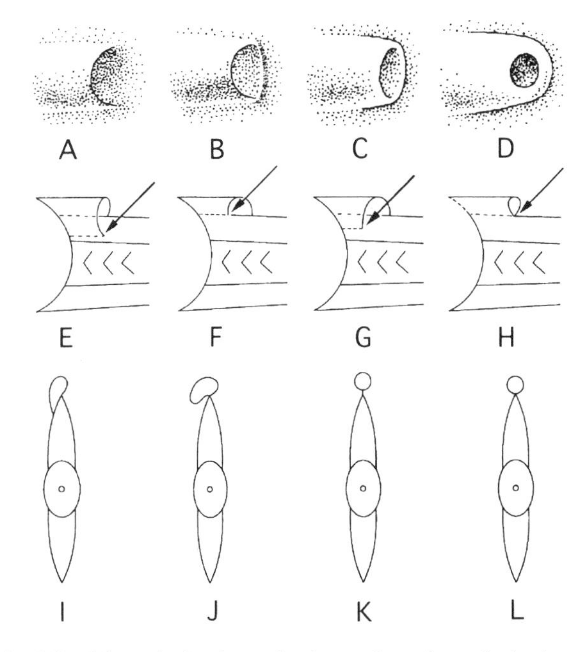

Fig. 3.5. Schematic drawings of major configurations of spiracle and vent tubes. *(Upper)* Left-lateral views of spiracular tubes that essentially face posteriorly: (A) inner (= centripetal) wall absent; (B) inner wall present as slight ridge; (C) inner wall free from body; and (D) inner wall free and formed such that aperture opens laterally instead of posteriorly. *(Middle)* Lateral views of the right side of supine tadpoles showing vent tubes (arrows indicate points of attachment of right wall): (E) right wall displaced dorsally; (F) right wall displaced anteriorly; (G) right wall displaced dorsally and anteriorly; (H) medial vent tube. *(Lower)* Schematic cross-sections through the base of the tail fin of supine tadpoles showing various placements of the vent tube: (I) right wall displaced dorsally, as in (E) above; (J) medial vent tube with lateral displacement; (K) medial vent tube with web between tube and fin; and (L) medial vent tube with both walls attached directly to ventral fin.

100 years ago (see Camerano 1890, 1892; Willey 1893), involve the position and orientation of the opening; apertures open medially (in line with the plane of the ventral fin) or dextrally (to the right of the plane of the ventral fin). Some authors (e.g., Cei 1980; Duellman and Hillis 1987; Kehr and Basso 1990) reported sinistral vent tubes in *Pleurodema, Gastrotheca,* and *Lysapsus,* respectively, and we have seen aperture variation within a single lot of *Ceratophrys cornuta.* Numerous other variations within each basic type are known (G. F. Johnston and Altig 1986), but no significance has been attached to the different configurations. We recommend that the character state (medial, dextral, sinistral) of the vent be determined relative to the position of the aperture, not the position of the tube. A vent tube that parallels the ventral margin of the fin may still have a dextral or sinistral aperture.

We know of no data supporting the notion that the vent tube is abdominal in origin but we assume that likely it is. All vents are medial during early development, and the ontogeny of vent tube formation has not been studied well. Those phyllomedusine hylids in which the entire vent tube and aperture is displaced to the right and not associated with the tail fin in any way have not been examined as embryos. The dextral condition in *Phyllobates lugubris* (Donnelly et al. 1990b) is not attained until later than normally expected (i.e., by stage 25; also see Lavilla and Langone 1991). Davies and Richards (1990) noted that until stage 32 the limb buds of *Nyctimystes dayi* are enclosed in a membranous sac associated with the vent flap (also see Van Dijk 1966; *Hemisus*). Variations on these themes are found among a number of rheophilous forms (see below).

A medial vent may or may not be associated with the ventral fin in several different ways (fig. 3.5). The tube may be long or short, free or attached directly to the fin, or attached via an intervening, fleshy web. The ventral wall may be shorter than, equal to, or longer than the dorsal wall. Differences in wall lengths produce distinctly shaped apertures that usually face in different directions. The right wall of dextral vent often is displaced anteriorly, dorsally, or more commonly anteriorly and dorsally. A ventral wall that is longer than the other walls causes the aperture to face in a number of directions, including dorsally. In some cases, dextral vents are completely separated from the tail fin and muscle (e.g., *Phasmahyla guttata*). In a number of suctorial forms (e.g., *Ascaphus, Heleophryne, Hyla bogotensis* group, *Telmatobufo*) a flap of tissue lies below the vent; these vent flaps take on several different shapes. In *Ascaphus,* the vent tube is attached to the dorsal surface of this flap, and the vent aperture is even with the distal margin of the flap. The margin of the aperture normally is smooth, but B. T. Clarke (1983) noted that in tadpoles of *Nannophrys ceylonensis* it was fluted and appeared similar to marginal papillae on the oral disc. We know of no data that indicate a correlation between vent tube morphology and position and tadpole ecology.

Transitory Embryonic Structures

Adhesive glands (fig. 3.3D), external gills, integumentary ciliated cells (visible in fig. 3.10), and hatching glands occur for various periods during embryology. The adhesive gland usually is visible between stages 18–25; it can be detected much earlier with proper techniques and may persist at least as a pigmented spot for several stages later in ontogeny. The term "sucker" should not be used in reference to the adhesive gland. The mode of attachment is via sticky secretions; there is no suction comparable to that found in the belly or abdominal sucker of gastromyzophorous tadpoles and the oral disc of suctorial forms. Hatchlings use the sticky secretions (Eakin 1963, *Pseudacris regilla*) of these transient glands to provide stabilization prior to the further development of the oral disc and tail which afford more coordinated adhesion and locomotion for active tadpoles. *Xenopus* and perhaps other hatchlings may hang from a mucous strand from this gland immediately after hatching (Bles 1905; also see Sive and Bradley 1996). Adhesive glands in *Xenopus laevis* first appear at about Gosner 18 (Nieuwkoop/Faber 15, early neurula), reach full development at stages 22–23 (35–36; see Drysdale and Elinson 1993), functionally degenerate by Gosner 25 (49), but persist as pigmented remnants posterolateral to the oral disc as late as Gosner 27–28. Lieberkund (1937) and Thiele (1888) surveyed the gross and histological structure of these glands in European tadpoles.

External gills (see chap. 5) are present in most tadpoles for a short period (Gosner 19–24) in early development. External gills appear as short nodules or long, branched fimbriae and their morphology may reflect the developmental environment. Løvtrup and Pigón (1968) discussed the influence of O_2 and CO_2 concentrations on their development, and there is some evidence of similar gill morphology among related species. Gills usually are on two visceral arches. Tadpoles of *Boophis, Mantidactylus* (Blommers-Schlösser 1979a) and a number of microhylids lack external gills.

Ciliated cells scattered throughout the epidermis of anuran embryos (Steinman 1968; see fig. 3.10) may have the following functions: removing microorganisms and impurities from the epidermis, transporting newly hatched embryos, providing respiratory ventilation of the body surface while in the egg as well as after hatching, and reducing the drag coefficient (Nachtigall 1982). Embryos move frequently within the egg jellies, and hatchlings move at surprising speeds via these cilia while lying on their side. Ciliated epidermal cells produce a cephalocaudal current along the embryo in the egg and after hatching (Kessel et al. 1974).

The hatching gland is a Y-shaped array of unicellular glands on the top of the head; the open end of the series faces anteriorly (Drysdale and Elinson 1993; Ohzu et al. 1987). Presumptive cells of the gland are present by the end of gastrulation, and differentiated cells start to form after neurulation is completed. These glands produce an enzyme that dissolves or weakens the egg jellies and facilitates hatching (Carroll and Hedrick 1974; K. W. Cooper 1936). Yoshizaki (1973, 1974, 1975, 1991), Yoshizaki and Katagiri (1975), and Yoshizaki and Yamamoto (1979) discussed the ontogeny of these glands and their enzymes (also see Gollman and Gollman 1993a; Noble 1926a).

Morphology and Ontogeny: Oral Apparatus

Terminology

Terminology used to describe certain features of the oral apparatus (fig. 3.6) is confusing and sometimes inappropriate. The terms "mouth" and "oral" have multiple usages, and clarification is required to avoid confusion and apparent contradictions. Anatomically, "mouth" refers specifically to the opening that forms by the invagination of the stomodeum, anterior growth of the archenteron, and eventual rupture of the oropharyngeal membrane (Cusimano-Carollo et al. 1962; Watanabe et al. 1984). The mouth provides access from the outside to the buccal or buccopharyngeal cavity. Use of "mouth" to refer to the oral disc is incorrect, and its use to refer to the buccal cavity is colloquial. We use "oral" as a general term to refer to the oral disc and its associated keratinized and soft structures that are situated adjacent and external to the mouth. Collective terms for these structures are oral apparatus and mouthparts. Restricting usage to these definitions means that papillae on the oral disc external to the mouth have incorrectly been called buccal (Cei 1980) or peribuccal (Lamotte and Lescure 1989) papillae. Likewise, papillae on the roof and floor of the buccal cavity confusingly have been called oral papillae (Wassersug 1980). In the latter case we recommend that "buccal" or "buccopharyngeal" are more appropriate terms to refer to those papillae and other structures on the roof and floor of the buccopharynx.

The infrequently used terms "aboral" (opposite, away, or distal from the mouth) and "adoral" (same, toward, or proximal to the mouth) can be interpreted in several ways. For describing relative locations of structures on the face of the oral disc, we prefer "anterior/posterior" relative to the longitudinal axis of the body and "proximal/distal" relative to a landmark (usually the mouth or jaw sheaths) and reserve aboral (= opposite) and adoral (= same) for denoting the back and face of the oral disc, respectively. Orientation of the oral disc relative to the longitudinal axis of the tadpole varies among tadpoles and is correlated with habitat and feeding mechanics; unfortunately, this variability complicates detailed descriptions. For example, the oral disc is positioned (see fig. 12.1 and other examples in chap. 12) ventrally (about 180°) in suctorial forms, anteroventrally (about 40°) in many tadpoles, terminally (90°) in some carnivores and some forms with reduced oral discs, and upturned or umbelliform (more than 90°) in certain surface feeders. As a result, the leading edge of the anterior labium may be anterior (oral disc ventral), dorsal (oral disc tèrminal), or posterior (oral disc umbelliform); and the trailing edge of the lower labium would be posterior, ventral, and anterior, respectively. In addition, a tadpole can attain many positions in the water. Understanding the orientation of the oral disc is an essential component of what we consider to be a satisfactory tadpole description and essential to interpreting data from other studies (e.g., feeding studies). To facilitate comparisons and refer to the location of structures of the oral disc without appearing contradictory, and certainly confusing, we have adopted an unspoken convention of thinking of the oral disc as if it were a published figure in face view with anterior (= dorsal, upper) at the top and posterior (= ventral, lower) at the bottom, regardless of the actual location of the disc on the tadpole or of the position of the tadpole in the water.

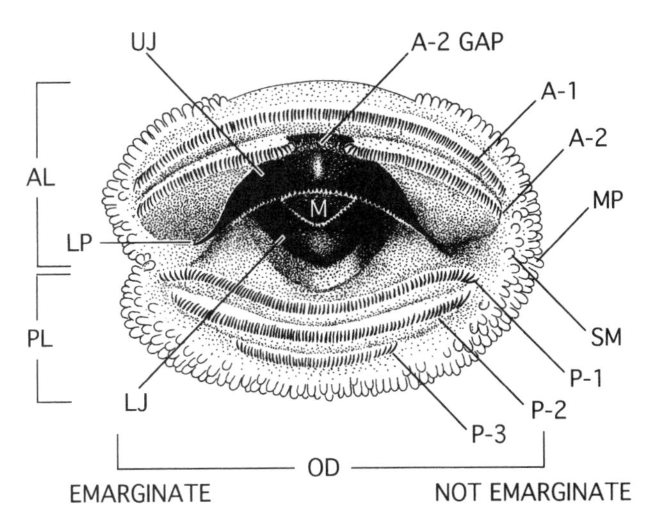

Fig. 3.6. Oral apparatus of a tadpole showing emarginate *(left side)* and not emarginate *(right side)* conditions of an oral disc. Abbreviations for descriptive terminology: AL = anterior (upper) labium; A-1 and A-2 = first and second anterior (upper) tooth rows; A-2 GAP = medial gap in second anterior tooth row; LJ = lower jaw sheath; LP = lateral process of upper jaw sheath; M = mouth; MP = marginal papilla; OD = oral disc; PL = posterior (lower) labium; P-1, P-2, and P-3 = first, second, and third posterior (lower) tooth rows; SM = submarginal papilla; and UJ = upper jaw sheath.

We strongly recommend that the term "denticle" be forever struck from the tadpole lexicon because of confusion resulting from its multiple usages for different structures of tadpoles and other organisms as well: labial teeth (e.g., Inger 1983), cusps on labial teeth (e.g., Faier 1972), serrations on jaw sheaths (e.g., Ruibal and Thomas 1988; M. H. Wake 1993a), and other small pointed structures. The preferred term is labial teeth and their structure and formation are much more complicated than the simple structure implied by "denticle" (e.g., Altig 1970; Boulenger 1892). By extension, the serrations on the head of an individual labial tooth are termed "cusps." Because a consistent terminology decreases the likelihood of confusion, and when coupled with long-term usage provides stability and understanding, a unique terminology (e.g., Van Dijk 1966) seems unwarranted or is premature pending better understanding of the distribution of morphological traits and their homology.

The jaws cartilages of tadpoles (chap. 4) are without basal articulation and significant adult derivatives and thereby differ from those of all other vertebrates. The sheaths that overlie the cartilaginous jaws of tadpoles are transitory and have a columnar histological structure different from that of similarly keratinized structures in birds, turtles, and some salamander larvae. In contrast, the beaks of other vertebrates (e.g., turtles and birds) are adult structures that overlie basally articulating jaws comprised of dermal bone. Because the sheaths that cover the jaws of tadpoles and the keratinized coverings (= rhamphotheca) of the jaws of other animals surely are not homologs, we believe they warrant a distinct term and recommend "jaw sheath." The so called jaw sheaths found in certain salamander larvae (e.g., *Ambystoma* and *Siren*) have not been examined histologically, but they appear to differ structurally from either of the previous two cases; although they overlie dermal bones, the specific bones involved are not entirely the same in larval salamanders as in other taxa. Also, the jaw sheaths of anuran larvae typically are not pointed and do not project, as is implied by the usual connotation of beak. Finally, the term "beak" has been used to denote the prominent medial convex projection of the margin of the upper jaw sheath of certain carnivorous tadpoles; this usage adds additional confusion, and we recommend incised as more appropriate descriptive terms for this jaw sheath morphology.

The conventional use of length to refer to the longer and usually longitudinal dimensions of a structure poses a conflict when describing jaw sheaths because their greatest dimension is transverse. For convenience we retain length for the transverse measurement. This dimension is usually measured or described without considering curvature because it is the only measure that can be obtained without dissection. Jaw length as appropriate for a mammal (= gape, the basal terminus of the jaw sheath in a straight line to the peak of the front curvature) has never been suggested as a measurement for a tadpole, again perhaps because it could not be measured on an intact specimen and because the jaw cartilages do not articulate basally. By retaining length for the longest but transverse dimension and width for the shorter, longitudinal dimension, jaw sheaths may be described and measured as follows: long/short (transverse measurement), wide/narrow (basal boundary of keratinization to serrated edge), and thick/thin (front, oral or adoral to back, buccal or aboral surface). Thickness has not been examined in many tadpoles.

The jaw sheath is only the keratinized part of the jaw covering and is usually pigmented. Descriptions should be based only on the sheath and should not include jaw and associated soft tissues. In those rare cases where the sheath is not pigmented (e.g., *Scaphiophryne*), determining the limits of the sheath is difficult. If the excised jaws of tadpoles are placed in water, the jaw sheaths will detach (soon if specimen not preserved, several days for preserved specimens) from their underlying cartilages, thereby facilitating detailed observations.

Oral Disc and Papillae

The oral disc of a typical tadpole is round to transversely elliptical. The poorly delimited upper (anterior) and lower (posterior) labia of a fully formed disc in nonsuctorial tadpoles are delineated by a fold that during closure (as usually occurs in preserved specimens) develops along a line that approximates the angles of the jaw cartilages. The degree to which parts of the upper and lower labia contribute to the entire disc determines the disc shape. For example, enlargement of lateral portions of the disc produce a bi-triangular shape, as seen in tadpoles of *Megophrys*. In tadpoles of the hylid *Scarthyla* (Duellman and De Sá 1988) and members of the *Scinax rostrata* group (e.g., McDiarmid and Altig 1990), the exceptionally short P-3 lies at the end of an armlike modification of the medial part of the posterior labium that rests in a gap in the ventral papillae. The lower labium commonly is larger than the upper labium and more of its area is free from the body wall. Tadpoles of species with many tooth rows have larger oral discs than those with fewer rows. The lower labium particularly may extend a considerable distance posterior to the last tooth row in some tadpoles (e.g., some *Ansonia*, some *Litoria*, and *Staurois*). The labia consist of loose connective tissue covered with an epidermis of varied thickness. Although the disc is more rigid in lotic than in lentic forms, the structural modifications that contribute to these differences in rigidity have not been determined. Although the cores of papillae and bases of tooth ridges sometimes are pigmented, the face of the disc usually lacks pigment, and mucous glands are absent. An inverted U-shaped depression or transverse slit commonly, perhaps always, occurs within the medial gap of the anterior tooth row adjacent to the upper jaw sheath. Neither the function nor development of this structure has been studied, and its occurrence in those species with a complete proximal, upper row is not known.

Most commonly the oral disc has all but its anterior margin free from the body wall. In tadpoles with a dorsal gap in the marginal papillae, the upper labium essentially is a continuation of the snout; in species with complete marginal papillae (e.g., suctorial forms) the entire margin is free. In some rheophilous *Litoria* from the Australopapuan region, the snout and anterior margin of the oral disc are separated by a crevice that appears to be covered by a thin membrane (personal observation). The function of this morphology is unknown but surely it signals a novel structure in stream-dwelling pelodryadine tadpoles as compared to stream-dwelling tadpoles of neotropical hylids. The back of the free portion of the oral disc, which is primarily in the lower labium, is typically smooth. In tadpoles of *Ceratophrys*, this surface is often fluted.

An umbelliform or upturned oral disc appears as a convergent trait in tadpoles of some arthroleptids, dendrobatids, hylids, mantelline rhacophorids, megophryids, and microhyline microhylids (fig. 9.5). Most umbelliform tadpoles occur in the backwaters of lotic systems. Tooth rows and jaw sheaths are reduced to absent in these forms, and large ridgelike papillae that project radially from the mouth are common. When present, marginal papillae are complete but greatly reduced in size. Morphological comparisons among umbelliform tadpoles have not been made, but cursory examinations show that both labia or primarily the lower labium form most of the disc. The bi-triangular disc of *Megophrys* is the most bizarre; when the tadpole submerges, the lateral, pointed flaps of the oral disc by their own elasticity fold medially; when the tadpole rises to the surface to feed, the flaps unfold to form a bi-triangular oral disc with the longer dimension oriented transversely. Even though functions other than feeding have been suggested for an umbelliform disc (e.g., stabilization among vegetation, float, etc.), observations on certain forms suggest to us that feeding on materials caught in the surface meniscus seems to be the most likely functional explanation for all umbelliform discs. The umbelliform disc tadpole of *Microhyla heymonsi* has a large, infolded semicircular structure at each corner of the mouth that appears derived from the lower labium.

The oral disc of some suctorial tadpoles does not fold shut, even in preserved specimens, and the discs of rheophilous forms commonly have lateral pleats (see emarginate below). These pleats presumably permit the seal between the disc and substrate to be maintained while the disc margin is expanding during the extension-retraction cycles of oral locomotion (C. L. Taylor and Altig 1995). Other folds and flexures commonly seen in preserved specimens, primarily along the disc margin, are caused by differential contractions of muscles and different types of connective tissues and may not occur in living tadpoles.

Emarginations, which are marginal indentations and not simply folds of a uniform margin, occur at various sites on the oral disc (fig. 3.6). Lateral emarginations are the most common (e.g., most *Bufo* and *Rana*) but emarginations may occur dorsally, ventrally, and or ventrolaterally. The disc often must be opened to distinguish an emargination from a pleat or fold.

Margins of the labia usually are defined by papillae that vary in density (number/distance), basal diameter, length, orientation, pigmentation, and shape (pointed vs. rounded tip). These marginal papillae occur in four basic patterns on the disc: complete (no gap), dorsal gap only (taxonomically and ecologically common), ventral gap only (rare), and dorsal and ventral gaps (most bufonids and a few hylids, mantellines, ranids, and rhacophorids). The condition of complete marginal papillae is most common in tadpoles that attach to rocky surfaces in fast-moving streams; it occurs uncommonly in a few nonrheophilous forms (e.g., *Anotheca*). The closely spaced, short, and numerous papillae presumably aid in forming a tight seal between the oral disc and the irregularities of the substrate. Van Dijk (1981) suggested that the ventral gap in the marginal papillae of bufonid tadpoles serves as a weirlike controlling device during feeding. Other explanations to account for gaps in marginal papillae have not been offered, but a large phylogenetic component seems obvious. In some species, the size of the ventral gap is associated with the length of the adjacent tooth row, but notions that these patterns may be developmentally linked have not been pursued.

We have noted obvious qualitative differences in the size (e.g., smaller in hylids than ranids) and density of marginal papillae among tadpoles but have found few published notices (e.g., I. Griffiths and De Carvalho 1965) of this variation. Generally the marginal papillae are relatively similar in size around the oral disc, but some ventral papillae may be greatly elongated (e.g., in *Hemisus,* some *Phrynobatrachus,* and some *Rana*). Elongate ventral papillae are rarely bifid (Channing 1978), and Van Dijk (1966) noted that the length of the long ventral papillae in *Phrynobatrachus* may decrease ontogenetically. Marginal papillae typically occur in one (uniserial) row but discs bordered by two (biserial) or several (multiserial) rows are not uncommon. Sometimes the bases of ventral papillae are slightly offset toward the back of the labium. The papillae forming a single (uniserial) row are properly aligned, but those that occur in two or more rows have their bases slightly offset and appear double, or the bases are in one series but alternate papillae project in different directions (also emulating a double row). These alternative patterns seemingly represent different solutions to the same problem and for clarity and understanding, they should be explained in descriptions.

Submarginal papillae, which occur on the face of the disc and sometimes grade centripetally from the marginal papillae, vary in density, distribution, shape, and size. The bases of submarginal papillae usually are circular but sometimes (e.g., *Nyctimystes dayi, Phasmahyla guttata,* and several umbelliform types) are elongate in a longitudinal or radial direction; S. J. Richards (1992) suggested that these types of papillae may be secretory. A patch of submarginal papillae commonly occurs near the lateral ends of the lower tooth rows, particularly in lotic tadpoles, and single rows of widely spaced, large papillae may occur distal to the first upper and last lower tooth rows in other forms. Some species of *Cardioglossa* and *Hyalinobatrachium eurygnatha* have a row of papillae that crosses the proximal face of the lower labium and is continuous with the lateral marginal papillae. In these situations, papillae lie in front of the recessed jaw sheaths. It is tempting to speculate that these papillae may be a vestige of a tooth ridge(s).

The oral disc and its presumed derivatives are reduced in various ways in tadpoles in several different lineages (e.g., some species groups of neotropical hylids, microhylids in general, pipids, and a few ranids) and feeding modes (see fig. 12.2). Some tadpoles have few (e.g., *Hyla marmorata* group) to no (e.g., *leucophyllata, microcephala,* and *parviceps* groups of *Hyla* spp., *Occidozyga* sp.) labial teeth but have normal or unusual (e.g., *Otophryne pyburni*) jaw sheaths. In these forms, the oral disc is reduced to differing degrees, typically to a U-shaped yoke around the mouth, and the marginal papillae appear as slightly differentiated mounds or are absent. The jaw sheaths, which sometimes are massive for the size of the tadpole, often are deeply recessed. During feeding, the snout is deformed and the jaws protruded to bring the sheaths in contact with the substrate. This configuration presumably is what prompted Lavilla (1990; *Hyla nana*) to describe what he called an "oral tube" that was visible only during feeding. The report (I. Griffiths 1963) of minute keratinized "denticles" in the oral apparatus of *Pseudhymenochirus* needs verification.

A few tadpoles other than microhylids, pipids, and rhinophrynids lack both labial teeth and at least keratinized jaw sheaths. Tadpoles of *Litoria subglandulosa* (Tyler and Anstis 1975) have complete marginal papillae, abundant papillae over the face of the disc, and small, unpigmented jaw sheaths. Six large papillae occur in a transverse row above, and a flat, white structure lies between the papillae adjacent to the midline. The tip of this structure is divided into 4–7 toothlike structures, each bearing 1–4 black, hairlike filaments, some of which are branched. Below this structure there are 2–6 papillae with lightly keratinized tips. Our unpublished data show some surprising similarities between the mouthparts of the tadpole of *Mantidactylus guttulatus* (mantelline rhacophorid; jaw sheaths absent) and those of *L. subglandulosa*. An unidentified *Boophis* has an immense oral disc that lacks keratinized structures and consumes exclusively clean-looking pieces of sand.

Species without keratinized mouthparts often have different or unusual soft structures. The paired, semicircular labial flaps of many microhylids presumably are homologs of the upper labium of a typical oral disc. These labial flaps, with or without papillate edges, are suspended in front of the mouth and separated by an inverted U-shaped medial notch (see Donnelly et al. 1990a for review). The flaps may be strongly semicircular, short and essentially without the medial notch (i.e., upper lip is transversely straight), excessively long and almost rectangular, or reduced to small protuberances around an extremely wide medial notch. A few microhylid tadpoles (e.g., *Nelsonophryne* from the New World and some *Microhyla* and *Scaphiophryne* from the Old World) have other structures that may be derived from the upper, lower, or both labia (Donnelly et al. 1990a). The unconventional microhylid tadpoles of *Nelsonophryne* (microhyline) and *Scaphiophryne* (scaphiophrynine with nonpigmented jaw sheaths) have oral discs with papillae. Except for vestigial structures that presumably are derived from typical jaw sheaths, the macrophagous carnivorous tadpole of *Lepidobatrachus* has an immense, slit-shaped mouth and no other mouthparts.

The superficial similarity of the soft mouthparts of rhinophrynids and pipids diminishes upon closer examination. Both *Rhinophrynus* and *Xenopus* have slight folds at the corners of the mouth that may represent vestiges of an oral disc (Orton 1943; Thibaudeau and Altig 1988), but the barbels of the two tadpoles are not homologous (Cannatella and Trueb 1988a). The single long barbel at each corner of the mouth of *Xenopus* has muscle fibers attached to an internal cartilaginous support and is mobile. The shorter and more numerous barbels in tadpoles of *Rhinophrynus* lack these characteristics; they also seemingly vary in number and length ontogenetically and certainly between cohorts at one site and between sites. Tadpoles of *Rhinophrynus* also have a single papilla-like projection at the symphysis of the lower jaw cartilages that probably is not a barbel.

Tooth Ridges and Labial Teeth

Transversely arranged tooth ridges form on the face of the oral disc during early ontogeny (Thibaudeau and Altig 1988; see Cherdantseva and Cherdantsev 1995). The number, length, position, shape (curved vs. straight), and spacing of these ridges on the disc vary interspecifically and ontogenetically. Darkly keratinized labial teeth that develop from mitotic sites in the tooth ridge are arranged linearly in one or more rows along the ridge. The tooth ridges of pond tadpoles are tall and have narrow bases and wide interridge valleys that are situated about equidistant between ridges; the ridges are quite flexible, and teeth generally can be removed easily. Rheophilous and particularly suctorial forms have shorter, flat-topped ridges with narrow interridge valleys that lie immediately behind the next proximal ridge. In all cases, the ridges become shorter and more closely spaced in the distal rows, and this is particularly pronounced in species with many tooth rows. Judged by the distribution of polarizing birefringence, the tooth ridge contains considerable connective tissue proximal and distal to a mitotic zone in the base of the tooth ridge that produces the labial teeth (figs. 3.7C, F).

Although known for some time (e.g., Schulze 1892; Weber 1898), the extrinsic musculature of the oral disc has not been well documented (e.g., Carr and Altig 1991; Gradwell 1968, 1972b, c; Noble 1929; Starrett 1973). Undoubtedly many unknown variations remain to be described, and several errors or omissions need to be clarified. Fibers of the m. mandibulolabialis (fig. 3.7G; see chap. 4) originate on the basal, ventrolateral surface of Meckel's cartilage. The m. mandibulolabialis superior serves the upper labium, and the m.m. inferior serves the lower labium (Carr and Altig 1991; McDiarmid and Altig 1990). When present, fibers of the muscle insert at the bases of the lateral quarters of the valleys between tooth ridges and at the bases of marginal papillae lateral to the tooth rows. Fibers of the m. mandibulolabialis usually do not extend distal to the most distal tooth row in either labium, although they do so in some suctorial *Litoria*. Examination of 24 species (Carr and Altig 1991) demonstrated little ontogenetic but considerable interspecific variation in these muscles. Typically, slips of both muscles are present, but in some species only the m. m. inferior has been detected. Contractions of these muscles apparently change the shape particularly of the lateral portions of the oral disc and of the tooth ridges. Distortion of the tooth ridges presumably rotates the labial teeth distally and effects a better contact with the substrate. Although the sequence

Fig. 3.7. Representative photomicrographs of labial teeth and jaw sheath of various tadpoles. (A) Multiserial tooth row (A-1) and biserial tooth rows (A-2 and A-3) in *Ascaphus truei* (Ascaphidae). (B) Tooth series in an intact tooth row of *Hyla chrysoscelis* (Hylidae). (C) Cross section (viewed with polarized light) of two posterior tooth rows *(mouth at bottom)* of *Hyla gratiosa* (Hylidae) showing two tooth series *(black structures)*, each surrounded by patches of connective tissue *(pale areas)* within the tooth ridge (P-2 and P-1) and with mitotic (tooth-forming) cells *(dark areas)* near their bases; slips of the mandibulolabialis muscle *(pale structures)* also visible *(lower left)*. (D) Labial teeth in nonnormal hillocks of tissue and on marginal papillae on oral disc of *Scaphiopus holbrookii* (Pelobatidae). (E) Lateral view of tooth series of about 15 teeth (series dissected from posterior tooth ridge, erupted tooth to *lower right*) of *Ascaphus truei* showing lengthy extensions of anterior base of each tooth sheath with associated connective tissue (white in polarized light). (F) Part of a posterior tooth row of *Ascaphus truei (distal perspective, mouth to right)* showing extensive connective tissue (white in polarized light) in tooth ridge. (G) P-2 and P-3 tooth rows of *Hyla gratiosa* (Hylidae) showing *(proximal perspective)* fibers of the mandibulolabialis inferior muscle (white in polarized light) inserting on tooth ridges and marginal papillae lateral to tooth rows. (H) Serrations on jaw sheath *(buccal surface to left)* of *Rana catesbeiana* (Ranidae). (I) Lateral view of five labial teeth *(distal surface or back to right)* from last posterior (multiserial) row of *Ascaphus truei.*

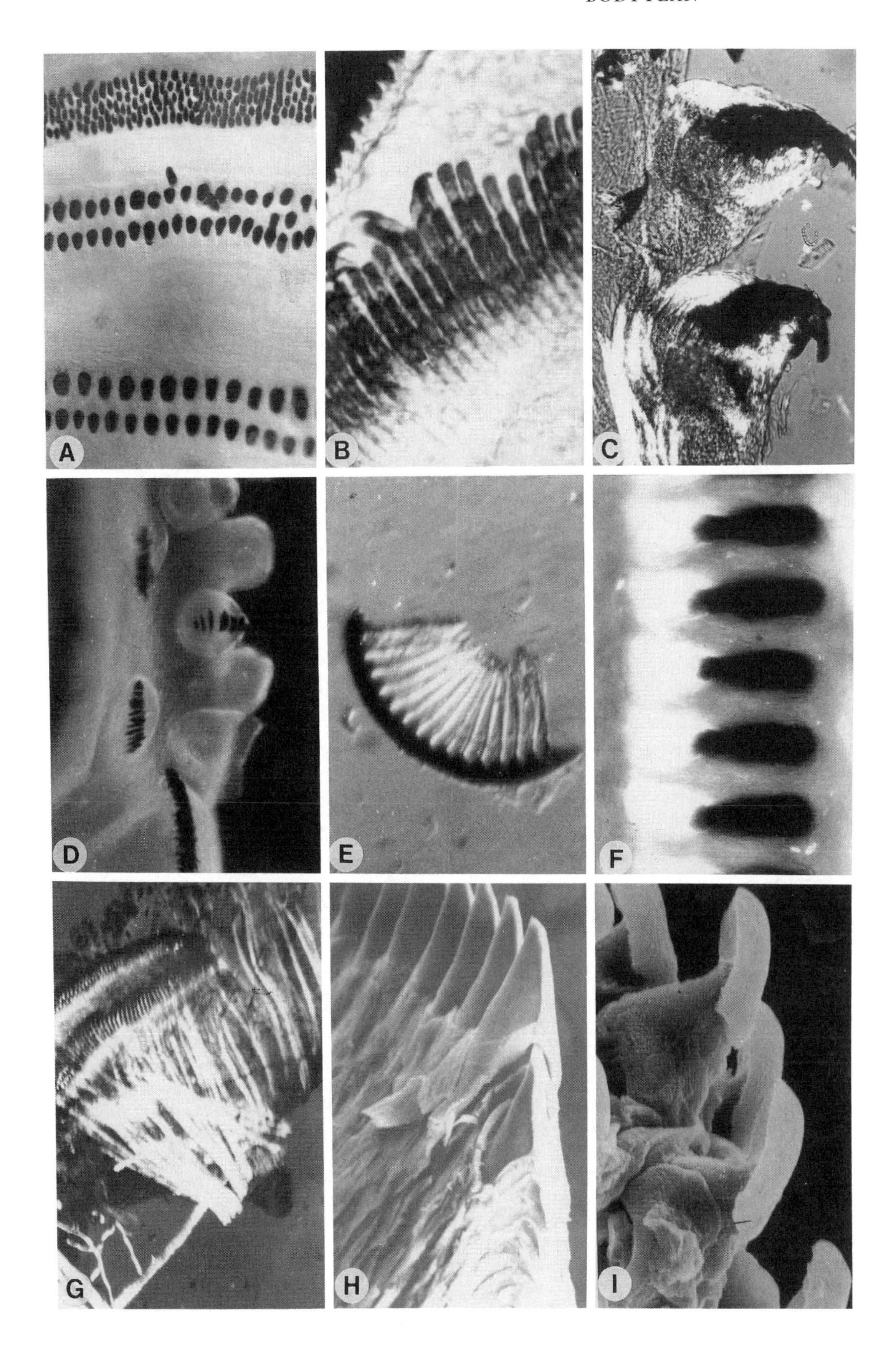
A
B
C
D
E
F
G
H
I

and timing of these muscle contractions during a feeding cycle are not clear, it seems that the muscles relax soon after maximum oral disc excursion and as closure begins (C. L. Taylor et al. 1996). The m. intermandibularis posterior (Gradwell 1972b) also may aid in tooth reorientation. Gradwell (1968) noted that the m. mandibulolabialis is innervated by the trigeminal nerve. We suspect that lymphatic spaces in and near the oral disc (chap. 5) also may function in the operation of some of the mouthparts.

The number and arrangement of tooth rows on the oral disc of tadpoles is species specific. The labial tooth row formula (LTRF) is a synoptic representation of this arrangement. A number of systems have been devised for numbering labial tooth rows and designating the rows with medial gaps. Any system is clear to an experienced user, but many systems are not obvious to a neophyte. To us, less cumbersome schemes seem to be more useful. Dubois (1995) and Dubois and Ohler (1994) present the most recent variations. Accordingly, we continue to use the fractional designation (e.g., Altig 1970) to specify the number and gross morphology of tooth row. This system accommodates all row configurations easily, does not use Roman numerals, has been used for quite a long time and is familiar to many workers, and takes less space than formulas that are written in spatial order. Rows on the anterior labium are numbered distal (labial margin) to proximal (mouth). The notation "A-1" denotes the first anterior (most distal from mouth) row, and designations extended sequentially through A-*n* to the row adjacent to the mouth. Rows on the posterior labium are numbered proximal to distal. The first row adjacent to the mouth is P-1, and more distal rows are numbered sequentially through P-*n*. Notations such as "PR-1" and "AR-1," denoting the tooth ridge for tooth row P-1 and A-1, can be extended for any anterior or posterior row if in developmental studies one wishes to refer to tooth ridges distinct from tooth rows. Rows with medial gaps are designated with parentheses, and rows that vary between individuals (gap present or absent) are placed within brackets. A gap in a tooth row is a physical break in the tooth ridge and therefore expressed in the tooth row; teeth occur only on tooth ridges. Even apparent atavistic teeth are positioned on hillocks of tooth ridge tissue (see below and fig. 3.7D). If the gap is quite narrow, as it often is in row P-1, staining (see chap. 2) may be necessary to discern if the ridge in fact is broken. The absence of teeth on part (or all) of a tooth ridge should be noted, but these breaks do not constitute gaps as defined here. Also, sharp bends in tooth rows caused by flexures of the disc in preservation are sometimes interpreted as gaps; one should open the oral disc manually to verify that a gap probably is not present in most of these cases.

The functional and developmental considerations of gaps in tooth rows have not been examined. These gaps may allow larger excursions of the jaw sheaths during feeding; adjacent tooth rows that have only narrow or no gaps perhaps signal smaller jaw excursions during feeding. We assume that most gaps are neither gained nor lost ontogenetically within rows and that in most cases gaps can be used to evaluate row homology.

In summary, a labial tooth row formula (LTRF) of 5(2–5)/3[1] indicates a tadpole with 5 upper tooth rows with medial gaps in rows A-2 through A-5 and 3 lower rows with or without a gap in P-1. An advantage is that this system apparently numbers homologous rows the same in different species and does so more often than other systems. In fact, no system will always number supposed homologs the same because of different patterns of ontogenetic and phylogenetic row alterations (Altig and Johnston 1989).

Some tadpoles, particularly of species of ranids and pelobatids, have accessory tooth rows situated in the lateral areas of the oral disc. The functional importance, developmental association, and homologous nature of these short rows compared to the more typical rows are not known. Two types of these accessory rows are distinguishable. In ranids, accessory rows essentially parallel the lateral ends of the posterior tooth rows; in pelobatids the rows tend to lie off the ends of and are oriented at nearly right angles to the posterior tooth rows. The formulation of Altig (1970) was modified by R. G. Webb and Korky (1977) by placing the number of accessory rows between solidi—5(2–5)/4/3[1]. Other than notations concerning ontogenetic change, few data (e.g., Bragg and Bragg 1958) illustrate the natural variations in tooth formulas. Echeverría and Filipello (1994) found that 77.4% of the tadpoles *Odontophrynus occidentalis* had one formula and the remaining 22.6% had several other formulas. The most deformities (26.7%) were in A-1, while P-1 showed the least (6.7%). Grillitsch and Grillitsch (1989) and C. L. Rowe et al. (1996, 1998a, b) reported similar data on ranid tadpoles.

Tadpoles of *Phrynohyas, Trachycephalus,* and some species of the *Hyla geographica* group share the uncommon feature of having A-1 with a wide median gap but with A-2 entire. In such cases, this row is probably not homologous with A-1 (but A-2 is) of other tadpoles. Certain structures on the oral disc of tadpoles with reduced mouthparts may be nascent tooth ridges without teeth. Bresler (1954) showed that the incidence of abnormal tooth rows (and jaw sheaths) increased with temperature, but all tooth rows were not affected equally. Jaw sheaths seemed to be less affected than tooth rows. Bresler and Bragg (1954) reported variations in tooth rows of five species in five genera, including *Spea bombifrons.* It is interesting that variation in tooth rows appears to be greater among individuals and populations of species of *Spea* than for almost any other tadpoles. Perhaps this is a reflection of the relatively warm aquatic environment and shorter developmental period characteristic of these species, but a phylogenetic influence cannot be ruled out. A review of tooth row variability in other pelobatids and an experimental test of the effect of different rearing temperature on tooth row variability would provide some insight. It also is tempting to speculate that this variation somehow is tied to the drastic ontogenetic changes that occur in cannibal morphotypes of this group.

Altig and Johnston (1989) proposed a balance value (i.e., the difference between the number of upper and lower tooth rows) to aid in understanding and comparing suspected functional differences in tooth row formulas among species.

They assumed that differences in balance values reflected differences in feeding mode. A tadpole with more tooth rows on the upper labium than on the lower labium has a positive imbalance (e.g., LTRF of 5/3, BV = +2); one with more rows on the lower labium than the upper has a negative imbalance (2/3, BV = −1); and one with equal number of rows on both labia is balanced (5/5, BV = 0). A balance value of −1 in the formula 2/3 is by far the most common. The extremes are +7 (10/3, *Boophis* sp.) and −11 (4/15, *Heleophryne* sp.), and not all intermediate values occur. The distribution of values shows that 88% of 627 species surveyed (51% had a LTRF of 2/3) had balance values between +2 and −2. A tadpole with a strong negative imbalance (e.g., LTRF 4/15; BV = −11) surely must use their labial teeth differently than one with a strong positive value (LTRF 10/3; BV = +7). A comparison of foraging behavior and feeding efficiency among tadpoles with the same total number of tooth rows (e.g., 12) but different balance values (LTRFs of 4/8 vs. 6/6 vs. 8/4) would provide needed data. It is interesting that the tadpole with the largest number of tooth rows (LTRF 17/21; *Hyla* sp.) has a nearly balanced formula.

Except for the pronounced ontogenetic changes characteristic of tadpoles of species with many tooth rows, the number and arrangement of labial tooth rows (LTRs) after about Gosner stages 25–26 are stable enough to be definitive of the species. A LTRF does not provide information about the position on or spacing between tooth rows on the face of the oral disc. Row spacing has not been examined, perhaps because the flexible oral disc makes it difficult to standardize the measurements, but differences apparently exist. An unidentified suctorial hylid tadpole from Papua New Guinea with a LTRF of 2/3 has an unusual arrangement of tooth rows on the lower labium; the space between P-1 and P-2 is over twice the distance between the other two rows. Large areas of the disc in certain tadpoles (e.g., *Duellmanohyla,* some *Taudactylus;* species with large oral discs and short tooth rows situated close to the jaw sheaths) are without teeth. Tadpoles of some centrolenids (personal observation) and suctorial forms of *Litoria* and *Staurois* (Inger and Wassersug 1990) have a large expanse of oral disc beyond the distal posterior row.

Although we have a poor appreciation of the nature of and limits to intraspecific variation, tadpole morphology at the generic level often is remarkably consistent. Interspecific variation in some taxa (e.g., most bufonids) is much less than in others (e.g., hylids, leptodactylids). Other than the cannibalistic morphotypes of *Spea,* true polymorphism, not just pronounced variation, is rare or poorly documented in tadpoles. Annandale (1917, 1918), Berry (1972), E. R. Dunn (1924), Khan and Mufti (1994), and M. A. Smith (1917) talked about notable differences within species, but the causes of the differences are unclear. I. Griffiths and De Carvalho (1965) used some of these cases as examples of dimorphisms that must be considered with caution when using larval features in systematic studies. Although the premise is valid, the underlying assumptions that the tadpoles were correctly identified and that the correct number of species was involved in all cases went unquestioned. Likewise, M. Kaplan (1994) suggested that variations of labial tooth rows in *Hyla minuta* prohibited the use of these features in phylogenetic studies; the problem here most certainly centers on the variation representing multiple species. A smattering of reports (e.g., Annandale 1917; Korky and Webb 1991; Lavilla 1984a, b; J. M. Savage 1960; M. A. Smith 1917) described apparent geographic variation in mouthparts, and a few others (Annandale 1917, 1918; Berry 1972; Caldwell 1982; E. R. Dunn 1924; Khan and Mufti 1994; M. A. Smith 1917) have documented notable examples of other types of morphological variation (e.g., coloration and pattern) within species. The paucity of data on both the nature of and limits to intraspecific variation raises questions about the reliability of some tadpole identifications. We frequently ask ourselves if observed differences are an indication of geographic variation within a species or a reflection of differences between species.

Tadpoles of species of *Spea* are notorious for having variable mouthparts beyond those induced by cannibalism (e.g., Bragg 1956; Bragg and Bragg; 1958; Bragg and Hayes 1963; Bresler and Bragg 1954; Hampton and Volpe 1963; Orton 1954; Potthoff and Lynch 1986). Yet, the relationship between a reported missing tooth row or patch of teeth and the presence of the underlying tooth ridge (see Bresler 1954) seldom has been explored. A missing tooth row may be the consequence of a missing tooth ridge (i.e., developmentally induced) or merely reflect the loss of teeth on the ridge (i.e., attributable to some postembryonic response). Too often, the condition is masked by an imprecise or incomplete description. If we are to make progress in understanding the nature of presumed geographic variation, this needs to be corrected!

Although intraspecific variation is poorly documented, the morphology of closely related species often is remarkably consistent. We have suggested (e.g., chap. 12) that this consistency occasionally signals the need for taxonomic reevaluation. Exceptions to this generic uniformity are known; the dramatically different tadpoles of *Litoria citropa* and *L. subglandulosa* (Tyler and Anstis 1975) and *Hyla chrysoscelis* and *H. avivoca* are examples.

In some tadpoles with an exceptionally short P-3 row, such as *Pseudacris triseriata* and *Pseudacris crucifer* (Gosner and Black 1957a), the third row is frequently absent, and a gap of equal size in the marginal papillae below the expected position of P-3 may be present. A LTRF of 2/3 is typical in most members of the *Rana pipiens* group, but a LTRF of 3/3 is not uncommon; the occurrence of the unconventional formulas of 2/4 and 3/4 in *Rana berlandieri* (Bresler and Bragg 1954) needs further study. Cannibal morphotypes of *Spea* have larger, differently shaped jaw sheaths and few to no teeth compared to the conspecific omnivorous morphs.

Aberrant and displaced tooth rows are not uncommon; parts of rows may be absent, rows may merge at any point or connect laterally, and teeth may form small circles within a row. Gosner (1959) attributed these alterations to wear, but this seems unlikely. Documentation of wear is not as common (e.g., Tubbs et al. 1993) as one might expect from structures that are being replaced continually. Temperature, nutrition, disease, and the substrates on which the teeth act may contribute to aberrant patterns. Cultured tadpoles often have higher incidences of abnormal tooth rows (and in specific patterns; Grillitsch and Grillitsch 1989), including a complete lack of keratinized structures, than field-collected specimens. Some taxa (e.g., *Bufo*) seem more likely to show such alterations than others (e.g., *Rana*). In 8 of 10 simultaneously field-collected specimens of *Heleophryne purcelli* aberrant teeth appear in the two proximal upper and three proximal lower rows. Apparently teeth were not shed as replacement teeth were produced, and the result was a remarkable array of long flexible strands of up to 40 teeth extending above the surface of the tooth ridge at the site where a single tooth would normally be. This condition surely prevented the teeth from working properly, but the tadpoles appeared healthy. Interpretation of some kinds of anomalies may provide hypotheses about the formation, evolution, and functions of mouthparts.

Each keratinized labial tooth (fig. 3.8) is derived from cells in the base of the tooth ridge and consists of three indistinct regions: a distal head, an intermediate body, and a basal, hollow sheath (Gosner 1959). A fracture plane in the body often is evident. At this point, the tooth head can break free of its body and expose the head of the next replacement tooth interdigitated in the sheath. The head of a tooth may be straight or variously curved and spoon-shaped (i.e., bluntly rounded and flattened; fig. 3.7I), thin with a sharp point (noncusped), or of various shapes with 2–18 terminal cusps. Even though the head often is flattened front to back, the body is the reverse—wide in lateral view and narrow in anterior view. Strands of connective tissue anchor the hollow sheath of the tooth within the tooth ridge; attachment is most prominent at the front (proximal to the mouth) base of each tooth. In most teeth, the front and back bases of the sheath in cross section are of comparable shape, but in suctorial forms, the front base of the sheath often extends considerably toward the mouth. This extension protrudes (fig. 3.7E) into the tissue of the tooth ridge proximal to the row and presumably acts as a brace to keep the tooth series and row from folding backward. The imprint of the nucleus of the original cell can be seen in the wall of the sheath of a tooth cleared in potassium hydroxide.

Tips of the labial teeth on both labia face the mouth—that is, in opposite directions. Usually two to three fully formed replacement teeth are successively interdigitated in the sheath of the preceding tooth below the erupted tooth. The amount and style of interdigitation appear to vary, but data are lacking. An erupted tooth and all its replacements are termed a tooth series (figs. 3.7B, E), and the population of tooth-forming cells is renewed continuously in a mitotic zone near the base of a tooth ridge (fig. 3.7C; Beaumont and Deunff 1958; Deunff and Beaumont 1959; Luckenbill 1964, 1965; Tachibana 1978). As the presumptive tooth cell moves toward the top of the tooth ridge, a replacement tooth slowly forms from the head toward the sheath. A tooth row comprises numerous tooth series aligned side by side within a tooth ridge. If the tissue of the tooth ridge is sufficiently transparent, the tooth series can be seen through

Fig. 3.8. Photomicrographs showing variations in labial teeth and their marks. (A) Short-cusped, moderately curved teeth of *Megaelosia goeldii* (Leptodactylidae). (B) Long-cusped, strongly curved teeth in single row/tooth ridge of *Hyla biobeba* (Hylidae). (C) Cusped, relatively straight tooth *(emergent in upper left)* and replacement series of *Leptodactylus chaquensis* (Leptodactylidae). (D) Noncusped (pointed), short-sheathed tooth of *Spea bombifrons* (Pelobatidae). (E) Negative cast of mark *(upper jaw sheath entered on left)* made on agar-based food by feeding tadpole of *Rana sphenocephala* (Ranidae). (F) Cusped, biserial teeth of *Bombina variegata* (Bombinatoridae). From Altig and Johnston (1989); reprinted by permission of *Herpetological Monographs*.

the tissue. What appears as a single long tooth with sequential light and dark areas are actually the replacement teeth stacked in a series. Sometimes an older tooth sticks to a newly erupted one and makes it appear longer than others in the row, and other times a tooth looks longer because of the difficulty of interpreting the extent of the soft tissue of the tooth ridge on the shaft. However, closely adjacent teeth are comparable in size and shape. Tooth size may vary in different parts of a row or in different rows (e.g., fig. 3.7A, I; Gosner 1959), but the usual pattern is a gradual shift to smaller size laterally within a row and distally among rows. Smaller specimens have smaller teeth than larger specimens. Meager data suggest that rows increase in length without changing tooth density; where and in what pattern teeth are added remains a mystery.

Most tadpoles have a single tooth row per ridge (= uniserial), but tadpoles with bi- (e.g., discoglossids) and even tri- or multiserial (fig. 3.7A) rows are known. Tooth densities of 30–100/mm, with higher counts in lotic tadpoles, are typical, and the number of cusps range from 0 (simple spike) to at least 18. Biserial sections in typically uniserial rows are not uncommon. The teeth of *Bombina orientalis* are uniserial at first appearance but become biserial during development. Noble (1929) noted biserial tooth rows in *Hoplobatrachus rugulosus,* but the text and adjoining figure leave doubt if these rows actually are biserial.

Tadpoles of *Xenopus laevis* develop true calcified teeth precociously (J. P. Shaw 1979). True teeth usually appear late in metamorphosis as the maxillae and premaxillae ossify, but in *X. laevis,* tooth buds are present at about stage 33 and start to calcify at stage 37. True teeth begin to form at about stage 36 in tadpoles of *Lepidobatrachus* (Starrett 1973).

Jaw Sheaths

Jaw sheaths are formed by the fusion of palisades of keratinized cells along their lateral margins (Kaung 1975; Kaung and Kollros 1976). Sometimes jaw sheaths have a striated appearance to at least their basal parts. As the serrated edges of the jaw sheaths wear away or break off, mitotic cells in the base of the sheaths continually produce new cells that keratinize in transit to the distal edge. Typically, jaw sheaths have a serrated edge, and the serrations vary in density, orientation (straight medially and angled laterally), presence (rarely absent, e.g., *Osteopilus brunneus* group; Noble 1929), shape (broad-based and short vs. narrow-based and long), and size. In a few cases, 1–3 enlarged serrations occur within

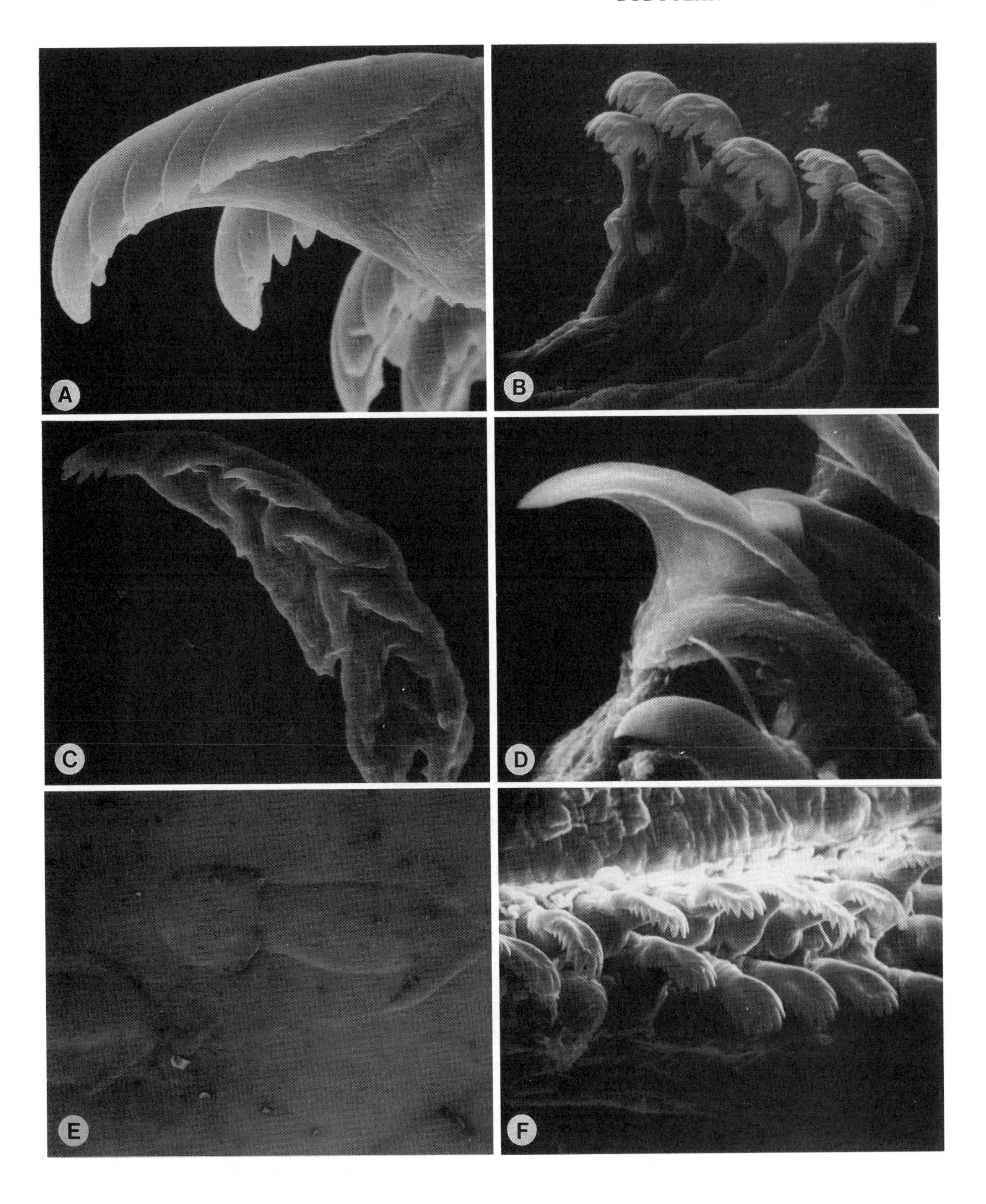
A
B
C
D
E
F

a normal graded series on the upper (e.g., *Plectrohyla ixil* and *P. matudai*) or both (e.g., *Leptodactylodon ventrimarmoratus*) sheaths. Individual serrations lacking a common sheath occur as transitory structures (e.g., Visser 1985, some young *Heleophryne*) or for the duration of larval life (e.g., Cei 1980, Ruibal and Thomas 1988, *Lepidobatrachus;* personal observation, *Mantidactylus lugubris;* Pyburn 1980, Wassersug and Pyburn 1987, *Otophryne*). Jaw sheaths with 30–80 serrations/mm are typical, and higher counts are known in lotic forms. In a few cases (e.g., some *Amolops* and *Boophis*), serrations on the edges are only 5–8/mm. Among suctorial forms of some *Boophis,* the sheaths appear to be composed of a series of fused columns; the irregular surfaces of these sheaths are in stark contrast to the smooth surfaces of typical forms. Some *Amolops* (Ranidae), *Ansonia* (Bufonidae), and *Litoria* (Hylidae) have one or both jaw sheaths divided medially; the functional significance of such arrangements is unknown. A comment by Menzies and Zweifel (1974) that older specimens of a species of *Litoria* with broken sheaths have a single sheath suggests ontogenetic changes, but this needs to be examined more closely. The upper and lower sheaths are usually similar in size, but the U-shaped lower sheath of *Ascaphus* is minute relative to the large, flattened, upper sheath. Tadpoles of *Heleophryne* lack one (upper) or both sheaths, and the upper one is absent in an unidentified tadpole of *Boophis* (personal observation). As with tooth rows and ridges discussed above, the apparent absence of jaw sheaths must be evaluated carefully. Absence of pigmentation does not corroborate the lack of keratinization (e.g., *Scaphiophryne*), so a judgment of absence should be based on an actual lack of a sheath on the labial cartilages. Jaw sheaths usually co-occur with labial teeth but may occur alone (e.g., hylid: *Hyla leucophyllata* group; ranid: *Occidozyga;* microhylid: *Scaphiophryne*). Rarely do tadpoles have labial teeth and no jaw sheaths (e.g., some *Heleophryne*). The upper sheath of some suctorial forms may be M-shaped, and some species (e.g., *Hyla pictipes*) appear to have a scouring surface on the front face of the upper sheath just proximal to the serrated edge. Based on the various accounts, the structures of the upper sheath of *Hoplobatrachus tigerinus* (surely confusion exists in distinguishing between the tadpoles of *Hoplobatrachus tigerinus* and *Hoplobatrachus rugulosus*) apparently are unique, although it cannot be determined whether one or two features are involved. Khan and Mufti (1994:28) stated that there is "A pair of long cylindrical thick papillae, tipped with keratinized plates . . . at the angles of the mouth opening. . . ." These structures, perhaps keratinized buccal papillae, do not appear associated with the upper jaw sheath. Kirtisinghe (1957) illustrated the lateral margins of the upper sheath with convexities but made no mention of them in the text. Khan and Mufti (1994) also noted that a ". . . median-buccal keratinized shield is observable through the mouth." This structure must be similar to the keratinized boss on the roof of the buccal cavity of *Spea* and some *Hyla.* Various jaw sheaths are included in fig. 3.9.

Shapes of jaw sheaths are difficult to describe because of their complex curved shapes. The curved jaw sheaths and trajectory of the cutting edge vary tremendously among taxa, but neither have been described well. Wide infra- or suprarostral cartilages accompanied by narrow keratinized sheaths occur, and the difference between this situation and one where narrow sheaths cover most or all of the cartilages needs to be qualified.

To visualize jaw sheath shapes better, imagine a rectangular piece of paper with the opposite short ends placed on a table so that the paper forms an arch that represents the typical smooth arc of the upper jaw sheath. The same concepts pertain to the lower sheath, although it is more difficult to simulate because the sheath is most often V-shaped. With proper pressures on and repositioning of the ends, the paper can be fashioned into various uniform (e.g., tall and narrow vs. short and wide) or nonuniform (e.g., less curvature in medial section) arches that effectively illustrate the known variations in the shapes of jaw sheaths. The facing edge, which equals the serrated edge of the sheath, can be cut to form various uniform or nonuniform (e.g., incised sheath with a medial convexity) borders. Also, various (uniform and nonuniform) serrations that usually grade from the largest medially to the smallest laterally can be cut into the facing edge. The ends of the arch can be modified in shape and lateral flexure to form the indistinctly differentiated lateral processes. Last, the roof (= front, oral, or adoral surface of sheath) and ceiling (= rear, back, aboral, or buccal surface of sheath) of the arch could be nearly flat and parallel, similar to the paper, or could curve to various degrees. The line of keratinization on the front face of the upper sheath typically describes a smooth, either abrupt or graded, arc, but the line of dark keratinization on the buccal face of the upper sheath often takes various shapes (Altig and Johnston 1989) with graded or abrupt margins. The lower sheath passes inside the upper sheath during a feeding stroke, and through scissorlike interactions serrations on the two sheaths cut or gouge surfaces for food removal. The degree of keratinization varies from weak (yellow to light brown) to strong (dense black); other than color differences, there is no system for even subjective evaluation of these differences. The cause or functional significance of light keratinization in sheaths that normally are heavily keratinized is unknown. Structures with melanic pigment are stronger than those that lack this pigment, but the strength and resistance to wear of sheaths that differ in degrees of keratinization have not been evaluated. Because of the way jaw cartilages articulate with their supportive structures, jaw sheaths are oriented and operate at several angles. A wide variety of sheath shapes can be seen in buccal view in Wassersug and Heyer (1988).

Other keratinized structures are associated with the oral disc and buccal cavity. A rounded, heavily keratinized boss occurs on the ceiling of the buccal cavity just posterior to the buccal base of the upper jaw sheath in tadpoles of *Spea.* Tadpoles of *Kassina senegalensis, Trichobatrachus robustus,* and several members of the *Rana pipiens* group, mostly with robust lower jaw sheaths, commonly have areas of thin keratinization lateral to the oral face of the lower jaw sheath. Although unclear, a comment by Annandale and Rao (1918) suggested a unique feature of the tadpole of *Euphlyctis hexadactylus:* "There is a deep groove, with its sides and base sometimes cornified, across the lower lip; the margin of the upper beak fits into this groove when the mouth is shut." This structure is not clear in their drawing.

Oral Ontogeny

Our understanding of oral ontogeny is based on relatively few studies of a small number of taxa and various periods of development (e.g., Agarwal and Niazi 1980; Cusimano-Carollo 1963, 1969, 1972; Cusimano-Carollo et al. 1962; Echeverría and Fiorito de Lopez 1981; Fiorito de Lopez and Echeverría 1984; Marinelli et al. 1985; Marinelli and Vagnetti 1988; Petersen 1922; Pyburn 1967; Tachihama et al. 1987; Thibaudeau and Altig 1988; Tubbs et al. 1993; our figs. 3.10–3.11). The assumption that oral development and the criteria used for staging tadpoles are tightly correlated is false. Because of this discordance and failure of investigators to accommodate for it, some studies show a more advanced tadpole with lesser developed mouthparts than a younger tadpole. The available data, although not always based on detailed and complete developmental series, suggest that interspecific differences in early embryology are primarily the result of timing and not pattern differences. Some events that occur slightly later (e.g., tooth row appearance) may involve several patterns, none of which is well understood. Unfortunately, all the species that have been studied are lentic forms with grossly similar mouthparts. Consequently, the stage notations in the following summary should be treated only as guides. Differences in the patterns of mouthpart formation or metamorphic atrophy and body development (e.g., Nodzenski and Inger 1990) are known, but we do not know whether changes in the developmental timing of the appearance or loss of mouthparts or the characters used in staging cause these patterns.

The stomodeum first appears as a dimple on the anteroventral part of the presumptive head at about stage 17. With continued development the stomodeum may become a more or less rounded, diamond-shaped, or triangular structure or appear as a transverse slit or T-shaped depression. As the head changes shape during development, the mouth eventually comes to lie ventrally, anteroventrally, terminally,

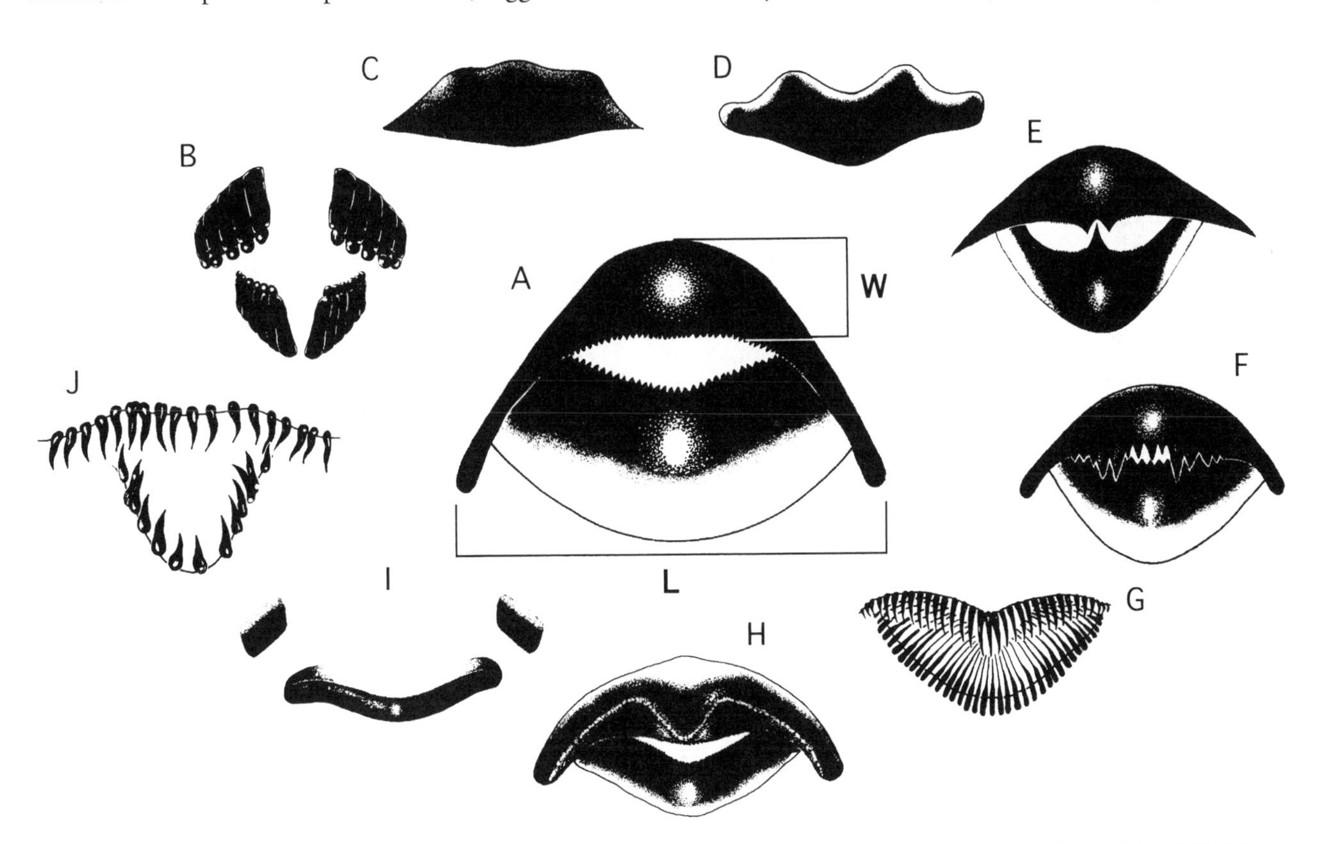

Fig. 3.9. Representative jaw sheaths of tadpoles. (A) Oral surface of upper and lower jaw sheaths of a typical tadpole showing standard width (W) and length (L) measurements. (B) Oral face of jaw sheaths of *Meristogenys orphnocnemis* (Ranidae); upper and lower sheaths divided and consisting of particularly large units. (C) Buccal face (cutting surface downward, lightly keratinized areas stippled) of upper jaw sheath of *Hyla femoralis* (Hylidae). (D) Buccal face (cutting surface downward, lightly keratinized areas stippled) of upper jaw sheath of *Rana sphenocephala* (Ranidae). (E) Oral face of incised sheaths of carnivorous tadpole of *Ceratophrys cornuta* (Leptodactylidae). (F) Oral face of jaw sheaths with hypertrophied serrations of *Plectrohyla ixil* (Hylidae). (G) Oral face of highly modified, long, spikelike derivatives of jaw sheaths of *Mantidactylus lugubris* (Mantellinae, Rhacophoridae). (H) Oral face of jaw sheaths showing M-shaped upper sheath of *Hyla pictipes* (Hylidae). (I) Oral face of jaw sheaths showing divided upper sheath of *Ansonia longidigita* (Bufonidae). (J) Spikelike, transitory (Gosner stages 23–25) "serrations" in species of *Heleophryne* (Heleophrynidae) that lack jaw sheaths (after description and illustration by Visser 1985).

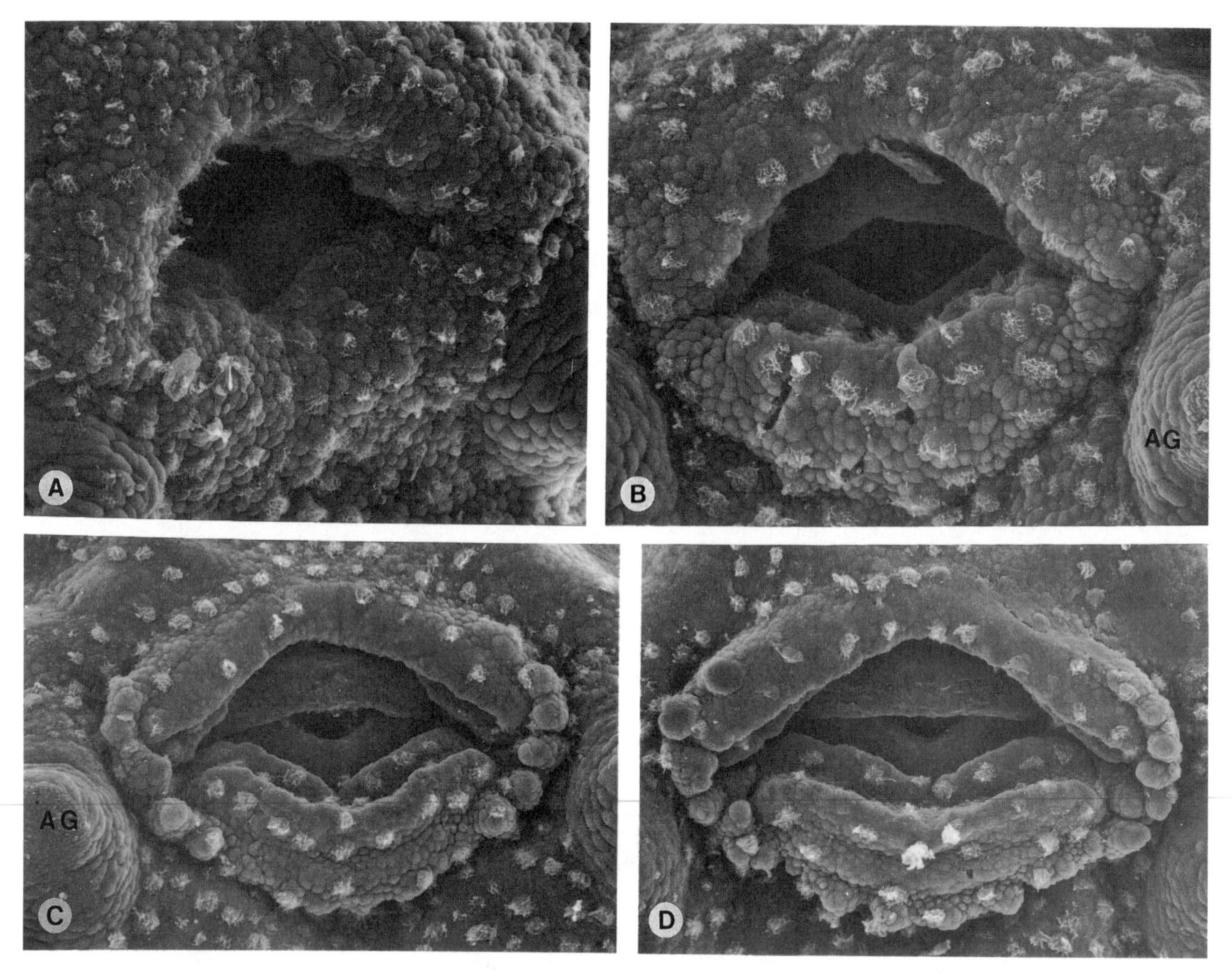

Fig. 3.10. Development of the oral apparatus of a typical tadpole with 2/3 LTRF, as illustrated by *Hyla chrysoscelis* (Hylidae). (A) Oral pad immediately after rupture of the oropharyngeal membrane, Gosner 21. (B) Labia distinct, jaw sheaths present, cleft at lateral margins separates presumptive tooth ridges for A-1 and P-2, Gosner 23. (C) All tooth ridges except P-3 distinct, lateral marginal papillae distinct, jaw sheaths more heavily keratinized, Gosner 24. (D) All components continuing development, early Gosner 25. Labial teeth erupt later in Gosner 25, medial parts of adhesive glands (AG) visible in lower corners of all panels, and ciliated cells with puffs of white threads at their centers visible on the epidermis. From Thibaudeau and Altig (1988), copyright © 1988, Wiley-Liss, Inc., a subsidiary of John Wiley & Sons, Inc.; reprinted by permission.

or slightly above the snout tip. In determining the location of the mouth one must take into account the actual rather than its apparent position as determined by arrangement of the oral disc. Subsequent increases in the depth of the stomodeum and growth of the archenteron eventually result in the rupture of the oropharyngeal membrane at about stage 21.

At about the same time, the oral pad, the composite anlage of both labia that surrounds the stomodeum, rises slightly above the surrounding surface. Ciliated cells that occur throughout the body ectoderm also occur here. Very soon after rising above the surrounding surface, the ventrolateral edges and then the anterolaterally and medioventral areas of the pad become undercut. If a lateral emargination is present on the disc, its beginning can be seen by stage 22, and nascent marginal papillae are present as indistinct crenulations. Most submarginal papillae do not form until after tooth ridges have appeared. The form of the jaw sheaths is apparent well before they start to keratinize at about stages 20–21 (Luckenbill 1964, 1965; Kaung 1975; Kaung and Kollros 1976). Usually, the lower sheath slightly lags the upper one in development.

Tooth ridges appear in a sequence (e.g., A-1, P-2, P-1, A-2, P-3; sequential LTRF of 1/0, 1/1, 1/2, 2/2, 2/3) during stages 22–24, but the species examined thus far do not differ significantly in LTRFs. The epithelial cells on the ridges are larger and appear slightly fuzzy compared with nonridge areas. Compared to more typical tadpoles, the oral structures of species with many tooth rows, especially suctorial forms, appear earlier in ontogeny (Nodzenski and Inger 1990), but the entire compliment of tooth rows may not appear until

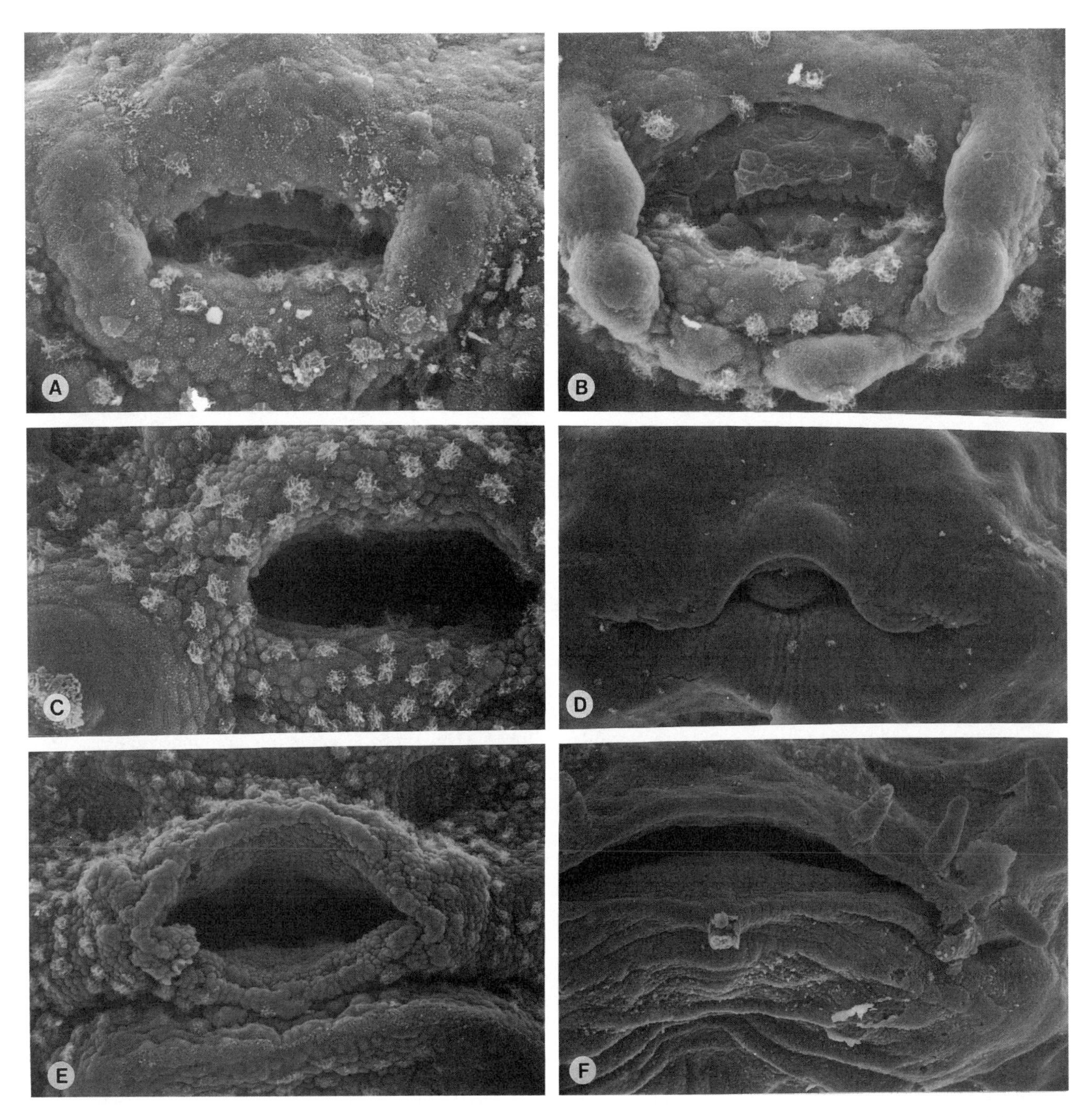

Fig. 3.11. Photomicrographs showing development of the oral apparatus in three tadpoles. *Top: Hyla sarayacuensis* (Hylidae), a tadpole with jaw sheaths, reduced oral disc, and no labial teeth. (A) Jaw sheath visible, marginal papillae starting to appear, Gosner 24. (B) Distinct serrations on jaw sheaths and marginal papillae are obvious, Gosner 25. *Middle: Gastrophryne carolinensis* (Microhylidae), a tadpole with oral flaps and no keratinized mouthparts. (C) Oral disc relatively undifferentiated, conical adhesive gland visible in lower left corner, Gosner 24. (D) Labial flaps beginning to differentiate, lower jaw (= infralabial prominence) visible as rounded projection in notch between labial flaps, Gosner 25. *Bottom: Rhinophrynus dorsalis* (Rhinophrynidae), a tadpole without keratinized mouthparts. (E) Oral disc undifferentiated, without jaw sheaths or barbels, linearly transverse adhesive gland visible below mouth, nares visible dorsal to mouth and still plugged with cells, Gosner 22. (F) Barbels beginning to differentiate around slitlike mouth, floor of the buccal cavity directly inside mouth visible because of slight downward flexure of lower jaw, medial papilla on lower jaw visible, Gosner 25. From Thibaudeau and Altig (1988), copyright © 1988, Wiley-Liss, Inc., a subsidiary of John Wiley & Sons, Inc.; reprinted by permission.

somewhat later in ontogeny compared with more typical tadpoles. Heyer (1985, *Hyalinobatrachium uranoscopa*) and Inger (1966, *Leptolalax gracilis*) noted cases where tooth rows are lost and the tooth ridges seemingly are transformed into a series of papillae. Some data (Altig and Johnston 1989) suggest several patterns of tooth row appearance once a formula of 2/3 (see above) has developed. Some tadpoles add rows proximally to the rows of the 2/3 formula on the anterior labium and distally on the lower labium. Others add rows distally on both labia. Teeth first appear near the medial parts of tooth ridges and then laterally. The sequence of tooth row appearance among ridges is the same as that for ridge formation. Hemispherical bulges in the epithelium can be seen where teeth are forming underneath, and the first teeth that appear may have a different form than subsequent ones (Thibaudeau and Altig 1988; Tubbs et al. 1993). From their first appearance and throughout tadpole ontogeny, as labial teeth are lost after wear or breakage, replacement teeth are continually formed from mitotic zones in the bases of the tooth ridges (e.g., Luckenbill 1964, 1965; Kaung and Kollros 1976). In most tadpoles, several replacement teeth are interdigitated in a series (figs. 3.6B, E); in a few cases replacement teeth apparently are absent (e.g., distalmost rows in *Ascaphus*).

Atrophy of the oral apparatus has not been well studied, but apparently it occurs in the approximate reverse order of ontogeny and more haphazardly. Presumably, mitoses that produce teeth cease and the last teeth are lost in patches. It is not known if a pattern to tooth loss within rows exists. The initiation, duration, and perhaps pattern of metamorphic atrophy varies among taxa and ecological types. For example, row P-3 (LTRF of 2/3) is the first to disappear, and A-1 usually is the last. Jaw sheaths are lost after all or most tooth rows. Ventrolateral marginal papillae are the last structures on the oral disc to atrophy.

Functional and Evolutionary Aspects of Tadpole Morphology

Too often we have indicated the paucity of information regarding the functional or evolutionary aspects of various larval structures; although bothersome, this reflects the inadequacies of our knowledge. Morphological features and their distributions among taxa are still being described, and often we have relatively little information on the development of such features and less on their functions. Until more of these holes in our knowledge are filled, probing discussions on the evolution of morphological structures in tadpoles will remain speculative.

Body Morphology

Exemplified by the hypotheses on ecomorphological guilds proposed by Altig and Johnston (1989), further observations suggest strong correlations among the following traits: behavior, body and fin configurations, ecology, metamorphic pattern (e.g., Nodzenski et al. 1989), and taxonomic position. Seven generalizations abstracted from Altig and Johnston's (1989) guild hypotheses involve associations among features of the body, eyes, fins, tail muscle, and mouthparts. (1) Lentic forms have less massive tail muscles than lotic forms, and the smallest muscles are associated with the largest fins. (2) Benthic forms have depressed bodies, dorsal eyes, and low fins whether in lentic or lotic environments. (3) Lentic (pond), nektonic (rarely lotic) forms have compressed (e.g., many hylids), depressed (e.g., microhylids and pipids, sometimes with a tail flagellum), or equidimensional (e.g., macrophagous feeders with jaw sheaths but lacking labial teeth, some *Hyla,* often with a tail flagellum) bodies and live in different parts of the water column. (4) Burrowers (e.g., centrolenids) and those tadpoles that live in confined spaces (e.g., bromeliad cisterns; *Hyla bromeliacia, H. dendroscarta*) are vermiform with depressed bodies, dorsal eyes, and low fins. (5) Semiterrestrial tadpoles have elongate bodies, narrow tail muscles with abbreviated fins, large eyes that bulge above the surrounding body surface, and hind legs that develop precociously (Drewes et al. 1989). (6) Tadpoles that live in slower reaches of streams resemble lentic-benthic forms except for a typical increase in the number of tooth rows; ranids make up a large component of this assemblage, and the number of upper tooth rows is often larger than the lower. (7) Lotic forms that use the oral disc to maintain position and feed in fast-moving water have complete marginal papillae and frequently have more tooth rows and higher tooth density; larger numbers of tooth rows presumably signal faster water habitats. The largest LTRF of 17/21 is from an unidentified *Hyla* from a high-energy stream in Venezuela. In these tadpoles the fins are low and the eyes are dorsal. There probably are two poorly understood patterns of streamlining.

Almost no data (see Gradwell 1971a, 1973, 1975b) exist on the actual functional proficiency of the basic morphological categories described above. The functions of other features (e.g., vent flaps, dextral vs. medial vent tubes and their many variations, tubular nares, naris size, and long, free spiracular tubes) have not been suggested or are seldom supportable by data. Based on observations of where tadpoles occur in the water column (e.g., Alford 1986b; S. L. Mitchell 1983; see Abelson et al. 1993 and Lancaster and Hildrew 1993) and their swimming abilities, associations between tail fin shapes and tail muscle configurations surely signal different capabilities. Unfortunately, the data on the swimming kinematics of tadpoles pertain only to a few species of lentic tadpoles (e.g., Hoff and Wassersug 1985, 1986; Wassersug 1989a; Wassersug and Hoff 1985).

Oral Morphology

The functional roles that have been proposed for oral papillae fall into two basic categories: chemosensory and tactile receptors and structures that control water flow (Van Dijk 1981), enhance attachment to substrates (Altig and Brodie 1972; Gradwell 1971a, 1975b), modify the shape of the oral disc during feeding, and manipulate food and substrate particles. As is usually the case, few data are available. The size and density of oral papillae in many cases seem to reflect lineages and habitats. For example, ranid tadpoles have larger papillae arranged more sparsely than hylids, and stream hylids have smaller, more densely arranged papillae than pond hylids. The number and prominence of marginal papillae varies concordantly with observed reductions in the size of the oral disc among tadpoles within the same lineage.

A comprehensive survey of labial tooth structure has not been made, and the few descriptions that have been published vary in many details (e.g., Altig 1973; Altig and Pace 1974; Faier 1972; Gosner 1959; Heron-Royer and Van Bambeke 1889; Hosoi et al. 1995; Inger 1983; Korky and Webb 1994; R. J. Nichols 1937). Little can be said about functional differences in tooth morphologies (e.g., Altig and Johnston 1989; Faier 1972; Gosner 1959; Inger 1983). Some suggested functions include acting as a broom, serving as a current generator, breaking up mucilaginous layers, sieving particles, combing strands into alignment for easier cutting (Altig and Johnston 1989), piercing plant cells, rasping food particles from a substrate (R. M. Savage 1952), holding food (Luckenbill 1965), attaching to a substrate (Altig and Brodie 1972; Gradwell 1975a, b), and trapping food (Tyler 1963). Few data exist to support any of these suggestions, and the immense variations remain enigmatic. Mertens (1960) and Nachtigall (1974) compared the morphologies of tadpole teeth to those of structures in other groups that presumably serve similar functions. Publications that contrast tadpole teeth to the snail radulae (e.g., Steneck and Watling 1982) provide other pertinent examples.

Likewise, only a few workers (e.g., Altig and Johnston 1989) have attempted to place the sizes, shapes, and serration patterns (e.g., fig. 3.6H) of jaw sheaths into a functional context. Even though the upper and lower jaw sheaths in a typical tadpole are similar in size and probably serve as gouging/biting/scraping structures, their striking differences in shape and the extensive array of morphologies across taxa suggest an innumerable variety of performance abilities. The immense, flattened upper sheath of *Ascaphus* is unique and must shave periphyton from various substrates during the retraction phase of the oral disc; the small lower sheath is probably inconsequential. The robust, narrowly arched jaw sheaths (i.e., short surface for actual biting) of semiterrestrial tadpoles presumably provide strength and may be essential to feeding on tough materials or unusual substrates characteristic of these thin (water film) habitats.

Except in the most general sense, the functional morphology of the oral apparatus of tadpoles is largely unknown. Extensive work has focused on the structural components and functional design of the food-trapping mechanisms in the buccopharyngeal area (e.g., Kenny 1969b, c; R. M. Savage 1955; Viertel 1984a, c, 1987; Wassersug 1972, 1980; Wassersug and Heyer 1988), but insufficient attention has been paid to the mechanics of the external feeding structures. Their small size and the rapid operation of these mouthparts pose several challenges to be overcome in designing a detailed study of the functional components of feeding by tadpoles. These problems are compounded by the procedural difficulty of producing standard food substrates for comparative testing. When these complications are considered in light of the enormous amount of interspecific variation, elucidating the functional and evolutionary aspects of the workings of the oral apparatus during feeding has proved to be especially difficult. Even so, a brief review is possible and worthwhile. The upper supra- and lower infrarostral cartilages (chap. 4) that are the jaws of a tadpole have almost no adult derivatives, are not articulated at their bases, and are operated in a unique mechanical fashion. The suprarostral cartilage articulates and is movable against the chondrocranium. The infrarostral cartilage articulates and is movable against Meckel's cartilage, which articulates and is movable against the chondrocranium. Therefore, two joints occur in the lower system and one in the upper. In addition, the suprarostral cartilage is not closed through muscular contraction; rather, this cartilage with its attached upper jaw sheath is closed through a tendon-and-pulley system via movements of the lower, infrarostral cartilage (Gradwell 1968, 1972b). This arrangement leads one to conclude that the jaws must operate together and that both sheaths are involved in food removal. Two sets of unpublished data (G. F. Johnston 1982, 1990) and other observations confirm that we do not understand the exquisite mechanics of this system. Carr and Altig (1992) also found interspecific differences of the rectus abdominis muscle that apparently correlate with feeding mode (figs. 3.3A–C). Among others factors, species, stage, position of the food in the water column, and density of agar-based food alter the following generalities. It also is important to remember that only a small group of ecologically similar tadpoles has been compared.

Feeding trials in the laboratory indicate that the lower sheath alone or the two sheaths acting together can remove agar-based food. In the latter case, the mark of the lower sheath on the agar is usually longer than the mark made by the upper sheath. Even though the two sheaths must move simultaneously, both do not have to contact the substrate. Substrate contact may be avoided by changing the angle that the tadpole takes relative to the substrate or perhaps by adjusting the longitudinal relationship of the two sheaths. The angle between the two infrarostral cartilages probably can change during a bite, and the depth of penetration of either jaw sheath can change during a bite. The lower jaw sheath may strike several times while the upper sheath takes but a single gouge from agar-based food. On agar that is too dense to penetrate, both jaws may strike repeatedly without closure. The labial teeth may leave no mark, leave a mark but show no movement, or show rasping movements with marks left on top of those made late in a feeding stroke by the jaw sheaths.

Insufficient information has contributed substantially to an inadequate understanding of how tadpole mouthparts work. A better appreciation of the following issues would add considerably to understanding their evolution: those factors important to the evolution and maintenance of the tadpole stage in the life history of a frog; a consensus on primitive or derived state of various larval traits; an appreciation of the full compliment of morphological variations and their functions; baseline data on geographic variation; and further facts on ontogenetic variation, developmental information, and the ecological significance of various larval morphologies. For these and other reasons, oral structures have been used infrequently in systematics analyses (e.g., Duellman and Trueb 1983, 1986; Grandison 1981; I. Griffiths 1963; I. Griffiths and De Carvalho 1965; Inger 1967; Kluge and Farris 1969; O. M. Sokol 1975) and debated less often. The lack of a clear phylogenetic hypothesis of some groups surely has exacerbated the situation, but cursory examinations suggest to us that tadpole morphology commonly is concordant with perceived taxonomic boundaries and perchance signals the need for closer examination of relationships (e.g., *Bufo debilis* group vs. other *Bufo, Hyla leucophyllata* group vs. other groups of hylids, *Scinax rostrata* vs. *S. rubra* groups, *Otophryne pyburni* vs. other microhylids). Boulenger (1897:110) made a knowing statement in this regard: "The structural differences which separate the genera and species in their tadpole condition reflect, on the whole, pretty accurately the system based on the perfect animals. . . ." Yet, even heuristic attempts to gain ideas on evolutionary patterns by overlaying larval traits on cladograms derived from analyses of adult traits have not been done (see chap. 4). Degree of commonness, at least as presently known (e.g., a 2/3 LTRF in many guilds and taxa), coupled with a bewildering array of morphological diversity, makes the situation appear chaotic. Even so, inclusion of larval traits in phylogenetic analyses of anuran groups is essential. We are not sure whether external or internal characteristics will prove to be more informative, so we recommend investigators consider both. A few examples adequately illustrate the need for much more work. J. D. Lynch (1971) considered a LTRF of 3/3 or greater as primitive within leptodactylid frogs, and Drewes (1984) noted that larger LTRFs belong to the more primitive taxa in hyperoliids. Noncusped teeth have been considered primitive by Noble (1931) and derived by Gosner (1959).

Although developmental and functional data may help define characters and their states, their inclusion in phylogenetic analyses should be with caution. We are convinced that understanding the developmental morphology of a character often provides insight into character homology. For example, determining the developmental history of a trait across lineages may enable one to distinguish between instances of what appears to be the independent occurrences of the same trait (homoplasy) and those that actually represent the occurrence of two traits (apparent homoplasy).

Altig and Johnston (1989) pointed out that patterns of early development and metamorphic atrophy were informative and suggestive of tooth row homology. Based on data from several sources (e.g., Dutta and Mohanty-Hejmadi 1984; Echeverría et al. 1987; Fiorito de Lopez and Echeverría 1984; Limbaugh and Volpe 1957; R. J. Nichols 1937; Thibaudeau and Altig 1988; Tubbs et al. 1993), tooth rows in species with a LTRF of 2/3 commonly appear in the developmental sequence of A-1, P-2, P-1, A-2, and P-3 (sequential formulas 1/0, 1/1, 1/2, 2/2, 2/3). Rows A-1 and P-2 are closely linked during development and may appear slightly reversed in time. Likewise, the first five rows to develop in species with LTRFs larger than 2/3 commonly appear in this sequence. In species with more than five tooth rows, subsequent rows apparently are added alternately on the upper and lower labia, but in at least four different patterns two of these are apparently common. In ranids, tooth rows are added proximally on at least the upper labium, whereas in ascaphids, heleophrynids, and hylids, additional rows are located distally on both labia. Species with many tooth rows (e.g., suctorial forms) attain the full compliment much later in ontogeny and retain tooth rows further into metamorphosis than species with fewer tooth rows (Menzies and Zweifel 1974; Nodzenski and Inger 1990). A thorough evaluation of row homology is hindered by the paucity of developmental series and the lack of understanding of how medial gaps in tooth rows develop and evolve.

The rows in a 2/3 formula are designated the prime formula. In species with a formula greater than 2/3, the pattern differences that were proposed to account for additional tooth rows require that the prime rows be renumbered once the entire compliment of tooth rows is present. Based on this observation, certain LTRFs should not exist. Of 124 species with fewer than 2/3 rows, 88 fit within the expected categories of 2/2 (loss of P-3), 1/2 (loss of A-2), 1/1 (loss of P-1), and 1/0 (loss of P-2) or 0/1 (loss of A-1). If homology is assumed, LTRFs such as 0/2, 2/1, and 1/3 (common in hyperoliids and African ranids) indicate the presence of other patterns. The genetics of tooth row formation, and for that matter of most other morphological characteristics of tadpoles, essentially is unknown. Fortman and Altig (1973) showed that F_1 crosses among species of *Hyla* from the southeastern United States resulted in hybrids that more closely resembled one parent than the other; seemingly many traits acted as simple dominants. Crosses of *Rana* sp. with LTRFs of 2/3 and 3/4 produced hybrids with a formula of 2/4 (Haertel and Storm 1970; also see Kawamura and Kobayashi 1960, Michalowski 1966). Although the relative sizes of various mouthparts remain nearly constant throughout large parts of larval ontogeny (Gates 1983), the patterns of growth of oral structures as they relate to feeding ecology are poorly known.

Summary

The biphasic life cycle of anuran amphibians comprises two quite different body plans on which selection can operate: tadpoles and adults. The products of this selection especially among the larval forms are amazing! As discoveries are made and papers are published, one is coerced into believing that the total morphological diversity of anuran tadpoles is nearly endless. Except for the pelagic areas in a continually flowing river or stream, there seems to be no freshwater habitat of any duration that does not harbor tadpoles. Even though we do not understand them well, numerous adaptations allow tadpoles to live and feed in these various habitats. The distinctions between ecological forces and phylogenetic restrictions remain muddled in part because of the shortage of phylogenetic hypotheses within major clades of advanced frogs.

Throughout much of tadpole diversity, a tadpole is a tadpole is a tadpole: a simple, depressed body shape, dorsal eyes, sinistral spiracle, marginal papillae with a dorsal gap, and 2/3 tooth rows describes many anuran larvae. Within this hazy construct, two patterns emerge: overall uniformity within a lineage or relative uniformity marked by a few striking variants. Tadpoles in the genus *Bufo,* a large taxon of more than 200 species with nearly worldwide distribution, feature an oppressively uniform morphology. The known tadpoles of centrolenid frogs are interesting but equally monotonous, and the morphological diversity among leptodactylid tadpoles is small relative to the number of taxa. Lineages with morphological uniformity among its tadpoles except for a few distinctive forms include dendrobatids (e.g., *Colostethus nubicola* and *Dendrobates pumilio* vs. much of the remainder), mantelline rhacophorids (*Mantidactylus lugubris* and *M. guttulatus* vs. most of the remainder), the megophryid-pelobatid line (e.g., *Megophrys* vs. most of remainder), microhylids (e.g., *Microhyla heymonsi, Nelsonophryne, Otophryne,* and *Scaphiophryne* vs. most of the remainder), and ranids (e.g., *Rana* sensu lato vs. *Amolops, Conraua, Indirana, Nannophrys,* and *Occidozyga*). Invariably, the deviants form a small percentage of the whole. Endotrophic development of several kinds also is sprinkled throughout many of the above groups (see chap. 7) as if by random experimentation. These generalized patterns will only become more refined and noteworthy with clearer phylogenetic hypotheses for the pertinent families. Whether external or internal characteristics of tadpoles will eventually prove to be more informative in these phylogenetic analyses requires further study. In the meantime, we recommend inclusion of both types.

In 1980, Jay Savage articulated Starrett's Rule; in so doing, he (Savage 1980b:1183) provided a tiny point of focus to the above discussion: ". . . most specialized and uniquely modified tadpoles develop into ordinary frogs whereas the most bizarre and distinctive frogs have ordinary, generalized tadpoles." We imagine that if Jay were given another opportunity to comment on the generality of Starrett's Rule, he might produce the first corollary: When you get no respect as a tadpole and have grown weary of herpetologists constantly scrutinizing your oral disc, give up and become an endotroph. In this light, perhaps the only encouraging observation that we can make is that tadpole biologists should be elated that members of the genus *Eleutherodactylus* do not have tadpoles.

ACKNOWLEDGMENTS

B. Grillitsch offered valuable comments on the terminology of tadpole mouthparts, A. Channing supplied photographs for our consideration for this chapter, and T. Kahn and K. Spencer effectively transformed some of our ideas into illustrations. Thanks to each of you for your help.

4

ARCHITECTURE
Cranial and Axial Musculoskeleton

David Cannatella

Semina limus habet virides generantia ranas
Et generat truncas pedibus, mox apta natando
Crura dat, utque eadem sint longis saltibus apta,
Posterior superat partes mensura priores.

Ovid, *Metamorphoses,* book XV (Pattist 1956)

Ev'n slime begets the frog's loquacious race:
Short of their feet at first, in little space
With arms, and legs endu'd, long leaps they take
Rais'd on their hinder part, and swim the lake,
And waves repel: for Nature gives their kind,
To that intent, a length of legs behind.

Translation by John Dryden in Garth (1961)

Introduction

Ovid's epic poem *Metamorphoses* chronicles miraculous changes of all sorts—a boy is transformed into an eft for offending Ceres, a statue carved by Pygmalion becomes a woman, the eyes of the dead musician Argus are made to adorn a peacock's train, and tadpoles metamorphose into frogs. The Roman poet's view of tadpoles as ephemeral stages is not unexpected, for even amphibian biologists do not fully appreciate the significance of tadpoles as free-living organisms. Tadpoles use the transient source of energy available in aquatic primary production, which is not available to adults (Wassersug 1975); the larval stage provides the advantage of rapid growth in unstable environments. Being a tadpole means that cranial morphology is canalized into unique feeding structures at the expense of all else. The body cavity contains mostly gut. The tail is emphasized over the limbs as the instrument of locomotion. The reproductive tissue lies outside of the body cavity, so reproduction (and thus neoteny) is not possible until metamorphosis, when the tail is resorbed, the limbs develop, and the body cavity enlarges. Thus, except for sex, tadpoles engage in most of the same activities as adult frogs—feeding, respiration, and locomotion—but in very different ways. This chapter emphasizes the structure and function of the musculoskeleton as it relates to these activities.

Anuran Phylogeny and the First Tadpole

What did the first tadpole look like? Given a phylogeny, some inferences are possible. L. S. Ford and Cannatella (1993) reviewed the status of anuran phylogeny and presented trees based on mostly adult and a few larval morphological characters (fig. 4.1; alternative phylogenies are discussed at the end of the chapter). The four most basal lineages of living Anura are *Ascaphus truei, Leiopelma,* Bombinatoridae, and Discoglossidae. These taxa have the Type 3 tadpole of Orton (1957). Pipanura (fig. 4.1) is composed of Mesobatrachia and Neobatrachia. Mesobatrachia includes Pipoidea, with Type 1 tadpoles, and Pelobatoidea, with Type 4 tadpoles. The tadpoles of Neobatrachia are Type 4 except for those of Microhylidae, which are Type 2. If Orton's (1953, 1957) four tadpole types are mapped as unordered character-states onto the phylogeny, it can be inferred that the ancestor of Anura had a Type 3 tadpole (jaw sheaths, labial teeth, medial spiracle), the ancestor of Pipanura had a Type 4 tadpole (jaw sheaths, labial teeth, sinistral spiracle), and the Type 1 tadpoles (no jaw sheaths or labial teeth, paired spiracles) and Type 2 tadpoles (no jaw sheaths or labial teeth, medial spiracle) were independently derived from Type 4 (fig. 4.2). This is similar to a proposal by O. M. Sokol

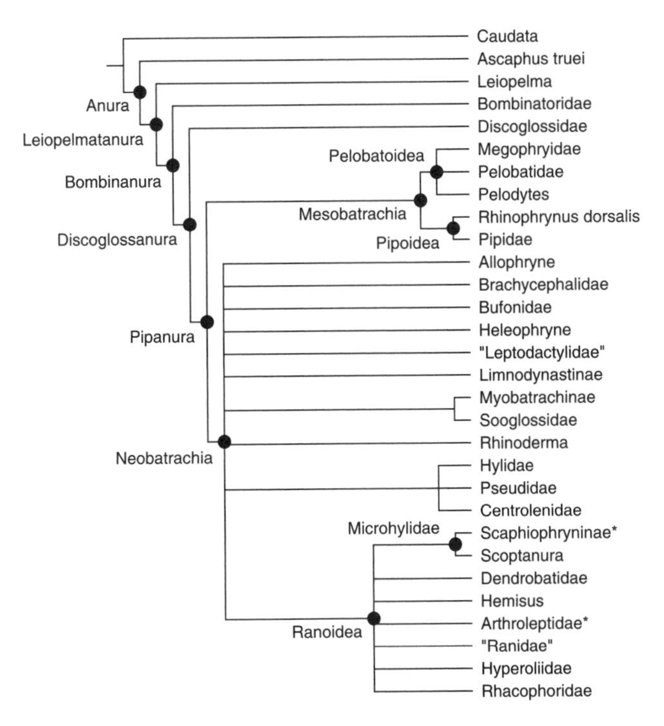

Fig. 4.1. Phylogenetic relationships among the major taxa of Anura. The dots represent node-based names. Symbols: asterisk = metataxon, names in quotation marks = nonmonophyletic taxa, and dots = node-based names (L. S. Ford and Cannatella 1993).

(1975) and in stark contrast to that of Starrett (1973), which is discussed in the section Evolution of Orton's Tadpole Types.

Although the ancestor of Anura can be inferred to have a Type 3 tadpole, in itself this is not very informative. Orton's tadpole types are defined primarily by the presence/absence of jaw sheaths and labial teeth, and the number and position of the spiracles. The taxa possessing the plesiomorphic Type 3 larva (fig. 4.2) are diverse. *Ascaphus truei* has a stream tadpole, and *Leiopelma* has larvae with "direct" development (chaps. 2 and 7). More typical pond tadpoles are present in *Bombina* and *Discoglossus* and in many other groups, but their widespread phylogenetic distribution does not guarantee that the pond tadpole morphology is plesiomorphic for Anura. It is generally assumed that the specialized morphologies of the *Ascaphus* tadpoles and *Leiopelma* larvae are derived from some pond tadpole ancestor (I. Griffiths 1963), but phylogenetic analysis of larval characters does not demand this interpretation. Certain features of the *Ascaphus* tadpole (e.g., reduced muscular process of the palatoquadrate) might be either plesiomorphic for Anura or apomorphic for *Ascaphus*. What can be said with confidence is that the ancestor of Bombinanura (figs. 4.1–4.2) had a pond tadpole, but this ancestral larva is quite removed phylogenetically from the typical pond larvae of *Rana catesbeiana* and *R. temporaria* on which much of this chapter is based. It is doubtful that the first anuran tadpole was structurally or functionally similar to the pond tadpoles of *Rana*, and the reader is cautioned not to extrapolate structure or function from *Rana* to other species. Instead, it is hoped that this chapter will stimulate descriptive and functional studies of other taxa, so that a more complete phylogenetic synthesis of tadpole structure and function may emerge.

General Anatomy

This chapter provides an overview of the skeleton and muscles of the chondrocranium, jaws, hyobranchium, and trunk. The general anatomy of each region is described, and intertaxic variation is discussed in a phylogenetic context when possible. Following the general comparative anatomy are sections treating changes during metamorphosis, the functional anatomy and evolution of gill irrigation mechanisms, and phylogenetic aspects of tadpole morphology. Only recently (e.g., Haas 1993, 1996b, 1997a) have phylogenetic analyses using primarily larval musculoskeletal characters appeared.

An ideal general description of tadpole anatomy would be based on a plesiomorphic pond tadpole such as *Bombina* or *Discoglossus*. Unfortunately, the literature on these species is not rich enough to permit this, although that situation is changing with papers such as those by Haas (1997) and Schlosser and Roth (1995). The descriptive cranial anatomy herein is based primarily on *Rana temporaria* (De Jongh 1968; Edgeworth 1935; Pusey 1938), and the functional aspects are based on *Rana catesbeiana* (Gradwell 1968, 1969a, b, 1970, 1972b, c; Gradwell and Pasztor 1968). Most of the terminology for the cranium derives from Gaupp (1893, 1894, 1896), and that for muscles comes from Edgeworth (1920, 1935), with modifications by Haas (1997). Although Edgeworth's terminology has been almost standard for many years, recent work has altered the proposed homologies of muscles, especially in the branchial region. The description of metamorphosis is generally based on De Jongh (1968) and Ridewood (1897b, 1898a). The information for the vertebral column and associated muscles is derived from *Xenopus laevis* because the literature on that species is much more complete than for *Rana*. The most comprehensive summary of development of all organ systems is that of Nieuwkoop and Faber (1967) for *Xenopus laevis*. Unfortunately, there are no anatomical illustrations, but there is an extensive literature section. Staging systems used in the text include Gosner (1960), Kopsch (1952), and Nieuwkoop and Faber (1967).

In general, English anatomical terms are used if both English and Latin are available. If no English term is available, or if the English term is ambiguous in meaning, the Latin is used; synonyms are listed in parentheses. Edgeworth (1935) provided a table of muscle synonyms for the early literature.

The Chondrocranium

The first study of the frog chondrocranium apparently was by Dugès (1834), who described and figured the larval chondrocranium of *Pelobates fuscus;* other literature on tadpole crania is summarized in table 4.1. Intertaxic variation in the larval chondrocranium is discussed below and in Haas

(1993, 1995, 1996b, 1997a) and O. M. Sokol (1975, 1981b) The reader is referred to Cannatella (1985), Duellman and Trueb (1986), and Trueb (1973, 1993) for variation of bones among taxa.

The chondrocranium of tadpoles is a cartilaginous case that protects the brain and supports the sense organs and jaw apparatus. The chondrocranium initially appears as two pairs of cartilaginous elements, the parachordals and trabeculae (fig. 4.3). The parachordals appear as longitudinal rods flanking the anterior end of the notochord. The trabeculae form in the same plane but anterior to the parachordals; the notochord does not intervene between the trabeculae. The trabeculae fuse anteromedially to form a trabecular plate (ethmoid plate, planum trabeculae anterior). The space where the trabeculae have not yet fused is the basicranial (hypophyseal) fenestra, which disappears as the trabeculae coalesce to form the cranial floor (basii cranii). Fusion of the parachordals forms the basal plate (planum basale; fig. 4.3) as the floor of the neurocranium; the planum may be perforated by the notochord. Before the basicranial fenestra disappears, the trabeculae and basal plate fuse. The junction between the basal plate and cranial floor is indicated by the primary carotid foramina (foramina carotica primaria) through which the internal carotid arteries pass. The parachordals project posteriorly to the otic capsules and support the occipital arch, which is serially homologous with the ver-

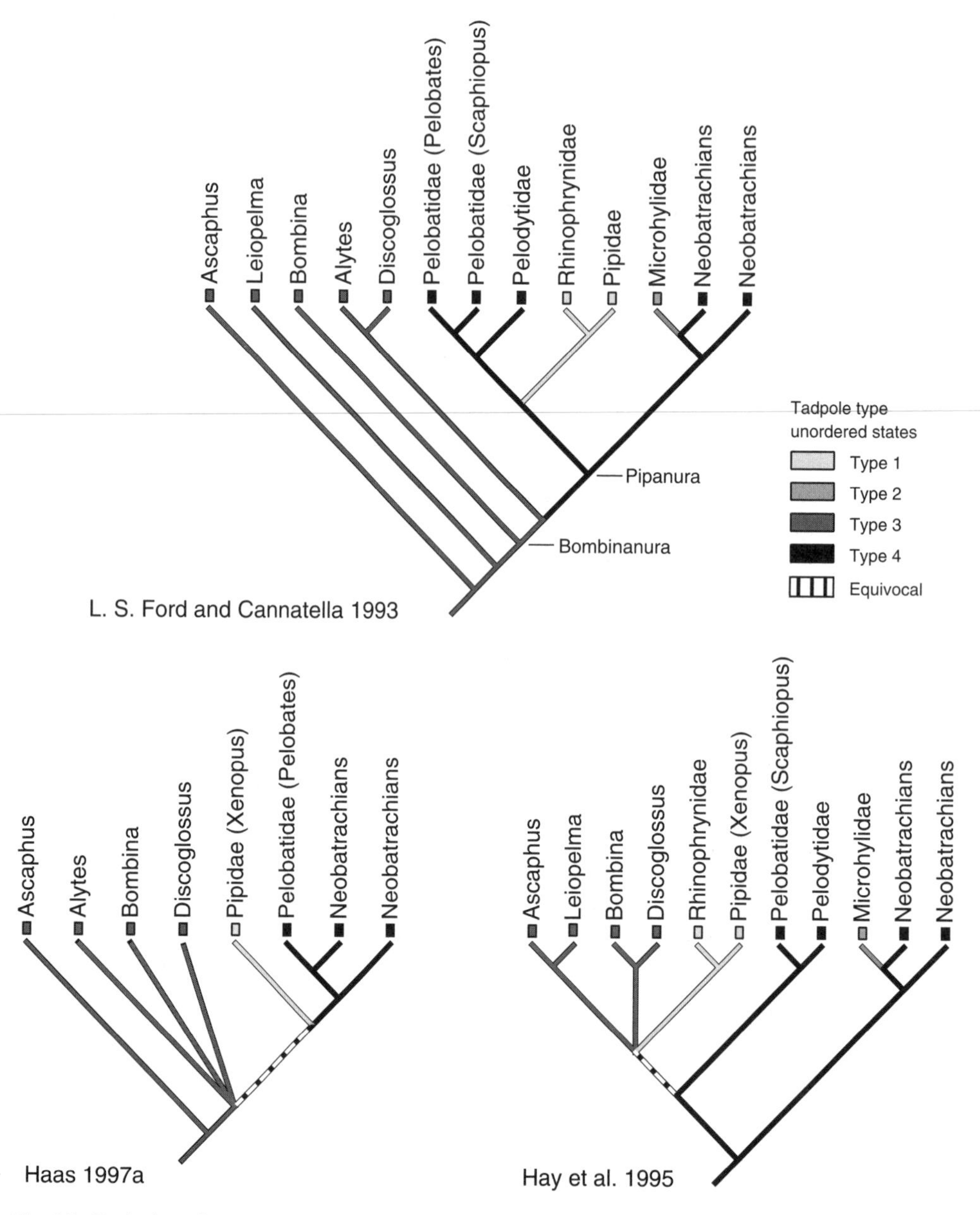

Fig. 4.2. Evolution of Orton's (1953) tadpole types under the assumption of unordered states, based on the phylogenies of L. S. Ford and Cannatella (1993), Haas (1997a), and Hay et al. (1995).

Table 4.1 Summary of literature concerning tadpole chondrocrania, muscles, and gill apparatus. Since this table was first compiled, similar compendia have appeared (Haas 1993, 1995, 1996b) that were of great help in updating the present table. All taxa treated in Haas (1996b) are also listed in Haas (1993). Some species names may be incorrect according to current taxonomy.

Taxon	Literature
Ascaphus truei	Gradwell 1971a, 1973; Haas 1995, 1996b, 1997a, b; Pusey 1943; Reiss 1997; Van Eeden 1951; Wassersug and Hoff 1982
LEIOPELMA	
Leiopelma archeyi	N. G. Stephenson 1951b
Leiopelma hochstetteri	N. G. Stephenson 1955
BOMINATORIDAE	
Bombina bombina	Goette 1875; Ridewood 1898b (as *B. igneus*); Severtsov 1969b
Bombina orientalis	Haas 1997a; O. M. Sokol 1975; Wassersug and Hoff 1982
Bombina variegata	Haas 1995, 1996b, 1997a
DISCOGLOSSIDAE	
Alytes obstetricans	De Beer 1985; Haas 1995, 1996b, 1997a; Krujitzer 1931; Magnin 1959; Pusey 1938; Ridewood 1898a, b; O. M. Sokol 1981b; Van Seters 1922; Wassersug and Hoff 1979, 1982
Discoglossus pictus	Kraemer 1974; Haas 1995, 1996b, 1997a; Pusey 1943; Ridewood 1898b; Schlosser and Roth 1995, 1997a, 1997b
Discoglossus sardus	Púgener and Maglia 1997; O. M. Sokol 1981b
MEGOPHRYIDAE	
Leptobrachium hasseltii	Ridewood 1898b
Megophrys montana	Haas 1996b; Krujitzer 1931; Ramaswami 1943
Megophrys nasuta	Haas 1995
Megophrys parva	Ramaswami 1943
Megophrys robusta	Ramaswami 1943
Megophrys stejnegeri	Haas 1995
PELOBATIDAE	
Pelobates fuscus	Dugès 1834; Haas 1997a; Luther 1914; Plasota 1974; Ridewood 1898b; Roček 1981; Schultze 1888, 1892; Severtsov 1969a
Pelobates syriacus	O. M. Sokol 1975, 1981b
Scaphiopus holbrookii	Haas 1995
Spea bombifrons	O. M. Sokol 1981b; Wassersug and Hoff 1979, 1982 (as *Scaphiopus*); J. J. Wiens 1989
PELODYTIDAE	
Pelodytes punctatus	Ridewood 1897b, 1898b; O. M. Sokol 1981a, b
PIPIDAE	
Hymenochirus boettgeri	O. M. Sokol 1959, 1962, 1977a
Pipa carvalhoi	O. M. Sokol 1977a
Pipa parva	Haas 1995; O. M. Sokol 1997a
Pipa pipa	W. K. Parker 1876 (as *P. monstrosa*); Ridewood 1897a, 1898b (as *P. americana*); Roček 1989, 1990; Roček and Veselý 1989
Silurana tropicalis	O. M. Sokol 1977a, b (as *Xenopus*)
Xenopus laevis	Dreyer 1915; Edgeworth 1930; Gradwell 1971b; Haas 1995, 1997a; Kotthaus 1933; W. K. Parker 1876 (as *Dactylethra capensis*); Paterson 1939a; Ramaswami 1941; Ridewood 1897a, 1898b; Sedra and Michael 1956a, 1957; Trueb and Hanken 1993; Wassersug and Hoff 1979, 1982; Weisz 1945a, 1945b
RHINOPHRYNIDAE	
Rhinophrynus dorsalis	Haas 1995, 1996b; Orton 1943; O. M. Sokol 1975, 1977a, 1981b; Swart and De Sá 1998; Wassersug and Hoff 1982
NEOBATRACHIA	
Heleophryne natalensis	Wassersug and Hoff 1979, 1982
Heleophryne purcelli	Ramaswami 1944; Van Der Westhuizen 1961
MYOBATRACHINAE	
Pseudophryne australis	C. M. Jacobson 1968
Pseudophryne bibronii	C. M. Jacobson 1968
Uperoleia lithomoda	Davies 1989
LIMNODYNASTINAE	
Limnodynastes peronii	Haas 1996b, 1997a
Limnodynastes tasmaniensis	W. K. Parker 1881
Mixophyes fasciolatus	Kesteven 1944
"LEPTODACTYLIDAE"	
Alsodes barrioi	Lavilla 1992a
Caudiverbera caudiverbera	W. K. Parker 1881; Reinbach 1939; Ridewood 1898b (as *Calyptocephalus gayi*); O. M. Sokol 1981b
Ceratophrys cornuta	Wild 1997
Ceratophrys cranwelli	Fabrezi and García 1994; Lavilla and Fabrezi 1992
Cycloramphus stejnegeri	Lavilla 1991
Eleutherodactylus coqui	Hanken et al. 1992; Schlosser and Roth 1997b
Eleutherodactylus nubicola	Lynn 1942
Lepidobatrachus laevis	Haas 1995, 1996b; Ruibal and Thomas 1988
Lepidobatrachus llanensis	Lavilla and Fabrezi 1992
Leptodactylus chaquensis	O. M. Sokol 1981b
Leptodactylus ocellatus	W. K. Parker 1881
Leptodactylus pentadactylus	Haas 1995, 1996b

Table 4.1 *continued*

Odontophrynus achalensis	Haas 1997a
Physalaemus pustulosus	Schlosser and Roth 1997b
Pleurodema bibronii	O. M. Sokol 1981b
Pleurodema borellii	Fabrezi and García 1994; Wassersug and Hoff 1982
Telmatobius bolivianus	Lavilla and De la Riva 1993
Telmatobius ceiorum	Fabrezi and Lavilla 1993
Telmatobius culeus	W. K. Parker 1881
Telmatobius jelskii	W. K. Parker 1881 (as *Cyclorhamphus culeus,* fide Ridewood 1898b)
Telmatobius laticeps	Fabrezi and Lavilla 1993
Telmatobius cf. *marmoratus*	Ridewood 1898b
Telmatobius pisanoi	Fabrezi and Lavilla 1993
PSEUDIDAE	
Pseudis paradoxa	W. K. Parker 1881; Ridewood 1898b
RHINODERMA	
Rhinoderma darwinii	Lavilla 1987
BUFONIDAE	
Bufo cf. *americanus*	W. K. Parker 1881
Bufo americanus	Wassersug and Hoff 1982
Bufo angusticeps	Barry 1956
Bufo bufo	Haas 1995, 1996b; W. K. Parker 1876; Ridewood 1898b (as *B. vulgaris*)
Bufo "lentiginosus"	Sedra 1951
Bufo melanostictus	Ramaswami 1940
Bufo regularis	Sedra 1950; Sedra and Michael 1956b, 1958, 1959
Bufo cf. *spinulosus*	W. K. Parker 1881
Bufo viridis	O. M. Sokol 1975
HYLIDAE	
Agalychnis callidryas	Haas 1995, 1996b
Anotheca spinosa	Haas 1996b; Wassersug and Hoff 1982
Cyclorana platycephala	Ridewood 1898b (as *Chiroleptes*)
Flectonotus goeldii	Haas 1996a
Gastrotheca espeletia	Haas 1996b
Gastrotheca gracilis	Fabrezi 1985; Fabrezi and Lavilla 1992
Gastrotheca cf. *marsupiata*	W. K. Parker 1881
Gastrotheca marsupiata	Haas 1996b
Gastrotheca orophylax	Haas 1996b
Gastrotheca peruana	Haas 1996b
Gastrotheca pseustes	Haas 1996b
Gastrotheca riobambae	Haas 1995, 1996b, 1997a
Hyla arborea	Haas 1996b; Ridewood 1898b; Severtsov 1969a
Hyla cinerea	Haas 1996b
Hyla lanciformis	De Sá 1988
Hyla microcephala	Haas 1996b
Hyla nana	Fabrezi and Lavilla 1992
Hyla pulchella	Fabrezi and Lavilla 1992; Lavilla and Fabrezi 1987
Hyla rosenbergi	Haas 1996b
Hyla squirella	O. M. Sokol 1981b
Phasmahyla guttata	Fabrezi and Lavilla 1992
Phyllomedusa boliviana	Fabrezi and Lavilla 1992
Phyllomedusa sauvagii	Fabrezi and Lavilla 1992
Phyllomedusa trinitatis	Haas 1996b; Kenny 1969a
Pseudacris crucifer	W. K. Parker 1881 (as *Acris pickeringi*)
Pseudacris regilla	O. M. Sokol 1981b (as *Hyla*)
Scinax acuminata	Fabrezi and Lavilla 1992
Scinax rubra	Haas 1996b
"RANIDAE"	
Amolops afghana	Ramaswami 1940, 1943 (as *Rana*)
Euphlyctis hexadactylus	Ramaswami 1940 (as *Rana*)
Hoplobatrachus tigerinus	Chacko 1965, 1976; Ramaswami 1940 (as *Rana*)
Occidozyga laevis	Ridewood 1898b (as *Oxyglossus*)
Rana catesbeiana	Gradwell 1968, 1970, 1972a, b; Starrett 1968
Rana clamitans	W. K. Parker 1881
Rana curtipes	Ramaswami 1940
Rana dalmatina	Kratochwill 1933 (as *R. agilis*)
Rana esculenta	Ridewood 1898b
Rana cf. *pipiens*	W. K. Parker 1881
Rana pipiens	Starrett 1968; Wassersug and Hoff 1979
Rana temporaria	De Beer 1985; De Jongh 1968; Gaupp 1893; Haas 1995, 1996b, 1997a; W. K. Parker 1871; Plasota 1974; Pusey 1938; Ridewood 1898b; O. M. Sokol 1975; Spemann 1898; Stöhr 1882
Rana "whiteheadi"	Ridewood 1898b

Table 4.1 *continued*

RHACOPHORIDAE	
Buergeria buergeri	Okutomi 1937
Philautus variabilis	Ramaswami 1938
Rhacophorus leucomystax	Ridewood 1898b
MICROHYLIDAE	
Breviceps adspersus	Swanepoel 1970
Dermatonotus muelleri	Lavilla 1992b
Gastrophryne carolinensis	Wassersug and Hoff 1982
Gastrophryne usta	Haas 1995, 1996b
Hamptophryne boliviana	De Sá and Trueb 1991
Hypopachus barberi	O. M. Sokol 1975, 1981b
Hypopachus variolosus	Starrett 1968 (as *H. alboventer*)
Microhyla ornata	Ramaswami 1940; Ridewood 1898b
Microhyla pulchra	Haas 1995, 1996b
Otophryne pyburni	Wassersug and Pyburn 1987 (as *O. robusta*)
Phrynomantis annectens	Gradwell 1974 (as *Phrynomerus*)
Uperodon systoma	Ramaswami 1940
DENDROBATIDAE	
Colostethus nubicola	Haas 1995, 1996b
Colostethus subpunctatus	Haas 1995, 1996b
Colostethus trinitatis	Ridewood 1898b (as *Phyllobates*)
Dendrobates auratus	De Sá and Hill 1998
Dendrobates tinctorius	Haas 1995, 1996b
Epipedobates anthonyi	Haas 1995, 1996b
Epipedobates boulengeri	Haas 1995, 1996b
Epipedobates tricolor	Haas 1995, 1996b
Phyllobates bicolor	Haas 1993, 1996b, 1997a

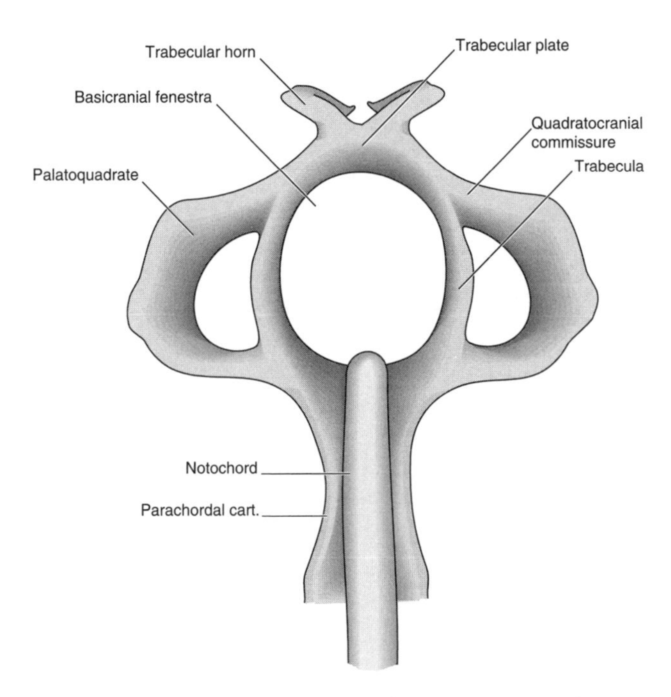

Fig. 4.3. Developing chondrocranium of *Rana temporaria* (7.5-mm larva) in dorsal view. Redrawn from Stöhr (1882).

tebral neural arches. The occipital arch fuses to the rear end of the skull.

Ethmoid Region

The anterior end of the primordial chondrocranium is elaborated to support the nasal capsules and the larval jaws. Two trabecular horns (cornua trabecularum) extend forward ventrally and anterolaterally from the ethmoid (= trabecular plate, fig. 4.3). The trabecular horns articulate with, or are fused to, a pair of suprarostral cartilages (cartilago labialis superior, superior rostral cartilage), the skeletal elements of the tadpole upper jaw. In other gnathostomes, the functional upper jaws are formed either by the palatoquadrate cartilage or by a series of dermal bones (e.g., maxillae and premaxillae). In this region two ligaments, the quadratoethmoid ligament (ligamentum quadratoethmoidale, l. cornu-quadratum mediale; Van Der Westhuizen 1961) and the lateral circumoral ligament (l. cornu-quadratum laterale, l. cornu-quadratum, l. circumoralis; O. M. Sokol 1981b) join the ethmoid region to the palatoquadrate cartilage; see details below under Circumoral Ligaments.

The nasal septum (septum nasi; fig. 4.4) arises from the trabecular plate and separates the nasal capsules. The nasal capsules of larvae usually are partially formed spheres. The roof of the nasal capsule (tectum nasi) is confluent posteriorly with the lamina orbitonasalis and is united to the septum nasi. The confluence of the trabecular horns joins the ventral part of the nasal capsule and forms the floor of the nasal capsule (solum nasi, trabecular plate, ethmoid plate; O. M. Sokol 1981b). The anterior wall of the nasal capsule is the anterior cupola (cupola anterior).

The nasal capsule is separated from the orbit by a wall, the lamina orbitonasalis (planum antorbitale). The lamina orbitonasalis is confluent with the tectum nasi anteriorly and the anterior end of the orbital cartilage (via the sphenethmoid commissure; see below) medially. The muscular process of the palatoquadrate is bound to the processus antorbitalis (Haas 1995) of the lamina orbitonasalis by the ligamentum tecti, which has two parts, the l. tecti superius and l. tecti inferius. These ligaments may be chondrified in some taxa to form the cartilago tecti ("cartilago tectum" of Sokol 1975) or commissura quadrato-orbitalis (Haas 1995). As metamorphosis begins, the cartilages and olfactory epithelium of the nasal region become elaborated into a highly folded saclike structure; details of these structures can be found in R. S. Cooper (1943), Denis (1959), Higgins (1920), Jurgens (1971), Rowedder (1937), and Swanepoel (1970).

Ossifications associated with the nasal region include several well-known dermal bones and some neomorphic elements. The nasals form part of the roof of the nasal capsules. The vomers are palatal bones that form the floor of the nasal capsule; these are absent primitively in frogs (Cannatella 1985) and secondarily lost in several taxa. The septomaxillae are generally small elements that are intimately associated with the nasolacrimal duct but develop independently of the nasal cartilages (De Jongh 1968; Jurgens 1971). Neomorphic elements, such as the internasals and prenasals in various taxa of casque-headed frogs (Trueb 1973), appear relatively late in ontogeny (Trueb 1985). Hanken and Hall (1984, 1988), N. E. Kemp and Hoyt (1969), Stokely and List (1954), and citations in Trueb (1985) provided ossification sequences in various species.

Diversity. The ligamentum tecti was figured for *Pelodytes* (O. M. Sokol 1981b) and *Xenopus laevis* (Sedra and Michael 1956a). Gradwell (1972b) illustrated it in *Rana catesbeiana* as the composite of the l. supraorbitale cranii and l. supraorbitalis ethmoidale. It is not mentioned in some taxa for which the chondrocranium is well described: *Ascaphus, Heleophryne,* or *Rana temporaria* (De Jongh 1968; Van Der Westhuizen 1961; Van Eeden 1951). Variation in the lateral circumoral ligament is discussed below under Circumoral Ligaments.

The ethmoid region of pipoids is difficult to compare with that of other frogs. For example, there is a single medial cartilaginous bar that supports the nasal septum. O. M. Sokol (1975) argued reasonably that this bar represents the fused trabecular horns, but Roček and Veselý (1989) considered this planum internasale to be nonhomologous with the trabecular horns. They also described several other features in which the ethmoid region of pipids differs from those of other frogs. The homology of ethmoidal structures was also discussed by Haas (1993, 1995).

Orbital Region

The lateral wall of the cranial cavity (cavum cranii) consists of an orbital cartilage (cartilago orbitalis; fig. 4.4) that arises from the posterior parts of the trabecular cartilages and the anterior ends of the parachordal cartilages. The orbital cartilage is joined to the floor of the braincase by several cartilaginous pillars; in some vertebrates the orbital cartilage arises as a discrete element, but it is not clear if this is so in frogs.

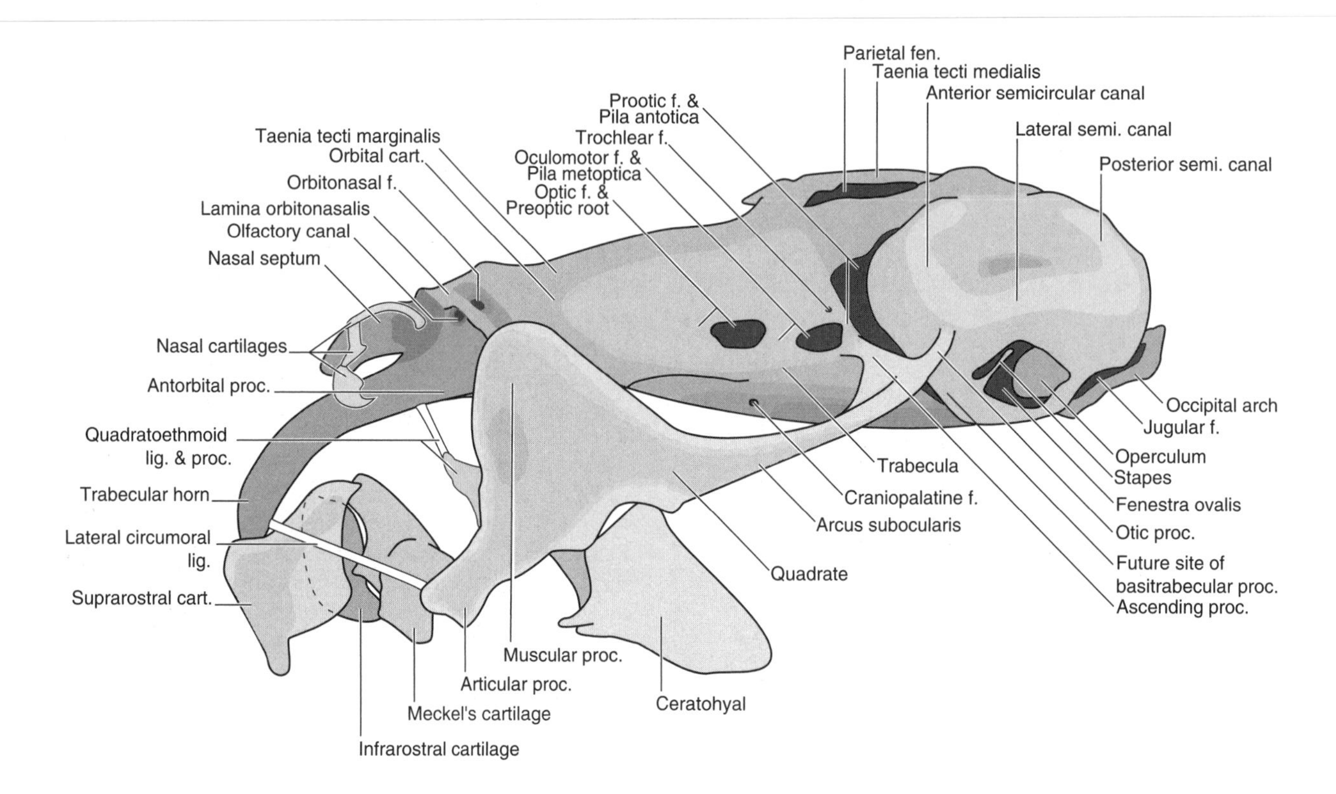

Fig. 4.4. Lateral view of the chondrocranium of a *Rana temporaria* tadpole (13 mm body length). Abbreviations: cart. = cartilage, f. = foramen, fen. = fenestra, lig. = ligament, and proc. = process. Redrawn from Pusey (1938).

The anteriormost of these pillars is the preoptic root forming the anterior margin of the optic foramen and attaching to the trabecula. In frogs and salamanders the preoptic root forms an extensive cartilaginous wall between the optic foramen and the sphenethmoid commissure; in caecilians (M. H. Wake and Hanken 1982) it is more of a strut. The pila metoptica grows upward from the trabecula and separates the optic foramen from the oculomotor (metoptic) foramen, the latter for the oculomotor nerve (III), ophthalmic artery, and pituitary vein. The pila antotica (pila prootica) arises from the parachordals and separates the oculomotor and prootic foramina. A fourth, more transitory strut, the pila ethmoidalis, forms the medial walls of the nasal capsules and olfactory canals (canales olfactorii, foramina olfactoria evehentia) and merges into the nasal septum (De Beer 1985). Posteriorly, the orbital cartilage joins the otic capsules via the taenia tecti marginalis. Anteriorly, the sphenethmoid commissure connects the orbital cartilage to the lamina orbitonasalis and the nasal capsule.

The orbitonasal foramen (foramen orbitonasalis; fig. 4.4) demarcates the lamina orbitonasalis from the orbital cartilage; the profundus nerve (ophthalmic branch of V) passes out of the orbit into the nasal cavity through this foramen. The optic foramen (for cranial nerve II) is the largest of the foramina of the orbital cartilage and is located posteriorly in the orbital cartilage. The oculomotor foramen (for cranial nerve III) is smaller and lies just posterior to the optic foramen. The tiny trochlear foramen (for cranial nerve IV) is located dorsal to the two aforementioned foramina near the taenia tecti marginalis. The prootic foramen (foramen prooticum) more or less separates the orbital cartilage from the otic capsule. The trigeminal (V) and facial nerves (VII; from the combined prootic ganglion), the abducens nerve (VI), and the lateral cephalic vein (vena capitis lateralis) pass through the prootic foramen.

Ossifications in this region include the palatines and sphenethmoid (orbitosphenoid). The palatines form along the lamina orbitonasalis and usually extend from the maxilla medially to near the midline. The sphenethmoid, usually an unpaired bone in adults, ossifies as paired elements along the anterolateral wall of the braincase anterior to the optic foramen.

Diversity. The pila antotica is absent in microhylid larvae (O. M. Sokol 1975) and *Rhinophrynus* (Trueb and Cannatella 1982). In the basal clades of frogs (*Ascaphus, Leiopelma, Discoglossus,* and *Bombina*) the prootic foramen is subdivided by a strip of cartilage called the prefacial commissure into the prootic foramen (for cranial nerve V) and palatine foramen (for cranial nerve VII; O. M. Sokol 1975). In taxa with the prefacial commissure, the trigeminal and facial ganglia are unfused, whereas in those lacking the commissure the ganglia are fused to form a prootic ganglion. As suggested by Sokol (1975), the lack of the commissure is a synapomorphy of Pipanura. Other aspects of variation were discussed by Haas (1993).

Braincase and Otic Capsules

The braincase of frog larvae and adults is open dorsally via the frontoparietal fenestra (fenestra frontoparietalis; fig. 4.5). The lateral rim of the fenestra is formed by strips of cartilage called taeniae tecti marginales, which are confluent with the orbital cartilages ventrally and with the tectum anterius anteriorly. Posteriorly, the taeniae roof the gap between the otic capsules and form the tectum synoticum. The frontoparietal fenestra may be subdivided by additional struts of cartilage, as in *Rana.* The taenia tecti transversalis is a transverse strut that partitions the parietal fenestra posteriorly from the frontoparietal fenestra anteriorly. The taenia tecti medialis, if present, extends from the taenia tecti transversalis to the tectum synoticum and subdivides the parietal fenestra into left and right parietal fenestrae. These tectal cartilages persist in adults.

The occipital arch forms the posterior end of the cranium; it is serially homologous with the neural arches of the vertebrae. Frogs have only one occipital arch; most other vertebrates have more than one (De Beer 1985). There are no pre-occipital arches in frogs (De Beer 1985), but see Kraemer (1974). The occipital arch is supported ventrally by the parachordals; early in ontogeny it fuses to the basal plate and otic capsules. The tectum posterius joins the halves of the occipital arch dorsally and fuses anteriorly to the tectum synoticum. The ventral part of the occipital arch bears the occipital condyles, which articulate with the cotyles of the atlas.

The otic capsules (capsulae auditivae; figs. 4.4–4.5) are subspherical structures that adjoin the posterior end of the braincase. Ontogenetically, they begin as hemispheres that are open medially. The hemispheres fuse to the parachordals and attach to the basal plate ventrally, the tectum synoticum dorsally, and the taeniae tecti marginales anterodorsally. Within these capsules the anterior and lateral semicircular canals (canalis semicircularis) appear and slightly later the posterior semicircular canals form. Each otic capsule develops a flattened crest, the crista parotica, that extends laterally from the outer wall of the lateral semicircular canal. The posterior part of the ear capsule forms the processus muscularis capsulae auditivae for muscle attachment. The otic ligament joins the anterior part of the crista parotica to the otic process (processus oticus) of the palatoquadrate (see below under Jaws).

The dorsal features and general size of the larval otic capsules are determined by the three semicircular canals of the inner ear. Chondrification and later ossification around these, especially the anterior and posterior canals, are evident as the epiotic eminences. Metamorphic changes in the inner ear are minimal; those of the middle ear are pronounced. The middle ear of adult frogs includes an opercularis system and usually a tympanic-columellar system; the differences in developmental timing of these two systems suggest that they are functionally independent (Hetherington 1988). Both of these communicate with the inner ear via the fenestra ovalis, a prominent opening in the ventrolateral wall of the otic capsule that is covered by a membrane in the tadpole. In the adult, the fenestra ovalis is largely occluded by two struc-

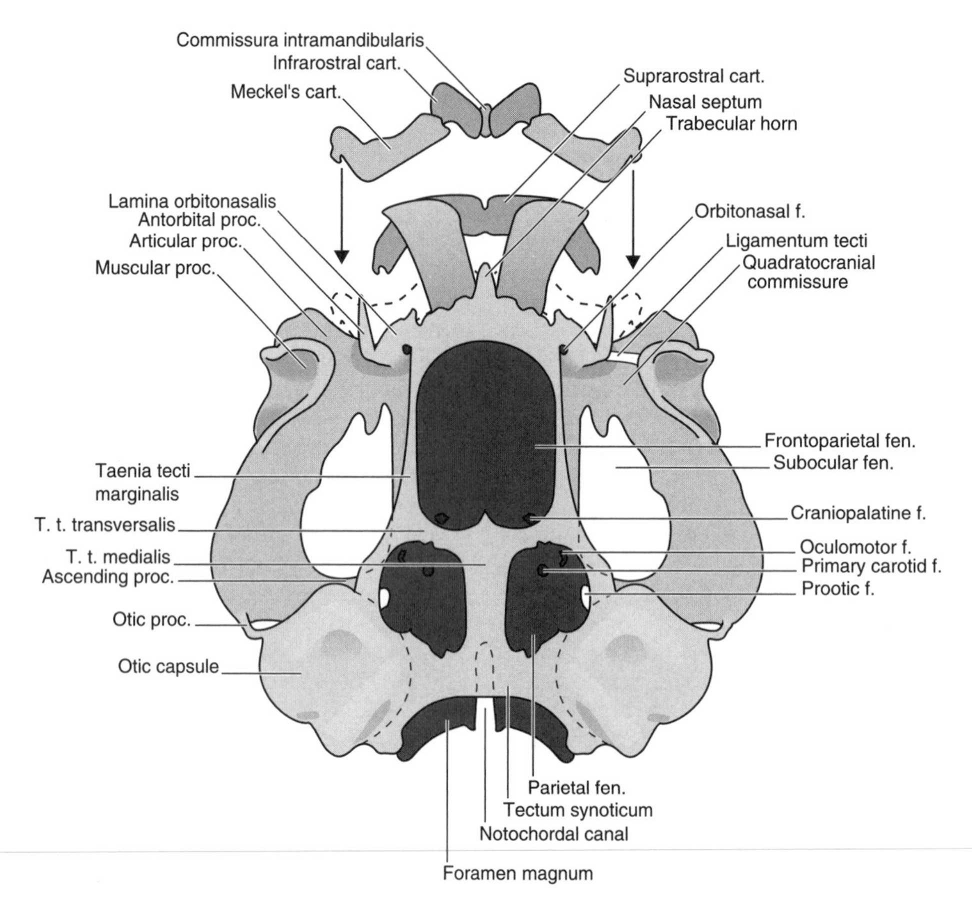

Fig. 4.5. Dorsal view of the chondrocranium of a *Rana temporaria* tadpole (13 mm body length). Abbreviations: cart. = cartilage, f. = foramen, fen. = fenestra, lig. = ligament, and proc. = process. Redrawn from Pusey (1938).

tures, the operculum (not to be confused with the homonymous flap of skin covering the larval gills) and the stapes (columella, plectrum). The operculum forms first from mesenchyme that accumulates over the surface of the membrane covering the fenestra ovalis (Eiselt 1942) at about the time of forelimb emergence before metamorphosis. A group of myoblasts near the suprascapula develops into the opercularis muscle and the levator scapulae superior. The opercularis muscle inserts onto the operculum at a raised point called the muscular process. Hetherington (1987) concluded that the opercularis system likely was functional at the time of forelimb emergence and first terrestrial activity. This is consistent with hypotheses concerning the reception of seismic or aerial signals by the opercularis system (Hetherington 1988).

The stapes appears at or just after metamorphosis as a footplate (pars interna) in the fenestra ovalis. Development proceeds distally to form the shaft (pars media), which later ossifies. The pars externa (extrastapes, extracolumella) forms more commonly as a separate chondrification, but in some taxa it forms as an extension of the shaft. During development of the stapes, the middle ear cavity forms from an evagination of the pharynx but maintains the pharyngeal connection in the adult via the openings of the Eustachian tube (ostia pharyngea). The skin overlying the distal end of the stapes thins to form the tympanum, and the cartilaginous tympanic annulus appears near the articulation of the palatoquadrate and Meckel's cartilage. The tympanic annulus migrates posteriorly with the palatoquadrate during metamorphosis and comes to lie slightly ventral and anterior to the stapes. Development of the columella and ear apparatus is discussed by Barry (1956), Sedra and Michael (1959), and Van Der Westhuizen (1961) and reviewed recently by Hetherington (1987, 1988).

The medial wall of the otic capsule is pierced by several holes that communicate nerves and vessels to the cavum cranii: one to three acoustic foramina for the statoacoustic nerve (acoustic or auditory nerve, VIII) are the most ventrally placed foramina. The anterior and posterior acoustic foramina transmit the anterior and posterior limbs of the acoustic ganglion, respectively. The intermediate acoustic foramen (foramen medium) transmits a small nerve from the ganglion (O. M. Sokol 1981b). Two or three perilymphatic foramina lie dorsal and posterior to the acoustic foramina; the superior perilymphatic foramen opens into the cavum cranii, and the inferior perilymphatic foramen opens into the condyloid fossa (see below). The perilymphatic ducts traverse the perilymphatic foramina and join the perilymphatic sac (saccus perilymphaticus, recessus scalae tympani). The perilymphatic sac is a diverticulum of the perilymphatic spaces located in the anterior part of the condyloid fossa; it communicates with the otic capsule by means of the perilymphatic ducts. The sac has a lateral aperture (De Beer 1985) over which is stretched the membrane of the round window (fenestra rotundum), which separates the perilymphatic sac from the middle ear cavity. The membrane of the round window dampens vibrations in the endolymphatic sac (Wever 1985). Witschi (1949) described diverticula of the lungs—bronchial columellae—in *Rana temporaria* that extended anteriorly to terminate at the round window; supposedly these transmit vibrations from the lung cavities to the inner ear (chap. 5). Van Bergeijk (1959) hypothesized that these diverticula had a hydrostatic function in *Xenopus laevis,* but Wassersug and Souza (1990) rejected this hypothesis.

The endolymphatic foramen is located anterior and dorsally near the taenia tecti marginalis. The endolymphatic duct, which expands to form the endolymphatic sac, passes through the endolymphatic foramen. The endolymphatic sac exits the cranial cavity and passes into the vertebral canal to form the sacs of chalky substance at the spinal ganglia (Wever 1985).

Posteriorly, the main openings in the skull (figs. 4.4–4.5) are the foramen magnum, inferior perilymphatic foramen, and jugular foramen (foramen jugulare, fissura metotica). The foramen magnum is the large medial opening through which the spinal cord passes into the cranium. The jugular foramen is actually a gap between the ventral part of the occipital arch and the otic capsule and is traversed by the glossopharyngeal (IX) and vagus (X) nerves. The jugular foramen and the inferior perilymphatic foramen open from the otic capsule into a wedge-shaped extracranial space, the condyloid fossa.

Ossifications associated with the braincase include the frontoparietals, parasphenoid, and in a few taxa (see below), the interfrontal and interparietal. The frontoparietals are paired elements that cover the frontoparietal fenestra and form the roof of the braincase. The parasphenoid is an unpaired medial element that forms the floor of the braincase, basis cranii, and, in nonpipoids, the otic capsules. The interfrontal is an unpaired element located between the anterior ends of the frontoparietals that is present only in some species of *Bombina* (Tschugunova 1981). The azygous interparietal is known only from *Pelobates* (Roček 1981); it fuses with the frontoparietals, which themselves are fused in adult *Pelobates.*

Ossifications associated with the otic capsules and occipital arch include the prootics, exoccipitals, and stapes. The prootic ossifies in association with the prootic foramen. The exoccipital forms in the occipital arch near the jugular foramen and occipital condyles. Coalescence of the exoccipital and prootic results in a usually solid otic capsule in the adult. The stapes is one of the last bones to ossify (Trueb 1985).

Diversity. *Caudiverbera caudiverbera* and *Ceratophrys cornuta* differ from other frogs in having a solid roof of the braincase so that there is no frontoparietal fenestra (Reinbach 1939; Wild 1997). Both the taenia tecti medialis and taenia tecti transversa are absent in mesobatrachians and *Ascaphus* (L. S. Ford and Cannatella 1993; Van Eeden 1951); Cannatella (1985) discussed the phylogenetic significance of the character. Evolutionary patterns in the number and position of acoustic foramina are unclear; early in ontogeny there is one foramen that later subdivides (De Beer 1985; Kraemer 1974; O. M. Sokol 1981b). However, it is uncertain whether homologous bundles of nerve fibers traverse the foramina in different taxa. Haas (1996b: table 3) summarized data on the acoustic foramina, and he (1995) noted the lack of data on intertaxic variation in the perilymphatic foramina. O. M. Sokol (1977a) reported "supraoccipital" ossifications in the occipital arch of *Pipa carvalhoi;* frogs are generally considered to lack supraoccipital ossifications.

Bronchial diverticula are present in *Rhinophrynus, Silurana tropicalis,* and *Xenopus laevis* (O. M. Sokol 1975, 1977a; Witschi 1949, 1950, 1955) but absent in *Bufo americanus* and *Pseudacris crucifer* (Hetherington 1987) and *Hymenochirus boettgeri* and *Pipa carvalhoi* (O. M. Sokol 1977a).

Jaws

The jaws of anuran tadpoles are unique in having suprarostral cartilages and infrarostral cartilages (figs. 4.4–4.5) that support the upper and lower keratinized sheaths but do not survive metamorphosis. The suprarostral cartilages are resorbed, and the infrarostral cartilages are incorporated into Meckel's cartilage. Other elements of the jaws include various circumoral ligaments, Meckel's cartilages, and the palatoquadrate.

Suprarostral Cartilages

The suprarostral cartilages attached to the trabecular horns (fig. 4.4) function as the movable upper jaw in tadpoles. The upper keratinized jaw sheath of tadpoles (when present) is supported by the suprarostral cartilage. In *Rana* the suprarostral cartilage arises as paired elements that fuse together (De Jongh 1968; O. M. Sokol 1981b) to form a medial corpus with lateral wings or alae. The suprarostral is connected dorsally to the trabecular horns by synovial joints or ligaments or is fused to the trabecular horns. The suprarostral disappears with metamorphosis, and its function as the upper jaw is assumed by the developing premaxilla and maxilla.

Diversity. In *Ascaphus,* the alae of the suprarostrals are separated from the corpus (Van Eeden 1951), producing a tripartite structure. In many other taxa, the suprarostral is divided into a lateral pars alaris and a medial pars corporis. The cornu trabeculae articulates with both the pars alaris and the pars corporis in bufonids, dendrobatids, hylids, and some leptodactylids. Haas (1995) offered that this conformation may be a synapomorphy of these taxa (= Bufonoidea?) or perhaps Neobatrachia; see also Haas (1993, 1996b).

The posterior end of the ala in some taxa (e.g., *Heleophryne, Megophrys,* and *Spea*) may have a collagenous or cartilaginous adrostral element associated with it (O. M. Sokol 1981b). The adrostral tissue mass is a collagen-rich connective tissue that is associated with the processus posterior dorsalis of the pars alaris. Haas (1995) described the details of its connection to the mandibulosuprarostral ligament and the lateral circumoral ligament (ligamentum cornuquadratum) in dendrobatids. He noted its presence (apomorphic) only in Type 4 tadpoles. I interpret its absence in microhylid larvae (Type 2) as a loss. Its absence in pipoids (Type 1) may also be a loss or may indicate that pelobatoids are more closely related to Neobatrachia than to pipoids, resulting in paraphyly of Mesobatrachia. Noble (1929) claimed that the absence of the suprarostral cartilage was characteristic of microhylid (his "brevicipitid") larvae. Starrett (1973) reported that the suprarostral was lacking in pipoid (Type 1) and microhylid (Type 2) larvae; this was linked to her view that the absence of jaw sheaths was pleiomorphic for anurans. O. M. Sokol (1975) pointed out that all frog larvae have suprarostral cartilages; in Types 1 and 2 larvae they are firmly ankylosed to the trabecular horns. The presence of keratinized jaw sheaths may be correlated with mobility of the suprarostral. The posteroventral process of the suprarostral (O. M. Sokol 1981b) is absent in Pipanura. *Alytes* and *Discoglossus* have long, distinctive, U-shaped corpora (O. M. Sokol 1981b). The lateral tip of the suprarostral plate in *Xenopus laevis* fuses with the quadratoethmoid cartilage to form the tentacular cartilage, which supports the tentacle (Trueb and Hanken 1993).

Palatoquadrate and Suspensorium

The larval palatoquadrate (pterygoquadrate; figs. 4.3 and 4.6) is a long flat strip of cartilage anchored to the neurocranium by four cartilaginous processes: ascending process (processus ascendens), basal process (processus basalis, processus pseudobasalis, hyobasal process), otic process (processus oticus), and the quadratocranial commissure (commissura quadratocranialis anterior).

The term suspensorium was used in its present sense by T. H. Huxley (1858) for the common attachment of the mandibular and hyoid arches to the skull (Pyles 1987). According to O. M. Sokol (1981b), the suspensorium is the posterior end of the palatoquadrate that lies at the anterior end of the otic capsule. It articulates with the otic capsules via the ascending and otic processes. The ascending process is a round bar that curves medially and fuses with the region of the pila antotica (see Diversity below). The otic process is the lateral part of the suspensorium; it is a thick plate that is curved almost vertically. The otic process is attached to the wall of the lateral semicircular canal near the crista parotica by the otic ligament; in some taxa this process chondrifies and is then called the larval otic process, which is considered homologous with the adult otic process (Van Der Westhuizen 1961; Van Eeden 1951). The larval otic process is destroyed during metamorphosis, but the cells released in the process rechondrify to form the adult process.

In most vertebrates the basal process of the palatoquadrate articulates with the basitrabecular process (fig. 4.4; a lateral projection of the trabecula) anterior to the palatine nerve. However, in most frogs the basitrabecular process is lost during ontogeny, and the basal process articulates or fuses with the neurocranium posterior to the palatine nerve (De Beer 1926). Because of the position of the process relative to the palatine nerve, Pusey (1938) rejected the homology of the two conditions; the newer process and its postpalatine connection have been termed the pseudobasal process (De Beer 1926) and pseudobasal connection (Pusey 1938). Van Der Westhuizen (1961) offered compelling evidence that the basal process simply has moved from the basitrabecular articulation to a more lateral position on the otic ledge to produce a postpalatine connection (commissure). Therefore, the basal and pseudobasal processes are homologous (Pyles 1987; Reiss 1997).

As one proceeds anteriorly, the palatoquadrate underlies the orbit and in this region is named the arcus subocularis (fig. 4.4). The wide gap that separates the palatoquadrate from the neurocranium is the subocular fenestra. The region anterior to the arcus subocularis is termed the quadrate (O. M. Sokol 1981b). The most prominent part of the quadrate is the muscular process (processus muscularis), a flattened triangular process that curves dorsally and medially and is the site of attachment for the orbitohyoideus and suspensoriohyoideus muscles. The muscular process is bound to the processus antorbitalis (Haas 1995) of the neurocranium by the ligamentum tecti (cartilago tecti or commissura quadrato-orbitalis when chondrified), which may be composed of two ligaments (see Ethmoid Region above), the ligamenti tecti superius et inferius. Anteromedially, the quadratocranial commissure (figs. 4.3 and 4.5) rises dorsally as a thick cartilaginous strip to join the quadrate to the floor of the braincase near the ethmoid plate. On the anterior margin of the commissure is the small triangular quadratoethmoid process (processus quadratoethmoidalis) to which the quadratoethmoid ligament (ligamentum quadratoethmoidale) is attached. A small processus pseudopterygoideus (Haas 1995) projects from the posterior part of the commissure into the fenestra subocularis.

The hyoquadrate process (processus hyoquadrati) is a condyle for the synovial articulation with the ceratohyal (figs. 4.4 and 4.7) and is located ventrally at the posterior part of the quadrate region. The hyoquadrate ligament attaches to a ridge posterior to the hyoquadrate process and binds the ceratohyal to the quadrate. The most anterior part of the quadrate is the stout articular process (pars articularis quadrati), which forms a synovial articulation, the adult jaw joint, with the posterior end of Meckel's cartilage.

During metamorphosis, the arcus subocularis, ascending process, hyoquadrate process, and muscular process are resorbed as the palatoquadrate rotates posteriorly (fig. 4.6). The quadratocranial commissure is partly destroyed, but its surviving part, with the quadratoethmoid process and processus maxillaris posterior of the lamina orbitonasalis, forms the pterygoid process (processus pterygoideus) of the adult. Most of the larval palatoquadrate disappears at metamorphosis, and some of its supportive function is assisted or taken over by several ossifications: maxilla, premaxilla, quadrate, quadratojugal, pterygoid, and squamosal.

The quadrate is an endochondral ossification that forms in the similarly named region of the palatoquadrate at the articular process, the point of articulation with the lower jaw. An arc of the premaxillae, maxillae, and quadratojugals takes

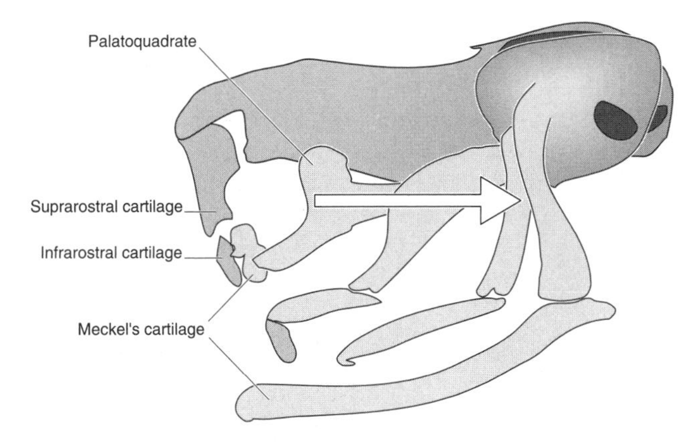

Fig. 4.6. Four arbitrary stages in the metamorphosis of the palatoquadrate of a generalized tadpole; lateral view. Arrow indicates direction of metamorphic shift of the jaw elements. After Sedra (1950).

over the role of the palatoquadrate in forming the upper jaw of the adult frog. The squamosal forms in association with the otic process and quadrate of the palatoquadrate cartilage. The pterygoid invests the adult pterygoid process, which forms in part from the quadratoethmoid process and the basal process.

Diversity. Several features of the palatoquadrate are highly modified in pipoids (O. M. Sokol 1977a; Swart and De Sá 1998). A novel structure, the ventrolateral process, is a synapomorphy of Pipoidea; it is present in *Pipa, Rhinophrynus, Silurana,* and *Xenopus* but not *Hymenochirus.* The muscular process and arcus subocularis are absent in *H. boettgeri* (O. M. Sokol 1962). The arcus subocularis is tenuously complete in *Xenopus laevis* (Trueb and Hanken 1993). In pipids, the otic process does not reach the anterior end of the otic capsule (O. M. Sokol 1975), and the ascending and otic processes are fused into a broad plate (O. M. Sokol 1977a).

The connection of the ascending process to the neurocranium is more dorsally placed (a "high" attachment) in basal taxa such as *Ascaphus* and *Bombina* and more ventrally placed ("low") in neobatrachians (O. M. Sokol 1975). In microhylids, the ascending process has a low attachment directly to the trabecula because the pila antotica is absent (O. M. Sokol 1975). The ascending process is absent in larvae of *Heleophryne, Otophryne,* and *Philautus* (see Haas 1995). The ascending process is lost during metamorphosis in all frogs except *Ascaphus truei* (Van Eeden 1951); Reiss (1997) disagreed with this interpretation, claiming that the persistent ascending process in *Ascaphus* is only part of the neurocranium. *Ascaphus truei* is also unique among Anura in retaining the "true" basal process in which the connection to the neurocranium is anterior to the palatine nerve (De Beer 1985; Pusey 1938). The condition in *Leiopelma* is somewhat intermediate (Pusey 1938; E. M. T. Stephenson 1951), and all other frogs surveyed have the pseudobasal process. Pyles (1987) summarized the issues regarding the homology of the pseudobasal/basal process. Reiss's (1997) intensive study of the larva of *Ascaphus* reached somewhat different conclusions about the phylogenetic significance of certain suspensorial characters than did L. S. Ford and Cannatella (1993), but because these characters are germane primarily to adults, a fuller critique of Reiss (1997) will not be undertaken here.

In *Leiopelma archeyi,* which has nidicolous development (Altig and Johnston 1989), the larval otic process remains unbroken through metamorphosis (N. G. Stephenson 1951b). *Heleophryne purcelli* has an apparently unique quadrato-otic ligament that attaches the rear end of the palatoquadrate to the floor of the otic capsule (Ramaswami 1944; Van Der Westhuizen 1961); this is distinct from the connection of the larval otic process to the otic capsules. Orton (1943) figured a cartilage of unknown homology in a *Rhinophrynus dorsalis* tadpole, which was identified as a free hyoquadrate process by O. M. Sokol (1975: fig. 12).

Ossification of the quadrate is known in only a few taxa of frogs (e.g., *Ascaphus truei*), and quadratojugals are absent in many taxa. The maxillae, premaxillae, pterygoids, and squamosals are present in almost all taxa (Trueb 1993).

Circumoral Ligaments

Four principal ligaments are associated with the jaws (fig. 4.4; see also fig. 4.16): (1) the quadratoethmoid ligament (ligamentum quadratoethmoidale, l. cornu-quadratum mediale; Schulze 1892), (2) lateral circumoral ligament (O. M. Sokol 1981b) (l. cornu-quadratum; l. cornu-quadratum laterale; l. circumorale; trabecular quadrate ligament; Pusey 1943), (3) mandibulosuprarostral ligament (l. rostrale superius cartilago Meckeli of Gradwell 1968), and (4) the suprarostral-quadrate ligament (l. rostrale superius quadrati; Gradwell 1968).

The quadratoethmoid ligament extends from a process (processus lateralis trabeculae; Haas 1995) on the posterolateral aspect of the trabecular horn to the quadratoethmoid process of the quadratocranial commissure. This ligament lies ventral and lateral to the internal naris. The lateral circumoral ligament passes from the tip of the trabecular horn to the anterolateral margin of the pars articularis quadrati. The l. mandibulosuprarostrale arises from the insertion of the m. levator mandibulae posterior superficialis on the dorsal surface of Meckel's cartilage and passes to the posterior dorsal process of the suprarostral or adrostral. The l. rostrale superius quadrati connects the posterior dorsal process of the suprarostral to the anterolateral margin of the pars articularis quadrati near the attachment of the lateral circumoral ligament.

Diversity. The quadratoethmoid ligament seems to be universally present. The mandibulosuprarostral ligament is absent in *Ascaphus,* microhylids, and pipoids (Haas 1995); in Type 4 tadpoles it attaches to the adrostral tissue mass rather than to the suprarostral cartilage as in discoglossids (Haas 1995). The lateral circumoral ligament attaches to the suprarostral rather than the trabecular horn in *Alytes, Ascaphus, Bombina,* and *Discoglossus* (O. M. Sokol 1981b). Pipids have lost the lateral circumoral ligament but *Rhinophrynus* apparently retains it in the discoglossoid condition (O. M. Sokol 1981b).

Lower Jaws

The lower jaws consist of paired Meckel's cartilages and paired infrarostral cartilages (infralabial cartilages, mentomeckelian cartilages) joined at the midline by connective tissue (fig. 4.5). The infrarostrals are placed posterior and slightly lateral to the suprarostrals. The two infrarostrals are connected by a slender cartilage, the commissura intramandibularis (fig. 4.5). In *Rana,* a second, condylar articulation is present ventral to the commissura, but it is not synovial. Although the infrarostrals have been termed mentomeckelian cartilages (O. M. Sokol 1975), the extent of these is much greater than the mentomeckelian bones (J. J. Wiens 1989).

Meckel's cartilages are generally stout and shallowly V-shaped; they are not associated with the lower jaw sheath. The medial end of Meckel's cartilage generally bears a cotyle for articulation with the infrarostral cartilages. The articulation with the pars articularis quadrati of the palatoquadrate is saddle-shaped and subterminal; thus, there is a posterolaterally directed retroarticular process on which the suspensorio-, hyo-, and quadratoangularis muscles insert (O. M. Sokol 1981b). The articulation with the palatoquadrate has a well-developed synovial capsule.

The bones associated with the lower jaw are the paired angulars (angulosplenials), articulars, dentaries, and mentomeckelians. The endochondral mentomeckelian bone ossifies in the medial ends of the larval infrarostral cartilage. The dentary and angular are elongate dermal bones that invest Meckel's cartilage. The angular is posterior and medial and the dentary anterior and lateral to Meckel's cartilage. The articular ossifies endochondrally in the posterior end of Meckel's cartilage at the jaw articulation.

Diversity. The infrarostrals of *Ascaphus* are splintlike elements (Van Eeden 1951). In pipoids, the shape of the lower jaw resembles the adult configuration—slender and gently curved. The angular ossifies earlier than in nonpipoid frogs (O. M. Sokol 1977a; Trueb 1985). The infrarostrals are fused medially in some microhylids (De Sá and Trueb 1991; O. M. Sokol 1981b) and apparently in *Hoplobatrachus tigerinus* (Ramaswami 1940). Fused infrarostrals (O. M. Sokol 1975) are a synapomorphy for Pipoidea. A symphysial joint later develops between the infrarostrals during metamorphosis, but mentomeckelian bones do not develop in Pipoidea. Separate, so-called sub-meckelian cartilages are known in *Heleophryne* (Van Der Westhuizen 1961), and *Alytes* (called paramandibularia; Van Seters 1922); these cartilages are not in the same position in *Alytes obstetricans* (personal observation) as in *Heleophryne.*

Buccal Floor Muscles

Most of the cranial muscles of premetamorphic larvae are found in the adult frog, albeit with different names at times. However, all muscles are transformed during metamorphosis as the larval fibers die and adult-type fibers develop from myoblasts (Takisawa et al. 1952a, b; Takisawa and Sunaga 1951). Gradwell and Walcott (1971) discussed the dual functional and structural aspects of these fiber types in the interhyoideus muscle. These muscles are summarized in table 4.2.

In a premetamorphic tadpole of *Rana temporaria* at Kopsch stage 25 (Kopsch 1952), the muscles of the buccal

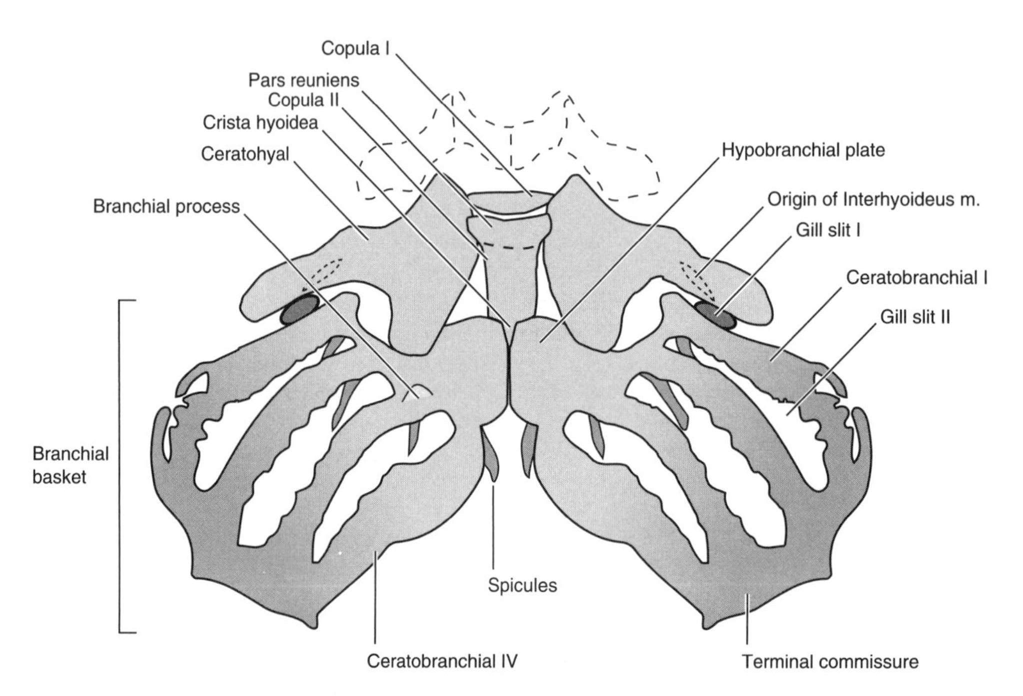

Fig. 4.7. Ventral view of the gill arch skeleton of a *Rana catesbeiana* tadpole. Symbols: dashed lines = lower jaw. After Gradwell (1972a).

Table 4.2 Origins and insertions of selected cranial muscles in a Gosner stage 35 tadpole of *Rana catesbeiana,* modified from Gradwell 1972b. Terminology is based on *Rana temporaria* (Edgeworth 1935) as modified by Haas (1997a); CB = ceratobranchial.

Muscle	Origin	Insertion
	Mandibular Group	
Levator mandibulae posterior superficialis	Dorsal part of palatoquadrate	Dorsal, medial part of Meckel's cartilage
L. m. p. profundus	Dorsal part of palatoquadrate	Lateral part of suprarostral cartilage
L. m. anterior	Dorsal part of ascending process of palatoquadrate; otic capsule	Dorsal, lateral part of Meckel's cartilage
L. m. externus	Anterior part of muscular process of palatoquadrate	Lateral part of suprarostral cartilage
L. m. a. articularis	Dorsal, lateral part of palatoquadrate	Dorsal, lateral part of Meckel's cartilage
L. m. a. subexternus	Anterior, medial surface of muscular process of palatoquadrate (appears at stage 40 in *Rana catesbeiana*)	Lateral part of suprarostral cartilage
L. m. a. lateralis	Anterior, medial surface of muscular process of palatoquadrate (appears at stage 39 in *Rana catesbeiana*)	Dorsal, lateral part of Meckel's cartilage
Intermandibularis anterior	Interconnects infrarostrals; no median raphe	—
Intermandibularis posterior	Median raphe	Medial part of Meckel's cartilage
Mandibulolabialis	Anterior part of Meckel's cartilage	Lower lip
	Hyoid Group	
Hyoangularis	Lateral part of ceratohyal	Retroarticular process of Meckel's cartilage
Quadratoangularis	Inferior, lateral part of muscular process of palatoquadate	Retroarticular process of Meckel's cartilage
Suspensorioangularis	Superior, lateral part of muscular process of palatoquadate	Retroarticular process of Meckel's cartilage
Orbitohyoideus	Lateral part of muscular process of palatoquadate	Lateral part of ceratohyal
Suspensoriohyoideus	Posterior, lateral part of muscular process of palatoquadate	Lateral part of ceratohyal
Interhyoideus	Median raphe	Lateral part of ceratohyal
Interhyoideus posterior	Median raphe	Palatoquadrate + otic process + diaphragm
	Branchial Group	
Levator arcus branchialis I	Lateral part of palatoquadrate	Lateral part of CB I
L. a. b. II	Otic process	Lateral part of CB I
L. a. b. III	Auditory capsule	Anterior, lateral part of CB II
L. a. b. IV	Auditory capsule	Posterior, lateral part of CB II
Constrictor branchialis I	Not present in *Rana;* see text	
C. b. II	Medial part of CB II	Terminal commissure of CB I and II
C. b. III	Medial part of CB II	Terminal Commissure of CB II and III
C. b. IV	Medial part of CB III	Terminal Commissure of CB II and III
Subarcualis rectus I, dorsal head	Posterior, medial part of ceratohyal	Medial part of CB I
S. r. II, ventral head	Posterior, medial part of ceratohyal	Medial part of CB II
S. r. II–IV	Medial part of CB II	Medial part of CB IV
Subarcualis obliquus	Crista hyoidea	Medial part of CB II
Transversus ventralis IV	Not present in *Rana;* see text	
Tympanopharyngeus	Auditory capsule	Lateral part of CB IV
Diaphragmato-branchialis IV	Medial part of diaphragm	Terminal commissure of CB II and III
Unnamed muscle (Cucullaris?)	Lateral part of CB IV	Scapula
	Spinal Group	
Geniohyoideus	Hypobranchial plate	Infrarostral cartilage
Rectus cervicis	Medial part of diaphragm; continuation of rectus abdominis	Medial part of CB II

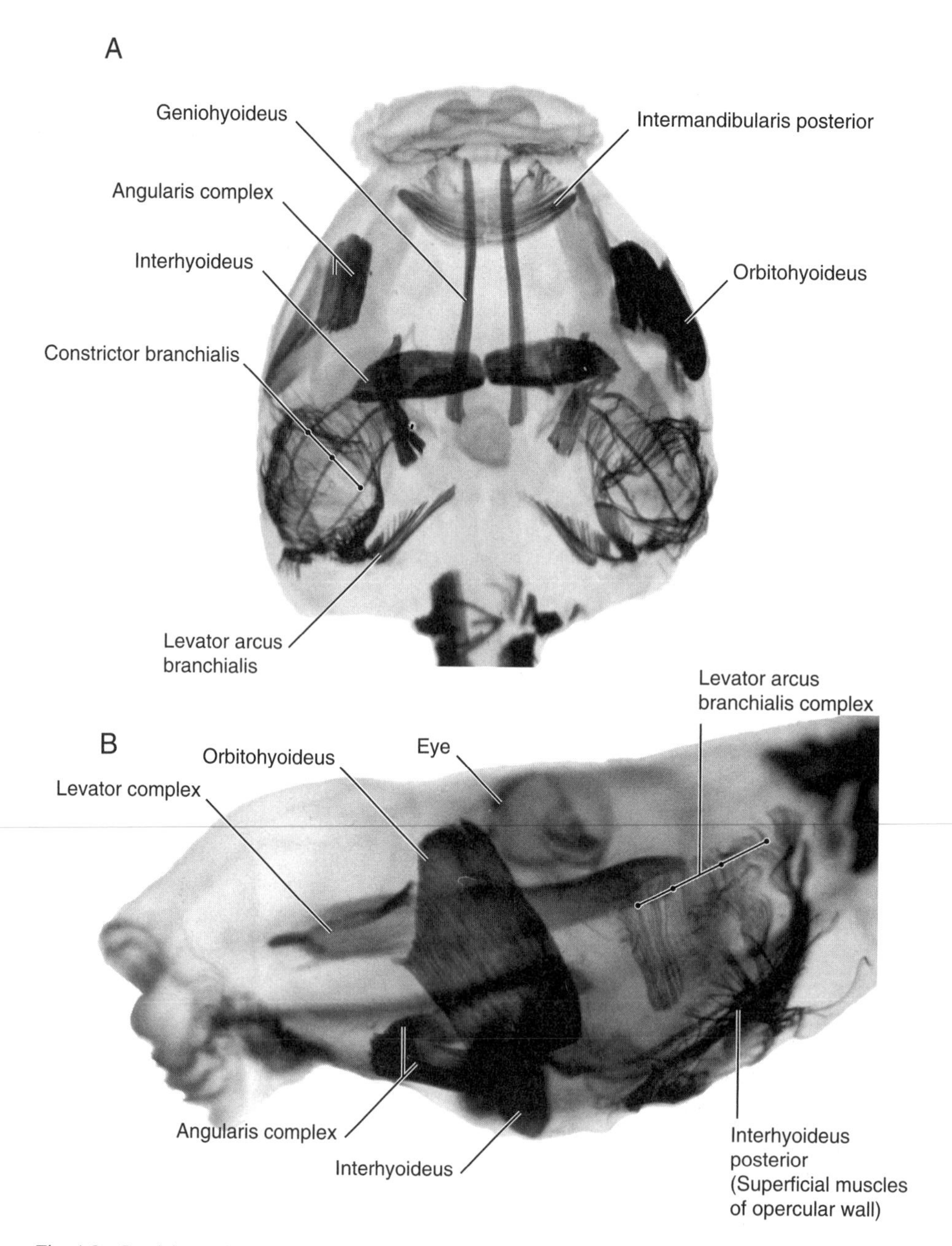

Fig. 4.8. Cranial muscles in a *Leptodactylus albilabris* tadpole (Gosner stage 28), as revealed by immunohistochemical staining of myosin protein (see Hanken et al. 1992). (A) Ventral view. (B) Left lateral view. Photographs by J. Hanken.

floor are simple (figs. 4.8–4.9). The mandibulolabialis is a tiny muscle that is attached to Meckel's cartilage near its articulation with the infrarostral. It inserts into the connective tissue of the lower lip (see below under Diversity). The mandibulolabialis is lost (De Jongh 1968) or perhaps incorporated into the intermandibularis (Sedra 1950) at metamorphosis. The intermandibularis anterior (submentalis) is a thin, ribbonlike, transverse muscle attached to the ventral side of the infrarostral and spans the symphysis; it develops just before and during metamorphosis. The intermandibularis (intermandibularis posterior; figs. 4.8–4.9) is a thin, flat, transverse muscle attached to the posterior half of Meckel's cartilage by a short aponeurosis. The left and right halves of the muscle curve posteriorly and ventrally to meet in a medial raphe. The interhyoideus (subhyoideus; figs. 4.8–4.9) is a transverse, ribbonlike muscle whose halves meet in a medial raphe. This muscle provides the force for the power stroke during gill irrigation; it inserts on an oblique crest on the ventral side of the ceratohyal near its lateral edge. The interhyoideus posterior (subbranchialis, constrictor colli) lies posterior to the interhyoideus; its halves join at a medial raphe and extend laterally to attach to the posterior palatoquadrate, the otic process, and the lateral part of the diaphragm. The larval diaphragm is the posterior wall of the sinus cervicalis, or pericardial space, surrounding the heart. In the adult frog these muscles are expanded to form a more or less continuous muscular floor for the buccal cavity.

The geniohyoideus is a paired, thin, elongate muscle (figs. 4.8–4.9) that originates on the hypobranchial plate and inserts on the infrarostral cartilage near the symphysis. In *Rana temporaria* it consists of a pars lateralis and pars medialis innervated by spinal nerves that comprise the hypoglossal nerve. The rectus cervicis muscle (sternohyoideus, diaphragmato-branchialis medialis), which extends from the branchial process of ceratobranchial II to attach to the larval diaphragm, is also supplied by the hypoglossal nerve. The tongue muscles proper, the genioglossus and hyoglossus, have not yet formed, but in *Rana temporaria* these differentiate as the forelimbs are emerging (De Jongh 1968).

Diversity. Differentiated portions of the intermandibularis form supplementary elements (Emerson 1976; Tyler 1971) in various taxa, but these are not apparent until near metamorphosis. Starrett (1973) reported that in microhylid tadpoles the intermandibularis posterior inserts on the basihyal (copula II) rather than a medial raphe. She also reported an infralabial retractor muscle that joins the infrarostral cartilage to Meckel's cartilage in microhylids. Presumably it retracts the lower jaw and closes the mouth. Starrett (1973) speculated that this muscle was homologous with the mandibulolabialis of Types 3 and 4 tadpoles. In several species a mandibulolabialis superior inserts onto the upper lip and a mandibulolabialis inferior inserts onto the lower lip (Carr and Altig 1991). The mandibulolabialis is apparently absent in some Type 1 larvae; it is present in *Gastrophryne carolinensis* and *Microhyla pulchra* and absent in *Hoplophryne rogersi* and *Phrynomantis* (= *Phrynomerus*) *annectens* (Gradwell 1974; Noble 1929).

Gradwell (1972b, c) pointed out that the interhyoideus posterior muscle (H3b in his terminology) occurs in *Pelobates, Pseudis, Rana catesbeiana* (contrary to Edgeworth 1935), and *R. fuscigula* but not in *Ascaphus, Phrynomantis,* pipoids, or *Scaphiopus.* Gradwell (1974) later illustrated and described an extensive interhyoideus posterior in *Phryno-*

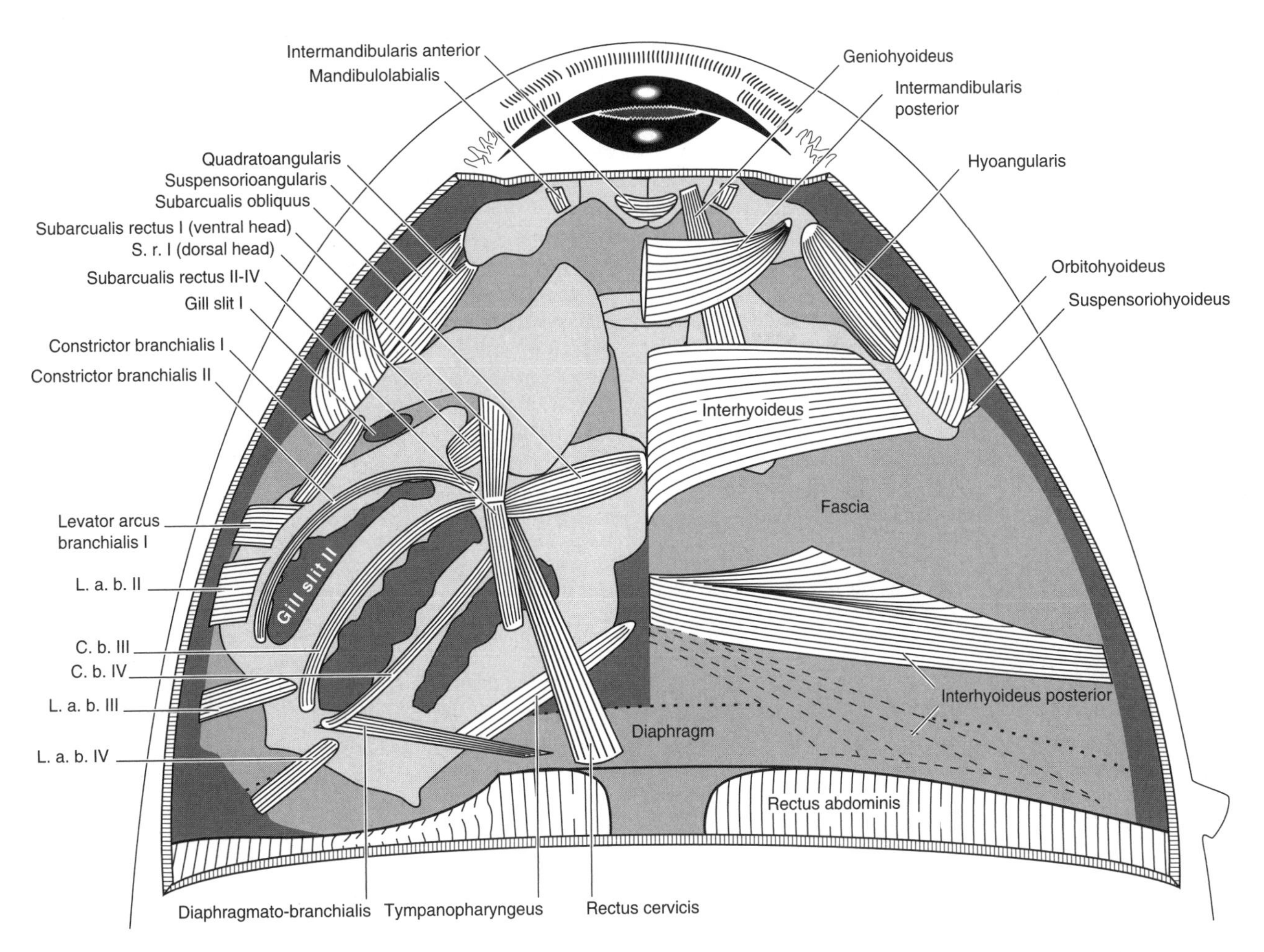

Fig. 4.9. Ventral view of the muscles of the jaws and branchial apparatus of an idealized tadpole, based on Gradwell (1972a), with additions.

mantis annectens. Noble (1929) mentioned that an extensive subbranchialis (interhyoideus posterior) covering the ventral wall of the branchial chamber in *Gastrophryne carolinensis* and *Microhyla pulchra* is absent in two species of *Hoplophryne.* O. M. Sokol (1975) reported that the interhyoideus posterior is absent in bufonids and leptodactylids and present in some hylids, microhylids, and some pelobatids.

Gradwell (1972b) also claimed that the superficial and deep layers of the interhyoideus posterior of *Rana catesbeiana* were identical to the subbranchialis and diaphragmatopraecordialis muscles (respectively) described by Schulze (1892) for *Pelobates.* Haas (1997a) pointed out that the diaphragmatopraecordialis muscle of *Pelobates* is a specialized part of the interhyoideus posterior. O. M. Sokol (1975; see fig. 4.9) illustrated the superficial and deep layers of the interhyoideus posterior of *Pelobates syriacus.* My comparison of the illustrations in Gradwell (1972b), Schulze (1892), and O. M. Sokol (1975) supports Gradwell's contention. Muscles with similar names were described by Noble (1929) in the suctorial tadpoles of *Amolops ricketti.*

Haas's (1997a) phylogenetic analysis indicates that the presence of an interhyoideus posterior is a synapomorphy of *Pelobates* + Neobatrachia, a clade that would render Mesobatrachia (fig. 4.1) paraphyletic. However, if the interhyoideus posterior is absent in *Scaphiopus* as claimed by Gradwell (1972b, c), and assuming that Pelobatidae (*Scaphiopus* + *Pelobates*) is monophyletic, then homoplasy in this character makes the putative synapomorphy for *Pelobates* + Neobatrachia ambiguous.

Noble (1929) described a pars locomotorius of the subhyoideus (interhyoideus) that inserts bilaterally into paired cutaneous flaps flanking the medial spiracle in *Hoplophryne.* Noble speculated that these flaps served as locomotor organs in the damp leaf crevices in which these nonaquatic larvae live. Edgeworth's (1935) table of muscle synonyms lists the constrictor colli muscle in frogs as a synonym of the subbranchialis, but his text does not mention the constrictor colli.

Jaw Levator Muscles

There are two morphological units of levator (adductor) muscles. One group consists of three levator muscles that originate from the posterior portion of the palatoquadrate and adjacent otic capsule and extend forward through the floor of the orbit medial to the muscular process of the palatoquadrate to insert onto the suprarostral and Meckel's cartilage. The second group originates from the anterior medial region of the muscular process and extends ventromedially to insert on Meckel's cartilage and the suprarostral. All of these muscles are innervated by the trigeminal nerve (V). The terminology of these muscles is particularly confused (table 4.3), and Starrett's use of the path of the mandibular branch (V_3) of the trigeminal nerve for distinguishing anterior, posterior, and externus groups of adductor muscles is not used here.

The most superficial muscle (fig. 4.10) in the first group is the levator mandibulae posterior superficialis. It originates on the dorsal surface of the palatoquadrate lateral to the ascending process and extends forward to insert on a rounded process on Meckel's cartilage lateral to its articulation with the infrarostral cartilage. The levator mandibulae posterior profundus lies just below the l. m. p. superficialis and originates on the dorsal surface of the palatoquadrate just below and slightly lateral to the origin of the l. m. p. superficialis. In *Rana temporaria* the l. m. p. superficialis and l. m. p. profundus are fused through the posterior half of their lengths (De Jongh 1968). The l. m. p. profundus inserts via a compound tendon on the ventrolateral edge of the suprarostral. This compound tendon also receives the insertion of the l. m. externus complex in *Rana temporaria* but apparently not in *R. catesbeiana* (Gradwell 1972b).

The deepest and most medial of the three muscles traversing the orbit is the levator mandibulae anterior (fig. 4.10). It originates from the anteroventral surface of the auditory capsule and the ventral aspect of the ascending process of the palatoquadrate; the area of origin is medial to that of the l. m. posterior superficialis and l. m. posterior profundus. Its fibers converge into a distinctive long tendon that inserts on the dorsal side of Meckel's cartilage just medial to the jaw joint. The l. m. anterior lateralis differentiates from the l. m. anterior during metamorphosis in *Rana temporaria* and at Gosner stage 39 in *R. catesbeiana* (Gradwell 1972b).

The second group of muscles arises from the muscular process of the palatoquadrate (fig. 4.10): the levator mandi-

Table 4.3 Comparisons of the terminology of jaw muscles in tadpoles (adapted from Starrett 1968) of Edgeworth 1935, Starrett 1968, Luther 1914, Sedra 1950, and the terminology for adults from Edgeworth 1935

Edgeworth	Starrett	Luther	Sedra	Edgeworth (Adult)
Levator mandibulae anterior	Adductor mandibulae anterior internus	A. m. pterygoideus	L. m. anterior	Levator mandibulae anterior
—	A. m. anterior longus	—	—	—
L. m. posterior superficialis	A. m. posterior superficialis	A. m. posterior longus superficialis	L. m. posterior superficialis	L. m. posterior
L. m. posterior profundus	A. m. posterior profundus	A. m. posterior longus profundus	L. m. posterior profundus	L. m. posterior
L. m. anterior subexternus	A. m. posterior subexternus	A. m. posterior subexternus	L. m. externus pars anterior	L. m. anterior subexternus
L. m. anterior articularis	A. m. posterior articularis	A. m. posterior articularis	—	L. m. anterior articularis
L. m. externus	A. m. externus superficialis	A. m. externus	L. m. externus pars posterior	L. m. externus
L. m. anterior lateralis	A. m. externus lateralis	A. m. posterior lateralis	L. m. anterior lateralis	L. m. anterior lateralis

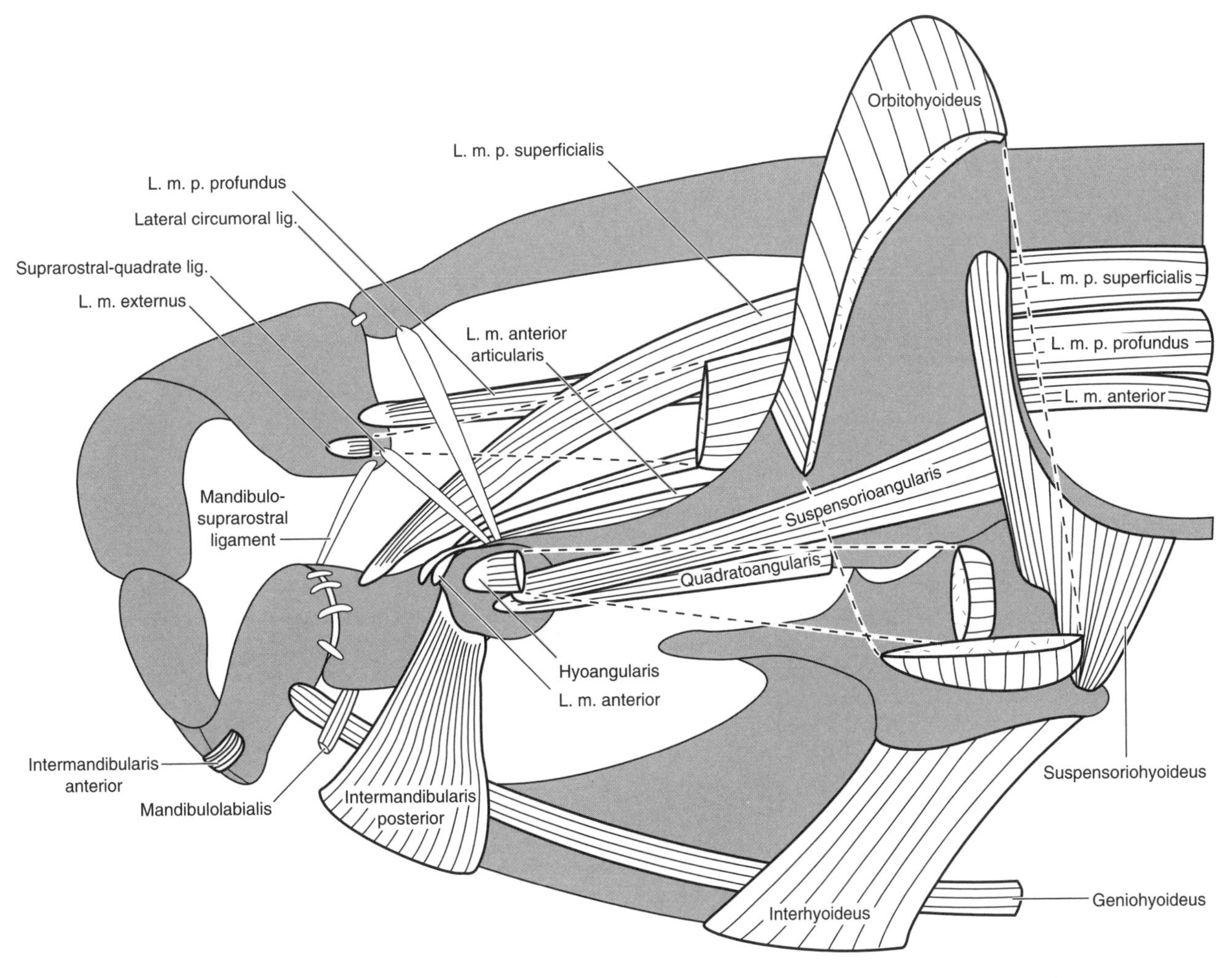

Fig. 4.10. Lateral view of muscles and ligaments of the jaw apparatus of the tadpole of *Rana catesbeiana;* the suprarostal and infrarostral cartilages have been forced apart slightly. Abbreviations: lig. = ligament and L.m.p. superficialis = Levator mandibulae pars superficialis. Redrawn from Gradwell (1972a).

bulae anterior articularis, levator mandibulae externus, and levator mandibulae anterior subexternus (l. m. subexternus of Haas 1996b). The l. m. anterior articularis arises from the medial aspect of the muscular process near its anterior edge. It extends medioventrally to insert on Meckel's cartilage via a short tendon medial to the jaw joint and just medial to the attachment of the tendon of the l. m. anterior.

The l. m. externus is lateral to the l. m. anterior subexternus, and the two muscles have a similar origin and insertion. They arise from the anteromedial surface of the muscular process just dorsal and anterior to the origin of the l. m. anterior articularis. They extend medioventrally, and their common tendon inserts into the tendon of the l. m. posterior profundus (see above), which inserts on the suprarostral. In *Rana temporaria* these two muscles are incompletely separated at Kopsch stage 25, separated at stage 26, and fused by stage 29 when the suprarostral has been resorbed. When the suprarostral disintegrates during metamorphosis, the tendons of the l. m. posterior profundus and the l. m. externus muscles shift to insert on Meckel's cartilage (De Jongh 1968). These muscles become part of the adult levator mandibulae complex.

Diversity. Little is known about variation in muscles of the mandibular and hyoid arches (e.g., Sedra 1950). The condition illustrated by Gradwell (1968) for *Rana catesbeiana* differs slightly from that for *R. temporaria.* Gradwell (1968: fig. 2) illustrated a l. m. externus pars posterior with a common insertion and origin with the l. m. posterior profundus. His l. m. externus pars anterior was identical in origin and insertion to the l. m. externus of Edgeworth (1935). Gradwell (1968) appeared to follow the terminology of Sedra (1950), who described a pars anterior inserting onto Meckel's cartilage and a pars posterior inserting into the profundus tendon in *Bufo regularis.* However, the two parts of the l. m. externus in *B. regularis* have a common origin on the muscular process, whereas in Gradwell's (1968) figure the origin of the pars posterior is the same as that of the l. m. posterior profundus. Perhaps the discrepancy was an oversight, because Gradwell (1972b: fig. 6) provided a modified drawing in which the pars posterior was no longer present, and he remarked that his pars anterior was homologous with the l. m. externus and that the l. m. subexternus (presumably his pars posterior) did not appear until stage 40 in *Rana catesbeiana.* The distinction between the l. m. externus and l. m. subexternus is particularly important because Starrett (1968) defined character states for the levator muscles based on the relationship of the mandibular branch of the trigeminal nerve (V_3) to these muscles.

The fates of the larval mandibular and hyoid muscles of *Bufo regularis, Rana temporaria,* and *Xenopus laevis* were discussed by Sedra (1950) and Sedra and Michael (1957). Metamorphosis in *Xenopus* appears to be very different from that in the other two taxa, but this may be caused by misidentifications of certain muscles. Sedra and Michael (1957) claimed to follow Edgeworth's (1930) muscle terminology for *Xenopus laevis,* but their treatment is problematic. For example, they named the largest and most superficial of the mandibular muscles the l. m. anterior and divided it into the l. m. anterior pars intermedius, pars lateralis, and pars medialis. The l. m. anterior pars medialis, on the basis of its topography and its long, slender tendon, corresponds to the l. m. anterior of *Rana temporaria.* The l. m. anterior pars intermedius corresponds to the l. m. posterior pars superficialis, and the l. m. anterior pars lateralis corresponds to the l. m. posterior pars profunda based on relative positions and points of insertion; compare Sedra and Michael (1957) to De Jongh (1968), Edgeworth (1935), and Gradwell (1971b, 1972b). Gradwell (1971b) called attention to this discrepancy, but unfortunately followed the terminology of Sedra and Michael (1956a, 1957).

In *Rana,* the l. m. posterior profundus inserts on the suprarostral cartilage. In pipids the suprarostral is fused to the trabecular horns, and the l. m. posterior profundus (misnamed the l. m. anterior pars lateralis by Sedra and Michael 1957) inserts on the cartilaginous base of the tentacle of *Silurana tropicalis* and *Xenopus laevis* (Gradwell 1971b; O. M. Sokol 1977a); this muscle was also called the m. tentaculi (Nieuwkoop and Faber 1967). In *Hymenochirus boettgeri,* which lacks tentacles, this slip inserts on the supralabial pad (O. M. Sokol 1977a), a transversely oval connective tissue pad that lies at the anterior edge of the ethmoid plate (O. M. Sokol 1962). Starrett (1973) tabulated differences in adductor muscles among Orton's four classes of tadpoles but more variation exists (O. M. Sokol 1975) than she reported.

Jaw Depressor Muscles

The hyoangularis, quadratoangularis, and suspensorioangularis, which are hyoid arch muscles innervated by the facial nerve (VII), insert on the retroarticular process of Meckel's cartilage. The suspensorioangularis is a fan-shaped muscle that extends from the lateral aspect of the base of the muscular process and inserts via a short aponeurosis on the retroarticular process of the lower jaw (figs. 4.9–4.10). The hyoangularis is cone-shaped and originates from the lateral aspect of the ceratohyal slightly ventral to the articulation of the ceratohyal and palatoquadrate cartilage. It inserts via an aponeurosis in common with the quadratoangularis just lateral to the insertion of the suspensorioangularis. The quadratoangularis originates on the ventrolateral aspect of the palatoquadrate, is bounded laterally by the bodies of the hyoangularis and suspensorioangularis, and inserts on a common aponeurosis with the hyoangularis. These three muscles form part of the depressor mandibulae of the adult.

Diversity. Starrett (1968) reported the occurrence of these muscles in many taxa, but these data are inconsistent with some other reports, including Starrett (1973). For example, pipoids have a single mandibular depressor muscle, the quadratohyoangularis, which is thought to represent fusion of the three muscles (O. M. Sokol 1962, 1977a); ontogenetically, it appears as a single muscle in *Xenopus laevis* (Nieuwkoop and Faber 1967). However, Starrett (1968) reported both the hyoangularis and quadratoangularis present in *Xenopus gilli* and *X. laevis.* The hyoangularis is said to be absent in *Ascaphus truei* (Pusey 1943; Starrett 1968) and *Discoglossus pictus* (Starrett 1968), but Starrett (1973) reported it present in Type 3 tadpoles.

Haas (1997a) noted that the origin of the suspensorioangularis from the palatoquadrate is behind the orbit in *Alytes, Ascaphus, Bombina,* and *Discoglossus* (state 0 of his character 23) but in front of the orbit in *Pelobates* and Neobatrachia (state 1). In pipids, he described the origin of the single angularis muscle as from both the quadrate and the ceratohyal and coded the state as missing. Sokol (1977a) provided greater detail on the origin of this muscle, but even so, the extreme distortion of the pipid palatoquadrate precludes a clear assignment to either state. In Haas's tree (1997a: fig. 15) with characters optimized using accelerated transformation (Swofford and Olsen 1990), state 1 is a synapomorphy of Pipanura. However, the delayed transformation optimization would suggest that state 1 is a synapomorphy of Pelobatoidea + Neobatrachia.

Hyobranchial Apparatus

The ceratohyals and branchial baskets (fig. 4.7) are the two major components of the hyobranchial skeleton. The ceratohyals and basibranchial elements (copulae and hypobranchial plates) are the pistons of the buccal pump that conveys currents of food-laden water to the gill filters and branchial food traps and irrigates the respiratory surfaces of the gills. The branchial baskets support the gill filters and gills. Water pumped posteriorly passes through the pharyngeal slits, opercular cavity, and finally, the spiracle(s).

Ceratohyals and Branchial Baskets

The ceratohyals are the main levers of the pump upon which muscle contraction acts. The ceratohyals lie anterior to the branchial baskets and are generally oriented transversely. The ceratohyals articulate with the palatoquadrates by means of the synovial hyoquadrate articulations, which are reinforced by a stout hyoquadrate ligament. The ceratohyals articulate with each other medially via copula II (posterior copula, copula posterior, basibranchial) and the interhyoid ligament (ligamentum interhyoideum, ligamentum interhyale), of which the posterior part is a mass of fibrous cartilage called copula I (anterior copula, copula anterior, basihyale). The copulae represent the medial basihyobranchial structures. The pars reuniens (fig. 4.7) is a medial mass of white fibrous tissue that binds the ceratohyals together medially and also separates copula I and copula II. The subtriangular space between copula II, the hypobranchial plate, and the ceratohyal is occupied by the intrahyoid ligament (ligamentum intrahyoideum; Gradwell 1972b). The crista hyoidea (urobranchial process) projects posteriorly from copula II and is the origin of the transversus ventralis II muscle. Within Amphibia, copula I is unique to tadpoles; the pars reuniens is believed to be homologous with the hypohyals of salamander larvae and copula II is believed to be homologous with the basibranchials and urobranchials (O. M. Sokol 1975). Together, the ceratohyals, copulae, ligaments, and parts of the hypobranchial plates form the major skeletal support of the buccal floor.

The branchial baskets are posterior to the ceratohyals. Each branchial basket (fig. 4.7) is composed of a subtriangular hypobranchial plate and four ceratobranchials. The hypobranchial plate is traditionally considered to be formed from two fused hypobranchial elements. However, Haas (1997a) argued that the plate is composed only of hypobranchial I, homologous with the similarly named urodelan structure. The four curved ceratobranchials are fused at their proximal and distal ends to form a bowl with three elongate slits; ceratobranchial I is most anterolateral. The proximal fusions are the proximal commissures (commissurae proximales) and the distal fusions are the terminal commissures (commissurae terminales). Near the anterior end of ceratobranchials II and III there is a prominence, the branchial process (processus branchialis), to which a number of branchial muscles attach (table 4.2; Gradwell 1972b). In *Rana catesbeiana* ceratobranchial I is fused to the hypobranchial plate by hyaline cartilage; the other three ceratobranchials are joined by connective tissue (Gradwell 1972b).

The spicules (spicula) are four elongate cartilaginous processes projecting dorsally and posteriorly from the ceratobranchials to support the ventral velum (chap. 5). Together the hypobranchials and ceratobranchials form the skeletal support for the gills and filter apparatus.

Diversity. Haas (1997a) provided detailed comparative accounts of the hyobranchial skeleton and muscles of *Alytes, Ascaphus, Bombina,* and *Discoglossus.* Ridewood (1898b) first pointed out that in *Alytes, Bombina,* and *Discoglossus,* copula II separates the two hypobranchial plates. This state is also present in *Ascaphus,* and the derived condition (hypobranchial plates in contact) is a synapomorphy of Pipanura. The fusion of hypobranchial I with ceratobranchial I is also a synapomorphy of Pipanura (Haas 1997a).

Pipa parva and *Xenopus* lack copula I and *Pipa* lacks copula II (Haas 1995; O. M. Sokol 1975). Copula I of microhylids is rodlike (Haas 1995). Ridewood (1897b) stated that there is no copula I in *Pelobates* or *Pelodytes,* but Roček (1981) reported copula I in *Pelobates fuscus.* Ridewood (1897a) termed copula II the basihyal but acknowledged the homology of copula II with the basibranchial and of copula I with the basihyal. In certain taxa the ceratobranchials are fused to the hypobranchial plates, and in other taxa the attachment is perichondrial or collagenous; the branchial baskets are bound to the copula by ligaments or perichondrium or may be fused in some taxa, such as suspension-feeding microhylids (Haas 1996b; O. M. Sokol 1975). Hymenochirines have only two of the four ceratobranchials (Menzies 1967; O. M. Sokol 1962). Certain ceratobranchials are also lost in the bromeliad-dwelling larvae of *Hyla picadoi* and *H. zeteki* (personal observation). Noble (1929) reported only one ceratobranchial, with a single branchial slit posterior to it, in two species of *Hoplophryne.* In most pipoid larvae and in some microhylid larvae, the filter plates, which are supported by the ceratobranchials, are chondrified. In many species, both ceratobranchials II and III bear a branchial process for muscle attachment at their proximal ends (Haas 1997a). Dendrobatids are unique in having free proximal ends of ceratobranchials II and III (Haas 1995). Haas (1993, 1995, 1997a) described several character states for the ceratobranchials in a wide variety of taxa.

Muscles Associated with the Ceratohyal

The orbitohyoideus and suspensoriohyoideus (figs. 4.9–4.10), which are hyoid muscles innervated by the facial nerve (VII), originate on the muscular process and insert on the ceratohyal. The orbitohyoideus is a very large, tapering muscle that originates along the curved upper margin of the muscular process. Most of the concave lateral surface of the process lacks muscle fiber attachments. The orbitohyoideus extends ventrally and slightly posteriorly to a fleshy insertion on the ventrolateral edge of the ceratohyal and conceals the origins of the hyoangularis, quadratoangularis, and suspensorioangularis. The suspensoriohyoideus muscle originates from a broad surface on the posterior edge of the muscular process and is partially concealed by the orbitohyoideus; it extends ventrally to insert just dorsal and slightly posterior to the orbitohyoideus. In the adult these muscles form part of the depressor mandibulae.

Diversity. Haas (1997a) compared the hyobranchial muscles of *Alytes, Ascaphus, Bombina,* and *Discoglossus.* His work indicates significant discrepancies in the literature treating these muscles. For example, the suspensoriohyoideus is reported absent in *Ascaphus* (Pusey 1943), but Haas (1997a) reported it as present. O. M. Sokol (1977a) reported the suspensoriohyoideus as absent in *Pipa carvalhoi* and present in *Hymenochirus boettgeri* and *Silurana tropicalis.* Gradwell (1971b) said it was present in *Xenopus laevis.* Starrett (1968) indicated that the suspensoriohyoideus was absent in *Rhinophrynus* and *X. laevis* but later (1973) reported it present in pipoids.

Muscles of the Branchial Baskets

The muscles associated with the ceratobranchials are complex; difficulties in assessing their homology within Anura and with other vertebrates have resulted in various sets of terminology. Clarification of the homologies would require extensive comparative work (see Fox 1965; Haas 1997a; Jollie 1982; Wiley 1979), but compared to the jaw muscles and chondrocranium, there are few comparative data on the muscles of the branchial baskets. Because of the general confusion regarding muscle identities, taxic diversity is included with the description of each muscle. These muscles are supplied by the glossopharyngeal (IX) and vagus (X) nerves. The functions of these muscles are discussed below under Gill Irrigation in *Rana catesbeiana.*

Levator Arcus Branchialis. Four thin, wide muscles comprise the levator arcus branchialis series or levatores arcuum branchialium (branchial levators; fig. 4.9). In *Rana temporaria* these are four discrete muscles that extend from the lateral and posterolateral margin of the ceratobranchial basket dorsally and attach to the posterior part of the palatoquadrate and adjacent otic capsule. This series was confusingly called the constrictor branchialis (for I–III) and levator arcus branchialis (for IV) by Sedra and Michael (1957) and the constrictor arcus branchialis by O. M. Sokol (1977b).

Levators I–III are fused in *Xenopus laevis;* Sedra and Michael (1957) incorrectly reported fusion of I–IV according to O. M. Sokol (1977b). In *Pipa carvalhoi* muscles I–II are fused, and *Hymenochirus boettgeri* lacks levators II and III, apparently related to the fusion (or lack) of ceratobranchials II–IV (O. M. Sokol 1977a). These muscles apparently have disappeared in adult *Rana* or possibly form the petrohyoid series.

Constrictor Branchialis. There are three constrictor branchialis muscles (I–III) in *Rana catesbeiana;* each muscle extends from the medial end of its own ceratobranchial along the anterior margin of the gill slit to the terminal commissure. These muscles are intimately associated with the afferent and efferent branchial arteries and gills. The constrictors apparently regulate the flow of water through the gill slits by changing the position of the interbranchial septa (O. M. Sokol 1977a) but do not function during regular phasic pumping (Gradwell 1971b, 1972c). No tadpole has a muscle extending along the length of ceratobranchial IV (O. M. Sokol 1975); Pusey (1943) claimed that constrictor IV was fused with levator IV.

The traditional seriation of the constrictors was challenged recently. Using whole-mount immunohistochemistry, Schlosser and Roth (1995) identified the innervation of four branchial arch muscles by the principal ramus of cranial nerve IX and three branchial trunks of cranial nerve X, respectively. They named these muscles constrictor branchialis I, II, and III. What was surprising was that their constrictors II and III corresponded in anatomical position to the traditional constrictors I and II as described for *Rana* by others. Their (new) constrictor branchialis I had been called the branchiohyoideus externus (= ceratohyoideus externus) because of its presumed homology to the urodelan branchiohyoideus, a muscle of the hyoid arch. The branchiohyoideus externus was reported in *Ascaphus, Bombina, Discoglossus* (Pusey 1943), and *Scaphiopus* (O. M. Sokol 1975) but not in *Pelobates* or any neobatrachian. Gradwell (1971a), without explanation, placed the branchiohyoideus correctly in his branchial rather than hyoid group.

Schlosser and Roth (1995) did not discuss the significance of their findings for issues of muscle homology. Haas (1997a) interpreted their implicit homologization to mean that the constrictor branchialis muscles had traditionally been misnumbered, and he argued strongly for adoption of the new labels.

Some observations support this new scheme: (1) the serial innervation of the four putative constrictor muscles by cranial nerve IX and three rami of cranial nerve X is consistent with this scheme, (2) the branchiohyoideus of salamanders is reportedly innervated by cranial nerve VII and thus is unlikely to be homologous with the "branchiohyoideus externus" of frogs (constrictor branchialis I, sensu novo), and (3) the distal end of the constrictor branchialis II and III (sensu novo) is closely placed to the proximal ends of levators II and III. The distal end of constrictor branchialis I (sensu novo) is also closely placed to levator I. This is in accord with Edgeworth's (1935) contention that the anuran constrictors and levators form from the dorsal portion of the four branchial muscle plates.

Other observations do not support the proposed homology but are not directly contradictory. (1) The constrictor branchialis I (sensu novo) extends between arches (i.e., ceratobranchial I to the ceratohyal) whereas each other constrictor lies along one ceratobranchial, and (2) the urodelan branchial depressors are believed to be homologous with the anuran constrictors. The urodelan depressor branchialis I lies only along ceratobranchial I and has no attachment to the ceratohyal (Haas 1997a: fig. 13), so if this muscle is homologous with the anuran constrictor branchialis I (sensu novo) then the attachments of the latter have shifted considerably. (3) The general placement of the urodelan depressors I, II, and III closely matches that of the three traditional anuran constrictors I, II, and III. (4) Each of the three classical anuran constrictors lies in intimate contact with an afferent branchial artery along the ceratobranchial; the constrictor branchialis I (sensu novo) has no such association. (5) Although the innervation of constrictor branchialis I (sensu novo) by nerve IX is consistent with that muscle belonging to the first branchial arch, it does not necessarily follow that the muscle is a constrictor branchialis per se.

These differences do not reject the proposed homology with constrictor branchialis I (sensu novo) but simply indicate that the anatomy of that muscle is somewhat modified. For example, Schlosser and Roth (1995) reported that the first branchial trunk of cranial nerve X innervates structures of the second branchial arch, including levator arcus branchialis II, constrictor branchialis II, subarcualis rectus II–IV, and transversus ventralis II (= subarcualis obliquus of Haas 1997a). However, their constrictor branchialis II actually lies along the first branchial arch in intimate contact with the first afferent branchial artery. The same applies to the second branchial trunk of cranial nerve X, which innervates the third branchial arch. One can infer that the constrictor muscles must have significantly shifted their points of attachment, assuming that the new homologies are correct. Accep-

Table 4.4 Homologies of muscles of the branchial baskets, as described by various authors for specified taxa

Edgeworth 1935 *Rana*	Pusey 1943 *Ascaphus*	Sedra and Michael 1957 *Xenopus*	Sokol 1975, 1977a *Pipa*	Haas 1996b *Gastrotheca*	Haas 1997a *Bombina*
Levator arcus branchialis I	Levator arcus brachialis I	Constrictores arcuum branchialium	Constrictores arcuum branchialium	Levator arcus branchialium I	Levator arcuum branchialium I
L. a. b. II	L. a. b. II	(fused)	(fused)	L. a. b. II	L. a. b. II
L. a. b. III	L. a. b. III	(fused)	(fused)	L. a. b. III	L. a. b. III
L. a. b. IV	L. a. b. IV	Levator arcuum IV	(fused)	L. a. b. IV	L. a. b. IV
Absent	Branchiohyoideus externus	Absent	Ceratohyoideus externus	Absent	Constrictor branchialis I
Constrictor branchialis I	Constrictor branchialis I	Subarcualis rectus II	Interbranchialis I	Constrictor branchialis I	C. b. II
C. b. II	C. b. II	Subarcualis rectus III	Interbranchialis II	C. b. II	C. b. III
C. b. III	C. b. III	Subarcualis rectus IV	S. r. II–IV	C. b. III	C. b. IV
Subarcualis rectus I	Subarcualis rectus I	Subarcualis rectus I	S. r. I	Branchiohyoideus internus	S. r. I (dorsal head)
S. r. II	Absent	Absent	Ceratohyoideus internus	Subarcualis rectus II	S. r. I (ventral head)
S. r. III + IV	Subarcuales recti IV and ?V	Absent	Absent	Subarcualis rectus IV	S. r. II–IV
Transversus ventralis II	Subarcualis obliquus II–IV and ?V	Transversus ventralis II	Subarcualis obliquus	Transversus ventralis II	Subarcualis obliquus
Diaphragmato-branchialis IV	Diaphragmato-branchialis	Not described	Diaphragmato-branchialis	Diaphragmato-branchialis	Diaphragmato-branchialis
Not described	Absent	Transversus ventralis IV	Tympanopharyngeus	Tympanopharyngeus	Absent
Absent	Transversus ventralis IV	Absent	Absent	Absent	Absent
Absent	Absent	Absent	Laryngeobranchialis	Absent	Absent

tance of this shift in attachment of constrictores branchiales II and III leads to the conclusion that the attachment of constrictor branchialis I is part of a common pattern. Given the nature of their whole-mount preparations, it is not clear whether Schlosser and Roth (1995) were able to see the branchial arch or the artery, or whether they recognized the significance of their muscle identifications.

I have followed the new arrangement of Schlosser and Roth (1995) as promoted by Haas (1997a). The traditional constrictors I, II, and III are simply renumbered II, III, and IV; and the branchiohyoideus externus muscle (found in only a few frog species) is renamed constrictor branchialis I (fig. 4.9; table 4.4).

Haas (1997a: fig. 13b) reported a short constrictor branchialis IV (sensu novo) in two specimens of *Ascaphus,* similar in appearance to that described by Pusey (1943). However, Haas described and illustrated it attaching to the base of ceratobranchial IV rather than ceratobranchial III as in Pusey (1943). Gradwell (1973) reported and illustrated the same muscle (as B2c) on ceratobranchial III of *Ascaphus.* Both Gradwell and Pusey reported that the muscle is attached to the incipient branchial process at the base of ceratobranchial III and inserted into soft tissue rather than onto the ceratobranchial. Haas (1997a) reported this muscle absent in *Alytes, Bombina,* and *Discoglossus.* Pusey (1943) noted its presence on ceratobranchial III in *Discoglossus.* O. M. Sokol (1975) erred in claiming that there was no constrictor (he termed these interbranchialis muscles) on ceratobranchial III in Types 1 and 3 tadpoles. Three constrictors were illustrated in *Xenopus laevis* (Sedra and Michael 1957) but were named subarcualis rectus II–IV. In reporting three muscles in *Xenopus,* Gradwell (1971b) followed Sedra and Michael's terminology but named them B2a, 2b, and 2c, as he did with the constrictor branchialis muscles of other nonpipid taxa that he studied.

Haas (1997a: fig. 13) also illustrated, but did not discuss, three branchial constrictors in *Odontophrynus* and *Pelobates.* The first of these in *Odontophrynus* is labeled constrictor branchialis II, but the last two are both labeled constrictor branchialis III, an error in labeling (Haas, personal communication). Assuming the last muscle is in fact constrictor branchialis IV, its origin is correctly shown on ceratobranchial III rather than IV as Haas (1997b) illustrated for *Ascaphus;* this does not mean that Haas's illustration of *Ascaphus* is incorrect. The attachments of the three constrictor muscles in *Odontophrynus* and *Pelobates* are almost identical to those in *Rana* (fig. 4.9).

Haas's (1997a) phylogenetic analysis indicated that the absence of the constrictor branchialis I (sensu novo) was a synapomorphy (character 19) for the clade Pipanura. However, Haas did not include data from *Scaphiopus* or *Spea,* and Sokol (1975) reported that these larvae have the muscle (I have verified this for *Spea bombifrons*). If one assumes that *Spea* is the closest relative of *Pelobates* (among the taxa Haas examined), then under parsimony optimization the muscle must have reappeared in *Spea.* Clearly, a greater range of taxonomic sampling is needed.

Subarcualis Rectus. The homology of the four subarcualis rectus muscles with those of other amphibians and fish is particularly troublesome (Jollie 1982; Wiley 1979). In anurans these muscles extend more or less longitudinally, connecting each gill arch to the one in front. Subarcualis rectus I connects the ceratohyal to the ceratobranchials, and subarcualis rectus II, III, and IV span the ceratobranchials.

Haas's (1997a) interpretation of subarcualis rectus I is followed here. This muscle has several conformations. In *Ascaphus* it has but one head. In some species (e.g., *Bombina*) there are two closely placed slips that connect the base of ceratobranchial to the ceratohyal. In other species the two slips are more distinct, and the deeper (viewed ventrally; fig. 4.9) dorsal head extends from the base of ceratobranchial I to the processus posterior hyalis on the ceratohyal. The ventral head stretches from the branchial process of ceratobranchial II to the ceratohyal. These heads are so distinct as to have been given different names (e.g., ceratohyoideus internus; table 4.4). Subarcualis rectus I is innervated by the principal ramus of cranial nerve IX (Schlosser and Roth 1975).

The remaining part of the muscle series, the subarcualis rectus II–IV, is a band of muscle extending across the proximal parts of ceratobranchials I–IV, as in *Ascaphus*. The compound nature of this muscle is suggested by the observation that in *Discoglossus* subarcualis rectus II–IV is innervated by the first, second, and (probably) the third branchial rami of cranial nerve X (Schlosser and Roth 1995). In other species (e.g., *Rana temporaria*) the muscle spans ceratobranchials II–IV (as subarcualis rectus III–IV) or only ceratobranchials III–IV (as subarcualis rectus IV; Haas 1997a).

The literature indicates general confusion about these muscles. The subarcualis rectus II, III, and IV of Sedra and Michael (1957) are actually constrictor branchialis muscles. O. M. Sokol illustrated a muscle in *Pipa* he termed the fused subarcualis rectus II–IV, but this appears to be the third branchial constrictor (1977a: fig. 6). Subarcualis rectus II–IV is absent in *Hymenochirus* (O. M. Sokol 1977a). Thus, it seems that there is no verified subarcualis rectus II–IV muscle in pipids. In *Rana* the subarcualis rectus muscles disappear at metamorphosis.

Subarcualis Obliquus. The subarcualis obliquus II (transversus ventralis II) is a stout muscle that extends from the crista hyoidea of copula II to the branchial process in *Rana* (Edgeworth 1935). Pusey (1943) argued, probably correctly, that Edgeworth's (1935) terminology was in error and that this muscle should be called subarcualis obliquus II. However, he pointed out that Edgeworth's transversus ventralis IV was correctly labeled. Edgeworth (1920) had earlier stated that transversus ventralis II was not present in any amphibian, so his (1935) listing of transversus ventralis II is curious.

Pusey (1943) also pointed out that only *Ascaphus* among frogs has subarcualis obliquus II–V. Muscles III–V are a fused series of slips that attach to the crista hyoidea and fan out to insert on the medial portions of the ceratobranchials and spiculum IV (Gradwell 1973; Pusey 1943). Other frogs, including *Alytes, Bombina,* and *Discoglossus,* have only subarcualis obliquus II, which Haas (1997a) referred to as the subarcualis obliquus. Haas (1997a) also speculated that subarcualis rectus I might actually represent subarcualis obliquus I, which is missing in frogs and salamanders. This muscle does not persist past metamorphosis.

Transversus Ventralis IV. Transversus ventralis IV is present in *Ascaphus truei* and salamanders (Gradwell 1973; Haas 1997a; Pusey 1943) but is not present in *Rana catesbeiana* (Gradwell 1972b). Haas (1997a) reported that in *Ascaphus* the origin is the proximal part of ceratobranchial IV and the insertion is a median connective tissue raphe anterior to the glottis. Edgeworth (1920, 1935) reported that a poorly developed transversus ventralis IV (hyopharyngeus in his terminology) disappeared during ontogeny in young larvae of *Rana temporaria;* he also (1935) reported it absent in *Pipa pipa* and *Xenopus laevis.* Sedra and Michael (1957) reported the muscle present in *Xenopus laevis,* and Gradwell (1971b) followed their usage. O. M. Sokol (1977a) considered the transversus ventralis IV of *Silurana tropicalis* and *Xenopus laevis* to be synonymous with the tympanopharyngeus of *Pipa.* Haas (1997a) believed that Sokol was correct in his interpretation, and I follow his opinion here; in his data matrix only *Ascaphus* has transversus ventralis IV. Noble (1929) reported a "hyopharyngeus" in *Hoplophryne,* but its topography does not correspond to the transversus ventralis IV of other taxa. The transversus ventralis IV does not survive metamorphosis.

Tympanopharyngeus. The tympanopharyngeus muscle, as described in *Rana catesbeiana* (Gradwell 1972b) and *Pelobates fuscus* (Schulze 1892), extends from the lateral aspect of ceratobranchial IV laterally to the tympanic region of the otic capsule. According to Sedra and Michael (1957) and Gradwell (1971b), in *Xenopus* the muscle extends from the medial aspect of ceratobranchial IV dorsal to the dilator laryngis muscle to meet the fibers of its companion in a medial raphe. In *Pipa* the tympanopharyngeus originates medial to the dilatator laryngis and inserts on the ventral surface of the posterior pharynx (O. M. Sokol 1977a). Gradwell (1972b: fig. 1) described and figured the tympanopharyngeus of *Rana catesbeiana* as extending from the otic capsule to ceratobranchial IV, essentially following Schulze (1892). Haas (1996b) gave the origin of the tympanopharyngeus in *Gastrotheca* as the pharyngeal wall and the insertion as the pericardium and indicated that it was closely associated with the levator arcus branchialis IV and the dilatator laryngis. The descriptions presented by O. M. Sokol, Sedra and Michael, Gradwell, and Haas are somewhat different. However, Haas (1997a) examined the muscle in a wide variety of frogs, and I accept his conclusions about the distribution of this muscle; the tympanopharyngeus is absent in *Alytes, Ascaphus, Bombina, Discoglossus,* and *Xenopus* and present in *Gastrotheca, Pelobates,* and other neobatrachians but not *Rana temporaria.* In his analysis, the presence of the muscle (character 16) is a synapomorphy of *Pelobates* + neobatrachians but is reversed in *Rana.*

Diaphragmato-branchialis. The diaphragmato-branchialis (IV) (diaphragmato-branchialis lateralis) extends anterolaterally from the larval diaphragm to the process at the distal end of ceratobranchial III; identification of this muscle in various taxa seems to be unproblematic. It is present in *Alytes, Ascaphus, Bombina, Discoglossus, Gastrotheca, Pelobates,* and *Rana catesbeiana* (Gradwell 1972b, 1973; Haas 1996b, 1997a; Pusey 1943; Schulze 1892). It is also present in *Hymenochirus* and *Pipa* (O. M. Sokol 1977a) but not reported in *Xenopus.* Although the muscle name is often followed by a "IV," it does not attach to the fourth arch. However, the muscle is innervated by the third branchial ramus of cranial nerve X, which supplies the fourth branchial arch. Pusey (1943) hypothesized this muscle to be homologous to the omoarcualis muscle of salamanders. Schulze (1892) described a diaphragmato-branchialis medialis in *Pelobates fuscus,* but Edgeworth (1935) and Noble (1929) considered this part of the rectus cervicis/rectus abdominis series. The diaphragmato-branchialis does not persist in adults.

Cucullaris. Gradwell (1971b) and Sedra and Michael (1957) described the cucullaris in *Xenopus laevis;* it extends from the lateral part of ceratobranchial IV to the scapula. In *Discoglossus* the muscle appears in a Gosner 33 tadpole (Schlosser and Roth 1997a). The cucullaris is a classical branchial arch derivative in tetrapods, but it is not described from any other frogs. Gradwell (1972b: fig. 1 and table 2) figured an unnamed branchial muscle with its origin on the lateral part of ceratobranchial and insertion on the scapula. This may be the cucullaris.

The Larynx and Its Muscles

In adult frogs (except pipids) the larynx consists of paired, crescent-shaped, movable arytenoid cartilages that are nestled within the cricoid ring (annulus cricoideus); in pipids the larynx is highly modified (Ridewood 1897a). In *Rana temporaria* the laryngeal muscles form before the arytenoid and cricoid ring chondrify during metamorphosis (Edgeworth 1920). Ridewood (1897a) noted that the larynx begins to appear in *Xenopus* at stages in which the hind legs are just appearing, and that in *Pipa pipa* the larynx is present in embryos of 12 mm SVL. Sedra and Michael (1957) noted that the primordia of the arytenoids are feebly developed in a Nieuwkoop/Faber 55 tadpole of *Xenopus laevis.* J. J. Wiens (1989) noted that the larynx of *Spea bombifrons* chondrifies rapidly at the end of metamorphosis. Trewavas (1933) discussed the laryngeal development of several taxa. It appears that the larynx develops earlier in *Pipa* and *Xenopus* than in other frogs.

The muscles of the larynx are derived from esophageal constrictor muscles rather than from the muscle of the gill arches (Edgeworth 1920); they are innervated by the laryngeal branch (ramus laryngeus) of the vagus nerve (X). Most of the laryngeal muscles develop after metamorphosis and are not discussed here. A few arise just prior to metamorphosis—the dilatator laryngis muscle arises from the edge of levator arcus branchialis IV and inserts on the arytenoid cartilage and the cricoid ring. The constrictor laryngis muscle consists of dorsal and ventral parts that arise from medial raphe and insert on the arytenoid cartilages (Sedra and Michael 1957). In *Discoglossus* the ramus laryngeus brevis innervates the dilatator laryngis and the r. laryngeus longus supplies the constrictor laryngis (Schlosser and Roth 1995). The dilatator and constrictor survive metamorphosis and bear the same name in the adult. The laryngeobranchialis muscle may be unique to *Pipa* (O. M. Sokol 1977a); it originates from the medial end of ceratobranchial III and inserts on the developing larynx.

Metamorphosis of the Hyobranchium

The hyobranchial apparatus does not undergo any appreciable change in shape until the beginning of metamorphosis (fig. 4.11). Metamorphosis of the apparatus generally begins at the time when the tadpole loses its jaw sheaths and the forelimbs break out. Metamorphosis will be described generally in three stages—the beginning of metamorphosis (Gosner 41–43), metamorphosis (stages 44–45), and newly metamorphosed (stage 46).

At the beginning of metamorphosis, the transversely oriented ceratohyals assume a more longitudinal orientation. The ceratohyals become larger and are in close medial contact when they reach their maximum size. The ceratobranchials are smaller. The spicules of the ceratobranchials disappear before the ceratobranchials. The cartilage of the ceratobranchials shrinks, and the clefts decrease in size. Copula I disappears and the pars reuniens gradually disintegrates. A thyroid fenestra or foramen initially forms as a depression near the medial margin of the hypobranchial plate and then perforates. The medial margin of this fenestra becomes rodlike and forms the incipient posteromedial process (thyrohyal). The hyoglossal and laryngeal sinuses enlarge. Meckel's cartilages begin to straighten and lose their sharp angulation; they lie in a gentle curve with the infrarostrals.

As metamorphosis proceeds, the ceratohyals orient yet more posteriorly and become slender (fig. 4.11D). They are united medially and lose their articulation with the palatoquadrate. In some taxa, new pieces of cartilage form along the anteromedial margin of the ceratohyal; these will form the anterior process of the ceratohyal. (This process of the metamorphosing ceratohyal is not homologous to the larval "anterior process" on the ceratohyal, which is resorbed; see De Jongh 1968.) The hypobranchial plates reduce, and the thyroid foramina enlarge. The distinction of the hypobranchial plates from copula II is less obvious but persists as a clear zone in the cartilage plate demarcating the posterior limits of copula II. The posteromedial processes form from the hypobranchial plate and do not represent the remnants of the fourth ceratobranchials (Ridewood 1897b). Lateral margins of the hypobranchial plate begin to form the anterolateral and posterolateral processes. The ceratobranchials are much smaller and break up, and the laryngeal and hyoglossal sinuses enlarge. The lower jaw is much larger and more U-shaped. The infrarostrals fuse to Meckel's cartilage giving the appearance of one element. The angulosplenial and later the dentary ossifies.

In the final and third stage of metamorphosis (stage 46) the ceratohyals are very slender and their posterior ends are fused to the otic capsule. They are no longer united medially,

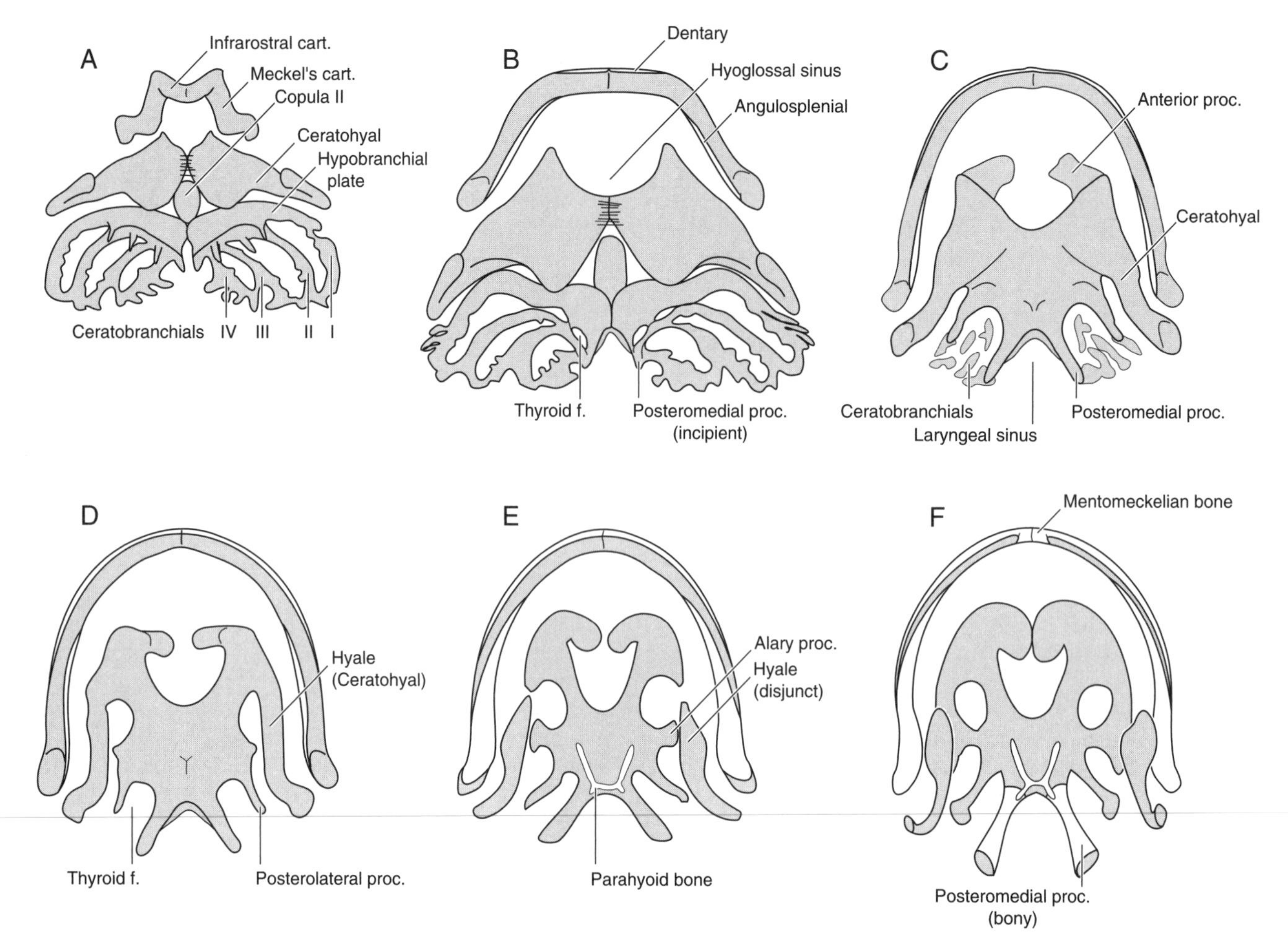

Fig. 4.11. Changes in the hyobranchial apparatus during metamorphosis in *Pelodytes punctatus*. (A) Dorsal view, tadpole with intact mouthparts, hind limbs 1 mm long; (B) dorsal view, metamorphosing tadpole with mouthparts absent and erupted forelimbs, tail 23 mm, hind limb 28 mm; (C) dorsal view, metamorphosing tadpole, tail 2.5 mm; (D) dorsal view, froglet with resorbed tail, 21 mm SVL; (E) ventral view, froglet with resorbed tail, 17 mm SVL; and (F) ventral view, adult frog, 37 mm SVL. Redrawn from Ridewood (1897b).

and resorption of cartilage results in copula II forming part of the margin of the large hyoglossal sinus. The body (corpus) of the hyoid generally is a flat plate (fig. 4.11F). The anterior processes of the ceratohyal, formed from autochthonous cartilage, become indistinguishable from the ceratohyal proper. The ceratobranchials have disappeared completely, and the thyroid foramen is a sinus in the margin of the hyoid plate. The posteromedial processes begin to ossify, and the anterolateral and posterolateral processes are well formed. The lower jaw attains the typical U shape of the adult, and Meckel's cartilage is invested by the angulosplenial and dentary bones; the mentomeckelian bones appear. Overall the appearance of the hyoid apparatus in a newly metamorphosed froglet is very similar to that of the adult. The general development of fate of the cranial muscles is summarized from Edgeworth (1935) in figure 4.12.

Diversity. W. K. Parker's (1871) account of development of the hyobranchial apparatus is riddled with inconsistencies (fide Ridewood 1897b). Copula I persists as an ossified element in adult *Hymenochirus* and *Pseudhymenochirus* (Cannatella and Trueb 1988a, b). In some pelobatoids (e.g., *Pelodytes*) fusion of the anterolateral process to the ceratohyal forms a "lateral foramen" (Ridewood 1897b) that is traversed by the glossopharyngeal nerve and the lingual branch of the carotid artery. The anterolateral process of *Alytes* is formed by adfusion of an autochthonous cartilage rather than from the lateral margins of the hypobranchial plate (Ridewood 1898a). In all pelobatoids and in *Rhinophrynus,* the middle part of the ceratohyal is resorbed in late metamorphosis (fig. 4.11) to produce a disjunct ceratohyal (Cannatella 1985; Ridewood 1897b). In contrast to Ridewood (1898a), Pusey (1943) claimed that the fourth spiculum becomes the thyrohyal of the adult. In pipoid larvae the anterior processes of the ceratohyals are greatly enlarged and underlie almost all of the buccal cavity (O. M. Sokol 1975).

The posteromedial processes of the hyoid are ossified in all frogs. A parahyoid bone forms late in metamorphosis; it is present primitively in Anura but lost independently in different lineages (Cannatella 1985). According to Ridewood (1898a) the parahyoid ossifies from a white fibrous precursor, apparently the pars reuniens. From the beginning it is a single bone (e.g., *Alytes* and *Pelodytes*) and is not formed from the fusion of two elements. Distinct endochondral ossifications, not homologous with parahyoid bones, form in the Bombinatoridae (L. S. Ford and Cannatella 1993). In adult hymenochirines the ceratohyals are ossified (Cannatella and Trueb 1988b).

Axial Skeleton and Muscles

Aspects of gross vertebral formation are perhaps best known in *Xenopus laevis* (Bernasconi 1951; Mookerjee 1931; Mookerjee and Das 1939; Nieuwkoop and Faber 1967; Trueb and Hanken 1993; D. B. Wake 1970), and this account is taken from that species. The account of the trunk muscles from Ryke's (1953) work on *X. laevis* is compared to *Bufo regularis* (Sedra and Moursi 1958).

Development

The eye and jaw primordia of the *Xenopus laevis* embryo form at Nieuwkoop/Faber 22–24, and the first signs of response to external stimuli occur at the same time and stage. The somites have differentiated into the dermatome, myotome, and sclerotome. The anterior somites develop more rapidly than the more posterior ones. The notochordal cells are not yet completely vacuolated. Prior to the appearance of the vertebrae, the axial skeleton consists of the notochord, which extends from the basal plate of the skull to the tip of the tail. The notochord is a flexible, turgid rod that in the mature premetamorphic tadpole consists of vacuolated cells surrounded by a tough notochordal sheath (Youn et al. 1980).

In Nieuwkoop/Faber 45–46, axial mesenchyme from the sclerotome migrates to a position between the myotomes and forms the myocommata which subdivide the dorsalis trunci muscle into myomeres. A mass of cells, the *Urwirbelfortsatz* (= primordial vertebral process), grows ventrally from the ventrolateral part of the myotome and later forms the ventral somatic muscles of the trunk. The *Urwirbelfortsatz* detaches from the dorsal somatic musculature in *Xenopus* at about Nieuwkoop/Faber 38, and the dorsal and ventral (*Urwirbelfortsatz*) muscle masses of the trunk develop independently. These dorsal and ventral muscle masses do not correspond to epaxial and hypaxial muscles. In gnathostome fishes epaxial and hypaxial muscles are divided by a horizontal septum at the level of the transverse processes of the vertebrae. The horizontal septum is fragmentary or completely missing in Anura and most tetrapods, but the distinction between epaxial and hypaxial muscles is maintained in the innervation by dorsal and ventral branches respectively of the ventral roots of the spinal nerves. The fascia lumbodorsalis (see below) forms in the position of the horizontal septum in *Xenopus* and separates epaxial from hypaxial mus-

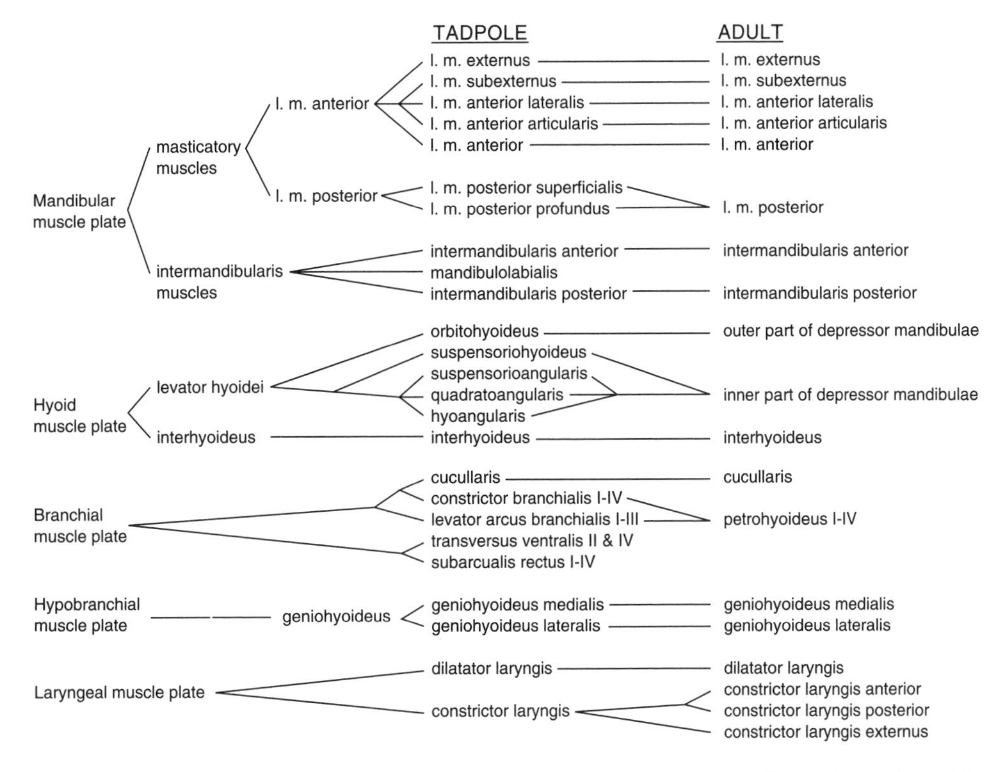

Fig. 4.12. Ontogenetic changes in the cranial musculature of *Rana temporaria*. Data from Edgeworth (1935) and Sedra and Michael (1957). Edgeworth's transversus ventralis II is the subarcualis obliquus of Haas (1997a). Abbreviations: l. m. = levator mandibulae.

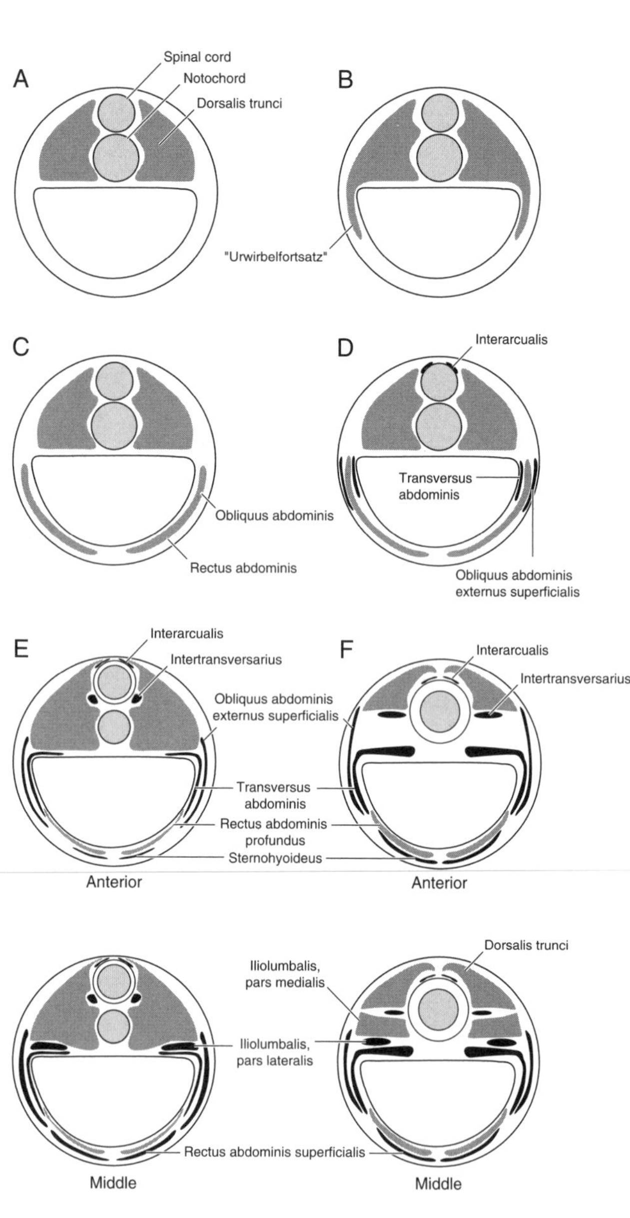

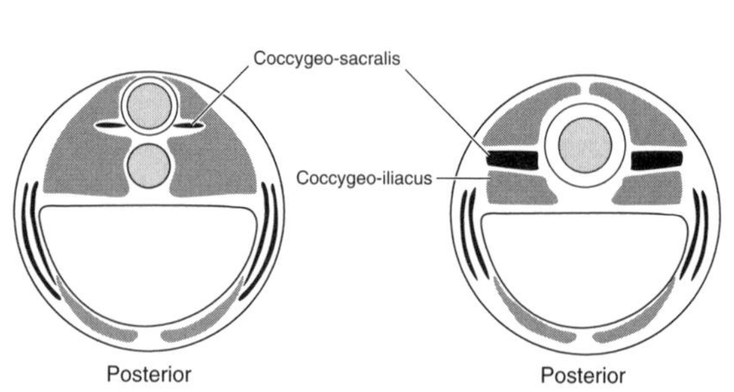

Fig. 4.13. Transverse sections of body wall muscles of *Xenopus laevis.* Consecutive stages in a premetamorphic tadpole: (A) presence of the myotome, (B) appearance of the *Urwirbelfortsatz* (= primordial vertebral process), (C) detachment of the ventral muscles, and (D) appearance of secondary muscles. (E) Sections from the anterior, middle, and posterior body regions of a metamorphosing tadpole. (F) Sections from the anterior, middle, and posterior body regions of a young postmetamorphic animal. Symbols: pale gray = notochord and spinal cord, medium gray = primary muscles, and black = secondary muscles. Redrawn from Ryke (1953).

cles, but because it forms late in ontogeny, Ryke (1953) did not regard it as a homolog of the horizontal septum.

Trunk myotome III spans the atlas and second vertebra. The first parts of the vertebrae to appear are the cartilaginous neural arches, which do so in an anterior-posterior sequence beginning at Nieuwkoop/Faber 48 (Gosner 27–28; Trueb and Hanken 1993). Although there are only 8 presacral vertebrae in *Xenopus laevis,* there are 12 neural arch rudiments. Vertebra IX forms the sacrum, and X–XII (postsacral) contribute to formation of the urostyle (coccyx) by fusion with the hypochord. The cartilaginous rudiments of the centrum begin to appear after neural arches X–XI.

Neural arches I–VIII and centra I–IV ossify at the same time as the first cranial elements, the frontoparietal and parasphenoid, at Nieuwkoop/Faber 55 (Gosner 35) in *Xenopus laevis* (Trueb and Hanken 1993). Vertebral ossification begins at Gosner 33 in *Hyla lanciformis* (De Sá 1988), stage 36 in *Spea bombifrons* (J. J. Wiens 1989), and just prior to stage 36 in *Hamptophryne boliviana* (De Sá and Trueb 1991). Ossification of the urostyle was investigated by Stokely and List (1955).

In pipids the ribs appear as distinct cartilaginous elements on presacral vertebrae II–IV and later fuse to the transverse processes of the vertebrae. The rib anlagen ossify at Nieuwkoop/Faber 57–58 (Gosner 40–41) at the same stage that all eight presacral centra and the sacrum ossify. By Nieuwkoop/Faber 64–65 (Gosner 44–45) the ribs fuse to the transverse processes, which themselves ossify at Nieuwkoop/Faber 64 (Gosner 44). The sacral diapophysis does not assume its expanded form until just after metamorphosis. The hypochord appears in cartilage at Nieuwkoop/Faber 60 (Gosner 41), ossifies at about Nieuwkoop/Faber 63 (Gosner 43), and fuses with the postsacral vertebrae to form the urostyle at Nieuwkoop/Faber 66 (Gosner 46). By this stage, the tadpole tail has largely been resorbed.

The myotomal trunk muscles can be grouped based on the time of development. The primary muscles develop directly from the myotomes, and secondary muscles develop from mesenchyme that has proliferated from the primary muscles closer to the time of metamorphosis. The primary somatic muscles of the adult frog trunk (fig. 4.13) are the coccygeo-iliacus, ileolumbalis pars medialis, longissimus dorsi, and rectus abdominis (also see fig. 3.3A–C). The secondary muscles (fig. 4.13) are the interarcuales, intertransversarii, obliquus abdominis externus superficialis, sternohyoideus, transversus abdominis, and perhaps the rectus abdominis superficialis.

The mass of primary myotomal muscle, called the dorsalis trunci (fig. 4.13A), is covered dorsally by a sheet of connective tissue, the fascia dorsalis. The first secondary muscles to differentiate from the dorsalis trunci are the interarcuales (fig. 4.13D), located between successive neural arches. Near the time of metamorphosis the intertransversarii (secondary) appear between the ossified transverse processes of the vertebrae.

At about the time of metamorphosis, the articulation between the ilium and now-ossified sacral diapophyses forms, and other related changes occur. The dorsalis trunci is separated into dorsal and ventral parts at the level of the transverse processes by a fascia lumbodorsalis (not homologous with the horizontal myoseptum). The ventral parts of myotomes IV–VIII thus form the ileolumbalis pars medialis (fig. 4.13F), which originates from the inner surface of the ilium and anterior margin of the sacral diapophysis and extends forward beneath the transverse processes of the vertebrae lateral to their centra. The coccygeo-iliacus (fig. 4.13F) is the continuation of this muscle mass posterior to the sacral diapophysis; it originates from the lateral surface of the urostyle and inserts on the medial aspect of the ilial wing. Both are primary muscles. The longissimus dorsi is the more dorsal part of the dorsalis trunci and is covered dorsally and laterally by the fascia dorsalis. The longissimus dorsi originates from the urostyle, extends anteriorly across the neural arches and transverse processes, and covers the interarcuales and intertransversarii. In adult *Xenopus* the interarcuales are fused to the longissimus dorsi. The entire muscle is divided into about nine segments by eight myocommata.

The ileolumbalis pars lateralis, a secondary muscle derived from the dorsalis trunci (fig. 4.13E), originates from the anterior tip of the ilium, and inserts onto connective tissue associated with the fourth vertebra. In adult *Xenopus* the ileolumbalis pars lateralis and ileolumbalis pars medialis are fused. Another new secondary muscle is the coccygeo-sacralis (fig. 4.13E), which is a caudad continuation of the intertransversarii. It extends from the posterior end of the sacral diapophysis to the lateral surface of the urostyle. The dorsal surface of the anterior half of the coccygeo-iliacus is covered by the coccygeo-sacralis, which precludes the attachment of the coccygeo-iliacus to the sacral diapophysis. In adult *Xenopus* these two muscles are fused.

The rectus abdominis (fig. 4.13C), derived from the ventrolateral part of the *Urwirbelfortsatz,* is the first of the ventral somatic muscles to develop. It is attached anteriorly to the ventral surface of the hyoid, pericardial sac, coracoid, and sternum and posteriorly to the epipubis or connective tissue in front of the pelvic girdle (see Carr and Altig 1992). During metamorphosis, the rectus abdominis differentiates a superficial layer, the rectus abdominis superficialis, and a ventromedial muscle, the sternohyoideus (rectus cervicis) (fig. 4.13E), which attaches to the hyoid anteriorly and to the dorsal surface of the developing sternum. Both of these are secondary somatic muscles. The remaining primary muscle is then termed the rectus abdominis profundus. In adult anurans, the r. a. superficialis and r. a. profundus are fused into a single rectus abdominis.

The obliquus abdominis (fig. 4.13C) is the dorsal part of the *Urwirbelfortsatz,* which in Anura represents the undifferentiated o. abdominis externus and o. abdominis internus. The o. abdominis differentiates from the o. abdominis externus superficialis and the transversus abdominis, which are both secondary muscles. According to Ryke (1953), the transversus abdominis is probably homologous with the pulmo-oesophageal muscle described by Beddard (1896) for *Pipa* and *Xenopus.* By the end of metamorphosis the obliquus abdominis has mostly disintegrated.

Some muscles of the girdles also are closely associated with the trunk. The latissimus dorsi and pectoralis (pars abdominalis) are derived from the ventrolateral parts of the myotomes near the developing shoulder girdle (Van Pletzen 1953). The iliacus externus develops between the iliac wing and muscles of the thigh, inserts near the crista femoris, and surrounds most of the ilial wing by metamorphosis (Ryke 1953). As *Xenopus* develops there is fusion of closely associated muscles. Ryke (1953) stated that muscles of postmetamorphic *Xenopus* are very similar to those of adult *Rana.*

Anatomy of Locomotion

Compared to the gill irrigation mechanism discussed below, locomotion in tadpoles is relatively simple. The apparatus is a trade-off between the demand for rapid metamorphosis and locomotory needs (Wassersug 1989a). Metamorphosis must be rapid because the locomotor performance of tadpoles is poor during this transition (Wassersug and Sperry 1977) and they are subject to intense selection through predation (S. J. Arnold and Wassersug 1978). The axial muscles of tadpoles, which are arranged in simple chevrons, provide almost all of the required forces for locomotion. The fibers are almost all white, which supports bursts of rapid acceleration but not sustained locomotion. The muscular control of locomotion is well established before the cartilaginous rudiments of the vertebrae appear in skeletogenous connective tissue septa into which the muscle fibers are already inserted. Compared to those of fish and other amphibians, the vertebrae do not ossify until just before metamorphosis, and then they are restricted to myotomes at the very base of the tail. Thus, the tadpole tail is very flexible, providing great maneuverability, but lacks the stiffness needed to store large amounts of potential energy. The limbs do not develop sufficiently to have any locomotor effect until just before metamorphosis; moreover, the forelimbs are covered by the opercular fold, which presumably reduces the amount of drag.

Swimming in *Xenopus laevis* larvae has been better studied than in other frog species (Wassersug 1996). Within two days of hatching, *Xenopus* larvae swim continuously and accelerate by increasing the amplitude of tail oscillation (Hoff and Wassersug 1986). In *Xenopus,* a higher velocity is initially achieved by incorporating more of the tail into the stroke; after the entire tail is engaged, greater speeds are attained by increasing the frequency of oscillation.

The kinematics of swimming is different in *Rana* and *Xenopus* (Wassersug 1989a). *Xenopus* tadpoles hang in the water and rapidly oscillate their tail tips (as do other obligate filter feeders such as microhylids and phyllomedusine hylids). The tadpole of *Rana,* like that of most other anurans, is more benthic and incorporates more of the tail into the stroke. Anatomical differences associated with these behavioral differences include the morphology of the postsacral spinal cord and the number of spinal nerves (K. Nishikawa

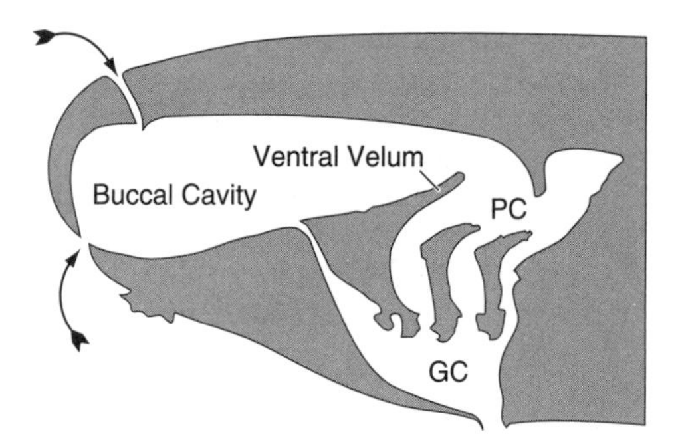

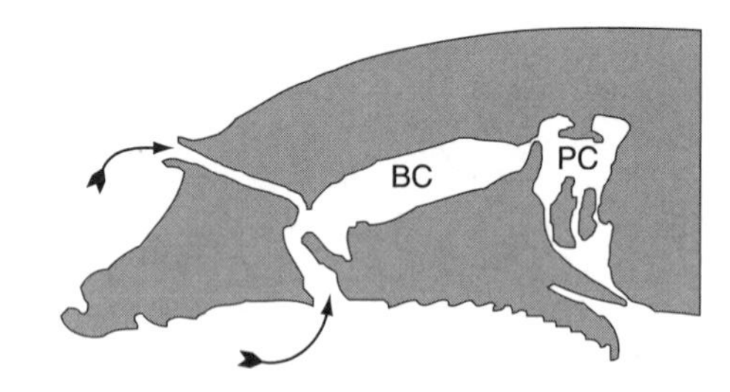

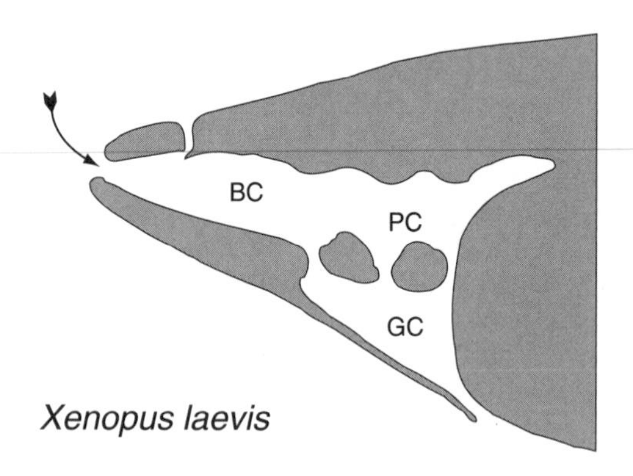

Fig. 4.14. Composite diagrams of sagittal and parasagittal sections through the heads of *Rana catesbeiana, Ascaphus truei,* and *Xenopus laevis.* Abbreviations and symbols: arrows = passage of water into the buccal cavity, BC = buccal cavity, GC = gill cavity, and PC = pharyngeal cavity. Redrawn from Gradwell (1971a, 1971b, 1972b).

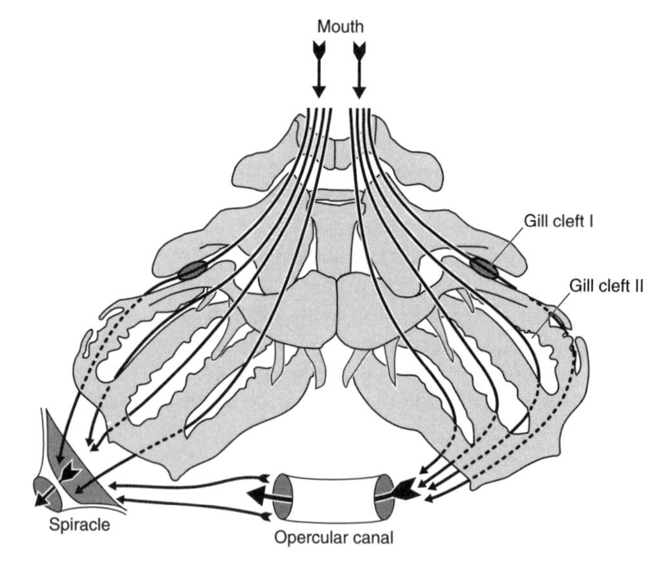

Fig. 4.15. Diagrammatic view of water flow relative to the visceral skeleton (dorsal view) of *Rana catesbeiana.* Upon entering the mouth, the water current is divided into left and right streams that come together just before the flow exits the spiracle. Redrawn from Gradwell (1972b).

and Wassersug 1988, 1989). However, too few taxa have been studied kinematically to permit broad generalizations about the relationship of neuroanatomy and locomotion in tadpoles (chaps. 6 and 9).

Gill Irrigation in *Rana catesbeiana*

The first important accounts of gill irrigation by Schulze (1888, 1892) for *Pelobates fuscus* remained the definitive works on the subject for many years. In a series of papers, Gradwell (1968, 1970, 1972b, c) and Gradwell and Pasztor (1968) described the anatomy of the gill irrigation mechanism in *Rana catesbeiana* as consisting of a buccal cavity (the principal one), pharyngeal cavity, and gill cavity. Each cavity functions at some time as the reservoir of a pump that moves water into the mouth, over the gills, and out through the spiracle.

Buccal, Pharyngeal, and Gill Cavities

The buccal cavity is demarcated from the pharyngeal cavity (fig. 4.14) by the ventral velum (anterior filter valve), a nonmuscular epithelial fold supported by the cartilaginous spicules of the ceratobranchials (Wassersug 1976a). The ventral velum extends laterally and dorsally to form the paired dorsal vela, which have no muscular or skeletal support. In most tadpoles the ventral velum produces strings of food-entrapping mucus from secretory columnar cells on its underside. Kenny (1969b, c) pointed out that rows of secretory columnar cells on the ventral velum were incorrectly identified as muscle fibers by R. M. Savage (1952). Together, the dorsal and ventral vela form a valve preventing backflow of water into the buccal cavity (Gradwell 1970).

In *Rana catesbeiana* the floor of the pharyngeal cavity has three gill clefts or slits. Additionally, endodermal pouch II is perforated to form an additional gill cleft, numbered I (figs. 4.7 and 4.15), which lies between ceratobranchial I and the ceratohyal. This gill cleft opens directly from the buccal cavity into the gill cavity and is not present in any discoglossoids or *Xenopus;* it is open in many neobatrachians, but Gradwell (1972b) and Haas (1997a) disagree on its condition in *Rana temporaria.*

The pharyngeal cavity of *R. catesbeiana* is separated from the gill cavity by three gill clefts corresponding to endodermal pouches III, IV, and V. The gill cavity is enclosed by the operculum, a fold of integument that grows back from the hyoid arch and is fused to the skin of the abdomen (Brock 1929), except at a sinistral opening called the spiracle (=

spout of Gradwell). The spiracle has no muscular or membranous valves in *Rana*. The left and right sides of the gill cavity are connected across the midline by the opercular canal through which water passes from the right side, joining the stream from the left, to exit the spiracle (fig. 4.15).

Within the gill cavity, there are four pairs of gills, each consisting of a series of about 7–15 highly vascularized tufts attached along a ceratobranchial in close association with branchial arteries and constrictor branchialis muscles II–IV (sensu novo); no muscle is associated with ceratobranchial IV. Additionally, the opercular cavity has a vascular lining, the membrana vasculosa opercularis (Gradwell 1969b). Gradwell (1972b) discussed the association of the first gill cleft with a well-vascularized operculum and the absence of the cleft in taxa with a poorly vascularized opercular lining. Haas (1997a) found the first gill cleft to be open (character 30) in *Pelobates* and all but one of the neobatrachians he examined.

Gradwell (1972a, c) reviewed the literature on gill irrigation and pointed out that De Jongh (1968), Kenny (1969c), and R. M. Savage (1952) recognized only a buccal pumping mechanism, whereas Kratochwill (1933) and Schulze (1892) recognized branchial and pharyngeal pumps in addition to the buccal pump. The movements associated with gill irrigation will be considered in several sections that overlap temporally: movements of the jaws, the buccal pump, the pharyngeal pump, and the branchial pump. Gradwell's (1968, 1972c) work was done on *Rana catesbeiana* tadpoles under light anesthesia in a supine (belly-up) position; it is unclear whether his results can be applied to tadpoles under more natural conditions. However, his extensive analysis employed muscle stimulation, denervation, electromyography, and pressure recordings. Gradwell's (1972c) account differs somewhat from his earlier report, and the latter one is used here.

Opening and Closing the Jaws

In general, jaw opening is effected by the jaw depressors: hyoangularis, suspensorioangularis, and quadratoangularis. However, opening of the upper jaw is accomplished not by muscles but by a coupling of the mandibulo-suprarostral ligament ("ligamentum rostrale superior cartilago Meckeli"; Gradwell 1968, 1972b). Jaw opening during irrigation occurs in three sequential phases: narrow opening, wide opening, and protrusion. During narrow opening of the mouth (fig. 4.16B), contraction of the quadratoangularis muscle causes Meckel's cartilage and the infrarostral to swing forward and outward from the buccal cavity, pulling open the lower jaw. Wider opening of the lower jaw is effected by additional recruitment of the hyoangularis muscle (fig. 4.16C). However, the suprarostral cartilage has not yet moved. During protrusion, additional contraction of the suspensorio-

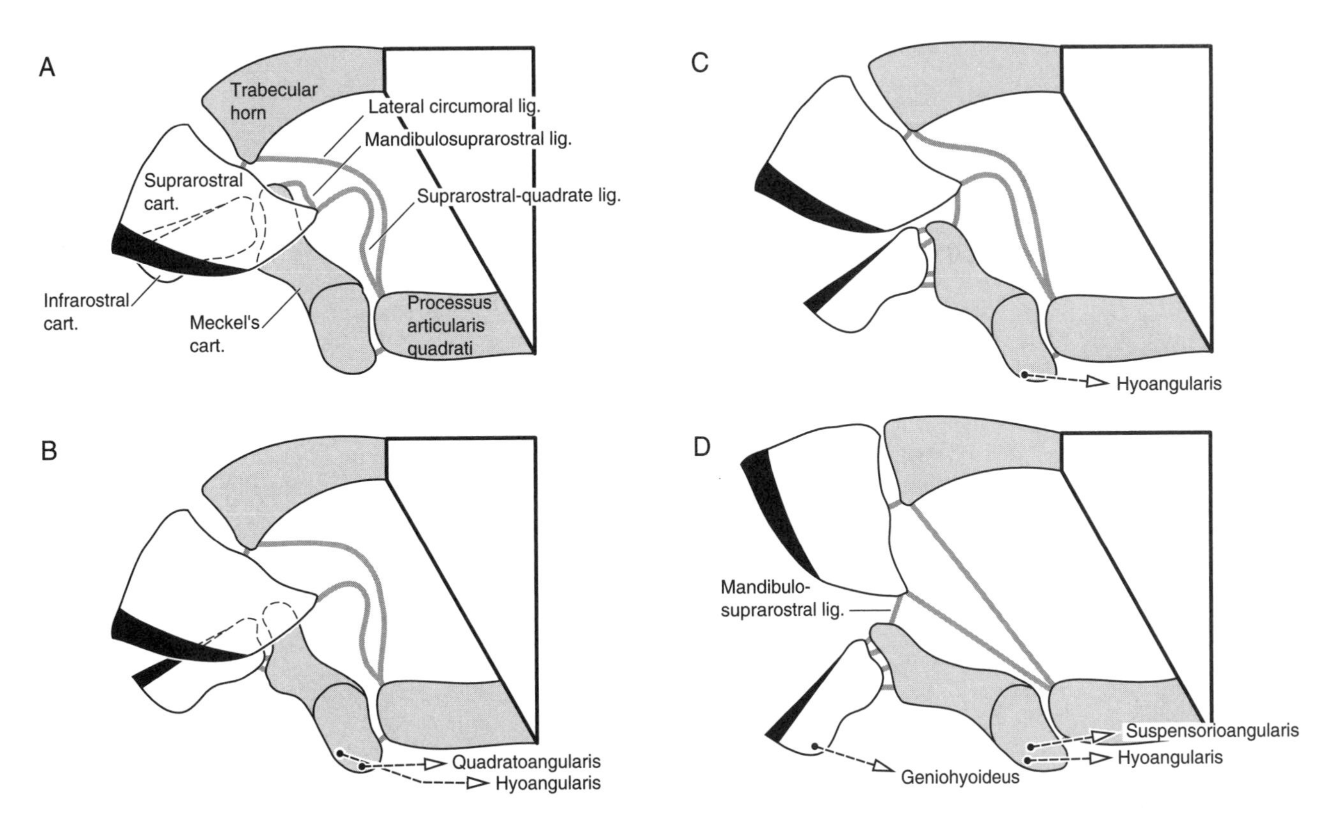

Fig. 4.16. Lateral views of the jaws of *Rana catesbeiana* during various phases of opening: (A) closed, (B) narrow opening, (C) wide opening, and (D) protrusion. Symbols: dashed arrows = direction of muscle action. Redrawn from Gradwell (1972b).

angularis causes yet further forward excursion of the lower jaw, and this takes up slack in the mandibulosuprarostral ligament, which relays the force to the suprarostral cartilage (fig. 4.16D). The lateral circumoral ligament (l. cornuquadratum laterale) and ligamentum rostrale superius quadrati prevent hyperabduction of the jaws.

The coupling protrudes upper and lower jaws at the same time, and the angle formed by the jaws becomes more obtuse. Although the geniohyoid muscle appears to be in a position to open the lower jaw, its role during rhythmic jaw opening is doubtful (see Starrett 1973). It may function during hyperinspiration.

In *Rana catesbeiana,* closing of the mouth occurs faster than the opening. Adduction of the upper and lower jaws takes place simultaneously. Closing during gill irrigation does not produce the close shearing of the upper and lower jaw sheaths as does feeding. In addition to elastic recoil, Gradwell (1968) suggested that closing may be effected by three muscles attached to Meckel's cartilage (i.e., levator mandibulae anterior, l. m. anterior articularis, and l. m. posterior superficialis) and two attached to the suprarostral cartilage (i.e., l. m. posterior profundus and l. m. externus, which is considered to have a pars anterior and pars posterior by Gradwell 1968). Gradwell (1972c) later retreated from this opinion because he was unable to detect electrical activity in the l. m. posterior superficialis, l. m. posterior profundus, l. m. anterior, and l. m. externus. The first three of these muscles lack pink, phasically contracting fibers, and it may be that these muscles function only during feeding and not during gill irrigation.

According to Gradwell (1972c), movements between the two jaws are tightly coupled by the ligaments. If the upper jaw is held closed, the lower jaw does not open; if the upper jaw is held open, the lower jaw cannot close. Similar restriction of movement applies to the upper jaw if the lower jaw is stabilized. If all of the levator muscles attached to the suprarostral cartilage are bilaterally cut, the upper jaw is still pulled closed by the ligamentum rostrale superius quadrati during closing of the lower jaw. Likewise, the lower jaw was closed by the ligament after the levators to Meckel's cartilage were cut bilaterally.

De Jongh (1968) considered musculoskeletal movements during feeding and irrigation to be essentially the same, but muscular contractions during feeding apparently have not been studied explicitly (Altig and Johnston 1989). Gradwell (1968, 1972a) suggested that the mandibulolabialis and intermandibularis posterior muscles function during feeding, perhaps to move the labial papillae and to bring the upper and lower jaws into a shearing position, respectively. Lesion of both muscles did not affect jaw movements during gill irrigation, and Gradwell (1972a) doubted De Jongh's (1968) contention that the intermandibularis posterior elevated the floor of the buccal cavity.

Buccal Pump

The buccal pump is the mechanism that draws water into the buccal cavity of the tadpole. Its piston is formed by the paired ceratohyal cartilages that underlie the floor of the buc-

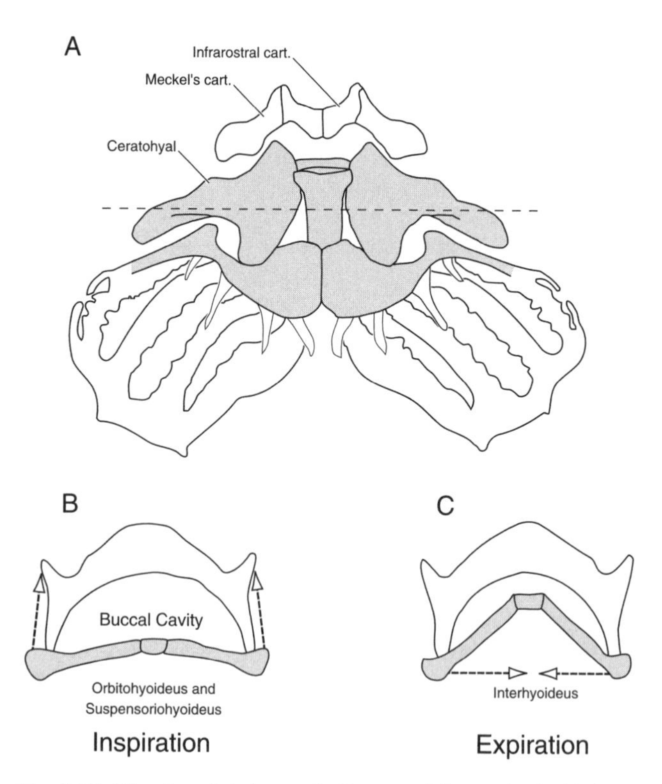

Fig. 4.17. The visceral skeleton of a *Rana catesbeiana* tadpole. (A) Dorsal view. (B) Inspiration occurs by depression of the buccal floor by contractions of the orbitohyoideus and suspensoriohyoideus. (C) Expiration occurs by contraction of the interhyoideus, which elevates the buccal floor. Symbols: dashed arrows = direction of muscle contraction, dashed line = position of the transverse sections in B and C, and gray elements = skeletal elements of the buccal floor. Redrawn from Gradwell (1968).

cal cavity. The ceratohyals are joined medially by the pars reuniens, copulae, interhyoid, and intrahyoid ligaments discussed earlier. These joints, plus the juxtaposition of the hypobranchial plates along the midline, permit bending along the longitudinal axis. The articulations between the anterior part of each hypobranchial plate and the posterior process of the ceratohyal, and between the hypobranchials and copula II, permit some bending in the transverse plane.

Most of the excursion about the elliptical hyoquadrate joint occurs such that the medial part of the ceratohyal moves dorsad and ventrad. Some movement also occurs such that the ceratohyal tilts anteroposteriorly. The lateral part of the ceratohyal bears a flat surface for the attachment of the hyoangularis, orbitohyoideus, and suspensoriohyoideus muscles.

Contraction of the stout interhyoideus muscle causes the ceratohyals to rotate upward around their articulations with the palatoquadrate and elevates the floor of the buccal cavity (expiration; fig. 4.17C). The water stream is divided into two parts, one to the right and one to the left of the pharynx, as verified by observing the flow of dye particles. The interhyoideus (and its principal antagonist, the orbitohyoideus) has both larval and "adult" fiber types. Larval fibers are histologically distinct in that they are thicker and have peripheral nuclei; these pink fibers, with "slow" contractile properties, many mitochondria, and phasic activity, are used primarily in the regular contractions of gill irrigation. The "adult" fibers are thinner, with centrally placed nuclei; these white fibers, of the "fast" type with fewer mitochondria and intermittent activity, are used primarily in hyperexpiration. The two types of fibers have different functions in the tadpole of *Rana catesbeiana* (Gradwell and Walcott 1971). It is the more anteriorly located pink fibers of the interhyoideus muscle that are active during gill irrigation; correspondingly, the anterior part of the buccal cavity, supported by the ceratohyals, is raised more than the posterior part. This configuration directs the water stream posteriorly into the pharynx. The white fibers of the interhyoideus do not function during pumping or hyperinspiration. The interhyoideus posterior consists only of white fibers; its contraction is sporadic and constricts the gill cavity, forcing water out through the spiracle.

Electromyographs and mechanograms (fig. 4.18; Grad-

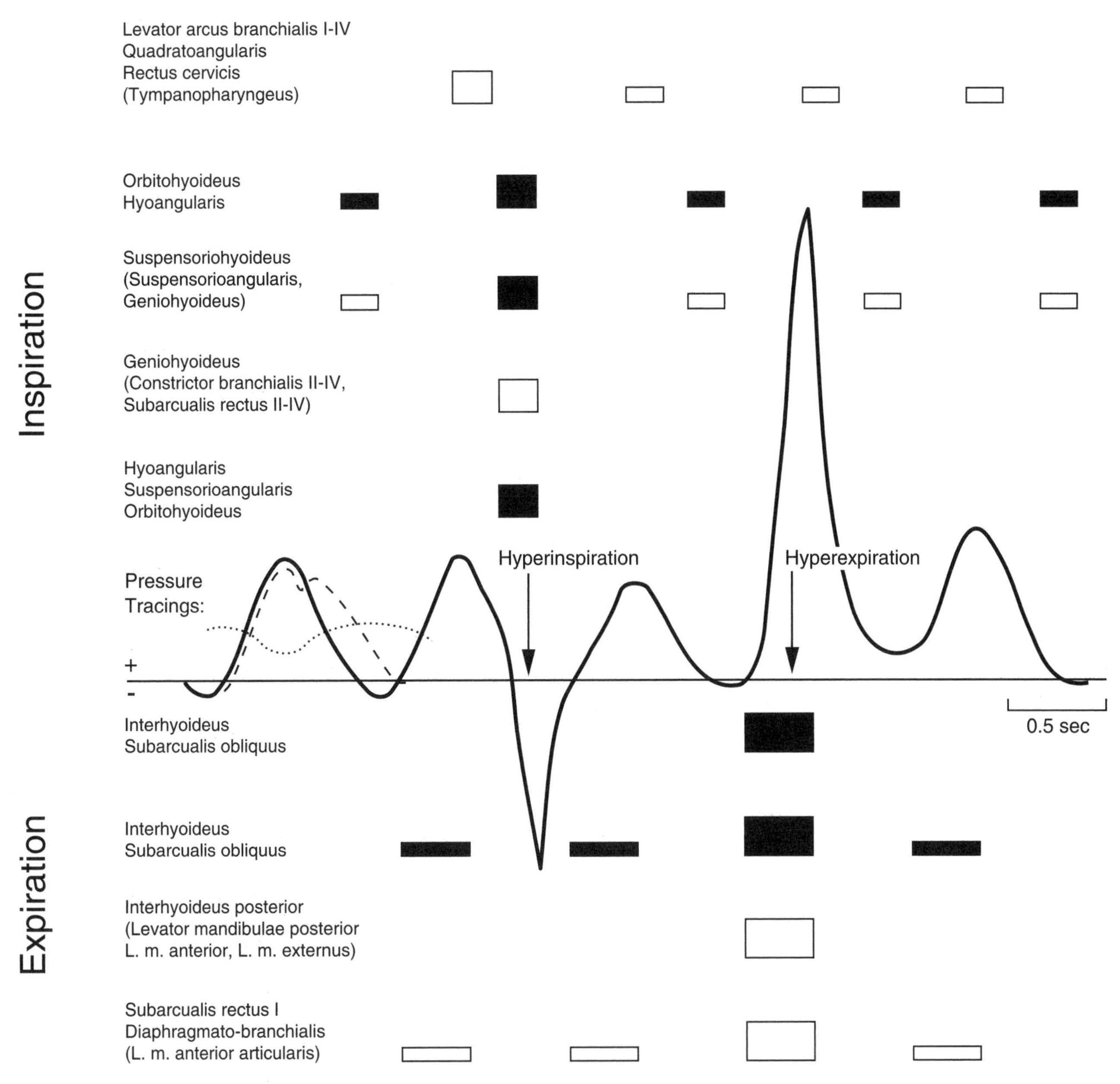

Fig. 4.18. Summary of muscle activity and hydrostatic pressures during gill irrigation in *Rana catesbeiana*. Muscle names in parentheses indicate that activity is postulated, not observed. Symbols: dashed line = typical cycle of pharyngeal pressure, dotted line = branchial cavity pressure, open rectangles = data from anatomy and direct observation, solid line = pressure tracing of buccal pressure for several cycles, and solid rectangles = data from electromyography. Redrawn from Gradwell (1972b).

well 1972c) indicate that both elastic recoil and contraction of the hyoidean depressors (orbitohyoideus and suspensoriohyoideus) depress the buccal floor for inspiration. The negative pressure draws water into the buccal cavity through the mouth. Sedra (1950) incorrectly judged that the hyoidean depressors lift the buccal floor. The suspensoriohyoideus also may help to stabilize the jaw joint during hyperinspiratory contraction of the hyoangularis muscle and to assist in closing gill cleft I (Gradwell 1968, 1972a). During inspiration in *Rana catesbeiana* and *R. fuscigula* (Gradwell 1972a) water flows in through the nares as well, and a narial valve prevents egress of water during expiration (Gradwell 1969a). Experiments using dye particles suggest that the left and right currents do not mix in the pharyngeal and buccal cavities (Gradwell 1972a) of these two species.

Gradwell (1968) found that denervation of the interhyoideus, orbitohyoideus, and suspensoriohyoideus muscles greatly reduced ceratohyal oscillations, but a slight rhythmic pumping was maintained. Bilateral denervation of the hyoidean depressors did not completely eliminate oscillatory buccal floor depression caused by the continued activity of the rectus cervicis and elastic recoil of the buccal floor. Denervation of the interhyoideus showed that buccal floor elevation persisted because of contraction of the subarcualis obliquus (transversus ventralis II) and subarcualis rectus I (Gradwell 1972c). Severing the ligamentum tecti (ligamentum supraorbitale cranii and l. supraorbitale ethmoidale) caused the muscular process of the quadrate to bend outward, suggesting that these ligaments stabilize the origins of the hyoidean depressors (Gradwell 1972c).

The oscillations of the buccal pump are mechanically linked with opening and closing of the mouth. When the ceratohyal was held against the buccal roof, contraction of the jaw depressors opened the mouth only slightly, and the jaws were not protruded as in normal mouth opening. When the mouth was held open the amplitude of ceratohyal excursion was reduced.

Pharyngeal Pump

Movements of the ceratobranchials during irrigation cause changes in the volume of the pharyngeal cavity; this constitutes the pharyngeal pump. Simultaneous with buccal floor elevation (expiration), anteromedial displacement of the ceratobranchials enlarges the pharyngeal cavity, and the gills are swept through the water in the gill cavity (fig. 4.19). This movement is effected by contraction of the subarcualis obliquus (transversus ventralis II), which is synchronized with that of the interhyoideus. When the ceratobranchials reach their forward limit, the buccal floor recoils passively, and the gills are swept posterolaterally, reducing the pharyngeal cavity. The spicules support the ventral velum against the buccal roof and prevent backflow into the buccal cavity. Movement of the ceratobranchials occurs both by elastic recoil and apparently by contraction of the rectus cervicis and the branchial levators I–IV, the apparent antagonists of the subarcualis obliquus.

Movements of the ceratobranchials and ceratohyals are synchronized to some degree by the subarcualis obliquus, which in addition to contraction apparently acts as a ligament joining the crista hyoidea of copula II to the branchial process of ceratobranchial II. Thus, during buccal floor elevation, the ceratobranchials are pulled forward and medially. The movements are assisted by subarcualis rectus I, which also assist in opening gill cleft I.

Branchial Pump

Substantial changes in branchial cavity volume are only occasional and occur during hyperexpiration (figs. 4.18–4.19). Contraction of the interhyoideus posterior muscle compresses the opercular chamber, and contraction of the subarcualis obliquus (= transversus ventralis II of Gradwell) and the normally inactive fibrillar parts of the interhyoideus muscle elevates the buccal floor. Water is thus forcefully expelled through the spiracle.

Pressure Dynamics

Inspiration begins with elastic recoil of the buccal floor from an elevated position. The buccal pressure drops (fig. 4.18), and as it approaches its minimum, the orbitohyoideus and hyoangularis contract, actively depressing the buccal floor and opening the mouth, respectively. Probably the suspensoriohyoideus and suspensorioangularis also contract at this time. Just after buccal pressure reaches its minimum, the interhyoideus and subarcualis obliquus contract to begin the expiration phase. The pharyngeal cavity is enlarged as the buccal floor is elevated and the ceratobranchials are displaced anteromedially. Probably the subarcualis rectus I and diaphragmato-branchialis assist as well. The mouth closes from elastic recoil, and buccal pressure increases greatly.

At the peak of buccal pressure, the branchial levators and rectus cervicis contract, constricting the pharyngeal cavity, closing the velar valve, and elevating the pharyngeal pressure. The resultant hydrostatic pressure is added to that just transmitted to the pharynx from the buccal cavity. Thus, water is pushed further posteriorly across the gill clefts. Immediately after the peak of buccal pressure, the inspiration phase begins again as the ceratohyals begin to recoil passively, the mouth opens, and buccal pressure drops. The pharyngeal pressure drops as well but lags behind the buccal pressure curve. At the onset of inspiration the pressure in the branchial cavity rises because of inflow generated by compression of the pharyngeal cavity; this is the time of greatest efflux from the spiracle. Most other workers have suggested incorrectly that spiracular outflow is greatest during expiration (Gradwell 1972c).

Evolutionary and Comparative Aspects of Gill Irrigation and Feeding

The mechanics of gill irrigation were studied in *Ascaphus truei* (Gradwell 1971a, 1973), *Bufo regularis* (Sedra 1950), *Hymenochirus boettgeri* (O. M. Sokol 1962), *Pelobates fuscus* (Schulze 1888, 1892), *Phrynomantis annectens* (Gradwell 1974), *Phyllomedusa trinitatis* (Kenny 1969c), various species of *Rana* (De Jongh 1968; Gradwell 1968, 1970, 1972b, c; Kratochwill 1933; Severtsov 1969a), and *Xenopus laevis* (Gradwell 1971b). Gradwell's studies of *Ascaphus, Phrynomantis, Rana,* and *Xenopus* provide consistent anatomical and functional comparisons of a species from each of the four tadpole types. The reader is cautioned against generalizing single-species accounts to all members of a particular tadpole type. Moreover, it is worth investigating whether the differences described herein exist at a more inclusive level for any tadpole type.

Ascaphus truei

The tadpole of *Ascaphus truei* (Type 3) has a large oral disc (see Noble 1929 for comparison with *Amolops ricketti,* another species with a large oral disc). The buccal pumping mechanism, powered by the orbitohyoideus, sucks water trapped between the substrate and the disc into the buccal cavity and generates negative pressure by which the sucker adheres to the substrate (fig. 4.14). A passive oral valve seals the mouth when sucker pressure becomes less than buccal pressure, thus maintaining the suction (fig. 4.20). Experimental lesions to the periphery of the oral disc or to the oral valve greatly decrease the effectiveness of the sucker mechanism.

Ascaphus tadpoles can "hitch" forward (but not laterally or backward) while adherent by alternating movements of the upper and lower lip (Altig and Brodie 1972; Gradwell 1971a; M. H. Wake 1993a). Backward movement of the lower jaw and lip by the quadratoangularis muscle opens the mouth and oral valve. Suction is thus reduced, and the friction of the lower labium against the substrate allows the snout to be pushed forward. During this phase the trunk and tail are abducted from the substrate, probably by contraction of the dorsalis trunci muscles. Closing of the lower

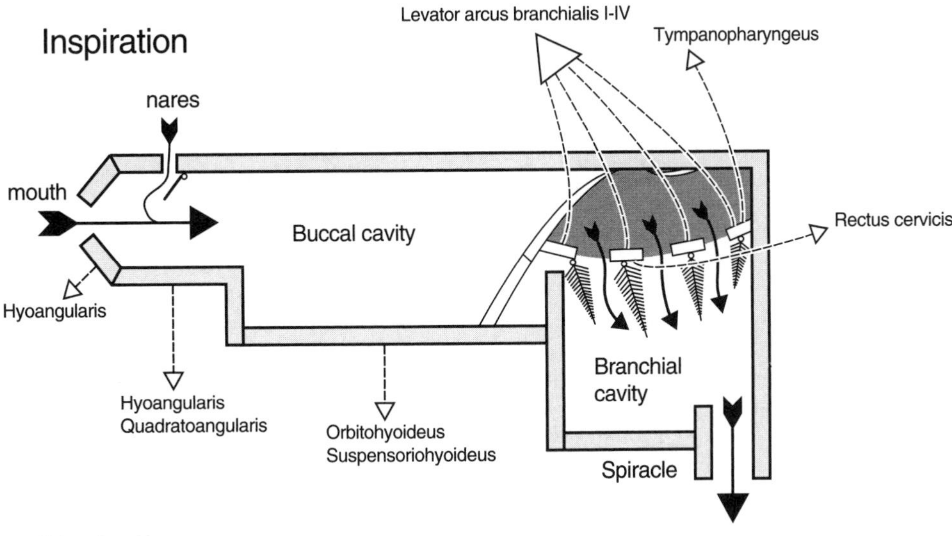

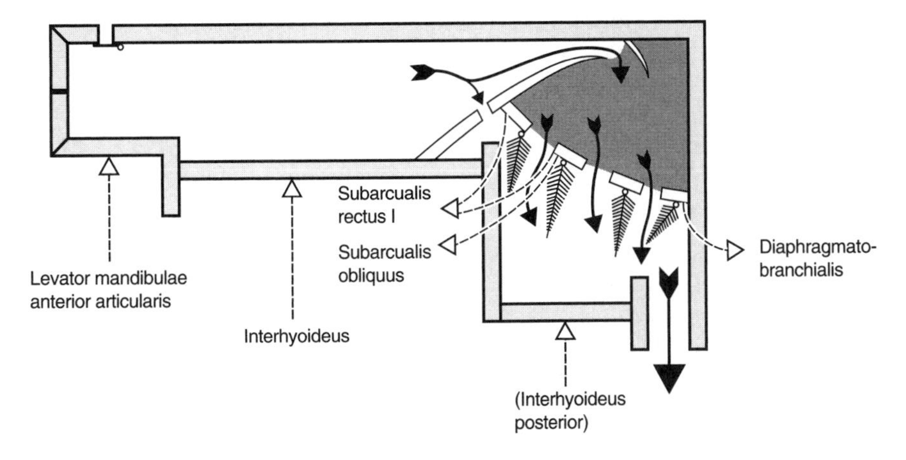

Fig. 4.19. Gradwell's model of the mechanics of gill irrigation in *Rana catesbeiana.* Symbols: dashed white arrows = direction of muscle action, gray area = pharyngeal cavity, and solid black arrows = water flow. The muscles in this model clearly have contracted or have electrical activity during gill irrigation (see fig. 4.14). The interhyoideus posterior muscle is enclosed with parentheses to designate its activity only during hyperexpiration. After Gradwell (1972b).

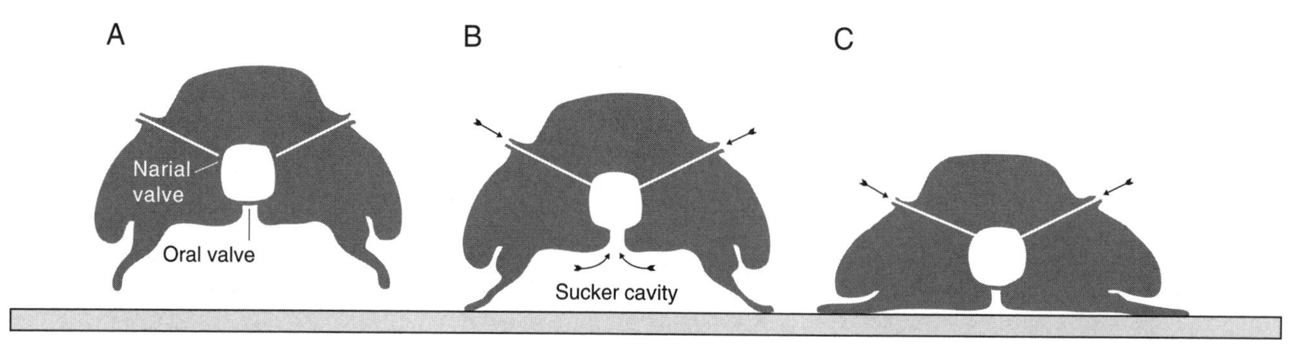

Fig. 4.20. Composite drawings of cross sections of the mouth and nares during oral disc engagement in *Ascaphus truei.* (A) Narial and oral valves closed during expiration. (B) During inspiration, water enters mostly through the mouth but some also enters through the nares; the periphery of the oral disc becomes sealed to the substrate and evacuation of water from the sucker cavity into the buccal cavity causes the oral disc to become engaged to the substrate. (C) During inspiration, reduction in buccal pressure draws in water through the nares, but the oral valve remains closed and adherence to the substrate is maintained. Redrawn from Gradwell (1971a).

jaw by adductor muscle contraction moves it forward, depression of the buccal floor increases the suction, and the body is pulled close to the substrate, possibly by the rectus abdominis muscles. The combination of suction pressure and friction of the lower lip may rasp algae from the substrate. Thus, "hitching" may also function simultaneously in feeding. However, the labial teeth do not serve in anchoring the tadpole to the substrate (Gradwell 1971a).

Tracings of buccal pressure and suction pressure are identical, and the frequency and amplitude of irrigation is similar in tadpoles whether or not suction is engaged. The flow of indicator particles demonstrates that when the sucker is engaged, water enters the buccal cavity only through the nostrils; when not engaged, water enters through both nostrils and mouth. Narial valves prevent reflux of water through the nostrils. As in *Rana,* there are both buccal and pharyngeal force pumps, slightly out of phase, with a velar valve that ensures the posterior flow of water over the gills. Pressure in the gill cavity fluctuates but is continuously positive, indicating a constant flow of water from the branchial cavity. Unlike *Rana,* there is no interhyoideus posterior muscle in the opercular wall, so there is no branchial pump. Gradwell (1971a) speculated that the single, midventral spiracular opening maintains a constant outflow and prevents detritus or parasites from entering the gill cavity.

Phrynomantis annectens

In contrast to *Ascaphus* and *Rana, Phrynomantis annectens* (Type 2) is a midwater filter feeder and lacks keratinized jaw sheaths and labial teeth. In the larval stage, the external nares have not yet opened (as in most microhylid taxa), so the only inlet is the mouth. As described for *Rana,* there is a jaw-ligament coupling such that both upper and lower jaws are protruded simultaneously. The intermandibularis anterior and mandibulolabialis muscles are absent. Unlike *Rana,* the lower jaw rotates in only one plane and the suspensorioangularis muscle is absent (Gradwell 1974).

Like *Ascaphus* and *Rana,* there are cyclic buccal and pharyngeal force pumps that operate slightly out of phase. Unlike *Rana,* the pharyngeal pump is small relative to the buccal pump, and the outflow of water from the branchial chamber is intermittent rather than continuous. Although a ventral velum is present, it is reduced and not completely free, and the dorsal velum assists substantially in maintaining unidirectional flow, unlike *Ascaphus* and *Rana.* A large interhyoideus posterior muscle can simultaneously constrict the buccal, pharyngeal, and gill cavities and may function in expelling detritus from the irrigation system. The flow of water from the midventral spiracle is intermittent, but because this tadpole is nektonic, there is no interference from the substrate.

Xenopus laevis

The tadpole of *Xenopus laevis* (Type 1) has a simpler jaw system than those described for *Ascaphus, Phrynomantis,* or *Rana* (fig. 4.14). Keratinized jaw sheaths and labial teeth are lacking, and the larva is obligately microphagous. Meckel's cartilage is elongate, as in the adult frog, and fused to the infrarostrals. The suprarostrals are fused to the trabecular horns. This results in a wide transverse slit for the mouth. Many of the jaw muscles appear to be fused. *Xenopus* larvae also lack gills and are obligate air breathers, so it is inappropriate to refer to buccal pumping as gill irrigation. Gradwell (1971b) stated that the buccal pump is not active until the larva finds suspended food particles. Phasic pumping then begins, and the larva suspends itself in the water column.

In *Xenopus* the mouth opens as inspiration begins, but the lower jaw is not protruded because of the nature of the jaw joint. Also, Meckel's cartilage is restricted to movement in the dorsoventral plane, like *Phrynomantis* but not as in *Ascaphus* or *Rana.* Correspondingly, the three depressors of the lower jaw are fused into a single quadratohyoangularis. Unlike the ligament-dependent linkage of *Rana,* the quadratohyoangularis muscle provides the only force to open the lower jaw during buccal depression. The upper jaw does not move during mouth opening and closing.

There is no ventral velum in *Xenopus,* and the combined mouth and throat cavity is called the buccopharyngeal cavity. Simultaneous recordings of pressure in the anterior and posterior parts of the cavity are perfectly in phase as expected considering the absence of a ventral velum; therefore, there is one functional pump. The posterior pressures are less than those recorded anteriorly, indicating a posterior flow of water. The interhyoideus muscle is the principal elevator of the buccal floor, but the subarcualis obliquus and the four branchial levators assist as well.

As buccopharyngeal compression begins, the mouth and narial valves are closed as in *Rana* and water passes posteriorly through the branchial clefts (there are no gills) and into the opercular cavity. The spiracles are paired and lateral with flaplike covers; water that enters the opercular chambers does not mix between left and right sides and passes quickly out of the spiracles.

Passive depression of the buccal floor contributes more to water intake than in other taxa, but contraction of the orbitohyoideus is important as well. The inspiration phase is much shorter than expiration. Because there is no ventral velum, the entire buccopharyngeal cavity fills with water, and reflux of water through the gill clefts is a possibility. However, depression of the buccopharyngeal floor causes the spiracular flaps to close, preventing reflux (Gradwell 1975c). Thus, water egress from the opercular cavities is intermittent, as in microhylids, but for a different reason. The constrictor branchialis muscles close the gill clefts, but they do not fire during normal phasic pumping. They do contract simultaneously with the interhyoideus muscle as shown when India ink is introduced into the buccopharynx. The mouth remains open, and the ink is regurgitated from the buccopharynx. There is no interhyoideus posterior to power a branchial pump.

Overall, the pumping apparatus of *Xenopus* is simplified compared to other tadpoles. Because of the fusion of suprarostral cartilage and shape of Meckel's cartilage, the number of movable skeletal joints and planes of movement is reduced. Several of the muscles are absent or fused to others. The ventral velum and a separate pharyngeal pump are absent. Gills and active muscular control of the gill cleft closing are absent, as is the interhyoideus posterior muscle and the branchial pump.

Comparative Ecomorphological and Functional Aspects of Buccal Pumping

Although the diverse ecologies of tadpoles have been appreciated, relatively little attention has been given to the relationship of ecology and the mechanics of buccal pumping. Wassersug and Hoff (1979) devised a model by which they estimated buccal volumes in a broad spectrum of tadpole guilds: *Rana catesbeiana,* 50 μl; *Rana sylvatica,* 10.55 μl; and *Xenopus laevis,* 14.7 μl. Negative allometry of body length with buccal volume among and within species suggests that feeding structures impose a size limit on how big a tadpole can get before it must metamorphose. The lever arm ratio, the ratio of the projected width of the lateral part of the ceratohyal (that part to which the buccal floor depressors attach) to the total width of the ceratohyal and the amount of rotation of the ceratohyal in the transverse plane about the hyoquadrate joint (assuming a contraction to 90% of resting length of the fibers of the orbitohyoideus) are two other morphological variables that correlate with feeding ecology. Buccal volume is strongly correlated with snout-vent length, but neither lever arm ratio nor ceratohyal rotation show significant correlation with size (Wassersug and Hoff 1979).

Wassersug and Hoff (1979) demonstrated that species they considered "microphagous suspension feeders" had small lever arm ratios (0.14–0.17) and largest angles of rotation (33.5°–45.0°); these included three species of microhylids, *Pachymedusa dacnicolor,* and *Xenopus laevis.* Species they classified as "macrophagous carnivores" are at the other end of the range of lever arm ratio (0.40–0.50) and rotation angles (14°–23°): *Anotheca spinosa* (oophagous), *Hymenochirus boettgeri, Scaphiopus holbrookii,* and *Spea bombifrons* (see Altig and Johnston 1989 for a different classification). Species termed "benthic, thigmotactic," such as *Heleophryne natalensis, Hyalinobatrachium fleischmanni,* and *Thoropa miliaris,* were less tightly clumped but distributed in the same half of the range for either lever arm ratio or angle of rotation. The middle part of the range consisted mostly of species with "typical" pond tadpoles. A regression of predicted buccal volume against snout-vent length (although not accounting for phylogenetic structure; Felsenstein 1985) was highly significant and indicated that both microphagous suspension feeders and carnivores had predicted buccal volumes that were larger than those expected from the regression model. In contrast, benthic larvae with suctorial discs had smaller buccal volumes than expected from the regression (Wassersug and Hoff 1979).

These patterns were explained by a consideration of feeding mechanics and prey characteristics (fig. 4.21). Carnivo-

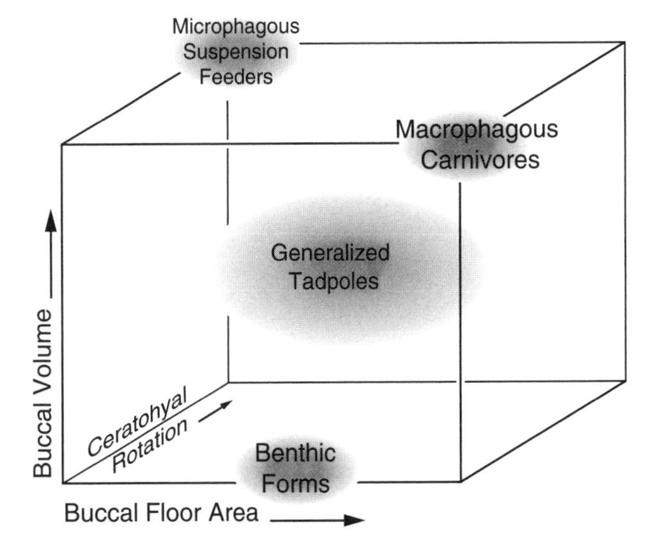

Fig. 4.21. Patterns in buccal pumping design of tadpoles related to the feeding ecology of tadpoles (see text). The size of the clouds is arbitrary. Redrawn from Wassersug and Hoff (1979).

rous tadpoles were predicted to require a large buccal volume to produce the single burst of suction needed to capture moving prey. In contrast, the feeding efficiency of microphagous forms depends on the quantity of water passed through the buccal cavity, and a larger buccal cavity will increase food intake. In both cases, buccal volume is increased, but in different ways. Macrophagous tadpoles had a short angle of ceratohyal rotation (stroke of the piston) and long lever arm combined with a large buccal floor area (bore of the cylinder). Conversely, midwater suspension feeders had a large degree of rotation and short lever arm, but a small buccal floor area (Wassersug and Hoff 1979).

Like the carnivores, the benthic larvae examined had long lever arms and little rotation of the ceratohyal, but the buccal floor, and thus buccal volume, was small. The large mechanical advantage of long lever arms generates large negative pressures at the suctorial disc for adhering to the substrate (e.g., *Ascaphus;* Gradwell 1971a, 1973). Because benthic larvae presumably feed on periphyton on the rocks, large buccal volumes needed for either elusive prey or to move large volumes of water with tiny food particles were not observed.

Tadpoles can effect a constant ingestion rate when food concentration is experimentally altered (Seale and Wassersug 1979). For suspension-feeding tadpoles, the rate of water pumping must be adjusted such that the gill filters do not become obstructed; this can be accomplished by either amplitude modulation (AM) or frequency modulation (FM) of the buccal pump (Wassersug and Hoff 1979). In frequency modulation the rate of pumping is adjusted, but each pump cycle transports the same volume of water (e.g., *Xenopus laevis*). In amplitude modulation the volume of the buccal cavity is altered so that more or less water is pumped in one cycle (e.g., *Rana sylvatica*). The type of modulation appears to be related to the feeding morphology. Forms with short lever arms are predicted to have more difficulty in modulating the amplitude of buccal floor displacement than do those with longer lever arms. *Xenopus laevis,* with a short lever arm, is a FM tadpole, and *R. sylvatica,* with a longer lever arm, is an AM tadpole (Seale and Wassersug 1979).

The muscles involved in modulating either amplitude or frequency of the buccal pump are primarily the orbitohyoideus (buccal floor depressor) and interhyoideus (elevator). Satel and Wassersug (1981) predicted that because macrophagous carnivores must generate a large suction pressure, they would have orbitohyoideus (OH) muscles that are large relative to the interhyoideus (IH), or a low IH/OH ratio. Microphagous larvae have dense gill filters that strain small particles and were predicted to require large buccal pressures and a large IH/OH ratio to force water through these filters.

Quantification of muscle cross-sectional areas supported these predictions (Satel and Wassersug 1981). The cross-sectional areas were isometric to body length. Obligate midwater feeders (i.e., five species of microhylids, *Agalychnis callidryas, Rhinophrynus dorsalis,* and *Xenopus laevis*) had the highest IH/OH values of 0.82), and macrophagous tadpoles such as *Anotheca spinosa, Hymenochirus boettgeri,* and *Spea bombifrons* had a mean IH/OH value of 0.26. Suctorial forms were expected to have strong buccal depressors for clinging to rocks. Slow-water stream forms such as *Hyla legleri* and *Rana boylii* had a lower mean value (0.30) than did pond generalists (0.44), but the three species of obligate, suctorial stream forms (*Ascaphus, Heleophryne,* and *Litoria*) had a mean value of 0.44, not different from generalists. The small sample sizes may be inadequate to detect significant differences.

These comparative studies were done before appropriate methods were available for analyzing correlations among continuous characters in an explicitly phylogenetic context (Harvey and Pagel 1991). Apparent correlations among traits surveyed in a range of taxa may be artificially high because a certain component of the covariation is attributable to phylogeny. It would be instructive to readdress these questions in an explicitly phylogenetic context with the appropriate methods. Reliable information on tadpole diet and ecology is also needed. Although macrophagous tadpoles have certain traits in common, the actual diet of these may vary greatly. Some are oophagous bromeliad dwellers (e.g., *Anotheca spinosa;* Lannoo et al. 1987), while others are open-water predators (*Hymenochirus boettgeri;* O. M. Sokol 1962). Other factors, such as physiological constraints (Feder et al. 1984a, b), merit consideration in generalizations concerning associations of diet, morphology, and function.

Phylogenetic Aspects of Tadpole Morphology

Caenogenesis and Pipoid Tadpoles

Frog tadpoles are caenogenetic (De Beer 1951; Gould 1977); they have deviated from the ancestral ontogeny in having larval specializations (nonterminal additions, such as keratinized jaw sheaths and labial teeth) and less obvious but equally important skeletal modifications (e.g., orientation of the palatoquadrate and presence of suprarostral and infrarostral cartilages). These larval feeding specializations provide a way of prolonging the larval life cycle (Slade and Wassersug 1975), and it may be that a reduction in food is a cue for metamorphosis (Wassersug 1986). At metamorphosis, these larval specializations are lost as the cranium of most tadpoles is extensively remodeled (Orton 1957). Metamorphosis in salamanders, the outgroup, is relatively gradual, but phylogenetic analysis indicates that the ancestor of Anura had drastic metamorphosis.

In Pipoidea, nearly all of the skull bones appear prior to the completion of metamorphosis whereas in other frogs several of the skull bones appear after metamorphosis (Trueb 1985; Trueb and Hanken 1993). It seems that the appearance of cranial ossifications is accelerated because metamorphosis in these pipoids is not as drastic as in other frogs (Wassersug and Hoff 1982) and less remodeling of the osteocranium is required. Pipoids have several peramorphic features (i.e., new morphologies or shapes that are added to the end of the ontogenetic sequence; Alberch et al. 1979). Examples of such features are the fusion of the frontoparietals in Pipoidea, fusion of the vomers in *Xenopus,* fusion of the parasphenoid to the sphenethmoid, expansion of the medial ramus of the pterygoid, and an elongate columella in Pipidae (Cannatella and Trueb 1988a, b). In contrast, some features are interpretable as paedomorphic (i.e., development has been arrested at an earlier stage in the ontogenetic trajectory): lateral line organs of the tadpoles retained in adults of Pipidae, the absence of eyelids and quadratojugals in Pipidae, and the absence of mentomeckelians in Pipoidea.

Pipids remain aquatic as adults (they are the most aquatic frogs), and lack of a transition to a terrestrial life may explain, in part, the less dramatic metamorphosis. Ossification begins earlier, and peramorphic features result from prolonged development. Although the extreme degree of filter feeding in most pipid larvae is apomorphic, pipids have reverted to an ontogeny reminiscent of salamanders by abandoning larval specializations associated with feeding.

Evolution of Orton's Tadpole Types

Starrett (1968, 1973), following Orton (1953, 1957), conceived of an evolutionary sequence (fig. 4.22) in which Type 1 was the most primitive, Type 2 was more derived, and Types 3 and 4 were the most derived, with Type 4 more specialized than Type 3. These assumptions of the primitiveness of pipoid larvae led to a historical controversy about higher level frog relationships because of incongruence of trees derived from larval data with trees based on adult data. Studies of adult morphology had concluded that "discoglossoid" frogs such as *Ascaphus, Bombina,* and *Leiopelma,* were the most primitive anurans (I. Griffiths 1963; Noble 1922). This "larva versus adult" controversy was comparable to the recent, but now obsolete, "molecules versus morphology" debates.

Starrett suggested that the plesiomorphic mode of feeding was similar to that of the Type 1 (pipoid) larvae. Water passing through the "beakless" mouth carries food particles, which are filtered out, and a steady stream of water exits the two spiracles. Feeding and respiration are the same actions.

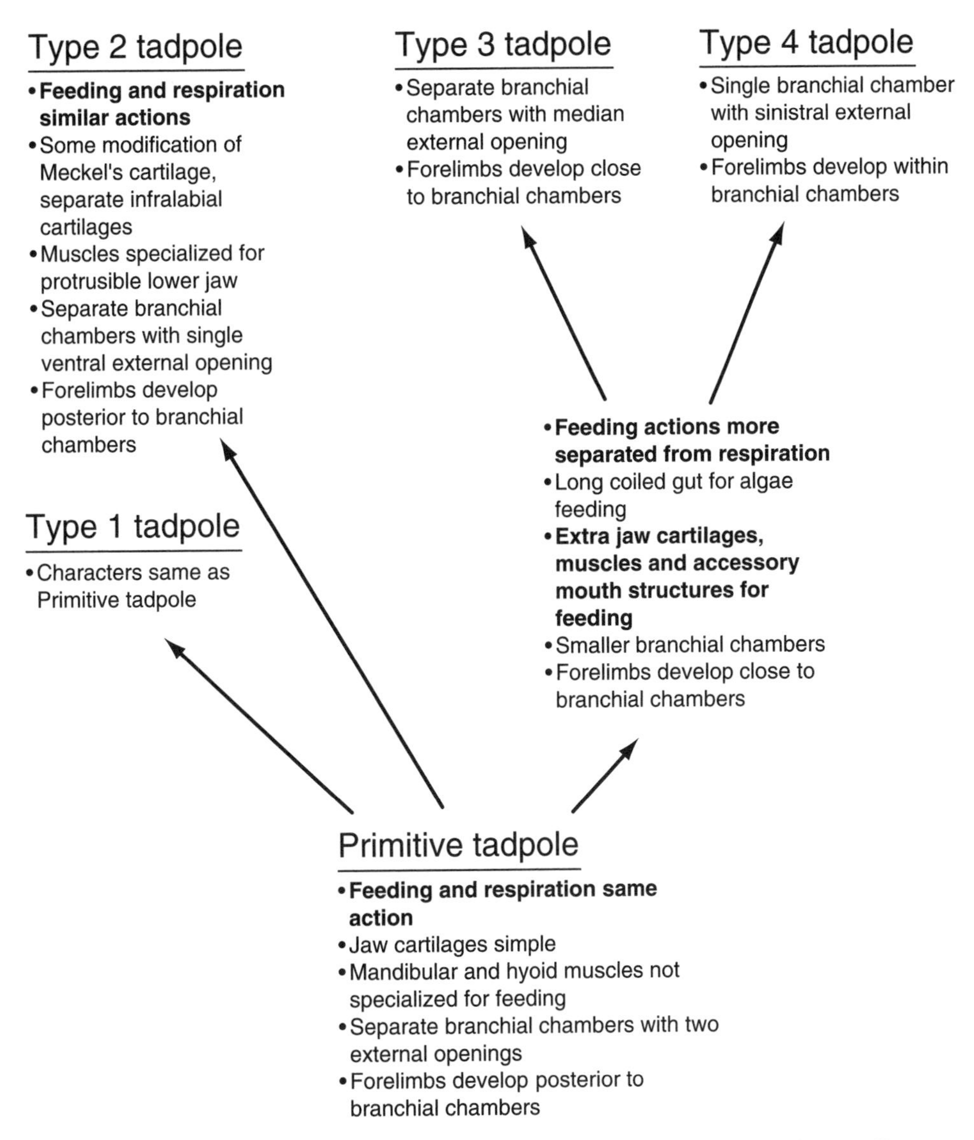

Fig. 4.22. Starrett's interpretation of the evolutionary sequence of Orton's (1953) tadpole types. Characters in boldface are discussed in the text. Redrawn from Starrett (1973).

The mouth is opened by the angularis complex and the geniohyoideus and closed by the adductor muscles.

She further argued that in the Type 2 (microhylid) tadpole some separation of feeding from respiration occurs. A strange, spoon-shaped lower lip is extended by the geniohyoideus and intermandibularis posterior. The lower lip and mouth are closed by the infralabial retractor, interhyoideus, and interhyoideus posterior. The evolution of keratinized jaw sheaths in Types 3 and 4 tadpoles contributes to the separation of feeding and respiration. The adductor muscles move the jaws in a rasping motion. The labial teeth and papillae are controlled by opposing actions of the geniohyoideus and adductor muscles. The addition of a joint between the infrarostral and Meckel's cartilage increases the range of motion of the lower jaw. The intermandibularis posterior muscle has abandoned its original function of assisting the interhyoideus in buccal floor elevation and functions in feeding only.

This evolutionary sequence suffers from two problems. One is the misinterpretation of larval anatomical and functional data, and the second is the lack of a phylogenetic context. Starrett's statement that feeding and respiration are the same action in pipoid larvae is incorrect. All pipoid larvae lack internal gills and are obligate air breathers (O. M. Sokol 1977a; Wassersug 1996); thus the movement of water through the buccal cavity serves no apparent (or a minor) respiratory function. *Xenopus laevis* tadpoles die if they cannot get air, even in highly oxygenated water (Wassersug 1992). Moreover, feeding in pipoid larvae is more diverse than was represented by Starrett (1973). *Rhinophrynus* larvae are known to be both carnivorous and microphagous (Satel and Wassersug 1981). *Silurana* and *Xenopus* larvae are obligately microphagous. *Hymenochirus boettgeri* is a predatory zooplanktivore and lacks gill filter plates completely (O. M. Sokol 1962), free-swimming *Pipa* larvae are macrophagous filter feeders (O. M. Sokol 1977a), and the direct-developing *Pipa* species (Trueb and Cannatella 1986) cannot feed as larvae. I suggest that the diversification of heterogenous feeding modes of pipoid larvae results from the complete decoupling (sensu Lauder 1981) of feeding from respiration. When the buccal pump is freed from the constraint of irrigating the gills, its mechanical parts are then able to evolve with the exploration of other trophic niches.

Some of Starrett's other statements (i.e., purported action of the geniohyoideus in opening the mouth and controlling the mouthparts, the muscles that close the mouth in microhylid tadpoles, the action of the intermandibularis in tadpoles with jaw sheaths, and a steady water stream exiting the spiracles in pipoid tadpoles) are not supported by the work of Gradwell (1968, 1971a, b, 1972b, c, 1973, 1974, 1975c). Other discrepancies were addressed by O. M. Sokol (1975).

Starrett's (1973) proposal assumed that the simpler musculoskeletal system (lack of jaw sheaths and labial teeth, fewer muscles and joints) of pipoid larvae is primitive for Anura, following an implicit assumption that structural simplicity is plesiomorphic. It is well established that determination of character polarity must be based on outgroup comparison (Wiley 1981) rather than preconceptions about the direction of morphological evolution. The structural simplicity argument is flawed because it assumes as fact the very hypothesis of morphological evolution that one is testing. It should be noted that although O. M. Sokol (1975) opposed many of Starrett's interpretations, his conclusions as well were often based on a priori notions of morphological transformation, including the assumption that fusions of cartilaginous elements are always evolutionarily derived.

Regardless of whether or not one accepts the structural simplicity argument, both outgroup comparison and the ontogenetic criterion (G. Nelson 1978) suggest a priori that certain "simple" characters of pipoid and microhylid larvae are plesiomorphic for Anura. Salamanders and caecilians lack labial teeth, and keratinized jaw sheaths are questionable in salamanders (perhaps only in sirenids), so outgroup analysis suggests their absence is plesiomorphic for frogs. These jaw sheaths and labial teeth appear later in ontogeny than do infrarostral and suprarostral cartilages, so by the ontogenetic criterion their absence is plesiomorphic. By either criterion, Starrett's assumption was correct.

However, such a priori hypotheses are tested by other data in the context of phylogenetic analysis. That is, the best-fitting phylogeny, based on all of the relevant data (Kluge 1989), will corroborate certain evolutionary transformations but reject others. Apparent conflicts in trees derived from different data sets are often resolvable if one analyzes the character data on which the trees are based (Miyamoto 1985). Starrett's hypothesis of the primitiveness of Type 1 larvae was not corroborated by phylogenetic analysis of other data. Analyses of primarily adult characters (Cannatella 1985; L. S. Ford and Cannatella 1993; J. D. Lynch 1973) indicated that the Type 3 larva is plesiomorphic, and that absence of jaw sheaths and labial teeth is derived within frogs and is an evolutionary loss. More recently, a phylogeny from larval morphological characters (Haas 1997a; fig. 4.2) also supported the placement of the Type 3 tadpole as plesiomorphic for Anura. In contrast, the phylogeny based on mitochondrial 12S and 16S nucleotide sequences (Hay et al. 1995; fig. 4.2) suggested that the Type 4 tadpole is plesiomorphic for Anura. However, a combined analysis by Kjer, Graybeal, and Cannatella (unpublished data) of morphological characters and this sequence data yielded a tree that supports that of L. S. Ford and Cannatella (1993). This analysis also indicated much of the incongruence is because of the rooting position of the molecular tree.

The continued use of Orton's tadpole types diverts attention from issues of morphological evolution to issues of definitions. L. S. Ford and Cannatella (1993) discussed the problem of the Type 2 (microhylid) tadpole, whose definition was strained after the discovery that tadpoles of at least one species of scaphiophrynine microhylid have jaw sheaths (Wassersug 1984, 1989b). None of the four tadpole types is homogeneous in structure or function, and the acceptance of Orton's groups as monolithic categories will only hinder our understanding of the tadpole evolution. Morphology evolves, and one would expect that definitions of the larval types will become fuzzy as more data are available.

Future Work

There is a resurgence of interest in the evolutionary morphology of tadpoles. New data are appearing in the form of anatomical descriptions, especially through the works of Rafael De Sá, Alexander Haas, and Esteban Lavilla. Equally important, interspecies comparisons are being analyzed phylogenetically both among closely related species (e.g., Haas 1993; Larson and De Sá 1998) and at higher levels. These phylogenetic data are being integrated into larger data sets of adult morphological and molecular characters. Detailed developmental studies, such as Schlosser and Roth's (1995) use of immunostained whole mounts, are providing interpretations on muscle innervation that challenge well-entrenched ideas of homology (Haas 1997a). When done in a phylogenetic context (Schlosser and Roth 1997a, b), these studies provide new insights into the evolution of ontogeny. What is lacking are detailed comparative functional and biomechanical studies coupled with careful natural history observations. Gradwell's studies of muscle function await verification and expansion. The relation of intra- and inter-specific morphological variation of life history and ecology (e.g., Pfenning 1992b; Wassersug and Hoff 1979) remains unappreciated. This genre of investigations, coupled with a phylogenetic perspective, would offer an integrated view of the evolution of this distinctive amphibian life-stage.

ACKNOWLEDGMENTS

I thank Richard Wassersug and the late Otto Sokol for numerous discussions about tadpoles; Wassersug in particular convinced me that tadpoles are interesting. Linda Trueb and Rafael De Sá provided preprints of pertinent literature. The manuscript was improved by reviews from Esteban Lavilla, Roy McDiarmid, and Ronald Altig. Alex Haas provided a copy of his unpublished dissertation and made valuable comments on the manuscript. James Hanken provided photographs of tadpoles with myosin staining. Preparation of the manuscript was supported in part by NSF grant 90-07485.

5

ANATOMY

Viscera and Endocrines

Bruno Viertel and Susanne Richter

The small, apparently insignificant, amphibian tadpole is a wondrous mechanism of organic complexity that is of profound significance in the study of basic biological problems . . .

Fox 1984:x

Introduction

An anuran larva is a nonreproductive, highly specialized stage within the complex, amphibian life cycle and not simply an extended embryonic phase that concludes with metamorphosis. A tadpole is a mosaic of organs that are either remodeled (e.g., exocrine pancreas and posterior alimentary tract; see Pretty et al. 1995; F. Sasaki and Kinoshita 1994; Yoshizato 1992) or degenerated during metamorphosis (e.g., blood-forming organs, filter apparatus, integumentary cells, lymphatic system, and pronephros), remain functionally quiescent (e.g., parts of the excretory and reproductive systems), or are functional throughout both larval and adult stages of the cycle (e.g., liver, lungs, pancreas, spleen, thymus, and thyroid, and the vascular system). The nonreproductive tadpole feeds and grows to produce a metamorph that has some probability of entering the breeding component of the population. In that context, much of the anatomy of a tadpole involves the gathering and processing of food, and many of the structures used in these activities, especially involving the buccopharyngeal area, are unique among vertebrates.

We summarize the anatomical diversity of the circulatory, digestive, respiratory, urogenital, and endocrine systems of tadpoles. Much of the data available on these subjects is derived from studies of only a few genera (e.g., *Bufo, Rana,* and *Xenopus*); when known, information on embryonic development and metamorphic changes is included. Additional information on some subjects is found in chapters 3, 4, and 6.

We used the staging tables of the original authors; translations of selected staging systems to that of Gosner (1960) are shown in table 2.1. The following staging systems are cited in this chapter and are presented by the author name(s) and the appropriate arabic or roman numeral(s): Cambar and Gipouloux (1956a, *Bufo bufo*), Cambar and Marrot (1954, *Rana dalmatina*), Cambar and Martin (1959, *Alytes obstetricans*), Gallien and Houllion (1951, *Discoglossus pictus*), Gosner (1960, all generalized anuran larvae), Nieuwkoop and Faber (1956, *Xenopus laevis*), Rossi (1959, *Bufo bufo*), Sedra and Michael (1961, *Bufo regularis*), Shumway (1940, *Rana pipiens*), and A. C. Taylor and Kollros (1946, *Rana pipiens*). The terminology we used in defining periods of development are as follows: embryonic = Gosner 1–24 (zygote to closure of operculum); larval or tadpole = Gosner 25–46; premetamorphic = Gosner 25–35 (growth of trunk and tail); prometamorphic = Gosner 36–41 (growth of limbs, thyroxine titers increase), and metamorphic climax = Gosner 42–46 (remodeling and degeneration of many larval structures, thyroxine titers reduced).

Circulatory System

The organs of the circulatory system are derived from mesoderm, and the heart and vitelline vessels are the first fully functional embryonic organs. The circulatory system transports materials between organs with external contacts (e.g., alimentary tract, lungs, gills, and skin) and those with internal functions (e.g., endocrine, excretory, muscular, nervous, and reproductive systems). The lymphomyeloid system is an immunological buffer between the external and cellular environments of the larva, especially for those organs that come into contact with an aquatic environment rich in microorganisms.

Heart

The histogenesis of the heart is similar in all vertebrates (Hirakow 1989; Hirakow et al. 1987, *Cynops pyrrhogaster;* Hirakow and Sugi 1990, *Gallus;* Volkov 1982, *R. ridibunda;* also Burggren and Fritsche 1997; Farrell 1997; Icardo 1997). Heart-forming potency is expressed only in the dorsal lip of the blastopore and in deep mesoderm between 30° and 45° lateral to the dorsal midline of the gastrula. Transplantation experiments show that heart mesoderm is established by a dorsalizing signal from the dorsal blastopore lip (Sater and Jacobson 1990, *Xenopus laevis*) well before the end of gastrulation. In contrast to the results of A. G. Jacobson and Duncan (1968, Caudata), Sater and Jacobson (1989, *X. laevis*) showed that the pharyngeal endoderm does not have an inductive effect and does not promote the onset of heart function.

In *Rana temporaria* (Shumway 1940), cells from the dorsal edge of the visceropleura form a single endocardial tube that is continuous with the endothelium of the ventral aorta by Shumway 18 (fig. 5.1A–G). The myocardial anlage originates from mesoderm at the ventral edge of the lateral plates (somatopleura and visceropleura; Cøpenhaver 1955 and DeHaan 1965). The dorsal part of the ventral edge folds around the endocardial tube from anterior to posterior. At Shumway 18 the heart is a longitudinal tube consisting of the conus arteriosus, ventricle, and sino-auricular complex (i.e., future sinus venosus and atrium). At this time, the pericardium develops as a folding of the more lateral somatopleura around the myocardium. Dorsally (Shumway 20) the myocardium attaches at the dorsal pericardium in the region between the conus arteriosus and the paired aortic trunks. The paired vitelline veins come into contact with the undivided auricular region and form the sinus venosus. Trabeculae form shortly after the heart begins to beat at Shumway 20. Up to Shumway 22, the embryonic heart tube is situated ventromedially and posterior to the pharynx. At Shumway 23 the right, left, and posterior distal conal valves and the septum coni differentiate from the endothelial lining of the conus arteriosus at the bifurcation of the aortic trunks. The endothelium grows rapidly compared to the myocardium and together with blood flow affects the epithelial folds. Cells migrating between the myocardium and endocardium stiffen the folds and form a valve. The conal valve changes sequentially (fig. 5.1E–G) from dorsal, to left, to ventral positions, and finally comes to lie near the ventricle at the proximal end of the conus (Shumway 25). At Shumway 24 an endothelial double-layered interauricular septum divides the atrium. By Shumway 25 the auriculo-ventricular valves are connected with the interauricular septum, the sinus venosus is distinct, trabeculae have developed in the ventricle, and the wall of the ventricle is spongy. The heart does not develop further until after metamorphosis. Bending and twisting of the heart tube is a typical feature in tetrapod heart development. In anurans the twisting of the ventricular region and the transitional region of ventricle and conus constitute essential changes, whereas the sinus venosus retains its position. The conus approaches the atrium (fig. 5.1H–I). In the fully developed larva, the heart is dorsal to and bounded laterally by the posterior branchial arches. In adults, the heart tilts more anteriorly than in the larva (fig. 5.1H–K; Benninghoff 1921, *Bombina variegata, Bufo bufo, Leptodactylus pentadactylus, R. catesbeiana, R. esculenta* complex, and *R. temporaria;* Ekman 1924, 1925, *R. temporaria;* Erdmann 1921, *Bombina* and *Rana;* Goette 1875, *Bombina;* Ison 1968; Nieuwkoop and Faber 1956, *X. laevis;* Prakash 1954 and Thomas 1972, *Hoplobatrachus tigerinus;* Rugh 1951).

Arteries and Veins

In *Xenopus laevis,* elongated cells appear in the lateral plate near the inner layer of skin at Nieuwkoop/Faber 31–32. They migrate from the lateral plate and engulf and surround the blood cells, and by Nieuwkoop/Faber 33–34 they are flat and in desmosomal contact with each other. At Nieuwkoop/Faber 35–36 these cells are found throughout the ventral side of the embryo, and by stage 37–38 they have extended toward the endoderm to form the blood vessels of the ventral intestinal tract (Mangia et al. 1970).

The formation of aortic arches and pulmonary arteries at the sixth aortic arch and the anatomical displacement and reduction of the aortic arches during metamorphosis have been studied extensively (figs. 5.2–5.3; see Delsol and Flatin 1972 and references therein). In contrast to the arterial system of the pharynx, little anatomical change occurs in the arterial system of the trunk and limbs. The subclavian (forelimbs), mesentrial (intestines and kidneys), and iliac arteries (hind limbs) branch sequentially from the dorsal aorta posterior to the pharynx. The dorsal aorta extends into the tail as the caudal artery (see Rhodin and Lametschwandtner 1993), and this vessel degenerates at metamorphosis.

Remodeling of the venous system occurs before and during metamorphosis primarily in the postcranial region (figs. 5.4–5.5). The medial posterior cardinal veins fuse to form an unpaired interrenal vein; the lateral posterior cardinal veins form and supply the opisthonephros with blood independently of arteries. As such, they represent the main drainage system for the caudal and iliac veins (i.e., the lateral pos-

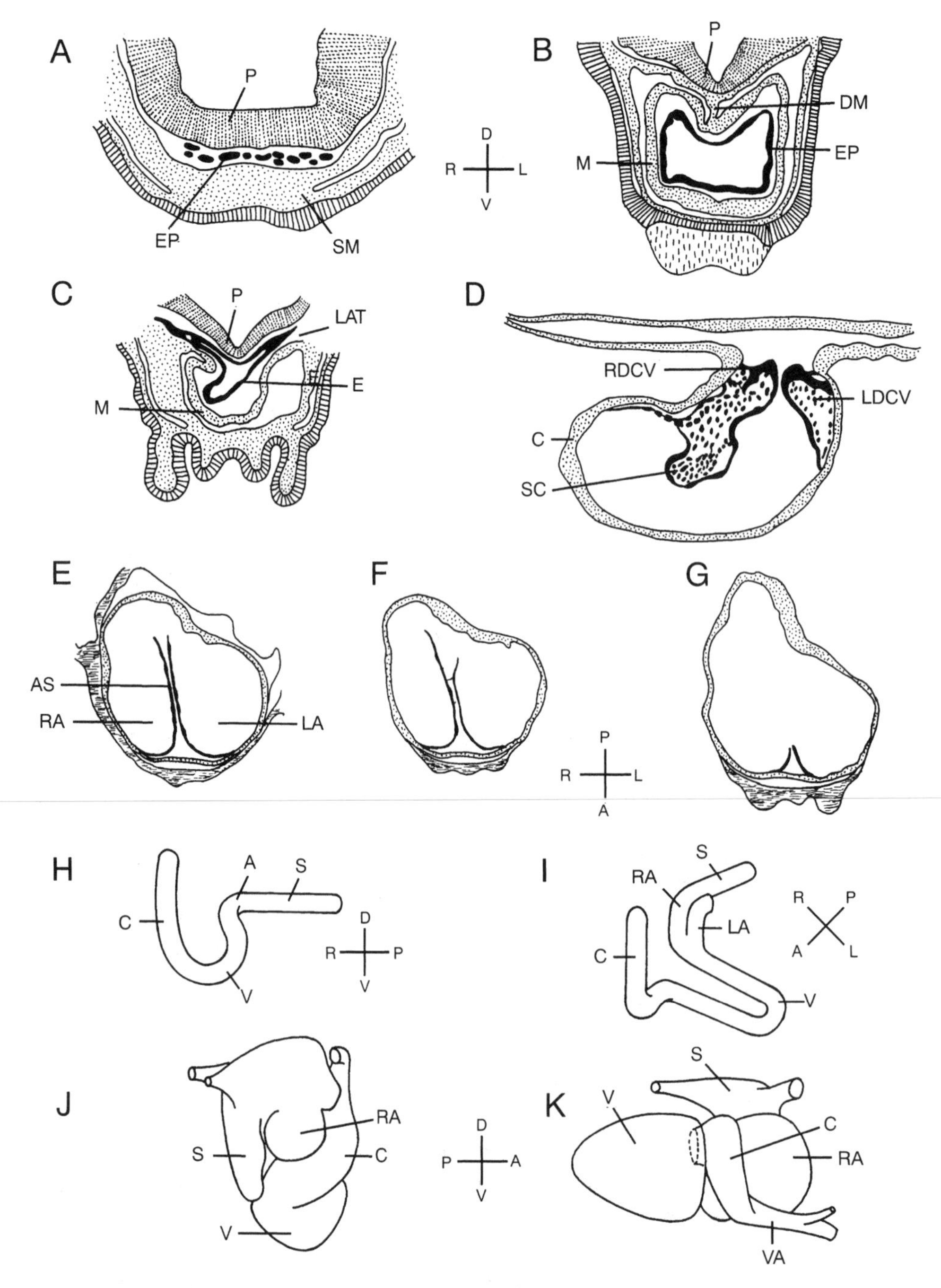

Fig. 5.1. Development of the heart of *Rana temporaria*. Cross sections of (A) plate of cells that will form the endocardium of the heart (Taylor/Kollros stage 17), (B) dorsal mesocardium (Taylor/Kollros 18), (C) endocardium continuous with the endothelial lining of the aortic arches (Taylor/Kollros 18), and (D) septum coni originating anteriorly as a continuation of the right distal conal valve (section through conus septum). Three horizontal sections (E–G) from dorsal to ventral, interauricular septum developing from the anterodorsal wall (Taylor/Kollros 24). Twisting of heart tube at early (H) and later (I) stages. Right lateral view of the heart at Taylor/Kollros 25 (J) and in the adult (K). Abbreviations: A = auricle, AS = interauricular septum, C = conus, DM = dorsal mesocardium, E = endocardium, EP = plate of cells that will form the endocardium of heart, LA = left auricle, LAT = left aortic arch, LDCV = left distal conal valve, M = myocardium, P = floor of pharynx, RA = right auricle, RDCV = right dorsal conal valve, S = sinus venosus, SC = septum coni, SM = splanchnic mesoderm, V = ventricle, and VA = ventral aorta. Orientational crosses: A = anterior, D = dorsal, L = left, P = posterior, R = right, and V = ventral. From Ison (1968); reprinted by permission of Taylor & Francis, Ltd.

terior cardinal veins are the afferent renal veins). One of the two omphalomesenteric veins forms capillaries in the liver and, passing caudally, drains the intestine via the newly formed portal vein; the left omphalomesenteric vein drains the stomach region via the gastric vein. The posterior vena cava arises anteriorly from the hepatic vein and posteriorly from the interrenal vein. A capillary network connects the abdominal and musculo-abdominal veins. Blood flows posteriorly via three ventral commissural vessels to the interrenal vein, then to the hepatic vein (i.e., the posterior vena cava), and from there to the sinus venosus. During metamorphosis, a blood sinus connects to the larval subintestinal vein (from the left omphalomesenteric vein) and forms the adult abdominal vein; blood flows anteriorly, opposite to that of the larval abdominal vein. Femoral and iliac veins flow into the anterior and only persistent ventral commissural vessel and thereby drain the hind limbs and tail. The head-pharynx region drains via the external (especially the filter apparatus) and internal jugular veins.

Blood

Turpen and Knudson (1982, *Rana pipiens*) described hematopoietic precursor cells in the late gastrula and early neurula (fig. 5.6). Cell aggregates in these plates of precursor cells differentiate to hematic cords of the blood islands and oval, primitive blood cells. Precursor cells identical with common pluripotent hematopoietic stem cells (PHSC; Broyles et al. 1981 and Cline and Golden 1979) migrate from the dorsal part of the lateral plates into the larval hematopoietic organs (Broyles 1981; Turpen et al. 1979; Turpen and Knudson

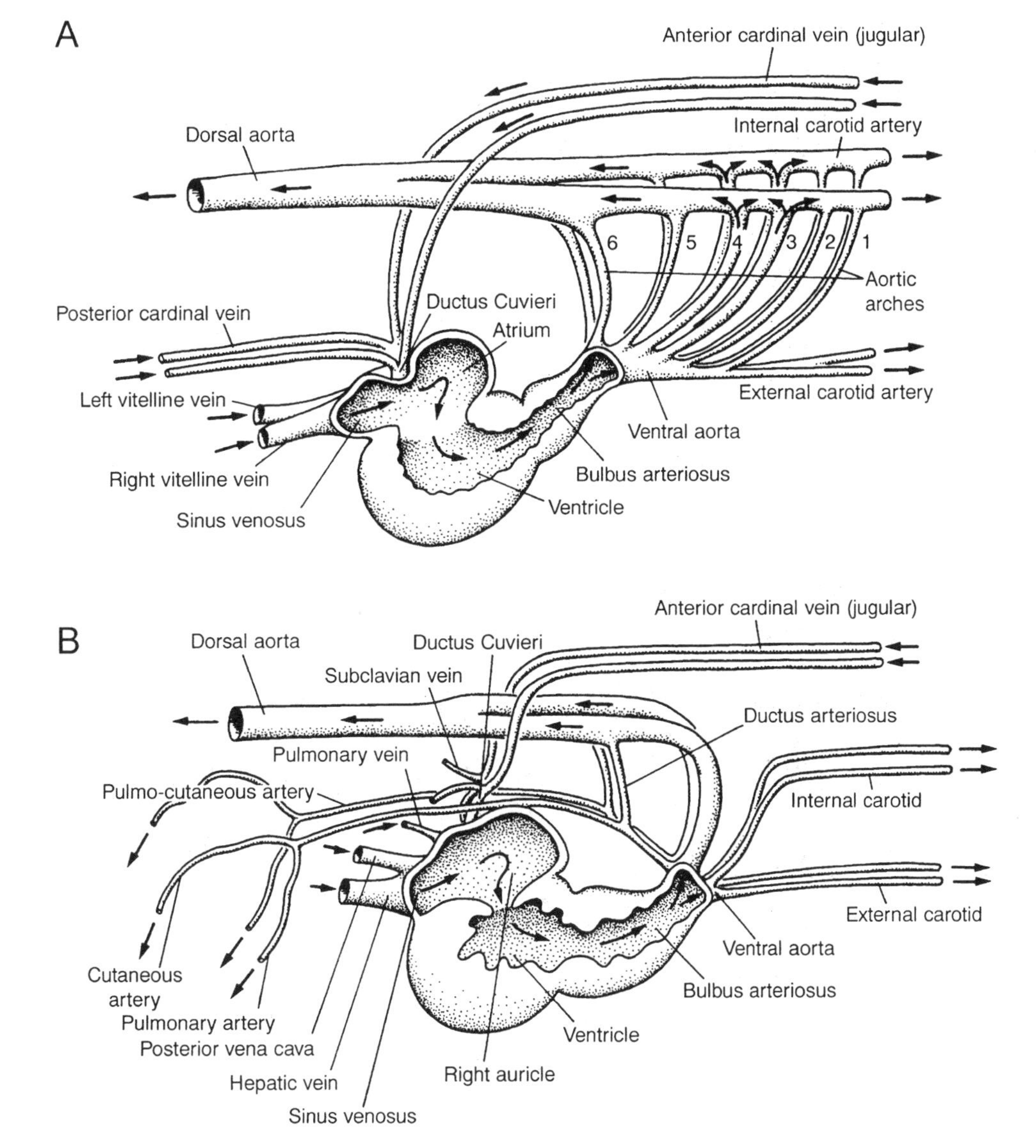

Fig. 5.2. Blood vascular systems of an early (A) and late (B) frog embryo viewed from the right side. From Rugh (1951); reprinted by permission of McGraw-Hill, Inc.

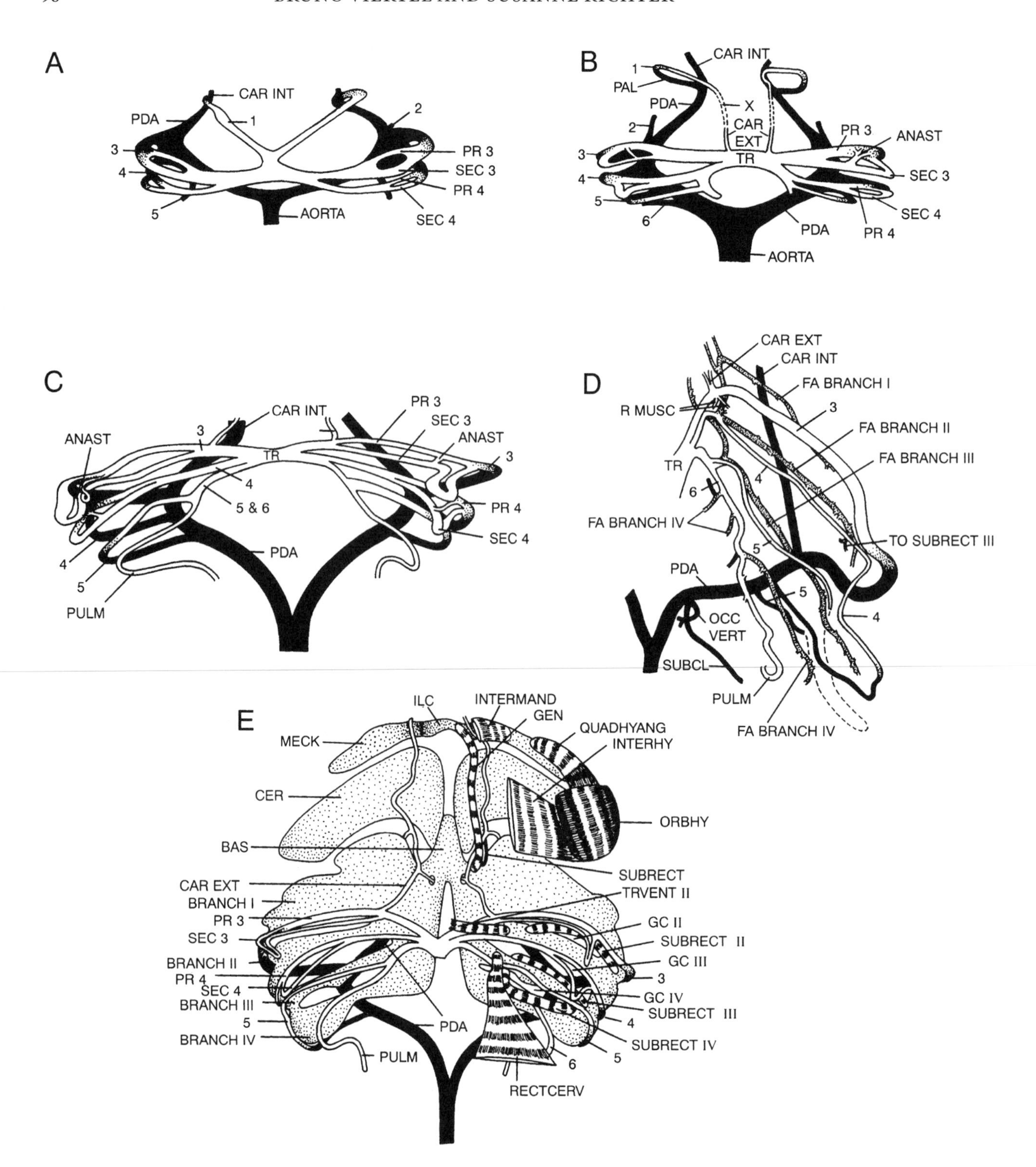

1982). Hemoproteins, first detectable in the anterolateral blood islands, differentiate in the lateral plates at Nieuwkoop/Faber 33–34. In subsequent stages, hematic cords extend into the posteroventral trunk region, and by Nieuwkoop/Faber 39 the anterior cords have disappeared. At Nieuwkoop/Faber 40 the subintestinal veins contain hemoproteins (Mangia et al. 1970). The intertubular liver and hepatic ducts, pro- and opisthonephros, and spleen are primary centers of hematopoiesis (Broyles et al. 1981, *R. catesbeiana;* G. Frank 1988, 1989a, b, *R. esculenta* complex; Maniatis and Ingram 1971a, b, c, *R. catesbeiana* (fig. 5.7). In *Rana pipiens* (Shumway 24–25) and other species, hematopoiesis begins

Fig. 5.3. Ventral views of the development of the aortic arches of *Xenopus laevis.* (A) In a 5-mm embryo, arches 1, 3, and 4 are complete; the other two have primary and secondary vessels. (B) In a 6-mm embryo, arches 3, 4, and 5 are complete; arch 1 is degenerating at point x; arches 2 and 6 are stumps. (C) In an 8-mm larva, secondary vessels of arches 3 and 4 loop backward into external gills; right side *(left in diagram)* is unusual in having two external gills on branchial arch 1, the second of which is supplied by an extra loop formed by the anastomosis between arch 3 and its secondary vessel. (D) Arrangement of aortic arches and arterial supply of filter apparatus *(stippled)* in a four-legged larva before completion of metamorphosis; note relative size of aortic arches; arch 5 is starting to degenerate *(dashed line).* (E) Reconstruction of aortic arches, branchial skeleton, and associated muscles in a 10-mm larva. Vessels supplying external gills are starting to atrophy, and arch 3 is wider than others; arteries to filter apparatus were omitted. Abbreviations: ANAST = anastomosis, AORTA = aorta, BAS = basihyal, BRANCH I–IV = branchial bars I–IV, CAR EXT = arteria carotis externa, CAR INT = arteria carotis interna, CER = ceratohyal, FA BRANCH I–IV = arteries of filter apparatus of branchial bars I–IV, GC II–IV = gill clefts I–IV, GEN = musculus geniohyoideus, ILC = inferior labial cartilage, MECK = Meckel's cartilage, INTERHY = musculus interhyoideus, INTERMAND = musculus intermandibularis, ORBHY = musculus orbitohyoideus, QUADHYANG = musculus quadrato-hyoangularis, RECTCERV = musculus rectus cervicis, SUBRECT I–IV = musculi subarcuales recti I–IV, TRVENT II = musculus transversus ventralis II, numerals 1–6 = aortic arches 1–6, OCCVERT = arteria occipito-vertebralis, PAL = arteria palatina, PDA = paired dorsal aorta, PR 3–4 = primary vessel of aortic arches 3–4, PULM = arteria pulmonalis, RMUSC = arteria carotis externa, ramus muscularis, SEC 3–4 = secondary vessel of aortic arches 3–4, SUBCL = arteria subclavia, and TR = truncus arteriosus. From Millard (1945); reprinted by permission of the Royal Society of South Africa.

in the pronephros (Broyles and Deutsch 1975 and Broyles and Frieden 1973, *R. catesbeiana;* K. L. Carpenter 1978, K. L. Carpenter and Turpen 1979, Foxon 1964, Hollyfield 1966, and Jordan 1933, *R. pipiens;* Meseguer et al. 1985, *R. ridibunda;* and Turpen et al. 1979, *Rana pipiens*), but the majority of larval erythrocytes is produced in the liver (Maniatis and Ingram 1971a, b, c, and Turpen et al. 1979, *R. catesbeiana*).

Salvatorelli (1970) distinguished four generations of erythrocytes in *Bufo bufo.* Cell diameters decrease in size from embryo to larva and increase at the beginning of metamorphosis. In embryos, the cells are 250–750 μm^2 (Cambar/Gipouloux III 10), 150–500 μm^2 in larvae (Cambar/Gipouloux IV 5), and 150–350 μm^2 at the beginning of metamorphosis (Cambar/Gipouloux IV 12). *Rana catesbeiana* (Taylor/Kollros X–XII; Broyles et al. 1981) has Type 1 cells (26.9 ± 0.03 [S.E.] μm; dominates in the kidney; contains Td-4) that are oblong to oval with an acentric nucleus, and Type 2 (23.50 ± 0.04 μm; dominates in the liver; contains Td-1, 2, and 3) cells are elliptical to round with the nucleus centered (Benbassat 1970; Hollyfield 1966; McCutcheon 1936). In contrast to V. I. Ingram (1972), Broyles et al. (1981) concluded that there were different erythrocytes lineages (Type I) emanating from the kidneys and (Type 2) the liver. The distribution of erythrocyte classes in the vessels varies individually within a species, and these cell survive for about 100 days in premetamorphic *R. catesbeiana* (Forman and Just 1976).

Spleen

In *X. laevis,* the spleen anlage forms at the dorsal mesentery near the anterior end of the stomach by Nieuwkoop/Faber 43 and is well-defined by stage 45–46. Blood corpuscles are visible at Nieuwkoop/Faber 47, lymphocytic differentiation is detectable at stage 48, and the organ is highly vascularized by stage 49. Large cells in the white pulp (lymphoid follicles) do not mature into medium or large lymphocytes before Nieuwkoop/Faber 50 or into small lymphocytes before stage 51 (Manning and Horton 1969; Turner and Manning 1972, 1974). At Nieuwkoop/Faber 50 "degenerating macrolymphocytes" are present in the boundary between the red and white splenic pulp.

The spleen anlage of *Rana pipiens* appears at Shumway 24–25 and contains erythrocytes, mesenchymal cells, and yolk platelets at Taylor/Kollros I. At Taylor/Kollros II the anlage consists of small masses of mesenchymal and elongated reticular cells interspersed with erythrocytes, large and medium lymphocytes, and a connective tissue capsule. At Taylor/Kollros III small lymphocytes, neutrophils, and granulocytes occur in the mesentery near the spleen, and the spleen begins to increase in size. The white pulp is formed during Taylor/Kollros VIII–IX. The red pulp merges smoothly into the white pulp as in the adult spleen (J. D. Horton 1971a, b).

The spleen is an important secondary lymphoid organ with significant reticuloendothelial functions. Phagocytic cells of the reticuloendothelial system of adults are situated in the red pulp and grouped around the white pulp (E. L. Cooper and Wright 1976, *Ascaphus truei;* and Diener and Nossal 1966, *Bufo marinus*). Macrophages prevent large particles and other cells from entering the white pulp. Turner (1969) ascertained that free macrophages of the body cavities have the same function in *X. laevis.*

Lymphatic and Reticuloendothelial Organs

The increased chance of infection caused by buccopharyngeal contact with the water that passes through it is countered by the lymphatic and reticuloendothelial systems (Aschoff 1924). The lymphatic, reticuloendothelial, and immune systems are closely related to each other and to the immune responses. Especially in amphibians, lymphoid and myeloid tissue appear together with macrophages and other reticuloendothelial cells (i.e., phagocytic cells) in the same organ. Lymph vessels drain body fluid from interstitial spaces back to the veins and to lymph glands, where it is filtered (fig. 5.8). The reticuloendothelial system consists of phagocytic macrophages in most organs and myeloid tissues, which differentiate into eosinophils, erythrocytes, macrophages, neutrophils, and thrombocytes. During ontogenesis, stem cells populate many structures (e.g., lymph glands, thymus, ventral cavity bodies, branchial and aortic arches, intertubular tissue of the kidneys, intestinal wall, liver, lungs, mesenteries, skin, and the walls of the bile and hepatic ducts; E. L. Cooper 1966 and Witschi 1956). E. L. Cooper (1967,

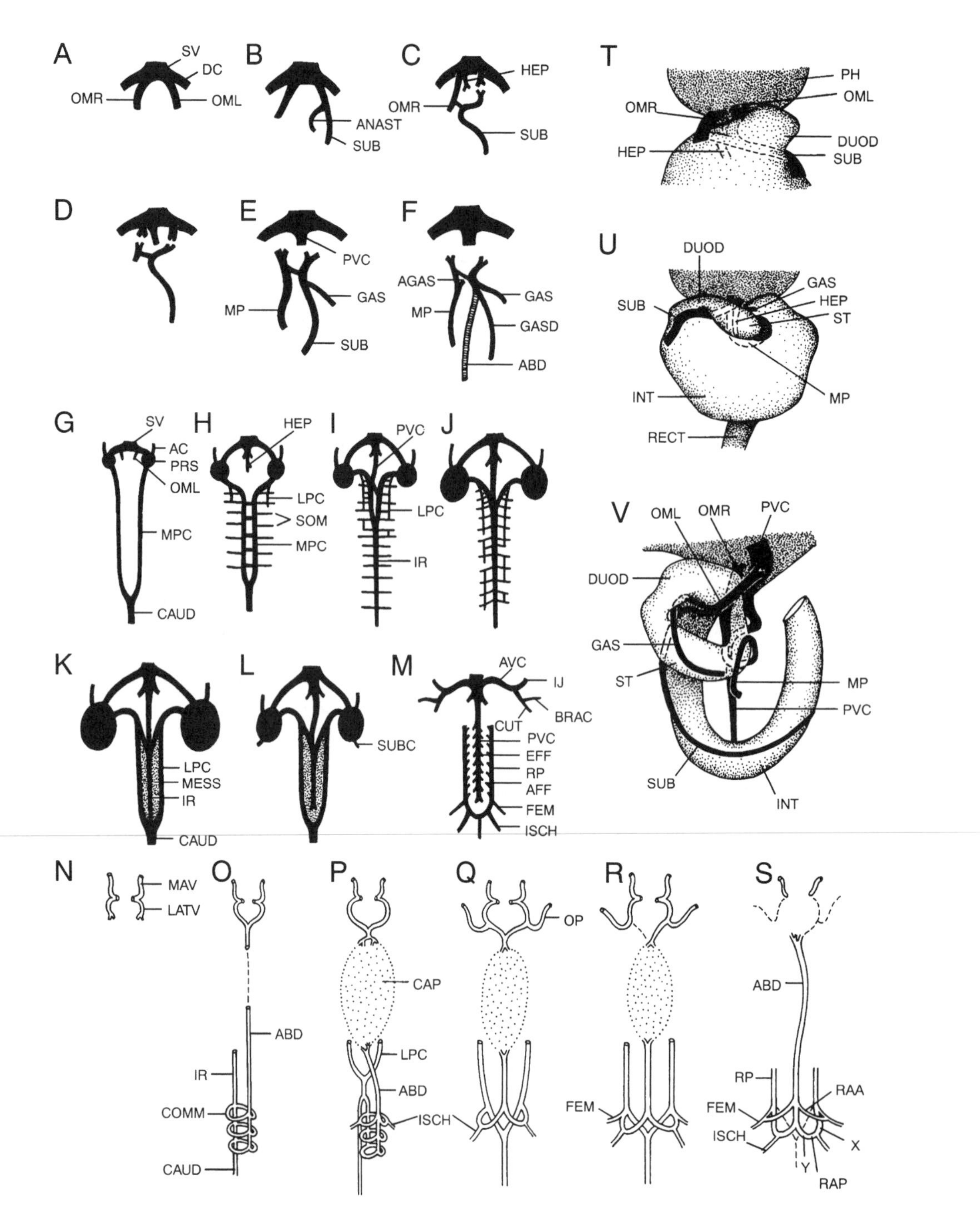

Fig. 5.4. Development of the venous system of *Xenopus laevis.* Diagrammatic representation (not to scale) of the development (ventral view) of the omphalomesenteric veins and hepatic portal system in (A) 5, (B) 6, (C) 6.5, (D) 7, (E) 8–40 mm larvae, and (F) metamorph. Diagrammatic representation (not to scale) of the development *(ventral view)* of the postcardinal veins and posterior vena cava in (G) 6, (H) 8, (I) 10, (J) 12, (K) 14, (L) 24-mm larvae, and (M) metamorph. Diagrammatic representation (not to scale) of the development *(ventral view)* of the abdominal and hind limb veins in (N) 7, (O) 10, (P) 18–24, (Q) 37-mm larvae, (R) early metamorph, and (S) adult. Reconstruction of the alimentary canal and its blood vessels in (T) 6.5, (U) 8, and (V) 24-mm larvae. Abbreviations: ABD = abdominal vein, AC = anterior cardinal vein, AFF = afferent vein, AGAS = anterior gastric vein, ANAST = dorsal anastomosis around gut, AVC = anterior vena cava, BRAC = brachial vein, CAP = capillary network, CAUD = caudal vein, COMM = commissural vessels between abdominal and caudal veins, CUT = cutaneous vein, DC = duct of Cuvier, DUOD = duodenum, EFF = efferent renal veins, FEM = femoral vein, GAS = gastric vein, GASD = gastroduodenal vein, HEP = hepatic vein, IJ = internal jugular vein, INT = intestine, IR = interrenal vein, ISCH = ischiadic vein, LATV = lateral vein, LPC = lateral postcardinal vein, MAV = muscularis abdominis vein, MESS = mesonephric sinus, MP = main portal vein, MPC = medial postcardinal vein, OML = left omphalomesenteric vein, OMR = right omphalomesenteric vein, OP = opercular vein, PH = pharynx, PRS = pronephric sinus, PVC = posterior vena cava, RAA = ramus abdominalis anterior (anterior anastomosis), RAP = ramus abdominalis posterior (posterior anastomosis), RECT = rectum, RP = renal portal vein, SOM = somatic veins, ST = stomach, SUB = subintestinal vein, SUBC = subclavian vein, and SV = sinus venosus. After Millard (1942); reprinted by permission of the Royal Society of South Africa.

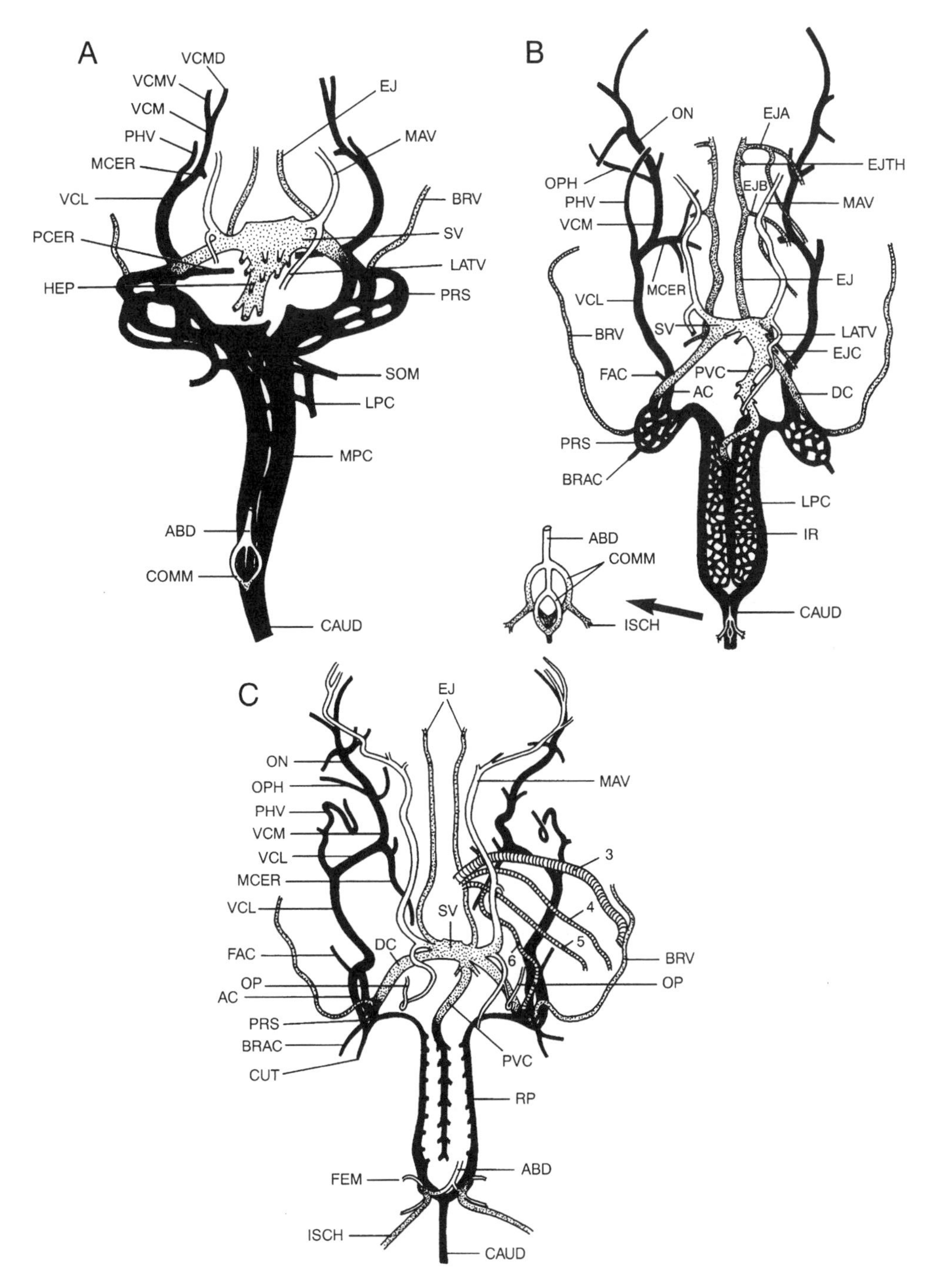

Fig. 5.5. Reconstruction of venous system (hepatic portal system not shown) of *Xenopus laevis.* (A) 8-mm larva, venous network of pronephros diagrammatic. (B) 24-mm larva, branches of right anterior cardinal and left external jugular veins shown in detail, venous network of pro- and opisthonephros diagrammatic. (C) Tadpole starting metamorphosis, branches of the external jugular vein not shown, left aortic arches included to indicate positions relative to veins. Abbreviations: ABD = abdominal vein; AC = anterior cardinal vein; BRAC = branchial vein; BRV = branchial vein; CAUD = caudal vein; COMM = commissural vessels between abdominal and caudal veins; CUT = cutaneous vein; DC = duct of Cuvier; EJ = external jugular vein; EJA, EJB, and EJC = branches of external jugular vein from branchial arches 2, 3, and 4; EJTH = branch of external jugular vein from thyroid gland; FAC = vena facialis; FEM = femoral vein; HEP = hepatic vein; IR = interrenal vein; ISCH = ischiadic vein; LATV = lateral vein; LPC = lateral postcardinal vein; MAV = muscularis abdominis vein; MCER = medial cerebral vein; MPC = medial postcardinal vein; numerals 3–6 = aortic arches 3–6; ON = vena orbitonasalis; OP = opercular vein; OPH = vena ophthalmica; PCER = posterior cerebral vein; PHV = pharyngeal vein; PRS = pronephric sinus; PVC = posterior vena cava; RP = renal portal vein; SOM = somatic veins; SV = sinus venosus; VCL = vena capitis lateralis; VCM = vena capitis medialis; VCMD = dorsal element of vena capitis medialis (= vena cerebralis anterior); and VCMV = ventral element of vena capitis medialis (= vena orbitonasalis). From Millard (1949); reprinted by permission of the Royal Society of South Africa.

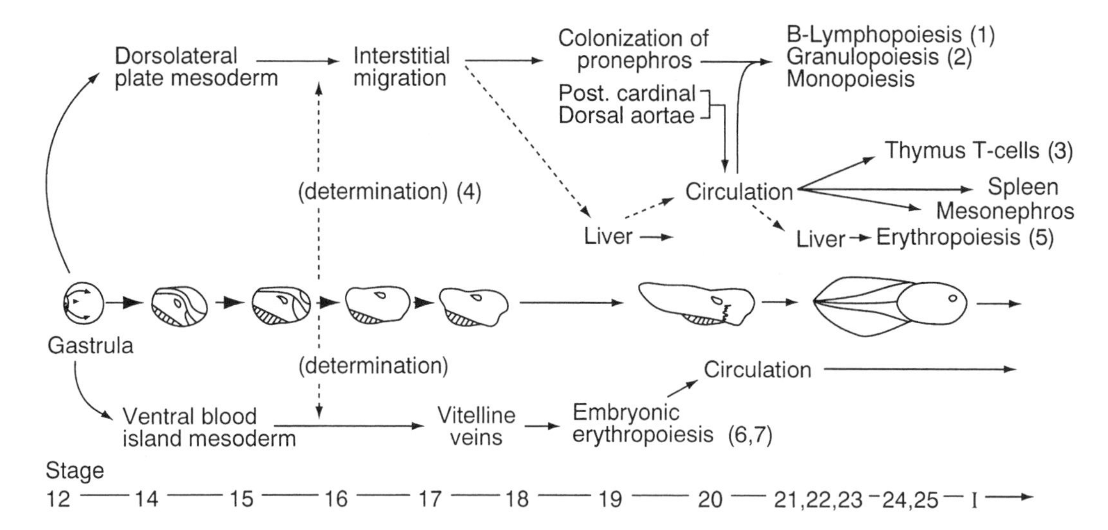

Fig. 5.6. Postulated relationships and migratory patterns of hematopoietic precursor cells during embryonic development in *Rana pipiens*. Dashed lines indicate events not documented experimentally. Stages 12–25 (embryonic period from gastrula to hatching) follow Shumway (1940) and stage I (beginning of larval period) is of Taylor and Kollros (1946). Numerals in parentheses refer to the following references: 1 = Zettergren et al. (1980), 2 = K. L. Carpenter and Turpen (1979), 3 = J. D. Horton (1971b), 4 = Fales (1935), 5 = Turpen et al. (1979), 6 = Hollyfield (1966), and 7 = Turpen et al. (1981a). From Turpen and Knudson (1982); reprinted by permission of Academic Press.

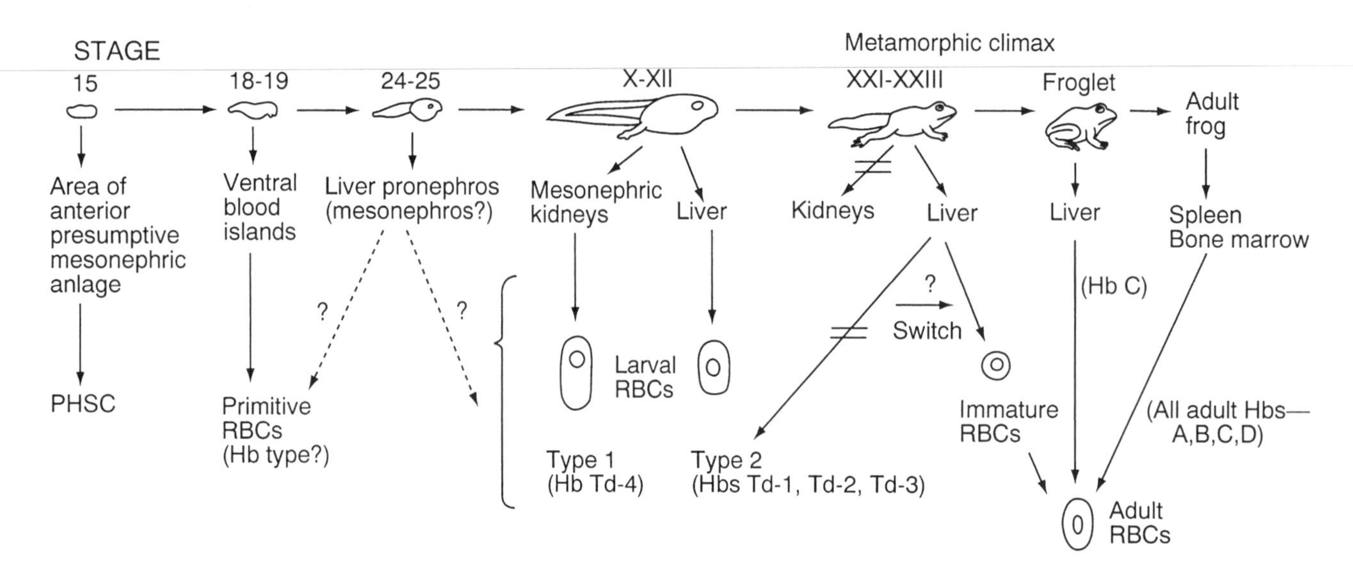

Fig. 5.7. Ontogeny of erythropoiesis and hemoglobin synthesis in *Rana pipiens* and *R. catesbeiana*. Arabic numerals 15–25 = Shumway stages and roman numerals X–XXIII = Taylor/Kollros stages. Abbreviations: Hb(s) = hemoglobin(s), PHSC = pluripotent hematopoietic stem cells, RBC = red blood cells, and Td-1 to Td-4 = hemoglobin globulin fractions. From Broyles (1981) and Turpen et al. (1979); reprinted by permission of Plenum Publishing Corporation and Academic Press.

1976), J. D. Horton (1971a, b), and Manning and Horton (1982) provide terminology for the lymphatic and endothelial systems.

The development of the trunci lymphatici laterales corporis occurs immediately after the cranial lymphatic hearts form at Gosner 18 (*Rana esculenta* complex). At Gosner 20 the lateral lymphatic sacs (vasa lymphatica caudale ventrale and dorsale) are visible. The hemolymphatic system has separated into blood vessels and lymphatic vessels at Gosner 24; lymphatic heart valves are present, and lymphatic vessels and lateral lymphatic sacs are present dorsally and around the eyes. By Gosner 25 (fig. 5.8) the vas lymphaticum caudale laterale (VLCL) and the trunci lymphatici laterales corporis (TLLC) have appeared and paired ductus thoracici are

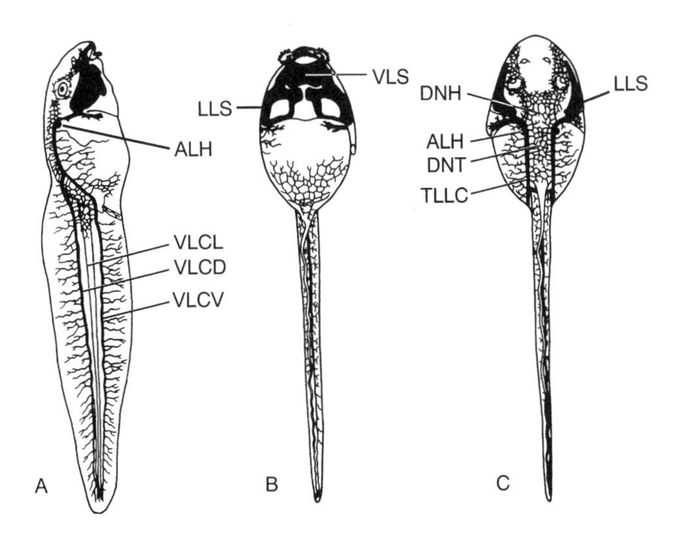

Fig. 5.8. Lymphatic vessels and spaces in *Rana esculenta.* (A) Lateral view. (B) Ventral view. (C) Dorsal view. Abbreviations: ALH = anterior lymph heart, DNH = dorsal net of head, DNT = dorsal net of trunk, LLS = lateral lymph sacs, TLLC = truncus lymphaticus lateralis corporis, VLCD = vas lymphaticum caudale dorsale, VLCL = vas lymphaticum caudale laterale, VLCV = vas lymphaticum caudale ventrale, and VLS = ventral lymph sac. After M. H. Hoyer (1905a).

differentiated. Caudal lymphatic hearts appear at Gosner 27. At Gosner 41 the paired ductus thoracici enlarge as sinusoidal spaces, and the lymphatic vessels of the trunk are reorganized as lymphatic sacs at stage 43. At the same stage one pair of cranial lymphatic hearts and two pairs of caudal lymphatic hearts develop. At Gosner 45, there are three caudal lymphatic hearts (M. H. Hoyer 1905a, b). Kampmeier (1915, 1922, 1958, 1969) supplied additional information on the anatomy of lymphatic vessels and hearts.

Thymus Gland

The thymus, a dorsal, ectodermal (J. D. Horton 1971b) derivative of the second visceral pouch (fig. 5.9A–C), is visible at Shumway 22–23 in *Rana pipiens.* At Shumway 23 the anlage consists of round and columnar epithelial cells with yolk platelets and scattered pigment. At Shumway 24–25 the thymic anlage lies ventrolateral to the auditory capsule and posterior to the eye and is attached to the epithelium of the first gill slit (Sterba 1950). At the same level, lateral branches of the dorsal aorta course more proximally or medially, while the anterior cardinal vein extends dorsally to the thymic buds. Posteroventral to the bud, the first efferent branchial artery extends to the branch of the dorsal aorta. At Shumway 25 the anlage, situated dorsolaterally at the level of the auditory capsule, increases in size and fibroblasts condense around it.

In *Rana pipiens* at Taylor/Kollros I (J. D. Horton 1971a, b), the thymus enlarges but remains attached to the visceral pouch. At Taylor/Kollros II growth and differentiation accelerate and the thymus begins to detach from the visceral pouch. In 14–19 day-old larvae, lymphocytes of all diameters are present in the medulla, although pale-staining irregular cells predominate (fig. 5.9A). Parts of the cortex are surrounded by connective tissue capsules, and blood capillaries begin to penetrate the dorsoventrally elongated mass. The increasing tendency of branching blood capillaries to push through the cortex increases the blood supply, and this invasion forms lobes in the gland. In 25–27 day-old larvae, concentric cytoplasmic striations found in a small number of these cells resemble those in Hassall's unicellular corpuscles in the cystic spaces of the medulla. Unicellular and multicellular Hassall's corpuscles are conspicuous in the medulla at Taylor/Kollros IV; from stages V–XVI, the organ increases in size to a maximum of about 1 × 2.5 mm (J. D. Horton 1971a, b). The persistence of the thymus throughout amphibian development suggests a favorable reaction to thyroxine or synthesis of a hormone that counteracts thyroxine (E. L. Cooper 1976).

Lymph Glands and Ventral Cavity Bodies

The term lymph gland (LM1, lymphomyeloid organ 1; dorsal or subthymic tonsils of *R. temporaria,* corpus lymphaticum of *R. esculenta,* dorsal cavity body of *R. temporaria,* dorsal gill remnant of *Bufo* and *Rana,* and glandula interposita or inclusa of *Hoplobatrachus tigerinus, Rana catesbeiana, R. occidentalis, R. ridibunda,* and *Rana* sp.; E. L. Cooper 1967) was first used by Witschi (1956) for *Rana catesbeiana* and, after J. D. Horton (1971a, b), it was used for *R. pipiens* and eventually for all anurans. These glands disappear at metamorphosis in *R. catesbeiana* and probably other closely related anurans. Authors who have described lymph glands in adults may have looked at a different lymphomyeloid organ (e.g., J. D. Horton 1971a, b).

In 9–12 day-old larvae of *Rana pipiens* (Taylor/Kollros II), groups of erythrocytes, lymphocytes and neutrophils occur lateral to the forelimb anlagen. Each lymph gland has close contact to similar cells in the pronephric region. Dorsally the lymph gland anlage extends into the anterior lymphatic space (division of the primary maxillary sinus; Kampmeier 1922) and contacts the dorsolateral wall of the opercular chamber ventrally. Structuring of the anlage into cellular cords enclosing sinusoids is complete in larvae of Taylor/Kollros II. Lymphocytes and reticular fibers occur in the glandular parenchyma, while granulocytes occur in blood sinuses lined by reticuloendothelial cells (phagocytic littoral cells; Baculi and Cooper 1968, *R. catesbeiana*).

The lymph glands grow rapidly until Taylor/Kollros III and now lie anterolateral to the forelimb anlagen. Lymphocytes, macrophages, and elongated reticular cells occur in the parenchyma; eosinophils, erythrocytes, lymphocytes, and neutrophils occur in the sinusoids. A connective tissue capsule surrounds the gland so that a lymphatic space appears. At their maximum size, lymph glands (Taylor/Kollros X–XII) appear reddish and lie anterodorsolateral to the base of the forelimb. They are connected laterally and medially to epithelium that limits and penetrates the anterior lymphatic space (J. D. Horton 1971a, b). Tadpoles of *Xenopus laevis* lack lymph glands (Manning and Horton 1969, 1982).

The ventral cavity bodies (LM2, lymphomyeloid organ 2) originate as invaginations of the epidermis of the opercu-

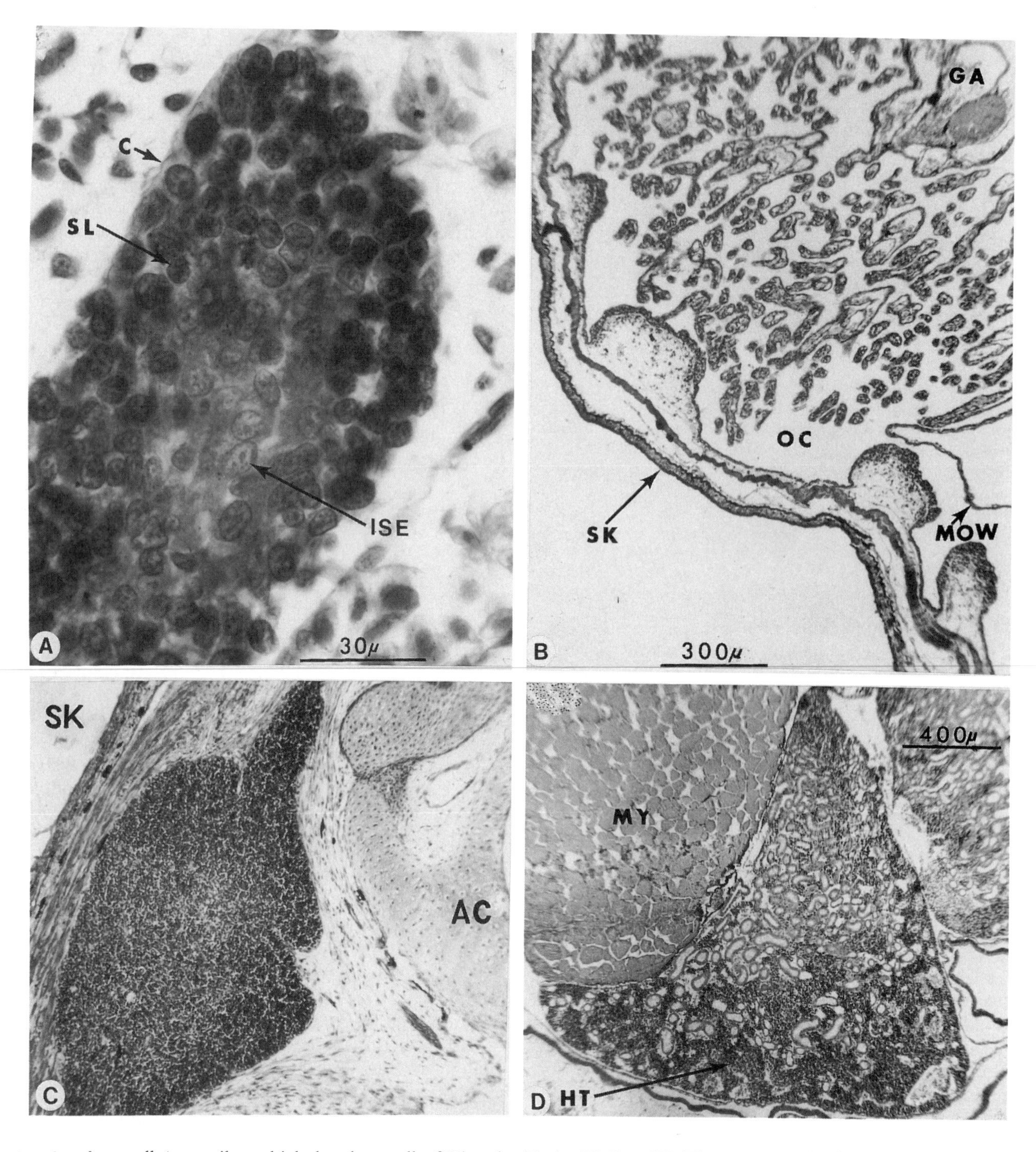

lar chamber wall (= peribranchial chamber wall of Viertel 1991). The first pair appears between days 16–19 in *R. pipiens* (Taylor/Kollros II) at the level of the second gill slit and lies in close contact with the filter apparatus and gill irrigation current. The bodies consist of nodular outgrowths underlain by connective tissue with numerous capillary branches. Lymphocytes and neutrophils occur in the capillaries and extravascularly in the surrounding tissue.

Up to four pairs of ventral cavity bodies have formed by Taylor/Kollros III. The most anterior lies medial to the posterior corner of the eye, and two of the posterior pairs are positioned half-way up the opercular wall (fig. 5.9B). Large round or oval cells that resemble hemocytoblasts (Cowden et al. 1968) or monocytes (Rouf 1969) are concentrated in the central region of the nodule connective tissue. Up to eight pairs of ventral cavity bodies may develop by Taylor/Kollros IV. Six pairs are in the lateral and ventral opercular chamber wall. Bodies positioned lateral to the heart and

Fig. 5.9. (A) Cellular details of the dorsal region of the right thymus of a 16-day-old *Rana pipiens* larva (Taylor/Kollros II). The cortex and medulla are beginning to differentiate; small lymphocytes can be recognized in the cortex; and paler staining, more irregularly shaped epithelial cells predominate in the central, medullary zone. From J. D. Horton (1971b). (B) Ventral branchial region of a 68-day-old *Rana pipiens* larva (Taylor/Kollros X) showing (low magnification) the location of a number of ventral cavity bodies projecting from the ventral and lateral opercular walls (peribranchial chamber walls) of the right side. The epithelia of all these nodules are lymphoid. The ventral cavity bodies shown here are diffusely populated (i.e., central connective tissue regions contain only a scattered array of leucocytes). Densely populated ventral cavity bodies are also found in leopard frog larvae at various stages of development. From J. D. Horton (1971a). (C) Right thymus of a 97-day-old *Rana pipiens* larva (Taylor/Kollros XXIII) showing (low magnification) the extent of this organ during metamorphosis. The ratio of medulla to cortex has increased since earlier larval life, and the densely staining cortical zone with its mass of small lymphocytes appears lobulate because of inwardly projecting connective tissue septa. From J. D. Horton (1971a). (D) The extent of the intertubular lymphomyeloid tissue of the right opisthonephros of a 68-day-old *Rana pipiens* larva (Taylor/Kollros X). The section passes through the more posterior part of the kidney whose ventral portion consists of a mass of hematopoietic tissue. From J. D. Horton (1971a). Abbreviations: AC = auditory capsule, C = capsule, E = epithelial cells (irregularly shaped), GA = gill arch–second branchial, HT = hematopoietic tissue, MOW = medial wall of opercular chamber, MY = myotomal muscle, OC = opercular chamber (peribranchial chamber), SK = skin, and SL = small lymphocytes. Reprinted by permission of the Wistar Institute Press and *American Zoologist.*

esophagus are not always paired. At Taylor/Kollros VIII–IX, a maximum of four pairs lie lateral to the heart and esophagus and others are at the ventral and lateral opercular wall (fig. 5.9B). The highest number of pairs of ventral cavity bodies (15) is reached between Taylor/Kollros X–XII, and by stage XIV this number has declined (*Rana pipiens*). A maximum of five pairs occur in the walls of the opercular cavity (E. L. Cooper 1967, *Rana catesbeiana;* J. D. Horton 1971a, b).

Tadpoles of *Xenopus laevis* have three pairs of ventral cavity bodies that resemble those of *R. catesbeiana.* One pair lies ventral to the posterolateral sides of the first branchial chamber between the first and second branchial arches and dorsal to the musculus subarcualis rectus II. The second pair lies along the musculus subarcualis III close to the point where the second branchial chamber opens into the opercular chamber near the spiracular openings. The third pair is situated at the cartilaginous base of the third branchial arch cartilage between the openings of the second and third branchial chambers near the spiracle. Based on their position, the procoracoid bodies of Sterba (1950, *Prokoracoidkörper*) appear identical to the first and second pairs of ventral cavity bodies (Manning and Horton 1969).

Pronephros and Opisthonephros

Neutrophils and blast cells occur in the pronephric intertubular tissue in *Rana pipiens* only after the onset of erythropoietic activity (Shumway 23). The complete cell repertoire includes eosinophils, various lymphocytes and neutrophils. Lymphocytes are present in the intertubular tissue of the opisthonephros.

By Taylor/Kollros III the percentage of intertubular lymphomyloid tissue has increased, especially granulocytes in the lateral and anterior regions of the pronephros. The pronephros and opisthonephros grow through Taylor/Kollros IV, and by stage VII the pronephros begins to decline. At Taylor/Kollros XVI the lymphomyeloid tissue of the pronephros consist mainly of granulocytes (fig. 5.9I; E. L. Cooper 1967, *R. catesbeiana,* and J. D. Horton 1971a, b). The opisthonephros continues to develop up to Taylor/Kollros XVI.

The interstices of the pronephros and opisthonephros and the mesenchymal sheath of the pronephric duct are granulopoietic (G. Frank 1988, *Rana esculenta* complex). Equal densities of eosinophils and heterophils are found in the pronephros and opisthonephros (F. R. Campbell 1970; Curtis et al. 1979; G. Frank 1988). G. Frank (1988) concluded that the opisthonephros has a lymphomyeloid interstitium. The only cells found in the mesenchymal sheath of the pronephric duct are eosinophils and heterophils of equal densities (G. Frank 1988). Tadpoles of *Xenopus laevis* lack the lymphomyloid tissue in the pronephros (Manning and Horton 1969).

At Taylor/Kollros III of *R. pipiens,* small lymphocytes and a few neutrocytes occur both intra- and extravascularly near the fibrous lamina propria and esophageal epithelium. After Taylor/Kollros III, lymphocytes accumulate in many areas of the small intestine and cloaca. Lymphomyeloid tissue with lymphocytes and granulocytes associate with capillaries and lymphatic spaces in the skin near the hind limb buds, and a pair of dorsal lymphomyloid accumulations near the myotomes are retained until prometamorphosis. Identical tissues are found in *Xenopus* near the limb anlagen (Nieuwkoop and Faber 1956).

Between Taylor/Kollros V and IX, lymphomyeloid tissues (gill remnant, LM4, lymphomyeloid organ 4; E. L. Cooper 1967) associated with the oral (lateral wall), pharyngeal (ventromedial wall), and dorsal branchial regions (ventral to thymus) appear. These typically paired structures are similar histologically to the lymphomyloid tissue of the esophagus. In the branchial region, two paired lymphatic accumulations appear at Taylor/Kollros V ventrolateral to the anterior parathyroid (visceral pouch 3) gland near the second gill cleft. A lymphatic space separates each pair from the anterior parathyroid. The posterior pair lies close to the posterior parathyroid glands (visceral pouch 4), is connected to the epithelium of the third gill cleft, and appears first in Taylor/Kollros VI. According to Sterba (1950) identical lymphatic accumulations are found in *Xenopus.*

Epithelial body is the term used by all other authors for lymphomyeloid organ 5 (LM5) of adults near the parathyroid glands. Often, the term epithelial body is used to describe the parathyroid glands together with the LM5 lying on top of them. These lymph tissues increase in size and form paired bodies as the larva develops. Two pairs lie opposite the tissue from which the jugular bodies of adults form during metamorphosis.

Small accumulations of lymphocytes and granulocytes appear from Taylor/Kollros X in and around the blood capillaries near the gills and branchial arches and in the posterior pharyngeal cavity at the level of, and medial to, the lymph gland. This tissue contacts the blood vessels in the region of the branchial arches and continues to grow until Taylor/Kollros XVI. Du Pasquier (1968, 1970, *Alytes obstetricans*) found large accumulations of lymphoid tissue in the esophagus and smaller lymphatic infiltrations in the gills, intestine, lungs, and mesentery. G. Frank (1988) showed the existence of granulopoietic tissue in the mesentery and in the mesenchymal coat of the bile and hepatic ducts in the larvae of the *R. esculenta* complex. The mesenteric artery and arterioles are surrounded exclusively by granulopoietic tissue, and smaller accumulations of granulocytes are attached to the bile duct.

Respiratory System

Tadpoles use the general body surface, lungs, and pharyngeal organs for gas exchange. The low capacitance coefficients of O_2 and CO_2 in water compared to air demand that a water breather ventilate 28 times as much as an air breather under the same conditions (Prosser 1973). The ontogenetically early differentiation of the lungs enables many tadpoles to be air breathers. Gas exchange is effected automatically with the locomotory movements of at least the tail (Wassersug 1989a) if the body surface is used for respiration. Under some hypoxic conditions, some tadpoles use aerial respiration.

Integument and Pigmentation

The epidermis of premetamorphic larvae consists of an outer periderm and an inner sensory layer (B. I. Balinsky 1981; Fox 1985; Gerhardt 1932; E. Marcus 1930). Numerous populations of cells make up the larval anuran epidermis, and considerable changes occur at metamorphosis (e.g., Kawai et al. 1994). Three epidermal layers occur during prometamorphosis and about six cell layers are present after metamorphosis (Fox 1977, 1984). Epidermal cells are connected by tight junctions and desmosomes and have little ultrastructural differentiation other than filamentous and fibrous structures (Fox 1977; Junqueira et al. 1984, *Pseudis paradoxa*). Dense arrays of tonofilaments are associated with abundant actin filaments near the surface membranes of cells of the sensory layer. Collagen in the basement lamella first appears in *Rana temporaria* at Cambar/Martin 25. An adepidermal space with lamellated bodies, the adepidermal membrane and collagen fibrils of the basement lamella form under the membrane. This system is reinforced by fibrils in the tail (Fox 1985).

Epidermal vascularization (subepidermal capillaries) is sparse in premetamorphic larva. Head, back, and tail have the smallest capillary net mesh size and are therefore considered to be the most active respiratory parts of the body surface (De Saint-Aubain 1982, *Bufo bufo* and *Rana temporaria*). In general, the role of skin in gas exchange is not as significant in larvae as one might expect from their large body surface. The capillary net becomes dense in prometamorphic larva. During metamorphic climax, cutaneous arteries and veins increase in diameter, subcutaneous arteries and veins become numerous, and a second venous network is developed in the form of anastomoses between the subcutaneous veins (De Saint-Aubain 1982, *Bufo bufo* and *Rana temporaria;* also see Strawinski 1956, *Rana esculenta complex*).

Merkel cells are found in *R. temporaria* from Cambar/Martin 33–36 and in *X. laevis* during Nieuwkoop/Faber 49–50. These small cells occur in the tail and body as well as in the tooth ridges of *R. japonica* and the barbels of *Xenopus*. Synaptic contacts with nonmyelinated nerves (Whitear 1983) suggest that barbels may be mechanoreceptive. Membrane-bound Merkel granules originating from the Golgi complex and located close to the synapses may contain various components that are released into the cytoplasm (Fox 1984; Fox and Whitear 1978; Ovalle 1979; Tachibana 1978). Mitochondria-rich cells (Type A cells of Hourdry 1974) in the tail and trunk (*Rana temporaria,* Cambar/Martin 33/34) have a high concentration of microridges on the apical pole and many mitochondria and may be stimulated by oxytocin (D. Brown et al. 1981). Type B cells Beaumont (1970) also have a high mitochondrial density and microvilli and may transport substances (Alvarado and Moody 1970; Maetz 1974) or have an osmoregulatory function. They resemble the chloride cells of fish (Whitear 1977), but it is unclear whether Type B cells are precursors of flask cells of adult amphibians (Fox 1985, 1986). *Stiftchenzellen* (= pin cells) found in the dorsal tail fin epidermis of tadpoles of *Rana* (Cambar/Martin 28–29) but absent in adults are probably chemosensory (Fox 1985, 1986; Whitear 1976). These pear-shaped cells with microvilli on the apical poles are thought to contact nerve fibers.

Mucous or goblet cells are known in *X. laevis* (Gartz 1970; Le Quang Trong 1974; Pflugfelder and Schubert 1965; Schneider 1957) and *R. temporaria* (Fox 1972, 1988). Leydig mucous cells (Leydig 1853, 1857; "clear cells" of Fox 1988) present during prometamorphosis are rich in tonofilaments and sparsely populated with granular ER and mitochondria. The mucous cells that Elias (1937), Le Quang Trong (1973), Poska-Teiss (1930), and Sato (1924) described as Leydig cells in the Bufonidae were classified as *Riesenzellen* (= alarm cells) by Fox (1988). The *Kugelzelle* (= ball cell or unicellular gland; Meyer et al. 1975a, b) occurs in high densities in *Ascaphus truei* until metamorphosis and uncommonly in *X. laevis* and *Hymenochirus boettgeri* (Fröhlich et al. 1977; Meyer et al. 1975a, b). Pfeiffer (1966) called these cells *Riesenzellen* in the Bufonidae. The mucus of some of these cells is acidic and extrusion is merocrine. *Riesenzellen,* originally described in *Bufo bufo* and *B. calamita* (Wenig 1913), that develop deep in the epidermis in young *Xenopus* are not homologous to Leydig cells (Fox 1988). The mucous cuticle is not an evaporation barrier in larvae or aquatic adults but probably functions as an ion trap in larvae, aquatic frogs, and fishes (Friedman et al. 1967). Epidermal melanophores up to 500 μm in diameter form a network between epidermal cells and are primarily involved in color change (Bagnara 1972, 1976; Nikeryasova and Golichenkov 1988). Various immigrant cells (e.g., mesenchymal macrophages, granulocytes, lymphocytes, leucocytes, and phagocytes) move to the epidermis but are not connected to the cells by desmosomes.

Although various cells are found in the dermis, there is a high concentration of connective tissue and capillaries. In the tail, muscle tissue penetrates deeply into the dermis. Melanocytes, or melanophores, like iridophores and xanthophores, are derivatives of the neural crest (Bagnara et al. 1979; B. I. Balinsky 1981). Melanophores are not connected to other cells by desmosomes and move amoebalike between epidermal and dermal cells. When the cells are in a dispersed state, numerous dendritic arms extend between the cells, and the pigment is brownish. In a contracted state, the cells are about a fifth of their dispersed size and are black. The average cell size is 300 μm, and melanosomes, organelles within the cells that contain melanin, measure 0.5–1.0 μm (J. D. Taylor and Bagnara 1972). Melanin is a complex polymer of tyrosine derivatives and proteins (Bagnara et al. 1979). There is a typical distribution of melanocytes for each species. Discoglossids form an adepidermal net of melanophores (G. Andres 1963).

Iridophores occur in large numbers in the body and less frequently in the tail. They are up to 300 μm in diameter and contain crystals of adenine, guanine, and hypoxanthine as reflecting stacks of platelets in organelles evidently derived from endoplasmic vesicles (Bagnara et al. 1979). In transmitted light, iridophores exhibit the refractive structural colors red, blue, and green depending on the position of the platelets relative to the light. In reflected light they are gold and silver (Bagnara 1976).

Xanthophores contain fat-soluble carotenoids obtained from the diet and pteridines synthesized by the cell (Bagnara et al. 1979) and stored in pterinosomes. Their color depends on the pattern of carotenoids and pteridines that often occur in the same xanthophore (Bagnara 1976; Bagnara et al. 1979). Xanthophores are rare in premetamorphic larvae. Erythrophores with carotenoid vesicles and perinosomes at the cell periphery are more numerous in adults than in larvae or juveniles.

Larvae generally adapt to light conditions by changing color primarily through dispersion, contraction, and migration of melanophores. Dermal melanophores surrounding the iridophores isolate them from the light and cause the larva to darken. The cells contract by withdrawing the dendritic processes and come to lie in the proximal dermis below the iridophores and xanthophores. Light transmission and reflection are particularly influenced by epidermal melanophores. The state of expansion of melanophores and xanthophores and the position of the platelets in iridophores is influenced by the melanophore-stimulating hormone (MSH) of the pars intermedia of the pituitary gland. In hypophysectomized larvae, melanosomes are contracted, platelets of the iridophores are dispersed, and xanthophore pteridine and carotenoid content is low. Thyroid hormones antagonize MSH, and catecholamines, epinephrine, estrogen, norepinephrine, pineal melatonin, progesterone, and testosterone also influence pigmentation (Fox 1985; also see Hayes and Licht 1995).

Dermal gland cells proliferate in the epidermis and migrate to the dermis. They are found in the body and at the base of the limbs in *X. laevis* (Nieuwkoop/Faber 57 onward), in the dorsal skin of *R. pipiens* at the beginning of prometamorphosis, and in *R. temporaria* at the beginning of metamorphic climax. Two types of secretory cells surrounded by myoepithelial-and mitochondria-rich cells on a basement membrane similar to the adepidermal membrane of the epidermis are included. Secretory cells of the mucous glands have a high density of granular ER and electronlucent vesicles; secretory cells of the granular glands have dense granules and microvilli at the apical pole. The ducts of the granular glands establish contact with the surface at Taylor/Kollros XVII, and the mucous gland ducts contact the surface in Taylor/Kollros XIX (Delfino 1977; Delfino et al. 1982).

Gills

Many authors consider the gills, branchial system, or branchial arch to consist of the dorsal filter plates (= gill filters), as well as the branchial food traps (= crescentric organs) and ventral gill tufts. The gill system is supported by branchial cartilage, which is a visceral arch, and most authors describe it as a branchial arch. The afferent and efferent branchial arteries and the musculus constrictor branchialis extend beside the arches.

McIndoe and Smith (1984a) discussed the respiratory function of filter plates and branchial food traps. They argued that the large netlike concentration of blood vessels on the rows of filter plates, as well as the placement in the irrigation current, indicate a respiratory function for these organs, but they recognized that certain features challenge this assumption. The epithelia are thick relative to the gill tufts, covered by mucus, and, as in the integument, the efferent blood stream is drained via venous vessels and has no connection with the heart via the branchial arteries. They suggested that the low PO_2 blood supply via the afferent branchial arteries probably means that the filter plates extract their own supply of O_2 from the surrounding irrigation current. McIndoe and Smith (1984a) argued that the gill tufts are the more significant organ in gaseous exchange.

The terms external and internal gills do not reflect the actual positions in the embryo. Prior to the closing of the opercular fold, all gills and their anlagen are external. The terms "outer" and "inner" gills generally apply only to Gosner 23–24. Apart from their position relative to the branchial arch, the fundamental difference lies in the fact that the proximal gills persist, whereas the distal (= outer) gills atrophy by the time the opercular fold closes. Viertel (1991) considered the terms persistent and transient gills to be more exact than inner and outer gills. Persistent gills develop along the ventral and transient epidermal gills along the ventrolateral parts of branchial arches I–IV. A gill tuft (fig. 5.10) consists of an epidermal covering, dermal connective tissue, areas occupied by individual endodermal-pharyngeal cells, and the blood vessels and mesodermal capillaries. Low temperatures and high O_2 pressures stimulate short gills while high temperatures and high CO_2 pressures stimulate the growth of longer gills (Løvtrup and Pigón 1968).

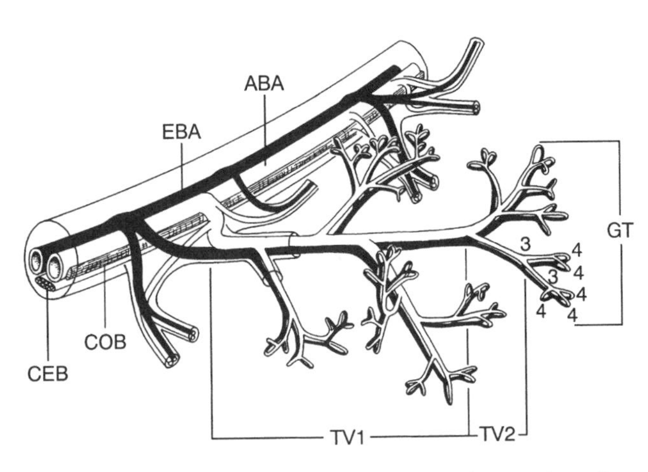

Fig. 5.10. Diagram of a gill tuft showing pattern of connections of afferent and efferent branchial arteries with primary and secondary tuft vessels. Abbreviations: numerals 3 and 4 = higher order branches, ABA = afferent branchial arteries, CEB = ceratobranchial cartilage, COB = coracobranchialis muscle, EBA = efferent branchial arteries, GT = gill tuft, TV1 = primary tuft vessels, and TV2 = secondary tuft vessels. From McIndoe and Smith (1984); reprinted by permission of Springer-Verlag.

Mitochondria-rich cells (Beaumont 1970) in the gill epidermis probably are responsible for the high adenosinetriphosphatase activity (Boonkoom and Alvarado 1971, *Rana catesbeiana*) linked to the active transport of ions. The epidermis also is permeable to O_2, CO_2, and water (Bentley and Baldwin 1980). Connective tissue is well formed at the base of the gill tufts, and the presence of endodermal cells from the visceral pouches shows that the gill tuft is part of the ectodermal-endodermal transition zone. In contrast, there is very little dermis and connective tissue in the distal areas of the gill tuft so that capillaries contact the sensory layer of the epidermis. Blood is provided by four aortic arches (= branchial arteries). The afferent branchial arteries supply blood from the heart, while three caudal efferent branchial arteries end blindly. Branches of the primary tuft vessels from the branchial arteries branch into secondary tuft vessels. Shunts bridge between arterial and venous capillaries in the gills and connect afferent with efferent tuft vessels (fig. 5.10; De Saint-Aubain 1981, 1985, *Rana temporaria* and *Bufo bufo;* McIndoe and Smith 1984a, *Litoria ewingi*). These authors assumed that the shunts maintain the flow of blood during the degeneration of gill tufts and capillaries. It seems likely that vasoconstriction of the shunts may regulate the amount of blood that flows through the capillaries and thus influence aquatic gaseous exchange. In species of the *R. esculenta* complex, there is surprisingly little vascularization of the gill tufts and filter plates. Gill regression begins when the operculum grows over the persistent gills.

Lungs and Trachea

Lung anlagen (Nieuwkoop/Faber 35–36, *Xenopus laevis*) appear as two ventrolateral pocketlike evaginations of the anterior foregut. These primordia originate from a transverse ridge that separates the buccopharynx from the intestinal tract, and a horizontal ridge divides the ventral primary liver cavity and the dorsal cavity of the trachea. At Nieuwkoop/Faber 39 the laryngeal anlage is visible as a horizontal ridge separating the tracheal cavity from the buccopharynx. The opening of the tracheal cavity to the esophageal and gastric cavity closes by Nieuwkoop/Faber 40. Laryngeal cartilages differentiate in Nieuwkoop/Faber 43, and the glottis perforates after stage 44 (B. I. Balinsky 1981; Nieuwkoop and Faber 1956, *X. laevis*). At Nieuwkoop/Faber 51–52 the lung epithelium folds, and connective tissue and blood vessels surround the lung sacs. The lungs expand posterodorsolaterally and by Nieuwkoop/Faber 54 extend to the pronephros.

In *Rana ridibunda* (Shumway 21), ventrolateral evaginations of the esophagus are visible anterior to the first nephrostome, and the esophagus closes soon thereafter. By Shumway 23 the lung sacs extend beyond the pronephros, and the laryngeal chamber is visible. The lateral and ventral walls of the esophagus form the aditus laryngis anlage. The musculus dilator laryngis opens the aditus laryngis at about Shumway 26, but laryngeal cartilages are not present at this stage (Müller and Sprumont 1972).

The inner walls of the lung sacs consist of a thin layer of squamous epithelium underlaid by thin, smooth muscles. The few blood vessels are not always in close anatomical contact with the smooth muscle layer; capillaries are sparse. The outer surfaces of the lungs and trachea are surrounded by the visceral pleura. Both these and the blood vessels are rich in melanocytes. Septa do not occur in premetamorphic larvae, but during and after metamorphic climax, septa develop as infoldings of the lung epithelium and the smooth muscle layer. Blood capillaries grow into the septa between the two smooth muscle walls. The smooth muscles thicken at the end of each septum (Atkinson and Just 1975; Goniakowska-Witalinska 1986). Strawinski (1956, *Rana esculenta* complex) found that the capillary network becomes denser during metamorphosis. Histological examination of the premetamorphic lung clearly illustrates its limited role in gas exchange. A nearly complete absence of capillaries, a small surface area, and a low capacity for self-ventilation suggest that the tadpole lung is a precursor of the adult lung. Lungs function as respiratory organs when the O_2 concentration of the water drops or when the buccopharynx and gills are blocked during intensive feeding. The primary function of lungs in premetamorphic larvae may be related to buoyancy.

The trachea begins posterior to the laryngeal chamber, which is a bilobed cavity connecting the glottis to the primary bronchi or lung buds (Rugh 1951). The laryngeal chamber is supported laterally by the laryngotracheal cartilages and surrounded ventromedially by the transversus and dilator laryngeus muscles (Sedra and Michael 1957). The tracheal epithelium of the developed larva is one cell thick (Nieuwkoop/Faber 44). A substance formed from fibrouslike bodies extruded by the epithelial cells lines the inner margin of the trachea and bronchus (Nieuwkoop/Faber 47, thickened at stage 48–49). Folds in the tracheal epithelium are penetrated by the fibrous substance. Epithelial cells are characterized by their small number of small, apically situated mitochondria. They also contain lipid granules, smooth vesicles, rough endoplasmic reticulum, Golgi complexes, and a high density of ribosomes in the matrix (Fox et al. 1970, 1972, *Xenopus laevis*). Fox et al. (1970) found no cilia in the tracheal system of *Xenopus,* but cilia occur in *R. temporaria* at metamorphic climax. It remains unclear whether the lamellar bodies in the lung sacs of adult *Bufo bufo* and *Hyla arborea* (Goniakowska-Witalínska 1984, 1986) are identical to the fibrouslike substance in larval *Xenopus.* The outer surface of the trachea is surrounded by connective tissue and blood vessels.

Digestive System

Most tadpoles are suspension feeders (Wassersug 1972), and the filter apparatus enables them to trap a wide range of particulate food. Tadpoles lacking keratinized mouthparts filter particles suspended in the water column. Tadpoles with keratinized mouthparts scrape or bite food from the substrate and filter the suspension. Many bottom feeders ingest periphyton, detritus and interstitial organisms from the sediment, and large quantities of indigestible material of different particle sizes (Seale et al. 1982; Seale and Wassersug 1979; Viertel 1990, 1992, 1996).

Buccopharynx

The buccopharynx consists of the buccal cavity (anterior buccopharynx) and pharynx (posterior buccopharynx). The pharynx develops as a filter apparatus by integrating anterior parts of the esophagus and lateral areas of epidermal ectoderm (Viertel 1991, 1996). Because of their position relative to the visceral skeleton and because the buccal cavity and pharynx cannot easily be distinguished from one another, these components are collectively referred to as the buccopharynx (fig. 5.11).

Wassersug (1976a, b) established the first comprehensive anatomical terminology and made broad comparisons (Wassersug 1980) among members of eight families (also Lannoo et al. 1987, *Osteopilus brunneus;* Viertel 1982, 1984b, *Alytes muletensis, A. obstetricans, Bombina variegata, Hyla arborea, Pelobates fuscus,* and central European *Bufo* and *Rana;* Wassersug and Heyer 1988, Leptodactylidae). Other investigators concentrated on the ontogenesis of the tongue (Hammerman 1965, 1969a, b, c, *R. catesbeiana* and *R. clamitans;* Helff and Mellicker 1941, *R. sylvatica*) or described the developmental changes and chemoreceptive characters of buccopharyngeal structures (Fiorito de Lopez and Echeverría 1989; Nomura et al. 1979a, b, *R. japonica*).

Authors from the beginning of the last century did not realize that they were describing a filter apparatus. They limited their investigations to individual components (Dugès 1834; Huschke 1826; Maurer 1888; W. K. Parker 1881; Rathke 1832), and Goette (1874, 1875) even rejected the notion that the "internal gills" functioned as a filter apparatus. Only Rusconi (1826) and Boas (1882) realized that they were investigating a filter apparatus. Informative descriptions by Kratochwill (1933, *R. dalmatina*), R. M. Savage (1952, *Bufo bufo* and *R. temporaria* and several microhylids), and Schulze (1888, *Pelobates fuscus*) form a bridge to the view of this organ that has been established in recent investigations.

The first descriptions of the surface of the branchial food traps examined by scanning electron microscopy (Wassersug and Rosenberg 1979) included members of nine families. Other papers that addressed the filter apparatus of specific species include the following: Das (1994, frogs from South India); O. M. Sokol (1981a, *Pelodytes punctatus*); Viertel (1984b, *Alytes muletensis, A. obstetricans,* and *Bombina variegata*); Wassersug (1984, compared *Scaphiophryne* with larval Types 2 and 4); Wassersug and Duellman (1984, egg-brooding hylid frogs), and Wassersug and Heyer (1983, leptodactylids living on wet rocky surfaces, *Cycloramphus duseni, Thoropa miliaris,* and *T. petropolitana*). The first investigations of the filter apparatus by transmission electron microscopy were by Viertel (1985, 1987, *Bombina variegata, Bufo bufo, R. temporaria,* and *X. laevis*).

The mouth, buccal cavity, pharynx, glottis, and esophagus are arranged serially only during the embryonic stages

Table 5.1 Abbreviations used in figures 5.11–5.16 and 5.25

ACI = aorta carotis interna, AFP = anlage of filter plates, AFPC = anlage of epidermal fold of peribranchial chamber (anlage of operculum), AGO = anlage of gonad, AL = anterior labium, AO + numerals = aortic arches (3, 4, and 6, latter is lung artery), AOC = arteria coronaria, APOP = anlage of postnarial papilla, ATE = anlage of tuba Eustachii, ATEG = anlage of transient epidermal gills, BI–BIV = branchial arches I–IV, BA = bulbus arteriosus, BAR = barbel, BC = buccal cavity, BFA (black dashed outline) = buccal floor arena, BFAP = buccal floor arena papillae or pustulations (shorter than twice the diameter of the base), BP = buccal pocket, BRA = buccal roof arena, BRAP = buccal roof arena pupillae, CC = ciliary cushion, CE = brain, CG = ciliary groove, CH = choanae, CHY = ceratohyale, CO = colon, CT = connective tissue, DU = duodenum, DV = dorsal velum, DVI–DVIII = dorsal velum I–III, EP = epidermis, ES = esophagus, ETZ = ectodermal-endodermal transition zone, EY = eye, FB = fat body, FL = forelimb, FP = filter plate, FPC = epidermal fold of peribranchial chamber (operculum), FPI–FPIV = filter plates I–IV, G = gill, GL = glottis, GS = gill slit with numerals, GZ = glandular zone, HL = hind limb, HM = hyoidean and mandibular muscles, HP = hypobranchial plate, HY = hyoid arch, IL = ilium, IN = internasal plate, IP = infralabial papillae, LA = left atrium, LFP = lateral filter plate, LI = liver, LJ = lower jaw, LP = lingual papillae, LRP = lateral ridge papillae, LTR = labial tooth row, LU = lung, MA = mandibular arch, MG = manicotto glandulare, MGH = musculus geniohyoideus, MGU = midgut, MLB = muculus levator branchiae, MNG = medial notch above glottis, MR = medial ridge, MSH = musculus subhyoideus, MST = musculus sternohyoideus, MT = musculus transversus, NA = external nares, NVP = narial valve projection, OC = oral cavity, OP = oral papillae, OPI = opisthonephros, OS = mouth, P = pancreas, PAP = prenarial arena pustulations, PCW = peribranchial wall, PD = pericardium, PET = peritoneum, PL = posterior labium, PONA= postnarial arena, POP = postnarial papillae, PRNA = prenarial arena, PRO = pronephros, PRP = prenarial papillae, RA = right atrium, RE = rectum, SC = secretory cell, SPL = spleen, SPM = spiracle, ST = stomadeum, SU = spiculum of hypobranchial plate, TA = tongue anlage, TEG = transient epidermal gills, TF = tail fin, TO = tractus olfactorius, TR = truncus arteriosus, UJ = upper jaw, VA = visceral arch, with numeral, VE = ventricle, VJE = vena jugularis externa, VP = prevelar papillae, and VV = ventral velum.

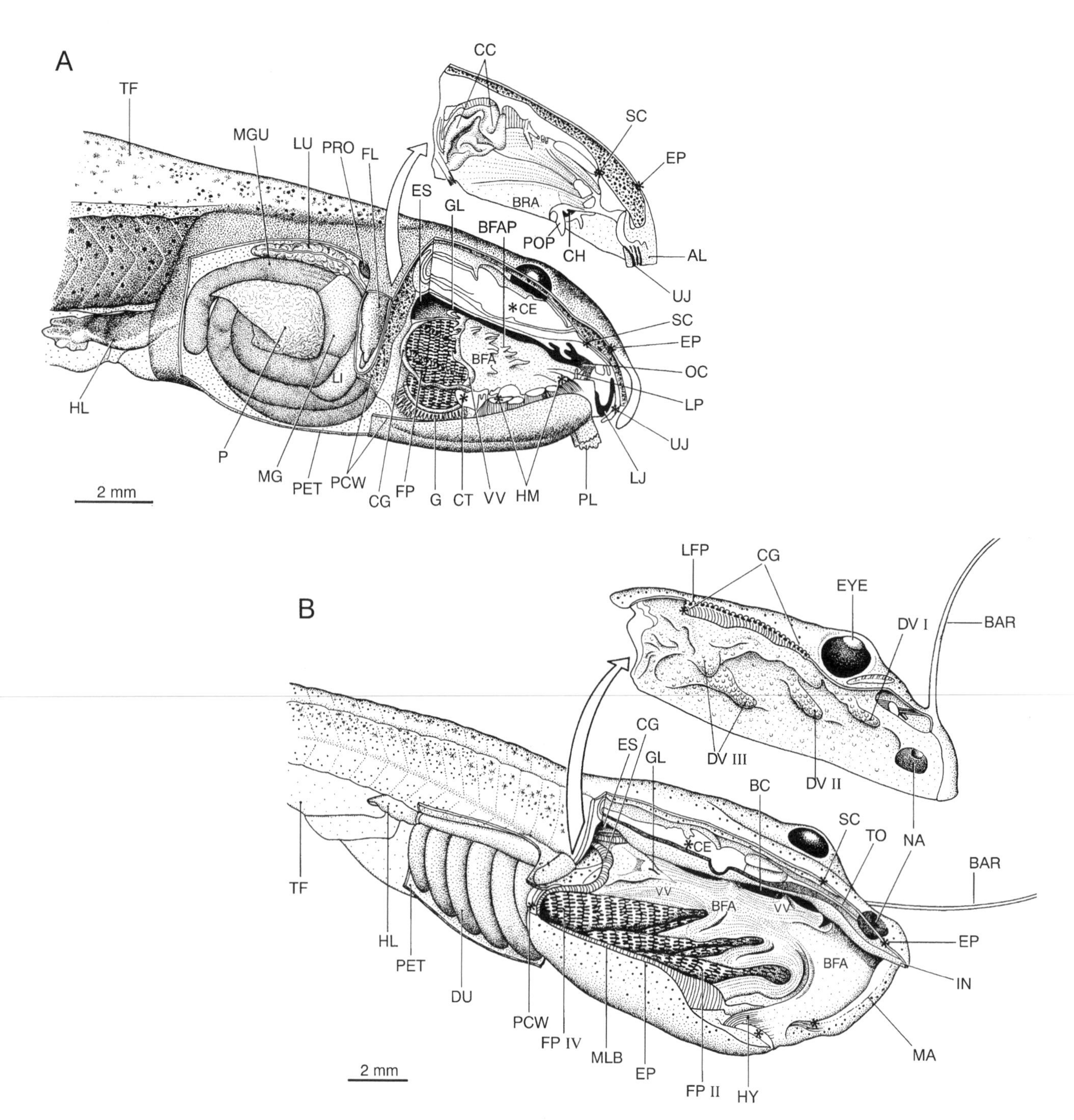

Fig. 5.11. Dorsolateral views of the alimentary tracts of (A) *Rana temporaria* (Orton's larval Type 4, Gosner stage 36) and (B) *Xenopus laevis* (Type 1, Nieuwkoop/Faber stage 54). Right, dorsal halves of the heads lifted and rotated and abdominal wall reflected. See table 5.1 for abbreviations. Redrawn from Viertel (1985, 1987); reprinted by permission of Springer-Verlag and Gustav Fischer Verlag.

of anuran larvae (figs. 5.11–5.14). In fully developed larvae, only the axis from the mouth to the ventral velum remains unaltered. The position of the pharynx in relation to the buccal cavity and longitudinal axis of the tadpole differs considerably among anuran larval types, and the position of the glottis relative to the buccopharynx is not the same in all larval types.

In pipoid tadpoles (larval Type 1), the pharynx is situated lateral to the posterior extension of the buccal floor whose limits are defined posteriorly by the glottis. The longitudinal axis of the pharynx is much shorter in all other larval types so that the buccal floor is widest at about the middle of the buccopharynx (fig. 5.11). Each side of the ventral velum forms an angle of 45° to the sagittal plane (*Rhinophrynus*) or extends posteriorly at an angle of 55°–70° (Types 2, 3, and 4; fig. 5.11A). Between a quarter and a third of the ventral pharynx is covered by the ventral velum. The branchial arches are more posteroventral than lateral to the buccal floor. The glottis of *Rhinophrynus* lies at the posterior extent of the buccal floor and, as is the case in the Pipidae, forms its posterior limit; it lies on the ventral velum and is connected to it medially. In microhylids, the glottis lies in the center of the buccal floor arena, while in larval Types 3 and 4, the glottis is posterior and more or less covered by the ventral velum between branchial arches 4. Ciliary epithelium surrounds the glottis, and although typical for the esophagus, is found nowhere else in the buccopharynx. This suggests that the glottis has no direct contact with the posterior buccal floor.

The morphology of the buccopharynx, its relations to the visceral skeleton and its musculature, and the role of the visceral skeleton as the pump for the irrigation of the filter apparatus (Gradwell 1970, 1971a, b, 1975b; Kenny 1969a; O. M. Sokol 1977a, 1981b; Wassersug and Hoff 1979) indicate how far the pharynx extends anteriorly. The positions

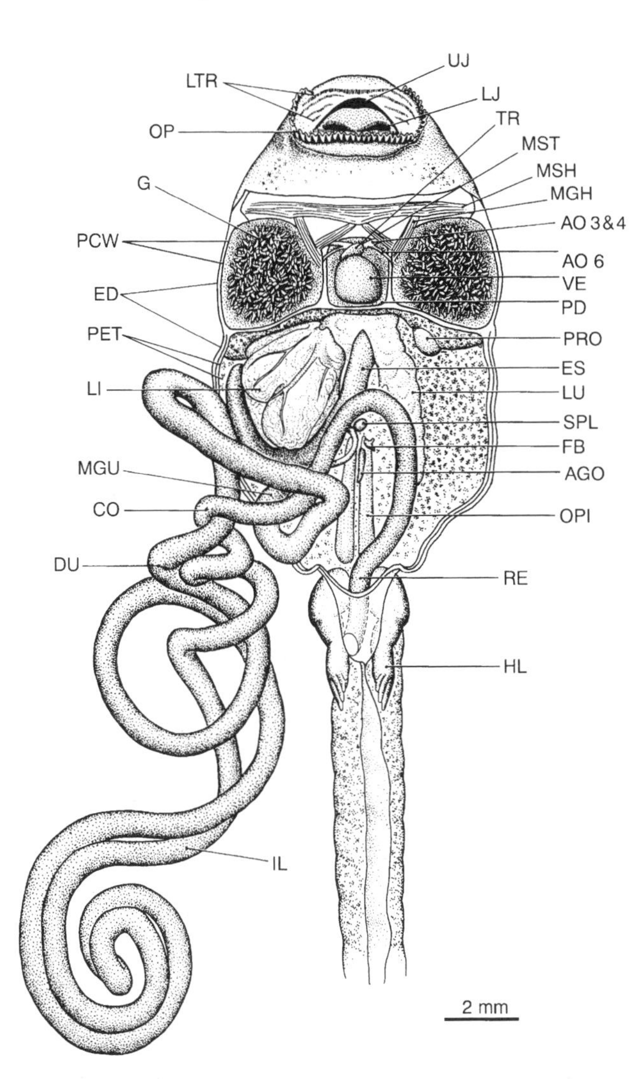

Fig. 5.12. *Rana temporaria* tadpole (Orton's larval Type 4, Gosner stage 36) in ventral view with ventral peribranchial wall, pericardium, and abdomen opened. See table 5.1 for abbreviations.

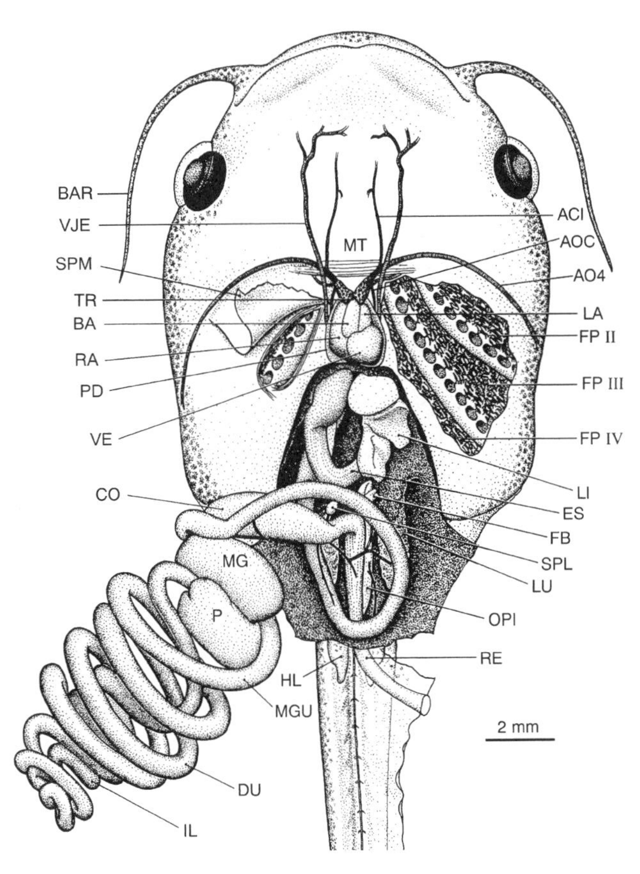

Fig. 5.13. *Xenopus laevis* tadpole (Orton's larval Type 1, Nieuwkoop/Faber stage 54) in ventral view dissected with peribranchial wall, pericardium, and abdomen opened. See table 5.1 for abbreviations.

of the hypobranchial plate, the ceratohyal, and the unpaired copula under the buccal floor show that the buccal floor is part of the anterior buccopharynx (figs. 5.11, 5.14C–D, and 5.15). Besides those cited specifically in this chapter, the following references include studies of the buccopharyngeal morphology of various tadpoles (Echeverría 1995, 1996, 1997; Echeverría et al. 1992; Grillitsch 1992; Inger 1985; Khan and Malik 1987; Pelaz and Rougier 1990).

The anterior buccopharynx of larval Types 3 and 4 has the most advanced structural organization of anuran larvae (figs. 5.15–5.16). Differentiation occurs in *Bufo bufo* between Gosner 25–28, and, beginning with the papillae, regression of the anterior buccopharynx starts at Gosner 41–42. Various authors have concentrated on the significance of the structures of the anterior buccopharynx for larval ingestion, systematic classification and understanding larval evolution. Wassersug (1980) and Wassersug and Heyer (1988) classified buccal structures according to family and genera and suggested evolutionary trends. Viertel (1982) wrote a key for the Central European species based on structures of the anterior buccopharynx. Wassersug (1980) linked the functional significance of the buccal papillae with suspension feeding. Although it is not yet possible to describe the exact function of each structure, a sieving apparatus may emerge via the interplay between anterior buccal structures and

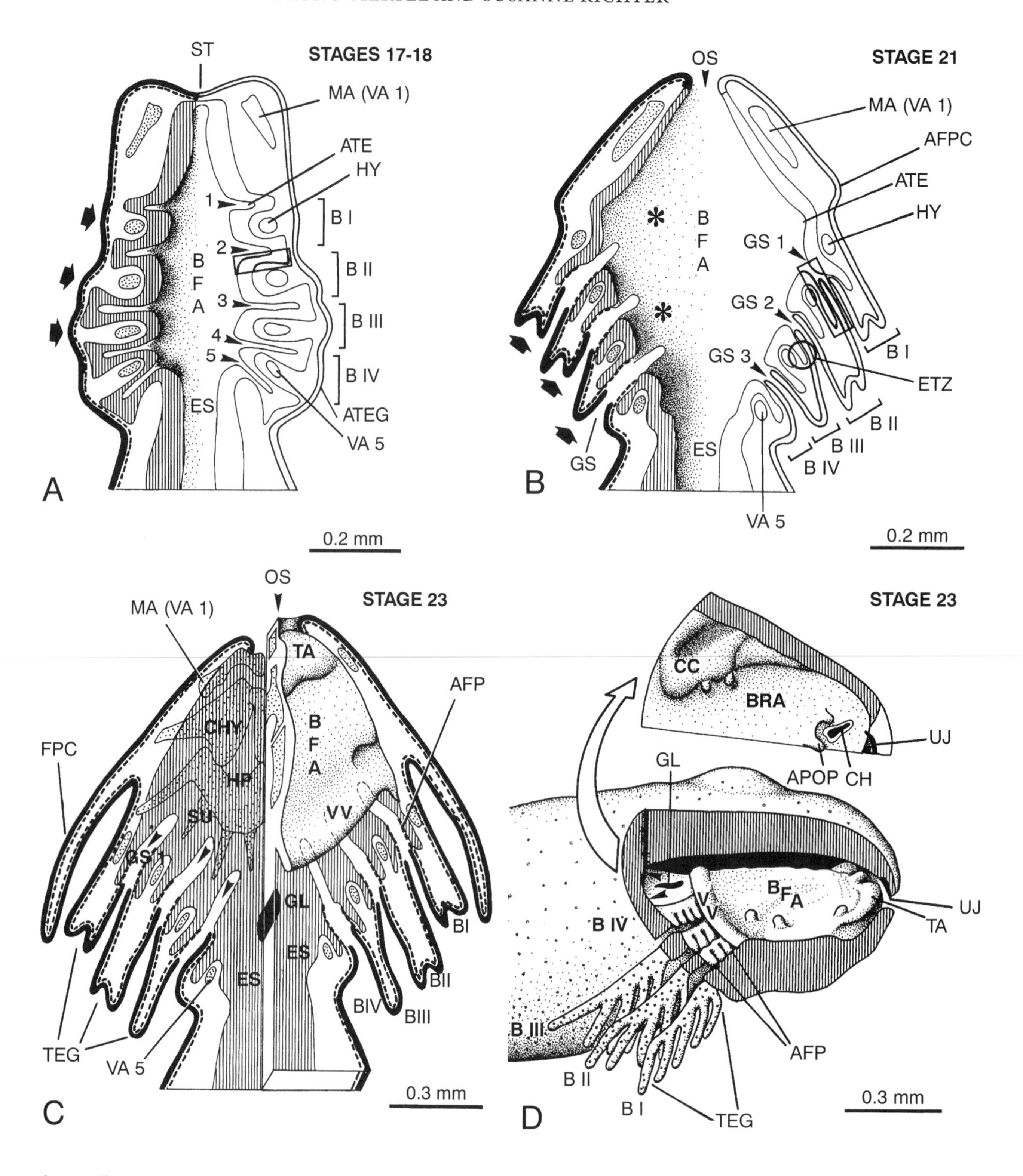

the ventilation movements of the entire buccopharynx (also Viertel 1991). The lingual papillae, buccal roof arena papillae, and buccal floor arena papillae may function as chemoreceptors (Hammerman 1969a, b, c; Helff and Mellicker 1941; Honda et al. 1992; Nomura et. al. 1979a, b).

The buccal floor begins posterior to the lower jaw sheath, and the tongue anlage is flanked on both sides by infralabial papillae (fig. 5.15). In the premetamorphic stages the tongue anlage forms a transverse bulge that changes in position and form. In prometamorphic stages it becomes cushion-shaped and extends along the longitudinal axis. Lingual papillae are differentiated on the tongue anlage during premetamorphic

Fig. 5.14. Ontogenesis of the anuran pharynx involving composite reconstructions of *Bombina variegata, Bufo calamita,* and *Rana temporaria* (Gosner stages 17–23). (A) Frontal section at the plane of stomodeum (generalized scheme of embryonic vertebrate pharynx, i.e., Gosner 17–18). (B) Frontal section at the level of transient epidermal gills and anlagen of filter plates (Gosner 21). Anlage of buccal floor arena arches upward between asterisks, visceral pouches 1–3 have broken through to form gill slits 1–3. Anlagen of transient epidermal gills are developing; sensory layer of epidermis overlying the endoderm of the visceral pouches (ectodermal-endodermal transition zone) forms the anlagen of the filter plates. (C) Left side, frontal section at the level of transient epidermal gills and anlagen of filter plates (same level as [B] but Gosner 23). The buccal floor arena is more strongly arched upward at the future sites of the mandibular and hyoid arches (= ceratohyal and hypobranchial plate). Epidermal fold of peribranchial chamber (= operculum) overgrowing the elongated transient epidermal gills. Right side, frontal section (plane higher than on left side) of transient epidermal gills, anlagen of filter plates, and esophagus. Ventral velum, buccal floor arena, and anlage of tongue reconstructed from SEM micrographs. Ventral velum is the posterior elongation of the buccal floor arena. (D) Lateral view of the anlage of the filter apparatus (Gosner 23). Right half of head lifted and reflected, oral cavity and peribranchial chamber opened, and epidermal fold of peribranchial chamber removed. Symbols: arrows = anlage of transient gills 1–5, arrow heads = direction of water movement, boxed areas = anlagen of filter plates, dotted lines = sensory layer of epidermis, heavy black lines = peridermis of epidermis, numerals = visceral pouches, stipple = visceral skeleton, and vertical lines = endoderm. See table 5.1 for abbreviations. From Viertel (1991); reprinted by permission of Springer-Verlag.

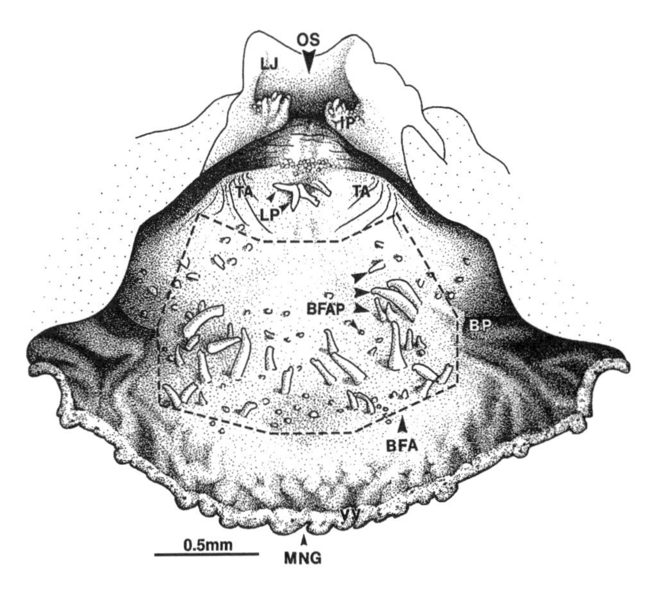

Fig. 5.15. The floor of the buccal cavity of a tadpole of *Bufo bufo* (Orton's larval Type 4, Gosner stage 36). See table 5.1 for abbreviations. From Viertel (1982); reprinted by permission of E. J. Brill.

stages (e.g., two in *Acris, Agalychnis, Anotheca, Osteopilus, Pelobates, Phyllomedusa, Ptychohyla, Rana esculenta* complex, and *Smilisca;* four in *Bufo, Gastrotheca,* central European brown frogs [= *Rana* spp.], *R. catesbeiana, R. clamitans,* and *R. sylvatica;* and 6–10 more or less fused together in a comblike structure in *Bombina, Discoglossus pictus, Scaphiophryne,* and *Thoropa*). In *Alytes cisternasi, A. muletensis,* and *A. obstetricans,* all authors report a number of small pustulations similar to those in *Ascaphus* and 0–2 lingual papillae. Wassersug and Heyer (1988) provided information on leptodactylids with 2, 3, or 4–11 lingual papillae; in several genera lingual papillae are absent, and by prometamorphic stages the surface of the tongue has fungiform papillae.

The buccal floor arena (fig. 5.15) lies posterior to the tongue anlage and is flanked by papillae. The arrangement, length, and number of these papillae differ among genera and species. The number of larger papillae (ca. 300–400 μm maximum length, except about 600 μm in *Crossodactylus*) varies from 10 (*Bufo*) to > 80 (*Ptychohyla leonhardschultzei*). Shorter pustulations (about 100 μm long) occur in roughly uniform numbers and are not confined to the buccal floor arena, and in many species (e.g., *Hyla arborea* and *R. lessonae*) these pustulations occur across the entire buccal floor as far as the buccal pockets. In this case they are termed prepocket papillae. The buccal pockets, situated lateral to the buccal floor arena, are slits in the buccal floor that form a bypass between the lumen of the buccopharynx and the peribranchial (= atrial or opercular) chamber. These pockets are not gill slits. The buccal pockets have lateral contact with the lat-

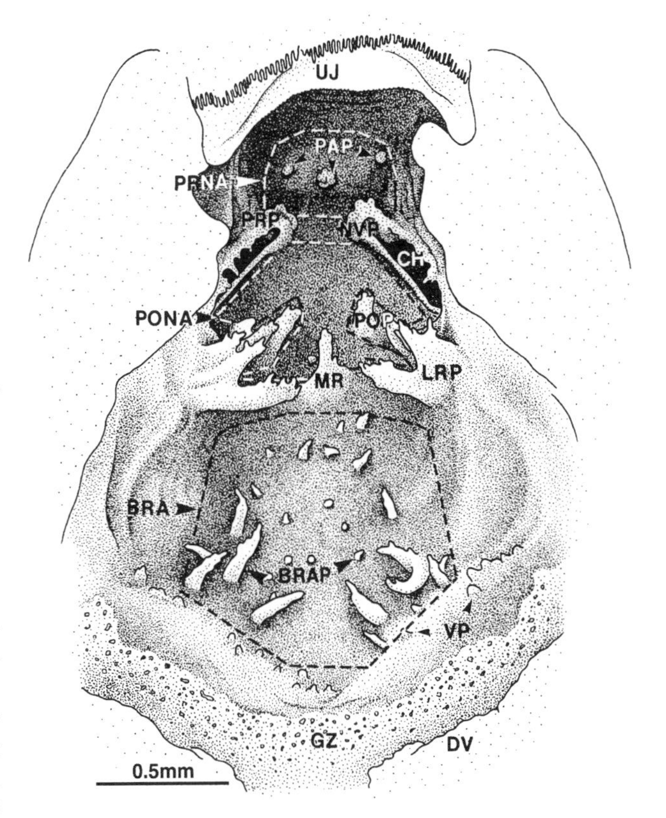

Fig. 5.16. The roof of the buccal cavity of a tadpole of *Bufo bufo* (Orton's larval Type 4, Gosner stage 36). See table 5.1 for abbreviations. From Viertel (1982); reprinted by permission of E. J. Brill.

eral areas of the ventral velum. The buccal floor terminates posteriorly at the ventral velum (Goette 1874 and R. M. Savage 1952, 1955; gill cover plate of Schulze 1888, 1892; and anterior filter valve of Kratochwill 1933). The ventral velum has a medial notch above the glottis that is particularly well developed in some species (e.g., *Hyla dendroscarta* and *H. phlebodes*). On the margin of the velum, particularly medially, there are several long, thin papillae (e.g., *Alsodes, Alytes obstetricans, A. muletensis, Atelognathus reverberii, Pelobates fuscus, Ptychohyla leonhardschultzei,* and *Thoropa*), short, broad-based papillae (e.g., European *Bufo* and *Rana, Crinia tasmaniensis, Gastrotheca, H. arborea, Leptobrachium hasselti,* and *Megistolotis lignarius*) or none (e.g., *H. ebraccata, H. rufitela, Osteopilus brunneus,* and *O. septentrionalis*).

The buccal roof (fig. 5.16) begins posterior to the upper jaw sheath and is divided into prenarial, postnarial, and buccal roof arenas. The posterior terminus of the buccal roof is the glandular zone (Kratochwill 1933; = dorsal food trap of Kenny 1969b) and the dorsal velum; (Goette 1874; R. M. Savage 1952, 1955; = posterior filter valves of Kratochwill 1933). A small number of pre-narial pustulations in the prenarial arena (e.g., *Alytes, Bombina, Bufo,* and *Pelobates*) are arranged in a U-shaped field with a posterior opening and are more or less fused together at the base (e.g., *Colostethus whymperi, Cyclorana australis, Hyla* spp., *Litoria alboguttata,* and *Osteopilus septentrionalis*). The internal nares (= choanae) lie between the prenarial and postnarial arena and are directed posteriorly in a V-shaped pattern (e.g., *Alytes, Bombina, Bufo, Hyla dendroscarta, Megaelosia goeldii,* and *Osteopilus*), directed anteriorly in an inverted V-shaped pattern (e.g., *Pelobates* and *Rana*), or are parallel (e.g., *Hyalinobatrachium fleischmanni, Hyla* spp., *Leptodactylus knudseni, L. wagneri,* and *Macrogenioglottus alipioi*). The choanae are flanked by the narial papillae and the narial valve projection. One to five postnarial papillae extend into the postnarial arena. In microhylids, *Hyla arborea, H. ebraccata, H. phlebodes, H. sarayacuensis,* and *Rhinophrynus* these papillae are not well developed and are absent in *Pelobates fuscus* and several leptodactylids. The lateral ridge papillae (to ca. 800 μm long) are situated posteroventral to the postnarial papillae. They occur individually (e.g., *Bufo, Pseudophryne bibronii,* and *Rana*) or form a flat planar structure fused only at the base (e.g., *Odontophrynus occidentalis, Paratelmatobius lutzii,* and *Pleurodema cinerea*) or along their entire length (e.g., *Alytes muletensis, A. obstetricans,* and *Bombina*). In some species, only one lateral ridge papilla occurs on either side (e.g., *Pelobates fuscus,* some leptodactylids), but there are usually 3–4. The medial ridge is situated between the lateral ridge papillae. The base is broad, and the terminus is shaped like a pointed lancet (e.g., *Anotheca spinosa, Bufo* spp., and *Hyla mixe*) or blunt (e.g., *Alytes* spp., *Discoglossus* spp., and *P. fuscus*). In some cases, the ridge is a flat structure with a wide base (e.g., *H. arborea, H. rufitela,* and *Thoropa*) with lateral or apical pustulations (e.g., *Alytes, Bombina,* and *Scaphiophryne*). The buccal roof arena is surrounded by papillae that may be short or long like the papillae of the buccal floor arena. The number of buccal roof arena papillae varies from ca. 10 large ones in *H. arborea* to 30–40 in *Crossodactylus*. There are no papillae in the buccal floor arena in *Adenomera marmorata* and *Heleophryne natalensis*. About 20 (*Bufo*) to over 100 (*Pelobates* and *Ptychohyla*) small pustulations less than 100 μm high are found on the buccal roof in all species and the postnarial arena in most species.

In *Rhinophrynus dorsalis* (larval Type 1), lingual papillae are absent, the surface anterior to the buccal pockets is large, the buccal floor arena papillae are arranged in a posteriorly directed arch, and papillae are absent on the ventral velum. The lateral ridge papillae, the medial ridge, and pre- and postnarial papillae of the buccal roof are absent, and the buccal roof is not divided into pre- and postnarial arenas. Buccal roof arena papillae and pustulations are present.

The buccal floor of microhylids (larval Type 2) is surrounded by papillae and lateral ridge papillae are present. Particularly well-developed prenarial papillae are found in *Microhyla berdmorei*, and the narial valve projection is very large in *M. heymonsi*. The papillae of the buccal floor and roof arenas are arranged in an arc in *M. heymonsi*, and prepocket papillae are present. Infralabial papillae have been described only for *M. heymonsi*. Lingual papillae and the medial ridge are absent in all microhylids.

In pipoid larvae (Type 1), the total width of the buccal floor arena is only a little larger than that of the oral orifice. It tapers just posterior to the mouth and extends posteriorly as a narrow strip between the two ventral vela and terminates at the glottis (fig. 5.11B). Its structural organization cannot be compared to that of Types 2 through 4. There is no tongue anlage before metamorphosis (see Channing 1984; Hammerman 1969a, b, c; Helff and Mellicker 1941; Lannoo et al. 1987; Viertel 1982, 1984b; Wassersug 1976a, b, 1980, 1984; Wassersug and Duellman 1984; Wassersug and Heyer 1983, 1988).

The filter plates and ciliary grooves are part of the posterior buccopharynx (i.e., filter apparatus; Gradwell 1972a, b, c, 1975b; Kenny 1969a, b; Kratochwill 1933; R. M. Savage 1952, 1955; Viertel 1984b, 1985, 1987; Wassersug 1976a, b, 1980; Wassersug and Rosenberg 1979). In larval Types 3 and 4, branchial food traps (= anterior filter valve of Kratochwill 1933) and ciliary cushions (Viertel 1985; = pressure cushions of Kratochwill 1933) are developed as parts of the posterior buccopharynx (figs. 5.11A, 5.17A, and 5.18A). Because of their functional significance (Gradwell 1970, 1972a, b, c), the ventral and dorsal vela are assigned to the filter apparatus (figs. 5.17A–E and 5.18). Although the persistent and transient gills are not involved in suspension feeding, they are important in the ontogenesis of the filter apparatus (fig. 5.17C). The anatomy of the filter apparatus in *Xenopus laevis* (larval Type 1) differs considerably from those of *Bombina variegata* (Type 3) and *Bufo bufo, B. calamita,* and *Rana temporaria* (Type 4).

In *Xenopus,* the ventral velum is paired (= pharyngobranchial tract of Weisz 1945a, b; filter valves of O. M. Sokol 1977a), and each half extends parallel to the longitudinal axis of the tadpole (figs. 5.11B and 5.19A–H); in larval Types 3 and 4, the single ventral velum is positioned transversely (figs. 5.11A, 5.14C, D, 5.15, and 5.17). Branchial food traps projecting ventrally from the velum via a transitional zone (Wassersug and Rosenberg 1979) probably are an integral part of the ventral velum in *Xenopus* (figs. 5.18A, G, and K and 5.20F–G). In larval Types 1, 3, and 4, four pairs of branchial arches are attached posteriorly to the ventral velum; the first has exclusively posterior filter plates, the second and third have filter plates on both sides, and the fourth has only anterior plates. In Types 3 and 4 only a rim (fig. 5.18G, K) separates the branchial food traps from the filter plates. The surface of the filter plates is enlarged by filter rows (fig. 5.18A–D). In *Xenopus laevis* the filter plates have a particularly large chondrified basal lamina (visceral skeleton; figs. 5.19B, G, and K) from which the filter row pedicels branch out into the filter rows. In most larvae of all four types the filter rows fold into primary and secondary side branches and thereby separate small filter niches from the filter canals. The transversely oriented dorsal velum in larval Types 3 and 4 is the dorsal equivalent of the ventral velum. In *Xenopus,* three pairs of dorsal vela (= velar vanes of Gradwell 1975c; pressure pads of O. M. Sokol 1977a) extend into the filter cavities between the filter plates (figs. 5.11B and 5.19A, B).

It is not clear whether the dorsal vela of larval Type 1 are homologous with the single velum of larval Types 3 and 4. In the latter, a glandular zone extends transversely anterior to the dorsal velum (Kratochwill 1933; = dorsal food trap of Kenny 1969a, b) and the posterolateral ciliary cushions. The ciliary grooves enclose the ciliary cushions (figs. 5.17A, E) on both sides posteriorly and continue into the esophagus. Ciliary cushions are absent in *Xenopus,* and the ciliary groove extends into the esophagus laterally (figs. 5.11B and 5.19A, D).

Tadpoles that are not suspension feeders (e.g., carnivores) have highly modified buccopharyngeal morphology. The visceral skeleton of the macrophagous-carnivorous tadpole of *Hymenochirus boettgeri* is small by comparison with that of other pipids (*Pipa* and *Xenopus*). A chondrified basal lamina of the filter plates and chondrified supporting filter row pedicels are lacking. The filter rows are flat and do not branch or enclose the filter canals. The surface of the filter plates is small, and ventral and dorsal vela are lacking (O. M. Sokol 1962, 1977a). The carnivorous larva of *Lepidobatrachus laevis* lacks many oral structures. There are some pustulations on the buccal floor and roof, flat narial valve projections and a tongue anlage. The ventral velum, branchial food traps and filter plates are lacking (Ruibal and Thomas 1988; Wassersug and Heyer 1988). All structures of the anterior buccopharynx, apart from the tongue anlage and the free ventral velum, are lacking in the arboreal, oophagous larva of *Osteopilus brunneus.* Filter plates with branching filter rows (primary and secondary side folds) and branchial food traps of the pitted type are present. In the closely related non-oophagous *O. septentrionalis,* all structures of the anterior buccopharynx are present, although they are not as fully developed as those of generalized hylid larvae (Lannoo et al. 1987). Wassersug and Duellman (1984) noted that the sequence of differentiation of the keratinized mouthparts, buccal projections and papillae, filter plates and gills, and ventral velum and branchial food traps in embryos of egg-brooding hylids (*Cryptobatrachus, Flectonotus, Gastrotheca* spp., *Hemiphractus,* and *Stefania*) relates to developmental mode, supply of yolk, and respiration. Some buccopharyngeal structures are lacking because the embryos do not pass through the late developmental stages (e.g., *Cryptobatrachus* and *Hemiphractus*) in which the anlagen of these structures are formed (= paedogenesis). Leptodactylids (e.g., *Cycloramphus duseni, Thoropa miliaris,* and *T. petropolitana*) that inhabit steep, wet rocky surfaces resemble stream-adapted larvae, but they share buccopharyngeal characteristics with arboreal or semiterrestrial larvae. They have fewer buccal papillae, no secretory ridges on the branchial food traps, and an exposed glottis (Wassersug and Heyer 1983).

Buccopharyngeal Histology and Ultrastructure

A basic scheme (squamous epithelium and bottle-shaped secretory cells, SC1) applies to the ventral and dorsal vela in *Xenopus.* The degree of differentiation appears to be advanced compared to larvae of Types 3 and 4 (figs. 5.20A–H). The SC1 are arranged in rows, and their secretions flow into secretory grooves and onto secretory ridges. The secretory grooves are bordered by secretory ridges and random cells of the squamous epithelium. The secretory grooves and ridges follow the course of the ventral velum. In the dorsal velum,

Table 5.2 Abbreviations used in figures 5.17–5.22 and 5.24

AC = apical cell, ACE = cerebral anlage, AFP = anlage of filter plates, APEG = anlage of persistent epidermal gills, AVV = anlage of ventral velum, BI–BIV = branchial arches I–IV, BFA = buccal floor arena, BFT = branchial food trap, BL = basal lamina, BLA = basal labyrinth, BP = buccal pocket, BRA = buccal roof arena, c = caudal, C = cilia, CA = cartilage, CC = ciliary cushion, CIC = ciliary cell, CL = capillary vessel, CT = connective tissue, d = dorsal, DV = dorsal velum, E = merocrine extrusions, ED = edge of filter plate III, ENC = endodermal cell, EP = epidermis, EPC = epidermal cell, ER = endoplasmic reticulum, ES = esophagus, ET = erythrocyte, EX = extrusion, EZ = zone of extrusion, FC = filter cavity, FN = filter niche, FP = filter plate, FPI–FPIV = filter plates I–IV, FPC = epidermal fold of peribranchial chamber (operculum), FR = filter row, G = gill, GL = glottis, GO = Golgi apparatus, GR = grana, GS = gill slit, IC = intercellular space, enlarged by fixation and dehydration, LFP = lateral filter plate, LV = lipid vacuole, M = mitochondrion, MF = middle fold, MRC = squamous epithelial cells with microridges, MV = microvilli, N = nucleus, NENC = nucleus of endodermal cells, NEPC = nucleus of epidermal cell, NSLC = nucleus of sensory layer cell, NC = nucleus of endothelial cell, NCT = nucleus of connective tissue cell, NSC = nucleus of secretory cell, NSQC = nucleus of squamous epithelial cell, OS = mouth, P = peripheral pustulation(s), PC = peribranchial chamber, PD = pericardium, PE = periderm, PET = peritoneum, PG = pigment granule, PI = pigment cell, PP = pustulation at base of filter row, PQ = cartilage of palatoquadrate, PS = primary side fold, R = rim, SC = secretory cell, SC1 = goblet cell filled with mucous vacuoles, SC2 = goblet cell after extrusion, SL = sensory layer, SP = secretory pit, SPC = supporting cell, SQ = squamous epithelium, SQC = squamous epithelial cell, SQE = squamous epithelium, SR = secretory ridge, SS = secondary side fold, TEG = transient epidermal gills, v = ventral, V = ventricle, VC = vacuole, VV = ventral velum, and YV = yolk vacuoles.

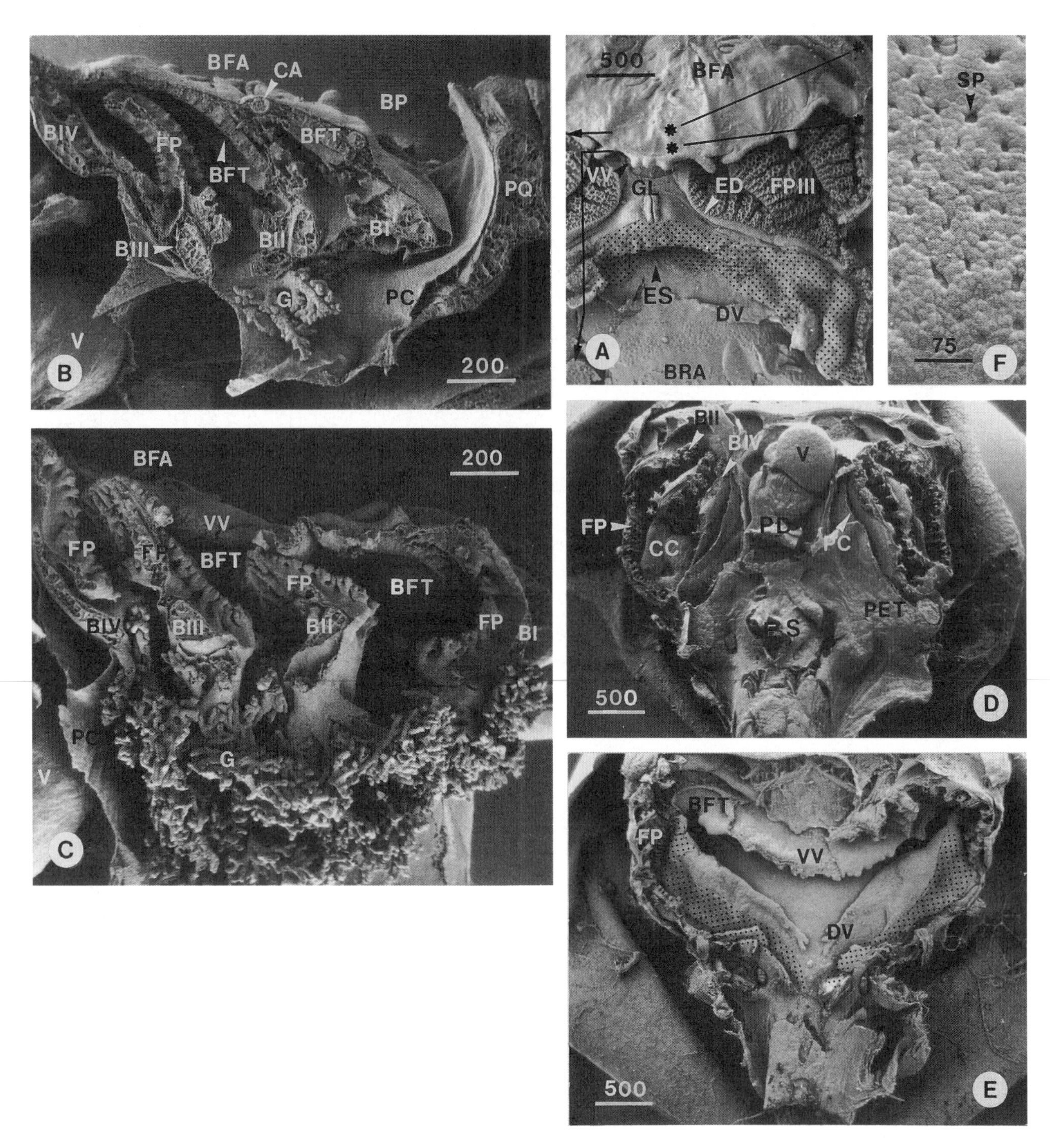

Fig. 5.17. Filter apparatus of *Rana temporaria* (Gosner stage 32). (A) Oral cavity opened showing buccal floor arena, buccal roof arena with glandular zone, and glottis; filter plates free; ciliary area of ciliary cushion *(stipple)* reaching the opening of the esophagus. Symbol: asterisks = planes of cross sections B and C. (B) Cross section of the right side of the buccal floor arena and filter apparatus. (C) Cross section of ventral velum, velar edge, and posterior filter apparatus on the right side. (D) Peribranchial chamber opened, posteroventral view; close connection of ciliary cushion and filter plates is distinct; wall of peribranchial chamber united with pericardium and peritoneum. (E) Proximal parts of filter apparatus removed, posteroventral view; position of ventral velum and dorsal velum forming a slit; ciliary cushion and ciliary groove *(stipple)* posterior to the dorsal velum. (F) Glandular zone of the buccal floor arena of *Bufo bufo* with mucus secreting secretory pits. Scale = micrometers. See table 5.2 for abbreviations. From Viertel (1985); reprinted by permission of Springer-Verlag.

24–36 SC1 cells secrete synchronously in a manner similar to that seen in the ventral velum of tadpole Types 3 and 4. The extrusion zones are prominent, and the secretory pits are not sunken. A transitional zone (fig. 5.14) lies between the dorsal velum and the ciliary groove. The ciliary groove of *X. laevis* does not extend between the filter plates (figs. 5.11B and 5.19A) as in tadpoles of Types 3 and 4. The ciliary groove is composed of ciliary cells and columnar (SC3) secretory cells in the center of the ciliary groove and pear-shaped (SC4) cells in the transition zone (figs. 5.19D–E). Goblet cells are absent. The ciliary groove is enclosed by squamous epithelium underlain by supporting cells in the transition zone.

The differences between the vela and secretory grooves and ridges and their positions and structure among Type 1 versus Types 3 and 4 tadpoles also apply to the structure of the filter plates. Multiple branches of the middle plate of *Xenopus* more frequently fold into primary and secondary side folds that greatly increase the surface area. As a result, filter niches are more numerous and the filter canals form a more ramified system than in larval Types 3 and 4 (figs. 5.18A–D, 5.19I–J, and 5.21C). The majority of organs in the filter apparatus are constructed according to a uniform principle: bottle-shaped glandular cells are embedded in a single-layered squamous epithelium supported by connective tissue. These secretory cells stand in groups and empty into secretory pits. This scheme applies to the dorsal and ventral vela, the glandular zone, and the branchial food traps (figs. 5.17E, 5.18G–J, 5.19C, and 5.20). Only in *Ascaphus, Rhinophrynus,* the Discoglossidae, and Pelobatidae is the pattern of distribution of the SC1, and therefore of secretory pits, as disparate as in the ventral margin of the velum. In *Rhinophrynus,* the epithelium forms wavelike folds. In microhylids and all Neobatrachia the secretory cells are arranged in rows and their secretion zones are situated on the exposed secretory ridges. In the Neobatrachia, SC1 and their secretion zones are always grouped together.

Filter plates, ciliary groove, and ciliary cushions in larval Types 3 and 4 are organized differently from that described for Type 1 larvae. Arched filter plates and the presence of middle and primary and secondary side folds increases the surface of the filter plates (figs. 5.18A–D, 5.19I–J, and 5.21C). Two overlapping layers of squamous epithelium are supported proximally by connective tissue and the visceral skeleton. Apical cells protrude from the epithelium and form the distal limits of the middle and side folds and also the lateral pustulations. The ciliary groove and ciliary cushions consist of ciliary cells and goblet cells (SC1 and SC2; figs. 5.18E–F and 5.24A, B). Mucus is present on the ventral velum, branchial food traps, filter rows, and ciliary cushions.

The ultrastructure of the ciliary groove, ventral vela, and dorsal vela in a Type 1 (*Xenopus laevis*) tadpole is summarized from Viertel (1985, 1987). The SC1 are taller than in larval Types 3 and 4, and abundant endoplasmic reticulum (ER) is found around the nucleus. The apical pole is densely filled with granules. The SC3 cells are rich in mitochondria from the middle of the cell to the apical pole where vacuoles and extrusions occur. The nucleus is situated at the cell base. The SC4 secretory cells are pear-shaped, rich in vacuoles, and densely filled with ER except at the apical pole, the surface of the secretory zone is small, and their nucleus is unusually large. The ciliary cells, apical cells, and squamous epithelial cells do not differ from those of larval Types 3 and 4.

Examples of Types 3 and 4 larvae include *Bombina variegata* and *Bufo bufo, B. calamita,* and *Rana temporaria,* respectively. Roughly bottle-shaped SC1 are rich in electron-dense granules and vacuoles. There are many mitochondria in the extrusion zone of the apical pole and many longitudinal microtubules. Microvilli maintain contact with the interstitial spaces, and a labyrinth of shallow folds forms the proximal limit of the cells. The shape of the cell changes considerably according to its vacuole content. Goblet cells, or goblet-cell-like mucous cells (SC2 secretory cells) lack microtubules, mitochondria and ER in the apical pole (fig. 5.21A). Mitochondria are found at the periphery of the cell, particularly proximally in the mucous zone. The presence of both filled and empty goblet cells indicates that mucus is being synthesized rapidly and extruded abruptly (fig. 5.21B).

Ciliary cells are funnel-shaped with a broad, flat, ciliated apical pole; the cells are rich in mitochondria and have an electron-dense cytoplasm caused by a high granular density (fig. 5.21A). The nucleus is basal. Squamous epithelial cells are especially flat in the filter plates, and more or less triangular in cross section in the branchial food traps and ventral velum (fig. 5.20). The nucleus is situated at the widest point and the cytoplasm is electron-dense. There is a single layer of vacuoles along the entire apical surface of the cell, and there is evidence of extrusion. Apical cells are papilla- or ridge-shaped with the nucleus situated at the cell base and surrounded by a thick layer of ER (fig. 5.21C). Mitochondria and the Golgi apparatus are located in the middle and apical parts of the cell. There is a high density of variously sized granules, a small number of vacuoles, and evidence of extrusion.

Ontogenesis of parts of the filter apparatus was described by Ekman (1913, *Bombina pachypus, Bufo bufo, Hyla arborea, R. esculenta,* and *R. temporaria*), Gegenbaur (1898–1901, *Bombina*), Gerhardt (1932, *H. arborea* and *Pelobates fuscus*), Goette (1874, *Bombina*), E. Marcus (1930, *R. arvalis*), and Naue (1890, *R. esculenta, R. temporaria,* and *P. fuscus*). The origin of the filter plates was the subject of detailed discussion and controversy (Ekman 1913; Gegenbaur 1898–1901; Gerhardt 1932; Goette 1874; Greil 1905; J. G. Kerr 1905; Mangold 1936; E. Marcus 1930; H. Marcus 1908; Maurer 1888; Naue 1890). Viertel (1991, *Bombina variegata, Bufo bufo, B. calamita, R. temporaria,* and *X. laevis*) questioned the homology of the filter apparatus with the pharynx, the evolution of anuran larvae and the function of the filter apparatus.

Morphogenesis and histogenesis of the filter apparatus is based on studies of *Bombina variegata, Bufo calamita,* and *Rana temporaria.* The anatomy of the anuran pharynx in Gosner 17–18 is comparable to that of other vertebrates, especially those of embryonic fish, salamanders, and caecilians. Visceral pouches 1–5 and the mouth are not perforated before Gosner 20 (fig. 5.14A). The anlagen of the transient

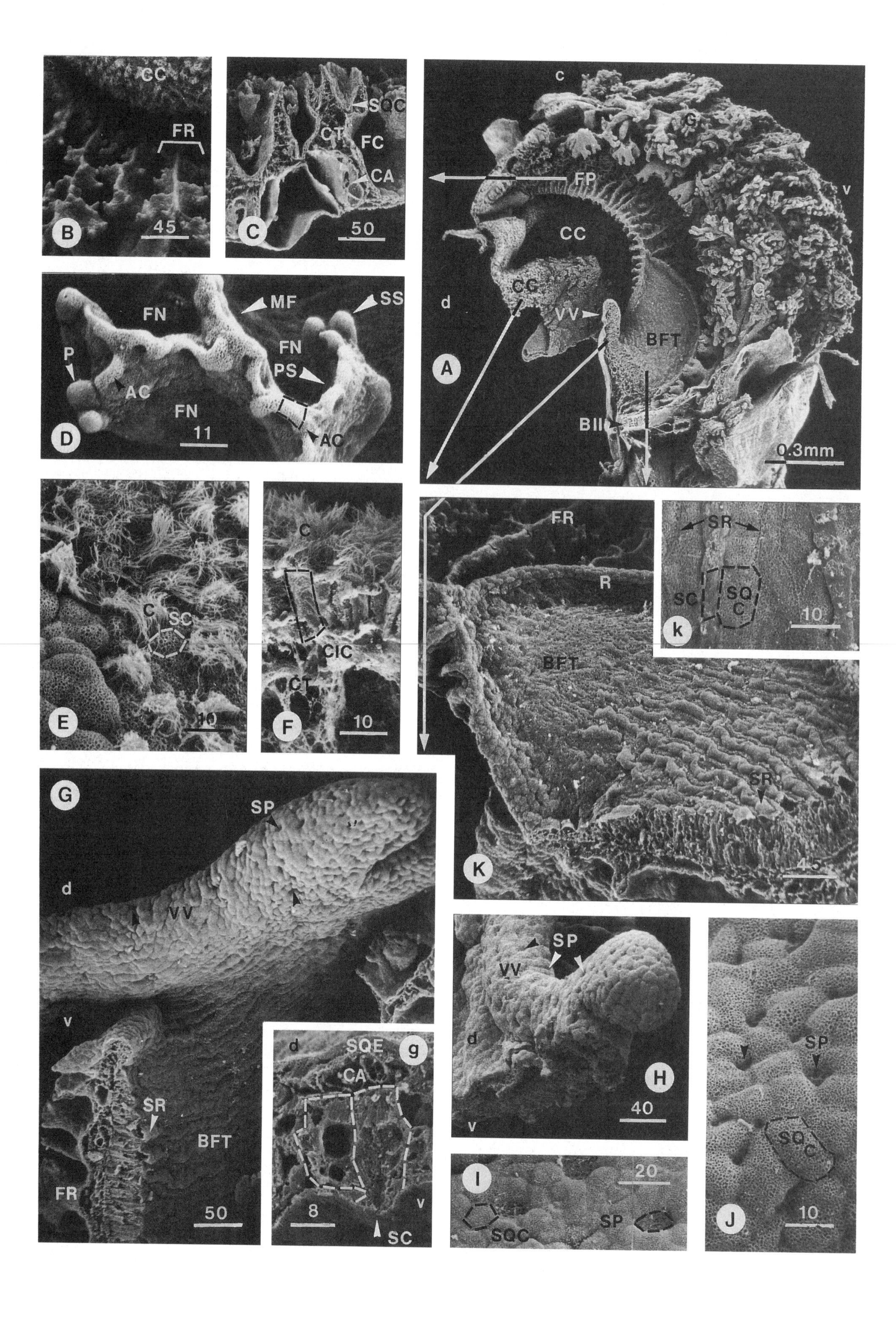

CC
FR
45
B
SQC
CT
FC
CA
50
C
c
G
FP
v
CC
CC
d
VV
BFT
A
BII
0.3mm
FN
MF
SS
P
FN
PS
AC
FN
AC
D
11
C
SC
10
E
C
CIC
CT
10
F
FR
R
SR
SC
SQ
C
10
k
BFT
SR
K
45
G
SP
d
VV
v
SR
BFT
FR
50
d
SQE
g
CA
8
v
SC
SP
VV
d
H
40
v
SP
SQ
C
J
10
I
20
SQC
SP

Fig. 5.18. Filter apparatus of *Rana temporaria* (Gosner stage 32). (A) Left branchial arch II, lateral view; note close position of filter rows and ciliary cushion and connection of branchial food trap and ventral velum. (B) Ciliary cushion touching the filter rows. (C) Section of filter rows. (D) Structure of a filter row. (E) Surface of ciliary cushion including ciliary and secretory cells. (F) Ciliary cushion opened, connective tissue supporting the ciliary cells. (G) Ventral velum and velar papilla; position of branchial food trap, filter plate, and ventral velum in posterior view; secretory ridges of branchial food traps changing to typical secretory pits of ventral velum. Inset g: Section of proximal ventral velum, cartilaginous clasp supports the thin squamous epithelium and a large vacuolized secretory cell. (H) Dorsal edge of ventral velum and velar papilla with a pattern of secretory pits. (I) Secretory pits of the velar edge. (J) Superficial secretory pits of the ventral side of the ventral velum surrounded by polygonal squamous epithelial cells. (K) Surface anatomy of branchial food trap. Inset k: Surface of branchial food trap showing cells on a secretory ridge. Scale = micrometers except millimeters in (A). See table 5.2 for abbreviations. From Viertel (1985); reprinted by permission of Springer-Verlag.

epidermal gills are visible as lateral protrusions. The buccal roof arena and buccal floor arena are still flat in Gosner 20–21, and visceral pouches 2–4 perforate from stages 20–21 (figs. 5.14B and 5.22A, B). Visceral pouches 1 and 5 do not perforate, and visceral pouch 1 later becomes the tuba eustachii. The transient gills reach their maximum size at Gosner 21 (fig. 5.14D) and regress during Gosner 24–25.

The distinction between endoderm (high density of yolk and lipid vacuoles) and ectoderm in the pharynx up to Gosner 23 (Viertel 1991) is corroborated by ontogenetic differentiation of the pharyngeal cells (i.e, increasing mitochondrial density) beginning at stage 23 (fig. 5.22C, D). The cell structure, cytogenesis, and site of the visceral pouches show that the buccal floor, buccal roof, dorsal and ventral vela, anlagen of filter plates, transitional zone, and branchial food traps are of endodermal origin (Viertel 1991).

The ventral velum at Gosner 22–24 starts as an archlike, transverse roll in the buccal floor anterior to visceral pouch 1 (figs. 5.14 and 5.22). In Gosner 24 the ventral velum covers the filter plates and gill slits which lie between them. The surface of the secretory pits, into which the fully differentiated SC1 open, is visible from this time onward. The region between the ventral velum and the branchial food traps is termed the transition zone (figs. 5.14B, C, and 5.22F). In principle, this zone, the branchial food traps, and the ventral velum do not differ histologically and cytologically. Ultrastructurally and histologically the anlagen of the filter plates are not different from the buccal floor (fig. 5.20). These anlagen mark the posterior limits of the pharynx and take up a progressively posterolateral position.

The endodermally derived ciliary cushions and ciliary grooves clearly are dorsolateral projections of the esophagus and not of pharyngeal origin (figs. 5.11A and 5.14D). Their surface anatomy and ultrastructure are the same as those of the esophagus. Ciliary and goblet cells in the epithelia of these organs do not occur in any other part of the buccopharynx. These cells, in contrast to the cells of the buccopharynx, are fully differentiated by Gosner 23.

Endoderm is represented largely in the proximal area of the filter plates and anterior to the branchial food traps. Apparently the endodermal anlagen of the filter plates are overlapped by a layer of epidermal cells proceeding from the anlagen of the persistent epidermal gills (fig. 5.22F). The single layer of cells and the absence of ciliary cells and Type A and B cells indicate that this is a sensory layer (fig. 5.22D). This zone, where the epidermal cells meet the endoderm (i.e., between the persistent epidermal gills and the filter plates), is termed the ectodermal-endodermal transition zone (fig. 5.22E).

The anlagen of the filter plates become progressively flatter and expand in a posterolateral direction. At Gosner 23, the anlagen of the middle folds of the filter rows are visible as protrusions on the surface of the filter plates. These increase in height from Gosner 24–25 and branch into the anlagen of the primary side folds. Secondary side folds are not differentiated by Gosner 25. After Gosner 23 some endodermal cells of the anlagen of the filter plates become oval or cube-shaped. The rate of growth of these cells does not increase until Gosner 25 when they become exposed as protrusions between the cells of the squamous epithelium. These are the anlagen of apical cells and continue to grow after Gosner 25.

Before Nieuwkoop/Faber 35–36, the pharyngeal development of *Xenopus laevis* (fig. 5.14A, B) is equivalent to the initial state in larval Types 3 and 4. From Nieuwkoop/Faber 38 the anlage of the ventral velum arches dorsally between the medial line and the visceral pouches. The anlagen of the filter plates, anterior visceral pouches and the gill slits change their position longitudinally and less so transversely. The extension of the pharyngeal longitudinal axis does not occur until Nieuwkoop/Faber 47. From Nieuwkoop/Faber 42–43 the anlage of the ventral velum has a serial arrangement of cells in three tiers, and secretory cells break through between the distal squamous epithelial cells. A full differentiation of the surface of the ventral velum takes place at Nieuwkoop/Faber 46. Secretory grooves into which SC1 flow are framed by secretory ridges and random squamous epithelial cells. The probable function of microtubules in the apical zone of secretory cells, which appear at Nieuwkoop/Faber 47, is to guide and transport vacuoles during extrusion (Berthold 1980). Mucus synthesis in the ventral velum is evident from Nieuwkoop/Faber 45, although no mucus is found on the anlagen of the filter rows in this stage.

The epidermis of the ectodermal-endodermal transition zone of *Xenopus* at Nieuwkoop/Faber 38 begins to penetrate into and overlay the visceral pouches (fig. 5.23). Judging by the distribution of yolk and the absence of ciliary cells, the sensory layer overlays the anlagen of the filter plates. In contrast to the case of *Bombina variegata*, *Bufo calamita*, and *Rana temporaria*, the sensory layer in *Xenopus* resembles squamous epithelium as does the epidermal mantle of the gills that are present ventrolaterally only during stages 38–47. From Nieuwkoop/Faber 42 the surface of the filter plates is enlarged by the dorsal arching of their anlagen and, from stage 47, by the appearance of the secondary side folds. The anlagen of the apical cells are formed by the endodermal

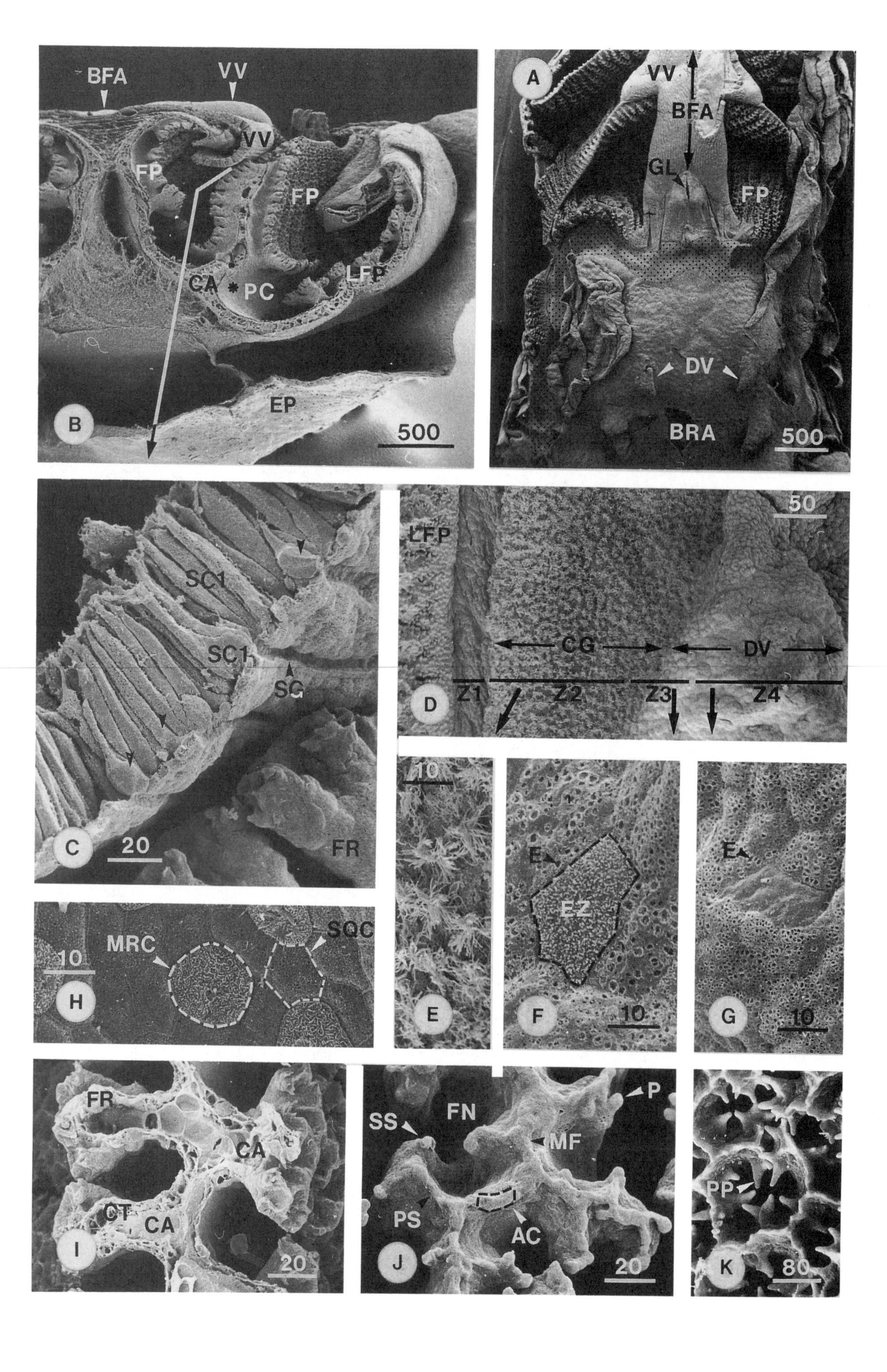

BFA
VV
VV
FP
FP
CA
PC
LFP
EP
B
500
A
VV
BFA
GL
FP
DV
BRA
500
SC1
SC1
SG
FR
C
20
50
LFP
CG
DV
Z1
Z2
Z3
Z4
D
10
E
E
EZ
F
10
E
G
10
MRC
SQC
10
H
FR
CA
CT
CA
I
20
FN
P
SS
MF
PS
AC
J
20
PP
K
80

Fig. 5.19. Filter apparatus of *Xenopus laevis* (Nieuwkoop/Faber stage 54). (A) Filter apparatus and buccal floor arena with views of ventral velum, filter plates, and dorsal vela. Symbol: stipple = ciliary groove. (B) Right side, anterior view of cross section at the plane of filter plate III and branchial cavities II and III with epidermis reflected. (C) Section of ventral velum showing different sizes of apical cell poles caused by different extrusion phases of mucus. (D) Ciliary groove and dorsal velum. Areas: Z1 = rim of squamous epithelium separating ciliary groove and lateral filter plate, Z2 = ciliary groove (also shown in E), Z3 = transitional zone of ciliary groove and dorsal velum (also shown in F), and Z4 = dorsal velum with smooth epithelial cells (also shown in G). (H) Epithelium at base of filter plate III; see asterisk in (B). (I–J) Morphology of filter row. (K) Filter rows of the ventral region with pustulations at the base of a filter row. Scale = micrometers. See table 5.2 for other abbreviations. From Viertel (1987); reprinted by permission of Gustav Fischer Verlag.

anlagen of the filter plates, break through the squamous epithelial cells of the sensory layer in Nieuwkoop/Faber 47, and come to lie between the squamous epithelial cells. Both cell types are supported by endodermal cells.

In *Bufo calamita,* ingestion begins prior to differentiation of the filter plates and before Type 1 secretory cells of the filter plates and ventral velum have become fully active (Viertel 1991). The steep rise in ingestion rate apparently results from the presence of goblet and ciliary cells that have differentiated by Gosner 23. The anlagen of the filter plates are brought continually closer to the ciliary cushions by ventilation movements of the buccopharynx in Gosner 23–24. This results in mucous contact between the goblet cells and anlagen of the filter rows. Threads of mucous that pass from dorsal to ventral in the lumen of the pharynx form a mucous entrapment system. Because a mechanism of transport of mucus-covered food particles is present in the ciliary cells of the ciliary cushions, trapped food can be moved to the esophagus. The ciliary cushions are the main producers of mucus for the functioning filter apparatus. Although the filter plates also produce mucus, their main function is to provide an extensive surface area and support for the distribution and deposition of mucus from the ciliary cushions. Mucus produced by one organ is deposited on the surface of another structure.

The larvae of *Bufo calamita* and *Rana temporaria* become vagile at Gosner 22, and after Gosner 23 the embryos actively search for and ingest food. This occurs prior to differentiation of the keratinized jaw sheaths and labial tooth rows, which precludes the scraping of attached foods. These young individuals depend upon filtration of small, planktonic, or sedimentary particles. Because the transition to a swimming stage in *B. bufo* embryos does not occur until Gosner 23, it is not surprising that ingestion activity begins at stage 24 rather than 23. In *B. bufo, B. calamita, R. temporaria,* and other anuran species with free-swimming larvae, there probably is strong selection to start feeding as soon as possible before the yolk material has been depleted. Early feeding allows tadpoles to proceed as early as possible to independent feeding and complement the diminishing yolk supply with a supplementary source of food. Perhaps the beginning of the larval phase and the end of embryonic development should be defined as that point at which ingestion first occurs and all larval organs are present at least as anlagen.

Ingestion of food by tadpoles of *Xenopus laevis* begins at Nieuwkoop/Faber 42, and by stage 46 all structures required for active food intake are present except the filter plates are not yet fully differentiated (Viertel 1991). Food particles are present in the esophagus of embryos by this stage, and chondrogenesis of the palatoquadrate and ceratohyals has been completed by this stage. According to Wassersug and Hoff (1979), these visceral elements, together with the connecting musculus orbitohyoideus, are responsible for the ventilation and irrigation of the filter apparatus. The synthesis of mucus on the ventral velum and filter rows does not begin until Nieuwkoop/Faber 45. At Nieuwkoop/Faber 46 the yolk reserves from the buccopharynx and the esophagus are exhausted, and those in the digestive tract are being depleted; the peribranchial chamber is closed. In other species, ingestion begins before the yolk reserves have been used up and before the peribranchial chamber closes. Beginning suspension feeding before the filter plates are enlarged and the keratinized mouthparts are developed in Type 4 tadpoles is an example of functional heterochrony (Viertel 1991, 1992, 1996).

The origin and evolution of the buccopharynx are now considered. Except for the ciliary cushions and groove, ciliary and goblet cells occur exclusively in the esophagus and larval stomach (Ueck 1967; figs. 5.24 and 5.25A, B). The presence and location of SC1 in the ventral velum, transition zone, branchial food trap, glandular zone, and dorsal velum suggest that the branchial food traps and transition zone are ventral differentiations of the ventral velum. The dorsal arching and arrangement in rows of the SC1 changes the branchial food traps of the pitted type (Type 3) into branchial food traps of the ridged type (Type 4). The filter plates and gills are epidermal-pharyngeal organs. The pharyngeal endoderm is the largest component of the mass of the filter plates, and epidermal ectoderm and mesoderm of the capillaries predominate in the gills.

The cell structure of the ciliary groove of *Xenopus laevis* is the same as that of the esophagus. Fox et al. (1972) and Kindred (1927) described ciliary cells in the esophagus that are identical to those of the ciliary groove from Nieuwkoop/Faber 44. Fox et al. (1970, 1972) described SC3 with apical vesicles in the esophagus. Goblet cells do not differentiate in the esophagus until the beginning of metamorphosis (Nieuwkoop/Faber 65; Fox et al. 1972) when the ciliary groove has already regressed. Fox et al. (1972) also corroborated the presence of microvillous stubs on the ciliary cells of the esophagus. Ciliary cells and SC3 are found only in the ciliary groove and esophagus. Their identical cellular composition and positions relative to each other show that the ciliary groove is an anterolateral projection of the esophagus.

The allocation of the ventral velum of *Xenopus* to the pharynx or buccal cavity is more difficult than with larval Types 3 and 4. Judging by its position relative to the visceral pouches (Nieuwkoop/Faber 38), the ventral velum is part of the buccal roof as in larval Types 3 and 4. Gradwell (1971a, b, 1975c) thought that the ventral velum of larval Types 3 and 4 originated similar to that of *Xenopus* by proliferation

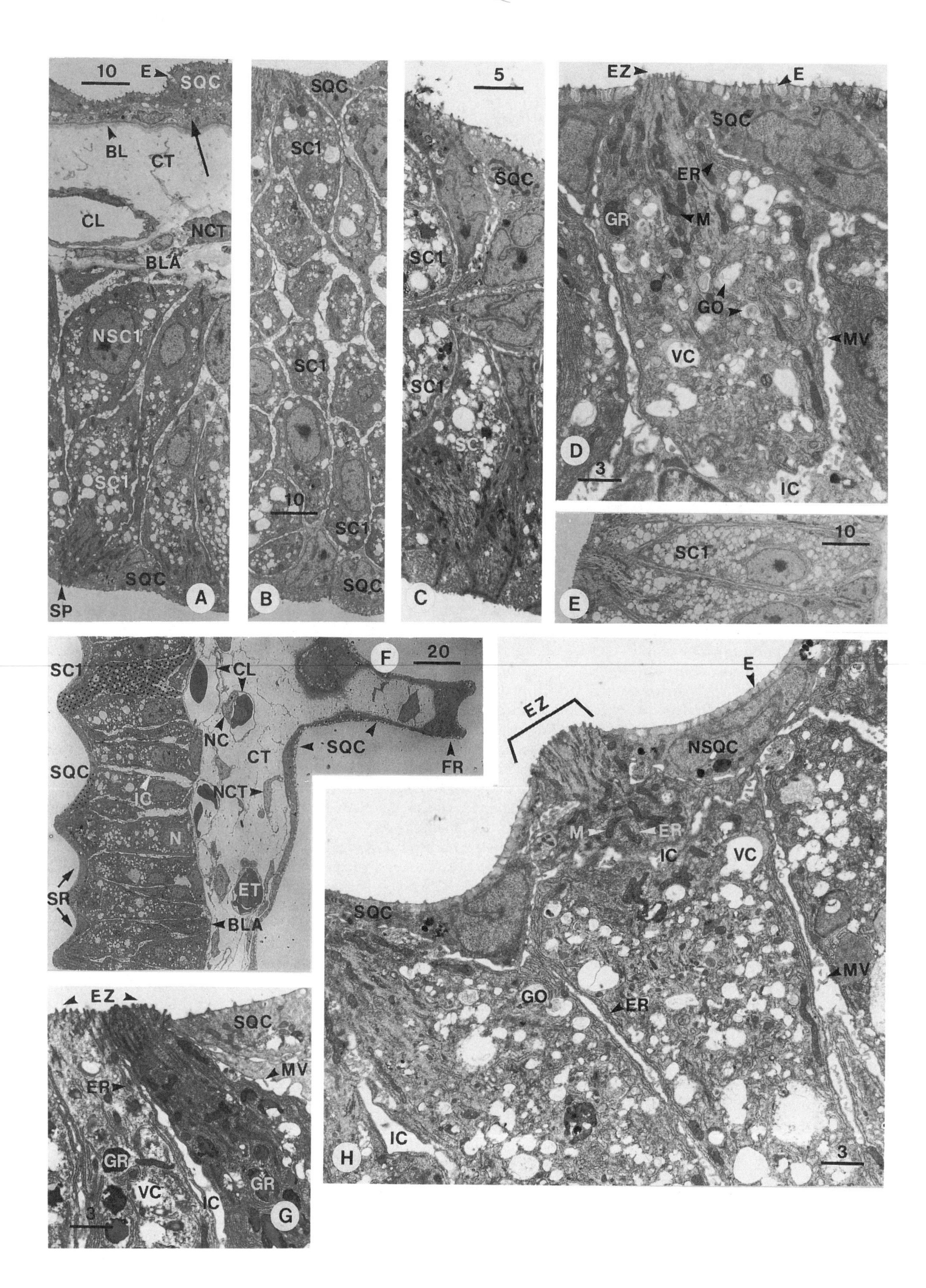

10
E
SQC
BL
CT
CL
NCT
BLA
NSC1
SC1
SQC
SP
A
SQC
SC1
SC1
10
SC1
SQC
B
5
SQC
SC1
SC1
SC
C
EZ
E
SQC
ER
GR
M
GO
VC
MV
D
3
IC
10
SC1
E
SC1
CL
NC
CT
SQC
F
20
FR
SQC
IC
NCT
N
ET
SR
BLA
EZ
SQC
MV
ER
GR
VC
IC
GR
3
G
E
EZ
NSQC
M
ER
IC
VC
SQC
GO
ER
MV
IC
H
3

Fig. 5.20. Cytology of the filter apparatus of *Rana temporaria* (Gosner stage 32). (A) Cross section of ventral velum anterior to the velar edge. (B) Cross section of the edge of ventral velum. (C) Longitudinal section of the edge of ventral velum. (D) Longitudinal section of the ventral side anterior to the edge of ventral velum. (E) Section of secretory cell. (F) Cross section of branchial food trap; secretory cells extending to the whole branchial food trap at SC1; squamous epithelial cells covering the secretory cells at SQC. (G) Cross section of branchial food trap. (H) Transverse section of branchial food trap of *Bufo bufo;* secretory cells exposed plainly. Scale = micrometers. See table 5.2 for abbreviations. From Viertel (1985); reprinted by permission of Springer-Verlag.

of the area between the halves of the velum in the oral cavity. Neither surface anatomy nor cell repertoire provides evidence of the origin of the ventral velum in *X. laevis.* The transition of endodermal cells to secretory cells SC1 is conspicuously similar to that in the ventral velum, transition zone, and branchial food traps of larval Types 3 and 4. The arrangement of SC1 in rows also is strikingly similar to the pattern in the ridged branchial food traps of Type 4 larvae. It is therefore probable that the branches of the ventral velum of *X. laevis* are homologous with the ventral velum and the branchial food traps of larvae of Types 3 and 4. The previous description of the filter plates in larval Types 3 and 4 also applies to *X. laevis.*

The term "pharynx" warrants further discussion. The site of the ceratobranchials shows that the anterior pharynx begins at the level of the buccal floor arena. The ectoderm that invaginates toward the anlagen of the filter plates does not belong to the pharynx any more than does the esophageal tissue that penetrates the pharyngeal roof to form the ciliary cushions and ciliary groove. These structures penetrate the embryonic pharynx and contribute to the formation of the filter apparatus. Relative to the origins and sites of the visceral skeleton, branchial nerves, and visceral musculature, the original embryonic pharynx does not differ from that of other vertebrate embryos (i.e., plesiomorphic vertebrate pharynx). It is a supporting structure for the invaginating epidermis and the esophageal endoderm; an ectodermal-esophageal filter apparatus has appeared.

If the anterior-posterior extension of the buccopharynx in *X. laevis* is a plesiomorphic character, then the buccopharynx may have evolved according to the following scenario. A fold morphologically and ultrastructurally similar to the ventral velum of larval Type 3, but whose location is similar to that of the Type 1 larva of *X. laevis,* limited the posterior extension of the buccal floor of the primitive anuran larva. This primitive morphology gave rise to two lineages. In one, the anatomical position of the buccopharynx was maintained, while the types of cell present in the ventral velum acquired their own pattern of differentiation and arrangement. This is the configuration found in xenoanuran larvae (Type 1; Starrett 1973). In a second lineage, the longitudinal axis of the mouth-buccal floor-glottis was shortened, especially posteriorly, while the transverse axis of the buccal floor was extended. Fewer cell types differentiated in the buccopharynx than in the xenoanuran larval type. This is the configuration in lemmanuran (Type 3 of Orton 1953) and acosmanuran (Type 4) larvae of Starrett (1973). These two lineages probably developed independently within different phylogenetic lines. Obviously the differentiation of a filter apparatus by anuran larvae is unique within the chordates and involved a specialized modification and association of the pharynx or *Kiemendarm* (i.e., gills and gut).

The common ancestor of all anuran larvae probably was a filter feeder with the following autapomorphous characters: an ectodermal-endodermal transition zone (filter plates) and an esophagus extending anterolaterally into the pharynx (ciliary groove and/or ciliary cushions). The position of the ventral velum in the tadpole of *X. laevis* is primitive and plesiomorphic within anuran larvae. Gradwell (1971a, b, 1975c) argued that the ventral velum in larval Types 3 and 4 was formed by proliferation of the area between the *Xenopus*-like ventral velum. This ancestral velum probably was characterized by the simple combination of squamous epithelial cells and SC1. If so, the complicated arrangement of SC1 and squamous epithelial cells in *X. laevis* suggests that the ventral velum was differentiated apomorphically. Accordingly, the ventral velum and branchial food traps of larval Types 3 and 4 are apomorphous with respect to their locations, and the branchial food traps of larval Type 4 are apomorphous relative to the more highly differentiated arrangement of SC1 (branchial food traps of the pitted type).

The epidermal covering of the anlagen of the filter plates (i.e., the ectodermal-endodermal transition zone of the visceral pockets; figs. 5.24 and 5.25) is common to *Bombina variegata, Bufo bufo, B. calamita, Rana temporaria, Xenopus laevis,* and probably all generalized larvae. Developmental association of two germ layers fits well into our knowledge of the vertebrate head, which is composed of ectodermal (neural crest) and endodermal elements, and especially the development of the vertebrate pharynx (splanchnocranium connected with the endodermal wall of the anterior gut).

Which characters of the anuran filter apparatus were derived within Lissamphibia (i.e., autapomorphic), and which characters are common to Lissamphibia and outgroups? Statements (B. I. Balinsky 1981; Hibiya 1982) that epidermis overlays mesodermal gill tissue in bony fish allow comparison to the anuran larval gills (B. I. Balinsky 1981; Fox 1986; Greven 1980; Lewinson et al. 1987). Because of certain synapomorphies (loss of sacculus vasculosus, presence of neurocranial endolymphatic sacs), the Dipnoi is considered to be the sister group of amphibians (Northcutt 1986; Rosen et al. 1981). Fox (1965) and Rauthner (1967) mentioned a deep penetration of gill epidermis that overlies the endoderm in the first visceral pouch of Dipnoi. The ectodermal-endodermal transitional zone of anuran larvae could be interpreted as the result of the continuing development of this synapomorphous character. If true, then the ancestor of anuran larvae had already developed a similar ectodermal-endodermal transitional zone of unknown differentiation.

Because of their uniformity (Lannoo et al. 1987; Wassersug 1980; Wassersug and Heyer 1988; Wassersug and Pyburn 1987), the surface structure and histology of filter

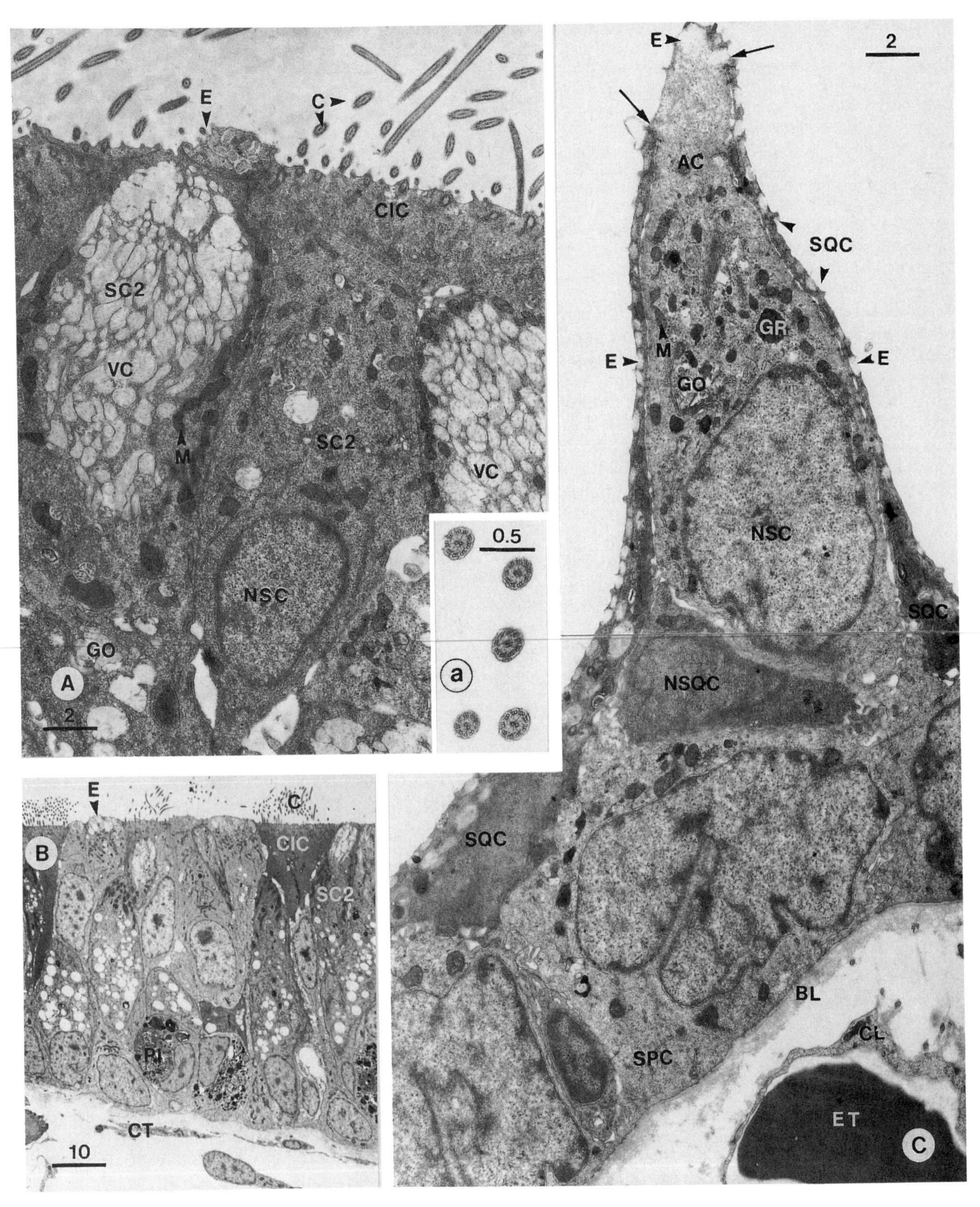

Fig. 5.21. Cytology of the filter apparatus of *Rana temporaria* (Gosner stage 32). (A) Different conditions of merocrine extrusion in the ciliary cushion. Inset a: Cilia on the ciliary cushion. (B) Cross section of ciliary cushion. (C) Cross section of primary side fold of filter row. Scale = micrometers. See table 5.2 for abbreviations. From Viertel (1985); reprinted by permission of Springer-Verlag.

plates do not lend themselves to cladistic analyses within anurans (O. M. Sokol 1975). Although quite different among anurans, the supporting cartilaginous system also lacks characters useful in a phylogenetic analysis. The esophageal derivation of ciliary grooves in *X. laevis* and of ciliary grooves and cushions in the other species is not found in Dipnoi, Urodela, or Gymnophiona and seems to be unique to anuran larvae. Ciliary grooves and ciliary cushions therefore are best considered as autapomorphic characters of anuran larvae, and their evolutionary origins should be considered carefully. Does the different structure of the ciliary groove in *Xenopus* compared with the ciliary grooves and ciliary cushions of *Bombina variegata, Bufo calamita, B. bufo,* and *R. temporaria* allow phylogenetic conclusions?

Goblet cells are the typical secretory cell of the ciliary grooves and ciliary cushions in these four species. In *Xenopus,* goblet cells do not occur in the ciliary groove (or esophagus), and Type 3 and 4 secretory cells are present (Viertel 1987). There are two alternative explanations. (1) The ciliary grooves of all the species studied were derived from a common anuran ancestral larva ("earlier stock" of Tihen 1965). This common character was modified structurally in different ways in *Xenopus* and other species, and the ciliary cushions were derived from the ciliary groove. (2) The ciliary grooves in *Xenopus* and other species (families) originated independently from the esophagus via convergence or parallelism. In this case the ciliary cushions would be derived from the esophagus.

The first alternative is supported by several other common anuran larval characters, such as the positions of the operculum (fold of peribranchial chamber) and spiracle(s) and the simplification of the mouthparts modified in different ways in *Xenopus* and tadpoles of larval Types 3 and 4 (Inger 1967; O. M. Sokol 1975; Tihen 1965). As for the ciliary grooves, these organs probably were inherited from a common ancestor and modified differently in the various larval types. As expected, corroboration of this alternative is found in the dendrograms of frog families (Inger 1967; Trueb 1973) and the cladistic analysis of Duellman and Trueb (1986); a ciliary groove was found in a common ancestor for all anurans and subsequently modified in *Xenopus* and other lineages. The alternative is corroborated by the first hypothetical phylogeny of frogs of Inger (1967) in which Pipidae and Rhinophrynidae (Type 1) on the one hand and all the other families (and larval types) on the other are derived independently from a common ancestor. In other dendrograms by Inger (1967), ciliary grooves would have been derived independently three times, and in the other hypothesized phylogenetic relationship (Duellman and Trueb 1986; Trueb 1973), four separate origins would be expected for the species studied here. If it is presumed that all anuran families developed ciliary grooves, then seven independent origins would be required. This high number of independent derivations suggests that the first alternative is much more probable than the second (see Ax 1984).

The ultrastructure and ontogenesis of the filter apparatus of Type 2 larvae (Microhylidae) also is in agreement with this discussion. Mucous cords are described by R. M. Savage (1952) and pressure cushions by Wassersug (1980); mucous cords are always associated with ciliary grooves. Whether the pressure cushions (Wassersug 1980) are ciliated or not is unknown. The surface anatomy of the filter plates is, in principle, identical with that of all other species (Wassersug 1980; Wassersug and Pyburn 1987). Based on a comparison of *Scaphiophryne* larva with the larvae of the Ranoidea (Type 4) and Microhylidae (Type 2), Wassersug (1984) concluded that the Type 2 larva was derived from Type 4 through developmental truncation. Wassersug viewed Ranoidea and Microhylidae as closely related taxa. If true, then the scenario above is also true for Type 2 larvae (Viertel 1991). The functional significance of these evolutionary features is the ability of anuran larvae to exploit a broad range of suspended and sedimented particles and, in larval Types 3 and 4, attached organisms. Most authors state that suspension feeding in tadpoles is a primitive character common to all anuran larvae (Inger 1967; Slade and Wassersug 1975; O. M. Sokol 1975). This feeding strategy must have had a tremendous influence on anuran evolution (Viertel 1991, 1992, 1996).

Digestive Tract and Diverticula

The foregut begins at the esophagus and includes the ciliary cushions and ciliary grooves that extend into the buccopharynx. The esophagus starts in the sagittal plane and then loops to the right where the manicotto (= manicotto glandulare, larval stomach) begins in the anterior bend of the second loop (I. Griffiths 1961, *Rana ridibunda;* figs. 5.12 and 5.26A, C, D). Mucus-producing goblet cells and ciliary cells are common in the foregut, and the ciliary cells are arranged in a species-specific pattern. In most species (e.g., I. Griffiths 1961, *Rana ridibunda,* Taylor/Kollros III; Rovira et al. 1993), cilia are still present in the left lateral loop of the esophagus. After a cilia-free section, cilia are found in most species anterior to the manicotto in two bands that extend at an angle of 90° to those of the ciliated esophageal loop. The lumen of the manicotto is ciliated. Posterior to the manicotto, two ciliated bands from the midgut (Ueck 1967) extend laterally. Because the manicotto lacks any muscular layer that enables peristalsis, the ciliary tracts likely promote food movement at points where accumulation might arise (see Naitoh et al. 1990 for information on motility of the larval intestine). The anatomical proportions of the foregut and the site of the liver and pancreas differ among genera (figs. 5.12 and 5.13). *Philautus gryllus,* in particular, deviates considerably from the general scheme.

The esophagus of suspension feeding anuran larvae is short with thin walls and longitudinal folds (fig. 5.26D). In most species it has a combination of ciliary and goblet cells (Barrington 1946; I. Griffiths 1961; Viertel 1992, *Bufo bufo, B. calamita,* and *Rana temporaria*); C. A. Brown et al. (1992) discussed vagal stimulation of the alimentary tract. Ciliary cells in *Xenopus* are developed by Nieuwkoop/Faber 44 (Fox et al. 1972; Viertel 1992; see Buccopharynx). Viertel (1987, 1992) did not find esophageal goblet cells in the ciliary groove of *X. laevis,* an anterior extension of the esophagus (Fox et al. 1972; Ueck 1967). Viertel (1987) describes bulb-shaped cells rich in ER and containing numerous hose-

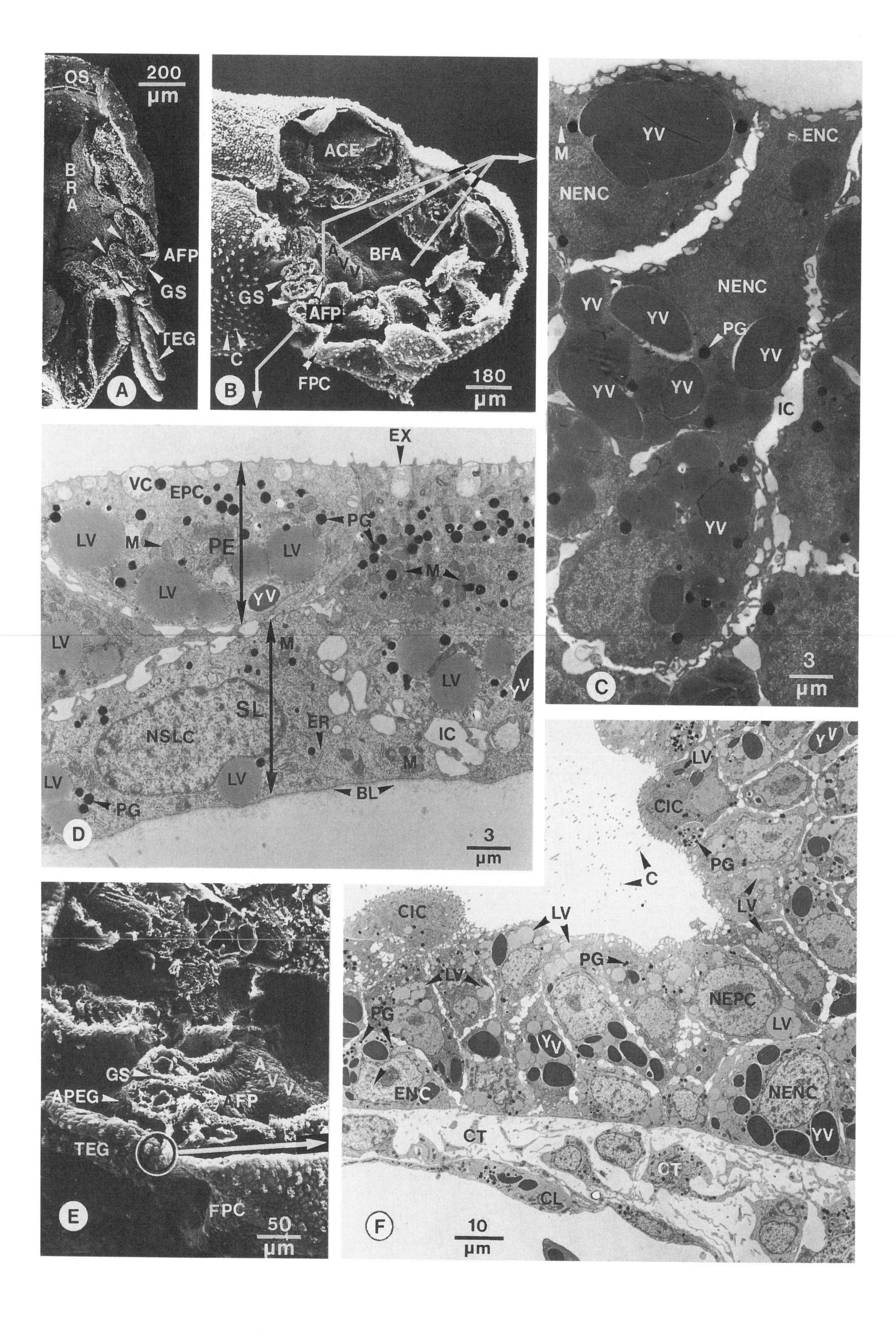

OS
200 µm
BRA
AFP
GS
TEG
A
ACE
BFA
AVV
GS
AFP
C
FPC
180 µm
B
YV
M
ENC
NENC
NENC
YV
PG
IC
3 µm
C
EX
VC
EPC
PG
LV
M
PE
YV
SL
NSLC
ER
IC
BL
3 µm
D
GS
AVV
APEG
AFP
TEG
FPC
50 µm
E
CIC
PG
C
LV
NEPC
ENC
NENC
CT
CL
10 µm
F

Fig. 5.22. Ontogeny of the filter apparatus of *Bufo calamita*. (A) Left side of buccal floor arena, anlagen of filter plates, and transient epidermal gills (Gosner stage 21). (B) Lateral view of anlage of filter plates, anlage of ventral velum, and transient epidermal gills (Gosner early 22). (C) Cross section of undifferentiated endodermal cells of the pharyngeal organs (Gosner 21). (D) Cross section of the ectoderm covering the transient epidermal gills (Gosner 22) with mitochondria-rich, Type B cells. (E) Site of anlage of filter plates, transient epidermal gills, anlage of persistent epidermal gills, and anlage of ventral velum (Gosner late 22). (F) Cross section of proximal base of transient epidermal gills with underlying endodermal cells (Gosner 22). Scale = micrometers. See table 5.2 for abbreviations. From Viertel (1991); reprinted by permission of Springer-Verlag.

shaped vacuoles in the ciliary groove; these are probably the vesicular cells of Fox (1981) and Fox et al. (1970, 1972).

The esophagus of the carnivorous larva of *Hymenochirus boettgeri* is long and forms diverticula. Strong fibrous cords of connective tissue develop posteriorly. Goblet cells are the dominant cell type in the anterior epithelium, whereas ciliary cells are found posteriorly almost to the exclusion of all other cell types. The esophagus evidently is elastic and secretes large quantities of mucus, which permit large prey to be swallowed whole (Ueck 1967). I. Griffiths (1961) described an abrupt histological transition from esophagus to manicotto in species in the Bufonidae, Discoglossidae, Hylidae, Leptodactylidae, Pelobatidae and Pipidae; members of the Ranidae and Rhacophoridae have a gradual histological transition.

The claim that the manicotto glandulare (Lambertini 1929; figs. 5.11A, 5.26A, and 5.27) is merely a storage-stomach (J. M. Dodd 1950) that lacks digestive functions or a precursor of the frog stomach (Süssbier 1936) has been disproved (Ueck 1967). Branched tubular glands lie beneath a ciliated epithelium, and large quantities of tubules thicken the wall. The tubular glands empty into the manicotto lumen through an irregular system of gaps rather than ducts (fig. 5.26E). Blood capillaries surround and penetrate an outer generative layer that forms the distal limit of the tubular gland tissue. A thin layer of muscles lies on top of the capillaries and generative layer (I. Griffiths 1961, *Rana ridibunda*). There is a single layer of circular muscle cells and a single layer of longitudinal muscle cells in *Xenopus laevis* (Ueck 1967), but the muscularis mucosae and submucosa are not present. Otherwise, the anatomy of *X. laevis* is similar to that of *R. ridibunda*.

Anatomical and histological comparisons of the manicotto reveal no fundamental differences among members of nine families, although a few species do not fit the general scheme (I. Griffiths 1961). Manicotto glands in *Alytes obstetricans* are found only dorsolaterally; a muscular foregut typhlosolis (tonguelike evagination of the gut wall) with ciliary cells and lined with ciliary tracks is situated ventrally. Goblet cells are highly concentrated in the manicotto epithelium. *Bombina variegata* lacks manicotto glands (Wilczynska 1981). In *Philautus gryllus* (fig. 5.27A–C; I. Griffiths 1961), the anatomical deviation is even greater, and the anatomy of microhylid larvae is more complex than in other forms.

Ribosomes, vesicles, and parallel arrays of smooth ER occur in actively secretory manicotto gland cells. The high density of ER forces the mitochondria to the base and periphery of the cell. Enzymatic granules usually found in small numbers in the middle of the cell suggest that mucus production is not the exclusive function of gland cells.

Gland development is dependent upon the nutritional state of the tadpoles. Nutritional deprivation reduces the total diameter of the tubules. The gland cells are cuboidal or slightly cylindrical with small microvilli. By contrast, gland cells of well-fed tadpoles are flattened to cuboidal with well-developed microvilli. The number of lipid vacuoles (neutral fats) increases with the intake of fat and decreases with the absence of dietary fat. They disappear in starved animals after 36 days. The cytoplasm of starving cells becomes denser while basophilic granules increase in size and number. This suggests that lipid vacuoles accumulate nutritive substances. Eating starch increases the size and number of lipid vacuoles more than eating dried pancreas tissue and yolk. The cytology of starved tadpoles that start feeding again soon approximates that of well-fed tadpoles. This indicates that the gland cells of the manicotto glandulare are functionally linked to the digestive process and that the basophilic granules are involved in protein digestion. Cytological comparison with the gastric parietal or oxynic cell of frogs suggests that the larval gland cell is an ion transporting cell. This cell produces and secretes electron-dense granules that can be distinguished from the pepsin granules of the *Magenhauptzelle* (= stomach main cells) of adult *Xenopus*.

The pH of the manicotto is below 3, and lipase activity was described by I. Griffiths (1961, *Rana temporaria* and *Xenopus laevis*). Hydrochloric acid kills microorganisms in the food, denatures protein to enable enzymatic breakdown, and dissolves salts from organic substances. The transformation of hemicellulose into sucrose in the cell walls of algae is especially important to suspension-feeding larvae. Because food passes through the manicotto quickly, the hydrochloric acid does not concentrate in the manicotto. Assuming that aggressive pepsin is not present, these protection mechanisms are sufficient. Inokuchi et al. (1991) found a pepsinlike enzyme in larval *R. catesbeiana* that may be cathepsin E, and Carroll et al. (1991) discussed pepsin in anuran larvae. I. Griffiths (1961) found evidence of proteolytic activity in early prometamorphic *Xenopus laevis* and *Rana temporaria* and tryptic activity in *R. ridibunda* up to Taylor/Kollros XX. Pepsin is present in the stomach of metamorphic *Xenopus laevis* at Nieuwkoop/Faber 63 (Fox 1984), beginning with Taylor/Kollros XXII in *R. ridibunda* (I. Griffiths 1961), and also in metamorphic *R. pipiens* (Kuntz 1924). Although pepsin or cathepsin may be present, many factors argue against a resorptive function of the manicotto (Ueck 1967), including the lack of direct contact between the food and the gland cells, the nonresorption of tryptophan blue, and the lack of alkaline phosphatase. Cytological comparisons of the gland cells with the resorbing cells of the intestine shows that the latter are more densely concentrated than gland cells, contain Golgi bodies, and metabolize low molecular weight substances.

Based on the systematic diversity of species studied, it is assumed that the manicotto glandulare is plesiomorphic for tadpoles and probably was present in their common ances-

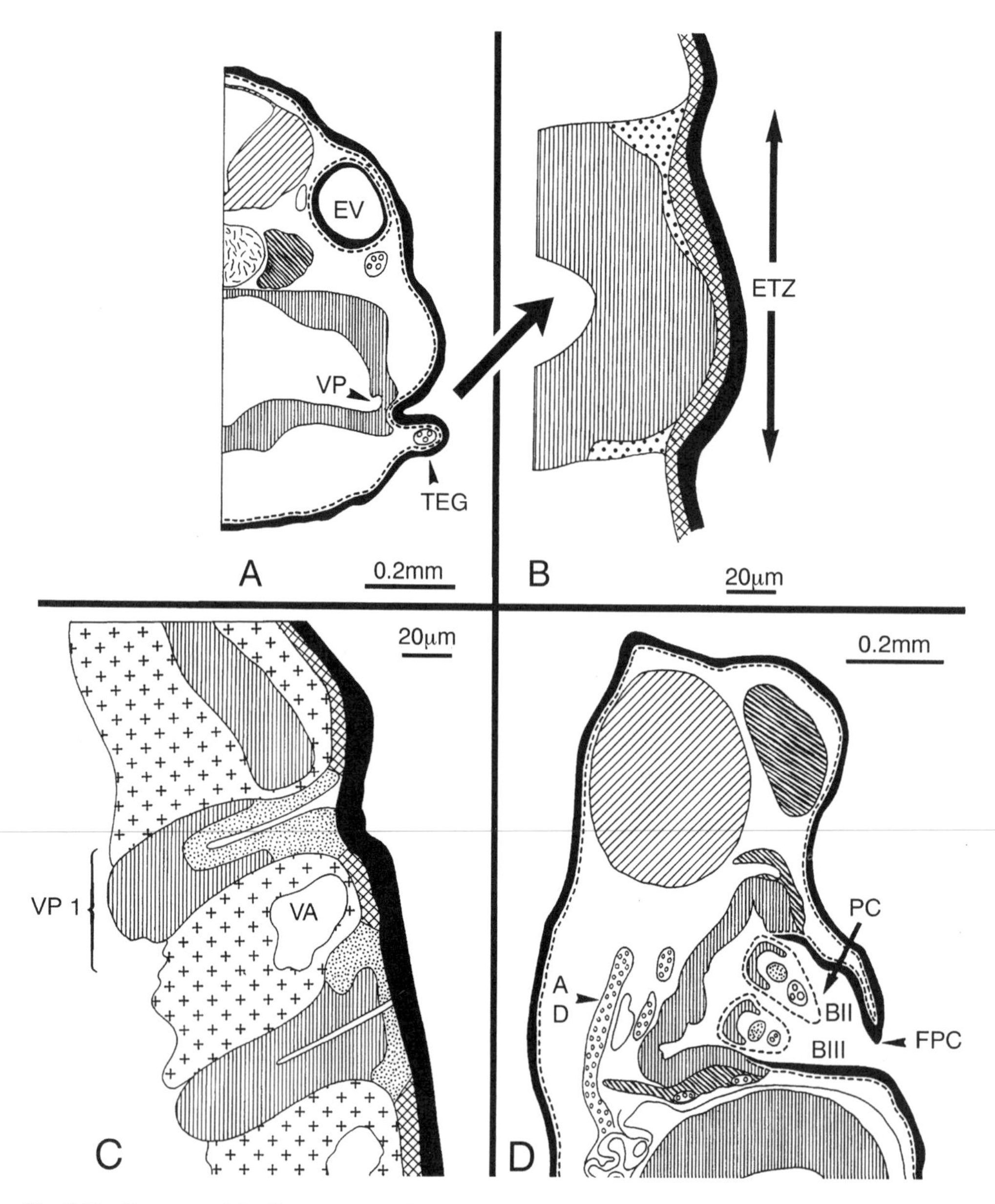

Fig. 5.23. Ontogeny of the filter apparatus of *Xenopus laevis*. (A) Cross section at the plane of the visceral pockets and the anlagen of transient epidermal gills (Nieuwkoop/Faber stage 38). (B) Magnified cross section from area in (A) of visceral pocket showing the sensory layer and endoderm. (C) Frontal section of left side of pharynx showing sensory layer and endoderm (Nieuwkoop/Faber 38). (D) Paramedial section of the head region, peribranchial chamber opened, anlagen of filter plates sectioned (Nieuwkoop/Faber 42). Schematics of TEM section of anlagen of filter row, yolky endoderm overlapped by the sensory layer of the epidermis: (E) an early hatchling (Nieuwkoop/Faber 40/41) and (F) an early tadpole (Nieuwkoop/Faber 45). Abbreviations: AD = aorta dorsalis, BII–III = branchial arches I to III, CT = connective tissue, CTC = connective tissue cell, ENC = endodermal cell, ETZ = ectodermal-endodermal transition zone, EV = ear vesicle, FPC = epidermal fold of peribranchial chamber (operculum), PC = peribranchial chamber, SL = sensory layer, SLC = sensory layer cell, TEG = transient epidermal gills, VA = visceral arch, VP = visceral pocket, VP1 = visceral pocket 1, and YV = yolk vacuoles. From Viertel (1991); reprinted by permission of Springer-Verlag.

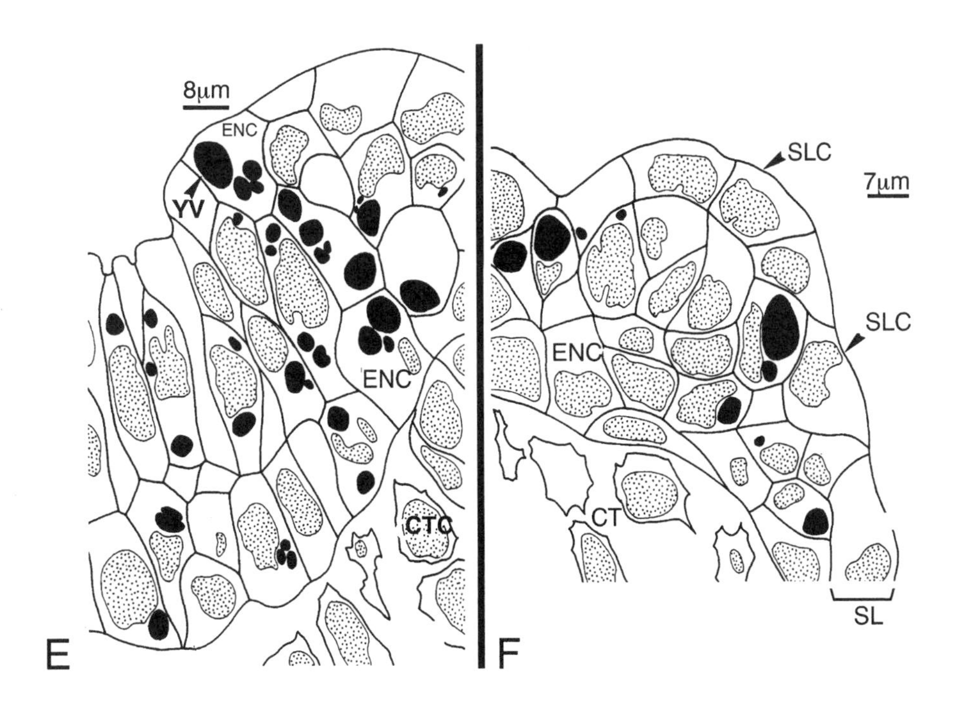

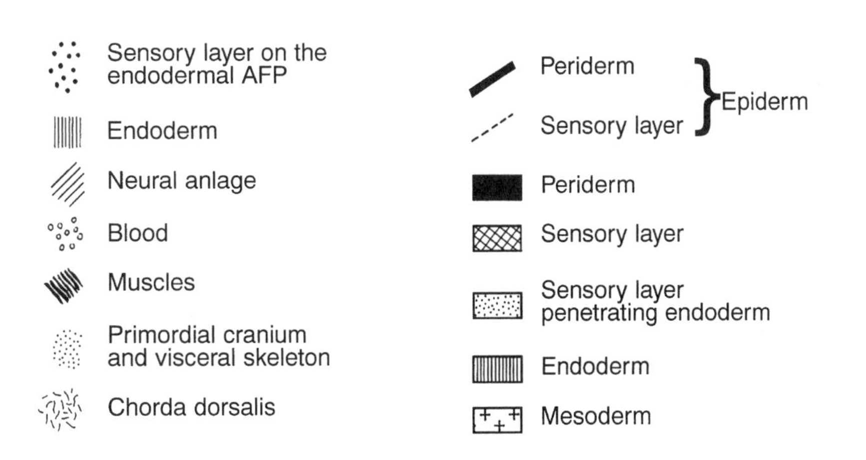

tor. Presumably the manicotto of carnivorous tadpoles is an apomorphous modification of the manicotto glandulare of the omnivorous suspension feeder (I. Griffiths 1961; Ueck 1967). The manicotto epithelium of the carnivorous larva of *Hymenochirus boettgeri* lacks ciliary cells and has a thick mucous layer containing acid phosphatase that protects the manicotto from ingested food. Branched tubular glands in the anterior part of the manicotto have cells similar to those in *X. laevis* and *R. temporaria,* but glands are absent in the posterior region. The posterior section of the manicotto is larger than any other part of the foregut (Ueck 1967) and functions as a storage area; this is functionally similar to the foregut in other animals that swallow and store their food whole. Like *Xenopus, H. boettgeri* lacks both muscularis mucosae and submucosae, and the longitudinal and circular muscles (Ueck 1967) are poorly developed. Other carnivorous forms develop a thick muscular coat, although there are no ciliary cells and few crypts (I. Griffiths 1961, *Lepidobatrachus laevis, Occidozyga laevis,* and *O. lima*). I. Griffiths (1961) suggested that the manicotto is a grinding chamber in these species.

The branched tubular glands differentiate at an early stage with the development of deep folds in the manicotto epithelium. Short tubules appear and then branch (Kopsch 1952, *Rana temporaria*). The synthesis of hydrochloric acid and the extrusion of lysosomes disprove the conclusion of Barrington (1946) that the ciliary cells, goblet cells, and *Schleimköpfchenzellen* (= mucous-headed cell) contribute to the formation of the gland cells. The lectin histochemistry of the manicotto epithelium is fundamentally different from that of the tubular glands (Ishizuya-Oka and Shimozawa 1990). I. Griffiths (1961, *Alytes obstetricans* and *Rana ridibunda*)

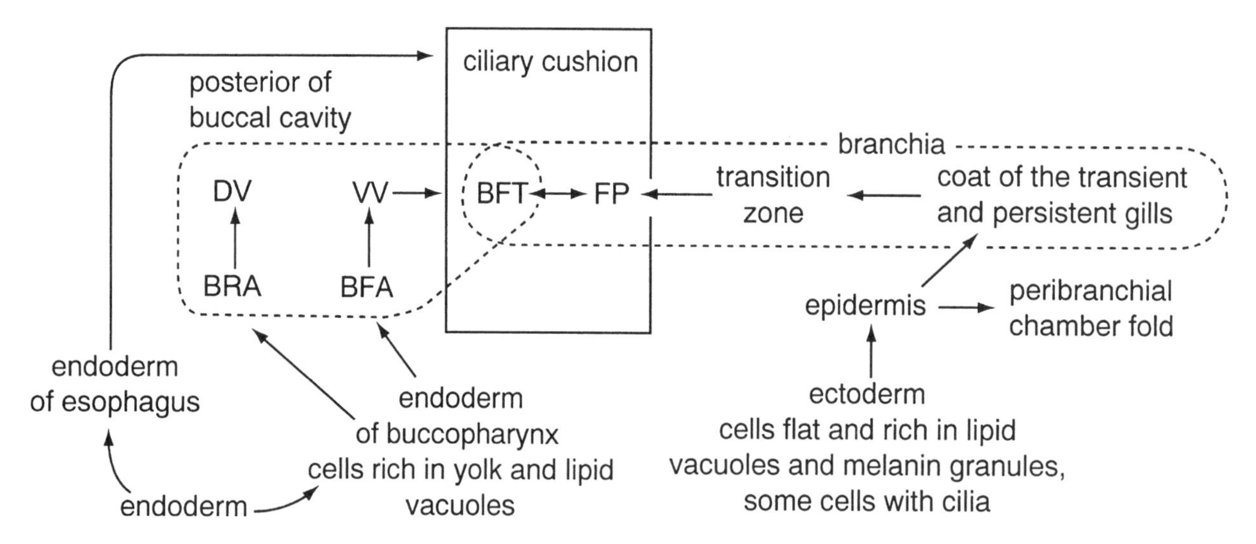

Fig. 5.24. Origin and morphogenesis of the filter apparatus. The filter plates (ectodermal-endodermal transition zone) and epidermal gills form one branchia. The parts of different origins (i.e., ciliary cushions, branchial food traps, and filter plates) are shown as a functional unit for mucus entrapment and ciliary transport. Symbols: arrows indicate change in morphogenesis, the box encloses functional units for production and transport of mucus, and dashed lines enclose unit of one organ. See table 5.2 for abbreviations. From Viertel (1991); reprinted by permission of Springer-Verlag.

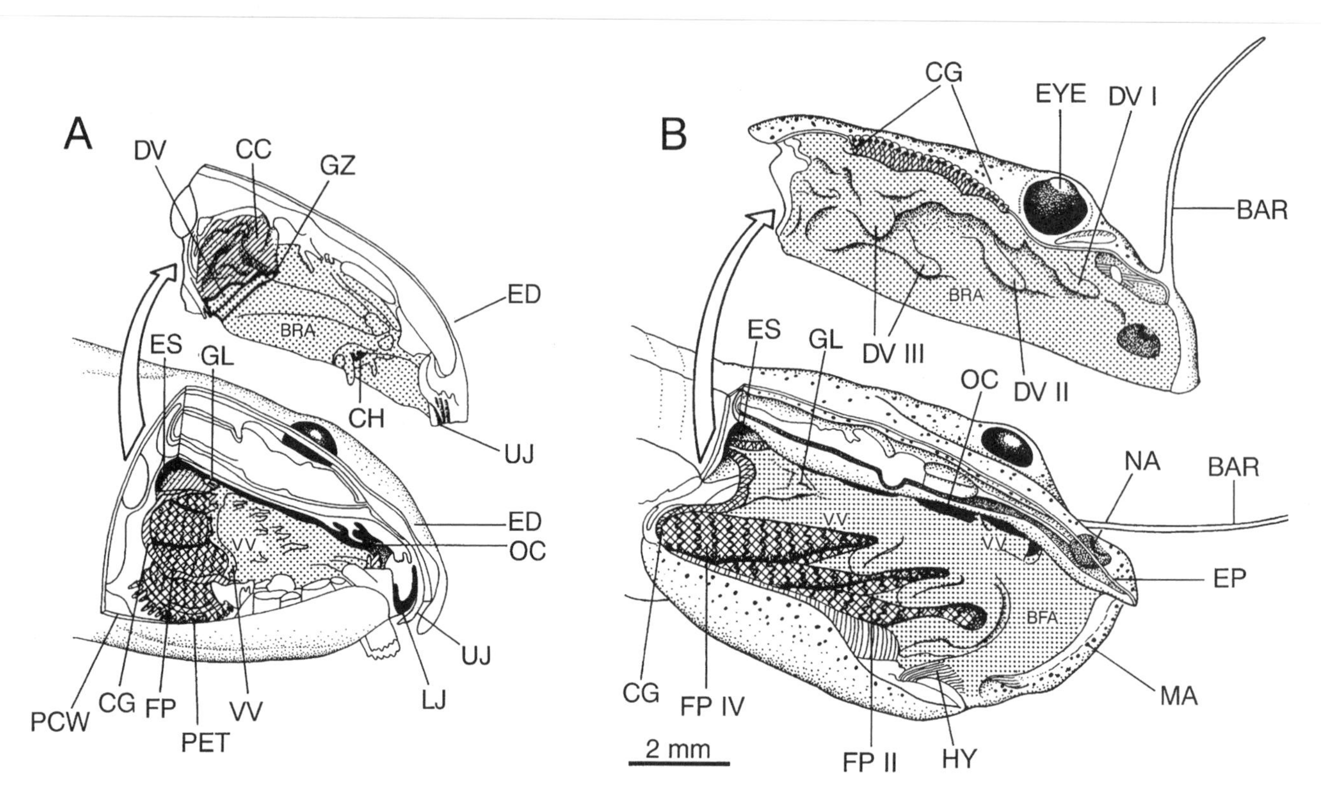

Fig. 5.25. Dorsolateral aspects of the buccopharynx of tadpoles. Right, dorsal halves of heads dissected and lifted. (A) *Rana temporaria* (Orton's larval Type 4, Gosner stage 36). (B) *Xenopus laevis* (Type 1, Nieuwkoop/Faber stage 54). Symbols: crosshatch = ectodermal-endodermal transition zone, stipple = endoderm, and diagonal lines = epithelium of esophagus (ciliary cushions, ciliary grooves). See table 5.1 for abbreviations. From Viertel (1991); reprinted by permission of Springer-Verlag.

suggested that the glandular part of the manicotto originates from elements of the ventral pancreatic anlage. Pancreatic tissue is found in the submucosa of the Dipnoi and Agnatha (Barrington 1942, 1957; J. G. Kerr 1919). Cells identical to those of pancreas cells in the manicotto apparently are not of pancreatic origin (Fox et al. 1972; Ueck 1967). The lytic regression of the gland cells and the neogenesis of a stomach during metamorphic climax show that the manicotto glandulare is a typical larval organ (see Carver and Frieden 1977 and Janes 1934 for further discussion of metamorphic regression of the gut).

Anuran larvae lack a pylorus, so the midgut begins immediately posterior to the manicotto and ends at the posterior opening of the hepatopancreatic duct (fig. 5.26A, C). This

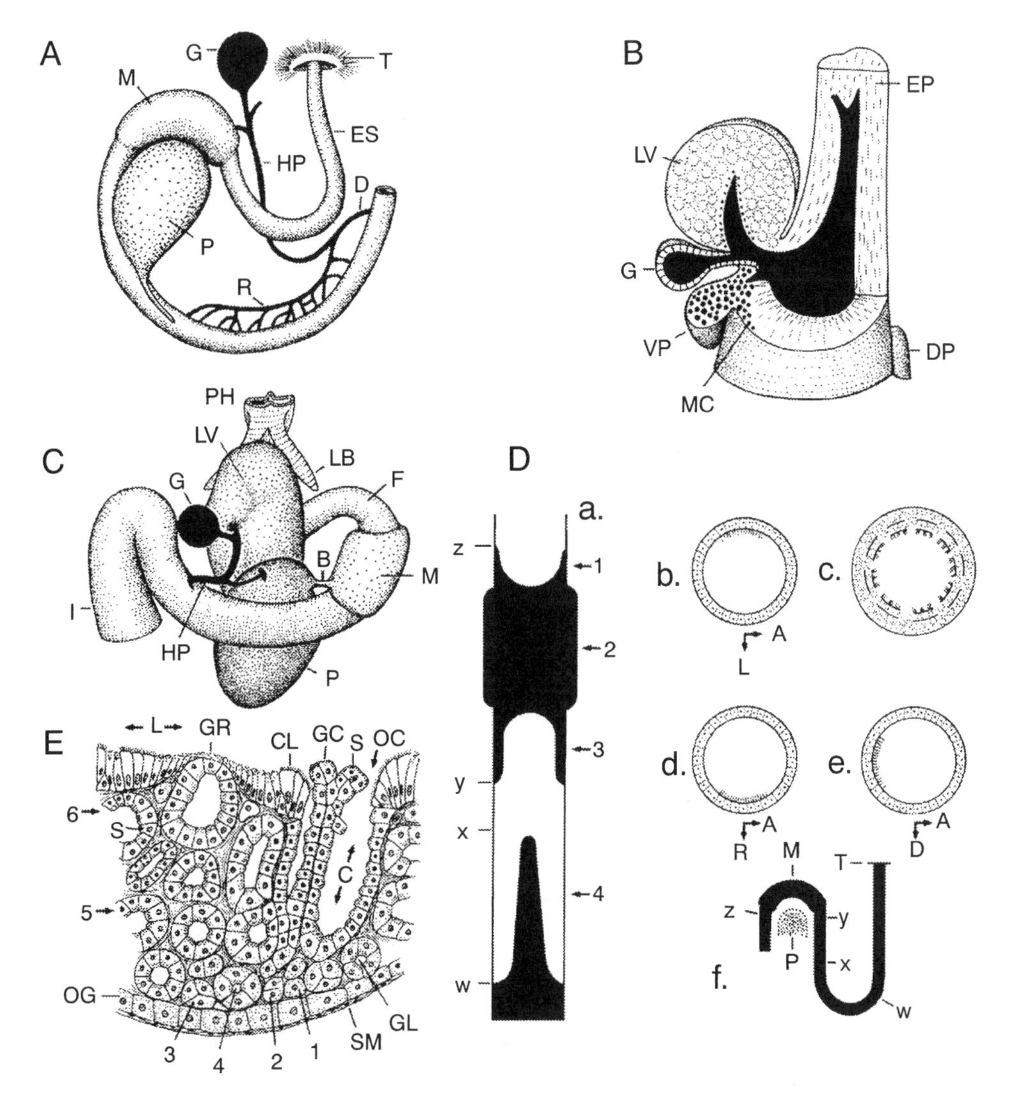

Fig. 5.26. Form and function of the foregut in *Rana ridibunda.* (A) Ventral view. (B) Foregut reconstruction; left, anterior half of gut tube removed (Shumway stage 22). (C) Ventral view of foregut (Shumway 25). (D) Ciliary patterns of the larval foregut: (a) Ventral view of foregut opened along ventral midline and edges reflected laterally, ciliary tracts in black, (b–e) diagrammatic cross sections at levels 1–4 in (a), and (f) lateral view of foregut showing topographic relationship of the planes w, x, y, and z in (a). (E) Manicotto of a larva (Taylor/Kollros stage V). Numerals 1–6 = successive stages in the development of manicotto glands. Abbreviations: B = manicotto-pancreatic bridge, C = manicotto crypt, CL = ciliated border of crypt opening, D = distal opening of hepatopancreatic duct, DP = dorsal pancreas, EP = esophageal plug, ES = esophagus, F = foregut, G = gallbladder, GC = gland cord, GL = gland lumen, GR = glandular rupture-point in the epithelium, HP = hepatopancreatic duct, I = intestine, L = gut lumen, LB = left lung bud, LV = liver, M = manicotto, MC = manicotto cells, OC = opening into manicotto crypt, OG = outer generative layer, P = pancreas, PH = pharynx, R = recurrent hepatopancreatic ductules, S = secretory granules, SM = serosa and attenuated muscle layer, T = transverse septum, and VP = ventral pancreas. Orientation marks as in fig. 5. 1. After I. Griffiths (1961); reprinted by permission of The Zoological Society of London.

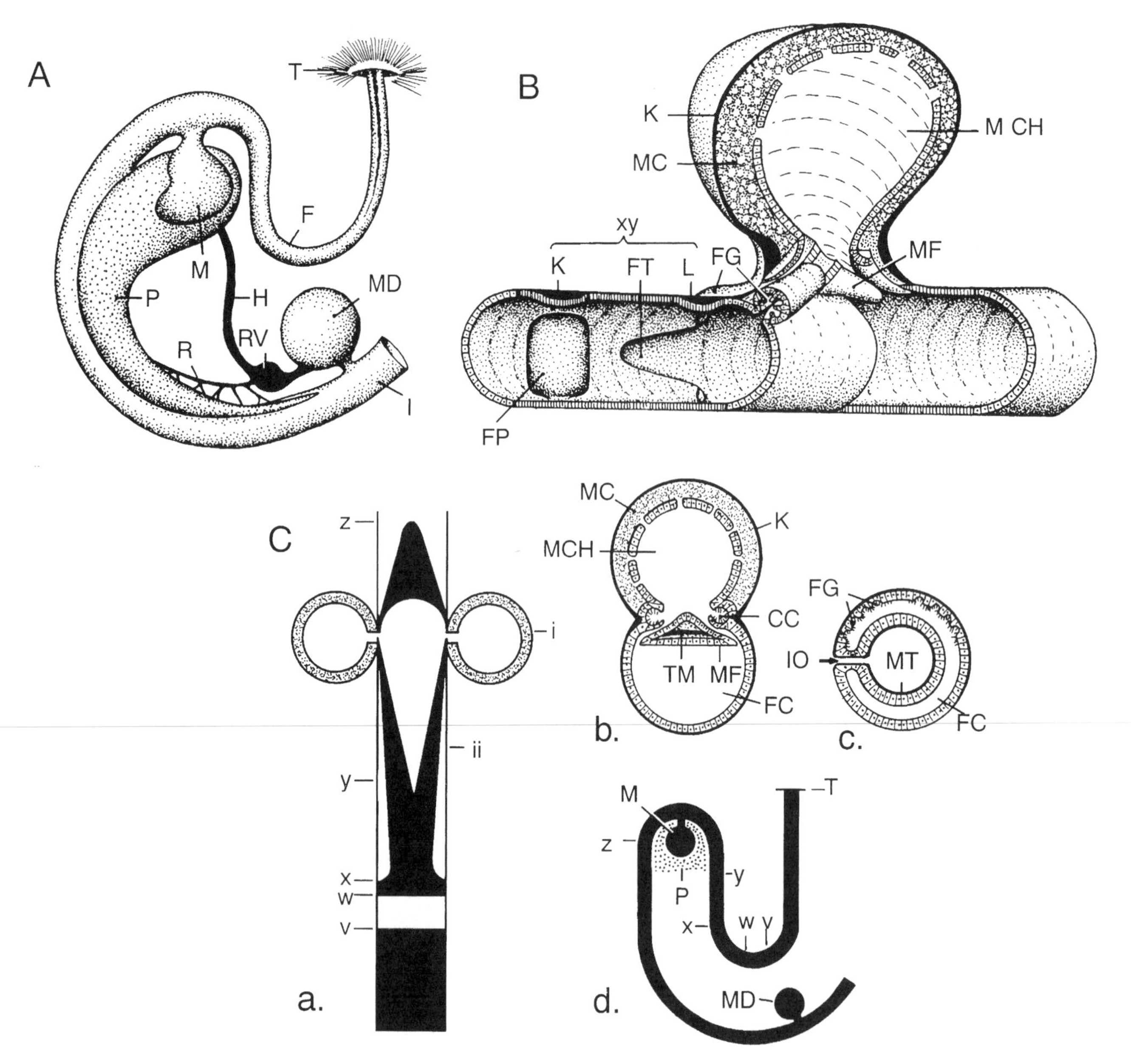

Fig. 5.27. Form and function of the foregut in *Philautus gryllus.* (A) Ventral view of foregut (Taylor/Kollros stage V). (B) Lateral view of foregut, half of manicotto capsule and parts of foregut wall removed (Taylor/Kollros V). Region marked as xy is that shown in (C) a and d. (C) Ciliation pattern of larval foregut: (a) Dorsal view of foregut opened along the dorsal midline and edges reflected laterally; ciliary tracts in black and manicotto tissue stippled, (b) diagrammatic cross section at "i" in (a), (c) diagrammatic cross section at "ii" in (a), and (d) lateral view showing topographic relationships of points indicated as v, w, x, y, and z in (a). Abbreviations: CC = lateral ciliated canals, F = foregut, FC = foregut cavity, FG = ciliated collecting grooves, FP = foregut plug, FT = foregut tongue, H = hepatopancreatic duct, I = intestine, IO = lateral opening of foregut tongue, K = muscular capsule of the manicotto, L = gut lumen, M = manicotto, MC = manicotto cells, MCH = manicotto chamber, MD = midgut diverticulum, MF = manicotto flap, MT = muscle of foregut tongue, P = pancreas, R = recurrent hepatopancreatic ductules, RV = reservoir, T = transverse septum, and TM = transverse muscles. After I. Griffiths (1961); reprinted by permission of The Zoological Society of London.

part of the second gut loop descends posterior to the manicotto and rises again anteriorly. The duodenum begins at the orifice of the hepatopancreatic duct and extends dorsally to form a loop that descends on the right and rises on the left. N. E. Kemp (1949, 1951) discussed how body shape influences the coiling pattern of the intestine (see Ichizuya-Oka and Shimozawa 1990 for development of gut connective tissue). Two spirals are formed, and a fibrous or muscular infolding of the intestinal wall (i.e., typhlosolis; Ueck 1967; see Rovira et al. 1993; Sesama et al. 1995) enlarges the anterior secretory and posterior resorptive surface of the intestine, controls clearance time of food through the gut, causes a narrowing of the intestinal passage, and slows mixing of the pancreatic enzymes with the food mass. The orifice of the hepatopancreatic duct lies next to the anterior end of the typhlosolis. The lack of secretory digestive granules in the duodenal cells suggests that this part of the intestine does not produce its own digestive enzymes but resorbs substances metabolized by pancreatic enzymes (Fox et al. 1970). Typical duodenal cells are columnar, mitochondria-rich, microvillous, ciliated cells (Fox 1981; Fox et al. 1970; Jorquera et al. 1982; Ueck 1967, *Xenopus laevis, Rhinoderma darwinii,* and *R. rufum*). Calciform cells have been observed in *R. rufum.* Concentrations of lipid vacuoles are high during embryonic and early premetamorphic stages (Jorquera et al. 1982). The yolk and lipid vacuoles and the cilia in *Xenopus* disappear by Nieuwkoop/Faber 65.

The boundary between the duodenum and the ileum is indistinct. The ileum extends ventrally in a left-handed outer spiral and then from posterior to anterior in 4–8 counterclockwise coils (figs. 5.11–5.13). There are no diverticula, villi, or caeca in the ileum; and the muscularis is thin. The diameter of the intestinal tube and the thickness of the epithelium decreases posteriorly from 70 to 10 μm. Goblet and microvillous adsorptive cells are present, and endoreduplication is common (Ueck 1967). The ileum resorbs metabolites digested by pancreatic enzymes, and the microvillous cells probably secrete various enzymes (Ueck 1967). High catalase activity found (Dauça et al. 1980, *R. catesbeiana*) in the "anterior intestine" probably relates to the ileum. Enzyme activity decreases during metamorphic climax and resumes after metamorphosis.

The colon (= rectum, coprodaeum) comprises the last two loops of the intestinal spiral (Ueck 1967, *X. laevis;* figs. 5.11B and 5.13) and can be distinguished from the ileum by its greater diameter. Intense enzyme activity and the presence of microorganisms in a characteristic layer near the epithelial cells suggest the symbiotic breakdown of plant cell walls. Symbionts are not found when larvae are fed only yolk (Ueck 1967), and Piavaux (1972) found no evidence of cellulases in the intestinal tract.

The rectum is positioned longitudinally and dorsally. Its cuboidal, microvillous, ciliary cells (Nieuwkoop/Faber 47 *X. laevis,* Fox et al. 1972) indicate resorptive and transport functions, especially water (figs. 5.11–5.13). Microvillous cells rich in mitochondria, smooth vesicles, and ribosomes are found exclusively in the anterior rectum in *X. laevis* (Nieuwkoop/Faber 47); during stages 54–58, they are present in the entire rectum (Fox et al. 1972; Ueck 1967).

The rectum, bladder, and ureters lead into the cloaca. In contrast to the intestine from the foregut to the colon, the cloaca is not exclusively endodermal in origin (B. I. Balinsky 1981; Monayong Ako'o 1981). Mesoderm contributes to the nephric ducts. In early larvae (Nieuwkoop/Faber 47), the urodaeum, the dorsal orifices of the nephric ducts, and the posterior proctodaeum are visible. The urodaeum is situated at the posterior margin of the coelom, and the proctodaeum is beyond the splanchnic wall of the splanchnotome (Van Dijk 1959, *Ascaphus truei*). In *A. truei* (Nieuwkoop/Faber 51) a ventral thickening of the urodaeum indicates the development of the bladder anlage. This grows ventrally, so that an externally unpaired and internally divided, saclike extension is present by Nieuwkoop/Faber 56. The bladder remains externally paired at stage 58, and both parts have a common connection to the ventral urodaeum (Van Dijk 1959).

The liver and gallbladder (fig. 5.26A, B, C) lie between the two foregut loops and the midgut loop. The liver surrounds the manicotto and lies ventral and posteroventral to the trachea and esophagus. A large gallbladder is in the anterior region of the liver, often extending beyond the liver in many species. The short neck of the gallbladder and the ductus cysticus lead into the bile duct, which courses laterally along the liver margin and crosses the liver posteriorly. At the pancreas, the bile duct follows the course of the two pancreatic ducts for a short distance before all three ducts merge to form the hepatopancreatic duct.

The development of the liver begins as a ventromedial diverticulum of the midgut anlage between the heart and duodenum anlagen (Nieuwkoop/Faber 35–36, *Xenopus laevis*). The pocketlike primary hepatic cavity corresponds with the gut cavity and has an outlet into the submesodermal space. Anterior diverticular folds appear and narrow all but the posterior region of the hepatic cavity (Nieuwkoop/Faber 37–40). At the same time, the area connected to the midgut narrows and forms the bile duct (Nieuwkoop/Faber 41). The remaining parts of the primary hepatic cavity form the main hepatic ducts that transport bile from the liver to the gallbladder. Remnants of the primary liver cavity form the gallbladder and ductus cysticus (B. I. Balinsky 1981; Nieuwkoop and Faber 1956).

Folds of the anterior diverticulum break into strands of cells corresponding to the primary hepatic cavity that are traversed by blood vessels and the sinuses of the vitelline veins (B. I. Balinsky 1981). The basic histological organization of the larval (and adult) liver consists of a parenchymatic frame of hepatocytes with an inner surface lined by endothelial and Kupffer cells. The perisinusoidal space (= space of Disse) lined by endothelial cells lies between the hepatocytes and the endothelial cells.

Precursor cells of liver parenchyma are arranged in strands, rich in yolk and lipid vacuoles, held together by desmosomes, and often undergo mitosis. These cells form lobules, and the enclosed intercellular space becomes a bile capillary. The cells increase in size, there is a marked increase in cristate mitochondria (yolk mitochondria), and some rough ER is present (Gosner 20–23). The yolk content is exhausted by Gosner 24, but lipid droplets are present.

Liver tissue remains firm during and after metamorphosis, and the spaces of Disse contain numerous microvilli and large nucleoli. Parenchymal cells of *Rana catesbeiana* change in fine structure in response to thyroid hormone (Cohen et al. 1978 and Fox 1984, *Bufo bufo* and *R. nigromaculata*). Peroxisome and microperoxisome densities increase in vitro under thyroxine influence, and catalase activity is increased by four times (Dauça et al. 1980; Dauça and Hourdry 1983). Hepatocyte cytology remains unchanged in *X. laevis* during metamorphosis (Fox 1984).

Columnar cells lining the lumen of the gallbladder have short microvilli, ribosomes, numerous smooth-walled vesicles, lipid vacuoles, a high density of mitochondria at the apical pole, and invaginate intercellular junctions. The wall of the bile duct is lined by ciliated and microvillous cuboidal cells and a small number of ciliary cells with microvillous stubs. A fuzzy coat suggests that secretion is probable (Toner and Carr 1968). Ciliary cells extend posterior to the nonciliated neck as far as the orifice of the hepatopancreatic duct (Fox et al. 1970, *X. laevis* at Nieuwkoop/Faber 47).

The pancreas (figs. 5.26A, C, and 5.27A, B) lies inside the second foregut loop with its distal region adjacent to the manicotto. The two pancreatic ducts pass to the left and right of the subintestinal vein and subsequently flow into the bile duct (Fox et al. 1970). In many species the hepatopancreatic duct branches distally into hepatopancreatic ductules and a distal hepatopancreatic orifice that marks the end of the midgut.

The pancreas develops from three endodermal anlagen. The first is a diverticulum at the connection between the primary hepatic cavity and the midgut (future orifice of the bile duct; Nieuwkoop/Faber 37–38, *X. laevis*). This diverticulum grows into two parts that surrounds the bile duct. The third part develops next to the dorsal midgut wall and, by Nieuwkoop/Faber 39, contacts the first anlage and separates from the midgut wall. The ventral pancreatic anlagen each differentiate into a pancreatic duct; the dorsal anlage does not. Differentiation of acinar cells starts at Nieuwkoop/Faber 41 in *X. laevis*. A single columnar cell layer lines the ducts (Fox 1984; Fox et al. 1970). Concentrically arranged pyramidal cells form acini and line the branching ducts. The apical pole of each cell projects into a common central lumen (Fox et al. 1970). In *Discoglossus* the exocrine pancreas does not differentiate until the β cells of the endocrine pancreas are visible (after Gallien/Houllion 28; Beaumont 1970).

It is not known if the larval exocrine pancreas synthesizes lipases, nucleases, peptidases, and proteases as in adults and other Amphibia (Karlson 1972). During metamorphic climax (Nieuwkoop/Faber 61 of *X. laevis*), 70%–80% of the total size and weight (Atkinson and Little 1972; Race et al. 1966 in Fox 1984) of the exocrine tissue degenerates. After metamorphosis the pancreas is reconstructed with a higher proportion of endocrine cells.

Urogenital System

The excretory and reproductive systems of anurans show striking morphological interrelationships and dependencies as a result of common mesodermal (i.e., intermediate plate = nephrotome or mesomere) origins and similar transport functions. The nephrotome, despite its irreversible determination in the late neurula (Bossard 1971, *Bombina variegata;* Burns 1955; Gipouloux and Hakim 1978, *Bufo bufo;* Hakim and Gipouloux 1978; *B. bufo;* Tung 1935, *Discoglossus pictus*), has few developmental options in its responses to the surrounding endoderm and chordamesoderm. The onset and character of development of the urogenital system are modified by the position of its blastema in the urogenital field. The spatiotemporal appearance of the reproductive, pronephric, and opisthonephric system must therefore be regarded as an aspect of regional differentiation in the larvae. The close morphological relationship of the excretory and reproductive systems is mainly based on embryonic propinquity, especially in males, where kidney tubules transport sexual products. The anatomy and development of the urinary and reproductive systems are treated in sequence below.

Urinary System: Anatomy

Most anuran larvae live in water, but even those larvae that develop outside of water occur in moist environments, and fluid exchange occurs between the larvae and the environment. The maintenance of acid-base balance, electrolytes, and fluids requires coordination between renal function and ion and water movement across the digestive system, gills, skin, and in later stages, the urinary bladder (Bentley 1971; Deyrup 1964). Tadpoles are good osmoregulators (Candelas and Gomez 1963, *Leptodactylus albilabris* and *Limnonectes cancrivorus;* M. S. Gordon and Tucker 1965, *Limnonectes cancrivorus;* R. M. Jones 1980, *Spea multiplicata;* McClanahan 1975; see chap. 8). The nitrogenous excretion rate is relatively high soon after hatching, declines during larval development, and increases again at metamorphosis. Larval anurans usually are ammoniotelic, and in response to enzymatic changes in the liver (J. B. Balinsky 1970; Cohen 1970, *Rana catesbeiana*), tadpoles shift to ureotely at metamorphic climax (Cohen 1966, *Hyla geographica*, 1970, *R. catesbeiana;* M. S. Gordon and Tucker 1965, *Limnonectes cancrivorus* Munro 1953, *Bufo bufo* and *R. temporaria;* Tahin et al. 1979, *Hyla geographica;* Zamorano et al. 1988, *Caudiverbera caudiverbera*). In the uricotelic *Phyllomedusa sauvagii* (Shoemaker and McClanahan 1982), *Scaphiopus couchii* and *Spea multiplicata* (R. M. Jones 1980), ammoniotelic *Pipa carvalhoi* (Tahin et al. 1979), and probably several leptodactylids (Candelas and Gomez 1963, *Leptodactylus albilabris;* Shoemaker and McClanahan 1973, *Leptodactylus bufonius*) and myobatrachids (A. A. Martin and Cooper 1972, *Geocrinia victoriana*), urea predominates in the premetamorphic stages (see Ashley et al. 1968, *Rana catesbeiana*). In *Xenopus*, the ammonia excretion rate remains almost unchanged throughout development (J. B. Balinsky et al. 1967, 1972; Munro 1953; see P. M. Wright and Wright 1996, *R. catesbeiana*).

The excretory system of anuran larvae likely evolved from a hypothetical holonephros (Price 1897) derived from the entire nephrotomal plate and with one nephron derived from each nephrotome. Differentiation of the nephrons progresses in an anteroposterior direction, and anterior nephrons are less organized than those that develop later. Thus, in anamniotes, a hypothetical uniform holonephros may have developed into a pro- and opisthonephros adapted to special developmental and functional conditions (Ciantar 1983; Fraser 1950; R. W. Hall 1904; Richter 1992; Van Den Broek et al. 1938; M. H. Wake 1979). Both types of kidneys are moderate-size structures lying retroperitoneally against the dorsal body wall (fig. 5.28A).

The paired, triangular pronephric kidneys (= head kidney) generally are located ventral to postotic somites III–V

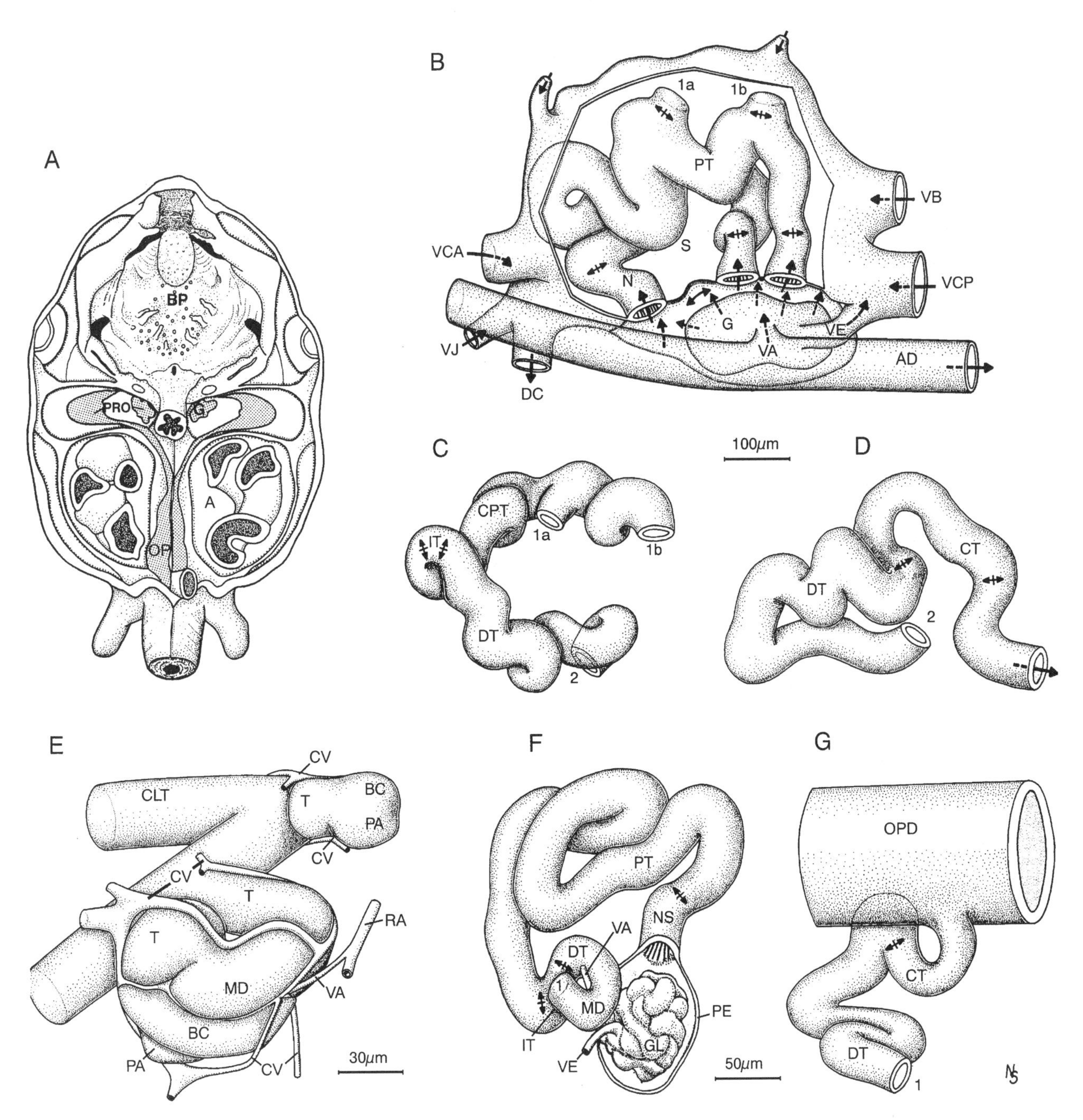

Fig. 5.28. Anatomy and development of the larval kidney. (A) Schematic representation of the configuration of the larval kidneys. (B–D) Pronephros of *Rana esculenta* (Gosner stage 27); 1a and 1b in (B) connect with 1a and 1b in (C), and 2 in (C) connects with 2 in (D). Development of opisthonephros of *R. esculenta.* (E) Developing nephrons showing renal vesicle (upper right) and S-shaped body *(lower right).* (F–G) Young opisthonephric nephron; 1 in (F) connects with 1 in (G). Arrows indicate direction of flow; double-headed arrows mark boundaries of tubular parts with an open lumen that connect elsewhere. See table 5.3 for abbreviations.

(Fox 1962b) or slightly anterior (M. I. Michael and Yacob 1974, *Hyla arborea savignyi*). This kidney begins to function at the muscular response stage (Cambar 1947a, *Rana dalmatina;* Rappaport 1955, *R. pipiens*) shortly after the pronephric duct joins the cloaca (Deyrup 1964; Jaffee 1954a, *R. pipiens;* Richter 1992, *R. esculenta*). This system functions during most of larval life and gradually disappears during metamorphic climax.

Each kidney consists of a coiled tubular complex, a large posterior cardinal sinus, interstitial hematopoietic tissue, and a glomus (figs. 5.28A and 5.29B). The cell structure of the tubular complex is similar among anurans (Ciantar

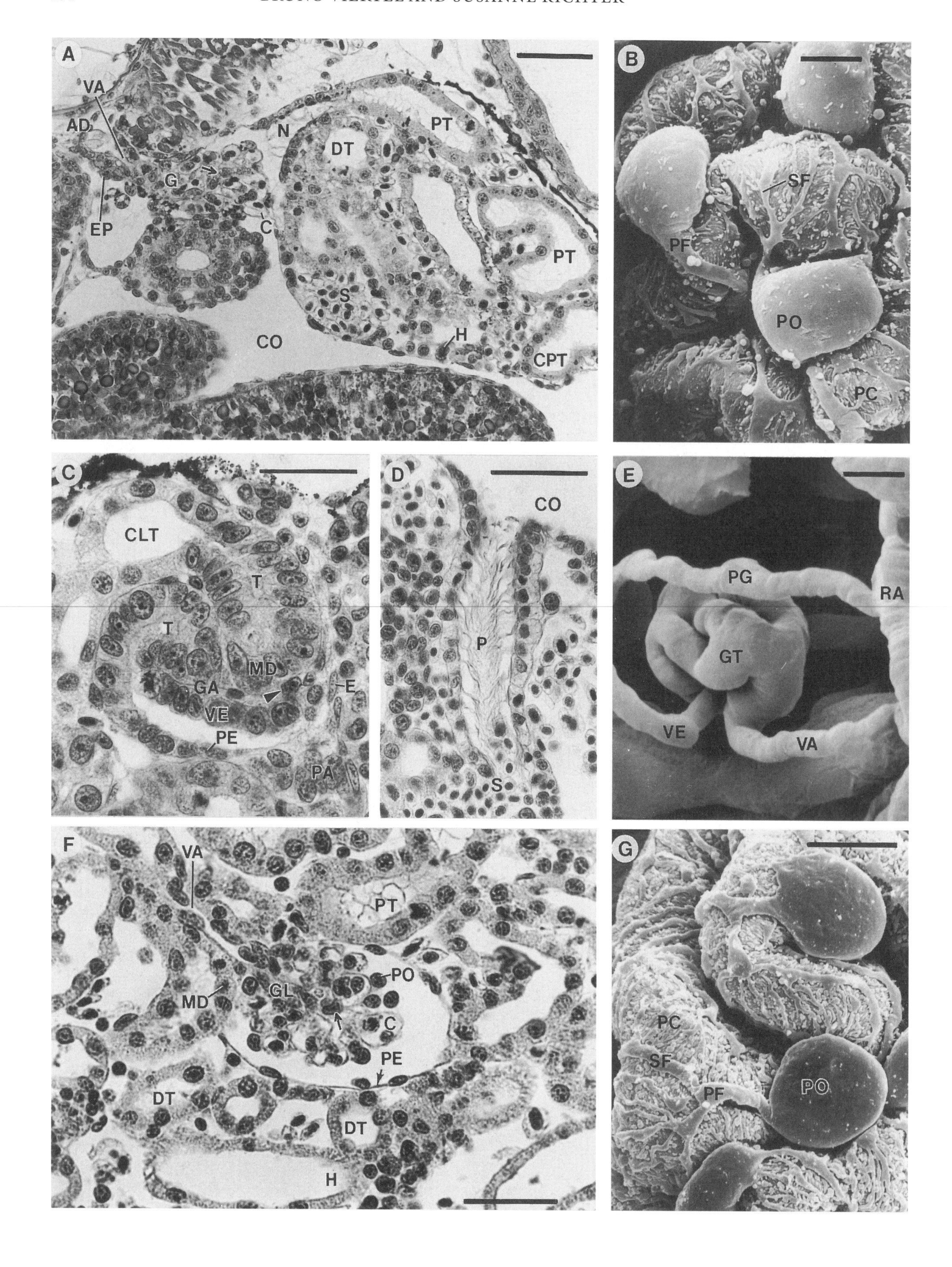
A
VA
AD
N
PT
DT
G
EP
C
PT
S
CO
H
CPT
B
SF
PF
PO
PC
C
CLT
T
T
MD
GA
E
VE
PE
PA
D
CO
P
S
E
PG
RA
GT
VE
VA
F
VA
PT
PO
MD
GL
C
PE
DT
DT
H
G
PC
SF
PF
PO

Fig. 5.29. Anatomy and development of the pro- and opisthonephros. Pronephros in *Rana esculenta:* (A) Glomus and tubular complex (Gosner 27); arrow indicates a mesangial cell; scale = 50 μm. (B) Visceral podocytic epithelium of the glomus; scale = 5 μm. Opisthonephros of *Rana esculenta* (C–D and F–G), and *Bufo bufo* (E): (C) S-shaped body; arrow indicates an invading mesangioblast; scale = 30 μm. (D) Peritoneal funnel; scale = 50 μm. (E) Immature glomerulus of *Bufo bufo;* scale = 20 μm. From Naito (1984); reprinted by permission of Niigata University School of Medicine. (F) Glomerulus and tubular complex; arrow indicates mesangial cells; scale = 40 μm. (G) Visceral podocytic epithelium of glomerulus; scale = 10 μm. See table 5.3 for abbreviations.

1983, *Xenopus laevis;* Fox 1963, 1970, 1971, *Rana temporaria;* Fox and Hamilton 1971, *X. laevis;* Gibley and Chang 1966, *R. pipiens;* Hurley 1958, *R. sylvatica;* Jaffee 1954a, *R. pipiens;* Kindahl 1938, *R. temporaria;* Matsurka 1935, *Bufo formosus;* M. I. Michael and Yacob 1974, *B. v. viridis, H. a. savignyi,* and *R. r. ridibunda;* Richter 1992, *R. esculenta;* Schluga 1974, *B. bufo* and *Pelobates fuscus*). Each tubular complex (figs. 5.28B–D, 5.29A) is composed of three nephrostomal segments that lead from the coelom, via nephrostomes, to three short proximal tubule segments. The cuboidal epithelium of the ciliated nephrostomal tubule segment possesses straight intercellular junctions and irregularly shaped mitochondria. The proximal segments unite to form a common proximal tubule segment. Micropinocytotic vesicles are visible between the microvilli of the apical cell membrane of the proximal tubule cells. Proximal tubular reabsorption, large numbers of cellular infoldings and lateral interdigitations, and the presence of enzyme systems have been demonstrated (Fox 1984, *Xenopus laevis;* Gérard and Cordier 1934a, b, *Alytes obstetricans, B. bufo, Discoglossus pictus,* and *R. temporaria;* Hah 1974, *Bombina orientalis;* Jaffee 1954b, 1963, *R. pipiens;* Piatka and Gibley 1967, *R. pipiens*). Golgi networks are scarce in *R. temporaria* (Fox 1970) but always present in *X. laevis* (Fox and Hamilton 1971). The common proximal tubule segment connects via an intermediate ciliated tubule segment that more distally becomes the connecting tubule segment. The distal tubule segment has a flat, striated epithelium where water is reabsorbed (Gérard and Cordier 1934a). The connecting tubule empties into the pronephric duct.

The glomus is a slightly elongated mass of capillaries and intercapillary cells within a spongy mesangial matrix and covered by a visceral podocytic epithelium (fig. 5.29B). Podocytes have well-defined, interdigitating primary, secondary, and tertiary processes and pedicels. The afferent vessel of the capillary tuft is derived from the dorsal aorta, and the efferent vessel is derived from the posterior cardinal vein (figs. 5.28B and 5.29A). Epithelioid-like cells around the afferent vessels may regulate arterial pressure (Richter 1992; Richter and Splechtna 1990, *Rana esculenta*). The space between the various tubule segments is occupied by sinusoids of the posterior cardinal vein that form an extensive vascular network around the tubules. The pronephric sinus plays an important role in hemopoiesis (K. L. Carpenter and Turpen 1979, *Rana pipiens;* Foxon 1964, G. Frank 1988, *R. esculenta;* Hollyfield 1966, J. D. Horton 1971b, *R. pipiens;* Latsis and Saraeva 1980, *R. temporaria;* Płytycz and Bigaj 1983; Turpen et al. 1981a, b; Turpen and Knudson 1982, *R. pipiens*).

The pronephric kidneys are drained by the pronephric ducts (= Wolffian ducts), which also serve as the vasa deferentia. The ducts originate from the anterior mesodermal duct primordium along with a minor contribution from the ectodermal cloacal diverticulum (Fox 1963; Fox and Hamilton 1964; Poole and Steinberg 1977, *Xenopus laevis;* Richter 1992, *R. esculenta*). The duct epithelium of polygonal cells joined by junctional complexes is underlain by a basal membrane and a fibrous extracellular matrix. Except for a few short microvilli, the lumen wall is almost smooth (Ciantar 1983, *X. laevis;* Delbos 1975, *R. dalmatina;* Fox and Hamilton 1971, *X. laevis*).

The oval, opisthonephric kidneys are generally asymmetrical and dissimilar in developmental rate and shape (Kindahl 1938, *R. temporaria;* Nodzenski et al. 1989, *Leptobrachium montanum, Leptolalax gracilis,* and *Megophrys nasuta*). These kidneys lie to either side of the dorsal mesentery, are covered ventrally by the adrenal glands, and start to function shortly before hind limb development (Richter 1992, *R. esculenta*). The opisthonephric kidneys are drained by the pronephric ducts (= Wolffian or opisthonephric ducts) and can be divided into lateral (primary nephrons), medial (younger nephron generations), and dorsomedial (nephric anlagen and opisthonephric blastema) zones. Malpighian corpuscles are always in the ventral zone. Ciliated peritoneal funnels restricted to the ventral kidney margin are numerous and prominent during larval life.

An opisthonephric nephron (figs. 5.28F–G and 5.29F) includes a neck segment at the urinary pole and an intermediate segment interposed between the thicker proximal and distal tubule segments. Both have cuboidal, ciliated epithelia. The proximal segment, which reabsorbs solutes, is comprised of cuboidal, granulated cells with a dense luminal brush border, and the epithelium has a permeability gradient (Gérard and Cordier 1934b, *Alytes obstetricans, Bufo bufo, Discoglossus pictus,* and *Rana temporaria*) that correlates with the pattern of tight junctions along the proximal epithelium in adult *R. esculenta* (Taugner et al. 1982). This segment shows alkaline phosphatase activity (Hah 1976, *Bombina orientalis*). The cuboidal cells of the distal tubule possess a smooth, apical plasma membrane and basal striations at Nieuwkoop/Faber 49. The connecting tubule segment and the collecting duct of larval and adult pipids are characterized, apart from the occurrence of light cells, by bottle-shaped cells (Bargmann 1937, 1978; Jonas and Röhlich 1970; Jonas and Spannhof 1971) differentiated from light cells at Nieuwkoop/Faber 49. Because their secretory material disappears in saltwater-acclimated tadpoles, these cells probably function in ion regulation (L. Spannhof and Jonas 1969). The cells contain a highly developed intercellular channel system filled with acid mucopolysaccharides (L. Spannhof 1956; L. Spannhof and Dittrich 1967). Each renal tubule is associated with a large oval Malpighian corpuscle (fig. 5.29F). The visceral epithelium of Bowman's capsule is composed of spherical podocytes (fig. 5.29G), and the cells of the parietal epithelium are flat. The afferent artery, surrounded by epithelioid-like cells (Richter 1992, *Rana esculenta*), branches into a large glomerular tuft directly from the renal artery. The efferent vessel gives rise to a rich capillary network ensheathing the tubular epithelium (Naito 1984, *R. catesbeiana;* Richter 1992, *R. esculenta*). Data are not available on the ultrastructure of the larval juxtaglomerular apparatus.

Nephrons are frequently innervated by myelinated nerve fibers. Occasionally, synaptic, vesicle-laden varicosities with dense granules are associated with both the arteriolar smooth muscle cells and the distal tubule cells (Tsuneki et al. 1984, *R. catesbeiana*). The nephrons are surrounded by the venous sinus of the renal portal system where waste material and excess water are removed (Engels 1935, *R. temporaria;* Millard 1949, *X. laevis;* Richter 1992, *R. esculenta*). Arterial blood flows from renal arteries branching from the dorsal aorta (Millard 1945, *X. laevis*). Lymphomyeloid (J. D. Horton 1971b, *R. pipiens;* Manning and Horton 1969, *X. laevis;* Płytycz and Bigaj 1983) and granulopoietic tissues are extensive (G. Frank 1988, *R. esculenta;* Meseguer et al. 1985, *R. ridibunda*) in the intertubular spaces.

Peritoneal funnels usually present in the opisthonephros (fig. 5.29D) open into the coelom, but their connection with Bowman's capsule is often lost before metamorphosis. In most larvae the peritoneal funnels open into the opisthonephric venous sinus, but in *Alytes obstetricans* and *Discoglossus pictus* (Gérard and Cordier 1933, 1934a) and *X. laevis* (De Waal 1973; Kunst 1936), they retain a connection with the nephrons throughout larval life. Peritoneal funnels are densely ciliated and carry liquid or particles from the coelomic cavity into either the veins or the nephrons (Gérard and Cordier 1933, 1934b).

The urinary bladder originates from a diverticulum extending anterolaterally from the cloaca at late prometamorphosis and then increases in developmental rate at the beginning of metamorphic climax (Nieuwkoop and Faber 1967, *Xenopus laevis,* Nieuwkoop/Faber 59; T. L. Powell and Just 1985, 1987, *Rana catesbeiana,* Taylor/Kollros XX–XXV, Gosner 41/42–46; Richter 1992, *R. esculenta,* Gosner 40). The bladder is suspended in the coelomic cavity by two strands of connective tissue and communicates with the cloaca through a wide orifice. The flat epithelium has a smooth luminal surface. A thin muscular layer develops within the bladder wall at metamorphic climax. The bladder accomplishes at least passive transport of water and nitrogenous excretion products and active sodium transfer. The major permeability barriers are the apical and lateral plasma membranes (DiBona et al. 1969, *Bufo marinus*). Net sodium transport and osmotic water flow across the bladder is stimulated by hormones (Leaf 1965). Water retention is lowest in young tadpoles and increases at metamorphic climax (Alvarado and Johnson 1966; Bentley and Greenwald 1970; T. L. Powell and Just 1987, *Rana catesbeiana*).

Urinary System: Development

The development of the paired pronephric kidneys is a progressive event beginning within an anteroposterior wave of tubulogenesis. The first pronephric anlage in *Xenopus* (Nieuwkoop/Faber 21; Richter 1992, *Rana esculenta,* Gosner 16–17) is a slight thickening of the intermediate plate. Tubulogenesis involves small, dorsolateral, metameric outgrowths of the somatic layer of the nephrotomes (fig. 5.29); these are initially blind, then lengthen, bend backward, and unite at their distal ends to form a longitudinal duct that differentiates into the tubular and pronephric ducts. The pronephric tubules increase in length and become highly convoluted. Cellular differentiation of the tubular complex and the pronephric duct occurs in the opercular-fold stage (*Rana esculenta,* Gosner 23; *Xenopus,* Nieuwkoop/Faber 33–34). The tubular complex consists of three ciliated nephrostomal tubule segments that open into the coelom via three nephrostomes (figs. 5.28B and 5.29A). In some species, a vestigial fourth nephrostome may be present (Fox 1962a, *R. temporaria;* Schluga 1974, *Bufo bufo* and *Pelobates fuscus*). As development proceeds, the nephrostomes eventually fuse into a single unit. The nephrostomal tubules join into three proximal segments that finally unite to a common proximal tubule segment. The latter continues as an intermediate, ciliated segment, and a nonciliated, distal portion leads into the connecting tubule segment that distally becomes the nephric duct. At the same time the posterior cardinal vein gives rise to a series of small capillaries that ramify among the coiling pronephric tubules; they rejoin and merge to open into the duct of Cuvier (*R. esculenta,* Gosner 22; *X. laevis,* Nieuwkoop/Faber 35–36, Millard 1949).

Table 5.3 Abbreviations used in figures 5.28–5.31

A = abdominal cavity, AD = aorta dorsalis, AX = auxocyte-like cell, B = basal lamina, BC = Bowman's capsule, BP = buccopharyngeal cavity, C = capillary of the glomular (glomerular) tuft, CE = coelomic epithelial cell, CI = inner epithelial mass, CLT = collecting tubule, CO = coelom, CPT = common proximal tubule segment, CS = surface epithelium, CT = connecting tubule segment, CV = capillary of the venous system, CX = cortex, D = degenerating auxocyte, DC = duct of Cuvier, DT = distal tubule segment, E = endothelial cell, ED = rudimentary efferent duct, EP = epitheloid cell, F = follicular cells, G = glomus, GA = glomerulus anlage, GC = gonial cell, GE = germ cell, GL = glomerulus, GS = secondary gonium, GT = glomerular tuft, H = interstitial hematopoietic tissue, IT = intermediate tubule segment, M = medulla, MD = macula densa, ME = mesenchymal tissue, MS = mesentery, N = nephrostomal tubule segment, NS = neck segment, O = oogonium, OC = ovarian cavity, OD = oocyte in diplotene stage, OE = early oocyte (leptotene–pachytene), OP = opisthonephros, OPD = opisthonephric duct, OT = opisthonephric tissue, P = peritoneal funnel, PA = anlage of the peritoneal funnel, PC = pedicels, PD = pronephric duct, PE = parietal epithelium of Bowman's capsule, PF = primary foot process, PG = primitive glomerular capillary intercalated between afferent and efferent vessel, PGC = primordial germ cell, PO = podocyte, PRO = pronephros, PT = proximal tubule segment, R = testicular rete, RA = renal artery, S = sinus of the posterior cardinal vein, SC = spermatocyte, SF = secondary foot process, SG = spermatogonium, ST = seminiferous tubule, T = tubule anlage, VA = vas afferens, VB = brachial vein, VCA = anterior cardinal vein, VCP = posterior cardinal vein, VE = vas efferens, and VJ = external jugular vein.

In the pronephros, the differentiation of the Malphighian corpuscle is independent of tubulogenesis (Gipouloux 1957; Gipouloux and Girard 1986). The visceral epithelium of the Malphighian corpuscle is formed by the fusion of three successively enlarging folds of the splanchnic layers of the nephrotomes (Richter 1992, *R. esculenta*). Blood vessels from the dorsal aorta and posterior cardinal vein eventually give rise to the glomular tuft and the afferent and efferent glomular vessels (Richter 1989, 1992, and Richter and Splechtna 1990, *R. esculenta*). In *Xenopus,* the glomular blood supply starts at Nieuwkoop/Faber 35–36 (*R. esculenta,* Gosner 19).

The entire pronephros of *Xenopus* becomes functional at Nieuwkoop/Faber 37–38 (*R. esculenta,* Gosner 24) and attains its maximum development at stage 56. In *R. temporaria,* pronephros size increases from Cambar/Marrot 29–47 (Nieuwkoop/Faber 41–55) and decreases after stage 49 (Fox 1962a). The pronephric nuclear population and the total pronephric tissue volume of *R. esculenta* increase with larval length (Fox 1961). In *Xenopus* the mitotic index of pronephric tubule cells is reduced after Nieuwkoop/Faber 51, although tubule length and lumen volume increase until stage 55 (Fox 1984). Chopra and Simnett (1969a, b, 1970a, b, 1971a, b) and Simnett and Chopra (1969) suggested that the pronephric growth rate in *Xenopus* is regulated by organ specific, antigenic mitotic inhibitors synthesized during the development of the pronephros.

The onset of pronephros regression apparently is variable among species but generally begins midway through prometamorphosis (Fox 1984). Histological changes in thyroid tissue and circulating thyroxine seem related to the degeneration process (M. H. I. Dodd and Dodd 1976; Fox 1971, *Rana temporaria;* Hurley 1958, *R. sylvatica;* M. I. Michael and Yacob 1974, *Bufo v. viridis, Hyla arborea savignyi,* and *R. r. ridibunda;* Richter 1992, *R. esculenta*). Reduction of thyroid activity with thyrostatics (Fox and Turner 1967, *R. temporaria* and *X. laevis;* Gardener and Peadon 1955, *Eleutherodactylus martinicensis;* Hurley 1958, *R. sylvatica*), thyroidectomy, hypophysectomy (Fox 1963), or the administration of prolactin (Vietti et al. 1973, *R. temporaria*) inhibits pronephric degeneration.

Metamorphic degeneration of the pronephros is uneven and variable in rate (Fox 1970, *R. temporaria*) among adjacent tubular cells. The tubules become somewhat swollen and the expanded lumens often contain necrotic tissue. The nephrostomes, which fused during the premetamorphic stages, are bordered by connective tissue (Nieuwkoop and Faber 1967, *X. laevis*). The blood vessels and the amount of hematopoietic tissue increase, and the Malphighian corpuscle becomes increasingly dense and smaller. Tubular cells contain large degeneration bodies in which mitochondria or other organelles undergo lysosomal hydrolysis. Synthesis of DNA ceases, while RNA synthesis, perhaps involved in the synthesis of lysosomes, proceeds unchanged (Fox 1970, 1971, *Rana temporaria* 1983). The brush border of the distal tubule segment, the cilia of the nephrostomal and intermediate segment, the tubular basement membrane together with its overlying collagen layer, and infoldings of the plasma membrane tend to persist almost to the final stage of tubule degeneration. Necrotic tubules first change into solid strands and are then ingested by leucocytes (Fox 1970, *R. temporaria;* Richter 1992, *R. esculenta*).

In *Xenopus,* the compact anlage of the pronephric duct begins to grow posteriorly (Nieuwkoop/Faber 26; *Rana esculenta,* Gosner 17–18) by their own independent growth along the lateral mesoderm (Cambar 1952a, and Cambar et al. 1962, *R. dalmatina;* Cambar and Gipouloux 1956b, *Bufo bufo;* Fox 1963; Richter 1992, *R. esculenta;* Vannini and Giorgi 1969, *B. bufo*). The cells of the lateral mesoderm and the duct tip possess filopodia (Ciantar 1983, *X. laevis* and Delbos 1975, *R. dalmatina*). Essentially, duct elongation appears to be a migratory phenomenon (Gipouloux and Cambar 1961, *B. bufo* and Overton 1959, *R. pipiens*) occurring by cell rearrangement. The migration of the anlage of the pronephric duct is stimulated and guided by the presence of morphogenetic factors, probably protein substances. These substances located on the cell surface of the lateral mesoderm (Cambar and Gipouloux 1973, *R. dalmatina*) create a predetermined pathway in space and time (Cambar and Gipouloux 1970a, *B. bufo, Discoglossus pictus,* and *R. dalmatina* and Tung and Ku 1944, *B. gargarizans* and *R. nigromaculata*). The direction of duct elongation is not influenced by the somites (Gipouloux and Cambar 1970a, *B. bufo* and *R. esculenta*). The rectal diverticula first appear in *Xenopus* at Nieuwkoop/Faber 32 (*R. esculenta,* Gosner 20) as small grooves in the dorsal proctodeal wall. They grow anteriorly but do not exert an inductive stimulus on the growing tips of the ducts (Cambar 1952b, *R. dalmatina*), which join the cloacal diverticula at Nieuwkoop/Faber 35–36 (*R. esculenta,* Gosner 22). The cellular differentiation of the duct epithelium is completed at Nieuwkoop/Faber 39 (*R. esculenta,* Gosner 23), and that of the rectal diverticula is completed stage 43 (*R. esculenta,* Gosner 25). The pronephric ducts between the pronephros and the opisthonephros begin to degenerate at Nieuwkoop/Faber 59 (*R. esculenta,* Gosner 41) and disappear at stage 60 (*R. esculenta,* Gosner 42). The posterior regions of the ducts, which belong to the opisthonephric anlage, thicken and the epithelium becomes stratified. At Nieuwkoop/Faber 58 (*R. esculenta,* Gosner 45), the wall consists of 2–3 cell layers enveloped by connective tissue.

A functional pronephros is essential for the development and maintenance of the pronephric duct (Cambar 1954, *R. dalmatina;* Vannini and Giorgi 1969, *B. bufo*). If the pronephros is extirpated, the anterior part of the duct reduces to a solid strand, while the posterior, tubular half functions as an inducer for the opisthonephric blastema (Cambar 1947b, *R. dalmatina*). This inductive stimulus and the required self-differentiation potential of the blastema are indispensable for the development of the opisthonephric tubules (Cambar 1948, *Alytes obstetricans, B. bufo, R. dalmatina,* and *R. esculenta;* Cambar and Gipouloux 1956c, *B. bufo;* Van Geertruyden 1946, 1948, *R. temporaria*). In *B. bufo* and *R. dalmatina* the inductive activity of the pronephric duct is attributed to extracellular substances (Delbos 1975 and Gipouloux and Delbos 1977). Both the inductive distance between the duct epithelium and the blastema (Cambar and Gipouloux 1956b, *B. bufo* and *Discoglossus pictus* and Gipouloux and Cambar 1970b, *B. bufo, D. pictus,* and *R. dalmatina*) and the duration of inductive potency (Cambar and Gipouloux 1970b, *B. bufo, D. pictus,* and *R. dalmatina*) depend on the pronephric duct and vary among species.

The opisthonephros develops in the posterior region of the intermediate plate where the nephrogenous cords develop from dissolved nephrotomes and gives rise to a set of distinct condensed cell aggregations that soon become renal vesicles (Gipouloux and Hakim 1976, *B. bufo* and *X. laevis;* P. Gray 1930, *Rana temporaria;* Richter 1992, *R. esculenta;* Schluga 1974, *B. bufo* and *Pelobates fuscus;* Vannini and Sabbadin 1954, *R. esculenta*). The renal vesicles are close to the epithelium of the nephric duct and receive an inductive signal from the latter. Their arrangement is not correlated with the segmentation of the myotomes, but there is a tendency for the primary units to develop at regular intervals. In *Xenopus,* the first renal vesicles, which are anlagen of primary nephrons, are present at Nieuwkoop/Faber 47 (*R. esculenta,* Gosner 25). Functional peritoneal funnels appear at Nieuwkoop/Faber 48 (*R. esculenta,* Gosner 26). The vesicles successively develop into tubules joining the pronephric duct, which becomes the opisthonephric duct. The remaining cells of the nephrogenic blastema proliferate, extend dorsomedially and give rise to new generations of nephrons. As new nephrons are added, the nephric duct and the primary units are pushed toward the lateral margin of the kidney (Arnauld and Cambar 1970 and Cambar and Arnauld 1970, *R. dalmatina;* P. Gray 1930, *R. temporaria*). Collecting ducts are formed from bulges in the wall of the nephric duct (Hirsch 1938; Richter 1992, *R. esculenta*). Younger nephric anlagen inductively stimulated by older nephrons join the epithelium of the collecting ducts at the dorsal base of the more differentiated nephrons.

Morphogenesis of the nephron starts with the accumulation of fibroblast-like cells (Ciantar 1983, *X. laevis;* Delbos 1975, *R. dalmatina;* Richter 1992, 1995, *R. esculenta*) arranged into hollow renal vesicles (fig. 5.28E) that probably require an extracellular matrix and a basement membrane for development of polarity (Ekblom 1989; Saxén 1987). The formation of the S-shaped body (figs. 5.28E and 5.29C) is generally attributed to two opposed invaginations caused by differential growth of the epithelium. The nephron anlage subsequently joins the uriniferous duct. Further elongations and cell differentiations of the epithelium lead to the establishment of tubular segments and to the formation of Bowman's capsule. The segment of the tubule attached to the visceral epithelium of Bowman's capsule becomes the macula densa (figs. 5.28F and 5.29F; Richter 1992, 1995, *R. esculenta;* Schluga 1974, *B. bufo* and *Pelobates fuscus*). The formation of the macula densa is correlated with the development of the glomerular capillary system (Richter 1992, 1995, *R. esculenta*). The glomerular tuft develops from capillary sprouts of interstitial arterial and venous vessels that grow into the anlage of the Bowman's capsule, merge, and form a conglomerated capillary network by the fusion of free-ending protrusions (fig. 5.29E; Ditrich and Lametschwandtner 1992, *Xenopus laevis;* Naito 1984, *R. catesbeiana;* Richter 1992 and Richter and Splechtna 1990, *R. esculenta*). The vessels are accompanied by immigrating mesenchymal cells distributed between the capillaries of the glomerular tuft and establish their position as mesangial, lacis, and epithelioid cells (Richter 1992, 1995; Richter and Splechtna 1990, *R. esculenta*).

Fig. 5.30. Schematic representation of the developing gonads (both sexes) and Bidder's organ. (A) Formation of genital ridge. Differentiation of gonads: (B) proliferation of coelomic epithelial cells of indifferent gonad and (C) cortex and medulla partially separated. Ovarian development: (D) cortex and medulla separated by mesenchymal tissue, (E) medullary tissue reduced and ovarian cavities differentiated, and (F) development of follicles. Testicular development: (G) cortex and medulla closely associated, (H) cortex separated into an inner epithelial mass and an outer epithelium (= surface epithelium), and (I) medullary mass differentiated into testicular rete and efferent ducts; formation of seminiferous tubules. Differentiation of Bidder's organ: (J) proliferation of somatic cells, (K) no definite cortico-medullary structure evident; germ cells scattered among somatic cells, (L) germ cells enlarge into auxocyte-like cells and somatic cells partially degenerate, (M) differentiation of auxocyte-like cells, and (N) auxocyte-like cells degenerate, and oocytes in meiotic stages. See table 5.3 for abbreviations. Redrawn from Tanimura and Iwasawa (1986, 1988); reprinted by permission of *Science Report of Niigata University* and Japanese Society of Developmental Biologists.

Peritoneal funnels develop as ventral outgrowths of the vesicle (fig. 5.29D). They either detach from the S-shaped body (Engels 1935, *Rana temporaria;* Gérard and Cordier 1933, 1934b, *B. bufo* and *R. temporaria;* P. Gray 1936 and Kindahl 1938, *R. temporaria;* Richter 1992, 1995, *R. esculenta;* Schluga 1974, *B. bufo* and *Pelobates fuscus*) or remain connected with the neck segment until metamorphosis begins (Gérard and Cordier 1933, 1934b, *Alytes obstetricans* and *Discoglossus pictus*). In the latter case, they separate from the tubules by the destruction of their proximal epithelium. In *Xenopus,* communication between the nephron and the peritoneal funnel persists throughout life (De Waal 1973; Kunst 1936). The peritoneal funnels usually join the renal veins and, clustered near the ventromedial margin of the kidney, unite at their distal ends with the ventral peritoneum (Kunst 1936).

At premetamorphosis the primary nephrons lose their connections with the pronephric duct and finally degenerate at metamorphic climax (P. Gray 1930, 1932; Hirsch 1938; Kindahl 1938, *Rana temporaria;* Richter 1992, *R. esculenta*). In *Xenopus,* degeneration starts at Nieuwkoop/Faber 55 (*R. esculenta,* Gosner 32), and the cellular structure of remaining nephrons changes at metamorphic climax. The number and metabolism of peroxisomes and catalase activity increases in the proximal tubule segment of *R. catesbeiana* at metamorphosis. No change in catalase activity or peroxisome number is detected in *Xenopus,* which remains aquatic and ammoniotelic throughout development (Dauça et al. 1982, 1983). The nephrons and the renal vascular system are reorganized in different zones within the kidney; each zone is composed of distinct nephron segments and special blood vessels, and the architecture of the metamorphic kidney now resembles the definitive adult kidney (Barch et al. 1966, *Bufo bufo;*

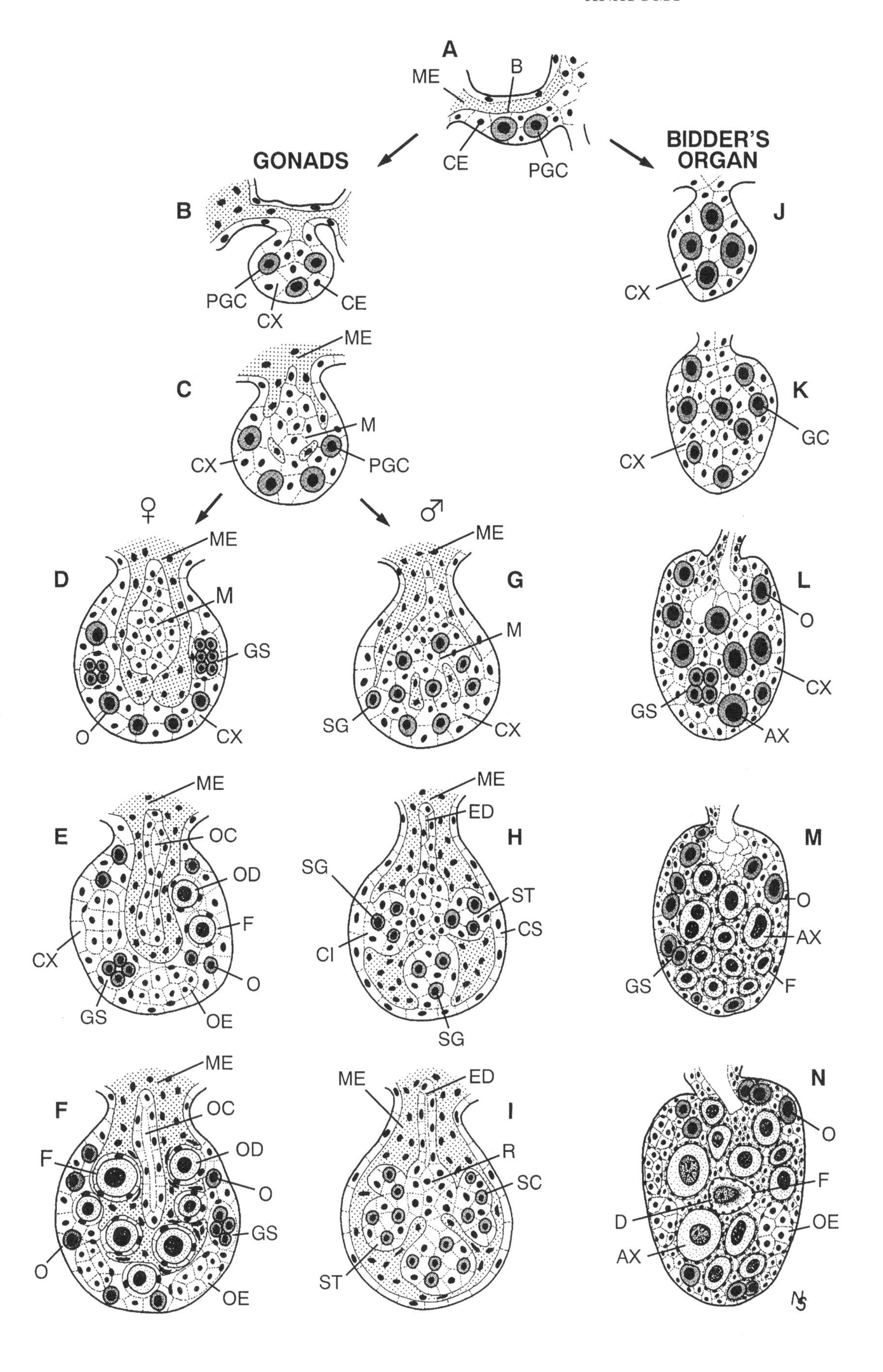
A
ME
B
CE
PGC
GONADS
BIDDER'S ORGAN
B
PGC
CX
CE
J
CX
C
ME
M
CX
PGC
K
GC
CX
♀
♂
D
ME
M
GS
O
CX
G
ME
M
SG
CX
L
O
CX
GS
AX
E
ME
OC
OD
F
CX
O
GS
OE
H
ME
ED
SG
ST
CS
CI
SG
M
O
AX
GS
F
F
ME
OC
F
OD
O
GS
O
OE
I
ME
ED
R
SC
ST
N
O
F
D
OE
AX

Bargmann et al. 1955, *X. laevis;* Berton 1964, *B. bufo, Hyla arborea, R. dalmatina,* and *R. esculenta;* Lametschwandtner et al. 1978, *Bombina variegata, B. bufo, R. ridibunda,* and *X. laevis;* Linss and Geyer 1964, *R. esculenta;* Morris and Campbell 1978, *B. marinus;* Ohtani and Naito 1980, *B. bufo;* Richter 1992, *R. esculenta*).

Reproductive System: Anatomy

The gonads of male and female larvae are paired structures that develop from sexually undifferentiated primordial gonads located near the presumptive kidneys. Gonadal differentiation in *Xenopus* starting at Nieuwkoop/Faber 49, Gosner 29 in *Rana nigromaculata* (Tanimura and Iwasawa 1988) and Gosner 26 in *R. ridibunda* (Ogielska and Wagner 1990) is completed in young frogs.

The primordial gonads appear as longitudinal bilateral thickenings covering the ventromedial region of the opisthonephros (fig. 5.30A) and are attached by the mesogonium to the inner body wall. The central region (= mesogonium) gives rise to the gonad proper, and the anterior region (= progonium) forms the fat bodies. The primordial gonad, differentiated from anterior to posterior, is composed of medullary and cortical regions covered with a basal lamina (Iwasawa and Yamaguchi 1984, *Xenopus laevis;* Lamotte et al. 1973, *Nimbaphrynoides occidentalis;* Lopez 1989, *Bombina orientalis;* Merchant-Larios and Villalpando 1981, *R. pipiens* and *X. laevis;* Ogielska and Wagner 1990, *R. ridibunda;* Tanimura and Iwasawa 1986, *Bufo japonicus formosus,* 1988, 1991, *R. nigromaculata;* Tanimura and Iwasawa 1989, *Rhacophorus arboreus*). A lack of ultrastructural differences between medullary and cortical cells of the primordial gonad indicates a cortical origin of the medullary tissue. In later stages, the mesogonium is invaded by blood vessels, nerves, and opisthonephric interstitial tissue. The primordial germ cells (= PGCs) located in the cortical region of the undifferentiated gonad form a primitive germinal epithelium together with surrounding prefollicular cells. The PGCs commonly become smaller after they enter the genital ridge in association with proliferation and digestion of yolk platelets (Lopez 1989, *Bombina orientalis;* Ogielska and Wagner 1990, *R. ridibunda;* Züst and Dixon 1975, 1977, *X. laevis*). Tight junctions between PGCs and prefollicular cells may prevent the passage of molecular substances between the two cell types. Prefollicular cells contain a flattened nucleus and few mitochondria (Delbos et al. 1984, *R. pipiens*).

Gonadal sex differentiation differs temporally among species but usually coincides with metamorphic climax. Female and male gonads are distinguishable by the number and size of germ cells and the amount of medullary tissue (figs. 5.30D–I; Ijiri and Egami 1975, *X. laevis;* Kobayashi 1975, *Rana nigromaculata;* Shirane 1986, *R. japonica* and *R. nigromaculata;* Tanimura and Iwasawa 1987, *Bufo japonicus formosus;* Zaccanti et al. 1977, *R. dalmatina, R. esculenta,* and *R. latastei*). Male gonads can develop in a differentiated, semidifferentiated, or undifferentiated mode. The undifferentiated and semidifferentiated types often express spontaneous early female sexuality (Tanimura and Iwasawa 1989, *Rhacophorus arboreus;* Takahashi 1971, *Rana rugosa;* Zaccanti et al. 1977, *Rana dalmatina* and *R. latastei*). In *Phyllomedusa sauvagii,* female gonadal differentiation initially displays a testicular trend (Rengel and Pisanò 1981).

Fig. 5.31. Gonadal differentiation of *Rana nigromaculata.* (A) Primordial gonad (Gosner stage 24), scale = 50 μm. Sexually indifferent gonads at (B) Gosner 26 and (C) Gosner 27; cortex and medulla clearly distinguished, direct contact *(arrows)* between cortical and medullary cells; scale = 20 μm. Rudimentary testis: (D) Gosner 30, cortical and medullary regions closely associated *(asterisks);* mesenchymal cells *(arrows)* and blood capillaries *(arrowheads)* are evident; scale = 40 μm; (E) Gosner 35, inner epithelial mass separated from surface epithelium, shows cordlike structure; scale = 50 μm; and (F) Gosner 46, rudimentary seminferous tubules *(arrows)* formed from the inner epithelial mass, and testicular rete connected to rudimentary seminiferous tubules *(arrowheads);* scale = 100 μm. Rudimentary ovary: (G) Gosner 30, cortex and medulla separated, mesenchymal cells *(arrows)* and blood capillaries *(arrowhead)* visible; scale = 40 μm; (H) Gosner 40, formation of ovarian cavity evident in medullary cell mass, primitive follicles *(arrows)* visible in cortex; scale = 100 μm; and (I) Many developing follicles present; scale = 100 μm. See table 5.3 for abbreviations. From Tanimura and Iwasawa (1988); reprinted by permission of the Japanese Society of Developmental Biologists.

Larval ovaries (figs. 5.30D–F and 5.31G–I) are larger than testes and have a slightly irregular outline. They usually consist of medullary and cortical tissue in which the proliferating gonial cells form several groups of oogonia and oocytes (Iwasawa and Yamaguchi 1984, *X. laevis;* Lamotte et al. 1973, *Nimbaphrynoides occidentalis;* Lopez 1989, *Bombina orientalis;* Ogielska and Wagner 1990, *R. ridibunda;* Tanimura and Iwasawa 1986, *Bufo japonicus formosus* 1988, 1991, *R. nigromaculata* 1989, *Rhacophorus arboreus*). The cortex is distinctly separated from the medulla by an acellular collagenous layer, and the medullary tissue has degenerated and lost its sex cord pattern. The cords, which are continuous with the rudiments of the rete ovarii system, dissolve. Small lumens formed in this way enlarge and finally fuse to the ovarian cavity. The medullary connective tissue thus becomes a simple epithelial layer lining the ovarian cavities.

Oogonia measure 17–25 μm in diameter and cytologically resemble PGCs except for the lack of pigment granules and abundant ER. Oogonia undergo periodic proliferation to replenish oogonial nests of about 16 gonial cells (Coggins 1973, *Xenopus laevis;* Lopez 1989, *Bombina orientalis;* Redshaw 1972, *X. laevis*). Intercellular bridges appear to be responsible for synchronous development during oogenesis (Coggins 1973, *X. laevis;* Ruby et al. 1970, *Rana pipiens*). The transition from primary to secondary oogonia (= primary oocytes) is characterized by a dark cytoplasm (Eddy and Ito 1971, *X. laevis*) and a change from lobed to round nuclei. Oogonia and primary oocytes contain germ plasm (Al-Mukthar and Webb 1971; J. B. Balinsky and Devis 1963; Coggins 1973; and Kalt 1973a, *X. laevis;* Delbos et al. 1971, *R. dalmatina;* Eddy and Ito 1971, *R. catesbeiana, R. clamitans,* and *R. pipiens;* Kress and Spornitz 1972, *R. esculenta* and *R. temporaria*).

Prophase of primary oocytes is initiated in larvae prior to metamorphosis (Al-Mukthar and Webb 1971; Coggins 1973; Jørgensen 1973; Lofts 1974; Lopez 1989; Ogielska and Wagner 1990; Van Gansen 1986). At metamorphic climax larval ovaries exhibit oogonia and leptotene, zygotene,

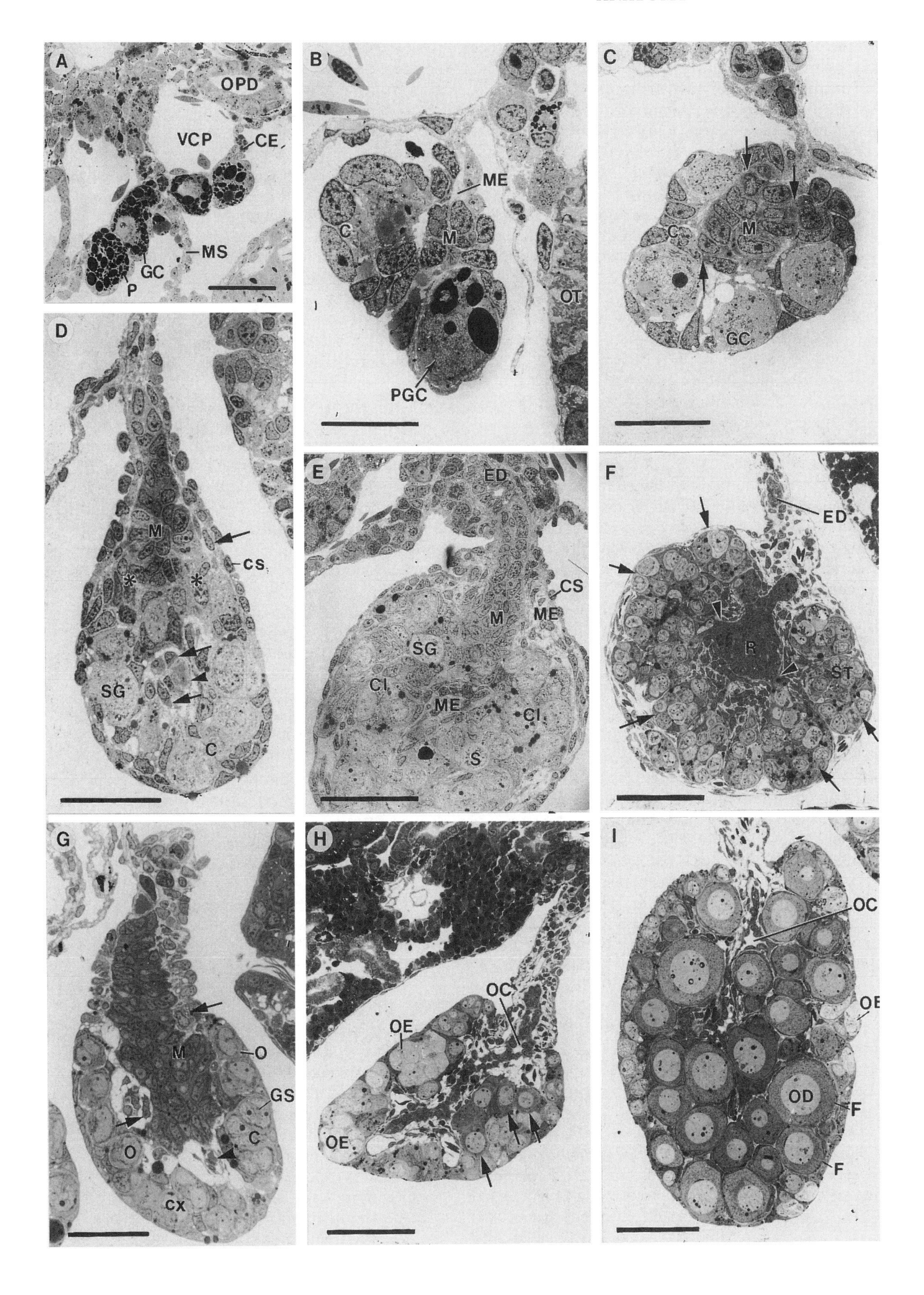
A
OPD
VCP
CE
GC
MS
P
B
ME
C
M
OT
PGC
C
M
C
GC
D
M
CS
SG
C
E
ED
CS
M
ME
SG
CI
ME
CI
S
F
ED
R
ST
G
M
O
GS
O
C
CX
H
OC
OE
OE
I
OC
OE
OD
F
F

pachytene, and diplotene oocytes. Near the end of leptotene, oogonial chromosomes attach to the inner surface of the nuclear membrane (= bouquet arrangement) at a site adjacent to the juxtanuclear mitochondrial aggregate, which frequently forms a cap over one end of the nucleus. The zygotene stage is characterized by the formation of short, axial chromosomes. Paired homologous chromosomes form synaptonemal complexes, and synapsis is completed in pachytene. Extrachromosomal rDNA, functioning in rRNA synthesis, increases, and multiple nucleoli are formed. The oocytes of a cell nest separate and become surrounded by follicular cells that differentiate into steroid-producing follicular epithelium (Redshaw and Nicholls 1971, *Xenopus laevis;* Saidapur and Nadkarni 1974, *Euphlyctis cyanophlyctis, Hoplobatrachus tigerinus, Limnonectes keralensis,* and *Uperodon systoma;* Thibier-Fouchet et al. 1976, *X. laevis*). During folliculogenesis, intercellular bridges disappear and oocytes begin to develop asynchronously. The cytoplasm and nucleus enlarge in diplotene oocytes, synaptonemal complexes are absent, and the chromosomes are in the lampbrush stage (Coggins 1973; Wischnitzer 1976, *X. laevis*). During early diplotene, the Balbiani body is formed by condensation of the juxtanuclear mitochondrial aggregate. In the late diplotene oocyte, dispersed germ plasm is stored in the cortex of the oocyte (J. B. Balinsky and Devis 1963, *X. laevis*).

In most anuran larvae, testicular differentiation starts at premetamorphosis (figs. 5.30G–I and 5.31D–F; Iwasawa and Kobayashi 1976, *Rana nigromaculata;* Iwasawa and Yamaguchi 1984, *X. laevis;* Lamotte and Xavier 1973, *Nimbaphrynoides occidentalis;* Lopez 1989, *Bombina orientalis;* Ogielska and Wagner 1990, *R. ridibunda;* Tanimura and Iwasawa 1986, *Bufo japonicus formosus,* 1988, 1991, *Rana nigromaculata,* 1989, *Rhacophorus arboreus*). The primitive germinal epithelium disintegrates to a simple peritoneal surface epithelium. The male gonial cells or spermatogonia, together with their surrounding follicular cells, are scattered throughout the compact medullary tissue, which becomes mixed with interstitial mesenchymal cells. In the center of the medulla, a network of tubules becomes increasingly distinct and finally differentiates into seminiferous tubules, which are attached to the opisthonephric tubules by the vasa efferentia. In *R. nigromaculata* (Tanimura and Iwasawa 1988) and *Rhacophorus arboreus* (Takasu and Iwasawa 1983; Tanimura and Iwasawa 1989), seminiferous tubules are derived from the inner layer of the cortical epithelium. Medullary cords near the hilus of the testicular primordium form the rete testis.

Spermatogenesis is virtually identical among species (Deuchar 1975; Kalt 1973a, b; J. B. Kerr and Dixon 1974; and Reed and Stanley 1972, *Xenopus laevis;* Iwasawa and Kobayashi 1976, *Rana nigromaculata;* Rastogi et al. 1983, *R. esculenta;* reviews by Grassé 1986 and Lofts 1974). Two types of spermatogonia are present in premetamorphic larvae; primary single spermatogonia are characterized by a lobed, pale nucleus with diffuse chromatin, and secondary spermatogonia are derived from proliferating primary spermatogonia. Secondary spermatogonia usually occur in clusters (= germinal cyst or nest) of cells of similar size and appearance that develop at a uniform rate (Fawcett et al. 1959, *X. laevis;* Rastogi et al. 1983, *R. esculenta*).

Meiosis usually starts in late metamorphosis with the transformation of secondary spermatogonia into primary spermatocytes. Leptotene spermatocytes occasionally observed in larval gonads (Iwasawa and Kobayashi 1976, *R. nigromaculata;* Kalt 1973a and Kalt and Gall 1974, *X. laevis*) have large nuclei with slightly condensed chromatin. Zygotene spermatocytes are characterized by a prominent Golgi complex, a few flattened vesicles, and partial synapsis and condensation of sister chromatids (= bouquet configuration). Synapsis is completed in the pachytene stage. Germ plasm disappears in pachytene spermatocytes (J. B. Kerr and Dixon 1974, *X. laevis*), and chromatin bivalents become separated in diplotene spermatocytes.

The Wolffian and Müllerian ducts of anurans transport gametes and, especially in males, nitrogenous wastes and water. Therefore, the anatomical structure of the genital ducts reflects the functions of water economy, egg protection, and nurturing of developing young (M. H. Wake 1979). The Wolffian ducts arise from the anterior region of each kidney and converge immediately behind the kidneys as they proceed to the cloaca. Wolffian ducts persist in both sexes to develop initially as primary nephric ducts. In females they retain their excretory function. In males they also serve as urogenital ducts with remarkable modifications, especially near the cloaca where a seminal vesicle develops in some species (Amer 1972, *Bufo regularis;* Beigl 1989, *Bombina variegata, Bufo bufo, Discoglossus pictus,* and *Rana esculenta;* Bhaduri 1953; Bhaduri and Basu 1957).

In young female anurans, the Müllerian ducts or oviducts usually arise at metamorphosis from tissue near the pronephric funnel (Richter 1992, *R. esculenta,* Gosner 43). They are attached to the dorsal body wall, extend posterolaterally to the Wolffian ducts, and connect to the cloaca. In larvae, Müllerian ducts are straight, thin-walled tubules, and in adults, they are extraordinarily coiled, especially during the breeding season. Each oviduct includes an ostium, a short, straight pars recta, and a long, coiled pars convoluta; the posterior sections are enlarged as the ovisacs (Amer 1972, *Bufo regularis;* Bhaduri 1953; and Bhaduri and Basu 1957; P. Horton 1983, *Rheobatrachus silus*). In males, Müllerian ducts either degenerate or remain vestigial and apparently functionless.

Bidder's organ, present in male and most female bufonids (Amer 1968–1969, *B. regularis;* Bhaduri 1953; Di Grande 1968, 1987; and Di Grande and Marescalchi 1987, *Bufo bufo;* Tanimura and Iwasawa 1986, 1987, *Bufo japonicus formosus;* Zaccanti and Vigenti 1980, *B. bufo*), is formed from the anterior mesogonial portion of the genital ridge. Because neither corticomedullary structure nor primitive gonial cavities develop, Bidder's organ, which retains the potentiality of an ovary in adults of both sexes, consists of a cortical layer of flattened cells, peripherally located gonial cells, a few oocytes, and large auxocyte-like cells that partially degenerate postmetamorphically (figs. 5.30J–N). Gonial cells and oocytes increase in number at metamorphic climax. The number and mitotic index of these cells is significantly higher in females (Tanimura and Iwasawa 1987, *B. j. formosus*). The fat body, derived from the progonium, is characterized by proliferating somatic cells that form fingerlike projections attached to the gonads. No data are available on the function of fat body in larvae, but in the adults, fat body mass and gonad mass are inversely correlated (Chieffi et al. 1980, *R. esculenta;* Jørgensen 1986 and Jørgensen et al. 1979, *B. bufo;* Pancharatna and Saidapur 1985 and Saidapur 1983, 1986, *Euphlyctis cyanophlyctis*), and fat body excision may (Chieffi et al. 1975, 1980, *R. esculenta* and Kasinathan et al. 1978, *Euphlyctis hexadactylus*) or may not (Kanamadi and Saidapur 1988, *E. cyanophlyctis*) result in testicular regression and spermatogenesis. In females, the fat body sustains vitellogenesis (Prasadmurthy and Saidapur 1987, *E. cyanophlyctis*).

Reproductive System: Development

Gonadal development starts with the accumulation of germ plasm at the vegetal pole of the fertilized egg. During cleavage the germ plasm is distributed among a few cells that become progenitors of the germ cell line (Bournoure 1937 and Bournoure et al. 1954, *Rana temporaria;* Buehr and Blackler 1970 and K. E. Dixon 1981, *Xenopus laevis;* Di Bernardino 1961; *R. pipiens;* Gipouloux 1962, 1975, *Discoglossus pictus, Phrynobatrachus natalensis, R. pipiens,* and *R. temporaria;* L. D. Smith 1965, 1966; *R. pipiens;* Wakahara 1978; Whitington and Dixon 1975, *X. laevis;* reviews by Beams and Kessel 1974 and Eddy 1975).

Presumptive primordial germ cells (= pPGCs), giving rise to all gametes, subsequently undergo a small number of cloning divisions during the migration to the genital ridge (Whitington and Dixon 1975, *X. laevis*). After gastrulation, these cells (= pPGCs) migrate from the embryonic gut to the root of the dorsal mesentery (Lopez 1989, *Bombina orientalis,* Gosner 22) and then laterally across the dorsal abdominal wall to the gonadal ridge (figs. 5.30A, B; Iwasawa and Yamaguchi 1984; Kamimura et al. 1976; and Züst and Dixon 1977, *X. laevis,* Nieuwkoop/Faber 41–46; Lopez 1989, *Bombina orientalis,* Gosner 23) where they differentiate into primordial germ cells (= PGCs). Between the early gastrula and differentiation from the endoderm, pPGCs divide twice (Dziadek and Dixon 1977 and Whitington and Dixon 1975, *X. laevis*). They synthesize DNA between Nieuwkoop/Faber 10–33 and RNA from the gastrula to early tail-bud-stage (Dziadek and Dixon 1975, 1977, *X. laevis*). The transcription of hnRNA, mRNA, and DNA is suppressed during Nieuwkoop/Faber 40–48 (Wakahara 1982). The cells remain mitotically quiescent between Nieuwkoop/Faber 40–48 after they enter the genital ridge; a second proliferative phase starts at about stage 48–52 (Ijiri and Egami 1975; Züst and Dixon 1977; Wakahara 1982). The pPGCs, averaging 30 μm in diameter, display polymorphism during migration from their endodermal position to the genital crest (Kamimura et al. 1980, *X. laevis*). The ultrastructure is similar in all anuran species examined thus far (Delbos et al. 1982, *Bufo bufo* and *R. dalmatina;* Ikenishi 1980 and Wylie and Heasman 1976a, b, *X. laevis*). Frequently pPGCs bear a single (15–30 μm long) pseudopodium at one end and a short (3–4 μm long) cellular process at the other extremity. Both are missing in PGCs situated at the genital ridge. The pPGCs show membrane adenylcyclase activity (Delbos et al. 1981, *R. esculenta* and *X. laevis*) and are separated from the small, polygonal somatic endodermal cells by a large intercellular space (Ikenishi 1982, *X. laevis*). The cells are rounded and contain many large yolk platelets, abundant amorphous lipid droplets, cortical granules, and germ plasm that is characterized by a mass of large, electron-dense mitochondria (Czovwska 1972; Ikenishi et al. 1974; Ikenishi and Kotani 1975; Kalt 1973a; and L. D. Smith and Williams 1979, *X. laevis;* Mohawald and Hennen 1971 and M. A. Williams and Smith 1971, *R. pipiens*). The round nucleus is, in contrast to the endodermal somatic nucleus, totipotent (L. D. Smith 1965, *R. pipiens* and Wylie et al. 1985, *X. laevis*) and contains a prominent nucleolus. Particles of β-glycogen supply the energy resource for migration (Delbos et al. 1981, *R. esculenta* and *X. laevis*). J. B. Kerr and Dixon (1974) suggested that germ plasm activated in germ cells between gastrulation and meiosis controls migration, mitosis, and meiosis.

During migration, pPGCs show three main types of differentiation: polarized pseudopodial projections containing microfilaments, accumulation of extracellular matrix around these projections, and junctions and desmosomes between the pPGCs and the substrate cells (Delbos et al. 1981, *R. esculenta* and *X. laevis;* Delbos et al. 1982, *B. bufo* and *R. dalmatina;* Heasman and Wylie 1978; Kamimura et al. 1980; Wylie and Heasman 1976a; and Wylie and Roos 1976, *X. laevis*). Migration of the pPGCs from the endoderm to the genital ridge likely is achieved by passive segregation of the pPGCs from other endodermal cells, by the fusion of lateral mesodermal plates to form the dorsal mesentery (Giorgi 1974 and Vannini and Giorgi 1969, *B. bufo*), and by the attraction exerted by the genital ridge on germ cells (Gipouloux 1970, *B. bufo, Discoglossus pictus, Hyla arborea, R. dalmatina, R. esculenta,* and *R. temporaria* and Gipouloux and Girard 1986). The basement membrane of the epithelial cells that line the migratory route of the pPGCs may be an additional factor. This is believed to regulate the movement of pPGCs from endoderm to mesentery by restricting those cells to the lateral plate (Heasman et al. 1985, *X. laevis*). The pPGCs also may be guided by diffusible substances released from the chordamesoderm (Giorgi 1974, *B. bufo;* Gipouloux 1964, 1970, *B. bufo, D. pictus, H. arborea, R. dalmatina, R. esculenta,* and *R. temporaria*) and neural crest cells (Gipouloux and Girard 1986, *B. bufo*). Diffusion of cAMP from the chordamesoderm is distributed along a concentration gradient that attracts pPGCs on their migratory pathway (Delbos et al. 1980, 1981, *R. esculenta* and *X. laevis;* Gipouloux et al. 1978, *X. laevis;* Ratiba et al. 1981, *R. dalmatina* and *X. laevis*). The cells move actively in a sequence involving elongation coupled with the extrusion of filopodia and broad blunt cell processes, waves of contraction, and ultimately by retraction of the trailing end of the cell (Heasman et al. 1977; Heasman and Wylie 1978; and Wylie et al. 1979,

X. laevis). They can adhere to mesentery cells via fibronectin and actin filaments produced by mesentery cells (Heasman et al. 1981 and Wylie et al. 1979, *X. laevis*).

Concomitantly with migration of the pPGCs, movement of peritoneal cells lateral to the mesentery to the germinal epithelium leads to the formation of paired, longitudinal, germinal ridges (figs. 5.30A and 5.31A; Lopez 1989, *Bombina orientalis,* Gosner 24; Ogielska and Wagner 1990, *R. ridibunda,* Gosner 25; Wylie and Heasman 1976a, b; Wylie et al. 1979, *X. laevis,* Nieuwkoop/Faber 45–50). The ridge can be divided into the anterior progonium, which gives rise to the fat body; the mesogonium, which forms the definitive gonads; and the posterior epigonium. The genital ridge soon acquires a gonadal medulla (figs. 5.30B–C and 5.31A–B) by the proliferation and displacement of germinal epithelial cells within the primordial gonad (Franchi et al. 1962; Iwasawa and Yamaguchi 1984, *X. laevis;* Merchant-Larios 1978 and Merchant-Larios and Villalpando 1981, *R. pipiens* and *X. laevis;* Tanimura and Iwasawa 1988, 1991, *R. nigromaculata,* 1989, *Rhacophorus arboreus*).

The onset and rate of development of the gonads is uneven and depends on the mode of differentiation (Iwasawa and Yamaguchi 1984, *Xenopus laevis;* Iwasawa and Kobayashi 1976; Kobayashi 1975; and Tanimura and Iwasawa 1988, 1991, *Rana nigromaculata;* Lamotte and Xavier 1973 and Lamotte et al. 1973, *Nimbaphrynoides occidentalis;* Lopez 1989, *Bombina orientalis;* Nodzenski et al. 1989, *Leptobrachium montanum, Leptolalax gracilis,* and *Megophrys nasuta;* Ogielska and Wagner 1990, *R. ridibunda;* Rengel and Pisanò 1981, *Phyllomedusa sauvagii;* Takahashi 1971, *R. rugosa;* Tanimura and Iwasawa 1986, 1987, *Bufo japonica formosus,* 1989, *Rhacophorus arboreus*). Sex differentiation essentially depends on a corticomedullary interaction mediated by elaboration and release of sex inductor substances (Burns 1955; Lofts 1974; Merchant-Larios 1978). Neither the genetic constitution nor the inductive role of germ cells influences sex determination (Di Grande 1968, 1987, and Di Grande and Marescalchi 1987, *B. bufo;* Shirane 1970, 1972, 1982, *R. japonica* and *R. porosa*). Other determining factors probably include somatic sex genes (Blackler 1970, *X. laevis*), sex hormones (Burns 1955; Lofts 1974; Merchant-Larios 1978; Shirane 1986, *R. japonica* and *R. nigromaculata*), and H-Y antigens (Engel and Schmid 1981, *B. bufo* and *X. laevis;* Wachtel et al. 1975, *R. pipiens* and *X. laevis*).

In *Xenopus* starting at Nieuwkoop/Faber 50 (Lopez 1989, *Bombina orientalis,* Gosner 31; Ogielska and Wagner 1990, *R. ridibunda,* Gosner 27), ovarian development (figs. 5.30D–F) obviously is controlled by estrogen and depends on age and not on metamorphic processes (Chang and Hsu 1987, *R. catesbeiana*). Oogonia, presumably the result of mitotic divisions of PGCs, located within the cortical area are arranged in oogonial nests surrounded by follicular cells (figs. 5.31H–I). The PGCs in the medullary region degenerate. The cortical epithelium and the central medulla are separated by an acellular collagenous layer or basal lamina (fig. 5.31G; Merchant-Larios 1978 and Tanimura and Iwasawa 1988, *R. nigromaculata*). Reduction of medullary tissue in *Xenopus* begins at Nieuwkoop/Faber 52 (Iwasawa and Yamaguchi 1984) and Gosner 28 in *R. ridibunda* (Ogielska and Wagner 1990). A series of cavities fuse to form the ovarian sac (figs. 5.31H–I), which remains incompletely developed at metamorphic climax. The medullary tissue becomes restricted to the lining of the cavities and to the rudimentary rete ovarii (K. L. Duke 1978). In *Xenopus,* oogonia enter meiotic prophase at Nieuwkoop/Faber 55 (fig. 5.31I; Al-Mukthar and Webb 1971, *Xenopus laevis;* Lopez 1989, *Bombina orientalis;* Ogielska and Wagner 1990, *R. ridibunda*). At metamorphosis most anuran ovaries contain oogonia and primary diplotene oocytes (Al-Mukthar and Webb 1971, *X. laevis;* Lopez 1989, *B. orientalis;* Ogielska and Wagner 1990, *R. ridibunda*). After metamorphosis, oogenesis and vitellogenesis (R. A. Wallace 1978; Wassermann and Smith 1978), and folliculogenesis (Guraya 1978) continue in a species-typical manner throughout adult reproductive life (Echeverría 1988, *Bufo arenarum;* Wagner and Ogielska 1990, *R. ridibunda;* Van Gansen 1986). In most temperate species the reproductive period of an individual lasts several years.

Testis formation (figs. 5.30G–I) begins with the incorporation of germ cells into the proliferating medullary tissue (Iwasawa and Yamaguchi 1984, *X. laevis* Nieuwkoop/Faber 50–58; Lopez 1989, *Bombina orientalis,* Gosner 31). Spermatogenesis occurs in clusters enclosed by follicular cells. In most anuran larvae, spermatogenesis begins late in metamorphosis. The duration of spermatogenesis in froglets depends on the duration of sexual maturation. Seminiferous tubules surrounding the germinal cysts are usually formed from the medullary tissue (Iwasawa and Yamaguchi 1984, *X. laevis;* Lofts 1974; Lopez 1989, *B. orientalis*). In *R. nigromaculata* (figs. 5.31E–F; Tanimura and Iwasawa 1988) and *Rhacophorus arboreus* (Takasu and Iwasawa 1983; Tanimura and Iwasawa 1989), seminiferous tubules develop from the inner region of the cortex. In *Xenopus,* formation of seminiferous tubules starts at Nieuwkoop/Faber 58 and in *B. orientalis* at Gosner 44 (Lopez 1989). It is completed after metamorphosis. Medullary cords near the gonadal hilus differentiate into the rete testis (fig. 5.31F). More anteriorly the cords give rise to the vas efferens which, in most anurans, connect with the anterior opisthonephric tubules to acquire open connections with the Wolffian ducts (figs. 5.31E–F). In some anurans, such as *Discoglossus pictus,* the efferent tubules join the Wolffian ducts directly (Beigl 1989). The follicle cells differentiate postmetamorphically into steroid-producing Sertoli cells (Bernardini et al. 1990, *X. laevis* and Guraya 1972, *R. pipiens;* review by Lofts 1974); interstitial cells develop into steroid-producing Leydig cells.

The Wolffian ducts initially develop as primary ducts draining the pro- and opisthonephros and, after connecting with the efferent tubules, also serve as urogenital ducts in males. Their posterior portions are either modified as seminal vesicles postmetamorphically or remain simple and undifferentiated (Amer 1972; Beigl 1989; Bhaduri 1953; Bhaduri and Basu 1957). The Müllerian ducts develop independently of the Wolffian ducts. Differentiation starts with the formation of the ostium abdominale developed by the closing of a peritoneal groove along an area of thickened epithelium at the site of the anterior pronephric funnel (Richter

1992, *R. esculenta,* Gosner 43). The lips of the groove close and the tube then grows posteriorly until it joins the cloaca. The Müllerian ducts are present in both males and females, although they typically degenerate in males (Bhaduri 1953; Bhaduri and Basu 1957; M. H. Wake 1979). Embryonic differentiation of both genital ducts is independent of hormonal influences, but as development proceeds they become sensitive to steroids secreted by the gonadal primordium. Estrogenic steroids are responsible for the differentiation of the oviduct and urogenital papilla and for jelly secretion (Gunasingh et al. 1982, *Euphlyctis hexadactylus*).

Bidder's Organ and Fat Body

Bidder's organ develops in male and female bufonids from the mesogonial section anterior to the definitive gonad. Differentiation starts soon after hatching with the proliferation of germ cells scattered among a few medulla-like cells (figs. 5.30J–N; Tanimura and Iwasawa 1986, *B. j. formosus;* see Petrini and Zaccanti 1998 for effects of various hormones and inhibitors). The germ cells enlarge and differentiate into gonial and auxocyte-like cells both enveloped by a layer of follicular cells. During larval development no sexual differentiation in the number of germ cells is found. Postmetamorphically the number of oocytes and secondary gonia are significantly higher in females (Tanimura and Iwasawa 1987, *B. j. formosus*).

The fat body composed only of somatic cells is derived from the progonial, germ cell-free part of the germinal ridge. In *Xenopus,* fat deposition starts at Nieuwkoop/Faber 51 (Ogielska and Wagner 1990, *R. ridibunda,* Gosner 29). Fat bodies are largely composed of adipose cells at the beginning of hind limb development (Lopez 1989, *B. orientalis,* Gosner 30; Nieuwkoop and Faber 1967, *X. laevis;* Tanimura and Iwasawa 1986, *B. j. formosus*).

Endocrine System

Regulation of the development, growth, and metamorphosis of tadpoles by the endocrine glands of endodermal or mesodermal origin is a fascinating aspect of larval biology. Except for the iodinated amino acid derivatives of the thyroid, endodermal glands produce polypeptides, while mesodermal glands synthesize steroid hormones. Endocrine glands of neural or neural crest origin consist of photoreceptor cells (e.g., dorsal pineal complex, ventral infundibulum, neurohypophysis anlage of pituitary gland) that evaginate from the diencephalon and produce oligopeptides, polypeptides, or, in the case of the pineal complex, a derivative of amino acid and polypeptides (chap. 6).

The dorsomedial adenohypophysis anlage, the ventromedial thyroid anlage, and the endocrine glands derived from the branchial arches all evaginate from the buccopharyngeal area. The endocrine pancreas originates from the midgut anlage. It is not clear whether the endocrine pancreas originates from an ancestrally diffuse gastrointestinal endocrine system, diffuse and already specialized endocrine cells of the intestinal wall, or the neural crest (Epple and Lewis 1973; Wessels 1968). The adrenal glands are of mesodermal and neural crest origin (Accordi and Grassi Milano 1990), while the gonadal primordia are of mesodermal origin.

Pituitary Gland

The stomodeal-hypophyseal anlage is visible at Nieuwkoop/Faber 20 in *Xenopus laevis.* The adenohypophyseal anlage segregates at Nieuwkoop/Faber 28, and by stage 41 the largest mass of the compact tissue lies anterior to the notochord. Differentiation of the pars intermedia begins at Nieuwkoop/Faber 47. Tuberal parts extend longitudinally in Nieuwkoop/Faber 51, and the entire hypophysis or pituitary gland (neurohypophysis and adenohypophysis) is well developed at stage 53 (Atwell 1918–1919; Eakin and Bush 1957; Nieuwkoop and Faber 1956).

The dislocation of the pars intermedia, pars nervosa, and medial eminence from late prometamorphosis to late metamorphic climax is similar in *Bufo americanus, Rana pipiens,* and *R. sylvatica.* The horizontal axis from the medial eminence to the pars nervosa assumes a perpendicular or transverse orientation, and the pars intermedia and adenohypophysis are drawn dorsally. By early metamorphic climax, the adenohypophysis loses direct contact with the medial eminence and pars intermedia (Etkin 1968; Hanke 1976).

Three types of pituitary cells occur in larvae (Batten and Ingleton 1987; Hanke 1976; Van Oordt 1974; figs. 5.32A–B): thyrotrops (Nieuwkoop/Faber 48), a smaller Type I basophil (stage 52), and adrenocorticotrops or ACTH cells. The A1 acidophils secrete prolactin, and A2 acidophils that secrete somatotropin are detectable after metamorphic climax. During metamorphic climax (Nieuwkoop/Faber 57) they are the most numerous cell type in the ventral half of the pars anterior.

Thyroid Gland

The thyroid begins to differentiate at Sedra/Michael 27–28 (Gosner 17–18) as a ventral diverticulum near the second visceral arch (Etkin 1936). At Sedra/Michael 35–36 (Gosner 20) the groove along the anterior anlage has disappeared. At Sedra/Michael 42 (Gosner 24) the anlage separates from the buccopharynx, contacts the basibranchial, and consists of two lobes bounded by a fibrous capsule and connected by an anterior, saddle-shaped structure. At Sedra/Michael 46 (Gosner 26–27) the lobes are separated, colloid-containing follicles are differentiated, and prefollicular cells are present. At Sedra/Michael 50 (Gosner 29) each lobe divides and comes into close contact with the hypobranchial plate. Vacuoles appear in the follicular cells, and a homogeneous basophilic colloid fills the follicular lumen. Beginning with Sedra/Michael 58 (Gosner 41) the lobes are about 1.5 times that at Sedra/Michael 57, the columnar cells are tall, and there are large chromatophobe droplets (Michael and Al Adhami 1974).

In *Xenopus laevis* (Nieuwkoop/Faber 33–34), the thyroid anlage is situated medially near the first visceral pouch. At Nieuwkoop/Faber 35–36 it grows posteroventrally as a thin strand of cells under the skin. The anlage thickens, splits into two lobes (Nieuwkoop/Faber 37–38), comes to lie near the branching point of the truncus arteriosus at stage 39, and is

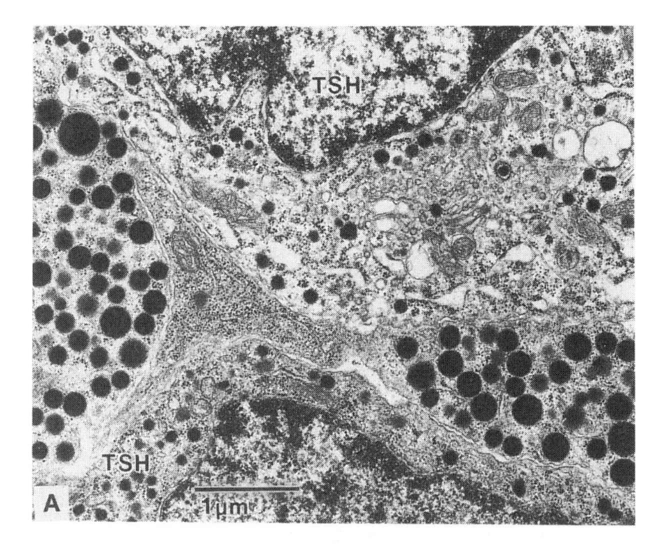

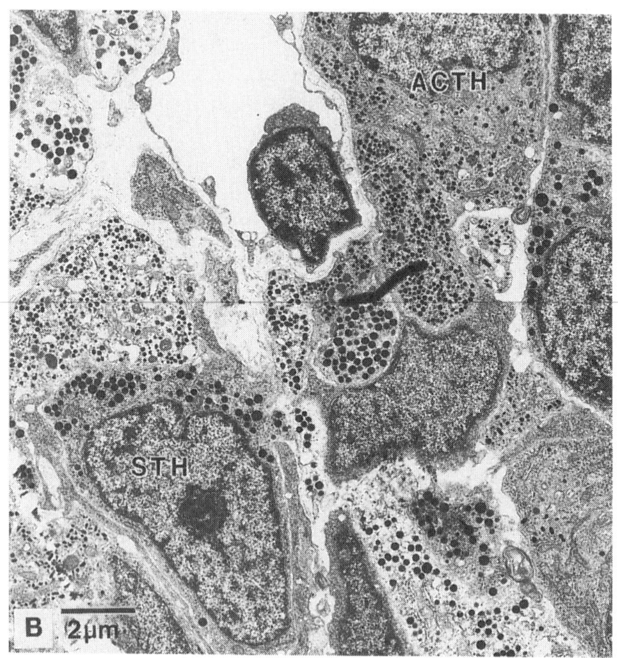

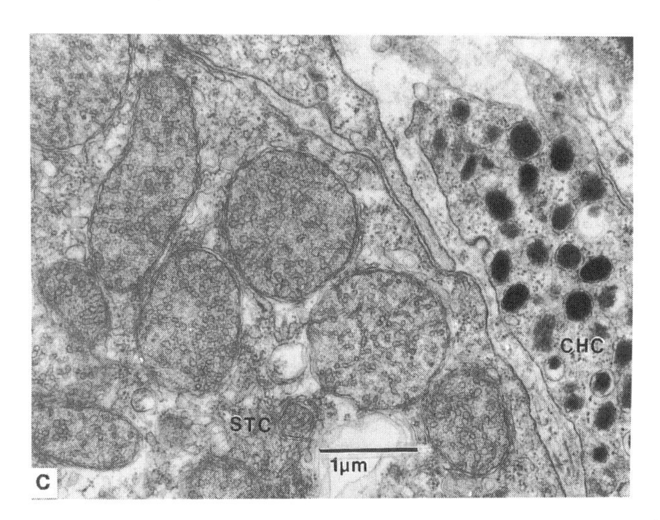

Fig. 5.32. (A–B) Adenohypophysis and (C) interrenal tissue and adrenal cortex of *Xenopus laevis* larva. Abbreviations: ACTH = cells that produce adrenocorticotropic hormone, Type III basophils, CHC = medullary chromaffin cell, STC = corticular steroidogenic cell, STH = somatotrophic cell (produces growth hormone), and TSH = cells that produce thyroid stimulating hormone, Type I basophils. Courtesy of Hanke.

connected with the epithelium of the buccopharyngeal floor by a solid cell strand (ductus thyreoglossus). This connection disappears by Nieuwkoop/Faber 40, and the thyroid forms two longitudinal lobes. At Nieuwkoop/Faber 41 the definitive thyroid gland appears as a paired organ lying ventral to each ventral velum and, by stage 43, on both sides of the hyoid arch. The ductus thyreoglossus disappears by Nieuwkoop/Faber 41. Follicles with colloid vesicles appear at Nieuwkoop/Faber 46–47 and are well developed and lined by mucous and cuboidal cells by stage 48–50; follicle mass and number increase, and the interfollicular connective tissue expands at about the same time. During early prometamorphosis, the flat follicle epithelium becomes cuboidal and then columnar (Jayatilaka 1978; Nieuwkoop and Faber 1956). Nanba (1972) outlined the development of the thyroid in *Rana japonica,* and Hayes (1995) and Hayes and Wu (1995) discussed the interdependence of corticosterone and thyroid hormones in the larvae of *Bufo boreas.*

Endocrine Pancreas

The islets of Langerhans develop from cells lining branching ducts or capillaries that extend through the exocrine pancreas tissue. The precursors of α-cells (= glucagon or A cells) are not visible before Gallien/Houllion 27–28 in *Discoglossus pictus* (Beaumont 1970) but are visible at Shumway 24 in *Rana pipiens* (Kaung 1981). The first α-granules are synthesized by the Golgi apparatus. At Gallien/Houllion 31 the α-cells are differentiated (Pouyet and Beaumont 1975, *Alytes obstetricans*). The β-cells (= insulin or B cells) are differentiated during limb-bud stage. At Taylor/Kollros II in *R. pipiens,* β-cells and insulin are already present and increase until Taylor/Kollros XVIII and XIX, respectively (Kaung 1983).

In *Xenopus laevis* the endocrine tissue starts to develop at Nieuwkoop/Faber 42 (Fox 1984), although islet cells are not differentiated before stage 48. Fox (1984) suggested that β-cells at Nieuwkoop/Faber 42 in *X. laevis* may be undifferentiated A cells or centroacinar cells of the adult pancreas. The endocrine pancreas of *Xenopus* metamorphs (after Nieuwkoop/Faber 61) consists of acini surrounded by α- and β-cells. After metamorphosis the endocrine cells occupy about 3% of the total pancreatic tissue, and there is a drastic increase in insulin production in prometamorphic and metamorphic climax larvae (from Taylor/Kollros XI; Buchan 1985; Frye 1964). After an initial decrease of β-cells during metamorphic climax (Taylor/Kollros XIX–XX), stasis in Taylor/Kollros XXI–XXV is followed by another decrease during later stages (Kaung 1983, *Rana pipiens*). During prometamorphosis and metamorphic climax, pancreatic weight and insulinlike immunoreactivity decreases (Hulsebus and Farrar 1985, *R. catesbeiana*).

Adrenal Gland

Steroidogenic cells of the adrenal glands are of mesodermal origin, and the chromaffin or catecholamine cells are modified neurons derived from the neural crest (Accordi and Grassi Milano 1990). Mesodermal cells proliferating from the dorsal root of the dorsal mesentery at the same level as the coeliacomesenteric artery extend to the opisthonephric anlage (Nieuwkoop/Faber 42–43, *X. laevis*). The interrenal primordium cells migrate dorsally to the ventral side of the aorta and fill the space posterior to the mesenteric artery between the two posterior cardinal veins. A medial peduncle forms ventrally in contact with the mesenteric dorsal root. When the two posterior cardinal veins unite to form the interrenal vein, the medial peduncle is split and the interrenal anlage is compressed between the interrenal vein and the ventral side of the aorta (Nieuwkoop/Faber 44–45). Between Nieuwkoop/Faber 46–50, two cell masses lie ventral to the aorta and dorsal to the interrenal vein along the medial face of the opisthonephric anlage. At Nieuwkoop/Faber 50 the precursors of chromaffin cells migrating from the sympathetic ganglia region penetrate the interrenal anlage. During Nieuwkoop/Faber 57–66 the adrenal tissue reaches its definitive position on the medioventral face of the opisthonephric kidneys (Accordi and Grassi Milano 1990; Grassi Milano and Accordi 1986; Nieuwkoop and Faber 1956). The position of the adrenal glands of *Bufo bufo* at Rossi XI changes similarly during metamorphic climax (Accordi and Grassi Milano 1977).

Precursors of chromaffin cells can be detected in low densities in *Bufo bufo* at the end of embryonic development (Rossi XXV). The number of chromaffin and cortical (steroidogenic) cells increases at the beginning of premetamorphosis (Rossi I–III) (fig. 5.32C). At the beginning of premetamorphosis, chromaffin cells are still in an embryonic state (Rossi IV) but are differentiated at the end of premetamorphosis (Rossi VI; Accordi and Grassi Milano 1977). In premetamorphic larvae, polygonal steroidogenic cells form the interrenal cords and enclose blood sinuses, and chromaffin cells form small disparate groups (Accordi and Grassi Milano 1990; Chester-Jones 1987; Grassi Milano and Accordi 1986). Two types of chromaffin cells have developed at the end of premetamorphosis. At the beginning of metamorphic climax (Rossi XII) the steroidogenic cells form thin cords that lie parallel to the surface of the gland and surrounded by connective tissue.

Acidophilic Stilling's or summer cells (Stilling 1898), probably related to steroidogenesis (Accordi and Grassi Milano 1990; Chester-Jones 1987), are found only in ranids and rhacophorids after metamorphosis and vary with season (Accordi and Cianfoni 1981; Grassi Milano et al. 1979; Grassi Milano and Accordi 1983, *Rana esculenta* complex, *R. montezumae,* and *Polypedates leucomystax*).

Corticosteroids are synthesized by the steroidogenic cells present at Nieuwkoop/Faber 54 in *Xenopus laevis* (Leloup-Hatey et al. 1990). Corticosterone concentration varies from about 15 ng ml^{-1} during premetamorphosis and at the end of metamorphic climax with a peak of 50 ng ml^{-1} at the middle of metamorphic climax. Aldosterone concentrations peak at the start and end of metamorphosis (Jolivet-Jaudet and Leloup-Hatey 1984; Leloup-Hatey et al. 1990).

Summary

Compared with other amphibians, larval and adult anurans are quite different, and these differences provide a unique opportunity to study a free-living and actively feeding developmental stage of a tetrapod. They develop larval organs in the sense of precursors or transitory stages of different graduations. These precursor organs (e.g., heart, artery, vein system, and lymphatic system) are not simply organs in waiting, they are highly adapted to aquatic life and are significant for larval survival. Anatomical characters of the heart and arteries are essential for a water and air breather. The lymphatic system, concentrated in the buccopharyngeal region, is the buffer of the filter-feeding larva against microorganisms ingested along with the irrigation current. Some of these organs and the endocrine glands are resorbed, rebuilt, and changed at metamorphosis. Branchiogene endocrine glands do not reach their definite position before metamorphosis. The skin is a respiratory, sensory, and mucus-producing organ, and few characters other than the lateral line are changed during metamorphosis. As is typical for precursor organs, only parts or certain anatomical features are especially significant for larval life and thus are changed during metamorphosis.

In contrast, the buccal structures, filter apparatus, and larval stomach (manicotto glandulare, an adaptation for suspension feeding) are almost totally resorbed at metamorphic climax. These organs are significant for special adaptations to aquatic life (e.g., suspension feeding) and are larval organs senso stricto even if a clear distinction between precursor organs and typical larval organs is not always possible. Aortic arches are typical adaptations to aquatic larval life that are reduced during metamorphosis, and some parts become the primary arteries of the lungs and heart. The larval excretory system consists of transitory and persistent parts that allow storage and transport of the gonadal products of the adult. The development of the kidneys begins with the differentiation of the pronephros (larval organ, sensu stricto), which is a transitional larval excretory organ. During larval life, the urinary function shifts to the opisthonephric kidney of the adult. The ducts of the opisthonephric kidneys are associated intimately with the reproductive system, and gonads develop from adjacent tissues; after the excretory ducts have developed, the reproductive system usually taps into them.

ACKNOWLEDGMENTS

The parts of this chapter written by B. Viertel were supported by the Ernst-Kalkhof Foundation. He thanks C. C. Bangnara (Utrecht), J. Bengassat (Jerusalem), H. Fox (London), G. Frank (Wien), W. Hanke (Karlsruhe), J. D. Horton (New Orleans), J. Meseguer (Murica), G. Pekny (Wien), J. B. Turpen (Omaha), and M. Ueck (Giessen) for kindly supplying illustrations. He also is indebted to B. Grillitsch and S. Richter for valuable bibliographical advice. He wishes to thank I. Galbraith (Wiesbaden) for his untiring work as a translator. He is indebted to K. Rehbinder for her drawings and M. Ullmann (Mainz) for her work in the photographic laboratory. He is grateful to the following publishers and copyright holders for permission to reproduce pictures and tables: Faber and Faber Ltd. Publishers (London); Taylor and Francis Ltd. (London); McGraw-Hill, Inc. (New York); Royal Society of South Africa (Cape Town); Academic Press (Orlando); Plenum Publishing Corporation (New York); Editrice Compositori Bologna (Bologna); Anne Orcet Livres Scientifiques (Pouilly Sur Loire); Cambridge University Press (Cambridge); American Society of Zoologists (Lawrence); Oxford University Press (Oxford); Springer-Verlag (Heidelberg); Alan R. Liss, Inc. (New York); Vaillant-Carmanne S. A. (Liège); Gustav Fischer Verlag (Jena); E. J. Brill (Leiden); Akademische Verlagsgesellschaft Geest und Portig K. G. (Leipzig); and the Zoological Society of London (London).

S. Richter is grateful to B. Viertel for his keen and continuous support. She is also grateful to A. Tanimura, H. Iwasawa, and I. Naito for their valuable and spontaneous help in providing illustrations. She is indebted to M. Stachowitsch for assistance in English translations and to S. Neulinger for preparing the illustrations.

6

INTEGRATION

Nervous and Sensory Systems

Michael J. Lannoo

The most refined methods of anatomical analysis cannot reveal the things that are of greatest significance for understanding the nervous system. Our primary interest is in the behavior of the living body, and we study brains because these organs are the chief instruments which regulate behavior.

Herrick 1948:5

Introduction

The nervous system, consisting of the brain and its cranial nerves and the spinal cord and its nerves (fig. 6.1), underlies all vertebrate behaviors. Vertebrates detect environmental cues filtered through their various senses and send these signals centrally via peripheral nerves. This information is processed by either relatively simple spinal reflexes or more complex brain circuitry, and commands are sent back through peripheral nerves to drive muscles and glands in some presumably adaptive manner. An adaptation is defined here as a feature that contributes to the survival of an individual or to the survival of its offspring (Liem and Wake 1985).

For most anurans, the tadpole neural circuitry is remodeled abruptly at metamorphosis. Sensory systems, muscle groups, and glands are reworked during the transition from aquatic, essentially limbless, swimming larvae to, in most cases terrestrial, quadrupedal, saltatory adults. During this remarkable transition, the nervous system must shift its ability to process and integrate the information received and the systems being affected between two very different types of organisms. This transition means that in addition to the usual and important questions of neural structure and function, the tadpole nervous system requires a further level of analysis; unlike that of most other vertebrates, the nervous system of the aquatic tadpole must be considered within the context of its rapid metamorphosis into the nervous system of a terrestrial frog. In this light, three questions concerning the neurobiology of tadpoles are addressed. (1) What portion of a tadpole's nervous system reflects what a tadpole actually uses? (2) Which systems must be reworked at metamorphosis? (3) What tadpole neural tissues are developed for later use by the adult? Also, interspecific variations in the behavior and ecology of tadpoles reflected in their neuronal organization and function must be considered.

The general neurobiology of anuran tadpoles is less well known than that of either anuran embryos or adults. For example, Spemann discovered the mesodermic induction of ectoderm to neuroectoderm in anuran embryos (Hamburger 1988). Adult anurans were used in the discoveries of the workings of the neuromuscular junction (Birks et al. 1960) and the rules underlying neuronal specificity within the visual system (e.g., Constantine-Paton et al. 1983; Reh et al. 1983; R. W. Sperry 1963). An excellent review of the neurobiology of amphibians, including adult structures, is provided by Wilczynski (1992). More recently, studies of tadpoles have provided an understanding of the growth of the cerebellum (e.g., Hauser et al. 1986a, b, and Uray

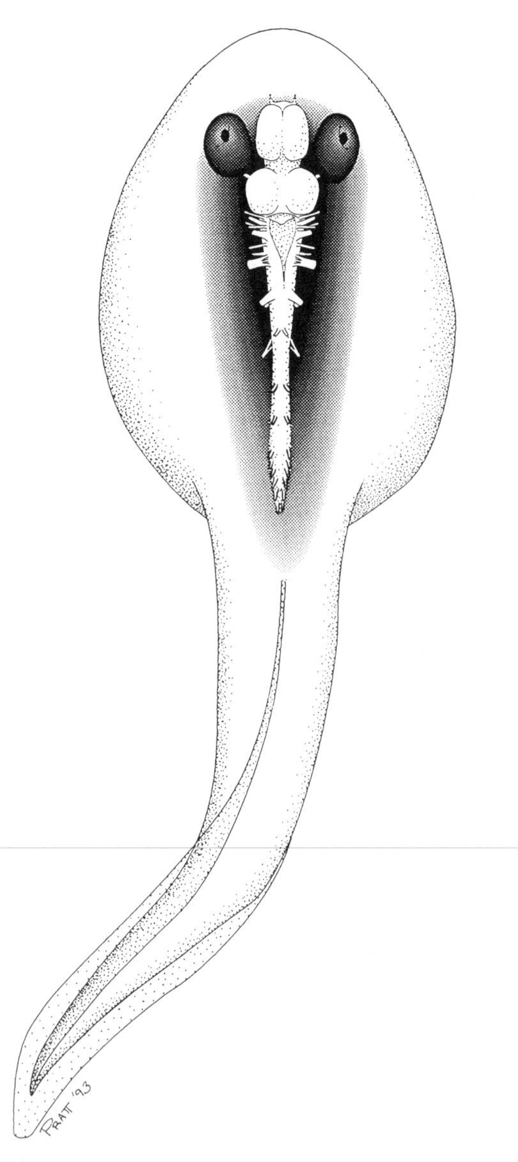

Fig. 6.1. Dorsal view of the brain and spinal cord in a generalized tadpole. Cranial and spinal nerves are shown as if they have been severed.

1985), the formation of and variation in the spinal cord (K. Nishikawa and Wassersug 1988, 1989; Nordlander 1984, 1986), and sexual differences within the nervous system (Gorlick and Kelley 1987).

The information contained in this chapter largely results from research on species of *Rana* and *Xenopus* with most other information based on species of *Bufo* and *Hyla*. The details of the organization of higher brain centers are known only for adults and are presented as a starting point for understanding the brain of tadpoles. To avoid the pitfalls of generalizing about "the tadpole," I point out known phylogenetic and ecological variations. I introduce the nervous system and the organization of the tadpole brain and then examine the organization and function of the spinal cord and brain, discuss the various sensory systems used by tadpoles, and summarize the organization of the tadpole nervous system with a brief consideration of the changes that it undergoes during metamorphosis.

An Overview

The nervous system can be divided into (1) peripheral (PNS) versus central (CNS) systems, with the CNS divided further into the brain and spinal cord (figs. 6.2–6.3); (2) somatic (external or voluntary) and visceral (internal or involuntary) systems; (3) sensory, integration, and motor systems; and (4) neurons and their more numerous supportive glial cells. Neural processing begins with information acquired through the sense organs. External sensory systems detect the types and intensities of energy present in the environment and translate this information into neuronal signals. Sensory systems act as filters for the reception of types of stimuli with adaptive value while presumably eliminating those that are less important. Internal sensory systems monitor the visceral environment and insure that the animal's physiological mechanisms are operating within their limits of tolerance.

The brain, which receives sensory information and transmits motor commands either through cranial (CN) or spinal (SN) nerves (figs. 6.2–6.3), is divided into a forebrain (prosencephalon), midbrain (mesencephalon), and hindbrain (rhombencephalon; fig. 6.4; also see D. Black 1917). The forebrain is divided into an anterior telencephalon and a posterior diencephalon (fig. 6.4), which function in receiving olfactory and terminal nerve information and integrating sensory inputs and motor outputs. One region of the diencephalon, the hypothalamus, regulates hormone production and involuntary autonomic responses. The midbrain consists of a dorsal, sensory tectum and a ventral, motor tegmentum (fig. 6.4). The midbrain integrates first-order (primary) visual inputs with higher order inputs from the lateral line, auditory, and vestibular systems. The hindbrain includes the cerebellum and medulla oblongata (fig. 6.4), and the combination of the mid- and hindbrains forms the brain stem. The area receives primary inputs from all of the sensory systems except the terminal nerve and olfaction, including vision, acousticolateral, gustatory, and general sensations, and sends motor commands to the muscles and glands of the head, cervical region, and viscera. The reticular formation of the brain stem is involved with integrating sensory inputs and with overall arousal. The cerebellum mediates motor learning and coordination.

The spinal cord is both a relay center and controller of reflexive behaviors and receives sensory information from specialized touch, temperature, pain, and postural receptor organs through its dorsal nerve roots. The spinal cord either directly provides the motor output to the muscles of the trunk, tail, and developing limbs (reflex behavior; see Shiriaev and Shupliakov 1986 for the anatomy of these connections) through its ventral roots or first relays these sensations to the brain and then transmits the brain commands back to the muscles. Within the brain stem and gray matter of the spinal cord, sensory centers tend to be dorsal, autonomic

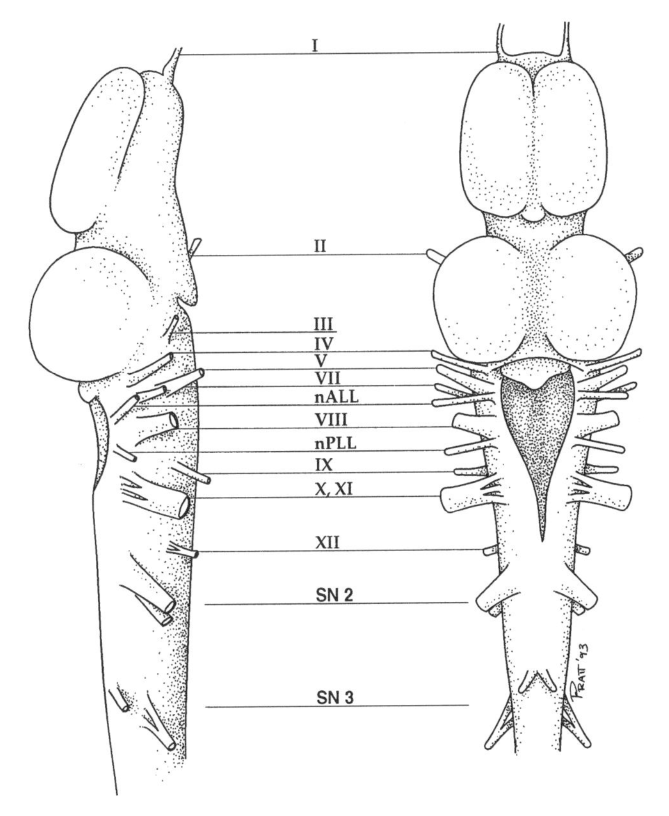

Fig. 6.2. Lateral *(left)* and dorsal *(right)* views of a generalized tadpole brain showing the positions of the cranial, lateral line, and selected spinal nerves. Cranial nerves (CN): I = olfactory (CN 0 is included with CN I), II = optic, III = oculomotor, IV = trochlear, V = trigeminal, VII = facial, VIII = acousticovestibular, IX = glossopharyngeal, X = vagus, XI = accessory, and XII = hypoglossal. Lateral line (LL) nerves: nALL = anterior division and nPLL = posterior division. Spinal nerves (SN): 2 and 3 are the anteriormost spinal nerves. From Stuesse et al. (1983) and Lowe (1987); reprinted by permission of Wiley-Liss, a subsidiary of John Wiley & Sons, Inc., and of Stuesse and Lowe.

centers are lateral and intermediate, and motor centers are ventral (fig. 6.5B).

In addition to neurons, the vertebrate brain contains glial cells of three basic types: astrocytes, ependymal cells, and oligodendrocytes. Glial cells are thought to provide the skeleton of the brain, to provide nutrition to neurons, and, in some specialized cases, to guide developing neurons as they migrate to their final position. Oligodendrocytes form the myelin sheaths that encircle neuronal axons.

The autonomic nervous system regulates involuntary body functions and insures that an organism is operating within its physiological limits. The sympathetic and parasympathetic components typically act antagonistically. The sympathetic system is energy expending and considered the "fight or flight" component; it halts gut peristalsis, tightens gut sphincters, dilates pupils, and increases heart rate (Pick 1970). The parasympathetic system is energy conserving and effects the opposite results. The most structurally visible features of the autonomic nervous system are the paired sympathetic trunks, each a series of ganglia that extend from anterior to posterior and parallel to the spinal cord (fig. 6.6; Taxi 1976). Sympathetic ganglia receive preganglionic fibers that travel through communicating rami from the spinal nerves (fig. 6.7; W. M. Davis and Nunnemacher 1974 and Schlosser and Roth 1995). About one quarter of the fibers in a spinal nerve are preganglionic parasympathetics (fig. 6.8; Vance et al. 1975). Preganglionic fibers enter the sympathetic trunk and may course either anteriorly or posteriorly

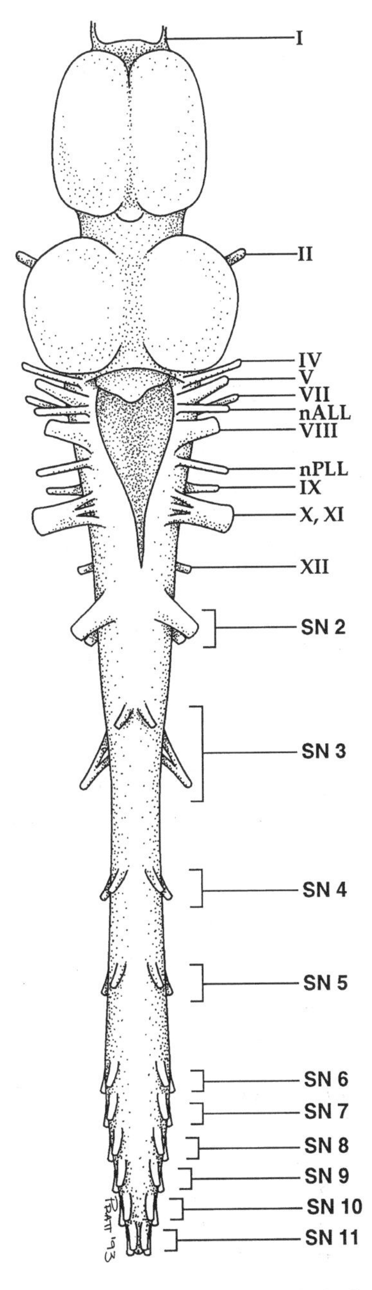

Fig. 6.3. Dorsal view of a generalized brain and spinal cord of a tadpole showing cranial and spinal nerves. Cranial nerves III and VI and the posteriormost 10–18 pairs of spinal nerves which innervate the tail segments have been omitted for clarity. Abbreviations as in fig. 6.2.

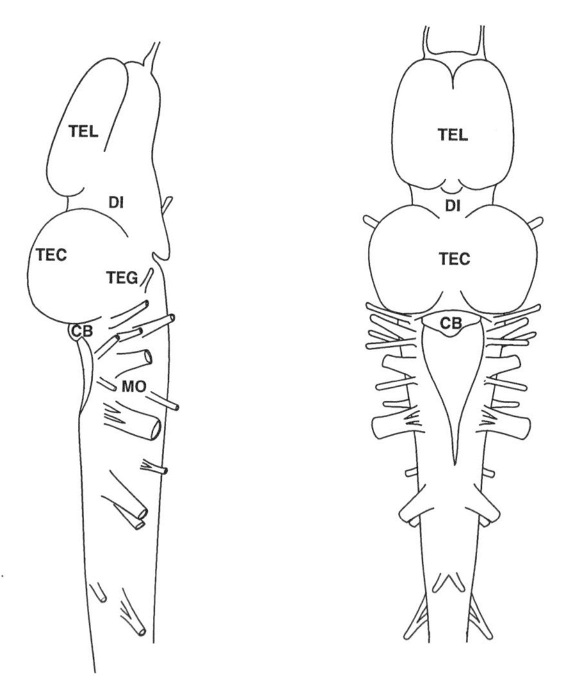

Fig. 6.4. Lateral *(left)* and dorsal *(right)* views of the tadpole brain. Abbreviations: Forebrain: TEL = telencephalon and DI = diencephalon; midbrain: TEC = tectum and TEG = tegmentum; and hindbrain: CB = cerebellum and MO = medulla oblongata.

through two or three ganglia prior to synapsing on postsynaptic neurons. Postsynaptic neurons leave the sympathetic trunk and follow the major arteries to visceral organs. Presynaptic neurons may also synapse distal to the sympathetic truck in visceral plexuses (figs. 6.6–6.7). For example, the solar plexus in anurans is formed by branches emanating from sympathetic ganglia 3, 4, and 5 that supply the stomach and adjacent parts of the alimentary canal (see C. A. Brown et al. 1992). The sympathetic ganglia and nerves of all vertebrates develop from neural crest cells. Neurons of the parasympathetic system leave the central nervous system either via cranial nerves or through the sacral parasympathetic system. Parasympathetic fibers from the laryngeus ventralis branch cranial nerve CN X provide parasympathetic innervation of the smooth muscles of the lungs, heart, and stomach.

Oka et al. (1989) observed parasympathetic neurons by backfilling CNs V, VII, IX, and X branches with cobalt lysine. Parasympathetic cell bodies form an almost continuous column through the brain stem dorsal to the corresponding motor neurons of the same cranial nerves. Heathcote and Chen (1991) detailed the development of the parasympathetic cardiac ganglion in *Xenopus laevis*. The sacral parasympathetic system functions in reproduction and for this reason may not be well developed in tadpoles.

Spinal Cord

The spinal cord occupies the neural canal within the vertebral column and is round or oval in cross section (figs. 6.7–6.8). Cell bodies (gray matter) are organized centrally into an H-shaped pattern with bilateral dorsal and ventral horns. Dorsal horns receive inputs from the sensory dorsal roots; ventral horns contain the motor neuron pools that supply the muscles of the trunk, tail, and limbs. Within the ventral horns, cells in both the medial and lateral motor pools tend to innervate the trunk and tail muscles, while additional cells from the lateral motor pool innervate the limb muscles (Ebbesson 1976; A. Roberts and Clarke 1982; Silver 1942). An intermediate gray area that contains preganglionic autonomic fibers is poorly understood in amphibians (Ebbes-

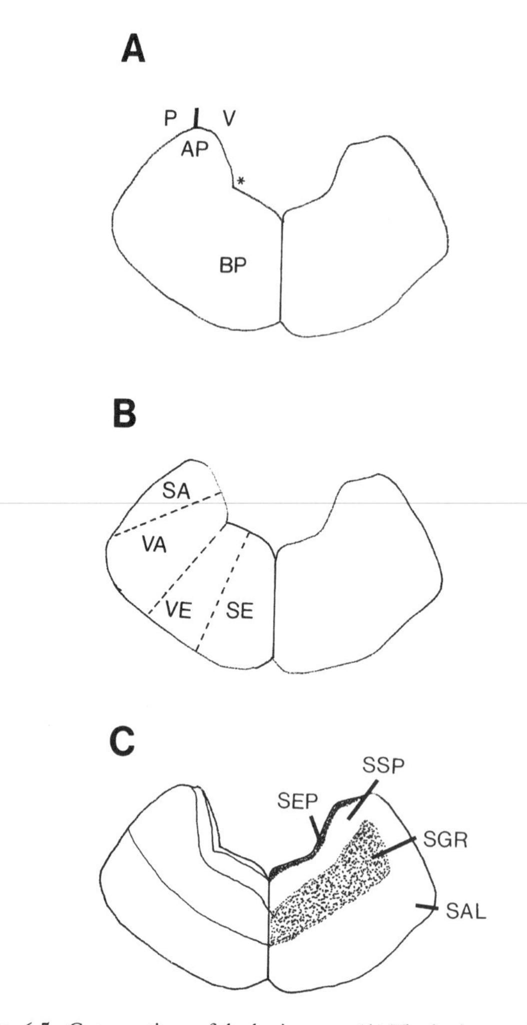

Fig. 6.5. Cross sections of the brain stem. (A) The brain stem has an inside (ventricular) and outside (pial) surface, and the dorsal, vertical alar plates are separated from the ventral, horizontal basal plates by the sulcus limitans in the lateral ventricular wall. (B) Cranial nerve and brain stem functional components are arranged into four longitudinal columns. (C) The cellular organization of the brain stem. Asterisk indicates position of sulcus limitans, and dotted lines separate longitudinal columns. Abbreviations: AP = alar plate, BP = basal plate, P = pial, SA = somatic afferents (sensory), SAL = stratum albumen (major area of fiber tracts), SE = somatic efferents, SEP = stratum ependymale cells, SGR = stratum griseum (major cell body layer), SSP = stratum subependymale cells, V = ventricular, VA = visceral afferents, and VE = visceral efferents (motor).

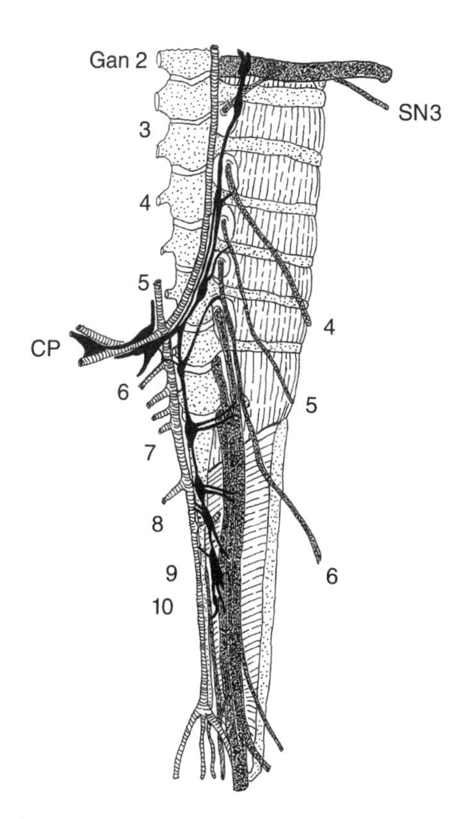

Fig. 6.6. The sympathetic trunk showing ganglia 2–10 and the coeliac (gut) plexus of an adult *Rana esculenta*. Ganglion 8 is split in the specimen drawn, and the ventral rami of spinal nerves 3–6 are indicated. Abbreviations: CP = coeliac plexus, Gan = ganglion, and SN = spinal nerve. Redrawn from Taxi (1976) with modifications of the original figure suggested by Dr. Taxi; reprinted by permission of Springer-Verlag and Taxi.

son 1976). Ascending and descending axons, organized in fiber tracts (white matter), surround the gray matter. Forehand and Farel (1982a, b), Holder et al. (1987), Nordlander (1987), Nordlander and Singer (1982a, b), Nordlander et al. (1991), A. Roberts (1976), A. Roberts and Patton (1985), A. Roberts and Taylor (1983), and J. S. H. Taylor and Roberts (1983) described the development of spinal neurons and fiber pathways in anuran larvae.

Variation in the morphology of the cord along its length depends on the type and number of its connections. The spinal cord, especially its white matter, is wider nearer the brain than at thoracic or posterior levels so that the cord tapers anteriorly to posteriorly (fig. 6.3). Imposed upon this tapering, the spinal cord is wider at the cervical and lumbar levels to accommodate the sensory and motor neurons that form the brachial and lumbar plexuses in the limb regions (K. Nishikawa and Wassersug 1988 and Sutherland and Nunnemacher 1981, *Hyla* and *Eleutherodactylus*). Lateral motor pool neurons increase in number with development of the limbs in the tadpole.

The spinal cord supplies each body segment with a pair of spinal nerves (figs. 6.7 and 6.9; K. Nishikawa and Wassersug 1988) numbered according to the vertebral level at which the pair exits. Proceeding down the cord, spinal nerves arise from the cord progressively more anterior than they exit the neural canal (fig. 6.9). Nerves exit the neural canal between the vertebrae, except for SN 10 and 11 (if present), which exit via foraminae in the coccyx. The morphology of the spinal cord and spinal nerves and their relationships to tail myotomes through metamorphosis have been studied by K. Nishikawa and Wassersug (1988) for both *Rana* and *Xenopus*. Spinal nerves range from 23 to 29 pairs primitively (e.g., ascaphids and discoglossids), and reductions have occurred independently at least seven times during anuran evolution. The posterior end of the spinal cord, termed the filum terminale, in *R. catesbeiana* and *R. pipiens* (Chesler and Nicholson 1985) contains a functional neuropile supported by a large number of glial cells (H. Sasaki and Mannen 1981).

Tadpoles have a ventral root in the position of SN 1 that is apparently homologous to CN XII in other vertebrates and is designated here as such. All other spinal nerves have a dorsal root with a large ganglion (M. R. Davis and Constantine-Paton 1983 and Wilhelm and Coggeshall 1981) and a ventral root that fuses with the dorsal root distal to the ganglion (fig. 6.7). Peripheral to this junction, a small branch of the spinal nerve passes dorsally to innervate the skin and muscles of the dorsal trunk. The large ventral branch of each spinal nerve innervates ventral and lateral skin muscles of the trunk and tail. The ultrastructural development of ventral roots has been studied by Nordlander et al. (1981). A ramus communicans containing preganglionic

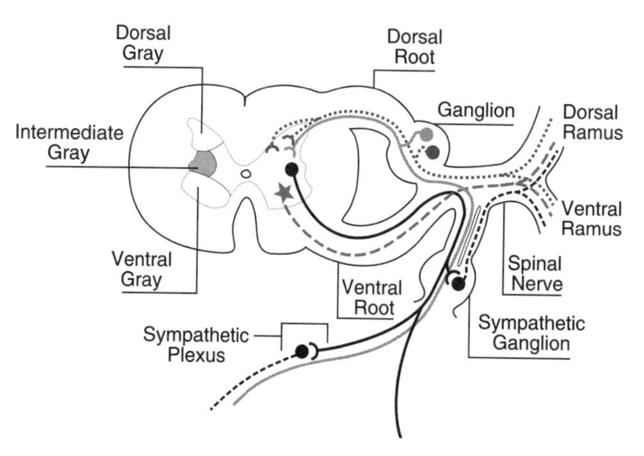

Fig. 6.7. A schematic representation of the organization of the spinal cord. The left side of the illustration shows dorsal and ventral horns composed of dorsal, intermediate, and ventral gray matter, and the white matter (fiber tracts) is peripheral. The right side shows the types of spinal fibers and their locations, including the sympathetic trunk and a sympathetic plexus. Symbols (dorsal to ventral): gray dotted = somatic afferents bringing general pain, pressure, temperature, and touch sensation from cutaneous receptors as well as afferents from muscle spindles through the dorsal ramus; gray solid = afferent fibers bringing visceral feedback; black solid = preganglionic sympathetic fibers to ganglia in the sympathetic trunk and visceral plexes; black dashed = postganglionic fibers to visceral organs to control autonomic functions; and gray dashed = somatic efferents to voluntary muscles through the ventral ramus. Redrawn from Noden and De Lahunta (1985); reprinted by permission of Williams & Wilkins Co., Noden, and De Lahunta.

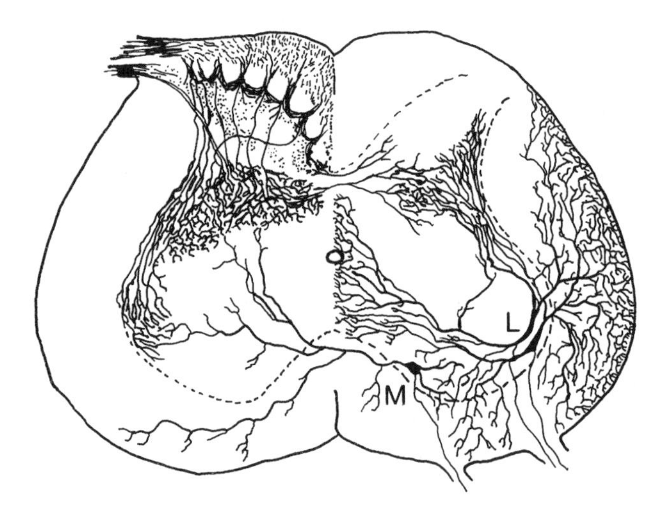

Fig. 6.8. Fiber types in the anuran spinal cord. The left side of the illustration shows dorsal sensory fibers and a dendritic tree from an axial motor neuron from the right side; on the right there are three motor neurons that send axons ventrally to the periphery. The ventromedial cell represents an axial motor neuron of the medial cell group (M), and the more lateral cells represent limb motor neurons from the lateral cell group (L). The transition between white matter and gray matter is shown by a dashed line. From Székely and Czéh (1976) after Székely (1976); reprinted by permission of Springer-Verlag and Székely.

sympathetic fibers extends from each ventral branch to the sympathetic nerve trunk.

Most anuran tadpoles have more than 20 pairs of spinal nerves (K. Nishikawa and Wassersug 1989), which reduce to about 10 pairs after metamorphosis. In *Rana,* SNs 2 and 3 join a connecting ramus to form the brachial plexus (e.g., Oka et al. 1989). Spinal nerves 7 and 8 form the lumbar plexus, and SN 7 gives off a branch (the iliohypogastric) to muscles of the lateral and ventral body wall. Spinal nerves 7 and 8 may fuse with SNs 9 and 10. The organization of neurons into nerve trunks from SNs 8 and 9 to the gluteus muscle in *Bufo marinus* adults was studied by D. R. Brown et al. (1989). The ventral branches of the spinal nerves between the brachial and lumbar plexuses give off branches to the lateral and ventral body wall musculature and to the skin. In anurans branches of SNs 9 and 10 form a plexus that innervates the bladder, cloaca, oviducts (in adult females), and lymph hearts.

Developmentally, prior to the formation of the spinal ganglia, peripheral sensation is provided by large, dorsal Rohon-Beard cells (fig. 6.10; Bixby and Spitzer 1982; Hughes 1957; Spitzer and Spitzer 1975). These cells are first present in the spinal cord at gastrulation (Lamborghini 1980), increase in numbers that peak in tadpoles, and are then reduced and eventually lost at metamorphosis (Decker 1976; Eichler and Porter 1981; Lamborghini 1987).

In the sensory spinal cord each nerve from the dorsal root courses centrally into the dorsal horn of the gray matter and divides into a medial and lateral division (e.g., Antal et al. 1980; see fig. 6.8 for general appearance of incoming sensory fibers). The lateral division consists of a small-diameter fiber that appears to terminate in the dorsal horn. The medial division consists of a large-diameter fiber that enters the dorsal funiculus and splits into ascending and descending collaterals. The ascending collateral courses cranially into the brain stem, where it continues into the stratum album (fig. 6.5C). Axons in this tract terminate in the hindbrain vestibular nuclei (spinovestibular fibers), the reticular formation (spinoreticular fibers; figs. 6.11–6.12), and in the diencephalon.

In anurans, cells from the dorsal root ganglion also project directly into the cerebellum (Székely et al. 1980). In the spinal cord these fibers ascend in the ipsilateral dorsal funiculus (Gonzalez et al. 1984; Joseph and Whitlock 1968; Van Der Linden and Ten Donkelaar 1987; Van Der Linden et al. 1988), enter the cerebellum, and synapse directly on Purkinje cells (Ebbesson 1976). In the cerebellum these dorsal spinocerebellar fibers mingle with fibers of the ventral spinocerebellar tract (a component of the ventral white matter). The anuran spinal cord differs from that of all other vertebrates except turtles (Künzle 1983) in having this direct projection from the dorsal root ganglia into the cerebellum. The function of this projection is not known.

In the ventral spinal cord, the dendrites of motor neurons

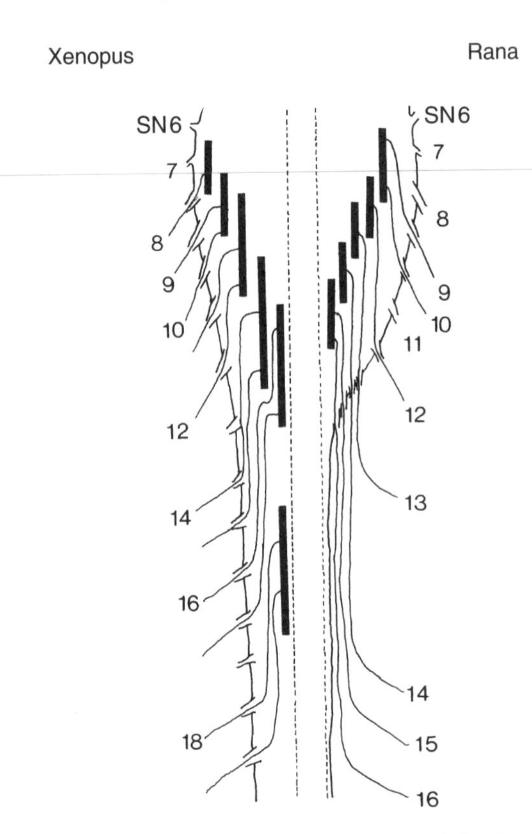

Fig. 6.9. Schematic view of the posterior spinal cords in *Xenopus laevis* and *Rana catesbeiana* tadpoles. The solid bars indicate the position of the cell bodies of motor neurons. Note that the spinal nerves (SN) exit the cord progressively more posteriorly than the position of their cell bodies and that the spinal cord extends more posteriorly in *Xenopus.* Redrawn from K. Nishikawa and Wassersug (1988); reprinted by permission of Wiley-Liss, a subsidiary of John Wiley & Sons, Inc., and of Nishikawa.

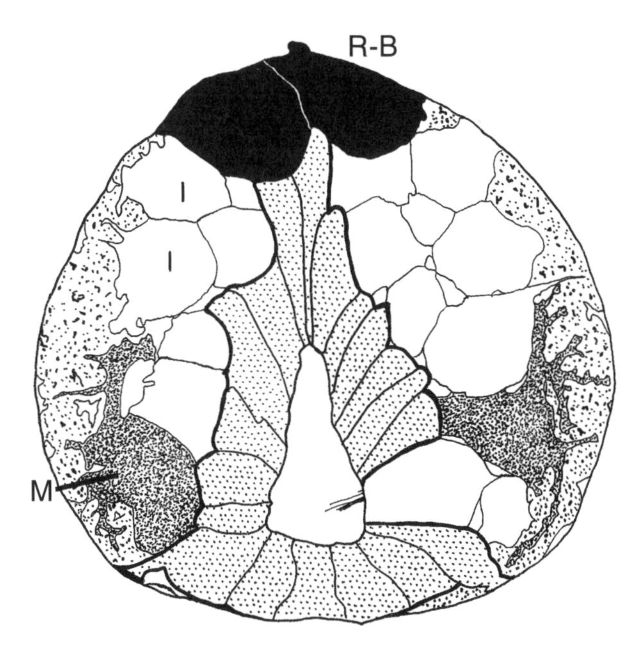

Fig. 6.10. The spinal cord in a young tadpole. Abbreviations and symbols: I *(no stipple)* = developing interneurons, M *(dark stipple)* = developing motor neurons, and R-B *(dorsomedial black)* = Rohon-Beard cells. The developing central canal *(central open space)* is delimited by germinal zones *(light stipple)*. Developing ascending and descending fiber tracts *(irregular stipple)* are located laterally. Redrawn from A. Roberts and Clarke (1982); reprinted by permission of The Royal Society of London and Roberts.

project laterally into the white matter, where they are contacted by axons descending from the brain stem (see fig. 6.8). The morphology of motor neuron dendrites has been studied extensively (Antal et al. 1986; Bregman and Cruce 1980; Rosenthal and Cruce 1985). Babalian and Shapovalov (1984) studied the synaptic interactions between single ventrolateral tract fibers on motor neuron dendrites.

The anuran spinal cord receives inputs from a variety of sources within the brain (Ten Donkelaar and De Boer–Van Huizen 1982). Afferents from the forebrain diencephalon travel in the ipsilateral thalamospinal tract (Ebbesson 1976). The bulk of this projection is to the two anterior spinal segments where neurons effect neck movements involved with fixing gaze in adult frogs. The presence and function of these projections in tadpoles is unknown. Vestibular neurons in the medulla also project to the spinal cord via the vestibulospinal tract, where they terminate within the dendritic fields of both medial and lateral motor neuron pools. Vestibular inputs influence posture and orientation (Stensaas and Stensaas 1971).

Reticulospinal pathways are poorly understood in anurans. Medial reticular cells project to the cervical spinal cord, and lateral reticular cells project to the lumbar spinal cord. Accumulating evidence suggests that reticulospinal neurons form part of the pattern generating circuits of a variety of behavioral patterns including swimming (Van Mier and Ten Donkelaar 1988). Corvaja and d'Ascanio (1981) investigated the spinal projections from the midbrain in *Bufo,* described the reticulospinal, rubrospinal, tectospinal, and trigeminospinal tracts; and characterized two smaller projections, one from the posterior ventral nucleus of the midbrain and one from the interstitial nucleus of the medial longitudinal fasciculus. A. Roberts and Alford (1986) described the role of descending neurons in producing fictive swimming in *Xenopus.*

In the ventral spinal cord, the number of motor neuron cell bodies is matched to the number of muscle motor units (McLennan 1988; Prestige 1967, 1973; Rubin and Mendell 1980; D.C. Sperry and Grobstein 1983). There is a developmental overproduction of motor neurons, and cells that do not establish connections or the proper type of connections with motor units eventually die (Ferns and Lamb 1987; Hughes 1961; Kett and Pollack 1985; Lamb 1981; McLennan 1988). The result appears to be an appropriate matching of motor neuron number and size with motor unit numbers and sizes (D. C. Sperry 1987). The organization of cervical spinal cord motor nuclei differs between anurans and caudates (D. B. Wake et al. 1988).

The development of spinal motor neurons and the mechanisms these cells use to match motor unit numbers is an active area of research (e.g., Farel and Bemelmans 1985, 1986; Lamb 1981; Lamb et al. 1989; McLennan 1988; D.C. Sperry and Grobstein 1983). Nordlander (1986) described the normal development of motor neurons in the tail of *Xenopus* tadpoles, and D.C. Sperry and Grobstein (1985) examined the effects of hormonal manipulation on lumbar motor neuron numbers and cell sizes in *Xenopus.* Van Mier et al. (1985) studied the development of motor neuron dendrites in first- and second-order cells within the ventral horn. C. L. Smith and Frank (1988a, b) and Van Mier and Ten Donkelaar (1988) examined the peripheral specificity of sensory connections in the developing spinal cord of *R. catesbeiana.*

Brain

Hindbrain (rhombencephalon)

The hindbrain consists of the medulla (caudal portion of the brain stem), isthmus (connection between the medulla and the midbrain), and cerebellum. The spinal cord grades into the medulla so that in gross appearance it is difficult to determine exactly where the spinal cord ends and the hindbrain begins (figs. 6.3–6.4).

The organization of the brain stem (hindbrain + midbrain) is generally the same as the spinal cord, with sensory elements located dorsally and motor elements ventrally (fig. 6.5). The vertically oriented sensory portion of the brain stem is termed the alar plate, and the horizontally oriented motor portion is termed the basal plate (fig. 6.5A). The division between these two areas is termed the sulcus limitans and is visible along the wall of the IVth ventricle (fig. 6.5A). The brain stem can be divided further into longitudinal columns that represent functional similarity (fig. 6.5B). The somatic motor column lies along the midline, somatic sensory portions lie along the lateral border, the visceral motor column is in a mediocentral position, and the visceral sensory

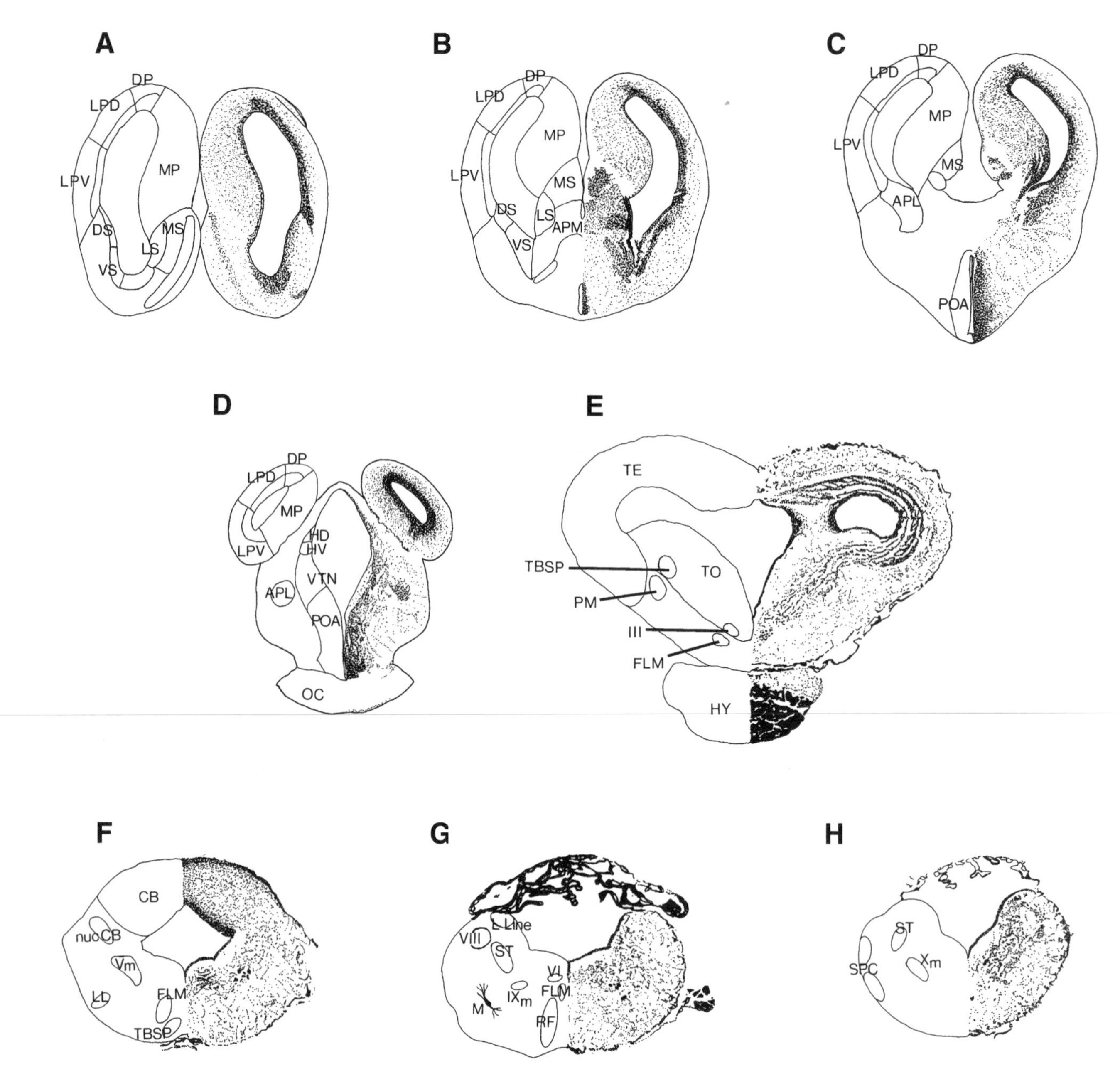

Fig. 6.11. A series of transverse sections through the tadpole brain. In each section, key nuclei and tracts are shown on the left, and gross histological appearance is shown on the right. Some structures may be located slightly anterior or posterior to the sections shown. Abbreviations: APL = lateral portion of the amygdala, APM = medial portion of the amygdala, CB = cerebellum, DP = dorsal pallium, DS = dorsal striatum, FLM = medial longitudinal fasciculus, HD = dorsal habenula, HV = ventral habenula, HY = hypothalamus, LL = lateral lemniscus, L Line = terminal area for primary lateral line afferents, LPD = lateral dorsal pallium, LPV = lateral ventral pallium, LS = lateral septum, M = Mauthner cell, MP = medial pallium, MS = medial septum, nuc CB = nucleus of the cerebellum, OC = optic chiasm of CN II, POA = preoptic area, PM = profundus mesencephali, RF = reticular formation, SPC = spinocerebellar tract, ST = solitary tract, TBSP = tectobulbar and tectospinal tract, TE = tectum, TO = torus, VS = ventral striatum, VTN = ventral thalamic nucleus, III = nucleus of CN III, VI = nucleus of CN VI, Vm = motor nucleus of CN V, VIII = terminal area for auditory and vestibular primary afferents, IXm = motor nucleus for CN IX, and Xm = motor nucleus for CN X. Redrawn from Nieuwenhuys and Opdam (1976); reprinted by permission of Springer-Verlag and Nieuwenhuys.

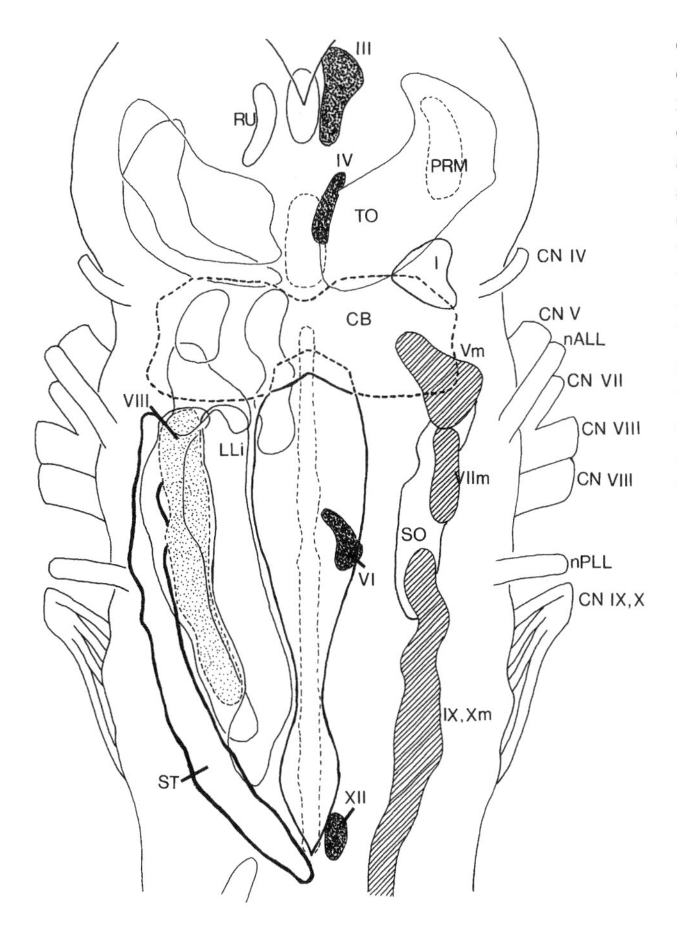

Fig. 6.12. Horizontal schematic section showing key brain stem nuclei through portions of the dorsal *(left)* and ventral *(right)* brain stem. Somatic motor nuclei are indicated by dark stipple, visceral motor neurons by hatching, and octaval region by light stipple. Abbreviations: CB *(heavy dashed line)* = cerebellum, CN IV = trochlear nerve, CN V = trigeminal nerve, CN VII = facial nerve, CN VIII = auditory/vestibular nerve, CN IX and X = common accessory and vagal nerve root, I = nucleus isthmi, LLi = lateral line region, nALL = anterior lateral line nerve, nPLL = posterior lateral line nerve, PRM = nucleus profundus, RU = nucleus ruber, SO = superior olive, ST = solitary tract, TO = torus, III = oculomotor nucleus, IV = trochlear nucleus, Vm = trigeminal motor nucleus, VI = abducens nucleus, VIIm = facial motor nucleus, VIII = octaval region *(light stipple),* IX and Xm = motor nuclei of the accessory and vagal spinal nerves, and XII = hypoglossal nucleus. Reprinted from Nikundiwe and Nieuwenhuys (1983); reprinted by permission of Wiley-Liss, a subsidiary of John Wiley & Sons, Inc.

column is laterocentral (fig. 6.5B). This guide can be used to determine the approximate locations of first- and second-order nuclei and their tracts (Nikundiwe and Nieuwenhuys 1983 and Opdam et al. 1976).

Histologically, four laminae within the brain stem are superimposed upon these organizational columns (fig. 6.5C). A medial lamina of cells termed the ependymal layer lines the fourth ventricle (figs. 6.5A, C). These small cells send a single neurite covered with numerous short appendages radially into more lateral laminae. Laterally adjacent to the ependymal layer, the subependymal layer is completely free of cells but contains the radial neurites of the ependymal cells in the brain stem. The stratum griseum contains the majority of the cell bodies. These cell bodies can be either large or small with large cells frequently grouped into clusters. The stratum albumen is the most peripheral and is a continuation of the peripheral fiber tracts of the spinal cord. Axons within the stratum albumen send collaterals or terminal axons into the stratum griseum. The stratum albumen also contains the terminations of ependymal cells and some cell groups embedded within the fiber layers. Among amphibians, the brain stem of anurans is considered to be more differentiated than that of caudates. Nikundiwe and Nieuwenhuys (1983) used cell staining techniques to characterize the nuclei of the brain stem in *Xenopus,* and these authors describe seven primary efferent (motor) nuclei, 13 primary afferent (sensory) nuclei, 7 reticular formation nuclei, and 15 "relay" nuclei (see Opdam et al. 1976 for an interpretation of the brain stem organization in *Rana*).

The brain stem receives or sends fibers from all of the cranial nerves except CN 0 (= terminal nerve) and CN I (= olfactory nerve) which project into the telencephalon. In the brain stem the division between the mid- and hindbrain occurs posterior to the level of CN IV (figs. 6.3). The somatic sensory column (SA, somatic afferents; fig. 6.5B; see longitudial organization above) is located dorsolaterally and receives inputs from—proceeding back to front—the posterior division of the lateral line nerve (nPLL), CN VIII, the anterior division of the lateral line nerve (nALL), and the sensory component of CN V (fig. 6.12). Within the acoustico-vestibulo-lateral line complex, the vestibular nucleus is most ventral, the two acoustic nuclei are central, and the lateral line nucleus is dorsal. The nuclei for the general and special visceral afferents (VA; fig. 6.5B) are located within the midlateral brain stem and consist of components of CNs V, VII, IX, and X. In general, these nuclei contain diffuse assemblages of cells and are not easily recognized (Nikundiwe and Nieuwenhuys 1983; Opdam et al. 1976). The nuclei for the general and special visceral efferents (VE; fig. 6.5B) occupy the mediocentral brain stem. The general visceral efferent is the parasympathetic innervation (carried by CNs III, IV, IX, and X). The special visceral efferent component consists of the neutrons projecting to muscles derived from branchiomeres (carried by CNs V, VII, IX, X, and XI). These visceral motor nuclei form a continuous column through the brain stem (Ebbesson 1976; Nikundiwe and Nieuwenhuys 1983; Oka et al. 1989; Opdam et al. 1976). The most medial brain stem column contains the somatomotor cell bodies (SE, somatic efferents; fig. 6.5B) that innervate the tongue (CN XII) and the extraocular eye muscles (CNs III, IV, and VI; fig. 6.12). Stuesse et al. (1983) reported that the hypoglossal nerve (CN XII) in *Rana pipiens* contains a dorsomedial and a ventrolateral subnucleus.

The reticular formation is located in the ventromedial portion of the brain stem (fig. 6.11G) and, although not shown completely here, projects from the midbrain through the hindbrain and into the cervical spinal cord. Its cells and fibers are diffuse. The reticular formation receives collaterals from brain stem nuclei and tracts involved in reflex arcs. The superior olive (fig. 6.12) is located within the basal plate and receives neurons from the dorsal octaval nucleus (CN VIII; Fuller and Ebbesson 1973) and the midbrain tegmentum (Wilczynski and Northcutt 1977). The olive sends fibers to the midbrain torus semicircularis (Nikundiwe and Nieuwenhuys 1983 and Opdam et al. 1976). The nucleus isthmi (fig. 6.12) may be involved in processing binocular inputs in adults and in tadpoles with overlapping visual fields. Most tadpoles do not have binocular vision and the function of this nucleus in tadpoles with lateral eyes is not clear. The nucleus isthmi is reciprocally connected to the midbrain tectum (Grobstein and Comer 1983; Udin et al. 1992; Wilczynski and Northcutt 1977).

Among fiber tracts the medial longitudinal fasciculus (figs. 6.11E, F, G) is a phylogenetically old feature of the vertebrate brain stem. This tract contains reticulospinal, vestibulomesencephalic, and vestibulospinal axons and mediates the vestibulo-ocular reflex. The lateral lemniscus (fig. 6.11F) carries second-order auditory and lateral line nerves from their primary nuclei to the midbrain torus semicircularis. (Wilczynski 1988; Will 1988). Senn (1972) detailed the development of these and other brain stem structures in *Rana temporaria.*

Mauthner neurons (fig. 6.11G) are the largest axons in the brain stem and appear to initiate the startle response of tadpoles, although other cells may also contribute (R. K. K. Lee and Eaton 1989). Mauthner neuron connections have been described in *Bombina, Bufo terrestris, Hyla cinerea, Kaloula pulchra, Polypedates leucomystax, Rana esculenta, Scaphiopus holbrookii,* and *Xenopus laevis* (Will 1986). Mauthner cell dendrites are contacted by fibers from the CNs VIII and V, lateral line neurons, and midbrain neutrons (Cioni et al. 1989; Will 1986). The spinal motor neurons that receive input from the Mauthner cells in larval amphibians are the earliest to develop (Blight 1978; Nordlander et al. 1985). Stefanelli (1951) reported that Mauthner cell degeneration is related to tail resorption at metamorphosis even in species with aquatic adults, while Moulton et al. (1968) and Will (1986) revealed the persistence of Mauthner cells in adult anurans.

The sensory and motor connections of CN XII (fig. 6.11H; Stuesse et al. 1983) of *Rana pipiens* originate from two nuclei located medially and laterally within the cauda medulla. Motor fibers from the medial nucleus project to the extrinsic tongue muscles in adults. Motor fibers from the lateral hypoglossal nucleus innervate the sternohyoid muscle. Sensory fibers arise mainly from the tongue, and in the brain stem, they travel posteriorly in the dorsolateral fiber tract to thoracic levels. The development of these motor and sensory connections in tadpoles probably depends on the extent of tongue development.

The central projections of the CN IX–X complex (fig. 6.12) were mapped by Stuesse et al. (1984; adult *Rana catesbeiana*) and Simpson et al. (1986; adult *Xenopus laevis*). Motor fibers of CN IX arise in the brain stem from a small ventrolateral nucleus, and those of CN X originate from a slightly more posterior nucleus (Stuesse et al. 1984). Afferents from CNs IX and X enter the dorsal roots of the respective nerves and descend in two tracts, either the solitary tract (fig. 6.12) or the spinal tract of CN V and the dorsolateral tract of the spinal cord. In *Xenopus,* CNs IX and X fibers are within the three roots of the IX–X complex and within the nPLL (Simpson et al. 1986). The most anterior of these IX–X roots (root 1) contains sensory fibers that terminate in the solitary tract and on lateral line efferents. Root 2 contains somatosensory fibers that terminate in the posterior medulla and anterior spinal cord and motor fibers from CNs IX and X. Root 3 contains motor fibers to the laryngeal muscles and efferents to the viscera. The results of these studies confirmed the same general conclusions of Rubinson and Friedman (1977) on *Rana catesbeiana, R. pipiens,* and *Xenopus muelleri.*

Lateral line, acoustic, and vestibular afferents terminate in different areas of the somatic sensory column (figs. 6.11–6.12; Fritzsch 1988; McCormick 1988; Will 1988; Will and Fritzsch 1988). Herrick (1948) and Larsell (1934) proposed that lateral line target neurons receive auditory fibers at metamorphosis. Fritzsch et al. (1984) and Jacoby and Rubinson (1983) did not support this hypothesis based on modern tract-tracing techniques. Lateral line fibers project into an intermediate nucleus, while auditory fibers project to a distinct nucleus located more laterally and ventrally (Fritzsch et al. 1984). According to Altman and Dawes (1983), some anterior fibers in the posterior division of the CN LL (nPLL) may be cutaneous afferents that project to the spinal cord in a manner similar to that of CN V. Within the lateral line system, second-order neurons project to the contralateral midbrain torus semicircularis (figs. 6.11–6.12) via the lateral lemniscus (fig. 6.11F).

Efferent neurons (not shown) project from the brain stem back to receptors and prevent hair cells from firing. These fibers originate from a CN VIII nucleus located ventrally and medially within the medulla (Fritzsch 1981a; Russell 1976). Primary afferents from the amphibian and basilar papillae (fig. 6.13, see auditory system) project into one or two nuclei at the dorsolateral edge of the alar plate (figs. 6.11–6.12; Altman and Dawes 1983; Fritzsch et al. 1988a, b; Jacoby and Rubinson 1983). Rubinson and Skiles (1975) suggested that primary auditory afferents also project directly to the superior olivary nucleus. Second-order projections from the primary auditory nucleus project to the superior olivary nucleus (Rubinson and Skiles 1975) and the midbrain torus semicircularis (fig. 6.12; Nikundiwe and Nieuwenhuys 1983; Opdam et al. 1976, *Rana;* Wilczynski 1988).

Sensory fibers from neurons innervating the semicircular canals, utricle, saccule, and lagena project into the medullary vestibular complex(fig. 6.13). Sensory vestibular fibers also project directly to the cerebellum (Altman and Dawes 1983). Efferents from the brain stem to the vestibular end organs arise from the same nucleus as lateral line efferents, and single efferent neurons may send axons to both the lateral line and vestibular systems (Claas et al. 1981). Based on second-order projections, the vestibular complex can be divided into anterior and posterior divisions (not shown). In *Rana*

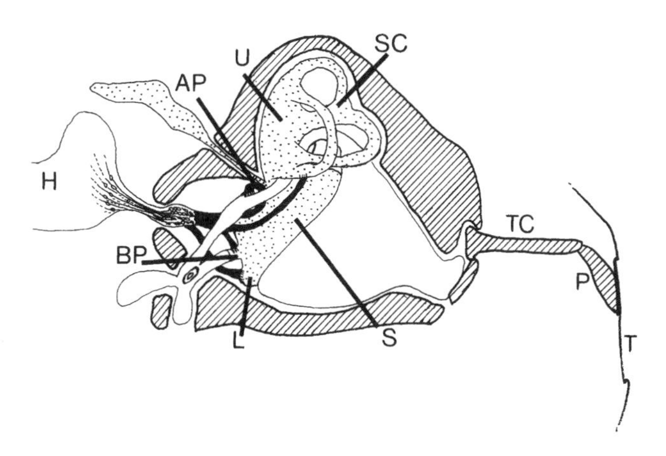

Fig. 6.13. The organization of the anuran ear in adults. In tadpoles, a bronchial columella pierces the dorsal aorta, and the plectrum and tympanic columella have not yet formed. Abbreviations: AP = amphibian papilla, BP = basilar papilla, H = hindbrain, L = lagena, P = plectrum, S = saccule, SC = semicircular canals, T = tympanum, TC = tympanic columella, and U = utricle. Redrawn from Capranica (1976); reprinted by permission of Springer-Verlag and Capranica.

pipiens the anterior portion projects to the nuclei of CNs III, IV, and VI and contributes to the vestibulo-ocular reflex; posterior efferents form the vestibulospinal pathway (Montgomery 1988).

The trigeminal nerve is composed of a motor nucleus that innervates the muscles of mastication (fig. 6.12), a midbrain (mesencephalic) sensory nucleus that mediates masticatory forces, a principle sensory nucleus, and a spinal sensory tract that is an anterior continuation of the spinal dorsolateral column (Nikundiwe and Nieuwenhuys 1983, *Xenopus;* Opdam et al. 1976, *Rana*). These last two trigeminal nuclei mediate somatosensation from the head.

The mesencephalic and motor trigeminal nuclei are involved in a reflex arc. Resistance to jaw closure in both tadpoles and adults is relayed through receptors to cells in the mesencephalic trigeminal nucleus (see Kollros and McMurray 1956 for cellular details of this nucleus) that project to motor trigeminal neurons and prevent further jaw closure. S. Lewis and Straznicky (1979) examined the development of these neurons. Motor trigeminal cells formed embryonically are responsible for buccal movements in both the tadpole and the adult. Alley and Barnes (1983) and Barnes and Alley (1983) found that the same cells that innervate the tadpole muscles innervate the adult muscles, despite the facts that adult muscles first arise at metamorphosis and that the tadpole and adult muscle groups subserve different feeding behaviors. These authors indicated that trigeminal motor neurons are recycled and respecified during metamorphosis. Omerza and Alley (1992) showed that about 80% of motor axons supplying adult muscle fibers originate from tadpole neuromuscular junctions.

Cerebellum

The cerebellum is perhaps the easiest neuronal structure to recognize because of its location (figs. 6.11F–6.12), highly regular cellular organization, and stereotyped inputs. The cerebellar cortex is composed of an external molecular layer, a thin layer of large Purkinje cells, and an internal granular cell layer (e.g., Herrick 1948; Larsell 1967; Sotelo 1976; Székely et al. 1980). Deep to the cortex, cerebellar nuclei (fig. 6.11F) serve as relay centers for afferent and efferent cerebullar fibers (Gonzalez et al. 1984; Montgomery 1988). The cerebellum receives inputs from the spinal cord, vestibular nuclei, lateral line system, and other hindbrain nuclei (Grover and Grüsser-Cornehls 1984; Larsell 1925; Sotelo 1976). Neurons in the dorsal root ganglia of the cervical and lumbar regions project directly to the cerebellum (Joseph and Whitlock 1968; Székely et al. 1980). These fibers terminate in a somatotopic pattern. A second type of spinocerebellar projection, arising from secondary cells in the spinal cord, is present in both dorsal and ventral spinocerebellar tracts (Van Der Linden et al. 1988) and terminates as mossy fibers in the cerebellar granular cell layer. Climbing fibers originate in the medulla at the location of the inferior olive. Efferents from the cerebellar nuclei project to the contralateral cerebellar nucleus and to the spinal cord. Cerebellar efferents also project bilaterally to the basal plate of the medulla and to the midbrain (Montgomery 1988).

The cerebellum functions in motor learning and coordination (Freeman 1965) and in anurans consists of the corpus, paired auricles, and a nodulus. The cerebellum does not fully mature until metamorphosis (e.g., Gona et al. 1982; Kollros 1981). The appearance of the corpus cerebellum in tadpoles is associated with the development of the tail musculature and the spinocerebellar tracts. The cerebellum auricles, which may receive and process lateral line inputs, are more fully developed in tadpoles than adults (Larsell 1967). The protracted development of the anuran cerebellum has made this structure a useful model for researchers interested in pattern formation within the nervous system (e.g., Hauser et al. 1986a, b; Uray 1985; Uray and Gona 1979, 1982; Uray et al. 1987, 1988).

Midbrain (mesencephalon)

The midbrain is the anterior continuation of the brain stem and is divided into a ventral tegmentum, which continues the ventral motor columns, and the dorsal tectum, which continues as the medullary sensory columns (figs. 6.4 and 6.11). The tegmental nuclei include CNs III and IV and nucleus ruber (fig. 6.12; Nikundiwe and Nieuwenhuys 1983). The nuclei of CNs III and IV (along with the hindbrain CN VI and perhaps CN V) send motor fibers to the extraocular muscles (Straka and Dieringer 1991). In the tegmentum these nuclei are located dorsomedially. Cranial nerve III emerges from the brain stem ventrally, and CN IV emerges dorsally. Both nuclei receive vestibular fibers from the medullary vestibular nucleus. The torus semicircularis receives second-order lateral line (Plassman 1980; Will et al. 1985) and second- and third-order auditory inputs (Feng and Lin 1991; Potter 1965), which it processes and sends to the tectum. The torus semicircularis also receives ascending inputs from the spinal cord (Ebbesson 1976), reticular formation, hypothalamus (Neary and Wilczynski 1977), and reciprocal connections from the contralateral torus (Nikundiwe and Nieuwenhuys 1983; review by Northcutt 1980). The nucleus ruber forms reciprocal connections with the cerebellum (Larson-Prior and Cruce 1992).

Three additional tegmental nuclei have been identified. The nucleus profundus (fig. 6.12) receives fibers from the superior olive (Rubinson and Skiles 1975), but its function is unknown. The optic nucleus receives bilateral connections projecting from each retina (Levine 1980). The mesencephalic nucleus of CN V contains large sensory neurons that mediate masticatory forces. More cells occur in the trigeminal mesencephalic nucleus of tadpoles than in adults and during metamorphosis a large number of these cells die (Kollros 1984, *Rana pipiens;* Kollros and Thiesse 1985, *Xenopus laevis*). The midbrain tectum is located dorsal to the tegmentum (figs. 6.4 and 6.11E) and receives first-order inputs from retinal ganglion cells and third-order inputs from the lateral line and auditory systems (Ebbesson 1970; Keating and Gaze 1970; Lowe 1986). The tectum integrates sensory inputs with motor commands (Elepfandt 1988a, b, lateral line system, *X. laevis*). Tectal inputs were considered by Wilczynski and Northcutt (1977), and Lowe (1987) showed that single cells within the tectum respond to both lateral line and visual inputs. Vestibular and auditory signals also reach the midbrain tectum. Auditory and lateral line information terminates in the torus semicircularis before being relayed to the tectum (Feng and Lin 1991; Pettigrew 1981; Wilczynski 1981). The torus is homologous with the mammalian inferior colliculus, and the tectum is a homolog of the superior colliculus.

The tectum is the best-studied structure in the anuran brain. Retinal ganglion cell axons of tadpoles project to the contralateral tectum in an ordered fashion, an arrangement that facilitates the study of neuronal connections (e.g., Constantine-Paton and Capranica 1976; Debski and Constantine-Paton 1990; Reh et al. 1983). The formation of order within the nervous system is itself an important question. Recent papers on this subject involving anurans include Constantine-Paton (1987), Constantine-Paton et al. (1983), Fujisawa (1987), Grobstein et al. (1980); Levine (1980), Montgomery and Fite (1989), Reh et al. (1983), and Straznicky and Tay (1982). Details of the salamander tectum in the context of visual behavior were presented by G. Roth (1987), and Lázár et al. (1991) demonstrated differences in the pattern of tectal lamination between *Rana esculenta* and *Xenopus laevis.* Anterior tectal projections in the tectothalamic tract and the postoptic commissure (Lázár et al. 1983; Vesselkin et al. 1971) will be considered with the forebrain. Posterior tectal efferents project to the nucleus isthmi (fig. 6.12) and in turn connect bilaterally back to the tectum (Grobstein and Comer 1983; Gruberg and Udin 1978; Udin and Fisher 1985). Other posterior tectal projections include tectobulbar (bulb refers to the medulla) and tectospinal fibers which may be responsible for generating or transmitting motor commands (Lázár et al. 1983). The pretectal nucleus lentiformis (not shown) receives contralateral retinal inputs as well as inputs from the tectum and tegmentum in *Rana pipiens* (Montgomery et al. 1985). Montgomery et al. (1982) examined the mesencephalic nuclei responsible for the optokinetic response in *R. pipiens* adults.

Forebrain

The telencephalon and diencephalon (fig. 6.4) integrate sensory inputs and initiate motor outputs. The telencephalon appears to integrate sensory information beyond that found in the thalamic nuclei (Ebbesson 1980; Northcutt 1981). The telencephalon receives anterior inputs from the olfactory bulbs (CN I) and terminal nerve (CN 0) and posterior inputs from the visual, auditory, and somatosensory systems by way of thalamic (diencephalic) relays (Kicliter 1979; Northcutt 1974; Wilczynski and Northcutt 1983a). The diencephalon includes the epithalamus, consisting of the habenula and the pineal gland; the hypothalamus, which controls homeostasis and drives such as hunger and thirst (fig. 6.11); and the thalamus, which functions primarily as a relay nucleus. Stehouwer (1987) described the effects of forebrain ablations on the behavior of tadpoles.

The terminal nerve (CN 0), of unknown function in anurans, is associated with the olfactory nerve (Herrick 1909; Hofmann and Meyer 1989; also Szabo et al. 1990). Hofmann and Meyer (1989) reported that the terminal nerve in *Bufo marinus* and *Xenopus laevis* projects to the forebrain medial septum, preoptic nucleus, and the hypothalamus. The olfactory bulbs (CN I) located distal to the anterior telencephalon are connected to the brain via olfactory tracts. The targets of these tracts are the main olfactory bulb and the accessory olfactory bulb (fig. 6.14; Scalia et al. 1991a, b). The main olfactory bulb, accessory olfactory bulb, and the anterior olfactory nucleus are interconnected ipsilaterally. The main olfactory bulb and anterior olfactory nucleus project to the contralateral medial wall of the telencephalon. Ipsilateral main olfactory bulb fibers also terminate near the origin of the brain stem medial longitudinal fasciculus. The accessory olfactory bulb projects to the lateral cortex of the contralateral telencephalon via the lateral forebrain bundle. The amygdala, a nucleus associated with the olfactory system (fig. 6.14), projects to two nuclei within the ipsilateral hypothalamus and a nucleus posterior to the ipsilateral nucleus isthmi (fig. 6.12; Kemali and Guglielmotti 1977). Scalia et al. (1991a, b) provided detailed descriptions of the main and accessory olfactory bulb projections in *Rana pipiens.* Jiang and Holley (1992) described the olfactory bulb outputs in *R. ridibunda,* and Burd (1992) and Byrd and Burd (1991) described olfactory development in *Xenopus laevis.*

The posterior telencephalon is divided into four areas around each lateral ventricle (figs. 6.11 and 6.14). The dorsal part of the posterior telencephalon is divided into medial and lateral pallial areas. The septal area lies ventral to the medial pallium, and the striatal area lies ventral to the lateral pallium and lateral to the septum. The amygdala lies in the posterior striatum. The septum and striatum are synaptic stations where olfactory fibers join with fibers from the diencephalic thalamus and the midbrain. The telencephalon connects with posterior brain areas through the medial and lateral forebrain bundles (fig. 6.14). The medial forebrain bundle contains ascending and descending connections that

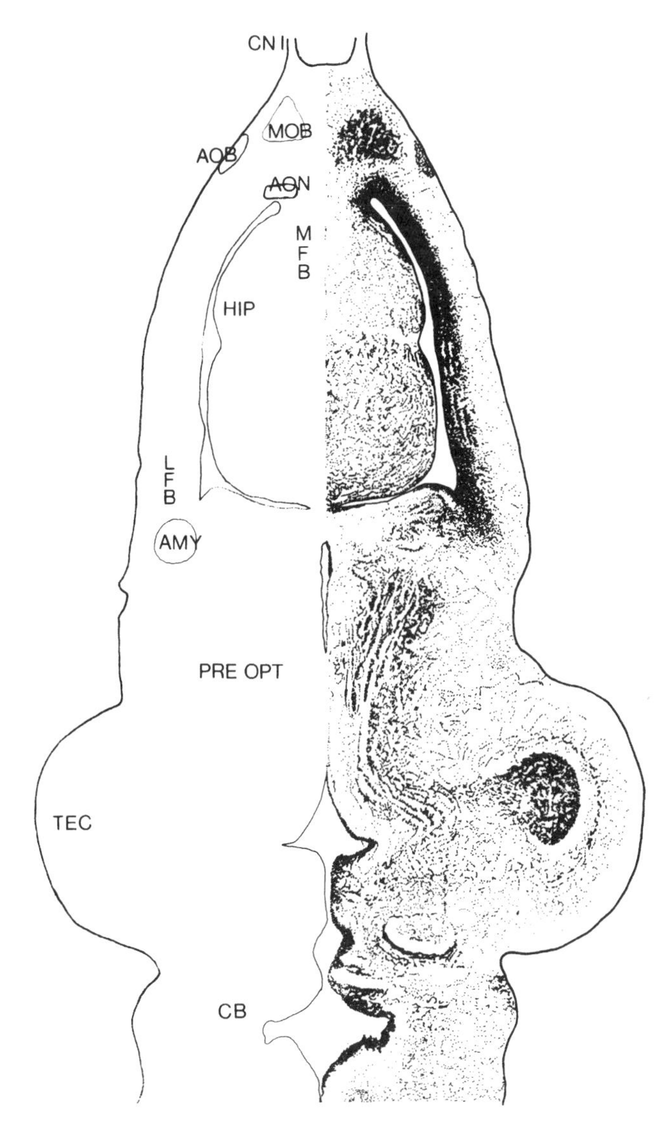

Fig. 6.14. Horizontal section of key forebrain nuclei. Locations of nuclei are shown on the left and gross histological appearance of the brain on the right. Abbreviations: AMY = amygdala, AOB = accessory olfactory bulb, AON = accessory olfactory nucleus, CB = cerebellum, CN I = cranial nerve I (olfactory, shown larger than life), HIP = hippocampus, MOB = main olfactory bulb, LFB = lateral forebrain bundle, MFB = medial forebrain bundle, PRE OPT = preoptic area, and TEC = tectum. Redrawn from Kemali and Braitenburg (1969) and Kemali and Guglielmotti (1987); reprinted by permission of Springer-Verlag.

join septal olfactory and limbic centers with the midbrain tegmentum and diencephalic preoptic and hypothalamic regions. Specific connections here include descending fibers from the septum that project to the ventral thalamus, preoptic area, and anterior hypothalamus. The lateral forebrain bundle connects the striatum with the posterior brain and contains afferents from thalamic nuclei, which in turn receive inputs from the midbrain torus and tectum (Wilczynski and Northcutt 1983a). Other lateral forebrain bundle fibers include inputs from the ipsilateral amygdala, the ventral thalamus, and adjacent preoptic areas. The striatum projects through the lateral forebrain bundle most heavily to the contralateral anterior endopeduncular nucleus, ventral thalamus, and preoptic area (Wilczynski and Northcutt 1983b).

Neary and Northcutt (1983) recognized two habenular nuclei within the epithalamus, two hypothalamic regions (preoptic area and infundibulum), three thalamic regions (dorsal, ventral, and tuberculum), and a transitional pretectal area in the diencephalon of adult *Rana catesbeiana*. The epithalamic habenula (fig. 6.11D, E) connects to the septum, hippocampus, hypothalamus, and thalamus (Kemali 1974; Kemali and Guglielmotti 1977) and is proportionally larger in animals with a well-developed forebrain. M. J. Morgan et al. (1973) described the left-right asymmetry in the habenulae of adult *R. temporaria*. The habenula plus the interpeduncular nucleus form the inner ring of the limbic system, which may be the highest correlative center in the anuran brain. Herrick (1948) suggested that this structure controls feeding.

The epithalamus also contains the pineal organ (epiphysis), which has a small projection (best developed in anurans within amphibians) to the dorsomedial surface of the epithalamus (fig. 6.15). These fibers contain the frontal and pineal tracts (parietal nerve) and enter the pretectal region in the posterior part of the epithalamus. In the lateral and ventral walls of the diencephalon, the thalamus forms a web of connecting fibers with all contiguous parts of the brain and serves as an important center for sensory correlation. For example, the optic nerves primarily project to the optic lobes of the midbrain, but along their course collateral fibers are given off to the thalamus (Fite et al. 1977) where they synapse with fibers from other sensory systems (Neary and Northcutt 1983). J. C. Hall and Feng (1987) considered the anatomy and physiology of the thalamic auditory region in adult *Rana pipiens* and presented evidence for parallel circuits.

Posteroventrally within the diencephalon, the bilobate hypothalamus (fig. 6.11E) is divided into preoptic and tuberoinfundibular regions. The preoptic nucleus is connected to the ventral thalamus and the ventral lobe of the hypophysis (pituitary). Neurons within the preoptic area and ventral hypothalamus respond to conspecific mating calls (Allison 1992, *Hyla cinerea*). The tuberoinfundibular nucleus is connected to the hypophyseal portal vessels. The general functions of the hypothalamus include activating the brain stem reticular formation; controlling the autonomic nervous system, metamorphosis, and pituitary gland; correlating gustatory, visceral, and olfactory inputs; and regulating feeding rates and water balance (Sarnat and Netsky 1981).

The hypothalamus and pre-optic area receive fibers from most regions of the forebrain, but telencephalic olfactory inputs and medullary gustatory inputs appear to predominate. Efferent fibers project to motor centers, the brain stem reticular formation, thalamus, forebrain, and olfactory limbic centers. The hypothalamus does not receive fibers from the major somatosensory pathways. In nonmammalian vertebrates, fibers of the optic tract that terminate directly in the

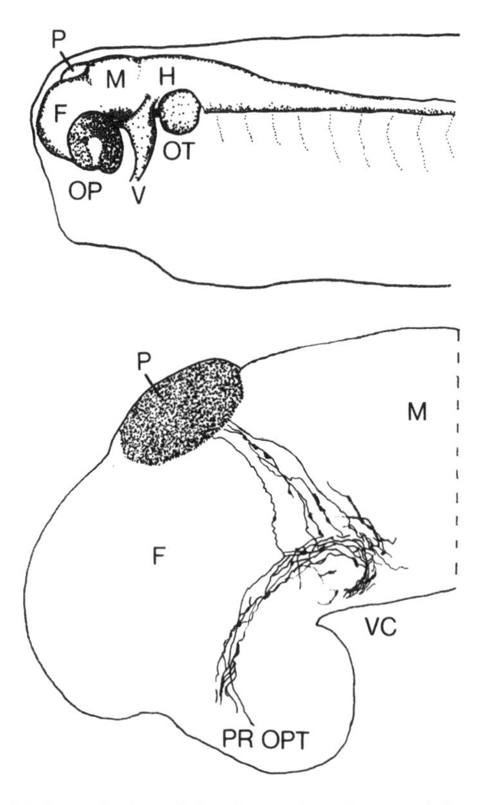

Fig. 6.15. The projection of pineal axons into the ventral diencephalon in embryonic anurans. Low-power diagram *(upper)* for orientation of brain regions within the embryo, and higher magnification *(lower)* to show features of the projections. Abbreviations: F = forebrain, H = hindbrain, M = midbrain, OP = optic placode, OT = otic placode, P = pineal gland, PR OPT = preoptic commissure, V = CN V, and VC = ventral commissure. Redrawn from R. G. Foster and Roberts (1982); reprinted by permission of Springer-Verlag and Foster.

preoptic area and hypothalamus may mediate color changes in the skin. The specific connections of the preoptic area of the hypothalamus (Neary and Northcutt 1983; fig. 6.14) that project to the pituitary are associated with the medial forebrain bundle, medial pallium, lateral amygdala, and infundibulum; see Fite (1985) for data on the pretectal area. The lateral forebrain bundle is the main afferent pathway for visual signals into the forebrain (Kicliter and Northcutt 1975). Information about the visual world, especially moving objects and changes in general illumination, is processed by telencephalic neurons.

The pituitary gland lacks neurons and is composed of glial astrocytes and pituicytes (Gerschenfeld et al. 1960). Within the pituitary, the neurohypophysis is located ventral to the hypothalamus and connected to it via a fiber tract called the infundibulum, and the adenohypophysis is located in a diverticulum in the roof of the mouth.

Sensory Systems

The sensory capabilities of generalized tadpoles include: audition and vestibular functions, mechanoreception through the lateral line system, olfaction, taste, vision, and miscellaneous senses (e.g., photodetection through the pineal complex and somatosensation—pain and postural sensation, temperature, and touch). For each system, the material will be presented in the sequence: receptor structure and function, peripheral innervation, first-order connections within the brain or spinal cord, development, and patterns of phylogenetic and ecological variation.

Auditory System

The anatomy, development, evolution, and physiology of the anuran auditory system have been well studied (e.g., Fritzsch et al. 1988a; Villy 1890; Wever 1985). The role of audition in spacing, courtship, and isolating mechanisms contributing to speciation is well known for adults (Fritzsch et al. 1988a; McCormick 1988; Ryan 1986, 1988; J. J. Schwartz and Wells 1983). It is important to realize that the mode of transfer of pressure waves (sounds) to the inner ear (Jaslow et al. 1988) is different in tadpoles than in adults. In adults, sound impinges on the tympanic membrane where pressure differentials are transformed into mechanical movements. The middle ear transmits these vibrations to the inner ear via (fig. 6.13) the plectrum (attaches to the tympanum), columella (the connecting bone), and operculum (links columella to the oval window, which in turn connects to the inner ear). In some anurans the columella has been lost (Trueb 1973, 1979). Motions of the oval window create movements of the perilymphatic fluid that generate movements of the endolymphatic fluid. The endolymphatic fluid stimulates the hair cells of the sensory neuroepithelia and the basilar and amphibian papillae.

Tadpoles hear despite lacking middle ear bones and a tympanum. Sound pressures are transmitted to the inner ear via a bronchial columella, which grows from the round window (a connection between the middle and inner ears) toward the ipsilateral bronchus and lung at about the time of opercular closure (Witschi 1949, 1956). In the process of growing, the columella pierces the dorsal aorta (see Boatwright-Horowitz and Simmons 1997). According to Capranica (1976) and Witschi (1949), this unusual arrangement may isolate the columella from other tissues and permit it a greater degree of movement. The columella attaches to the lung and bronchus, which vibrate in response to changes in water pressure.

The inner ear of tadpoles and adults contains two auditory receptive areas, the amphibian papilla and basilar papilla (fig. 6.13; E. R. Lewis and Lombard 1988). Amphibians are unique in this regard because most other vertebrates have one auditory receptive area epithelium (e.g., saccule in fish, basilar papilla in reptiles, and cochlea in birds and mammals; Wever 1973). The surface morphology of the amphibian papilla of *Rana catesbeiana* is described by E. R. Lewis (1976) and E. R. Lewis and Lombard (1988). The basilar and amphibian papillae are tuned to different frequencies which tend to be specialized to an animal's acoustic environment. The number of hair cell receptors within these papillae varies interspecifically and with size intraspecifically. Typically, large animals have large papillae that contain numerous hair cells

(Capranica 1976). The basilar papilla is the generalized auditory organ. Its hair cells receive only afferent innervation and respond to frequencies above 500 Hz. The amphibian papilla is sensitive to both low- and high-frequency stimuli.

During metamorphosis of *Bufo regularis,* the walls of the dorsal aorta close; the bronchial columella is resorbed; and the oval window, operculum, and tympanic columella form. At metamorphosis the cartilage for the plectrum is visible and the tympanic membrane is visible as a crescent-shaped structure (Capranica 1976).

Auditory fibers from the basilar and amphibian papillae travel in CN VIII (acousticovesticular nerve) and terminate in the auditory nucleus in the brain stem (Boord et al. 1970; Capranica and Moffat 1974; Larsell 1934; Matesz 1979; McCormick 1982). Acoustic fibers within CN VIII are organized by the frequency of the stimulus they carry. The auditory system in anurans has been considered in detail by Fritzsch et al. (1988a).

Shofner and Feng (1984) examined the developing basilar and amphibian papillae with both light and scanning electron microscopy and described an increase in the size of the sensory epithelium and the addition of hair cells as the animal grows. Capranica (1976), Fritzsch et al. (1988a, b), Sedra and Michael (1959), and Witschi (1949, 1956) detailed the changes in the auditory apparatus through larval development and metamorphosis in *Bufo* and *Rana.* Mudry and Capranica (1980, 1987), Mudry et al. (1977), Neary (1988), and Neary and Wilczynski (1986) described auditory pathways in the forebrain of adult *Rana.* See Fritzsch (1988) for a description of the evolution of the auditory system in anurans.

Vestibular System

The vestibular system consists of three orthogonally oriented semicircular canals plus a utricle, a saccule, and a lagena that function together in orientation and maintenance of body position in space (fig. 6.13; E. R. Lewis and Lombard 1988). As with the lateral line and auditory systems, the sensory mechanism is fluid (endolymphatic fluid in the vestibular system) motion across an epithelium of mechanoreceptive hair cells. The sensory epithelia of the three semicircular canals are contained in ampullary bulges within the canals, while utricular and saccular hair cells project into an otolithic membrane. The semicircular canals detect angular acceleration or rate of body turning and tilting and are oriented in orthogonal planes with respect to each other (e.g., E. R. Lewis and Lombard 1988; Precht 1976). Each of the semicircular canals emerges from the elongate utricle. The anterior vertical semicircular canal detects motion in an anterolateral-posteromedial plane, the posterior vertical canal detects motions in an anteromedial-posterolateral plane, and the horizontal canal detects horizontal movements. The anterior vertical canal of one side of the tadpole is oriented in the same plane as the contralateral posterior vertical canal. The utricle and saccule monitor linear acceleration and gravitational pull. The utricular macula is oriented horizontally and is crescent-shaped around a central structure termed the striola. The hair cells are polarized in a direction radial to the striolar curve and, taken together, are responsive to movements through 360°. Type I hair cells with long stereocilia that extend to the tip of the kinocilium are found along the striola. Type II hair cells with a long kinocilium and short stereocilia are found along the sides of the striola.

The saccular macula is oriented nearly vertically and faces posterolaterally. The striola is curved and the hair cells, considered together, are sensitive through 360°. Outside the striola, kinocilia are directed outward and inside they face inward. Most saccular hair cells resemble the utricular Type II cells. The posterior protrusion from the saccule is termed the lagena and has a ciliary pattern resembling the utricle. Afferent neurons are bipolar cells. Axons terminate in the vestibular nucleus (e.g., R. F. Dunn 1978). Hair cells also receive efferent fibers. Caston and Bricout-Berthout (1985) described connections of vestibular nerve to the horizontal semicircular canal in *Rana esculenta* and showed that the efferents both inhibit and facilitate the afferent response. These authors reported that neurons within the vestibular nucleus respond both to motion through the vestibular nerve and light pulses from the midbrain tectum. Van Bergeijk (1959) described a reflexive vestibular behavior that allows *Xenopus* tadpoles to maintain a stable position in the water column.

Lateral Line System

The lateral line system of aquatic anamniotes consists primitively of mechanoreceptive and electroreceptive end-organs (Bullock et al. 1983). Anurans retain mechanoreceptors (= neuromasts; Fritzsch 1981b) but have lost their electroreceptive subsystem. Mechanoreceptive neuromasts are sensitive to local water displacements (i.e., the large-scale movements of water molecules) rather than changes in water pressure (small local movements of water molecules that vibrate around a relatively fixed point) to which the auditory system is most sensitive. The lateral line system of tadpoles consists of three main lines each on the trunk and head (fig. 6.16A; Escher 1925; Kingsbury 1895; Lannoo 1987; Shelton 1970, 1971; M. Uchiyama et al. 1991). Tadpoles with midventral spiracles have a symmetrical arrangement of the neuromasts, but those with sinistral spiracles often have asymmetrical arrangements perhaps because of developmental interference caused by the spiracle (Lannoo 1987). Hair cells are the actual displacement detectors of neuromasts (fig. 6.16B; Flock 1965; Flock and Wersäll 1962; Jande 1966; I. J. Russell 1976; M. Uchiyama et al. 1990, 1991).

Hair cells contain a ciliary bundle (the "hairs") at their external surface that consists of a single kinocilium (fig. 6.16C) located at an edge of the bundle and numerous stereocilia. This anatomical polarization corresponds to a directional sensitivity. Displacements of the cilia in the direction of the kinocilium depolarize the cell while displacements in the opposite direction hyperpolarize the cell. Depolarizations increase the rate of firing of the cell, and hyperpolarizations prevent the cell from firing (Flock 1965; review by I. J. Russell 1976). Typically, adjacent hair cells are oriented oppositely so that a stimulus that depolarizes one hair cell will

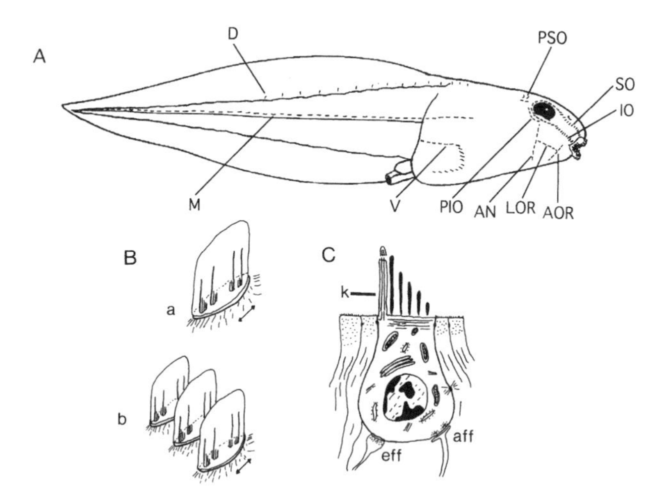

Fig. 6.16. The lateral line system of a generalized tadpole. (A) Named components and orientation of neuromasts in the lateral line system. Abbreviations: AN = angular, AOR = anterior oral, D = dorsal body, IO = infraorbital, LOR = longitudinal oral, M = middle body, PIO = posterior infraorbital, PSO = posterior supraorbital, SO = supraorbital, and V = ventral body line. (B) One primary neuromast (a) and three secondary neuromasts (b) forming a receptor unit (= stitch) with the polarization of hair cells as they project into the gelatinous copula. (C) A single hair cell cut through the kinocilium (k) and stereocilia bundle and showing the afferent (aff) and efferent (eff) innervation. Redrawn from Lannoo and Smith (1989); reprinted by Oxford University Press.

hyperpolarize the other. Neuromasts are innervated by multiple neurons, and hair cells that are oriented similarly are innervated by fibers of the same neuron.

Neuromasts are usually elongated, and their direction of maximum sensitivity corresponds to the long axis of the oval. Sensitivity in directions other than the main axis is proportional to the cosine of the angle from the main axis (Flock 1965). Unlike other tadpoles, pipoid larvae have round neuromasts; such structure makes their axis of sensitivity difficult to determine (Lannoo 1987). In contrast, aquatic pipoid adults have linear neuromasts (Shelton 1970, 1971). Typically, neuromasts within a line are oriented in the same direction (fig. 6.16A; Görner 1963; Lannoo 1987; Shelton 1970). Dorsal trunk neuromasts are most sensitive to displacements in a dorsoventral direction, and middle and ventral neuromasts are most sensitive to anterior-posterior displacements. On the head, supraorbital and infraorbital lines around the eye are most sensitive in directions tangential to the margin of the eye. As these lines run onto the snout, neuromasts change orientation to become anterior-posteriorly sensitive. The angular line is most sensitive to dorsoventral displacements while the two hyoid lines combine to be sensitive to both dorsoventral and anterior-posterior displacements (Lannoo 1987).

As generalized, pond-dwelling tadpoles grow, primary neuromasts formed embryonically divide to form secondary neuromasts (fig. 6.16B; Lannoo 1987; Russell 1976). All secondary neuromasts derived from a single primary neuromast are aligned in the same direction as the primary neuromast and compose a receptor unit (after Zakon 1984; previously termed stitch or plaque, Lannoo 1987, 1988; I. J. Russell 1976).

Hair cells within neuromasts are innervated by bipolar primary afferent neurons whose cell bodies are located in ganglia (Shelton 1970; Van Der Horst 1934). Neuromast afferents course through branches of either the anterior (nALL, head) or posterior (nPLL, trunk) lateral line nerve. Once in the brain, all lateral line afferents bifurcate to send one collateral anteriorly and one posteriorly within the medullary nucleus medialis (Altman and Dawes 1983; Boord and Eisworth 1972; Fritzsch et al. 1984; Jacoby and Rubinson 1983; Lowe and Russell 1982; Matesz and Székely 1978). Neuromasts also are innervated by efferent fibers whose cell bodies are located in the ventral medulla (Will 1982). Efferent axons course through the lateral line nerves. Single efferent cells may send axons to the lateral line, auditory, and vestibular systems (Claas et al. 1981). The efferent system appears to hyperpolarize hair cells, which prevents neuromast function and precludes overstimulation (e.g., during swimming; I. J. Russell 1976). The lateral line system may be responsible for maintaining the spacing between individuals in schools of *Bufo* tadpoles. Diminishing the sensitivity of neuromasts through exposure to streptomycin, which is toxic to hair cells, alters the orientation of individuals and the distance between tadpoles of *Xenopus laevis* (Lum et al. 1982). In adult *Xenopus,* Elepfandt (1988a, b) found that the oriented response to water displacements diminishes with the surgical ablation of neuromasts.

Neuromast receptor units vary in number of organs, number of hair cells per organ, and their position relative to the epidermal surface. These morphological features correlate well with the particular habitat of a tadpole (Lannoo 1987) and presumably are adaptive. Generalized pond tadpoles have a large number of units composed of three to five secondary neuromast organs that are positioned flush with the epidermis. In contrast, stream and arboreal tadpoles have fewer receptor units composed solely of primary neuromasts that are positioned below the epidermal surface. In general, reductions in the number of receptor units occur within lines rather than through the elimination of whole lines, although *Ascaphus* tadpoles appear to have lost their oral line (Lannoo 1987). Considerable variation in neuromast features is found within the genus *Osteopilus* (Hylidae). Older tadpoles of the bromeliad-dwelling *O. brunneus* (see Thompson 1996 for information on the habitat and ecology of these tadpoles) have receptor units composed of single (primary) neuromasts, while congeneric, pond-dwelling tadpoles (*O. dominicensis* and *O. septentrionalis*) have receptor units containing multiple secondary neuromasts (Lannoo et al. 1987).

The lateral line system develops embryonically from pre- and postoptic placodes (R. G. Harrison 1904; Lannoo and Smith 1989; S. C. Smith et al. 1988, 1990; Stone 1922, 1933; see J. F. Webb and Noden 1993). These placodes give rise to primordia that migrate in anterior and posterior directions to form the neuromast organs and their innervation (review by Lannoo and Smith 1989). Primordia from preotic placodes migrate anteriorly to form the cranial lateral line component, and postotic primordia migrate posteriorly to form trunk and tail components. At least three preotic and three postotic placodes give rise to individual nerve branches and their ganglia (Northcutt 1989; S. C. Smith et al. 1988). Neuromasts derived from preotic placodes are innervated by branches of the nALL, while neuromasts from postotic placodes are innervated by the nPLL. Winklbauer and Hausen (1983a, b, 1985a, b) developed and tested a model for the formation of the supraorbital neuromast line in *Xenopus.* They proposed that stem cells are allocated randomly to neuromasts as they are dividing. Each stem cell divides seven times before becoming terminal; terminal cells become hair cells. Therefore, the number of hair cells per neuromast depends on the numbers of stem and terminal cells allocated per neuromast. The generality of this model to other species, and in fact to other lines of neuromasts within *Xenopus,* has not been determined.

Olfactory System

The olfactory system of anurans is divided into two parallel subsystems termed the primary or main olfactory system (MOS) and the secondary or accessory olfactory system (AOS). In the MOS, the olfactory afferents innervate receptors located in the sensory epithelium of the nasal capsule and terminate on target neurons in the main olfactory bulb (MOB; fig. 6.14). The ultrastructure of the olfactory bulb and its neurons is described in K. H. Andres (1970) and Burton (1985). Neurons from the MOB project into the lateral cortex of the telencephalon (Kemali and Guglielmotti 1987; Northcutt and Royce 1975; Scalia 1976; Scalia et al. 1968, 1991a, b). Neurons from the AOS afferents form the vomeronasal nerve, which innervates olfactory receptors in the sensory epithelium of the vomeronasal organ and projects to the accessory olfactory bulb (AOB; fig. 6.14). Neurons in the AOB project to the amygdaloid nucleus on the lower surface of the telencephalon (Kemali and Guglielmotti 1987; Scalia 1976). The AOS is well developed in anurans and caecilians but poorly developed in most salamanders. The degree of development of the AOS is positively correlated with the development of Jacobson's organ, which is absent in neotenic salamanders. The amygdaloid nucleus is correspondingly well developed in anurans and caecilians, as is the striatum (see Forebrain).

The sensory epithelia of the MOS and AOS are similar to each other and to the nasal epithelium of other vertebrates (fig. 6.17). Cilia on the external surface of the olfactory epithelium originating from the olfactory neurons are interposed by an almost continuous sheet of microvilli (fig. 6.17). Below the surface of the epithelium, the olfactory neurons become grouped into olfactory filae containing hundreds of unmyelinated axons that form a compact olfactory nerve (Burton 1985). The main olfactory and accessory olfactory nerves course separately into the anterior portion of the olfactory bulb where their synapses form olfactory glomeruli. Each glomerulus is a spherical group of terminals from numerous primary olfactory neurons that contacts the dendrites of mitral cells, the principle postsynaptic neurons. Each mitral cell appears to contact more than one glomerular dendrite. The axons of the mitral cells form the efferent projections of the olfactory bulbs.

The olfactory system develops from an ectodermal placode in *X. laevis* (Klein and Graziadei 1983). Whether the

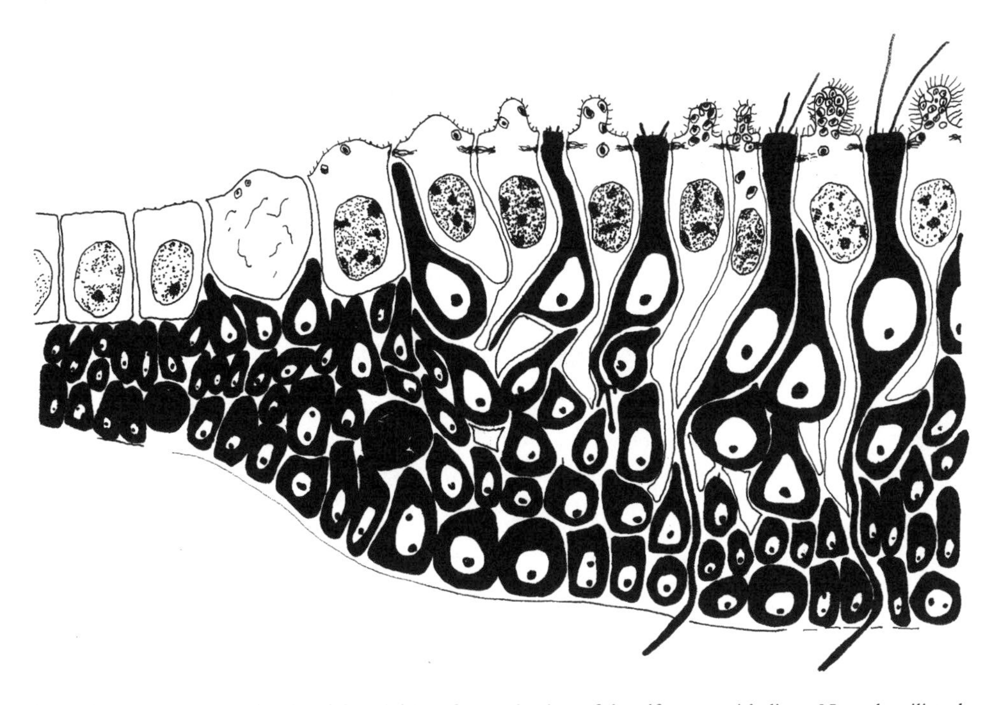

Fig. 6.17. Developmental series *(left to right)* and organization of the olfactory epithelium. Note the ciliated cells separating the dark olfactory neurons. From Klein and Graziadei (1983); reprinted by permission of Wiley-Liss, a subsidiary of John Wiley & Sons, Inc., and of Klein.

olfactory system of pipoids, with their aquatic tadpole and adult stages, is representative of all tadpoles remains to be determined. Likewise, the embryonic differentiation of olfactory placode tissue into principal and accessory systems remains to be described, although there is likely phylogenetic and ecological variation in the olfactory system of tadpoles. There are known variations in the structure of the internal nares and associated skin folds (Wassersug 1976a). Pond, stream, and arboreal tadpoles probably are sampling fundamentally different olfactory environments and using their olfactory systems to mediate different behaviors. For example, pond tadpoles appear to recognize kin via olfaction (chap. 9).

Gustation

In adult anurans, taste buds are located on papillae or protrude only slightly from the mucosal surface and are scattered over the floor and the roof of the buccal cavity and the dorsal surface of the tongue. Taste buds in the roof and floor of the buccal cavity are nonpapillary, while those on the tongue occur on fungiform papillae. Each taste bud papillary disc, which can contain up to 700 receptor cells depending on the species, is innervated by 5–10 myelinated nerve fibers (C. B. Jaeger and Hillman 1976). Taste buds are innervated by CNs VII and IX. The facial nerve innervates the roof and floor of the buccal cavity, while the glossopharyngeal nerve innervates the tongue. To my knowledge, taste buds have not been described in tadpoles but are assumed to be present.

Visual System

The visual system of anurans is well known (Fite 1976). What is not known is the extent that tadpoles, some of which are nocturnally active, rely on visual cues. The extraocular eye muscles of tadpoles appear to be the same as those in adult frogs but less developed (Nowogrodzka-Zagórska 1974). Adult frogs also have levator bulbi muscles innervated by CN V which lift the eyes back into position after they have assisted in pushing food into the esophagus.

Amniotes focus images on the retina by changing the shape of the lens. Amphibians and fishes focus by changing the distance between the lens and retina—proximally for far vision and distally for near vision. Because the refractive index of water and the vitreous humor are similar, tadpoles have less need for accommodation than do adults. The cornea is divided into an inner and an outer layer that fuse at metamorphosis (reviewed by Kollros 1981). The retina receives and processes light energy with two basic receptor types, three kinds of interneurons, and ganglion cells that project into the brain (figs. 6.18–6.19). The vertebrate retina has a number of features that are unusual from the perspectives of mechanical design and neural processing. First, the sensory surfaces of the photoreceptors face backward so that light must pass through the ganglion cells and their axons and the interneuron layer before reaching the receptors. Second, photoreceptors are stimulated by darkness; light hyperpolarizes and inhibits rods and cones. Third, the neurotransmitter released from the photoreceptors hyperpolarizes one

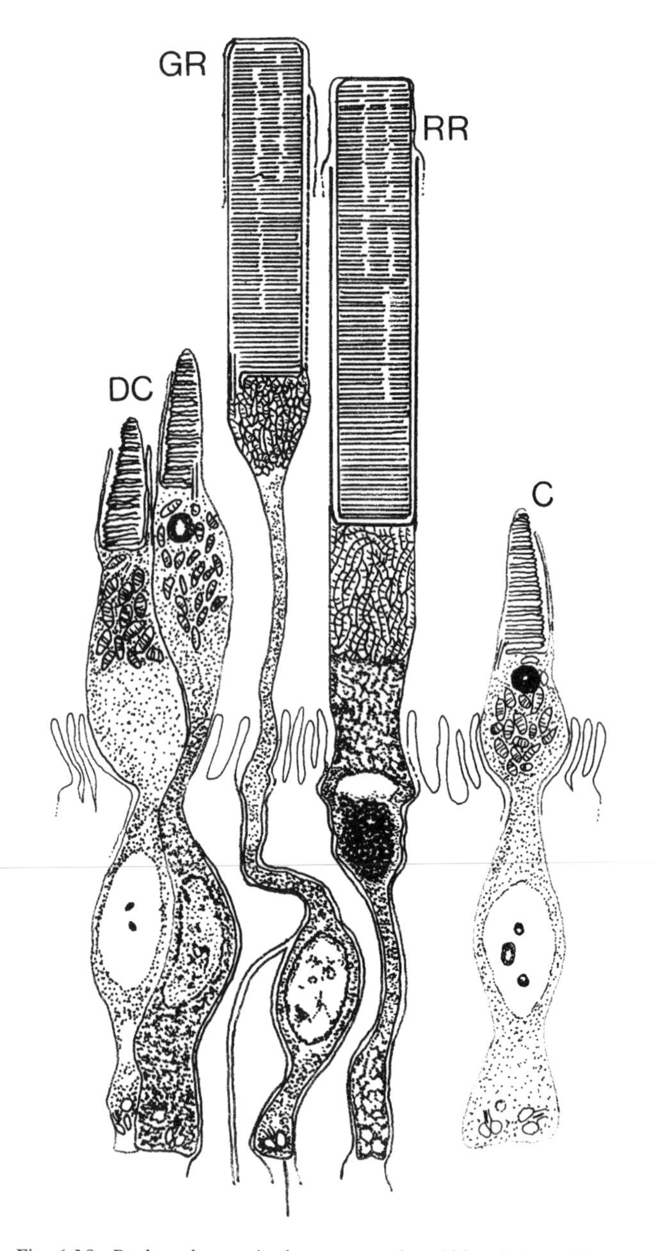

Fig. 6.18. Rods and cones in the anuran retina. Abbreviations: C = single cone, DC = double cone, GR = green rod, and RR = red rod. Redrawn from Donner and Reuter (1976); reprinted by permission of Springer-Verlag and Donner.

bipolar cell type while depolarizing the second bipolar cell type. Fourth, photoreceptors carry and process information without generating action potentials.

Rods and cones are composed of three segments (fig. 6.18). An outer distal segment consists of a series of stacked disks containing the photoreceptor pigment (Corless and Fetter 1987 and Corless et al. 1987a, b, *Rana pipiens*). The inner segment contains many mitochondria, the nucleus, and ribosomes. The synaptic terminal occurs proximally and transmits the receptor signal to the interneurons, which is the first step in sending visual information to the brain. The anuran retina contains green (sensitivity peak at 433 nm) and red rods (502 nm; fig. 6.18; Donner and Reuter 1976; J. Gordon and Hood 1976; W. R. A. Muntz 1962a, b; W. R. A. Muntz and Reuter 1966). Green rods are sensitive to violet and blue light and are known to occur only in anurans and caudates. Green rods are important in light adaptation and provide inputs that contrast with those from cones. Red rods appear important in dark adaptation. Anurans possess a dichromatic cone system of color vision. Single cones have a peak frequency sensitivity between 575 and 580 nm, and double cones are most sensitive at 502 nm (Hailman 1976). As photons excite these visual pigments they become bleached; a chemical change in the permeability of the receptor membranes affects the transmission of neural signals. In tadpoles, adult retinal pigments are preceded by their corresponding 3-dehydroretinal pigments (i.e., porphyropsin rather than rhodopsin; Donner and Reuter 1976). The developmental shift in the spectral preference in tadpoles was studied by R. G. Jaeger and Hailman (1976). Amphibians usually lack a macula, an area of high visual acuity. For a discussion of the intricate biochemistry underlying phototransduction, see Donner and Reuter (1976) and J. Gordon and Hood (1976).

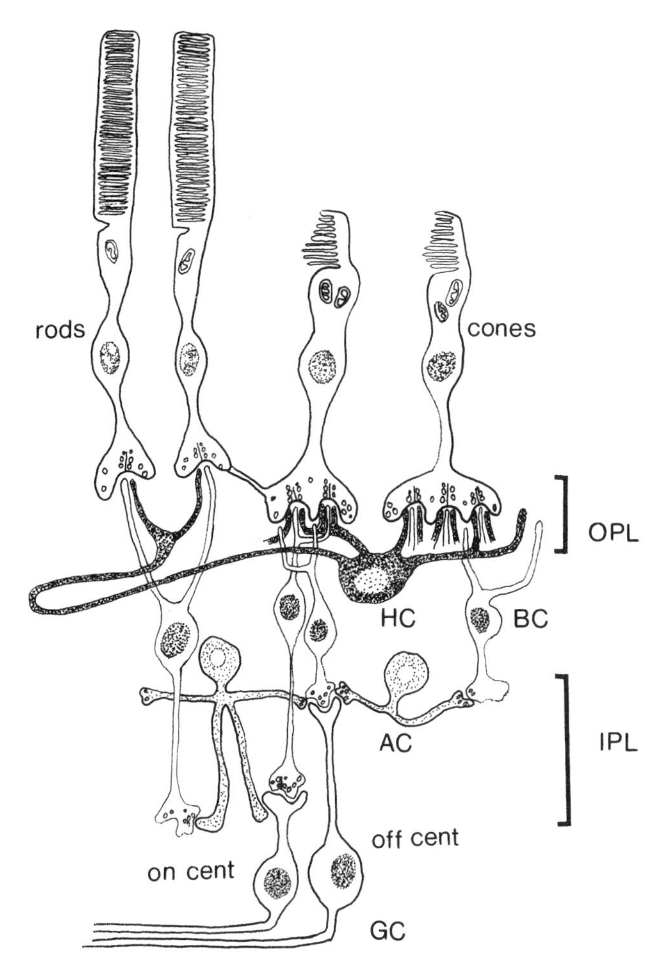

Fig. 6.19. Connections of anuran photoreceptors within the retina. Abbreviations: AC = amacrine cell, BC = bipolar cell, HC = horizontal cell, GC = ganglion cell, IPL = inner plexiform layer, off cent = off-center ganglion cell, on cent = on-center ganglion cell, and OPL = outer plexiform layer. Redrawn from Bailey and Gouras (1985). Copyright 1985 by Elsevier Science Publishing Co., Inc.; reprinted by permission of Elsevier, Bailey, and Gouras.

Rods and cones differ in their sensitivities. In general, rods are able to detect lower light levels than cones but are poor at spatially and temporally resolving this light. This is because the large photoreceptor outer pigment is more likely to be hit by photons, but this size compromises spatial resolution. Also, rod photoreceptor pigment stacks are less polarized and therefore more sensitive to reflected light, and rods converge onto bipolar neurons. The slower response of rods to photostimulation compared with cones limits their temporal resolution. In the retina the majority of synapses occur in the inner (IPL) and outer plexiform layers (OPL; fig. 6.19). The thin IPL is where rods and cones synapse, and the thick OPL is where the amacrine, bipolar, and ganglion cells synapse (Dowling 1976; J. Gordon and Hood 1976). Rods and cones send information to the ganglion cells either directly through bipolar interneurons or indirectly via amacrine, bipolar, and horizontal interneurons. Information from the photoreceptors is sent to either on-center or off-center bipolar cells (fig. 6.19). Bipolar cells are connected by inhibitory horizontal cells. When one cell is excited it inhibits the surrounding cells of the same polarity (e.g., on-center cells inhibit surrounding on-center cells). This arrangement increases contrast and sharpens edges. Bipolar cells connect to ganglion cells directly or indirectly through amacrine cells (fig. 6.19), which mediate interactions between the on-center and off-center systems and have several other specialized functions (Donner and Reuter 1976).

Ganglion cells transmit information from the retinal cells centrally into the brain through CN II—the optic nerve and tract. The retina (technically part of the brain) in *Rana pipiens* contains about 450,000 ganglion cells, 5–7 times as many bipolar cells and 2–3 times as many photoreceptors. Several types of ganglion cells are classified on the basis of the types of information they carry; B. D. Frank and Hollyfield (1987a, b) described seven classes in *R. pipiens.* Stirling and Merrill (1987) detailed the functional morphology of a ganglion cell type that transmits the off-center response. Gaze and Grant (1992) described the patterns of retinal cell death during ontogeny of *Xenopus laevis.* The retina projects principally to the contralateral midbrain tectum, although a small ipsilateral projection that appears at metamorphosis may also exist in adults (M. Schütte and Hoskins 1993). Other retinal fibers project bilaterally to the diencephalic thalamus and pretectal visual centers (these nuclei in turn project to the tectum). The brain also sends neurons centrifugally to the retina. This small projection (30–40 neurons, *Rana*) originates from the telencephalic lamina terminalis near the olfactory nuclei (fig. 6.14) and courses through the preoptic area and into the optic tract to the retina. While the number of afferent fibers in the optic tract increases during ontogeny, the number of efferents remains constant (H. Uchiyama et al. 1988).

The literature on the visual orientation and feeding behavior of anurans has been dominated by studies on adult frogs (Ingle 1976; Maturana et al. 1960). From a phylogenetic perspective (review by Grüsser and Grüsser-Cornehls 1976), the retinas of adult ranids respond to cues from stationary objects, those of bufonids respond to cues from moving objects, and hylids respond to intermediate stimuli. These visually guided behaviors correspond to ecological activity patterns (sit-and-wait vs. mobile predators) and result from differing proportions of retinal neuronal classes. Interestingly, the tadpoles of the species reviewed by Grüsser et al. (1976; *Bufo bufo, Osteopilus septentrionalis, Rana esculenta,* and *R. pipiens*) were all generalized pond dwellers (Orton 1944). This begs the question of whether the retinal morphology of these tadpoles reflects their common lifestyle or the more variable lifestyles of their respective adults.

Retinal cell number increases with growth in tadpoles. Beach and Jacobson (1979a, b) and P. Grant et al. (1980) examined the patterns of cell proliferation in the retina of *Xenopus.* S. Grant and Keating (1986) studied the metamorphic and postmetamorphic maturation of the retina and the midbrain optic tectum. Dunlop and Beazley (1981) described the increasing number of ganglion cells in tadpoles through adults in *Heleioporus eyrei,* where a visual streak is present in adults but not tadpoles; Bork et al. (1987) described the growth of optic fibers in the retina of *Xenopus,* and Cullen and Webster (1979) described the metamorphic changes that occur in the myelin sheath of optic nerve axons in *Xenopus.*

Miscellaneous Senses

Adult frogs, and apparently tadpoles, are sensitive to pain and touch through free nerve endings in the epidermis and dermis (reviews by Catton 1976 and Spray 1976). Cutaneous nerves, among the 60–80 myelinated neurons arising from each spinal nerve, can be grouped into two functional categories: those less than 3 μm, which are also the slowest conducting fibers (0.8–5 m/s) that transmit temperature and pain impulses, and those greater than 5 μm with conduction velocities greater than 7 m/s that transmit phasic and tonic pressure information. In addition to small fibers with slow conduction velocities, pain and temperature receptors are characterized by a long latency from stimulus to discharge and by an action potential of long duration.

The pineal organ (epiphysis) is a small projection attached to the dorsomedial surface of the epithalamus (fig. 6.15) which, within amphibians, is best developed in anurans. These fibers enter the pretectal region in the posterior portion of the epithalamus. The cytology of the pineal is described by Van de Kamer (1965). This organ functions in detecting features of ambient light, such as seasonal variation in photoperiod and perhaps the strength and inclination of sunlight, that mediate physiological functions such as breeding and overwintering (Bagnara 1965; Binkley et al. 1988). Locomotor activity is altered by a sudden drop in light intensity detected by the pineal organ (R. G. Foster and Roberts 1982). Hendrickson and Kelly (1971) described the development of this structure. Frogs (D. H. Taylor and Ferguson 1970) and tadpoles (Justis and Taylor 1976) can orient relative to a shoreline when allowed only pineal input.

In most anuran tadpoles and adults, the peripheral portion of the pineal organ is termed the frontal organ (= stirnorgan; reviewed by Adler 1976). The skin with reduced pigmentation overlying the frontal organ is called the brow spot. Eldred et al. (1980) showed that the central projections of the frontal organ are similar to the ordinary pineal projections and are widespread, including the amygdala, pretectal region, and the central gray of the mesencephalon and diencephalon.

Muscle spindles are stretch receptors that provide information about the location and position of the limbs, trunk, and tail. Spindles are reflexively connected to motor neurons through a pathway termed the stretch reflex, which serves to maintain muscles at their functionally appropriate length (e.g., E. G. Gray 1957; reviewed in Ottoson 1976); Gans and De Gueldre (1992) presented some data on the physiology of larval muscles.

Metamorphosis

The changes that occur in the nervous system of anurans during metamorphosis and the effects of thyroxine in driving these changes are summarized in Hughes (1976) and Kollros (1981). Some conclusions can be drawn about how the nervous system accommodates the tadpole and adult stages. First, the tadpole nervous system comprises features (e.g., lateral line system and posterior spinal cord) that are exclusively larval traits lost at metamorphosis. Second, the tadpole has nervous elements (e.g., limb motor neurons and the corpus of the cerebellum) that apparently are not important in tadpoles but will become functional in adults. Third, the tadpole nervous system includes cells (e.g., motor trigeminal cells) that perform one function in tadpoles and are reworked at metamorphosis to perform a new function in adults. Finally and perhaps most remarkably, a majority of the cells, especially in higher order nuclei, are not known to change during metamorphosis. These cells may function appropriately within the demands of both the tadpole and adult worlds.

The brains of tadpoles are not all alike. Variation appears to be common even within the limited number of tadpoles that have been studied. Orton's (1944) tadpole types offer a starting point for comparative neurobiologists interested in the ecological and phylogenetic bases (but see chap. 4) of neuronal variation in these animals, and the diversity described in chapter 12 provides the mileposts for such investigations. Despite enormous efforts by many researchers to provide basic information about the organization and functioning of the anuran nervous system, we are only now beginning to understand how tadpoles integrate nervous system development and differentiation, and how it is altered to fit the demands of different environments. Learning by tadpoles and how this is modified at metamorphosis is poorly studied (e.g., Munn 1940a, b; Punzo 1991; Strickler-Shaw and Taylor 1991; Youngstrom 1938).

Summary

There are two significant generalizations that can be drawn from a review of the nervous and sensory systems of anuran tadpoles. Few taxa have been examined, but there are many variations exhibited even among this small number. Future studies of a more ecologically and phylogenetically diverse assemblage will most certainly turn up other interesting patterns. The second generality arises from the profound and abrupt remodeling of the neural circuitry at metamorphosis (also chap. 5). In a life cycle that includes a profound metamorphosis from an aquatic larva to a terrestrial adult, questions about adult influences on tadpole nervous system morphology must be considered in addition to the more obvious questions of neural structure and function in tadpoles. Several conclusions can be drawn. First, tadpole nervous systems have elements (e.g., lateral line system) that typically function only in tadpoles. Second, tadpole nervous systems have other elements (e.g, limb motor neurons) that are not important to tadpoles but are important in adults. Third, other elements (e.g, trigeminal motor neurons) are reworked at metamorphosis to perform new functions in adults. Fourth, the majority of neurons are not known to change during metamorphosis. This latter uniformity is remarkable when one considers the differences between a swimming, herbivorous tadpole and a hopping, carnivorous adult frog. Finally, the brains of tadpoles are not all alike; there is no such thing as "the tadpole brain" even within the limited number of species that have been studied. Despite enormous efforts by many researchers, we are now only beginning to understand tadpole nervous systems.

ACKNOWLEDGMENTS

I express my greatest respect and appreciation to those whose work I have cited here and apologize to those whose work I have inadvertently slighted either through oversimplification or omission. I thank D. Pratt, who produced the original artwork, and S. Lannoo, who redrew the figures borrowed from other sources. Acknowledgments to R. J. Wassersug, K. Nishikawa, K. Hoff, and D. Townsend (my ex-labmates in Richard Wassersug's lab at Dalhousie University) and to R. Altig, R. McDiarmid, J. Eastman, and B. Bock for their comments on earlier drafts of this chapter. This chapter is contribution number 452 from the Iowa Lakeside Laboratory and was supported by NIH Grant NS30702-01 and a grant from the Ball State University Office of Research and Sponsored Programs.

7

ENDOTROPHIC ANURANS

Development and Evolution

Giselle Thibaudeau and Ronald Altig

While it might seem strange to focus on a kind of evolution that fails to affect the morphology of the adult, it is precisely because of this fact that direct developing forms offer such superb experimental systems to examine the modifications possible within ontogenies that achieve the same end by different trajectories.

Raff 1987:6

Introduction

The quote from Raff (1987) concerns the development of sea urchins, but the conspicuous parallels during the evolution of direct development in sea urchins and endotrophic anurans are significant. Endotrophic anurans obtain their entire developmental energy from parental sources, most commonly vitellogenic yolk (Altig and Johnston 1989), and the developmental patterns of these anurans deviate in many ways from those of typical anurans. The evolutionary history of the various forms of endotrophic development and the pertinence of understanding these developmental modes relative to anuran systematics still are debated (see chap. 11 and Bogart 1981). Lutz (1948:29) made a fascinating projection: "As long as anuran taxonomy continues to be based exclusively on adult morphology they will be left aside in establishing phylogenetical relationships." Many papers pertinent to understanding the evolution of endotrophy in anurans involve ancillary subjects or deal with other organisms (table 7.1)

Altig and Johnston (1989) defined the following six developmental guilds of endotrophs: (1) viviparous (after exhaustion of vitellogenic yolk, fetus in the oviduct feeds on oviducal materials to complete a modified development before birth as a froglet), (2) ovoviviparous (embryo completes a modified development in the oviduct via only oogenic energy sources and is birthed as a froglet; see Blackburn 1994 for opinions on terminology), (3) direct developer (embryo completes highly modified development via oogenic energy sources in an oviposited egg that is not intimately associated with a parent's body and hatches as a froglet), (4) paraviviparous (embryo completes a modified development via oogenic energy sources in a site other than the reproductive tract of the mother and is "born" as a froglet), (5) exoviviparous (embryo develops via oogenic energy sources in a terrestrial egg before the hatchling moves to a site usually in or on the male parent's body and a froglet eventually is birthed from that site), and (6) nidicolous (eggs oviposited terrestrially, and embryo develops from oogenic energy sources to produce various sorts of free-living, nonfeeding larvae). The nidicolous and direct development modes are more common than other endotrophic guilds. Diversity is highest among the nidicolous group where various developmental patterns provide a range of morphotypes; at the unmodified end of this continuum an entire tadpole morphotype is produced. Near the other end of the continuum, there are various forms of embryos with progressive losses of typical tadpole features.

If endotrophy occurs in a presently recognized genus, usually all species involved are in one guild; exceptions include *Eleutherodactylus* (telmatobine leptodactylid, 2 endotrophic guilds), *Gastrotheca* (hemiphractine hylid, 1 endotrophic and 1 exotrophic guild; Matthews 1957), *Mantidactylus* (mantelline rhacophorid, 1 endotrophic and several exotrophic guilds), and *Megophrys* (megophryid, 1 endotrophic and 1 exotrophic guild). Among microhylids, endotrophy occurs in 6 of 10 subfamilies and is pervasive in 4 of these; all brevicipitine (Africa) and asterophrynine and genyophrynine (Australopapua) frogs have direct development. All cophylines (Madagascar) are nidicolous with similar morphologies, although embryos of *Stumpffia* seem to have a different developmental trajectory. These observations should not be used to imply anything about the evolution of endotrophy within these lineages.

The characterization of a taxon as endotrophic often is based solely on the presence of a small clutch of large, usually nonpigmented ovarian eggs. Based on known relationships between egg size, pigmentation, and the occurrence of tadpoles, this criterion should be employed cautiously. Even though more than 1,000 species in more than 90 genera and 11 families are involved and data have appeared over a span of more than 225 years (e.g., Bavay 1873; Bello y Espinosa 1871; Fermin 1765; Spallanzani 1785), developmental data on endotrophic anurans generally are sparse to absent. There has been little standardization in terminology, staging systems, and information presented. The majority of this information is derived from *Eleutherodactylus* (telmatobiine leptodactylid), *Gastrotheca* (hemiphractine hylid—even though the most commonly studied species in fact produces tadpoles) and *Nectophrynoides* (bufonid). The only complete staging table of an endotroph is based on *E. coqui* (Townsend and Stewart 1985), and considerable data on other *Eleutherodactylus* make the developmental patterns of this genus the best understood. Salthe and Mecham (1974) suggested that direct development of *Eleutherodactylus* may represent the most advanced type of endotrophy, although this idea may only reflect the paucity of data for other taxa and our understanding of the evolution of the various forms of endotrophy. Primarily through the work of Eugenia Del Pino (e.g., 1975, 1989a, 1996; Del Pino et al. 1992) and her colleagues, the unique development of hemiphractine hylids, based primarily on the genus *Gastrotheca,* is also well documented.

The first part of this chapter is a collation of information on the development of endotrophs (tables 7.2–7.3; figs. 7.1–7.2) based on a common terminology and the staging systems of Townsend and Stewart (1985; direct developers) and Gosner (1960; nidicolous larvae). In table 7.4 we standardize stages of Townsend and Stewart (1985) with three other systems involving *Eleutherodactylus.* Townsend and Stewart (1985:432, table 3) compared stages and developmental times of eight other species of *Eleutherodactylus* with those of *E. coqui.* Stages of *Gastrotheca riobambae* (Del Pino and Escobar 1981) can be standardized against Gosner stages (table 7.5). In species of *Gastrotheca* that do not have tadpoles, a modified staging scheme is needed at least for

Table 7.1 Summary of ancillary subjects, often pertaining to organisms other than amphibians, that could have a notable impact on various studies of endotrophic anurans

Subject	Reference
Developmental regulation	Alberch 1982, 1985, 1987
Evolution of viviparity	Amboroso 1968
Evolution of viviparity and egg retention	Andrews and Rose 1994
Hybridization of ovo- and viviparous lizards	Arrayago et al. 1996
Developmental vs. evolutionary rates	Arthur 1982
Fertilization mode and parental care	Beck 1998
Parental care in *Eleutherodactylus*	Bourne 1997, 1998
Froglet transport in *Eleutherodactylus*	Diesel et al. 1995
Reproductive energy allocation/viviparous lizard	Doughty and Shine 1998
Multigene families regulate development	Dressler and Grusse 1988
Ontogeny/phylogeny relationships	Fink 1982
Reproductive strategies in heterogeneous habitats	Giesel 1976
Regulatory genes and development	Gluecksohn-Waelsch 1987
Evolution of viviparity in fishes	Guillette 1987
Evolutionary loss of feeding larvae in an echinoderm	Hart 1996
Evolution of larval types in invertebrates	Havenhand 1993
Developmental energy use in echinoids	Hoegh-Guldberg and Emlet 1997
Maternal-fetal oxygen transport in lower vertebrates	Ingerman 1992
Evolution of viviparity	Mathies and Andrews 1996
Immunological/endocrinological problems in viviparity	Medewar 1953
Maternal body volume constrains viviparity	Qualls and Shine, 1995, 1996
Costs of viviparity in scorpions	L. R. Shaffer and Formanowicz 1966
New hypothesis for evolution of viviparity	Shine 1995
Loss of larval types in invertebrates	Strathmann 1978
Evolution of direct development in salamanders	D. B. Wake and Hanken 1996
Evolutionary significance of fetal maintenance	M. H. Wake 1977
Phylogenesis of direct development and viviparity	M. H. Wake 1989
Evolutionary scenario of viviparity	M. H. Wake 1993b
Evolution of oviducal viviparity and egg retention	M. H. Wake 1993c

Table 7.2 Alphabetical list of families and genera including species (number of species in parentheses) known or suspected of being endotrophs. More detailed data for some taxa appear in table 7.3.

Arthroleptidae
  Arthroleptinae: *Arthroleptis* (12)
Bombinatoridae: *Barbourula* (2)
Brachycephalidae: *Brachycephalus* (2) and *Psyllophryne* (1)
Bufonidae: *Altiphrynoides* (1), *Andinophryne* (3), *Bufo* (part, 3 out of 217 species may be endotrophs, at least under some circumstances) *Crepidophryne* (1), *Didynamipus* (1), *Frostius* (1), *Laurentophryne* (1), *Metaphryniscus* (1), *Nectophryne* (2), *Nectophrynoides* (5), *Nimbaphrynoides* (2), *Oreophrynella* (6), *Osornophryne* (6), *Pelophryne* (8), *Rhamphophryne* (8), *Truebella* (2), and *Wolterstorffina* (2)
Dendrobatidae: *Colostethus* (part, 3 out of 90)
Hylidae
  Hemiphractinae: *Cryptobatrachus* (3), *Flectonotus* (5, feeding occurs under some circumstances in some species), *Gastrotheca* (part, 27 out of 43 species), *Hemiphractus* (5), and *Stefania* (7)
Leiopelmatidae: *Leiopelma* (3)
Leptodactylidae
  Leptodactylinae: *Adenomera* (part, unknown number out of 6) and *Vanzolinius* (1)
  Telmatobiinae: *Adelophryne* (2), *Barycholos* (2), *Cycloramphus* (1 out of 25 species), *Dischidodactylus* (2), *Eleutherodactylus* (517), *Euparkerella* (4), *Eupsophus* (8), *Geobatrachus* (1), *Holoaden* (2), *Ischnocnema* (4), *Phrynopus* (20), *Phyllonastes* (4), *Phyzelaphryne* (1), and *Zachaenus* (3)
Megophryidae: *Megophrys* (part, 1 out of 24 species)
Microhylidae
  Asterophryinae: *Asterophrys* (1), *Barygenys* (7), *Callulops* (13), *Hylophorbus* (1), *Mantophryne* (3), *Pherohapsis* (1), *Xenobatrachus* (17), and *Xenorhina* (6)
  Brevicipitinae: *Balebreviceps* (1), *Breviceps* (13), *Callulina* (1), *Probreviceps* (3), *Spelaeophryne* (1)
  Cophylinae: *Anodonthyla* (4), *Cophyla* (1), *Madeccassophryne* (1), *Platypelis* (9), *Plethodontohyla* (14), *Rhombophryne* (1), and *Stumpffia* (6)
  Genyophryninae: *Aphantophryne* (3), *Copiula* (5), *Choerophryne* (1), *Cophixalus* (28), *Genyophryne* (1), *Oreophryne* (24), and *Sphenophryne* (24)
  Melanobatrachinae: *Melanobatrachus* (1) and *Parhoplophryne* (1)
  Microhylinae: *Adelastes* (1), *Albericus* (3), *Arcovomer* (1), *Gastrophrynoides* (1), *Hyophryne* (1), *Kalophrynus* (10), *Myersiella* (1), *Phrynella* (1), *Synapturanus* (3), and *Syncope* (2)
Myobatrachidae
  Limnodynastinae: *Kyarranus* (3) and *Philoria* (1)
  Myobatrachinae: *Arenophryne* (10), *Assa* (1), *Bryobatrachus* (1), *Geocrinia* (7), *Myobatrachus* (1), and *Rheobatrachus* (2)
Pipidae
  Pipinae: *Pipa* (part, 1 out of 7 species)
Ranidae
  Petropedetinae: *Anhydrophryne* (1) and *Arthroleptella* (6)
  Raninae: *Batrachylodes* (8), *Ceratobatrachus* (1), *Discodeles* (5), *Palmatorappia* (1), *Phrynodon* (1), *Platymantis* (42), and *Taylorana* (2)
Rhacophoridae
  Mantellinae: *Mantidactylus* (part, 6 and 7 in each of two species groups out of 61 species)
  Philautinae: *Philautus* (87, unknown number are endotrophs)
Rhinodermatidae: *Rhinoderma* (2)
Sooglossidae: *Nesomantis* (1) and *Sooglossus* (2)

later stages. We focus on developmental, anatomical, and physiological aspects of endotrophy that have been largely ignored in previous summaries of anuran reproductive biology (e.g., W. C. Brown and Alcala 1983; Crump 1974; Duellman 1985; Salthe and Duellman 1973; Salthe and Mecham 1974; Tyler 1985). Suffice it to say that endotrophs typically produce fewer, larger eggs than similar-size frogs that produce tadpoles. Discussions of breeding modes (e.g., Del Pino 1980a; Del Pino and Escobar 1981; Duellman and Gray 1983; Duellman and Maness 1980), the evolution and morphology of the brood pouches of egg-brooding hylids (e.g., Del Pino 1980b; Del Pino et al. 1975; R. E. Jones et al. 1973), and parental care (e.g., Taigen et al. 1984; Townsend 1986, 1996; Townsend et al. 1984) are omitted. Taxonomic names used here are from D. R. Frost (1985) and may differ from those found in the published papers. Citations that pertain to specific species are in table 7.3.

The second part of this chapter is a heuristic discussion of developmental trajectories and related topics concerning the evolution of endotrophy; parts of chapter 11 also are pertinent.

Development of Endotrophic Anurans

Because the direct-developing *Eleutherodactylus coqui* has been studied extensively (e.g., Townsend and Stewart 1985), data for this species are summarized first, and data for other endotrophs are added in comparison. The study by Lynn (1942; *Eleutherodactylus nubicola*) provided the most anatomical details of an endotroph. Elinson et al. (1990) summarized considerable information on endotrophic larvae including culturing methods. The incomplete information on endotrophs is obvious in the following text.

Gametes and Fertilization

Eggs of endotrophs usually are larger than those of similar-size frogs that produce tadpoles and are laid in various sites with sufficient moisture (e.g., Estrada 1987; Pérez-Rivera and Nadal 1993). Parental care is common. Much of the research on oogenesis and egg components (i.e., yolk density, meiotic mechanisms, RNA components) of endotrophic eggs was based on hemiphractine hylids. The 3-mm eggs of *Gastrotheca riobambae* (produces tadpoles) have lower ribosomal gene amplification, fewer nucleoli (about 300), less

Table 7.3 Summary of breeding biology of endotrophic anurans. Emg = ecomorphological guild (Altig and Johnston 1989): DD = direct developer, EV = exoviviparous, NI = nidicolous, OV = ovoviviparous, PV = paraviviparous, and VI = viviparous; Mod = mean ovum diameter, mm; Clutch = eggs/clutch; Dev_t = developmental time, days; Site = developmental site: UO = under objects, IB = in burrow; ES = elevated sites; OV = oviducal-ovoviviparous, IS = in stomach of female, BS = in back skin of female, OB = on back skin of female, IP = in partial or complete dorsal pouch of female, VI = oviducal-viviparous; VS = in male vocal sac, ID = in dorsal pouch of male, and OM = on back of male; Hatch = hatchling SVL, mm; and Reference = a pertinent reference.

Taxon	Emg	Mod	Clutch	Dev_t	Site	Hatch	Reference
Arthroleptidae, Arthroleptinae							
Arthroleptis							
crusculum	DD	3.0	—	15	UO	—	Guibé and Lamotte 1958
poecilonotus	DD	5.0	20–25	20–30	UO	—	Lamotte and Perret 1963b
wahlbergi	DD	2.5	11–30	28	UO	6.0	Wager 1986
Bufonidae							
Altiphrynoides malcomi	NI	2.9	11–31	100	UO	4.4	H. H. Wake 1980
Didynamipus sjoestedti	DD	2.3	—	18	UO	—	Grandison 1981
Laurentophryne parkeri	DD	2.0	—	16	UO	—	Grandison 1981
Nectophryne afra	NI	—	22–36	—	UO	—	Scheel 1970
Nectophrynoides torneiri	OV	3.5	9–37	60–90	OV	—	Orton 1949
Nimbaphrynoides occidentalis	VI	6.0	4–35	270	VI	—	Lamotte and Xavier 1973
Oreophrynella nigra	DD	3.0	8–35	—	UO	4.0	McDiarmid and Gorzula 1989
Pelophryne brevipes	NI	—	5–20	—	UO	—	Alcala and Brown 1982
Dendrobatidae							
Colostethus							
chalcopis	NI	2.7	1–4	—	UO	—	Kaiser and Altig 1994
stephani	NI	2.0	—	3.8	UO	—	Juncá et al. 1994
Hylidae, Hemiphractinae							
Cryptobatrachus fuhrmanni	PV	4.0	20–30	—	OB	—	Ruthven 1915
Flectonotus							
fitzgeraldi	NI	3.0	—	—	IP	—	Duellman and Gray 1983
goeldii	NI	4.0	—	26	IP	—	Weygoldt and De Carvalho e Silva 1991
ohausi	NI	3.0	—	—	IP	—	Heyer et al. 1990
pygmaeus	NI	3.0	5–11	—	IP	—	Duellman and Maness 1980
Gastrotheca							
ceratophryes	PV	8.0	10	—	IP	—	Del Pino 1980b
christiani	PV	4.0	12	—	IP	—	Del Pino 1980b
ernestoi	PV	6.0	30	—	IP	—	Del Pino 1980b
ovifera	PV	8.0	36	—	IP	—	Del Pino 1980b
testudinae	PV	4.0	48	—	IP	—	Del Pino 1980b
weinlandi	PV	9.0	14	—	IP	—	Del Pino 1980b
Hemiphractus scutatus	PV	10.0	10	—	OB	—	Duellman and Maness 1980
Stefania scalae	PV	9.0	—	—	OB	—	Duellman and Maness 1980
Leiopelmatidae							
Leiopelma							
archeyi	EV	5.0	4–12	84–105	OM	—	Archey 1922
hamiltoni	EV	5.5	7–10	133–147	OM	—	Bell 1978
hochstetteri	NI	5.5	10–22	—	UO	—	N. G. Stephenson 1955
Leptodactylidae, Leptodactylinae							
Adenomera							
hylaedactyla	NI	2.6	4–25	—	IB	—	Heyer and Silverstone 1969
marmorata	NI	—	8–10	—	IB	—	Heyer et al. 1990
Leptodactylidae, Telmatobiinae							
Eleutherodactylus							
abbotti	DD	—	21	—	UO	—	A. Schwartz and Henderson 1991
atkinsi	DD	3.5	42–94	—	UO	—	A. Schwartz and Henderson 1991
audanti	DD	—	22	—	UO	10	A. Schwartz and Henderson 1991
augusti	DD	—	6–25	—	UO	—	Valett and Jameson 1961
cooki	DD	4.0	25	—	UO	—	A. Schwartz and Henderson 1991

Table 7.3 *continued*

Taxon	Emg	Mod	Clutch	Dev$_t$	Site	Hatch	Reference
coqui	DD	3.5	3–45	17–26	ES	—	Townsend and Stewart 1985
cornutus	DD	5.0	45	> 30	UO	—	Heatwole 1962
dimidiatus	DD	—	24–60	—	—	—	A. Schwartz and Henderson 1991
fowleri	DD	8.0	12	—	—	—	A. Schwartz and Henderson 1991
gossei	DD	—	32	—	—	—	A. Schwartz and Henderson 1991
hedricki	DD	—	14–32	—	—	7.0	A. Schwartz and Henderson 1991
inoptatus	DD	—	20–35	17–20	UO	—	A. Schwartz and Henderson 1991
jasperi	OV	3.3	1–6	33	OV	—	M. H. Wake 1978
johnstonei	DD	3.7	—	13	UO	—	Ovaska 1991
martinicensis	DD	4.5	29–40	13–14	UO	—	L. Adamson et al. 1960
nasutus	DD	5.0	13–20	28	UO	—	Lynn and Lutz 1946b
nubicola	DD	3.0	26–74	26	UO	6.0	Lynn 1942
parvus	DD	4.0	49–86	—	UO	—	Lutz 1944a
patriciae	DD	—	18–24	—	—	—	A. Schwartz and Henderson 1991
planirostris	DD	2.0	3–26	13–20	UO	4.8	A. Schwartz and Henderson 1991
portoricensis	DD	—	4–27	18–20	UO	—	Gitlin 1944
thorectes	DD	—	—	3	—	—	A. Schwartz and Henderson 1991
varians	DD	3.7	30	≈15	—	—	A. Schwartz and Henderson 1991
varleyi	DD	—	—	4	—	3.7	A. Schwartz and Henderson 1991
wetmorei	DD	—	10	—	—	—	A. Schwartz and Henderson 1991
Eupsophus							
roseus	NI	—	—	—	UO	—	Formas and Vera 1980
taeniatus	NI	—	—	—	UO	—	Cei and Capurro 1958
vittatus	NI	—	—	—	UO	—	Formas and Vera 1980
Zachaenus							
parvulus	DD	6.0	30	17	UO	—	Lutz 1944b
Microhylidae, Brevicipitinae							
Breviceps							
adspersus	DD	—	—	42	IB	—	De Villiers 1929b
Microhylidae, Cophylinae							
Anodonthyla							
boulengeri	NI	2.0	23–30	16–20	ES	—	Blommers-Schlösser 1975
Cophyla							
phyllodactyla	NI	3.0	—	—	ES	—	Glaw and Vences 1992
Madecassophryne							
truebae	—	4.0	18	—	UO	—	Glaw and Vences 1992
Platypelis							
grandis	NI	4.0	90	35	ES	—	Blommers-Schlösser 1975
Plethodontohyla							
notostica	NI	3.0	120	28	ES	—	Blommers-Schlösser 1975
Rhombophryne							
testudo	NI	—	—	—	IB	—	Glaw and Vences 1992
Stumpffia							
grandis	NI	1.5	20	—	UO	—	Glaw and Vences 1992
Microhylidae, Genyophryninae							
Cophixalus							
parkeri	DD	3.5	18	65–100	UO	—	Simon 1993
Oreophryne							
annulata	DD	3.5	3–9	—	ES	—	W. C. Brown and Alcala 1983
Microhylidae, Microhylinae							
Calophrynus							
pleurostigma	NI	—	—	16	UO	—	Berry 1972
Myersiella							
microps	DD	—	—	28	UO	—	Izecksohn et al. 1971
Synapturanus							
salseri	NI	—	4	—	IB	—	Pyburn 1975
Syncope							
antenori	NI	—	—	—	ES	—	Krügel and Richter 1995
Myobatrachidae, Limnodynastinae							
Kyarranus							
sphagnicolus	NI	3.3	45–52	30	UO	—	De Bavay 1993
Philoria							
frosti	NI	4.0	50–170	70–126	UO	—	Littlejohn 1963
Myobatrachidae, Myobatrachinae							
Arenophryne							
rotunda	DD	5.5	—	73	IB	—	J. D. Roberts 1984

Table 7.3 *continued*

Taxon	Emg	Mod	Clutch	Dev_t	Site	Hatch	Reference
Assa							
darlingtoni	EV	2.5	10	60	ID	–	Ehmann and Swan 1985
Myobatrachus							
gouldii	DD	7.4	–	–	IB	–	J. D. Roberts 1981
Rheobatrachus							
silus	PV	–	26–40	45	IS	–	Tyler and Davies 1983
Pipidae, Pipinae							
Pipa							
pipa	PV	6.0	40–114	77–136	BS	–	F. Schütte and Ehrl 1987
Ranidae, Petropedetinae							
Anhydrophryne							
rattrayi	DD	2.2	11–19	26	UO	–	Warren 1922
Arthroleptella							
hewitti	DD	3.0	20–40	20–31	UO	3.1	De Villiers 1929c
lightfooti	DD	4.5	5–8	7–10	UO	≈ 4.0	Wager 1986
Ranidae, Raninae							
Ceratobatrachus							
guentheri	DD	5.0	–	–	IB	–	specimens
Platymantis							
guentheri	DD	2.2	8–47	–	UO	–	Alcala and Brown 1957
hazalae	DD	3.4	6–8	45–49	UO	–	Alcala 1962
meyeri	DD	3.2	15–40	39	UO	6.4	Alcala 1962
Rhacophoridae, Philautinae							
Philautus							
lissobrachius	–	3.0	17–19	38–41	ES	–	Alcala and Brown 1982
schmackeri	–	2.8	6–15	–	ES	–	Alcala and Brown 1982
Rhinodermatidae							
Rhinoderma							
darwinii	EV	–	20–30	–	VS	–	Jorquera et al. 1972
Sooglossidae							
Sooglossus							
gardineri	DD	1.8	65–12	–	UO	–	specimens

18 and 28S rRNA, and slower developmental rates (2 weeks to complete gastrulation) than those of *Xenopus laevis* (14 h; Del Pino et al. 1986). The degree of rRNA amplification is generally low in the slow-developing hemiphractine hylids, but *Flectonotus pygmaeus* (Del Pino and Humphries 1978; Macgregor and Del Pino 1982) has about 2,000 meiotic oocyte nuclei with low levels of rRNA amplification and the embryos develop more rapidly. This results from about 11 rounds of oogonial mitoses occurring in ovarian chambers or cysts (Del Pino and Humphries 1978) in which nuclear division occurs without cytokinesis; such a process produces fewer eggs with more nutrients and other components (e.g., high concentrations of rRNA, C = 1.7×10^{-12} g) necessary for endotrophic development. The nuclei eventually segregate into a peripheral band of large nuclei and a central group of smaller ones. The central group and all but one peripheral nucleus disintegrate, and that peripheral nucleus, perhaps the one with the highest DNA content, remains. The amount of nuclear DNA present may also affect developmental rates. The mean Au/nucleus for 9 species of *Eleutherodactylus* is 10.7±2.2 (7.9–15.2) versus 13.2±4.6 (4.5–26) for a number of other species of frogs (Goin et al. 1968). Mantellines have 8.3–13.3 pg of DNA per diploid nucleus (Blommers-Schlösser 1981), and development in all members of two species groups is thought to be endotrophic.

Oocyte maturation (De Albuja et al. 1983) is assumed to be mononucleate (Del Pino 1989b; Nina and Del Pino 1977, *Eleutherodactylus;* Gall 1968, *E. johnstonei;* Nina and Del Pino 1977, *E. unistrigatus* (have many large nucleoli in the germinal vesicle). Del Pino (1989b) suggested that oogenesis of *Eleutherodactylus* may be modified to allow accumulation of RNA and proteins to accommodate faster development of the large eggs.

Sperm motility is high in low tonicity conditions, and sperm entry is limited to about 10% of the surface of an *Eleutherodactylus* egg (Elinson 1987b). Polyspermy may occur with normal development, and all nuclear events of the first cell cycle occur near the animal pole (Gitlin 1944; Goin 1947; Sampson 1904). In *Gastrotheca riobambae* a translucent spot containing the single nucleus is visible at ovulation (Del Pino et al. 1986). Cleavage is assumed to be holoblastic and unequal (Elinson 1987b), although meroblasty has been suggested (Del Pino and Escobar 1981 and Del Pino et al. 1975, *G. riobambae*) or reported (De Villiers 1929b, c, *Arthroleptella* and *Breviceps;* Inger 1966, *Philautus hosii;* Warren 1922, *Anhydrophryne rattrayi*). Warren (1922) reported that the animal pole of *Anhydrophryne* was syncytial. Subsequent stages through neurulation probably follow the typical anuran pattern (Del Pino and Elinson 1983; Elinson 1987b). Gastrulation in *Kyarranus sphagnicolus,* an unmodified nidic-

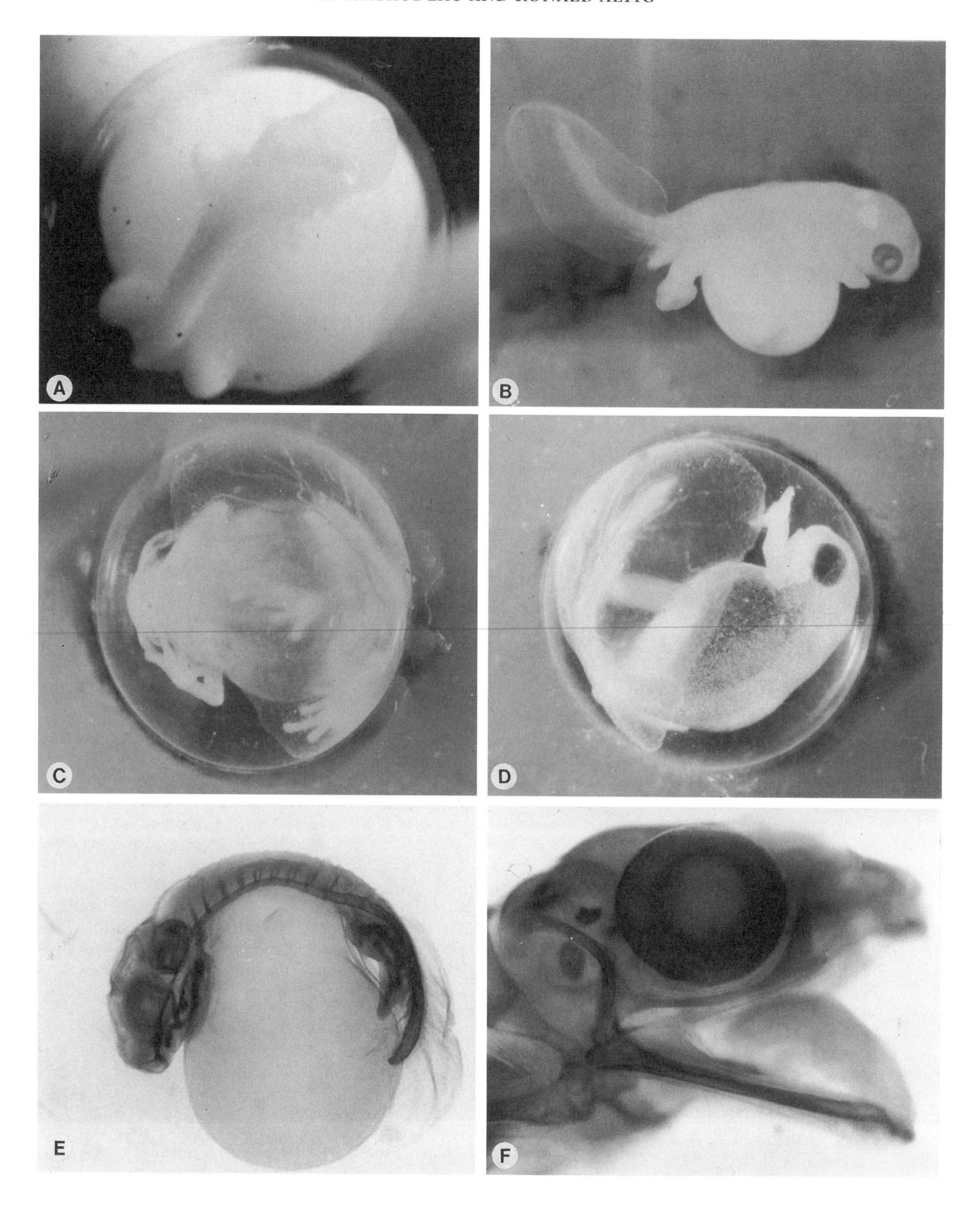
A
B
C
D
E
F

Fig. 7.1. Embryos and hatchling of *Eleutherodactylus coqui*. Stages are from table of Townsend and Stewart (1985). (A) Embryo, dorsal view, head upward, limbs buds visible, Townsend/Stewart 4. (B) Embryo, lateral view, removed from egg capsules, Townsend/Stewart 6. (C) Embryo, ventral view, within egg capsules, Townsend/Stewart 12. (D) Embryo, lateral view, within egg capsules, Townsend/Stewart 12. (E) Embryo, stained with a monoclonal antibody for Type II collagen, most dark areas are cartilage, Townsend/Stewart 9. (F) Hatchling, lateral view of head, stained for cartilage.

olous form, resembles that in *Rana* (J. A. Moore 1961), although the eventual neural plate appears narrower and shorter.

Early Embryology

Hemiphractine hylids follow a novel pattern of early development. Although *Gastrotheca riobambae* eventually produces a tadpole (Del Pino and Elinson 1983; Del Pino and Loor-Vela 1990; Elinson and Del Pino 1985), an embryonic disc forms as a result of gastrulation. The embryo proper is derived from this disc in a pattern similar to that seen in birds and reptiles (see Gatherer and Del Pino 1992). Involution is limited, and the primitive gut is limited to a small cavity at one end of the embryo. In most genera the yolk

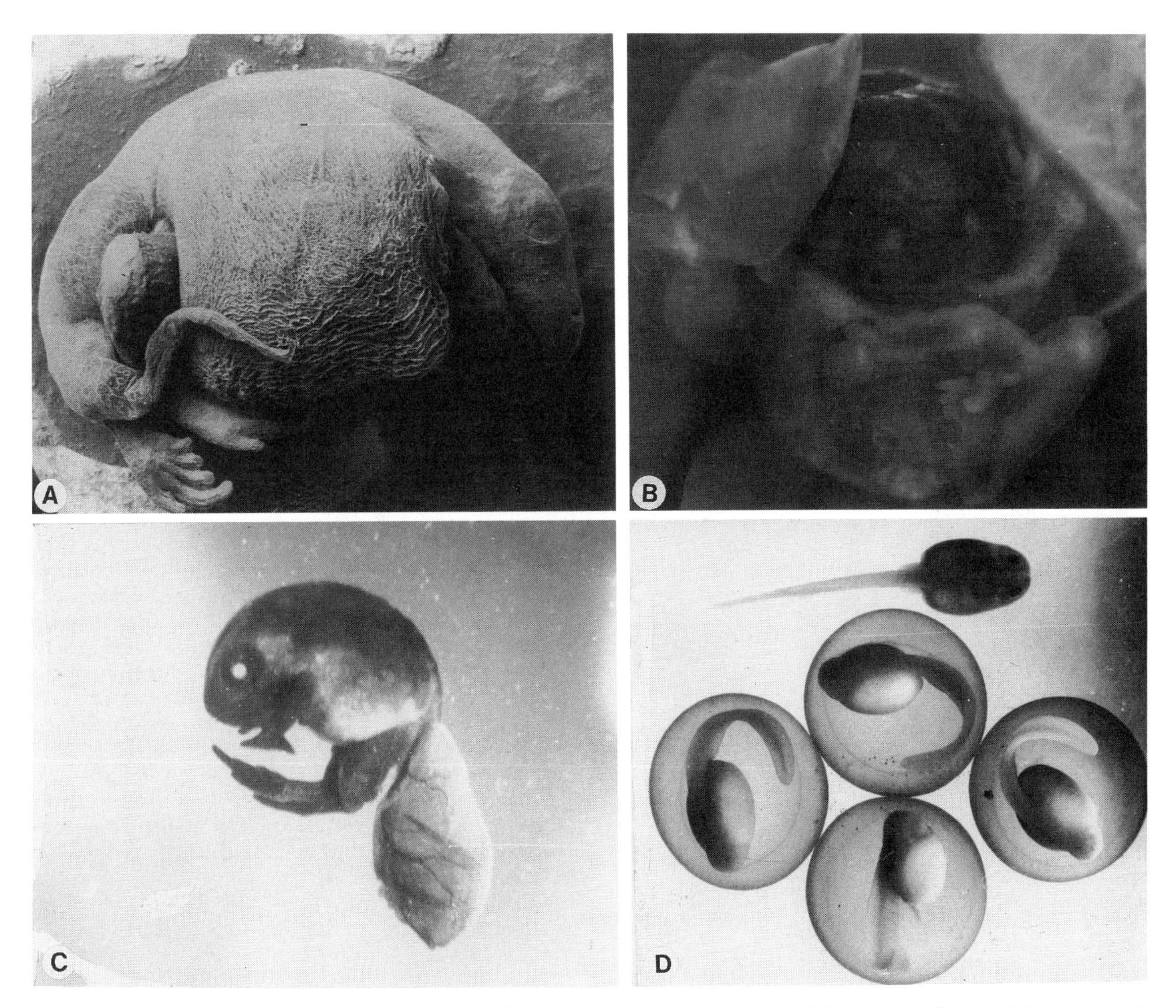

Fig. 7.2. Embryos of different endotrophic frogs. Stages are from the table of Townsend and Stewart (1985), except in (D). (A) *Ceratobatrachus guentheri* (Ranidae, Raninae), lateral view, Townsend/Stewart 13. (B) *Gastrotheca* sp. (Hylidae, Hemiphractinae), ventral view, bell gills visible, Townsend/Stewart 15. (C) *Copiula fistulans* (Microhylidae, Genyophryninae), removed from egg capsules, lateral view, Townsend/Stewart 14. (D) Embryos and nidicolous tadpole of *Plethodontohyla inguinalis* (Microhylidae, Cophylinae), intracapsular embryos, Gosner stage 20; nidicolous tadpole, Gosner 27.

Table 7.4 Staging systems of Lynn (1942, *Eleutherodactylus nubicola*), Lynn and Lutz (1946b, *E. nasutus*), and Sampson (1904, *E. martinicensis*) standardized against that of Townsend and Stewart (1985, *E. coqui*)

Townsend/Stewart	Lynn/Lutz	Lynn	Sampson
1	—	—	—
2	—	24	I
3	I	23.5	—
4	—	—	—
5	—	23	II
	—	22	III
	—	21	IV
6	II	19	—
7	—	18	V
8	—	17	VI
9	III	16	VII
—	—	14	VIII
10	—	13	—
11	—	—	—
12	—	12	—
	—	11.5	IX
	—	11	X
	—	10.5	XI
	—	10	XII
13	IV	9	—
14	—	6	XIII
15	—	—	—
Hatch	V	hatch	XIV

Table 7.5 Comparison of the staging system of Del Pino and Escobar (1981) for *Gastrotheca riobambae* with that of Gosner (1960)

Event	Gosner	Del Pino/Escobar
Fertilization	1	1
Gray crescent	2	—
2 cell	3	2
4 cell	4	3
8 cell	5	4
16 cell	6	—
32 cell	7	5
Midcleavage	8	—
Late cleavage	9	6
Dorsal lip	10	7
Midgastrula	11	8
Late gastrula	12	9
Neural plate	13	10
Neural folds	14	11
Rotation	15	—
Neural tube	16	12
Tail bud	17	—
Muscle response	18	13
—	—	14
Heart beat	19	15
Gill circulation	20	16
Clear cornea	21	18
Fin circulation	22	—
Operculum begins	23	19
Operculum continues	24	20
Operculum closes	25	21
Limb $< 0.5 \times$ diameter	26	—
Limb $\geq 0.5 \times$ diameter	27	—
Limb $\geq 1 \times$ diameter	28	20
Limb $\geq \times$ 1.5 diameter	29	21
Limb $= 2 \times$ diameter	30	22
Toe development	31–37	23–25

mass remains roughly spherical throughout development, but in *Ceratobatrachus* (personal observation) and *Platymantis* (Alcala and Brown 1957; Atoda 1950), the yolk starts to partition along the ventral midline, most pronounced anteriorly at the beginning of the process, at about Townsend/Stewart 6. The tail extends anteriorly in this sagittal depression, and the invagination increases progressively until by about Townsend/Stewart 13 the yolk mostly resides as two ventrolateral masses. Moury and Hanken (1995) is the sole paper that addresses neural crest migration in a direct developing frog.

Body, Limbs, and Tail

The lateral body wall appears between the limbs at Townsend/Stewart 7 as a small convex, pigmented border. This border grows ventrally to surround the yolk mass through Townsend/Stewart 12 (Goin 1947, *Eleutherodactylus planirostris;* Lynn 1942, *E. nubicola;* Lynn and Lutz 1946a, *E. guentheri*). The four limb buds appear in *Eleutherodactylus* embryos soon after the neural groove closes in Townsend/Stewart 4 and before gill buds appear. Derived from lateral plate mesoderm, the rounded buds lie lateral to and seemingly are not connected to the presumptive trunk. By Townsend/Stewart 5, limb buds have increased in size and visibly connect with the trunk. Elbow and knee joints appear at Townsend/Stewart 6, and foot paddles have formed by Townsend/Stewart 7. Digits begin to differentiate in Townsend/Stewart 8, and during stages 9–13 the limbs and toes elongate and toe pads eventually form. The front legs usually are folded medially below the head, while the hind limbs are flexed ventrally or ventrolaterally along the body. Elinson (1994) showed that proper leg development of *E. coqui* may be controlled by a systemic factor and an influence of thyroid hormone could not be demonstrated (also Lynn and Peadon 1955).

Limb buds of other taxa may appear simultaneously (e.g., *Arthroleptis, Eleutherodactylus, Leiopelma, Nectophrynoides tornieri, Philautus,* and *Pipa*), the front legs may develop earlier than the hind legs (e.g., *Breviceps fuscus*), or the hind legs may appear earlier than the front legs (e.g., *Anhydrophryne rattrayi*). When larvae of *Flectonotus* are released from the pouch, their hind legs are comparable to those of a tadpole at Gosner 39. Alcala (1962) noted that the advanced embryo of *Platymantis guentheri* with large front limbs and no hind limbs or tail reported in Alcala and Brown (1957) was malformed.

Except for hind limb development, the relatively unmodified nidicolous larvae at advanced stages (e.g., Gosner 30–36) closely resemble much younger larvae of exotrophs (e.g., 21–23) in proportions and shape. The hind limb buds are proportionately and absolutely huge for the size of the larva (Juncá et al. 1994; Kaiser and Altig 1994).

In *Eleutherodactylus coqui,* a tail bud appears at Townsend/Stewart 4, the tail elongates and membranous fins develop during stages 6–10. The fins begin to regress after Townsend/Stewart 12 and are entirely resorbed within 2 days of hatching. The tail curls ventrally between the hind legs and along the venter or laterally along the side of the embryo, usually to the left. In other species, the tail and fins are distorted so as to appear somewhat like a mushroom whose cap covers the posterior hemisphere of the embryo. A section of the tail fin at about midlength appears firmly adherent to the inner surface of the egg jelly in *E. inoptatus* (personal observation). Other species of *Eleutherodactylus* (e.g., *E. hazelae* and *E. meyeri*) and frogs in several other guilds (e.g., *Assa, Ceratobatrachus,* several *Gastrotheca, Myobatrachus, Nectophrynoides tornieri*) have small tails with fins small to absent. Embryos of *Discodeles opisthodon* lack a tail, and there appear to be no data to support the statement of Rostand (1934) that the tail of *Nectophrynoides* is hollow. The tail fins of direct developers are thin and nonpigmented, and the blood vessels are obvious. Whether or not there is a sufficient increase in vascularity to suggest a specific respiratory function for the tail is debatable. De Villiers (1929b) found no specialized vascularization in the fin of *Arthroleptella* and concluded that gas exchange is enhanced only by increased surface area. Warren (1922) stated that embryos of *Anhydrophryne rattrayi* developing from smaller eggs had larger tails than embryos developing from larger eggs. Even though Townsend and Stewart (1985) intimated potential ecological, developmental, and systematic correlations among the number and size of gills and tail morphology, definitive data are lacking.

Integument and Egg Tooth

Most endotrophic eggs and embryos are nonpigmented and appear white, yellow, or orange. Those of *Pipa pipa* and some nidicolous forms are exceptions, and Noble (1917) noted that early embryos of *Cryptobatrachus fuhrmanni* are black; based on some of our observations, we suggest that this coloration is a preservation artifact. Embryos of *Eleutherodactylus coqui* and probably most other taxa begin to deposit pigment on the head and along the dorsal midline by Townsend/Stewart 5. In subsequent stages, chromatophores differentiate laterally but do not go beyond the limits of the body wall. By Townsend/Stewart 12 all but a sliver of yolk is enclosed by the body wall, and the entire yolk mass is pigmented by stage 13. Pigment is most intense dorsally and laterally, and characteristic patterns on the head and legs are evident by Townsend/Stewart 14. As a generality, the adult pattern is present by hatching in direct developers but appears later in other endotrophic froglets. The tail muscle and fins are typically nonpigmented.

In species of *Eleutherodactylus,* a single keratinized egg tooth forms on the skin adjacent to the premaxillary symphysis late in Townsend/Stewart 12. The conical tooth projects vertically from its insertion and terminates in a single or bifid point (Hardy 1984; Lutz 1944a; Noble 1926a). The single point of the conical egg tooth of *E. urichi* from Trinidad projects anteriorly. Differences in the number of points and the projecting direction suggest the employment of different motions during hatching or differences in the characteristics of the egg jelly. Except for the ranid *Discodeles opisthodon* (Alcala and Brown 1957; Boulenger 1886), an egg tooth has not been reported from members of any other genus. Advanced embryos of *Stefania goini* have a small, triangular, keratinized tubercle in the middle of the upper jaw that Duellman and Hoogmoed (1984) thought might be a remnant of the upper jaw sheath. Based on the location of the tubercle and the age of the embryos, this structure may be an egg tooth. An unidentified leptodactylid embryo (R. W. McDiarmid, personal communication) has an apparently analogous keratinized structure on the symphysis of the protruding lower jaw.

Mouthparts

No vestiges of tadpole mouthparts are known in *E. coqui* or other *Eleutherodactylus,* although a detailed developmental study of the mouth area is lacking. Vestiges of oral structures are present in some hemiphractine hylids (Wassersug and Duellman 1984), *Arthroleptella* (Lambiris 1988a, b), and *Nimbaphrynoides occidentalis* (Lamotte and Xavier 1972); the latter authors suggested that the mouthparts in fetuses of *Nectophrynoides* were somehow used in feeding in utero. *Arthroleptella* has infra- and suprarostal cartilages (Lambiris 1988a, b; Van Dijk 1966), and the occurrence of mouthpart remnants in this genus agrees with Elinson's ideas (1990) that the typical lack of mouthparts results from the absence of the proper inducing sources. Embryos of *Cycloramphus stejnegeri* have an almost complete oral apparatus, and those of *Rhinoderma* have nascent tooth ridges on the lower labium (Cei 1987). In direct-developing species of *Gastrotheca,* pigmented mouthparts form but as development proceeds they undergo changes associated with metamorphosis and produce the mouth configuration of a frog (Wassersug and Duellman 1984). *Nectophrynoides tornieri* has a ridge at the place the lower labium would form in an exotrophic tadpole, and the sides of the upper jaw overhang the lateral parts of the lower jaw (Orton 1949). Most nidicolous cophyline larvae have reduced, soft mouthparts that resemble those of the exotrophic scaphiophrynine *Scaphiophryne;* larvae of *Stumpffia* do not have such mouthparts and mouth development proceeds more directly toward that of a frog.

Elinson (1990) suggested that the lack of mouthparts results from the absence of the inducing cartilages or ectodermal competence, or both, but it is obvious that detailed, comparative developmental studies are needed to understand the developmental patterns more effectively. Adhesive glands apparently are lacking in all nonnidicolous endotrophs and often are nonfunctional in nidicolous forms (e.g., Cei and Capurro 1958; Formas and Pugín 1978; Formas and Vera 1980, *Eupsophus roseus* and *E. vittatus;* Alcala and Brown 1957, *Pelophryne brevipes*). The less-modified embryos of *Philautus* (Alcala and Brown 1982) and *Philoria* (Littlejohn 1963) have functional adhesive glands.

Gills, Operculum, Spiracle, and Respiratory Systems

Pharyngeal slits apparently are absent in all endotrophs. One or two pairs of small gills with fimbriae present as not much more than terminal knobs and bumps are present for less than one-third of development in *E. coqui,* and the number of gills (0–2) and arches involved vary among species. Gill buds with circulating blood appear at Townsend/Stewart 5, reach maximum extent by stage 7, and regress by stage 9. Most *Nectophrynoides* lack gills, but *N. tornieri* has two pairs.

Atoda (1950) and Boulenger (1886; see Duellman and Trueb 1986; Noble 1927) thought that a pleated area (= abdominal sacs of Alcala 1962) on the lateral body wall between the legs of *Ceratobatrachus, Discodeles,* and *Platymantis* serve as a respiratory surface. Our observations of the development of *Ceratobatrachus guentheri* agree with those of Noble (1925b, 1927) who suggested that these pleats are simply wrinkles in the lateral body wall that develop as the yolk reserve is depleted.

The unique bell gills of hemiphractine hylids exist in several configurations. Not long after neurulation is complete, two pairs of anlagen are visible immediately behind the presumptive head (Del Pino et al. 1975; Noble 1927). The anlagen do (e.g., *Gastrotheca*) or do not fuse to produce the 1–2 pairs of functional gills attached to the first or second branchial arch, or both (Duellman and Gray 1983). The gills develop into concave, vascularized discs or bell-shaped cups, lie in intimate contact with the egg jelly (Del Pino 1980b; Del Pino et al. 1975), cover 50%–100% of the embryo, and the right and left sides may overlap. Analogies to placental extraembryonic membranes have been suggested (e.g., Noble 1917; Valett and Jameson 1961). One or two gill stalks, each containing efferent and afferent vessels, extend between the top of the bell and the first and second aortic arches of the embryo. Noble (1917) and I. Spannhof and Spannhof (1971) noted that the strands of single-stranded gills actually are composed of two tightly wrapped strands. Striated muscle cells may pull the gill into the gill chamber at birth (Noble 1917), although Del Pino et al. (1975) noted that circulation to the gills ceases soon after birth and that the gills are resorbed within 24 hr in *Gastrotheca riobambae.* Embryos cultured in Ringer's solution retain gills. The development of the aortic arches is relatively simple in *Eleutherodactylus coqui,* and the vascular network of vessels extending around the yolk mass are based on one or two large vessels that enter near the branchial arches (see Lynn 1940b).

Opinions concerning the operculum and spiracle (e.g., presence, configuration, homology among endotrophs and between endotrophs and exotrophs, and terminology) in endotrophs are confusing. By definition, a spiracle cannot form if the operculum is absent or does not form to a similar extent as it does in typical tadpoles. Several authors stated that an operculum is absent in *Eleutherodactylus,* but lateral folds (= dermal folds, branchial or gular fold) continuous with the skin of the head and sides of the body cover the base of the forelimb (Lynn 1942). These folds remain separated and never enclose the entire limbs or produce a spiracle. Sampson (1904) stated that these folds were not homologous with the operculum of typical tadpoles. Warren (1922) noted that in *Anhydrophryne* a fold grows forward over each limb from behind, but he did not discount that these folds were homologs of the true operculum even though shifted in position. *Leiopelma hochstetteri* (Bell 1978, 1985) and *Nectophrynoides* (Orton 1949) have an operculum, and *L. archeyi* and *L. hamiltoni* have a small fold at the base of the legs. Embryos of *Platymantis* (Alcala 1962; Alcala and Brown 1957) and *Ceratobatrachus guentheri* have a membrane that hides the front legs until late in development. Either the front legs develop later than the hind legs or an operculum covers the front legs of *Anhydrophryne rattrayi* and *Arthroleptella lightfooti,* and the front legs of *Breviceps adspersus* are hidden "under the skin" (Wager 1986). At least some cophyline larvae have bilateral, slitlike spiracles contrary to statements by Blommers-Schlösser (1975).

A number of modified nidicolous species lack a spiracle. A spiracle is absent in most egg-brooding hylids but opercular folds remain open and then regress and close by the time of birth. Del Pino and Escobar (1981) suggested that these openings represent a partial development of the spiracle. A spiracle forms in *Flectonotus, Philoria, Pipa, Rhacophorus,* and *Rhinoderma.*

The respiratory use of the small gills that only persist for a short period in *Eleutherodactylus* is questioned. The large tail of some direct developers may serve as a respiratory organ (see Townsend and Stewart 1985). The walls of the pouches of egg-brooding hylids and the integumentary-incubating crypts of *Pipa pipa* may allow gas exchange with the embryos (Del Pino et al. 1975; Noble 1925b; I. Spannhof and Spannhof 1971). Lutz (1944b) suggested that the large surface area of the yolk of *Zachaenus parvulus* may serve in respiration and afford protection from predators.

Sensory and Nervous Systems

Optic placodes are evident by Townsend/Stewart 4 in *E. coqui.* The lens and iridial pigment appears by Townsend/Stewart 6, and the iris is black by stage 10. Retinal layers form in Townsend/Stewart 9, and by stage 13 the pupil darkens and the iris attains the adult coloration. Auditory vesicle diverticula form at Townsend/Stewart 3 and soon lose their connections with surface ectoderm. These vesicles exist as single chambers until the saccule and utricule form at Townsend/Stewart 9. Semicircular canals are apparent by Townsend/Stewart 13, and the eustachian tube and tympanic cavity appear somewhat later.

The endolymphatic calcium deposits (ECD; otocyst of Chibon 1960; otoliths of Hughes 1962) are present in many direct developing anurans. These deposits, important to calcification of the skeleton, may be homologous with $CaCO_3$ deposits in the endolymphatic sacs (Dempster 1930; Etkin 1964) of typical tadpoles. The ECD appears as small spots at Townsend/Stewart 6 and develops into quadrangular patches by stage 8. Extensions from the ECD form during Townsend/Stewart 9–12, join at the midline by late in stage 12, and fade by stage 15. An ECD has been documented in *Ceratobatrachus* and other *Eleutherodactylus* (*antillensis, guentheri, johnstonei, martinicensis, nasutus,* and *portoricensis*); in *Ceratobatrachus,* similar deposits are arranged along the dorsolateral extent of the spinal column.

Olfactory organs open to the pharynx at Townsend/Stewart 6 and olfactory pits appear apparent by stage 9. Neuromasts are absent in *Eleutherodactylus nubicola* (Lynn 1942) and uncommon in other nonnidicolous endotrophs. Embryos of *Arthroleptella* have neuromasts on the head (De Villiers 1929b; Lambiris 1988a, b). The neural tube is closed by Townsend/Stewart 3 in *Eleutherodactylus coqui,* primary brain divisions are present by stage 6, and all parts of the brain are present in correct proportions by stage 9 (also Hughes 1959; Lynn 1940b). Schlosser and Roth (1997b) studied the modified development of the peripheral nervous system of *E. coqui.*

Digestive, Urogenital, and Reproductive Systems

The primitive gut separates the early embryo from the large yolk mass, ends blindly anteriorly, and is open posteriorly via a persistent blastopore. Rupture of the oropharyngeal membrane during Townsend/Stewart 9 allows communication between the buccal cavity and the exterior. Rapid absorption of yolk by the endodermal cells lining the gut causes the digestive tract to become convoluted and filled with yolk. Only the anterior and posterior ends remain hollow throughout embryogenesis, and the intestine never forms a coil typical of tadpoles. Warren (1922) stated that the gut of *Anhydrophryne* develops first at the anterior and posterior ends. Jorquera et al. (1982, *Rhinoderma* spp.) and Teran and Cerasuolo (1988, *Gastrotheca gracilis*) also studied larval gut development. Vilter and Lugand (1959) noted precocious development of the alimentary tract in *Nectophrynoides* and suggested that the small fetus "drinks" a nutritive substance secreted by the oviducal epithelium. Labial filaments found on these embryos may be specializations to aid in the stimulation and absorption of oviducal secretions (Lamotte and Xavier 1973; M. H. Wake 1980, 1982).

Three pairs of convoluted tubules form the pronephros, and a straight tube leading to the Wolffian duct forms the mesonephros at Townsend/Stewart 6. By stage 9 the mesonephros has enlarged, and by stage 13 the pronephros is nonfunctional (Lynn 1942; also see Meseguer et al. 1996; Selenka 1882). The urinary bladder diverticulum from the cloaca and the liver and pancreas diverticula from the gut are present by stage 6. In the next few stages, the liver becomes lobed and the gallbladder is formed. The gonadal ridge appears between Townsend/Stewart 6–13. Chibon (1962, *E. martinicensis*) and Lamotte and Xavier (1973) and Lamotte et al. (1973, *Nimbaphrynoides occidentalis*) discussed gonad development.

Cei (1980:281) commented on the development of *Batrachyla taeniata* as follows: "After 15–20 days [when embryos are intracapsular] solid whitish wastes can be seen in the tail." Cei speculated that this storage was similar to that of a cleidoic egg. He did not state if the observed tadpoles were alive or preserved, and if this white material is different from the white precipitate that commonly occurs in preserved tadpoles, further investigations certainly are warranted.

Skeletogenesis

The sequence of postcranial ossifications in *Eleutherodactylus* is similar to that of *Rana,* although these bones ossify along with cranial ossifications (Hanken and Summers 1988; Lynn 1940a) that are highly modified from the norm. The first ossifications to appear in *Eleutherodactylus* are in the appendicular skeleton at Townsend/Stewart 12. The ischium, sternum, and episternum ossify during metamorphosis in *Rana* but show no ossification at hatching in *Eleutherodactylus.* The angular, mentomeckelian, and septomaxilla appear later in development than in exotrophic tadpoles. An embryo of *Eleutherodactylus* slightly before hatching is very similar to a ranid tadpole at the completion of metamorphosis, although they each arrived at this state by different routes. Cranial elements associated with larval mouthparts are absent. Swanepoel (1970, *Breviceps adspersus*) specifically examined the ontogeny of the chondrocranium and nasal sacs, and Haas (1996) reported heterochronic changes in the development of the head of hemiphractine hylids.

Physiology

Burggren (1985) suggested that conditions of O_2, CO_2, and pH within aquatic egg masses are not as severe as has been suggested (e.g., R. M. Savage 1935). Bradford and Seymour (1985, 1988a, b), Gollman and Gollman (1993a), A. A. Martin and Cooper (1972), Seymour and Bradford (1995), Seymour et al. (1991a, b, 1995), Seymour and Roberts (1991), and Shoemaker and McClanahan (1973) supplied pertinent data on various facets of intracapsular metabolism and development of terrestrial eggs. Callery and Elinson (1996) examined the urea-cycle enzymes during ontogeny of *Eleutherodactylus coqui,* and Alcocer et al. (1992) determined that ureotelism was the prevailing mode of nitrogen excretion in larvae of *Gastrotheca riobambae.*

Endotrophs take 13 days to 9 months to hatch, and the developmental period of *Eleutherodactylus coqui* ranges from 17.1 days at 24.6°C to 26.3 days at 21.1°C (Townsend and Stewart 1985, 1986). A decrease of 1°C increased developmental time by 2.5 days. Their Q_{10} of 3.92 is higher than any of 15 species of temperate-region frogs except *Ascaphus.* This large value suggests that direct developers have a greater sensitivity to environmental temperatures and reduced tolerance to thermal deviations. Wager (1986) noted that *Breviceps adspersus* takes about 2 months to hatch and *Anhydrophryne rattrayi* and *Arthroleptis wahlbergi* take 4 weeks.

Eleutherodactylus coqui can develop in water of low tonicity from fertilization through hatching (Elinson 1987b). Wager (1986) noted that *Anhydrophryne* larvae drowned if placed in water, and Lynn (1948) found that embryos of *E. ricordi* raised in water grew more slowly than those in air unless the jelly was removed. The development of embryos of *Gastrotheca riobambae* (Del Pino et al. 1975) in water or Ringer's depended on the stage of development. Heart rate is about 12 beats/min at 30°C in *Platymantis pelewensis* at about Townsend/Stewart 7 (Atoda 1950); Burggren et al. (1990) presented data on the developmental changes in cardiac physiology of *E. coqui.* Larvae of *Batrachyla* start development in what appears to be a nidicolous mode but spend most of their development as exotrophic tadpoles.

The thyroid gland differentiates by Townsend/Stewart 13. Tail degeneration is stimulated by thyroid hormones (Hughes 1966; Lynn and Peadon 1955), while limb initiation (but not completion) is independent of thyroxine in contrast to the condition in typical tadpoles. The lack of effects of thyroidectomy (Hughes 1966; Lynn and Peadon 1955) suggests that development is thyroid independent, although Elinson (1990) noted that there may be thyroid hormones stored in the egg. Goiterogenic properties of phenylthiourea affected pigmentation in *E. ricordi* more severely than in other tadpoles (Lynn and Sister Alfred de Marie 1946). Oviducal secretions of *Nimbaphrynoides occidentalis* are maintained during pregnancy by corpus luteum-progesterone mediation until the last two months of gestation (Xavier 1977). Brink (1939, *Arthroleptella*) and Lynn (1936, 1948, *Eleutherodactylus*) addressed growth and function of the thyroid gland, and other references that pertain to endocrinology, development and metamorphosis include Exbrayat and Hraouibloquet (1994), Hourdry (1993), Hughes and Reier (1972), D. H. Jennings (1994), and B. E. Morgan et al. (1989).

Evolution of Endotrophy

Endotrophy was attained through modifications of the normal amphibian developmental program (also see chap. 11). We do not imply that all lineages with similar development necessarily were produced by the same evolutionary pathway, but certain common denominators are present. The fact that these alterations have occurred repeatedly with similar results suggests provocative ideas about potential evolutionary and genetic mechanisms (see Fang and Elinson 1996). All observed "intermediates" would not have had to occur as a progression; those that appear to be intermediates between typical amphibian development and the more derived conditions should suggest probable sequences of changes and promote an understanding of developmental constraints. Others have used this reasoning to illustrate sequential alterations of breeding modes of amphibians (e.g., Grandison 1978; Heyer 1969, 1973a; M. H. Wake 1982) and other taxa (e.g., Raff 1987; Strathmann 1978, 1993; Strathmann et al. 1992). In some circumstances the intermediates seemingly would be at a disadvantage by not expressing the appropriate phenotype for either option. In the case of anuran development, an exotrophic tadpole, a profoundly modified developmental pattern as in *Eleutherodactylus,* and many cases that appear as intermediate morphologies along the nidicolous continuum all produce a froglet. Because of the dual life-history pattern of anurans, the developmental exclusion of features unique to tadpoles causes elaborate changes in larval developmental patterns with no effect on the adult. Although the understanding of development of endotrophs is based on rather few data involving a small assemblage of taxa, we suggest several ideas that differ from current dogma.

Genetic on-off signals (e.g., morphogen production and distribution, cellular competence and induction, pattern formation) influence the sequence of events or pattern of development. The period during which an event(s) occurs reflects the rate at which a developmental event or process occurs; Genetic mechanisms are of prime importance in considering sequence of developmental events, while abiotic factors, particularly temperature, often are more instrumental in immediate considerations of altering developmental period. Alterations in developmental sequence were instrumental in the evolution of endotrophy, while alterations in developmental period probably had a minor influence on these unusual developmental patterns. Even cursory comparisons show that endotrophy is not simply a shortened or telescoped sequence of events. Developmental periods of typical anurans from egg to metamorph range from 9 days (e.g., Forge and Barbault 1977, *Bufo pentoni*) to 3 or more years (e.g., Metter 1967, *Ascaphus truei*) with inconsequential changes in general developmental sequence; both extremes exceed the known developmental periods of endotrophs, and in the examples given, differences in sequence reflect habitat (Nodzenski and Inger 1990) and not development. A common staging table will work in either case.

A reasonably consistent developmental sequence in early stages of all exotrophs eventually results in a wide diversity of morphologies (e.g., the intricate oral apparatus) because of developmental modifications later in ontogeny. Because most endotrophs lack the oral apparatus and have less variation in tail morphology compared with exotrophic tadpoles, the morphological variation in endotrophs is lower than in exotrophs. Conversely, the multiple times that endotrophy has evolved in a number of lineages suggest that (1) the genetic mechanisms involved are either relatively simple, simply regulated, or likely to occur; (2) similar mechanisms operated repeatedly to produce similar developmental patterns; and (3) the mechanisms involved alterations in gene regulation rather than mutations of structural genes. It seems highly unlikely that homologous mutations of structural genes would occur repeatedly in the proper sequence in so many different lineages. If there has been a reversal from endotrophy to exotrophy (e.g., Wassersug and Duellman 1984), gene regulation rather than structural mutations seems a more likely mechanism. We do not imply that the process was not complicated; changes in the (1) development, genetics, and physiology of the egg and embryo and (2) behavior, morphology, and physiology of the adult had to be properly orchestrated (e.g., Xavier 1977). Based on the observations that apparent intermediates between exotrophic and endotrophic development occur, that species of *Gastrotheca* may produce either tadpoles or froglets, and that a frog results from all such developmental variations, we reiterate that the examination of extant taxa will provide insight into the probable patterns and mechanisms and aid in illustrating the developmental alterations. Depending on the hierarchical level of developmental regulation that is postulated, all lineages would not necessarily pass through the same sequence of events. These presumed intermediate species should be placed in a sequence only for heuristic purposes, and they should not be viewed as incomplete attempts (Lutz 1948) at terrestrial development. In hopes of stimulating thoughts on the subject and collection of pertinent materials, we continue this discussion under the following three topics: (1) two general pathways based on the presence or absence of oviposition, (2) characteristics of exotrophs that seem to imitate conditions found in early stages of nidicolous and direct developing endotrophs, and (3) changes within the various forms of endotrophy.

Oviposition and Endotrophy

Although the changes most certainly occurred in concert, we suggest that various alterations in oviposition were a fundamental factor in the evolution of endotrophy. Without egg retention (= lack of oviposition), the associated changes in both adult reproductive systems that allow ovovivipary and vivipary could not have arisen. Even so, the ovoviviparous forms (e.g., *Eleutherodactylus jasperi* is more similar to direct developers compared to some species of *Nectophrynoides,* which are more similar to nidicolous larvae) do not follow similar embryological patterns. If so, one would predict that these conditions arose differently; the condition in *Eleutherodactylus* may have involved the retention of a direct-developing egg, while in *Nectophrynoides* the changes involved the retention of an egg that would develop at some point along the nidicolous continuum. Although these differences do not suggest it, ovovivipary and vivipary also could have evolved in sequence once oviposition was suppressed, or at least a series of ovoviviparous-like conditions (i.e., vary as to length of time retained) may have been intercalated into the evolution of vivipary.

The other forms of endotrophy involve a common oviposition pattern and modified developmental patterns. Hemiphractine hylids have highly modified breeding behaviors (i.e., fertilization followed by positioning of eggs) and a unique embryology. Exovivipary, which essentially involves nidicolous larvae that are intimately tended by a parent, could easily be derived from the normal nidicolous types by modification of either parental (e.g., *Rhinoderma*) or larval (e.g., *Assa,* some *Leiopelma,* and some *Sooglossus*) behavior. Negative geotaxis, perhaps modulated by a chemotactic response to material from the adult skin, would suffice in the latter case.

Even though we do not have enough details of even morphological (much less genetic) development to make astute predictions about the possible evolutionary steps leading to the larval types along the nidicolous continuum, the developmental regulation that produced these forms is surely less complex than that which produced direct development. At least for the present discussion, we will consider nidicoly as a potential precursor to direct development. A postulate of where on the nidicolous continuum that direct development might have arisen will be tenuous at best. In summary, hemiphractine hylids possess highly modified breeding biologies and strangely different developmental patterns that certainly evolved via mechanisms different from all other endotrophs. Other endotrophic forms may have similar developmental stages, and in the next section we present scenarios whereby particularly nidicoly and then direct development might have arisen from exotrophic development.

Exotrophs Approaching Endotrophy

Environmentally influenced variations in development are important, but they only alter developmental period and not developmental sequence. Regardless of the embryonic period, exotrophic embryos typically hatch within a narrow range of stages centered at about Townsend/Stewart 22. From hatching until the tadpole stage at Gosner 25 (about 3% of developmental time and about four developmental stages), the hatchling is particularly vulnerable, locomotes slowly over short distances by epidermal cilia, and stabilizes itself with a transitory adhesive gland. Yolk reserves support development for about the first 22 stages, and perhaps 80% of the developmental time commonly remains after yolk exhaustion; this period primarily involves isometric growth with minor developmental change. A second period of development (i.e., metamorphosis) comes at the end of larval ontogeny. We first suggest ways in which exotrophic development might be altered without requiring changes in developmental sequence.

Initial developmental changes of exotrophs would affect the period, and therefore the stage, to hatching. The embryo could remain intracapsular during part or all of the time that is normally spent as a hatchling (Lamotte and Lescure 1977). "Late hatching" commonly accompanies oviposition out of water, on the soil surface near water (e.g., *Chirixalus idiootocus, Rana grayi*), on leaves (e.g., centrolenids, some hyline and phyllomedusine hylids), and underground or under objects (e.g., some *Leptodactylus, Pseudophryne;* see Alcala 1962). The development that ensues during the extended intracapsular period enhances the probability of survival and therefore the selection for oviposition out of water if egg mortality is low (only 8.1% survival in *Philoria;* Malone 1985). If reduced oxygen concentration is the hatching stimulus (Petranka et al. 1983), the increased oxygen concentration in an aerial environment may allow an extended intracapsular period by allowing the threshold concentration of oxygen to be reached later in development. Higher oxygen concentrations also may cause the hatching glands to develop and secrete later than normal. Embryos that hatch "late" often have enlarged gills (e.g., Lamotte and Lescure 1977) presumably to accommodate the larger size and advanced development while confined in the jelly. Embryos of *Rana grayi* may reach at least Gosner 25 prior to hatching (personal observation), and embryos of *Geocrinia victoriana* may arrest at stage 26 for 4 months if conditions are not amenable to hatching (A. A. Martin and Cooper 1972). Downie and Weir (1997) found that the response of tadpoles of *Leptodactylus fuscus* that were artificially induced to arrest differed in growth and metamorphic response. The longer the tadpoles stayed in the foam, the more slowly they grew when released and a smaller proportion eventually metamorphosed. The mean metamorphic size of those that arrested for longer periods was greater than those that left the foam earlier, but the variability was greater. Based on oviposition sites on land and vegetation out of water, embryos of *Dendrophryniscus minutus* and *Tachycnemis seychellensis* may also have the capability to remain intracapsular for longer periods than normal. Researchers need to study the feedback interactions that may exist among remaining yolk reserves, stimulation of hatching, and developmental stage at hatching. We argue that no additional energy or alteration of development would be needed to make these minor developmental adjustments common to nonaquatic eggs, and the individual can reach Gosner 25 (i.e., a tadpole) in this manner.

If particulate food is not present, the tadpoles of *Bufo periglenes* metamorphose at a smaller size but at the same time as siblings that feed (Crump 1989b; also see Inger 1992b, *Rhacophorus gauni;* McDiarmid and Altig 1990, *Bufo haematiticus;* Weygoldt 1989, *Flectonotus goeldii*). Seemingly there are no data on the possibility of amphibian larvae absorbing dissolved organic material (DOM; see Manahan 1990), and bacterial populations were not monitored in Crump's experiments. The ovum of *B. perilgenes* is large and lightly pigmented and the clutch size is small for a *Bufo* of about 50.0 mm snout-vent length. Altig and Johnston (1989) classified *B. periglenes* as an exotroph because they considered exotrophy to be the primitive condition and feeding to be the likely scenario followed in most natural cases. In the present context, one also could consider *B. periglenes* as a facultative endotroph residing at the absolute terminus of the nonmodified end of the developmental continuum of nidicolous anurans (Altig and Johnston 1989). Therefore endotrophic development is possible without developmental modifications in species whose eggs are well within the size range and presumed energy content of exotrophs. *B. periglenes* has sufficient energy in the egg, a function of maternal reproductive physiology, so that the resulting offspring can survive to metamorphosis without feeding, if that scenario is encountered in the environment. The benefits of this facultative form of development are obvious, and one has to wonder if other species that develop rapidly in habitats with apparently low productivity can respond similarly. Also, we suggest caution in subjectively equating the absence of visible algal growths with limited food availability. At least in dystrophic, tropical pools, the detrital/fungal/bacterial component is likely more important as a food source than photosynthetic algae. Similar cases occur in highly ephemeral desert pools.

Changes within Endotrophs

The next step toward endotrophy would be to change the facultative nonfeeding type (e.g., *B. periglenes*) to an obligatory nonfeeding type (e.g., *Anodonthyla, Philoria, Phrynodon,* and *Plethodontohyla*). That is, an entire tadpole morphotype that does not feed is produced, but one must be aware of the danger of generalizing based on incomplete knowledge of the development of frogs with large eggs. A case in point: Duellman and Gray (1983) showed that the tadpole of *Flectonotus fitzgeraldi* did not feed and metamorphosed in 5–20 days; they predicted that the tadpole of other *Flectonotus* would respond similarly. Weygoldt (1989) reported that the tadpoles of *F. goeldii* feed readily and take 23 days to metamorphose. Also, the fetus of *Nimbaphrynoides occidentalis* feeds from uterine secretions (Vilter and Lugand 1959), and the embryos of *Rhinoderma darwini* may feed from the lining of the vocal pouch of the male (Goicoechea et al. 1986). We need to know more about comparative visceral development among members of various developmental guilds. For example, does the development of the intestine of *Nectophrynoides* spp. differ from that of *Eleutherodactylus* spp.?

The next step might involve the first changes in developmental sequence. Because structures lost by individuals in the intermediate and modified parts of the nidicolous continuum occur only in embryos or tadpoles, the adult phenotype is not affected. Considerably less than an entire tadpole morphotype can produce a frog via metamorphosis, and development can proceed toward metamorphosis from an organism that resembles various stages of a typical exotrophic hatchling. A tadpole morphotype is mandatory only if feeding is mandatory because of developmental energetics but not for the production of a froglet. Whether eggs that develop into nidicolous larvae have more yolk or a higher yolk density than eggs of exotrophs is not known, but some increased amount certainly is expected. Patterns of phylogenetic and ontogenetic gain or loss (e.g., Alberch and Gale 1983; Thibaudeau and Altig 1988) suggest that modifications of development that involve losses may occur in the reverse order of ontogeny of typical species. We have reliable information on the ontogeny of typical embryos but lack these details on a diversity of nidicolous forms. If these data were available, we should be able to discern common mechanisms among lineages and make hypotheses about the development of various features or groups of features controlled by a single inductive cascade (e.g., Elinson 1990). Early interruption of the cascade would prohibit the formation of all subsequent structures because of their developmental interdependence. Within the confines of little data, we speculate that a sequence of expected loss may be adhesive glands, external gills, oral apparatus, coiled gut, lateral line, and spiracle. Conversely, Juncá et al. (1994) and Kaiser and Altig (1994) could not find concordance among character losses of nidicolous larvae within the same genus of dendrobatids (see E. N. Arnold 1994). The nidicolous embryos of *Arthroleptella hewitti* provide another illustration. Probable vestiges of marginal papillae at the corners of the mouth and neuromasts are present, and infralabial and supralabial cartilages, which are probably instrumental in the inductive cascade of mouthparts (Elinson 1990), are present. Detailed studies of the developmental pattern and developmental regulation of features of representative types of embryos are needed.

As indicated above, the early appearance of large limb buds and the usual lack of a number of larval features in both nidicolous and direct developers support a very tenuous assumption that nidicolous-like patterns occurred during the evolution of direct developers. For the evolution of direct development, (1) an increase in energy, by some combination of increases in egg size or yolk density, or both, beyond that which was posulated as being needed to enter the nidicolous grade of endotrophy from exotrophy would be required (see Digression below), and (2) the first profound changes, something more than developmental truncation, in developmental timing could have occurred. The most obvious alteration concerns precocial limb and head formation, and the early appearance of these structures signals a highly modified cascade of developmental events. While remembering that stage designations have no relevance to size or real time per stage, limb buds appear at Gosner 26 in typical tadpoles—shortly after the tadpole becomes free-swimming. In direct developers, limb buds appear immediately after neurulation. Comparisons of timing among developmental modes may be frivolous other than as an indication of degree, but limbs begin to develop perhaps 10%–20% (much more for head development) and 9–10 Gosner stages earlier in relative developmental time than in typical exotrophs. Large, heavily vascularized tail fins presumably serve as a respiratory surface, but no data are available to support this idea; embryos of many species, often ones larger than others with large tails, have small tails. External gills of direct developers never approach the size or complexity of those of many exotrophic hatchlings.

Retention of eggs by a direct developing species may result in ovoviviparity and seems like a relatively minor shift in development, but the similarity of development and morphology between embryos of ovoviviparous forms and the modified end of the nidicolous guild (not direct-developing forms) suggests that the evolution of ovoviviparity and direct-development are independent events or that the constraints of ovoviviparity are different from those for direct development. Changes in embryonic physiology, especially respiration, adjustments in breeding behavior of the adults to accomplish internal fertilization, and alterations in maternal endocrinology to repress ovipositional stimuli would be involved. Even so, the small number of ovoviviparous frogs signals either strong, negative selective pressures or constraints. In most circumstances the energetic cost of carrying embryos may be too extreme over too long a period, especially if being gravid retards preparation for subsequent broods. The transition from ovoviviparity to nonplacental viviparity (see M. H. Wake 1978) is perhaps a smaller step, although some sort of allometry of the feeding and digestive structures would have to occur to allow intrauterine feeding.

Digression on Egg Size and Composition

The lack of information concerning the relationships between egg composition and size is a strong deterrent to a thorough understanding of reproductive biologies. Egg size is commonly invoked as an important modifier of anuran development, and the largest anuran eggs are laid by endotrophs. Numerous studies show that larger eggs provide greater fitness, even for up to eight years later (Semlitsch et al. 1988), but it seems that no one knows what it is about a big egg that is good. A big egg is assumed to contain more energy, but (1) variations in yolk density (yolk weight/volume) and (2) other important components influenced by vitellogenic physiology (e.g., mRNA, polypeptides translated by the mother) have not been taken into account. Egg size alone is a poor predictor of endotrophy of any kind; there is a 25% overlap in the range of endo- and exotrophic eggs sizes. Exotrophic eggs range from about 0.8 (e.g., *Dendrophryniscus*) to at least 4.0 mm (e.g., *Ascaphus truei*), and endotrophic eggs range from about 1.8 (*Sooglossus gardineri*) to 10.0 mm (*Gastrotheca ceratophryes, G. weinlandi,* and *Hemiphractus scutatus*). *Geocrinia rosea,* with an egg diameter of about 2.4 mm, develops directly in about 6–8 weeks, while *G. victoriana* has a typical tadpole that develops normally from an egg 3.1 mm in diameter (A. A. Martin and Cooper 1972). Based on these observations and the case of *Bufo periglenes,* we contend that the increase in available energy does not increase linearly with increases in egg size, and an increase of energy beyond that contained in some average egg is not the prime prerequisite of endotrophic development. Even through Salthe and Duellman (1973) suggested that selection favors increased ovum size in small frogs and that only small and medium-size frogs are able to enter fully terrestrial reproductive zones, there must also be some correlation of egg composition and the likelihood for the evolution of endotrophy within that small size group. Even if direct development and small adult size are correlated, one must consider these two factors on development and eventual morphology (e.g., Hanken 1983, 1984; Hanken and Wake 1993).

Based on the geometry of a sphere, does a 5-mm egg contain an increase of exactly 27% of all components relative to a 4.5-mm egg? Doubling egg diameter equals an 8-fold volume increase, whereas doubling the diameter of a large egg is a much larger absolute amount but a smaller percentage increase than for a small egg. No one seems to know how the composition of energy stores change with egg size and what patterns might be most advantageous under any given circumstance. How does metabolic regulation of yolk consumption differ among developmental modes (e.g., A. A. Martin and Cooper 1972)? Answers to these types of questions likely would demonstrate alterations of maternal oogenic physiology.

The amount and distribution of yolk affects cleavage. The membrane-bound, protein-matrix yolk platelets of various sizes and shapes distributed throughout the egg persist at least through neurulation. The transcription of the zygotic genome and the degradation of the platelets as an energy source do not start immediately after fertilization. Eggs vary intra- and interspecifically in yolk density, and there are more and larger platelets toward the vegetal pole compared with the animal pole (= moderately telolecithal). This dispersion produces an increasing yolk gradient from animal to vegetal pole, and an increase in yolk density would increase the slope of the gradient. The position of the third cleavage plane (first horizontal plane) above the equator is determined by this yolk gradient. We assume that an increase in the amount of yolk would shift the curve of the yolk gradient toward the animal pole. We suggest that the latitudinal position of the third cleavage furrow above the equator reflects the shape of the yolk gradient, and that this cleavage plane probably occurs at a point of similar yolk concentration (at least in eggs of about the same size). If so, species that have a high yolk density, whether exotrophic or endotrophic, compared to a less dense egg would have the third cleavage plane at a higher latitude. This circumstance would cause the formation of smaller micromeres and larger macromeres than in a companion species with lower yolk density. Also, all anurans, exo- and endotrophs, may not practice holoblasty of all cleavage furrows (e.g., Alcala 1962; De Villiers 1929b; Elinson and Del Pino 1985; Warren 1922), perhaps because of yolk amount or distribution. If so, this further increases the size and decreases the number of macromeres. The embryonic disc of *Gastrotheca* that resembles the meroblastic blastodisc of amniotes is a paramount example (Elinson and Del Pino 1985).

Multicellularity is a primary result of cleavage, and differences in cleavage patterns (not including holoblastic vs. meroblastic conditions), and therefore blastomere number and size, probably have no developmental consequences. The results of these patterns can be used to determine egg composition. Embryos of many direct developers and certain exotrophs are elevated above the yolk mass. Eggs of exotrophs with extraordinarily small micromeres (i.e., high-latitude third cleavage) and elevated embryos (e.g., centrolenids, *Heleioporus*) may share other characteristics with endotrophs.

Are there other factors in a large egg that might make a larger embryo? The rather small intraclutch and certain intraspecific variation in egg size may reflect something as simple as water content rather than energy content, and measures of weight/volume (and not simply diameter) or chemical analyses are needed to clarify this idea. The length of the neural groove is the first constraint on eventual size by influencing the starting length of the presumptive body. Eggs with larger diameters provide a longer distance between latitudes across the animal pole to the opposite side, although the amount of increase per unit change in size decreases as egg size increases. If the factor(s) that results in a larger egg is coupled with a faster growth rate, then the benefits attributed to this larger embryo should persist throughout life.

We do not know how much energy it takes to make a direct developing hatchling of a given size, but we question if the generally larger eggs of direct developers are in fact necessary for production of a froglet. We suggest that a large part of the yolk in direct developers is for posthatching growth to an appropriate size to feed—not for expenditure on the formation of a froglet per se. This assumes that endotrophy is no more energetically expensive than normal development. When an endotrophic froglet hatches, there is always considerable yolk that remains. Similar yolk reserves in hatchlings of exotrophic tadpoles support what is essentially a hatched embryo until it grows to a feeding stage. Some part(s) of the feeding or digestive system of direct developers, structural or enzymatic, may not be capable of handling extraneous food at the time of hatching (e.g., Toloza and Diamond 1990a, b). We know of no data pertinent to this hypothesis (see Twombly 1996; Wassersug 1986, 1996). Also, the froglet may simply be too small to find suitable-size food items. Vestigial-winged *Drosophila* are too large for hatchling *Eleutherodactylus coqui* (Elinson et al. 1990) that are about 6.0 mm SVL (mouth gape = 1.5 mm; distance between jaw articulations 3.3 mm). Adult *Sooglossus gardineri* (Mitchell and Altig 1983) commonly eat mites, and hatchlings (about 1.6 mm SVL) may also be able to consume mites.

We ask if one could observe normal development after the removal of the amount of yolk that normally remains after hatching of a direct developer and replace it with an inert material to maintain egg conformation? We know of no data that suggest the kinds of relationships among egg size (or other egg qualities), yolk remaining at hatching, and time and growth before first feeding. The absolutely smallest egg that can produce a froglet directly may be dictated by embryological mechanics and scaling functions associated with locomotion, feeding, and dehydration, and the 1.6-mm egg of *S. gardineri* is probably near the lower limit.

Changes in Egg Jelly

Although not a direct modifier of endotrophic development, certain qualities of the egg jellies had to be modified in association with the various routes out of the aquatic environment. Suggested functions for the layers of mucopolysaccarhide jelly around amphibian eggs include: protection against predators, bacteria, fungi, dehydration, and thermal shock; enhancing solar radiation; attachment to a substrate; and restriction of sperm entry to a specific zone and inhibition of polyspermy. Few correlations among phylogenetic relationships of frogs, their reproductive ecologies (e.g., J. A. Moore 1949b), breeding modes, and egg jelly configurations have been suggested. The eggs of phyllomedusine hylids and *Chirixalus* hatch explosively, and touching one may cause numerous others to hatch in a crackling fury. This explosive hatching suggests quite a different osmotic regime in these eggs compared with aquatic eggs, which become increasingly flaccid as development proceeds. Exiting the aquatic environment would produce additional demands on the functions of egg jelly (e.g., maintaining turgidity to protect the embryos from trampling by parents and avoid the loss of surface area for respiratory exchange by collapse of the jelly, retaining spherical egg conformation for proper development in the absence of water buoyancy, protecting the egg from desiccation without inhibiting gas exchange, and aiding attachment to the damp surface of the mother's back). The outer jelly layer of direct developing eggs often is tough and elastic (eggs bounce when dropped), while the ovum will rupture under its own weight if the jelly is removed in air. The thick, proximal layer is a stiff gel. Especially in hemiphractine hylids that carry exposed eggs, the outer jelly becomes brown and parchmentlike.

Exovivipary and Paravivipary

The site of oviposition, the sex and behavior of the attendant parent, and activities of newly hatched larvae were important parameters used to delimit exoviviparous and paraviviparous guilds (Altig and Johnston 1989). While accepting that a discussion of such behavioral modifications of both embryos and adults is entirely speculative, we suggest that initial stages in the evolution of exovivipary and paravivipary were driven primarily by selection on male behavior. If tactile or visual stimuli are important in breeding, the male probably has an increased likelihood of perceiving these clues based on the amplectic positions of most anurans. Experimental protocols should involve the responses of blinded and anosmic frogs to real and dummy egg masses. In the evolution of paravivipary, a change in ovipositional behavior of the male would position the eggs upon the back of the female (e.g., *Hemiphractus*) with little or no alterations of the back morphology or behavior of the female. A photograph (collection of R. W. McDiarmid) of an *Atelopus cruciger* with two eggs stuck to the lower left quadrant of its back suggests that no great modification of the egg membranes or maternal skin must occur to allow at least initial cohesion of the eggs and skin (see Bourquin 1985 and Pérez et al. 1992a, b). The evolution of brood pouches could then proceed. Perhaps the ovipositional loop of *Pipa* (Rabb and Rabb 1960) is an aborted attempt to go to the surface to breath; seemingly the female would have to initiate the trip to the surface, but the male could interrupt the attempt via steering. Apparently behavioral changes primarily in males accomplish internal fertilization in *E. coqui* and egg positioning in *Gastrotheca*.

Understanding how hormonal profiles are modified in these complex forms of breeding biologies needs additional research. Hormone-mediated behaviors of males, females, and embryos must be orchestrated, and studies of similar behaviors in exotrophs (e.g., *Alytes* and dendrobatids) should be informative. One would suspect that some behaviors operate only for short periods and in the proper temporal and reproductive contexts. Do male *Assa* and *Rhinoderma* not eat their young because they recognize embryos for what they are via sight or olfaction, or are the responses to what appears as a potential food item inhibited? In other cases, excitation (e.g., brooding by either sex; Townsend and Moger 1987) must occur. The behavior of hatchlings must be modified to allow them to mount the back of the male at hatching and remain there (e.g., *Dendrobates* spp., *Leiopelma* spp., and *Sooglossus sechellensis*) or enter a pouch (e.g., *Assa*). Males that carry embryos aid adhesion or movements of the young by skin secretions (G. J. Ingram et al. 1975). The gaseous and nutrient interactions of parents and larvae need considerable investigations (e.g., De Pérez et al. 1992a, b; Wells 1980), and the potential for immunological interactions between adults and embryos should be investigated.

Summary

As noted by E. R. Dunn (1924), anuran adults and larvae live in different worlds so that the significant changes of metamorphosis allow independent selection on the two life stages. In a functional context, one can consider two partially overlapping genomes within each amphibian egg. Largely different sections of this composite genome must be deployed sequentially in order to produce larval features during development and adult features at metamorphosis. Within this single genomic program, the different developmental patterns are produced by alterations in gene expression, and alterations in either section of the genome must not adversely affect the other stage (M. J. Rose 1982). The relative uniformity of most endotrophic developmental patterns suggests that the regulatory changes that were employed were relatively similar in many cases, but the trajectory of the development of hemiphractine hylids certainly is grossly different. At a time when the phylogenetic relation-

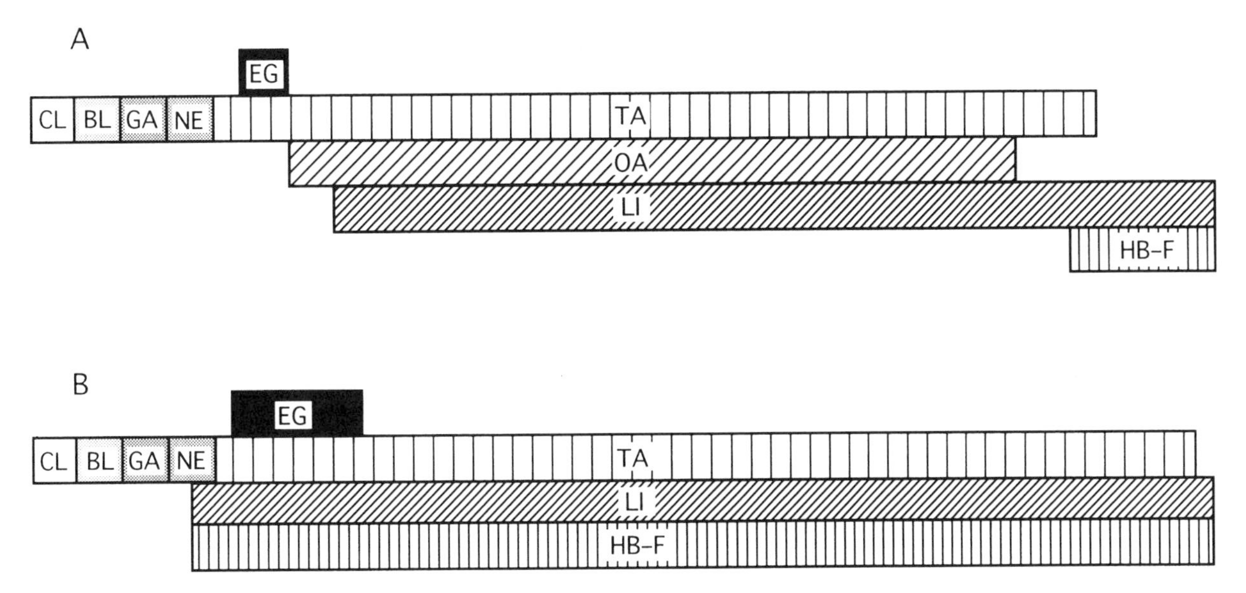

Fig. 7.3. Hypothetical scheme of the sequential expression of various batteries of developmental and regulatory genes. (A) Typical exotrophic tadpole. (B) Direct developer. The fact that some characters may be controlled within an inductive cascade is ignored in this demonstration. Abbreviations: BL = blastulation, CL = cleavage, EG = external gills, GA = gastrulation, and HB-F = formation of the head and body morphology of a frog, LI = limbs, NE = neurulation, OA = oral apparatus, and TA = tail. Lengths of blocks indicate relative time periods of expression.

ships of egg-brooding frogs were not known at the familial level, Noble (1917) noted that a common mode of embryology was a strong signal.

Endotrophy offers a means of shifting life history dramatically (e.g., Raff and Kaufman 1983); and depending on the lineage being examined, some of the developmental changes require that such shifts occur relatively early in development. Raff et al. (1987) suggested that such situations are the result of heterochronies that occur over relatively short evolutionary intervals (and see Alberch 1985; Alberch et al. 1979; Fink 1982). This suggestion does not imply that the basis of the changes is not genetic or that they are not genetically fixed, nor does it imply monophyly among endotrophs or identical regulatory changes in all lineages. A hypothesis involving heterochronic changes gains support from the occurrence of similar energetic-developmental constraints in other groups. Such comparisons are most pertinent when made among groups that include developmental patterns that produce both free-living larvae and direct developers. The discussion by Raff (1987) on the evolution of direct development in sea urchins is an intriguing comparison. Diagrams similar to those of Raff (1987) present speculative ideas on operative sequences of major gene groups of a direct developer compared to a typical tadpole (fig. 7.3). We suggest that mutations in higher level regulatory genes (= homeobox, hox), perhaps coupled with gene duplication, allowed for major steps in developmental patterns to be made quickly. Direct development in anurans results from a suppression of probable gene families that govern typical larval traits (e.g., larval chondrocranium and hyobranchium; all oral features of tadpoles) and an advancement in timing of other cascades (e.g., adult head formation; leg ontogeny) that result in early initiation and formation of a typical frog morphotype usually without a tadpole morphotype. We reiterate that endotrophic development is not a simple shortening or telescoping of typical development, and a thorough understanding of the evolution of endotrophy will come only through an understanding of the developmental regulation that causes the retiming of the various events.

In summary, the occurrence of frog eggs that develop without the quintessential tadpole suggested by general dogma is intriguing to almost everyone. While considering the outlandish idea that a male frog would selectively swallow embryos into his vocal sac and later burp up baby frogs, perhaps Lynn's (1961:157) pun marks our amazement the best: "There can be no doubt that this is a case in which the sudden arrival of quintuplets left the father speechless." Surely it is only the relative rarity of finding endotrophic embryos that is responsible for the large gaps in our knowledge. Because of the divergent developmental pattern compared with typical frogs, direct-developing anurans, with *Eleutherodactylus coqui* being the most commonly studied species, are prime subjects for the study of developmental processes (Elinson 1987a; Elinson et al. 1990). Comparisons among several nidicolous forms with increasing degrees of divergence from the norm also would be a viable option. Because of the multiple occurrences of endotrophy throughout the Anura, we must remain cautious about making premature generalizations.

Additional data are needed on the development, endocrine controls (Hourdry 1993), genetics, and physiology of endotrophic anurans as related to reproductive behavior (e.g., Townsend et al. 1981). Both classical and molecular approaches will be productive (see Hanken et al. 1997). Additional details of egg composition as a function of size among species and developmental modes need to be illustrated in light of the effects on development and the energetics relative to breeding strategies. Transplant experiments between developmental types (Elinson 1990), hybridization of exotrophic and endotrophic species of *Gastrotheca,* and manipulation of egg components would be informative. Surely one is not going to produce an endotroph from an exotroph, but the possibility of generating a developmental variant similar to *Bufo periglenes* discussed above might not be totally out of the question.

ACKNOWLEDGMENTS

In alphabetical order, D. Auth, A. Channing, R. C. Drewes, W. E. Duellman, M. Cherry, M. Hero, D. L. Jameson, A. J. L. Lambiris, W. Magnusson, H. Marx, R. W. McDiarmid, J. I. Menzies, R. A. Nussbaum, J. D. Roberts, L. Rodriguez, F. Slavens, R. Stocks, P. Tolson, and J. Vindum contributed or loaned valuable specimens for use in this study or assisted our efforts in other ways. W. J. Diehl, R. P. Elinson, J. Hanken, and D. S. Townsend contributed substantially to earlier drafts. K. Adler and J. R. Mendelson III provided citations and other pertinent information, R. Rappaport commented on certain of the ideas presented above, and J. Hanken kindly supplied several of the photographs used in this chapter.

8

PHYSIOLOGY

Coping with the Environment

Gordon R. Ultsch, David F. Bradford,
and Joseph Freda

In the Amphibia, as well as in other groups of animals, it often seems that the larvae receive less attention than the adults.

R. M. Savage 1962:24

Introduction

Early scientific interests in the study of tadpoles centered mainly upon developmental biology and date back to at least the morphological investigations of Jan Swammerdam (e.g., 1737–1738). Physiological studies, also inspired by an inquisitiveness about developmental processes, are generally dated more recently, usually from the seminal work of Wilhelm Roux (1888), who tested the preformation theory of development. He destroyed one blastomere of the two-cell stage of a frog with a hot needle and concluded that the remaining cell could not form a complete embryo. This conclusion was later proven to be incorrect (McClendon 1910); frogs have radial and indeterminate cleavage, and the effects Roux saw were caused by not separating the dead cell from the live one. Nevertheless, Roux is usually considered the "father of experimental embryology." In a sense, Roux can also be considered to have been the initiator of studies on the general physiology of tadpoles (see B. I. Balinsky 1981; Browder et al. 1991; Carlson 1988; J. A. Moore 1972).

Interests in nondevelopmental aspects of the physiology of tadpoles were insignificant until the last two to three decades. An increase in the number of papers on tadpole physiology during that period parallels a general proliferation of studies of the comparative physiology and physiological ecology of lower vertebrates. In this chapter we concentrate on those aspects of physiology that are especially relevant to the ecology of tadpoles, particularly respiration physiology, thermal relations, and ion and water balance.

Discussions of amphibian energetics by Seale (1987) are applicable to several subjects presented below. Staging notations are from A. C. Taylor and Kollros (1946) in Roman numerals and Gosner (1960) parenthetically in Arabic numerals (and see table 2.1).

Respiratory Physiology

It only has been within about the last 20 years that an appreciable amount of research has been done on the respiration physiology of tadpoles. As with other vertebrates, tadpoles need to take up oxygen, eliminate carbon dioxide, and maintain an appropriate acid-base status. Beyond this rather generic assertion, further generalizations are risky because of the diversity of microhabitats in which tadpoles are found and the variety of gas exchange organs that are present. Furthermore, most studies of gas exchange and acid-base balance have been done on North American species, and the majority of those have been on the bullfrog (*Rana catesbeiana*). Finally, there are apparently physiological clines (e.g., hemoglobin functions) occurring in this wide-ranging frog that require caution about drawing general conclusions even about this single species. This review, therefore, deals largely with ranid tadpoles, but the reader must bear in mind the impressive variety of tadpoles, particularly those in the tropics (e.g., Duellman 1970; Inger 1985), about which we have practically no physiological information.

Hemoglobin Function

Hemoglobin (Hb) greatly increases the O_2 carrying capacity of blood. Reports of the O_2 capacity of bullfrog tadpole blood range from 7.8 (McCutcheon 1936) to only about 3 vol % (Pinder and Burggren 1983). McCutcheon (1936) first reported that the P_{50} of tadpole (bullfrog) Hb was considerably lower than that of an adult, although there is not enough information to state this as a general case for anurans. A low P_{50} is presumably adaptive for an animal that is primarily a water breather, especially if it lives in vegetated areas that are prone to seasonal and diurnal hypoxia (Nie et al. 1999; Ultsch 1973, 1976a). McCutcheon (1936) also reported a negative Bohr shift in tadpole Hb contrary to the positive Bohr effect found in most animals, including adult bullfrogs. A shift to the left of the O_2 dissociation curve is adaptive if an animal lives in a hypoxic environment that is also hypercarbic. This shift enhances O_2 loading from the hypoxic environment, and hypoxic aquatic environments are often hypercarbic (Ultsch 1973, 1976a). However, the presence or absence of a negative Bohr shift in tadpole Hb solutions is dependent upon the buffer solution used (Aggarwal and Riggs 1969; K. W. K. Watt and Riggs 1975), and more important, the Bohr shift of whole blood is positive (Johansen and Lenfant 1972; Pinder and Burggren 1983). Thus, the potential ecological adaptive value of a negative Bohr shift is apparently not realized in whole animals.

McCutcheon's (1936) finding of an ontogenetic increase in P_{50} has been corroborated for Hb solutions (reviews by Broyles 1981; Riggs 1951; Sullivan 1974) and for whole blood (fig. 8.1; Hazard and Hutchison 1978; Johansen and Lenfant 1972; Pinder and Burggren 1983). As is the rule in blood from other species, the P_{50} of whole tadpole blood is considerably higher than that of stripped Hb solutions because of the presence of cofactors such as nucleoside triphosphates (Aggarwal and Riggs 1969; Sullivan 1974). However, the increase in P_{50} from tadpole to adult is not related to organic phosphates, which decrease with metamorphosis (Hazard and Hutchison 1978; Johansen and Lenfant 1972; Pinder and Burggren 1983) and is caused by larval Hb being abruptly replaced with adult Hb at about Taylor/Kollros XXII–XXIV (Gosner 44–45; fig. 8.2; Just and Atkinson 1972; Moss and Ingram 1968b; Theil 1970). The hemoglobins of both larval and adult bullfrogs are composed of several components (Just and Atkinson 1972; Moss and Ingram 1968a; Wise 1970). Of the four components (I–IV) of bullfrog tadpole blood (K. W. K. Watt and Riggs 1975), components I and II predominate in young tadpoles and have higher O_2 affinities than components III and IV, which predominate in older tadpoles. Because the lungs of young bullfrog tadpoles are less developed than those of older ones

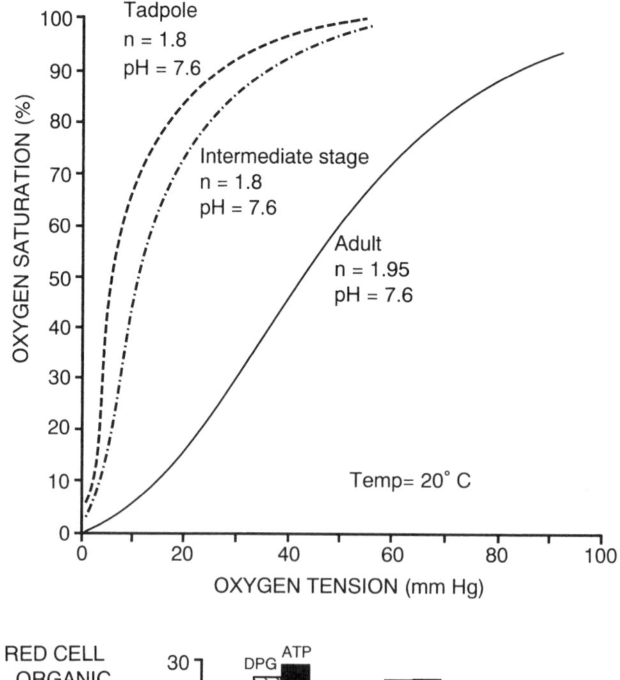

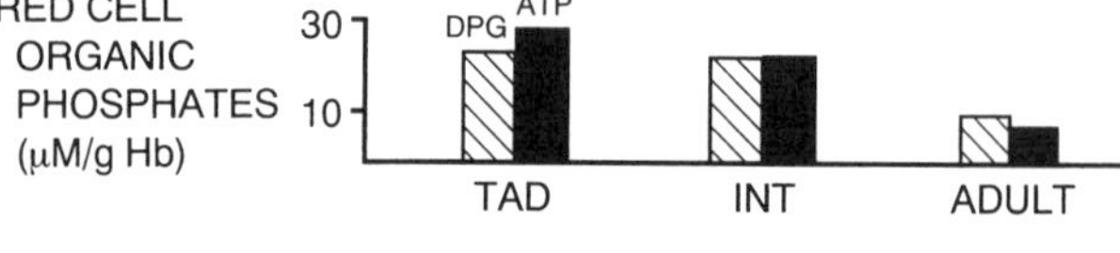

Fig. 8.1. Oxygen dissociation curves and erythrocyte organic phosphate concentrations of tadpoles (TAD), intermediate stages (INT), and adults of *Rana catesbeiana;* n = calculated Hill coefficient. After Johansen and Lenfant (1972).

(Atkinson and Just 1975), the higher affinities could be an adaptation to aquatic hypoxia for a stage that is more dependent upon aquatic O_2 uptake.

There seems to be much less flexibility in Hb function in tadpoles than in adult frogs. Pinder and Burggren (1983) subjected bullfrog tadpoles and adults to 28 days of combined aerial and aquatic hypoxia (70–80 mm Hg; 20°–23°C) and compared changes in respiratory properties of the blood to normoxic controls. There were no changes in hematocrit, RBC count, Hb concentration, mean corpuscular Hb concentration, O_2 capacity, Bohr effect, Hill's coefficient, or intraerythrocytic concentrations of nucleoside triphosphates (ATP + GTP, and 2,3-DPG); a slight decrease in P_{50} from 9.2 to 7.0 mm Hg was discounted as physiologically unimportant. In contrast, adults showed significant increases in hematocrit, a doubling of O_2 capacity, and a decrease of 11 mm Hg in the P_{50}. The authors concluded that the adult responded to hypoxia by altering the respiratory properties of the blood, while tadpoles enhanced the efficacy of the gas exchange organs by changes in their morphology (Burggren and Mwalukomo 1983; see below). The fact that the responses appeared in adults, which do not experience hypoxia because they breathe air, but not in tadpoles, which live in an aquatic environment much more prone to hypoxia, is curious. Perhaps adaptations for adults are related to underwater hibernation as a compensation for their high P_{50} relative to that of tadpoles, but because these experiments were done at 20°–23°C, further experiments at low temperatures are needed to test this hypothesis.

A final comment deals with the caveats mentioned above. Essentially all of the studies of Hb function have used bullfrogs, and the sources of the animals have varied. A number of studies (particularly earlier ones) involving ontogenetic comparisons used tadpoles from one area and adults from another, mixed sources of animals, or used animals with unknown latitudinal origins (e.g., McCutcheon 1936; Moss and Ingram 1968a, b; Riggs 1951). Significant physiological clines relating to respiration and acid-base balance have been demonstrated in other ectotherms (e.g., turtles, Ultsch et al. 1985), and similar clines may exist in anurans. For example, Moss and Ingram (1968a) noted that tadpoles from different suppliers frequently showed differences involving the major Hb components, and Aggarwal and Riggs (1969) found that both the proportions of the Hb components and their amino acid compositions varied with the origin of tadpoles. Therefore one must largely restrict conclusions from studies to date about Hb function to bullfrogs and then be cautious about particulars.

Oxygen Uptake

Gatten et al. (1992) reviewed the literature on metabolic rates of amphibian larvae (see also Løvtrup and Werdinius 1957) and concluded that only at 20°C was there enough data to permit generalizations. They found that the routine metabolic rate was essentially the same for anuran and salamander larvae and that the pooled estimate of the metabolic rate of amphibian larvae was about 50% higher than that of adult salamanders and about 33% lower than that of adult frogs. The intermediate position of the amphibian larvae might be partially caused by the relatively low metabolic rates of salamanders, combined with (at least for anuran larvae) the high water content and high gut volumes of tadpoles relative to the adults.

A number of studies address the relationship between stage of development in tadpoles and $\dot{V}O_2$, but it is risky to generalize. The majority of studies were done under conditions that would not result in determinations of routine or standard $\dot{V}O_2$ (e.g., in respirometers where animals were in vessels that were shaken continuously; Fletcher and Myant 1959; Funkhouser and Foster 1970; Funkhouser and Mills 1969; Sivula et al. 1972; Wills 1936) or in a flow-through respirometer where the excurrent PO_2 (therefore also the PO_2 about the animal) was hypoxic (Etkin 1934). Most studies did not correct for the effects of the large increase in body size as development proceeds, for changes in the proportion of body mass as water, or for changes in the relative contribution of the gut and its contents. A number of studies did not give the tadpoles access to air. Feder (1982) recognized and avoided most of these objections in his study of the relationships of body size, developmental stage, and $\dot{V}O_2$ in four species of tadpoles at 25°C; he concluded that dry mass accounted for most of the variation (59%–90%) in $\dot{V}O_2$, at least up to metamorphic stages (Taylor/Kollros XX–XXV; Gosner 42–46). Feder (1981) also found that 91% of the variance in $\dot{V}O_2$ could be attributed to dry mass in *Xenopus laevis* tadpoles at 20°C. Among species, there is some indication of adaptive differences in the metabolic rates of larvae, but data are not plentiful. *Bufo* tadpoles, for ex-

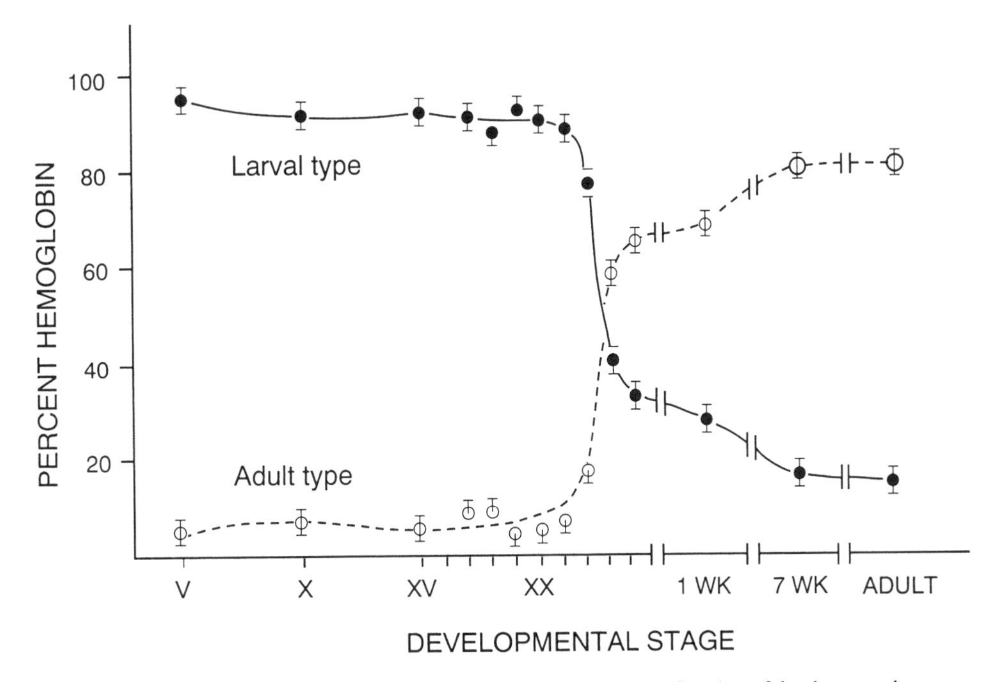

Fig. 8.2. Relative proportions of larval- and adult-type hemoglobins as a function of developmental stage (Taylor/Kollros) of *Rana catesbeiana*. After Just and Atkinson (1972).

ample, have relatively high metabolic rates even when considering their small size (Feder 1982; Noland and Ultsch 1981), which could be considered adaptive if a high metabolic rate results in a faster developmental rate that would allow them to escape from a drying pond.

Gatten et al. (1992) concluded that the average Q_{10} for anuran larvae was 2.14 over the temperature range of 15°–25°C but did not generalize about Q_{10} values over lower temperature ranges. It would be advantageous for the Q_{10} to increase considerably as temperature falls, particularly at temperatures associated with overwintering (i.e., a "metabolic depression" to considerably below that expected from the usual value for Q_{10} of 2–3 observed at higher temperatures; see Ultsch 1989). A high Q_{10} would indicate a metabolic depression that would be adaptive to both a reduced food supply (if overwintering tadpoles feed) and the potentially severe hypoxia that can occur in northern ponds during the winter. There is some indication that metabolic depression occurs in tadpoles at low temperature (as indicated by a high Q_{10}; Bradford 1983 and G. E. Parker 1967), but the case for a metabolic depression associated with low temperature, hypoxia, or both, is not compelling. There is a metabolic depression during starvation in tadpoles (Bradford and Seymour 1985), and adult anurans and other amphibians reduce their metabolic rates 70%–82% while estivating (Seymour 1973; Van Beurden 1980; Whitford and Meltzer 1976; review by Pinder et al. 1992). Therefore it is likely that there is an increase in Q_{10} in tadpoles at low temperatures (also see Thermal Relations).

Among vertebrates that are exclusively water breathers (e.g., fishes, although the following discussion is also relevant for air-breathing vertebrates), it is typical for $\dot{V}O_2$ to be independent of water PO_2 over some ecologically relevant range from the PO_2 of air-saturated water downward. Such animals are called metabolic O_2 regulators, as contrasted with metabolic O_2 conformers whose $\dot{V}O_2$ declines steadily with declining ambient PO_2. The range of naturally occurring PO_2 over which metabolic O_2 regulation can occur is one measure of the degree of adaptation to hypoxia. The PO_2 at which regulation is no longer possible, and $\dot{V}O_2$ starts to decrease with further lowering of ambient PO_2, is termed the critical O_2 tension (P_c). The lower the P_c of an animal, the better adapted it is to hypoxic water. At ambient O_2 tensions below the P_c, the relationship between $\dot{V}O_2$ and PO_2 is usually close to linear. The slope of this linear relationship between $\dot{V}O_2$ and PO_2 below P_c has been termed the O_2 transport capacity of the animal (Beckenbach 1975) or the whole-body O_2 conductance (J. T. Duke and Ultsch 1990); the latter was named in analogy to the thermal conductance of endotherms.

Changes in metabolic rate and whole-body O_2 conductance can effect changes in P_c (fig. 8.3). Consider an animal that is a metabolic O_2 regulator with a given standard metabolic rate (SMR) and a P_c of 100 mm Hg (point A, fig. 8.3). If the animal can double its O_2 conductance, P_c will reduce by half to 50 mm Hg (point B, fig. 8.3) without any change in the SMR. Alternatively, if the animal can reduce its SMR by one-half, P_c will still reduce by half to 50 mm Hg (point D, fig. 8.3) without any change in O_2 conductance. If the animal can reduce its SMR by half and double its O_2 conductance simultaneously, then it can lower its P_c by a factor of 4 (25 mm Hg, point C, fig. 8.3). Assuming that these are the maximal possible adjustments in SMR and O_2 conductance, the range of potential adjustments in P_c resulting from all combinations of changes in SMR and O_2 conductance is defined by area ABCD.

Decreases in SMR (metabolic depression) of ectotherms can be induced by a number of factors, including diving (in air-breathing animals), lowering temperature, torpor and hibernation, and hypoxia or anoxia (reviews by Hochachka 1988 and Ultsch 1989). Mechanisms normally involve biochemical regulation of rate-limiting enzymes (review by Hochachka and Somero 1984). Increases in the whole-body O_2 conductance can be attributed to any mechanisms that can favorably alter the rate-limiting step in the transport of O_2 from the external environment to the cell. Potential O_2-exchange limiting mechanisms have been grouped into those involving the movement of O_2 across the surface of the external gas exchanger (diffusion limitations), delivery of blood to the external gas exchanger (perfusion limitations), or circulatory limitations between the external gas exchanger and the cells (convective limitations; reviews by Feder and Burggren 1985; Piiper 1982; Piiper and Scheid 1977).

As mentioned above, some evidence indicates that metabolic depression may occur in tadpoles in response to low temperatures during overwintering. Most tadpoles that overwinter are ranids, which during warmer seasons can use their lungs to supplement aquatic O_2 uptake from hypoxic water. However, ice cover during the winter would force the tadpoles to rely exclusively upon dissolved O_2, and a metabolic depression would be highly adaptive. Whether a metabolic depression in response to diving occurs in tadpoles is open to question and may depend on the importance of air breathing to a given species. Feder (1981) found that *Xenopus laevis* tadpoles reduced $\dot{V}O_2$ by 20%–40% during forced submergence at 25°C relative to controls with access to air, while there was no effect of forced submergence on the $\dot{V}O_2$ of *Rana berlandieri* larvae. In later studies on the same two species, there was no decrease in the $\dot{V}O_2$ during forced submergence of either species relative to those with access to air (Feder 1983a; Feder and Wassersug 1984). Thus the more usual finding is no change in total $\dot{V}O_2$ with submergence at near-normoxic PO_2. Considering the relatively small contribution of pulmonary respiration to total $\dot{V}O_2$ in normoxic water (Burggren and West 1982; Feder 1983a; Feder and Wassersug 1984), this is not a surprising result. However, in hypoxic water (less than about 70 mm Hg), the aquatic $\dot{V}O_2$ of *Rana berlandieri* and *Xenopus laevis* tadpoles is much reduced in forcibly submerged animals, and access to air does not raise the total $\dot{V}O_2$ to control (normoxic water with access to air) levels (Feder 1983a; Feder and Wassersug 1984). These findings suggest that the set point for aquatic $\dot{V}O_2$ is lowered. The possibility that such a controlled response occurred is supported by the observation that if *Rana berlandieri* tadpoles were exercised in water with a PO_2 of 44–70 mm Hg, the total $\dot{V}O_2$ averaged 80% of control values and

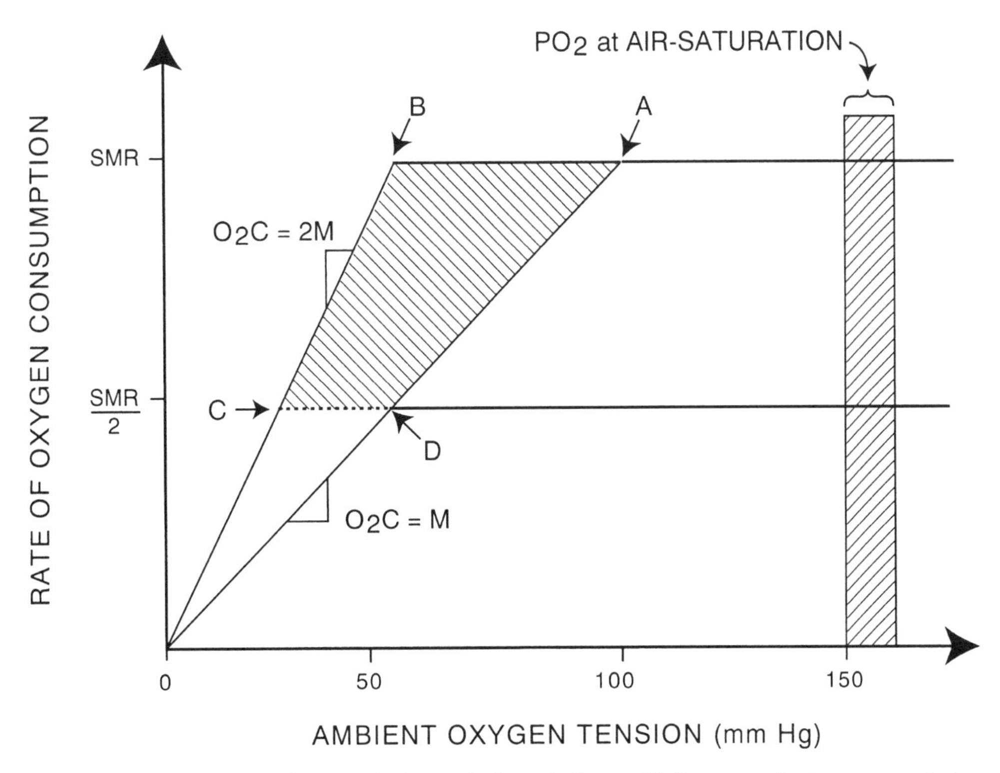

Fig. 8.3. Theoretical effects of changes in the standard metabolic rate (SMR, expressed as O_2 consumption) and whole-body O_2 conductance (O_2C, calculated as the slope M of the left portion of the relationship between O_2 consumption and ambient O_2 tension) on the critical O_2 tension (P_c). A given P_c can be affected by changes in SMR and by changes in O_2C. For example, P_c at 100 mm Hg (A) can be reduced to 50 mm Hg by either halving the SMR (D) or doubling the O_2C (B). The PO_2 of air-saturated water is shown as an approximate range for 0°–40°C. See text for details.

reached as much as 200%, indicating that the ability to take up O_2 at control levels was still present (Feder 1983a).

Whether tadpoles can increase their whole-body O_2 conductance for aquatic respiration has not been studied directly, but indirect evidence suggests they can. A number of short- and long-term strategies are presumably available to them, including cutaneous capillary recruitment (review by Feder and Burggren 1985), adaptive changes in the morphology of the skin and gills (see below), and modifications of hemoglobin function. If tadpoles can alter their whole-body O_2 conductance (e.g., as in the salamander *Siren lacertina;* J. T. Duke and Ultsch 1990), those individuals within a species living in routinely hypoxic environments may have lower P_c values than those from routinely normoxic environments.

The effect of declining water PO_2 on aquatic $\dot{V}O_2$ is of interest even if tadpoles can make up a decrement in aquatic $\dot{V}O_2$ by pulmonary respiration, because surfacing increases predation risk (see below). Thus, a low P_c will be ecologically adaptive if water is the only source of O_2 (determined experimentally by forced submergence), and one might expect to find the lowest values for P_c among species that frequent hypoxic waters. In North America these situations generally confront species that breed in permanent ponds. Conclusions drawn from the relevant data (table 8.1) should be tentative. One qualitative conclusion that seems fairly well supported is that the P_c of tadpoles is higher than that of fishes (Ott et al. 1980; Ultsch et al. 1980b, 1981a) and lower than that of most aquatic salamanders (J. T. Duke and Ultsch 1990; Ultsch 1976b; Ultsch and Duke 1990), indicating that the efficacy of tadpole mechanisms for breathing water are intermediate between the two. There is a trend toward a lowering of P_c with falling temperature, which could be expected because of the decreased SMR (fig. 8.3), but data are sparse. The P_c would be expected to increase with body size as the surface-to-volume ratio falls in tadpoles if the skin is an important gas exchange site (see below). This was found to be the case by Feder (1983b) for resting *Bufo woodhousii*, but the results are mixed for *B. terrestris* and *Rana sphenocephala* (Noland and Ultsch 1981). Moreover, Feder (1983b) found no relationship between body size and P_c in *R. berlandieri*, but Crowder et al. (1998) found the P_c of *R. catesbeiana* tadpoles to be relatively constant with body size through Taylor/Kollros XVI (table 8.1). More data relevant to this point for species with late lung development would be especially interesting.

Gatten et al. (1992) called attention to the paucity of physiological data on anuran larvae in their review. They reported that the ability of tadpoles to increase mass-specific aerobic rates of O_2 consumption (i.e., metabolic scope) is

Table 8.1 Estimates of critical oxygen tension (P_c, mm Hg) of various water-breathing tadpoles relative to developmental stage (Taylor/Kollros 1946), body mass (g), and temperature (°C)

Taxon	Stage	Mass	Temperature	P_c	Reference
Bufo terrestris					
	II–VII	0.030–0.058	22	32	Noland and Ultsch 1981
	IX–XIV	0.088–0.260	22	30	Noland and Ultsch 1981
	II–XII	0.028–0.080	32	39	Noland and Ultsch 1981
	IX–XIV	0.010–0.255	32	52	Noland and Ultsch 1981
Bufo woodhousii					
	—	0.002–0.022[a]	25	25–60[b]	Feder 1983b
Pseudophryne bibronii					
	I–II	—	12	29–33	Bradford and Seymour 1988b
Rana berlandieri					
	I–XV	—	25	38[c]	Feder 1983a
Rana catesbeiana					
	I–IV	2.52	23	29.2	Crowder et al. 1998
	V–VII	4.35	23	31.6	Crowder et al. 1998
	VII–X	9.00	23	35.3	Crowder et al. 1998
	XI–XIII	13.64	23	33.4	Crowder et al. 1998
	XIV–XVI	10.62	23	36.5	Crowder et al. 1998
	XXVII–XIX	15.88	23	51.1	Crowder et al. 1998
	XX	13.05	23	64.1	Crowder et al. 1998
	XXI	11.75	23	76.7	Crowder et al. 1998
Rana muscosa					
	—	2.76–4.78	4	15	Bradford 1983
Rana pipiens					
	—	2.9–6.0	26	35	Helff and Stubblefield 1931
Rana sphenocephala					
	I–V	0.04–0.37	22	20	Noland and Ultsch 1981
	IX–XIV	1.29–2.95	22	34	Noland and Ultsch 1981
	I–V	0.39–1.10	32	49	Noland and Ultsch 1981
	IX–XIV	1.27–2.62	32	32	Noland and Ultsch 1981
Xenopus laevis					
	I–XX	0.010–0.136[a]	25	75–85[b]	Feder and Wassersug 1984

[a]Dry mass.

[b]Estimated from increases in lactate concentration and visual inspection of graphs.

[c]Estimated from increases in lactate concentration after 6 hours submergence.

largely size-independent at 20 and 25°C, compared to the usual dependence of mass-specific routine metabolic rate on body mass. The differing relationships between body mass and metabolic rate for resting and activity $\dot{V}O_2$ means that one cannot make a general statement about the amount by which a tadpole can increase its metabolic rate with activity from general equations for the two situations. Such a comparison can be made for lizards (activity $\dot{V}O_2$ about 6 times the resting $\dot{V}O_2$; Bennett and Dawson 1976) and salamanders (average increase about 5 times; Gatten et al. 1992). On a species-by-species basis, the average increase of 12 times in adult anurans indicates an efficient oxygen transport capacity. Apparently tadpoles are less capable of increasing their $\dot{V}O_2$ than adults, with factorial increases of about 3 times, but this is a preliminary conclusion based on only two studies (Feder 1983b; Quinn and Burggren 1983).

The ability to increase $\dot{V}O_2$ may be of little ecological necessity because most tadpoles are not continuously active. Even when stimulated to high rates of activity when escaping predators, the usual strategy is to dash away and hide. Such burst activity requires only a few seconds at most, and because an instantaneous acceleration is involved, these activities are most efficiently powered by available stores of ATP and other high-energy compounds such as phosphocreatine. The regeneration of these compounds can then be accomplished aerobically without the need for lactate production. Gatten et al. (1984), for example, found that tadpoles of three species had little lactate production after 30 sec of intense swimming at 4°–27°C; tail muscle phosphocreatine fell to nil in the one species in which they measured it (*Rana catesbeiana*).

While it is likely that tadpoles rarely accumulate lactate in the field as a result of activity, they do have a capacity for lactate anaerobiosis, which may be used when water becomes hypoxic. Anaerobiosis during hypoxia would be especially important among tadpoles that overwinter, when ice cover prevents air access (e.g., several species of ranids, Collins and Lewis 1979; review by Pinder et al. 1992), and in species that do not overwinter but have poorly developed lungs (e.g., bufonids).

The role of anaerobiosis during overwintering among amphibians, larval or adult, has not been studied. In laboratory experiments, adult frogs are intolerant of anoxia relative to other aquatic ectotherms, even at temperatures as low as

5°C (review by Pinder et al. 1992). Yet Friet and Pinder (1990), in a preliminary study of hibernacula of adult *Rana catesbeiana* in Nova Scotia, found that the PO_2 of the water within 50 cm of the bottom was as low as 5 mm Hg, with a sustained PO_2 of less than 20 mm Hg for more than 8 weeks. Their findings suggest that bullfrogs are much more tolerant of severe hypoxia in the field than would have been deduced from experimental studies. Whether this conflict is because of different origins of animals, acclimation to progressive hypoxia, or some other effect remains to be seen. Because tadpoles tend to be more tolerant of low PO_2 at low temperatures than adults (Bradford 1983), it is likely that tadpoles are also more capable of anaerobiosis and perhaps of tolerating the resultant lactic acidosis. However, the physiological mechanisms that permit overwintering among tadpoles have not been studied.

There have been several studies of activity metabolism in tadpoles, including the role of anaerobiosis. While burst activity seems to be largely aerobic in tadpoles (Bennett and Licht 1974; Gatten et al. 1984), continuous exercise can lead to appreciable lactate accumulation. For example, Bennett and Licht (1974) found that 30 sec of rapid swimming by bullfrog tadpoles at 20°C increased the resting lactate concentration by only 33% but that 10 min of slow swimming raised lactate concentrations by 270%; by exercising the tadpoles to exhaustion, Quinn and Burggren (1983) induced bullfrog tadpoles to accumulate lactate to levels twice as high as those in the Bennett and Licht (1974) study.

Quinn and Burggren (1983) conducted a thorough analysis of the consequences of exhaustive exercise in bullfrog tadpoles and adults. They found that blood and muscle lactate concentrations at rest were about the same in both stages and that muscle lactate concentration increased by 5–6 fold in both after exercise. Adult blood lactate concentration peaked within 1 min after exercise, but it took 30 min to peak in the larvae, which indicates a slower release to the blood, reconversion in situ, or both, rather than by export to the liver. Adults showed the usual increase in $\dot{V}CO_2$ that results from lactic acidosis when body bicarbonate stores are titrated, but this was not the case for larvae (fig. 8.4), or at least it appeared so when the first measurements were taken 45 min after exercise. However, the persistent low value for the respiratory quotient (fig. 8.4) suggests that body CO_2 stores were being replenished (i.e., repayment of a "CO_2 debt"). This in turn suggests that there may have been a postexercise peak in $\dot{V}CO_2$ that had dissipated by the time of the first $\dot{V}CO_2$ measurement. Tadpoles may be able to eliminate excess CO_2 more rapidly than adults because they are smaller and because they can use lungs, gills, and skin for CO_2 elimination. By having multiple sites for CO_2 exchange, the tadpoles may also be less susceptible to respiratory acidosis during exercise.

The importance of lungs to tadpoles in hypoxia is obvious, but they are also important during activity. Stamina increases with access to air even when swimming in normoxic water, and hypoxia decreases stamina more in lungless *Bufo americanus* tadpoles than in the lunged tadpoles of *Rana berlandieri* and *Xenopus laevis* and (Wassersug and Feder 1983).

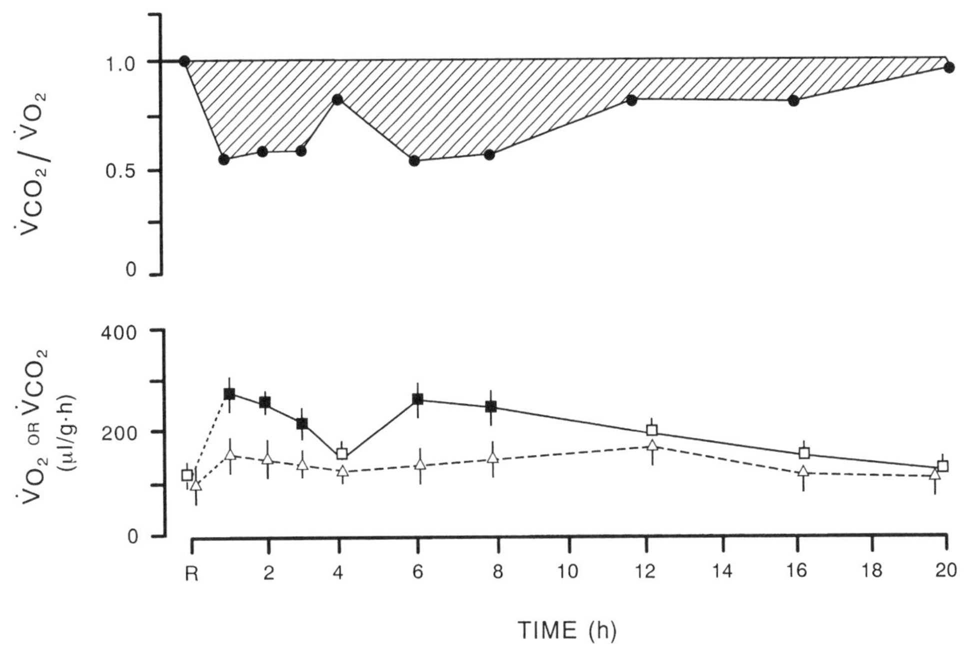

Fig. 8.4. Respiratory responses of tadpoles of *Rana catesbeiana*. *(Upper)* Respiratory exchange ratio (R = $\dot{V}CO_2/\dot{V}O_2$) at rest and after exhaustive exercise at 23°C; the hatched area indicates a sustained decrease in R. *(Lower)* Means ± S. E. of O_2 consumption *(squares)* and CO_2 elimination *(triangles)* under the same conditions; solid symbols indicate values that increase significantly from resting levels. After Quinn and Burggren (1983).

However, the ecological importance of the ability to accumulate lactate is most likely related to hypoxia tolerance. From this viewpoint it is interesting that Feder (1983b) found that *Bufo woodhousii* tadpoles swimming to exhaustion accumulated no more lactate than those at rest in water with a PO_2 of 19 mm Hg. The effects of prolonged elevations of lactate during chronic hypoxia have not been studied.

Gas Exchange Partitioning

Tadpoles, like many aquatic salamanders, may have several sites of gas exchange (e.g., gills, skin, and lungs) and therefore may exchange gases with air, water, or both. The extent to which gas exchange is partitioned among sites and between media depends on a host of factors, including the presence or absence of lungs, body size, respiratory gas concentrations in the water (at the moment and over time), temperature, stage of development, level of metabolism, and microhabitat (e.g., Jia and Burggren 1997a, b). To understand the complexities of gas exchange in tadpoles, one must consider the morphological characteristics of the gas exchangers and their vascular supply. A number of species, such as all bufonids studied to date and some torrent-dwelling species (Feder 1981; Nodzenski et al. 1989; Wassersug and Heyer 1983), have rudimentary lungs or do not develop lungs until metamorphosis. These species are likely more dependent on cutaneous gas exchange than those with lungs. Microhylids (R. M. Savage 1952; Wassersug and Pyburn 1987), some leptodactylids (Wassersug and Heyer 1988), and some pipids (Malvin 1988) have a reduced development of the respiratory portions of the gills, which likely lowers the ability of the gills to exchange gases with water. Ranids have all three sites functional from an early age (R. M. Savage 1952). Because bullfrogs and other ranids are most commonly studied, the discussions herein are heavily weighted toward that group.

Probably the only universally important gas exchange organ among tadpoles is the integument. Both terrestrial and aquatic tadpoles possess the typical gas-permeable skin of amphibians, and all amphibians exchange gases, especially carbon dioxide, through it. The degree of gas exchange cannot be predicted directly from morphology. For example, the importance of the skin of tadpoles as a gas exchanger for oxygen has been questioned by De Saint-Aubain (1982, 1985) and McIndoe and Smith (1984b); they noted that cutaneous vascularization is sparse relative to that of aquatic salamanders and adult anurans. However, the skin can be an important site of O_2 uptake (see below). There also may be no coupling between the degree of skin vascularization and the presence or absence of lungs. De Saint-Aubain (1982) found no morphological differences in the cutaneous vascular system between the tadpoles of *Bufo bufo* and *Rana temporaria,* although the latter has well-developed lungs and the former has only rudimentary lungs (R. M. Savage 1952) and presumably is dependent upon aquatic respiration. Perhaps the small size of *Bufo* larvae negates any need for an increased cutaneous O_2 conductance.

In considering the potential of the skin as a gas exchanger, Fick's law of diffusion is appropriate: $\dot{M} = D(S)(dP)/l$, where $\dot{M}$ is the rate of gas transfer, D the diffusion coefficient, S the effective surface area for gas exchange, dP the partial pressure gradient across the skin, and l the distance over which diffusion must occur (water [or air]-to-blood barrier). It has been argued that gas exchange across the integument of amphibians is primarily diffusion-limited (Clemens and Feder 1990; review by Feder and Burggren 1985) and that $\dot{M}$ can be regulated only by varying dP (e.g., changing the internal gas pressures or moving to a more favorable respiratory microenvironment). However, Burggren and his coworkers argued that at a given dP, $\dot{M}$ can be regulated by changing the effective values of either S or l. They proposed that the effective cutaneous exchange area can be varied by capillary recruitment and derecruitment (Burggren and Moalli 1984; Feder and Burggren 1985; also Feder et al. 1988) and that the apparent thickness of the diffusion path can be decreased by convection that reduces the thickness of the boundary layer (Burggren and Feder 1986; Feder and Pinder 1988); this layer is potentially important in limiting cutaneous gas exchange (Booth 1990; Pinder and Feder 1990).

Whether such cutaneous regulatory changes occur, and whether they are effective in altering cutaneous $\dot{V}O_2$, is an open question. For example, the cutaneous $\dot{V}O_2$ of the skin of adult *Rana pipiens* in water with access to air fell as water PO_2 fell (Vitalis 1990), which implies that mechanisms available for increasing cutaneous $\dot{V}O_2$ during progressive hypoxia were not effective. Because the frogs could breathe air, it could be argued that the shortfall in cutaneous $\dot{V}O_2$ was simply made up by an increase in pulmonary O_2 uptake, which negated any need for cutaneous regulatory mechanisms to be brought into play. In contrast, the much larger and predominantly skin-breathing salamander *Cryptobranchus alleganiensis* can maintain its cutaneous $\dot{V}O_2$ over the same range of water PO_2 (Ultsch and Duke 1990) that the frogs in the study by Vitalis could not (or did not). Therefore, among the amphibians in which the question of the regulation of cutaneous $\dot{V}O_2$ in water has been addressed, there is no definitive answer; data on tadpoles are lacking.

Apparently changes in the rate of gas transfer, particularly O_2 uptake, can occur among aquatic amphibians over both short (Feder et al. 1988; Malvin and Hlastala 1986a, b, 1988) and long terms (J. T. Duke and Ultsch 1990), but mechanisms and their control remain conjectural (Malvin and Riedel 1990; Pinder et al. 1990). The only study involving tadpoles (Burggren and Mwalukoma 1983) showed that chronic exposure of bullfrog tadpoles to hypoxia resulted in a thinning of the skin, a doubling of capillary mesh density, and a reduction of the blood-water barrier by one-half.

Anurans develop external gills prior to hatching and retain them for several days after hatching. The external gills are replaced by the internal gills coincident with the development of the operculum; the internal gills are lost by Taylor/Kollros XXIII (Gosner 45; Malvin 1988). In bullfrogs between Taylor/Kollros V and XX (Gosner 30–42), the gills are highly vascularized and remain a constant proportion of body mass; by Taylor/Kollros XXII (Gosner 44) the importance of the gills is diminished as resorption starts and the gill capillaries move deeper into the epithelium (Atkinson and Just 1975).

It is adaptive for aquatic amphibians to have variable perfusion patterns that are related to environmental O_2 tensions. Tadpoles in well-oxygenated water exhibit cardiorespiratory synchrony when at rest (Wassersug et al. 1981c). In severely hypoxic water, when the lungs are the primary organ of O_2 uptake, O_2 would be lost to the water if gill perfusion was inevitable. Most, if not all, gilled aquatic salamanders have shunts that either direct blood from the ventral to the dorsal portions of the aortic arch before it enters the gill lamellae or bypass the respiratory portion of the lamellae by flowing directly from the afferent to efferent branchial arteries (C. L. Baker 1949; Darnell 1949; Figge 1936; Malvin 1988; Malvin and Boutilier 1985). While the physiological data are not as extensive as for salamanders, some tadpoles possess gill shunts (De Saint-Aubain 1981, *Bufo bufo* and *Rana temporaria;* McIndoe and Smith 1984b, *Litoria ewingi*), and tadpoles likely exert control over the degree of shunting (De Saint-Aubain 1985). As with salamanders (Figge 1936; Malvin 1985a, b), the proportioning of blood flow between the respiratory portions of the gills and the shunts is presumed to be controlled by selective vasoconstriction, vasodilation, or both, perhaps under the influence of circulating catecholamines.

As with external gills among other amphibians (Bond 1960), the degree of development of anuran internal gills is responsive to the concentrations of dissolved respiratory gases (Løvtrup and Pigón 1968). Bullfrog tadpoles in Taylor/Kollros XV–XX (Gosner 40–42) exposed to a water PO_2 of 70–80 mm Hg for 4 weeks increased the size and number of gill filaments; chronic hyperoxia (387–430 mm Hg) had no effect (Burggren and Mwalukoma 1983). Although tadpoles in hypoxic water in that study had access to a gaseous phase, the gas had the same PO_2 as the water, so it is uncertain how great the morphological response of the gills would have been had the tadpoles had access to normal air, especially because they were at a stage where the lungs are well developed. Nevertheless, the presence of a morphological response capability was evident.

The lungs of bullfrog larvae gain weight from Taylor/Kollros V–XV (Gosner 30–40); after that, lung absolute weight remains relatively unchanged, but their relative weight increases as body weight falls (Atkinson and Just 1975). These authors reported the degree of lung septation to be minor at Taylor/Kollros X (35) and increasing considerably by stage XXII (44) as tail resorption progresses. This does not mean that the lungs are not functional in early stages, because even early-stage bullfrog tadpoles surface occasionally to breathe and routinely have gas-filled lungs (Crowder et al. 1998; Just et al. 1973; also Orlando and Pinder 1995).

The ability to breathe air is not crucial to development or survival in bullfrog larvae through Taylor/Kollros XXI (Gosner 43) when submerged in normoxic water at 20°C, but later stages will drown (Crowder et al. 1998; Just et al. 1973). It appears that all stages breathe air (Crowder et al. 1998; Just et al. 1973). The utility of breathing air is obvious in hypoxic water, and the microhabitat of ranid tadpoles can become quite hypoxic (Nie et al. 1999; Noland and Ultsch 1981). The reason tadpoles breathe air in normoxic water may be to keep the lungs inflated; if they did not breathe air, the lung gases would be lost as O_2 and then N_2 followed the diffusion gradients out of the lung. This situation is similar to that of diving insects that use a gas bubble as a diffusion gill (K. Schmidt-Nielsen 1983). Lungs dissected from premetamorphic bullfrog tadpoles that have been living in normoxic water with access to air float (Crowder et al. 1998; also Just et al. 1973), but if the animals are denied air access for a day, the dissected lungs sink, which indicates that most or all of the lung gases have been lost. This reasoning is supported by the observation of West and Burggren (1982) that bullfrog tadpoles in Taylor/Kollros XVI–XIX (Gosner 40–41) did not ventilate their lungs during 1 hr of hyperoxia (about 600 mm Hg), while ventilation (albeit low-level at 1/hr) did occur in normoxia. A possible explanation is that the lungs did not deflate in hyperoxic water; had they done so, the tadpoles probably would have breathed. West and Burggren (1983) also showed that pulmonary stretch receptors must be in the lungs. The loss of lung gases has also been suggested by Feder and Wassersug (1984); *Xenopus laevis* tadpoles in normoxic water became negatively buoyant after 10–63 min of submergence, and a negative hydrostatic pressure of −1 atm would not cause these tadpoles to release an air bubble. Air would be released from the lungs of tadpoles kept in normoxic water with access to air if they were similarly tested.

Burggren and Mwalukoma (1983), in the studies of morphological changes in hypoxia-acclimated *Rana catesbeiana* tadpoles discussed above, found an increased lung volume and cava density in response to hypoxia. Hyperoxia had no effect on tadpoles, and neither hypoxia nor hyperoxia affected adults. Because the animals were breathing both air and water of the same PO_2, it is not certain what the changes in lung structure in the tadpoles would have been if they had been kept in hypoxic water with access to normoxic air. The fact that adults showed no changes may be caused by their higher degree of lung development; that is, 70–80 mm Hg may not be a severe enough hypoxia to present a significant respiratory challenge. Tadpoles of *Discoglossus pictus* exposed to 426 and 710 mm Hg O_2 for 15–30 days in both the air and the water phase had greatly suppressed lung development (De Quiroga et al. 1989). The degree of suppression was greatest in earlier stages and least in newly metamorphosed animals, which led to the conclusion that the effect was primarily one of delayed lung development caused by the excess aquatic oxygen supply. These authors suggested that the reason they found a response, while Burggren and Mwalukoma (1983) did not, was that the PO_2 used in the latter study was not high enough to trigger a response; it is also possible that the difference in responses was species-specific.

Finally, it is not known if the lungs are merely accessory respiratory mechanisms that enable a tadpole to live in water with a PO_2 below the P_c determined during forced submergence experiments (i.e., the "ecological P_c" is less than the "experimental P_c"; see Nie et al. 1999), or if the lungs can be sufficient in an environment where the water is chronically and severely hypoxic. Salamanders that live in such environments (e.g., *Amphiuma, Siren*) have well-developed lungs and can inhabit areas where the water remains virtually anoxic for months (Ultsch 1976a). Some lunged tadpoles can tolerate extended hypoxia (e.g., Lannoo et al. 1987, *Osteopilus brunneus*), but whether they can tolerate continuous anoxia is unknown (see Costa 1967).

Aquatic tadpoles lacking or with poorly developed lungs partition gas exchange between the skin and the buccopharyngeal epithelia, with the main exchange surface in the latter being the gills. Bufonid tadpoles, which are essentially lungless, have gills with well-developed respiratory lamellae. It seems to be a general rule that while there are aquatic tadpoles with reduced or absent lungs and others with reduced or absent respiratory lamellae, there are none in which both of these respiratory structures are reduced or absent. Such might be the case if cutaneous respiration is sufficient and the tadpole is unlikely to encounter hypoxia. If such a tadpole exists, one might find it in relatively cool torrent habitats and the tadpole would be small, or both, and therefore have a favorable surface-to-volume ratio for integumentary gas exchange. *Ascaphus* tadpoles inhabit cool mountain streams, and while they have reduced lungs, the respiratory portion of their gills is not diminished (Gradwell 1971a, 1973). *Bufo* tadpoles, which are uniformly small regardless of adult size (Werner 1986), have rudimentary lungs and functional gills. Tadpoles that obtain oxygen exclusively through the skin are not known, but their nonexistence would not seem to be because it is physiologically impossible for them to exchange gases solely in this way. Some large aquatic salamanders are (Ultsch and Duke 1990, *Cryptobranchus*) or can be (Shield and Bentley 1973, *Siren*) predominantly skin-breathers in normoxic water. No attempt has been made to partition gill and cutaneous oxygen uptake in any tadpole that lacks a significant pulmonary capability. Such an experiment on *Bufo* tadpoles would be particularly interesting because the sparse cutaneous vasculature (De Saint-Aubain 1982, *Bufo bufo*) suggests that the skin is an inefficient gas exchanger.

The relative roles of the skin and gills in gas exchange of lunged tadpoles in water of near-normoxic PO_2 and with access to air have been studied. Burggren and West (1982) used *Rana catesbeiana* tadpoles and four-year-old adults, both from Massachusetts; the larvae were of early (Taylor/Kollros IV–V; Gosner 29–30), intermediate (XVI–XIX; 40–41), and late (XXIII–XXIV; 45) stages. Late stages had lost their gills, so measurements were only of pulmonary and cutaneous exchange; early stage tadpoles were presumed to be mainly aquatic breathers and were not permitted access to air. This constraint made their pulmonary component 0%, which is probably an underestimate even in water of high PO_2. The authors stated that intermediate-stage larvae are obligate air breathers at 20°C. However, bullfrog tadpoles from Alabama submerged in aerated water at 20°C at Taylor/Kollros V–XII (Gosner 30–37) can develop to at least stage XXI (43) without access to air (Crowder et al. 1998). These contrasting results suggest an important effect of latitude of origin or developmental PO_2 history of the tadpoles.

Burggren and West's (1982) study of bullfrog tadpoles showed that the major proportion of the $\dot{V}CO_2$ was through the skin in all stages, as is the usual case in amphibians, and that the cutaneous $\dot{V}O_2$ was dominant over branchial $\dot{V}O_2$ in the premetamorphic stages (fig. 8.5). This effect was not caused by an inefficiency of the gills, which extracted about 60% of the O_2 from the water passing over them; this rate compares favorably with the gills of teleost fishes (but see West and Burggren 1982, where the O_2 extraction by bullfrog gills was only 43%). In addition, the high extraction rates were paired with high gill irrigation rates of about 95/min. Experimentally induced increases in gill irrigation rates did not raise the branchial $\dot{V}O_2$, which indicates that O_2 exchange across the gills was not diffusion-limited. Also, the predominance of cutaneous $\dot{V}O_2$ is unaffected by increases in total $\dot{V}O_2$, at least when the increases are caused by increases in temperature, and remains at about 60%–80% from 15°–33°C for both *Rana berlandieri* and *R. catesbeiana* (Burggren et al. 1983). One explanation for the high cutaneous $\dot{V}O_2$ may be that a significant proportion of the tadpole tissues (and therefore $\dot{V}O_2$) are peripheral (e.g., large tail surface) and cutaneous diffusion may suffice for those tissues (Burggren and West 1982; De Saint-Aubain 1985).

The most ecologically relevant aspect of gas exchange partitioning in tadpoles is between media and not among sites. Having an ability to use air as an O_2 source and CO_2 sink allows tadpoles to inhabit microenvironments from which they might otherwise be excluded by intolerable water hypoxia, hypercarbia, or both. A number of studies have dealt with aerial and aquatic gas exchange (especially O_2 uptake) as a function of developmental stage, water PO_2, temperature, and degree of activity. Several studies of *Rana catesbei-*

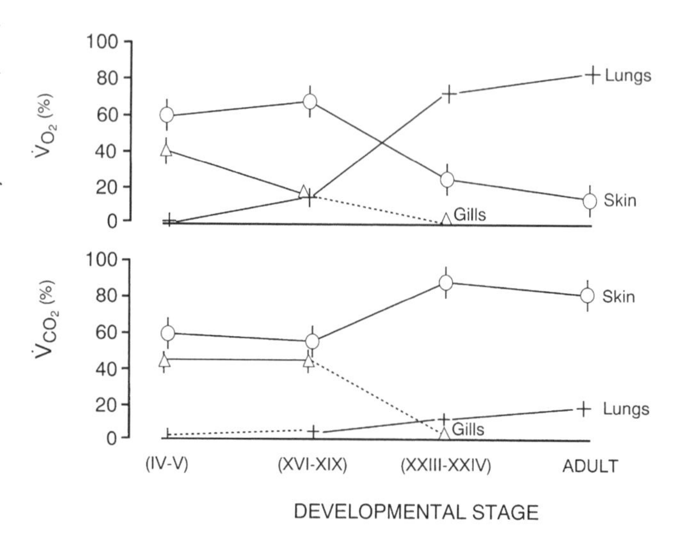

Fig. 8.5. The relative contributions of gills, skin, and lungs to gas exchange in tadpoles and adults of *Rana catesbeiana* at 20°C. *(Upper)* O_2 uptake ($\dot{V}O_2$). *(Lower)* CO_2 elimination ($\dot{V}CO_2$). Tadpoles and metamorphosing individual staged according to Taylor and Kollros (1946). Dotted lines connect to presumed values. After Burggren and West (1982).

ana tadpoles related aerial and aquatic respiration. West and Burggren (1982) studied Taylor/Kollros XVI–XIX (Gosner 40–41) tadpoles in water of 140, 82, 43, and 21 mm Hg (see Crowder et al. 1998). There were progressive increases in the frequency of lung ventilation with increasing hypoxia (1/hr at 140 mm Hg to 51/hr, with much variability, at 21 mm Hg) with an indication that tidal volume also increased. Average gill irrigation frequency and heart rate were unaffected by hypoxia (also *Xenopus laevis* larvae, Feder and Wassersug 1984; see Tang and Rovainen 1996), but there was a transient decrease in both the frequency and amplitude of gill ventilations in hypoxic water after an air breath. Marian et al. (1980) found an increase in surfacing frequency with a decrease in water PO_2 in *Euphlyctis cyanophlyctis,* but in contrast to the studies just mentioned, they also found an increase in the frequency of gill ventilation as hypoxia increased. West and Burggren (1983) concluded that both lung inflation (via stretch receptors) and the PO_2 of lung gas affect the gill irrigation minute-volume. They suggested that the decrease in gill irrigation immediately after an air breath could be adaptive in preventing diffusive O_2 loss to the hypoxic water.

Feder and Wassersug (1984) examined aerial and aquatic $\dot{V}O_2$ in *Xenopus laevis* larvae, which lack gill filaments and presumably use the enlarged gill filters, buccopharyngeal epithelia, and skin for aquatic gas exchange. In water with more than 100 mm Hg of O_2, the aerial component of $\dot{V}O_2$ was a relatively minor 17%, as was the case in bullfrog tadpoles (Burggren and West 1982). At a water PO_2 of 75–85 mm Hg, the frequency of lung-breathing started to increase from control values of about 2/hr to a high of 45/hr at 25–50 mm Hg. When the water PO_2 became very low, there was an indication of a reduction in overall $\dot{V}O_2$, demonstrating that the lungs were not able to make up the shortfall in the aquatic $\dot{V}O_2$; lactate did not increase. There was much scatter in the data, so a suggestion that these results indicate a depression of metabolic rate in hypoxic water must remain tentative. In addition, Wakeman and Ultsch (1976), in similar studies with three species of salamanders ranging from completely to partially aquatic, found that total $\dot{V}O_2$ could be maintained in all three species down to a water PO_2 of 20 mm Hg. Whether a metabolic suppression (or a controlled reduction) occurs among tadpoles in severely hypoxic water probably will also be dependent upon such factors as degree of lung development and temperature.

It is difficult to draw general conclusions about the extent of partitioning without considering other variables. For example, the developmental stage has significant effects on partitioning related to differences in the relative development of lungs and gills and the large increases in tadpole body size between hatching and the late premetamorphic stages (Burggren and Doyle 1986). The gill-lung-skin partitioning of $\dot{V}O_2$ in bullfrog tadpoles was relatively independent of temperature from 15°–33°C in spite of increases in metabolic rate (Burggren and Pinder 1983). The relative importance of air breathing will depend upon species-specific differences in the efficacy of exchange mechanisms, and these vary greatly. For example, *Osteopilus brunneus,* a tropical hylid, is an obligate air breather in its natural environment and does not irrigate its buccal cavity (Lannoo et al. 1987).

Although Burggren and West (1982), Burggren and Pinder (1983), and Feder and Wassersug (1984) have shown that air breathing is infrequent and that pulmonary respiration is a minor component of total $\dot{V}O_2$ for tadpoles in normoxic water, it is clear that an increased dependence on air breathing will occur in hypoxic water. In order to minimize predation risk, one would expect tadpoles to use one or more of a variety of adaptations, including selection of high-PO_2 habitats, unpalatability, minimizing exposure time during breathing, acclimatization to hypoxic habitats (see below), and adaptive behavior patterns associated with air breathing. The selection of habitats with high PO_2 would reduce the dependence upon air breathing and would be adaptive if all other factors were equal. In habitats where the distribution of dissolved O_2 is heterogeneous, PO_2 is normally highest in open (i.e., lacking dense vegetation) water (Ultsch 1973, 1976a; Wakeman and Ultsch 1976). Open water is not a good place for tadpoles to hide or find food, and when they do surface, they are much more likely to be detected by predators (Feder 1983c).

Noland and Ultsch (1981) studied the relationships among temperature tolerance, hypoxia tolerance, and habitat selection in lungless tadpoles of *Bufo terrestris* and lunged tadpoles of *Rana sphenocephala.* Water temperature and dissolved O_2 (DO) were measured from March to August at nine sites where one or both species occurred, and an attempt was made to sample all microhabitats at each site. After these measurements, each microhabitat at each site was sampled to determine if either of the species was present. They found a trend toward a lowering of the critical thermal maximum (CT_{max}) with increasing stage of development (also *Bufo woodhousii,* Sherman 1980). *Bufo* tadpoles had a CT_{max} about 2°C above that of *Rana* at all stages; higher acclimation temperatures resulted in higher values of CT_{max}. Critical O_2 tensions (P_c) of submerged larvae were relatively low at 20–52 mm Hg and tended to increase with temperature but showed no clear relationship with stage or between species. Several conclusions can be drawn from the field data (fig. 8.6). (1) The range of temperature and DO was similar at sites containing *Bufo* and those with *Rana* (some sites had both species). (2) *Bufo* tended to live at higher temperatures (median of 30°C) than *Rana* (median of 24°C), largely because of their occurrence in shallow water and temporary ponds. (3) Eighty-two percent of *Bufo* larvae were found at a DO of more than 50% air saturation, and 92% were found above their P_c. (4) Only 37% of the *Rana* larvae were found in water with a DO more than 50% of air saturation, while 38% were found below their P_2. From these data, Noland and Ultsch concluded that *Bufo* breed in temporary ponds partly because such ponds tend to have both a high DO and high temperature relative to permanent ponds. The high DO is obviously adaptive to a tadpole that lacks lungs, while both high DO and high temperature will increase the already relatively high metabolic rate of *Bufo* larvae, speed their development, and increase the probability of metamorphosis before pond drying (see also Dupré and Petranka 1985; Pat-

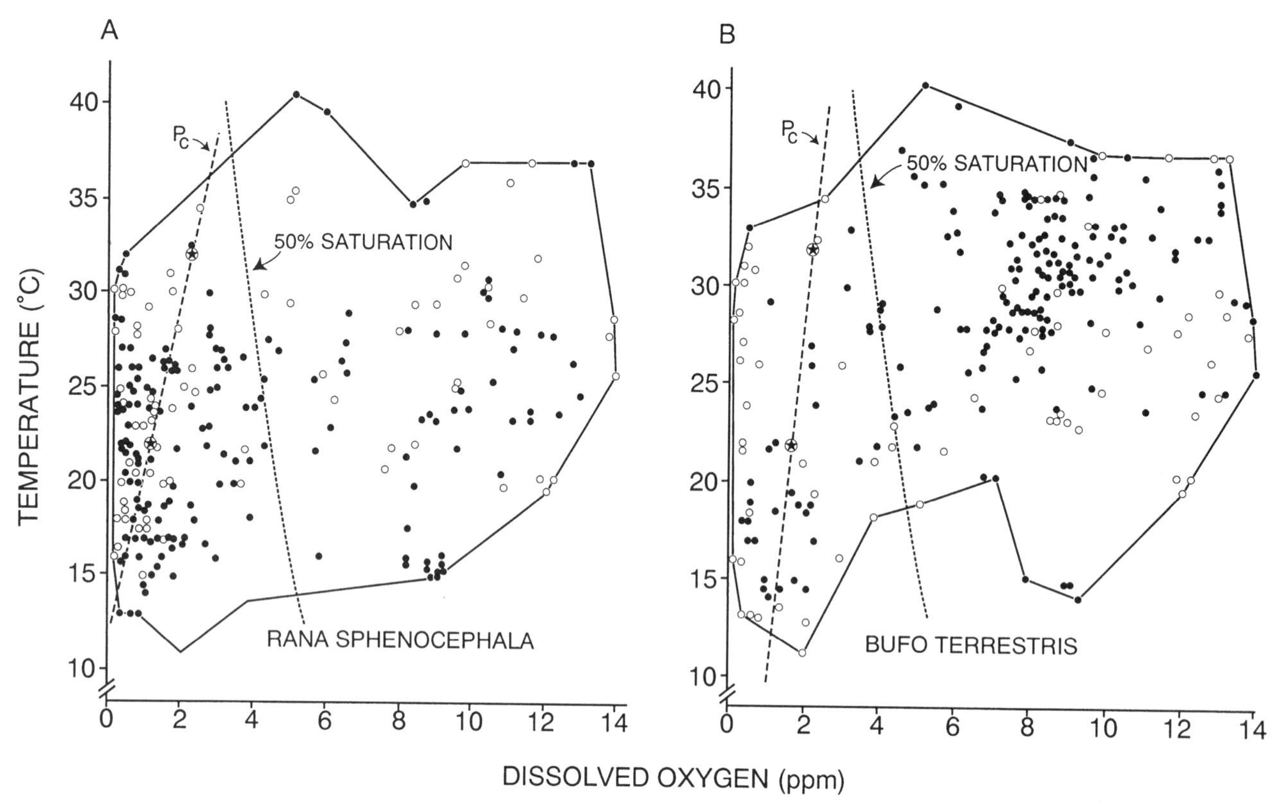

Fig. 8.6. Dissolved oxygen-temperature polygons for some aquatic habitats in Alabama. These areas represent the sum of possible combinations of dissolved oxygen and temperature within the habitats. The presence of tadpoles (*Rana sphenocephala* and *Bufo terrestris*) at a location is indicated by dots and their absence by circles. The critical O_2 tension (P_c) for each species and 50% saturation level of dissolved oxygen are both indicated as a function of temperature. After Noland and Ultsch (1981).

terson and McLachlan 1989). *Rana,* on the other hand, breed in more permanent ponds that will tend to have lower DO levels, and the tadpoles compensate by breathing air during the lengthy developmental period. Nie et al. (1999) also studied the effects of ambient PO_2 and temperature on the distribution of tadpoles (*Rana catesbeiana*) in Alabama. They found that during the summer, the tadpoles in one of the study ponds were frequently found in water with a PO_2 below their P_c in spite of the lack of predatory fishes and the fact that 56% of the pond bottom had water with a PO_2 greater than the P_c of the tadpoles. Tadpoles caught in submerged traps in the hypoxic areas drowned, which indicates that air breathing was obligate in those microenvironments.

If a tadpole lives in shallow water to ensure being in a high DO environment or breathes air in hypoxic water, it is exposing itself to predation. Predation is probably the major cause of mortality of tadpoles, especially in permanent ponds, and may be a major selective force that has resulted in many anurans breeding in temporary (fish are absent) ponds (see chaps. 9 and 10). Moreover, predation may determine anuran community structure (Britson and Kissell 1996; Cortwright and Nelson 1990; Heyer et al. 1975; Woodward 1983; see Axelsson et al. 1997) and can eliminate entire populations of tadpoles (Cortwright and Nelson 1990; Sexton and Phillips 1986; D.C. Smith 1983; see Barreto and Moriera 1996). Some anurans that breed in temporary ponds have relatively unpalatable larvae (e.g., R. A. Griffiths 1986; Reading 1990; Voris and Bacon 1966; Wassersug 1971; Wilbur 1987; some *Bufo*), which would allow them to be exposed in shallow, well-oxygenated waters, but in general temporary-pond breeders have tadpoles that are more palatable than those of permanent-pond breeders (see chap. 9). Among anurans that breed in permanent ponds, tadpoles of those with extended developmental periods may also be less palatable than those with shorter developmental periods. Bullfrog tadpoles, while apparently not absolutely unpalatable, will be avoided by some predators if there are other anuran larvae available (Kruse and Francis 1977; Woodward 1983).

The relationships between behavior and air breathing have not been extensively studied. The decreasing frequency of air breathing with increasing water PO_2 (Crowder et al. 1998, Wassersug and Seibert 1975; reviews by Feder 1984, Burggren 1984) indicates that pulmonary respiration is an accessory means of gas exchange in tadpoles. Using lungs only when necessary should lower the predation risk associated with breathing at the surface (Feder 1983c). In addition, when a tadpole does surface to breathe air, the breath is taken quickly and followed by a rapid descent to minimize exposure time. A few reports of synchronous air breathing among aggregating tadpoles are known (e.g., R. W. McDiarmid, personal communication, *Leptodactylus;* Altig and Christensen 1981, *Rana heckscheri;* M. S. Foster and McDiarmid 1983, Stuart 1961, *Rhinophrynus dorsalis;* N. D. Richmond 1947, *Scaphiopus holbrookii*); this activity has been interpreted as an antipredator behavior among fishes (Kramer and Graham 1976).

Tadpoles lacking functional lungs are not necessarily unable to benefit from atmospheric oxygen. "Air gulping" is a common behavior in fishes in hypoxic water (Ultsch 1989) that can increase blood O_2 transport (Burggren 1982). Even in severely hypoxic water the upper few millimeters will be oxygenated, and if an aquatic animal can irrigate its gills with this water, it can use atmospheric O_2 indirectly. Small fishes with upturned mouths are well adapted for this behavior (Kramer 1987; W. M. Lewis 1970). *Bufo woodhousii* tadpoles appear to use the latter strategy, as they move to the surface in hypoxic water but apparently do not gulp air (Wassersug and Seibert 1975).

Carbon Dioxide Elimination and Acid-Base Balance

The majority of CO_2 elimination of all amphibians is cutaneous, and tadpoles in normocarbic water would not be expected to be an exception (reviews by Burggren 1984, Burggren and Just 1992). This supposition was true for Taylor/Kollros XVI–XIX (Gosner 40–41) bullfrog tadpoles in a flow-through respirometer, where 98% of the $\dot{V}CO_2$ was aquatic (57% cutaneous, 41% branchial) and only 2% was pulmonary even though 17% of the $\dot{V}O_2$ was pulmonary (Burggren and West 1982). As with aquatic $\dot{V}O_2$, aquatic $\dot{V}CO_2$ partitioning in resting tadpoles is not affected by developmental stage (and therefore body mass) throughout the premetamorphic stages. The dominant role of the skin in CO_2 elimination that is characteristic of the adult is assumed when a larva enters metamorphic climax at Taylor/Kollros XX–XXV (Gosner 42–46; Burggren and West 1982).

There is no reason to believe that the partitioning of $\dot{V}CO_2$ between air and normoxic water would change significantly with an increase in pulmonary respiration. If the water becomes hypercarbic, a tadpole would have to increase its plasma PCO_2 in order to maintain an outward diffusion gradient for aquatic CO_2 elimination. Even a relatively slight environmental hypercapnia would be a potentially severe acid-base challenge. The plasma PCO_2 of tadpoles is normally very low and closer to that of fishes than to that of other amphibians (reviews by Heisler 1986; Toews and Boutilier 1986; Ultsch 1996; see Busk et al. 1997). According to the Henderson-Hasselbalch relation, an increase in blood PCO_2 of 3 mm Hg in a tadpole such as *Rana catesbeiana* represents about a doubling from normal values. This would cause much more of an effect on plasma pH than would be the case for aquatic salamanders with a normal plasma PCO_2 2–6 times that of a tadpole (Ultsch 1987, 1996).

Normal acid-base variables among tadpoles resemble those of fishes that live in normoxic water (Heisler 1986; Ultsch 1996; see Ultsch and Jackson 1996), although the data are limited (Erasmus et al. 1970–1971; Helff 1932; Just et al. 1973). Perturbations of acid-base status can be expected from the same sources as with fishes: exogenously from water acidification and environmental hypercarbia and endogenously from accumulation of acidic metabolites, especially lactate.

The reactions of tadpoles to acidic precipitation and acid mine drainage are discussed below. The effects of water hypercarbia on tadpoles have received almost no attention. Hypercarbia elicits an increased dependence on pulmonary respiration in aquatic salamanders (Wakeman and Ultsch 1976), as was found for the late stages of bullfrog tadpoles (Infantino 1989); increasing water PCO_2 up to 12 mm Hg had little effect on rates of gill or lung ventilation in bullfrog tadpoles in Taylor/Kollros IV–XV (Gosner 29–40). A ventilatory response had developed by Taylor/Kollros XVI–XIX (Gosner 40–41), when the frequency of breathing doubled and the frequency of gill irrigation reduced by half; these changes would be consistent with a pulmonary hyperventilation that would partially compensate for a respiratory acidosis and a reduction in aquatic $\dot{V}O_2$ that might slow the rate of CO_2 entrance. For tadpoles of earlier stages with small lungs and for lungless species, this strategy is unavailable. Extracellular pH compensation would have to involve increases in plasma concentrations of HCO_3^-, a mechanism common in fishes (reviews by Heisler 1986, Ultsch 1996) but lacking in the aquatic salamanders *Amphiuma* and *Siren* (Heisler et al. 1982), which often inhabit hypercarbic water. These salamanders allow the extracellular pH to fall during hypercapnia but defend intracellular pH by elevating intracellular concentrations of HCO_3^-, partly by exchanging Cl^- for HCO_3^- from the environment.

With no field data on the upper limits of PCO_2 in tadpole microenvironments, the level of respiratory acidosis that might exist in natural populations is unknown. Tadpoles may lack effective mechanisms for coping with high CO_2 tensions and therefore avoid hypercarbic water.

The accumulation of lactate should present an acid-base challenge, and it is clear that lactate does accumulate both during exercise and severe hypoxia (see above). However, as with most other sources of acidosis among tadpoles, there are no data on mechanisms of acid-base compensation to the acidosis that accompanies lactate formation. In aquatic turtles, lactate accumulation results initially in both metabolic and respiratory acidosis, as HCO_3^- will be titrated at a greater rate than the animal can eliminate the resultant CO_2 (review by Ultsch 1989). In tadpoles, because of their multiple and efficient modes of CO_2 elimination and their relatively small body size (and therefore favorable surface-to-volume ratio for cutaneous CO_2 elimination), a significant contribution of respiratory acidosis is not to be expected when lactate accumulates; this would ameliorate the pH disturbance. In bullfrog tadpoles, lactate can be eliminated by the gills, skin, and kidney, although elimination is not important relative to gluconeogenesis in processing lactate that results from exhaustive exercise (Quinn and Burggren 1983). These paths of elimination might be important during chronic lactic acidosis induced by long periods of anaerobiosis such as might occur during overwintering in hypoxic water.

Thermal Physiology

Of all the physical parameters in the aquatic environment, temperature is perhaps the most dramatic in its effect on the physiology, ecology, and behavior of anuran larvae. For example, environmental temperature dramatically affects the time taken to reach metamorphosis, which can be critical to the survival of an individual faced with a drying habitat or the onset of winter. Temperature also affects differentiation and growth rates, body size at metamorphosis, mechanisms of gas exchange, rates of energy metabolism, and undoubtedly many other physiological parameters documented in other ectothermic vertebrates. Moreover, the limits of temperature tolerance and temperature-dependent life history traits of anuran larvae are generally related to the geographic distribution of the species.

The thermal environment of all anuran larvae is variable in space and time, often extremely so in shallow water exposed to intense insolation. For example, tadpoles of *Rana catesbeiana* may encounter seasonal temperatures ranging from near 0° to over 35°C, and near-shore temperatures may increase by more than 20°C in less than 6 hr (Lillywhite 1970; Menke and Claussen 1982). Water temperature adjacent to tadpoles of some species in temporary ponds may regularly rise during the day to 42°C (H. A. Brown 1969; Dunson 1977). Anuran larvae have generally accommodated to variation in the thermal environment by (1) possessing the physiological ability to function over a wide range in body temperature, (2) behaviorally regulating body temperature, and (3) physiologically adjusting to changes in environmental temperature (i.e., acclimatization, or in the laboratory, acclimation). On an evolutionary scale, variation in the thermal environment has resulted in natural selection for differences among anuran larvae in a number of temperature-dependent physiological or life history traits.

Temperature Tolerance Limits

The limits of temperature tolerance of an animal are commonly described by a critical thermal maximum (CT_{max}) and critical thermal minimum (CT_{min}). These usually are defined as the temperatures at which "locomotor activity becomes disorganized and the animal loses its ability to escape from conditions that will promptly lead to its death" (Cowles and Bogert 1944). The CT_{max} for most anuran larvae, when compared at the same stage of development and at intermediate acclimation temperatures, is between about 38° and 42°C (table 8.2). Extreme values are several degrees higher (44.9°C, *Bufo marinus;* fig. 8.7; table 8.2), and CT_{max} is apparently often approached in nature. For example, *Cyclorana cultripes, C. platycephala,* and *Litoria rubella* occur in Australian pools at 39.2°C (Main 1968), *Limnonectes cancrivorus* at 42°C (Dunson 1977), *Scaphiopus couchii* at 39°–40°C (Mayhew 1965), *Pseudacris triseriata* at 36.5°C (Hoppe 1978), *Osteopilus septentrionalis* suspected at 42°C (H. A. Brown 1969), and a number of Australian species at 35°–45°C (Tyler 1989).

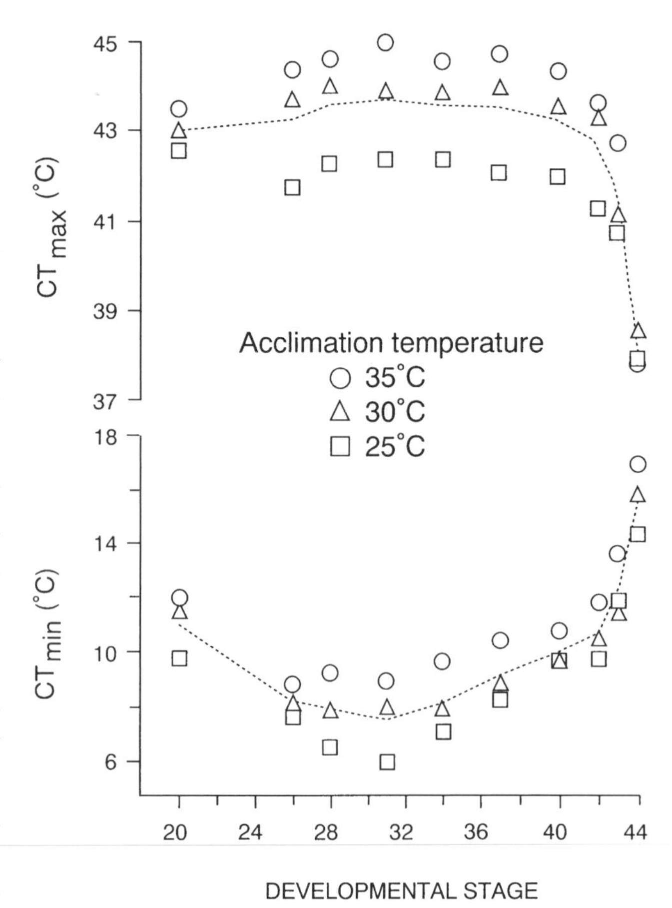

Fig. 8.7. Critical thermal maxima (CT_{max}) and critical thermal minima (CT_{min}) of larvae of *Bufo marinus* at three acclimation temperatures and various stages of development. Dotted lines connect mean responses of larvae averaged over all acclimation temperatures. Staging follows Gosner (1960). After Floyd (1983).

CT_{max} of tadpoles appears to be related to the natural environmental temperatures to which the animals are exposed. Such a pattern is well established for anuran embryos (Bachmann 1969; J. A. Moore 1949b). The highest reported CT_{max} values for tadpoles occur in species that typically inhabit shallow ponds exposed to intense insolation (e.g., *Bufo marinus* and other species, *Gastrophryne carolinensis,* and *Limnonectes cancrivorus*). Among two species of hylids and two species of pelobatids (table 8.2), lethal temperature is directly correlated with the maximum temperatures likely to be experienced in the field (H. A. Brown 1969). In *Ascaphus truei,* a species inhabiting cool streams, CT_{max} may be exceptionally low, as tadpoles always avoid temperatures above 22°C in a thermal gradient (De Vlaming and Bury 1970). *Pseudacris triseriata* in Colorado has a significantly higher CT_{max} at low elevation (e.g., 38.6°C at 1530 m, Taylor/Kollros XX [Gosner 42]) than at high elevation (37.5°C at 2710 m; Hoppe 1978). Likewise, in *R. sylvatica* heat tolerance decreases with both increasing elevation and latitude

Table 8.2 Preferred body temperatures (PBT, as mean selected temperature) and CT_{max} and CT_{min} for anuran larvae at various stages and acclimation temperatures. In studies including more than one acclimation temperature or larval stage, an intermediate acclimation temperature and stage closest to Gosner (1960) stage 39 were chosen. Values in parentheses are the extreme means of CT_{max} or CT_{min} observed at any larval stage or acclimation temperature. FTP = final thermal preferendum, (i.e., acclimation temperature equals PBT; Reynolds and Casterlin 1979) and RT = room temperature.

Species	Stage	Acclimation Temperature (°C)	PBT	CT_{max}	CT_{min}	Reference
ASCAPHIDAE						
Ascaphus truei	2 yr	10	12.2	(>22)	(<0)	De Vlaming and Bury 1970
BUFONIDAE						
Bufo americanus	39–40	FTP	33.3	—	—	Dupré and Petranka 1985
	35–40	22.5	≈32	—	—	Beiswenger 1978
	39	20	—	40.3 (42.0)	—	Cupp 1980
B. boreas	35–40	22.5	≈31	—	—	Beiswenger 1978
	40	—	—	36–38	—	Karlstrom 1962
B. canorus	40	—	—	36–38	—	Karlstrom 1962
B. exsul	—	—	—	41–42	—	Straw 1958
B. hemiophrys	35–40	22.5	≈31	—	—	Beiswenger 1978
B. marinus	39	27	28.2	—	—	Floyd 1984
	40	30	—	43.4 (44.9)	9.5 (6.0)	Floyd 1983
	—	21	—	41.5	—	Heatwole et al. 1968
B. terrestris	—	25	—	43.3 (43.5)	—	Davenport and Castle 1895
	35–39	27	—	40.3 (41.7)	—	Noland and Ultsch 1981
B. woodhousii	26–41	—	—	42.5 (42.5)	—	Sherman 1980
	39	20	—	41.2 (42.8)	—	Cupp 1980
HYLIDAE						
Osteopilus septentrionalis	26–40	25	—	40.2[a] (41.5[a])	—	H. A. Brown 1969
Pseudacris regilla	26–40	20	—	35.0[a](36.1[a])	—	H. A. Brown 1969
P. triseriata	36–38	FTP	28.3	—	—	Dupré and Petranka 1985
	38	—	—	38.8	—	Hoppe 1978
	39	20	—	38.3 (40.1)	—	Cupp 1980
LEPTODACTYLIDAE						
Leptodactylus albilabris	—	21	—	38.8	—	Heatwole et al. 1968
MICROHYLIDAE						
Gastrophryne carolinensis	39	20	—	42.5 (43.3)	—	Cupp 1980
PELOBATIDAE						
Scaphiopus couchii	26–40	25	—	40.2[a] (41.4[a])	—	H. A. Brown 1969
S. holbrookii	—	18–20	—	37.5	8.5	Gosner and Black 1955
Spea hammondii	26–40	26.5	—	38.7[a] (40.5[a])	—	H. A. Brown 1969
RANIDAE						
Limnonectes cancrivorus	35–39	22–25	—	42.3[b] (43[b])	—	Dunson 1977
R. cascadae	39–40	21	28.1	—	—	Wollmuth et al. 1987
R. catesbeiana	40	21	31.8	—	—	Wollmuth and Crawshaw 1988
	41	24	29.2	—	—	Dupré et al. 1986
	39–40	26.7	19.3	—	—	Hutchison and Hill 1978
	—	RT	30	—	—	Lucas and Reynolds 1967
	32–40	15	—	38.8 (39.3)	1.7 (0.2)	Menke and Claussen 1982
R. clamitans	—	20	27	—	—	Workman and Fisher 1941
	32–38	20	—	39.4 (40.4)	—	Willhite and Cupp 1982
R. muscosa	31–41	4	—	—	< −2.0	D. F. Bradford, unpublished data
R. pipiens	—	20	20	—	—	Workman and Fisher 1941
	—	FTP	27.9	—	—	Casterlin and Reynolds 1979
	—	RT	30	—	—	Lucas and Reynolds 1967
	35–39	27	—	38.8 (40.0)	—	Noland and Ultsch 1981
R. sphenocephala	38–40	FTP	28.8	—	—	Dupré and Petranka 1985
R. sylvatica	—	20	17	—	—	Workman and Fisher 1941
	39	20	—	38.4 (39.6)	—	Cupp 1980
	33–36	22	—	37.9, 38.2	—	Manis and Claussen 1986

[a]Incipient lethal temperature (i.e., maximum temperature that 50% of the tadpoles can tolerate indefinitely).

[b]Lethal temperature, dead in 1–2 hr.

(Manis and Claussen 1986). In contrast, no significant difference in CT_{max} was found between populations of *Leptodactylus albilabris* at 15 and 570 m elevation in Puerto Rico (Heatwole et al. 1968).

The CT_{min} also appears to correlate with environmental temperatures, although it has been determined for tadpoles of only a few species (table 8.2). In *Rana muscosa,* a species in which tadpoles overwinter in frozen lakes, CT_{min} is below −2°C. The CT_{min} is near or below 0°C in *R. catesbeiana,* a temperate species that overwinters in some populations (Collins 1979), and *Ascaphus truei,* a species inhabiting cool temperate streams. In contrast, the tropical *Bufo marinus* has a much higher Ct_{min} (i.e., 6.0°C or greater and dependent on acclimation temperature and larval stage; fig. 8.7). Similarly, in the temperate species *Scaphiopus holbrookii,* which undergoes larval development entirely during the summer, CT_{min} is 8.5°C in larvae acclimated at 18°–20°C.

The temperature tolerance range of anuran larvae is large in comparison to most ectothermic vertebrates. In species in which both CT_{max} and CT_{min} have been determined, the temperature tolerance range for a single stage of development and acclimation regime is 34°C in *Bufo marinus,* 37°C in *R. catesbeiana,* and 28°C in *Scaphiopus holbrookii* (table 8.2). This range is even greater if acclimation effects are included. For example, *B. marinus* tadpoles at Taylor/Kollros VI (Gosner 31) can tolerate a temperature range of 39°C if acclimated prior to exposures to the low and high temperatures (fig. 8.7; Floyd 1983; also Abe and Godinho 1991).

Data are not available for the temperature tolerance limits of tropical forest species that experience relatively small variations in temperature. It might be expected that tadpoles of these species would have narrower temperature tolerance ranges than species exposed to wider temperature variation. However, among adult anurans, tropical species do not generally have narrower temperature tolerance ranges in comparison to temperate forms, although their temperature tolerance limits are higher than those for temperate species (Brattstrom 1968).

The typically higher CT_{max} of tadpoles relative to adults is a striking difference in the thermal relations of larval and adult anurans. This is the case for *Bufo americanus, B. exsul, B. marinus, B. woodhousii, Gastrophryne carolinensis, Rana catesbeiana, R. pipiens, R. sylvatica,* and *Scaphiopus holbrookii* (Cupp 1980; Floyd 1983; Hathaway 1927; Sherman 1980; Straw 1958; compare data in table 8.2 with Brattstrom 1968). The only apparent exceptions to this pattern are *Pseudacris triseriata,* in which larval CT_{max} appears to be similar to adult CT_{max} (Hoppe 1978), and *B. canorus,* in which larval CT_{max} is lower than adult CT_{max} (Karlstrom 1962). The finding that *R. pipiens* tadpoles die more quickly at high temperature than adults (Orr 1955) may have been caused by differences in body size. The generally higher CT_{max} of tadpoles than adults may reflect the need for tadpoles to minimize the time to complete metamorphosis and escape inimical environmental conditions and aquatic predators (see below). Alternatively, this difference may reflect the greater likelihood for tadpoles than adults to become trapped in warm surroundings. Adults on land continually lose heat by evaporation (Tracy 1976) and often have the option to move to alternative thermal microenvironments (Lillywhite 1970).

Behavioral Thermoregulation

Both larval and adult anurans regulate body temperature behaviorally by moving or changing posture among available thermal microenvironments (Brattstrom 1962, 1970). When placed in a thermal gradient in the laboratory, anuran larvae invariably select a certain temperature or temperature range more frequently than others. The arithmetic mean of these selected temperatures is often defined as the preferred body temperature (PBT; Reynolds and Casterlin 1979). Many biochemical and physiological processes, such as active metabolic rate, maximum sustained speed, growth rate, food conversion efficiency, and learning and memory capabilities, are optimal at or near the PBT (Huey and Stevenson 1979; Hutchison and Maness 1979). Among anuran larvae, important processes known to change as a function of temperature include differentiation rate, growth rate, body size at metamorphosis, mechanisms of gas exchange, and metabolic rate (see below). Clearly, temperature regulation allows maximization or control of differentiation and growth rates (Smith-Gill and Berven 1979). For example, in *Pseudacris ornata, Rana clamitans,* and *R. sylvatica,* differentiation and growth rates increase with increasing temperature until an inhibitory temperature well below CT_{max} is reached (fig. 8.8;

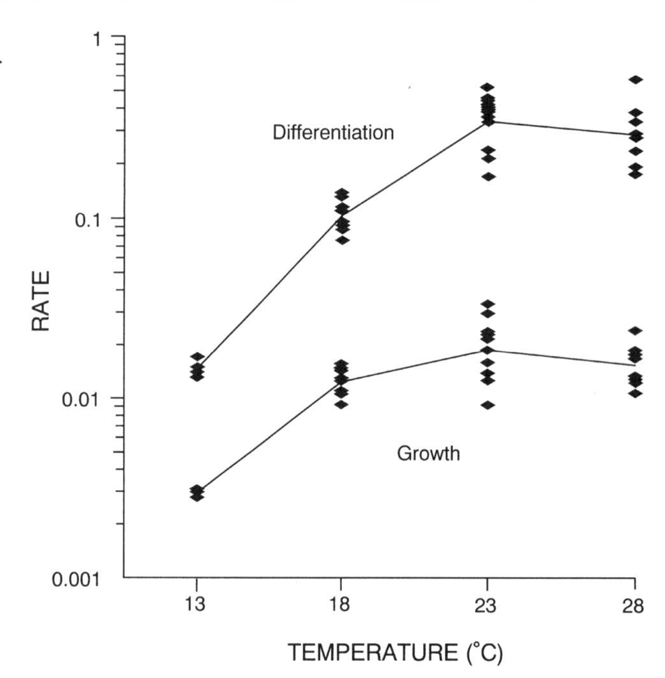

Fig. 8.8. Temperature dependence of differentiation rate and growth rate for individual *Rana pipiens.* Growth rate is computed as the regression coefficient of body volume$^{0.33}$ on days up to Taylor/Kollros XVII (Gosner 40). Differentiation rate is computed as the regression coefficient of stage on days through Taylor/Kollros XX (Gosner 42). Lines connect geometric means at each temperature. After Smith-Gill and Berven (1979).

table 8.2; Berven 1982; Berven et al. 1979; Harkey and Semlitsch 1988; Smith-Gill and Berven 1979). Temperature selection may be critical to a tadpole faced with a deteriorating aquatic environment, the end of the growth season, or an increased risk of predation. Behavioral temperature regulation also may serve to minimize exposure to deleterious temperatures and conditions. For example, *Pseudacris ornata* larvae have fewer deformities and greater average survival when raised at intermediate temperatures of 20° and 25°C than when raised at 10° and 30°C (Harkey and Semlitsch 1988). In overwintering species, avoidance of the lowest temperatures may minimize the probability of entrapment in ice that may lead to death by freezing or anoxia (Bradford 1984b).

A temporary elevation in PBT, termed behavioral fever, occurs in all major groups of ectothermic vertebrates (Kluger 1978). Behavioral fever can be induced in tadpoles by injection of killed *Aeromonas hydrophila,* a common pathogen in fishes, amphibians, reptiles, and mammals (Casterlin and Reynolds 1977). This response has an adaptive value in some ectotherms by increasing survival rate after infection with live *A. hydrophila* (Kluger 1978). However, behavioral fever in tadpoles may alter their behavior and increase their susceptibility to predators (Lefcort and Eiger 1993).

The mechanisms of heat exchange involved in behavioral thermoregulation differ between larval and adult anurans because only adults can thermoregulate on land where radiation and evaporation are important mechanisms of heat exchange (Lillywhite 1970; Tracy 1976). In water, conduction and convection predominate as heat exchange mechanisms. Consequently, the body temperature of a tadpole is determined almost entirely by the temperature of the surrounding water. Neither tadpoles nor frogs are known to possess any physiological means of controlling heat flux when in water (Brattstrom 1979). Nevertheless, solar radiation may contribute directly to the heat input of tadpoles in shallow water (see below).

The PBT of most anuran larvae studied is between 28° and 32°C (table 8.2). These values are roughly 10°C below the CT_{max} for each species, although PBT does not appear to be well correlated with CT_{max}. There is some evidence for a weak correlation between natural environmental temperature and PBT. For example, in *Rana clamitans, R. pipiens,* and *R. sylvatica* in the northeastern United States, tadpole PBT is directly correlated with the beginning date of the breeding season and the northernmost distribution of the species (J. A. Moore 1949a; Workman and Fisher 1941). Moreover, the two species with the lowest known PBT (table 8.2; *Ascaphus truei* and *R. sylvatica*) inhabit the coldest environments (also see Miranda et al. 1991).

The precision of temperature selection varies considerably among species and larval stages. For example, *Pseudacris triseriata* and *Rana sphenocephala* behave as weak thermal selectors, especially early in larval development, whereas *Bufo americanus* behaves as a strong thermal selector throughout development (Dupré and Petranka 1985). For the developmental stages showing the most precise temperature selection in these species, the standard deviation for selected temperature in groups of tadpoles is 2.2°C in *P. triseriata,* 2.1°C in *R. sphenocephala,* and 1.6°C in *B. americanus.* The corresponding value in *R. catesbeiana,* in which selected temperatures were determined for individuals rather than groups, is 0.7°C (Wollmuth and Crawshaw 1988).

In contrast to CT_{max}, the PBT of tadpoles shows no general pattern in comparison to PBT for adults. *Rana cascadae* tadpoles have higher PBTs than adults (Wollmuth et al. 1987), *R. pipiens* tadpoles appear to have a similar PBT to adults (Casterlin and Reynolds 1979), *R. catesbeiana* have similar or lower PBTs than adults (Hutchison and Hill 1978; Lillywhite 1970; Lucas and Reynolds 1967; Wollmuth and Crawshaw 1988), and *Ascaphus truei* tadpoles have lower PBTs than adults (Claussen 1973; De Vlaming and Bury 1970). Data are not available for tropical forest anurans that experience a narrow temperature range or live in thermally homogeneous habitats. It is possible that some species in these conditions do not behaviorally regulate body temperature like some tropical forest-dwelling lizards (Huey and Webster 1976).

Ontogeny of Temperature Tolerance and Regulation

The temperature tolerance limits of anuran larvae typically change considerably during ontogeny. The most striking and consistent change is a decrease in CT_{max} near the completion of metamorphosis (about Taylor/Kollros XXI–XXV [Gosner 43–46]; fig. 8.7), as in *Bufo americanus, B. marinus, B. terrestris, B. woodhousii, Gastrophryne carolinensis, Pseudacris triseriata, Rana sphenocephala,* and *R. sylvatica* (Cupp 1980; Floyd 1983; Hoppe 1978; Noland and Ultsch 1981; Sherman 1980). The only exception is *R. catesbeiana,* in which this pattern was evident at only one of two acclimation temperatures (Menke and Claussen 1982). Prior to metamorphosis, CT_{max} remains fairly stable over a broad range of stages in most species, although significant changes are sometimes apparent (above references; Gosner and Black 1955).

Ontogenetic changes in the CT_{min} has received little attention. In *Bufo marinus* the change in CT_{min} with stage is pronounced and is largely the opposite of the change in CT_{max} (fig. 8.7). In determinations of CT_{min} for three groups of stages of *R. catesbeiana,* CT_{min} was near 0°C and varied little among the stages (Menke and Claussen 1982; see Dainton 1988, 1991).

Based on these limited data, the temperature tolerance range of anurans (i.e., CT_{max} to CT_{min}) appears to be reduced during the final stages of metamorphosis, rather than shifted to a lower level, as postulated by Cupp (1980) and Floyd (1983). The characteristic reduction in CT_{max} and the apparent reduction in temperature tolerance range at the end of metamorphosis may be consequences of biochemical and morphological rearrangements occurring at this time (Sherman 1980). Floyd (1983) noted that the times of least tolerance to temperature extremes in *B. marinus* coincide with the times of greatest morphological and biochemical rearrangement early and late in larval development.

The thermal preference of tadpoles also changes during larval development. Throughout most of development, PBT generally increases with larval stage (e.g., De Vlaming and

Bury 1970; Dupré et al. 1986; Dupré and Petranka 1985; Floyd 1984; Hutchison and Hill 1978; Lucas and Reynolds 1967; Wollmuth and Crawshaw 1988; Wollmuth et al. 1987; *Ascaphus truei, Bufo americanus, B. marinus, Pseudacris triseriata, Rana cascadae, R. catesbeiana, R. pipiens,* and *R. sphenocephala*). However, in *R. cascadae* and *R. catesbeiana* (Oregon), PBT declines substantially during final metamorphic stages (Wollmuth and Crawshaw 1988; Wollmuth et al. 1987), whereas it appears to remain high at this time in *B. americanus, P. triseriata, R. catesbeiana* (Oklahoma), and *R. sphenocephala* (Dupré et al. 1986; Dupré and Petranka 1985; Hutchison and Hill 1978).

Precision of temperature selection generally increases throughout larval development, along with the increase in PBT (e.g., *Bufo americanus, Pseudacris triseriata, Rana catesbeiana* [Kentucky], and *R. sphenocephala;* Dupré et al. 1986, Dupré and Petranka 1985). Standard deviations of selected temperature in groups of tadpoles of these species range from about 4°–8°C in early developmental stages to about 2°–3°C in later stages. In *R. catesbeiana,* in which individual tadpoles were monitored rather than groups, PBT and the precision of temperature regulation also generally increased during development (Wollmuth and Crawshaw 1988). However, both precision and PBT decreased at the end of development (i.e., Taylor/Kollros XXII–XXV [Gosner 44–46]). The standard deviation for selected temperature in this study varied among stages from 0.7°–4.6°C.

The adaptive significance of ontogenetic changes in PBT is not clear. Floyd (1984) argued that ontogenetic changes in PBT of *B. marinus* are correlated with changing thermal characteristics of the natural microhabitats, and PBT is not coupled to ontogenetic change in temperature tolerance. Dupré and Petranka (1985) argued that the general increase in PBT during development may reflect greater selection pressure for rapid development at the end of the developmental period; this is a time when larvae become increasingly susceptible to habitat deterioration (Wilbur 1980) and predation (S. J. Arnold and Wassersug 1978; Wassersug and Sperry 1977), and the growth season may be ending. Finally, ontogenetic changes in PBT may spatially segregate age or size classes and thereby potentially reduce competition or cannibalism, or encourage the formation of size-specific, cooperative aggregations (Alford and Crump 1982; Dupré and Petranka 1985). Although age segregation of *R. cascadae* occurred in the field, groups were in such close proximity that neither cannibalism nor competition were likely to have been affected (Wollmuth et al. 1987).

Temperature Acclimation

Over a period of hours to days in a laboratory setting, tadpoles accommodate to changes in the thermal environment by physiologically adjusting to new temperatures (= temperature acclimation; Hutchison and Maness 1979). When such adjustments occur in the natural environment, the process is called acclimatization (K. Schmidt-Nielsen 1983). Temperature acclimation has been discussed extensively for adult amphibians and for ectotherms in general (e.g., Duellman and Trueb 1986; Hutchison and Dupré 1992; Hutchison and Maness 1979; and references therein). In all amphibian larvae examined (table 8.2), temperature tolerance limits are significantly affected by the temperature of acclimation. For example, an increase in acclimation temperature of *Bufo marinus* from 25° to 35°C resulted in an increase in CT_{max} of about 2.5°C and an increase in CT_{min} of about 1°C (fig. 8.7). Among five other species (*Bufo, Gastrophryne, Pseudacris, Rana*), an increase in acclimation temperature from 10° to 30°C resulted in an increase in CT_{max} of about 1°–3°C (Cupp 1980), and an increase in acclimation temperature from 15° to 35°C in three species (*Osteopilus, Scaphiopus,* and *Spea*) resulted in an increase in lethal temperature by about 4°C (H. A. Brown 1969). Such effects are probably essential for surviving temperature extremes in natural conditions, as a number of species have been observed in the field thriving in water at temperatures near their CT_{max} (see above). The effect of acclimation temperature on PBT is much less distinctive than its effect on temperature tolerance limits. In *Ascaphus truei* and *Bufo marinus,* there is no effect of temperature acclimation (De Vlaming and Bury 1970; Floyd 1984), whereas a complex interaction was found among acclimation temperature, stage of development, and PBT in *Rana cascadae, R. catesbeiana,* and *R. pipiens* (Hutchison and Hill 1978; Lucas and Reynolds 1967; Wollmuth et al. 1987). The time required to acclimate completely to a new temperature has not been documented for anuran larvae. In adult anurans about two to four days are required for CT_{max} and CT_{min} to stabilize after a large change in acclimation temperature (Brattstrom 1968; Duellman and Trueb 1986). Possibly as little as 24–36 hr are required in larvae, and this time has been used as the minimum required for measuring the "final thermal preferendum" (= the temperature ultimately selected regardless of previous thermal experience; Casterlin and Reynolds 1979; Dupré and Petranka 1985; Reynolds and Casterlin 1979).

In addition to acclimation, which is a gradual process, tadpoles of *Rana catesbeiana* undergo heat and cold "hardening" (Menke and Claussen 1982). In this process tadpoles exposed briefly to near-lethal temperatures become more tolerant to subsequent exposure to these temperatures within 90 min.

Other Influential Factors

A number of factors (e.g., time of day, season, light intensity, photoperiod, and nutritional state) other than stage of development and thermal history (i.e., acclimation) influence thermoregulatory behavior or temperature tolerance in anuran larvae (Duellman and Trueb 1986; Hutchison and Maness 1979). The magnitude of the effects of these parameters on PBT and temperature tolerance limits generally is small in comparison to the effects of thermal history.

Diel and seasonal cycles in temperature tolerance and PBT have been found in many ectothermic vertebrates, although it is not established whether all cycles are truly endogenous (Hutchison and Maness 1979). Among anuran larvae a diel pattern in PBT was observed in *Rana clamitans* and *R. pipiens* but not in *Ascaphus truei* (Casterlin and Reynolds 1979; De Vlaming and Bury 1970; Willhite and Cupp

1982). Seasonal differences in PBT have been found in *R. catesbeiana* and *R. pipiens* (Lucas and Reynolds 1967; Wollmuth and Crawshaw 1988). In tadpoles of *R. catesbeiana* at Taylor/Kollros XI–XVI (Gosner 36–40), PBT was above 30°C during summer and in the mid-20s during winter. In contrast, earlier stage tadpoles selected the same temperature throughout the year. Although the cues underlying this seasonal difference in later stage tadpoles of *R. catesbeiana* are unknown, a seasonal difference in PBT would function to minimize rate of differentiation at a time inappropriate for rapid development and metamorphosis (Wollmuth and Crawshaw 1988).

Although tadpoles of many species are positively or negatively phototaxic (Duellman and Trueb 1986; Wollmuth et al. 1987), daytime PBT does not differ significantly from nighttime PBT of *Rana pipiens* (Casterlin and Reynolds 1979). Many positive phototaxic responses may occur in anticipation of increasing water temperatures that follow exposure to sunlight (Beiswenger 1977; Duellman and Trueb 1986). Photoperiod affects the PBT of second-year but not first-year tadpoles of *Ascaphus truei;* PBT increases when photoperiod increases (De Vlaming and Bury 1970). Photoperiod affects CT_{max} and CT_{min} in *B. marinus* tadpoles in a way that appears to increase survival under potentially dangerous thermal conditions (Floyd 1985). At high acclimation temperatures, a long photoperiod increases CT_{max}, and at low acclimation temperature, a short photoperiod reduces CT_{min}. The magnitude of these effects in *B. marinus* is a small fraction of the changes in CT_{max} and CT_{min} caused by thermal history.

Nutritional state affects PBT in many ectothermic vertebrates, including adult anurans. The increase in PBT that occurs when the animal is digesting food (Lillywhite et al. 1973) presumably maximizes rate of digestion and conserves energy when digestion is not occurring. This pattern may be irrelevant for many anuran larvae, which are virtually never without food in the digestive tract. No data are available to assess the effect of nutritional state on PBT in tadpoles. Nutritional state affects thermal tolerance limits in *Bufo americanus* and *B. marinus* by decreasing the range of tolerance during food deprivation (Cupp 1980; Floyd 1985). However, the magnitude of the reduction in both species is small in comparison to the effects of thermal history on temperature tolerance limits.

Thermoregulation and Aggregation Behavior

Although selection by tadpoles of a preferred body temperature is evident in the laboratory, a host of factors (e.g., light intensity, food availability and quality, dissolved oxygen, population density, presence of predators, and tadpole social interactions) other than temperature may affect microhabitat selection in the field (Beiswenger 1975, 1977; Lemckert 1996; Noland and Ultsch 1981; Wollmuth et al. 1987). Nevertheless, temperature is often the dominant factor determining microhabitat selection, and thermoregulatory behavior has been documented in a number of studies.

Tadpoles in many temperate ponds typically lie offshore at night in warmer, deeper water and migrate to shallows during the daytime as water temperature increases. Tadpoles commonly select the warmest temperatures available (e.g., Beiswenger 1977, 1978; Bradford 1984b; Brattstrom 1962; Mullally 1953; Noland and Ultsch 1981; Tevis 1966; Wollmuth et al. 1987; *Bufo boreas, B. canorus, B. hemiophrys, B. punctatus, B. terrestris, Rana cascadae,* and *R. muscosa*). Overwintering *R. muscosa* select the warmest temperature below the thermocline in frozen lakes (i.e., 3°–4°C; Bradford 1984b). For some anurans (e.g., Beiswenger 1977, 1978; Brattstrom 1962; Noland and Ultsch 1981, *Bufo americanus, Pseudacris crucifer, Rana boylii,* and *R. sphenocephala*) surface waters become sufficiently warm that they are avoided or no longer sought.

A comparison of the thermal relations of *Bufo terrestris* and *Rana sphenocephala* in the field and laboratory demonstrates relationships among tadpole temperature tolerance limits, tadpole thermoregulatory behavior, and the thermal characteristics of the environment (Noland and Ultsch 1981). Tadpoles of *B. terrestris* typically select the warmest temperatures available in shallow water, which often results in body temperatures above 30°C. In contrast, *R. pipiens* avoids shallow water when temperatures become warm, and body temperatures are typically well below 30°C. These differences correlate with a higher CT_{max} in *B. terrestris* than in *R. pipiens* (table 8.2). Also, the body temperature of *B. terrestris* in the field more frequently is near the CT_{max} than *R. pipiens.* The conspicuous behavior that maximizes body temperature in *B. terrestris,* a species that frequently breeds in shallow, temporary waters, is viewed as a strategy for shortening developmental time and thus increasing the probability of completing metamorphosis before the habitat dries. In contrast, *R. pipiens* usually inhabits more permanent habitats than *Bufo* and does not develop as rapidly. Moreover, *Rana* habitats typically have lower ambient temperatures than *Bufo* habitats.

Dense aggregations of a few to thousands of anuran larvae frequently occur (see chap. 9). Although a number of factors are involved in the formation and maintenance of these aggregations, water temperature is often the dominant factor (Beiswenger 1975, 1977, 1978). For example, in *Bufo americanus,* dispersed tadpoles are activated in the morning by increasing light intensity and move to shallow water on sunny days where they accumulate in warm microhabitats (Beiswenger 1975). Once tadpoles reach sufficient density in the lighter and warmer microhabitats, feeding and social activities draw them into dense aggregations. At night as water temperatures decrease in the shallows, tadpoles return to warmer, offshore water. Beiswenger (1975) argued that aggregating is a way for tadpoles to exploit food more efficiently, engage in certain social activities, and seek suitable temperatures more effectively. In contrast, Brattstrom (1962, 1970) and Wollmuth et al. (1987) argued that aggregations in *B. boreas, Pseudacris crucifer, P. regilla, Rana boylii,* and *R. cascadae* occurred because of, and to augment, the thermal characteristics of the local microhabitat. Clearly, dense aggregations in many species are often observed in the warmest microhabitats in a pond.

Brattstrom (1962) provided experimental evidence that

tadpoles aggregate, in part, because they absorb more solar radiation than the surrounding water, which results in more heat being absorbed in water containing aggregations. Beiswenger (1977) described observations both supporting and refuting this argument. The use of white porcelain pans in which much of the radiation may have been reflected by the white substrate is a limitation in evaluating the data in Brattstrom's (1962) paper. Further experiments are needed in which the radiative properties of the substrate in the test system better represent those in the field.

Temperature, Differentiation, and Growth Rate

The influence of temperature on differentiation and growth rates probably is the most dramatic effect of environmental temperature on the ecological physiology of anuran larvae. This influence determines larval body size and length of the larval period (Reques and Tejedo 1995; Smith-Gill and Berven 1979), which in turn are directly related to survival and reproductive success (Berven 1982; Berven and Gill 1983; Berven et al. 1979; Collins 1979; Crump 1989a; Harkey and Semlitsch 1988; D.C. Smith 1987). Typically, an increase in temperature results in faster rates of development and growth, shorter time to metamorphosis, and smaller body size at metamorphosis.

A model that predicts the timing of metamorphosis based on differentiation rate has been proposed and tested in the laboratory and field by Smith-Gill and Berven (1979). In *Rana pipiens,* differentiation rate (defined as the slope of the stage-age relationship) is strongly correlated ($r^2 = 0.95$) with the length of the larval period, whereas growth rate is considerably less correlated ($r^2 = 0.51$). Growth rates of *R. pipiens* and *R. sylvatica* larvae in some circumstances show no significant relationship to length of the larval period. Growth parameters are often poor predictors of timing of metamorphosis largely because growth and differentiation are affected differentially by environmental factors such as temperature and population density. Also, growth rate is functionally dependent upon differentiation rate, whereas the converse is not true. This is because growth and developmental rates are both functions of circulating hormone levels, which are in turn dependent upon the stage of differentiation reached. Nevertheless, growth rate frequently is related to the duration of larval development (e.g., Wilbur and Collins 1973) because conditions favorable for differentiation often are also favorable for growth (Smith-Gill and Berven 1979).

Temperature is a major factor determining larval differentiation and growth patterns, although other factors such as density, food abundance and quality, and water quality are clearly important (Berven 1982; Berven and Gill 1983; Harkey and Semlitsch 1988; Pandian and Marian 1985b; Smith-Gill and Berven 1979; Viparina and Just 1975; Wilbur 1980; Wilbur and Collins 1973). As temperature increases, rates of differentiation and growth both increase until an inhibiting temperature is reached (Berven 1982; Berven et al. 1979; Harkey and Semlitsch 1988; Smith-Gill and Berven 1979). Such an increase is predictable based on Arrhenius enzyme kinetics. For example, a temperature increase from 13° to 23°C results in about a 10-fold increase in differentiation rate and a 6-fold increase in growth rate in *Rana pipiens* (fig. 8.8; Smith-Gill and Berven 1979). Similar patterns are also found in *R. clamitans* and *R. sylvatica* (Berven 1982; Berven et al. 1979).

Environmental temperature has a pronounced effect on length of the larval period by virtue of its influence on differentiation and growth rates (Smith-Gill and Berven 1979). In some species, reducing the time to metamorphosis to a minimum is critical because tadpoles must complete metamorphosis prior to drying of the habitat or onset of other inimical conditions. Behavioral temperature selection resulting in maximal differentiation rate would be expected (e.g., Brattstrom 1962; Sherman 1980; Wollmuth et al. 1987). The differential effect of temperature on differentiation and growth rates promotes emergence from ponds during the warm part of the year, particularly if ponds become overheated during pond drying (Etkin 1964). The same mechanism inhibits emergence during the inappropriate cold season.

In some species and environments, temperature may be sufficiently low to make overwintering of larvae obligatory (Berven et al. 1979; Collins 1979; Collins and Lewis 1979; Smith-Gill and Berven 1979). Overwintering occurs in a number of anurans at high elevation and latitude where average temperatures are lower and the growing season is shorter than at lower elevation and latitude. For example, *Rana catesbeiana* larvae in Louisiana may complete metamorphosis in a few months, whereas individuals in Michigan may not metamorphose until their third summer (Collins 1979). *Ascaphus truei* tadpoles in subalpine sites in Washington require five summers to complete larval development (H. A. Brown 1990b). In *R. sylvatica* in montane environments, a one-month difference in oviposition results in nearly a year difference in completion of metamorphosis because larvae from eggs oviposited later must overwinter (Berven et al. 1979). At extreme elevations and latitudes, low temperatures that prevent metamorphosis may be the factor limiting the geographic range of a species (Smith-Gill and Berven 1979; Uhlenhuth 1921).

Amphibian larvae are typically larger when raised at low temperature, a phenomenon that appears to be general for larval ectotherms (Berven 1982; Berven and Gill 1983; Berven et al. 1979; Collins 1979; Etkin 1964; Harkey and Semlitsch 1988; Kollros 1961; Smith-Gill and Berven 1979; Uhlenhuth 1919). For example, *R. pipiens* raised at 13° and 23°C in the laboratory differ in body size by about 3-fold at all stages, including metamorphic climax (fig. 8.9; Smith-Gill and Berven 1979). In *R. sylvatica* in the laboratory and in *R. clamitans* in both the field and laboratory, a 10°C decrease in temperature results in about a 2-fold increase in body size at metamorphic climax (Berven 1982; Berven and Gill 1983; Berven et al. 1979). Larger size at lower temperature can be attributed to the greater effect of temperature on differentiation rate than on growth rate (fig. 8.8; Berven et al. 1979; Smith-Gill and Berven 1979). Thus, larvae at low temperatures gain more mass and are larger at any given stage, including metamorphic climax, than at higher tem-

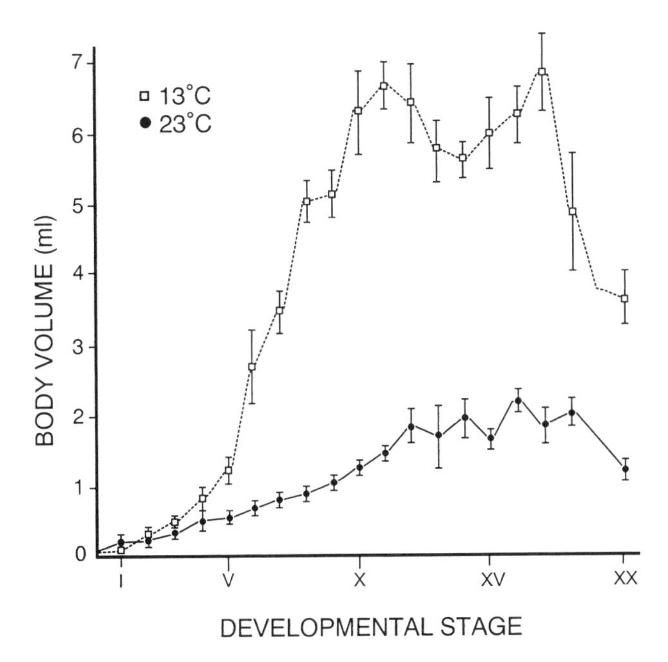

Fig. 8.9. Body size of larvae of *Rana pipiens* reared at 13° and 23°C through metamorphic climax (Taylor/Kollros XX, Gosner 42). Vertical bars were not defined in original reference. After Smith-Gill and Berven (1979).

peratures. A secondary consequence of larger body size in *Hoplobatrachus tigerinus* appears to be increased energetic efficiency in converting tadpole tissue into frog tissue (Pandian and Marian 1985b).

The differential effects of temperature on rates of differentiation and growth are evident in the field as well as the laboratory. In *Rana clamitans* larvae that had not completed metamorphosis by September, differentiation ceased whereas growth continued (Smith-Gill and Berven 1979). Eventually the larvae ceased both differentiation and growth as temperatures dropped to below 10°C. In the following spring, growth commenced earlier, at a lower temperature, than did differentiation. Subsequently, these larvae metamorphosed at a larger size than those that both grew and differentiated at faster rates during the previous summer. Larvae with the slowest growth and differentiation were largest at metamorphic climax (Berven et al. 1979; Smith-Gill and Berven 1979).

Clinal variation in larval development and morphology associated with temperature differences has both environmental and genetic components. In *Rana clamitans* (lowland and montane populations) and *R. sylvatica* (lowland, montane, and tundra populations), the relatively cold environments at high elevation and high latitudes shorten the available breeding seasons, slows tadpole developmental rates, prolongs the larval periods, and increases the sizes of larvae at all stages in comparison to the lowland sites (Berven 1982; Berven and Gill 1983; Berven et al. 1979). Complementary field and laboratory studies (including reciprocal transplant and breeding experiments) demonstrated that this clinal variation is primarily induced by environmental effects of temperature, rather than genetic or maternal effects. Nevertheless, significant genetic differences were found. For example, in *R. clamitans* genetic differences were found in the temperature at which differentiation and growth rates were maximal, in temperature sensitivity of differentiation and growth rates, and in stage-specific growth rates. These genetic differences were counter to the observed clinal variation. For example, in cold-temperature experiments designed to simulate mountain-top conditions, montane tadpoles developed faster and completed metamorphosis earlier and at a smaller size than did lowland tadpoles. Thus, natural selection in *R. clamitans* has favored the shortest possible larval periods given the ambient temperature constraints. This pattern is viewed as a case of counter-gradient selection, in which natural selection favors genotypes that minimize the range of phenotypes potentially induced by the environments along a gradient. A similar counter-gradient pattern exists between environmental temperature and developmental rate in anuran embryos (H. A. Brown 1967; J. A. Moore 1949b).

Temperature Effects on Other Processes

Energy metabolism measured by rate of oxygen consumption of anuran larvae increases approximately exponentially as a function of temperature, a pattern typical of most ectothermic animals. Q_{10} values are 1.6–2.2 in *Rana berlandieri* (Burggren et al. 1983; Feder 1985), 1.9–2.4 in *R. catesbeiana* (Burggren et al. 1983; Feder 1985), 1.8–1.9 (Noland and Ultsch 1981) and 2.6 (G. E. Parker 1967) in *R. sphenocephala*, 1.2–1.8 in *Bufo terrestris* (Noland and Ultsch 1981), 1.3–2.0 in *Xenopus laevis* (Feder 1985; Hastings and Burggren 1995), and 3.2 in *Limnodynastes peronii* (E. Marshall and Grigg 1980b). An overall Q_{10} of 2.14 between 15° and 25°C has been derived from data for both anuran and caudate larvae (Gatten et al. 1992).

Thermal acclimation of O_2 consumption occurs in larvae of three of the four species examined: *Rana berlandieri, R. pipiens,* and *X. laevis* (Feder 1985; G. E. Parker 1967). The findings are similar to those for many adult anurans and caudates (Feder 1985; Feder et al. 1984a) and many fishes and invertebrates (Bullock 1955) in which thermal acclimation affects rate of oxygen consumption. Commonly, ectothermic animals reared at cool temperatures have higher metabolic rates at a given temperature than animals reared at warmer temperatures (Prosser and Brown 1961; K. Schmidt-Nielsen 1983). Interestingly, tadpoles of the Australian leptodactylid *Limnodynastes peroni* show no thermal acclimation of metabolism at any temperature (E. Marshall and Grigg 1980b), and one has to ask if this trait may be a common feature of the larvae and adults of Australian anurans.

Temperature affects many other physiological variables that are ecologically relevant in a wide variety of ectothermic vertebrates. Examples are digestive efficiency and rates (Warkentin 1992a, b), survival of disease, maximum locomotor acceleration and velocity (e.g., Lefcort and Blaustein 1995), endurance, active metabolic rate (see Toloza and Diamond 1990a, b), and learning and memory capabilities (Huey and Stevenson 1979; Hutchison and Maness 1979; Kluger

1978). Presumably, many of these variables are likewise affected by temperature in anuran larvae.

Osmoregulatory Physiology

Our current knowledge of ion and water balance in tadpoles has come from three principle areas: comparative physiology, properties of adult and larval frog skin, and acidification of freshwater habitats by acidic deposition. The latter causes mortality of amphibian larvae (e.g., Ling et al. 1986) primarily through the disruption of Na^+ and Cl^- regulation. Studies exploring the mechanisms of acid water toxicity have revealed much basic information about ion regulation in tadpoles (e.g., Pierce 1985; Pierce and Harvey 1987; Pierce et al. 1984). Although osmoregulation in tadpoles has been studied from these different perspectives, our overall knowledge still remains incomplete in comparison with our knowledge of gas exchange in tadpoles or osmoregulation in fishes. Thus far, ion regulation in tadpoles appears to be similar to ion regulation in teleost fishes. Significant differences between tadpoles and teleosts include an active uptake of ions across the skin, gills that are less developed morphologically, and metamorphosis providing a gradual transition to air breathing. Many interesting details regarding ion and water balance in tadpoles await discovery. In this section, our current state of knowledge is presented and promising areas for research are suggested.

Body Composition

The ionic composition of the plasma of tadpoles, like most freshwater vertebrates, is markedly hyperosmotic and hyperionic relative to the surrounding environment. Ions of Na^+, Cl^- and HCO_3^- (table 8.3) are the major constituents responsible for 50%, 30%, and 15% of the osmotic pressure of plasma respectively (Just et al. 1977). The concentration of Na^+ in bullfrog tadpoles is much more dilute than in typical teleost fishes (90 and 150 mEq/liter, respectively). In the crab-eating frog (*Limnonectes cancrivorus*), Na^+ and Cl^- are also the most important ionic constituents, but their concentrations are about 50% higher than in bullfrogs acclimated to similar conditions (table 8.3; M. S. Gordon and Tucker 1965). *Limnonectes cancrivorus* inhabits brackish water ponds and both tadpoles and adults can survive prolonged exposure to full-strength seawater. The plasma osmolarity of spadefoot toad (*Spea hammondii*) tadpoles is also higher than that of bullfrog tadpoles, and Funkhouser (1977) hypothesized that this is an adaptation for survival in ephemeral ponds that exhibit high salinities caused by evaporation (also see Ferrari 1998). Little is known about the plasma ion concentrations of other species of tadpoles; as this information becomes available, it will be interesting to see if the concentrations of plasma electrolytes are correlated with the concentrations of salts in their habitats.

The osmotic pressure of tadpole plasma increases steadily throughout development and is highest during the later stages (Degani and Nevo 1986; Funkhouser 1977; J. E. Richmond 1968). Osmolarities of 160 mOsmol/liter in premetamorphic stages of *Rana catesbeiana* tadpoles increased to 259 mOsmol/liter at the end of metamorphic climax (J. E. Richmond 1968). Just et al. (1977) reported a much smaller increase (181 to 198 mOsmol/liter) during development of *R. catesbeiana*. *Spea hammondii* tadpoles also experienced a similar increase in plasma osmolarity at metamorphosis (Funkhouser 1977). This increase is presumably because of the increase in concentrations of plasma electrolytes resulting from a characteristic dehydration that occurs at metamorphosis (Just et al. 1977). This phenomenon is in keeping with the general concept that aquatic animals benefit from lower plasma ion concentrations because of a reduction of osmotic and ionic gradients between the animal and the water. Hematocrit remains relatively stable during development and averages about 30% in *R. catesbeiana* tadpoles (Just et al. 1977).

Few measurements of osmotic concentration and composition of tadpole blood have been reported for species other than large ranids. Because of this limitation, many investigators have reported ion and water concentrations as whole-body concentrations, and large interspecific and intraspecific variations exist (table 8.4). Caution must be taken when interpreting experiments that measure whole-body concentrations of ions and water because several factors influence their levels. The recent feeding history must be considered because a large percentage of the mass of a tadpole is ingested material in the gut, and the ionic content of the food can directly influence the mass-specific ion content of tadpoles. The mass of the gut contents also can have an indirect effect on mass-specific ion content by influencing body mass. In addition, body ion and water content changes dramatically

Table 8.3 Ionic composition, osmotic pressure and hematocrit of the plasma of tadpoles

Species	Ionic composition (mEq/liter)				Osmotic Pressure (mOsm/liter)	Hematocrit (%)	Reference
	Na^+	K^+	Cl^-	HCO_3			
Limnonectes cancrivorus[a]	153	5	99	—	272	—	M. S. Gordon and Tucker 1965
Rana catesbeiana	90	4	68	25	70–195	—	Alvarado and Moody 1970
	96–109	3–4	57–60	8–24	1–200	30	Just et al. 1977
	—	—	—	—	—	20–23	S. C. Brown et al. 1986
	—	—	—	—	60–259	—	Funkhouser 1977
Spea hammondii	—	—	—	—	190–290	—	Funkhouser 1977

[a]*Limnonectes cancrivorus* was acclimated to 100% freshwater.

Table 8.4 Ion content and percent body water of five species of *Rana* tadpoles

Species of *Rana*	Ion Content (μMol/g dry body mass)				Body water (%)	Reference
	Na^+	K^+	Ca^{++}	Cl^-		
catesbeiana	573–1080	339–355	—	—	—	Freda and Dunson 1984
	573	—	—	—	—	Freda and Dunson 1986
clamitans	369	—	—	—	—	Freda and Dunson 1984
	60–70[a]	30–35[a]	45–55[a]	30–40[a]	—	McDonald et al. 1984
muscosa	—	—	—	—	88–91	Bradford 1984b
pipiens	1365–1815	507	—	—	—	Freda and Dunson 1984
sylvatica	400–800	300–600	—	300–825	87–96	Freda and Dunson 1984

[a]units = μEq/g wet mass.

as a tadpole develops. Freda and Dunson (1984) reported the body ion and water concentrations of wood frog (*Rana sylvatica*) tadpoles every week from one week posthatching until metamorphosis. From days 7–14, body sodium concentrations increased from 400 to 800 μEq/g dry mass. This presumably occurred from a combination of active uptake across the skin and gills and the initiation of feeding. Body water also increased from 80% to 96%. The net result was that tadpoles gained weight very rapidly early in development because of the uptake of water. For example, dry mass increased by only 20%, while wet mass increased by 300%. This rapid increase in body mass early in development has ecological and physiological significance. Many predators (e.g., newts and salamander larvae; Brodie and Formanowicz 1983; Crump 1984) of tadpoles are gape-limited, that is, they feed on what they can fit in their mouths and swallow. Tadpoles should be most vulnerable to predation during early stages of development, and a rapid volume increase after hatching would tend to reduce predation. Throughout the remainder of development of *R. sylvatica,* body Na^+ concentration and body water gradually declined. Body Na^+ concentration has been shown to be linked with tadpole size for other species. For *R. catesbeiana,* body Na^+ concentration is negatively correlated with dry mass (body Na^+ concentration = 255 g − 0.485, $r^2 = 0.94$; Freda and Dunson 1986). Similarly, Bradford (1984a) reported that the concentrations of body water and solute are negatively correlated with body mass, and Zamachowski (1985) found that percent body water declined as *R. temporaria* tadpoles grew. The mechanisms causing this relationship are not known, but if the extracellular fluid space shrinks during growth and metamorphosis, body water and solute concentrations would be expected to decline. A second possibility is that as tadpoles grow, they contain a higher percentage of tissues (e.g., bone) that might reduce mass-specific concentrations of water. McDonald et al. (1984) showed that body Ca^{++} concentrations nearly triples in *R. clamitans* during premetamorphic and metamorphic stages. This presumably occurs as a result of limb development and increased calcification of the skeleton.

Species comparisons are also confounded by environmental conditions (e.g., temperature, salinity, and pH), which can have large effects on body ion and water composition (see below), and intraspecific variation in body ion and water concentrations. For example, Freda and Dunson (1984; see Pehek 1995) reported that *R. clamitans* tadpoles collected from an acidic bog and acclimated in the laboratory to water with a pH of 5.8 had a body Na^+ concentration of 369 μEq/g dry mass. McDonald et al. (1984) reported a body Na^+ concentration of 60 μEq/g wet mass (= 600 μEq/g dry mass, assuming 90% water) for *R. clamitans* tadpoles collected from an undescribed pond (see M. Uchiyama and Yoshizawa 1992). Conclusions about any general effects of pH require further studies because the different species that have been tested have not been of the same age or size or maintained under comparable conditions (see Bradford et al. 1994; Schmuck et al. 1994; Verma and Pierce 1994).

Water Balance

As discussed in the previous section, the body fluids of most tadpoles are hyperosmotic and hyperionic relative to the freshwater habitats in which they live. These osmotic and ionic gradients are the driving force for the continual osmotic uptake of water and diffusive leak of plasma ions. As with freshwater fishes, tadpoles remain in ion balance by excreting copious quantities of dilute urine and by actively transporting ions across external epithelia.

Few direct measurements of unidirectional exchanges of water in tadpoles have been reported. This is surprising because the body water concentration of tadpoles changes during growth and development and is strongly influenced by temperature (Bradford 1984b). Water may be taken up across the gills, skin, or gut. Tadpoles drink the external medium, probably during the course of feeding. Alvarado and Moody (1970) reported drinking rates (*Rana catesbeiana*) of 0.014 ml/g·h (see Territo and Smits 1998 for a discussion of physiological calculations based on various measures of body mass), of which 0.006 ml/g·h is absorbed. S. C. Brown et al. (1986) reported preliminary measurements of urine production to be 0.006 ml/g·h, and Mackay and Schmidt-Nielsen (1969) and B. Schmidt-Nielsen and Mackay (1970) reported urine flow in *R. clamitans* at about 0.02 ml/g·h. Because the mass of these animals did not change during the measurements, urine flow rate approximated the uptake of water across the gills, skin, and gut. Detailed studies of water uptake and loss across the gills, skin, gut, and kidney remain to be done.

Ion Regulation

Tadpoles continuously lose Na^+, Cl^- and other ions because of the large concentration gradient between their body fluids and the water. In addition, ions are lost in the urine. We are not aware of studies that quantify the relative importance of these different loss routes. Several studies reported loss rates of Na^+ and Cl^- for *Rana catesbeiana* and *R. clamitans* (Alvarado and Moody 1970; Dietz and Alvarado 1974; Freda and Dunson 1984; McDonald et al. 1984). Uptake of Na^+ and Cl^- generally exceeds loss so the animals remain in positive ion balance. Possible routes of active uptake of ions include the gills, skin, gut, and buccopharyngeal epithelium. Initially uptake was thought to occur only at the gills because Na^+ and Cl^- was lost from solutions used to perfuse the gills, but uptake was not increased if the perfusate was allowed to contact the skin (Alvarado and Moody 1970). In addition, studies of isolated larval skin did not detect active uptake or a transepithelial potential (TEP; Alvarado and Moody 1970). In support of the notion of the lack of dermal transport, active transport and TEP across the skin abruptly appear late in metamorphosis and are correlated with an increase in skin Na^+-K^+-ATPase (Boonkoom and Alvarado 1971; Kawada et al. 1969; R. E. Taylor and Barker 1965). This conclusion has been challenged because stripping the skins from the underlying musculature caused considerable stress to the preparation. In addition, no specific precautions were taken to prevent edge damage. Cox and Alvarado (1979; see Alvarado and Cox 1985) developed techniques for the study of isolated larval skin that minimized dissection and edge damage. They found that larval skin actively transports Na^+ inward at a rate of 5%–10% of that of adult skin. Their work suggests that the basal membrane in the skin of larval and adult anurans is similar. The difference in the rate of Na^+ transport may be caused by the absence of amiloride-inhibitable Na^+ channels in the apical membrane of larval skin. The addition of nystatin (forms cation-selective channels in lipid bilayers) to the outer bathing solution stimulated Na^+ influx to rates comparable to untreated adult skins (Cox and Alvarado 1983; also Zamorano and Salibián 1994).

The many factors that influence the active uptake of Na^+ include temperature, external ion concentration, water hardness, and pH. Interspecific differences may occur, but conclusive data are lacking. The uptake mechanisms for Na^+ and Cl^- display typical saturation kinetics. The Km and V_{max} for Na^+ uptake in *Rana catesbeiana* tadpoles were 0.2 mEq/liter and 0.25 μEq/g·h (Dietz and Alvarado 1974). Freda and Dunson (1986) reported a Km and V_{max} for Na^+ transport in *R. catesbeiana* to be 0.28 mEq/liter and 13.4 μEq/g·h on a dry mass basis (= 1.61 μEq/g·h; assuming 90% body water). The chloride transporter was saturated over the range of 0.2–3.0 mEq/liter and the V_{max} for Cl^- was 0.10–0.15 μEq/g·h (Dietz and Alvarado 1974). The location and type of cells responsible for Na^+ and Cl^- transport have not been identified. Na^+-K^+-ATPase activity in microsomal fractions of gill homogenate correlates well with the intensity of Na^+ transport (Boonkoom and Alvarado 1971). Na^+-K^+-ATPase has also been measured in the skin of *Rana catesbeiana* tadpoles (Kawada et al. 1969).

The transport of Na^+ and Cl^- are independent. For example, acetazolamide inhibits the branchial influx of Cl^- without affecting the influx of Na^+ or the efflux of Na^+ or Cl^- (Dietz and Alvarado 1974). The counter ions for Na^+ transport are thought to be H^+ and NH_4^+, while Cl^- is exchanged for HCO_3^-. These findings bring up another important point. In addition to their function in salt and water balance, the gills of tadpoles are also important sites of acid-base regulation and NH_4^+ excretion. Dietz and Alvarado (1974) reported net NH_4^+ excretion in *Rana catesbeiana* tadpoles of 260 nEq/g·h. McDonald et al. (1984) reported net losses of NH_4^+ for *Rana clamitans* of 634 nEq/g·h. In addition, NH_4^+ excretion in postmetamorphic animals was 12% of that in premetamorphic animals, indicating that the well-characterized switch from ammoniotely to ureotely had been completed. In the terrestrial *Bufo bufo,* this switch starts at the time when forelimbs are developed under the operculum and accelerates with forelimb emergence but does not occur in the aquatic *Xenopus laevis* (Munro 1953).

Effects of Temperature

When tadpoles are exposed to cold temperatures (e.g., 4°C), they rapidly increase in mass because of the uptake of water. The water content of *Rana muscosa* tadpoles increased by 6% after 1 month of exposure to 4°C, after which water content remained stable for 7 months (Bradford 1984a). Between 7 and 12 months of exposure, water content again increased slightly. During the first month of exposure, the osmotic pressure of the peritoneal fluid dropped by 14%, approached control values after 7 months, and then again declined at 12 months. It appears that water uptake is accommodated by the extracellular fluid volume (ECF; Bradford 1984a). In a related study, S. C. Brown et al. (1986) studied the weight loss in cold-acclimated (5°C) tadpoles after return to warmer temperatures. *Rana catesbeiana* tadpoles transferred to 11° and 18°C lost 7% and 10% of body weight after 5 days, while plasma Na^+ concentration increased by 28% and 21%, respectively. Daily treatment with ovine prolactin or growth hormone prevented weight loss or the concomitant increase in plasma Na^+ concentration. Neither propylthiouracil nor arginine vasotocin had an effect. These authors concluded that the changes in plasma solutes reflected changes in ECF volume. The adaptive significance, if any, of cold-induced edema in larval or adult anurans is not known. It would be interesting to know if water gain was simply because of an inhibition of osmotic regulatory mechanisms at low temperature, or if it is a specific physiological adaptation to winter conditions.

Effects of Salinity

Frogs occur in aquatic habitats that range widely in salinity. Many habitats are extremely dilute, such as soft-water bogs, rainwater filled temporary ponds, and water in granitic basins. *Rana catesbeiana* tadpoles can be reared in distilled water (Alvarado and Moody 1970). Similar to frogs and fishes, exposure to distilled water or very dilute solutions stimulates the sodium transport system. Elevated V_{max} for Na^+ transport and an unchanged Km (Dietz and Alvarado 1974; Freda and Dunson 1986) may reflect an increase in the total number of transport sites. Freda and Dunson (1986) also found that exposure to high Na^+ concentrations increased Km and depressed V_{max}. These changes in the characteristics of the transport mechanism would help maintain Na^+ balance in dilute water and reduce energetic costs of ion regulation in water with high concentrations of Na^+. It is not known whether branchial or skin permeability is adjusted in response to water salinity.

Some amphibians occur in highly saline habitats such as ponds and streams influenced by saltwater spray or tidal movements and desert ponds as they evaporate. The majority of field studies have focused on the distribution of adults with reference to salinity (Christman 1974; Ruibal 1959, 1962). Many species have been found in brackish salt marshes, and it appears that species vary in their tolerance to salinity. A few specialized species (*Bufo viridis* and *Limnonectes cancrivorus*) can survive in 40% and full-strength seawater, respectively (M. S. Gordon 1962; M. S. Gordon and Tucker 1965; M. Uchiyama et al. 1990). Very few studies of the distribution of tadpoles with regard to habitat salinity have been reported.

The best-known euryhaline amphibian is the crab-eating frog (*Limnonectes cancrivorus*) of southeastern Asia. The adults and tadpoles may be found in brackish water ponds and mangrove swamps. The tadpoles can survive exposure to full-strength seawater (M. S. Gordon and Tucker 1965; M. Uchiyama et al. 1990; M. Uchiyama and Yoshizawa 1992) and have been found in ponds ranging from 16% to 75% seawater (M. S. Gordon and Tucker 1965; Dunson 1977). However, M. S. Gordon and Tucker (1965) could not induce tadpoles to metamorphose in the lab if water salinity was greater than 20% seawater, and the largest tadpoles were observed in the most saline ponds. These observations suggest that *L. cancrivorus* tadpoles cannot metamorphose at high salinities, but this suggestion awaits more rigorous verification. The adults are osmoconformers and are able to survive in highly saline waters by elevating plasma urea, which raises plasma osmotic pressure and thereby eliminates the driving force for the dehydration that would occur in saltwater. Larvae are osmoregulaters; if water salinity is increased from 100% freshwater to 100% saltwater (a more than 10-fold increase in osmotic pressure), plasma osmotic pressure only doubles (M. S. Gordon and Tucker 1965). Over all salinities (0%–100% seawater), 90%–100% of the osmotic pressure of plasma is caused by Na^+, K^+, and Cl^-. Similar to marine teleost fishes, these tadpoles osmoregulate in seawater by drinking the hyperosmotic medium and by excreting a small amount of isosmotic urine. Although it has never been verified, the tadpoles probably excrete salts by an extrarenal pathway such as Na^+ transporters in the skin and gills. M. S. Gordon and Tucker (1965) also found that tadpoles become osmoconformers as froglets (Taylor/Kollros XXV; Gosner 46). In 80% seawater, starting at Taylor/Kollros XX (Gosner 42), plasma osmotic pressure increased and plasma Na^+, K^+, and Cl^- concentrations did not change. This pattern was presumably the result of the initiation of urea production. More detailed experimental data are clearly needed on the physiology of this interesting species.

Effects of Low pH

Concern over acidic precipitation has stimulated a great deal of recent research on the effects of acidic water on amphibians. Lowered rain pH over the northeastern United States, southeastern Canada and Europe has been implicated in the depression of alkalinity and pH of many freshwater habitats. Acidic mine drainage associated with coal and metal strip mining can also seriously impact freshwater habitats (Kinney 1964; Porges et al. 1986). Although anthropogenic acidification has received much attention, it is generally not realized that many freshwater habitats are acidic because of naturally occurring humic acids or the biological activity of *Sphagnum* (reviews by Gorham et al. 1985; Kilham 1982; see Rosenberg and Pierce 1995). Naturally acidic waters have been described throughout the world: major branches of the Amazon and Congo rivers (Marlier 1973); boreal peat bogs of North America, Europe, and Asia (Gorham et al. 1985); many waters of the eastern and Gulf Coastal plain of the United States (Gosner and Black 1957b) and the southwest cape of South Africa (Picker 1985; Picker et al. 1993); heathlands of Great Britain (Beebee and Griffin 1977); swamps of Malaysia (D. S. Johnson 1967); and ponds of northern Australia (T. E. Brown et al. 1983). Detailed discussions of the distribution of acidic waters in the United States, the relative tolerance of different species, and the ecological effects of acidic waters are reviewed by Freda (1986) and Pierce (1985). Tadpoles are more tolerant of low environmental pH than embryos but less so than adults (Freda 1986). Amphibians as a group are much more tolerant than fishes. The primary toxic effect of low pH is disruption of Na^+ and Cl^- balance (Freda and Dunson 1984; McDonald et al. 1984). Acute exposure to lethal pH inhibits the active uptake of Na^+ and Cl^- while causing a massive stimulation in passive Na^+ and Cl^- loss. Influx of Na^+ is linearly related to pH and 100% inhibition occurs near pH 4.0. The loss of Na^+ increases exponentially just above the lethal pH. In all species tested, death occurs when body Na^+ concentration is reduced by about one-half. The increased efflux of Na^+ and Cl^- at low pH is quantitatively responsible for the reductions in body ions, and variation in the rate of loss defines intra- and interspecific variation in acid tolerance (Freda and Dunson 1984). The inhibition of Na^+ influx may be caused by direct competition of H^+ and Na^+ for the carrier sites (which may not be the same for the two ions), as increasing the concentration of Na^+ in acidic water elevates Na^+ uptake (Freda and Dunson 1986). Ion loss presumably occurs across the skin and gills. In fishes it is thought that Na^+ loss at low pH results from the leaching of calcium from the branchial epithelium, which opens up tight junctions and compromises the structural integrity of the gill epithelium (Freda and McDonald 1988; also see Stiffler 1996).

Chronic exposure to sublethal pH generally causes a 20% decline in body Na^+ concentration and stimulates compensatory mechanisms to help maintain ion balance (Freda and Dunson 1984, 1985, 1986). The influx of Na^+ in *Rana pipiens* tadpoles was first depressed but then exceeded control values during seven days of exposure to pH 4. 5. Efflux of Na^+ was initially stimulated but later declined to levels just above controls (Freda and Dunson 1984). Similarly, McDonald et al. (1984) reported that Na^+ influx and efflux in *R. clamitans* in closed chambers where the water was initially of pH 4.0–4.2 returned to control values after 7 hr of exposure. Although the inhibition of influx is not significant during lethal exposures, restoration of influx during sublethal exposures is important in maintaining ion balance. Freda and Dunson (1986) showed that exposure to low pH exposure acts similarly to other salt-depleting conditions (e.g., distilled and soft water) by increasing the V_{max} for Na^+ transport. The reduction in sodium efflux that occurs after prolonged exposure has not been studied but presumably involves a reduction of Na^+ permeability of the gills, skin, or both.

Summary

To understand the ecological physiology of tadpoles as a group, it is essential to examine phylogenetically diverse taxa and diverse environmental conditions. At present, most of the knowledge about tadpole ecological physiology is based on very limited subsets of taxa and environmental settings. For example, most ecophysiological studies have been done with temperate zone species, rather than tropical species, yet the greatest diversity of taxa are tropical. Even within the temperate zone, the preponderance of work has been limited to members of the Bufonidae and Ranidae. In part, this is because relatively few investigations have been conducted on amphibian larvae in comparison to adult amphibians and other vertebrate groups. Also, these studies have often sought to understand how amphibian larvae differ from adults and other vertebrates rather than to understand how tadpole biology differs among species and environments. Investigators have tended to work with local species or species easily maintained in the laboratory.

Future ecophysiological studies should ask questions concerning the interaction of environmental parameters and the physiology of tadpoles in environmental settings that are poorly known. Studies of ontogenetic differences would be helpful (e.g., Burggren 1984; Burggren and Fritsche 1997). Examples may be temperature relations of tadpoles in tropical forests with little fluctuation in temperature and gas exchange of tadpoles inhabiting bromeliads. To the extent possible, such studies should be designed to compare phylogenetically disparate taxa and different settings in order to distinguish between phylogenetic and environmentally induced patterns. We should also attempt to understand the physiology of tadpoles in the context of the many considerations faced by the individual (e.g., obtaining food, avoiding abiotic and biotic hazards, and maintaining an internal environment within tolerable limits; Feder 1992). For example, behavioral temperature regulation can expedite development, and air gulping can compensate for low oxygen levels in the water. In contrast, the former may lead to stranding, and both activities may render an individual more susceptible to predation (e.g., Lefcort and Eiger 1993). Our challenge is to understand the trade-offs between the physiology-driven processes and other considerations. This task should be interesting, although perhaps difficult, because many amphibians can tolerate wide variations in some physiological parameters (e.g., body water, temperature, and osmolyte concentration) in comparison to other vertebrates, especially birds and mammals (Feder 1992). This is exemplified for tadpoles of many species by their relatively wide temperature tolerance ranges (table 8.2). Also, many aspects of the physiology of tadpoles change during larval development. This apparently wide tolerance of some physiological conditions combined with ontogenetic changes in physiology may allow tadpoles considerable flexibility in accommodating to the spatial and temporal variations in their abiotic and biotic environments.

ACKNOWLEDGMENTS

We thank T. Dietz, A. Riggs, and R. J. Wassersug for editing parts of this chapter. Additional comments were provided by W. Burggren and M. Feder. Bradford acknowledges the environmental Science and Engineering Program at the University of California, Los Angeles, for support during the early stage of preparation of this chapter.

9

BEHAVIOR

Interactions and Their Consequences

Karin vS. Hoff, Andrew R. Blaustein, Roy W. McDiarmid, and Ronald Altig

I find studying the behaviour of animals in their natural surroundings a fascinating hobby.

Tinbergen 1968:iii

Introduction

A tadpole is an energy-gathering, growing, nonreproductive larval stage in the biphasic anuran life cycle. Various authors (e.g., Wilbur and Collins 1973) have modeled the impacts that different aquatic environments have on tadpole growth and feeding, and growing to the largest size in the shortest period remains a driving force for selection. Optimizing growth enhances the probability of completing metamorphosis and reaching the reproductive stage. Forms of paedomorphosis that result in a larval individual capable of reproduction occur in salamanders but are unknown in anurans. As a result, behaviors usually associated with sexual activity and species recognition (e.g., courtship displays) among reproductive adults are absent in larvae. Instead, the behavioral repertoire of tadpoles is restricted to activities that augment development, growth, and survival to metamorphosis. Aggregations of tadpoles in nature are relatively common (see Duellman and Trueb 1986; Lescure 1968; Lyapkov 1996; P. J. Watt et al. 1997) and may be interpreted broadly as ecophysiological, metamorphic, feeding, and social responses. Relative to aggregative behavior, we focus primarily on feeding and social interactions.

Three basic subject areas provide the organization for this chapter: (1) aspects of general maintenance (feeding, respiration, thermoregulation, and locomotion); (2) dispersion and social behaviors (habitat selection, social interactions, and kin recognition); and (3) sensory abilities (vision, cutaneous senses, olfaction, and chemical communication) and learning. Tadpole behavior is not well studied, and we point out specific areas that need attention.

All notations of developmental stages follow the system of Gosner (1960; see table 2.1 and fig. 2.1).

General Maintenance

Feeding

Tadpoles usually are considered highly specialized, filter-feeding herbivores (e.g., Duellman and Trueb 1986). Most ingest planktonic material from the water column, obtain organic materials from pond sediments, or scrape material (i.e., aufwuchs, periphyton) from submerged substrates. Given the diverse and sometimes bizarre morphology of their feeding structures, the variety of sizes and kinds of food items recorded in their guts, the absence of cellulase for digesting plant materials, the diversity of microhabitats occupied, and our lack of basic knowledge of their diets determined from conventional gut content analyses convinces us that tadpoles are better thought of as opportunistic omnivores or detritivores.

Stimuli emanating from food sources that initiate locomotor or feeding responses (e.g., P. H. Harrison 1990) also have not been studied well. Systematic and comparative evaluations of the food habits of tadpoles are uncommon (but see Diaz-Paniagua 1985; Heyer 1973b; Inger 1986 and references therein) and generally not very informative because many ingested items pass undamaged through the gut, while other soft-bodied organisms and bacteria are not detected. Such observations have led some authors to suggest that the real sources of food for tadpoles may be dissolved inorganic and organic nutrients, bacteria, and viruses (Heyer 1973b; Inger 1986). The contribution of these small items to tadpole diets remains unstudied in nature, and that part of the material that a tadpole ingests that is actually digested is not known.

Items reported from the guts of tadpoles include sand, detritus, viruses, bacteria, small unicellular organisms (protists), algae (diatoms, filamentous green, euglenophytes), plant fragments of assorted sizes, pollen grains, fungi, various kinds of small animals (annelids, cladocerans, copepods, gastrotrichs, insects, nematodes, rotifers, tardigrades, water mites), anuran eggs, and heterospecific and conspecific tadpoles (Costa and Balasubramaniam 1965; Diaz-Paniagua 1985; J. D. Harrison 1987; Heyer 1973b; Inger 1986; Lajmanovich 1994; Sabnis and Kuthe 1980; Sekar 1992). Tadpoles also are known to feed on fecal material (Steinwascher 1978a) and sundry types of carrion (= necrophagous group of Beiswenger 1975). In all cases, caution must be exercised in interpreting feeding habits and their possible correlations with morphology. A. Haas (personal communication) recently found intact ephemeropteran larvae in an unidentified *Boophis* tadpole from Madagascar. This situation was unexpected in a suctorial tadpole that lives in very fast water and maintains position with an exceptionally large oral disc.

Research has focused on evidence of food selection from several approaches: stomach content analysis (e.g., Jenssen 1967); comparative aspects of buccal anatomy related to feeding (e.g., Viertel 1985; Wassersug 1980); comparative mechanical restrictions on feeding behavior based on anatomy and limited experimentation (e.g., Wassersug and Hoff 1979); feeding dynamics of a small number of species in the laboratory (e.g., Seale 1980); and relationships between feeding modes and microhabitat use, oral morphology, diet, and phylogenetic groups (e.g., Altig and Johnston 1989; Diaz-Paniagua 1985; Heyer 1973b; Inger 1986). Warkentin (1992a, b) examined the effects of temperature and illumination on microhabitat use and feeding rates of *Rana clamitans* tadpoles.

Most tadpoles feed by producing currents that carry particles into the buccal cavity and across food entrapment surfaces. A tadpole opens its mouth (see fig. 4.16), depresses the floor of the buccal cavity to draw in water, closes the mouth, and then elevates the floor of the buccal cavity to pump water across food entrapment surfaces and out through the spiracle(s) (e.g., De Jongh 1968; Gradwell 1972b, c; Kenny 1969b, c; Severtsov 1969a; Wassersug 1972).

Tadpoles occur in countless aquatic habitats, feed at many sites (benthic, midwater, surface) throughout the water column, and have characteristic morphologies and behaviors. To demonstrate this ecological and morphological diversity and to provide some background for reading this chapter, we illustrate two somewhat hypothetical larval communities, one typical of a forest pond (fig. 9.1) and the other of a forest stream (fig. 9.2). We selected our examples from the Neotropical region because that is a fauna with which we are familiar. It also is broadly representative of faunas in other tropical regions, which are ecologically and taxonomically diverse compared to those in comparable habitats in temperate zones. More detailed illustrations of these and other tadpole morphotypes are shown in figure 12.1, and their associated mouthparts are shown in figure 12.2.

Heyer (1973b) described the microhabitat use, feeding and swimming behavior, and diet of pond-dwelling tadpoles of 17 species in a dry forest site in Thailand, and Inger (1986) and Inger et al. (1986a) reported a comparable study for 12 stream and 4 nonriparian tadpoles of 16 frog species from a wet forest site in Borneo. Pond tadpoles of the African frog *Xenopus laevis* are typically positioned head downward at about 45°, while those of *Agalychnis spurrelli* (Duellman and Trueb 1986) and other phyllomedusine hylids usually position themselves upward at about 45°. Suspension-feeding microhylid tadpoles usually float horizontally near the surface or in midwater, and tadpoles of *Rhinophrynus dorsalis* usually swim horizontally.

The buccal pumping mechanism is similar in at least all rasping tadpoles, but species differ in the efficiency of entrapment based on the sizes (i.e., 0.126 in Wassersug 1972 to greater than 200 μm in Seale 1980) of particles that are captured. These differences in efficiency have been correlated with variations in skeletal elements of the pump (Wassersug 1972; Wassersug and Hoff 1979) and density of the gill filters (Wassersug 1972). Tadpoles without keratinized mouthparts (e.g., *Gastrophryne* and *Xenopus*) cannot graze on plant material that is too large to fit into their mouths, and tadpoles lacking ultraplanktonic entrapment surfaces (e.g., *Hymenochirus*) feed on items large enough to be retained without entrapment surfaces.

Wassersug and Hoff (1979) noted specific behavioral correlates with buccal pump design. Deflection of the floor of the cavity by a lever action of the ceratohyal causes the buccal cavity to act as a hydraulic pump. The area of the buccal cavity floor, the length of the lever, and the amount of deflection determine maximum buccal volume and maximum suction force (negative pressure) generated by the pump. Microphagous tadpoles tend to have large buccal volumes and short lever arms and regulate feeding rate by frequency modulation (FM) of pumping frequency (e.g., *Xenopus laevis*). Tadpoles with relatively longer lever arms vary buccal volumes and pressures by amplitude modulation (AM) of the ceratohyal deflection (e.g., *Rana sylvatica*). Suctorial forms that rely on buccal suction to adhere to surfaces (e.g., *Ascaphus*) have large lever arms and small buccal volumes to generate high pressures. Macrophagous tadpoles have large volumes and long lever arms.

The validity of morphological modeling to predict feed-

Fig. 9.1. An ecological diorama of a hypothetical composite tadpole community in and near a neotropical pond. The tadpoles of six species with distinct morphotypes that might occur there are illustrated exhibiting typical behaviors (e.g., feeding, swimming, hiding, etc.). Top center: *Hyla bromeliacia* (Hylidae), an arboreal (Type 1) tadpole that occurs in the central axil of a bromeliad. Top right: *Hyla microcephala* (Hylidae), a midwater macrophagous form with reduced mouthparts searching for small crustaceans among stems of aquatic plants. Lower right: *Scinax staufferi* (Hylidae), a nektonic midwater form with high fins that scrapes food from leaves and stems and other underwater surfaces. Right center: *Rhinophrynus dorsalis* (Rhinophrynidae), a filter feeding tadpole that swims in large, midwater schools. Left center: *Leptodactylus pentadactylus* (Leptodactylidae), an omnivorous benthic tadpole that often is carnivorous on hatchlings and small tadpoles. Lower left: *Bufo marinus* (Bufonidae), a black, benthic, rasping tadpole that sometimes aggregates.

ing behaviors is supported by studies of feeding dynamics in *Rana sylvatica* and *Xenopus laevis* (Seale et al. 1982 and Seale and Wassersug 1979; see feeding dynamics below) and also is consistent with the suggestion (Wassersug and Seibert 1975) that the buccal pump design of tadpoles reflects feeding more than respiratory needs (see Respiration below).

Polis and Myers (1985) tabulated 14 species of frogs whose tadpoles ate conspecific tadpoles or eggs, and other species have been added to the list (e.g., *Osteopilus septentrionalis,* Crump 1986; also see table 10.2). Crump (1986) distinguished between opportunistic cannibalism (such as oophagy and necrophagy, in which the prey does not attempt to evade the predator, and cannibalism, in which conspecifics are attacked, killed, and eaten (e.g., Summers and Amos 1997; also see Caldwell and De Araújo 1998 and Ńguenga et al. 1997). In their review of arboreal tadpoles, Lannoo et al. (1987) tabulated species known to consume frog eggs; other hylids that breed in phytotelmata (i.e., bromeliads and tree holes), including *Phrynohyas resinifictrix* (R. W. McDiarmid, personal observation), *Osteocephalus* cf. *leprieurii* (R. W. McDiarmid, personal observation), *O. oophagus* (Jungfer and Schiesari 1995); *Anotheca spinosa* (Jungfer 1996) and *O. planiceps* (L. Haugen, personal communication) can be added to that list. *Chirixalus eiffingeri* (Rhacophoridae) nests in tree holes and broken or cut bamboo stems and has oophagous tadpoles (Ueda 1986; Kam et al. 1996). Tadpoles of *Philautus* sp. also found in a tree hole in Thailand apparently feed on frog eggs, likely their own (Wassersug et al. 1981a).

Fig. 9.2. An ecological diorama of a hypothetical composite tadpole community in and near a neotropical stream. Likely occurring representative morphotypes are shown in different microhabitats and exhibiting typical behaviors (e.g., feeding, swimming, hiding, etc.). Top right: *Thoropa miliaris* (Leptodactylidae), a semiterrestrial tadpole that lives on the wet surface where rivulets of water flow over a vertical rock faces. Lower right: *Cochranella granulosa* (Centrolenidae), a fossorial tadpole that buries within leaf packs in still water. Right center: *Otophryne pyburni* (Microhylidae), a psammonic tadpole buried in sand in shallow, slow-moving water. Left center: *Phasmahyla guttata* (Hylidae), a neustonic tadpole feeding from the surface film in shallow water with an umbelliform oral apparatus. Lower left: *Hyla armata* (Hylidae), a suctorial tadpole that attaches to rocks in fast water. Top left: *Eleutherodactylus ridens* (Leptodactylidae), embryos of a direct developing species that places its eggs among fallen leaves that often accumulate in the axils of understory plants. Top center: *Phrynohyas resinifitrix* (Hylidae), an arboreal (Type 2), omnivorous, often times oophagous, tadpole that occurs in tree holes.

Cannibalism may be more common than reported, especially among tadpoles of frogs that breed in small, ephemeral ponds and phytotelmata (e.g., bromeliads and tree holes) that are subject to unpredictable drying or where resource shortages and overcrowding may be common and survivorship low (Bragg 1957, 1965; Crump 1986, 1990, 1992; Downie 1990b; see chap. 10). Cannibalistic tadpoles most commonly eat conspecifics of different sizes or developmental stages. Large tadpoles may eat eggs (Crossland 1997; Crossland and Alford 1998; Heusser 1970a) and hatchlings (Crump 1983), and premetamorphic tadpoles may prey on metamorphosing individuals (Bragg 1957; Crump 1986). Heusser (1970a) speculated that conspecific oophagy may be yet another reason for synchronous oviposition, and this argument can be extended to synchronous metamorphosis. Reports of tadpoles eating conspecific tadpoles in the field have not always made it clear whether the prey were already dead before they were eaten. However, conspecific tadpole cannibalism has been observed in the field (Crump 1990, *Hyla pseudopuma*); Petranka and Thomas (1995) argued that explosive breeding in *Rana sylvatica* reduced egg and tadpole cannibalism.

In laboratory studies, Crump (1986) allowed tadpoles of *Osteopilus septentrionalis* to metamorphose in aquaria with emergent surfaces to reduce the likelihood of drowning. The presence of these surfaces did not eliminate cannibalism by younger tadpoles in the same aquaria, and metamorphosing tadpoles (Gosner 42–43) in both experimental treatments (i.e., those that were not eaten by conspecifics) climbed up the vertical glass walls. All metamorphosing froglets survived in the control treatments (i.e., those without younger conspecifics). It seems likely that cannibalism, at least in this species, is active predation by younger (Gosner 36–39) tadpoles, perhaps siblings. Crump (1986) speculated that metamorphosing tadpoles were more vulnerable to predation because they had diminished locomotor capacity compared to either tadpoles without forelegs or froglets without tails. Wassersug and Sperry (1977) also demonstrated that anurans in the middle of metamorphosis (four legs and a tail) do not swim as fast as tadpoles younger than Gosner 42 or hop as far as froglets older than stage 45. It may be that tadpoles prey preferentially on conspecific tadpoles when they are most vulnerable; this is the same stage at which they are also preferentially preyed upon by other species (e.g., *Thamnophis,* S. J. Arnold and Wassersug 1978).

Some tadpoles (e.g., *Hymenochirus* and *Lepidobatrachus*) are specialized for consuming large food items (Wassersug and Hoff 1979), and some are actively predaceous on heterospecifics. The hypertrophied jaw serrations of the tadpoles of *Otophryne pyburni* (Wassersug and Pyburn 1987) and *Mantidactylus lugubris* (R. Altig, personal observation) superficially appear to be specializations for carnivory but are probably used to sort particles during filtration processes. Wassersug and Pyburn (1987) suggested that the tadpole of *Otophryne pyburni* feeds by passive aspiration caused by water flowing past the aperture of the elongate spiracle tube.

An amazing example of feeding behavior and parent-offspring communication has been documented in the Central American frog, *Dendrobates pumilio.* After 10–12 days of development, tadpoles hatch from eggs (clutches range from 5 to 20 eggs; mean of 10 clutches = 10.9) laid and fertilized on horizontal dry leaves (Weygoldt 1980). The female parent transports a single tadpole on her back to a suitable bromeliad and releases the tadpole in a leaf axil with water. Periodically (1–9 day intervals), she returns, inspects the leaf axil holding her tadpole, and deposits an unfertilized egg which the tadpole eats. When a tadpole of *D. pumilio* detects (see Sensory Abilities, Vision) a frog visiting the bromeliad axil in which it lives, it begins a special tail-vibrating "dance" that can be seen by the visiting frog. Instead of swimming with the usual tail undulations, the tadpole creates noticeable circular movements beneath the water surface by stiffening and rapidly vibrating its tail. If the visitor is the female parent coming to feed her tadpole, she backs into the axil and the tadpole swims against her hindquarters to stimulate her to oviposit nutritive (unfertilized) eggs. The tadpoles of *D. pumilio* appear to be obligate egg feeders, and attempts to raise them on other foods have failed (Weygoldt 1980). This remarkable behavior characterized by posthatching parental care, communication between a tadpole and its female parent, and the use of nutritive eggs to feed ones offspring was unknown among amphibians when Weygoldt published his extraordinary observations with captive frogs. This behavior was later observed in *D. pumilio* in the field in Panama (Brust 1993) and likely is characteristics of all eight species in the group (e.g., Jungfer 1985, *D. speciosus;* Zimmermann and Zimmerman 1981, *D. histrionicus*).

Another remarkable case of tadpole-mother communication that also is associated with deposition of nutritive eggs was described by Jungfer (1996) for *Anotheca spinosa.* Tadpoles of this species normally are found in tree holes, and in the laboratory they show a peculiar behavior when the female parent enters the basin containing her larvae. The normally inactive tadpoles swim around the mother, touching her with their snouts and "sucking" slightly on her skin; just seconds before nutritive eggs are extruded, the tadpole movements become faster and more hectic, sometimes including bites on the extruded cloaca of the mother (Jungfer 1996). Usually, the eggs are eaten immediately by tadpoles (fig. 9.3A).

Other species of bromeliad and tree-hole breeding frogs also may rely on deposition of eggs to provision their tadpoles. Are the eggs laid in a bromeliad that already contains tadpoles provided by the female parent of the tadpoles? Are the eggs nutritive and fertile (i.e., capable of developing into tadpoles) or nonfertile and provided only as a food source for the tadpoles? What is the role of the male frog in these situations, and is there any communication between the tadpoles and the parent(s)? The answers to these and other intriguing questions are beginning to appear. In their description of *Osteocephalus oophagus,* Jungfer and Schiesari (1995) noted that females, clasped by a male, returned at intervals of about five days to bromeliads that contain their tadpoles and deposit eggs that are eaten by the tadpoles (fig. 9.3B). If no tadpoles are present the eggs develop. The tadpoles apparently are oophagous and starve if eggs are not provided. The tadpoles of *Chirixalus eiffingeri* (Kam et al. 1996) and *Mantella laevigata* may be other candidates for this type of nutrition.

Different types of parent-offspring communication, other than those described for some oophagous tadpoles, may be more common than we think. Tadpoles of *Leptodactylus bolivianus* move toward the source of small waves made by rhythmic movements of their mother, and by her movements, she leads the tadpoles to other sites (Wells and Bard 1988; see Downie 1996). When the frequency of rain slows and puddles begin to dry, male *Pyxicephalus adspersus* construct channels between puddles and lead tadpoles through the interconnecting channel to other parts of the system that contain more water (Kok et al. 1989). These reports of parent-tadpole communication are exciting observations and provide a fertile arena for future investigation.

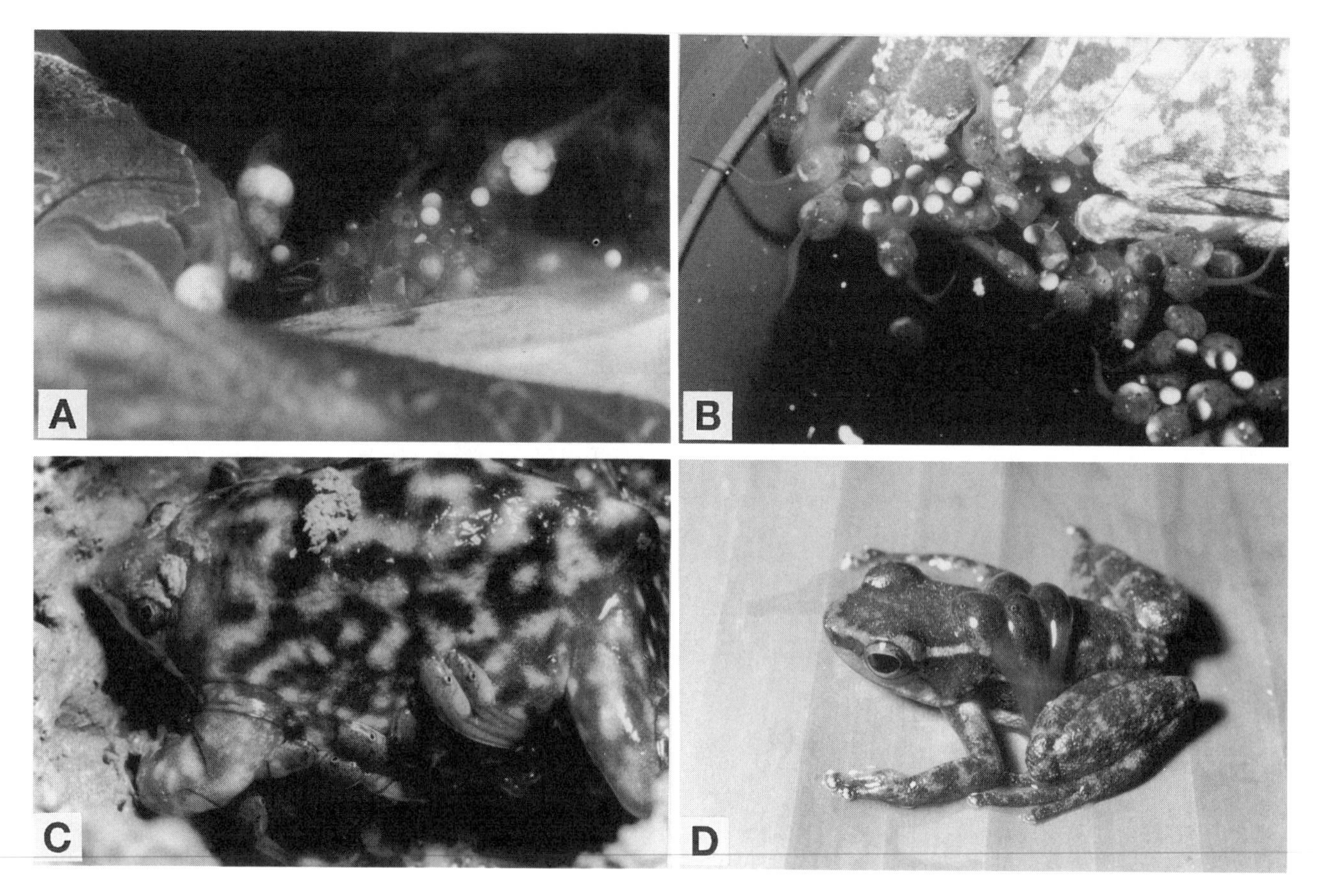

Fig. 9.3. Examples of parent-offspring communication and behavior include the provisioning of nutritive eggs to tadpoles and transport of larvae from egg deposition sites to aquatic habitats. (A) Three tadpoles of *Anotheca spinosa* (Hylidae) eating nutritive eggs oviposited by their mother (see Jungfer 1996). (B) Tadpoles of *Osteocephalus oophagus* (Hylidae) feeding on eggs laid by an amplexing pair, and the female is the parent. (C) Tadpoles of *Hemisus marmoratus* (Hemisotidae) on and beneath the attending female parent in a nest cavity in Camoé National Park, Ivory Coast (see Rödel 1996 and text for discussion). (D) Four tadpoles atop a nurse frog of *Colostethus nexipus* (Dendrobatidae) from the Rio Cenepa drainage in Amazonas, Peru.

Two other interesting examples apparently involve communication between the tadpole and parent. Wager (1929) reported maternal care in *Hemisus marmoratus.* Females remain in a hollowed-out cavity atop the developing eggs until they hatch. Newly hatched tadpoles are lively and "indefatigable wrigglers." Based on excavations of burrows and other circumstantial evidence, Wager (1929:129) commented that ". . . it seems certain therefore that the mother burrows towards the water making a tunnel down which the tadpoles wriggle in a squirming mass until they reach the water." These observations were repeated without qualification by Wager (1965, 1986) and have been repeated in the general literature (e.g., Duellman and Trueb 1986). On reviewing all the published reports, Van Dijk (1985) rejected Wager's scenario and noted that others had reported hatched larvae swarming over the back of an attending female and on the back of another frog, presumably a female. Van Dijk suggested that female *Hemisus* carry their tadpoles to the water. M.-O. Rödel (personal communication) reported tadpoles following a female as she constructed an open channel from the nest to water; during this period some of the tadpoles were attached to the female. Other observations indicate that as ponds fill female *Hemisus* open flooded nest chambers to allow tadpoles to escape. Circumstantial evidence (e.g., nests more than 100 m from water without any indication of channels, tadpoles in rock pools above the level of the surrounding savannah) suggested to M.-O. Rödel (personal communication) that maternal behavior (digging channels or transporting larvae) of *Hemisus marmoratus* may vary depending on nest site topography and the surrounding habitat. Although larvae have been observed on female *Hemisus marmoratus* in the nest cavity (fig. 9.3C) and in open channels, no one has actually observed a female *Hemisus* carrying a tadpole across land.

Larval transport is well known in dendrobatid frogs (personal observations; fig. 9.3D) and a few other species (Duellman and Trueb 1986). Larval transport in dendrobatids usually involves the male, except in those species where females provision the larvae with nutritive eggs (see above). Wells (1978) reported that male *Dendrobates auratus* may use visual or tactile cues to determine when tadpoles can be carried from the nest site (usually in leaf litter) to a suitable aquatic site (bromeliad or tree hole). When tadpoles become more active just prior to hatching, attending males assume a distinctive posture on the clutch with back arched and ankles pressed together to form a trough. Tadpoles attempted to wriggle onto the male's back but often fell off. In *D. auratus* only one or a few tadpoles are carried at a time; in other species most of a clutch may be transported. Stebbins and Hendrickson (1959) described larval transport in *Colostethus subpunctatus* noting that attachment is by sticky mucus produced by the nurse frog and not by the larval mouth. They described the release of larvae but had no information on how the tadpoles got onto the male's back. Additional data on parent-larva communication prior to transport could be obtained from cultured frogs. Active wriggling by the hatchling larvae seems to be an obvious parallel in all of these cases and probably serves as a visual or tactile cue to initiate the appropriate parental behavior. This seems to be a fertile area for future investigation.

Under certain unknown and seemingly uncommon circumstances, tadpoles that occur in temporary sites with bare soil substrates form large expanses of tightly spaced feeding depressions. In the sedimentary literature (see Dionne 1969; T. D. Ford and Breed 1970; Maher 1962; Opatrny 1973), these depressions have been called "tadpole holes" or "tadpole nests" for well over 100 years. J. H. Black (1974, *Bufo americanus* and *Pseudacris streckeri;* see J. H. Black 1971) noted these holes once in six years of observing about 30 pools, and one of us (R. Altig, personal observation) has seen these holes only at two sites during the same summer (*Bufo terrestris* and *Pseudacris triseriata*) over 26 years of field observations. It is as if the tadpoles are deliberately seeking specific material beneath the soil surface. The tail of a feeding tadpole vigorously beating against the bottom produces a circular, hexagonal, or pentagonal berm about the feeding point that delimits a depression about twice the length of the tadpole (fig. 9.4). These holes are not ripple interference marks made simply by tadpoles swimming (Dionne 1969).

The film that develops on the surface of the water apparently is a bountiful place to feed (e.g., Danos et al. 1983; Goldacre 1949), but the circumstances that promote this feeding mode is not known. Tadpoles with umbelliform oral discs (see chaps. 3 and 12) specialize in feeding from this film (fig. 9.5). Because of the orientation of the disc, they feed in a slightly tail-down but upright attitude relative to the surface. Typical benthic and nektonic tadpoles commonly feed from the surface film in ponds and slow-moving streams. These tadpoles either turn vertically or roll over on their backs (e.g., species of *Bufo, Hyla, Rana,* and *Scaphiopus*) when feeding from the surface film. Hatchlings commonly hang from the surface film (see Wangersky 1976), and one could speculate that they gain some of their first nutrition from the surface just as they begin feeding.

The details of feeding behavior and optimal foraging are known for only a few species. Feeding dynamics have been investigated mostly in tadpoles of species that are obligate suspension feeders (e.g., Schoonbee et al. 1992, *Xenopus laevis*). Suspension clearance rates and buccal pumping rates are relatively easy to monitor in the laboratory. Seale et al. (1982) examined the relationships among food concentration, buccal pumping rate, and ingestion rate in *Xenopus*

Fig. 9.4. Tadpole holes (depressions in the muddy bottom) formed by feeding tadpoles of *Bufo terrestris* (Bufonidae), photographed in June in Jackson County, Mississippi.

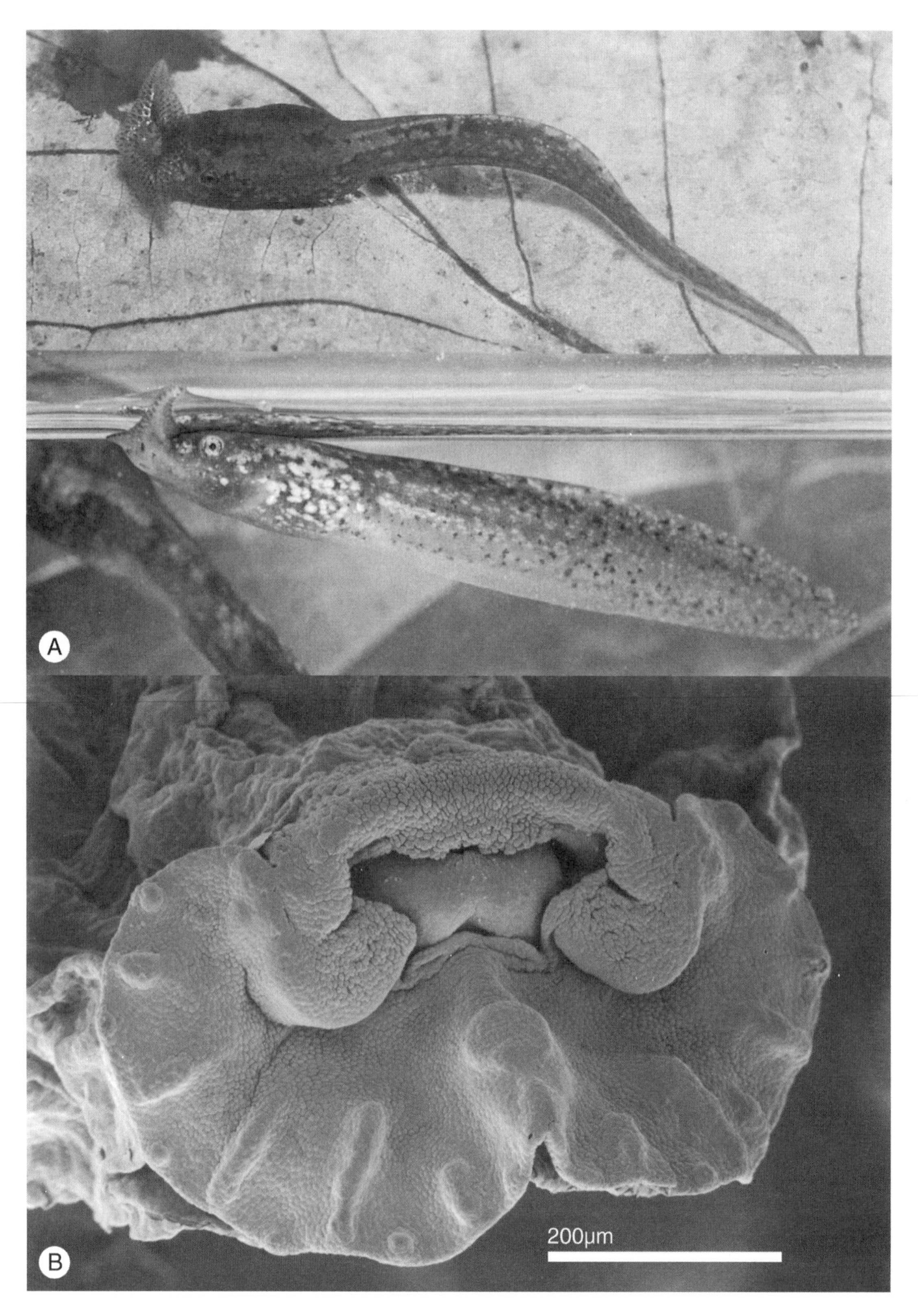

Fig. 9.5. Tadpoles of several unrelated species have umbelliform oral discs to harvest particulate material in the surface film. (A) Tadpole of *Megophrys montana* (Megophryidae) feeding on the surface film. (B) Scanning electron micrograph of the oral disc of *Microhyla heymonsi* (Microhylidae). Maximum width of oral disc is 2.5 mm.

laevis and fitted the data to either a modified Monod (Michaelis-Menten) or an energy optimization model. Tadpole feeding dynamics resemble those of other suspension-feeding organisms and provide for both high energy return to the tadpole and stability of populations of food organisms. *Xenopus laevis* tadpoles, presented with concentrations of food (yeast) ranging from 5×10^{-3} to 5×10^{2} mg/liter, increased feeding rates as food concentrations increased until a maximum ingestion rate was attained. Seale et al. (1982) speculated that the maximum ingestion rate was determined by gut packing or by saturation of the gill filters. The low food concentrations (below about 0.1 mg/liter) at which tadpoles cease to feed is presumed to reflect an autoregulatory mechanism that avoids energy loss during attempts to harvest scarce resources. Feeding cessation also provides a concentration refugium for food items so that scarce items are not eliminated. Work with a population of *Rana catesbeiana* in a natural pond showed that phytoplankton populations responded reasonably well and their density stabilized at the concentrations that were predicted by the laboratory experiments (i.e., after algal blooms, the concentration gradually decreased and stabilized at about the refugium concentration predicted in the laboratory; Seale 1980). Viertel (1990, 1991) showed that tadpoles at early stages of development may have lower threshold concentrations (the lowest concentration at which feeding starts) than older tadpoles.

Farlowe (1928) assumed that tadpoles were indiscriminate suspension feeders and used intestinal contents as an index of the algal constituents of ponds. Jenssen (1967, *Rana clamitans*) and Hendricks (1973, *Rana pipiens*) found that foregut contents of tadpoles and water column constituents were indistinguishable. Tadpoles also graze on periphyton and epiphytic algae (Dickman 1968; see L. M. Johnson 1991; Kamat 1962; and Sahu and Khare 1988 for other studies of tadpole feeding habits), epibenthic algae (Calef 1973), and their own fecal pellets (Steinwascher 1978a). The issue of selection among these patchy food sources has not been addressed, but C. L. Taylor et al. (1995) showed that tadpoles can choose among foods of different nutritional qualities.

During an examination of ingestion rates of five larvae (*Bufo americanus, B. woodhousii fowleri, Pseudacris crucifer, Rana catesbeiana,* and *R. sylvatica*), Seale and Beckvar (1980) assumed that ingestion rates were indicative of food preference. They demonstrated some differences among species that were fed on single algal cultures but found no differences with *R. catesbeiana* when six species of green and blue-green algae were offered. In all cases, tadpoles appeared to adjust feeding rate to food volume or biomass (see Seale and Wassersug 1979). Tadpoles of *Pseudacris crucifer* had a much higher threshold concentration (= food concentration below which feeding ceases) than did those of *B. w. fowleri* or *R. catesbeiana* when these species were fed a filamentous blue-green alga (*Anabaena spherica*). These findings suggest that the concentration range over which food is available may differ among tadpoles and may affect selection of food types (e.g., Anholt and Werner 1995). Studies of the microspatial distribution of potential tadpole foods in ponds must accompany future laboratory selection experiments. Ponds that experience intense algal blooms (e.g., Seale 1980) may not present opportunities for selection among food types.

Many authors have suggested that microhabitat use and temporal partitioning (e.g., Degani 1982; Heyer 1976a; Inger 1986; Löschenkohl 1986; chap. 10) reduce interspecific competition for food among tadpoles. Laboratory measurements of growth rates in *Hyla gratiosa* versus *Hyla femoralis* indicate that one species may decrease the growth rate in another and that this effect is variable among sibships (Travis 1980b). Although Steinwascher (1978b) detected both chemical interference and behavioral exploitation competition in laboratory experiments with *Rana sphenocephala,* behaviors associated with exploitation competition have not been rigorously examined. Larger tadpoles probably push smaller tadpoles away from food patches, although this would most likely occur only with particulate food sources and at high tadpole densities relative to food availability (Travis 1980b; Werner 1994). Also, a tadpole of one feeding mode (e.g., rasps and swims actively) may physically interfere with a tadpole of another feeding type (e.g., suspension feeder remains quiescent). Chemical interference may have no obvious behavioral correlates beyond preferential associations or avoidance (see Social Behavior below). R. A. Griffiths (1991) noted that competition mediated by nutrient depression may occur between species in the absence of physical interactions.

Respiration

Tadpoles respire through the skin, gills, and lungs (Respiratory Physiology, chap. 8). Most work on tadpole respiratory behavior has focused on the rates of ventilation with different oxygen availabilities, development of lung use, partitioning air and water breathing, trade-offs between respiration and feeding, and the relationship between respiration and locomotor performance (e.g., Watkins 1997). Tadpoles that use both gills and lungs respond to changes in dissolved oxygen (DO) concentration by varying the rate of air breathing. To breath air, tadpoles characteristically swim to the surface, quickly fill the buccal cavity with air, and rapidly swim down again (= "bobbing" of Wassersug and Seibert 1975; S. Wong and Booth 1994). Bobbing rates are consistent with the state of lung development in tadpoles of *Bufo woodhousii, Pseudacris triseriata,* and *Rana pipiens;* these forms showed a significant negative correlation between DO concentration and bobbing rate once DO was below a critical value. The bobbing rate of *Spea bombifrons* was negatively correlated with DO throughout the range of DO concentrations used. Air bubbles taken into the buccal cavity may provide oxygen by absorption through the operculum, filter apparatus, and walls of the buccal cavity. The tadpoles of *Bufo woodhousii,* essentially lungless until metamorphosis, may not take atmospheric air into their buccal cavities but may come to the oxygen-rich surface water for longer periods of time to take in oxygen through the skin (chap. 8).

Feder (1983a) noted increases in lung ventilation as a response to aquatic hypoxia in *Rana berlandieri,* and West and

Burggren (1983) found the converse; forced lung ventilation per se (as well as the resulting increase in PO_2) contributed to the suppression of gill ventilation in *Rana catesbeiana,* although the lungs contributed relatively little to gas exchange at early developmental stages. As tadpoles approach metamorphosis, the lungs increase in importance because of increased oxygen demands, and aquatic respiratory behavior (gill ventilation rate) in *Rana pipiens* increased with DO concentration in tadpoles at Gosner 41 (Wassersug and Seibert 1975). Differential ability to use aerial and aquatic oxygen resulted in interspecific differences in locomotor stamina (time to fatigue) under conditions of hypoxia (Wassersug and Feder 1983). Comparisons of lungless *Bufo americanus* versus *R. berlandieri* and *Xenopus laevis* (which use both lungs and gills; see Pronych and Wassersug 1994) showed that air-breathing tadpoles could moderate the reduction in stamina seen in all species in hypoxic water. Breathing air also reduces stamina because of increased buoyancy (*X. laevis,* Wassersug and Feder 1983; Campeny and Casinos 1989), increases exposure and risk of predation (Feder 1984), and reduces growth rate because of the metabolic cost of surfacing to breathe (Pandian and Marian 1985c).

Ontogenetic changes in buoyancy were described for five species (*Bufo americanus, Hyla chrysoscelis/versicolor, Pseudacris triseriata maculata, Rana pipiens,* and *R. sylvatica*) at hatchling, larval, and early metamorphic stages by Gee and Waldick (1995). Hatchlings were negatively buoyant and sessile; their lungs were not inflated. All but *R. pipiens* were neutrally buoyant through most larval stages and active in midwater; *R. pipiens* larvae were negatively buoyant and active or sedentary near the bottom. All species were negatively buoyant and benthic in metamorphic stages. Gee and Waldick (1995) showed that the tadpole lung has a hydrostatic role and some species were able to regulate precisely their buoyancy to compensate for environmental changes. Campeny and Casinos (1989) examined buoyancy in tadpoles of the midwife toad, *Alytes obstetricans* (see Feder and Wassersug 1984, Pronych and Wassersug 1994, and Snetkova et al. 1995).

Suspension-feeding tadpoles use the buccal pumping mechanism for both food and oxygen uptake. Feder et al. (1984b) found that tadpoles of *Xenopus laevis* increased buccopharyngeal respiration in hypoxic water at the expense of feeding when prevented from breathing air; food ingestion decreased even though buccal pumping rate increased, and tadpoles likely actively suppressed mucus secretion to reduce food entrapment rate. Wassersug and Murphy (1987) noted that aerial respiration promoted growth in obligate suspension feeding *Xenopus.* Feder et al. (1984b) speculated that the evolution of air breathing in aquatic vertebrates occurred as the result of selection for the complete dedication of the buccopharyngeal surfaces to feeding.

Thermoregulation

Although tadpoles often aggregate in warmer areas of ponds and sometimes follow changing patterns of sunlight (Brattstrom 1962; Schley et al. 1997), the mechanism for temperature selection has not been identified (also see Behavioral Thermoregulation in chap. 8). There have been numerous speculations on its adaptiveness, but there is no consensus that any factor(s) (e.g., developmental changes, energy metabolism, gas exchange, osmoregulation, predator avoidance) adequately explains the observed behavioral temperature selection in any species. Most of the hypotheses reviewed by Brattstrom (1979) remain untested. Lefcort and Eiger (1993) discussed how a fever caused by a pathogen influenced tadpole behavior and may influence pathogen transmission.

Cupp (1980) found interspecific, geographical, and ontogenetic differences in the critical thermal maxima (CT_{max}; *Bufo woodhousii, Gastrophryne carolinensis, Rana catesbeiana,* and *R. sylvatica,* but not in *Pseudacris triseriata*). Of those that showed changes in the CT_{max}, the highest temperature was tolerated by tadpoles in Gosner 30–40. Dupré and Petranka (1985) noted similar ontogenetic trends in final thermal preferendum (FTP; mean temperature selected by tadpoles in thermal gradients) of several tadpoles including *Pseudacris triseriata;* the FTP peaked sharply at premetamorphic stages and did not track precisely the development of CT_{max} noted by Cupp (1980) and Sherman (1980). Dupré and Petranka (1985) speculated that behavioral temperature selection reflects changes in thermal sensitivity of cutaneous cold-sensitive neurons (also see Crawshaw et al. 1992; Hammerschlag et al. 1989; Wollmuth and Crawshaw 1988; Wollmuth et al. 1987).

Indeed, the many ontogenetic changes in the amphibian nervous system (Kollros 1981) include changes in the temperature at which static impulse frequency occurred in cutaneous cold-sensitive neurons (*Rana catesbeiana;* Dupré et al. 1986). Tadpoles at younger stages were not as selective of temperature as older tadpoles, even though developmental trends in static impulse frequency of individual neurons and behavioral temperature selection were similar. Static impulse frequency usually occurred at somewhat lower temperatures than those that were behaviorally selected (Dupré et al. 1986), and the development of additional warm-sensitive neurons may also influence temperature selection. Such neurons have not yet been identified. Other aquatic ectotherms (including salamander larvae) modulate temperature selection in response to hypoxia (Dupré and Wood 1988), but this has not been examined in anuran larvae.

Locomotion

Anuran larvae are not shaped like any other aquatic vertebrate. Tadpoles have globose bodies, no lateral fins, and a relatively long, muscular, flexible tail that comprises most of the body length (Hoff and Wassersug 1985; Wassersug 1989a; chap. 3). This shape suggests that tadpoles swim differently from more typical aquatic vertebrates, such as fishes, and often has led to the assumption that tadpole swimming is relatively ineffective and uncontrolled. Recent studies have shown that many aspects of the swimming movements of tadpoles do not differ greatly from those of typical subcarangiiform fishes, such as trout (Wassersug and Hoff 1985).

In the sense of large fishes that swim continuously (e.g., tuna), steady swimming in tadpoles does not exist. Although tadpoles have impressive endurance while swimming against a slow current in a circulating flow tank (i.e., more than 90 min for a *Bufo* tadpole; Wassersug and Feder 1983), reported observations of tadpole swimming generally fall into two categories: rapid bursts to evade predators, or very slow swimming between patches of food or in midwater aggregates (see Social Interactions below). Occasionally tadpoles swim for more than a few seconds without being actively pursued (e.g., aggregations of tadpoles following patches of sun across a pond; Brattstrom 1962, R. J. Wassersug, personal communication). Even in these cases, steady swimming of undisturbed tadpoles is best characterized as a series of short swimming bouts. Dudley et al. (1991) provided information on the effects of shape and limb formation on drag forces during swimming in *Rana catesbeiana* tadpoles, and Boothby and Roberts (1995) examined the effects of tactile stimulation on escape swimming responses in *Xenopus laevis* hatchlings. Parichy and Kaplan (1995) examined the functional consequences of developmental plasticity on locomotor performance on hatchlings. Denver (1997) addressed the concept of hormonal changes caused by environmental stress and their effects on metamorphosis. This form of plasticity has seldom been considered when discussing variations in metamorphic timing and its effects on behavior such as locomotion.

For this presentation, we have combined steady and periodic swimming and consider only swimming bouts that last at least three locomotor cycles (tail beats) and show little acceleration or deceleration. Assisted by high-speed cinematography and computer-assisted frame-by-frame analyses, laboratory studies of swimming have focused on quantitative descriptions of swimming kinematics. These descriptions were compared to the swimming of other aquatic vertebrates (fishes: reviews by P. W. Webb 1975, 1984a, b). Oxner et al. (1993) presented a mathematical model of body kinematics in swimming tadpoles.

Tadpoles use lateral undulation of the body to propel themselves through the water (Hoff and Wassersug 1985; Wassersug and Hoff 1985). Waves of bending initiated anteriorly are propagated to the end of the tail (Hoff 1986a) exactly as in other undulating aquatic animals (see P. W. Webb 1975). For lentic bufonid and ranid tadpoles, a linear relationship exists between the frequency of the tail beat and the swimming velocity. When velocity is measured in body lengths per second, swimming of all sizes of free-swimming tadpoles up to just before metamorphosis can be described by the same regression equation. Chovanec (1992) compared swimming velocity in *Bombina bombina, Bufo bufo, Hyla arborea,* and *Rana dalmatina* at early (26–28), middle (30–33) and late (39–40) Gosner stages. Species differed in velocity and activity; at intermediate and late stages of development, *Bufo* and *Hyla* had high activity with low cruising speeds, especially *Hyla,* which showed the lowest cruising velocities. *Rana* usually swam in short bursts with high velocities, and *Bombina* showed intermediate patterns. Chovanec (1992) evaluated these patterns with regard to predation by dragonfly naiads (see Sensory Abilities, chemical communication, later in this chapter). In the Wassersug and Hoff studies, the highest steady velocity was about 12 body lengths/sec.

The maximum amplitude of a tadpole tail beat (measured at the tip) increases gradually as swimming speed increases to an asymptotic value of about 0.25 body lengths. Tadpoles with relatively longer tails (e.g., *Rana clamitans* or *R. septentrionalis* compared to *B. americanus* or *Rana catesbeiana*) had a relatively lower maximum amplitude, but all amplitudes were well within the range of steady swimming recorded for many species of fishes. Because of the odd shape of pond-type tadpoles (i.e., heavily weighted toward the front end as compared to many streamlined fishes), a head wobble is apparent during swimming. Subsequent study (Hoff and Wassersug 1985) has shown that these standard kinematic measures that relate to efficiency and control (e.g., maximum amplitude of the traveling wave at all points along the body) are similar among tadpoles of several bufonid and ranid species and subcarangiiform (i.e., troutlike) fishes. Indeed, the apparent head wobble is in large part an illusion created by the snout moving from side to side. The pivot point for this motion is on the sagittal line between the otic capsules (Wassersug and Hoff 1985) and not more posteriorly at the center of the body mass. This suggests that tadpoles have some mechanism for adjusting that point of minimum amplitude (i.e., minimum lateral displacement) and may invest considerable energy in maintaining stability during swimming. It is reasonable to expect that the point of minimum lateral displacement would be in the vicinity of the semicircular canals so that equilibrium is not effected by the side-to-side movements of the head. Wassersug (1992) showed that a tadpole can bend its body vertically to initiate upward and downward movements.

The mechanical efficiency of tadpole swimming, measured as the relationship between tail beat frequency and swimming speed, falls near the middle of the range for all fish species described (Hoff and Wassersug 1985). In a more telling comparison of the distance traveled per tail beat among the "average" tadpole, the "average" fish, and ambystomatid salamander larvae, Hoff et al. (1989) showed that anuran tadpoles traveled farther per tail beat than did salamander larvae but not quite as far as the "average" fish. The great similarity between tadpoles and many fish species suggests that mechanical design constraints on moving through water by lateral undulations are very strong. Although tadpoles do not cruise often or quite as well in all regards as the best cruising fishes, the mechanical similarities are more remarkable than the differences.

Tadpoles that maintain position in midwater (e.g., *Xenopus laevis;* Hoff and Wassersug 1986) present a different mechanical issue. Some aspects of their swimming mechanics are consistent with the argument that maintenance of stability (lack of side-to-side movements) of the head is conducive to increased feeding efficiency in obligate suspension feeders. Up to about 6 body lengths/sec, *Xenopus* tadpoles do not use their entire tail for propulsion; they use only the flagellum at the end of the tail while holding position (swimming down against their own buoyancy) and gradually recruit greater lengths of the tail to increase swimming speed. The bending wave on the tail is not initiated at the base of the tail until the tadpoles are swimming moderately rapidly (6 body lengths/sec); at these speeds the tadpoles probably cease feeding. Restricting lateral movement to the end of the tail, combined with a midventral keel-like extension of the tail fin onto the body, probably contribute greatly to maintaining anterior stability (Hoff and Wassersug 1986). At higher swimming speeds, *Xenopus* tadpoles increase the tail beat frequency of the entire tail to increase swimming speed much like more benthic species (Wassersug and Hoff 1985); at these speeds their heads have the characteristic tadpole wobble.

Burst swimming used for predator evasion is another mode of swimming. Feder (1983c) reported that if *Rana berlandieri* tadpoles pursued by turtles could extend the chase (in some cases for 11 sec or more), the likelihood of their evading predation would increase. Those experiments were conducted in a large open pool with no cover for the tadpoles and may illustrate a threshold for the attention or the sensory ability of the turtle rather than any measure of the capacity of anuran tadpoles to evade predators. Increasing plant density and cover (= habitat heterogeneity) has significant positive effects on the survival of *Bufo terrestris* in the presence of visual-hunting odonate naiads and giant water bugs (Babbitt and Jordan 1996; also see Brönmark and Edenhamm 1994 and Stauffer and Semlitsch 1993 for behavioral responses of tadpoles to fish). Tadpoles observed under more natural conditions (i.e., with cover sites or bottom substrate) make short dashes followed by periods of immobility (Caldwell et al. 1980). During these predator-avoiding dashes, tadpoles move with speeds comparable to fishes of the same size range (2–10 cm total length; P. W. Webb 1978). In studies of electrically induced startle responses, Hoff (1986b, 1988b) found that maximum velocity in a short dash was correlated with the proportional and absolute amounts of axial muscle among tadpoles (i.e., *Bufo americanus, Rana catesbeiana, R. clamitans, R. septentrionalis,* and *Xenopus laevis*) that varied in size and shape. Although the performances of size-matched tadpoles were broadly overlapping, the performance ranks correlated with the muscle mass; *Rana septentrionalis* was the fastest tadpole with 26% of its mass as axial muscle and *Xenopus laevis* (3.5–4.0 cm TL) was the slowest with 16% of its mass as axial muscle. Most fishes have greater amounts of axial muscle compared to any tadpole examined (Hoff 1988b). This disparity suggests that muscle mass alone is not what permits comparatively good burst swimming performance; overall body shape may be an important factor. The maximum absolute velocity recorded (Hoff 1986a, 1988b) was 104 cm/sec for a large tadpole, and the highest relative velocity was 29 body lengths/sec for a small tadpole (about 2 cm long). The range in locomotor performance was very large for all species, and many individuals did not respond by swimming rapidly. Maximum electrically inducible speeds may not be indicative of the utility of speed alone to evade predators. Even at these surprisingly high maximum speeds, it is clear that most tadpoles and small fishes would still be consumed. Their speeds are inadequate when compared to the recorded velocities of feeding events by relatively large, potential predators of tadpoles (i.e., lunging and gape-and-suck feeding in several fish species). The ability to evade a predator likely depends on the sensory capabilities of the predator and prey and the complexity of the habitat and availability of cover, as well as the locomotor abilities of a tadpole.

The neuromuscular control of locomotion in tadpoles is not completely understood. Variations in the distribution of peripheral nerves in the tails of *Rana* and *Xenopus* (K. Nishikawa and Wassersug 1988, 1989) may relate to patterns of muscle recruitment that vary with swimming speed. *Xenopus* has a long spinal cord and many nerves branching (presumably from different motor neurons) along its entire length that apparently innervate individual (at least few) muscle segments (see chap. 6). *Rana* tadpoles have a cauda equina branching (at the point where the spinal cord ends in the adult) into large nerves traversing (and possibly innervating) many muscle segments. The patterns of innervation are correlated with the observed ability of *Xenopus* to initiate waves of bending anywhere along the tail as they increase the portion of the tail used with increases in swimming speed, rather than increasing the frequency of the tail beat, as was shown for tadpoles of *Bufo* and *Rana* (Hoff and Wassersug 1986). *Rana* tadpoles seem to use their tails in larger blocks or functional units (half the tail or the whole tail) and initiate waves of bending only at the base or about at the middle of the tail.

Waves of bending pass posteriorly as the animal moves forward. It has been presumed that these traveling waves of bending are accompanied by waves of muscle contraction (e.g., J. Gray 1936), but Blight (1976) suggested that at least three patterns of muscle contraction can produce traveling waves of bending. Electromyographic studies of slow swimming in several amphibians (Blight 1976, larval newts; A. Roberts et al. 1984, anuran hatchlings; and Stehouwer and Farel 1980, 1981, tadpoles) and other aquatic vertebrates (Grillner 1974, dogfish; Grillner and Kashin 1976, carp) showed that ipsilateral axial muscles can contract serially to produce a wave of muscle activity that travels with the bending wave. Examination of other swimming activities (e.g., starting, stopping, very slow and very fast swimming) in *Rana catesbeiana* showed that traveling waves of bending can be produced when contralateral muscles are acting simultaneously and also can be transmitted passively (i.e., without muscle activity in the posterior tail) if the wave is initiated anteriorly. It is not clear whether this range of locomotor mechanisms is specific to *Rana catesbeiana* or to anurans, or if it is a general feature of vertebrates that use undulatory swimming.

L. Muntz (1975) described five stages in the behavior of *Xenopus* larvae relative to trunk myogenesis: nonmotile (Nieuwkoop/Faber 20–22; myotomic muscles start to differentiate), premotile (stage 22–24; first striated fibrils visible, contractions possible), early flexure (stage 24–27; reflexes and peripheral nerves present), early swimming (stage 28–33), and free swimming (stage 32–46; coordinated swimming possible, muscle cells uninucleate). Muscle cells did not become multinucleate until metamorphosis (Nieuwkoop/Faber 48–50). The behavioral development of leg muscles was similar but, as expected, occurred at later stages: nonmotile (Nieuwkoop/Faber 48–52; differentiation minimal, nerves present), premotile (stage 53–54; limb trembles and cell striations forming), motile (stage 55–58; limb can make stepping movements, cells striated and multinucleate), and fully functional (stage 60–63). Hughes and Prestige (1967) studied the behavior of the hind limb during development.

Some tadpoles move in ways other than by swimming. Suctorial tadpoles occur in a number of families and have large oral discs (e.g., Altig and Brodie 1972; *Ascaphus truei*). The large disc allows the tadpole to move about and feed on rock substrates in high-energy streams without loosing contact (C. L. Taylor and Altig 1995; C. L. Taylor et al. 1996). Having the belly modified into a sucker (e.g., Ranidae: *Amolops, Huia,* and *Meristogenys;* Bufonidae: *Atelophryniscus, Atelopus,* some *Bufo*) apparently allows inhabitation of faster currents (Altig and Johnston 1989), but it is the extension-retraction cycles of the oral disc that provide the locomotor force. Like suctorial and other rheophilous tadpoles, those with abdominal suckers can move across the substrate by extension-retraction cycles of the oral disc (= oral hitching of Altig and Brodie 1972). Although some stream tadpoles can swim powerfully (e.g., Formas 1972; *Telmatobufo australis*), there are no tadpoles that continually occupy the water column in fast flowing water like some fishes do. Some nonsuctorial tadpoles that live in flowing water escape the current by hiding within crevices, rocky substrates, or leaf packs. These tadpoles often do not have the typical morphology associated with stream inhabitants; they are often fusiform (= vermiform) with less robust tail muscles and low fins. It is unlikely that tadpoles of this morphotype swim against currents (Wassersug and Heyer 1983 and citations therein; R. Altig, personal observation).

Tadpoles of the family Centrolenidae are stream dwellers with fusiform bodies, long tails and low fins, poorly developed eyes, and highly vascularized and lightly pigmented skin. Villa and Valerio (1982) reported collecting specimens in soft, spongy earth of a stream bank, in moist earth beneath a plant 3 m from the stream, and in leaf packs in the stream. They called them "fossorial" and speculated that the long tails were useful in pushing the fusiform body (= burrowing) into mud and leaf packs. We have collected centrolenid tadpoles of several species and most often found them in dense leaf packs and decaying organic material along streams, sometimes above the waterline. The tadpole of *Staurois natator* (Inger and Tan 1990; Inger and Wassersug 1990) is a morphologically similar burrowing tadpole from streams in the lowlands of Borneo. In addition to being strongly convergent on the centrolenid morphology, this tadpole has a large, uninflated lung (does not breathe air) and, like centrolenid tadpoles, probably relies heavily on cutaneous respiration; in this respect it differs from other elongate tadpoles.

Other elongate tadpoles live in bromeliads and do very little swimming (table 3 in Lannoo et al. 1987). The tadpole of *Osteopilus brunneus* has a tail to body ratio of 3.9 and probably is the most elongate tadpole known. Lannoo et al. (1987) showed that *O. brunneus* tadpoles swim very little, ineffectively (i.e., they have to beat their tails twice as fast as a *Rana* or *Xenopus* larva to achieve the same velocity), and with excessive lateral motion indicative of low kinematic efficiency. Three explanations had previously been offered to account for the very long tail of *O. brunneus* tadpoles (and certain other arboreal, bromeliad dwelling forms as well): cutaneous respiration, locomotion in a viscous medium, and an agitator to aerate the water. Lannoo et al. (1987) reviewed each of these postulates and found them somewhat unsatisfactory. As a result, they offered a fourth: the tail of *O. brunneus* serves as a static postural organ to keep the head directed upward and nearer the water surface. They suggested that the lungs serve for buoyancy and assist in this behavior. Pulmonary respiration is important to the tadpole of *O. brunneus,* and this posture saves the tadpoles energy by them not having to swim to the surface. Also a heads-up position could provide the longest tadpoles with an advantage in getting to eggs laid by females at the water surface (see previous oophagy discussion).

Oldham (1977) described terrestrial locomotion in the tadpoles of two West African frogs, *Chiromantis rufescens* and *Leptopelis hyloides* (Hyperoliidae). *Leptopelis* deposits eggs in a "nest" dug up to 6 cm beneath the surface in moist, soil close to water; there the eggs hatch and the tadpoles wait for rain. When the nest area floods, frantic wiggling by the group of tadpoles creates a water-filled cavity. The roof of the cavity eventually collapses and the tadpoles "burrow" to the surface. The tadpoles have unusually long tails with many muscle bundles per linear unit and are able to reach water by wiggling like eels in very shallow water across the substrate (Oldham 1977). In contrast, *Chiromantis rufescens* builds foam nests on vegetation, usually above water; when the nest disintegrates, tadpoles drop into the water and carry on as typical pond larvae. Sometimes, the nest is not above water and the tadpoles have to make their way across wet soil to water. The tail of *C. rufescens* tadpoles is relatively shorter and less muscular. In a different scenario, Caldwell and Lopez (1989) reported that tadpoles of *Leptodactylus mystaceus* also make foam in their terrestrial nests to avoid dehydration.

Under experimental conditions designed to duplicate hatching conditions, *Leptopelis hyloides* larvae traveled further and faster and always moved on their bellies while propelling themselves with lateral undulations. Tadpoles of *Chiromantis rufescens* used less flexure in wiggling and moved more rapidly by landing on their sides and "flipping" with tail movements much like a fish out of water, and some larvae of *Mantidactylus* spp. also move effectively in this manner (R. Altig, personal communication). *Leptopelis* larvae were able to survive and wiggle toward water longer than *Chiromantis* in the same trial conditions. Both species showed some ability to modify the rate of movement with changes in surface moisture and relative humidity, and *Leptopelis* showed some ability to move directionally, perhaps through geotaxis. The sinuous wiggling locomotion of *Leptopelis* larvae was correlated with the longer, more muscular tail and seemed to work better under the conditions tested. Oldham (1977) suggested that the probability of reducing aquatic predation on the defenseless egg stage and having a relatively large and well-advanced tadpole arrive in the pond constituted the survival value of terrestrial eggs. In these experiments, the tadpoles of *Bufo maculatus* used for comparison had no ability to move on land. Wager (1986) reported that tadpoles of a related species from South Africa, *Leptopelis natalensis,* after hatching and emerging, moved toward water as a group.

Another pattern of locomotion characteristic of larvae with long, low-finned tails, well-developed eyes and hind limbs, cryptic coloration, and unusual mouthparts (see chap. 12) we call "tail flipping." This type of movement is associated with tadpoles called "subaerial" by Wassersug and Heyer (1983) and "semiterrestrial" by Altig and Johnston (1989). In most other situations, we would give priority to the earlier name, but for obvious reasons we use "semiterrestrial" here. These tadpoles use their tails to flip around on wet, exposed rocks in streams, in films and trickles of water flowing over rocks or from seepages, and on stems and leaves and between axils of plants. Wassersug and Heyer (1983) examined the tadpoles of *Cycloramphus* and *Thoropa* (Leptodactylidae) and noted their similar morphologies (e.g., elongate and relatively finless tail, depressed body and branchial baskets, compressed and deep jaw sheaths). These tadpoles are obligate air breathers found on vertical rock surfaces where water flow is negligible. Some of these have a flattened abdomen (a possible precursor to a belly sucker) presumably to increase surface area for cohesion and as an adjunct to lung ventilation.

Drewes et al. (1989) described the tadpole of *Arthroleptides martiensseni* (Ranidae) and commented on the striking morphological and ecological convergence between their *Arthroleptides* tadpoles and those of *Cycloramphus* (Heyer 1983a, b) and *Thoropa* (Cocroft and Heyer 1988). *Arthroleptides martiensseni* tadpoles were collected from vertical or steep faces of smooth rocks in the stream but outside the main current; many tadpoles were in fissures. These authors noted the large eyes of these tadpoles and found that the hind legs of this species developed much more precociously than those of tadpoles of *Bufo* or *Rana.* When disturbed, tadpoles moved haphazardly by tail flipping, sometimes for a distance of 1–2 cm per flip.

The obvious convergences on the semiterrestrial morphology in tadpoles from several families are notable. Tadpoles of *Nannophrys ceylonensis* (Ranidae; Kirtisinghe 1958) also are morphologically similar to those of *Arthroleptides, Cycloramphus,* and *Thoropa* and occur in shallow sheets of water in seeps. The strikingly similar tadpoles of *Indirana beddomei* were described by Annandale (1918) as skipping rapidly over damp rocks.

Fig. 9.6. Semiterrestrial tadpole of *Petropedetes parkeri* (Ranidae). Tadpoles were found climbing on slime-covered rock faces along a road cut through a forest in Cameroon, West Africa.

Blommers-Schlösser and Blommers (1984) noted that tadpoles of the Malagasy frogs in the *Mantidactylus pulcher* group can move from one axil of *Pandanus* to another by violent wiggling; they also have elongate, depressed bodies and low fins on a long tail. Lawson (1993) reviewed the natural history of the tadpoles of the Cameroon species of *Petropedetes* (Ranidae). In general, they are elongate tadpoles with long tails, well-developed hind legs that apparently develop early in the larval stages and small heads. Lawson (1993) noted that eggs are laid on the undersides of leaves on low shrubs and trees along rock and pebble banks of seasonal streams. On hatching, the tadpoles squirm along the branches for some time before dropping to the ground below. In his figure 19, Lawson (1993) showed tadpoles of *P. cameronensis* on stems and a leaf and a tadpole of *P. parkeri* with well-developed hind limbs and a long tail sitting on a leaf; *P. parkeri* tadpoles were found crawling on slime-covered, vertical rock faces covered by trickling water (fig. 9.6), and *Petropedetes* tadpoles were found at several localities in the Korup National Park actively moving on the forest floor hundreds of meters from the nearest water. Kunte (1998) observed ecologically similar tadpoles in India that were probably *Indirana leithii.* These tadpoles tried to escape the predatory strikes of the snake *Ahaetulla nasuta* by hopping into cracks and depressions in the surrounding rocks.

Tadpoles of *Hoplophryne rogersi* and *H. uluguruensis* (Noble 1929) live in the damp crannies between leaves of wild bananas or within split bamboo stems and have a bilateral pair of fingerlike, fleshy projections on the ventrolateral margin of the body at the posterior end of the buccopharynx. These projections move upon contraction of the subhyoideus muscle (chap. 3), and Noble suggested they give an added "kick" to assist the tadpoles as they wiggle from one moist axil to another. It is interesting that the tail of a well-formed embryo is longer than the head and body and wraps across the ventral surface of the developing embryo; hind limb buds are clearly visible in these tadpoles before they hatch but external gills are not (Noble 1929). Recently hatched larvae of *H. uluguruensis* have well-developed, inflated lungs that extend two-thirds the length of the body cavity; their size implies that pulmonary respiration is important for these tadpoles (Noble 1929).

It is interesting to speculate about the origin and adaptive nature of patterns of tail-flipping/semiterrestrial locomotion. Undoubtedly, the precursor of this behavior can be found in more typical tadpoles but little information is available. M. A. Smith (1916) noted that the carnivorous tadpoles of *Occidozyga* and *Phrynoglossus* lie quietly for long periods, seldom move with their tails, and push themselves about with their precocial hind legs. Perhaps other examples are in the literature but have not been evaluated in this regard. Our understanding of the evolutionary interplay between the breeding ecology of most frogs and the feeding and locomotion of their tadpoles is elementary. A cursory review of the literature on more specialized patterns of breeding (Duellman and Trueb 1986), especially on frogs that deposit eggs in relatively small and potentially ephemeral habitats such as phytotelmata, suggests fruitful areas of investigation. The convergent evolution of similar morphology (e.g., body shape, tail length, oral structures), physiology, locomotion, and feeding in different lineages in similar habitats is striking. What are the trade-offs and patterns of convergences between feeding and locomotion in these different habitats in which tadpoles find themselves? For example, what is the evolutionary history of the independent trends toward increasing tail length, dependency on pulmonary respiration and development of semiterrestrial/tail-flipping locomotion, and changes toward macrophagy as exemplified by carnivory, cannibalism, and oophagy? Surely, more tadpole watching will provide insight and answers to some of these questions and bring other morphological and behavioral correlates to light; we encourage such investigative efforts.

There is considerable information on the structure and function of hatching glands (chap. 3), but few observations have been made across species to identify potential differences in the timing and actual hatching process and behavior of the embryo (e.g., Bradford and Seymour 1985; Gollman and Gollman 1993a; Kothbauer and Schenkel-Brunner 1981; Lutz 1944a; Noble 1926a; Yoshizaki and Yamasaki 1991). Thrashing about by embryos may enhance hatching of other embryos in a clutch of eggs with high water turgor. Physical features like pH (e.g., Dunson and Connell 1982) and oxygen tension (e.g., Petranka et al. 1983) also influence hatching; one of us (R. Altig, personal observation) has seen several cases where the surface-film eggs of *Hyla cinerea* became encased in a mat of algae in a particularly eutrophic pool. These embryos died after reaching Gosner 25 because they could not escape the jelly membranes.

Behavior of hatchlings is largely unknown, but specific activities that have not been studied must occur in those frogs that have larval transport or some other form of larval brooding. Hatchlings of some endotrophs (e.g., G. J. Ingram et al. 1975, *Assa;* Brauer 1898, *Sooglossus sechellensis*) and exotrophs with larval transport (e.g., De Pérez et al. 1992a, dendrobatids; Inger and Voris 1988, Inger et al. 1986b, *Limnonectes microdisca*) must respond by negative geotaxis, pheromones, or a combination of these and other cues to arrive at the proper place in or on the parent.

Dispersion and Social Behavior

Habitat Selection

Numerous physical (e.g., distance from shore, oxygen concentration, substrate qualities, water depth and flow rate, site duration, and temperature; discussed by O'Hara 1981) and biological (e.g., presence and distribution of vegetation, other tadpoles, other organisms, and the phenology of all organisms) factors influence the spatial and temporal distribution of tadpoles among microhabitats. Tadpoles may select habitats because of attraction to or avoidance of conspecifics and predators.

Temperature often influences the behavior and ecology of tadpoles (review by Duellman and Trueb 1986; chaps. 8 and 10). Temperature influences tadpole activity patterns, growth and development, metabolic rate, and the timing of metamorphosis (Kollros 1961; Smith-Gill and Berven 1979; Wilbur and Collins 1973). Rapid growth and development are especially important for species that live in ephemeral habitats, such as in deserts where pools dry up quickly or at high elevations where ponds may freeze (A. R. Blaustein 1988; Hokit and Blaustein 1995; O'Hara 1981; Skelly 1996). In nature, the gathering of tadpoles (e.g., *Bufo, Hyla, Pseudacris, Rana,* and *Scaphiopus*) in the warmest thermal gradients suggests selection of favorable temperatures for rapid growth (e.g., Ashby 1969; Beiswenger 1972; Bragg 1968; Brattstrom 1962, 1963; C. C. Carpenter 1953; Mullally 1953; O'Hara 1981; Tevis 1966). This assumption is supported by experimental studies (e.g., Beiswenger 1972; Beiswenger and Test 1966; De Vlaming and Bury 1970; Herreid and Kinney 1967).

Few experimental tests have demonstrated that tadpoles selectively respond to key features of the substrate. Altig and Brodie (1972) showed that tadpoles of *Ascaphus truei* preferred smooth rocks above 55 mm in diameter. J. A. Wiens (1970, 1972) showed that *Rana aurora* and *R. cascadae* tadpoles could be conditioned to prefer certain substrates, although the two species displayed opposite preferences in identical experimental regimes. *Rana aurora* tadpoles reared over featureless or square-patterned substrates showed no preference for striped or square-patterned substrates, but larvae reared on striped substrates preferred striped-patterned habitats. The preference for striped substrates was established during the first 14–17 days in the striped habitat. This preference was retained after isolation from the substrate and could be reestablished in both young and old tadpoles. *Rana cascadae* tadpoles reared in featureless or striped-pattern environments showed no preference for either square-patterned or striped substrates, but tadpoles raised in a square-patterned environment showed a significant preference for that regime. J. A. Wiens (1972) suggested that the different responses were because of differences in the substrates of the habitats in which these tadpoles are found. *Rana aurora* typically breeds in shallow ponds prone to summer drying and in areas where permanent ponds have overflowed (Storm 1960; J. A. Wiens 1970). Branches, and stems of emergent plants that are characteristic of these habitats cast a pattern of linear shadows on the bottom. The stripes that larval *R. aurora* responded to may resemble the linear substrates and shadows found in their natural habitat.

Methods similar to those of J. A. Wiens (1970, 1972) were used to test habitat selection in *Kaloula pulchra* (Punzo 1976) and *Bufo americanus, Rana clamitans,* and *R. sylvatica* (O'Hara 1974). *Kaloula* tadpoles reared in striped habitats preferred stripes, those reared in square-patterned habitats preferred squares, and those reared in featureless habitats displayed no preferences. Punzo (1976) suggested that tadpoles imprint on the habitat in which they were reared and that they select this type of habitat in nature. In contrast to studies by Wiens and Punzo, the rearing regime had no influence on the habitat choices of *B. americanus, R. clamitans,* and *R. sylvatica* (O'Hara 1974).

Results obtained for *Rana cascadae* tadpoles by J. A. Wiens (1972) and by O'Hara (1981) are contradictory. O'Hara (1981) tested the responses of tadpoles reared on a smooth tank bottom versus natural substrates (e.g., sand, gravel, and rock). Tadpoles of *Bufo boreas, Pseudacris regilla,* and *R. cascadae* preferred fine-grained over coarse-grained substrates. Tadpoles from different populations and wild-caught tadpoles versus those reared in the laboratory environment behaved similarly. *Rana cascadae* tadpoles were influenced by the rearing regime in the study by Wiens, whereas in O'Hara's experiments they were not. One must be cautious about interpreting results of experiments concerning habitat selection when the environment is quite artificial. Comparing experiments that used different techniques is difficult; these results may reveal real ecological differences in nature (see O'Hara 1974, 1981, and J. A. Wiens 1970 for discussions).

Tadpoles may use visual, tactile, and olfactory cues to distinguish among substrate types in field and laboratory tests (O'Hara 1981). Waringer-Löschenkohl (1988) showed that the presence of vegetation, as well as other species, caused a significant shift in the distribution of European tadpoles from the bottom of aquaria toward the water surface. Experimental results (Pfennig 1990a, b) suggested that tadpoles of *Spea multiplicata* select their habitat by using diet-based environmental cues; tadpoles reared on similar diets aggregated with one another; he also argued that larvae have increased growth and survival when they were restricted to their natal environment. J. A. Hall et al. (1995) found that larvae of *Spea intermontana* did not prefer only cues of their natal environment but were able to react preferentially to novel stimuli (larvae raised on different diets) in forming associations.

Social Interactions

Many aggregations of larval anurans form as a result of social attraction toward conspecifics or because of abiotic parameters. Through field observations and experimentation in the laboratory and field, the various modes of tadpole aggregative behavior have been assessed and classified (e.g. Bragg 1965; Caldwell 1989; Downie 1990a; Duellman and Trueb 1986; Lescure 1968; Wassersug 1973). Analyses of the geometric structure of tadpole aggregations (e.g., L. C. Katz et al. 1981; Potel and Wassersug 1981; Wassersug et al. 1981b), the sensory bases by which individuals maintain contact with conspecifics (e.g., A. R. Blaustein and O'Hara 1982a; Waldman 1985a; Wassersug et al. 1981b) and the selective pressures that may have influenced aggregation formation (e.g., A. R. Blaustein 1988; Waldman and Adler 1979; Wassersug 1973) have been investigated and discussed. Studies of the roles of competition and predation in aquatic communities in which tadpole aggregations are a significant component have provided important insight into how aquatic communities are structured (e.g., Morin 1981; Travis 1980b; Wilbur 1972, 1984; Woodward 1982, 1983; chap. 10).

Several authors (e.g., Beiswenger 1975; Bragg 1965; Caldwell 1989; Wassersug 1973) categorized tadpole aggregations qualitatively; Bragg (1965) described tadpole aggregations as being asocial or social and suggested that asocial groups form through association with environmental features (e.g., food patches or temperature gradients) that are attractive to individuals. Social aggregations form because of the attraction of individuals to conspecifics. Because of the difficulty in distinguishing between social versus asocial factors, Wassersug (1973) suggested a broad, functional classification of two types: a simple, taxic aggregation in response to some physical feature (e.g., temperature, light, current) was equivalent to the asocial aggregation of Bragg (1965), and a biosocial aggregation (= school) was equivalent to the social aggregate of Bragg (1965). Either type of aggregate may be polarized (tadpoles oriented in same direction) or nonpolarized. Tadpole schools are biosocial aggregates that display some synchronized movement patterns, but the degree of polarization within a school varies. Also, some tadpoles form slow-moving, polarized aggregations within which individuals may lie on the bottom of the body of water or be suspended motionless in midwater. Beiswenger (1975) also proposed a functional classification of tadpole aggregations focused on moving (streams and schools) and stationary (feeding and metamorphic) patterns.

Because of the difficulty in assessing whether aggregations observed in nature were simple or biosocial, M. S. Foster and McDiarmid (1982), O'Hara (1981), Wassersug (1973) and Wassersug and Hessler (1971) conducted a series of controlled laboratory experiments to test rigorously how and if tadpoles of various species were socially attracted to one another. These studies consisted of placing one tadpole of a particular species within each of four sections of a glass or plastic tray (M. S. Foster and McDiarmid 1982; Wassersug 1973; Wassersug and Hessler 1971) or an aquarium (O'Hara 1981). Each section was separated from the others by water-tight, transparent partitions so that tadpoles in each section had visual but not chemical contact with one another (see M. S. Foster and McDiarmid 1982 for results of chemical contact with the same design). Each section was marked into four equal quadrants (fig. 1 of Wassersug and Hessler 1971). If visually mediated social interactions among test individuals were random, then it was assumed that tadpoles would not associate in one particular quadrant over another. If tadpoles were positively attracted toward one another, they would be found most often in the quadrants where they could be as close to one another as possible. If tadpoles were avoiding one another, tadpoles would be found more often in the quadrant farthest from other individuals. The tadpoles of at least 10 species have been tested with these methods with variable results (see M. S. Foster and McDiarmid 1982 for a critical evaluation of these methods).

Tadpoles known to aggregate in nature generally preferred to associate in the quadrant closest to conspecifics (e.g., O'Hara 1981, Wassersug 1973, *Bufo boreas;* O'Hara 1981, *Rana cascadae;* Wassersug 1973, *Rhinophrynus dorsalis* and *Xenopus laevis;* Wassersug and Hessler 1971, *Xenopus laevis*). Tests of species that probably only occasionally form aggregations in nature, such as the ranids *Rana boylii, R. catesbeiana,* and *R. pipiens,* displayed a random distribution among the quadrants (Wassersug 1973). There was one discrepancy between the results obtained by O'Hara (1981) and Wassersug (1973). Wassersug (1973) found *Pseudacris regilla* tadpoles exhibited no biomutual attraction whereas O'Hara (1981) showed that they were mutually attracted to one another. *P. regilla* tadpoles form relatively loose, intermittent aggregations in nature and their aggregating tendencies seem to differ between populations (A. R. Blaustein 1988; O'Hara 1981). The different testing apparati used by Wassersug and O'Hara may have accentuated the apparently discrepant results. Of the ranids tested, only *Rana cascadae* tadpoles displayed a clear-cut biosocial attraction toward conspecifics. In nature, *R. cascadae* tadpoles are one of the most social of the ranids and form small, cohesive aggregations in ponds and lakes (A. R. Blaustein 1988; Hokit and Blaustein 1997; O'Hara 1981). Tadpoles of *Spea* do not display an attraction for conspecifics (Wassersug 1973) even though they form aggregations in nature (also see J. A. Hall et al. 1995). As Wassersug (1973) pointed out, these tadpoles may require more than visual cues in order to form aggregations or, as exemplified by *Pseudacris regilla* tadpoles, the tendency to form aggregation may vary among populations.

Based upon laboratory experiments and field observations, Wassersug (1973) described two basic modes of tadpole group formation (= schooling): the *Xenopus* mode and the *Bufo* mode, with comments on a third mode made up of schooling tadpoles that occur in ephemeral breeding sites. He judged that tadpoles within different families generally fit into one of these basic modes of schooling behavior. In the *Xenopus* mode, schools consist of clusters of strongly polarized individuals in midwater, and individuals avoid contact with one another. The *Bufo* mode is characterized by schools being polarized or not and hundreds to thousands of relatively slow swimming individuals in frequent contact that "appear as dense black mats in shallow water or near the bottom" (Wassersug 1973:285). The ephemeral group was characterized by highly motile, strong swimmers that occur in very active, closely packed schools; these tadpoles are often omnivorous in terms of food particle size and quality. Several species of pelobatid frogs (e.g., *Scaphiopus* and *Spea*), *Rhinophrynus dorsalis, Pyxicephalus adspersus,* and many Australian species fit this mode. Many schooling species remain to be categorized. Tadpoles of *Rana heckscheri* (Altig and Christensen 1981) move about in immense schools of several cohorts; tadpoles in front are feeding and those behind are moving and coming to the surface to breathe. A large school can be heard by the crackling noise caused by so many tadpoles surfacing. In laboratory tests, there was evidence of social facilitation in feeding. Tadpoles of several species of *Leptodactylus* aggregate in nonpolarized schools that extend from the bottom to the top of the water, and these tadpoles also are stimulated to follow small waves made by their attendant mother. Likewise, the nonfeeding tadpoles of *Nectophryne afra* follow vibrations caused by movements of the attendant male (Scheel 1970). Tadpoles of *Aubria subsigillata* form tight football-shaped clusters that appear to roll across the bottom (Schiøtz 1963). Tadpoles of *Pyxicephalus adspersus* promptly move through channels dug by the attendant males from drying parts of the habitat to remaining water (Kok et al. 1989). Tadpoles of *Pyxicephalus* may school with those of *Schismaderma* (R. Altig, personal observation; also see R. A. Griffiths and Denton 1993).

Caldwell (1989) suggested that it may be difficult to place tadpole schooling behavior in one of Wassersug's modes because of considerable variation within a species and potentially different modes displayed by other species. Because similarities in the behaviors that result in groups may occur in species that are not closely related, Caldwell (1989) suggested that these behaviors should be designated by types rather than by taxonomic names. According to her scheme, Type I behavior is illustrated by tadpoles that form loosely aggregated schools in shallow areas or on the bottom of ponds but rarely in midwater. These aggregations may or may not be polarized, and movement of the group is amoeboid and slow (e.g., *Bufo* and possibly some *Leptodactylus*). Type II behavior (e.g., *Xenopus,* some species of *Phyllomedusa* and some microhylids), is characterized by well-organized schools in midwater, and individuals are not in contact with one another. Movement is slow in these strongly polarized schools. Type III behavior (e.g., *Hyla geographica,* some ranids and rhacophorids) involves well-organized, polarized schools in the shape of a sphere; and tadpoles are always in physical contact with one another. These schools are found throughout the water column, near the edge or on the bottom. Individuals continuously moving toward the center of the school creates a rolling effect.

Comparable data on the structural organization and sensory bases of fish schooling behavior (e.g., Partridge 1982; Partridge and Pitcher 1980; review by Pitcher 1986 and E. Shaw 1978) are not available for anuran larvae. Analysis of the geometric structure and sensory bases of schooling (L. C. Katz et al. 1981; Wassersug et al. 1981b) suggested that schooling behavior of *Bufo woodhousii* and *Xenopus laevis* tadpoles is similar to that of fish. In the absence of other cues, both *Bufo* and *Xenopus* larvae exhibited two features of fish schools—oriented parallel to and showed preferential bearing to their nearest neighbors. In spite of these similarities, most tadpole schools do not exhibit the mobility displayed by fish schools. Compared to those of fish, most tadpole schools are relatively stationary and individuals in schools are usually at random distances from their nearest neighbors. A preferred distance to neighbors displayed by fish is rarely observed in tadpole schools (Wassersug et al. 1981b). The mobile schooling behavior of *Hyla geographica* is an exception to the generalization that tadpole schools are relatively stationary (Caldwell 1989); the same is true for schools of *Rhinophrynus dorsalis* (Stuart 1961; R. W. McDiarmid, personal observation). Tadpoles of certain species, such as *Phyllomedusa vaillanti,* may be evenly spaced when schooling in midwater (Branch 1983; J. P. Caldwell, personal communication). *Bufo* and *Xenopus* tadpoles differ in some aspects of their schooling behavior. The bearing relationship of one individual changed with illumination in *Xenopus* but not in *Bufo.* Although *Bufo* tadpoles in schools seem to be generally weakly polarized (Breden et al. 1982), they polarize more extensively in the field than in the laboratory (Wassersug et al. 1981b)

Like some fishes (e.g., Radakov 1973), tadpoles in schools may orient toward individuals of similar sizes (e.g., Alford and Crump 1982; Branch 1983; Breden et al. 1982). The magnitude of and the factors involved in size sorting varies among species. *Phyllomedusa vaillanti* larvae in Brazil (Branch 1983) formed temporary schools of mixed size classes that eventually segregated into schools composed of different size classes. Unequal locomotor capabilities of size classes, social attraction of like-sized individuals, and preferences for specific water depths may contribute to sorting by size in these schools (Branch 1983). *Rana sphenocephala* tadpoles of different size classes may live in different habitats, and both interference and exploitative competition may contribute to intraspecific size sorting (Alford and Crump 1982). Small *R. sphenocephala* tadpoles may be displaced by large ones. Although *Bufo woodhousii* larvae seem to associate with individuals of the same size, the magnitude of size sorting is small (Breden et al. 1982). Both large and small individuals may be found near one another.

Group living has numerous benefits, including the possibility of avoiding, detecting, and deterring predators, enhancing foraging efficiency, accruing benefits in competitive interactions, and increasing the efficiency of thermoregulation (Alexander 1974; Bertram 1978; Hamilton 1971; Pitcher 1986; Pulliam and Caraco 1984; G. C. Williams 1964). Tadpoles that form aggregations may gain many of these benefits (A. R. Blaustein 1988; Brodie and Formanowicz 1987; O'Hara 1981; Waldman 1982a; Wassersug 1973). For example, stirring of the substrate by groups of toad tadpoles results in suspending food that larvae can filter from the water (e.g., Beiswenger 1972, 1975; Wilbur 1977b). Tadpoles that form groups could be more efficient than individuals in the early detection and avoidance of predators (O'Hara 1981). Theoretical, observational, and experimental evidence suggest that as prey group-size increases early predator detection and probability of escape increases (e.g., S. J. Arnold and Wassersug 1978; Hamilton 1971; Kenward 1978; G. V. N. Powell 1974; Pulliam 1973; Siegfried and Underhill 1975; R. J. Taylor 1976, 1979; Treisman 1975; Vine 1973). Tadpole groups may efficiently swamp potential competitors. O'Hara (1981) and Olson (1988) occasionally observed aggregations of western toad (*Bufo boreas*) tadpoles displacing small groups of Pacific treefrog (*Pseudacris regilla*) tadpoles. Intraspecific competition may be mediated by aggressive behavior between individual tadpoles; apparent aggressive "butting" has been observed in *Bufo americanus* tadpoles (Beiswenger 1975; D.C. Smith 1990 and references therein; Waldman 1982a).

Groups of tadpoles may thermoregulate more efficiently than individuals. An aggregation of black tadpoles may efficiently absorb solar radiation that raises the temperature of the surrounding water (Caldwell 1989; Guilford 1988; Wassersug 1973). O'Hara (1981) repeatedly found temperature differentials of 2–3°C inside and outside of *B. boreas* aggregations; he ascribed these to solar heat absorption and possibly metabolic activity. Also, tadpoles may accrue inclusive fitness by associating with related individuals. The concept of inclusive fitness (Hamilton 1964a, b, c) predicts, all else being equal, that individuals will cooperate with or aid related individuals over unrelated individuals. Although not required, Hamilton suggested that an ability to discriminate between kin and nonkin (= kin recognition) could facilitate this so-called kin selection (Maynard Smith 1964) process.

Kin Recognition and Kin Association

Tadpoles of many species are socially attracted toward and form schools with conspecifics in nature, and tadpoles in schools may orient toward individuals of similar sizes. Also, based on numerous laboratory experiments that controlled for body size, tadpoles of several species seemingly prefer to associate with relatives over unrelated individuals regardless of body size (A. R. Blaustein 1988; A. R. Blaustein et al. 1987a; A. R. Blaustein and Waldman 1992; Waldman 1991; also see Cornell et al. 1989; Dawson 1982; O'Hara and Blaustein 1985; Waldman 1982a, 1986a). It seems likely, therefore, that tadpole schools of certain species are composed primarily of related individuals.

Investigators have used variations of two basic methods, laboratory choice tests or laboratory open-field tests, to determine whether or not a tadpole can discriminate between kin and nonkin. Aggregation choice tests of tadpoles of the western toad (*Bufo boreas*), Pacific treefrog (*Pseudacris regilla*), and the red-legged (*R. aurora*), Cascades (*Rana cascadae*), and spotted (*R. pretiosa*) frogs have been conducted in the laboratory (review by A. R. Blaustein 1988; A. R. Blaustein and Waldman 1992; A. R. Blaustein and Walls 1995) after test animals were reared in one of four regimes (summarized in A. R. Blaustein and O'Hara 1986a): (1) kin only, (2) kin and nonkin (mixed-rearing regime), (3) in isolation from an early embryonic stage, and (4) nonkin only. Association preferences were measured by allowing test individuals to associate near members of two stimulus groups within a rectangular tank (A. R. Blaustein and O'Hara 1986a). Stimulus compartments that were separated from the main portion of the tank by a screen were placed on opposite ends of the tank, and various numbers of tadpoles that varied in genetic affinities were placed in each compartment. In most tests, the time out of 20 min that tadpoles spent in the half nearest one stimulus group was compared with the time that would be expected if associations were random. In other experiments, the time that tadpoles spent in the third of the tank closest to one stimulus group, the third closest to the other stimulus group, or the time spent in the middle third of the tank was compared with a random expectation. Results obtained by A. R. Blaustein and O'Hara are difficult to compare directly with similar tests of kin recognition done by Cornell et al. (1989, *R. sylvatica*) and Fishwild et al. (1990, *Pseudacris crucifer, R. pipiens,* and *R. sylvatica*) because rearing and testing protocols were different.

Waldman (1981, 1984, 1985b) and Waldman and Adler (1979) conducted open-field laboratory experiments to test kin recognition in *B. americanus* and *R. sylvatica* tadpoles (also see Rautio et al. 1991). After being reared under a variety of regimes, members of two sibships were dyed different colors, equal numbers of the two groups were placed in a laboratory pool simultaneously, and the distances from each tadpole to its nearest sibling and nearest nonsibling were recorded. Waldman (1985a, 1986a) also used choice tests in a Y-maze to assess kin recognition in *B. americanus* tadpoles. In some of these tests, tadpoles were simultaneously presented with water flowing from two containers, each holding members of different sibling groups. In other tests, tadpoles were simultaneously presented with water flowing from a container holding siblings and one holding dechlorinated tap water. Dawson (1982) used both choice tests and open-field tests to investigate kin recognition in *B. americanus* tadpoles. Like in experiments of Cornell et al. (1989), the test apparatus used by Dawson was much smaller than the one used by A. R. Blaustein and O'Hara. The methods and rearing regimes used by Dawson (1982) differed somewhat from those used by A. R. Blaustein and O'Hara and Waldman, so again direct comparisons are difficult.

Significant differences in how tadpoles discriminated between kin and nonkin exist among species (table 9.1), and the different methods used by the various investigators cannot fully explain the observed differences. Differences in discrimination between species were apparent even when identical rearing and testing regimes were used. For example, the three ranids tested and reared identically by A. R. Blaustein and his colleagues displayed significant differences in how they discriminated between kin and nonkin (table 9.1).

Table 9.1 A summary of the results of kin recognition experiments involving various anuran tadpoles. ENC = experiment not conducted.

Species	Discriminates Siblings From Nonsiblings	Identifies Half-Siblings	References
Bufonidae			
Bufo americanus	Yes[a]	Yes[b]	Dawson 1982; Waldman 1981
B. boreas	Yes	Yes[c]	A. R. Blaustein et al. 1990; O'Hara and Blaustein 1982
Hylidae			
Pseudacris crucifer	No	ENC	Fishwild et al. 1990
P. regilla	No	ENC	O'Hara and Blaustein 1988
Ranidae			
Rana aurora	Early stages only	ENC	A. R. Blaustein and O'Hara 1986b
R. cascadae	Yes[a]	Yes[d]	A. R. Blaustein and O'Hara 1981, 1982a
R. pipiens	No	ENC	Fishwild et al. 1990; O'Hara and Blaustein 1981
R. pretiosa	No	ENC	O'Hara and Blaustein 1988
R. sylvatica	Yes	Yes[e]	Cornell et al. 1989; Fishwild et al. 1990; Waldman 1984

[a]Corroborated in field experiments (O'Hara and Blaustein 1985; Waldman 1982a).

[b]Preference for full siblings over paternal half-siblings; maternal half-siblings not distinguished from full siblings (only half-sibling categories tested).

[c]All categories of full and half-siblings have been tested, but results indicate only preferences for full siblings over paternal half-siblings and for maternal half-siblings over nonsiblings.

[d]Can discriminate between all categories of full and half-siblings and between half-siblings and nonsiblings (details in text).

[e]Preference for paternal half-siblings over nonsiblings; experiments not conducted on other half-sibling categories.

Rana cascadae discriminated between kin and nonkin after being reared under a variety of regimes. *Rana aurora* tadpoles discriminated only between kin and nonkin in early larval stages and only after rearing with full siblings. *Rana pretiosa* tadpoles failed to discriminate between kin and nonkin regardless of how they were reared (A. R. Blaustein 1988). Similarly, *R. sylvatica* tested by Fishwild et al. (1990) discriminated between kin and nonkin whereas *Pseudacris crucifer* and *R. pipiens,* tested and reared identically to *R. sylvatica,* did not. Using choice tests and a laboratory apparatus slightly modified from that used by O'Hara and Blaustein (1981), J. A. Hall et al. (1995) investigated the ability of *Spea intermontana* tadpoles to discriminate between kin and diet-based (= environmental) cues in forming associations. They found no evidence of kin-based associations for individuals raised on the same diet, and mixed results between kin and nonkin groups raised on different diets. They concluded that if kin discrimination occurs, environmentallly derived cues may be the basis for such discrimination and that those cues may be external to the natal environment of the tadpoles.

Rana cascadae and *R. sylvatica* seem to have the most acute kin recognition abilities. The rearing regime had little influence on the discrimination abilities of *R. cascadae* tadpoles. Results of testing mixed-reared tadpoles show that individuals preferred to associate with unfamiliar siblings over unfamiliar nonsiblings (O'Hara and Blaustein 1981). They also preferred to be nearest an unfamiliar stimulus group composed of siblings only, over a familiar group containing 50% siblings and 50% nonsiblings (O'Hara and Blaustein 1981). Also, tadpoles reared in isolation subsequently associated preferentially nearest unfamiliar siblings over unfamiliar nonsiblings (A. R. Blaustein and O'Hara 1981, 1982). These results suggest that familiarity with other individuals is not required for kin recognition, but siblings reared in mixed-rearing regimes failed to discriminate between familiar siblings and familiar nonsiblings. Tadpoles reared with siblings could assess the relative composition of sibling groups to some degree; individuals discriminated between a stimulus group composed of 50% kin and 50% nonkin and one containing 100% nonsiblings by preferring to associate nearest the former group (A. R. Blaustein and O'Hara 1983).

Embryos may respond to cues that emanate from the jelly derived from the mother's oviduct (A. R. Blaustein and O'Hara 1982a; Waldman 1981). Tests of *Rana cascadae* tadpoles that were reared after the jelly had been removed or after the jelly had been removed and replaced with jelly from a nonkin egg mass indicate that tadpoles were unaffected by these manipulations (A. R. Blaustein and O'Hara 1982a). *Rana cascadae* tadpoles preferred to associate with full siblings over either maternal or paternal half-siblings and with half-siblings (either maternal or paternal) over nonsiblings (A. R. Blaustein and O'Hara 1982a). Because full siblings were chosen over maternal half-siblings, and paternal half-siblings were chosen over nonsiblings, maternal cues are not necessary for kin recognition. Because maternal half-siblings were chosen over paternal half-siblings, maternal cues may exert a stronger influence than paternal cues. *Rana cascadae* tadpoles showed a positive attraction to kin rather than a repulsion to nonkin (A. R. Blaustein and O'Hara 1987).

Waldman (1984) conducted open-field laboratory tests of *Rana sylvatica* tadpoles that had been reared with siblings or in mixed-rearing regimes. Tadpoles reared only with siblings displayed a smaller, mean nearest-neighbor distance to siblings than to nonsiblings. The importance of familiarity is difficult to interpret from these experiments. Mixed-reared tadpoles distinguished between familiar siblings and familiar nonsiblings, but individuals reared in mixed regimes had closer contact with siblings than nonsiblings; this could have influenced their preferences (Waldman 1984). The tadpoles showed a tendency to associate more closely with familiar siblings than with unfamiliar siblings. Cornell et al. (1989) showed that *R. sylvatica* did not discriminate between familiar and unfamiliar siblings but did discriminate between unfamiliar paternal half-siblings and unfamiliar nonsiblings. These results suggest that neither maternal cues nor prior association with individuals are necessary for recognition in *R. sylvatica* tadpoles.

The discrimination system of *Rana aurora* tadpoles differs significantly from those illustrated by *R. cascadae* and *R. sylvatica.* Only those *R. aurora* tadpoles reared with siblings and tested in early developmental stages associated with kin (A. R. Blaustein and O'Hara 1986b; A. R. Blaustein et al. 1993). This tendency disappeared as tadpoles developed.

The larvae of *Bufo americanus* and *B. boreas* showed both similarities and differences to each other and to the ranids. The ontogeny of *B. boreas* kin recognition was influenced by the rearing regime; tadpoles reared with kin preferentially associated with kin over nonkin in laboratory choice tests (O'Hara and Blaustein 1982), but individuals reared in mixed-rearing regimes and with nonkin only displayed random association with respect to siblings and nonsiblings. Even when preferences for siblings were fully established after prolonged rearing with siblings, short-term exposure to nonsiblings nullified these preferences (O'Hara and Blaustein 1982).

The use of methods and sample sizes identical to those used by A. R. Blaustein and O'Hara (1982a, *Rana cascadae*) in their investigation of half-sibling recognition in *Bufo boreas* tadpoles (A. R. Blaustein et al. 1990) makes comparisons between these studies especially meaningful. *Bufo boreas* tadpoles preferentially associated with full siblings over paternal half-siblings and with maternal half-siblings over nonsiblings (A. R. Blaustein et al. 1990). They did not discriminate between full siblings and maternal half-siblings or between paternal half-siblings over nonsiblings. These results suggest that a maternal component is necessary for discrimination, which contrasts with the data on *Rana cascadae* and *R. sylvatica* tadpoles. Unlike *R. cascadae* tadpoles, *Bufo boreas* tadpoles did not discriminate between maternal and paternal half-siblings. The role of maternal cues in *B. boreas* kin recognition may be relatively complex (see A. R. Blaustein et al. 1990).

Bufo americanus tadpoles reared only with siblings or in isolation associated significantly nearer their siblings than nonsiblings (Waldman 1981; Waldman and Adler 1979). Tadpoles reared with siblings in early development and then exposed to siblings and nonsiblings in later development retained their close association with siblings. However, tadpoles that were reared with both siblings and nonsiblings in early development and exposed to siblings only in later development did not preferentially associate with familiar siblings over familiar nonsiblings (Waldman 1981). This suggests the possibility of a sensitive period in early development during which *B. americanus* tadpoles familiarize themselves with other individuals (Waldman 1981). Maternal cues seem to be important in *B. americanus* kin recognition because maternal half-siblings were not distinguished from full siblings, yet paternal half-siblings were distinguished from full siblings (Waldman 1981).

In laboratory Y-maze experiments, *Bufo americanus* tadpoles preferred unfamiliar siblings rather than familiar nonsiblings (Waldman 1986a). These data provide further evidence that familiarity is not the sole basis for kin recognition in *B. americanus.* In other Y-maze tests, tadpoles chose water lacking conspecific chemical cues over water containing nonsiblings (Waldman 1985a). Thus, avoidance of nonsiblings, rather than attraction to siblings, could be the basis for kin association in *B. americanus* tadpoles. Tests of *B. americanus* kin recognition (Dawson 1982) showed that tadpoles generally associated with siblings over nonsiblings in choice tests but failed to do so in open-field tests.

Results of laboratory kin recognition tests of similar methodologies with tadpoles of *Bufo americanus* (Waldman 1982a) and *Rana cascadae* (O'Hara and Blaustein 1985) have been corroborated in field experiments. Larvae of *R. cascadae* from six populations were reared in pure sibling groups or in mixed-rearing regimes. Equal numbers of tadpoles from two unrelated sibships were dyed different colors and released together in ponds they naturally inhabit. For several days, single aggregations were censused for sibship composition. Aggregations consisted of tadpoles primarily from one of the two color groups (controls revealed that color did not affect aggregation behavior). Tadpoles raised in pure sibling and in mixed-rearing regimes associated in groups composed principally of siblings. Unlike the case of laboratory tests, familiar siblings were distinguished from familiar nonsiblings.

Results of field experiments on *Bufo americanus* tadpoles generally corroborate the laboratory results (Waldman 1982a). As with *Rana cascadae, B. americanus* tadpoles could discriminate between familiar siblings and familiar nonsiblings in the field, but they could not do so in the laboratory (Waldman 1982a). These data suggest that field experiments probably are a more sensitive measure of kin recognition than are laboratory experiments (A. R. Blaustein 1988).

Two ranids (Fishwild et al. 1990, *Rana pipiens;* O'Hara and Blaustein 1988, *R. pretiosa*) and two hylids (Fishwild et al. 1990, *Pseudacris crucifer;* O'Hara and Blaustein 1988, *Pseudacris regilla*) failed to discriminate between kin and nonkin in laboratory choice tests. These results do not negate the possibility that these species may discriminate kin from nonkin under different conditions. Tadpoles tested in the laboratory may not have been motivated to display discrimination behavior (A. R. Blaustein et al. 1987b); kin recognition may be manifested differently in different populations, may be a polymorphic trait within populations (A. R. Blaustein 1988; A. R. Blaustein et al. 1987b), and may change ontogenetically (A. R. Blaustein and O'Hara 1986b).

Studies of *Bufo americanus* (Dawson 1982; Waldman 1985a), *B. boreas, Rana aurora* (A. R. Blaustein and O'Hara, unpublished data) and *Rana cascadae* (A. R. Blaustein and O'Hara 1982b) illustrate that kin recognition in tadpoles is mediated by water-borne chemical cues. Kin recognition persists after metamorphosis in at least *R. cascadae* (A. R. Blaustein et al. 1984) and *R. sylvatica* (Cornell et al. 1989; Waldman 1989); and based on the experimental regimes used, it does not seem as if water-borne chemical cues were used by metamorphs to discriminate between siblings and nonsiblings.

Pfennig (1990b) suggested that kin association in spadefoot toad tadpoles resulted from habitat selection based on diet-based environmental cues rather than from social preferences. Tadpoles with similar diets aggregate in the laboratory. In nature, because of the spatial and temporal proximity of siblings within a pond, individuals eating similar foods are more inclined to be related and therefore may form aggregations. Work by Pfennig implies that "kin recognition" behavior and kin association in tadpoles of some species may be an artifact of habitat selection (also see arguments by J. A. Hall et al. 1995). O'Hara and Blaustein (1982) posed a similar argument concerning habitat selection for kin association in *Bufo* tadpoles, and Grafen (1990) suggested that all cases of vertebrate kin recognition can be more consistently explained as species recognition. A. R. Blaustein et al. (1990) addressed this point in detail; and for some species of anuran larvae, the proposal by Grafen (1990) may be correct (A. R. Blaustein et al. 1990).

A. R. Blaustein and O'Hara (1986b) suggested at least two important parameters associated with the tendency for a tadpole to display kin recognition behavior: dispersal characteristics and aggregation behavior. These parameters relate to the probability of whether a tadpole will have the opportunity in nature to interact with kin. Species whose larvae randomly disperse (see Augert and Joly 1994) from their oviposition sites would have a relatively low probability of interacting with relatives; A. R. Blaustein and O'Hara (1986b) predicted that larvae of such species would not exhibit kin-selected behaviors. Similarly, in species whose larvae do not socially interact (form cohesive aggregations) with conspecifics, kin-selected behaviors may have little opportunity to evolve.

In nature, the patterns of aggregation formation and dispersal characteristics differ among the anuran larvae tested for kin recognition. *Bufo americanus, B. boreas,* and *Rana cascadae* form persistent, compact, social aggregations (A. R. Blaustein 1988). *Rana cascadae* aggregations are small and usually composed of a number (less than 100) much smaller than the clutch size (A. R. Blaustein 1988). *Rana sylvatica* tadpoles may aggregate similarly in certain ponds but not in others (DeBenedictis 1974; Hassinger 1972; Waldman

1984), and their aggregations are usually composed of hundreds of thousands of individuals from numerous sibships (Waldman 1984). Although little is known about the larval ecology of *Rana aurora*, there is evidence that intermittent, temporary, aggregations may form in early larval stages (A. R. Blaustein, unpublished data; Calef 1973). *Rana pretiosa* aggregations probably form only sporadically (C. C. Carpenter 1953; R. K. O'Hara and Blaustein, unpublished data). *Pseudacris regilla* tadpoles occasionally form small, loose, aggregations (O'Hara 1981); tadpoles of *Pseudacris crucifer* and *Rana pipiens* are not known to aggregate (Fishwild et al. 1990). *Bufo* aggregations are generally much larger than aggregations of *Rana* tadpoles, are often extremely dense (O'Hara and Blaustein 1982; Waldman 1982a; Wassersug 1973), and may be composed of hundreds of thousands of individuals from numerous sibships.

Small group size and low dispersal from sites of oviposition make it likely that aggregations of *Rana cascadae* tadpoles are composed primarily of kin. Because of these behaviors, *R. cascadae* tadpoles appear to have sufficient opportunity to interact with siblings and display a kin recognition system that would be efficient in their natural habitats (A. R. Blaustein et al. 1987b; O'Hara 1981; O'Hara and Blaustein 1985). Potential exists for mixing with nonsiblings, especially because egg masses may be laid communally (A. R. Blaustein 1988; O'Hara and Blaustein 1981). If there are benefits to aggregating with kin, tadpoles of *R. cascadae* must be able to discriminate between siblings and nonsiblings; this ability must be resistant to modification after exposure to nonsiblings. Laboratory and field experiments indicate that the kin recognition system of *R. cascadae* is established early in development and is generally not altered by exposure to nonkin (A. R. Blaustein and O'Hara 1981, 1982a, b, 1983; O'Hara and Blaustein 1981, 1985). In certain ponds *Rana sylvatica* tadpoles may disperse rapidly and far from sites of oviposition, whereas in other ponds dispersal may be more limited and tadpoles may avoid one another (DeBenedictis 1974; Hassinger 1972; Waldman 1984).

In ponds where the tadpoles aggregate, there is some opportunity for the formation of kin groups. At times, large, single, polarized, schools break up into smaller, feeding groups of several hundred to several thousand individuals (Waldman 1984). Although most schools of the tadpoles of *R. sylvatica* consist of individuals from many sibships, these smaller units may represent sibling groups (Waldman 1984). *Rana sylvatica* tadpoles display a kin recognition system that generally is not influenced by exposure to nonsiblings. As in *R. cascadae,* a system that is resistant to modification after exposure to nonsiblings would be required because of communal egg laying and exposure to nonkin during larval stages (Waldman 1982b, 1984).

Dispersal from sites of oviposition in both *Bufo americanus* and *B. boreas* tadpoles is high and members of different sibships intermingle (Beiswenger 1972; O'Hara 1981; O'Hara and Blaustein 1982; Waldman 1982b). The kin recognition system of *B. americanus* is generally consistent with its larval ecology and behavior. Large aggregations may subdivide into smaller schools (Beiswenger 1972, 1975; Waldman 1982a) that may represent sibling cohorts (Waldman 1982a). Mixing with nonsiblings is frequent in active larvae (Beiswenger 1972; Waldman 1982a), and sibling preferences, once established, are resistant to modifications (Waldman 1981). The larval behavior of *B. boreas* tadpoles may preclude kin association; extremely large aggregations form from individuals from numerous clutches, and the feeding behavior of the individuals in aggregation entails frequent moving, intermingling, and splitting off. The formation of discrete kin associations may be difficult even if such behavior conferred fitness benefits (O'Hara and Blaustein 1982). Maintaining cohesive sibling groups would be difficult even if tadpoles retained a capability to recognize siblings after encounters with nonsiblings. As in *B. americanus,* smaller schools split off from the main aggregation, and these may be sibling cohorts.

The failure of *Pseudacris crucifer, P. regilla, Rana pipiens,* and *R. pretiosa* to discriminate between kin and nonkin is generally consistent with their larval ecology (discussed by A. R. Blaustein 1988). Because these species either do not aggregate or form aggregations only intermittently, there is little opportunity for siblings to interact. Too little is known about the larval ecology of *R. aurora* to speculate whether there is opportunity for kin to interact (A. R. Blaustein and O'Hara 1986b). These larvae are the only ones known to have a kin recognition system that changes ontogenetically.

It is not necessary for a tadpole to aggregate with kin to gain the benefits of group living, but those individuals that preferentially associate with kin may accrue additional benefits through an increase in inclusive fitness (Alexander 1974; Hamilton 1964a, b). Kin recognition potentially enhances the formation of and maintenance of kin groups, which is important if there are benefits to associating with kin. Tadpoles living in groups composed primarily of kin could warn relatives if they release an alarm substance when they are attacked (Hews and Blaustein 1985; Hrbâçek 1950; Waldman 1986b). This is a more likely benefit of kin association in *R. cascadae* tadpoles (Hews and Blaustein 1985). Injured *R. cascadae* tadpoles release a chemical cue that causes others to elicit an escapelike response (Hews and Blaustein 1985). In this situation, kin would be warned because individuals in *R. cascadae* aggregations are presumably mostly kin, and kin recognition may be maintained by kin selection. Unpalatability and noxiousness are probably important antipredator adaptations in some species of tadpoles (Brodie et al. 1978; Formanowicz and Brodie 1982; Wassersug 1973; Werschkul and Christensen 1977). The evolution of distastefulness or aposematic coloration has often been explained by a kin selection model (e.g., Benson 1971; Fisher 1930; Harvey and Greenwood 1978). If members of a single kin group aggregate with each other and a predator that samples one or more distasteful siblings learns to avoid others in the group, then a gene for distastefulness could increase in frequency through kin selection. This scenario has been invoked to explain the advantage of associating with kin in toad tadpoles (Waldman and Adler 1979; Wassersug 1973).

Social behavior influences growth and development. For example, in some insects, larval growth and development is enhanced in groups composed of certain proportions and combinations of given genotypes (Bhalla and Sokal 1964; Lewontin 1955; Sokal and Huber 1963; Sokal and Karten 1964). Social behavior may be especially important in the growth and development of tadpoles. Those that develop at high elevations where the growing season is short or where habitats are ephemeral may be under intense selection for rapid growth. Some data suggest that tadpole growth and development is influenced by the proportion of related individuals to unrelated individuals within a population or group (Hokit and Blaustein 1994, 1997; Jasieński 1988; Shvarts and Pyastolova 1970; D.C. Smith 1990; Waldman 1986b).

Interference competition has been documented in anuran larvae (chap. 10), and kin recognition may allow tadpoles to direct intraspecific competition toward nonkin when resources are limited. Kin recognition could also direct cannibalism toward nonkin or, in some cases, toward kin (A. R. Blaustein et al. 1987b; Pfennig et al. 1993; Walls and Blaustein 1995). Because kin recognition persists for at least some time after metamorphosis (A. R. Blaustein et al. 1984; Cornell et al. 1989; Waldman 1989), kin recognition may also play a role in adult anurans (e.g., balance inbreeding with outbreeding; see Bateson 1983).

When tadpoles form kin groups, the benefits may be directed primarily toward kin even if the signals mediating the response are not kin-specific (see Waldman 1986b). For example, alarm substances released by a tadpole injured by a predator would become diluted as they diffuse away from the aggregation (Waldman 1986b). Those tadpoles in closest proximity to the injured individual would benefit most directly. In the case of *Rana cascadae,* kin would benefit most because they are in closest proximity to one another (O'Hara and Blaustein 1985). Similarly, chemicals affecting growth and development may be favored by natural selection because they are behaviorally directed toward kin (Waldman 1986b).

In summary, it is known that tadpoles of several species of anurans can discriminate between kin and nonkin. Kin selection may have played an important role in the evolution and maintenance of kin recognition in certain species. The tadpoles of those species that did not discriminate in laboratory tests may discriminate under different conditions. Perhaps, kin recognition and kin association were important in the evolutionary past under different environmental and social regimes, and what we now see are remnants of behaviors that persisted. Finally, in light of evidence that newly metamorphosed ranids discriminate between kin and nonkin (A. R. Blaustein et al. 1984; Cornell et al. 1989), enhanced optimal outbreeding may be a potential function of kin recognition (sensu Bateson 1983). We need much more empirical information to derive specific conclusions about the adaptive value of kin recognition and kin association in anuran larvae (A. R. Blaustein et al. 1990).

Sensory Abilities and Learning

Anurans frequently are used as model organisms for investigations of spinal reflexes, and considerable literature exists on behaviors with simple triggers (e.g., visually elicited flight and feeding; see review by Ingle 1976). Comparatively little is known about the sensory capabilities of tadpoles (chap. 6) and how these relate to specific behaviors. This deficiency simply may be because tadpoles are aquatic and small, and thereby somewhat less tractable for many types of manipulation. Work to date suggests that the sensory mechanisms of tadpoles differ from adults in ways that are consistent with the large ecological differences between the two life stages.

The sensory abilities, especially pertaining to vision, of tadpoles and their ability to modify their behavior by learning is a poorly understood field. How much they use vision during various behaviors (e.g., escape and feeding) is not known. Also, only speculations can be made on the effects that the variations in eye positions (i.e., dorsal vs. lateral; see chaps. 3 and 12) and the amount of corneal and lenticular protrusion has on the vision field. The large, laterally placed eyes with curved corneas and prominent lenticular protrusion of a nektonic hylid tadpole must afford visual abilities that differ greatly from the small, dorsal eyes with more flattened corneas and little lenticular protrusion of a bufonid tadpole.

Vision

Tadpoles are nearsighted. Retinoscopic measurements of resting refractive state of larval toads (*Bufo americanus*) show a myopic defocus of about −275 Diopter (D) in air and −30 D in water for young tadpoles to about −4 D in water for tadpoles just before metamorphosis (Mathis et al. 1988). These measurements are similar to measurements on larval salamanders (myopic with a resting refractive state of about −3 D; Manteuffel et al. 1977) and unlike measurements on adult amphibians (hyperopic with a defocus of about 8 D, or nearly emmetropic; Manteuffel et al. 1977; Mathis et al. 1988). It is therefore unlikely that tadpoles, even in the clearest water, use vision to detect anything at great distances or with much precision. They are sensitive to light levels (Laurens 1914), and some species (Dunlap and Satterfield 1985, *Rana pipiens*) select the brightest background regardless of preconditioning or track patches of sun across a pond (although this could be thermal sensitivity; Brattstrom 1962). These responses may be similar to phototactic behavior characteristic of many other animals including salamanders (Anderson 1972) and a few anurans (e.g., Altig and Brodie 1972, *Ascaphus truei*). Other species (e.g., *Osteopilus brunneus*) flee from intense light presumably to avoid overheating (Lannoo et al. 1987), and the phototactic preferences of tadpoles change ontogenetically (R. G. Jaeger and Hailman 1976).

Vision also may be important in determining spatial arrangements in stationary schools in some species. Wassersug et al. (1981b) found that schools of *Xenopus* were more organized (i.e., nearest neighbors were more nearly parallel or antiparallel) in the light than in the dark. Tadpoles of *Rana catesbeiana* can move their eyes (Stehouwer 1988), and *Xenopus* tadpoles track movements with their eyes and turn their bodies to follow moving visual patterns (i.e., optokinetic response, P. Grobstein, personal communication). Auburn and Taylor (1979) showed that *R. catesbeiana* larvae oriented to polarized light cues indoors if they had been trained outdoors. Tadpoles were acclimated in an outdoor tank with a deep end at a particular compass heading. Presented with indoor patterns of polarized light, these animals oriented to the compass heading (deep end, outside) inferred from the pattern of polarized light expected at that time of day. Goodyear and Altig (1971) studied orientation of bullfrogs during metamorphosis, and Justis and Taylor (1976) demonstrated that extraocular photoreception could be used in compass orientation of larval bullfrogs.

Many amphibians, including anurans, have extraoptic photoreception via the pineal body that may have some function in orientation (review by Adler 1970); we know of no studies of pineal-mediated behavior in tadpoles. Leucht (1987) found that *Xenopus* tadpoles changed pigmentation to match the background differently when weak magnetic field fluctuation was added to the normal light cue.

Cutaneous Senses and Chemical Communications

Tadpoles have well developed lateral line systems (Lannoo 1987; chap. 6) that presumably detect disturbances in the water. Lum et al. (1982) noted that stationary schools of *Bufo americanus* and *Xenopus* are significantly nonrandom even in the dark. It is likely that lateral line information plays an important part in school structure, but the critical experiment (e.g., severing the lateral line nerves) has not been done.

Tadpoles are extremely sensitive to touch, and light stroking or prodding of the skin elicit struggling or escape response from both embryos and larvae. These responses may be mediated by Rohon-Beard cells, the terminals of which are in the epidermis (Fox et al. 1978) and rapidly adapt to light stroking (A. Roberts and Hayes 1977). No studies have addressed the details of this behavior. The specific activity of cold-sensitive neurons in the skin is correlated with behavioral temperature selection (Dupré et al. 1986)

Tadpoles of several species preferentially associate with siblings and the mechanism for sibling recognition is presumably chemical (review in A. R. Blaustein et al. 1993; see Aggregations above). Aversive reactions to predators and apparently conspecifics under some conditions are also well documented.

The presence of predators often results in tadpoles spending more time in refugia (Petranka et al. 1987; Semlitsch and Gavasso 1992) or otherwise alters tadpole biology (e.g., R. A. Griffiths et al. 1998; Kupferberg 1998; Rödel and Linsemair 1997), and controlled predation studies have highlighted the survival value of different patterns of swimming and activity among tadpoles. Chovanec (1992) compared the susceptibility of tadpoles of four European frogs (*Bombina bombina, Bufo bufo, Hyla arborea,* and *Rana dalmatina*) to predation by dragonfly naiads (*Aeshna cyanea*). Differential predation was attributed to different swimming speeds and tail flexion, activity patterns, and habitat use among species. Chovanec (1992) suggested that the lateral eyes and greater visual field of *Hyla* accounted for lower predation in that species. Azevedo-Ramos et al. (1992) examined predation by aeshnid naiads (*Gynacantha membranalis*) on tadpoles of the hylid frogs *Hyla geographica, Osteocephalus taurinus, Phyllomedusa tarsius,* and *Scinax rubra* from near Manaus, Brazil. They found a significant negative correlation between average distance moved per unit of time and survival time, and a positive relationship between rapidity of movement independent of distance covered and survival time. Species (e.g., *Scinax rubra*) that were motionless for longer periods and then moved quickly, suffered less predation. Tadpole immobility is an effective deterrent to sight hunting predators but potentially results in reduced foraging efficiency.

Skelly and Werner (1990) showed that tethered naiads of predatory dragonflies (*Anax junius*) caused a 41% decrease in activity of *Bufo americanus* kept in small containers, and the tadpoles avoided the side of the container inhabited by the odonate naiads. Werner (1991) established that in the presence of *Anax* naiads, small larvae of *Rana catesbeiana* and *R. clamitans* also reduced activity and sought retreats, but *R. catesbeiana* became the superior competitor because it reduced its foraging less than did *R. clamitans.* Werner and Anholt (1996) showed that such activity led to a decrease in growth rate, presumably because of reduced foraging activity in smaller tadpoles; they reported that these behavioral responses were mediated through chemical modalities, with detection occurring either directly from the presence of caged predators or indirectly through exposure to water that previously held the predator. Apparently, tadpole responses are size dependent because larger *R. catesbeiana* tadpoles continued to forage in the presence of caged *Anax* naiads and had higher growth rates because they had access to the resources made available by the depression of foraging activity of smaller tadpoles (compensatory reduction in competition) as a result of the naiad presence. In the presence of *Anax* naiads, growth rates of the larger bullfrog tadpoles increased 1.26-fold and a positive effect was detected both on the fraction of the population that metamorphosed and the mass of the metamorphs. Such behaviorally indirect effects acting through body size and activity levels of the species can have a major and variable effect on community dynamics (chap. 10).

Such results suggest that maintenance activity (e.g., feeding and swimming) as well as temporal and spatial use of microhabitats in the presence of predators may be mediated by sensitivity to chemicals emitted by the predator (also see Kiseleva 1984, 1992, 1997; Kiseleva and Manteifel 1982; Lefcort 1996; Manteifel 1995; Manteifel and Zhushev 1996; see D. J. Wilson and Lefcort 1993 for effects of predator diet on the alarm response). The presence of species-specific alarm substances is suggested by the occurrence of alarm responses in *Bufo boreas* to conspecifics damaged by giant water bugs but no such response to damaged heterospecific tadpoles (Hews 1988).

Chemical communication is implicated in intraspecific and possibly interspecific interference competition. Steinwascher (1978b, 1981) documented changes in growth rate and survivorship in mixed-size (one species) and two-species assemblages. One species can inhibit the growth of other species, and large tadpoles may inhibit the growth of smaller conspecifics (chap. 10). Behavioral changes such as reduced feeding activity were not documented in these studies.

Learning

Most behavioral work on anuran larvae has concentrated on immediate responses to a variety of sensory stimuli (e.g., Munn 1940a, b). Other studies considered the effects of early experience on habitat preference (Punzo 1976; see Aggregations above) and simple avoidance learning (W. J. Hoyer et al. 1971; Punzo 1991). Both types of studies illustrate the ability of anuran larvae to modify somewhat their behavior by assimilating new information, an ability previously thought to be the exclusive province of long-lived animals in heterogeneous environments. Punzo (1991) noted that learning of simple, active, dark-cued, shock avoidance (i.e., light turned off before electric shock) in some tadpoles that school (e.g., *Rana heckscheri*) was increasingly enhanced as group size increases to about 10 individuals. Nonschooling tadpoles have not been tested, but other studies suggest that schooling and nonschooling species may have different learning abilities. Hoff (1988a) showed that escape directions relative to the positions of nearest neighbors differed between schooling and nonschooling species. In particular, shock-induced startle responses of nonschooling *Rana septentrionalis* failed to reflect the changing position of the conspecifics around them (i.e., there were many collisions), while schooling tadpoles of *Bufo americanus* never collided with their conspecifics during startle behavior. The pattern of avoidance directions of the *R. septentrionalis* tadpoles was identical to that of adult anurans avoiding present or recently seen obstacles (Hoff 1989, *R. catesbeiana;* Hoff and Ingle 1988, *R. pipiens*). That is, they avoided the positions previously held by their nearest neighbors but not those positions currently held by those neighbors. This observation suggests that the short-term spatial memory used by adult anurans to form a map of potential escape directions while their eyes are focused on near-field activities such as feeding (Hoff and Ingle 1988) may be present in nonschooling tadpoles (which deal only with stationary objects in their environment). This ability may be absent or suppressed in schooling tadpoles that must update spatial information continuously during avoidance behavior.

Strickler-Shaw and Taylor (1990, 1991) and D. H. Taylor et al. (1990) showed that sublethal doses of lead inhibited learning in tadpoles of *Rana clamitans.* Steele et al. (1991) showed that tadpoles do not avoid lead-rich water.

Escape behavior of tadpoles of *Rana sylvatica* changes ontogenetically, and R. M. Brown and Taylor (1995) suggested that these changes are correlated with poorer swimming performances of younger and older tadpoles relative to those of intermediate stages. During stages 29–37, tadpoles had a low probability of deviating from a straight escape trajectory, while younger (< 29) and older (> 37) tadpoles showed a greater propensity to turn and maneuver at sharp angles from the initial trajectory.

Summary

Although there is much that is not known about the mechanical aspects of the behavior of anuran larvae, the rudiments of behaviors associated with feeding, respiration, thermoregulation, and locomotion are understood and documented. The same cannot be said of the social behavior of these animals. Empirical investigations suggest that there is as much or more variety in the details of social behavior as there is in any aspect of tadpole behavior. Although the ecological context of some behaviors, such as preferential kin association, is understood for some species, the understanding of general patterns must await more experimentation.

ACKNOWLEDGMENTS

Kate Spencer, whose artistic skills are considerable, designed and produced the two drawings for this chapter; we greatly appreciate her willingness to work with us. Karl-Heinz Jungfer loaned slides and provided information on the tadpoles of *Anotheca spinosa* (fig. 9.3A), *Osteocephalus oophagus* (fig. 9.3B), and *Megophrys montana* (fig. 9.5A). Mark-Oliver Rödel and Karsten Grabow shared their knowledge of *Hemisus marmoratus* biology with us and K. Grabow loaned the slide reproduced in figure 9.3C. Jonathan A. Campbell loaned the slide of *Petropedetes parkeri* shown in figure 9.5. To each of these colleagues we express our thanks.

10

ECOLOGY

Resource Use, Competition, and Predation

Ross A. Alford

An adequate picture of the instability of life in a pond, and of the ups and downs of tadpole life, cannot be gained by compressing the events into one colorless summary. The risks, the successes, the disasters overtaking whole populations are fundamental to the ecology of the animal, and for their understanding detailed descriptions of specific instances are essential.

R. M. Savage 1962:24

Introduction

As noted in the quote above, R. M. Savage (1962) recognized that the ecology of tadpoles is a complex and variable reflection of most of the processes that govern the dynamics of populations and communities. During the 1960s and 1970s several pioneering studies took advantage of this idea by using larval amphibians as models for examining ecological and evolutionary processes (see fig. 10.1). Continuation of this trend into the 1980s and 1990s led to a rapid expansion of the tadpole literature. Anuran larvae were used in studies ranging from the genetics of growth (also see Rist et al. 1997) and development to the organization of complex communities and ecosystems. Seale's (1987) synthesis of the literature on amphibian energetics is applicable to various subjects below, and Wilbur (1980) summarized the state of tadpole ecology at the start of the 1980s. In this chapter, I provide a similar summary, and point out new areas of inquiry that have arisen from work done primarily since 1979.

All staging notations are from the system of Gosner (1960; also see table 2.1).

Tadpole Habitats and Communities

General Features

The generalized anuran life history, as defined by Wassersug (1974, 1975), includes a free-living, primarily herbivorous larval stage. Because anuran larvae are specialized for growth and must transform before reproduction (Slade and Wassersug 1975; Wassersug 1974; Wilbur 1980), they cannot form stable long-term ecological associations. Some larvae exploit relatively simple, predictable habitats, such as water-holding epiphytes, large nut husks, and large snail shells (Caldwell 1993; Grandison 1980; Inger 1985; Lannoo et al. 1987; Orton 1953; Starrett 1973) in tropical areas. Others develop in more complex permanent aquatic habitats as temporary invaders in established communities (e.g., Faragher and Jaeger 1998; Sinsch 1997; Werner and McPeek 1994). These situations can lead to increased competition and predation pressure from permanently resident species (Bradford 1989; Heyer et al. 1975; Kats et al. 1988; Wassersug 1974; Werner 1986).

Most anuran larvae inhabit temporary habitats that range from depressions in the trunks of fallen trees in tropical forests (Starrett 1973) to large pools in boreal habitats (Koskela 1973). Unpredictable temporal and spatial distributions and cyclic patterns of nutrient availability are common features of these habitats. New patches of habitat may appear and disappear on a timescale of a few years. Existing patches often fill and dry in response to broadly seasonal patterns that vary between sites and years. Mineral nutrients and high-quality detritus are likely to be most available early in each episode of filling and drying (Bärlócher et al. 1978; McLachlan 1981a, b; Osborne and McLachlan 1985; Wassersug 1975). The cycle of filling and drying, combined with fluctuations in resources, leads to complex changes in the quality and quantity of available habitat (Wilbur 1987). These changes preclude the establishment of stable climax communities and may facilitate invasion by transient organisms such as anuran larvae.

Ecological Interactions

It is likely that the interaction of biotic and abiotic factors influences tadpole ecology (Dunson and Travis 1991; fig. 10.1). During their lives, tadpoles of most anuran species

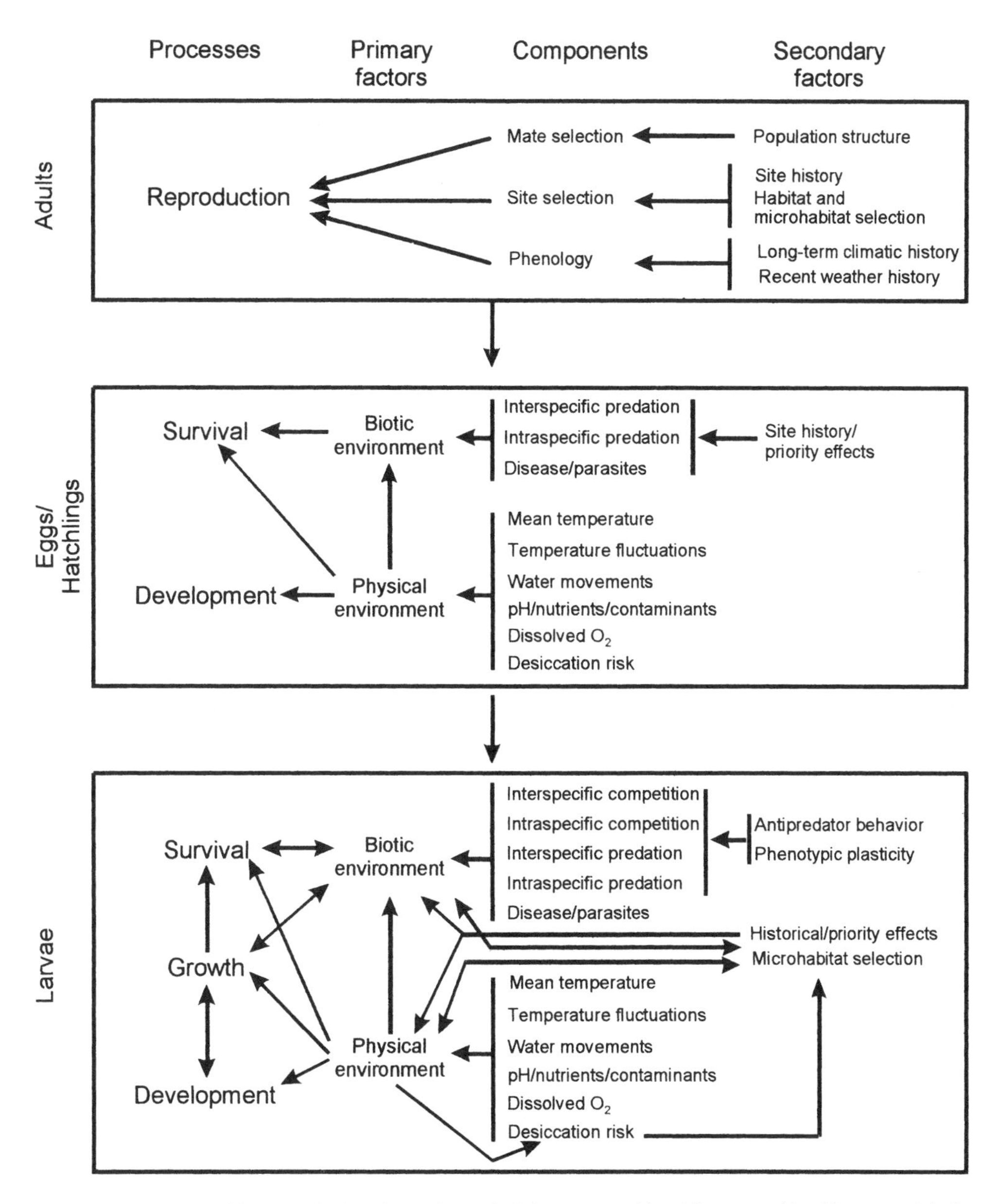

Fig. 10.1. Summary of factors affecting the ecology of adults, eggs, and hatchlings (combined because of similar processes), and tadpoles of anuran amphibians. Four major processes control transitions through and between life history stages, and these processes are influenced by primary factors in or of the biotic environment. While secondary factors are reflected primarily in the reproductive ecology and behavior of adults, they can affect directly the biotic environment of a larva (e.g., microhabitat selection). Arrows indicate interactions and their directions, which need not be directly causal. For example, developmental differences are larval responses to components of the environment and influence the duration of larval period and thus the degree and type of exposure to the physical environment.

increase in mass by factors of 50 or more. They may experience local densities of several hundred individuals/m² or 25⁺/liter (table 10.1). Most of the habitats occupied by tadpoles are used by more than one species and contain herbivores of a variety of other taxa, many of which may compete with tadpoles (Morin et al. 1988). Most also contain predators (Cooke 1974; C. L. Rowe et al. 1994), and the roles of predator and competitor may switch through time (L. Blaustein and Kotler 1993; L. Blaustein and Margalit 1994; Hearnden 1992; Petranka et al. 1998). The phenology and reproductive site preferences of adult frogs control the exposure of tadpoles to competitors and predators.

The general type of reproductive habitat chosen (temporary or permanent water, surrounding habitat type) has a strong influence on the entire developmental strategy followed by many species. Reproduction in permanent water decreases the risk of drying before development is complete, generally increases the diversity of competitors and predators present, and decreases the availability of food. Species spawning in permanent water tend to have prolonged larval periods and metamorphose at sizes larger than temporary-pond spawners (Patterson and McLachlan 1989). In part this may be influenced by adult reproductive parameters; egg size and clutch size can depend on the type of reproductive habitat used by a species (Tejedo and Reques 1994a). Woodward (1987a) found that species breeding in temporary ponds had relatively smaller numbers of relatively larger eggs than permanent pond species. Also, differences in relative egg and clutch sizes among breeding aggregations suggested that inexperienced females might produce fewer, smaller eggs and visit less optimal sites (Woodward 1987b).

Larvae of species that spawn early in the history of a temporary pool experience different ecological conditions from those that reproduce later (Wilbur and Alford 1985). Species that reproduce explosively produce a single cohort of larvae that initially are very similar ecologically. Members of such a cohort may diverge into a variety of size classes as growth and development proceed (Wilbur 1984). Species with extended reproductive periods are likely to spawn several times during a season and produce distinct larval cohorts with different body sizes that may function as separate "ecological species" (Alford 1986a; Alford and Crump 1982; Polis 1984). In some species, this ecological separation between different-size conspecifics reaches the point where larger individuals become predators on smaller ones (Crump 1983). Larvae of early cohorts may also prey on conspecific eggs of late cohorts, and large larvae may also prey on small larvae or eggs of other species (Banks and Beebee 1987; Hearnden 1992). Tadpoles are thus likely to encounter a variety of predators and both intra- and interspecific competitors. Their ecological interactions often will depend not simply on the identity of the species they interact with but

Table 10.1 The mean, standard deviation, minimum, and maximum numbers (listed in order vertically) of tadpoles per liter and tadpoles per square meter in field samples from a temporary pond in northern Florida in 1981. Number of samples and pond area (m²) are given below each date. Expanded from Alford 1986b.

1 March		8 March		23 March		5 April		13 April		17 April		19 April	
7		12		6		11		10		10		12	
1400		1470		612		264		80		50		20	
liter	m²	liter	m²	liter	m²	liter	m²	liter	m²	liter	m²	liter	m²
Bufo woodhousii													
0.00	0.0	0.00	0.0	0.00	0.0	6.26	998.2	0.64	83.5	4.82	603.0	0.75	67.9
0.00	0.0	0.00	0.0	0.00	0.0	12.40	2294.0	1.35	171.4	14.91	1864.0	1.78	147.0
0.00	0.0	0.00	0.0	0.00	0.0	0.00	0.0	0.00	0.0	0.00	0.0	0.00	0.0
0.00	0.0	0.00	0.0	0.00	0.0	38.30	7659.8	4.07	488.9	47.26	5907.8	6.11	488.9
Pseudacris ocularis													
0.00	0.0	0.28	25.5	0.23	20.4	0.49	68.5	0.90	120.2	0.80	105.9	0.87	83.2
0.00	0.0	0.37	32.6	0.22	18.2	0.45	60.6	0.80	107.6	0.79	113.5	1.39	126.9
0.00	0.0	0.00	0.0	0.00	0.0	0.00	0.0	0.00	0.0	0.00	0.0	0.00	0.0
0.00	0.0	1.16	101.9	0.58	40.7	1.43	183.3	2.24	325.9	2.53	366.7	4.55	387.1
Pseudacris ornata													
0.56	58.2	0.35	23.8	0.16	17.0	0.03	5.6	0.12	18.3	0.22	26.5	0.11	11.9
1.08	95.9	0.48	21.0	0.09	8.3	0.07	13.2	0.20	32.5	0.27	31.9	0.12	13.6
0.00	0.0	0.00	0.0	0.00	0.0	0.00	0.0	0.00	0.0	0.00	0.0	0.00	0.0
2.94	264.8	1.75	61.1	0.29	20.4	0.20	40.7	0.62	101.9	0.89	101.9	0.33	40.7
Rana sphenocephala													
0.43	49.5	0.61	40.7	0.52	47.5	0.11	18.5	0.60	83.5	0.50	71.3	2.81	337.8
0.50	45.3	0.91	43.4	0.61	61.3	0.15	26.5	0.56	77.1	0.48	70.7	2.66	362.8
0.00	0.0	0.00	0.0	0.00	0.0	0.00	0.0	0.00	0.0	0.00	0.0	0.21	20.4
1.36	122.2	3.06	122.2	1.55	163.0	0.41	81.5	1.88	244.5	1.30	183.3	8.01	1161.2

on their own body size and that of others (Alford 1986a; Werner 1986; Wilbur 1988).

The direct ecological interactions experienced by tadpoles occur through a variety of mechanisms. Intra- and interspecific competition can be caused by exploitation: the simple depletion of resources (Kuzmin 1995). The intensity of exploitative competition depends on the degree of resource overlap between species, and the degree of overlap can change evolutionarily or behaviorally in response to local conditions (Alford 1986a). Negative effects of density can also be caused by interference—some tadpoles either reduce the access of others to preferred resources (R. M. Savage 1952; Steinwascher 1978b) or release allelopathic chemicals or symbionts (Goater 1994; R. A. Griffiths et al. 1993; Licht 1967; Steinwascher 1979b; also see Tejedo and Reques 1994b). In most populations, the negative effects of density are probably produced by a combination of interference and exploitation mechanisms (Alford 1994; Berven and Chadra 1988; Semlitsch and Caldwell 1982; Steinwascher 1979b; this chapter). The effects of either of these mechanisms can be mediated by larval genotype (Alford 1989b; Newman 1988b, 1995; Travis 1980a, b, 1983a, b; Travis et al. 1985a), body size (Alford and Wilbur 1985; Ebenman 1988; Kupferberg 1997a; Werner and Gilliam 1984), species-specific competitive ability, environmental heterogeneity and microhabitat differentiation (Alford 1986a; Diaz-Paniagua 1987; Waringer-Löschenkohl 1988), and behavior (Alford 1986a; Morin 1986).

Tadpoles also interact with other types of herbivores. A variety of adults and larvae of herbivorous, aquatic insects, crustaceans, and zooplankton co-occur with tadpoles and use similar resources. Few studies have examined either the impact of tadpoles and other aquatic herbivores on their shared resources or possible interphyletic competitive interactions involving tadpoles. Alford (1989c) demonstrated experimentally that tadpole densities affected the growth rates of larval newts, which prey on zooplankton, and suggested that this was probably an indirect effect of competition between tadpoles and zooplankton. Morin et al. (1988) demonstrated that interphyletic competition can influence the growth of tadpoles. Brönmark et al. (1991) found that larval *Rana temporaria* negatively affected the growth of snails, while snails apparently facilitated the growth of *Rana*, possibly by altering the composition of the algal assemblage. Interphyletic interactions have also been demonstrated by L. Blaustein and Margalit (1994) and Kupferberg (1997b).

Interactions of tadpoles with predators have recently been studied in some detail. These interactions are generally complex and not described well by simple species-specific coefficients of predation rate. Many predators of tadpoles, such as fish and salamanders, are gape limited. Larger tadpoles swim faster (Wassersug and Hoff 1985) and may be more capable of escaping predators (Feder 1983c). Risk of predation is thus often size-specific, either decreasing monotonically with increasing tadpole body size (S. J. Richards and Bull 1990a, b) or increasing to a maximum and then decreasing (Brodie and Formanowicz 1983; Wilbur 1988). Tadpoles may also alter predation risk by behaving differently when predators are present. Behavioral changes often involve alterations in microhabitat selection (Formanowicz and Bobka 1989; Hews 1995; Horat and Semlitsch 1994; Petranka et al. 1987; Semlitsch and Reyer 1992a) or total activity level (Chovanec 1992; Hews 1988; S. P. Lawler 1989; Skelly and Werner 1990).

The ultimate effects of ecological conditions on tadpoles are often difficult to determine. Changes in tadpole body size relative to age, stage, or metamorphosis (e.g., Newman and Dunham 1994) and survival and developmental rates are relatively easy to measure. Difficulties arise in interpreting how these responses affect the population biology of species throughout their life cycle. Although it is possible to detect some patterns of relationship between adult body size and size at metamorphosis (Patterson and McLachlan 1989; Pough and Kamel 1984), most anuran larvae have considerable flexibility in their rates of growth and development (Wilbur and Collins 1973). This means that single measures of tadpole performance such as metamorphic body mass can be misleading; tadpoles from two populations may metamorphose at the same size after very different larval periods. Alternatively, two populations may have similar larval periods but very different masses at metamorphosis. Berven (1990) and D.C. Smith (1987) showed that increased metamorphic mass can increase survival to first reproduction. John-Alder and Morin (1990) demonstrated that jumping ability in metamorphic *Bufo woodhousii* is positively correlated with body mass. Blouin (1992b) demonstrated genetic correlations between larval life-history characters and aspects of terrestrial morphology. Egg deposition in environments with relatively high predation risk may be advantageous if predation reduces competition sufficiently to increase the body size and thus the survival of metamorphs. These complications do not affect the validity of tadpoles as model organisms for studies of ecological processes, but it is difficult to relate studies of tadpole ecology to the population biology of frogs throughout their life cycles.

Resource Use and Effects on Resources

Kenny (1969b, c) and Wassersug (1972) described the basic morphology and mechanics of feeding in anuran larvae (chaps. 2, 3, and 4). Most tadpoles have keratinized jaw sheaths surrounded to some extent by rows of labial teeth, which in turn are surrounded by labial papillae. The jaw sheaths and teeth remove material from substrates, and the jaw sheaths also chop larger pieces of material into sizes that fit into the mouth. Food and respiratory water are transported through the mouth, buccal cavity, and gills by buccal pumping. Food scraped into suspension from a substrate or ingested directly from the water column is removed from suspension by branchial filaments and entrapped on branchial mucus.

Species differ in their filtration rates and abilities to ingest particles of different sizes, but most are very efficient at extracting a wide variety of particle sizes from the incoming water. Many kinds of tadpoles can ingest particles that may range from below 1 μ to as large as hundreds of micrometers (Kenny 1969b, c; Seale et al. 1982). Seale and coworkers

(Seale 1982; Seale and Beckvar 1980; Seale et al. 1982; Seale and Wassersug 1979) investigated the feeding dynamics of several species of tadpoles. Laboratory trials (Seale and Beckvar 1980) compared the abilities of *Bufo americanus, B. woodhousii fowleri, Pseudacris crucifer, Rana catesbeiana,* and *R. sylvatica* to ingest suspended blue-green algae. All five species ingested cultured algal cells (*Anabaena spherica*) at similar maximum rates. Ingestion rates of *B. w. fowleri* and *R. catesbeiana* were constant when the density of algal cells exceeded about 1×10^7 μm³/ml³. *Pseudacris crucifer* reached its peak ingestion rate at a concentration about 2 times greater. Filtration rate decreased as food concentration increased but at a rate that kept ingestion rate constant. *Rana catesbeiana* larvae ingested monocultures of six species of algae of widely varying sizes and morphologies at similar maximum rates, suggesting that these tadpoles are relatively unselective feeders. Seale (1982) compared the filtration ability of an obligate suspension feeder (*Xenopus laevis*) to those of facultative filter feeders (*Rana pipiens* and *R. sphenocephala*). She found that *Xenopus* tadpoles regulated their filtration and ingestion rates by altering pumping rate, while tadpoles of *Rana* altered their pumping volume. *Xenopus* were able to reach maximum ingestion rates at a lower particle concentration then *Rana*. Her results suggest that there may be some trade-off between pure filter-feeding efficiency and the ability to feed by scraping particles from the substrate. Viertel (1990) examined suspension feeding at low algal concentrations by larvae of *Bufo bufo, B. calamita, Rana temporaria,* and *Xenopus laevis.* He found that *Rana* filtered algae at very low concentrations, while *Bufo* filter-fed at low concentrations only in later developmental stages. All species were more efficient at higher concentrations of algae, and all but *Xenopus* scraped food from the substrate in addition to filter feeding. Viertel (1992) found that the feeding rates of four species responded differently to different particle sizes and that responses also changed ontogenetically within species. In general, feeding was initiated at lower particle concentrations when particles were larger (see Kupferberg et al. 1994).

Some authors have used oral and body morphology to suggest likely foods and feeding behaviors for tadpoles. Orton (1953) suggested that tadpoles fell into seven major ecological groups: arboreal, carnivorous, direct development, generalized lentic, mountain stream, nektonic, and surface-feeding. Orton's work was expanded by Starrett (1973). Inger (1986, discussed below) recognized five ecological types based on morphology and diet. Altig and Johnston (1989) synthesized the available information on feeding morphology and behavior of tadpoles. They proposed the terms endotrophic, to describe species that gain their nutrition entirely from initial oogenic processes of the mother, and exotrophic, to describe species that consume food during development. They also proposed an ecomorphological classification that includes 6 types of endotrophic larvae and 18 types of exotrophic larvae. This scheme should be of considerable use in developing hypotheses about ecological interactions among tadpoles. At least one species of tadpole (Crump 1989b; *Bufo periglenes*) seems to switch facultatively between exotrophic and endotrophic nutrition.

The contributions of Kenny, Seale, and Wassersug greatly advanced our knowledge of the mechanics and rates of tadpole feeding, and the classification of Altig and Johnston (1989) allows more specific suggestions about likely feeding mechanisms based on morphology; data on food usage and conversion efficiency are relatively scarce. Several studies have examined the ability of tadpoles to use different food types. Some researchers used morphological features to predict diet composition, while others directly examined growth and developmental rates on a variety of foods. Most tadpoles are primarily herbivorous, but many are also capable of feeding on carrion and a few are carnivorous. R. M. Savage (1952) examined the gut contents and clearance rates of tadpoles throughout development. The guts of *Bufo bufo* and *Rana temporaria* contained a wide variety of food ranging from nektonic algae to detritus. He concluded that long-term observations of feeding preferences and growth efficiencies would be necessary to establish the value of each food source. Digestive clearance times for *R. temporaria* averaged 6.25 hr at 17°C, while *B. bufo* cleared their guts in 4.75 hr at the same temperature. Some algal cells and fragments of higher plants apparently passed intact through the guts of both species. Pavignano (1989) also found identifiable material in feces and suggested that fecal samples could be used instead of gut contents to study tadpole diets. Tadpoles must lack the ability to digest cellulose and rely on mechanical rupture of plant cell walls to extract the contents (R. M. Savage 1952).

Steinwascher (1979a) raised *Rana clamitans* tadpoles under a variety of conditions in the laboratory and fed them powdered rabbit pellets. In some treatments, this food was incorporated into gelatin-agar pellets, which forced tadpoles to scrape their food into suspension before ingestion. The concentration of food in pellets was either high or low; this presumably varied the energy required to extract a similar amount of food from the pellets. The tadpoles exhibited two basic feeding modes: scraping food from the pellets and filtering suspended food and fecal material from the water column. At low food levels, the proportion of time spent filtering food from the water column increased. The concentration of food in pellets had little effect on tadpole growth rates (fig. 10.2). At high food levels, tadpoles spent more time scraping and grew more rapidly on more concentrated pellets. Under these conditions, scraping was a more efficient feeding mode for *R. clamitans* than filtering, and scraping from concentrated pellets was more efficient than scraping from more dilute pellets.

Altig and Kelly (1974) measured gut lengths and areas and examined gut organic contents of 13 species of tadpoles. Relative gut lengths varied from 1.43 times body length in *Ascaphus truei* to 8.08 times body length in *Scaphiopus couchii.* They suggested that species feeding primarily on animal material should have relatively shorter guts, while those feeding on plant material and detritus should have relatively longer guts. Reexamining their data shows no correlation between the ashed weight/mm³ of gut contents (a measure of inorganic material, and thus detritus, in the diet) and relative gut length ($r^2 = 0.018$; $p = 0.63$). Diet is not well known for

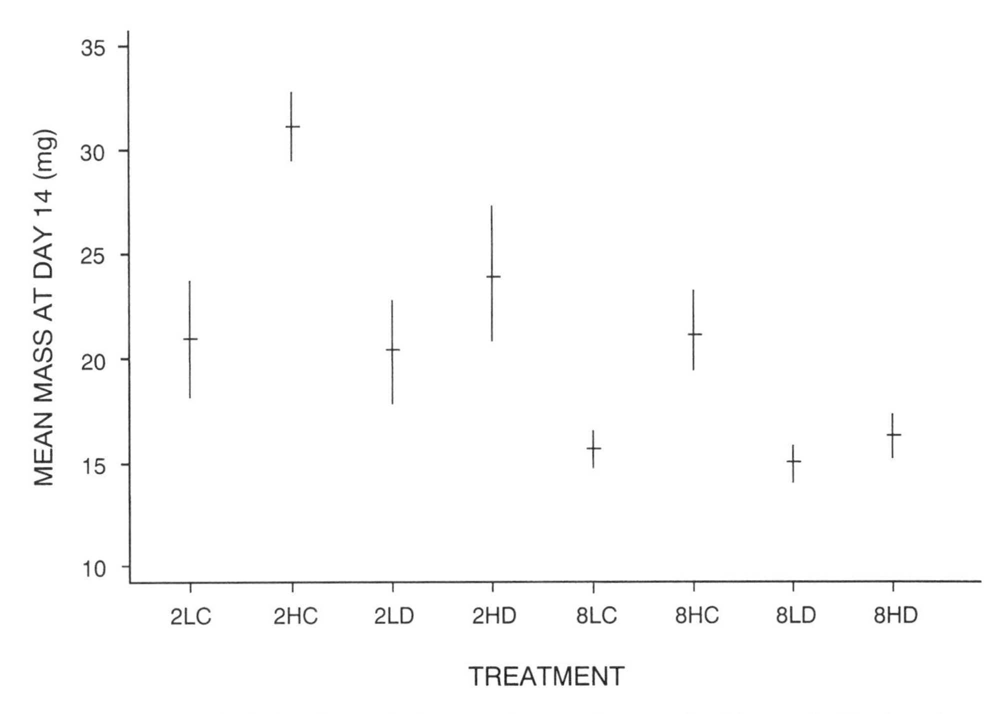

Fig. 10.2. Growth of tadpoles of *Rana clamitans* reared (see x-axis) at two densities, two food levels, and two degrees of food dispersion. Treatments are 2 or 8 individuals per container, high (H) or low (L) food level, and concentrated (C) or dispersed (D) food. Vertical bars = 95% confidence limits of the means. Data from Steinwascher (1979a).

most of the species they examined, and they presented no data on the composition of gut contents, so drawing firm conclusions from their data is difficult.

An alternative approach to understanding the significance of differences in relative intestine length appeared in Horiuchi and Koshida (1989), Nodzenski et al. (1989), and Yung (1904). Horiuchi and Koshida (1989) showed that the intestines of *Rhacophorus arboreus* tadpoles that were fed plant material from hatching were about 1.5 times as long on day 15 as larvae fed an animal diet. Adding cellulose to the animal diet produced longer intestines. Intestines of tadpoles switched from an animal diet to a plant diet partway through the larval period responded by becoming longer and reaching maximum lengths similar to those of tadpoles always fed plant material. This study showed that relative intestine length can be related to the proportion of plant material in the diet. It also indicates that interspecific comparisons of tadpoles from field collections should be made cautiously because there may be considerable phenotypic plasticity for relative intestine length. Nodzenski et al. (1989) examined the development of the viscera of pelobatid tadpoles and found that different organs grow at different rates after tadpoles hatch; in particular, the gut and pancreas experience growth spurts shortly after hatching.

Percent assimilation and clearance times were measured by Altig and McDearman (1975) for tadpoles of five species fed rabbit pellets. Mean percent assimilation of organic matter for four species ranged from 53.8% to 85.7%. The fifth species, *Rana catesbeiana*, showed considerably lower digestive efficiencies of 7.8% at 22°C to 24.9% at 30°C and the fast clearance times ranged from 29 to 101 min at 22°C. Digestive efficiencies may have been higher in this study than they are in nature because processing of rabbit pellets may rupture cell walls and expose a greater proportion of their contents for easy absorption than would be true for intact algal cells. Dash and Mishra (1989) examined digestive efficiency as part of a study on the growth and metamorphosis of *Polypedates maculatus*. Tadpoles may sometimes feed on cryptic items; animals apparently scraping the surface of macrophytes, for example, may actually be feeding on epiphytic diatoms (Kupferberg et al. 1994). Even though larval diets and other rearing factors have been studied by L. E. Brown and Rosati (1997), Carmona-Osalde et al. (1996), Culley (1991), Culley et al. (1977), and Culley and Sotairidis (1983) in the laboratory and by Castanet and Caetano (1995) in the field, dietary requirements and metabolic efficiency of tadpoles need further attention.

A few studies have used growth and developmental rates of laboratory populations to compare the relative value of feeds. Li and Lin (1935) raised larval *Kaloula borealis* at a density of 1/liter (274/m^2). They found that tadpoles fed animal material (egg yolk and macerated pork) grew more rapidly and survived at a greater rate than those fed algae or a hay infusion. Nagai et al. (1971) examined the relative values of several diets for larval *Bufo bufo*. Food mixtures included vegetable material only, vegetable and animal material, animal material only, and a cannibal diet of ground conspecifics. Unfortunately, the cannibal tadpoles were fed only one-fifth as much dry weight per day. Tadpoles on the animal and mixed animal/vegetable material diets reached the greatest metamorphic sizes, while cannibals showed the greatest conversion efficiency. This experiment established that *Bufo* tadpoles benefit considerably from the addition of animal material to their diet and that cannibalism or scavenging from dead conspecifics is an efficient feeding mode. Larval *Bufo stomaticus* and *Hoplobatrachus tigerinus* respond with optimal growth rates and size at metamorphosis on a diet combining animal and vegetable material (Dash and Mahapatro 1990; Hota and Dash 1986; Mohanty and Dash 1986), as do larval *Xenopus laevis* (L. E. Brown and Rosati 1997). Ahlgren and Bowen (1991) found that rates of survival and growth of larval *Bufo americanus* fed on detritus depended on the source and particle size of the material. Pollen was not a good source of nutrition for larval *Pseudacris triseriata feriarum* (Britson and Kissell 1996).

The effect of cannibalism on the growth of larval *Hyla pseudopuma* was investigated experimentally by Crump (1990). These opportunistic cannibals and predators of conspecific and heterospecific eggs and hatchlings in the field were raised individually in laboratory containers on a 1:1 (dry weight) mixture of ground tadpoles and artificial feed. The tadpoles used as food were either *H. pseudopuma, Osteopilus septentrionalis,* or *Scinax staufferi.* In one experiment, each *H. pseudopuma* tadpole was fed 20 mg of one of the types of food every 2 days, and in the other experiment, the quantity of food was adjusted so that each tadpole received food with equal energy content. In both experiments, tadpoles given food containing conspecifics reached significantly greater mass at metamorphosis than tadpoles fed heterospecifics. Crump suggested that cannibalism is a particularly valuable source of animal food for tadpoles and discussed the implications of cannibalism for species population biology and evolution; she pointed out that species that engage in cannibalism should have mechanisms that reduce the probability of consuming close relatives.

In experiments with artificial feeds, Steinwascher and Travis (1983) demonstrated that growth rates of larval *Rana clamitans* do not respond to changes in the ratio of protein to carbohydrate, while larval *Hyla chrysoscelis* grow much better at high protein:carbohydrate ratios. Test and McCann (1976) noted that larval *Bufo americanus* aggregate and switch from scraping substrates to filter feeding in response to high concentrations of protozoans and suggested that protozoans are a higher quality resource than periphyton. In general, the results of these studies suggest that tadpoles grow more rapidly when animal material occurs in their diets.

Many tadpoles may supplement their diets with animal protein through predation or scavenging on conspecific and heterospecific eggs and tadpoles (Crump 1983, 1990, 1992; Magnusson and Hero 1991; Tejedo 1991). Crump (1983) suggested that cannibalism might be particularly common in species that reproduce in ephemeral ponds where it can serve as a means of rapidly consolidating resources in a high-quality form. Larvae of at least 30 species are known to eat conspecific eggs, tadpoles, or metamorphs (table 10.2; Crump 1992). Three of these are *Spea,* which develop specialized cannibal morphotypes under appropriate environmental conditions (Pfennig 1990a). Larvae of at least 15 species eat normal conspecific eggs, while two eat unfertilized nutritive eggs provided directly by the mother. At least 12 species have unspecialized tadpoles that are known to consume conspecific tadpoles, and one species consumes conspecific metamorphs. At least three species are known to cannibalize both eggs and smaller conspecific larvae, while some species (e.g., Heusser 1970b, *Rana temporaria*) prey on eggs of both their own and other species.

Interspecific predation among anurans, particularly predation by tadpoles on the eggs and hatchlings of other species, may prove to be very common. The predators in this situation gain nutrition while reducing or eliminating potential competition or predation. Hota and Dash (1983) attempted an experiment on competition between *Bufo melanostictus* and *Hoplobatrachus tigerinus,* but the experiment failed because the *Rana* larvae consumed many of the *Bufo* in most of their treatments. Tejedo (1991) found that predation by larval *Pelobates cultripes* and *Pelodytes punctatus* strongly affected rates of embryonic survival in *Bufo calamita.* Magnusson and Hero (1991) tested 14 species of tadpoles with eggs of 6 species of frogs; every species tested consumed at least some heterospecific eggs. Crossland and Alford (1998) found that larvae of all six species of Australian frogs they tested attempted to consume eggs of *Bufo marinus* when they were available, despite the fact that *Bufo* eggs were highly toxic to them. Crossland (1997) found that larval *Limnodynastes ornatus* consumed eggs of all five other species they were tested with, as well as cannibalizing conspecific eggs. He also found that the competitive effects of larval *Limnodynastes ornatus* on *B. marinus* were reduced when *B. marinus* entered ponds as eggs after *L. ornatus.* This occurred because consumption of *Bufo* eggs killed larval *L. ornatus,* reducing their numbers and their competitive effects on *Bufo.* Larval *Rana amurensis* preyed on eggs and hatchlings of *Bombina orientalis* (R. H. Kaplan 1992). Petranka et al. (1998) found that larval *Rana sylvatica* consume eggs of the salamander *Ambystoma maculatum,* which is a predator on *Rana* tadpoles. An alternative method for feeding on animal matter is that of *Rhacophorus viridis,* which eat their own foam nests after hatching (Tanaka and Nishinara 1987). Foam nests may also serve to protect eggs from cannibalism or interspecific predation in some species (Downie 1990b; Magnusson and Hero 1991; personal observation).

Estimates of the rate of occurrence of opportunistic cannibalism and cannibal morphotypes are available for a few species, but in general little is known about the importance of cannibalism and carnivory in the diets of generalist tadpoles. In a review of the evolution of cannibalism and cannibal morphotypes in amphibians, Polis and Myers (1985) concluded that there was little evidence for genetic polymorphisms and much evidence for switches to cannibalism in response to environmental cues. Pfennig (1990a) demonstrated that the development of cannibal morphotypes in *Spea multiplicata* is controlled by the availability of animal food and (Pfennig 1992b) that diet switches can cause

Table 10.2 Cannibalism in anuran larvae. Abbreviations: CL = cannibal-morph larvae, E = eggs, L = larvae, M = metamorphs prior to tail resorption, and NE = special nutritive eggs.

Species	Predator	Prey	Reference
Anotheca spinosa	L	NE, L	Jungfer 1995
Bufo bufo[a]	L	L	Nagai et al. 1971
B. marinus[a]	L	E	Hearnden 1992
Chacophrys pierotti[a]	L	E, L	Polis and Myers 1985
Dendrobates pumilio[a]	L	NE	Brust 1993
Euphlyctis cyanophlyctis[a]	L	L	Polis and Myers 1985
Hoplobatrachus tigerinus	L	L	Polis and Myers 1985
Hyla arborea[a]	L	E	Heusser 1970b
H. boans[a]	L	E	Magnusson and Hero 1991
H. geographica[a]	L	E	Magnusson and Hero 1991
H. pseudopuma	L	E	Crump 1983, 1990
H. zeteki[a]	L	E	Polis and Myers 1985
Hymenochirus boettgeri[a]	L	L	Polis and Myers 1985
Lechriodus fletcheri	L	E	Polis and Myers 1985
Leptodactylus pentadactylus	L	L	Heyer et al. 1975
Limnodynastes ornatus[a]	L	E, L	Crossland 1997
Litoria rubella[a]	L	E	Crossland 1997
Osteocephalus oophagus	L	E	Jungfer and Schiesari 1995
O. taurinus[a]	L	E	Magnusson and Hero 1991
Osteopilus brunneus[a]	L	E	Polis and Myers 1985
O. septentrionalis[a]	L	M	Crump 1986
Phyllomedusa vaillanti[a]	L	E	Magnusson and Hero 1991
Pyxicephalus adspersus	L	L	Polis and Myers 1985
Rana sylvatica[a]	L	L	Bleakney 1958
R. temporaria[a]	L	E	Heusser 1970b
Rhinophrynus dorsalis	L	L	Polis and Myers 1985
Scaphiopus holbrookii	CL	L	Polis and Myers 1985
Spea bombifrons	CL	L	Polis and Myers 1985
S. hammondii	CL	L	Polis and Myers 1985
S. multiplicata	CL	L	Pfennig 1990a

[a]Not in table 12.1 of Crump 1992.

S. multiplicata to change morphotypes. Pfennig suggested (1992b) that the proximate cause of switches may be thyroxine obtained from the bodies of prey. Pfennig (1992a) showed that the frequency of cannibal and noncannibal morphotypes also appears to be regulated by pond duration and that different ponds have different frequencies of the two morphotypes. This result was extended by Pfennig and Frankino (1997), who showed that both *S. multiplicata* and *S. bombifrons* are less likely to develop cannibal morphotypes when they are raised with groups of siblings. The occurrence of cannibal morphotypes in *Spea* may be a threshold character mediated by underlying genetic variation in liability. Predisposition to cannibalism in species lacking polyphenisms is also likely to be underlain by quantitative genetic variation. These questions deserve further investigation using quantitative genetic techniques similar to those recently applied to salamander paedomorphosis (Harris 1987; Semlitsch and Wilbur 1989).

Relatively little information is available on the diets of tadpoles; Culley and Sotairidis (1983), Horseman et al. (1976), Leibovitz et al. (1982), and G. A. Marshall et al. (1980) supply various data concerning tadpole nutrition derived from culturing of *Rana catesbeiana*. Field studies of tadpole diets are not common (e.g., Belova 1964; Costa and Balasubramanian 1965; Farlowe 1928; Lajmanovich and Faivovich 1998; Sekar 1992; Sin and Gavrila 1977; also see Ahlgren and Bowen 1991; Płytycz and Klimek 1992; Syuzyamova and Iranova 1988) and often not very informative. After examining gut contents of *Rana clamitans* before and during metamorphosis, Jenssen (1967) concluded that the larvae are unselective continuous filter feeders. The larvae ceased feeding at stage 42 (forelimb eruption) and did not feed again until tail resorption was complete. Hendricks (1973) found that 97% of the food items in the gut contents of *Rana pipiens* larvae were free-floating forms. By examining the diets of larval *Acris crepitans* from two ponds with differing phytoplankton assemblages, L. M. Johnston (1991) concluded that the abundance of algae in tadpole guts were related to their abundance in the environment and that tadpoles from the pond with more abundant phytoplankton grew more rapidly. Johnston's data also suggest that *A. crepitans* are selective feeders; comparisons of proportional similarity showed that the diets of animals were 25% and 67% similar to the composition of the algal assemblage from the two ponds but were 82% similar to each other. C. L. Taylor et al. (1995) demonstrated experimentally that larval *Rana sphenocephala* and *Bufo woodhousii* selected agar blocks with lower agar concentrations and containing food with more animal material.

Dietary information may suggest ecological relationships among species. The diets of seven species living in Spain were related to morphology and microhabitat by Diaz-

Paniagua (1985, 1989). Proportional similarities (Feinsinger et al. 1981) between pairs of five species varied from 0.4 to 0.7. Relative intestine length varied among species and was short in detritivores. Relative intestine length generally increased with development to about Gosner 36 and then remained constant or decreased. Pavignano (1990) concluded that the diets of larval *Hyla arborea* and *Pelobates fuscus* showed large overlaps consistent with a high degree of competition. Twelve of 16 species in a Bornean rain forest (Inger 1986) occupied stream microhabitats. Inger divided these inhabitants into 3 groups based on food particle size: macrophages—2 species that fed on relatively large particles (mean size > 0.12 mm) of mostly vascular plant fragments; mesophages—3 species that fed primarily on intermediate sized particles (0.069 mm < mean size < 0.08 mm); microphages—7 species feeding on small particles (mean size < 0.05 mm). He found broad overlaps in types of food eaten among species within feeding types. Some of this apparent lack of dietary differentiation is probably because of the low statistical power of tests caused by a small sample size (maximum N = 3 per species). Inger suggested that much of the dietary differentiation he observed was because of phyletic differences in feeding behavior, microhabitat preference and morphology rather than to immediate competitive pressures.

Trenerry (1988) examined the diets of four species of tadpoles from a rain forest stream in northern Queensland, Australia, during wet and dry seasons. The suctorial tadpoles of *Litoria nannotis* and *Nyctimystes dayi* inhabit riffles and other fast-moving water in the same areas of the creek. Their diets were 89% similar in the wet season and 85% similar in the dry season (proportional similarity index; Feinsinger et al. 1981). The large dietary overlap suggests that they compete or occupy different microhabitats, and food may not be a limiting resource. Tadpoles of two pool species, *Litoria genimaculata* and *Mixophyes schevilli,* had diets that were 61% similar in the wet season and 68% in the dry season. The diets of the pool and riffle species were 65% similar on average.

Parasites and commensals of tadpoles have not been studied extensively, and this field of research would seem to be worthwhile in light of the current interest in the declining and deformed amphibian phenomena. I present only examples of citations on the subject. The numbers, developmental stages, and species of enteric protozoans vary among tadpoles and depend on the developmental stage of the host (Affa'a 1979, 1983; Affa'a and Amiet 1994; Hegner 1922; Oku et al. 1979; Schorr et al. 1991; Wessenberg 1978), but the significance of these presumed commensals to the welfare of the host has not been determined (see Beebee and Wong 1992a, b). Affa'a and Amiet (1994) presented an engaging summary of the biology of nyctotherans within the frog fauna of Cameroon. Studies of other parasites include Desser et al. (1986, myxozoan), Hird et al. (1981, *Aeromonas*), Nyman (1986, *Aeromonas*), Papernal and Lainson (1995, microsporidian), and Poinar (1988, nematodes). Studies such as Aho (1990) on parasite communities relative to the ecology and phenology of the tadpoles would be of interest. M. L. Adamson (1981a, b) discussed the epidemiology and seasonal changes in populations of a helminth in wild tadpoles, and Goater (1994) and Goater and Vandenbos (1997) discussed the effects of parasites on larval life history of tadpoles; see Eberhard and Brandt (1995) and Schmutzer (1977) for other information on the interactions of tadpoles and parasites. Thrall (1972) reported that there were no bacteria in the guts of tadpoles of *Rana catesbeiana,* while others (e.g., Battaglini and Boni 1967; R. Altig, personal communication) report numerous bacteria.

Because tadpoles often occur at relatively high densities, usually have broad diets, and have high rates of ingestion and clearance, they may be the major primary consumers in some ecosystems, particularly temporary ponds. Tadpole grazing can strongly alter the standing crop and species composition of algae. Dickman (1968) showed that grazing by caged *Rana aurora* tadpoles reduced the standing crop of filamentous green algae to 1%–2% of levels found in ungrazed controls. This altered the species composition of the algal community. Lamberti et al. (1992) found that tadpoles of *Ascaphus truei* (5/m^2) could reduce periphyton biomass by 98% and chlorophyll *a* by 82% in streams in Washington, a response that was about equal to the total effects of grazing by small invertebrates. Seale (1980) showed that tadpoles, feeding primarily by filtration from the water column, significantly altered the distribution of nitrogen from suspended to dissolved compounds. Tadpoles also shifted the community composition of algae in a pond from blue-green algae to green algae. Morin et al. (1990) demonstrated that *Hyla andersonii* tadpoles reduced the standing crop of periphyton in artificial ponds. In some habitats, tadpoles may alter algal species composition through feeding and by promoting nutrient cycling (Kupferberg 1997a; Osborne and McLachlan 1985; Weigmann 1982). Tadpoles may also simply decrease algal biomass without affecting species composition (Holomuzki 1998). The feces of tadpoles can have greater nutritive value than the food the tadpoles ingest (Steinwascher 1978a). Tadpole feces can form a nearly continuous layer several centimeters deep in a Carolina bay (Harris et al. 1988) at the end of summer (personal observation). If the tadpole populations observed by Trenerry (1988) in a northern Queensland creek were processing food at rates similar to those measured by Altig and McDearman (1975), in late summer his pool 2 would receive roughly 1000 g of tadpole feces per day. It seems likely that in some systems tadpole grazing may strongly affect the composition of the plant community and that tadpole feces may be an important source of fine particulate organic matter for detritivores of many taxa. This possibility deserves further investigation.

Growth, Development, and Survival

A host of factors has been suggested as potential influences on the rates of growth, development, and survival of tadpoles in nature. Many species may grow exponentially (Alford and Jackson 1993). Growth rates may be influenced by competition, environmental temperature, and other aspects of environmental quality such as dissolved oxygen and pH (see Bradford et al. 1994; Cummins 1989; Dunson and Connell 1982; Picker et al. 1993; Saber and Dunson 1978). Competition could be of several types, including intra- and intercohort competition within species, interspecific competition within the tadpole "guild" (cf. Root 1967), competition with other vertebrates such as larval fish, and competition with members of other phyla. The mechanism of competition could be exploitation (usually defined as a resource demand:supply ratio greater than 1:1) or interference. Interference competition may be behavioral or allelopathic, which could be caused by chemicals, parasites, or pathogens. Rates of development may respond directly to environmental influences such as dissolved oxygen, pH, and temperature (Hayes et al. 1993; chap. 8). Endocrine feedback systems may also lead to responses to extrinsic factors such as photoperiod (e.g., M. L. Wright et al. 1990) or to intrinsic cues such as growth rate or body size thresholds (chap. 11, and below). Survival is directly affected by predation and may be influenced strongly by growth rate; mortality rates of tadpoles may decrease with increasing size, and faster-growing tadpoles often metamorphose earlier than slower-growing individuals and thus experience lower cumulative mortality rates as larvae. Survival is also dependent on biotic and abiotic environmental factors—the composition of the assemblage of potential predators and the probability of environmental changes such as drying of ephemeral ponds.

A model of influences on development, growth, and survival of tadpoles proposed by Wilbur and Collins (1973) stimulated a great deal of work. They noted that the developmental rates and metamorphic sizes of individuals of the same species often vary widely within a given physical environment. They suggested that this might be caused by an evolutionary advantage conferred on species that are able to adjust their developmental rate in response to their growth rate. Adjustment would allow individuals to alter the proportion of total growth between egg and first reproduction that occurs in the aquatic environment. If aquatic conditions were favorable, allowing rapid larval growth, tadpoles would slow their developmental rate and metamorphose at a large body size. If aquatic conditions were unfavorable, tadpoles would accelerate their developmental rate, metamorphose at a species-specific minimum body size, *b,* and continue growing in the terrestrial stage. Collins (1979) modified their suggested feedback process slightly (fig. 10.3).

One problem with the Wilbur-Collins model is the lack of a specific mechanism for the feedback of growth rate on developmental rate. Crump (1981) found that larval *Pseudacris crucifer* raised at high density and fed ad libitum had lower mass specific energy contents and metamorphosed at smaller sizes than tadpoles raised at low density. She suggested that the accumulation of energy reserves, in the form of body fat, might affect differentiation rate (McCarrison 1921). A direct mechanism that might allow feeding rate, and thus growth rate, to control differentiation rate was suggested by Wassersug (1986). He found that a hormone in the prostaglandin E group (PGE_2) is secreted into oral mucus by some tadpoles and transported into the stomach with entrapped food particles. Inhibition of the development of adult stomach features is caused by PGE_2 (Wassersug 1986). Higher quantities of PGE_2 are ingested at higher feeding rates. Wassersug suggested this should slow the development of the stomach and could also act to inhibit the operation of the hormonal mechanisms promoting development of other features. Mobbs et al. (1988) investigated this hypothesis with isolated tadpole tails in vitro and found no evidence for PGE_2 inhibition of thyroid-hormone induced metamorphosis. They pointed out, however, that PGE_2 might not act directly on tail tissue or that some other component of oral mucus may regulate metamorphosis. Delidow (1989) examined the effects of a variety of growth hormones on larval *Rana* and found differential effects on tail and body growth. He suggested that differentiation rate was not simply regulated by thyroxine.

A number of studies have examined predictions of the Wilbur-Collins model, while others have proposed alternative models for the control of growth and development in tadpoles. In their original paper, Wilbur and Collins made several predictions about the responses of growth and developmental rates of individual tadpoles to ecological conditions and the responses of tadpole and metamorph populations to ecological conditions. They predicted that populations in stable, predictable environments would tend to maximize the proportion of growth accomplished in the aquatic environment. Most individuals would metamorphose at or near the species-specific maximum body size, *c,* after widely varying larval periods. Conversely, they suggested that populations in unstable, unpredictable environments such as temporary ponds should show greater variance in both larval period and metamorph body size as individuals adjusted their differentiation rates in response to local conditions. Within populations, they suggested that at low-density/high-resource availability, most individuals would grow at rates greater than *g* and thus would delay metamorphosis until reaching maximum size, *c.* At higher densities/lower resource availabilities, the first individuals to reach minimum metamorphic size would be growing slowly and would metamorphose at or near *b.* Their departure would make resources available to the remaining tadpoles and allow them to grow more rapidly and delay their metamorphosis until reaching slightly larger body sizes. This process would continue until the growth rates of remaining individuals were sufficient to allow them to metamorphose at maximum size.

Wilbur and Collins (1973) did a preliminary test of one prediction of their model: individuals should grow more rapidly when released from density stress. They selected larval *Rana sylvatica* in 3 size classes (< 100, 101–200, 201–400 mg body mass) from populations that had been growing at high densities (806/m^2; 5.33/liter) and placed them in pens at low density (13.4/m^2; 0.089/liter). All of the new populations began growing rapidly. Wilbur and Collins interpreted this as support for their model. Collins (1979) used the model to predict how the size-frequency distribution of metamorphs emerging from a population of *Rana catesbeiana* larvae should change over time. He suggested that the availability of resources present after fast-developing individuals metamorphosed should lead to populations that obey the Wilbur-Collins model by showing a pattern of increasing metamorph mass with increasing larval period. The *R. catesbeiana* populations he studied appeared to follow this pattern.

Pfennig et al. (1991) detected the opposite pattern in four natural populations of *Spea multiplicata*. They varied food levels and the duration of artificial ponds to examine the causes of this correlation. Age and size at metamorphosis were negatively related. Larval period and size at metamorphosis were positively correlated in long duration, low-food ponds and were negatively correlated in high-food ponds. They suggested that negative correlations are likely to be common in nature because as ponds age, the animals in them are likely to experience diminishing resource availabilities. Kehr and Adema (1990), Reques and Tejedo (1995), and Tejedo and Reques (1992) also discuss correlations between size at metamorphosis and length of larval period in the context of the Wilbur-Collins model.

Several studies examined the effects of decreases in population density late in the larval period on rates of growth and differentiation in tadpoles. Semlitsch and Caldwell (1982) reared *Scaphiopus holbrookii* tadpoles at three densities in aquaria. On days 7, 15, 28, and 56 of their experiment, they removed 10 individuals from each density treatment and reared them alone. They found that length of the larval period was affected by density and by the timing of release from density stress, but the effect of timing of release was independent of density. This suggests that after density release, all individuals shifted to similar developmental programs. Their experiment demonstrated that differentiation rate can respond to changes in ecological conditions during the larval period. Travis (1984) used larval *Hyla gratiosa* to determine whether low food level alone could produce a positive correlation between mass at metamorphosis and length of larval period within a cohort and to examine the effects of competitive release. He reared individual tadpoles in 500 ml of water. Two sets of 40 individuals were fed at constant rates of 1 and 2X per individual. In 2 other sets of 40 (the "release" treatments), the food rations of metamorphosed animals were divided among those remaining. Larvae fed at the 1X rate had more variable larval periods than larvae fed at 2X. Larval periods of animals in the release treatments were no more variable than those animals fed at constant relative rates per individual. Metamorphic masses of individuals in the low-food, release treatment were greater on average than those of animals in the low, constant food treatment and were also more variable. Because the release treatments did not appear to affect length of larval period, Travis proposed that differentiation rate might be fixed early in the larval period.

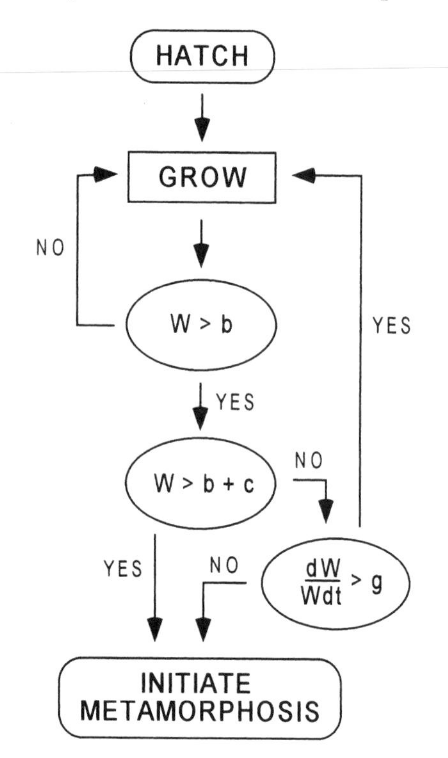

Fig. 10.3. The Wilbur-Collins model relating growth and development rates of tadpoles. Abbreviations: b = minimum size for metamorphosis; c = difference between minimum and maximum body size for metamorphosis; g = size-specific growth rate below which metamorphosis is initiated, above which metamorphosis is delayed; and W = body mass. After Collins (1979).

The laboratory experiments discussed thus far examined responses to competitive release but did not address the prediction of the Wilbur-Collins model that animals in deteriorating environments should increase their differentiation rates and metamorphose more rapidly than animals in constant environments. Alford and Harris (1988) used larval *Bufo woodhousii* to simultaneously examine the effects of constant, increasing, and decreasing rates of food supply on growth and differentiation rates. We raised tadpoles alone and supplied food at two relative rates. On days 12, 18, and 30 of the experiment, we switched one treatment from the low food regimen to the high food regimen and one treatment from high to low food. We found that differentiation rate was not fixed early in development; individuals switched from high to low food levels as late as day 18 had larval periods nearly identical to individuals raised at constant low food. Individuals switched from low food level to high food level at days 12 and 18 reached metamorphic masses nearly identical to individuals raised at constant high food after very similar larval periods. Individuals switched from low to high food on day 30 reached metamorphic masses very similar to constant high-food individuals but took longer to do so. Differentiation rate thus responded to both increases and decreases in growth rate in a manner consistent with the Wilbur-Collins model. Hensley (1993) performed a series of experiments with *Pseudacris crucifer* larvae that were designed similarly to that of Alford and Harris (1988). Hensley found that differentiation rate in this species responded to growth rate in accordance with the Wilbur-Collins model until about Gosner 35, at which time it became fixed. Beck (1997) and Leips and Travis (1994) have continued this line of research by focusing on when in the larval period developmental rate loses its ability to adjust to different growth conditions. They demonstrated that the stage at which fixation occurs varies among species. Beck (1997) summarized adaptive explanations that have been proposed to explain this variation. Although it has also been suggested that this variation may be nonadaptive, to my knowledge no studies have examined the possibility that it may be explained by simple functional morphological constraints. That is, if the development of the forelimbs constrains the filtering and food collecting apparatus in any way, the potential increase in growth rate achievable if resource levels increase must decrease. Perhaps developmental rates become fixed at or after the point at which the trade-off between forelimb size and filtering rate becomes unfavorable for additional growth. This possibility would be fairly easy to examine experimentally.

Crump (1989a) examined whether tadpoles of *Hyla pseudopuma* increased their developmental rates when water level decreased. She raised tadpoles at a single density under constant high water levels, constant low water levels, and decreasing water levels. Tadpoles raised in decreasing water levels had the shortest larval periods and metamorphosed at sizes similar to tadpoles raised at constant low water levels. Tadpoles raised at constant high water levels were significantly larger at metamorphosis than either of the other groups and metamorphosed after intermediate larval periods. Tejedo and Reques (1995) showed that *Bufo calamita* larvae metamorphosed more rapidly from ponds with shorter duration. These studies again demonstrated that developmental rate can respond to ecological deterioration in the direction predicted by the Wilbur-Collins hypothesis. Although most tadpoles appear to have at least some ability to respond as predicted by Wilbur and Collins, larval salamanders may not. Beachy (1995) found that developmental rate of larval *Desmognathus ochrophaeus* did not respond to changes in growth rate at any point during the larval period.

The Wilbur-Collins model is an abstract model concerned with suggesting how relative rates of growth and development should be controlled within a population experiencing a common physical environment. Smith-Gill and Berven (1979) developed a predictive model for larval period and size at metamorphosis in populations of *Rana pipiens* exposed to different thermal regimes. They found that differentiation, as expressed by position along the Taylor/Kollros staging series (A. C. Taylor and Kollros 1946), was strongly correlated with size and time. This result might have been predicted; the Taylor/Kollros staging series was developed by choosing the morphological benchmarks for each stage deliberately so that stage increased linearly with time in a laboratory population of *Rana pipiens*. Smith-Gill and Berven (1979) also found that the relationship between size and developmental stage depended on environmental temperature; developmental rate decreased faster than growth rate as temperature decreased so that larvae grown at lower temperatures grew more in each stage and metamorphosed at larger body sizes than those reared at higher temperatures. Berven et al. (1979) applied these concepts to explain differences between montane and lowland populations of *Rana clamitans*. Pandian and Marian (1985a) developed a similar model that predicts metamorphic size based on the feeding rates and stage-specific growth rates of *Hoplobatrachus tigerinus* larvae raised under a variety of regimens involving density, food quality and quantity, and temperature.

Werner (1986) developed a model (see chap. 11) that suggests that the timing of metamorphosis should be determined by trade-offs between growth and mortality rates in the aquatic and terrestrial environments. In a stable population, this would lead to metamorphosis at a body size that minimized the ratio of mortality rate to growth rate at each size. Further development of this type of model by Ludwig and Rowe (1990) and L. Rowe and Ludwig (1991) has increased the model's realism by allowing for constraints on the timing of first reproduction. These models are not incompatible with the Wilbur-Collins model or any other model thus far proposed for the proximate control of growth and metamorphosis in anurans. These models provide a framework for broader interspecific comparisons (e.g., Patterson and McLachlan 1989; Pough and Kamel 1984).

The remainder of this section concentrates on studies that have examined aspects of the growth and survival of single species of tadpoles in a variety of natural and artificial systems.

Several of these studies originally suggested that tadpole mortality rates decreased as tadpoles grew. Petranka (1985) examined data from several studies and showed that the mortality rate (proportion of the population dying/time) of tadpoles is usually constant over the larval period. Herreid and Kinney (1966) studied populations of *Rana sylvatica* tadpoles at four small ponds in central Alaska. All populations experienced nearly constant proportional mortality from oviposition through metamorphosis. They noted considerable predation by larval diving beetles (*Dytiscus* sp.), whose growth patterns may enable them to prey at nearly constant rates on growing tadpoles (Brodie and Formanowicz 1983). Augert and Joly (1993) found that larval *Rana temporaria* from two adjacent sites differed in size at metamorphosis. They suggested that this difference was caused by environmental rather than genetic effects. Calef (1973; *Rana aurora* larvae in Marion Lake, British Columbia, Canada, 1969–1970) counted the numbers of egg masses deposited and followed the resulting tadpole populations through metamorphosis. In both years, the tadpoles died at a relatively constant rate, and about 5% survived to metamorphosis. Mean body size at each census was strongly correlated to the number of degree-days since hatching of the cohort. Caging experiments established that food was not limiting. Most mortality appeared to be caused by a variety of invertebrate (e.g., odonate naiads) and vertebrate (e.g., rainbow trout, *Oncorhynchus mykiss*) predators. This wide range of predator sizes may explain the apparent lack of a size refuge from predation. Relatively constant rates of mortality were also found by Riis (1991) in populations of larval *Rana dalmatina* and *R. temporaria*.

Predation also appeared to be the major cause of mortality in two populations of *Rana catesbeiana* larvae (Cecil and Just 1979) near Lexington, Kentucky. For two years, mortality rates appeared to be constant but growth was exponential. Absolute population densities ranged from $7/m^2$ shortly after hatching to less than $1/m^2$ nearly 1 year later. Many invertebrate and vertebrate predators were present. Licht (1974) measured growth and survival of *R. aurora* and *R. pretiosa* in a temporary pond and a river in British Columbia, Canada. Differences in percent survival to hatching, an average of 71% of *R. pretiosa* and 92% of *R. aurora,* were caused by choice of oviposition sites; *R. pretiosa* deposited eggs near the margins of water bodies where they were vulnerable to desiccation caused by small fluctuations in water level. Both species showed relatively constant mortality rates throughout development. Cumulative survival from egg to metamorph was less than 1% in the pond and about 5% in the river site. Licht suggested that most tadpole mortality was caused by predation. When predation is absent, tadpoles can survive to metamorphosis at high rates. Seigel (1983) estimated that *R. sylvatica* in a highly ephemeral pond experienced only 37.5% mortality from egg to metamorph.

D. C. Smith (1983) presented one of the most comprehensive studies of growth and survival of tadpoles under natural conditions. He studied populations of the chorus frog, *Pseudacris triseriata,* which breed in ephemeral rocky pools on the shorelines of two small islands in Lake Superior, just off the coast of Isle Royale National Park, Michigan. *Pseudacris triseriata* required ephemeral pools with intermediate size and duration for development. Larger, more permanent pools harbor populations of dragonfly naiads (*Anax junius*) that eliminate tadpoles from them, while very small, ephemeral pools were either scoured out by rainfall or dried before tadpoles could metamorphose. Removal of *A. junius* increased tadpole survival in large pools. Variation in the persistence of intermediate-size pools caused tadpole survival rates to vary from 0 to 100%. Tadpole densities in these pools were high (mean initial density = 0.95/liter) and often increased as tadpoles grew and pool size decreased. There was strong evidence for effects of density on the growth rates and size at metamorphosis of tadpoles. Experimental manipulations of density and food level showed that competition was occurring.

Schmuck et al. (1994) studied the effects of density on growth and metamorphic performance as well as associated physiological responses. Tadpoles of *Hyperolius viridiflavus* responded to increases in nitrogenous waste from metabolic processes or evaporation by increases in the ornithine cycle (increased urea synthesis; also Shoemaker and McClanahan 1973). In contrast, tadpoles of *H. marmoratus taeniatus* tolerated high ammonia levels (also Candelas and Gomez 1963; Harpur 1968). The deterioration of natal pond conditions also increased the postmetamorphic responses of *H. v. nitidulus* and *H. v. ommatostictus.*

Newman (1987) conducted a study that used natural ponds but allowed a rigorous analysis of patterns. He monitored the growth and survival of *Scaphiopus couchii* tadpoles in 82 ephemeral desert ponds over a 3-year period. In 16 ponds, predation was the major cause of *S. couchii* mortality, and desiccation was the major cause of death in 49 ponds; only eight ponds produced metamorphs. High initial densities of eggs led to slow growth and mortality through desiccation. Supplementing the available food in high-density ponds increased individual growth rates, allowed some animals to metamorphose, and demonstrated that resources were in limited supply.

Berven (1990) studied the population biology of terrestrial and aquatic stages of *Rana sylvatica* at two ponds in Maryland over a 7-year period. Enclosure of the ponds with drift fences allowed complete censuses of frogs as they arrived at and departed from the ponds. Survival from egg to departing metamorph varied from 0.36% to 8.0% ($\bar{x}$ = 4.5%) at one pond and 0.09% to 1.7% ($\bar{x}$ = 0.95%) in the other. There was strong correlational evidence that larval growth and survival in each pond depended on larval density. Spearman rank correlations for the two ponds were −0.77 and −0.90 between the proportion surviving to metamorphosis and number of eggs deposited, −0.77 and −0.94 between mean snout-ischium length of metamorphs and number of eggs, and 0.77 and 0.83 between mean larval period and number of eggs. Cohorts developing in years of high adult reproductive effort survived at lower rates, reached smaller metamorphic sizes, and spent longer in the aquatic environment than cohorts in years with lower egg input. Juveniles produced from larger, faster developing larvae were more likely to survive to reproductive age, reproduced earlier, and were larger and more fecund as adults. Berven suggested that the consistently lower larval survival rates in one pond may have been caused by larger numbers of predators in the pond that dried less often. Larval density, predation, and pond duration appeared to control the growth and survival rates of larval *Bombina variegata* at 46 Swiss ponds studied by Barandun and Reyer (1997). Skelly (1996) showed that larvae of *Pseudacris crucifer* and *P. triseriata* are affected by density, predators, and pond permanence, while L. Blaustein and Margalit (1995) demonstrated that the distribution of larval *Bufo viridis* among pools was correlated with the availability of filamentous algae.

While the studies cited above have produced useful estimates of growth rate and survival of tadpoles under natural conditions, most allow many possible interpretations of the observed patterns. Growth and survival of tadpoles have been studied rigorously in nearly natural systems by two additional approaches. The first approach is to subdivide natural populations in pens and manipulate factors such as density and predators. The second approach is to construct artificial communities in large outdoor ponds or tanks.

Brockelman (1969) placed pens in two ponds to examine the responses of larval *Bufo americanus* to conspecific density, food level, and the presence of predators. Larval period increased and metamorph body size decreased with increasing density, and the addition of supplemental food reduced but did not eliminate the effects of increased density. In one pond, his experimental pens were heavily colonized by predators, and low survival was not related to initial density. In the other pond, mortality was lower and strongly related to density; about 45% survived to metamorphosis at an initial density of 67 tadpoles/m^2 (= 0.4/liter), and survival decreased to about 32% at 178 tadpoles/m^2 (= 1.18/liter) and

14% at 471 tadpoles/m^2 (= 3.12/liter). The natural tadpole density in the second pond was higher, and his estimate of cumulative larval survival was also higher—about 5% of tadpoles surviving to metamorphosis.

The pen technique was also used by Wilbur (1976) in his study of density effects on the survival, growth, and development of *Rana sylvatica* tadpoles. He reared tadpoles at initial densities ranging from 100 (= 0.66/liter) to 1612/m^2 (= 10.66/liter) in pens placed in a natural pond. Proportion surviving to metamorphosis decreased exponentially with increasing density. Mean body mass at metamorphosis also decreased exponentially with density, with a lower asymptote of about 150 mg reached at an initial density of 403 tadpoles/m^2 (= 2.66/liter). The distribution of body mass at metamorphosis was not significantly different from a normal distribution. Wilbur suggested that this arose as a result of individual larvae following the Wilbur-Collins model of growth and differentiation. Skelly (1995b) used pen experiments to show that larval *Pseudacris crucifer* are not excluded from permanent ponds by competition with *Rana clamitans* larvae.

Travis (1983b), Travis et al. (1985a), and Travis and Trexler (1986) used populations raised in field enclosures to examine how ecological and genetic factors affect the growth and development of several species of tadpoles. *Hyla gratiosa* larvae of 7 sibships reared in enclosures at an initial density of 1.6 animals/liter (235/m^2) survived to 5 weeks of age at rates varying from 1% to 22% (Travis 1983b). The enclosures excluded larger predators but were colonized by a variety of invertebrate predators. In each enclosure, percent surviving was positively correlated with mean body size. This suggests that faster growing populations suffered less mortality from predation than slower growing populations. Travis et al. (1985b) further examined the influences of size-limited predation on *H. gratiosa*. They raised tadpoles of 2 sibships at initial densities of 21, 42, and 84/m^2 (0.055, 0.11, and 0.22/liter) with predators (naiads of *Tramea lacerta*, larval *Ambystoma opacum*, or both) present or not. The experiment started 2 weeks after egg deposition. Survival to day 10 of the experiment varied from 0 to 78% with a median of 31%. Initial tadpole density did not affect survival to day 10. On average, 47% of tadpoles survived to day 10 when both predators were absent. Both predators affected survival equally by each reducing survival by 24% compared to treatments where both predators were absent. The predators did not influence each other; survival when both predators were present was exactly as predicted if there was no interaction between the sources of mortality. The two sibships grew at different rates, and the faster growing sibship had consistently greater survival than the slower growing sibship.

Using an ecologically different species, Travis and Trexler (1986) examined the interaction between influences of general environmental harshness and density on the growth and survival of tadpoles of *Bufo terrestris* of three initial size classes. In 1 experiment, they placed tadpoles belonging to 3 size classes in the same enclosures used by Travis et al. (1985a) at densities of 42, 84, and 168/m^2 (0.11, 0.22, and 0.44/liter). Initial body size explained a significant amount of the variability in survival to metamorphosis; tadpoles with the largest initial sizes survived better than medium-sized tadpoles, which did better than the smallest ones. The initial medium-sized group metamorphosed at the same body size as the initially large group.

Petranka (1989) used manipulations of a seminatural pond, experiments in field enclosures and laboratory experiments to examine whether chemical interference competition among tadpoles occurs outside the laboratory. His study focused primarily on larval *Rana sphenocephala*. He introduced hatchling tadpoles to a 10 × 19 m pond devoid of tadpoles (initial density about 416/m^2; 1.4/liter). He assayed local crowding effects by growing hatchling tadpoles in water from locally crowded and uncrowded areas of the pond. Early in the history of the population, when local densities were high, water from aggregations significantly inhibited the growth of assay tadpoles. Diffuse inhibitory effects were assayed by raising individual tadpoles in situ in enclosures that afforded various degrees of exposure to possible inhibitory substances. These tadpoles showed no evidence for growth inhibition over 10 days of growth. He also collected 13 water samples from 11 breeding ponds and assayed them for inhibitory effects on the growth of *R. sphenocephala*. Two of the samples inhibited *Rana* growth: one from a pond with high densities of *Gastrophryne carolinensis* and *Hyla chrysoscelis* tadpoles and one from a pond with a relatively high biomass/unit area of overwintering *R. sphenocephala* tadpoles. Petranka concluded that chemical interference may be relatively uncommon in natural populations but may be important when aggregation or pond drying produces high local densities. Much remains to be learned about chemical interference; for recent perspectives on this subject and the role that the unpigmented alga *Prototheca* might play, see papers by G. C. Baker and Beebee (1997), Beebee (1995), R. A. Griffiths (1995), and Petranka (1995).

Field enclosures were used by Skelly (1991) to demonstrate that antipredator behavior by larval *Hyla versicolor* has a cost in reduced growth rate. He placed larval *Ambystoma tigrinum* in mesh enclosures inside pens in a pond. *Hyla versicolor* larvae in pens with *Ambystoma* larvae grew more slowly than those in pens with empty predator enclosures. Laboratory experiments showed that the overall activity levels of tadpoles decreased when predators were present. Tadpoles of many species decrease activity when predators are detected (Hokit and Blaustein 1995; Lefcort 1996, 1998). Changes in activity levels are often accompanied by shifts in microhabitat use (Anholt et al. 1996; Hews 1995; Kupferberg 1998; Manteifel 1995). Antipredator responses can be depend upon the size of the tadpole (Anholt et al. 1996; Bridges and Gutzke 1997), the relatedness of tadpoles in groups (Hokit and Blaustein 1995), and the previous diet of the predator (Laurila et al. 1997).

Plastic bags floating in ponds were used as field enclosures by R. H. Kaplan (1992) for experiments that demonstrated that changes in maternal investment as reflected by egg size can alter hatchling morphology and vulnerability to predation in *Bombina orientalis*.

Field enclosure experiments offer the benefits of increased control but often suffer from high variability among replicates. The use of artificially assembled experimental communities is the next level of increased control and decreased realism. Morin (1981, 1986) first used this technique to study tadpole ecology. The artificial ponds used in his studies, like most others who have employed this approach, are cattle watering tanks that contain about 1,000 liters of water. Tanks usually are covered with lids made from screening to prevent colonization by aquatic insects and insect larvae. In a typical experiment, tanks are filled with water one to two weeks before hatchling tadpoles are added. Leaf litter, supplementary nutrients, and a standardized set of zooplankton and phytoplankton are added sometime after the ponds are filled. This technique establishes a set of extremely similar but totally independent communities that can be manipulated by selective additions or removals of larval anurans and other species. The advantages and drawbacks of this approach to experimental ecology are discussed by Hairston (1989), R. G. Jaeger and Walls (1989), Morin (1989), and Wilbur (1989).

Morin (1986) examined the effects of intraspecific competition and predation by the newt *Notophthalmus viridescens* on the growth, larval period, and survival of *Pseudacris crucifer* tadpoles. Three densities of hatchling tadpoles (31, 62, and 124/m^2; 0.1, 0.2, and 0.4/liter) were combined with 0, 2, or 4 adult newts per pond. Only newt density affected survival of tadpoles; it declined from 73% with newts absent to 61% and 47% at medium and high newt densities, respectively. The growth rate, length of larval period, and mass at metamorphosis of tadpoles were primarily controlled by initial tadpole density; tadpoles at higher densities grew more slowly and emerged at smaller sizes after longer larval periods. Tadpoles avoided newts by switching from foraging in exposed sites to foraging in hidden, benthic sites. This apparently did not affect their growth rates.

Artificial ponds were used by A. H. Roth and Jackson (1987) to examine the effects of pond size on survival and predation rates of larval *Hyla cinerea*. They established pools of about 0.09, 1.5, and 3 m^2 and allowed insects to colonize some of them for 3 months. Over the remainder of the year, they performed 3 experiments in which larval *H. cinerea* were stocked at a constant density of 43/m^2 (0.43/liter) and followed survival of the tadpoles through metamorphosis. Most predation occurred in the first week of larval life, and predation rates increased significantly as pool size increased. They suggested that greater predation pressure in large pools might be one factor causing selection of small pools as reproductive sites.

Fauth (1990) manipulated the occurrence of adult *Notophthalmus viridescens* and the crayfish *Cambarus bartonii* in 375-liter artificial ponds and examined the effects on the growth and development of larval *Hyla chrysoscelis* at a density of 200 individuals per pond (130/m^2; 0.53/liter). The experimental treatments included neither predator present, only *N. viridescens* present (two per pond), only *C. bartonii* present (two per pond), and two predators of each species present per pond. Each treatment was replicated eight times. Four replicates were destructively sampled on days 5 and 6 of the experiment, and those that remained were destructively sampled on day 30. In ponds with no predators, an average of 90% of *Hyla* tadpoles survived the first 5–6 days of the larval period, and 43% survived to day 30. When *N. viridescens* only were present 21% of *Hyla* survived to day 5/6, and none survived to day 30. Predation by *C. bartonii* followed a different pattern; 78% of *Hyla* exposed only to crayfish survived to day 5/6, and 14% survived to day 30. When both predators were present, an average of 29% of *Hyla* survived to day 5/6 and 1% survived to day 30. Despite their lower density, *Hyla* tadpoles exposed to either predator or to both together were smaller on day 5/6 than tadpoles not exposed to predators. Tadpoles survived to day 30 in reasonable numbers only when predators were absent or *C. bartonii* was the only predator; data (table 1 of Fauth 1990) indicate that mean tadpole size at day 30 in these treatments did not differ significantly. These results can be explained by antipredator behavior by *H. chrysoscelis* tadpoles. Presence of either predator elicited avoidance behavior that reduced the growth rates of tadpoles in the first days of life. Avoidance behavior elicited by the presence of both predators simultaneously may have reduced mortality caused by either predator more than the avoidance elicited by only one species of predator. Fauth concluded that the effects of the predators on the growth and survival of *H. chrysoscelis* were not additive and that their final effects at the end of the larval period could not be predicted from short-term studies near the beginning of the larval period.

The artificial pond technique allows for controlled experiments under seminatural conditions but still leaves some variation uncontrolled. For example, because artificial ponds are largely self-sustaining systems, most tadpoles feed on primary production that occurs in the ponds. Rates of primary production, and thus of food supply, are likely to differ between ponds given different experimental treatments. Additionally, unbiased intermediate censuses and measurements are difficult to conduct; artificial ponds present the same sampling problems that arise in natural ponds. It is difficult to follow individuals, because that would require that they be individually marked. Fine control of factors like temperature and photoperiod is also difficult. To look in more detail at one or a few controlled factors, many studies have been conducted under laboratory conditions.

Early studies of tadpole populations in the laboratory usually were concerned with examining factors controlling growth and differentiation rates from a mechanistic rather than an ecological perspective. R. M. Savage (1952) observed that large *Bufo* tadpoles behaviorally kept smaller individuals away from preferred food sources. He suggested that this might lead to inhibition of the growth of smaller individuals. Adolph (1931) showed that growth rates of *Rana pipiens* and *R. sylvatica* declined with increasing density and that the growth rates of tadpoles raised alone decreased as the size of the rearing container decreased (see Guyetant 1977a, b; Hourdry and Guyetant 1979; Lametschwandtner et al. 1982; Streb 1967). He attributed the effects of crowding to "stress," which increased metabolic rate and decreased growth efficiency of tadpoles. An experiment by Lynn and Edelman (1936) suggested that stress caused by physical encounters between individuals might decrease growth rates under crowded conditions. They reared *Rana sylvatica* tadpoles in an aquarium separated into a series of compartments by mesh screen that allowed water circulation and vision between compartments. They used a wide range of densities (7.8–121.0/liter) and supplied food ad libitum. Tadpoles in the more crowded compartments grew more slowly, experienced higher rates of mortality, and metamorphosed later than tadpoles in less crowded compartments. The feeding regimen eliminated food as a possible cause of these effects. Water circulation and visual communication between compartments should have reduced any effects of diffusible substances or visual disturbance. The effects of water-borne substances may not have been eliminated, as after initial mixing the circulation of any substances produced would have been by diffusion, and appreciable concentration gradients could be maintained. Rugh (1934) reached a similar conclusion with a similar experiment on *R. pipiens.*

Gromko et al. (1973) suggested that reduced growth in larval *R. pipiens* reared under crowded conditions might be caused by adrenocortical responses to social stress. Although perhaps a different phenomenon is involved, Downie (1994a, b) investigated the developmental arrest of *Leptodactylus* tadpoles in a foam nest caused by the failure of the nest to be inundated. John and Fenster (1975) raised *R. pipiens* tadpoles in aquaria containing 20 individuals (3.5/liter). Four aquaria were bare, and four contained partitions that partially subdivided them into 10 connected chambers. Mean body volume at metamorphosis in the undivided aquaria (1.88 cm^3) was significantly lower than mean size in divided aquaria (2.34 cm^3). They suggested that this was caused by stress induced when animals come into physical contact with others and that spatial complexity might be important in controlling tadpole growth rates. Increasing spatial complexity would decrease encounter rates and thus the effects of stress. John and Fusaro (1981) showed that the size of growth chambers could dramatically affect growth rates of larval *R. pipiens* fed ad libitum; species that normally inhabit small volumes of water (e.g., Caldwell 1993) need to be studied in this regard.

Other work on growth rates of individuals in laboratory populations has concentrated on biological or chemical factors that may inhibit tadpole growth. These mechanisms fall into three general categories: simple fouling of water with metabolic wastes, production of or infestation with parasites that increases with density, and production of inhibitory chemical substances. The first mechanism usually is accepted as potentially important in every study but is seldom measured in detail. Some studies have eliminated fouling by using flow-through systems. Most studies reduce fouling by changing all or part of the water in rearing containers on a regular schedule.

Several authors have suggested that growth is inhibited by living cells or parasites. Water conditioned by large *R. pipiens* tadpoles reduced the growth rates of other *R. pipiens* (C. M. Richards 1958). Richards concluded that the inhibitory effects of conditioned water were caused by cells of an unknown nature that were present in the feces of the tadpoles that conditioned the water. One large *Rana pipiens* tadpole growing in 3 liters of water, replaced daily, could prevent smaller *R. pipiens* tadpoles in the same container from growing (S. M. Rose 1959). This suggested that tadpoles which gain a size advantage early in growth can maintain and enlarge that advantage through inhibition of slower growing individuals. In a later study, S. M. Rose (1960) cultured groups of 14, 16, 37, and 53 tadpoles in 12 liters of water with food provided ad libitum. The mean mass attained by all tadpoles in a container declined as density increased, but the 14 largest individuals from each container had the same mean mass at all densities. This suggests that inhibitory relations between large and small individuals are asymmetrical. Several other experiments by S. M. Rose produced similar results. Water conditioned by large tadpoles inhibited the growth of small tadpoles even after the large animals were removed. This inhibitory effect was removed by heating the water to 60°C or greater. Rose did not present detailed evidence regarding the nature of the inhibitory substance but agreed with C. M. Richards (1958) that it was cellular. Nakata et al. (1982) proposed a mathematical model of tadpole growth inhibition that explains results obtained by Rose in terms of growth rates and inhibitor substance concentrations.

Steinwascher (1978b, 1979b) identified cells interacting with larval *Rana clamitans* and *R. sphenocephala* as a yeast, *Candida humicola.* He demonstrated that low concentrations of *C. humicola* cells enhanced the growth of all tadpoles, but higher concentrations decreased the growth rates of smaller tadpoles without affecting growth rates of larger individuals. Steinwascher suggested that this host-parasite interaction might be selectively advantageous to larger tadpoles because it increased their growth rates by reducing the growth rates of smaller individuals and thus their demands on resources. More recently, Pelaz (1987) found cells similar to the ones found by Richards and Rose in water conditioned by larval *Rana ridibunda.* He identified the cells as green algae of the genus *Chlorella* but was unable to show that they exerted any inhibitory effect on tadpole growth. Beebee (1991) and R. A. Griffiths et al. (1993) reported the results of experiments on interference between anuran larvae that is mediated by a recently described, nonpigmented alga *Prototheca richardsi* (A. L.-C. Wong and Beebee 1994); for further discussion of this topic, see recent exchanges among Beebee (1995), Petranka (1995), and R. A. Griffiths (1995).

A case for chemical rather than cellular inhibitory substances was made by Akin (1966), who experimentally reared *R. pipiens* tadpoles with and without access to their own feces to reduce the rate of ingestion of any inhibitory cells. Both groups of tadpoles grew at the same rate, which suggested that the inhibitory substance was not contained in the feces and thus was not cellular. Akin also assayed the effects of water conditioned by other aquatic species on the growth and mortality rates of *R. pipiens* tadpoles. Inhibitory effects decreased with increasing phylogenetic distance between *R. pipiens* and the conditioning species. Akin proposed that inhibitors are chemical compounds that act primarily to inhibit the growth of conspecifics with allelotoxic effects being largely incidental. This interpretation was contradicted by Licht (1967), who examined the effects of water conditioned by *Bufo woodhousii* and *R. pipiens* tadpoles on the growth of many other species. Licht found no evidence for decreasing effects with increasing phylogenetic distance and attributed the inhibitory effects of conditioned water to parasitic cells shed in the feces of tadpoles.

To summarize, growth inhibition is common in laboratory populations of tadpoles. It can have several causes, including exchange of parasitic organisms and release of chemical compounds. It probably does not occur on a large spatial scale in natural populations but may be important where tadpoles are locally dense, in hatchling aggregations, and in habitats such as small pools and larger pools in the final stages of drying.

Many recent studies have used tadpoles as models of the operation of intraspecific competition in general. These usually have concentrated on exploitation competition where the rate of resource supply is not sufficient to meet maximum demand. In exploitation competition, individuals prevent others from using the resource by consuming it rather than by defending it. Many studies have concentrated on competition among individuals reared at a variety of densities on identical food supplies. This varies the rate of food supply per individual and hence the degree of exploitation competition. Unfortunately, without control treatments that vary density while holding rate of supply per individual constant, such experiments cannot distinguish effects caused by resource limitation from effects caused by interference. In general, interference competition, which may be caused by any of the inhibitory effects just discussed, can increase the effects of density observed in such experiments.

Wilbur (1977a) investigated the effects of density on laboratory populations of larval *Bufo americanus*. The variance of body mass among individuals reared in isolation and fed ad libitum increased as they grew. There was no correlation between the rates of growth and development of individuals. Other *Bufo* were raised at densities of 10, 20, 40, 80, 160, and 320 per 0.08 m^2 in 2-liter pans with the same ration of food supplied to all densities. Instantaneous mortality rates were not related to initial density, and larval growth and development rates declined with increasing density. This led to the apparently paradoxical result that despite density-independent instantaneous mortality rates, the probability of survival to metamorphosis was strongly correlated to initial density. At high initial density, low growth and differentiation rates mean that individuals are exposed to equivalent instantaneous chances of mortality for longer periods, and thus a greater proportion of individuals die before metamorphosis. Dash and Hota (1980, *Hoplobatrachus tigerinus*) and A. Sokol (1984, *Litoria ewingi*) found similar patterns in comparable experiments.

In order to separate the effects of exploitative and interference competition, Wilbur (1977b) raised larval *Rana sylvatica* in the laboratory at 1, 2, 4, and 8/0.0110 m^2 (250-ml dish) and 1, 2, 4, and 8X then 6X multiples of a ration of ground rabbit food. This allowed a set of comparisons of growth and development rates between tadpoles raised at equivalent rates of food supply per individual over a range of population densities. If the effects of density were caused only by exploitation, populations given the same rate of supply per individual should have grown and developed at the same rate. Figure 10.4 summarizes his results for mean body mass at day 29–30 at densities of 1, 2, and 4 individuals and feeding rates of 1, 2, and 4X (highest density and feeding regime omitted because the highest food level changed during the experiment). His results are difficult to interpret. When a single individual was raised alone, its mass declined with increasing food level. At density 2, mass increased when food went from 1 to 2X, then declined. Only at density 4 did mass increase consistently with increasing food level. At density 2, mass increased when food went from 1 to 2X and then declined. Only at density 4 did mass increase consistently with increasing food level. The broken lines connecting treatments with equivalent rates of food supply per individual suggest that there was no consistent effect of density independent of food supply. The hypothesis most consistent with these results is that at lower densities and higher feeding rates, growth was depressed by fouling effects caused by excess food.

A series of papers by Steinwascher (1978a, b, 1979a, b) explored the ecological implications of competition among tadpoles in laboratory systems. He found that tadpoles can shift feeding modes as resource availability changes, that larger individuals may monopolize higher quality resources, and that some of the effects of interference competition may actually be caused by host-parasite interactions involving a yeast, *Candida humicola*. Steinwascher (1978a) demonstrated that allowing access to feces increased the growth rate of larval *Rana catesbeiana*. He suggested that the primary function of coprophagy was to increase the length of time food was resident in the digestive tract and that it might also allow some microbial digestion. The food value of feces was lower than the value of a similar quantity of original food, but filtering feces from suspension required less energy than scraping solid food from a substrate. When he fed *R. catesbeiana* solid food at low rates, larger individuals fed primarily by filtration of feces and forced smaller individuals to expend additional energy in scraping food from the substrate.

Semlitsch and Caldwell (1982) examined the effects of density on the growth and development of larval *Scaphiopus holbrookii*. In one experiment, total food ration was held constant, while individual density varied from 3 to 30 individuals per container (1.2–12.0/liter; 0.01–0.1/cm^2). This experiment produced a complex set of results; growth rate decreased and length of larval period increased as density increased. In combination, these two trends led a reduction in mass at metamorphosis as density increased from 1.0 to 4.8 individuals per liter and then increased with increasing density. This increase was caused by developmental rate slowing faster than growth rate as density increased. They suggested that this was caused by early metamorphosis at small sizes by faster growing individuals in the high-density treatments, which allowed remaining individuals to take advantage of the increase in resource availability by growing more rapidly while reducing their rates of development. The remaining individuals emerged at large sizes after an extended larval period as proposed by Wilbur and Collins (1973). This experiment did not separate effects caused by exploitation from effects of interference.

A complex experiment performed by Berven and Chadra (1988) evaluated the influences of egg size, density, and food level on tadpoles of *Rana sylvatica*. They raised *Rana* tadpoles hatching from large and small eggs at densities of 2, 4, or 8 individuals per pan (size unspecified). These egg size and density treatments were completely crossed with three food supply rates. The absolute quantities of food supplied were altered over the experiment but remained in the ratio 1:2:4. They found that size at metamorphosis decreased and length of larval period increased with decreasing food per tadpole. They found no effect of density independent of food level; interference competition was apparently minimal or absent. It is difficult to interpret this lack of interference competition because they presented no information about the absolute densities per unit area or volume used in their experiment. Egg size affected metamorphic parameters only at higher food levels and lower initial densities. Kehr (1989) experimentally manipulated food level and density of *Hyla pulchella* tadpoles in unreplicated aquaria. The tadpoles exhibited effects of both interference and exploitation competition; at constant low feeding rates per animal, larval period increased with increasing density while body length remained constant. When feeding rate per individual decreased with increasing density, length at metamorphosis decreased and larval period increased.

I recently conducted a laboratory experiment designed to separate the effects of interference and exploitation competition on the growth of larval *Limnodynastes ornatus*. I raised tadpoles at densities of 1, 2, and 4 individuals in 10 × 7 cm circular plastic containers holding 250 ml of water and completely crossed these densities with feeding rates of 1, 2, and 4X multiples of a basic food ration consisting of a 1:3 mixture by weight of finely ground tropical fish flakes and alfalfa pellets. I weighed all tadpoles in the experiment on day 10 of development (fig. 10.5). The design of this experiment was similar to that of Wilbur (1977b), but its outcome was different. In *L. ornatus* the effects of interference and exploitation competition appear to be nearly equivalent, particularly at lower densities. For example, the performance of isolated individuals raised at a 1X ration is nearly identical to the performance of individuals raised in pairs with a ration of 4X. The effects of interference competition increased less when density increased from 2 to 4 individuals per container. These results suggest that the intensity of interference competition may increase nonlinearly with density and perhaps reaches an asymptote beyond which adding individuals does not increase the intensity of competition. This could be true if interference were caused by stress, and the asymptote would be the point at which hormonal systems were maximally perturbed by disturbance. This could also be true if

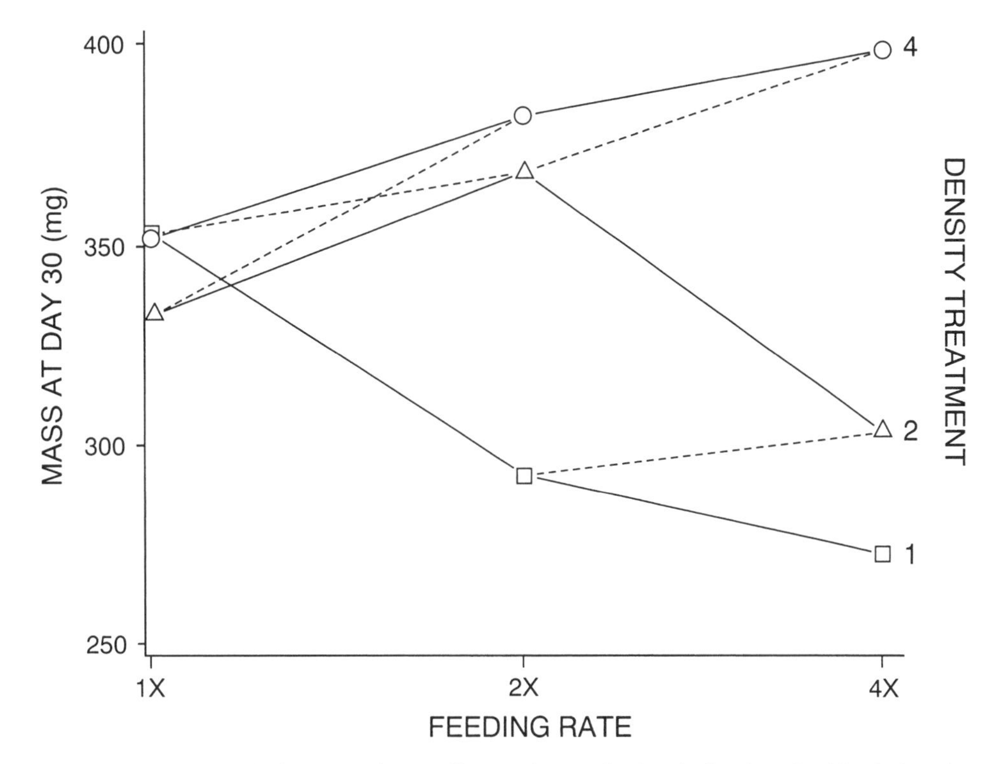

Fig. 10.4. Responses in body mass at day 30 of larvae of *Rana sylvatica* raised at three densities (1, 2, and 4 per container) and three food levels (1X, 2X, and 4X). Dashed lines connect means of treatments expected to have the same mean mass if interference competition were absent, and solid lines connect means of replicates raised at the same density. Data from Wilbur (1977b).

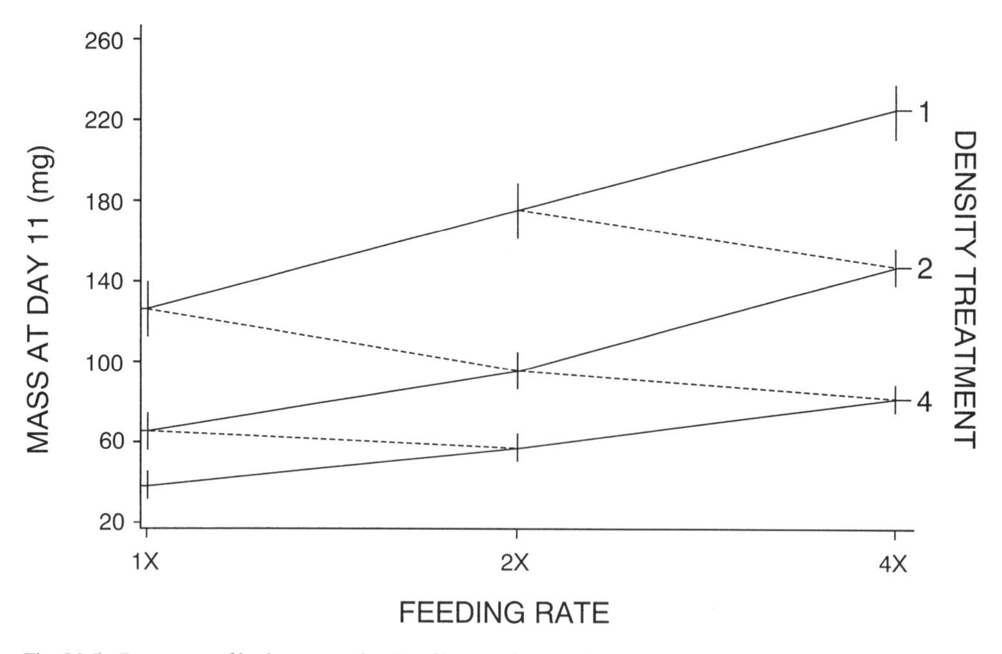

Fig. 10.5. Responses of body mass at day 11 of larvae of *Limnodynastes ornatus* raised at three densities (1, 2, and 4 per container) and three food levels (1X, 2X, and 4X). Dashed lines connect means of treatments expected to have the same mean mass if interference competition were absent, and solid lines connect means of replicates raised at the same density. Vertical lines are Gabriel simultaneous 95% comparison limits for means.

interference was caused by physical interactions between individuals, and a maximum of aggressive encounters might be reached at intermediate densities beyond which individual activity could decline. I obtained similar results in an experiment with larval *Bufo marinus* (Alford 1994). Other laboratory experiments that have examined intraspecific competition include Hota and Dash (1981), Mahapatro and Dash (1987), Mishra and Dash (1984), and Murray (1990).

Parichy and Kaplan (1992b) examined effects of maternal investment on growth and survival of larval *Bombina orientalis* in the laboratory and found that tadpoles from smaller eggs performed worse than those from larger eggs in low-quality environments, but egg size had no consistent effects in higher quality environments. They suggested that these results are consistent with the hypothesis that frogs produce eggs of a variety of sizes as a form of "bet hedging." In a second paper (Parichy and Kaplan 1995), they showed that larvae from larger eggs have higher sprint swimming speeds and may thus be more capable of avoiding predation. In contrast, Tejedo (1992) found no trade-off between egg size and number in *Bufo calamita* and suggested that increases in both the size and number of eggs with increasing female body size must be caused by some other factor, perhaps "phylogenetic constraints." Egg size may not always be related in a simple way to ecological properties of tadpoles; Williamson and Bull (1989) found that variation in egg size within clutches of *Crinia signifera* was related to size at hatching but not to time to hatching or feeding. Clutches with larger average egg sizes both developed more rapidly and produced larger hatchlings. They also found effects of clutch on size and hatching time that were unrelated to mean egg size; these effects were of about the same magnitude as the effects that were related to egg size.

Newman (1988a, b) investigated genetic and environmental influences on the growth and development of *Scaphiopus couchii*. In a laboratory experiment, Newman (1988b) demonstrated that developmental rate has a high heritability and that the ranking of performance of families can depend on the competitive regime. In another set of experiments, Newman (1988a) showed that five full sibships differed in developmental rate in the laboratory. He raised the same sibships in experimental ponds with long and short durations and showed that similar differences persisted; the sibship with the slowest rate of development experienced greater mortality in short-duration ponds but reached greater sizes in long-duration ponds.

Blouin (1992a) raised tadpoles at different food levels (*Hyla cinerea* and *H. gratiosa*) and temperatures (*H. cinerea, H. gratiosa,* and *H. squirella*). He compared species mean norms of reaction of mass at metamorphosis and length of larval period across levels of food and temperature and found that the mean reaction norms of *H. cinerea* and *H. gratiosa* were parallel across levels of food. Evidence of differences in slope suggested species-specific differences in responses to temperature variation. Blouin suggested that selection may have favored different plastic responses to temperature in the three species. Blouin (1992b) showed that some aspects of metamorph morphology in *H. cinerea* are genetically correlated with larval life history characters, so that selection on length of larval period, for example, might also affect size-adjusted head width at metamorphosis.

Breden and Kelly (1982) were interested in the extent to

which social interaction affected developmental synchrony in *Bufo americanus* tadpoles. They hypothesized that the aggregative behavior often shown by *B. americanus* might facilitate metamorphic synchrony and perhaps allow them to swamp terrestrial predators upon emergence. They held the rate of food supply and the density of tadpoles per unit volume constant. Tadpoles were raised in isolation in individual compartments in aquaria that allowed visual and chemical but not physical interaction and in open aquaria that allowed full interactions. They found that tadpoles raised alone metamorphosed at the smallest sizes, partially interacting tadpoles metamorphosed at intermediate sizes, and fully interacting tadpoles metamorphosed at the largest sizes. Tadpoles in the partially interacting treatment had the shortest mean length of larval period (33.8 days), followed by the fully interacting (36.6 days) and the isolated treatment (37.3 days). Variance of larval period among replicates was greatest in the fully interactive followed by the partially interactive and isolated treatments. This is the reverse of the order predicted by their hypothesis. They suggested that their results were caused by competition; faster growing tadpoles inhibited the growth of smaller individuals until the larger animals metamorphosed, which allowed smaller individuals to grow more rapidly later in their larval periods. If faster growing individuals tend to delay metamorphosis until they reach larger sizes, this might explain their results.

The hypothesis that female mate choice should lead to superior larval performance was evaluated by Woodward (1987d; see Woodward and Travis 1991). He captured 12 female and 16 male *Scaphiopus couchii* and mated each female with a large and a small male. Within females, tadpoles sired by larger males survived to day 34 of the experiment at a higher rate than tadpoles sired by smaller males. No other larval or metamorph characters were affected by paternal size; there was no evidence that females would have chosen large males if allowed their choice of mates, so his results must be interpreted cautiously. Woodward et al. (1988) carried out a more elaborate set of experiments that allowed females to mate with the male they had amplexed with in nature. Six female *Pseudacris crucifer* were captured in amplexus. Each female was mated with her original mate, a male which had entered amplexus with another female, one unmated male ("bachelor") from the upper half of the male body size distribution, and one bachelor from the lower half of this distribution. Tadpoles were raised individually. Offspring of naturally mated males grew slightly but significantly faster than offspring of bachelors. Offspring of large bachelors grew significantly faster than offspring of small bachelors. Male attributes other than body size had stronger effects on growth rate. They concluded that under some ecological conditions, females may enhance their reproductive success by choosing larger males as mates.

Woodward (1987c) investigated the possible effects of interactions between cohorts of larval *Bufo woodhousii* in laboratory containers. He raised hatchlings in 1.5-liter containers alone, in pairs, or in pairs involving tadpoles from a cohort about 20 days older. Food supply rate did not appear to be limiting for hatchlings raised alone or in pairs; both treatments followed indistinguishable growth trajectories. Tadpoles reared with larger individuals grew more slowly and reached minimum metamorphic size later than tadpoles reared alone or with members of the same cohort. The presence of a larger tadpole also reduced survivorship of hatchlings. This suggested that females reproducing later in the season face selective disadvantages caused by competition from earlier cohorts.

How does intraspecific competition among close relatives compare with competition among unrelated individuals? In general, simple ecological theory would suggest that the intensity of exploitative competition should increase with increasing relatedness among competitors because closely related individuals should have more nearly identical niches than distantly related individuals. Interference competition through allelopathy might be less intense among relatives than nonrelatives because an effective allelotoxin should not negatively affect the individual producing it and thus might be expected to have weaker effects on close relatives of the producer. At least five studies have examined some aspect of this question in tadpoles. Shvarts and Pyastolova (1970) found that larval *Rana arvalis* grew better when raised in mixed-sibship groups than when raised in single-sibship groups. Travis (1980b) included a mixed-sibship treatment in an experiment examining interspecific competition between *Hyla femoralis* and *H. gratiosa* (also see Gollman and Gollman 1993b; Semlitsch and Schmiedehausen 1994). Jasieński (1988) examined competition within and among sibships of *Bombina variegata*. Tadpoles from eight sibships were raised in two experimental treatments: groups of eight individuals from one sibship with one replicate per sibship and eight mixed groups of eight individuals with one from each sibship. Density was 50/liter, and food was supplied ad libitum. This feeding regimen should result in any competition being caused by interference rather than exploitation. Mean body mass at day 43 of single-sibship groups was greater than mixed groups. Variation in body mass was much greater within mixed groups than within single-sibship groups. The largest individuals in mixed groups were larger, and the smallest individuals were smaller than in single-sibship groups. Jasieński's results are consistent with the hypothesis that interference competition occurred in his beakers and that it acted less strongly among sibs than among nonsibs. D.C. Smith (1986, 1990) experimentally manipulated the genetic composition of populations of *Pseudacris triseriata* in small natural pools. He found that in small ponds with high initial densities, groups of animals from single sibships grew more quickly and survived to metamorphosis at higher rates than mixed-sibship groups and suggested that these results were probably caused by relatedness-mediated interference. Hokit and Blaustein (1997) showed that the mass of *Rana cascadae* larvae reared in mixed groups was skewed toward smaller body sizes compared to tadpoles reared in single sibships, which suggests that competition may have been more intense.

In summary, the experiments reported to date suggest that interference competition tends to be weaker within groups of siblings than within groups of unrelated individuals. It is not clear whether exploitation competition among siblings is stronger than exploitation competition among nonrelatives. This should be explored further with experiments designed to examine both types of competition simultaneously.

Three additional topics on the ecology of tadpole growth and development deserve mention: responses to environmental temperature, dissolved oxygen and pH (see chap. 8). Three general themes are apparent in the literature on responses to temperature and dissolved oxygen. One set of studies (e.g., Floyd 1985; Harkey and Semlitsch 1988; Herreid and Kinney 1967; Licht 1971; Zweifel 1977) examined the effects of temperature on growth and development and related these to conditions encountered in nature; thermal tolerances as limitations on the reproductive biology and range of species are emphasized. Another group of studies examines how tadpoles behaviorally respond to small-scale thermal or oxygen concentration variation by orienting themselves in gradients (e.g., Beiswenger and Test 1966; Bradford 1984b; Lucas and Reynolds 1967; Noland and Ultsch 1981; Wassersug and Seibert 1975). The third set of studies looked at how thermal preferenda change over time through acclimation (Hutchison and Hill 1978; E. Marshall and Grigg 1980a, b) or during development (Dupré and Petranka 1985; Wollmuth and Crawshaw 1988; Wollmuth et al. 1987). All of these studies are useful as examinations of the effects of thermal and oxygen constraints on tadpole growth and development.

The effects of environmental pH on growth and development of tadpoles are of interest because of the natural acidity of some wetlands (Gosner and Black 1957b), and because acidification by human activities may have a major impact on frog populations (e.g., Henrickson 1990). Recent studies concentrated on North Temperate areas where the problem of acidification is perceived as most severe (e.g., Andrén et al. 1988; Cummins 1989; Gascon and Bider 1985; Padhye and Ghate 1988; Pierce and Harvey 1987; Pierce et al. 1984). Pierce (1985) reviewed acid tolerance studies and concluded that amphibians in general were relatively acid tolerant, with larval mortality usually increasing only below pH 4. Some species were very sensitive and would thus be particularly vulnerable to anthropogenic acidification. Warner et al. (1991, 1993, discussed below) demonstrated that alterations in pH had complex effects on competition between *Hyla femoralis* and *H. gratiosa* tadpoles, and photoperiod (M. L. Wright et al. 1990) may act similarly. Changes in environmental acidity may also affect the interactions of species with their predators, sometimes reducing the vulnerability of tadpoles to size-limited predators by decreasing predator growth rates, as Kiesecker (1996) demonstrated for the interaction between larvae of *Pseudacris triseriata* and *Ambystoma tigrinum*. Responses to increased acidity can depend on larval size or developmental stage (Rosenberg and Pierce 1995) so that the effects of transient increases may depend on their timing relative to the larval life span of affected species.

Interactions with Predators

Field data on survival rates of tadpoles suggest that predation is the major source of mortality. Predation begins at the egg stage (Villa et al. 1982) and continues through metamorphosis (S. J. Arnold and Wassersug 1978; Wassersug and Sperry 1977). The presence of predators has been suggested as a limiting factor in pond use by many anurans (Bradford 1989; Kats et al. 1988; Sexton and Phillips 1986; Woodward 1983). Predation may affect the outcome of other interspecific interactions in tadpole communities (Morin 1981). Tadpoles are vulnerable to a wide range of invertebrate and vertebrate predators. Their vulnerability can be influenced by tadpole behavior, body size, coloration, genotype, habitat preferences, and palatability.

Wassersug (1971) persuaded several human volunteers to taste eight species of tadpoles from Costa Rica and documented considerable variation in the palatability of tadpoles. The tail of most species was least disagreeable to the volunteers, the skin second, and the body most disagreeable. *Bufo marinus* tadpoles were rated the least palatable. Wassersug suggested that this is in accord with the known toxicity and relatively low predation rate on bufonid tadpoles. This result agrees with experiments performed by myself and students on *B. marinus* tadpoles in Queensland, Australia; all species of freshwater fish we have tested either will not taste them or immediately spit them out after tasting them (personal observation). These tadpoles appear to be distasteful or harmful to some (e.g., larval *Dytiscus* sp. and the odonates *Hemianax papuensis* and *Orthetrum caledonicum*) but not all invertebrate predators; naiads of the cosmopolitan odonate *Pantala flavescens* consume them avidly (personal observation).

Licht (1969) found that ovarian eggs of four *Rana* species were palatable to larval *Ambystoma gracile, Gasterosteus aculeatus,* and *Salmo clarkii.* Ovarian eggs of *Bufo valliceps* were toxic to most potential predators (Licht 1968). Walters (1975) documented variation in feeding rates of salamanders on anuran eggs, hatchlings, and swimming larvae. She suggested that *Rana clamitans* eggs and hatchlings were distasteful to salamanders. Kruse and Francis (1977) also suggested that palatability varied among tadpoles of different species. They found that tadpoles of four species were eaten by three species of fish but that larval *Rana clamitans* were taken only by *Micropterus salmoides* and only when alternatives were not abundant. They suggested that tadpoles from more permanent habitats should be less palatable. A similar result was obtained by Grubb (1972), who found that *Gambusia* ate more eggs of temporary-pond frog species than of permanent-pond species.

Formanowicz and Brodie (1982) showed that tadpoles of *Bufo americanus, Pseudacris crucifer, Rana clamitans, R. palustris,* and *R. sylvatica* in Gosner 38–42 were vulnerable to predation by larvae of the diving beetle *Dytiscus verticalis.* Vulnerability decreased at metamorphosis (Gosner 45–46), probably because of the development of glands secreting distasteful substances. Brodie and Formanowicz (1987) demonstrated that hatchling (22–25) stages of *B. americanus* were unpalatable to newts (*Notophthalmus viridescens*) and sucking (*Belostoma* sp.) and chewing (*Anax junius*) invertebrates. Predators exposed to unpalatable hatchlings eventually avoided more palatable intermediate stage (Gosner 30–33) tadpoles. Brodie and Formanowicz suggested that this learned avoidance protects later stages from predation without continued production of defensive compounds. Sheratt and Harvey (1989) investigated the effects of absolute and relative density of tadpoles of *Phyllomedusa trinitatis* and *Physalaemus pustulosus* on rates of predation and prey selectivity by larvae of the odonate *Pantala flavescens*. The consistent preference of *Physalaemus* over *Phyllomedusa* independent of absolute or relative density might reflect a difference in palatability between the species. Denton and Beebee (1991), Henrickson (1990), Peterson and Blaustein (1991), and Reading (1990) found that *Bufo* eggs, embryos, or larvae were less palatable than other species.

The effect of tail coloration on tadpole predation by odonate naiads (*Anax junius*) was investigated by Caldwell (1982). Tadpoles of *Acris crepitans* and *A. gryllus* have black, mottled, or clear tails. The clear morphotype was more common in permanent water containing fish, while the black tail pattern was most common in temporary pools where *A. junius* was the dominant predator. She suggested that this resulted from natural selection for crypsis in the presence of fish and for deflection of predator strikes away from the body when odonates were the dominant predators. The hypothesis that tadpole tails may be lost to predators with little loss in fitness was supported by an experiment conducted by Wilbur and Semlitsch (1990). They experimentally damaged the tails of *Rana catesbeiana* and *R. sphenocephala* tadpoles and raised them in pond enclosures and in the laboratory. Removing any part of the tail of *R. catesbeiana* tadpoles slowed their rates of growth and development in field enclosures. Growth rate over a 36-day period was reduced by about 33%. A second experiment with *R. catesbeiana* in which tadpoles were reared in the laboratory showed no effect of any degree of tail damage on growth or development. There was no effect of tail damage on survival rates of tadpoles in either experiment. Tail damage also did not affect survival rates of *R. sphenocephala* exposed to predation by adult *Notophthalmus viridescens*. Figiel and Semlitsch (1991) demonstrated that tail damage needed to be severe (75% lost) to increase rates of crayfish predation on larval *Hyla chrysoscelis* in 20-liter containers. Semlitsch (1990) also found that 75% tail damage increased the rate of predation by dragonfly naiads (*Tramea lacerta*) on *H. chrysoscelis* larvae. In some species tail damage may have greater effects. Parichy and Kaplan (1992a) showed that *Bombina orientalis* that hatch from larger eggs are less affected by tail injury, as are larvae raised at higher food levels. Tail injury tended to increase the larval period and decrease survival to and size at metamorphosis in this species.

The influence of body size on tadpole vulnerability to predation has received considerable attention. In general, mechanical limitations should cause most predators to prefer prey within a limited range of body sizes. This is less true for some predators, such as odonate naiads, than for gape-limited predators like fish and salamanders. Size limitation of predators may mean that tadpoles are exposed to differing predation risks from differing suites of predators as they grow. The ecological implications of size-specific predator-prey interactions were discussed in detail by Persson (1988), Polis (1988), and Wilbur (1988). M. K. Moore and Townsend (1998) discussed the effects of abiotic factors on these interactions. Nearly all predators that have been examined in the laboratory show some evidence for size-selectivity (table 10.3). For many predators, decreasing predation rate with increasing prey size may be caused by a combination of several factors. The operation of the mouth or mouthparts may limit prey taken to sizes less than the maximum gape, faster burst swimming rates may enable large prey to escape more effectively (Huey 1980; S. J. Richards and Bull 1990b), and larger and later stages may develop glands that secrete protective substances (Formanowicz and Brodie 1982). Laboratory experiments on prey selection may or may not be realistic simulations of nature; prey may be forced into closer contact with predators than they would normally allow, and outcomes may depend on exact numbers of prey exposed to predators and on when data are taken.

Encounter rates between tadpoles and predators are influenced by the behavior of both predator and prey. Tadpole behavior may change in response to extrinsic cues having no relation to predatory interactions and may also respond to cues related to predation. Some behaviors of tadpoles may actually facilitate predation. Ideker (1976) suggested that aggregation of larval *Rana berlandieri* near shore increased the rate of predation by grackles. Feder (1983c) demonstrated that both swimming and air-breathing increased the distance from which painted turtles (*Chrysemys picta*) would attack *Rana berlandieri* tadpoles. Moving tadpoles were attacked from distances as great as 175 cm, while the maximum distance for attack on nonmoving tadpoles was 30 cm. Air breathing increased rates of attack on tadpoles but did not increase their escape ability and thus increased their overall rate of loss to turtle predation. When behavior increases predation risk, this increase must be offset by gains in fitness from other sources. Such trade-offs might prove to be a fertile field for further investigation.

Many species of tadpoles also show behavioral responses to predators that decrease their vulnerability to predation. Tadpoles of *Ascaphus truei* altered their feeding behavior in response to nonvisual cues from four predators (Feminella and Hawkins 1994). Caldwell (1989) suggested that *Hyla geographica* tadpoles form schools to reduce individual vulnerability to predation, and Rödel and Linsenmair (1997) demonstrated that swarms of larval *Phrynomantis microps* form in response to visual or olfactory cues of predator presence. Morin (1986) suggested that *Pseudacris crucifer* tadpoles altered their microhabitat use in response to predation by *Notophthalmus viridescens*. This change may have been a response to chemicals released by newts or other tadpoles or might have been a reaction to other sensory cues. Some chemical cues eliciting avoidance behavior may originate in the bodies of tadpoles. Hews and Blaustein (1985) demonstrated that larval *Bufo boreas* responded to an extract of

Table 10.3 Summary of the effects of predator and prey body size on predation rates on tadpoles. C = constant, D = decrease, I = increase, — = not measured. Predator taxa: A = bird, B = beetle adult, b = beetle larva, F = fish, H = hemipteran or heteropteran adult, N = odonate naiad, S = salamander adult, and s = salamander larva.

Prey	Predator	Effect of Prey Size Increase	Effect of Predator Size Increase	Reference
Bufo arenarum	H, *Belostoma oxyurum*	D	I[a]	Kehr and Schnack 1991
B. bufo	S, *Triturus vulgaris*	D[b]	—	Henrikson 1990
	B, *Dytiscus lapponicus*	I then D[b]	—	Henrikson 1990
	B, *Rhantus exoletus*	D[b]	—	Henrikson 1990
	N, *Aeshna* sp.	I then D[b]	—	Henrikson 1990
	N, *Leucorrhinea dubia*	D[b]	—	Henrikson 1990
	H, *Notonecta glauca*	I then D[b]	—	Henrikson 1990
B. marinus	N, *Pantala flavescens*	D	—	Heyer et al. 1975
	F, *Lates calcarifer*	D	—	K. L. Lawler and Hero 1997
Crinia signifera	N, *Hemicordulia tau*	D	—	S. J. Richards and Bull 1990b
Hyla chrysoscelis	F, *Lepomis macrochirus*	D	I	Semlitsch and Gibbons 1988
	N, *Tramea lacerta*	D	—	Semlitsch 1990
H. gratiosa	s, *Ambystoma talpoideum*	D	—	Caldwell et al. 1980
H. pseudopuma	B, *Rhantus guticollis*	D	—	Crump 1984
	N, *Sympetrum nigrocentrum*	D	—	Crump 1984
	N, *Aeshina* sp.	C	—	Crump 1984
Limnodynastes ornatus	F, *Lates calcarifer*	C	—	K. L. Lawler and Hero 1997
Litoria ewingi	N, *Hemicordulia tau*	D	—	S. J. Richards and Bull 1990b
Osteocephalus taurinus	F, *Pyrrhulina* sp.	D	I	Gascon 1989b
	N, Aeshnidae	D	—	Gascon 1989b
Physalaemus pustulosus	N, *Pantala flavescens*	D	—	Heyer et al. 1975
	F, *Lebiasina panamensis*	I or D[c]	—	Hews 1995
Pleurodema borellii	A, *Pitanqus sulphuratus*	I	—	Crump and Vaira 1991
Pseudacris crucifer	b, *Dytiscus verticalis*	D	I[a]	Brodie and Formanowicz 1983
	s, *Ambystoma jeffersonianum*	D	—	Brodie and Formanowicz 1983
	H, *Lethocerus americanus*	D[d]	—	Brodie and Formanowicz 1983
Pseudophryne bibronii	N, *Hemicordulia tau*	D	—	S. J. Richards and Bull 1990b
Rana areolata	N, *Tramea lacerta*	D	—	Travis et al. 1985b
	H, *Notonecta indica*	I then D	—	Cronin and Travis 1986
	H, *Notonecta undulata*	D	—	Cronin and Travis 1986
R. arvalis	S, *Triturus vulgaris*	D[b]	—	Henrikson 1990
	N, *Dytiscus lapponicus*	I then D[b]	—	Henrikson 1990
	B, *Rhantus exoletus*	D[b]	—	Henrikson 1990
	N, *Aeschna* sp.	I then D[b]	—	Henrikson 1990
	N, *Leucorrhinea dubia*	D[b]	—	Henrikson 1990
	H, *Notonecta glauca*	I then D[b]	—	Henrikson 1990
R. clamitans	B, *Dytiscus verticalis*	D	I	Formanowicz 1986
R. sylvatica	N, *Anax junius*	D	C	Brodie and Formanowicz 1983
	b, *Dytiscus verticalis*	D	I[a]	Brodie and Formanowicz 1983
	H, *Belostoma* sp.	D	—	Brodie and Formanowicz 1983
R. warschewitschii	N, *Libellula herculea*	D	—	Caldwell 1994
	N, *Megaloprepus caerulatus*	D	—	Caldwell 1994
Smilisca phaeota	N, *Pantala flavescens*	D	—	Heyer et al. 1975

[a]Increasing predation concentrated on larger prey.

[b]Statistical significance not determined.

[c]Depended on microhabitat.

[d]Not statistically significant, $P > 0.05$.

chemicals from injured conspecifics by increasing their overall activity level and avoiding areas containing the substance. *Bufo boreas* tadpoles captured by the water bug *Lethocerus americanus* release chemicals that act as alarm substances (Hews 1988). Other *B. boreas* tadpoles increased activity rates and avoided areas where tadpoles had been injured. *Bufo* tadpoles did not respond to feeding by *L. americanus* on tadpoles of other species. In another experiment, Hews (1988) showed that this avoidance behavior was effective at decreasing the vulnerability of *B. boreas* to predation by naiads of *Aeshna umbrosa*. Other chemical cues may originate in predators. Petranka et al. (1987) demonstrated that *Hyla chrysoscelis* tadpoles exposed to water conditioned by the fish *Lepomis cyanellus* spent more time in refuges than tadpoles in unconditioned water. Petranka and Hayes (1998) showed that larval *Bufo americanus* and *Rana sylvatica* responded to water conditioned by the predatory odonate *Anax junius* by reducing activity levels and moving away from the source of the odor. Larval *B. americanus* responded identically to conspecific tissue extracts, while *R. sylvatica* larvae ignored this cue. Altig and Christensen (1981), Pfeiffer (1966, 1974), R. J. F. Smith (1977), and Summey and Mathis (1998) are other pertinent references; others are cited below.

Tadpole microhabitat use can determine the degree of exposure of tadpoles to predators and thus influence predation rate. Shifts in microhabitat use are one method some tadpoles may use to decrease predation rates (Morin 1986). Ghate and Padhye (1988) suggested that larval *Microhyla ornata* are particularly vulnerable to predation by *Gambusia affinis* because *Microhyla* are stationary midwater feeders and thus have a high encounter rate with foraging *Gambusia*. Naiads of the odonate *Hemicordulia tau* require tactile stimulation to elicit predatory strikes against *Xenopus laevis* tadpoles (S. J. Richards and Bull 1990a). If *Xenopus* were able to avoid the immediate vicinity of *Hemicordulia* naiads, they would be largely immune to predation by them. Semlitsch and Reyer (1992a) demonstrated that both activity levels and microhabitat use of two species (*Rana esculenta* and *R. lessonae*) responded to the presence of predators (and see Babbitt and Jordan 1996; Babbitt and Tanner 1997; Kupferberg 1998; Skelly 1995a, b; D.C. Smith and Van Buskirk 1995). Formanowicz and Bobka (1989) found that *Hyla versicolor* and *Rana sylvatica* tadpoles preferred to remain in areas of experimental pools with complex structural features. Larval *Dytiscus verticalis* also preferred these areas when they were placed in pools with no prey. When both predators and prey were present, all microhabitats were used with no apparent preference. The authors suggested that this results from interactions between the behavioral responses of predators and prey to one another. The prey shift from the preferred microhabitat and thus reduce the capture rate of predators. The predators then expand their foraging range into less preferred microhabitat and reduce the advantage of habitat shifts of prey. The advantage of habitat shifts may not be eliminated entirely because more dispersed prey force predators to expend more time in searching and thus may reduce overall predation rates (Formanowicz and Bobka 1989). Habitat shifts might also be more effective against predators that forage with very different efficiencies in different habitat types.

At least some detection of potential predators occurs by olfaction; Semlitsch and Gavasso (1992) found that adding scents of predators caused larval *Bufo bufo* and *B. calamita* to alter their microhabitat use. It is interesting that both species responded equally to all predator scents and to a control scent, almond extract. Semlitsch and Gavasso suggested that tadpoles may exhibit generalized behavioral responses to any novel olfactory stimulus, which could be advantageous in uncertain pond habitats. Manteifel (1995) also found that tadpoles avoided novel scents. In addition to affecting encounter rates with predators, tadpole behavior can also affect capture rates when encounters occur. Woodward (1983) performed pair-wise experiments comparing the vulnerability of temporary pond (*Scaphiopus couchii* and *Spea multiplicata*) and permanent pond (*Rana catesbeiana* and *R. pipiens*) tadpoles to predation by belostomatids, larval corydalids, notonectids, three species of aeshnid dragonfly naiads, larval *Ambystoma tigrinum*, and *Gambusia affinis*. All predators showed a strong bias toward eating the temporary pond tadpoles. Woodward suggested that this might have been caused by the much greater frequency of movements he documented for the temporary pond tadpoles; movements attract the attention of visually orienting predators and bring tadpoles into range of less active predators. Chovanec (1992) also concluded that rates of movement were an important determinant of the vulnerability of tadpoles of *B. bufo*, *Hyla arborea*, and *R. dalmatina* to dragonfly naiads (*Aeshna cyanea*). S. P. Lawler (1989) found that predation rates of newts (*Notophthalmus viridescens*) on tadpoles of *Bufo woodhousii*, *Hyla andersonii*, *H. versicolor*, and *Pseudacris crucifer* were correlated with levels of activity shown by the species in the absence of predators; she suggested that baseline activity levels may be more important than responses to predator presence in determining vulnerability to predation.

Skelly (1991), Skelly and Werner (1990), and Werner (1991) showed that predator avoidance behavior carries costs as well as benefits. Skelly and Werner (1990) found that *Bufo americanus* larvae raised in the laboratory in containers with restrained naiads of *Anax junius* reduced their levels of activity and metamorphosed at smaller sizes than animals raised without predators present. Werner (1991) found that the presence of restrained *A. junius* naiads altered the outcome of competition between larval *Rana catesbeiana* and *R. clamitans* in laboratory containers. When predators were absent, both species grew at similar rates and appeared to have equal competitive ability. When restrained predators were present, both species reduced levels of activity and altered their use of the available space, and *R. clamitans* grew to much smaller sizes. Skelly (1991) demonstrated similar effects with *Hyla versicolor* tadpoles in field pens.

The presence of fish seems to be a major factor controlling pond use by tadpoles. Fourteen amphibian species used three farm ponds in Missouri prior to the colonization of two ponds by fish; after colonization, tadpoles of most species were eliminated (Sexton and Phillips 1986). They suggested that the sunfish *Lepomis cyanellus* had the greatest impact on tadpole diversity. At Railroad Pond (Seale 1980), tadpole diversity declined from 11 to 2 species (*Rana catesbeiana* and *R. pipiens*) after invasion by fish. Bradford (1989) surveyed lakes in the Sierra Nevada of California and concluded that introduced fish had eliminated *Rana muscosa* from many of them. Donnelly and Guyer (1994) concluded that fishes were responsible for a substantial proportion of the difference they observed among years in recruitment into the terrestrial juvenile stage at a site in Costa Rica.

Kats et al. (1988) examined eight species of tadpoles and seven species of salamanders for defenses against fish predation. They classified tadpoles by usual habitat (temporary pond species—*Pseudacris crucifer*, *P. triseriata*, and *Rana sylvatica;* permanent pond species—*R. catesbeiana* and *R. clamitans*), including those that used both habitats (*Acris gryllus*, *Bufo americanus*, and *Hyla chrysoscelis*) and suggested that exclusive temporary pond species should seldom encounter fish in nature. Palatability trials with sunfish (*Lepomis cyanellus* or *L. macrochirus*) demonstrated that all temporary-pond species were palatable to fish, but the permanent-pond species and *B. americanus* were unpalatable. The only species that regularly occur in permanent ponds and are palatable were *A. gryllus* and *H. chrysoscelis.* None of the temporary-pond species or the unpalatable species spent significantly more time in refuges when exposed to water conditioned by fish. Tadpoles of *A. gryllus* spent 86% of their time in refuges when fish were absent and 91% when fish-conditioned water was present, but the increase was not statistically significant. Larval *H. chrysoscelis* significantly increased time spent in refuges from 49% when water was not fish-conditioned to 82% when fish chemicals were present. Kats et al. (1988) suggested that the patterns were consistent with evolution of defenses against fish predation by species regularly exposed to it in nature. They also suggested that small-scale geographic comparisons between populations exposed to differing predators might provide interesting information on the evolution of defenses in tadpoles.

I demonstrated that genetic variation for factors affecting vulnerability to predation exists in *Hyla chrysoscelis* (Alford 1986b). I reared 4 sibships of tadpoles, the offspring of a factorial cross of 2 males with 2 females, for 23 days. I then exposed groups of 9 tadpoles to predation by male and female broken-lined newts (*Notophthalmus v. dorsalis*) in 24 × 30 × 12 cm plastic trays. Male and female parents had significant effects on tadpole growth at day 23. Male parent, and the three-way interaction of male parent, female parent, and newt sex had significant effects on the numbers of tadpoles remaining in the trays after 30, 60, and 120 min of exposure to predation. The least vulnerable families were not those with the largest bodies; this suggested that some genetically determined factor other than size had affected predation rate. This could have been a difference in behavior or in levels of distasteful secretions. Semlitsch and Gibbons (1988) induced body size variation in larval *H. chrysoscelis* by rearing five sibships at three densities. When these were exposed to predation by bluegill sunfish (*Lepomis macrochirus*), both tadpole body size and sibship controlled vulnerability to predation. In contrast, Semlitsch (1990) found no effects of sibship on the vulnerability of *H. chrysoscelis* tadpoles to predation by larval *Tramea lacerta.*

D. C. Smith and Van Buskirk (1995) demonstrated that competition and predation interacted with the phenotypes of larval *Pseudacris crucifer* and *P. triseriata* in predictable ways. Both species showed some alterations to their phenotype when raised in the presence of predators. *P. crucifer,* which normally occurs with predators, responded less than *P. triseriata,* which does not. McCollum and Leimberger (1997) and McCollum and Van Buskirk (1996) showed that larval *Hyla chrysoscelis* and *H. versicolor* raised in the presence of predators developed different morphology and color patterns than those raised alone. The changes were such that predation rates would be likely to decrease. Van Buskirk et al. (1997) examined the selective pressures exerted by predatory dragonfly (*Anax*) naiads on larval *Pseudacris triseriata* and phenotypic responses by tadpoles to that pressure. They found that predation favored larvae with shallow, narrow bodies, high tail fins, and a wide tail muscle. Tadpoles exposed to *Anax* during development acquired higher tail fins and wider tail muscles despite not showing greater activity, which suggests that this response may reflect morphological plasticity triggered by exposure to predator cues. Warkentin (1995) demonstrated that late-embryonic *Agalychnis callidryas* respond to snake attack by hatching early, dropping into water, and avoiding the predator. The evolution of tadpole defenses against predation is an area that deserves greater attention in future work. The influence of predation on species interactions among tadpoles will be considered later in this chapter.

Composition and Ecology of Species Assemblages

Composition and Ecology

Species composition and habitat use in tadpole assemblages can be looked at in two ways. One approach compares ponds by examining which species co-occur and which do not and the factors that explain patterns of species co-occurrence. The results of this approach are often difficult to interpret in terms of tadpole ecology; the spatial distribution of tadpoles among ponds and their temporal patterns of occurrence result from the spatial and temporal distribution of reproductive effort by adult frogs, and adults may respond to many factors in addition to the ecological requirements of their larvae. For example, spawning site use may be constrained by adult predation risk, predation or other mortality risk to eggs, or the quality of surrounding habitat for postmetamorphic juveniles. Any of these factors, or many others, may cause tadpoles to occur in habitats that are not necessarily "optimal" for their growth and development but reflect compromises between the ecological requirements of tadpoles and other stages in the life history. Bradfield (1995) undertook a series of experiments and observations to examine this hypothesis with larvae of two species of rainforest tree frogs. Larval *Litoria genimaculata* naturally occur only in stream pools, and larval *L. xanthomera* occur in pools on the forest floor. Using transplant experiments, Bradfield showed that both species grew faster and survived at higher rates in forest pools than in stream pools. This suggests that *L. genimaculata* are constrained to occur in streams by the requirements of other life history stages or by phylogenetic history. The second approach to the study of tadpole assemblages is to examine and attempt to explain the spatial and temporal distribution of tadpoles within ponds. This corresponds to the usual notion of a community study in which ecological relations among two or more co-occurring species are examined. Both approaches depend to some extent on examining patterns of reproductive habitat use by adult frogs.

Tadpoles have little control over the general habitat type (permanent or temporary, still or running water) they occupy. This is determined by the choice of spawning sites by adults, which determines the range of species with which tadpoles may potentially interact. The timing of frog reproduction determines the temporal distribution of tadpoles within sites. This affects the interactions each cohort of tadpoles experiences with other cohorts of the same species, with other species of anurans, and with other vertebrates and invertebrates.

The most diverse tadpole faunas occur in wet tropical regions. Detailed studies of spawning habitat use in the tropics include that of Crump (1974), who examined the spatial and temporal use of reproductive habitats by 53 species of aquatic-breeding anurans in a relatively aseasonal area at Santa Cecilia, Ecuador. She studied 8 breeding sites used by 26 species. The maximum number of species breeding at any site was 16 and most sites were used by fewer species. Six of the 8 sites she studied were used by 7–13 species. The 2 most similar sites shared 56% of their species. Duellman (1978) further analyzed the tadpole communities at Santa Cecilia and found that no more than 10 species occurred at any site simultaneously. He examined microhabitat use and identified three broad habitat types: vegetation-choked areas, pelagic areas, and pond bottoms. A maximum of 7 species occurred in any microhabitat.

Aichinger (1987) examined temporal and spatial use of seven reproductive sites by frogs in the seasonal wet tropics of Peru. A total of 34 anurans bred at his aquatic sites during 1 year of monitoring. Five of his sites were temporary and dried for long periods during the winter months. Twenty or more species of males called at four of the sites, but the maximum number observed at any time was 12 at a permanent pond site and eight at a temporary puddle site. Most species were reproductively active mainly or exclusively in the wet season. No species reproduced at all seven sites. Magnusson and Hero (1991) examined the distribution of tadpoles of 18 species among water bodies in a rain forest area in Amazonian Brazil; they found that site use appeared to be dictated by predation pressure exerted by anuran egg predators. Gascon (1991b) examined the phenology and reproductive habitat use of 25 species of frogs at 53 aquatic sites in rain forest in Brazilian Amazonia. He found considerable variation in the degree of reproductive seasonality shown by species and that activity by more seasonal species tended to be strongly affected by rainfall. He used discriminant function analysis to determine the environmental factors that appeared to be responsible for determining the spatial distribution of the 11 most common tadpoles. A variety of factors were implicated, including the size and depth of water bodies, vegetation cover, and dissolved oxygen. Gascon did not find evidence for discrete structured assemblages of co-occurring species; he concluded that each species responds individually to the environment in selecting sites for reproduction and subsequent larval development. Gascon (1991a) examined the detailed responses of *Leptodactylus knudseni* to seasonal and annual variation in rainfall. Moreira and Lima (1991) also found that rainfall was an important factor controlling reproduction and recruitment in four species of Amazonian leaf-litter frogs, and Donnelly and Guyer (1994) reached a similar conclusion for an eight-species assemblage of hylids in Costa Rica.

In the seasonally dry tropics, Heyer (1973b) studied the tadpole communities of a series of ponds in Thailand. Ten species called at these sites and larvae of eight species were found. Six species started calling near the onset of the wet season, while two species delayed for 1–2 months after the onset of heavy rains. Species composition of the larval assemblages in his pools ranged from three to eight and was not correlated in any obvious way with differences among the pools. He concluded that the species composition of the larval assemblage at most ponds was largely caused by chance. In a similar study, Heyer et al. (1975) studied larval species assemblages in Costa Rican tropical wet forest and concluded that the timing of oviposition and site selection by many species appeared to minimize larval exposure to predation.

Several students and I are currently studying the use of reproductive habitats by frogs in seasonally dry eucalypt woodland in the Australian tropics near Townsville, Queensland. We have observed that adult frog density near ponds sometimes decreases temporarily after the first light wet-season rains, probably because frogs disperse to forage. Adult density increases dramatically, and most species begin calling after the first heavy rains of the wet season. A few species are attracted to very temporary habitats, such as flooded paddocks, creek overflows, and roadside ditches. These species also reproduce at less ephemeral sites such as dams and permanent water holes. For all but the most ephemeral habitats, the species composition of larval assemblages early in the wet season does not appear to follow a strong pattern. The most reproductively catholic species is the introduced *Bufo marinus,* which reproduces at temporary or permanent water in any month of the year when there is rainfall. Larval *B. marinus* are distasteful to most fishes and to some dragonfly naiads (notably *Orthetrum caledonicum,* the most abundant species in many habitats) and larval dytiscid beetles. This low vulnerability to predators may minimize selection for specialization by site or season.

Reproductive timing and site choice have also been studied in detail in temperate anurans. Detailed analyses of reproductive timing and sites are available for many temperate species but fewer entire species assemblages. Some of these are at least as diverse as the seasonal tropical communities; H. M. Wilbur (personal communication) observed 18 species of anurans calling at a single pond in the Sandhills region of North Carolina. Most of these species breed in the two months after the first heavy rains in the spring. Utsunomiya et al. (1983) examined the timing and spatial distribution of reproduction by five species of *Rana* in a wooded mountain stream in Okinawa, Japan. They observed a greater than normal degree of pattern; none of the species overlapped in reproductive habitat use. Four used mutually exclusive habitats. The fifth overlapped with one of the first four in habitat but differed in season. Strijbosch (1979) documented reproductive habitat selection by six frogs in fen pools in The Netherlands. He found that ranids were the most selective species and bufonids the least selective. Most species avoided very oligotrophic or acidic sites. Thirty-three ponds in desert habitats in New Mexico (Woodward 1983) had six anuran species present. Four species (one *Bufo,* three *Scaphiopus*) occurred primarily in temporary ponds, while two *Rana* used only permanent ponds. He suggested that this dichotomy was caused primarily by differing vulnerabili-

Table 10.4 Records of frogs calling in a seasonal tropical eucalypt woodland near Townsville, Queensland, Australia. Perm = permanent water, Temp = temporary water.

	Timing in Wet Season							
	Early		Mid-		Late		Post-	
Taxon	Temp	Perm	Temp	Perm	Temp	Perm	Temp	Perm
Bufo								
marinus	X	X	X	X	X	X	X	X
Crinia								
deserticola	X	X	X	X	X	X		
Cyclorana								
brevipes	X		X					
novaehollandiae	X		X					
Limnodynastes								
ornatus	X		X		X			
tasmaniensis	X	X	X					
Litoria								
alboguttata	X	X	X	X	X	X		
bicolor	X	X	X	X	X	X		
caerulea	X	X	X	X	X	X		
fallax	X	X	X	X	X	X		
gracilenta	X	X	X	X		X		
inermis		X		X		X		
nasuta	X	X	X	X				
rothi	X	X	X	X				
rubella	X	X	X	X				
Uperoleia								
lithomoda	X		X					
mimula	X		X					

ties of larvae to predation; the temporary-pond species were more vulnerable to permanent-pond predators than were the two *Rana* species. Semlitsch et al. (1993) examined the factors governing the reproductive phenology of salamanders in South Carolina. Their study is remarkable for the large number of marked individuals used and the availability of data collected over an extended period.

More detailed studies of the cues governing reproductive habitat selection have been performed for many individual species. Of 75 rock pools on the west coast of Sweden (Andrén and Nilson 1985), *Bufo calamita* reproduced in 31 of them. The rate of use by toads was positively correlated with flatness of shoreline, height above sea level, and distance from the sea. Use by toads was negatively correlated with maximum water depth and pool size. The correlation with shoreline flatness may primarily reflect adult toad preferences; flatter shorelines make better calling sites and simplify entry and exit from the water. The other correlations probably primarily reflect larval habitat requirements; increasing distance from and elevation above sea level reduce water salinity, while small, shallow pools reach greater temperatures which allow faster growth and development. They suggested that choice of pool size was a compromise between avoidance of drying and attainment of maximum growth and development rates.

Some species appear to be capable of evaluating the state of the aquatic community in small pools and adjusting their oviposition behavior accordingly. Adult *Hyla chrysoscelis* in North Carolina (Resetarits and Wilbur 1989; see Petranka et al. 1994) discriminated among experimental pools for calling and oviposition based on the presence of aquatic predators and competitors, including *H. chrysoscelis* belonging to earlier cohorts. The experiment by Resetarits and Wilbur (1989) was criticized by Ritke and Mumme (1993), who suggested (among other things) that their experimental ponds were too small and close together, and that salamanders in one of their predator treatments may have consumed large numbers of eggs before they could be counted. Even so, R. W. McDiarmid (personal communication) observed a similar pattern of oviposition by *Smilisca phaeota* on the Osa Peninsula, Costa Rica, and *Hyla pseudopuma* also appear to be able to choose oviposition sites based on the condition of the habitat. Crump (1991) showed that they are capable of discriminating based on water depth and the presence of potential egg predators. Frogs may select reproductive habitats based on the presence or absence of other frog species. Gascon (1992b) showed that adult *Pipa arrabali* greatly decrease the survival of eggs and larvae of *Osteocephalus taurinus;* he suggested that the species may coexist because *O. taurinus* choose to spawn in sites in which *P. arrabali* are absent. Pearman and Wilbur (1990) used computer simulations to show that increased patchiness of oviposition may increase population stability. Anurans are not always successful in choosing breeding habitats; Tevis (1966) documented repeated reproductive failures, most caused by drying of aquatic habitats, in a population of *Bufo punctatus,* and Laurila and Aho (1997) showed that *Rana temporaria* did not avoid depositing eggs in rock pools containing predatory fish.

Most of the studies cited so far concentrated on choices of calling and spawning sites by adult frogs. Characteristics of the chosen sites and spawning phenology affect the ecology of tadpoles by defining which species of competitors and predators are likely to be present and whether the habitat may dry before the end of the larval period. Species that spawn very early in the history of a temporary pool may gain an advantage for their larvae from the flush of primary productivity that arises from the high levels of allochthonous nutrients available early in the history of such pools. Their larvae may also gain size refuges from gape-limited predators and gain a size-mediated advantage in competition with other herbivores. Larvae of early spawners in permanent water bodies may gain similar advantages against the larvae of fish and insects (e.g., Majecki and Majecka 1996) that use these sites. Several disadvantages are balanced against the advantages; spawning before temporary pools have completely filled may increase the risk of desiccation for tadpoles, and spawning early in the summer season in permanent water may lead to slow growth and development caused by low water temperatures. Such slow growth can increase the period of exposure to size-limited predators and decrease size-mediated competitive ability. With such complex possibilities, it is not surprising that frog reproduction follows a variety of phenological patterns.

Blair (1961) summarized the reproductive phenology over a four-year period of seven species of anurans that bred at several sites near Austin, Texas. One species bred primarily in winter, four in summer, and two in spring and autumn. Six species called for more than 1 month. The calling activity and initiation of first calling of all species was in response to heavy rains. After initial calling, some species continued to call any time that it rained or even with no rain, while others limited calling activity to periods of heavy rainfall. The sequence of calling and spawning by the species varied considerably among years. Temporal separation of calling and spawning activity of adult frogs inhabiting a temporary pond in Mexico was documented by J. R. Dixon and Heyer (1968); they suggested that this phenology enforces temporal habitat segregation and resource-use differences in the tadpole assemblage. Seasonal occurrence of tadpoles and salamander larvae was documented at a site in Maryland over a 2-year period by Heyer (1976a). He found that the phenology and density of some species was very similar between years, while considerable pattern changes of other species in the same pond produced large changes in the total species assemblage. The same pattern continued over a further two years of monitoring at the same site (Heyer 1979).

Brook (1980) identified three seasonal patterns of calling and spawning by frogs in Victoria, Australia: frogs with 4–6 week reproductive seasons occurring at roughly the same time each year regardless of precipitation history, frogs with short reproductive seasons (2–3 days) apparently initiated by heavy rainfall, and frogs with extended reproductive seasons. The frogs breeding for 4–6 weeks began calling at different times depending on the genus; *Litoria, Philoria,* and *Uperoleia* initiated calling in early summer, while *Pseudophryne* started in late summer. Eggs or amplectant pairs occurred in more than 1 month in 20 of 33 species.

Diaz-Paniagua (1986, 1988) studied the reproductive phenology of seven anuran species at Donaña, Spain, over four years. The heaviest rains usually fell during the autumn and early winter months. *Pelobates* and *Pelodytes* had short reproductive seasons associated with heavy winter rains. Four other species bred over longer periods during early spring, and *Rana perezi* had an extended reproductive season during summer. Temporal overlap in pond occupancy of larvae varied considerably among years; in 1978–1979, the heaviest rains arrived in mid-winter, and most species reproduced almost simultaneously. Temporal overlap varied less among the other three years of the study.

Timed tape recorders were used (W. Martin 1991) to examine details of the calling phenology of frogs in seasonally dry eucalypt woodland in the Australian tropics near Townsville, Queensland (table 10.4; see Pearman 1995 and C. L. Rowe and Dunson 1995). He used daily calling data to construct similarity matrices for calling patterns among species within sites. Classification analyses of these matrices revealed only two clearly delineated phenologies; *Cyclorana brevipes* and *C. novaehollandiae* call early in the wet season in response to heavy rainfall, while all other species have extended calling seasons with peaks in activity often corresponding to heavy rains. Although the classification analysis did not reveal clear phenological groups within the second category, Mantel tests comparing the similarity matrices among sites showed high degrees of correlation, implying that interspecific similarities in calling pattern remained constant between sites. This suggests that the large group of species with extended calling seasons contains a complex set of subgroups with slightly different phenologies.

The longest series of data available on reproductive phenology for a single species (Terhivuo 1988) involves locations and dates of spawning by *Rana temporaria* in Finland between 1846 and 1986, with an extremely detailed supplemental data set for 1983. Terhivuo found that the date of the onset of spawning correlates well with temperature at any site. Populations in southwestern Finland (60°N) spawn earliest, and spawning proceeds north and east as temperatures increase. The northernmost populations (70°N) spawn an average of 35 days later than the southernmost ones. Unfortunately, data include only the onset of spawning, not its duration. Semlitsch et al. (1996), in their analysis of 16 years of detailed data from a natural pond, showed that late spawners may sometimes gain a strong advantage: *Scaphiopus holbrookii* and *Gastrophryne carolinensis* were most successful when they spawned after the pond had filled and dried earlier in the season, leading to low densities and diversities of competitors and predators.

Larval anuran species assemblages are frequently complex but often less complex than adult species assemblages in the same geographical region. Duellman (1978) reports 10 species of tadpoles found in a single habitat simultaneously in an aseasonal tropical region with at least 53 species of aquatic-breeding anurans, and Hero (1991) found a similar number in Brazil. In Madagascar, Blommers-Schlösser and Blommers (1984) found 15 species in one brook, and R. Altig (personal communication) collected 13 species within an area of about 2 m^2 in a stream in Madagascar. Nearly as many species have been found simultaneously in seasonal tropical and temperate habitats with much lower adult anuran species richness. This apparent low limit on the number of simultaneously occurring species may be caused in part by competition. Temporary pools are relatively unpredictable in space and time and contain a relatively unpredictable set of competitors and predators. This may limit the degree of specialization in resource use attainable by tadpoles. Because most tadpoles are relatively unspecialized filter-feeders, most species in a pond may be potential competitors. Differentiation of site selection and reproductive phenology by adult frogs may serve to decrease the levels of competition experienced by larvae to tolerable levels.

Density and Resource Use

Relatively few studies describe or explain the composition of tadpole species assemblages in the field. Although R. M. Savage (1952) did not set out to produce a community study, the detailed studies of the autecology of two species in the same ponds can be interpreted as one. Heyer (1973b, 1974) examined species distributions among and within ponds at a seasonal tropical location in Thailand, examined gut contents and relative gut lengths (Heyer 1973b), and calculated niche breadths and overlaps (Heyer 1974). His quantitative microhabitat use and niche measurements are based entirely on species occurrence in sweep net samples taken as an indicator of microhabitat usage. He found a broad range of overlap values, as might be expected from a community containing 8 species and thus 28 possible species pairs. He examined the food particle sizes used by species pairs showing the greatest spatial overlap and found little indication that food might be partitioned based on particle size. He suggested that differences in feeding mode overlooked by his relatively coarse-grained sampling technique, for example surface-film versus midwater versus bottom feeding, might separate some of the more overlapping species (see Bowen 1980). Heyer (1973b) also reported a series of qualitative observations on microhabitats that include the observation that different cohorts of some species may occupy different microhabitats.

Heyer (1976a) studied habitat use and partitioning by tadpoles at sites in Maryland and on Barro Colorado Island, Panama. He defined habitat partitioning as including use of breeding ponds by adult frogs, microhabitat use by tadpoles within ponds, and temporal occurrence within ponds. As discussed earlier, I would suggest that habitat partitioning of the first type is better viewed as a consequence of adult breeding biology rather than as a component of tadpole ecology. Heyer examined microhabitat use at his Maryland site by dividing the available habitat into surface film, midwater, and bottom. Heyer sampled tadpoles of six species during 1974 and 1975. All species for which sample sizes were reasonably large showed evidence for microhabitat preference. The microhabitat of *Pseudacris crucifer* larvae changed as they grew; larger larvae occurred relatively more often in deeper water. Indices of microhabitat overlap between species generally changed markedly between years. Only three species pairs maintained consistent overlaps in both years. Heyer suggested that overlap values for most species pairs indicated that competition was unlikely. He also noted that most species experienced very different environments in the two years. The study on Barro Colorado Island (Heyer 1976a) concentrated on examining species occurrences in a series of smaller streambed pools. Most pools contained only one species of tadpole. This might be caused by temporal or spatial partitioning of reproductive habitat by adults or by a variety of other factors including interspecific egg predation or its avoidance. Heyer (1979) summarized the results of two additional years of monitoring tadpole habitats at the Maryland site. He found considerable between-years variation in habitat use, population size, and recruitment. He suggested that this variation resulted from complex interactions among causative factors including adult reproductive pattern, food resources, intra- and interspecific competition, physical-climatic factors, and predation.

Turnipseed and Altig (1975) looked simultaneously at the density and age structure of *Acris gryllus, Hyla avivoca,* and *H. cinerea* in a 4,000 m^2 farm pond in Mississippi. They took weekly samples over 17 weeks starting on 14 April. Two samples were taken along each of four sections of the pond bank with a quadrat sampler placed at the waterline. Captured tadpoles were counted, staged, and weighed. Their collection methods allowed a more quantitative examination of spatial and temporal variation in habitat use than previous studies. They found that *A. gryllus* made up 95.7% of the tadpoles in the area sampled, and the spatial distribution of *Acris* tadpoles showed a consistent bias away from one area of the pond. The density varied through time, indicating some combination of changes in their propensity to remain near the shore and changes in total population size. The stage composition of the *Acris* population changed as cohorts developed and new cohorts were added by reproduction. The populations of both *Hyla* species were apparently more evenly distributed spatially but showed temporal patterns similar to *Acris,* which suggest that they were responding to the same environmental stimuli.

In 1980 and 1981, I used a circular quadrat sampler to examine the size structure and spatial distribution of populations of tadpoles in a pond in northern Florida. In March 1980, only larval *Rana sphenocephala* were present, but they were divided into three size classes that were distributed differently in the pond (Alford and Crump 1982). An experiment in an aquarium showed that habitat use differences among the size classes persisted in a simplified environment. In 1981, I sampled the pond from just after hatching of the first species of frogs that used it until it dried (Alford 1986a). Larval *Pseudacris ornata* and *R. sphenocephala* were present throughout the sampling period. *Bufo americanus* and *Pseudacris ocularis* appeared after sampling began. All species persisted until the pond dried, and *B. americanus, Pseudacris ocularis,* and *P. ornata* produced metamorphs. I selected sampling locations haphazardly within several habitat types and sampled most areas of the pond which was at most 60 cm deep (table 10.1; see Wild 1996). I found *R. sphenocephala* at densities of up to 3.1/liter early in the season and at densities as high as 8.0/liter as the pond began to shrink. Tadpoles of *B. americanus,* which often form dense aggregations, reached densities as high as 47.3/liter on 17 April. I measured a number of physical and biotic environmental parameters, including plant cover, substrate type, temperature and water depth. Tadpoles of *B. americanus, Pseudacris ocularis,* and *R. sphenocephala* showed differences in spatial use between individuals of different sizes within sampling dates, which suggest that habitat use changes ontogenetically and that cohorts of different ages may function as separate "ecological species" (Polis 1984; Werner and Gilliam 1984). Kehr (1997) also showed stage-specific shifts in microhabitat use by *P. ocularis.*

In my study, aspects of the spatial distribution of all four species responded to one or more environmental parameters. Spatial distributions of some species pairs were correlated, which suggest that habitat use might, in part, be caused by behavioral responses of species or cohorts to one another. I evaluated this hypothesis with a laboratory experiment in an environment devoid of asymmetrical spatial features. *Rana sphenocephala* tadpoles tended to be overdispersed with respect to other species. This may have been caused by other species avoiding *Rana* which were the largest tadpoles. Overall, this study suggested that tadpole habitat use is species-specific, changes ontogenetically within species, and may be plastic with tadpoles responding to short-term factors such as the presence of larger tadpoles. This complexity is not surprising given the great environmental variability encountered by temporary-pond tadpoles of each species during their evolutionary history.

Diaz-Paniagua (1987) and Löschenkohl (1986) studied tadpole habitat use in Europe. Löschenkohl examined microhabitat, resource use, and seasonal differences among five species in the same pond. Four species (*Hyla arborea, Pelobates fuscus, Rana dalmatina,* and *R. ridibunda*) overlapped seasonally. The fifth, *Bufo bufo,* reproduced earlier in the year, and most *Bufo* tadpoles had metamorphosed before the other species bred. The four seasonally overlapping species had very similar gut contents but differed in their distributions among microhabitats. Waringer-Löschenkohl (1988) examined differences in the use of three depth strata in experimental aquaria. Aquaria were established with different population densities, vegetation, and species composition. The results showed that the species differed in microhabitat use within the aquaria. Tadpoles of *B. bufo* always preferred the bottom and those of *H. arborea* preferred the surface. Tadpoles of *R. dalmatina* preferred the bottom when alone but altered this preference when other species were present, and *P. fuscus* preferred the bottom when alone but was distributed more widely when *Hyla* were present. Tadpoles of *R. ridibunda* preferred the bottom of a bare aquarium, but this preference disappeared when the aquarium contained macrophytes.

Diaz-Paniagua (1987) defined five vegetation zones within ponds: (1) grass, (2) short floating vegetation, (3) emergent vegetation, (4) submergent vegetation, and (5) no vegetation. Samples from 16 ponds showed that all tadpoles avoided zone 5. *Hyla meridionalis* tadpoles were found primarily in zone 4 and some were in zone 3. Tadpoles of *Bufo bufo* were found mostly in zones 1 and 2. *Bufo calamita, Discoglossus galganoi,* and *Rana perezi* tadpoles were distributed widely in zones 1 through 4. Diaz-Paniagua suggested that differences in zone use were related to differences in feeding method among species.

Degani (1986) found tadpoles of *Bufo viridis, Hyla arborea, Pelobates syriacus,* and *Rana ridibunda* in a 40 × 60 m temporary pond in Israel and examined their microhabitat use in aquaria. He recorded the proportion of time spent in seven depth strata and in areas with and without macrophytes. All species except *B. viridis* spent 60%–70% of their time near macrophytes, while *Bufo* spent only 45% of their time there. *Bufo* and *Rana* tadpoles spent most of their time at or near the bottom, *Hyla* tadpoles were distributed nearly evenly among depth zones, and *Pelobates* tadpoles were bimodally distributed between the surface or near the bottom. Degani used an experiment to determine whether these microhabitat preferences affected tadpole growth rates. He followed the growth of each species in three cages placed near the surface and three on the bottom. Results agreed with the habitat preference study; *Bufo* and *Rana* grew best in cages on the bottom, while *Hyla* and *Pelobates* grew equally well at either depth. Barreto and Moreira (1996) demonstrated differences in microhabitat use among the larvae of six species of frogs breeding in a pond in central Brazil.

All of the studies above were carried out in still water environments, many in highly productive temporary pools with relatively high nutrients. A few studies have examined rain forest stream environments, which are generally thought to be relatively low in nutrients and rates of primary production (Bishop 1973). Inger (1969) commented on tadpole habitat use by frog communities along small rain forest streams. Odendaal et al. (1982) documented differences in habitat use between *Crinia riparia* and *C. signifera* in streams of South Australia. In laboratory experiments they showed that the habitat preference of *C. signifera* shifted to reduce overlap when *C. riparia* was present. Inger et al. (1986a) documented differences in microhabitat use among 29 species of tadpoles collected in streams in primary rain forest in Borneo. They identified four taxonomically heterogeneous groups: leaf-pack, pothole, riffle-shingle-open-pool, and side-pool pothole species. Variation in species composition was greater between than within sites over all of their sampling. Species composition was least variable in the leaf-pack habitat. They suggested that little evidence for competition existed and that most differences in microhabitat use probably were because of differences in adaptations for coping with the physical environment.

Trenerry (1988) examined habitat use by four species of tadpoles living in a tropical rain forest stream in northern Queensland, Australia. He divided the habitat into pools, riffles, and runs. *Litoria genimaculata* and *Mixophyes schevilli* primarily lived in pools, and *L. nannotis* and *Nyctimystes dayi* inhabited only riffles. Runs were not regularly censused but appeared to be inhabited at low densities by pool species. Both riffle species were suctorial, while the pool species had typical lotic body forms. Population densities reached relatively high levels in summer months. In a pool measuring about 10 × 6 m with a mean depth of 40 cm, he estimated a population of 62±14 *M. schevilli* (1/m^2; 0.0025/liter) and 916±258 *L. genimaculata* (15.3/m^2; 0.04/liter) in late summer. Hawkins et al. (1988) presented similar information for tadpoles of *Ascaphus*.

Interspecific Competition and Predation

Field Studies

Field studies cited above have shown that, regardless of habitat, species usually overlap broadly in diet and microhabitat use and that tadpoles often reach high densities. Many tadpole habitats show a great deal of year-to-year and site-to-site variation in seasonal availability, duration, and size. Species composition of tadpole assemblages varies similarly. This spatial and temporal variation probably reduces the degree of permanent niche differentiation among species. The combination of high population densities and relatively low niche differentiation suggests that interspecific competition may be relatively common in tadpole assemblages. Interspecific competition has been studied by the same general approaches used for intraspecific competition; populations have been monitored in the field, experiments have been performed in field enclosures and large-scale artificial communities, and experiments have been performed in laboratory systems.

Few field experiments have examined interspecific competition alone. Dumas (1964) used a field experiment to examine possible competitive exclusion between *Rana pipiens* and *R. pretiosa*. The percentages and numbers surviving contradict the numbers that he introduced, so the following is my interpretation of his methods and results. He removed all tadpoles from 4 small natural ponds and then introduced 400 *Rana pretiosa* to 1 pond, 400 *R. pipiens* to another, and mixed groups with 200 individuals of each species to the other 2 ponds. Nine weeks later, 44% of *R. pretiosa* and 39% of *R. pipiens* had survived in the single-species ponds. In the mixed-species ponds, survival of *R. pipiens* was 86% and 84%, while *R. pretiosa* survived at rates of 38% and 52%, respectively. He suggested that these results indicated that *R. pipiens* was capable of expanding its range and displacing *R. pretiosa*.

Wilbur (1977a) examined the relative strengths of intraspecific and interspecific competition by raising *Bufo americanus* alone and with larval *Rana palustris* in 61 × 61 × 244 cm pens placed in a pond. *Bufo* alone were at densities of 67, 101, 134, 202, 336, and 605/m^2 (0.44, 0.67, 0.88, 1.33, 2.22, and 4.00/liter). *Bufo* raised with *Rana* were always present at a constant density (100 per pen; 67/m^2; 0.44/liter). *Rana* densities were 0, 17, 34, 67, 134, and 202/m^2 (0, 0.11, 0.22, 0.44, 0.88, and 1.33/liter). Each treatment was replicated twice. A reanalysis of his means for treatments with 100 *Bufo* larvae raised with 50, 100, and 200 *Bufo* and 50, 100, and 200 *Rana* suggests that there was no effect of density (F_2,6 = 0.08, P = 0.922), competitor species (F_2,6 = 0.02, P = 0.900) or the interaction of density and competitor species (F_2,6 = 0.02, P = 0.980) on mass at metamorphosis of *Bufo*. There were also no significant effects on arcsine-transformed proportion survival. When these treatments are compared to the treatment with 100 *Bufo* raised alone, there is some evidence for reductions in proportion surviving and mean mass at metamorphosis, but the reductions caused by *Rana* appear to be equivalent to reductions caused by *Bufo*.

Wiltshire and Bull (1977) used a field enclosure experiment to examine competition between *Pseudophryne bibronii* and *P. semimarmorata,* which have narrow zones of geographic overlap in southeastern Australia. They evaluated the hypothesis that the observed geographic distribution might be maintained in part by competitive exclusion. They collected eggs of both species from allopatric populations, hatched them in the laboratory, and introduced hatchlings into experimental enclosures (30 × 30 × 130 cm) placed in a pond in the range of *P. bibronii*. Tadpoles were raised at 40 or 80 individuals per enclosure (116 or 232/m^2; about 1.5 or 3.0/liter). At density 40 there were 3 treatments: *P. bibronii* alone, *P. semimarmorata* alone, and 20 of both species. At density 80, there were 3 similar treatments that were repeated with and without supplemental food. Survival of *P. semimarmorata* was lower than survival of *P. bibronii* in all treatments. When both species were present without food supplementation, no *P. semimarmorata* survived to metamorphosis. With supplemental food, the presence of *P. bibronii* reduced the mass of *P. semimarmorata* metamorphs. Wiltshire and Bull suggested that under their experimental conditions *P. bibronii* outcompeted *P. semimarmorata* and that larval competition might partially explain the narrow geographic overlap between the species (also see Kupferberg 1997b).

Interactions between another species pair with primarily allopatric distributions were examined by Odendaal and Bull (1983). *Crinia signifera* typically occupies creeks with slow flow rates, and *C. riparia* lives in faster moving water. They measured growth and survival of both species in enclosures placed in a creek in the range of *C. signifera,* in the range of *C. riparia,* and in an overlap zone. Tadpoles were captured in the field and relocated to cages measuring 100 × 50 × 40 cm high. Densities placed in the experimental cages ranged from 124 to 222/m^2 (0.62–1.11/liter). Some cages were shielded from stream flow while others were exposed. In general, both species grew and survived similarly in shielded cages; but in unshielded mixed cages, *Crinia riparia* grew and survived better than *C. signifera*. The results from the shielded and single-species cages indicated that *C. signifera* could survive in areas normally occupied by *C. riparia,* but the results from the mixed, unshielded cages show that tadpoles of *C. signifera* are at a competitive disadvantage when *C. riparia* occur with them in those areas. Odendaal and Bull suggested that competition was unlikely to be caused by exploitation but was more likely a result of interspecific avoidance behavior. Tadpoles of *C. riparia* were more active in laboratory experiments and were more capable of maintaining position in flowing water. Their movements may disturb *C. signifera* when both species occur in high flow-rate environments and reduce their ability to remain settled on favorable substrates. This could result in *C. signifera* being swept away by currents or could simply reduce their growth and survival rates by forcing them to expend greater proportions of their energy in returning to sheltered sites.

An experiment by Woodward (1982) examined potential competitive interactions between temporary and permanent pond tadpoles from a desert anuran community. He excavated an array of 124 ($1 \times 1 \times 0.4$ m) pools in a flat field; each pond contained 227 liters of water. He used these ponds to perform pair-wise competition experiments involving tadpoles of *Bufo woodhousii* and *Rana pipiens, Scaphiopus couchii* and *Spea multiplicata, R. pipiens* and *Scaphiopus couchii,* and large and small *R. catesbeiana* and *Scaphiopus couchii.* Total tadpole densities were between 50 and 200 individuals per pond (50–200/m^2; 0.22–0.88/liter). Tadpoles of *Spea multiplicata* eliminated *R. pipiens,* apparently via predation. *B. americanus* and *Scaphiopus couchii* had negative effects on one another, with the effect of *Scaphiopus* on *Bufo* stronger than the reverse. *Scaphiopus couchii* also negatively affected *R. pipiens* while *R. pipiens* had only a slight affect on *S. couchii. R. catesbeiana* of both sizes had negative effects on *S. couchii* while being affected rather little themselves. Woodward suggested that most of the competitive effects he observed were because of exploitation competition. His results also indicated that in most cases, permanent pond species could be excluded from temporary ponds by competitively superior temporary pond species. The tadpoles of *R. catesbeiana,* the only permanent pond species able to hold its own in competition with *Scaphiopus,* might be excluded from temporary ponds by its normally extended larval period.

Morin and Johnson (1988) examined competition between *Pseudacris crucifer* and *Rana sylvatica* in artificial pond communities. They raised tadpoles of *Pseudacris* alone at 83 and 166/m^2 (0.15 and 0.30/liter), *Rana* alone at 83 and 166/m^2, and *Pseudacris* and *Rana* mixed 1:1 at a total density of 166 and 332/m^2 (0.30 and 0.60/liter). These treatments allowed an independent examination of the effects of intraspecific and interspecific competition and compared the interspecific effects of both species with each other. Phytoplankton standing crops were examined at intervals, and water from the ponds were assayed for inhibitory effects on the growth of *Pseudacris* tadpoles in the laboratory. The mean mass at metamorphosis of both species decreased when density changed from 83 to 166/m^2. The effect on *Rana* was about twice as great as that on *Pseudacris.* The species differed greatly in their interspecific effects; *Pseudacris* had very little effect on *Rana* at either density, while *Rana* strongly affected most responses of *Pseudacris* at both low and high densities. The interspecific effect of *Rana* on *Pseudacris* was greater than the intraspecific effect of a similar number of *Pseudacris. Rana* tadpoles had a greater effect on the standing crop of periphyton than did *Pseudacris* tadpoles. The assays for growth inhibitors suggested that interference competition through inhibition did not occur. They suggested that the greater effects of *Rana* were likely because of their greater body size and effect on the availability of resources.

Trenerry (1988) used a reciprocal-transplant enclosure experiment to evaluate the potential for competition between larval *Litoria genimaculata* and *Nyctimystes dayi* in tropical rain forest streams. Tadpoles of *L. genimaculata* normally occur in slow-flowing pools, and *N. dayi* tadpoles are suctorial and primarily occupy riffles. Both species are found in runs with intermediate flow rates. He measured and staged tadpoles, placed them in 3.5-liter plastic jars with screen wire sides, and anchored the jars in the stream. Nine jars were placed in a pool and nine in a riffle: 3 containing 6 *L. genimaculata* (390/m^2; 1.7/liter), 3 containing 6 *N. dayi,* and 3 containing 3 tadpoles of each species. After 34 days, he measured the mean growth increments of tadpoles and the number surviving in each jar. Both species showed significant increases in body length only when they were alone in their normal microhabitat. The data also suggested that mortality of each species was lower in its normal microhabitat. Results indicate that *L. genimaculata* and *N. dayi* are sufficiently specialized that they perform poorly in each other's microhabitats. The fact that presence of the other species caused them both to perform poorly in their normal microhabitats indicates that this specialization may be reinforced by competition when they co-occur. Penned populations in the field are commonly used to examine competitive relationships. For example, R. A. Griffiths (1991) found that *R. temporaria* increased the larval period and decreased the growth rate, mass at metamorphosis, and survival to metamorphosis of *B. calamita* even when the species were not allowed physical contact (see Bardsley and Beebee 1998). Likewise, R. A. Griffiths et al. (1991) found that the growth of larval *B. calamita* was inhibited by co-occurring *R. temporaria,* and by *R. temporaria* feces alone.

The effects of pond drying and competition between tadpoles of *Bufo calamita* and *Hyla arborea* and either *Rana esculenta, R. lessonae,* or *R. ridibunda* were examined by Semlitsch and Reyer (1992b). *Rana esculenta* is a hybrid species that is a sexual parasite on the two nonhybrids. In outdoor artificial ponds containing 1,100 liters of water, they used a 3 (*Rana* species) $\times$ 2 (interspecific competitors present/absent) $\times$ 2 (ponds dry/do not dry) factorial design. Each tank contained 40 tadpoles of one *Rana* species. Half of the tanks also contained 35 *Hyla* and 100 *Bufo.* Half of the tanks in each combination of *Rana* species/competition treatment dried completely in 60 days; remaining tanks had a constant water level. The tadpoles of *R. ridibunda* survived at low rates in all treatments. There was a significant interaction between the effects of species and drying regime on the proportion of tadpoles that metamorphosed in 60 days. Several other interactions involving species identity that appear in their figures 1–5 were not statistically significant. Competition and drying regime also had significant main effects on percentage of individuals metamorphosing and on mass at metamorphosis. Days to metamorphosis was significantly affected only by competition. Tadpoles of *R. esculenta* performed better under poorer conditions than either of the nonhybrid species, while the nonhybrid *R. lessonae* outperformed *R. esculenta* under favorable conditions. Semlitsch and Reyer suggested that environmental heterogeneity may be an important factor in maintaining populations of all the species necessary for the stability of this hybridogenetic complex (see Schmidt 1993).

Laboratory Studies

Compared to the large number of laboratory studies of intraspecific competition, relatively few have examined interspecific effects. Some experiments, such as those of Akin (1966) and Licht (1967), have already been discussed with reference to growth inhibition. Smith-Gill and Gill (1978) examined interference competition between *Rana pipiens* and *R. sylvatica* and fed the tadpoles boiled lettuce ad libitum. The effects of conspecific density on *R. sylvatica* were curvilinear and *R. pipiens* had little effect on *R. sylvatica*. Conversely, the effect of *R. sylvatica* density on *R. pipiens* was greater than the conspecific effect of *R. pipiens*. The interspecific effect of *R. sylvatica* on *R. pipiens* was also curvilinear (i.e., increased more rapidly than density). They also found evidence for positive effects of increasing density at lower densities. They concluded that the intensity and even the sign of interspecific competition coefficients may depend on density and that extrapolation from single-density experiments to predict interactions over a wide range of densities is unlikely to result in realistic predictions. It would be interesting to perform a similar experiment under conditions where exploitation competition was likely to be dominant as it usually is in nature (Petranka 1989).

Travis (1980b) examined the responses of *Hyla gratiosa* larvae belonging to six sibships to competition with *H. femoralis* larvae from a single sibship. Six of 7 experimental treatments included 20 individuals from 1 *H. gratiosa* sibship and 20 *H. femoralis* sibships, and the final one included 4 *H. gratiosa* from each of 5 of the sibships plus 20 *H. femoralis*. The initial densities in all treatments were equal with 40 individuals in a plastic pan holding 6 liters of water (555/m^2; 6.66/liter). His analysis indicated that *H. gratiosa* sibships differed in their competitive effects on *H. femoralis* whose mass at day 47 of the larval period varied from means of 125–185 mg depending on the *H. gratiosa* sibship with which they were competing. The *H. gratiosa* families also differed in their responses to the combination of inter- and intraspecific competition. The treatment that included *H. gratiosa* from a mixture of families showed responses intermediate between, but more variable than, the single family groups.

The laboratory experiment of Alford (1989b) was designed to determine whether variation in reproductive phenology of adults might interact with genetic variation for factors determining competitive ability in larval *H. chrysoscelis*. I designed a 2 × 2 × 3 factor experiment in which the first 2 factors were male and female parentage and the third was competitive regimen. Crossing two males with each of two females gave four sibships of tadpoles. I raised these in 3 competitive regimens: individuals alone in plastic boxes containing 250 ml of water (83/m^2; 4/liter), 1 *H. chrysoscelis* with 4 *R. catesbeiana* introduced simultaneously (415/m^2; 20/liter), and 1 *Hyla* with 4 *Rana* introduced on day 11 of the *Hyla* larval period. These treatments simulated variation in the relative timing of reproduction of *Hyla* and *Rana* similar to that encountered in monitoring of field populations. I supplied food to all containers at the same daily rate and evaluated the effects of competition on the *H. chrysoscelis* tadpoles by weighing them on day 41 of their larval periods. The presence of *R. catesbeiana* significantly reduced the mass of *Hyla* tadpoles. The four *H. chrysoscelis* sibships responded differently to competition, and the offspring of different females differed significantly. Offspring from different male parents differed significantly, but their ranking from large to small changed depending on whether competing *Rana* were introduced simultaneously or later. These results suggest that selection caused by phenological variation may be a factor leading to the retention of genetic variation in the competitive ability of tadpoles because different suites of characters may be advantageous depending on when competitors reproduce.

In another case (personal observation), I examined competitive effects of tadpoles of *Limnodynastes ornatus* and *Litoria inermis* from Australia on larvae of the exotic *Bufo marinus*. I raised tadpoles in 10-cm diameter × 7-cm deep plastic containers holding 250 ml of water and supplied food at 1 and 6X. As part of the experiment, I used 5 experimental treatments to examine interspecific competition: 1 *Bufo* tadpole raised alone with 1X food, 1 *Bufo* raised alone with 6X food (density 127/m^2; 4/liter), 1 *Bufo* tadpole raised with 5 *L. inermis* (762/m^2; 24/liter), 1 *Bufo* with 5 *L. ornatus,* and 6 *Bufo* tadpoles. There were large differences in relative effects of each species on *Bufo* growth rate. Tadpoles of *L. inermis* had very little effect on growth rates of *Bufo,* which grew at nearly the same rate in competition with five *L. inermis* as they did when raised alone. Intraspecific competition from *Bufo* had an intermediate effect, while competition from *L. ornatus* had the greatest effect; *Bufo* raised with *L. ornatus* grew very slowly and all died before metamorphosis. These results are consistent with what is known of the basic biology of tadpoles of each species. *Limnodynastes* are specialized for rapid growth in extremely ephemeral pools; they hatch from relatively large eggs with large yolk supplies that allow them to reach large sizes within 2 days posthatching. They are voracious feeders and behaviorally interfere with the access of other tadpoles to food sources by butting and can reach metamorphosis in as little as 13 days (personal observation). All of these characteristics suggest that *L. ornatus* tadpoles should be strong competitors. Tadpoles of *L. inermis,* which normally occur in semipermanent or permanent dams and water holes, hatch from much smaller eggs than *L. ornatus* and at a much smaller sizes, grow more slowly posthatching, and appear to be less aggressive (personal observation). *Bufo marinus* are intermediate in all characters except aggression while feeding; they appear to be as aggressive as *L. ornatus,* but because they are usually smaller after hatching and their relative size decreases with the passage of time, *L. ornatus* larvae can outcompete them for macroscopic food items. Semlitsch (1993) examined competition between tadpoles of species of the hybridogenetic *Rana esculenta* complex and showed that interspecific competition exists, with the hybrid *R. esculenta* a somewhat stronger competitor than *R. lessonae.*

Competition of Tadpoles with Other Organisms

Although it seems likely that tadpoles and many other groups of primarily herbivorous aquatic organisms overlap in their food requirements, very little information is available on competition outside the tadpole guild. In her laboratory study of interspecific growth inhibition, Akin (1966) found that water conditioned by other taxa did not inhibit the growth of *Rana pipiens* tadpoles. Morin et al. (1988) examined the responses of *Hyla andersonii* tadpoles to competition from *Bufo woodhousii fowleri* and an assemblage of aquatic insects. Twelve artificial ponds containing *H. andersonii* tadpoles at a density of 83/m^2 (0.15/liter) were used. Six ponds also contained *B. w. fowleri* tadpoles (83/m^2). They allowed aquatic insects to colonize three of the ponds with both tadpole species and three with only *H. andersonii*. Both *Bufo* and insects reduced the mean metamorphic mass of *Hyla* tadpoles. The combined effects of *Bufo* and insects were similar to the effects of either group alone. Competition had no effect on larval period or survival rate of *Hyla*. By measuring the standing crop of periphyton on glass slides on a single date, they found that insects reduced the standing crop and presumably the availability of periphyton for tadpole consumption. They suggested that in natural temporary ponds lacking vertebrate predators, competition with insects may be at least as important as competition within the tadpole guild. Competition of tadpoles with other taxa has also been shown by Alford (1989c, newt larvae), L. Blaustein and Margalit (1994, 1996; mosquito larvae), Brönmark et al. (1991, snails), and Kupferberg (1997b, insect larvae). These results strongly indicate that such effects should be considered in future synthetic studies of tadpole ecology.

Interactions between Competition and Predation

Competition and predation can have strong effects on the development, growth, and survival of tadpoles. In natural systems, tadpoles are commonly exposed to both of these factors, which may often interact. Predation obviously affects competitive interactions by altering the density of competing species. Differences in palatability or predator avoidance among species may interact with differences in competitive ability; species with better antipredator mechanisms may suffer less from interspecific competition as predators reduce the numbers of their competitors. The intensity of competition may affect predation rates and cumulative risks of predation by changing the time periods that prey spend at sizes that are vulnerable to particular predators (reviewed by Sih et al. 1985). The complexity of these possible effects makes it necessary to experimentally evaluate their importance.

DeBenedictis (1974) examined competition between larval *Rana pipiens* and *R. sylvatica* in pens placed in a pond (101–806 individuals/m^2; 0.67–5.36/liter) over a 2-year period. In the first year, density did not appear to limit growth or survival of either species in any experimental treatment. He suggested that this indicated that the experimental densities were below the carrying capacity of the environment. In the second year, there were strong effects of density and evidence for interspecific competition with inter- and intraspecific competition having about equal effects. Predators (fish, leeches, and odonate naiads) reduced survival and increased metamorph size of both species. Many of the results of DeBenedictis are difficult to interpret because of highly variable replicates.

Artificial pond communities generally exhibit less variation among replicates than pond enclosures and therefore allow complex experiments with interpretable results. Interactions between competition and predation were first examined in artificial pond communities by Morin (1981). He established 16 experimental ponds containing adult broken-striped newts (*Notophthalmus viridescens dorsalis*) at densities of 0, 2, 4, or 8 per pond (0, 1.1, 2.2, or 4.4/m^2). Five days later he added hatchlings of *Pseudacris crucifer, Rana sphenocephala,* and *Scaphiopus holbrookii* at densities of 0.2, 0.3, and 0.1/liter (110, 165, and 55/m^2), respectively. The community was completed 36 days later with the addition of *Bufo terrestris* hatchlings at a density of 0.3/liter (165/m^2). Morin found that *Scaphiopus* had very high survival to metamorphosis in the absence of predators ($\bar{x}$ = 93%). *Bufo* also survived at a relatively high rate when predators were absent ($\bar{x}$ = 40%). Morin attributed the low survival of *Pseudacris* when predators were absent ($\bar{x}$ = 4%) to interspecific competition. The survival rates of *Bufo* and *Scaphiopus* declined with increasing newt density and reached means of 8% and 18%, respectively, at the highest newt density. Survival of *Pseudacris* increased to 28% at a newt density of 1.1/m^2 and remained constant at higher newt densities. Mean body mass at metamorphosis of *Pseudacris* increased from 72 mg at a newt density of 1.1/m^2 to 211 mg when 4.4 newts/m^2 were present. Data from his table 1 suggest that total metamorph biomass produced per pond declined from about 93 g when newts were absent to about 46 g when 4.4 newts/m^2 were present; this suggests that some competition may have been caused by interference. Alternatively, this result could have been caused by increased survival and competition from larval *Rana* that did not metamorphose during the period covered by his analyses.

Morin (1981) demonstrated that predation could alter the competitive hierarchy in a temporary-pond system but suffered from some lack of realism by including only a single predator. In a second overlapping experiment, Morin (1983) manipulated the abundance of a second predator, larval tiger salamanders (*Ambystoma tigrinum*). Newts were established at the same densities (0, 2, 4, and 8 per pond) used in the first experiment. Two additional treatments were added: one with four adult *N. v. dorsalis* per pond and four *A. tigrinum* per pond and one with only four *A. tigrinum* per pond. Hatchling anurans were added in the sequence and densities previously outlined in Morin (1981). Twenty-two days after the introduction of *Bufo terrestris,* Morin added *Hyla chrysoscelis* and *H. gratiosa* both at densities of 83/m^2 (0.15/liter). Larval *A. tigrinum* removed nearly all tadpoles from all ponds they occurred in. The effects of newts on the tadpole metamorph assemblage were similar to those reported in Morin (1981), although the observations were over a longer period. Inclusion of data on *H. chrysoscelis* and

H. gratiosa showed that total metamorph biomass did not decrease consistently with increasing newt density, which indicates that competition was probably caused by exploitation of resources. Correlational analysis of interspecific competition suggested that it was strongly asymmetrical, with *B. terrestris, H. chrysoscelis, R. sphenocephala,* and *S. holbrookii* negatively affecting the growth and survival of *H. gratiosa* and *P. crucifer,* while not being affected by them. Morin suggested that *H. gratiosa* and *P. crucifer* may depend on reduction of the density of competing species by predators for their persistence in anuran communities. He also discussed the possibility that some of the asymmetry in competitive effects may have been caused by the phenology of introduction of species.

In both of his experiments, Morin (1981, 1983) could infer competition only through retrospective correlations of metamorph abundances with performance. Wilbur et al. (1983) addressed this problem with an artificial pond experiment by manipulating the abundances of adult newts and tadpoles simultaneously. Newt densities were 0, 4, or 8 per pond (0, 2.2, or 4.4/m^2), and *Bufo woodhousii fowleri* tadpoles were introduced to all ponds at a constant density of 165/m^2 (0.3/liter). Three other species were introduced to low-, medium-, and high-density ponds as follows: *Bufo terrestris* (55, 110, or 220/m^2; 0.1, 0.2, or 0.4/liter), *Rana sphenocephala* (28, 55, or 110/m^2; 0.05, 0.1, or 0.2/liter), and *Scaphiopus holbrookii* (55, 110, or 220/m^2; 0.1, 0.2, or 0.4/liter). Total tadpole densities in the ponds were: low = 138/m^2 (0.25/liter), medium = 275/m^2 (0.5/liter), and high = 550/m^2 (1.0/liter). *Bufo* species were competitively dominant at medium and high densities when newts were absent. *Scaphiopus* had low survival rates, which was attributed to their inability to feed on the filamentous algae that dominated the ponds. *Rana* appeared to be strongly affected by interspecific competition at the high density. When newts were present, they eliminated all *S. holbrookii.* Newts at 2.2/m^2 greatly reduced survival of both *Bufo* species, and at 4.4/m^2 they eliminated all *Bufo* from 8 of 9 ponds. Newts also greatly reduced the survivorship of *R. sphenocephala,* but some *Rana* survived in most ponds with newts. Surviving *Rana* benefitted from reduced inter- and intraspecific competition and metamorphosed at mean body masses at least twice as large as *Rana* from ponds without predators. Wilbur et al. (1983) suggested that *R. sphenocephala* was the only species capable of attaining a size refuge from newt predation and that predation might be the dominant organizing factor in many tadpole assemblages.

Other studies focused primarily on interactions between competition and predation as influences on tadpole ecology. Van Buskirk (1988) examined whether effects of multiple predators can be regarded as additive. He studied predation by dragonfly naiads on a tadpole assemblage in artificial ponds by independently varying the densities of *Tramea carolina* (0 or 9.9/m^2; 0 or 0.019/liter) and *Anax junius* (0 or 2.2/m^2; 0 or 0.004/liter). The tadpole assemblage consisted of *Bufo americanus* (138/m^2; 0.26/liter), *Pseudacris crucifer* (39/m^2; 0.07/liter), *P. triseriata* (50/m^2; 0.09/liter), and *Rana sphenocephala* (83/m^2; 0.16/liter). The composition of the metamorph assemblage was affected strongly by predators. *Rana* were almost eliminated from ponds with no predators, but about 10% survived in ponds with one species of predator. *Rana* survival decreased to 2% when both predators were present. *Bufo* and *P. crucifer* were affected similarly by *A. junius,* which reduced their survival rates from 74% and 62%, respectively, to 30%. *Tramea carolina* affected *P. crucifer* and *Bufo* differentially with 0% and 35% surviving when *Tramea* were the only predators. When both predators were present, their effects on *Bufo* appeared additive, with about 19% of *Bufo* reaching metamorphosis. The survival of *P. triseriata* was affected strongly by both dragonfly species by declining from 62% when predators were absent to about 10% when either predator was present and 0% in the presence of both. Predation benefitted surviving *B. americanus* and *P. triseriata* by increasing their growth rate and decreasing their larval periods; it did not affect the performance of the other tadpole species. Van Buskirk concluded that the effects of predators on the tadpole assemblage were essentially additive and allowed the effects of both species together to be predicted from the effects of each one alone. This offers some hope for the eventual construction of models of species interactions in complex communities.

Cortwright (1988) examined interactions among predators and the effects of predators on their prey in 1 × 2.5-m pens placed in a permanent pond. His predators were adult *Ambystoma opacum* and larval *A. jeffersonianum,* and *A. maculatum.* The presence or absence of *A. opacum* (0 or 4.4/m^2) was crossed with 3 initial densities (17, 34, or 67/m^2) of hatchling *A. jeffersonianum.* Other species were added at equal initial densities to all pens: hatchling *A. maculatum* (34/m^2), hatchling *Hyla chrysoscelis* (92/m^2), hatchling *Rana sylvatica* (200/m^2), and year-old overwintering *Rana clamitans* (12/m^2). Cortwright found that the yearling *R. clamitans* were not vulnerable to salamander predation presumably because of their large body size. *Hyla chrysoscelis* were virtually eliminated from all pens with a maximum survival rate of 0.2%. The fate of *R. sylvatica* depended on the presence of *A. opacum;* when these were present, *R. sylvatica* were almost eliminated. When *A. opacum* were absent, about 10% of *R. sylvatica* survived to metamorphosis. There was some evidence that surviving *R. sylvatica* benefitted from reductions in conspecific density when the initial density of *A. jeffersonianum* was higher. Cortwright and Nelson (1990) report on additional results from this experiment.

Wilbur and Fauth (1990) examined competition and predation in a two-predator, two-prey system in artificial ponds. Their experiment crossed the presence and absence of *Anax junius* naiads (two per pond) and adult newts (*Notophthalmus viridescens;* two per pond) with the presence and absence of tadpoles of *Bufo americanus* and *Rana palustris,* both at 276/m^2 (0.5/liter). When predators were absent, there was significant intra- and interspecific competition. The level of competition was reduced when predators lowered survival rates. The survival of both species was reduced in ponds with one predator and one prey species but to a species-specific extent; *Rana* were more strongly affected by *Anax* than were *Bufo.* When one or two predators and both prey were present, the prey survived at higher rates than when they were alone with the predator; they appeared to share the rise of predation so that each species benefitted from the presence of the other. When both predators were present with one prey species, the prey suffered about additive mortality from both predators—the survival rate of prey in two-predator, one-prey systems could be predicted from their survival in one-predator, one-prey systems. Wilbur and Fauth examined their results further with several models of multispecies interactions. Fauth and Resetarits (1991) reported on the results of an experiment in which newts (*Notophthalmus viridescens*) and sirens (*Siren intermedia*) preyed on a five-species tadpole assemblage. They found that newts acted as keystone predators, altering the composition of the tadpole assemblage, while sirens were nonselective. The predators interacted, apparently through predation by sirens on newts. Peacor and Werner (1997) found complex interactions between competition, predation, and behavioral antipredator mechanisms in experimental systems containing two size classes of larval *Rana catesbeiana*, larval *Rana clamitans*, and two odonate predators.

These field experiments demonstrate that competition and predation interact in a complex manner in determining the success of tadpoles. Some species that have effective antipredator mechanisms (e.g., *Pseudacris crucifer*) may benefit unambiguously from predation on competing species. Surviving individuals of many species may benefit from increased growth rates and body sizes at metamorphosis caused by reductions in the levels of inter- and intraspecific competition. The extent of these benefits may depend on interactions between freeing of resources by the thinning of competitors and the costs of antipredator mechanisms such as reductions in rates of movement (Van Buskirk and Yurewicz 1998). Whether the advantages outweigh the disadvantages of decreased survivorship for species with poor defenses probably varies from time to time and place to place depending on the densities of competitors and on the presence of other mortality sources such as pond drying. Some of these questions are addressed in the next section.

Effects of Environmental Uncertainty

Environmental uncertainty has been featured prominently throughout this chapter. Reproductive site choice by frogs, and thus possible tadpole community composition, is influenced by a complex interaction between environmental quality, environmental uncertainty, life history, predation risk, and resource availability. Tadpoles are capable of responding to alterations in the environment, both physical and biotic, by altering their behavior and physiology and changing their relative rates of growth and development. As Morin and Johnson (1988) pointed out, competition between species of tadpoles is often asymmetrical and may depend on relative body sizes. Relatively recently, a number of field experiments have attempted to evaluate how environmental uncertainty of several types effects ecological relations within and between species of tadpoles.

Alford and Wilbur (1985) were interested in determining whether the outcome of competition was affected by small changes in priority of reproduction by *Bufo americanus* and *Rana sphenocephala* in artificial pond systems. We crossed three *Rana* treatments (absent, 100 [55/m^2, 0.1/liter] introduced on day 0, and 100 introduced on day 6) with three *Bufo* treatments (absent, 500 [275/m^2, 0.5/liter] introduced on day 0, and 500 introduced on day 6). For each species, these treatments yielded ponds with no interspecific competition and ponds with interspecific competitors introduced simultaneously, 6 days earlier, and 6 days later. The treatment with neither species introduced was used in a separate experiment (Wilbur and Alford 1985; see below). Both species performed best in the absence of interspecific competition. Both species also performed better when they were introduced on day 6 of the experiment than when they were introduced on day 0. The effect of *Rana* on *Bufo* appeared to increase when they were introduced earlier than *Bufo* and to decrease when they were introduced later than *Bufo*. This result was consistent with the idea that the effects of interspecific competition are size-specific, with larger animals having greater competitive effects. *Rana* actually performed better when *Bufo* preceded them into ponds and worse when *Bufo* followed them into ponds. This may have been caused by a reduction in the number of *Bufo* competitor days experienced by *Rana* when *Bufo* preceded them into the ponds. The performance of tadpoles can depend on the absolute fine timing of reproduction by adult frogs and on the relative timing of reproduction by competing species. L. Blaustein and Margalit (1995) have demonstrated similar effects of small changes in priority on the interactions between larval *Bufo viridis* and mosquitoes (*Culiseta longiareolata*).

The study of Alford (1989a) was designed to examine the effects of short-term variation in reproductive phenology in greater detail than had been revealed in the preceding experiment (Alford and Wilbur 1985). I used 9 experimental treatments to examine the responses of tadpoles of *Bufo americanus* and *Rana palustris* to 7 relative timings of introduction of the other species that spanned a range of 11 days preceding to 11 days following. All treatments included both species at the same initial densities of 55 *Rana*/m^2 (0.1/liter) and 274 *Bufo*/m^2 (0.5/liter). Growth and survival of both species failed to respond to the timing of introduction of the other species. Correlations between growth rates and number of metamorphs indicated that both species responded to their own density, but there was little evidence of interspecific competition. These species frequently co-occur in North Carolina ponds, and I suggested that the lack of interspecific competition might reflect co-adaptation to avoid competition and that the effects of phenological variation on species interactions depend on the species being examined.

Effects of short-term variation in reproductive phenology were also explored by Gascon (1992a) and S. P. Lawler and Morin (1993). Gascon examined competitive relations between the early reproducing *Osteocephalus taurinus* and the later-reproducing *Epipedobates femoralis* and *Phyllomedusa tomopterna* in Amazonian Brazil. He conducted three experiments in 45-cm diameter × 19-cm deep basins placed outdoors in the forest. Neither of the late-breeding species affected the early species, but the early species affected both of the late-breeding species. The relative timing of introduction altered the effects of *O. taurinus* on *P. tomopterna;* they were reduced when both species were introduced simultaneously. Relative timing did not alter the competitive effect of *O. taurinus* on *E. femoralis.* S. P. Lawler and Morin (1993) used artificial pond experiments to determine how *Bufo woodhousii* and *Pseudacris crucifer* larvae responded to 0, 7, and 14 day lags in timing of reproduction. They found that the species responded asymmetrically; *Pseudacris* were affected more strongly by *Bufo* when *Bufo* were introduced 7 days before them, while *Bufo* were not affected by *Pseudacris.* They concluded that the usual pattern in nature, in which *Pseudacris* reproduce very early, probably maximizes the success of this competitively inferior species.

Competition may be direct, through the use of the same resources or behavioral or chemical interference, or indirect, through alterations to the physical or biological structure of the environment. Wilbur and Alford (1985) demonstrated that tadpoles may affect other species through persistent effects on the environment. We examined the effects of pond history and competition on the performance of *Hyla chrysoscelis* larvae. *Hyla* were introduced at a density of 274/m² (0.5/liter) to the artificial ponds used in a separate experiment (Alford and Wilbur 1985, discussed above) 80 days after the ponds were initially filled. They were also raised in a set of ponds that had been filled only 15 days previously and had never been occupied by other tadpoles. This led to *H. chrysoscelis* tadpoles experiencing ponds with 10 different histories: 15- and 80-day-old ponds that had never been occupied by tadpoles, 80-day-old ponds that had been occupied by *Rana sphenocephala* or *Bufo americanus* for 66 or 72 days, and 80-day-old ponds that had been occupied by both *Rana* and *Bufo* for various combinations of 66 and 72 days. Most *Bufo* had completed metamorphosis before any *Hyla* were introduced to the ponds; effects of *Bufo* on *Hyla* thus were caused by environmental alterations rather than direct competitive effects. Growth and survival of *H. chrysoscelis* was greatest in 15-day-old ponds. Eighty-day-old ponds that had never been occupied by other tadpoles produced the next best *Hyla* performance. The presence of *Bufo* or *Rana* decreased the performance of *Hyla* metamorphs, which was also influenced by the order in which *Bufo* and *Rana* had been introduced.

Indirect effects between temporally separated species were examined further by Morin (1987), who introduced *Pseudacris crucifer* tadpoles to experimental ponds at 55, 110, and 220/m² (0.1, 0.2, and 0.4/liter). These were crossed with densities of the predatory newt *Notophthalmus viridescens* at 0, 2, or 4 per pond (0, 1.1, or 2.2/m²). Forty-three days later, about 1 week before *Pseudacris* began to metamorphose, Morin added hatchling *Hyla versicolor* to all ponds at a density of 110/m² (0.2/liter). Newts had little effect on the growth and survival of *Pseudacris,* but increasing newt density strongly decreased the survival to metamorphosis of *Hyla.* When newts were absent, the total froglet biomass exported from ponds was relatively constant (about 75 g). The proportion of that biomass attributed to *Hyla* depended strongly on the initial density of *Pseudacris;* although all *Pseudacris* completed metamorphosis early in the *Hyla* larval period, there was a significant negative correlation between *Pseudacris* biomass and *Hyla* biomass exported from ponds. *Hyla versicolor* growth and survival rates also decreased with increasing densities of *Pseudacris.* Morin pointed out that this pattern is consistent with a hypothesis of preemptive exploitation competition for resources; a limited total quantity of nutrients is available when the pond fills and the export in metamorphs of early breeding species may reduce the availability to later reproducers. He also noted that *Pseudacris* appeared to be competitively superior to *Hyla* and that their temporal separation might not preclude competitive exclusion under some conditions.

Warner et al. (1991) examined interactions among the effects of density, long-term reproductive priority, and pH on the larval biology of *Hyla femoralis* and *H. gratiosa* in outdoor tanks. All remaining *H. gratiosa* were removed before the introduction of *H. femoralis* 108 days after *H. gratiosa* had been introduced. They found that *H. gratiosa* larvae had longer larval periods and lower metamorphic body sizes when densities were higher. Lower environmental pH increased the effects of intraspecific density on *H. gratiosa. Hyla femoralis* responded similarly, but the effects of pH did not interact with the effects of density. *H. femoralis* introduced to ponds that had previously contained *H. gratiosa* performed better than ones in ponds with no previous tadpole occupants. Warner et al. (1991) suggested that this facilitation might have been caused by effects of *H. gratiosa* on resources; their grazing may have changed the abundances of algae, favoring types with better nutritive quality, or may have enhanced nutrient cycling or reduced the prevalence of some toxic substance.

The consequences of variation in reproductive timing within a species were examined by Morin et al. (1990). They introduced a second cohort of *Hyla andersonii* tadpoles at a density of 55/m² (0.1/liter) to 12 artificial ponds used in an experiment discussed earlier under competition with other taxa (Morin et al. 1988). The second cohort was introduced 34 days after the first. Their results indicated that the second cohort experienced longer larval periods and reduced growth rates and survival to metamorphosis when compared to the first cohort. Mean masses at metamorphosis were similar between the cohorts except when interspecific competitors and predators were excluded; in that treatment, the earlier cohort reached metamorphic masses nearly twice as great as the later cohort. Hearnden (1992) demonstrated similar effects in Australian populations of *Bufo marinus,* with the additional factor that when a second cohort was introduced as eggs, the earlier cohort preyed upon them, adding predatory reductions in survival to competitive effects.

The preceding experiments showed that the reproductive phenology of frogs can affect ecological interactions among tadpoles. Variation in reproductive phenology has the potential to alter the outcome of competition between some species. Variability in the persistence of ephemeral ponds is probably another aspect of environmental uncertainty that is important in tadpole ecology. Wilbur (1987) examined interactions between competition, pond drying rates and predation in experimental pond communities. He hypothesized that predation may benefit some tadpoles in temporary habitats by reducing interspecific and intraspecific competition by allowing survivors to grow more rapidly and leave ponds before they dry. He tested this hypothesis with an artificial pond experiment including three rates of pond drying: constant water depth of 50 cm, drained after 50 days, or drained after 100 days. He also included 2 predation regimes of adult *Notophthalmus viridescens dorsalis* (absent or present at a density of 2.2/m^2 [0.004/liter] and low and high tadpole densities (totals of 242 and 970/m^2; 0.44 and 1.76/liter). The tadpoles included in each pond were *Bufo americanus* (110 or 441/m^2; 0.2 or 0.8/liter), *Hyla chrysoscelis* (69 or 276/m^2; 0.125 or 0.5/liter), *Rana sphenocephala* (28 or 110/m^2; 0.05 or 0.2/liter) and *Scaphiopus holbrookii* (36 or 143/m^2; 0.065 or 0.26/liter). The composition of the metamorph assemblage was influenced significantly by predation, drying time, density, and their two-way interactions. *Scaphiopus* appeared to be the dominant competitor. *Rana* negatively affected the growth of all species except *Scaphiopus. Bufo* and *Hyla* appeared to have negative effects only on themselves. At low initial density, newts eliminated nearly all tadpoles. At high density, similar total numbers of tadpoles metamorphosed regardless of newt presence, but the composition of the metamorph assemblage changed; newts reduced the dominance of *Scaphiopus* and allowed other species to metamorphose in greater numbers. This effect was more pronounced in ponds that dried. With predators absent, only *Scaphiopus* emerged from ponds that dried in 50 days, but with predators present, both *Bufo* and *Scaphiopus* metamorphosed. Similarly, only *Bufo* and *Scaphiopus* emerged in any numbers from ponds that dried in 100 days when predators were absent, but the presence of predators allowed significant numbers of *Hyla* to metamorphose as well. There was evidence that tadpoles responded directly to pond drying by initiating metamorphosis earlier. Wilbur interpreted his results as indicating that competition and desiccation probably dominate the dynamics of populations exploiting very ephemeral habitats, with predation possibly acting as a moderating influence. The importance of predation should increase as habitat duration increases.

Harris et al. (1988) documented the temporal patterns of abundance of adult and larval newts (*N. v. dorsalis*) in a pond in North Carolina over a 3-year period. We showed that the timing of newt arrival and departure and the duration of newt presence varied among years. This variation was not well correlated with variation in the tadpole assemblage (personal observation). All of the experimental studies cited thus far which examined predation did not explicitly allow for the fact that predators typically follow their own phenological patterns. I conducted an artificial pond experiment (Alford 1989c) to examine the effects of predator phenology and phenological variation. My experiment was combined with the one reported by Wilbur (1987). The same four anuran species were introduced at the same two densities. I established 4 predation regimes that included 2 extreme treatments in which adult *Notophthalmus* at a density of 2.2/m^2 (0.004/liter) were absent or continuously present. Two additional regimes mimicked the extremes in newt departure time observed by Harris et al. (1988)—newts initially present but removed from ponds on 1 April and newts initially present but removed on 13 and 14 May.

A correlation analysis suggested that interspecific competition followed a hierarchy with *S. holbrookii* being the dominant competitor followed in order by *R. sphenocephala, B. americanus,* and *H. chrysoscelis.* The presence of predators for any length of time altered the composition of the metamorph assemblage. The anurans that reproduced earliest (*Rana* and *Scaphiopus*) were affected most negatively. When newts persisted for longer periods, they also affected later reproducers. The abundance and contribution to metamorph biomass of *Bufo* declined when newts persisted until 13–14 May, while *Hyla* increased. The relatively abundant *Hyla* in this predation regime appeared to have competitive effects on the other species, which suggests that interspecific competition may not be completely described by the simple hierarchy postulated earlier. Forcing newts to remain in the ponds through the summer reduced the abundance and biomass of *Hyla* by increasing the relative contributions of the other species to the metamorph assemblage. *Bufo, Rana,* and *Scaphiopus* tadpoles appeared to reach size refugia after which newt removal did not alter their survival patterns. There was strong correlational evidence for negative effects of increased tadpole density on the performance of newt larvae. Because the primary resource of tadpoles is plant material and newt larvae feed primarily on zooplankton, this negative effect probably was caused by diffuse competition through two intermediate levels of the trophic web; tadpoles altered the algal community, which altered the zooplankton community, which affected the performance of newt larvae. This experiment confirms that incorporating predation into models of temporary pond and other communities requires some knowledge of predator phenology in addition to predator abundance.

Summary

The available information on the ecology of tadpoles has increased greatly since Wilbur's (1980) review. Tadpoles have proved to be a valuable model system for investigation of many aspects of population and community ecology. In the course of these investigations, it has become clear that a full understanding of the ecology of any tadpole assemblage requires data of several types. First, there must be information on reproductive habitat choice and phenology of the frog species inhabiting that region. Choices of reproductive habitats by frogs set the range of possible tadpole assemblages that may develop. Reproductive phenology can predetermine the outcomes of interactions between tadpoles and their competitors and predators. Second, information on the reproductive habitat use and phenology of potential predators is necessary for the same reasons. I suggest that information on potential competitors belonging to other phyla will also be needed. Third, it is necessary to have information on tadpole, predator, and probably interphyletic competitor densities. Finally, information on resource availability, habitat structure, behavioral responses of tadpoles to competitors and predators, and the previous history of the habitat may be needed. A lack of any of this information may lead to incorrect predictions about the course of interactions in any tadpole assemblage. Extrapolation from ecological interactions experienced by tadpoles to their effects on the population biology of frogs requires much more information on the biology of metamorph and juvenile frogs than is currently available for nearly all species. This complex picture is far from the optimistically simple views of community ecology that prevailed in the early 1970s (see May 1974). It is also far from the nondeterminism espoused by some authors (see Strong 1983). Rather, it appears that tadpole assemblages, and probably complex assemblages of other species, are highly deterministic but that a great deal of information is needed to unravel the sources of that determinism.

I identified several areas that are in particular need of further investigation. To fully understand the autecology of tadpoles, detailed studies on their foods, feeding rates, and assimilation efficiencies must be initiated. Such studies also would aid in understanding the role of tadpoles in communities. Although the evidence is sparse, tadpoles may provide a major source of fine particulate organic matter for detritivores. Studies of interphyletic interactions between tadpoles and other aquatic herbivores would also aid in examining this connection.

Despite the comparatively great amount of work that has examined the relative importance of interference and exploitation as competitive mechanisms in tadpoles, some good areas for investigation in these fields remain. Combining the quantitative genetic approach with ecological experiments designed to separate interference and exploitation may lead to some insight into the evolution of these competitive mechanisms. The extreme competitive mechanism of carnivory or cannibalism, which often appears to be facultative in tadpoles, also should be looked at more extensively with quantitative genetic analysis. Interactions between tadpole behavior, inter- and intraspecific competition, and predation risk also have been explored only preliminarily.

I suspect the reader will have arrived independently at a list of research needs that differs from mine. This should emphasize the fact that although much more is known about tadpole ecology in 1998 than was known in 1980, much more remains to be discovered.

ACKNOWLEDGMENTS

I would like to thank S. J. Richards for his considerable assistance in reading and commenting on drafts of the manuscript and for access to his comprehensive reprint collection. R. Altig and R. W. McDiarmid provided immense assistance in keeping up with developments in the field during the prolonged gestation of the book and made many invaluable editorial comments and corrections. I would also like to thank L. J. Alford and M. N. Hearnden for comments on the manuscript and my students K. S. Bradfield, M. R. Crossland, W. Martin, and M. P. Trenerry for discussions and access to their data. Support for my research was provided by several Australian Research Council grants, the Council of Nature Conservation Ministers, and the Cooperative Research Centre for Tropical Rainforest Ecology and Management.

11

THE ANURAN TADPOLE

Evolution and Maintenance

Reid N. Harris

The course of Darwinian evolution is thus seen as determined . . . not only by the type of selection, not only by the frequency of mutation, not only by the past history of the species, but also by the nature of the developmental effects of genes and of the ontogenetic process in general.

J. Huxley 1942:555

Introduction

The complex life cycle is often viewed as an adaptation that allows a species to exploit two or more different ecological environments (Wassersug 1975; Wilbur and Collins 1973). Adult anurans are terrestrial carnivores, but their tadpoles usually are aquatic herbivores with many specialized traits that are not retained by the adult. The ability to exploit different environments effectively may require separate or uncorrelated developmental programs of larval and postlarval stages. Such independence allows each stage to evolve separate adaptations and to respond to environmental changes in separate environments (Alberch 1987; Inger 1967; also see Rieger 1994).

This review first summarizes ecological adaptations of tadpoles, including timing of metamorphosis as an adaptation, and the genetics that underpin them. Two questions relate to the origin of the anuran complex life cycle: (1) What factors promote the evolution of divergent life histories? (2) What is the degree of phenotypic and genetic correlation across life-history stages? Finally, the maintenance of the complex life cycle is reviewed and two questions are discussed: (1) What factors have led to the maintenance of the tadpole stage? (2) What factors have led to abandonment of the tadpole stage (e.g., a shift to direct development; Garstang 1922; H. B. Shaffer 1984; Sweet 1980; see Moran 1994 and Rieger 1994)? These questions are related because successful larval adaptations can promote the retention of a free-living larval stage.

The Tadpole Stage as an Adaptation

The aquatic environment of most anuran larvae presents an array of opportunities and liabilities. A recently filled temporary pond may be an environment of low predation pressure and high primary productivity, although neither characteristic has been strongly demonstrated. Wassersug (1975) and Wilbur (1980, 1987) argued that the anuran complex life cycle is an adaptation to capitalize on this favorable growth environment.

There are mortality factors associated with the pond environment that can act as agents of selection. Predation, inter- and intraspecific competition, parasitism, and pond drying cause mortality and alter aspects of individual quality, such as larval period and body size. By causing differential mortality among genotypes, these ecological factors are agents of natural selection. Travis (1983a, 1984) and Wilbur (1972, 1980, 1987) argued that fitness can be maximized in temporary pond environments by maximizing growth rate. Thus, the evolutionary hypothesis of interest is that the larval stage is where adaptations to maximize growth rate are expected. Evidence consistent with this hypothesis would include the presence of nonadditive genetic variation for growth rate and a response to natural selection on genotypes that differ in growth rate. Trade-offs between growth rate and other traits, such as competitive ability or antipredator mechanisms, might also be expected (Alford 1986a, 1989a, b). If the growth rate maximization hypothesis is correct, selection has favored growth over other traits to the degree that any conflict among functions exist.

Rapid growth rate allows a tadpole to reach a size refuge from many gape-limited predators and achieve a large body size at metamorphosis. Also, larvae quickly attain the minimum size needed for metamorphosis and have the option to metamorphose if the pond begins to dry or if predation pressures increase. Body size at metamorphosis can be negatively correlated with age at first reproduction and positively correlated with size at first reproduction (Berven and Gill 1983; D.C. Smith 1987). Thus, a high larval growth rate is also of potential value in terrestrial survival, fecundity, mating success, and age at first reproduction. Predator avoidance and resistance to parasitism are other important traits that can be disassociated from growth rate and may be involved in phenotypic or genotypic trade-offs with growth rate.

Morphologically, tadpoles appear primarily adapted for feeding and therefore achieving maximal growth. A large proportion of a tadpole's biomass (50%) is accounted for by its digestive tract (Wassersug 1974). Many of the morphological variations of buccopharyngeal (buccal filters) and oral (papillae, labial teeth) features are correlated with the mode of feeding (e.g., carnivorous, suspension feeding, scraping; Altig and Johnston 1989; Orton 1953; Wassersug and Heyer 1988). In general, carnivores have reduced branchial filters, buccal papillae, and labial teeth, whereas suspension feeders have the most elaborate buccopharyngeal structures. Suctorial and surface feeding forms have major morphological features associated with feeding (Altig and Johnston 1989; Orton 1953).

Growth and Metamorphosis

Growth Models

Growth rate is a critical adaptation in anuran tadpoles, and I will review the basic growth models that workers have used to study tadpole growth. Growth is a physiological process that can be mathematically modeled (M. J. Katz 1980; Laird et al. 1965); the presentation below follows Laird et al. (1965). Unrestrained exponential growth can be represented by

$$\lim_{\Delta t \to 0} \frac{1}{W}\frac{\Delta W}{\Delta t} = k \qquad (1)$$

where W is size, t is time, and k is a constant. This equation models the case of unlimited cell division. Exponential growth is restrained and the degree of restraint can be modeled by the coupled equations

$$\frac{dW(t)}{dt} = \gamma W(t) \qquad (2)$$

and

$$\frac{d\gamma}{dt} = -\alpha\gamma. \qquad (3)$$

Here, γ is the size-specific growth rate, and α is the exponential decay of the size-specific growth rate. Physiologically, γ can represent the rate of cell division; α can represent cell death, slowing of division rates, and the truncation of growth as tissue differentiation occurs. The solution of the coupled equations is

$$W(t) = W_0 \operatorname{Exp}\left\{\frac{\gamma_0}{\alpha}\left[1 - \operatorname{Exp}(-\alpha t)\right]\right\}, \qquad (4)$$

where W_0 is initial size and γ_0 is initial size-specific growth rate. The specific growth rate γ exponentially decays as described by

$$\gamma = \gamma_0 \operatorname{Exp}(-\alpha t). \qquad (5)$$

This model provides an important framework for a discussion of evolutionary and ecological questions (Travis 1980a; Wilbur and Collins 1973). First, it delineates three ways in which final size can be affected: (1) variation in initial size (W_0), (2) variation in size-specific growth rate (γ), and (3) variation in the exponential decay of the size-specific rate (α). All parameters can be affected intrinsically by genotype and extrinsically by environmental factors. In addition, variation in initial size (W_0) could result from variation in egg sizes, including yolk content, or from genetically or environmentally caused variation in embryogenesis. The parameter α can be thought of as the rate of change of the size-specific growth rate and may increase or decrease throughout the larval period. Thus, differences in size-specific growth rate, γ, can be based on differences in γ_0, α, or both (Travis 1980a).

Based on developmental events, the previous equations predict that growth will be size specific; this leads to interactions among the parameters. Differences in initial size among individuals or sibships will be translated into differences in dW/dt. An increase in final size at metamorphosis can be achieved by beginning at a larger size that can be maintained by a constant size-specific growth rate and a constant damping parameter (α). A large size at metamorphosis can also be achieved by beginning growth earlier and at a higher size-specific rate (γ_0). Early growth may be a conservative trait maximized by selection, but the damping parameter, α, may be plastic and underlain by a large amount of genetic variation (Travis 1980a). Rapid early growth allows the tadpole to escape gape-limited predators and to reach the size suitable for metamorphosis if the pond begins to dry or predation pressure increases. Tadpoles are under selection for rapid growth; thus, a variable α would seem maladaptive.

Maintenance of genetic variation for the damping parameter, α, may be explained by relatively weak selection on late growth rate as compared to early growth rate in the tadpole stage (Travis 1980a). Once a tadpole achieves the minimal size necessary to metamorphose, the selective pressure on growth rate may relax because of the option of metamorphosis. If resources decline or predation pressure remains high, then selection would favor the metamorphic decision (Wilbur and Collins 1973). A negative genetic correlation between growth rate and length of larval period (Travis et al. 1987; discussed below) is another explanation for the persistence of genetic variation for later growth rate.

A term for developmental rate (Alford and Harris 1988; Hensley 1993; Leips and Travis 1994; Smith-Gill and Berven 1979; Wilbur and Collins 1973) is lacking in the above equations. Wilbur and Collins (1973) predicted that growth rate and developmental rate should be dissociable (i.e., developmental rate should be responsive to changes in growth rate). In an experimental study, *Bufo woodhousii fowleri* tadpoles did slow their developmental rate in response to an enhanced growth rate (Alford and Harris 1988). In *Scaphiopus couchii,* the degree of correlation of developmental and growth rate varied among sibships. Sibships that had a rapid developmental rate could not alter it in response to an enhanced opportunity for growth (e.g., a long-duration pond). This implies a "tight link between growth and development" (Newman 1988a). Other sibships characterized by a weaker correlation between growth and developmental rates had a slower overall developmental rate and were able to take advantage of enhanced growth opportunities. Temperature affects growth and developmental rate differentially in tadpoles of *Rana sylvatica* (Smith-Gill and Berven 1979). Experimental evidence indicates that growth and developmental rates are separate biological entities and are dissociable in anuran larvae. A quantitative model that incorporates the growth parameters discussed above with developmental parameters is needed.

Models of Metamorphic Timing

The ability to time the change between life-history stages in a way that will enhance fitness is a key adaptation of any species that has more than one life-history stage. Wilbur and Collins (1973) proposed a model that predicts the timing of amphibian metamorphosis. The cornerstone of their hypothesis is that growth rate (GR) and developmental rate (DR) are dissociated (see chaps. 9 and 10). The degree to which dissociation is possible represents an adaptation; the degree to which dissociation is not possible represents constraint. Growth rate is the change in body size over time. Developmental rate is the appearance of new features (biochemical, morphological, behavioral) in the organism over time (Gosner 1960; A. C. Taylor and Kollros 1946). The Wilbur-Collins model predicts that if the growth rate of the larva increases, developmental rate is retarded to capitalize on the improved growth conditions. If growth rate decreases, developmental rate is predicted to increase to escape a deteriorating environment. According to the model, the ability to dissociate growth rate and developmental rate can occur throughout the larval period after a minimum metamorphic size is attained.

Travis (1984) presented a fixed-rate model that predicts that growth rate and developmental rate are dissociated only during the early parts of the larval period. Developmental rate is set toward the beginning of the larval period, and subsequent changes in growth rate change only size at metamorphosis. Hensley (1993) modified the Wilbur-Collins model to take into account a fixed developmental rate late in the larval period. Changes in growth rate that occur late in development affect size at metamorphosis and not developmental rate. The proportion of the larval period that is sensitive to changes in growth rate is the fundamental difference between the Wilbur-Collins and fixed-rate models. This sensitive period may be variable. Indeed, K. A. Berven (personal communication) has evidence that populations of *Rana sylvatica* differ in the fraction of the larval period that is responsive to changes in growth rate. This difference suggests that variation among populations and species could be correlated with selective environments or represent different "solutions" to maximizing size at metamorphosis.

Alford and Harris (1988) argued for an experimental manipulation to evaluate adequately the Wilbur-Collins or fixed-rate models because different correlative patterns are consistent with either model. For example, Collins (1979) stated that a positive relationship between length of larval period and mass at metamorphosis of field-collected metamorphs would support the Wilbur-Collins model. As the first larvae metamorphose, more resource per capita is left. Larvae delay metamorphosis to take advantage of the increased growth rate. This assumes that resource level is constant or increasing in the pond. Resource level can decline during the season if primary productivity decreases or if primary productivity is constant, and increasingly larger tadpoles consume more. In this case, a negative correlation would be generated even if the larvae are conforming to the Wilbur-Collins model.

The fixed-rate model also can generate either a positive or negative correlation of length of larval period and size at metamorphosis, assuming some variation in developmental rate is present. Larvae with a fixed rapid developmental rate metamorphose early, freeing resources for larvae that have a slow developmental rate. This situation would lead to a positive correlation. If resources are declining over time, a negative correlation would be produced. Variation in developmental rate can be caused by underlying genetic variation or by a response to competition early in the larval period. In summary, since either model can generate a positive or negative correlation of length of larval period and size at metamorphosis, an experimental analysis is required to ascertain how individuals are making metamorphic decisions.

Three experiments that tested larval responses to increases and decreases in food level address predictions of the Wilbur-Collins and fixed-rate models. Alford and Harris (1988) exposed tadpoles of *Bufo woodhousii fowleri* to eight experimental treatments. Four treatments initially had low food environments and therefore low growth rates, and four initially had high food environments and high growth rates. Three of the initially low growth rate treatments were switched to high growth rate treatments, and three of the initially high growth rate treatments were simultaneously switched to low growth rate treatments. All of the switched treatments were consistent with the Wilbur-Collins model, and five of the six switched treatments were inconsistent with the fixed-rate model. In general, larvae that initially were growing slowly and then switched to a rapid growth rate delayed metamorphosis until a large body size was reached. Larvae that were initially growing rapidly either metamorphosed at a minimum size if switched to a slow growth rate before the minimum size was attained or initiated metamorphosis if the switch occurred after the minimum size was attained. Larvae that switched from high to low food late in the larval period did not accelerate metamorphosis relative to the control of constant high food. The fixed differentiation rate set by an early differentiation rate, in general, was not supported, but differentiation rate appeared to be set late in the larval period.

Hensley (1993) used *Pseudacris crucifer* tadpoles in a similar design to that of Alford and Harris (1988). Changes in food level prior to Gosner 35–37 affected both size at metamorphosis and length of larval period. After those stages, changes in food level did not affect length of larval period. Mass at metamorphosis was not affected if food level was decreased late in the larval period but increased if food level was increased any time during the larval period. Hensley (1993) suggested that metamorphosis was "initiated" at Gosner 35–37 by increasing titers of metamorphic hormones; Hensley incorporated a point of fixed developmental trajectories to the Wilbur-Collins model to improve its fit to his data.

Leips and Travis (1994) also studied effects of changes in resource availability on timing of and body size at metamorphosis. Tadpoles of *Hyla gratiosa* (breeds in temporary ponds) and *H. cinerea* (breeds in permanent ponds) were raised at different food levels at two temperatures. Length of larval period tended to be affected by changes in food level in the first 60% of the larval period; the complex responses of metamorphic size are discussed below. *Hyla gratiosa* displayed greater plasticity in responses to changes in food level, although it did not have a greater period of sensitivity to changes in food level. These results tend to support the Wilbur-Collins model as modified by Hensley (1993), although Leips and Travis (1994) argued that too much of the larval period was insensitive to changes in growth rate (about the later 40%) for the Wilbur-Collins model to be supported strongly.

Recent studies that have tested models of anuran metamorphosis have also supported the general result that developmental rate can respond to changes in growth rate early in the larval period but not late in the larval period (Audo et al. 1995; Beck 1997; Newman 1994; Tejedo and Reques 1994). In contrast, the developmental rate of larval plethodontid salamanders appears not to respond to changes in growth rate at any point in the larval period (Beachy 1995, *Desmognathus ochrophaeus;* B. E. O'Laughlin and R. N. Harris, unpublished data, *Hemidactylium scutatum*).

Werner (1986, 1988) expanded the Wilbur-Collins formulation from an emphasis on aquatic growth rate to a consideration of aquatic and terrestrial growth and mortality rates. Werner predicted that metamorphosis should occur when the ratio of size-specific mortality to size-specific terrestrial growth is less than the same ratio in the aquatic phase. Finding the appropriate measurement of terrestrial growth and survival is a problem with testing the Werner model. Aquatic larvae are unable to know or predict what terrestrial growth conditions are likely to be in any given season, and the Werner model would postulate that selection has adjusted metamorphic timing with regard to the long-term environmental conditions as they relate to aquatic and terrestrial growth and survival. In order to test the model, estimates of long-term aquatic and terrestrial growth and survival are necessary. The number of generations needed in the "long term" estimate is unclear but would be related to the heritability of the "timing of metamorphosis" character and to selection intensity on that character. Neither of these parameters has been estimated, but their estimation would be an important contribution.

L. Rowe and Ludwig (1991) extended Werner's model explicitly to include the period of time between metamorphosis and reproduction. Two amphibian cases are modeled: explosive breeding that occurs at one time during a season, and prolonged breeding with an advantage to earlier within-season reproduction. In contrast, the Wilbur-Collins and Werner models assume that amphibians can reproduce continually over infinite time. Time constraints in the L. Rowe and Ludwig (1991) model mean that variation in metamorphic size is positively correlated with variation in size at reproduction and therefore reproductive output (i.e., smaller individuals at metamorphosis do not have time to catch up in body size before reproductive season commences). They assume that predation is the sole source of mortality and consider only small differences in aquatic and terrestrial size-specific mortality rates. Their general conclusion is that time constraints lead to optimal sizes at metamorphosis that vary with time to reproduction. Whether this relationship is positive or negative depends on how mortality and growth rates compare in aquatic and terrestrial habitats. "Intrapopulation variation in size at metamorphosis [is expected] even when larval growth and mortality rates are stable throughout a season" is an important prediction that contrasts with that of the Wilbur-Collins model. Consider the case of higher growth and mortality rates in the terrestrial habitat. Larvae that hatch later may perceive a decreased time until reproduction and in which to achieve reproductive size. They may have to metamorphose at a smaller size to achieve the higher terrestrial growth rate even if the mortality risks are higher. Given that the terrestrial juvenile period of amphibians is variable (up to several years), it remains to be seen whether a positive correlation of size at metamorphosis and size at first reproduction will be a general finding (D. C. Smith 1987). If this correlation does not exist, then the predictions of L. Rowe and Ludwig do not follow. Neither the models of Werner nor L. Rowe and Ludwig explicitly considers changes in size-specific growth rates during a season, although it should be possible to incorporate such changes into their model. Changes in recent growth history are the cornerstone of the Wilbur-Collins model.

Other models of metamorphic timing are based on staging series. Pandian and Marian (1985a) and Smith-Gill and Berven (1979) proposed a reliance on developmental rate to predict timing of metamorphosis and found that differential rate is a good predictor of metamorphic timing. Staging series used in both studies are intended explicitly to reflect changes in form over time leading to metamorphic climax. Thus, extrapolation to the end point of a developmental stage versus time relationship should predict when metamorphosis occurs, if environmental conditions within a population remain constant. Indeed, both studies found that different population densities and temperatures could alter the relationship of developmental rate and growth rate, a prediction made by Wilbur and Collins.

The claim that growth rate is functionally dependent on developmental rate (Smith-Gill and Berven 1979) is not supported in the sense that variation in developmental rate has not been shown to affect changes in growth rate. The hypothesis that an underlying developmental process exists that effects changes both in size and stage, however, has not been disproved. As the biology of growth and development becomes better known, this hypothesis can be evaluated. White and Nicoll (1981) discussed evidence that prolactin and thyroid hormones are antagonistic. A switch in concentration from growth hormones (i.e., prolactin) to a rising concentration of thyroid hormones (TH) with the dominance of TH could generate an interaction between growth and development. Wassersug (1986) noted that prolactin increases mucus production in tadpoles, and mucus contains prostaglandin, which slows development of an adult stomach.

The hypothesis that developmental rate is adjusted for changes in growth rate is predicted by both the Wilbur-Collins and Werner models. The adaptation proposed by Wilbur-Collins and Werner is subject to some constraints. Developmental rate is not responsive to changes in growth rate throughout the entire larval period. This was suggested by Wilbur-Collins and by the different results obtained by Alford and Harris (1988), Audo et al. (1995), Hensley (1993), Leips and Travis (1994), and Travis (1984). Seemingly, changes in growth rate have no effect on developmental rate beyond a certain developmental stage. An extreme case may be when metamorphosis is initiated by a surge in thyroxine. Any change in growth rate at that time, no matter how dramatic, is unlikely to affect that process. Hensley (1993) suggested that at least by Gosner 38, *Pseudacris crucifer* initiates metamorphosis and that developmental rate is fixed at that point. The verbal model of Leips and Travis (1994), which tries to unite the models of Wilbur-Collins and Smith-Gill-Berven, assumes that resources are allocated between growth and development differently in early and late larval stages. At early ages, resources are mainly allocated to development: "development has the primacy." At later stages, the allocation pattern changes such that fewer resources are allocated to development and that "the proportionate allocation of resources to growth . . . is total after the developmental trajectory is fixed." Presumably they mean that after the developmental trajectory is fixed, any additional resources are allocated to growth. They derive predictions from this model which they state "accommodates all of the extant data" relevant to amphibian responses to changes in food level. The main prediction of the model is an asymmetry in response to increasing and decreasing food levels at different larval ages. Earlier increases in food level affect body size less than do later increases because of an increasing proportional allocation of resources to growth. Earlier decreases in food level affect body size more than do later decreases because of the compensatory effect of a greater proportional allocation of resources to growth late in the larval period.

I propose an alternative graphical model (fig. 11.1) that posits that energy allocated to development increases late in the larval period as metamorphosis is initiated. This model also distinguishes the proportion of energy allocated to growth and development from the absolute amount of energy allocated to growth and development. At low food levels, individuals devote proportionally more resources to development and less to growth in order to metamorphose at the minimum size in the shortest time. At high food levels,

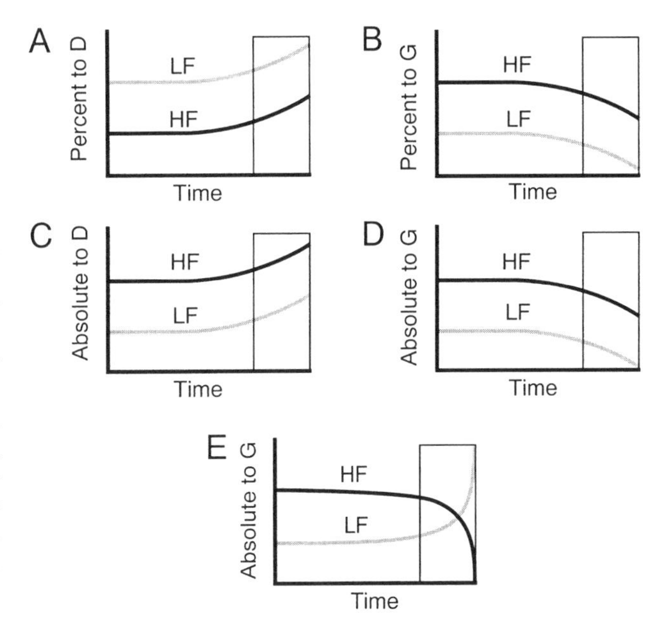

Fig. 11.1. Graphical model of energy allocation during development and growth of larval amphibians. (A) The proportion of energy allocated to development by larval anurans. (B) The proportional amount of energy allocated to growth. (C) The absolute amount of energy allocated to development. (D) The absolute amount of energy allocated to growth. (E) The absolute amount of energy allocated to growth if food level changes after metamorphosis has been initiated. The rectangle at the right side of each graph denotes that period of time from the initiation of metamorphosis until metamorphic climax. Abbreviations: D = development, G = growth, HF = high food condition, and LF = low food condition.

individuals devote proportionally fewer resources to development and more to growth in order to take advantage of favorable conditions for growth (figs. 11.1A, B). Individuals are able to devote absolutely more energy to both development and growth at high food conditions (figs. 11.1C, D). It follows that individuals at low food conditions metamorphose at a smaller size after a long larval period, and individuals at high food conditions metamorphose at a larger size in a short time. Upon the initiation of metamorphosis, suggested to be between Gosner 35–38 (Hensley 1993), developmental rate is fixed and begins to consume a greater proportion of energy because of morphologic and other changes associated with metamorphosis induced by increasing titers of thyroxine. Proportional and absolute amounts of energy devoted to growth decline. The observation that individual growth curves tend to level off or drop toward metamorphic climax supports the idea that less energy is devoted to growth, presumably in favor of an increased allocation to development.

Changes in the food level prior to fixation of developmental rate will have predictable results; increases in food level will lower the proportion of energy allocated to development but allow a greater absolute amount of energy to be allocated to development. A greater proportion and absolute amount of energy is allocated to growth; length of larval period will decrease and size at metamorphosis will increase. Decreases in food level will increase the proportion of energy allocated to development but a small absolute amount of energy will be available for development. Proportionally and absolutely, little energy is allocated to growth. Length of larval period increases and size at metamorphosis decreases relative to individuals that experience constant high food.

Changes late in the larval period occur after the developmental trajectory is fixed; the absolute amount of energy devoted to development also is fixed. A change to high food conditions means that a small but increasing amount of energy is committed to finishing metamorphosis. A very large proportion of the increased resources can be devoted to growth, which results in a large increase in size at metamorphosis (fig. 11.1E). A change to low food implies that a large and increasing amount of energy is required to complete the fixed developmental trajectory. Essentially, all energy is devoted to development and none to growth, with perhaps some negative growth required to complete metamorphosis on the rapid developmental trajectory (fig. 11.1E). A large decrease in body size is possible. No asymmetries in response to increasing and decreasing food levels at different ages are predicted.

The evidence from three studies that have investigated larval responses to increases and decreases in food level, in addition to supporting the modified Wilbur-Collins model as presented by Hensley (1993), also supports the graphical model of resource allocation discussed above. Recall that Leips and Travis (1994) predicted asymmetrical effects from changing food levels at different times, whereas the graphical model does not. Alford and Harris (1988) found no asymmetry from changing food levels at different times. All groups that received increased food at different times were not significantly different from each other in mass at metamorphosis. Two groups that received decreased food, including the last group to be switched, metamorphosed at the same mass. Hensley (1993) also found no asymmetry in responses. Early increases and decreases in food levels had their greatest effect on size at metamorphosis. Leips and Travis (1994) presented evidence that responses were asymmetric for *Hyla cinerea* but not for *H. gratiosa.* Body size (SVL) at metamorphosis did not differ significantly for all groups that received increased food levels at different times in both species. Two groups of *H. gratiosa* that received decreased food, including the last group to be switched, metamorphosed at the same size. Earlier decreases in food level had more of an effect than late decreases in *H. cinerea.* No asymmetry of responses was found in recent studies by Beck (1997) on *H. squirella* and Tejedo and Reques (1994b) on *Bufo calamita.*

Metamorphosis does not seem to be possible until a minimum size is attained. A panadaptationist might suggest that metamorphosis at a very small size is better than death in a drying pond. Metamorphosis at a small size can decrease terrestrial performance (Emerson 1986; Pough and Kamel 1984; Taigen and Pough 1985). A developmental constraint may exist because adult organs may not form below a certain size. The maximum size that a species attains at metamorphosis is very likely an adaptation to average opportunities for growth and mortality rates experienced on land. However, mapping metamorphic trait values on anuran phylogenies would be valuable as a means of assessing the relative role of selection and evolutionary history.

Ecological Genetics of Larval Traits

Several investigators (e.g., Newman 1988a, b; Travis et al. 1987) provided a coherent picture of various aspects of the genetics of growth rate and possible trade-offs with other fitness traits in larval anurans. They used the breeding designs of quantitative genetics (Falconer 1989) to assay for degree and type of genetic variation present and for genetic correlations of fitness traits. For a full-sib design, eggs are collected from single, isolated matings; egg masses collected in the field will not guarantee that all larvae are full sibs. In a mating aggregation of *Bufo americanus* or *Rana sylvatica,* for example, it is possible that one female's eggs may be fertilized by two or more males (personal observation). A full-sib design can provide considerable information relatively easily, but several sources of variation in the response are confounded into the effect of family: additive genetic variation (necessary for a response to selection), dominance variation, and maternal effects. The inability to isolate maternal effects is a major problem because egg size variation could be caused by nongenetic factors (e.g., feeding history of the mother) and lead to a spurious estimate of the amount of genetic variance present in a population. Frogs could be raised through one or more acclimation generations to eliminate nongenetic maternal effects, but this usually is impractical given the long generation times. In general, a full-sib design would be appropriate as a pilot study to estimate variances for use in a statistical power calculation to determine the number of replicates to use in further studies.

Half-sib designs and partial diallel designs allow partitioning of the effect of family into additive genetic, nonadditive genetic, and environmental components of variation. In the half-sib design, females are mated to a number of males. The effect of male parent on offspring fitness is free of any maternal effects and is used as a measure of additive genetic variation. Unless an acclimation generation is used, the estimate of nonadditive genetic variation is confounded with nongenetic maternal effects. In a partial diallel design, all males are each mated to all females and vice versa. The external fertilization of anurans facilitates a partial diallel design. For practical reasons, it may be possible to use relatively few parents. For example, an 8 females × 8 males design with 5 replicates would require 320 experimental units (e.g., pools or cages). The amount of maternal effect can be estimated by comparing the magnitude (sums of squares) of the male effect with that of the female effect. The male effect is an estimate of additive genetic variation; the female effect is an estimate of additive genetic variation plus maternal effects. Thus, maternal effects can be estimated by subtraction. Partial diallel designs also allow for estimation of nonadditive genetic effects (male × female interaction).

In order to interpret the nature of genetic variation for a particular trait, the ecological context of the population must be considered (Travis et al. 1985a). Any given value of a genetic parameter can be caused by several different evolutionary scenarios. For example, an estimate of zero genetic variance can be explained by strong directional selection or by a low mutation rate coupled with weak directional selection. Conversely, an ecological study alone cannot determine whether a trait is adaptive without a knowledge of genetic underpinnings. For example, the large body sizes of montane species is largely caused by the low temperature environment of the larvae (Berven et al. 1979). Studies of the ecological genetics of frogs have progressed to the point where both kinds of information are present and credible scenarios of past selection can be presented.

Genetics of Growth Rate

Smith-Gill and Berven (1979), Travis et al. (1987), and Wilbur and Collins (1973) argued that growth rate is an important measure of larval fitness under almost any imaginable environmental regime and suggested that growth rate is a valid measure of competitive ability. Individuals that grow rapidly are the winners of competition; individuals that grow slowly are the losers. Slow growth places larvae at an increased risk of predation for longer periods of time and at an increased risk of dying in a drying pond (Newman 1989; Wilbur 1987).

In an early experiment in the ecological genetics of anuran larvae, Travis (1980b) raised six full-sib families of *Hyla gratiosa* in competition with *H. femoralis* under laboratory conditions. Significant variation among the families of *H. gratiosa* with regard to early growth rate (τ) and length of larval period suggested possible genetic variation for those traits. Maternal effects were also suggested as the potential cause of variation among families. Individual tadpoles were considered as observations (degrees of freedom) in the statistical analysis, which led to a nonconservative analysis.

The suggestion of genetic variation for growth rate was initially troubling because if the trait were so important to an individual's fitness, then growth rate should be maximized by selection, and genetic variation for growth rate should be exhausted. A resolution was suggested by the finding that early growth rate and the length of larval period were negatively correlated, as were early growth and size at metamorphosis (Travis 1980b). Individuals that initially grew rapidly metamorphosed early at a small size. Individuals that grew slowly early in the larval period metamorphosed later at a large size. This led to a positive correlation between size at metamorphosis and length of larval period. A trade-off between minimizing time spent in the pond (reducing risk of predation and desiccation) and maximizing size at metamorphosis (important for terrestrial survival and later reproduction) could maintain genetic variation for growth rate. The combination of correlated traits evolves to a joint optimum value in the population (Via and Lande 1985), which is not the individual optimum of either trait.

The presence of variation in growth rates between populations of *H. gratiosa* was confirmed by the results of four breeding episodes spread over three years (Travis 1983a). Full-sib families differed in growth rate in three out of four bouts of breeding, and the difference among families tended to be consistent over two larval densities. No consistent relationship between average egg sizes and average growth rates was found. Thus, the effects of nongenetic maternal effects on significant among-family variation in growth rate were discounted. Within families of interacting individuals, size at metamorphosis and length of larval period were positively correlated. Among families, size at metamorphosis and length of larval period were negatively correlated. This means that rapidly growing families tend to develop quickly and metamorphose rapidly. Individuals of more slowly growing families on average metamorphose at a smaller size, but given proper environmental conditions, such as a constant or increasing food supply, they could attain large sizes at metamorphosis.

In summary of full-sib studies, variation in growth rate and usually in length of larval period exists among sibs in natural populations (Travis 1980a, b, 1983a). However, variation in size at metamorphosis is not always found. The differences among sib groups could have been caused by maternal effects, additive genetic variation, or nonadditive genetic variation. Experiments including half-sib or partial diallel designs could distinguish among these effects.

In a half-sib design with *Pseudacris triseriata*, differences in growth rate, length of larval period, and size at metamorphosis were found among families in the laboratory (Travis 1981b). Differences in growth rate among families appeared due largely to nongenetic maternal effects. In another study involving a half-sib design using *H. gratiosa*, Travis (1980a) found genetic variation in growth rate and length of larval period, but not in size at metamorphosis. By the mid 1980s, genetic and nongenetic bases had been demonstrated for growth rate, and species-specific differences had been found. Progress would be made when more elaborate breeding designs and ecological information were combined to infer the history of natural selection.

At least three extensive breeding designs (partial diallels or chain block designs) have been published. Travis et al. (1987, *Pseudacris crucifer*) and Newman (1988b, *Scaphiopus couchii*) found no or little additive genetic variation but significant nonadditive genetic variation for growth rate. By pooling maternal and additive genetic variation, Travis et al. (1987) showed that their results were similar to those reported earlier; the implication was that nonadditive genetic effects for growth rate were important in earlier studies (Travis 1980a, b, 1983a, b). Blouin (1992b) found a positive heritability for larval growth rate in *Hyla cinerea* but did not explicitly test for additive versus nonadditive genetic variation. In a partial diallel experiment, evidence of significant additive genetic variation for growth rate was found in *H. chrysoscelis* (Alford 1986a, 1989a). Species-specific differences in the distribution of genetic variation for growth rate remain to be fully discovered and explained.

Genetics of Length of Larval Period

In full-sib experiments, Travis (1980a) found significant differences among sibships in length of larval period. Płytycz et al. (1984) reported differences among maternal half-sibs in length of larval period. In more complex designs, little additive genetic variation for length of larval period was found, but significant nonadditive variation was present in the populations of *Pseudacris crucifer* (Travis et al. 1987). Working with *Scaphiopus couchii,* Newman (1988b) found a predominance of additive genetic variation for length of larval period, or the equivalent trait, developmental rate. Berven (1987) reported high levels of additive genetic variation for development time in lowland and montane populations of *Rana sylvatica.* Blouin (1992b) found a positive heritability for length of larval period in *H. cinerea.*

Genetics of Size at Metamorphosis

In full-sib designs, variation among sibs in size at metamorphosis—independent of variation caused by growth rate—was found by Travis (1983a) in *H. gratiosa.* This contrasted with the results from a more limited study (Travis 1980a) in which no variation in size at metamorphosis was found. Two partial diallel studies gave mixed results: Travis et al. (1987) found significant additive genetic variation for size at metamorphosis, but Newman (1988b) did not. Blouin (1992b) found a positive heritability for size a metamorphosis but did not test for nonadditive variation. Woodward (1986) found no genetic variation among sib groups that differed by male parent in size at metamorphosis but did find that offspring from larger sires were larger than offspring from smaller dames 30 days after metamorphosis. Berven (1987) found no additive genetic variation for size at metamorphosis in lowland populations of *R. sylvatica* but high levels of genetic variation in montane populations.

In summary, variation among full sibs exists routinely in the fitness traits of growth rate, length of larval period, and size at metamorphosis. Not enough information on the actual genetic architecture of these traits exists, but the partial diallel designs of Alford (1986a, 1989a), Blouin (1992b), Newman (1988a, b), and Travis et al. (1987) show important progress in this area. The pattern of existing underlying variation (additive, nonadditive, maternal) is expected to differ depending on past selective environment of the population (see discussion below).

Correlation among Characters

As phenotypic characters, growth rate and length of larval period invariably are inversely correlated. The correlations of growth rate with size at metamorphosis and length of larval period with size at metamorphosis can be negative, positive, or zero and depend on environmental conditions. Travis et al. (1987) reported a positive additive genetic correlation between growth rate and length of larval period, but Newman (1988b) found a negative correlation between size at day 5 (early growth) and length of the larval period for *Scaphiopus couchii.* This result should be viewed cautiously because Newman detected no additive genetic variation for size at day 5. Blouin (1992b) found no additive genetic correlation between growth rate and length of the larval period in *Hyla cinerea.* Travis et al. (1987) reported a positive additive genetic correlation between growth rate and size at metamorphosis and between length of larval period and size at metamorphosis; they found a negative nonadditive correlation between growth rate and length of larval period and between size at metamorphosis and length of larval period. A positive, nonadditive correlation existed between growth rate and size at metamorphosis. Newman (1988b) detected no other significant additive genetic correlations and did not estimate nonadditive correlations because none of his characters showed nonadditive variation. Blouin (1992b) found a positive genetic correlation between length of larval period and size at metamorphosis but no correlation between growth rate and either larval period or juvenile growth rate. Berven (1987) described a negative additive genetic correlation between developmental time and size at metamorphosis in montane populations of *Rana sylvatica* but no correlation in lowland populations.

Antagonistic pleiotropy (M. J. Rose 1982) between growth rate and the reciprocal of length of larval period was a key finding of Berven (1987) and Travis et al. (1987). Selection will act to maximize growth rate and minimize length of larval period under many ecological conditions. Genes that lead to this desirable combination of traits will be fixed quickly and will not contribute to genetic variation for either trait. Genes that "code" for a slow growth rate and a rapid development, or vice versa, will not be fixed but will come to joint equilibrium in the population. This effect can explain the existence of genetic variation for either trait (table 11.1A; Falconer 1989). As Travis et al. (1987) noted, neither growth rate nor length of larval period in this population had much additive genetic variation that needed to be explained. Berven (1987) also invoked the explanation of antagonistic pleiotropy to explain the maintenance of high levels of additive genetic variation for development time and

Table 11.1 Heuristic models that explore the outcome of selection on larval and adult traits. (A) Depiction of a model by Falconer (1989) and (B–D) extensions of that model. Fitness values for genes affecting larval (horizontal axis) and adult (vertical axis) traits are shown in left panel of each section. Genetic combinations that are fixed, selectively eliminated, or remain neutral are shown in right panel.

A. Correlated selection on larval and adult stages: no selection for coupling or uncoupling of stages, and the life cycle is complex. Genes affecting the same trait in the larva and adult in favorable and unfavorable ways are given fitness values of +1 and −1, respectively.

Fitness Values				Genetic Combinations that Are		
				Fixed	Eliminated	Neutral
	−	0	+	+/+	−/−	+/−
−	−2	−1	0	+/0	−/0	−/+
0	−1	0	1	0/+	0/−	0/0
+	0	1	2			

B. Selection on the adult stage, uncoupling of stages, and no evolution in the larval stage; the life cycle is complex. Genes that affect a trait in the adult in favorable and unfavorable ways are given fitnesses of 1 and −1, respectively. Genes that affect the trait in the larva in any way are selected against (fitness of −1). There is selection for uncoupling (+1) and against coupling (−1) of the trait in larva and adult.

Fitness Values				Genetic Combinations that Are		
				Fixed	Eliminated	Neutral
	−	0	+	0/+	+/+	+/0
−	−3	0	−3	0/0	−/+	−/0
0	0	1	0		+/−	0/−
+	−1	2	−1		−/−	

C. Selection on the adult stage in a species with direct development. Genes that affect the trait in the adult in favorable (+) and unfavorable (−) ways are given fitnesses of 1 and −1, respectively. Genes that affect the trait in the embryo in any way are selectively neutral. There is no selection for uncoupling or against coupling of the trait in larva and adult.

Fitness Values				Genetic Combinations that Are		
				Fixed	Eliminated	Neutral
	−	0	+	+/+	+/−	+/0
−	−1	−1	−1	0/+	0/−	0/0
0	0	0	0	−/+	−/−	−/0
+	1	1	1			

D. Selection on the adult stage for uncoupling of stages and against any current or past correlated evolution of the embryo; the life cycle is now complex after a period of direct development. Genes that affect the adult in favorable (+) and unfavorable (−) ways are given fitness values of 1 and −1, respectively. Genes that affect the larva in any way are selected against (fitness of −1). There is selection for uncoupling and against coupling of the trait in larva and adult as in B.

Fitness Values				Genetic Combinations that Are		
				Fixed	Eliminated	Neutral
	−	0	+	0/+	+/+	+/0
−	−3	0	−3	0/0	+/−	−/0
0	0	1	0		0/−	
+	−1	2	−1		−/+	
					−/−	

size at metamorphosis in montane populations of *Rana sylvatica*.

The work of Travis et al. (1987) included a principal component analysis (PCA) of the phenotypic, additive, nonadditive, and environmental correlation matrices for growth rate, length of larval period, and size at metamorphosis. For the phenotypic variation, the first principal component that maximized the amount of variation explained also reflected a gradient between large size at metamorphosis and rapid growth to small size and slow growth. Length of larval period did not come into play in the PCA but was inversely related to growth rate and positively correlated with size at metamorphosis. Tadpoles that grew rapidly achieved a maximum size in a short time and metamorphosed.

Environmental correlations were also analyzed with PCA. The first principal component described an axis with slow growth, long larval periods, and large sizes at metamorphosis at one extreme, and rapid growth, short larval period, and small sizes at metamorphosis at the other. These environmental effects can be produced by temperature variation and some types of competitive environments. The PCA on the nonadditive correlations were of considerable interest because theoretical population genetics predicts that the outcome of past selection is likely to be reflected in these correlations (Travis et al. 1987). Favorable genetic combinations should be protected from recombination by inversion that would cause linkage or gametic disequilibrium and be reflected by an estimate of nonadditive genetic variation. The first principal component described an axis with rapid growth, short larval period, and large size at metamorphosis at one extreme. This was contrasted with slow growth, a long larval period, and small size at metamorphosis. In many ecological situations, this mirrors a selection gradient from traits that are adaptive to traits that are not.

Summary

Estimates of genetic variation can be coupled with ecological information to construct probable scenarios of past selection, adaptation, and constraint. For example, Travis et al. (1987) reported nonadditive genetic variation for length of larval period, whereas Newman (1988b) did not. Although Berven (1987) and Blouin (1992b) did not estimate nonadditive variation, they found additive genetic variation for a correlate of larval period length. These differences are related to ecological factors. The desert ponds containing *Scaphiopus* that Newman studied were arguably more variable in duration than the ponds studied by Travis, Berven, and Blouin. Newman (1988b) documented a negative correlation between developmental rate and plasticity in growth rate. Rapid developers did not delay metamorphosis if growth rate increased and could not take advantage of the increased growth rate. The combination of slower development and plasticity in growth rate may be favored in pools of longer duration. Spatial and temporal variations in selection means that no particular developmental rate is favored, and no dominance variation, and therefore no dominance covariance structure, is expected in *Scaphiopus.*

The presence of additive variation for the inverse of the length of the larval period in *Rana sylvatica* was explained in two ways (Berven 1987). In lowland populations, size at metamorphosis is canalized. Large size at metamorphosis is correlated with earlier age at first reproduction. For short-lived colonizing species such as lowland wood frogs, age at first reproduction should be minimized. Larvae are selected to attain a large size at metamorphosis regardless of development time, which would be under weak selection. In montane populations, the inverse of larval period length shows a negative genetic correlation with size at metamorphosis. Early metamorphosis that allows terrestrial growth or late metamorphosis that allows a large body size at metamorphosis will lead to a large juvenile body size. This may enhance overwintering survival in the rigorous montane environment. In both environments, large size at metamorphosis is favored by selection, but the different genetic architectures that have evolved suggest that multiple adaptive solutions are possible (Harris et al. 1990).

Quantitative genetics, when combined with ecological information, has yielded important insights into adaptation and constraint of larval anuran fitness traits. Although not enough studies have been conducted to allow any broad generalizations, it seems that (1) half-sib and diallel breeding designs are necessary because nonadditive genetic variation is an important quantity to measure, (2) genetic correlations must be estimated in order to evaluate hypotheses about adaptation and constraint, and (3) ecological information must be combined with genetic data to generate a credible analysis of past selection pressures.

Origin of a Complex Life Cycle

Anurans possess morphologically and ecologically divergent larval and postlarval stages, each with an array of specializations. Larval specializations include keratinized teeth, a buccal pump mechanism, a long coiled intestine, and caudal locomotion. Many of these specializations are related to feeding. Adult specializations include lack of a tail, fused postsacral vertebrae, an elongated ilium, a radius fused to the ulna, and a tibia fused to the fibula. These adaptations are related to saltation. At metamorphosis, the larval tail degenerates; and the locomotory, digestive, and nervous systems are largely or completely replaced (Alberch 1987; Fox 1981; Kollros 1981). In contrast, the shortened presacral vertebral column remains constant across stages.

How can such divergent life-history stages evolve? The origin of a free-living larval stage can be explained with the phylogenetic null hypothesis, i.e., anurans evolved from ancestors that had a complex life cycle. The ancestral larval to postlarval size versus shape trajectory is thought to have included a minor metamorphosis (Alberch 1987; Bolt 1977; Eaton 1959; Noble 1925a; Orton 1953; Wassersug and Hoff 1982; fig. 11.2A). I define a minor metamorphosis as minimal variation between many traits in the larva and postlarva. Compared with frogs, salamanders have a minor metamorphosis (Harris 1989; Reilly and Lauder 1990). A minor metamorphosis seems to have occurred in the dissorophids (Bolt 1977, 1979), the presumed ancestors to modern amphibians, but the question of anuran origin is still unresolved. Three scenarios for the evolution of divergent phenotypes in larval and postlarval anurans are (1) the larval stage evolved specializations before the postlarval stage, (2) the postlarval stages evolved specializations before the larval stage, or (3) specializations occurred in both stages simultaneously.

Selection for specializations in the adult stage will cause allometric and other changes in the larval stage unless a state of independence or lack of correlation exists between stages. Change in a size-shape trajectory in the adult stage, in either elevation or slope, will affect the larval trajectory (fig. 11.2B). Adaptive evolution in the adult stage would produce correlated responses in the larval stage that may not be adap-

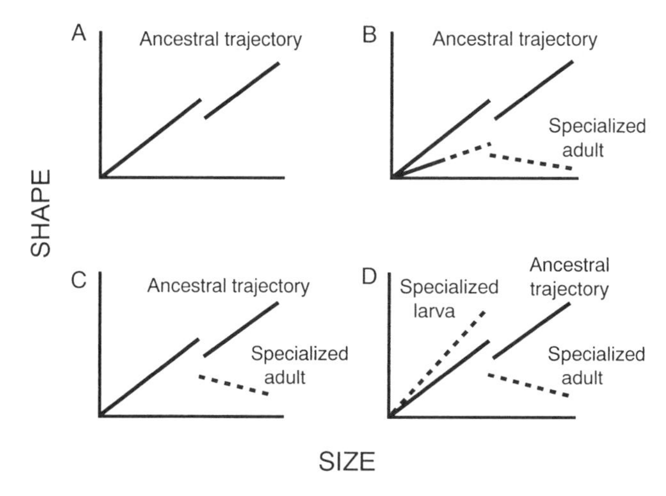

Fig. 11.2. Graphical model to illustrate the evolution of divergent life history stages in amphibians. (A) Ancestral size-shape trajectory that leads to little phenotypic variation between stages. Metamorphosis occurs at the break in the trajectory. (B) Selection for a specialized adult reduces the slope of the larval trajectory if the developmental programs are correlated. Metamorphosis occurs at the breaks in the trajectories. The dashed lines in the late larval and adult trajectories indicate new shapes that are not characteristic of previous ontogenetic stages. (C) Selection for a specialized adult when the developmental programs are uncoupled. The larva of the specialized adult follows the ancestral larval trajectory. Metamorphosis occurs at the breaks in the trajectories. The dashed line in the adult trajectory indicates a new shape that is not characteristic of previous ontogenetic stages. (D) Once uncoupling occurs, the larva also can diverge from the ancestral trajectory. Metamorphosis occurs at the breaks in the trajectories. The dashed lines in the adult and larval trajectories indicate new shapes that are not characteristic of previous ontogenetic stages.

tive. Similarly, adaptive evolution in the larval stage would produce correlated responses in the adult stage that may not be adaptive. To the degree that maladaptive morphologies are produced, selection should occur simultaneously for an uncoupling of stages (Raff 1987). A partial uncoupling could speed the rate of specialization in both stages. As each stage comes to sit on different "adaptive peaks," selection for specialization and uncoupling would stop. Developmentally, uncoupling of stages may occur by compartmentalization of embryonic cells. One group of cells differentiates into the larval structure, while another group of cells remains undifferentiated. At metamorphosis, the previously undifferentiated cell line produces the adult structure (Alberch 1987).

Is it possible to select simultaneously for specialization in one stage and for uncoupling of stages? Simultaneous selection for two traits eventually leads to a negative correlation between the traits (Falconer 1989). In Falconer's model (table 11.1A), the two traits would be the same trait in both the larval and postlarval stages. A negative correlation arises because pleiotropic genes that affect the trait in the larval and postlarval stages as desired (defined here as $+_{\text{larval}}/+_{\text{adult}}$) will come to fixation. Also, genes that affect both traits in the undesired direction (−/−) will be selectively eliminated, which means that much of the remaining covariation is caused by genes that affect one trait favorably (+/−) and the other trait unfavorably (−/+) and are selectively neutral. Thus, the heritability of the traits in the larval and postlarval stages simultaneously becomes zero and will not respond to joint selection (Falconer 1989; M. J. Rose 1982).

The two traits under selection in the evolution of the protoanuran life history are a hypothetical trait (e.g., characteristics of the digestive tract) in one life-history stage and a uncoupling of that trait in the other stage. Decoupling is considered a trait in this model, and uncoupled traits are defined as traits that are genetically and developmentally uncorrelated. Evolution of the trait in one stage will not affect evolution of that trait in another stage. I extend the heuristic model by Falconer (1989) to include the case of selection for specialization of an adult trait and for uncoupling of the trait in the larva and in the adult.

I define total fitness of the genotype as the genotypic effect on the trait in the adult plus its effect on the same trait in the larva plus its effect on the correlation between the trait in the larva and in the adult. It is assumed that any correlated evolution (allometry) in the larval stage is maladaptive (fig. 11.2B). This means that an early larva that begins to develop specializations of an adult frog instead of maintaining the ancestral larval form is selected against. In this example, the frog larva would begin to develop a carnivorous digestive tract before metamorphosis. Thus, we can assign genes that affect the adult trait in the desired direction (+) a fitness of 1, genes that affect the adult trait in the undesired direction (−) a fitness of −1, and genes that cause evolution of the larval trait in any direction and therefore away from its adaptive peak (+ or −) a fitness of −1. A correlation between traits is assigned a fitness of −1, and no correlation between traits is assigned a fitness of 1 (table 11.1B). For example, the 0/+ genes have a fitness of 2 (+1 for affecting the adult trait in the desired direction and +1 for causing no correlation between the trait in the larva and adult) and will move toward fixation. The +/+ and −/+ genes have a fitness of −1 and the +/− and −/− genes have a fitness of −3. Genes with negative fitness values will tend toward selective elimination. Combinations of −/0, +/0, and 0/− are neutral.

This heuristic model has two outcomes: (1) selection will uncouple larval and adult traits, and (2) neither stage will evolve because + and − genes that are fixed or neutral balance each other. The 0/− combination is neutral and will not be selected against; it limits evolution of the adult in the desired direction. Simultaneous selection for adult specialization and uncoupling leads only to uncoupling. Once uncoupling has evolved, presumably via compartmentalization of embryonic cells, mutations can arise that would allow the evolution of adult and larval specializations. The simple model does suggest that simultaneous selection for evolution in one stage and uncoupling is not likely to be effective. The same result occurs when simultaneous selection occurs for larval specialization and uncoupling; uncoupling evolves but neither stage changes.

Alternatively, direct development is a developmental mode that potentially can facilitate postlarval specializations and uncoupling (Jägersten 1972). Direct development removes a larva from many of the factors of external selection present in the environment of a free-living larva (Lutz 1948). Selection for adult specialization would produce correlated larval morphologies that may be maladaptive, but such morphologies would be exposed to weak selection. The genes that coded for adult specializations and larval evolution, +/+, −/+, and 0/+, would tend toward fixation (table 11.1C). The neutral combinations of −/0, +/0, and 0/0 would covary. Adult evolution occurs, but the embryo does not evolve (fig. 11.1C).

Should selection again favor an aquatic larva, the embryo (now tadpole) would be exposed to selection. With selection for uncoupling of larval and postlarval stages, the +/+ and −/+ genes would be eliminated rapidly (table 11.1D). The loss of + genes that affect adult specialization in the desired direction would lead to regression of the adult specialization. After selection, only the +/0 and −/0 genes would contribute to the covariance because the 0/− genes had been selectively eliminated.

What are the differences between simultaneous selection for a change in adult specialization and uncoupling in species with a complex life cycle (table 11.1B) as opposed to a species that secondarily evolved a complex life cycle from direct development (tables 11.1C, D)? The simple models predict that in both cases the 0/+ genes will tend toward fixation and that the evolution of uncoupling of traits has occurred. One difference is that the 0/− genes are neutral in the simultaneous selection case but eliminated selectively in the direct development scenario. Selection for uncoupling after direct development allows a greater initial specialization in the adult stage. This may make it more probable that a complex life cycle with a major metamorphosis will evolve from an ancestor with direct development. This simple heuristic model needs to be developed quantitatively in order to explore further the evolution of divergent life history stages.

The model suggests an alternative pattern to the one proposed for plethodontid salamanders. Based on work with *Eurycea,* Alberch (1987) proposed that compartmentalization of larval and adult development programs may have allowed a suppression of the larval structures by direct developers. If true, compartmentalization preceded direct development. The model presented here for protoanurans suggests that direct development preceded selection for compartmentalization.

Is there any evidence that direct development did or did not facilitate the evolution of divergent life stages in anurans? Can direct development be considered primitive in frogs? I argue that if direct development is primitive, then an analysis of *Leiopelma,* which retains many primitive traits, or a fossil ancestor should reveal a larval size-shape trajectory that retains primitive, perhaps salamander-like traits. If direct development is derived, then *Leiopelma* and other species with direct development should have a larval size-shape trajectory that retains some trace of a tadpolelike trajectory. Most workers have considered a complex life cycle to the primitive in anurans (Duellman 1989; Noble 1925a; Orton 1953; Szarski 1957; but see Bogart 1981). The argument for primitiveness is that most anurans have complex life cycles and that most fish and salamanders also have complex life cycles. However, commonness of a trait cannot be used to establish a character polarity. For example, most mammals are placentals, but it is generally concluded that the placenta is a derived character. The life cycles of the coelacanth or lungfishes are not helpful in determining what life cycle is primitive in frogs. The crossopterygian *Latimeria* has direct development and the lungfishes have complex life cycles.

Fossil evidence does not seem conclusive in determining whether a complex life cycle is primitive in frogs. Bolt (1977) suggested that the Dissorophoidea is ancestral to the three modern groups of Lissamphibia. The evidence includes similarities in the pedicellate teeth of both groups. Adult and juvenile remains of the dissorophids have been found. Bolt proposed a paedomorphic origin of the modern amphibians based on their similarity to juvenile dissorophids. It cannot be concluded that dissorophids had a free-living larval stage because larval remains have not been found. Protoanurans are known from the fossil record, but larval stages of these transitional forms are not known.

The Recent frog genus that retains the most primitive traits may be *Leiopelma,* and it has direct development (exovivipary of Altig and Johnston 1989) to various degrees. Can it be determined if direct development is primitive in *Leiopelma?* If direct development is derived in *Leiopelma,* then the prehatching stages should have some trace of a tadpole trajectory, although extensive ontogenetic repatterning associated with direct development may have occurred. It is possible that the tadpole trajectory may be truncated relative to a free-living tadpole. If direct development is primitive, then the trajectory of the prehatching stages of *Leiopelma* should begin as a primitive (perhaps Urodela-like) trajectory, then diverge toward the adult trajectory (fig 11.2B). The literature has contradictory reports on the trajectory of *Leiopelma.* Eaton (1959) and N. G. Stephenson (1951a, 1955) stated that *Leiopelma* does not resemble a tadpole at any stage during its development, and N. G. Stephenson commented that prehatching *Leiopelma* resemble caudate larvae. Larval *Leiopelma* have a medially expanded ceratohyal, which would suggest a derived direct development (Wassersug 1975). Further analysis is needed in the form of a series of complete size-shape trajectories for a variety of traits in *Leiopelma.* Even if *Leiopelma* turns out to have a derived direct development, the hypothesis that direct development can facilitate a phenotypic divergence between life stages remains testable should suitable fossil material be found.

The argument that a complex life cycle cannot be derived from direct development seems to be false as a general rule. Some members of the hemiphractine genus *Gastrotheca* apparently evolved a complex life cycle from ancestors with direct development (Duellman 1989; Wassersug and Duellman 1984). It is unclear if a primitive complex life cycle reappeared or if the complex life cycle is in some way derived. If Duellman's phylogenetic hypothesis is supported, it suggests that several pathways are possible in the evolution of anuran developmental modes.

I stress that the suggestion of direct development being

a primitive feature of "protoanurans" is a hypothesis based on a simple genetic model (sensu Falconer 1989) of the evolution of morphologically divergent life-history stages. A broad analysis of morphological transitions among a variety of metamorphic and direct developing anurans is needed. Combining this analysis with a well-supported phylogenetic scheme will be essential in resolving whether direct development is a primitive or derived trait in the Anura.

Maintenance of the Complex Life Cycle

A food-rich aquatic habitat, a higher probability that free-living larvae will escape predation or parasitism better than embryos of direct developing species, and the increased opportunity for kin selection in free-living larvae compared to embryos in an egg mass can be associated with high larval growth and survival and therefore might promote the retention of the larval stage. It has been suggested that the modern tadpole may be an adaptation to exploit abundant but ephemeral resources found in temporary pools (Wassersug 1975; Wilbur 1980; Wilbur and Collins 1973). Indeed, the species diversity of anurans is higher in temporary ponds than in permanent ponds in the southeastern United States (Wilbur 1987; personal observation). Free-living larvae may have enhanced survival in the presence of predators because they can attain a size refuge from predators and alter their activity level to avoid detection by predators (Skelly 1990, 1991). Although the increased opportunity for kin selection caused by behavior of tadpoles has not been suggested as a reason for the maintenance of a complex life cycle, behavioral adaptations would have the effect of increasing the fitness of individuals or family groups that retain the complex life cycle. Kin selection also acts on antipredator responses (chap. 9).

Although some have asserted that temporary ponds provide a flush of primary productivity as they fill and recently fallen leaves and organic matter become saturated and release nutrients (Wassersug 1975; Wilbur and Collins 1973), little limnological evidence supports their claim. Nutrients surely are released upon pond filling (Osborne and McLachlan 1985); but the rate of that release, and the relationship between the release rate and primary productivity available to tadpoles, is debatable. In fact, larval resources might be at their highest level as the pond dries. At that time nutrients become concentrated, and a flush of primary productivity might occur. Before the hypothesis that a complex life cycle is an adaptation to exploit a flush in primary productivity can be addressed critically, studies on the phenology of primary productivity and its availability to tadpoles in temporary ponds are needed.

Conversely, Wassersug (1975) argued that the larval stage might be abandoned when ponds do not have a predictable pulse in productivity. Tropical ponds may lack a predictable pulse in productivity, and that may have been one selective pressure for direct development in the tropics. Data on the phenology of productivity of tropical ponds also are lacking, so a meaningful comparison with temperate ponds is impossible at this time. The hypothesis that tropical and temperate ponds differ in mean phenologies of primary productivity cannot be adequately addressed with current information.

Predation pressure on tadpoles may be less in recently filled ponds than in permanent ponds (Wilbur 1980, 1987). Upon filling, a temporary pond may be a relatively predator-free environment. Fish would be absent, but salamanders and other predators might be important. In western Virginia, newts (*Notophthalmus viridescens*) migrate to small montane ponds in time to eat hatchlings of the wood frog (Gill 1978, *Rana sylvatica;* personal observation). Individuals of a species of *Siren* (Caudata: Sirenidae) emerged from a dried pond bed in the southeastern United States within two days of refilling (personal observation) and began eating tadpoles. Predaceous insects may be scarce in a recently filled temporary pond, but predators of small tadpoles, such as diving spiders, may colonize quickly. Unless salamanders and insects are much less efficient predators than fish, it is not clear that recently filled temporary ponds are characterized by lower predation pressure than permanent ponds. More study is needed to compare predation on tadpoles and embryos in temporary ponds, permanent ponds, and terrestrial habitats. If predation pressures vary systematically among temporary ponds, permanent ponds, and terrestrial settings, then predation can be invoked as an evolutionary factor that selects for larval habitat. Pond size is an important variable; extremely small temporary pools (e.g., the size of cow hoof print) will not contain fish or salamanders but will retain water long enough for some species of tadpoles to metamorphose (e.g., *Gastrophryne carolinensis* and *Physalaemus pustulosus*).

Direct development may be favored when predation or parasitism is higher on free-living tadpoles than on direct developers (Duellman and Trueb 1986; Lutz 1948; Magnusson and Hero 1991). Parental care can interact with direct development to deter predation and lower the incidence of parasitism (McDiarmid 1978). Critical data on these points are lacking and a need exists for further investigation. Direct development is more prevalent in the tropics and might be related to the intensity of aquatic predation. Average predation rates in temporary ponds in tropical and temperate zones may be so different that a comparison would be profitable. Vermeij (1976) compared predation on gastropods in the Pacific and Atlantic Oceans, and both prey and predators in the Pacific have evolved stronger defenses and predatory abilities, respectively.

An increased opportunity for kin selection may favor the retention of a complex life cycle. An individual's inclusive fitness is composed of its individual fitness and its effect on the fitness of other related individuals (Hamilton 1964a). Kin selection among embryos in a terrestrial egg mass might lead to the evolution of a communally produced toxic substance. However, I argue that the opportunity for kin selection is greater among free-living tadpoles because the behavioral repertoire is greater. Options such as schooling behavior and release of alarm chemicals are not available to embryos restricted to an egg mass. Enhanced behavioral possibilities can increase an individual's inclusive fitness by allowing individuals to aid related individuals. Kin selection has played an important role in the larval biology of many species (chap. 9), and this may lead to an overall increase in fitness to the individual with a complex life cycle in environments that allow it.

Complex life cycles have been stable over long periods of evolutionary time (Bolt 1977; Wassersug 1975). A theoretical model by Istock (1967) suggested that such stability is unexpected, thus creating Istock's paradox. A critical assumption of Istock's model was that no genetic correlation exists between traits in both stages thereby allowing independent evolution. This leads to the suggestion that stages will differ consistently in rates of adaptation. Genotypes associated with a reduction or elimination of the less adapted stage will be favored. The assumption of genetic correlation has been evaluated in only two studies with anurans (Blouin 1992b; Emerson et al. 1988) and is discussed below. Istock's paradox can be resolved with the following three explanations: functional constraints, fluctuating selection, and genetic correlations.

Functional constraints of tadpole morphology and physiology may lead inevitably to metamorphosis. Anuran tadpoles become less efficient feeders as they grow. Wassersug (1975) argued that the surface area-to-volume ratio of the filtering surfaces of a tadpole decreases as biomass of the tadpole increases, and this creates less efficient feeding at increasing size. This ratio creates an upper limit to the size of the branchial basket in tadpoles of all species. In order to overcome this constraint, evolution in the tadpole would have to proceed via a peak shift (an ontogenetic change in trophic morphology and behavior tantamount to a metamorphosis; S. Wright 1931).

Morphological analyses provide an explanation for the decreasing trophic efficiency hypothesis. Oral trophic structures show isometric growth or negative allometric growth with body size (Wassersug 1976a). In the latter case, feeding structures become relatively smaller as the tadpole increases in size. A positive allometry of filtering surface and body size would lead to a tadpole with an increasingly crowded body cavity. Wassersug (1975) suggested that there may be an upper limit to the size of the branchial basket; the branchial baskets of large *Pseudis paradoxa* tadpoles are no larger in absolute terms than those of *Rana catesbeiana* larvae.

Wassersug (1975) tested the hypothesis that large body size decreases the relative size of the filtering surfaces and thus decreases feeding efficiency. He found that 12 small *Rana pipiens* tadpoles cleared their guts more rapidly than three large tadpoles of the same combined weight. Wassersug (1975) did not measure the metabolic costs of feeding in small and large individuals. Large tadpoles may be less efficient feeders, but they may spend less energy feeding. If true, the net energy return per unit time may have been greater in the large tadpoles. Further experiments with a variety of species with the same and different feeding modes need to be conducted to test fully the hypothesis. For example, *R. pipiens* has keratinized mouthparts, which suggests that the species may not function primarily as a filter feeder (R. Altig, personal communication). Seale and Beckvar (1980) and Seale and Wassersug (1979) showed that tadpoles can adjust filtering rates to the concentration of suspended algae. How resource concentration would interact with relative size of the filtering surfaces to affect size-specific growth rate needs investigation.

A comparison of size-specific growth rates of ranid tadpoles tends to support the hypothesis that a large body size decreases the efficiency of suspension feeding. Werner (1986) reviewed data for several species of *Rana* and found that size-specific growth rates increase and then decrease over the larval period. The decrease in growth rate as body size increases late in the larval period is not associated with the typical depression in growth rate at metamorphosis. These data are consistent with a mechanistic explanation of decreasing efficiency of tadpole feeding as body size increases and with the hypothesis that larvae (see above) allocate greater proportions of energy to development late in the larval period. However, some species can grow exponentially throughout their larval period if resources are adequate (e.g., Alford and Harris 1988, *Bufo woodhousii fowleri*). Exponential growth implies a constant size-specific growth rate throughout the larval period (see eq. [1]).

Under some ecological conditions, aquatic growth as a large larva, with its relatively smaller filtering surfaces, may still carry a higher fitness than metamorphosis to a terrestrial stage at a smaller size. This would select for paedomorphosis or the relative lengthening of the larval stage. An unfavorable terrestrial environment would create such a selective regime. More studies that integrate the relationship of larval size to filtering efficiency, assimilation percentage (Altig and McDearman 1975), metabolic costs of feeding, and specific growth rates are needed. For example, tadpoles of species that metamorphose at a small body size (e.g., *Bufo*) apparently can maintain a constant size-specific growth rate while those of species that metamorphose at a large body size (e.g., *Rana*) cannot.

Functional constraints in the form of decreasing trophic efficiency with increasing larval body size can provide a partial resolution of Istock's paradox. The tadpole stage is thought to be an adaptation to exploit ephemeral but favorable growing conditions, although the isometry of filtering surfaces to body size may lead to a slowing of further growth (Wassersug 1976a). Wassersug's argument leads to the suggestion that the adult stage will displace the larval stage over evolutionary time because the rate of adaptation slows in the larval stage (i.e., larvae are as "adapted" as they can get because of allometric constraints) but not in the adult stage. This resolution rests on the assumption that large size in a tadpole is always adaptive. The existence of species that typically metamorphose at a small size, such as *Bufo*, suggests that metamorphosis at a small body size can be an adaptation. Furthermore, the rate of larval adaptation can proceed along numerous other axes, including the evolution of mechanisms of predator avoidance, increased competitive ability, and digestive efficiency.

Wassersug (1975) considered life-cycle evolution in dichotomous terms: either the population retains or eliminates a stage. It seems more likely that selective pressures will operate to increase or decrease the relative durations of a stage. If selection favored the larval stage for a long period, two outcomes are conceivable. One would be the elimination of the adult stage (paedomorphosis). As Wassersug noted, this would require a reworking of the tadpole body plan to make space for enlarged gonads, to alter amplexus or other mating behaviors, and to overcome a lack of emetic behavior that is required for oviposition and sperm extrusion (Naitoh et al. 1989). Paedomorphic tadpoles may not be an evolutionary possibility. This is a consequence of the high degree of divergence between life stages. The same phenomenon occurs in insects; the only truly larval reproduction in insects is associated with parthenogenesis (H. F. Nijhout, personal communication). Wassersug's question, "Why there are no paedomorphic tadpoles?" could be followed by the question, "Why are there no parthenogenetic tadpoles?"

The relative lengthening of the larval stage, resulting in a single breeding period shortly after metamorphosis (e.g., ephemeropteran insects; Istock 1984) is another possible evolutionary outcome of a highly adapted larval stage. This hypothesis could be tested by geographical comparisons. For example, one might expect that populations that occupy favorable adult environments but consistently use poor larval environments should have a relatively short larval period. Populations from the center of a geographical range may occupy equally favorable larval and adult habitats. Populations at the edge of the range likely exist in a poor larval or poor adult habitat. The goal would be to identify areas where the environment was limiting only on the larva or only on the adult. A geographical comparison of the relative durations of larval and adult periods would be of great interest. Indeed, Berven et al. (1979) found that montane populations of *Rana clamitans* have evidence of past selection for a shorter larval period relative to lowland populations. The montane ponds arguably are poorer larval environments than lowland ponds. The techniques of common garden (e.g., Berven et al. 1979) and reciprocal transplant experiments could be employed to test whether any observed differences have a genetic basis. Because of the long generation times of anurans, these experiments would be far more practical than a selection experiment.

Istock (1984) postulated another resolution of Istock's paradox: persistent genetic variation in length of larval period could preserve a complex life cycle. When ecological factors periodically favor lengthening the larval period via fluctuating selection, genetic variation for the length of the larval period is maintained. If length of the adult stage is inversely related to length of the larval stage, then such selection would concomitantly decrease the length of the adult stage. Fluctuating selection might alter the relative lengths of the two stages, but neither stage would be eliminated if the larval period were periodically favored. Fluctuating selection is one factor that leads to the maintenance of genetic variation (Hartl and Clark 1989), in this case for length of larval period.

The fluctuating-selection explanation could be addressed first by surveying populations for genetic variation for length of larval period. Such genetic variation has been found in some populations (Berven 1987; Blouin 1992b; Newman 1988b; Płytycz et al. 1984; Travis et al. 1987). Second, the assumption that the lengths of the larval and adult periods are inversely related could be examined. Genetic correlation between lengths of the two stages is of interest. If longevity is constant among individuals, then the length of larval and adult stages necessarily are negatively correlated, but longevity in the larval and adult stages might be positively genetically correlated. Individuals with longer larval periods might also live longer after metamorphosis. If there is no genetic correlation between length of larval stage and length of adult stage then persistent or fluctuating variation in length of larval period will have little effect on evolution in the adult stage and therefore persistence of a complex life cycle. This has not been estimated in anurans, and its estimation in a variety of species would be of interest.

The occurrence of genetic correlations for life-history traits across stages provides a third resolution of Istock's paradox. The possibility of antagonistic pleiotropy (M. J. Rose 1982) could explain the maintenance of a complex life cycle. If the genotype that codes for a complex of adaptive adult traits does not code for the most favorable complex of adaptive larval traits, then genetic variation for those traits and the concomitant complex life cycle could be maintained. A polymorphism would be achieved as two solutions come to an equilibrium: less adapted larvae and more adapted adults versus more adapted larvae and less adapted adults. If genetic correlations across life-history stages exist, it would be unlikely or impossible for one stage to evolve at a faster rate. This hypothesis could be tested with a conventional quantitative genetic analysis, such as a half-sib design that follows tadpoles into adulthood and measures trait values for both stages (Falconer 1989). Additive genetic correlations between larval and adult life-history traits could be estimated (e.g., Emerson et al. 1988). Selection experiments on larval and adult traits with an assay for correlated selection responses of larval and adult traits also could be used. The long generation time of many anuran species would make selection experiments problematic. Species with generation times of one year would be good candidates for a test of the antagonistic pleiotropy hypothesis. As argued above, anuran evolution probably has been characterized by a genetic or developmental independence between stages. Such independence would allow divergence of stages and would allow each stage to adapt to different environments. If true, the antagonistic pleiotropy hypothesis is unlikely to provide a resolution to Istock's paradox. Viability selection acts on the larval stage before viability selection or fecundity selection acts on the adult. Whether this could create a differential rate of adaptation in favor of the larval stage is unclear because such difficult models have not been constructed (Hartl and Clark 1989).

Experimental Approaches to the Evolution of the Complex Life Cycle

Quantitative Genetics

A central characteristic of the anuran complex life cycle is a phenotypic divergence between larval and postlarval stages. I have argued that uncorrelated developmental programs are necessary for such divergence to have evolved. Support for this hypothesis would be a quantitative genetic analysis that estimates no genetic correlations between a trait in the larval stage and the same trait in the adult stage. An analysis of this kind has not been performed, although genetic correlations between larval life-history traits and postlarval morphological traits have been estimated. Emerson et al. (1988) showed that time to metamorphosis and growth rate of larvae were not correlated with hind limb shape of metamorphs, and Blouin (1992b) demonstrated that growth rate of larvae was not correlated with head width. These results suggest that larval life-history traits and postlarval hind limb traits are able to evolve independently. The generality of these results require further study.

Estimations of the genetic correlation between values of the same trait in larval and postlarval stages are avenues for future study. One such trait is body shape in the tadpole and in the metamorph. For example, if selection is imposed for a narrower tadpole, will there also be a response to that selection in the postlarval stage? Lack of an additive genetic correlation would suggest that independent evolution is possible. Estimation of genetic correlations between traits in larval and postlarval salamanders would be an interesting companion study. I hypothesize that correlations in salamanders will be large, while correlations in frogs will not be significant. If true, this would help to explain the far greater phenotypic divergence between stages in anurans.

Lack of genetic correlations between stages would also explain why paedomorphosis is common in caudates and not in anurans. The ability of a species to reproduce in a larval morphology (paedomorphosis) is inversely related to the degree of phenotypic divergence between stages. A lack of developmental flexibility (e.g., loss of possibility of paedomorphosis) has adverse macroevolutionary consequences (higher extinction rates) in benthic marine invertebrates (Valentine and Jablonski 1983). Such a pattern is unstudied in anurans.

Developmental Biology

An alternative to the quantitative genetic methodology is rooted in descriptive histology and developmental biology. Using this approach, Alberch (1987) identified two developmental pathways that led to the formation of the larval and postlarval epibranchial in the salamander *Eurycea*. The unique formation of the epibranchial cartilage during the embryonic and larval periods is a primitive pattern. During metamorphosis, larval chondrocytes can undergo cell death or remain static. The compartmentalization of the precursors to larval chondrocytes (neural crest cells) into prospective larval and adult cells is the derived pattern. While some neural crest cells differentiate into the larval epibranchial, others remain quiescent. At metamorphosis, undifferentiated cells develop into the postlarval epibranchial and the larval chondrocytes die. The derived compartmentalized development allows for independent evolution of larval and postlarval stages.

Anurans also are expected to have compartmentalized developmental programs (Alberch 1987). The greater divergence between stages in anurans perhaps is explained by a greater number of compartmentalized developmental pathways. A research program involving the marking of larval cells in a variety of species and assaying their fates across metamorphosis would be of great interest.

Summary

The complex life cycle of anurans is characterized by a phenotypic divergence between larval and postlarval stages. Selection for specialization of one stage in protoanurans may have been accompanied by selection for uncoupling of traits in the other stage. Genetic and developmental independence is expected between stages that allow independent evolution. Direct development may have facilitated the origin of a divergent complex life cycle by allowing greater initial adult specialization. The divergent larval stage (tadpole) seems to be specialized for growth. Adaptations that promote rapid growth and provide an optimal timing of metamorphosis are expected and have been examined by many workers.

The evolutionary stability of a complex life cycle has been questioned (Istock 1967), and several solutions have been discussed. Once highly divergent life cycles evolve, elimination of the adult stage may be impossible without parthenogenesis. A parthenogenetic solution has been exploited by insect groups (H. F. Nijhout, personal communication) but not by anurans. However, the relative lengths of the anuran larval and adult stages can be adjusted. In anurans, the length of larval period is under genetic control, but genetic control of adult longevity is unknown. By identifying larval and adult environments that differ in quality, it should be possible to test the hypothesis that length of larval period has been adjusted accordingly. For example, a consistently lush larval habitat surrounded by a harsh terrestrial environment would select for paedomorphosis in salamanders but could select for a long larval period and short adult life span in frogs.

Further work is needed in several areas. Morphometric and phylogenetic analyses are needed in a variety of species to assess phenotypic correlations of traits across metamorphosis. The case of *Leiopelma* is of particular interest in that it is unclear if the early size-shape trajectory of the embryo resembles a tadpole (fig. 11.2D), an ancestral salamander-like larva (fig. 11.2C), or neither (fig. 11.2B). Quantitative genetic analyses coupled with ecological information are invaluable for testing hypotheses about current selection on anurans. Such analyses also can test the hypothesis that tadpole and adult traits are genetically uncorrelated. Developmental biology will become increasingly important in addressing the general issue of uncoupling of life-history stages (Alberch 1987). The combination of morphological, genetic, ecological, developmental, and phylogenetic approaches surely will clarify our understanding of life-history evolution in anurans.

ACKNOWLEDGMENTS

I thank R. Alford, J. Hanken, S. Marks, and G. Wyngaard for helpful comments on the manuscript. James Madison University provided me with a research leave during which I developed the ideas presented in this chapter.

12

DIVERSITY

Familial and Generic Characterizations

Ronald Altig and Roy W. McDiarmid

Tadpoles are ". . . highly specialized, and seemingly unlimited variation shows little organization on preliminary study."

Orton 1953:68

Introduction

Orton (1953) clearly recognized the impediments that commonly are encountered when attempting to identify anuran tadpoles. If it were only the inherent morphological traits characteristic of the larvae of different species that were holding back our understanding of tadpole diversity, we would have made considerable progress over the past 45 years. Unfortunately, the tadpoles of many species remain to be identified, and we have only meager understanding of the nature of ontogenetic, geographic, and metamorphic variation (e.g., Cei and Crespo 1982; Gollman and Gollman 1991, 1995, 1996; Inger 1966; Korky and Webb 1996; Lavilla 1984b; J. M. Savage 1960). Our ability to recognize and identify new forms progressively improves with each description, but many factors, including a chronic lack of detail in many descriptions, inadequate collections, and too few workers, hamper progress. Intangible insights come from looking at many tadpoles of many species, and through such exposure one begins to develop an understanding of their morphological diversity. Having a readily available summary of the known morphological diversity of tadpoles also would be beneficial. At times, interspecific variation seems to be slight or absent; whether such variation simply is not recognized (e.g., *Bufo*) or poorly understood (many taxa) remains to be shown. Phenotypic plasticity, especially of pigmentation and fin shape, is confusing. Conspecific tadpoles collected from turbid versus clear water vary tremendously in color, and cultured tadpoles often have aberrant mouthparts compared to wild-caught individuals of the same species. Tadpoles of the same species may differ in shape in still versus flowing water (R. D. Jennings and Scott 1993; Van Dijk 1966), and ontogenetic changes in color and oral morphology are occasionally profound. Improper fixation and storage (chap. 2) tend to complicate all other issues. Examination of the small, complex oral apparatus of a tadpole provides the major characteristics for species identifications but often is frustrating. The discovery of a simple method for properly preserving a tadpole with its oral disc wide open and without distortion would facilitate consistent observations and be a major advance. In spite of many gains, we still find ourselves in the realm of the "preliminary study" mentioned by Orton (1953).

Because research on tadpoles is increasing rapidly, the need for properly identified tadpoles is becoming more common. Researchers are realizing that sampling tadpoles usually is an efficient, viable means of assessing local biodiversity. Tadpoles are present in aquatic habitats for longer periods than breeding adults and are often more easily collected. H. B. Shaffer et al. (1994) presented useful ideas in this regard, but Wassersug (1997b) raised some interesting counterpoints. The importance of understanding ontogenetic variation and patterns of development has been discussed previously (chaps. 2 and 3), and the need for collections of ontogenetic series is even more apparent. Developmental data provided an understanding of the differences between the dual lateral spiracles of pipids and rhinophrynids and those of *Lepidobatrachus* (Leptodactylidae; Ruibal and Thomas 1988; see Nieuwkoop and Sutasurya 1976). Even so, morphological data of tadpoles have seldom been gathered or examined to answer question about character homology. The use of larval data in systematic studies has been debated, and tadpole features have been used to various degrees (e.g., Diaz and Valencia 1985; Donnelly et al. 1990a; Duellman and Trueb 1983; I. Griffiths 1963; I. Griffiths and De Carvalho 1965; Inger 1967; Kluge 1988; Kluge and Farris 1969; Kluge and Strauss 1985; Noble 1926b; O. M. Sokol 1975; Starrett 1973; Yang 1991). Disagreements relative to the use of tadpole characters in systematic analyzes seemingly result from incomplete understanding of character states and their distributions among taxa (e.g., Wassersug 1989b). When appropriate traits have been investigated, issues relative to character homology, functional capacities, and both geographic and ontogenetic variations can often be confronted.

We reviewed the morphological diversity of tadpoles in chapter 3, and Harris discussed the evolution of the tadpole as a life-history stage in chapter 11. In this chapter we summarize the known larval morphology of pertinent taxa, and discuss the limitations of current knowledge. Two data sets are needed before meaningful advances in understanding the interactions of tadpole ecomorphology and phylogeny can be made. We must have a more detailed phylogeny of the families and genera of anuran amphibians and a better understanding of the origin and morphology of the primitive tadpole (see chaps. 4 and 11). Once this has been achieved, we should be able to examine the distribution of larval traits among the several clades of modern anurans and speculate on their evolution. The modulators of growth and metamorphosis and their combined effects on tadpole ecology (e.g., Wilbur and Collins 1973) need to be considered, as do minute differences in developmental patterns, if we are to expand the current notions of character homology and better understand the evolution of larval traits. We must be able reliably to distinguish between those features that are more labile in response to ecological pressures (i.e., result in ecomorphological convergences among lineages) and those that hold reliable phylogenetic information. For example, morphology of the chondrocranium (chap. 4 and citations therein), position of the front limbs relative to the buccopharyngeal structures (Starrett 1973), and patterns of formation and position of the spiracle(s) might be expected to carry strong phylogenetic signal, while certain features of the oral apparatus more likely reflect ecologically pertinent selection. In the same light, the distribution of some oral features among lineages seems to indicate lineage-specific constraints.

From our perspective the majority of tadpoles fall within boundaries that typically coincide with the phylogenetic patterns derived from adult frogs, but the extremes in ecological and morphological diversity among anuran larvae (figs. 12.1–12.2) seem truly astounding. Some of this diversity is captured in the composite of microhabitats and morpho-

Table 12.1 Guide to select literature on tadpole faunas and regional treatments of groups. References with keys are indicated with asterisks.

Europe
- General: Boulenger 1892*, E. N. Arnold and Burton 1985*, Nöllert and Nöllert 1992*, Berninghausen 1997*
- Austria: Grillitsch et al. 1983*, Lower Austria
- Belgium: De Witte 1948*
- France: Angel 1946*
- Germany: Berninghausen 1994*, Brauer 1961*, Engermann et al. 1993*
- Italy: Lanza 1983*
- Portugal: Serra and Albuguerque 1963*
- Russia: Terent'ev and Chernov 1965*
- Spain: Salvador 1985*
- Turkey: Başoğlu and Özeti 1973

Africa
- East Africa: Schiøtz 1975, hyperoliids
- South Africa: Van Dijk 1966*, Wager 1986
- West Africa: Rödel 1996, Schiøtz 1967, hyperoliids
- Cameroon: Mertens 1938
- Madagascar: Blommers-Schlösser and Blanc 1991, Glaw and Vences 1992*
- Natal: Lambiris 1988a*, b*
- Zimbabwe: Lambiris 1989*

Asia
- Southeast Asia; Bourret 1942*
- Borneo: Inger 1966*, 1985*
- China: C.-C. Liu 1940, 1950
- Japan: Okada 1931, Maeda and Matsui 1989
- Java: Schijfsma 1932
- Malaysia: Berry 1972
- Pakistan: Khan 1982a*
- Philippines: Inger 1954
- Russia: Terent'ev and Chernov 1965*
- Taiwan: Chou and Lin 1997*
- Thailand: M. A. Smith 1916a, 1917

Australia/New Guinea
- Australia: Watson and Martin 1973, Tyler 1989*; Queensland: Retallick and Hero 1988*, rainforest streams in one area

North America
- Canada: Orton 1952*, genera, Altig 1970*; British Columbia: Corkran and Thoms 1996*; Alberta: A. P. Russell and Bauer 1993*
- United States: Orton 1952*, genera, Altig 1970*, Wright 1929*; Eastern United States: A. H. Wright and Wright 1949*; Florida: Fanning 1966*, Stevenson 1976*; Louisiana: Siekmann 1949*; North Carolina: Travis 1981a*; Oregon: Corkran and Thoms 1996*; Washington: Corkran and Thoms 1996*

Middle America
- General: Duellman 1970, hylids
- Costa Rica: J. M. Savage 1968*, dendrobatids; 1980a*, Lips and Savage 1996*
- Honduras: L. D. Wilson and McCranie 1993*
- Mexico: Altig and Brandon 1971*, genera, Altig 1987*
- Panama: J. M. Savage 1968*, dendrobatids

South America
- Argentina: Cei 1980, 1987; Kehr and Williams 1990
- Brazil: Hero 1990*, Manaus; Heyer et al. 1990, Boracéia
- Chile: Cei 1962
- Ecuador: Duellman 1978*, Santa Cecilia/Upper Amazon Basin

types depicted in figures 9.2–9.3. The characterization of tadpoles by taxa presented in this chapter pinpoints some of the difficulties associated with species identifications, illustrates gaps in the data, provides a point of departure, and lists pertinent references to improve the likelihood of proper identification. In our experience, larval morphology is often concordant with accepted systematic treatments (e.g., *Bufo veraguensis, Hyla leucophyllata,* and *H. parviceps* groups) and in other instances likely signals the need for adjustment (e.g., species in the *Scinax rostrata* group and *Scarthyla*).

The literature that has proven useful for identification of tadpoles generally is scattered. In this chapter we review the sources of major keys and other large works that include considerable tadpole information (table 12.1). As we have noted previously, making correct identifications of tadpoles often is difficult. In table 12.2, we list citations that correctly identify previously misidentified tadpoles and in table 12.3 we provide identifications and references to tadpoles that were described originally without a name. Mouthparts of tadpoles are admittedly odd looking structures without an obvious orientation, and they are sometimes printed upside down; known examples include *Pseudis paradoxa* (Dixon et al. 1995); *Hyla zeteki* (Duellman 1970); *Bufo bufo, Kaloula rugifera,* and *Rana japonica* (C.-C. Liu 1940); and *Microbatrachella* sp. (Wager 1986). The tadpole described as *Centrolene buckleyi* by Rada de Martinez (1990) either is misidentified or deviates strongly (e.g., complete marginal papillae and many more than 2/3 tooth rows) from the conservative centrolenid morphotype; her description of *Scinax rostrata* also seems in error.

The primitive tadpole presumably was some sort of benthic inhabitant of a pond-type environment, perhaps ephemeral in nature. It likely fed by rasping food materials from submerged surfaces with keratinized mouthparts. Bogart (1981) and others have presented alternative ideas (see chap. 11). As previously noted, suggestions regarding the evolution of specific characteristics among anuran larvae are rare and usually weakly defined. Given the state of knowledge, the concepts of ecomorphological guilds (Altig and Johnston 1989) or similar designations of ecologically functional larval groups (Van Dijk 1972) at the moment appear to be the best summaries of morphological diversity among lineages.

Familial and Generic Characterizations

Understanding the diversity of tadpole morphology (figs. 12.1–12.2; also see chaps. 2 and 3) is a requisite of successful identifications. Appreciating how those morphologies are distributed across taxa and which tadpoles have been described in each group also is invaluable. Because of gaps in our knowledge and the general conservativeness of tadpoles, some tadpoles are more difficult to characterize, and therefore identify, than others. These obstacles may be at the generic (e.g., all centrolenids, most dendrobatids, *Hyla* vs. *Osteocephalus, Scaphiopus* vs. *Spea*) or specific levels of organization (e.g., species of *Bufo* and the *Rana pipiens* complex, Hillis 1982; Scott and Jennings 1985; Wassersug 1976c).

Table 12.2 Identifications of previously misidentified tadpoles. Names are arranged alphabetically by initial genus and species.

Original Identification Citation	Corrected Identification Citation
Bufo cavifrons Korky and Webb 1973	*Bufo macrocristatus* Mendelson 1997
Chirixalus vittatus Pope 1931	*Chirixalus doriae* Heyer 1971
Chiromantis rufescens Guibé and Lamotte 1958	*Phlyctimantis leonardi* Schiøtz 1975
Colostethus haydeeae Rivero 1976	*Hyla platydactyla* La Marca 1985
Colostethus subpunctatus Cei and Roig 1961	*Colostethus semiguttata* J. Faivovich, personal communication
Gephyromantis methueni Razarihelisoa 1973b	*Mantidactylus bicalcaratus* Blommers-Schlösser and Blanc 1991
Heterixalus arnoulti Razarihelisoa 1979	*Heterixalus betsileo* Blommers-Schlösser 1982
Hyla loquax[a] Duellman 1970	Unidentified *Rana* R. Altig, personal observation
Hyla siopela Caldwell 1974	*Hyla celata* Toal and Mendelson 1995
Hyla sp. Stuart 1948	*Plectrohyla teuchestes* Duellman and Campbell 1992
Hyperolius picturatus Ahl 1931	*Rana albolabris* Schiøtz 1967
Kassina senegalensis Lamotte and Zuber-Vogeli 1956	*Kassina maculata* Schiøtz 1967
Leptobrachium hasselti Wassersug 1980	*Leptobrachium hendricksonii* Inger 1983
Leptobrachium gracilis[b] Inger 1966	*Leptobrachella mjobergi* Inger 1983
Mantidactylus alutus Arnoult and Razarihelisoa 1967	*Heterixalus betsileo* Blommers-Schlösser 1979a
Mantidactylus brevipalmatus Razarihelisoa 1973a	*Mantidactylus aerumnalis* Blommers-Schlösser and Blanc 1991
Mantidactylus ulcerosus Razarihelisoa 1969	*Mantidactylus curtus* Blommers-Schlösser 1979a
Nyctimystes montana H. W. Parker 1936	*Nyctimystes cheesmanae* Davies and Richards 1990
Nyctimystes dayi Czechura et al. 1987	*Litoria nyakalensis* S. J. Richards 1992
Microhyla borneensis H. W. Parker 1934	*Kalophrynus* sp. Inger 1966
Otophryne robusta Wassersug and Pyburn 1987	*Otophryne pyburni* Campbell and Clarke 1998
Paratelmatobius lutzi Heyer 1976b	Unknown W. R. Heyer, personal communication
Philautus variabilis Annandale 1919	*Nyctibatrachus pygmaeus* Pillai 1978
Pseudacris brachyphona Green 1938	Probably *Rana pipiens* R. Altig, personal observation
Pternohyla fodiens Duellman 1970	*Rana forreri* R. Altig, personal observation
Ptychadena oxyrhynchus Power 1927b	*Ptychadena anchietae* Van Dijk 1966
Rana hosii Van Kampen 1923	Unknown Inger 1966
Rana [= *Limnonectes*] *macrodon* Flower 1899	Unknown Inger 1966

(continued)

Table 12.2 *continued*

Rana [= *Limnonectes*] *macrodon* Inger 1956	*Limnonectes microdisca* Inger 1966
Rana microdisca palavanensis Inger 1956	*Limnonectes paramacrodon* Inger 1966
Rana nicobariensis M. A. Smith 1930	Unknown Inger 1966
Rana pleskei[c] Annandale 1917	Perhaps *Scutiger mammata* Boulenger 1919
Rena vertebralis Hewitt 1927	*Rana umbraculata* Van Dijk 1966
Rhacophorus leucomystax Inger 1956	*Polypedates macrotis* Inger 1966
Staurois natator Mocquard 1890	Unknown Inger 1966
Staurois latopalmatus Inger 1966	*Rhacophorus bimaculatus* Inger 1985
Rhacophorus bimaculata Inger 1985	*Rhacophorus gauni* Inger and Tan 1990

[a]See J. C. Lee 1996 for correct description.

[b]Toothless.

[c]See C.-C. Liu 1950.

Table 12.3 Identification of tadpoles originally described without a name. Names and citation are presented in alphabetical order by initial genus and species.

Original Identification Citation	Corrected Identification Citation
Amolops sp. A Inger 1966	*Meristogenys phaeomerus* Inger and Gritis 1983
Amolops sp. B Inger 1966	*Meristogenys poecilus* Inger and Gritis 1983
Rhacophorus sp. B Inger 1966	*Rana luctuosa* Inger 1985
Unknown Pyburn 1980	*Otophryne pyburni* Wassersug and Pyburn 1987
Unknown hylid Stuart 1948	*Rana maculata* Hillis and De Sá 1988
Unknown E. H. Taylor 1942	*Rhinophrynus dorsalis* Orton 1943
Unknown Inger and Wassersug 1990	*Staurois natator* Inger and Tan 1990

The apparent lack of differentiating characteristics among species in some genera or inadequate study of certain groups sometimes forces workers to key tadpoles only to the species rank. In this regard considerable work is needed on enigmatic taxa such as *Bufo,* all centrolenids and dendrobatids, *Leptodactylus,* some ranids, and *Telmatobius.* Attempts to diagnose larvae above the species level with morphological (e.g., Lavilla 1985, 1988) or anatomical (e.g., Viertel 1982) data are uncommon, and ecomorphological guilds (Altig and Johnston 1989) do not reflect phylogenetic relationships.

In this chapter we provide the first summary of familial and generic characterizations of tadpoles based on a common terminology. The taxonomic arrangement of D. R. Frost (1985), updated through 1996, was followed and includes most of the ranoid arrangement proposed by Dubois (1987, 1992; see comments by Inger 1996). Representative, but usually not exhaustive, citations concerning each genus are provided. Subfamilies are included primarily to separate groups of genera.

The Gosner system (1960) of staging was used throughout. After each family, the number of genera and species, geographic range, major ecomorphological divisions of Altig and Johnston (1989), and other characteristics are given; for monogeneric families or subfamilies, the comparable information is presented only under the higher category. Character states common to all genera within a family or subfamily are not repeated at the generic level. The state of a character for which data are lacking is indicated with a dash. Morphological traits are given only for exotrophic (i.e., feeding; see chap. 7 for endotrophic forms) forms; genera, all of whose species are entirely endotrophic, are included only to provide a complete listing (see also chap. 7).

Character abbreviations used in the compilation are **CO, CO**mposition (family/subfamily: number of genera/number of species; genus: number of species); **GR, G**eographic **R**ange; **EMG, E**co**M**orphological **G**uild; **LTRF, L**abial **T**ooth **R**ow **F**ormula (see chap. 3; ontogenetic and other variations not included); **OA, O**ral **A**pparatus (typical oral disc with labial teeth and jaw sheaths regardless of configurations: position descriptor); **MP, M**arginal **P**apillae (distribution: arrangement); **SUP, SU**bmarginal **P**apillae; **DEM, D**isc **Em**argination; **NA, Na**res; **VT, V**ent **T**ube; **EP, E**ye **P**osition; **SP, Sp**iracle; **UJ, U**pper **J**aw sheath (width notation, descriptor of shape of serrated edge); **LJ, L**ower **J**aw sheath (width notation, descriptor of shape of serrated edge); **DF, D**orsal **F**in (general height descriptor, anterior site of origin, tip descriptor); **BS, B**ody **S**hape (dorsal view/lateral view); **CP, C**olor and **P**attern; **TL, T**otal **L**ength/Stage; **NO, No**tes; and **CI, Ci**tation(s). Generalized drawings of representative morphotypes appear immediately before the pertinent taxon, and data on the specimens used in the preparation of these drawings follow.

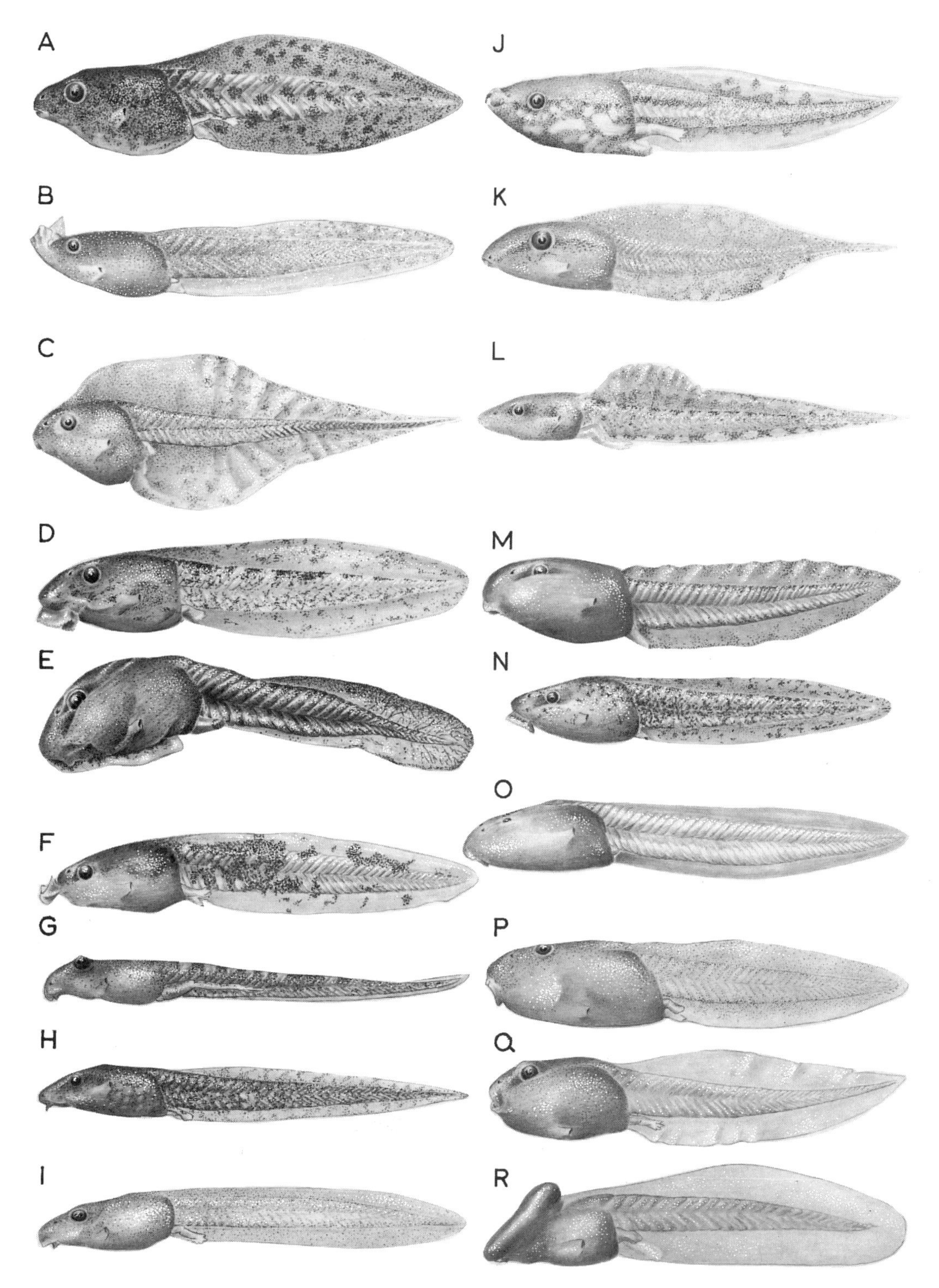

Fig. 12.1. Representative body shapes and fin configurations among anuran tadpoles. (A) *Rana palmipes* (Ranidae, clasping). (B) *Megophrys montana* (Megophryidae, neustonic). (C) *Scinax nebulosa* (Hylidae, nektonic). (D) *Hyla rivularis* (Hylidae, suctorial). (E) *Atelopus ignescens* (Bufonidae, gastromyzophorous). (F) *Colostethus nubicola* (Dendrobatidae, neustonic). (G) *Thoropa petropolitana* (Leptodactylidae, semiterrestrial). (H) *Leptopelis hyloides* (Hyperoliidae, benthic). (I) *Hyla bromeliacia* (Hylidae, arboreal Type 1. (J) *Gastrophryne carolinensis* (Microhylidae, suspension feeder). (K) *Hyla microcephala* (Hylidae, macrophagous). (L) *Occidozyga lima* (Ranidae, macrophagous). (M) *Hyla lindae* (Hylidae, suctorial). (N) *Duellmanohyla uranochroa* (Hylidae, adherent). (O) *Cochranella griffithsi* (Centrolenidae, fossorial) (P) *Anotheca spinosa* (Hylidae, arboreal Type 2). (Q) *Ceratophrys cornuta* (Leptodactylidae, carnivorous Type 1). (R) *Stephopaedes anotis* (Bufonidae, arboreal Type 3). From Altig and Johnston (1989), after Duellman and Trueb (1986); original drawings by Linda Trueb. Reprinted by permission of *Herpetological Monographs* and Trueb.

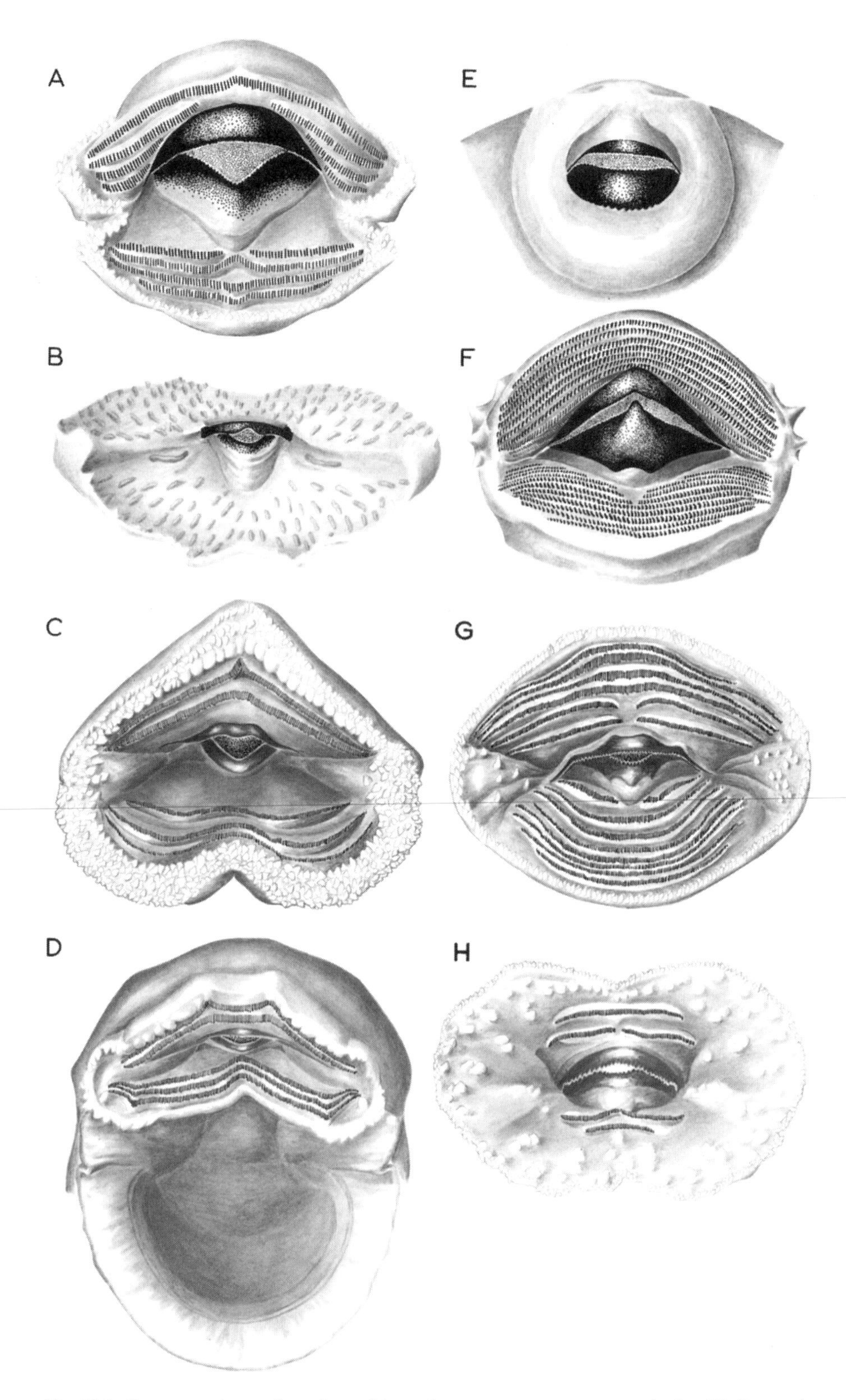

Fig. 12.2. Representative configurations of the oral apparatus among anuran tadpoles. (A) *Rana palmipes* (Ranidae, clasping), (B) *Megophrys montana* (Megophryidae, neustonic), (C) *Hyla rivularis* (Hylidae, suctorial), (D) *Atelopus ignescens* (Bufonidae, gastromyzophorous), (E) *Hyla microcephala* (Hylidae, macrophagous), (F) *Ceratophrys cornuta* (Leptodactylidae, carnivorous Type 1), (G) *Hyla lindae* (Hylidae, suctorial), and (H) *Duellmanohyla uranochroa* (Hylidae, adherent). From Altig and Johnston (1989) after Duellman and Trueb (1986); original drawings by Linda Trueb. Reprinted by permission of *Herpetological Monographs* and Trueb.

ALLOPHRYNIDAE: CO: 1/1. **GR:** Guyana–Brazil. Tadpole unknown.
Allophryne: **CO:** 1. **GR:** Guyana–Brazil. Tadpole unknown.

ARTHROLEPTIDAE: CO: 6/77. **GR:** sub-Saharan Africa. **EMG:** exotroph: lotic: adherent, neustonic, suctorial; lentic: benthic; endotroph. **LTRF:** several variations between 0/0 and 9/11. **OA:** typical: anteroventral, terminal, umbelliform. **MP:** absent, medium to wide dorsal gap, complete: uni- and multiserial. **SUP:** absent, single row distal on lower labium, oblong papillae arranged radially to mouth. **DEM:** absent, lateral. **NA:** nearer snout than eye. **VT:** medial. **EP:** dorsal, lateral. **SP:** sinistral. **UJ:** wide, smooth arc to V-shaped, sometimes serrations coarse and medial ones may be enlarged. **LJ:** similar to upper sheath. **DF:** low, originates on tail muscle posterior to the dorsal tail-body junction, tip round to pointed. **BS:** oval to elongate/depressed. **CP:** uniformly dark to mottled. **TL:** 29–80/36.

Arthroleptinae: CO: 2/52. **GR:** sub-Saharan Africa.

Arthroleptis: **CO:** 35. **GR:** sub-Saharan Africa. **EMG:** endotroph.

Cardioglossa: **CO:** 17. **GR:** western Africa. **EMG:** exotroph: lentic: benthic. **LTRF:** 0/0. **OA:** typical: anteroventral. **MP:** wide dorsal gap: uniserial, long and widely spaced ventrally. **SUP:** single row distal on lower labium. **DEM:** absent. **NA:** small, nearer snout than eye. **VT:** medial. **EP:** dorsal. **SP:** sinistral. **UJ:** wide, smooth arc, coarsely serrate. **LJ:** wide, coarsely serrate. **DF:** low, tip rounded. **BS:** oblong/depressed. **CP:** uniformly dark. **TL:** 29/36. **NO:** ventral fin originates on tail muscle. **CI:** Lamotte (1961).

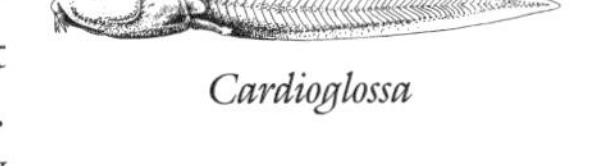
Cardioglossa

Astylosterninae: CO: 4/25. **GR:** western Africa.

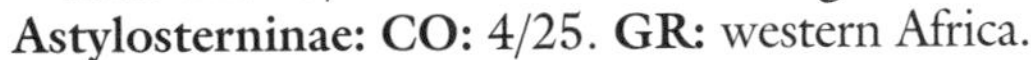

Astylosternus: **CO:** 12. **GR:** Sierra Leone–Zaire. **EMG:** exotroph: lotic: adherent. **LTRF:** 0/1, 3(2–3)/3(1–2). **OA:** typical: almost terminal, anteroventral. **MP:** medium to wide dorsal gap: uniserial. **SUP:** row distal to distal most lower row. **DEM:** lateral, absent. **NA:** nearer snout than eye, small medial tubercle at margin. **VT:** dextral. **EP:** lateral. **SP:** sinistral. **UJ:** narrow to wide, smooth arc with single large, medial serration. **LJ:** exceptionally wide, V-shaped, coarsely serrate. **DF:** low, tip pointed. **BS:** oblong/depressed. **CP:** uniformly dark. **TL:** 80/37. **NO:** ventral fin originates posterior to origin of dorsal fin; body with lateral sacs of various degrees of development, function unknown (Amiet 1970, 1971). **CI:** Amiet (1971); Lamotte and Zuber-Vogeli (1954b).

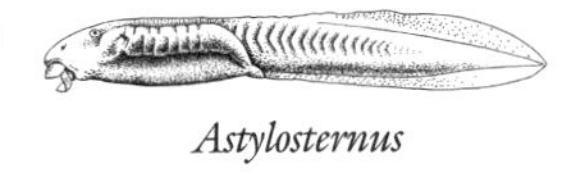
Astylosternus

Leptodactylodon: **CO:** 11. **GR:** Cameroon–Nigeria. **EMG:** exotroph: lotic: neustonic. **LTRF:** 0/0, but with ridges that suggest a 0/1–2 arrangement. **OA:** typical: umbelliform. **MP:** absent, margin crudely crenulate. **SUP:** oblong papillae arranged radially to mouth and scattered throughout disc. **DEM:** absent. **NA:** small, about equidistant between snout and eye. **VT:** dextral. **EP:** lateral. **SP:** sinistral. **UJ:** massive, all serrations large, especially laterally. **LJ:** massive, similar to upper sheath. **DF:** tip rounded. **BS:** oval/depressed. **CP:** uniformly dark, mottled on tail. **TL:** 50/36. **NO:** a tooth ridge–like ridge occurs immediately distal to the lower jaw sheath; body with lateral sacs of various degrees of development, function unknown (Amiet 1970, 1971). **CI:** Amiet (1970).

Leptodactylodon

Scotobleps: **CO:** 1. **GR:** Nigeria–Gabon. Tadpole unknown; see Lawson (1993).

Trichobatrachus: **CO:** 1. **GR:** Nigeria–Equatorial Guinea. **EMG:** exotroph: lotic: suctorial. **LTRF:** 9(7–9)/11(1–2). **OA:** typical: ventral. **MP:** complete: multiserial. **SUP:** absent. **DEM:** lateral. **NA:** equidistant between eye and snout. **VT:** dextral. **EP:** dorsal. **SP:** sinistral. **UJ:** medium, strongly V-shaped. **LJ:** medium, V-shaped. **DF:** tip rounded. **BS:** oblong/depressed. **CP:** uniformly dark. **TL:** 91/36. **NO:** large subbranchial muscle along anterior two-thirds of body gives ventral silhouette a distinctive shape; body with lateral sacs of various degrees of development, function unknown (Amiet 1970, 1971). **CI:** Mertens (1938).

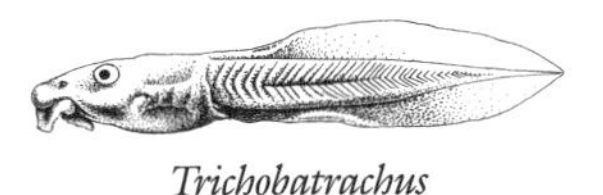
Trichobatrachus

ASCAPHIDAE: CO: 1/1. **GR:** northwestern United States–adjacent Canada. **EMG:** exotroph: lotic: suctorial. **LTRF:** 2–3/9–11(1). **OA:** typical: suctorial. **MP:** wide dorsal gap, but anterior lip fleshy and crenulate: uniserial. **SUP:** multiserial distal to distalmost lower row. **DEM:** absent but folds present. **NA:** nearer eye than snout, prominent tubular extensions. **VT:** medial. **EP:** dorsal. **SP:** midventral on chest, no dorsal wall. **UJ:** large, platelike lying almost parallel with substrate, sometimes with medial break. **LJ:** minute, U-shaped. **DF:** low, originates at dorsal tail-body junction, tip broadly rounded. **BS:** oval/depressed. **CP:** rocklike mottling, tail tip typically white bordered by black. **TL:** 60/36. **NO:** prominent small skin glands present ventrally; LTRF A-1 and most distal posterior row multiserial.

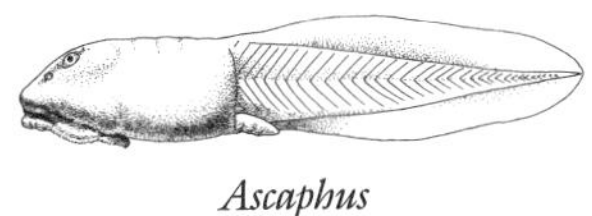
Ascaphus

Ascaphus: **CO:** 1. **GR:** northwestern United States–adjacent Canada. **CI:** Altig (1970); H. A. Brown (1990a); Metter (1964, 1967); Stebbins (1985).

BOMBINATORIDAE: CO: 2/8. **GR:** Eurasia–Philippines–Borneo

Barbourula: **CO:** 2. **GR:** Philippines–Borneo. **EMG:** Tadpole unknown, perhaps endotrophic.

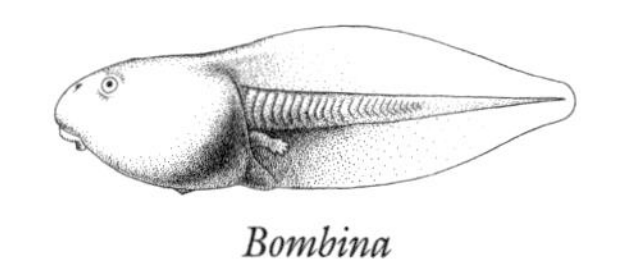

Bombina

Bombina: **CO:** 6. **GR:** Eurasia. **EMG:** exotroph: lentic: benthic. **LTRF:** 2/3[1], biserial. **OA:** typical: anteroventral. **MP:** complete: uniserial. **SUP:** row ventrolaterally. **DEM:** lateral. **NA:** nearer eye than snout. **VT:** medial. **EP:** dorsal. **SP:** low on left side. **UJ:** medium, straight area medially, coarsely serrate. **LJ:** narrow, V-shaped, coarsely serrate. **DF:** low, originates near dorsal tail-body junction, tip rounded. **BS:** oval/depressed. **CP:** melanophores form network pattern in skin. **TL:** 30–38/36 **CI:** Boulenger (1892); Lanza (1983).

BRACHYCEPHALIDAE: CO: 2/4. **GR:** Brazil. **EMG:** endotroph.

Brachycephalus: **CO:** 2. **GR:** Brazil. **EMG:** endotroph.

Psyllophryne: **CO:** 2. **GR:** Brazil. **EMG:** endotroph.

BUFONIDAE: CO: 33/381. **GR:** widespread except Australia. **EMG:** exotroph: lentic: benthic; lotic: benthic, suctorial, gastromyzophorous; endotroph. **LTRF:** 2/3, 2(2)/3, 2(2)/3(1), 2(2)/2. **OA:** typical: anteroventral, rarely ventral and suctorial. **MP:** wide dorsal and ventral gaps, complete: uni- and biserial, wide dorsal gap only. **DEM:** lateral, absent. **NA:** often large, often not round and with medial papilla, sometimes small and rounded, closer to eye than snout or reverse. **VT:** medial, dextral. **EP:** dorsal. **SP:** sinistral. **UJ:** usually narrow and broadly arched. **LJ:** usually narrow and broadly V-shaped. **DF:** low, originates near or posterior to dorsal tail-body junction, tip round to slightly pointed. **BS:** oval to round/depressed. **CP:** typically uniformly black; tail muscle sometimes bicolored or banded. **TL:** 15–40/36.

Altiphrynoides: **CO:** 1. **GR:** Ethiopia. **EMG:** endotroph.

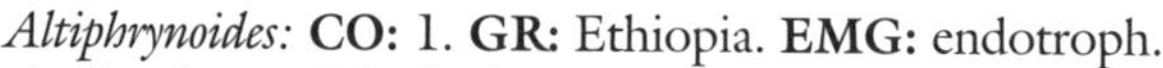

Andinophryne: **CO:** 3. **GR:** Ecuador–Colombia. **EMG:** endotroph.

Ansonia

Ansonia: **CO:** 19. **GR:** India–Borneo–Philippines. **EMG:** exotroph: lotic: clasping, suctorial, gastromyzophorous. **LTRF:** 2/3. **OA:** typical: ventral. **MP:** small, with medium to wide dorsal gap, complete: uniserial. **SUP:** row(s) distal to P-3, few laterally. **DEM:** lateral, absent. **NA:** small, nearer eyes than snout. **VT:** medial. **EP:** dorsal. **SP:** sinistral. **UJ:** narrow to medium, divided, absent. **LJ:** narrow, very open V-shaped. **DF:** low, originates on tail muscle or near dorsal tail-body junction, tip bluntly pointed. **BS:** oval/depressed. **CP:** uniformly dark, with middorsal stripe. **TL:** 18–35/36. **NO:** large variations in oral morphology within genus; anterolateral terminus of abdominal sucker of *A.* "sucker" of Inger (1992a) has fingerlike projection. **CI:** Inger (1966, 1985, 1992a).

Atelophryniscus: **CO:** 1. **GR:** Honduras. **EMG:** exotroph: lotic: gastromyzophorous. **LTRF:** 2/3. **OA:** typical: ventral. **MP:** wide dorsal and smaller ventral gaps: uniserial ventrally, biserial dorsally. **SUP:** few anterolaterally. **DEM:** absent. **NA:** small, nearer eye than snout. **VT:** medial, not attached to ventral fin. **EP:** dorsal. **SP:** sinistral. **UJ:** narrow, without serrations, V-shaped. **LJ:** narrow, V-shaped. **DF:** low, originates slightly posterior to dorsal tail-body junction, tip rounded. **BS:** oval to round/depressed. **CP:** black tail muscle boldly banded; other white marks on body. **TL:** 26/37. **NO:** ventral fin originates on tail muscle; resembles tadpoles of *Atelopus.* **CI:** McCranie et al. (1989).

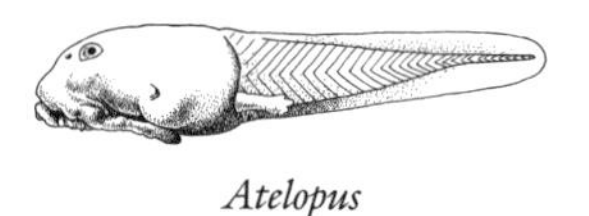

Atelopus

Atelopus: **CO:** 50. **GR:** Costa Rica–Bolivia. **EMG:** exotroph: lotic: gastromyzophorous. **LTRF:** 2/3. **OA:** typical: ventral. **MP:** wide ventral gap: uniserial. **SUP:** absent. **DEM:** lateral. **NA:** nearer eye than snout. **VT:** medial. **EP:** dorsal. **SP:** sinistral. **UJ:** narrow, smooth arc. **LJ:** narrow, open U-shaped. **DF:** low, originates near or slightly distal to dorsal tail-body junction, tip rounded. **BS:** oval to round/depressed. **CP:** uniformly black, may have white blotches or bands on tail. **TL:** 15–20/36. **CI:** Duellman and Lynch (1969); Gascon (1989a); Lindquist and Hetherington (1988); J. D. Lynch (1986); Starrett (1967).

Bufo

Bufo: **CO:** 217. **GR:** widespread. **EMG:** exotroph: lentic: benthic; lotic: benthic, suctorial, gastromyzophorous; endotroph. **LTRF:** 2(2)/3(1), 2(2)/3, 2/3, 2(2)/2(1). **OA:** typical: anteroventral. **MP:** wide dorsal and ventral gaps, complete: uniserial. **SUP:** few laterally, absent. **DEM:** lateral, absent. **NA:** large and usually not round, often with me-

dial papilla. **VT:** medial, dextral. **EP:** dorsal. **SP:** sinistral. **UJ:** narrow, smooth arc, angled laterally. **LJ:** narrow, open V-shaped. **DF:** low, originates at dorsal tail-body junction, tip rounded. **BS:** oval-round/depressed. **CP:** typically uniformly black, tail muscle sometimes bicolored or banded. **TL:** 15–35/36. **NO:** small size is common; morphologically conservative and species characteristics poorly documented; no consistently recognized morphological differences among numerous species groups except in *debilis* group; Crump (1989b) reported facultative endotrophy in *B. periglenes.* **CI:** Altig (1970); Berry (1972); McDiarmid and Altig (1990); Zweifel (1970).
Bufoides: **CO:** 1. **GR:** India. Tadpole unknown; see Pillai and Yazdani (1973).
Capensibufo: **CO:** 2. **GR:** southwestern South Africa. **EMG:** exotroph: lentic: benthic. **LTRF:** 2(2)/3. **OA:** typical: anteroventral. **MP:** wide dorsal and ventral gaps: uniserial. **SUP:** absent. **DEM:** lateral. **NA:** large, nearer eye than snout. **VT:** medial. **EP:** dorsal. **SP:** sinistral. **UJ:** medium, broad medial convexity. **LJ:** medium, open U-shaped. **DF:** low, originates at dorsal tail-body junction, tip round. **BS:** oval/depressed. **CP:** uniformly dark. **TL:** 35/36. **CI:** Power and Rose (1929); Wager (1986).
Crepidophryne: **CO:** 1. **GR:** Costa Rica. Unknown developmental trajectory.
Dendrophryniscus: **CO:** 3. **GR:** Ecuador–Peru–Guianas–Brazil. **EMG:** exotroph: lentic: benthic, arboreal. **LTRF:** 2/3, 2(2)/3. **OA:** typical: anteroventral. **MP:** wide dorsal and ventral gaps: uniserial. **SUP:** absent. **DEM:** lateral. **NA:** large, nearer eye than snout. **VT:** medial. **EP:** dorsal. **SP:** sinistral. **UJ:** medium, angled at dorsolateral points. **LJ:** medium, open V-shaped. **DF:** low, originates at dorsal tail-body junction, tip rounded. **BS:** oval/depressed. **CP:** uniformly dark. **TL:** 10–13/36. **CI:** Izecksohn and Da Cruz (1972).
Didynamipus: **CO:** 1. **GR:** Cameroon. **EMG:** endotroph.
Frostius: **CO:** 1. **GR:** Brazil. **EMG:** endotroph.
Laurentophryne: **CO:** 1. **GR:** Zaire. **EMG:** endotroph.
Leptophryne: **CO:** 2. **GR:** Malaysia. **EMG:** exotroph: lentic: benthic. **LTRF:** 2(1–2)/3. **OA:** typical: anteroventral. **MP:** wide dorsal gap: uniserial. **SUP:** absent. **DEM:** lateral. **NA:** large like *Bufo* and equidistant between eye and snout. **VT:** medial. **EP:** dorsal. **SP:** sinistral. **UJ:** narrow, smooth arc. **LJ:** narrow, open U-shaped. **DF:** low, originates at dorsal tail-body junction, tip rounded. **BS:** oval/depressed. **CP:** uniformly black. **TL:** 15/34. **NO:** lack of gap in marginal papillae of lower labium is unusual. **CI:** Berry (1972); Inger (1985).

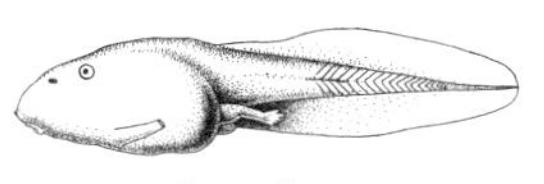

Leptophryne

Melanophryniscus: **CO:** 11. **GR:** Brazil–Paraguay–Argentina. **EMG:** exotroph: lentic: benthic. **LTRF:** 2/3[1]. **OA:** typical: anteroventral. **MP:** wide dorsal and ventral gaps: uniserial. **SUP:** absent, few laterally. **DEM:** lateral, absent. **NA:** nearer eye than snout. **VT:** medial. **EP:** dorsal. **SP:** sinistral. **UJ:** medium, smooth arc, coarsely serrate. **LJ:** medium, open U-shaped. **DF:** low, originates near dorsal tail-body junction, tip round. **BS:** oval/depressed. **CP:** uniformly dark, tail muscle bicolored, fins speckled. **TL:** 19/36. **NO:** figure in Prigioni and Arrieta (1992) shows lateral gaps in marginal papillae. **CI:** Prigioni and Langone (1990); Starrett (1967).
Mertensophryne: **CO:** 2. **GR:** Zaire–Tanzania. **EMG:** exotroph: arboreal. **LTRF:** 1/2. **OA:** typical: anteroventral. **MP:** wide dorsal gap: uniserial. **SUP:** absent. **DEM:** lateral. **NA:** nearer snout than eye, within the fleshy, hollow "crown" that encircles the eyes and nares. **VT:** medial. **EP:** dorsal. **SP:** sinistral. **UJ:** wide, almost straight edge, coarsely serrate. **LJ:** similar to upper sheath. **DF:** low, originates near dorsal tail-body junction, tip broadly rounded. **BS:** oblong/depressed. **CP:** uniformly and lightly pigmented. **TL:** 13/30. **NO:** see *Stephopaedes.* **CI:** Grandison (1980).
Metaphryniscus: **CO:** 1. **GR:** Venezuela. **EMG:** endotroph.
Nectophryne: **CO:** 2. **GR:** Cameroon–Zaire. **EMG:** endotroph.
Nectophrynoides: **CO:** 5. **GR:** Tanzania. **EMG:** endotroph.
Nimbaphrynoides: **CO:** 2. **GR:** Liberia–Ivory Coast. **EMG:** endotroph.
Oreophrynella: **CO:** 6. **GR:** Venezuela–Guyana–Brazil. **EMG:** endotroph. **CI:** McDiarmid and Gorzula (1989).
Osornophryne: **CO:** 5. **GR:** Colombia–Ecuador. **EMG:** endotroph.
Pedostipes: **CO:** 6. **GR:** India–Borneo. **EMG:** exotroph: lentic: benthic. **LTRF:** 2(2)/3(1). **OA:** typical: ventral. **MP:** wide dorsal and ventral gaps: blunt and uniserial. **SUP:** absent. **DEM:** lateral. **NA:** nearer eye than snout. **VT:** medial. **EP:** dorsal. **SP:** sinistral.

UJ: medium, smooth arc, coarsely serrate. **LJ:** medium, open V-shaped. **DF:** low, originates near dorsal tail-body junction, tip round. **BS:** oval/depressed. **CP:** uniformly dark. **TL:** 16/36. **CI:** Inger (1966, 1985).

Pelophryne: **CO:** 7. **GR:** Philippines–China. **EMG:** endotroph.

Peltophryne: **CO:** 9. **GR:** Greater Antilles. **EMG:** exotroph: lentic: benthic. **LTRF:** 2(2)/3, 2/2. **OA:** typical: anteroventral. **MP:** wide dorsal and ventral gaps: uniserial. **SUP:** row lateral to lower tooth rows. **DEM:** lateral. **NA:** nearer eye than snout. **VT:** medial. **EP:** dorsal. **SP:** sinistral. **UJ:** medium, smooth arc with slight medial indentation. **LJ:** medium, open U-shaped. **DF:** low, originates near dorsal tail-body junction, tip rounded. **BS:** oval/depressed. **CP:** uniformly dark. **TL:** 21/36. **CI:** Ruiz Garcia (1980).

Pseudobufo: **CO:** 1. **GR:** Borneo. Tadpole unknown.

Rhamphophryne: **CO:** 9. **GR:** Panama–Peru. **EMG:** endotroph.

Schismaderma

Schismaderma: **CO:** 1. **GR:** Tanzania–Zaire–South Africa. **EMG:** exotroph: lentic: nektonic. **LTRF:** 2/3. **OA:** typical: anteroventral. **MP:** wide dorsal and ventral gaps: small and uniserial. **SUP:** absent. **DEM:** lateral. **NA:** nearer snout than eye. **VT:** medial. **EP:** dorsal. **SP:** sinistral. **UJ:** medium, long low convexity in margin; Wager (1986): upper sheath lacks serrations. **LJ:** narrow, open U-shaped. **DF:** low, originates near dorsal tail-body junction, tip rounded. **BS:** oval/depressed. **CP:** uniformly dark. **TL:** 35/36. **NO:** unique semicircular flap across body behind eyes; commonly schools in midwater/surface at least during the day. **CI:** Wager (1986).

Spinophrynoides: **CO:** 1. **GR:** Ethiopia. **EMG:** exotroph: lentic: benthic. **LTRF:** 2(2)/3. **OA:** typical: anteroventral. **MP:** wide dorsal and ventral gaps: uniserial. **SUP:** –. **DEM:** –. **NA:** –. **VT:** –. **EP:** –. **SP:** –. **UJ:** –. **LJ:** –. **DF:** –. **BS:** –. **CP:** –. **TL:** –. **CI:** Grandison (1978); M. H. Wake (1980).

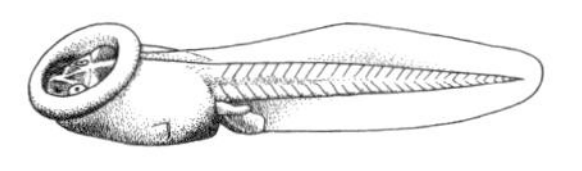

Stephopaedes

Stephopaedes: **CO:** 2. **GR:** Tanzania–Mozambique. **EMG:** exotroph: arboreal. **LTRF:** 2(2)/2. **OA:** typical: anteroventral. **MP:** wide dorsal gap: uniserial, some appear branched and small gaps between papillae. **SUP:** absent. **DEM:** absent. **NA:** nearer eye than snout. **VT:** medial. **EP:** dorsal. **SP:** sinistral. **UJ:** wide, edge nearly straight. **LJ:** absent or not keratinized. **DF:** low, originates near dorsal tail-body junction, slight arc about midlength, tip broadly rounded. **BS:** oval/depressed. **CP:** uniformly pale. **TL:** 20/37. **NO:** fleshy "crown" encircles nares and eyes; see *Mertensophryne.* **CI:** Channing (1978, 1993).

Truebella: **CO:** 2. **GR:** Peru. **EMG:** endotroph. **CI:** Graybeal and Cannatella (1995).

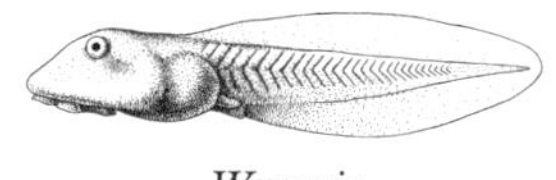

Werneria

Werneria: **CO:** 4. **GR:** Togo–Cameroon. **EMG:** exotroph: lotic: suctorial. **LTRF:** 2/3. **OA:** typical: ventral, suctorial. **MP:** wide dorsal gap: biserial. **SUP:** absent. **DEM:** lateral. **NA:** small, nearer snout than eyes. **VT:** medial. **EP:** dorsal. **SP:** sinistral. **UJ:** narrow. **LJ:** narrow, open V-shaped. **DF:** low, originates at dorsal tail-body junction. **BS:** oval/depressed. **CP:** uniformly dark. **TL:** 27/36. **CI:** Mertens (1938); Perret (1966).

Wolterstorffina: **CO:** 2. **GR:** Cameroon–Nigeria. **EMG:** endotroph.

Cochranella

CENTROLENIDAE: CO: 3/103. **GR:** Mexico–Brazil–Bolivia. **NO:** generic and specific differences poorly known.

Centrolene: **CO:** 33. **GR:** Honduras–Peru–Venezuela. **EMG:** exotroph: lotic: burrower. **LTRF:** ≤ 2/3. **OA:** typical: anteroventral. **MP:** wide dorsal gap: uniserial. **SUP:** absent. **DEM:** lateral. **NA:** small, about equidistant between eye and snout. **VT:** medial, dextral. **EP:** dorsal: tiny and near sagittal line. **SP:** sinistral. **UJ:** narrow to medium, smooth arc, sometimes coarsely serrate. **LJ:** narrow to medium, U- or V-shaped. **DF:** low, originates on tail muscle to near tail-body junction, tip rounded. **BS:** oval/strongly depressed. **CP:** reddish in life because of skin vascularization, uniformly tan in preservative. **TL:** 40–50/36. **CI:** J. D. Lynch et al. (1983); Mijares-Urrutia (1990).

Cochranella: **CO:** 44. **GR:** Nicaragua–Bolivia–Brazil. **EMG:** exotroph: lotic: burrower. **LTRF:** ≤ 2/3. **OA:** typical: anteroventral. **MP:** wide dorsal gap: uniserial. **SUP:** absent. **DEM:** lateral. **NA:** small, about equidistant between eye and snout. **VT:** medial, dextral. **EP:** dorsal: tiny and near sagittal line. **SP:** sinistral. **UJ:** narrow to medium, smooth arc, sometimes coarsely serrate. **LJ:** narrow to medium, U- or V-shaped. **DF:** low, originates on tail muscle to near tail-body junction, tip rounded. **BS:** oval/strongly depressed. **CP:** reddish in life because of skin vascularization, uniformly tan in preservative. **TL:** 40–50/36. **CI:** Duellman (1978).

Hyalinobatrachium: **CO:** 26. **GR:** Mexico–Bolivia–Tobago–Guyana–Brazil. **EMG:** exotroph: lotic: burrower. **LTRF:** ≤ 2/3. **OA:** typical: anteroventral. **MP:** wide dorsal gap: uniserial. **SUP:** absent. **DEM:** lateral. **NA:** small, about equidistant between eye and snout. **VT:** medial, dextral. **EP:** dorsal: tiny and near sagittal line. **SP:** sinistral. **UJ:** narrow to medium, smooth arc, sometimes coarsely serrate. **LJ:** narrow to medium, U- or V-shaped. **DF:** low, originates on tail muscle to near tail-body junction, tip rounded. **BS:** oval/strongly depressed. **CP:** reddish in life because of skin vascularization, uniformly tan in preservative. **TL:** 40–50/36. **CI:** Duellman (1978); Heyer (1985).

DENDROBATIDAE: CO: 8/164. **GR:** Nicaragua–Bolivia–Brazil. **EMG:** exotroph: lentic: benthic, arboreal; lotic: benthic, neustonic; endotroph. **LTRF:** number of variations from 0/0 to 2/3. **OA:** typical: anteroventral, terminal, umbelliform. **MP:** medium to wide dorsal gap: uni- and biserial. **SUP:** few lateral, absent. **DEM:** lateral, absent. **NA:** nearer eye than snout or reverse. **EP:** dorsal. **SP:** sinistral. **UJ:** narrow to wide, smooth arc. **LJ:** narrow to wide, U- or V-shaped. **DF:** low, originates near dorsal tail-body junction, tip rounded to slightly pointed. **BS:** oval to rounded/depressed. **CP:** uniform or muted variations. **TL:** 12–68/36. **NO:** generic and specific differences poorly understood.

Aromobates: **CO:** 1. **GR:** Venezuela. **EMG:** exotroph: lotic: benthic. **LTRF:** 2(2)/3(1). **OA:** typical: anteroventral. **MP:** wide dorsal gap: uniserial on upper labium, biserial on lower labium. **SUP:** few ventrolaterally. **DEM:** lateral. **NA:** nearer eye than snout. **VT:** dextral. **EP:** dorsal. **SP:** sinistral. **UJ:** massive, finely serrated edge a uniform arc. **LJ:** large, moderately V-shaped. **DF:** low, originates near dorsal tail-body junction, tip rounded. **BS:** globular/depressed. **CP:** uniformly dark. **TL:** 68/37. **CI:** Myers et al. (1991).

Colostethus: **CO:** 85. **GR:** Costa Rica–Peru–Brazil. **EMG:** exotroph: lentic: benthic, arboreal, neustonic; lotic: neustonic; endotroph. **LTRF:** 2(2)/3[1], 2/2, 0/0. **OA:** typical: anteroventral, umbelliform, largely absent in endotrophs. **MP:** medium dorsal gap: uniserial. **SUP:** row throughout extent of marginal papillae, absent, few laterally and distal to P-3. **DEM:** lateral. **NA:** nearer eye than snout. **VT:** medial. **EP:** dorsal. **SP:** sinistral. **UJ:** medium to wide, smooth arc. **LJ:** medium to wide, open V-shaped. **DF:** low, originates near dorsal tail-body junction, tip rounded. **BS:** oval/depressed. **CP:** muted tones, dark line from nostril to snout common. **TL:** 15–30/36. **NO:** *C. nubicola* group has umbelliform oral disc and jaw sheaths but no labial teeth. **CI:** Kaiser and Altig (1994); Lescure (1984); J. M. Savage (1968).

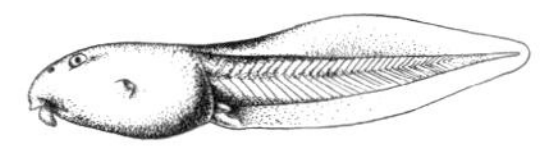

Colostethus; C. trinitatis

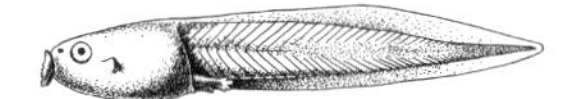

Colostethus; C. flotator

Dendrobates: **CO:** 26. **GR:** Nicaragua–Peru–Brazil. **EMG:** exotroph: lentic: benthic, arboreal. **LTRF:** 1/1, 2(2)/3[1]. **OA:** typical: anteroventral. **MP:** wide dorsal gap: uniserial, sometimes narrow ventral gap. **SUP:** absent. **DEM:** lateral. **NA:** nearer snout than eye. **VT:** medial. **EP:** dorsal. **SP:** sinistral. **UJ:** medium to very wide, smooth arc. **LJ:** medium to wide, open U-shaped. **DF:** low, originates near dorsal tail-body junction, tip rounded to slightly pointed. **BS:** oval/depressed. **CP:** uniform and muted variations. **TL:** 15–25/36. **CI:** J. M. Savage (1968); Silverstone (1975).

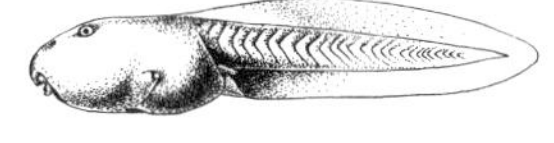

Dendrobates

Epipedobates: **CO:** 24. **GR:** northern South America. **EMG:** exotroph: lentic: benthic, arboreal. **LTRF:** 2(2)/3[1–2]. **OA:** typical: anteroventral. **MP:** wide dorsal gap: uniserial. **SUP:** few laterally, absent. **DEM:** lateral. **NA:** nearer snout than eye. **VT:** medial. **EP:** dorsal. **SP:** sinistral. **UJ:** narrow to medium, smooth arc. **LJ:** narrow to medium, open U-shaped. **DF:** low, originates near dorsal tail-body junction, tip rounded. **BS:** oval/depressed. **CP:** uniform and muted variations. **TL:** 15–20/36. **CI:** J. M. Savage (1968); Silverstone (1975).

Mannophryne: **CO:** 9. **GR:** Venezuela–Tobago. **EMG:** exotroph: lentic: benthic, arboreal. **EMG:** exotroph: lentic: benthic, arboreal, neustonic; lotic: neustonic; endotroph. **LTRF:** 2(2)/3[1], 2/2, 0/0. **OA:** typical: anteroventral, umbelliform, largely absent in endotrophs. **MP:** medium dorsal gap: uniserial. **SUP:** row throughout extent of marginal papillae, absent, few laterally and distal to P-3. **DEM:** lateral. **NA:** nearer eye than snout. **VT:** medial. **EP:** dorsal. **SP:** sinistral. **UJ:** medium to wide, smooth arc. **LJ:** medium to wide, open V-shaped. **DF:** low, originates near dorsal tail-body junction, tip rounded. **BS:** oval/depressed. **CP:** muted tones, dark line from nostril to snout common **TL:** 15–30/36. **NO:** see *Colostethus.* **CI:** J. R. Dixon and Rivero-Blanco (1985).

Minyobates: **CO:** 9. **GR:** Venezuela–Ecuador. **EMG:** exotroph: lentic: benthic. **LTRF:**

2(2)/3. **OA:** typical: anteroventral. **MP:** wide dorsal gap: uniserial. **SUP:** absent. **DEM:** absent, lateral. **NA:** nearer eye than snout. **VT:** medial. **EP:** dorsal. **SP:** sinistral. **UJ:** wide, smooth arc. **LJ:** wide, open U-shaped. **DF:** low, originates near dorsal tail-body junction, tip rounded. **BS:** oval/depressed. **CP:** uniformly dark. **TL:** 12–18/36. **CI:** J. M. Savage (1968); Silverstone (1975).

Nephalobates: **CO:** 5. **GR:** Venezuela. **EMG:** exotroph: lentic: benthic. **LTRF:** 2(2)/3. **OA:** typical: anteroventral. **MP:** wide dorsal gap: uniserial. **SUP:** absent. **DEM:** absent, lateral. **NA:** nearer eye than snout. **VT:** medial. **EP:** dorsal. **SP:** sinistral. **UJ:** wide, smooth arc. **LJ:** wide, open U-shaped. **DF:** low, originates near dorsal tail-body junction, tip rounded. **BS:** oval/depressed. **CP:** uniformly dark. **TL:** 12–18/36. **NO:** former *Colostethus albogularis* group. **CI:** La Marca and Mijares-U (1988).

Phyllobates: **CO:** 5. **GR:** Costa Rica–Colombia. **EMG:** exotroph: lotic: benthic. **LTRF:** 2(2)/3. **OA:** typical: anteroventral. **MP:** wide dorsal gap: uni- and biserial. **SUP:** absent. **DEM:** lateral. **NA:** nearer eye than snout. **VT:** medial. **EP:** dorsal. **SP:** sinistral. **UJ:** wide, smooth arc. **LJ:** wide, open V-shaped. **DF:** low, originates near dorsal tail-body junction, tip rounded. **BS:** oval/depressed. **CP:** uniformly dark. **TL:** 15–25/36. **CI:** Donnelly et al. (1990b); Lescure (1976); J. M. Savage (1968).

DISCOGLOSSIDAE: **CO:** 2/9. **GR:** Europe. **EMG:** exotroph: lentic: benthic. **LTRF:** 2/3(1), biserial. **OA:** typical: anteroventral. **MP:** complete: uniserial. **SUP:** absent. **DEM:** lateral. **NA:** nearer snout than eye. **VT:** medial. **EP:** dorsal. **SP:** low on left side. **UJ:** narrow, smooth arc. **LJ:** narrow, open U-shaped. **DF:** low, originates near dorsal tail-body junction, tip rounded. **BS:** oval/depressed. **CP:** chainlike melanophore pattern in skin. **TL:** 35–55/36.

Alytes: **CO:** 4. **GR:** Europe. **EMG:** exotroph: lentic: benthic. **LTRF:** 2/3(1), biserial. **OA:** typical: anteroventral. **MP:** complete: uniserial. **SUP:** absent. **DEM:** lateral. **NA:** nearer snout than eye. **VT:** medial. **EP:** dorsal. **SP:** low on left side. **UJ:** narrow, smooth arc. **LJ:** narrow, open U-shaped. **DF:** low, originates near dorsal tail-body junction, tip rounded. **BS:** oval/depressed. **CP:** chainlike melanophore pattern in skin. **TL:** 35–55/36. **CI:** Boulenger (1892); Lanza (1983).

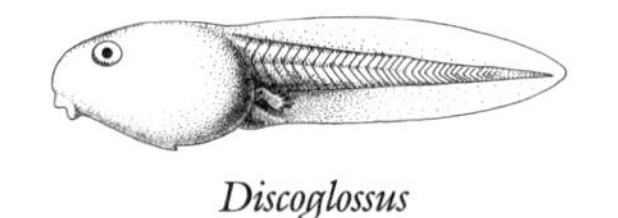

Discoglossus

Discoglossus: **CO:** 5. **GR:** Europe–Syria. **EMG:** exotroph: lentic: benthic. **LTRF:** 2/3(1), biserial. **OA:** typical: anteroventral. **MP:** complete: uniserial. **SUP:** absent. **DEM:** lateral. **NA:** nearer snout than eye. **VT:** medial. **EP:** dorsal. **SP:** low on left side. **UJ:** narrow, smooth arc. **LJ:** narrow, open U-shaped. **DF:** low, originates near dorsal tail-body junction, tip rounded. **BS:** oval/depressed. **CP:** chainlike melanophore pattern in skin. **TL:** 35–55/36. **CI:** Boulenger (1892); Lanza (1983).

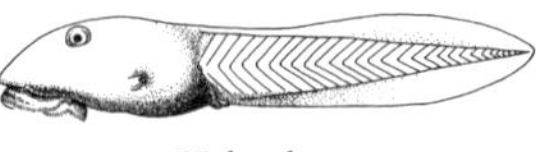

Heleophryne

HELEOPHRYNIDAE: **CO:** 1/5. **GR:** South Africa. **EMG:** exotroph: lotic: suctorial. **LTRF:** 4/14–15(1). **OA:** typical: ventral, upper jaw sheath absent, sometimes lower also. **MP:** complete: biserial. **SUP:** rows below distal most lower row. **DEM:** absent. **NA:** equidistant between eye and snout, slightly tubular. **VT:** medial. **EP:** dorsal. **SP:** sinistral. **UJ:** absent. **LJ:** when present: medium, open V-shaped. **DF:** low, originates on tail muscle, tip rounded. **BS:** oblong/depressed. **CP:** rock/lichenlike mottling, yellow/gold often in background color in life. **TL:** 50–85/36. **NO:** fewer than 14 tooth rows on lower labium common; hypertrophied, isolated jaw serrations in the position of jaw sheaths of small tadpoles are soon lost (Visser 1985); large subbranchial muscle present ventrolaterally along anterior two-thirds of body gives ventral silhouette a distinctive shape.

Heleophryne: **CO:** 5. **CI:** Channing (1984); Channing et al. (1988); Van Dijk (1966).

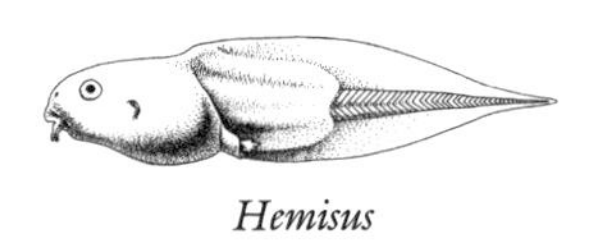

Hemisus

HEMISOTIDAE: **CO:** 1/8. **GR:** sub-Saharan Africa. **EMG:** exotroph: lentic: nektonic. **LTRF:** 5(3–5)/4(1), 6(3–6)/3. **OA:** typical: anteroventral. **MP:** wide dorsal gap: uni- and biserial, some ventral papillae exceedingly long. **SUP:** absent, few laterally. **DEM:** absent. **NA:** small, closer to snout than eye. **VT:** dextral. **EP:** lateral. **SP:** sinistral. **UJ:** medium to wide, strong medial convexity. **LJ:** medium, open U-shaped. **DF:** low, originates near dorsal tail-body junction, tip pointed. **BS:** oval/compressed. **CP:** stripes on tail, distal third dark, basal connective tissue sheath visible. **TL:** 40–60/36.

Hemisus: **CO:** 8. **CI:** Lambiris (1988a, b); Van Dijk (1966); Wager (1986).

HYLIDAE: **CO:** 41/743. **GR:** Holarctic–Neotropic–Australopapua. **EMG:** exotroph: lentic: several guilds; lotic: several guilds; endotroph. **LTRF:** many variations between 0/0 and

17/21. **OA:** typical: anteroventral, ventral, terminal. **MP:** dorsal gap, complete, dorsal and ventral gaps: variable, often bi- or multiserial. **SUP:** highly variable. **DEM:** absent, lateral. **NA:** round, variable in size, closer to snout than eye or reverse, often rimmed and countersunk. **VT:** dextral, medial. **EP:** lateral, dorsal. **SP:** sinistral, low on left side, ventral. **LJ:** highly variable. **DF:** highly variable. **BS:** oval to oblong/depressed, equidimensional and compressed. **CP:** highly variable. **TL:** 20–60/36. **NO:** marginal papillae always smaller and more densely arranged than ranids, the most commonly confused taxa; few taxa with 1–3 anterior rows with medial gaps distal to nonbroken rows.

Hemiphractinae: CO: 5/64. **GR:** Panama–South America.

Cryptobatrachus: **CO:** 3. **GR:** Colombia. **EMG:** endotroph.

Flectonotus: **CO:** 5. **GR:** Brazil–Venezuela–Trinidad. **EMG:** endotroph; exotroph: arboreal. **LTRF:** 0/0. **OA:** typical but lacking labial teeth; anteroventral. **MP:** dorsal gap: uniserial. **SUP:** absent. **DEM:** absent. **NA:** nearer snout than eye. **VT:** medial. **EP:** dorsal. **SP:** ventral. **UJ:** medium, wide arc. **LJ:** narrow, wide U-shaped. **DF:** low, originates near dorsal tail-body junction, tip round. **BS:** oblong/depressed. **CP:** lightly pigmented. **TL:** 18/31. **NO:** exotrophic larvae are morphologically similar to those of nidicolous endotrophs. **CI:** Duellman and Gray (1983); Heyer et al. (1990); and Weygoldt (1989).

Gastrotheca: **CO:** 44. **GR:** Panama–Brazil. **EMG:** exotroph: lentic: benthic; endotroph. **LTRF:** 2(2)/3(1). **OA:** typical: anteroventral. **MP:** medium dorsal gap: uniserial. **SUP:** ventrolaterally. **DEM:** absent. **NA:** equidistant between snout and eye. **VT:** dextral. **EP:** dorsal. **SP:** sinistral. **UJ:** medium, smooth arc. **LJ:** wide, open V-shaped. **DF:** low, originates near dorsal tail-body junction, tip rounded. **BS:** oval/depressed. **CP:** uniformly dark. **TL:** 40–60/36. **CI:** Wassersug and Duellman (1984).

Hemiphractus: **CO:** 5. **GR:** Panama–Bolivia. **EMG:** endotroph.

Stefania: **CO:** 7. **GR:** Venezuela–Guyana. **EMG:** endotroph.

Hylinae: CO: 26/488. **GR:** Eurasia and North, Central, and South America.

Acris: **CO:** 2. **GR:** eastern North America. **EMG:** exotroph: lentic: benthic. **LTRF:** 2(2)/2. **OA:** typical: anteroventral. **MP:** wide dorsal gap: uniserial. **SUP:** few laterally. **DEM:** absent. **NA:** nearer snout than eye. **VT:** dextral. **EP:** dorsal. **SP:** sinistral. **UJ:** medium, wide smooth arc. **LJ:** medium, U-shaped. **DF:** low, originates near dorsal tail-body junction, tip pointed. **BS:** oval/depressed. **CP:** dorsum of tail muscle banded; tail tip usually black. **TL:** 40–50/36. **NO:** spiracular tube long. **CI:** Altig (1970); Orton (1952); A. H. Wright and Wright (1949).

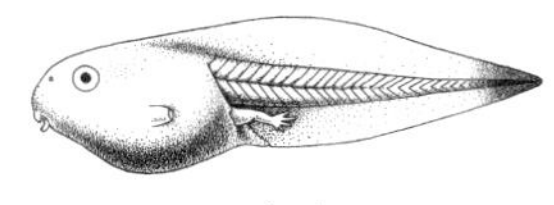

Acris

Anotheca: **CO:** 1. **GR:** Mexico–Panama. **EMG:** exotroph: lentic: arboreal. **LTRF:** 2(2)/2. **OA:** typical: nearly terminal. **MP:** complete: uniserial. **SUP:** few laterally. **DEM:** absent. **NA:** about equidistant between snout and eye. **VT:** dextral. **EP:** dorsal. **SP:** sinistral. **UJ:** wide, long shallow arc, finely serrate. **LJ:** wide, open V-shaped. **DF:** low, originates near dorsal tail-body junction, tip rounded. **BS:** oval/depressed. **CP:** uniformly purplish brown. **TL:** 45/38. **NO:** gut not coiled. **CI:** Robinson (1961); E. H. Taylor (1954).

Aparasphenodon: **CO:** 2. **GR:** Brazil–Venezuela. **EMG:** lentic: benthic. **LTRF:** 2(1)/5–6(1). **OA:** typical: anteroventral. **MP:** wide dorsal gap: multiserial. **SUP:** small group ventrolaterally. **DEM:** absent. **NA:** –. **VT:** medial. **EP:** lateral. **SP:** sinistral. **UJ:** –. **LJ:** –. **DF:** low, originates anterior to the dorsal tail-body junction, tip pointed. **BS:** ovoid in dorsal view. **CP:** dark brown with irregular dark spots; tail muscle banded dorsally. **TL:** 47/36. **CI:** H. R. de Silva (personal communication).

Aplastodiscus: **CO:** 1. **GR:** Brazil. **EMG:** exotroph: lentic: benthic. **LTRF:** 2(2)/4(1). **OA:** typical: anteroventral. **MP:** medium dorsal gap: uniserial. **SUP:** absent. **DEM:** absent. **NA:** nearer eye than snout. **VT:** dextral. **EP:** lateral. **SP:** sinistral. **UJ:** medium, straight medial edge. **LJ:** medium, U-shaped. **DF:** low, originates at dorsal tail-body junction, tip pointed. **BS:** oval/depressed. **CP:** distal one-third of tail darkened; anterior part of tail muscle with lateral stripe. **TL:** 68/38. **CI:** Caramaschi et al. (1980).

Argenteohyla: **CO:** 1. **GR:** Argentina–Paraguay. **EMG:** exotroph: lentic: benthic. **LTRF:** 2(2)/4(1). **OA:** typical: anteroventral. **MP:** medium dorsal gap: biserial. **SUP:** absent. **DEM:** absent. **NA:** nearer eye than snout. **VT:** medial. **EP:** dorsal. **SP:** sinistral. **UJ:** medium, smooth arc. **LJ:** medium, open V-shaped. **DF:** somewhat arched, originates on body, tip pointed. **BS:** oval/depressed. **CP:** uniformly brown. **TL:** 42/31. **CI:** De Sá (1983).

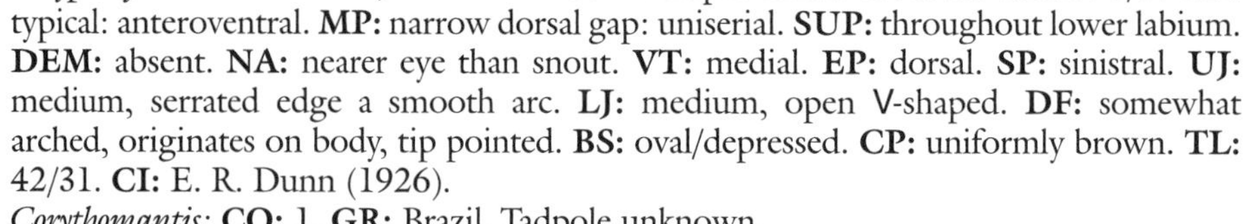

Calyptahyla: **CO:** 1. **GR:** Jamaica. **EMG:** exotroph: lentic: arboreal. **LTRF:** 1/0. **OA:** typical: anteroventral. **MP:** narrow dorsal gap: uniserial. **SUP:** throughout lower labium. **DEM:** absent. **NA:** nearer eye than snout. **VT:** medial. **EP:** dorsal. **SP:** sinistral. **UJ:** medium, serrated edge a smooth arc. **LJ:** medium, open V-shaped. **DF:** somewhat arched, originates on body, tip pointed. **BS:** oval/depressed. **CP:** uniformly brown. **TL:** 42/31. **CI:** E. R. Dunn (1926).

Corythomantis: **CO:** 1. **GR:** Brazil. Tadpole unknown.

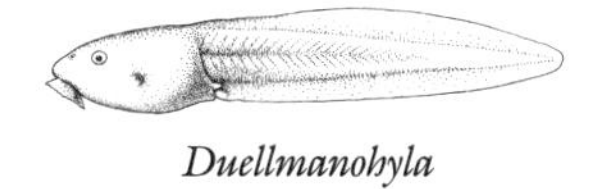

Duellmanohyla

Duellmanohyla: **CO:** 8. **GR:** Mexico–Costa Rica. **EMG:** exotroph: lotic: adherent. **LTRF:** 3(1,3)/3[1], 2(2)/3, 2(2)/2. **OA:** typical: ventral. **MP:** complete: small and uniserial. **SUP:** scattered throughout disc. **DEM:** midventral. **NA:** small, nearer snout than eye. **VT:** dextral. **EP:** dorsal. **SP:** sinistral. **UJ:** wide, coarsely serrate, smooth arc to slightly indented medially. **LJ:** medium, coarsely serrate, V-shaped. **DF:** low, originates near dorsal tail-body junction, tip rounded. **BS:** oval/depressed. **CP:** largely brown, some mottling on tail. **TL:** 33/26. **NO:** tooth rows much shorter than transverse diameter of oral disc. **CI:** J. A. Campbell and Smith (1992); Duellman (1970).

Hyla; H. bromeliacia

Hyla; H. picadoi

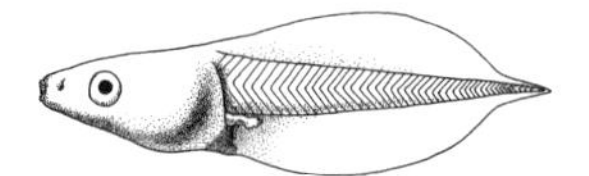

Hyla; H. sarayacuensis

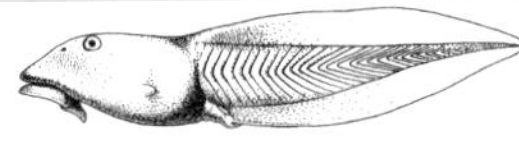

Hyla; H. armata

Hyla: **CO:** 289. **GR:** North, Central and South America–Eurasia. **EMG:** exotroph: lotic: arboreal, benthic, adherent, clasping, suctorial; lentic: benthic, nektonic, arboreal. **LTRF:** many variations between 0/0 and 17/21; 2/3 most common. **OA:** typical: terminal, anteroventral, ventral; reduced to various degrees. **MP:** dorsal gap, dorsal and ventral gaps: uni-, bi-, and multiserial. **SUP:** few to many laterally, distinct rows above upper and below lower tooth rows, absent. **DEM:** absent. **NA:** various positions. **VT:** dextral, medial. **EP:** dorsal, lateral. **SP:** sinistral. **UJ:** many variations. **LJ:** many variations. **DF:** very low to very high, originates anterior to, near (most common), or posterior to dorsal tail-body junction, tip pointed to round. **BS:** oval to oblong/depressed, compressed or equidimensional. **CP:** variable. **TL:** 20–60/36. **NO:** a number of species groups have distinctive morphologies. **CI:** Caldwell (1974); Duellman (1970, 1978); Duellman and Altig (1978); Duellman and Fouquette (1968); Duellman and Trueb (1989); Peixoto and Da Cruz (1983); J. M. Savage (1980b).

Nyctimantis: **CO:** 1. **GR:** Ecuador. Tadpole unknown.

Osteocephalus: **CO:** 14. **GR:** Guianas–Venezuela–Amazon Basin–Argentina. **EMG:** exotroph: lentic: benthic, arboreal. **LTRF:** 2(2)/5(1), 3(2–5)/5(1). **OA:** typical: anteroventral. **MP:** wide dorsal gap: biserial. **SUP:** few laterally. **DEM:** absent. **NA:** equidistant between eye and snout. **VT:** dextral. **EP:** dorsal. **SP:** sinistral. **UJ:** narrow, wide smooth arc. **LJ:** narrow, V-shaped. **DF:** low, originates near dorsal tail-body junction, tip rounded. **BS:** oval/depressed. **CP:** uniformly dark. **TL:** 49/37. **CI:** Duellman (1978); Duellman and Lescure (1973); Trueb and Duellman (1970).

Osteopilus: **CO:** 3. **GR:** Greater Antilles–Florida. **EMG:** exotroph: lentic: benthic, arboreal. **LTRF:** 2(2)/4(1). **OA:** typical: anteroventral, ventral. **MP:** dorsal gap: uniserial. **SUP:** absent. **DEM:** absent. **NA:** nearer eye than snout. **VT:** dextral, medial. **EP:** lateral, dorsal. **SP:** sinistral. **UJ:** medium, edge forms smooth arc. **LJ:** medium, deep medial indentation. **DF:** low, originates near dorsal tail-body junction, tip slightly pointed. **BS:** oval/depressed. **CP:** uniformly dark or pale. **TL:** 37/41. **CI:** Duellman and Schwartz (1958); Lannoo et al. (1987).

Phrynohyas: **CO:** 5. **GR:** Mexico–Argentina. **EMG:** exotroph: lentic: nektonic, benthic, arboreal. **LTRF:** 4(1–2,4)/6(1), 3(1,3)/5(1), 2(2)/4[1]. **OA:** typical: anteroventral. **MP:** medium dorsal gap: biserial. **SUP:** few ventrolaterally. **DEM:** absent. **NA:** nearer eye than snout. **VT:** medial. **EP:** lateral. **SP:** sinistral. **UJ:** medium, edge describes a smooth arc, coarsely serrate. **LJ:** medium, open U-shaped. **DF:** medium, originates on body, tip pointed. **BS:** oval/compressed. **CP:** tail muscle striped. **TL:** 30–38/36. **NO:** unusual LTRF shared with *Trachcephalus;* Pyburn (1967) shows a medial constriction in the upper jaw sheath; Zweifel (1964) shows a medially broken upper sheath; apparently random breaks, not gaps as noted above, common in lower rows. **CI:** Pyburn (1967); Zweifel (1964).

Phyllodytes: **CO:** 7. **GR:** Trinidad–Brazil. **EMG:** exotroph: lentic: benthic, arboreal. **LTRF:** 2(2)/3. **OA:** typical: anteroventral. **MP:** wide dorsal gap: uni- and biserial. **SUP:** few laterally. **DEM:** absent. **NA:** nearer snout than eye. **VT:** medial. **EP:** dorsal. **SP:** sinistral. **UJ:** medium, smooth arc. **LJ:** medium, open U-shaped. **DF:** low, originates near dorsal tail-body junction, tip rounded. **BS:** oval/depressed. **CP:** uniformly dark. **TL:** 29/36. **CI:** Bokermann (1966); F. M. Clarke et al. (1995); Giarretta (1996).

Plectrohyla: **CO:** 16. **GR:** Mexico–Guatemala. **EMG:** exotroph: lotic: clasping. **LTRF:** 2(2)/3. **OA:** typical: ventral. **MP:** complete: uni- and biserial. **SUP:** row above A-1 and below P-3, scattered laterally. **DEM:** absent. **NA:** small, nearer snout than eye. **VT:** dextral. **EP:** dorsal. **SP:** sinistral. **UJ:** wide, uniform arc, uniform graded series of serrations, hypertrophied serrations in specific sections. **LJ:** wide, open U-shaped. **DF:** low, originates near dorsal tail-body junction, tip round. **BS:** oval/depressed. **CP:** uniformly dark, some mottling on tail. **TL:** 35–43/34. **CI:** J. A. Campbell and Kubin (1990); Duellman (1970).

Pseudacris: **CO:** 13. **GR:** North America. **EMG:** exotroph: lentic: nektonic, benthic. **LTRF:** 2(2)/3[1]. **OA:** typical: anteroventral. **MP:** medium dorsal gap: uni- and biserial, sometimes narrow ventral gap. **SUP:** few laterally. **DEM:** absent. **NA:** nearer eye than snout. **VT:** dextral. **EP:** lateral. **SP:** sinistral. **UJ:** narrow to medium, wide arc, sometimes medial zone straight. **LJ:** medium, V-shaped. **DF:** medium to high, originates near dorsal tail-body junction, on body, tip pointed. **BS:** oval/compressed. **CP:** muted mottling, stripes or bicolored on tail. **TL:** 25–45/36. **CI:** Altig (1970); Duellman (1970); A. H. Wright and Wright (1949).

Pternohyla: **CO:** 2. **GR:** southern Arizona–Mexico. **EMG:** exotroph: lentic: benthic. **LTRF:** 2(2)/3. **OA:** typical: anteroventral. **MP:** dorsal gap: uniserial dorsolaterally, biserial elsewhere. **SUP:** few laterally. **DEM:** absent. **NA:** small, equidistant between eyes and snout. **VT:** dextral. **EP:** lateral. **SP:** sinistral. **UJ:** wide, edge a smooth arc. **LJ:** medium, open U-shaped. **DF:** high, originates on body, tip somewhat pointed with slight flagellum. **BS:** oval/compressed. **CP:** pale dorsolateral stripes. **TL:** 38/36. **CI:** R. G. Webb (1963).

Ptychohyla: **CO:** 10. **GR:** Mexico–Panama. **EMG:** exotroph: lotic: benthic, suctorial. **LTRF:** 4(1,4)/7(1), 4(4)/7(1), 4(4)/6(1). **OA:** typical: nearly ventral. **MP:** complete: biserial. **SUP:** many laterally. **DEM:** absent. **NA:** small, nearer snout than eye. **VT:** dextral. **EP:** dorsal. **SP:** sinistral. **UJ:** medium, edge a smooth arc. **LJ:** medium, V-shaped. **DF:** low, originates near dorsal tail-body junction, tip pointed to rounded. **BS:** oval/depressed. **CP:** dark, some minor mottling, especially on tail. **TL:** 33–39/25–31. **CI:** J. A. Campbell and Smith (1992); Duellman (1970).

Scarthyla: **CO:** 1. **GR:** Peru. **EMG:** exotroph: lentic: nektonic. **LTRF:** 2(1–2)/2(1). **OA:** typical: nearly terminal. **MP:** wide dorsal and ventral gaps: uniserial. **SUP:** absent. **DEM:** absent. **NA:** nearer snout than eye. **VT:** dextral. **EP:** lateral. **SP:** sinistral. **UJ:** medium, prominent arch, coarsely serrate. **LJ:** wide, open U-shaped, coarsely serrate. **DF:** low, originates posterior to dorsal tail-body junction, tip pointed. **BS:** oval/depressed. **CP:** tail fins with prominent serial blotches that look like incomplete bands. **TL:** 28/36. **NO:** see *Scinax.* **CI:** Duellman and De Sá (1988).

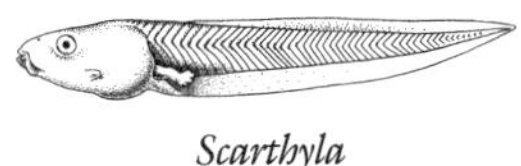

Scarthyla

Scinax: **CO:** 81. **GR:** Mexico–Uruguay. **EMG:** exotroph: lentic: nektonic, benthic. **LTRF:** 2(2)/3[1]. **OA:** typical: anteroventral, almost terminal. **MP:** narrow dorsal gap: uni- and biserial; *rostrata* group: ventral gap ≥ extent of short P-3 that sits upon an armlike derivative of the lower labium. **SUP:** few laterally. **DEM:** absent. **NA:** nearer eye than snout. **VT:** dextral. **EP:** lateral. **SP:** sinistral. **UJ:** wide, smooth arc or with medial convexity, sometimes coarsely serrate. **LJ:** medium to wide, V- to open U-shaped. **DF:** low, originates near dorsal tail-body junction, tip pointed, or quite high, originates anterior to dorsal tail-body junction, tip pointed, sometimes with flagellum. **BS:** oval/depressed or compressed. **CP:** lightly pigmented, eye line common. **TL:** 20–30/36. **NO:** see *Scarthyla;* at least *rostrata* and *rubra* groups with distinctive morphology of the oral apparatus. **CI:** Duellman (1970); Hero and Mijares-Urrutia (1995); Heyer et al. (1990); McDiarmid and Altig (1990).

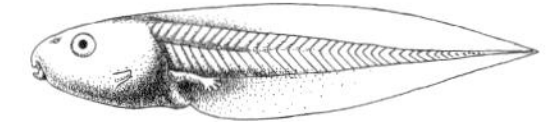

Scinax; S. sugillata

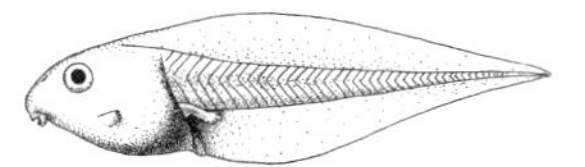

Scinax; S. elaeochroa

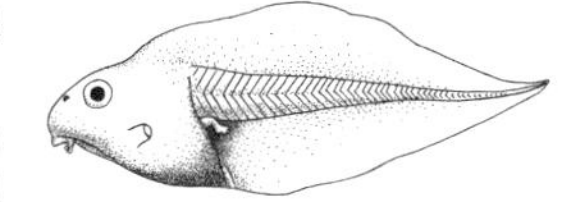

Scinax; S. staufferi

Smilisca: **CO:** 6. **GR:** southern Texas–northern South America. **EMG:** exotroph: lentic: benthic; lotic: adherent. **LTRF:** 2(2)/3. **OA:** typical: anteroventral, ventral. **MP:** dorsal gap, complete: uni- and multiserial. **SUP:** many laterally. **NA:** nearer eye than snout. **VT:** dextral. **EP:** lateral, dorsal. **SP:** sinistral. **UJ:** medium to wide, edge a smooth arc or indentation medially. **LJ:** medium to wide, open U-shaped. **DF:** low, slightly arched, originates near dorsal tail-body junction or onto body, tip pointed. **BS:** oval/slightly depressed. **CP:** uniformly brownish; pale dorsolateral stripe on body; dorsum of tail muscle lightly banded. **TL:** 22–40/30–40. **CI:** Duellman (1970).

Sphaenorhynchus: **CO:** 11. **GR:** Trinidad, Orinoco and Amazon basins of northern South America. **EMG:** exotroph: lentic: benthic. **LTRF:** 2(2)/3(1). **OA:** typical: anteroventral. **MP:** wide dorsal gap: uniserial. **SUP:** few laterally and ventrolaterally. **DEM:** absent. **NA:** small, nearer snout than eye. **VT:** medial. **EP:** lateral. **SP:** sinistral. **UJ:** wide, slight medial convexity. **LJ:** wide, very open U-shaped. **DF:** low, originates near dorsal tail-body junction, tip pointed. **BS:** oval/depressed. **CP:** stripes below eye and on tail. **TL:** 57/35. **CI:** Bokermann (1974); Da Cruz and Peixoto (1980a); Heyer et al. (1990).

Tepuihyla: **CO:** 7. **GR:** eastern Venezuela. **NO:** no tadpoles known for the *O. rodriguezi* group that was recently elevated to a genus distinct from *Osteocephalus.*

Trachycephalus: **CO:** 3. **GR:** Brazil–Ecuador. **EMG:** exotroph: lentic: nektonic. **LTRF:** 4(1–2,4)/6(1). **OA:** typical: anteroventral. **MP:** medium dorsal gap: biserial. **SUP:** patch ventrolaterally. **DEM:** absent. **NA:** small, nearer snout than eye. **VT:** medial. **EP:** lateral. **SP:** sinistral. **UJ:** medium, forms smooth arc. **LJ:** medium, open V-shaped. **DF:** high, originates on body, tip pointed. **BS:** oval/compressed. **CP:** body and tail striped. **TL:** 38/34. **NO:** shares uncommon LTRF with *Phrynohyas.* **CI:** McDiarmid and Altig (1990).

Triprion: **CO:** 2. **GR:** western and southern Mexico. **EMG:** exotroph: lentic: benthic. **LTRF:** 2(2)/3. **OA:** typical: anteroventral. **MP:** medium dorsal gap: biserial. **SUP:** few laterally. **DEM:** absent. **NA:** nearer eye than snout. **VT:** dextral. **EP:** lateral. **SP:** sinistral. **UJ:** wide, smooth arc. **LJ:** wide, broadly U-shaped. **DF:** low, originates at dorsal tail-body junction, tip rounded. **BS:** oval/slightly depressed. **CP:** uniformly brown. **TL:** 27/30. **CI:** Duellman (1970).

Xenohyla: **CO:** 2. **GR:** Brazil. **EMG:** exotroph: lentic: nektonic. **LTRF:** 2(2)/3(1). **OA:** typical: anteroventral. **MP:** medium dorsal gap: uniserial. **SUP:** absent. **DEM:** absent. **NA:** nearer tip of snout than eye. **VT:** –. **EP:** lateral. **SP:** sinistral. **UJ:** medium, with moderate medial convexity. **LJ:** wide, broadly U-shaped. **DF:** high, originates directly behind plane of eye, tip with prominent flagellum. **BS:** oval/compressed. **CP:** bold bands on tail, prominent eye stripe. **TL:** 32/33. **CI:** Izecksohn (1997).

Pelodryadinae: CO: 4/143. **GR:** Australia–New Guinea.

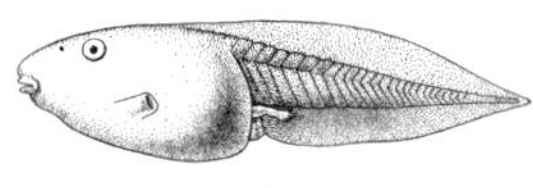

Cyclorana

Cyclorana: **CO:** 12. **GR:** Australia. **EMG:** exotroph: lentic: benthic. **LTRF:** 2(2)/3(1). **OA:** typical: anteroventral. **MP:** wide dorsal gap: biserial. **SUP:** absent. **DEM:** absent. **NA:** nearer snout than eye. **VT:** dextral. **EP:** dorsal. **SP:** sinistral. **UJ:** wide, smooth arc. **LJ:** wide, V-shaped. **DF:** low, originates near dorsal tail-body junction, tip pointed. **BS:** oval/depressed. **CP:** unicolored to faintly mottled or blotched. **TL:** 40–70/42. **CI:** Watson and Martin (1973).

Litoria: **CO:** 104. **GR:** Australia–New Guinea. **EMG:** exotroph: lentic: benthic; lotic: benthic, suctorial. **LTRF:** 2/3, 2(2)/3[1], 2(2)/2. **OA:** typical: anteroventral, ventral. **MP:** narrow to wide dorsal gap: uni- and biserial; complete: biserial. **SUP:** absent, few laterally, many laterally. **DEM:** absent. **NA:** nearer eye than snout and reverse. **VT:** dextral. **EP:** lateral, dorsal. **SP:** sinistral. **UJ:** narrow to wide, smooth arc, slight medial convexity. **LJ:** narrow to wide, open U-shaped. **DF:** low, originates near dorsal tail-body junction, tip pointed to rounded. **BS:** oval/depressed or compressed. **CP:** variable. **TL:** 30–60/36. **NO:** species groups not well defined (e.g., Barker and Grigg 1977; King 1981; Tyler 1968; Tyler and Davies 1978). **CI:** Watson and Martin (1973).

Nyctimystes: **CO:** 24. **GR:** northern Australia–New Guinea. **EMG:** exotroph: lotic: suctorial. **LTRF:** 2/3. **OA:** typical: ventral. **MP:** complete: uniserial. **SUP:** radially oriented, elongate papillae across lower labium distal to P-3. **DEM:** absent. **NA:** nearer snout than eye. **VT:** medial. **EP:** dorsal. **SP:** sinistral. **UJ:** wide, medial modification for suctorial existence. **LJ:** wide, open U-shaped. **DF:** low, originates posterior to dorsal tail-body junction, tip round. **BS:** oval/depressed. **CP:** muted mottling. **TL:** 33/36. **CI:** Davies and Richards (1990); Tyler (1989).

Pelodryas: **CO:** 3. **GR:** Australia. **EMG:** exotroph: lentic: benthic. **LTRF:** 2(1–2)/3. **OA:** typical: anteroventral. **MP:** medium dorsal gap: biserial. **SUP:** absent. **DEM:** absent. **NA:** equidistant between eye and snout. **VT:** dextral. **EP:** lateral. **SP:** sinistral. **UJ:** medium, smooth arc, coarsely serrate. **LJ:** medium, open V-shaped, coarsely serrate. **DF:** low, originates near or slightly posterior to dorsal tail-body junction, tip pointed. **BS:** oval/depressed. **CP:** tail striped. **TL:** 41/41. **NO:** see *Litoria.* **CI:** Tyler et al. (1983).

Phyllomedusinae: CO: 6/48. **GR:** Mexico–Bolivia.

Agalychnis: **CO:** 8. **GR:** Mexico–Ecuador. **EMG:** exotroph: lentic: suspension-rasper, nektonic, arboreal. **MP:** medium dorsal gap, complete: biserial. **SUP:** absent. **DEM:** absent. **NA:** nearer snout than eye. **VT:** dextral. **EP:** lateral. **SP:** low on left side without medial wall. **UJ:** medium to wide, smooth arc, slight medial convexity. **LJ:** medium, open U-shaped. **DF:** low, originates near dorsal tail-body junction, tip rounded. **BS:** oval/compressed. **CP:** lightly pigmented. **TL:** 30–60/36. **NO:** ventral fin often higher than dorsal fin. **CI:** Donnelly et al. (1987); Duellman (1970); Hoogmoed and Cadle (1991).

Hylomantis: **CO:** 2. **GR:** Brazil. **EMG:** exotroph: lentic. **NO:** similar to tadpoles of *Phasmahyla* (O. L. Peixoto, personal communication).

Pachymedusa: **CO:** 1. **GR:** western Mexico. **EMG:** exotroph: lentic: suspension-rasper. **LTRF:** 2(2)/3(1). **OA:** typical: anteroventral. **MP:** wide dorsal gap: uniserial. **SUP:** few bordering marginal papillae. **DEM:** absent. **NA:** nearer snout than eye. **VT:** dextral. **EP:** lateral. **SP:** low on left side without medial wall. **UJ:** medium, small serrations on smooth, broad arc. **LJ:** medium, open V-shaped. **DF:** medium, originates at dorsal tail-body junction, tip pointed. **BS:** oval/compressed. **CP:** bluish gray in life, tan in preservative. **TL:** 45/34. **CI:** Duellman (1970).

Phasmahyla: **CO:** 4. **GR:** Brazil. **EMG:** exotroph: lentic: neustonic. **LTRF:** 1/2(1). **OA:** typical: umbelliform. **MP:** complete, absent: minute. **SUP:** scattered, some large and oriented radial to mouth. **DEM:** dorsal, dorsal and ventral. **NA:** equidistant between eye and snout. **VT:** dextral. **EP:** lateral. **SP:** low on left side without medial wall. **UJ:** medium, prominent medial convexity. **LJ:** medium, open V-shaped. **BS:** oval/compressed. **DF:** low, originates near or anterior to dorsal tail-body junction, tip pointed. **CP:** body uniformly pale, tail similar or with distal portion darker. **TL:** 39/36. **NO:** ventral fin extends onto abdomen and higher than ventral fin; vent tube not associated with ventral fin. **CI:** Da Cruz (1980, 1982); Heyer et al. (1990).

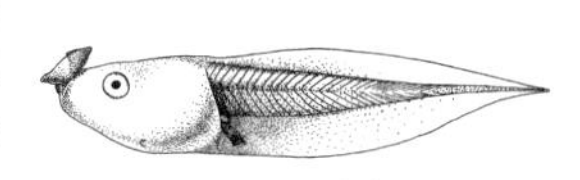
Phasmahyla

Phrynomedusa: **CO:** 5. **GR:** Brazil. **EMG:** exotroph: lentic: suspension-rasper. **LTRF:** 2/3(1), 2(2)/3(1). **OA:** typical: anteroventral. **MP:** complete: bi- or multiserial. **SUP:** few to many laterally. **DEM:** absent. **NA:** nearer snout than eye. **VT:** dextral. **EP:** lateral. **SP:** low on left side without medial wall. **UJ:** medium, smooth arc. **LJ:** medium, open V-shaped. **DF:** low, originates near dorsal tail-body junction, tip pointed. **BS:** oval/compressed. **CP:** uniformly pale. **TL:** 30–46/28–38. **NO:** ventral fin often higher than dorsal fin. **CI:** Da Cruz (1982); Heyer et al. (1990); Izecksohn and Da Cruz (1976).

Phyllomedusa: **CO:** 28. **GR:** Costa Rica–Argentina. **EMG:** lentic: suspension-rasper. **LTRF:** 2(2)/3/(1). **OA:** typical: anteroventral. **MP:** medium dorsal gap: uni- and biserial, sometimes with ventral gap. **SUP:** few laterally. **DEM:** absent. **NA:** nearer snout than eye. **VT:** dextral. **EP:** lateral. **SP:** low on left side without medial wall. **UJ:** medium, smooth arc with straight area medially. **LJ:** medium, V-shaped. **BS:** oval/compressed. **DF:** low, originates anterior of dorsal tail-body junction, tail pointed with flagellum. **CP:** uniformly pale; diffuse yellow, orange to black area common in anterior part of lower fin. **TL:** 30–50/33–37. **NO:** ventral fin higher than dorsal fin and extends anteriorly onto abdomen. **CI:** Cei (1980); Da Cruz (1982); Duellman (1970).

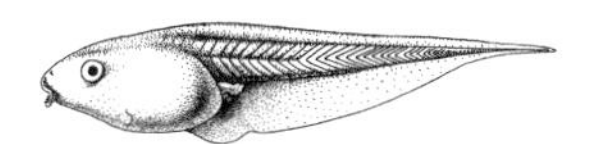
Phyllomedusa

HYPEROLIIDAE: CO: 19/232. **GR:** sub-Saharan Africa–Madagascar. **EMG:** exotroph: lentic: benthic, nektonic; lotic: benthic. **LTRF:** variations between 0/0 and 5/3; 1/3 most commonly. **OA:** typical: anteroventral, terminal. **MP:** medium to wide dorsal gap, sometimes narrow ventral gap: uni- and biserial. **SUP:** absent, few laterally, throughout extent of marginal papillae. **DEM:** absent. **NA:** often small, usually nearer snout than eye. **VT:** dextral, medial. **EP:** lateral, dorsal. **SP:** sinistral. **UJ:** narrow to wide and massive, smooth arc or medial convexity, coarse serrations in some. **LJ:** narrow to wide, U- or V-shaped. **DF:** low, originates near dorsal tail-body junction, tip pointed to rounded or very high, originates anterior to dorsal tail-body junction, tip pointed as flagellum. **BS:** oval/depressed and compressed. **CP:** variable—usually rather uniform, may have striped tail or tip black. **TL:** 25–85/36.

Hyperoliinae: CO: 12/161. **GR:** Africa–Madagascar–Seychelle Islands.

Acanthixalus: **CO:** 1. **GR:** Cameroon–Zaire. **EMG:** exotroph: lentic arboreal. **LTRF:** 4(2–4)/3. **OA:** typical: subterminal. **MP:** wide dorsal and narrow ventral gaps: biserial. **SUP:** abundant centripetally around lower labium. **DEM:** absent. **NA:** nearer snout than eye. **VT:** medial. **EP:** dorsal. **SP:** sinistral. **UJ:** wide, large medial convexity. **LJ:** wide,

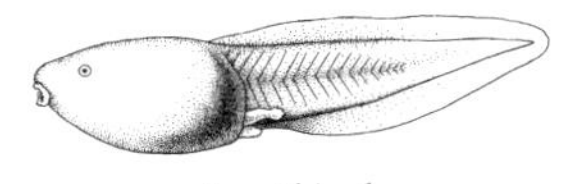
Acanthixalus

open U-shaped. **DF:** low, originates near dorsal tail-body junction, tip rounded. **BS:** oval/depressed. **CP:** uniformly dark. **TL:** 52/33. **CI:** Lamotte et al. (1959a).

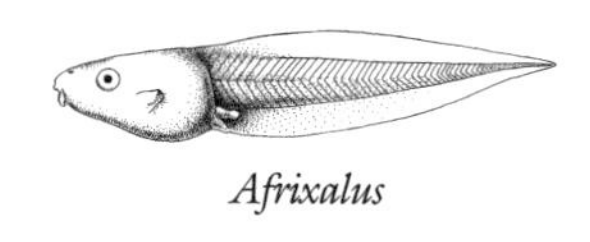

Afrixalus

Afrixalus: **CO:** 28. **GR:** sub-Saharan Africa. **EMG:** exotroph: lentic: benthic. **LTRF:** 0/0, 0/1. **OA:** typical: nearly terminal. **MP:** wide dorsal gap: uni- and biserial ventrally, large, widely spaced. **SUP:** few midventrally. **DEM:** absent. **NA:** very small, nearer snout than eye, or reverse. **VT:** dextral. **EP:** lateral. **SP:** sinistral. **UJ:** wide, smooth arc or with medial convexity, serrations often coarse. **LJ:** wide, smooth arc. **DF:** low, originates near dorsal tail-body junction, tip pointed. **BS:** oblong/depressed. **CP:** uniformly dotted. **TL:** 30–65/36. **NO:** ecomorphological guild poorly known, may in fact be occur in midwater or near the surface. **CI:** Drewes et al. (1989); Pickersgill (1984); Schiøtz (1967, 1975).

Alexteroon: **CO:** 1. **GR:** Cameroon. **EMG:** exotroph: lotic: benthic. **LTRF:** 2(2)/3(1). **OA:** typical: anteroventral. **MP:** wide dorsal gap: uniserial. **SUP:** single row from tip of A-1. **DEM:** absent. **NA:** nearer snout than eye. **VT:** dextral. **EP:** dorsal. **SP:** sinistral. **UJ:** medium, broad arc with medial area straight, coarsely serrate. **LJ:** medium, open U-shaped. **DF:** low, originates near dorsal tail-body junction, tip slightly pointed. **BS:** oval/depressed. **CP:** uniformly brown, margins of fins largely unpigmented. **TL:** 29/33. **CI:** Amiet (1974b).

Arlequinus: **CO:** 1. **GR:** Cameroon. **EMG:** exotroph: lentic: benthic. **LTRF:** 1/3. **OA:** typical: anteroventral. **MP:** medium dorsal gap: uniserial. **SUP:** many throughout the extent of the marginal papillae. **DEM:** absent. **NA:** nearer snout then eye. **VT:** dextral. **EP:** lateral. **SP:** sinistral. **UJ:** medium, smooth arc. **LJ:** medium, open U-shaped. **DF:** low, originates near dorsal tail-body junction, tip pointed. **BS:** oval/depressed. **CP:** black stripe the length of dorsal half of tail muscle and body bordered above and below with pale stripes; dorsum of tail muscle white. **TL:** 43/36. **NO:** adult pattern appears before metamorphosis. **CI:** Amiet (1976).

Callixalus: **CO:** 1. **GR:** Rwanda–Zaire. Tadpole unknown.

Chlorolius: **CO:** 1. **GR:** Cameroon-Gabon. **EMG:** exotroph: lentic: benthic. **LTRF:** 1/3(1). **OA:** typical: anteroventral. **MP:** wide dorsal gap: uniserial. **SUP:** single row closely adjacent to marginal papillae laterally and ventrally. **DEM:** absent. **NA:** nearer snout than eye. **VT:** dextral. **EP:** lateral. **SP:** sinistral. **UJ:** wide, slight convexity medially, coarsely serrate. **LJ:** medium, V–shaped. **DF:** low, originates near dorsal tail-body junction, tip pointed. **BS:** oval/depressed. **CP:** uniform body pigmentation with pale lines from snout to eyes; distal half of tail lightly pigmented. **TL:** 38/36. **CI:** Amiet (1976).

Chrysobatrachus: **CO:** 1. **GR:** Zaire. Tadpole unknown.

Cryptothylax: **CO:** 2. **GR:** Cameroon–Zaire. **EMG:** exotroph: lentic: benthic. **LTRF:** 1/3. **OA:** typical: anteroventral. **MP:** wide dorsal gap: uniserial. **SUP:** single row extending from A-1. **DEM:** absent. **NA:** nearer snout than eye. **VT:** –. **EP:** dorsal. **SP:** sinistral. **UJ:** medium, serrated edge describes a smooth arc. **LJ:** narrow, open U-shaped. **DF:** low, originates near dorsal tail-body junction, tip pointed. **BS:** oblong/depressed. **CP:** body brown, tail faintly striped. **TL:** 22/26. **CI:** Lamotte et al. (1959b).

Heterixalus: **CO:** 7. **GR:** Madagascar. **EMG:** exotroph: lentic: benthic. **LTRF:** 1/3(1). **OA:** typical: anteroventral. **MP:** medium dorsal gap: uniserial. **SUP:** few laterally and row inside marginal papillae of lower labium. **DEM:** absent. **NA:** nearer snout than eye. **VT:** medial. **EP:** lateral. **SP:** sinistral. **UJ:** medium, wide, smooth arc, coarsely serrate. **LJ:** medium, open V-shaped. **DF:** low, originates near dorsal tail-body junction, tip rounded. **BS:** oval/compressed. **CP:** uniform body, blotched tail. **TL:** 35/36. **CI:** Arnoult and Razarihelisoa (1967); Blommers-Schlösser (1982); Blommers-Schlösser and Blanc (1991).

Hyperolius: **CO:** 116. **GR:** sub-Saharan Africa. **EMG:** exotroph: lentic: benthic. **LTRF:** 1/3[1], 2/3. **OA:** typical: anteroventral. **MP:** wide dorsal gap: uni- and biserial, ventral papillae sometimes elongate. **SUP:** absent. **DEM:** absent. **NA:** aperture margin sometimes papillate. **VT:** dextral. **EP:** lateral. **SP:** sinistral. **UJ:** wide, smooth arc or slight medial convexity, serrations coarse. **LJ:** wide, open U-shaped. **DF:** low, originates near dorsal tail-body junction, tip pointed. **BS:** oval/depressed. **CP:** variable: unicolored or striped tail, tail tip black, fin margins dark. **TL:** 30–50/36. **NO:** P-3 usually much shorter than P-2. **CI:** Amiet (1979); Channing and De Capona (1987); Lamotte and Perret (1963c); Schiøtz (1967, 1975).

Kassinula: **CO:** 1. **GR:** Zambia–Zaire. Tadpole unknown.
Nesionixalus: **CO:** 2. **GR:** Sao Tomé. Tadpole unknown.

Kassininae: CO: 5/21. **GR:** sub-Saharan Africa.

Kassina: **CO:** 13. **GR:** sub-Saharan Africa. **EMG:** exotroph: lentic: nektonic. **LTRF:** 1/2(1) or 1/3. **OA:** typical: anteroventral. **MP:** dorsal gap: wide; ventral gap: narrow or absent. **SUP:** numerous laterally. **DEM:** absent. **NA:** closer to snout than eye. **VT:** medial. **EP:** lateral. **SP:** sinistral. **UJ:** massive, smooth but prominent arc. **LJ:** massive robust, very widely V-shaped, at least medially. **DF:** very high, originates well anterior to dorsal tail-body junction, tip pointed. **BS:** oval/compressed. **CP:** clear parts of tail fins often red or orange. **TL:** to 83/36. **NO:** keratinized area on lower disc below jaw sheath sometimes present. **CI:** Largen (1975); Schiøtz (1975); Wager (1986).

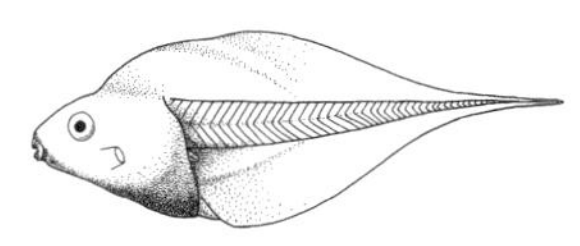
Kassina

Opisthothylax: **CO:** 1. **GR:** Cameroon–Zaire. **EMG:** exotroph: lotic: benthic. **LTRF:** 0/0. **OA:** typical: reduced. **MP:** wide dorsal gap: uniserial and large. **SUP:** absent. **DEM:** absent. **NA:** tiny, nearer the snout than the eye. **VT:** dextral. **EP:** dorsal. **SP:** sinistral. **UJ:** very narrow, smooth arc. **LJ:** wider than upper, open V-shaped. **DF:** low, originates near dorsal tail-body junction, tip slightly pointed. **BS:** oval/depressed. **CP:** body brown, fins and tail muscle mottled; throat dark. **TL:** 26/36. **CI:** Amiet (1974a).

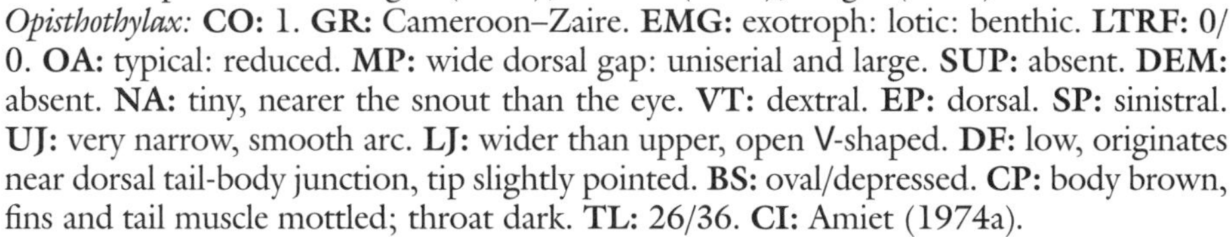

Paracassina: **CO:** 2. **GR:** Ethiopia. **EMG:** exotroph: lentic: nektonic. **LTRF:** 1/3(2). **OA:** typical: anteroventral. **MP:** wide dorsal gap, very narrow ventral gap where P-3 positioned. **SUP:** many laterally. **DEM:** absent. **NA:** –. **VT:** medial. **EP:** lateral. **SP:** sinistral. **UJ:** massive, serrated edge describes smooth arc. **LJ:** massive, open V-shaped. **DF:** tall, highly arched, originates well anterior of the dorsal tail-body junction, tip pointed. **BS:** oval/compressed. **CP:** uniformly dark on body, tail dark in distal third. **TL:** 54/36. **NO:** keratinized area posterolateral to lower jaw sheath. **CI:** Largen (1975).

Phlyctimantis: **CO:** 4. **GR:** Ivory Coast–Tanzania. **EMG:** exotroph: lentic: benthic. **LTRF:** 1/3[1–2]. **OA:** typical: anteroventral. **MP:** wide dorsal gap: biserial. **SUP:** numerous throughout extent of marginal papillae. **DEM:** absent. **NA:** nearer snout than eye. **VT:** medial. **EP:** dorsal. **SP:** sinistral. **UJ:** wide, serrated margin describes a fairly uniform arc. **LJ:** medium, open U-shaped. **DF:** low, originates near dorsal tail-body junction, tip pointed. **BS:** oval/depressed. **CP:** uniformly pale with blotches. **TL:** 49/34. **NO:** P-3 exceptionally short. **CI:** Guibé and Lamotte (1958); Schiøtz (1975).

Semnodactylus: **CO:** 1. **GR:** South Africa. **EMG:** exotroph: lentic: nektonic. **LTRF:** 1/3(1). **OA:** typical: anteroventral. **MP:** dorsal gap: biserial. **SUP:** few laterally. **DEM:** absent. **NA:** nearer snout than eye. **VT:** dextral. **EP:** lateral. **SP:** sinistral. **UJ:** wide, coarsely serrate. **LJ:** massive, open U-shaped. **DF:** high, originates on body, tip pointed. **BS:** oval/compressed. **CP:** distinct dark stripe on tail muscle. **TL:** 58/36. **NO:** P-3 exceptionally short. **CI:** Wager (1986).

Leptopelinae: CO: 1/49. **GR:** sub-Saharan Africa. **EMG:** exotroph: lentic: benthic. **LTRF:** 3–5/3[1], usually 2–n rows on upper labium broken medially. **OA:** typical: anteroventral. **MP:** wide dorsal gap: uniserial dorsally, biserial ventrally. **SUP:** absent. **DEM:** absent. **NA:** small, nearer snout than eye. **VT:** dextral. **EP:** dorsal. **SP:** sinistral. **UJ:** wide, medial indentation. **LJ:** wide, V-shaped. **DF:** low, originates near dorsal tail-body junction, tip pointed. **BS:** oval/depressed. **CP:** uniformly dark. **TL:** 49/36.

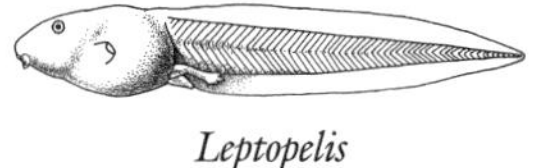
Leptopelis

Leptopelis: **CO:** 49. **CI:** Drewes et al. (1989); Lamotte and Perret (1961b); Lamotte et al. (1959b); Schiøtz (1967, 1975).

Tachycneminae: CO: 1/1. **GR:** Seychelle Islands. **EMG:** exotroph: lotic: benthic. **LTRF:** 3(2–4)/3. **OA:** typical: anteroventral. **MP:** wide dorsal gap: uniserial midventrally. **SUP:** –. **DEM:** absent. **NA:** very small, nearer to snout than eye. **VT:** dextral. **EP:** dorsal. **SP:** sinistral. **UJ:** wide, broad arch. **LJ:** medium, V-shaped. **DF:** low, originates distal to body. **BS:** oval/depressed. **CP:** uniformly pale brown, venter clear in preservative. **TL:** 44/27.

Tachycnemis: **CO:** 1. **GR:** Seychelle Islands. **CI:** specimens.

LEIOPELMATIDAE: CO: 1/3 **GR:** New Zealand. **EMG:** endotroph.

Leiopelma: **CO:** 3. **GR:** New Zealand. **EMG:** endotroph.

LEPTODACTYLIDAE: CO: 48/850. **GR:** southern United States–South America–Caribbean. **EMG:** exotroph: lentic: various guilds; lotic: various guilds; endotroph. **LTRF:** many variations between 0/0 and 9/9, 2/3 most common. **OA:** typical: anteroventral or all keratinized mouthparts absent. **MP:** medium to wide dorsal gap, complete, narrow ventral gap: uni- and biserial. **SUP:** absent, few laterally or dorsolaterally. **DEM:** absent, lateral. **NA:** nearer snout than eye or reverse. **VT:** medial, dextral or sinistral. **EP:** dorsal. **SP:** sinistral except in *Lepidobatrachus* (dual/lateral). **UJ:** variable, narrow to wide, usually a smooth arc, incised or not. **LJ:** variable, narrow to wide, U to V-shaped. **DF:** low, originates near, slightly anterior to or well posterior to the dorsal tail-body junction, tip rounded to pointed. **BS:** round, oval to oblong/depressed. **CP:** variable but uniformly dark is common. **TL:** 15–130/36.

Ceratophryinae: CO: 6/33. **GR:** South America.

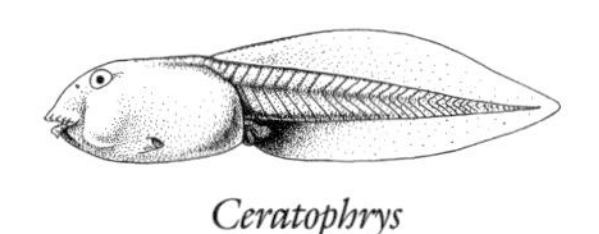

Ceratophrys

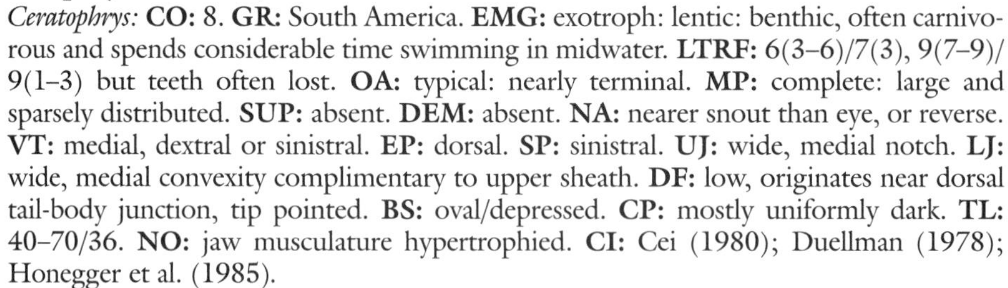

Ceratophrys: **CO:** 8. **GR:** South America. **EMG:** exotroph: lentic: benthic, often carnivorous and spends considerable time swimming in midwater. **LTRF:** 6(3–6)/7(3), 9(7–9)/9(1–3) but teeth often lost. **OA:** typical: nearly terminal. **MP:** complete: large and sparsely distributed. **SUP:** absent. **DEM:** absent. **NA:** nearer snout than eye, or reverse. **VT:** medial, dextral or sinistral. **EP:** dorsal. **SP:** sinistral. **UJ:** wide, medial notch. **LJ:** wide, medial convexity complimentary to upper sheath. **DF:** low, originates near dorsal tail-body junction, tip pointed. **BS:** oval/depressed. **CP:** mostly uniformly dark. **TL:** 40–70/36. **NO:** jaw musculature hypertrophied. **CI:** Cei (1980); Duellman (1978); Honegger et al. (1985).

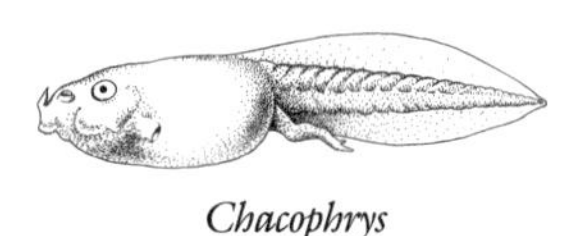

Chacophrys

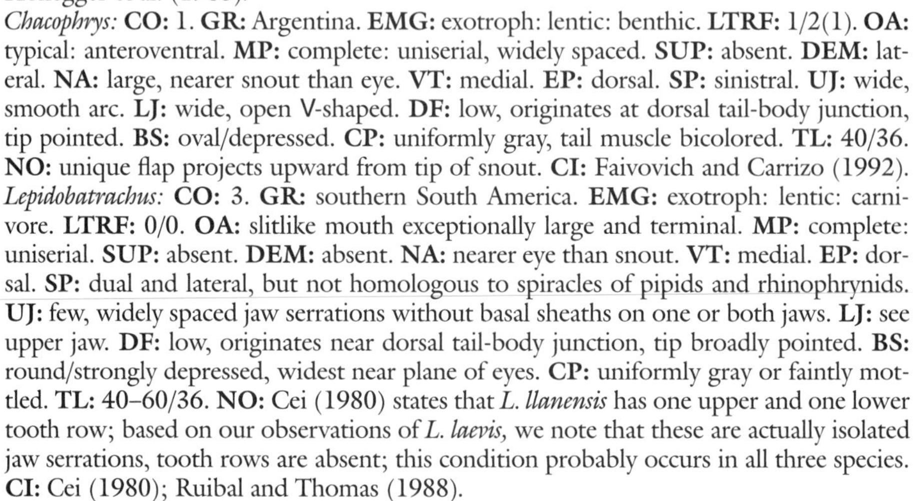

Chacophrys: **CO:** 1. **GR:** Argentina. **EMG:** exotroph: lentic: benthic. **LTRF:** 1/2(1). **OA:** typical: anteroventral. **MP:** complete: uniserial, widely spaced. **SUP:** absent. **DEM:** lateral. **NA:** large, nearer snout than eye. **VT:** medial. **EP:** dorsal. **SP:** sinistral. **UJ:** wide, smooth arc. **LJ:** wide, open V-shaped. **DF:** low, originates at dorsal tail-body junction, tip pointed. **BS:** oval/depressed. **CP:** uniformly gray, tail muscle bicolored. **TL:** 40/36. **NO:** unique flap projects upward from tip of snout. **CI:** Faivovich and Carrizo (1992).

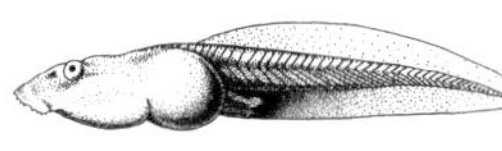

Lepidobatrachus

Lepidobatrachus: **CO:** 3. **GR:** southern South America. **EMG:** exotroph: lentic: carnivore. **LTRF:** 0/0. **OA:** slitlike mouth exceptionally large and terminal. **MP:** complete: uniserial. **SUP:** absent. **DEM:** absent. **NA:** nearer eye than snout. **VT:** medial. **EP:** dorsal. **SP:** dual and lateral, but not homologous to spiracles of pipids and rhinophrynids. **UJ:** few, widely spaced jaw serrations without basal sheaths on one or both jaws. **LJ:** see upper jaw. **DF:** low, originates near dorsal tail-body junction, tip broadly pointed. **BS:** round/strongly depressed, widest near plane of eyes. **CP:** uniformly gray or faintly mottled. **TL:** 40–60/36. **NO:** Cei (1980) states that *L. llanensis* has one upper and one lower tooth row; based on our observations of *L. laevis,* we note that these are actually isolated jaw serrations, tooth rows are absent; this condition probably occurs in all three species. **CI:** Cei (1980); Ruibal and Thomas (1988).

Macrogenioglottus: **CO:** 1. **GR:** Brazil. **EMG:** exotroph: lentic: benthic. **LTRF:** 2(2)/3(1). **OA:** typical: anteroventral. **MP:** wide dorsal gap: uniserial. **SUP:** absent. **DEM:** lateral. **NA:** equidistant between eye and snout. **VT:** dextral. **EP:** dorsal. **SP:** sinistral. **UJ:** very narrow, wide, smooth arc, serrations lacking. **LJ:** very narrow, open U-shaped, serrations laterally. **DF:** low, originates at dorsal tail-body junction, tip pointed. **BS:** oblong/depressed. **CP:** brownish-gray, reticulated with nonpigmented areas, dorsum of tail muscle banded, edges of fins with black areas. **TL:** 35/30. **CI:** Abravaya and Jackson (1978).

Odontophrynus: **CO:** 8. **GR:** eastern and southern South America. **EMG:** exotroph: lentic: benthic, lotic: benthic. **LTRF:** 2(2)/3[1]. **OA:** typical: anteroventral. **MP:** medium dorsal gap: uniserial. **SUP:** absent. **DEM:** absent. **NA:** nearer eye than snout. **VT:** dextral. **EP:** dorsal. **SP:** sinistral. **UJ:** wide, smooth arc. **LJ:** wide, open U-shaped. **DF:** low originates near dorsal tail-body junction, tip pointed. **BS:** oval/depressed. **CP:** body dark, tail uniform to blotched. **TL:** 35–45/36. **CI:** Caramaschi (1979); Cei (1980).

Proceratophrys: **CO:** 12. **GR:** Brazil–Paraguay. **EMG:** exotroph: lentic: benthic. **LTRF:** 2/3(1), 2(2)/3(1). **OA:** typical: anteroventral. **MP:** wide dorsal gap: uniserial. **SUP:** absent, few anterolateral. **DEM:** lateral, two ventrolateral. **NA:** large, not round, nearer eye than snout. **VT:** dextral. **EP:** dorsal. **SP:** sinistral. **UJ:** wide, smooth arc. **LJ:** wide, open V-shaped. **DF:** low, originates near tail-body junction, tip rounded. **BS:** oval/depressed. **CP:** tail muscle weakly banded, body and fins speckled. **TL:** 33/36. **CI:** Da Cruz and Peixoto (1980b); Izecksohn et al. (1979).

Hylodinae: CO: 3/28. **GR:** eastern and southern South America.

Crossodactylus: **CO:** 9. **GR:** southern South America. **EMG:** exotroph: lotic: benthic.

LTRF: 3(1,3)/5, 2(2)/3[1]. **OA:** typical: anteroventral. **MP:** medium dorsal gap: uniserial. **SUP:** absent. **DEM:** absent. **NA:** nearer eye than snout. **VT:** dextral. **EP:** dorsal. **SP:** sinistral. **UJ:** wide, uniform arc and coarsely serrate. **LJ:** wide, open U-shaped, coarsely serrate. **DF:** low, originates near dorsal tail-body junction, tip rounded. **BS:** oval/depressed. **CP:** body uniformly dark, tail and fins with dark specks. **TL:** 52/37. **CI:** Caramaschi and Kisteumacher (1989); Caramaschi and Sazima (1985); Cei (1980).

Hylodes: **CO:** 15. **GR:** eastern South America. **EMG:** exotroph: lotic: benthic. **LTRF:** 2(2)/3[1], 2(2)/4. **OA:** typical: anteroventral. **MP:** medium dorsal gap: uniserial. **SUP:** absent, few laterally. **DEM:** lateral. **NA:** nearer eye than snout. **VT:** dextral. **EP:** dorsal. **SP:** sinistral. **UJ:** wide, smooth arc but coarsely serrate, medial serration hypertrophied. **LJ:** wide, open U-shaped, coarsely serrate. **DF:** low, originates near dorsal tail-body junction, tip pointed. **BS:** oval/depressed. **CP:** uniformly dark. **TL:** 36/25. **CI:** Bokermann (1966); Heyer et al. (1990); Sazima and Bokermann (1982).

Megaelosia: **CO:** 4. **GR:** Brazil. **EMG:** exotroph: lentic: benthic. **LTRF:** 2(2)/3(1). **OA:** typical: anteroventral. **MP:** wide dorsal gap: small and uniserial. **SUP:** few laterally near emargination and below P-3. **DEM:** lateral. **NA:** small, nearer snout than eye. **VT:** dextral. **EP:** dorsal. **SP:** sinistral. **UJ:** wide, edge describes smooth arc. **LJ:** wide, V-shaped. **DF:** low, originates near dorsal tail-body junction, tip pointed. **BS:** oval/depressed. **CP:** uniformly dark. **TL:** 128/25. **CI:** Giarretta et al. (1993); Heyer et al. (1990).

Leptodactylinae: CO: 10/121. **GR:** southern Texas–Chile–Antilles.

Adenomera: **CO:** 6. **GR:** South America east of Andes. **EMG:** endotroph. **NO:** De la Riva (1995), based on observations in Bolivia, noted that all species of *Adenomera* are not entirely endotrophic; if the larvae of some forms gain access to free water, they can develop as typical lentic-inhabiting, leptodactylid tadpoles and resemble the tadpoles of some *Leptodactylus* spp., LTRF 2(2)/3(1). **CI:** De la Riva (1995); Heyer and Silverstone (1969).

Edalorhina: **CO:** 2. **GR:** Colombia–Peru. **EMG:** exotroph: lentic: benthic. **LTRF:** 2(2)/3(1). **OA:** typical: anteroventral. **MP:** dorsal gap: biserial. **SUP:** few dorsolaterally. **DEM:** absent. **NA:** equidistant between eye and snout. **VT:** dextral. **EP:** dorsal. **SP:** sinistral. **UJ:** narrow, wide uniform arc. **LJ:** medium, open U-shaped. **DF:** low, originates near dorsal tail-body junction, tip rounded. **BS:** oval/depressed. **CP:** uniformly pale brown. **TL:** 14/25. **CI:** De la Riva (1995); Schlüter (1990).

Hydrolaetare: **CO:** 1. **GR:** Colombia–Peru. Tadpole unknown.

Leptodactylus: **CO:** 50. **GR:** southern Texas–Mexico–South America–West Indies. **EMG:** exotroph: lentic: benthic. **LTRF:** 2/3, 2(2)/3[1]. **OA:** typical: anteroventral. **MP:** medium to wide dorsal gap: uniserial, narrow ventral gap rarely. **SUP:** absent. **DEM:** lateral, absent. **NA:** nearer snout than eye. **VT:** medial. **EP:** dorsal. **SP:** sinistral. **UJ:** narrow to medium, wide smooth arc. **LJ:** narrow to medium, U- or V-shaped. **DF:** low, originates near dorsal tail-body junction, tip rounded to pointed. **BS:** oval/depressed. **CP:** often uniformly dark. **TL:** 25–60/36. **NO:** tadpoles of 4–5 species groups can be recognized by features more subtle than presented here. **CI:** Cei (1980); Heyer (1973a); Sazima and Bokermann (1978).

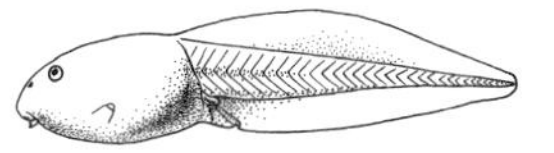
Leptodactylus; L. melanonotus

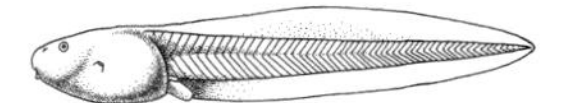
Leptodactylus; L. pentadactylus

Lithodytes: **CO:** 1. **GR:** Venezuela–Ecuador–Peru. **EMG:** exotroph: lentic: benthic. **LTRF:** 2(2)/3(1). **OA:** typical: anteroventral. **MP:** wide dorsal gap. **SUP:** absent. **DEM:** absent. **NA:** nearer snout than eye. **VT:** medial. **EP:** dorsal. **SP:** sinistral. **UJ:** medium, smooth arc. **LJ:** narrow, wide, open U-shaped. **DF:** low, originates near dorsal tail-body junction, tip rounded. **BS:** oval/depressed. **CP:** uniformly dark. **TL:**47/34. **NO:** large lower labium extends far beyond lower tooth row. **CI:** Lamar and Wild (1995); Regös and Schlüter (1984).

Paratelmatobius: **CO:** 3. **GR:** Brazil. **EMG:** exotroph: lotic or lentic: benthic. **LTRF:** 2(2)/3(1). **OA:** typical: anteroventral. **MP:** wide dorsal gap: uniserial midventrally. **SUP:** absent. **DEM:** lateral, absent. **NA:** nearer eye than snout. **VT:** dextral, medial. **EP:** dorsal. **SP:** sinistral. **UJ:** narrow, edge a smooth arc. **LJ:** medium, open U-shaped. **DF:** low, originates near dorsal tail-body junction, tip bluntly pointed. **BS:** oval/depressed. **CP:** uniformly pale brownish, belly lightly pigmented, fins nonpigmented to brownish, tail muscle blotched. **TL:** 26–32/37. **CI:** Cardoso and Haddad (1990); Giarretta and Castanho (1990).

Physalaemus: **CO:** 37. **GR:** Mexico–South America. **EMG:** exotroph: lentic: benthic.

LTRF: 2(2)/2, 2/3(1), 2(2)/3(1). **OA:** typical: anteroventral. **MP:** wide dorsal gap: uniserial, sometimes ventral gap. **SUP:** few laterally. **DEM:** lateral. **NA:** equidistant between snout and eye. **VT:** dextral. **EP:** dorsal. **SP:** sinistral. **UJ:** wide, smooth arc. **LJ:** medium, open V-shaped. **DF:** low, originates near dorsal tail-body junction, tip pointed. **BS:** oval/depressed. **CP:** uniformly dark, mottled. **TL:** 23/38. **NO:** some (e.g., *P. petersi*) with prominent, white glands on body. **CI:** Barrio (1953); Heyer et al. (1990).

Pleurodema: **CO:** 12. **GR:** southern South America. **EMG:** exotroph: lentic: benthic. **LTRF:** 2(2)/2, 2(2)/3[1]. **OA:** typical: anteroventral. **MP:** wide dorsal gap: uniserial, ventral gap sometimes. **SUP:** rare laterally. **DEM:** lateral. **NA:** nearer eye than snout. **VT:** medial. **EP:** dorsal. **SP:** sinistral. **UJ:** medium, smooth arc. **LJ:** medium, U-shaped. **DF:** low, originates near dorsal tail-body junction, tip rounded. **BS:** oval/depressed. **CP:** uniformly dark, tail muscle mottled. **TL:** 25/36. **CI:** Cei (1980); Leon-Ochoa and Donoso-Barros (1970); Peixoto (1982).

Pseudopaludicola: **CO:** 8. **GR:** southern South America. **EMG:** exotroph: lentic: benthic. **LTRF:** 2(2)/3. **OA:** typical: anteroventral. **MP:** wide dorsal and narrow ventral gaps: uniserial. **SUP:** row(s) ventrolaterally. **DEM:** lateral. **NA:** nearer snout than eye. **VT:** dextral. **EP:** dorsal. **SP:** sinistral. **UJ:** medium, wide smooth arc. **LJ:** medium, wide U-shaped. **DF:** low, originates near or anterior to dorsal tail-body junction, tip pointed. **BS:** oval/depressed. **CP:** uniformly dark. **TL:** 22/36. **CI:** Barrio (1953); Cei (1980).

Vanzolinius: **CO:** 1. **GR:** Ecuador–Peru. **EMG:** exotroph: lentic: benthic. **LTRF:** 2/3. **OA:** typical: anteroventral. **MP:** dorsal gap: uniserial. **SUP:** –. **DEM:** lateral. **NA:** nearer eye than snout. **VT:** medial. **EP:** dorsal. **SP:** sinistral. **UJ:** narrow. **LJ:** narrow. **DF:** low, originates near dorsal tail-body junction, tip rounded. **BS:** oval/depressed. **CP:** uniformly tan or brown. **TL:** 25/30. **CI:** Duellman (1978).

Telmatobiinae: CO: 29/668. **GR:** central to southern South America.

Adelophryne: **CO:** 2. **GR:** Guiana Shield. **EMG:** endotroph.

Alsodes: **CO:** 11. **GR:** Chile–Argentina. **EMG:** exotroph: lotic: benthic. **LTRF:** 2(2)/3(1). **OA:** typical: anteroventral. **MP:** wide dorsal gap: uniserial. **SUP:** few dorsolaterally and row(s) distal to P-3. **DEM:** lateral. **NA:** equidistant between eye and snout, nearer eye than snout. **VT:** dextral. **EP:** dorsal. **SP:** sinistral. **UJ:** wide, smooth arc. **LJ:** wide, open V-shaped. **DF:** low, originates distal to dorsal tail-body junction, tip broadly rounded to pointed. **BS:** oval/depressed. **CP:** uniformly dark, few small spots on tail. **TL:** 60–75/36. **CI:** Diaz and Valencia (1985).

Atelognathus: **CO:** 7. **GR:** Chile–Argentina. **EMG:** exotroph: lentic: benthic. **LTRF:** 2(2)/3(1). **OA:** typical: anteroventral. **MP:** medium dorsal gap: uniserial. **SUP:** few laterally. **DEM:** lateral. **NA:** equidistant from eye and snout. **VT:** dextral. **EP:** dorsal. **SP:** sinistral. **UJ:** wide, medial zone of serrate edge straight. **LJ:** wide, open V-shaped. **DF:** low, originates near dorsal tail-body junction, tip pointed. **BS:** oval/depressed. **CP:** brown, sparsely pigmented, belly transparent. **TL:** 57/36. **CI:** Cei (1964, 1980).

Atopophrynus: **CO:** 1. **GR:** Colombia. Tadpole unknown.

Barycholos: **CO:** 2. **GR:** Ecuador–Brazil. **EMG:** endotroph.

Batrachophrynus: **CO:** 2. **GR:** Peru–Bolivia. **EMG:** exotroph: lentic: benthic. **LTRF:** 2(2)/3(1). **OA:** typical: anteroventral. **MP:** wide dorsal gap: biserial. **SUP:** throughout extent of marginal papillae. **DEM:** absent. **NA:** –. **VT:** medial. **EP:** dorsal. **SP:** sinistral. **UJ:** wide, smooth arc. **LJ:** wide, open U-shaped. **DF:** low, originates at dorsal tail-body junction, tip broadly rounded. **BS:** oval/depressed. **CP:** uniformly dark, small spots on fins. **TL:** 67/26. **CI:** Sinsch (1990).

Batrachyla: **CO:** 3. **GR:** Chile–Argentina. **EMG:** exotroph: lentic: benthic. **LTRF:** 2(2)/3(1). **OA:** typical: anteroventral. **MP:** wide dorsal gap: uniserial. **SUP:** dorso- and ventrolaterally. **DEM:** lateral. **NA:** equidistant between eye and snout. **VT:** dextral. **EP:** dorsal. **SP:** sinistral. **UJ:** medium, smooth arc. **LJ:** medium, open V-shaped. **DF:** low, originates near dorsal tail-body junction, tip rounded. **BS:** oval/depressed. **CP:** uniformly dark, tail speckled. **TL:** 27–37/33. **NO:** early stages develop like nidicolous endotrophs. **CI:** Cei (1980); Cei and Capurro (1958); Formas and Pugín (1971).

Caudiverbera: **CO:** 1. **GR:** Chile. **EMG:** exotroph: lotic: benthic. **LTRF:** 2(2)/3(1). **OA:** typical: anteroventral. **MP:** medium dorsal gap: uniserial. **SUP:** few laterally. **DEM:** lateral. **NA:** nearer snout than eye. **VT:** dextral. **EP:** dorsal. **SP:** sinistral. **UJ:** narrow, edge describes smooth arc. **LJ:** narrow, open U-shaped. **DF:** low, originates at dorsal tail-body junction, tip pointed. **BS:** oval/depressed. **CP:** pale green in life, distal third of tail dark. **TL:** 95/37. **NO:** Cei (1962) shows a LTRF of 3(2)/3(1) and P-2 and P-3 with nonmedial gaps. **CI:** Diaz and Valencia (1985); Jorquera and Izquierdo (1964).

Crossodactylodes: **CO:** 3. **GR:** Brazil. **EMG:** exotroph: lentic: arboreal. **LTRF:** 2(2)/3. **OA:** typical: anteroventral. **MP:** wide dorsal gap: uniserial. **SUP:** absent. **DEM:** lateral. **NA:** nearer snout than eye. **VT:** dextral. **EP:** dorsal. **SP:** sinistral. **UJ:** medium, small but prominent medial convexity, basal margin also with convexity. **LJ:** medium, open V-shaped. **DF:** low, originates near dorsal tail-body junction, tip round. **BS:** oblong/depressed. **CP:** uniformly dark. **TL:** 19/35. **CI:** Peixoto (1981).

Cycloramphus: **CO:** 25. **GR:** Brazil. **EMG:** exotroph: lotic: semiterrestrial; endotroph. **LTRF:** 2/3, 2(1–2)/3(1), 2/3(1), 2(1)/3(1). **OA:** typical: anteroventral, ventral. **MP:** wide dorsal gap: uniserial. **SUP:** absent. **DEM:** absent. **NA:** see notes. **VT:** medial. **EP:** dorsal. **SP:** sinistral. **UJ:** medium, strongly arched. **LJ:** medium, strongly arched. **DF:** low, originates well posterior of dorsal tail-body junction, tip pointed. **BS:** oblong to oval/depressed. **CP:** uniform to mottled, tail muscle may be slightly banded. **TL:** 20–40/36. **NO:** Heyer et al. (1990) reported nares absent in *C. boraceiensis* but seem apparent in figure; Heyer (1983a, b) did not locate a spiracle on *C. boraceiensis* or *C. valae;* may have parts of abdominal wall expanded as a free flap. **CI:** Heyer (1983a, b).

Dischidodactylus: **CO:** 2. **GR:** Venezuela. **EMG:** endotroph.

Eleutherodactylus: **CO:** 515. **GR:** North, Central, and South America–Caribbean. **EMG:** endotroph.

Euparkerella: **CO:** 4. **GR:** Brazil. **EMG:** endotroph.

Eupsophus: **CO:** 8. **GR:** Chile–Argentina. **EMG:** endotroph.

Geobatrachus: **CO:** 1. **GR:** Colombia. **EMG:** endotroph.

Holoaden: **CO:** 2. **GR:** Brazil. **EMG:** endotroph.

Hylorina: **CO:** 1. **GR:** Chile–Argentina. **EMG:** exotroph: lentic: benthic. **LTRF:** 2(2)/3(1). **OA:** typical: anteroventral. **MP:** wide dorsal gap: uniserial. **SUP:** few centripetally arranged throughout extent of marginal papillae. **DEM:** lateral. **NA:** equidistant between snout and eyes. **VT:** dextral. **EP:** dorsal. **SP:** sinistral. **UJ:** wide, serrated edge almost transversely straight. **LJ:** wide, open U-shaped. **DF:** somewhat arched, originates near dorsal tail-body junction, tip broadly rounded. **BS:** oval/depressed. **CP:** uniform with dark spots throughout tail. **TL:** 51/38. **CI:** Barrio (1976); Formas and Pugín (1978).

Insuetophrynus: **CO:** 1. **GR:** Chile. **EMG:** exotroph: lotic: benthic. **LTRF:** 2(2)/2(1–2) or less. **OA:** typical: ventral, jaw sheaths and labial tooth rows reduced. **MP:** medium dorsal gap: uniserial, papillae very small. **SUP:** absent. **DEM:** absent. **NA:** small, nearer eye than snout. **VT:** dextral. **EP:** dorsal. **SP:** sinistral. **UJ:** small, weakly developed but serrate. **LJ:** similar and U-shaped. **DF:** low, originates at dorsal tail-body junction, tip broadly rounded. **BS:** oval/depressed. **CP:** greenish brown in life, irregularly mottled. **TL:** 61/37. **NO:** lower tooth rows sometimes absent, one drawing of oral variation shows complete marginal papillae, lateral line visible. **CI:** Diaz and Valencia (1985); Formas et al. (1980).

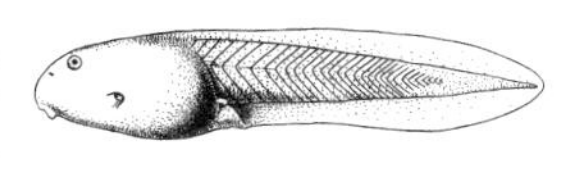

Insuetophrynus

Ischnocnema: **CO:** 3. **GR:** Amazonia–Brazil. **EMG:** endotroph.

Limnomedusa: **CO:** 1. **GR:** Brazil–Argentina. **EMG:** exotroph:. **LTRF:** 2(2)/3(1). **OA:** typical: anteroventral. **MP:** medium dorsal gap: uniserial. **SUP:** numerous along all marginal papillae. **DEM:** lateral. **NA:** nearer eye than snout. **VT:** medial. **EP:** dorsal. **SP:** sinistral. **UJ:** wide, uniform arc. **LJ:** medium, open U-shaped. **DF:** low, originates near dorsal tail-body junction, tip pointed. **BS:** oval/depressed. **CP:** body uniform, tail boldly spotted. **TL:** 42/34. **CI:** Cei 1980; Gehrau and De Sá (1980).

Phrynopus: **CO:** 20. **GR:** Andean South America. **EMG:** endotroph.

Phyllonastes: **CO:** 4. **GR:** Peru–Ecuador. **EMG:** endotroph.

Phyzelaphryne: **CO:** 1. **GR:** Brazil. **EMG:** endotroph.

Scythrophrys: **CO:** 1. **GR:** Brazil. Tadpole unknown.

Somuncuria: **CO:** 1. **GR:** Argentina. **EMG:** exotroph: lotic: benthic. **LTRF:** 2(2)/3(1). **OA:** typical: anteroventral. **MP:** wide dorsal and smaller ventral gaps: biserial. **SUP:** absent. **DEM:** lateral. **NA:** equidistant between eye and snout. **VT:** dextral. **EP:** dorsal. **SP:** sinistral. **UJ:** wide, smooth arc, or perhaps a slight medial concavity in margin. **LJ:** wide, open U-shaped. **DF:** low, originates near dorsal tail-body junction, tip rounded. **BS:** oval/depressed. **CP:** body dark, tail blotched. **TL:** 46/36. **CI:** Cei (1980).

Telmatobius: **CO:** 35. **GR:** Ecuador–Argentina. **EMG:** exotroph: lentic: benthic; lotic: benthic. **LTRF:** 2(2)/3(1). **OA:** typical: anteroventral. **MP:** wide dorsal gap: uni- and biserial. **SUP:** few dorsolaterally, laterally and distal to P-3. **DEM:** lateral. **NA:** equidistant between eye and snout. **VT:** dextral. **EP:** dorsal. **SP:** sinistral. **UJ:** wide, edge a smooth arc. **LJ:** wide, open U-shaped. **DF:** low, originates near dorsal tail-body junction, tip rounded to pointed. **BS:** oval/depressed. **CP:** uniform to muted mottling. **TL:** 60–100/37. **CI:** Diaz and Valencia (1985); Lavilla (1984a, 1985, 1988).

Telmatobufo: **CO:** 3. **GR:** Chile. **EMG:** exotroph: lotic: suctorial. **LTRF:** 2/3(1). **OA:** typical: ventral. **MP:** complete: uniserial. **SUP:** multiple rows centripetally to all marginal papillae except at lateral margins of disc. **DEM:** lateral. **NA:** nearer eye than snout. **VT:** dextral, vent flap underlies tube. **EP:** dorsal. **SP:** sinistral. **UJ:** medium, open V-shaped. **LJ:** medium, open V-shaped, "internal borders smooth" (Diaz et al. 1983). **DF:** low, originates well posterior of dorsal tail-body junction, tip broadly pointed. **BS:** oval/depressed, may narrow appreciably posterior to oral disc. **CP:** dark brown, orange protuberances over body dorsum. **TL:** 107/40. **CI:** Diaz et al. (1983); Formas (1972).

Thoropa

Thoropa: **CO:** 5. **GR:** Brazil. **EMG:** exotroph: lotic: semiterrestrial. **LTRF:** 2(1–2)/3(1). **OA:** typical: ventral. **MP:** wide dorsal gap: uniserial. **SUP:** absent. **DEM:** absent. **NA:** nearer snout than eye. **VT:** medial. **EP:** dorsal. **SP:** sinistral. **UJ:** medium, strongly arched and coarsely serrate. **LJ:** wide, strongly arched. **DF:** low, originates on tail muscle well posterior to dorsal tail-body junction, almost absent in some, tip rounded. **BS:** oblong/depressed. **CP:** uniform dark to mottled; tail muscle banded. **TL:** 20–30/36. **NO:** some have parts of lateral margins of abdominal wall expanded as a free flap; nidicolous endotrophs may be involved (Bokermann 1965). **CI:** Caramaschi and Sazima (1984).

Zachaenus: **CO:** 3. **GR:** Brazil–Argentina. **EMG:** endotroph.

MEGOPHRYIDAE: CO: 9/79. **GR:** northern India–China–Philippines. **EMG:** exotroph: lotic: clasping, neustonic, benthic; endotroph. **LTRF:** 0/0 to 7/7. **OA:** typical: anteroventral, ventral; umbelliform: oral disc bi-triangular. **MP:** narrow dorsal gap, complete, absent: uniserial. **SUP:** scattered near mouth, few laterally, some arranged radially to mouth. **DEM:** absent, dorsal and ventral. **NA:** equidistant between eye and snout, nearer eye than snout or reverse. **VT:** dextral, medial. **EP:** dorsal, lateral. **SP:** sinistral but well below longitudinal axis. **UJ:** absent, medium to wide, small medial convexity or not. **LJ:** absent, medium to wide, U- or V-shaped. **DF:** low, originates near or posterior to dorsal tail-body junction, tip rounded to pointed. **BS:** oval to elongate/depressed. **CP:** muted mottling to unicolored. **TL:** 25–75/36.

Atympanophrys: **CO:** 1. **GR:** China. Tadpole unknown.

Brachytarsophrys: **CO:** 1. **GR:** Burma–China–Thailand. Tadpole unknown.

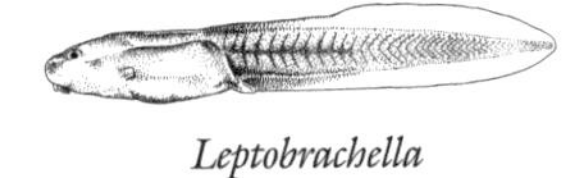
Leptobrachella

Leptobrachella: **CO:** 7. **GR:** Borneo–Natuna Islands. **EMG:** exotroph: lotic: benthic. **LTRF:** 0/0. **OA:** typical: ventral. **MP:** narrow dorsal gap: uniserial. **SUP:** scattered near mouth. **DEM:** dorsal and ventral. **NA:** nearer snout than eye, one medial projection. **VT:** dextral. **EP:** dorsal. **SP:** sinistral but well below longitudinal axis. **UJ:** medium, small medial convexity. **LJ:** medium, open U-shaped. **DF:** low, especially in proximal third, originates near dorsal tail-body junction, tip round. **BS:** oblong/depressed. **CP:** uniformly dark. **TL:** 30–45/36. **CI:** Inger (1983, 1985).

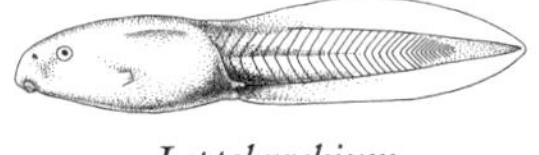
Leptobrachium

Leptobrachium: **CO:** 10. **GR:** Indonesia–China–Philippines. **EMG:** exotroph: lotic: benthic. **LTRF:** 3(1–3)/4(1–3) to 7(2–6)/6(1–5). **OA:** typical: anteroventral to ventral. **MP:** complete: uniserial. **SUP:** few laterally. **DEM:** absent. **NA:** equidistant between eye and snout, 1–3 medial projections. **VT:** dextral. **EP:** dorsal. **SP:** sinistral but well below longitudinal axis. **UJ:** wide, smooth arc. **LJ:** wide, wide V-shaped. **DF:** low, originates near dorsal tail-body junction, tip pointed to rounded. **BS:** oval to oblong/depressed. **CP:** uniformly dark. **TL:** 30–50/36. **NO:** LTRF very variable, integumentary glands sometimes present, submarginal papillae commonly have labial teeth. **CI:** Berry (1972); Inger (1983, 1985).

Leptolalax: **CO:** 7. **GR:** Burma–China–Borneo. **EMG:** exotroph: lotic: suctorial, clasping. **LTRF:** 5(2–5)/3(1–2). **OA:** typical: ventral. **MP:** narrow dorsal gap or complete: uniserial. **SUP:** scattered above and below mouth, probably rudimentary tooth ridges. **DEM:** dorsal and ventral. **NA:** nearer snout than eye. **VT:** dextral. **EP:** dorsal. **SP:** sinistral but well below longitudinal axis. **UJ:** wide, smooth arc. **LJ:** wide, open U-shaped. **DF:** low, particularly in first fourth, originates distal to dorsal tail-body junction, tip rounded. **BS:** oblong/very depressed. **CP:** uniformly brown. **TL:** 25–50/36. **NO:** integumentary glands dorsolaterally. **CI:** Inger (1966).

Megophrys: **CO:** 25. **GR:** Southeast Asia–China–Philippines–Greater Sundas. **EMG:** exotroph: lotic: neustonic; endotroph. **LTRF:** 0/0. **OA:** umbelliform: bi-triangular. **MP:** slight crenulations to none. **SUP:** scattered throughout oral disc, some arranged radially to mouth. **DEM:** absent. **NA:** nearer snout then eye. **VT:** medial. **EP:** lateral. **SP:** sinistral but well below longitudinal axis. **UJ:** absent, minor keratinization. **LJ:** minor keratinization. **DF:** low, originates near or posterior to dorsal tail-body junction, tip pointed. **BS:** oblong/depressed. **CP:** belly striped in some. **TL:** 30–50/36. **CI:** Inger (1985); C.-C. Liu and Hu (1961); Ye and Fei (1996).

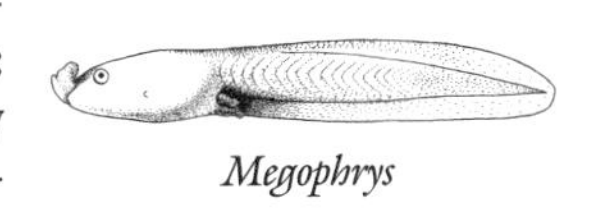

Megophrys

Ophryophryne: **CO:** 3. **GR:** Vietnam–China. Tadpole unknown.

Scutiger: **CO:** 24. **GR:** India–China. **EMG:** exotroph: lotic: benthic, clasping. **LTRF:** 5(2–5)/4(1–4), 5(2–5)/6(1–5), 6(2–6)/6(1–5), 7(2–7)/7(1–6). **OA:** typical: anteroventral. **MP:** narrow dorsal gap: uniserial, dorsal gap absent: uniserial. **SUP:** few laterally. **DEM:** absent. **NA:** nearer snout than eye. **VT:** dextral. **EP:** dorsal. **SP:** sinistral but well below longitudinal axis. **UJ:** very wide, smooth arc, very coarsely serrate. **LJ:** wide, wide U-shaped, coarsely serrate. **DF:** low, originates near dorsal tail-body junction, tip rounded. **BS:** oval/depressed. **CP:** uniformly dark. **TL:** 60–75/36. **NO:** C.-C. Liu (1950) showed labial teeth on submarginal papillae. **CI:** C.-C. Liu (1950).

Vibrissaphora: **CO:** 1. **GR:** southern China. **EMG:** exotroph: lotic. **LTRF:** 6(2–6)/4(1), 6(2–6), 6(1–5), A-1 exceptionally short. **OA:** typical: anteroventral. **MP:** narrow dorsal gap: uniserial. **SUP:** few ventrolaterally. **DEM:** absent. **NA:** nearer snout than eye. **VT:** –. **EP:** dorsal. **SP:** sinistral but well below longitudinal axis. **UJ:** very wide, narrow U-shaped, very coarsely serrate. **LJ:** very wide, wide U-shaped, coarsely serrate. **DF:** low, originates near dorsal tail-body junction, tip rounded. **BS:** oval/depressed. **CP:** uniformly dark body, fins with coarse blotches. **TL:** –. **CI:** C.-C. Liu and Hu (1961).

MICROHYLIDAE: CO: 65/327. **GR:** North and South America, Africa, India–Korea–Austropapuan region. **EMG:** exotroph: lentic: suspension feeder, psammonic; endotroph. **LTRF:** 0/0. **OA:** keratinized structures absent (except in *Otophryne* and *Scaphiophryne*); any resemblance to an oral disc absent (except in umbelliform *Microhyla, Nelsonophryne, Otophryne,* and *Scaphiophryne*), oral apparatus usually involves semicircular to straight-edged oral flaps pendant over terminal mouth, few species with umbelliform oral disc. **MP:** not applicable except in *Nelsonophryne* and some endotrophs. **SUP:** not applicable except perhaps in *Nelsonophryne.* **DEM:** not applicable. **NA:** usually absent until metamorphosis but positioned nearer snout than eye. **VT:** medial, dextral. **EP:** lateral, dorsal. **SP:** midventral on chest, abdomen or near vent, appears sinistral in different configurations in *Otophryne* and *Stereocyclops.* **UJ:** usually absent but *Scaphiophryne* has unpigmented jaw sheaths, *Otophryne* has hypertrophied serrations without a basal sheath. **LJ:** as upper jaw. **DF:** low, originates near dorsal tail-body junction, sometimes with flagellum. **BS:** round to oval/depressed. **CP:** dorsum uniform, sometimes with sagittal line, venter sometimes mottled or striped. **TL:** 15–35/36.

Asterophryinae: CO: 8/49. **GR:** New Guinea.

Asterophrys: **CO:** 1. **GR:** New Guinea. **EMG:** endotroph.
Barygenys: **CO:** 7. **GR:** New Guinea. **EMG:** endotroph.
Callulops: **CO:** 13. **GR:** New Guinea. **EMG:** endotroph.
Hylophorbus: **CO:** 1. **GR:** New Guinea. **EMG:** endotroph.
Mantophryne: **CO:** 3. **GR:** New Guinea. **EMG:** endotroph.
Pherohapsis: **CO:** 1. **GR:** New Guinea. **EMG:** endotroph.
Xenobatrachus: **CO:** 17. **GR:** New Guinea. **EMG:** endotroph.
Xenorhina: **CO:** 6. **GR:** New Guinea. **EMG:** endotroph.

Brevicipitinae: CO: 5/19. **GR:** eastern and southern Africa.

Balebreviceps: **CO:** 1. **GR:** Ethiopia. **EMG:** endotroph.
Breviceps: **CO:** 13. **GR:** southern Africa. **EMG:** endotroph.
Callulina: **CO:** 1. **GR:** Tanzania. **EMG:** endotroph.
Probreviceps: **CO:** 3. **GR:** Tanzania–Zimbabwe. **EMG:** endotroph.
Spelaeophryne: **CO:** 1. **GR:** Tanzania. **EMG:** endotroph.

Cophylinae: CO: 7/36. **GR:** Madagascar.

Anodonthyla: **CO:** 4. **GR:** Madagascar. **EMG:** endotroph.

Cophyla: **CO:** 1. **GR:** Madagascar. **EMG:** endotroph.
Madecassophryne: **CO:** 1. **GR:** Madagascar. **EMG:** endotroph.
Platypelis: **CO:** 9. **GR:** Madagascar. **EMG:** endotroph.
Plethdontohyla: **CO:** 14. **GR:** Madagascar. **EMG:** endotroph.
Rhombophryne: **CO:** 1. **GR:** Madagascar. **EMG:** endotroph.
Stumpffia: **CO:** 6. **GR:** Madagascar. **EMG:** endotroph.

Dyscophinae: CO: 2/9. **GR:** Madagascar and Southeast Asia.

Calluella: **CO:** 6. **GR:** Southeast Asia. **EMG:** exotroph: lentic: suspension feeder. **LTRF:** 0/0. **OA:** straight-edged upper lip without medial notch above terminal mouth. **MP:** absent **SUP:** absent. **DEM:** not applicable. **NA:** absent until metamorphosis **VT:** medial. **EP:** lateral. **SP:** midventral below gut **UJ:** absent **LJ:** absent **DF:** low, originates near dorsal tail-body junction, tip pointed. **BS:** oval/depressed. **CP:** pale brown, translucent, distal part of tail dark. **TL:** 24/31. **NO:** ventral fin with lobe near body, hangs in midwater at a 45° angle. **CI:** C.-C. Liu and Hu (1961); M. A. Smith (1917).

Dyscophus: **CO:** 3. **GR:** Madagascar. **EMG:** exotroph: lentic: suspension feeder. **LTRF:** 0/0. **OA:** straight-edged upper lip without medial notch above terminal mouth. **MP:** absent **SUP:** absent. **DEM:** not applicable. **NA:** absent until metamorphosis **VT:** medial. **EP:** lateral. **SP:** midventral below gut **UJ:** absent **LJ:** absent **DF:** low, originates near dorsal tail-body junction, tip pointed. **BS:** oval/depressed. **CP:** uniformly pale melanic pigment, sometimes more intense on distal part of tail. **TL:** 36/36. **CI:** Blommers-Schlösser (1975); Blommers-Schlösser and Blanc (1991); Glaw and Vences (1992); Pintar (1987).

Genyophryninae: CO: 8/89. **GR:** Australia–New Guinea–Philippines.

Albericus: **CO:** 3. **GR:** New Guinea. **EMG:** endotroph.
Aphantophryne: **CO:** 3. **GR:** New Guinea. **EMG:** endotroph.
Choerophryne: **CO:** 1. **GR:** New Guinea. **EMG:** endotroph.
Cophixalus: **CO:** 28. **GR:** New Guinea–northeastern Australia. **EMG:** endotroph.
Copiula: **CO:** 5. **GR:** New Guinea. **EMG:** endotroph.
Genyophryne: **CO:** 1. **GR:** New Guinea. **EMG:** endotroph.
Oreophryne: **CO:** 24. **GR:** Philippines–Celebes–New Guinea. **EMG:** endotroph.
Sphenophryne: **CO:** 24. **GR:** New Guinea–northeastern Australia. **EMG:** endotroph.

Melanobatrachinae: CO: 3/4. **GR:** India–Tanzania.

Hoplophryne: **CO:** 2. **GR:** Tanzania. **EMG:** exotroph: lentic: arboreal, probably spends considerable time out of water. **LTRF:** 0/0. **OA:** absent to transversely straight upper lip above terminal mouth. **MP:** not applicable. **SUP:** not applicable. **DEM:** not applicable. **NA:** absent. **VT:** medial. **EP:** dorsal. **VT:** medial. **SP:** midventral on belly. **UJ:** absent. **LJ:** absent. **DF:** low, originates near dorsal tail-body junction, tip pointed. **BS:** oval/depressed. **CP:** uniform pale pigmentation. **TL:** 30/36. **NO:** fingerlike projection at anterolateral extent of abdomen. **CI:** Noble (1929).

Melanobatrachus: **CO:** 1. **GR:** southern India. **EMG:** endotroph.
Parhoplophryne: **CO:** 1. **GR:** Tanzania. **EMG:** endotroph.

Microhylinae: CO: 29/106. **GR:** North and South America and Southeast Asia.

Adelastes: **CO:** 1. **GR:** Venezuela. **EMG:** endotroph.

Altigius: **CO:** 1. **GR:** Peru. **EMG:** exotroph: lentic: suspension feeder. **LTRF:** 0/0. **OA:** immense, paired, quadrangular oral flaps pendant over terminal mouth. **MP:** not applicable. **SUP:** not applicable. **DEM:** not applicable **NA:** absent. **VT:** dextral. **EP:** lateral, fully visible from below. **SP:** midventral, near vent. **UJ:** absent. **LJ:** absent **DF:** low, originates near dorsal tail-body junction, tip pointed. **CP:** generally dark with middorsal stripe. **TL:** 59/36. **CI:** Wild (1995).

Arcovomer: **CO:** 1. **GR:** Brazil. Tadpole unknown.

Chaperina: **CO:** 1. **GR:** Borneo. **EMG:** exotroph: lentic: suspension feeder. **LTRF:** 0/0. **OA:** small semicircular oral flaps with smooth edges pendant over terminal mouth. **MP:** not applicable **SUP:** not applicable. **DEM:** not applicable. **NA:** absent. **VT:** medial. **EP:** lateral. **SP:** midventral on abdomen. **UJ:** absent. **LJ:** absent. **DF:** low, originates near dorsal tail-body junction, tip pointed. **BS:** rounded/depressed. **CP:** uniformly black. **TL:** 25/42. **CI:** Inger (1956, 1966, 1985).

Chiasmocleis: **CO:** 15. **GR:** Panama–South America. **EMG:** exotroph: lentic: suspension feeder. **LTRF:** 0/0. **OA:** small semicircular oral flaps with smooth edges pendant over terminal mouth. **MP:** not applicable. **SUP:** not applicable. **DEM:** not applicable. **NA:** absent. **VT:** medial. **EP:** lateral. **SP:** midventral on lower abdomen. **UJ:** absent. **LJ:** absent. **DF:** low, originates at dorsal tail-body junction, tip pointed with flagellum. **BS:** rounded/depressed. **CP:** uniformly pale. **TL:** 17/36. **CI:** Schlüter and Salas (1991).

Ctenophryne: **CO:** 2. **GR:** Colombia–Brazil. **EMG:** exotroph: lentic: suspension feeder. **LTRF:** 0/0. **OA:** straight-edged upper lip without oral flaps above terminal mouth. **MP:** not applicable. **SUP:** not applicable. **DEM:** not applicable. **NA:** absent. **VT:** medial. **EP:** lateral. **SP:** midventral. **UJ:** absent. **LJ:** absent **DF:** low, originates at dorsal tail-body junction, tip round. **BS:** rounded/depressed. **CP:** uniformly pale. **TL:** 12/25. **CI:** Schlüter and Salas (1991).

Dasypops: **CO:** 1. **GR:** Brazil. **EMG:** exotroph: lentic: suspension feeder. **LTRF:** 0/0. **OA:** labial flaps reduced without medial notch. **MP:** not applicable. **SUP:** not applicable. **DEM:** not applicable. **NA:** absent. **VT:** medial. **EP:** lateral. **SP:** midventral near vent. **UJ:** absent. **LJ:** absent. **DF:** low, originates at dorsal tail-body junction, tip rounded without flagellum. **BS:** oval/depressed. **CP:** uniformly dark, belly spotted and banded. **TL:** 52/37. **CI:** Da Cruz and Peixoto (1978).

Dermatonotus: **CO:** 1. **GR:** Brazil–Paraguay. **EMG:** exotroph: lentic: suspension feeder. **LTRF:** 0/0. **OA:** semicircular oral flaps with smooth edges pendant over terminal mouth. **MP:** not applicable **SUP:** not applicable. **DEM:** not applicable. **NA:** absent. **VT:** medial. **EP:** lateral. **SP:** midventral below vent. **UJ:** absent. **LJ:** absent. **DF:** moderately high, originates at dorsal tail-body junction, tip rounded as flagellum. **BS:** rounded/depressed. **CP:** uniformly dark. **TL:** 36/35. **NO:** tail sheath (see text) present. **CI:** Cei (1980); Lavilla (1992b); Vizotto (1967).

Elachistocleis: **CO:** 4. **GR:** Panama–Argentina. **EMG:** exotroph: lentic: suspension feeder. **LTRF:** 0/0. **OA:** semicircular oral flaps with smooth edges pendant over terminal mouth. **MP:** not applicable **SUP:** not applicable. **DEM:** not applicable. **NA:** absent. **VT:** medial. **EP:** lateral. **SP:** midventral near vent. **UJ:** absent. **LJ:** absent. **DF:** low, originates near dorsal tail-body junction, tip pointed. **BS:** rounded/depressed. **CP:** uniformly pale, tail muscle striped. **TL:** 15–22/36. **CI:** J. D. Williams and Gudynas (1987).

Gastrophryne: **CO:** 5. **GR:** United States–Costa Rica. **EMG:** exotroph: lentic: suspension feeder. **LTRF:** 0/0. **OA:** semicircular oral flaps with smooth edges pendant over terminal mouth. **MP:** not applicable **SUP:** not applicable. **DEM:** not applicable. **NA:** absent. **VT:** medial. **EP:** lateral. **SP:** midventral near vent. **UJ:** absent. **LJ:** absent. **DF:** low, originates near dorsal tail-body junction, tip pointed. **BS:** rounded/depressed. **CP:** uniform dorsally, sagittal stripe, belly mottled or blotched. **TL:** 35/36. **CI:** Donnelly et al. (1990a); C. E. Nelson and Altig (1972); C. E. Nelson and Cuellar (1968); Orton (1952); A. H. Wright and Wright (1949).

Gastrophrynoides: **CO:** 1. **GR:** Borneo. **EMG:** exotroph: lentic: arboreal (probably still feeds as a suspension feeder). **LTRF:** 0/0. **OA:** greatly reduced straight-edged oral flap above terminal mouth. **MP:** not applicable **SUP:** not applicable. **DEM:** not applicable. **NA:** absent. **VT:** medial. **EP:** lateral. **SP:** midventral on gut. **UJ:** absent. **LJ:** absent. **DF:** low, originates near dorsal tail-body junction, tip round. **BS:** rounded/depressed. **CP:** uniformly dark. **TL:** –. **CI:** tentative identification based on Inger (1985:38).

Glyphoglossus: **CO:** 1. **GR:** Burma–Thailand. **EMG:** exotroph: lentic: suspension feeder. **LTRF:** 0/0. **OA:** straight-edged upper lip without oral flaps above terminal mouth. **MP:** not applicable **SUP:** not applicable. **DEM:** not applicable. **NA:** absent. **VT:** medial. **EP:** lateral. **SP:** midventral below gut. **UJ:** absent. **LJ:** absent. **DF:** low, originates near dorsal tail-body junction, tip pointed. **BS:** oval/depressed. **CP:** pale brown, translucent, distal part of tail dark. **TL:** 40/36. **NO:** M. A. Smith (1917) reports data on the nares as if they were present, hangs in midwater at a 45° angle. **CI:** M. A. Smith (1917).

Hamptophryne: **CO:** 1. **GR:** northern and western Amazon Basin–Guianas–Bolivia. **EMG:** exotroph: lentic: suspension feeder. **LTRF:** 0/0. **OA:** semicircular oral flaps with smooth to slightly crenulate edges pendant over terminal mouth. **MP:** not applicable **SUP:** not applicable. **DEM:** not applicable. **NA:** absent. **VT:** medial. **EP:** lateral. **SP:** midventral near vent. **UJ:** absent. **LJ:** absent. **DF:** low, originates near dorsal tail-body junction, tip pointed. **BS:** rounded/depressed. **CP:** dark, tail striped. **TL:** 32/36. **CI:** Duellman (1978); Schlüter and Salas (1991).

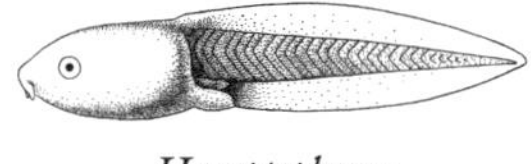

Hamptophryne

Hyophryne: **CO:** 1. **GR:** Brazil. Tadpole unknown.

Hypopachus: **CO:** 2. **GR:** southern United States–Costa Rica. **EMG:** exotroph: lentic: suspension feeder. **LTRF:** 0/0. **OA:** semicircular oral flaps with scalloped–papillate edges pendant over terminal mouth. **MP:** not applicable **SUP:** not applicable. **DEM:** not applicable. **NA:** absent. **VT:** medial. **EP:** lateral. **SP:** midventral near vent. **UJ:** absent. **LJ:** absent. **DF:** low, originates near dorsal tail-body junction, tip pointed. **BS:** rounded/depressed. **CP:** dark reticulations, lightly striped tail. **TL:** 21/36. **NO:** tadpoles described by E. H. Taylor (1942) as *H. alboventer* likely belong to a species of *Gastrophryne.* **CI:** C. E. Nelson and Cuellar (1968); Stuart (1941).

Kalophrynus: **CO:** 10. **GR:** southern China–Java–Philippines. **EMG:** endotroph

Kaloula: **CO:** 10. **GR:** Korea–Sri Lanka. **EMG:** exotroph: lentic: suspension feeder. **LTRF:** 0/0. **OA:** straight-edged upper lip without oral flaps above terminal mouth. **MP:** not applicable. **SUP:** not applicable. **DEM:** not applicable. **NA:** absent. **VT:** medial. **EP:** lateral. **SP:** midventral near vent. **UJ:** absent. **LJ:** absent. **DF:** low, originates near dorsal tail-body junction, tip pointed. **BS:** rounded/depressed. **CP:** uniform, tail striped, terminal part of tail may be unpigmented. **TL:** 25–30/36. **CI:** Kirtisinghe (1958); C.-C. Liu (1950).

Metaphrynella: **CO:** 2. **GR:** Malaysia–Borneo. **EMG:** exotroph: lentic: suspension feeder. **LTRF:** 0/0. **OA:** dorsoterminal, no oral flaps, 2–3 papillae at lower jaw. **MP:** not applicable. **SUP:** not applicable. **DEM:** not applicable. **NA:** absent. **VT:** medial. **EP:** dorsal. **SP:** midventral on chest. **UJ:** absent. **LJ:** absent. **DF:** low, originates slightly posterior of dorsal tail-body junction, tip round. **BS:** oval/depressed. **CP:** uniformly brown. **TL:** 32/36. **NO:** Berry (1972) reports open nares from stage 26 onward. **CI:** Berry (1972).

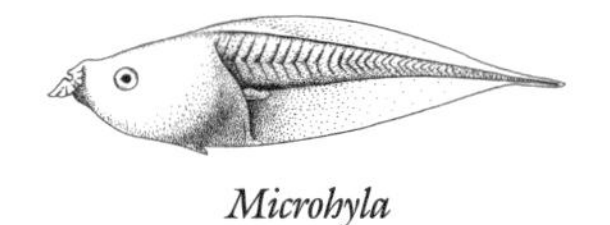

Microhyla

Microhyla: **CO:** 24. **GR:** Sri Lanka–Japan–China–Southeast Asia. **EMG:** exotroph: lentic: suspension feeder. **LTRF:** 0/0. **OA:** small to large semicircular oral flaps with smooth edges pendant over terminal mouth, umbelliform taxa have oral disc with radially arranged submarginal papillae, and formed almost entirely of lower labium. **MP:** not applicable. **SUP:** not applicable. **DEM:** not applicable. **NA:** absent. **VT:** medial. **EP:** lateral. **SP:** midventral on belly or near vent. **UJ:** absent. **LJ:** absent. **DF:** low, originates near dorsal tail-body junction, tip pointed with or without a flagellum. **BS:** round to oval/depressed. **CP:** dark, belly sometimes mottle or banded, tail striped. **TL:** 20–30/36. **CI:** Chou and Lin (1997); Heyer (1971); Inger (1985); Inger and Frogner (1979).

Myersiella: **CO:** 1. **GR:** eastern Brazil. **EMG:** endotroph.

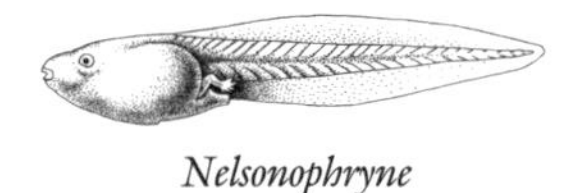

Nelsonophryne

Nelsonophryne: **CO:** 2. **GR:** Costa Rica–western Ecuador. **EMG:** exotroph: lentic: suspension feeder. **LTRF:** 0/0. **OA:** labial flaps absent, upper lip derivative forms large lateral flaps, small lower labium. **MP:** absent, although some papillae midventrally resemble marginal papillae. **SUP:** absent. **DEM:** not applicable. **NA:** absent. **VT:** medial. **EP:** lateral. **SP:** midventral on gut. **UJ:** absent. **LJ:** absent. **DF:** low, originates near dorsal tail-body junction, tip rounded. **BS:** oval/depressed. **CP:** uniformly dark. **TL:** 44/34. **CI:** Donnelly et al. (1990a).

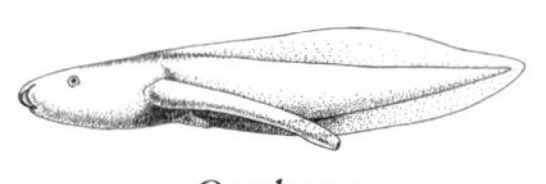

Otophryne

Otophryne: **CO:** 3. **GR:** northern South America. **EMG:** exotroph: lotic: psammonic. **LTRF:** 0/0. **OA:** oral disc derivatives reduced to lateral, vertically oriented flaps, nascent tooth ridges present below lower jaw but no teeth present. **MP:** not applicable. **SUP:** not applicable. **DEM:** not applicable **NA:** absent **VT:** medial. **EP:** lateral. **SP:** appears sinistral but starts development midventrally, spiracular tube very long and free from body wall. **UJ:** isolated, hypertrophied jaw serrations without basal sheath. **LJ:** similar to upper jaw. **DF:** low, originates near dorsal tail-body junction, tip rounded. **BS:** oblong/depressed. **CP:** uniformly pale brown. **TL:** 36/36. **CI:** J. A. Campbell and Clarke (1988); Pyburn (1980); Wassersug and Pyburn (1987).

Phrynella: **CO:** 1. **GR:** Malaysia–Sumatra. **EMG:** endotroph.

Ramanella: **CO:** 8. **GR:** India–Sri Lanka. **EMG:** exotroph: lentic: suspension feeder. **LTRF:** 0/0. **OA:** smooth-edged, entire, small oral flap slightly pendant over terminal mouth. **MP:** not applicable **SUP:** not applicable. **DEM:** not applicable. **NA:** absent. **VT:** medial. **EP:** lateral. **SP:** midventral near vent. **UJ:** absent. **LJ:** absent. **DF:** low, originates near dorsal tail-body junction, tip rounded. **BS:** rounded/depressed. **CP:** uniform. **TL:** 27/32. **NO:** Kirtisinghe (1958) reports open nares. **CI:** Kirtisinghe (1958).

Relictovomer: **CO:** 1. **GR:** Panama–Colombia. Tadpole unknown.

Stereocyclops: **CO:** 1. **GR:** Brazil. **EMG:** exotroph: lentic: suspension feeder. **LTRF:** 0/0.

OA: semicircular oral flaps with smooth edges pendant over terminal mouth. **MP:** not applicable **SUP:** not applicable. **DEM:** not applicable. **NA:** absent. **VT:** medial. **EP:** lateral. **SP:** midventral but opens to left of the plane of the ventral fin along with vent. **UJ:** absent. **LJ:** absent. **DF:** low, originates near dorsal tail-body junction, tip pointed. **BS:** round/depressed. **CP:** uniform (see note). **TL:** –. **NO:** I. Griffiths and De Carvalho (1965) discuss polymorphisms of a middorsal stripe and spiracular tube length; tail sheath (see text) present. **CI:** I. Griffiths and De Carvalho (1965).

Synapturanus: **CO:** 3. **GR:** Colombia–Guianas–Brazil. **EMG:** endotroph.

Syncope: **CO:** 2. **GR:** eastern Ecuador–Peru. **EMG:** endotroph.

Uperodon: **CO:** 2. **GR:** India–Sri Lanka. **EMG:** exotroph: lentic: suspension feeder. **LTRF:** 0/0. **OA:** smooth-edged entire (no medial notch) oral flap pendant over terminal mouth. **MP:** not applicable **SUP:** not applicable. **DEM:** not applicable. **NA:** absent. **VT:** medial. **EP:** lateral. **SP:** midventral near vent. **UJ:** absent. **LJ:** absent. **DF:** low, originates near dorsal tail-body junction, tip pointed. **BS:** oval/depressed. **CP:** uniform or mottled with striped tail. **TL:** 28/36. **NO:** Bhaduri and Daniel (1956) and Mohanty-Hejmadi et al. (1979) report open nares. **CI:** Kirtisinghe (1957).

Phrynomerinae: **CO:** 1/5. **GR:** sub-Saharan Africa. **EMG:** exotroph: lentic: suspension feeder. **LTRF:** 0/0. **OA:** no keratinized mouthparts, entire (without medial notch) oral flap with smooth edges pendant over terminal mouth. **MP:** not applicable. **SUP:** not applicable. **DEM:** not applicable. **NA:** absent. **VT:** medial. **EP:** lateral. **SP:** midventral, below vent. **UJ:** absent. **LJ:** absent. **DF:** low, originates near dorsal tail-body junction, flagellum present. **BS:** round/depressed. **CP:** uniformly pale, tail fins may be darker. **TL:** 37/36.

Phrynomantis: **CO:** 5. **CI:** Lamotte (1964); Wager (1986).

Scaphiophryninae: **CO:** 2/10. **GR:** Madagascar

Paradoxophyla: **CO:** 1. **GR:** Madagascar. **EMG:** exotroph: lentic: suspension feeder. **LTRF:** 0/0. **OA:** semicircular oral flaps with smooth edges pendant over terminal mouth. **MP:** absent. **SUP:** absent. **DEM:** not applicable. **NA:** absent. **VT:** medial. **EP:** lateral. **SP:** midventral, on abdomen. **UJ:** absent. **LJ:** absent. **DF:** low, originates near dorsal tail-body junction, tip pointed. **BS:** round/depressed. **CP:** dark dorsally with pale spot on snout and over lungs. **TL:** 28/36. **NO:** looks like a microhyline tadpole. **CI:** Blommers-Schlösser and Blanc (1991); Glaw and Vences (1992).

Scaphiophryne: **CO:** 9. **GR:** Madagascar. **EMG:** exotroph: lentic: suspension feeder, may rasp some materials, sometimes carnivorous. **LTRF:** 0/0. **OA:** small oral disc with papillae and unpigmented, serrated jaw sheaths, labial teeth lacking. **MP:** complete. **SUP:** few dorsally and ventrally. **DEM:** absent. **NA:** absent. **VT:** medial. **EP:** dorsal. **SP:** midventral at center of belly. **UJ:** wide, uniform arc, unpigmented. **LJ:** wide, open U-shaped, unpigmented. **DF:** low, originates near dorsal tail-body junction, tip pointed. **BS:** oval/depressed. **CP:** very lightly pigmented, translucent. **TL:** 30/36. **NO:** only microhylid tadpole with fully formed jaw sheaths (see *Otophryne*). **CI:** Blommers-Schlösser (1975); Blommers-Schlösser and Blanc (1991); Glaw and Vences (1992); Wassersug (1984).

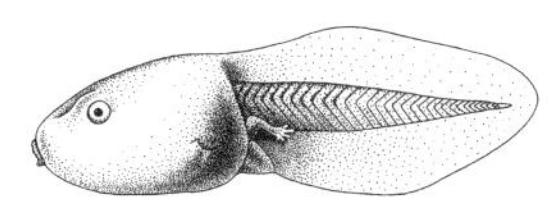

Scaphiophryne

MYOBATRACHIDAE: **CO:** 23/118. **GR:** Australia–New Guinea. **EMG:** exotroph: lentic: benthic, carnivore, nektonic; lotic: benthic, clasping; endotroph. **LTRF:** many variations between 0/0 and 8/3. **OA:** typical: anteroventral, ventral. **MP:** complete, medium to wide dorsal gap, narrow to wide ventral gap. **SUP:** absent, throughout extent of marginal papillae, below P-3, few laterally. **DEM:** absent, lateral. **NA:** nearer eye than snout or reverse. **VT:** dextral, medial. **EP:** dorsal. **SP:** sinistral. **UJ:** narrow to wide, smooth arc. **LJ:** narrow to wide, U- to V-shaped. **DF:** low, originates near dorsal tail-body junction, high and originates anterior to dorsal tail-body junction, tip rounded to pointed. **BS:** oval/depressed. **CP:** uniform to various muted patterns. **TL:** 15–70/36.

Limnodynastinae: **CO:** 11/51. **GR:** Australia–New Guinea.

Adelotus: **CO:** 1. **GR:** eastern Australia. **EMG:** exotroph: lentic: benthic. **LTRF:** 3(2–3)/3(1). **OA:** typical: anteroventral. **MP:** medium dorsal gap: uniserial. **SUP:** throughout extend of marginal papillae. **DEM:** absent. **NA:** nearer snout than eye. **VT:** dextral. **EP:** dorsal. **SP:** sinistral. **UJ:** narrow, wide smooth arc. **LJ:** narrow, open V-shaped. **DF:** low, originates near dorsal tail-body junction, tip rounded. **BS:** oval/depressed. **CP:** uniform. **TL:** 33/33. **CI:** Watson and Martin (1973).

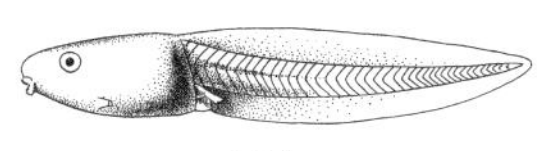

Adelotus

Heleioporus

Heleioporus: **CO:** 6. **GR:** Australia. **EMG:** exotroph: lentic: benthic. **LTRF:** 8(2–8)/3(1), 6(2–6)/3(1), 5(3–5)/3(1), 5(2–5)/3(1). **OA:** typical: anteroventral. **MP:** medium dorsal gap: uniserial. **SUP:** row below P-3. **DEM:** lateral. **NA:** nearer eye than snout. **VT:** medial. **EP:** dorsal. **SP:** sinistral. **UJ:** medium, smooth arc. **LJ:** medium, open U-shaped. **DF:** low, originates near dorsal tail-body junction, tip rounded. **BS:** oval/depressed. **CP:** uniformly dark. **TL:** 30–45/36. **CI:** Davies (1991); A. K. Lee (1967).

Kyarranus: **CO:** 3. **GR:** eastern Australia. **EMG:** endotroph.

Lechriodus: **CO:** 5. **GR:** eastern Australia–New Guinea. **EMG:** exotroph: lentic: carnivore. **LTRF:** 6(2–6)/3(1). **OA:** typical: anteroventral. **MP:** wide dorsal gap: biserial. **SUP:** throughout extent of marginal papillae. **DEM:** lateral. **NA:** nearer snout than eye. **VT:** medial. **EP:** dorsal. **SP:** sinistral. **UJ:** medium, wide smooth arc. **LJ:** medium, open V-shaped. **DF:** low, originates near dorsal tail-body junction, tip broadly rounded. **BS:** oval/depressed. **CP:** uniformly dark. **TL:** 33/34. **CI:** Watson and Martin (1973).

Limnodynastes: **CO:** 12. **GR:** Australia–New Guinea. **EMG:** exotroph: lentic: benthic; lotic: benthic. **LTRF:** 6(2–6)/3(1), 5(2–5)/3(1), 4(3–4)/3(1), 4(2–4)/3(1), 2(2)/3(1). **OA:** typical: anteroventral. **MP:** medium dorsal gap: biserial. **SUP:** few laterally. **DEM:** lateral. **NA:** nearer snout than eye or reverse. **VT:** medial. **EP:** dorsal. **SP:** sinistral. **UJ:** medium, smooth arc. **LJ:** medium, open U-shaped. **DF:** low, originates near dorsal tail-body junction, tip pointed. **BS:** oval/depressed. **CP:** uniform to muted mottling. **TL:** 35–70/36. **CI:** Tyler et al. (1983).

Megistolotis: **CO:** 1. **GR:** northwestern Australia. **EMG:** exotroph: lotic: clasping. **LTRF:** 6(3–6)/3. **OA:** typical: anteroventral. **MP:** narrow dorsal gap: uniserial ventrally and biserial dorsolaterally. **SUP:** few laterally and below lower tooth rows. **DEM:** absent. **NA:** nearer eye than snout. **VT:** medial. **SP:** sinistral. **EP:** dorsal. **UJ:** medium, edge describes smooth arc. **LJ:** medium, open V-shaped. **DF:** low, originates near dorsal tail-body junction, tip rounded. **BS:** oval/depressed. **CP:** uniformly dark. **TL:** 52/42. **CI:** Tyler et al. (1979).

Mixophyes: **CO:** 6. **GR:** eastern Australia. **EMG:** exotroph: lotic: benthic; lentic: benthic. **LTRF:** 6(2–6)/3(1). **OA:** typical: anteroventral. **MP:** complete: uniserial. **SUP:** few laterally. **DEM:** lateral. **NA:** nearer eye than snout. **VT:** dextral. **SP:** sinistral. **EP:** dorsal. **UJ:** medium, smooth arc. **LJ:** medium, very wide U-shaped. **DF:** low, originates near dorsal tail-body junction, tip rounded. **BS:** oval/depressed. **CP:** uniformly dark. **TL:** 60–70/37. **NO:** 1–5 accessory rows may be present anterior of lateral ends of lower rows. **CI:** Davies (1991); Watson and Martin (1973).

Neobatrachus: **CO:** 10. **GR:** Australia. **EMG:** exotroph: lentic: benthic. **LTRF:** 3(2–3)/3(1). **OA:** typical: anteroventral. **MP:** dorsal gap: uniserial. **SUP:** absent. **DEM:** lateral. **NA:** nearer eye than snout. **VT:** dextral. **EP:** dorsal. **SP:** sinistral. **UJ:** medium to wide, smooth arc. **LJ:** medium to wide, open V-shaped. **DF:** low, originates near dorsal tail-body junction, tip rounded. **BS:** oval/depressed. **CP:** uniformly dark. **TL:** 40–50/36. **CI:** Davies (1991); Watson and Martin (1973).

Notaden: **CO:** 4. **GR:** Australia. **EMG:** exotroph: lentic: benthic. **LTRF:** 3(2–3)/3(1), 2(2)/2(1). **OA:** typical: anteroventral. **MP:** wide dorsal and ventral gap: uniserial. **SUP:** absent. **DEM:** absent. **NA:** nearer eye than snout. **VT:** dextral. **EP:** dorsal. **SP:** sinistral. **UJ:** medium, smooth arc. **LJ:** medium, open U-shaped. **DF:** low, originates near dorsal tail-body junction, tip rounded. **BS:** oval/depressed. **CP:** uniform. **TL:** 25/24. **CI:** P. Slater and Main (1963); Tyler et al. (1983).

Philoria: **CO:** 1. **GR:** northeastern Australia. **EMG:** endotroph.

Rheobatrachus: **CO:** 2. **GR:** northeastern Australia. **EMG:** endotroph.

Myobatrachinae: **CO:** 12/67. **GR:** Australia–New Guinea.

Arenophryne: **CO:** 1. **GR:** Western Australia. **EMG:** endotroph.

Assa: **CO:** 1. **GR:** northeastern Australia. **EMG:** endotroph.

Bryobatrachus: **CO:** 1. **GR:** Tanzania. **EMG:** endotroph.

Crinia: **CO:** 14. **GR:** Australia–New Guinea. **EMG:** exotroph: lentic: benthic; lotic: benthic. **LTRF:** 2(2)/3(1), 2/3(1), 2(2)/2. **OA:** typical: anteroventral. **MP:** wide dorsal gap and narrow to wide ventral gap: uniserial. **SUP:** absent. **DEM:** lateral, absent. **NA:** nearer snout than eye. **VT:** dextral. **EP:** dorsal. **SP:** sinistral. **UJ:** medium wide smooth arc. **LJ:** medium, open U-shaped. **DF:** low, originates near dorsal tail-body junction, tip rounded. **BS:** oval/depressed. **CP:** dark, minor mottling. **TL:** 15–20/36. **CI:** Littlejohn and Martin (1965); Watson and Martin (1973).

Geocrinia: **CO:** 7. **GR:** southern Australia. **EMG:** exotrophic: lentic: benthic. **LTRF:** 0/0, 2(2)/3[1]. **OA:** typical: anteroventral. **MP:** wide dorsal and narrow ventral gaps: biserial. **SUP:** absent. **DEM:** absent. **NA:** nearer eye than snout. **VT:** dextral. **EP:** dorsal. **SP:** sinistral. **UJ:** medium, smooth arc with medial part straight. **LJ:** medium, open U-shaped. **DF:** low, originates near dorsal tail-body junction, tip round. **BS:** oval/depressed. **CP:** uniformly dark. **TL:** 22/30. **CI:** Littlejohn and Martin (1964); Watson and Martin (1973).

Metacrinia: **CO:** 1. **GR:** southwestern Australia. Tadpole unknown.

Myobatrachus: **CO:** 1. **GR:** Western Australia. **EMG:** endotroph.

Paracrinia: **CO:** 1. **GR:** eastern Australia. **EMG:** exotroph: lentic: nektonic. **LTRF:** 2(2)/2. **OA:** typical: anteroventral. **MP:** medium dorsal gap: uniserial. **SUP:** absent. **DEM:** absent. **NA:** nearer eye than snout. **VT:** dextral. **EP:** dorsal. **SP:** sinistral. **UJ:** medium, smooth arc. **LJ:** narrow, open U-shaped. **DF:** high, originates anterior to dorsal tail-body junction, tip pointed. **BS:** oval/depressed. **CP:** uniformly pale. **TL:** 34/34. **CI:** Watson and Martin (1973).

Pseudophryne: **CO:** 10. **GR:** eastern Australia. **EMG:** exotroph: lentic: benthic. **LTRF:** 2(2)/3[1]. **OA:** typical: anteroventral. **MP:** wide dorsal gap and narrow ventral gap: uniserial. **SUP:** absent. **DEM:** lateral. **NA:** nearer eye than snout. **VT:** dextral. **EP:** dorsal. **SP:** sinistral. **UJ:** medium, smooth arc. **LJ:** medium, open U-shaped. **DF:** low, originates near dorsal tail-body junction, tip pointed. **BS:** oval/depressed. **CP:** uniformly dark. **TL:** 20–25/36. **CI:** Watson and Martin (1973).

Spicospina: **CO:** 1. **GR:** southwestern Australia. Tadpole unknown. **CI:** J. D. Roberts et al. (1997).

Taudactylus: **CO:** 6. **GR:** eastern Australia. **EMG:** exotroph: lotic: benthic. **LTRF:** 0/0, 2(2)/3(1). **OA:** typical: anteroventral, ventral. **MP:** complete: uniserial. **SUP:** absent. **DEM:** lateral. **NA:** nearer eye than snout. **VT:** dextral. **EP:** dorsal. **SP:** sinistral. **UJ:** very narrow, smooth arc, coarsely serrate. **LJ:** very narrow, open U-shaped, coarsely serrate. **DF:** low, originates near dorsal tail-body junction, tip round. **BS:** oval/depressed. **CP:** uniformly dark. **TL:** 15–25/36. **CI:** Retallick and Hero 1998; Watson and Martin (1973).

Uperoleia: **CO:** 23. **GR:** northeastern Australia–New Guinea. **EMG:** exotroph: lentic: benthic. **LTRF:** 1/3, 2(2)/3[1]. **OA:** typical: anteroventral. **MP:** wide dorsal and ventral gaps: uniserial. **SUP:** absent. **DEM:** absent. **NA:** nearer eye than snout. **VT:** dextral. **EP:** dorsal. **SP:** sinistral. **UJ:** wide, smooth arc with slight medial convexity. **LJ:** medium, open U-shaped. **DF:** low, originates near dorsal tail-body junction, tip slightly rounded. **BS:** oval/depressed. **CP:** uniformly dark. **TL:** 31/36. **CI:** Tyler et al. (1983).

PELOBATIDAE: CO: 3/11. **GR:** Europe–North America. **EMG:** exotroph: lentic: benthic, carnivore. **LTRF:** 4/4 to 6/6, A-1 usually very short. **OA:** typical: anteroventral, almost terminal. **MP:** narrow dorsal gap: uniserial. **SUP:** absent, few laterally. **DEM:** absent. **NA:** nearer snout then eye. **VT:** medial. **EP:** dorsal. **SP:** low on left side, sometimes almost midventral. **UJ:** wide, smooth arc, unless a carnivore and then with a medial convexity. **LJ:** wide, smooth arc, unless a carnivore and then with a medial concavity. **DF:** low, originates near dorsal tail-body junction, tip bluntly pointed. **BS:** oblong/depressed. **CP:** uniformly dark or pale, depending on environmental conditions. **TL:** 20–80/36. **NO:** cannibal morphotypes of *Spea* have highly modified keratinized mouthparts and jaw musculature.

Pelobates: **CO:** 4. **GR:** Europe. **EMG:** exotroph: lentic: benthic. **LTRF:** 4(2–4)/5(1–3). **OA:** typical: anteroventral. **MP:** narrow dorsal gap: uniserial. **SUP:** row throughout extent of marginal papillae. **DEM:** absent. **NA:** nearer snout then eye. **VT:** medial. **EP:** dorsal. **SP:** low on left side. **UJ:** medium, smooth arc. **LJ:** wide, open U-shaped. **DF:** low, originates near dorsal tail-body junction, tip bluntly pointed. **BS:** oval/globular. **CP:** muted mottling. **TL:** 40–60/36. **CI:** Boulenger (1892); Lanza (1983).

Scaphiopus: **CO:** 3. **GR:** North America. **EMG:** exotroph: lentic: benthic. **LTRF:** 6(4–6)/6(1), 4(3–4)/4(1–2). **OA:** typical: anteroventral. **MP:** narrow dorsal gap: uniserial. **SUP:** few laterally. **DEM:** absent. **NA:** nearer snout then eye. **VT:** medial. **EP:** dorsal. **SP:** low on left side. **UJ:** narrow, smooth arc. **LJ:** medium, open U-shaped, basal part may appear striated. **DF:** low, originates near dorsal tail-body junction, tip bluntly

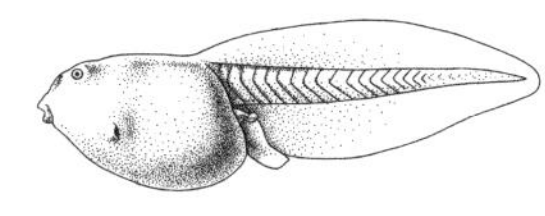

Scaphiopus

pointed. **BS:** oval/depressed. **CP:** uniformly dark. **TL:** 20–35/36. **CI:** Altig (1970); A. H. Wright and Wright (1949).

Spea: **CO:** 4. **GR:** North America. **EMG:** exotroph: lentic: benthic, facultative carnivore. **LTRF:** 4(2–4)/4(1–3), 4(3–4)/4(1–2). **OA:** typical: almost terminal. **MP:** narrow dorsal gap: uniserial. **SUP:** few laterally, absent. **DEM:** absent. **NA:** nearer snout then eye. **VT:** medial. **EP:** dorsal. **SP:** low on left side. **UJ:** typical: wide, smooth arc; carnivore: medial convexity. **LJ:** typical: wide, V-shaped; carnivore: medial indentation, basal part may appear striated. **DF:** low, originates near dorsal tail-body junction, tip bluntly pointed. **BS:** oval/globular. **CP:** uniformly dark or pale. **TL:** 40–80/36. **NO:** usually has keratinized knob on the roof of the buccal cavity, LTRF variable. **CI:** Altig (1970); A. H. Wright and Wright (1949).

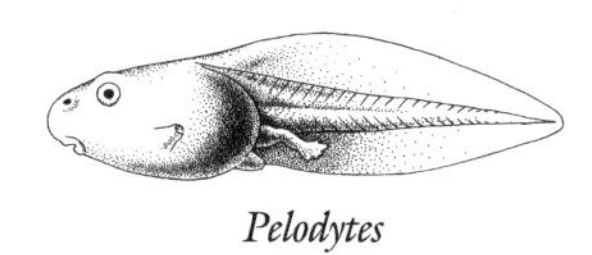

Pelodytes

PELODYTIDAE: CO: 1/2. **GR:** southwestern Europe–Caucasus. **EMG:** exotroph: lentic: benthic. **LTRF:** 4(3–4)/5(1–3). **OA:** typical: anteroventral. **MP:** dorsal gap: wide. **SUP:** absent. **DEM:** lateral. **NA:** nearer snout than eye. **VT:** medial. **EP:** dorsal. **SP:** sinistral. **UJ:** narrow, uniform arc. **LJ:** narrow, wide U-shaped. **DF:** medium, originates near dorsal tail-body junction. **BS:** oval/depressed. **CP:** uniformly dark. **TL:** 26/36.

Pelodytes: **CO:** 2. **CI:** Boulenger (1892); Lanza (1983).

PIPIDAE: CO: 5/28. **GR:** South America and sub-Saharan Africa. **EMG:** exotroph: lentic: suspension feeder, carnivore; endotroph. **LTRF:** 0/0. **OA:** oral disc or keratinized mouthparts absent, slitlike mouth, terminal or nearly so, sometimes with large barbels, highly derived protrusible mouthparts of microphagous carnivore. **MP:** not applicable. **SUP:** not applicable. **DEM:** not applicable. **NA:** much nearer snout than eye. **VT:** medial. **EP:** lateral. **SP:** dual/lateral. **UJ:** not applicable. **LJ:** not applicable. **DF:** low, originates near dorsal tail-body junction, tip pointed, often with flagellum that undulates continually. **BS:** oval/depressed. **CP:** uniformly lightly pigmented, often translucent. **TL:** 30–60/36. **NO:** low ventral fin extends onto abdomen, often as a pronounced lobe, most resemble rhinophrynid tadpoles.

Pipinae: CO: 1/7. **GR:** South America. **EMG:** exotroph: lentic: suspension feeder; endotroph. **LTRF:** 0/0. **OA:** absent, mouth transverse and slitlike. **MP:** not applicable. **SUP:** not applicable. **DEM:** not applicable. **NA:** near mouth. **VT:** medial. **EP:** lateral. **SP:** dual/lateral. **UJ:** not applicable. **LJ:** not applicable. **DF:** low, originates near dorsal tail-body junction, tip pointed as a flagellum. **BS:** oval/depressed. **TL:** 40–50/36. **NO:** barbels absent, ventral fin extends onto abdomen as a lobe.

Pipa: **CO:** 7. **GR:** Panama–northern South America. **CI:** Gines (1958); F. Schütte and Ehrl (1987); O. M. Sokol (1977a).

Xenopodinae: CO: 4/21. **GR:** sub-Saharan Africa.

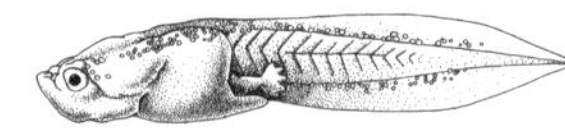

Hymenochirus

Hymenochirus: **CO:** 4. **GR:** Cameroon–Zaire. **EMG:** exotroph: lentic: carnivore. **LTRF:** 0/0. **OA:** absent, protrusible, rounded opening, opens dorsally when at rest. **MP:** not applicable. **SUP:** not applicable. **DEM:** not applicable. **NA:** nearer mouth than eye. **VT:** medial. **EP:** lateral. **SP:** dual/lateral. **UJ:** not applicable. **LJ:** not applicable. **DF:** low, originates near dorsal tail-body junction, tip bluntly pointed **BS:** oval/depressed. **TL:** 12–21/34. **CI:** O. M. Sokol (1959, 1962).

Pseudhymenochirus: **CO:** 1. **GR:** western Africa. **EMG:** exotroph: lentic: carnivore. **LTRF:** 0/0. **OA:** protrusible, rounded opening, opens dorsally when at rest. **MP:** not applicable. **SUP:** not applicable. **DEM:** not applicable. **NA:** small, nearer snout than eye. **VT:** medial. **EP:** lateral. **SP:** dual/lateral. **UJ:** not applicable. **LJ:** not applicable. **DF:** low, originates posterior to dorsal tail-body junction, tip rounded. **BS:** oval/depressed. **CP:** uniformly pale. **TL:** 19/34. **NO:** I. Griffiths (1963) reported fine "denticles" on the mouthparts. **CI:** Lamotte (1963).

Silurana: **CO:** 2. **GR:** western Africa. **EMG:** exotroph: lentic: suspension feeder. **LTRF:** 0/0. **OA:** single barbel at corners of terminal, slitlike mouth. **MP:** not applicable. **SUP:** not applicable. **DEM:** not applicable. **NA:** near snout, transversely elliptical. **VT:** medial. **EP:** lateral. **SP:** dual/lateral. **UJ:** not applicable. **LJ:** not applicable. **DF:** very low, originates near or posterior to dorsal tail-body junction, tip pointed as a constantly undulating flagellum. **BS:** oval/depressed. **CP:** uniformly pale. **TL:** 35–60/36. **NO:** barbel with extrinsic musculature and cartilage support, ventral fin extends onto abdomen as a rounded lobe; floats head-downward in water. **CI:** O. M. Sokol (1977a).

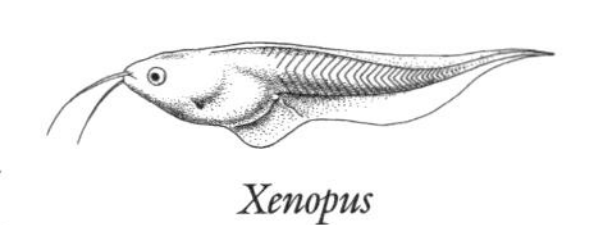
Xenopus

Xenopus: **CO:** 14. **GR:** sub-Saharan Africa. **EMG:** exotroph: lentic: suspension feeder. **LTRF:** 0/0. **OA:** single barbel at corners of terminal, slitlike mouth. **MP:** not applicable. **SUP:** not applicable. **DEM:** not applicable. **NA:** near snout, transversely elliptical. **VT:** medial. **EP:** lateral. **SP:** dual/lateral. **UJ:** not applicable. **LJ:** not applicable. **DF:** very low, originates near or posterior to dorsal tail-body junction, tip pointed as a constantly undulating flagellum. **BS:** oval/depressed. **CP:** uniformly pale. **TL:** 35–60/36. **NO:** barbel with extrinsic musculature and cartilage support, ventral fin extends onto abdomen as a rounded lobe; floats head-downward in water. **CI:** Arnoult and Lamotte (1968); Nieuwkoop and Faber (1967); Rau (1978); Vigny (1979).

PSEUDIDAE: CO: 2/5. **GR:** South America. **EMG:** exotroph: lentic: benthic, nektonic. **LTRF:** 2/3[1], 2(2)/3. **OA:** typical: anteroventral. **MP:** medium dorsal gap: biserial. **SUP:** few laterally, absent. **DEM:** absent. **NA:** nearer snout than eye. **VT:** medial, sinistral. **EP:** lateral. **SP:** sinistral. **UJ:** medium to wide, medial edge arced or straight. **LJ:** medium, open U-shaped. **DF:** high, originates well anterior of dorsal tail-body junction or low, originates well anterior of dorsal tail-body junction, tip rounded to pointed. **BS:** oval/compressed. **CP:** often brightly patterned and prominent ontogenetic change. **TL:** 34–230/36.

Lysapsus: **CO:** 3. **GR:** South America east of Andes. **EMG:** exotroph: lentic: benthic. **LTRF:** 2(2)/3. **OA:** typical: anteroventral. **MP:** medium dorsal gap: biserial. **SUP:** few laterally. **DEM:** absent. **NA:** nearer snout than eye. **VT:** sinistral. **EP:** lateral. **SP:** sinistral. **UJ:** wide, medial edge straight. **LJ:** medium, open U-shaped. **DF:** low, originates well anterior of dorsal tail-body junction, tip pointed. **BS:** oval/compressed. **CP:** entire tail boldly banded. **TL:** 34/37. **NO:** tail tip black, black bar at base of tail muscle and fin, perhaps only case reported of a sinistral vent tube being characteristic of a species. **CI:** Kehr and Basso (1990).

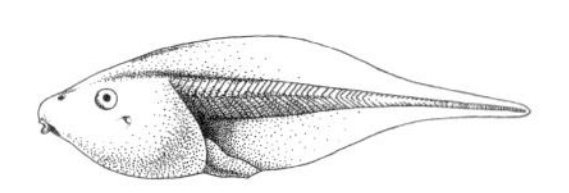
Pseudis

Pseudis: **CO:** 2. **GR:** South America. **EMG:** exotroph: lentic: nektonic. **LTRF:** 2/3(1). **OA:** typical: anteroventral. **MP:** medium dorsal gap: biserial. **SUP:** absent. **DEM:** absent. **NA:** nearer snout than eye. **VT:** medial. **EP:** lateral. **SP:** sinistral. **UJ:** medium, smooth arc. **LJ:** medium, open U-shaped. **DF:** high, originates near or anterior to plane of eyes, tip rounded. **BS:** oval/compressed. **CP:** young tadpole boldly banded, older ones unicolored dark. **TL:** 230/36. **NO:** metamorphic atrophy of fins precedes that of tail myotomes. **CI:** De Sá and Lavilla (1997); Dixon et al. (1995); Kenny (1969a).

RANIDAE: CO: 45/635. **GR:** widespread. **EMG:** exotroph: lentic: several guilds; lotic: several guilds; endotroph. **LTRF:** many variations between 0/0 and 9/8, 2/3 and n/3 most common. **OA:** typical: anteroventral, ventral to almost terminal, oral disc reduced and LTRF absent. **MP:** wide dorsal gap, dorsal and ventral gaps: uni- and biserial. **SUP:** absent, few laterally. **DEM:** lateral, absent. **NA:** nearer eye than snout or reverse. **VT:** dextral, medial. **EP:** dorsal, lateral. **SP:** sinistral. **UJ:** narrow to wide, smooth arc, medial convexity. **LJ:** narrow to wide, U- to V-shaped. **DF:** low, originates near dorsal tail-body junctions, very low, originates on tail muscle, tip pointed to round. **BS:** oval/depressed. **CP:** mostly muted tones of mottling. **TL:** 30–100/36.

Petropedetinae: CO: 13/94. **GR:** Africa.

Anhydrophryne: **CO:** 1. **GR:** South Africa. **EMG:** endotroph.

Arthroleptella: **CO:** 3. **GR:** South Africa. **EMG:** endotroph.

Arthroleptides: **CO:** 2. **GR:** Tanzania–Kenya. **EMG:** exotroph: lotic: semiterrestrial. **LTRF:** 3(1–3)/3(1). **OA:** typical: ventral. **MP:** wide dorsal gap: biserial ventrally. **SUP:** absent. **DEM:** absent. **NA:** small, nearer snout than eye. **VT:** medial. **EP:** dorsal. **SP:** sinistral. **UJ:** wide, strongly arched, medial convexity. **LJ:** similar to upper sheath. **DF:** very low, originates on tail muscle, tip round. **BS:** oval/depressed. **CP:** mottled. **TL:** 30/36. **NO:** eyes bulge noticeably, hind legs develop precociously relative to front legs and general development. **CI:** Drewes et al. (1989).

Cacosternum: **CO:** 7. **GR:** southern Africa. **EMG:** exotroph: lentic: benthic. **LTRF:** 2(2)/3(1), 3(2–3)/3, 4(2–4)/3(1). **OA:** typical: anteroventral. **MP:** wide dorsal gap: uniserial. **SUP:** few in row laterally. **DEM:** lateral, absent. **NA:** equidistant or nearer eye than snout. **VT:** dextral. **EP:** dorsal. **SP:** sinistral. **UJ:** wide, medial convexity, coarsely serrate. **LJ:** wide, open V-shaped. **DF:** low, originates near dorsal tail-body junction, tip pointed to rounded. **BS:** oval/depressed. **CP:** uniformly dark. **TL:** 28–50/36. **CI:** De Villiers (1929a); Power (1927a); Van Dijk (1966).

Dimorphognathus: **CO:** 1. **GR:** Cameroon–Gabon. Tadpole unknown.
Ericabatrachus: **CO:** 1. **GR:** Ethiopia. **EMG:** Tadpole unknown.
Microbatrachella: **CO:** 1. **GR:** South Africa. **EMG:** exotroph: lentic: benthic. **LTRF:** 3(2–3)/3(1). **OA:** typical, anteroventral. **MP:** wide dorsal and narrow ventral gaps: uniserial. **SUP:** few laterally in a row. **DEM:** absent. **NA:** nearer snout than eye. **VT:** dextral. **EP:** dorsal. **SP:** sinistral. **UJ:** medium, medial convexity. **LJ:** medium, open U-shaped. **DF:** low, originates at dorsal tail-body junction, tip pointed. **BS:** oval/depressed. **CP:** brown dorsally, white ventrally. **TL:** 25/36. **CI:** Van Dijk (1966); Wager (1986).
Natalobatrachus: **CO:** 1. **GR:** South Africa. **EMG:** exotroph: lotic: benthic. **LTRF:** 5(2–5)/3. **OA:** typical, anteroventral. **MP:** wide dorsal gap: biserial. **SUP:** absent. **DEM:** lateral. **NA:** nearer snout than eye. **VT:** dextral. **EP:** dorsal. **SP:** sinistral. **UJ:** wide, prominent medial convexity. **LJ:** medium, open V-shaped. **DF:** low, originates at dorsal tail-body junction, tip rounded. **BS:** oval/depressed. **CP:** pale gray. **TL:** 41/36. **CI:** Van Dijk (1966); Wager (1986).
Nothophryne: **CO:** 1. **GR:** Malawi–Mozambique. Tadpole unknown.
Petropedetes: **CO:** 7. **GR:** Cameroon–Sierra Leone. **EMG:** exotroph: lotic: suctorial, semiterrestrial. **LTRF:** 3(1–3)/3(1), 3(3)/3(1), 4/5, 7(5–7)/6(1). **OA:** typical: ventral. **MP:** wide dorsal gap: uniserial, large folds in lateral part of disc. **SUP:** two rows below distal most lower tooth row. **DEM:** lateral. **NA:** nearer eyes than snout. **VT:** medial. **EP:** lateral. **SP:** sinistral. **UJ:** wide, smooth arc, broken into two pieces connected by a narrow part. **LJ:** wide, open V-shaped. **DF:** low, originates near dorsal tail-body junction, tip pointed. **BS:** oval/depressed. **CP:** uniformly dark. **TL:** 35/36. **NO:** upper rows much longer than lower rows in *P. natator.* **CI:** Lamotte et al. (1959a); Lamotte and Zuber-Vogeli (1954b).
Phrynobatrachus: **CO:** 67. **GR:** sub-Saharan Africa. **EMG:** exotroph: lentic: benthic. **LTRF:** 1/2, 1/3, 1/4(1), 2(2)/2. **OA:** typical: anteroventral. **MP:** medium dorsal gap: uniserial, ventral papillae sometimes elongate. **SUP:** row throughout lower labium. **DEM:** lateral, absent. **NA:** nearer snout than eye. **VT:** medial. **EP:** dorsal. **SP:** sinistral. **UJ:** wide, prominent medial convexity. **LJ:** wide, open U-shaped. **DF:** low, originates near dorsal tail-body junction, tip pointed. **BS:** oval/depressed. **CP:** dark. **TL:** 20–35/36. **CI:** Lamotte and Dzieduszycka (1958); Van Dijk (1966).
Phrynodon: **CO:** 1. **GR:** Cameroon–Fernando P. **EMG:** endotroph.
Poyntonia: **CO:** 1. **GR:** South Africa. **EMG:** exotroph: lentic: benthic. **LTRF:** 1/2 or 1(2)/2. **OA:** typical. **MP:** wide dorsal gap **UP:** lateral patch. **DEM:** absent. **NA:** small, about equidistant between eye and snout. **VT:** dextral. **EP:** dorsal. **SP:** sinistral. **UJ:** wide, forms shallow arch with a medial convexity in serrated margin. **LJ:** V-shaped. **DF:** low, originates posterior of dorsal tail-body junction, tip rounded. **BS:** oval/depressed. **CP:** pale brown with dark speckles. **TL:** 32/26. **NO:** skin glands prominent throughout dorsum, margin of vent tube slightly scalloped. **CI:** Channing and Boycott (1989).

Raninae: CO: 32/541. **GR:** widespread.
Amolops: **CO:** 24. **GR:** northern India–China–Southeast Asia–Indonesia. **EMG:** exotroph: lotic: gastromyzophorous. **LTRF:** 4–8/3–4, usually with three complete distal rows on upper labium and a medial gap in first lower row. **OA:** typical: suctorial. **MP:** wide dorsal gap. **SUP:** absent. **DEM:** lateral. **NA:** closer to eye than snout. **VT:** medial. **EP:** dorsal. **SP:** sinistral. **UJ:** medium, wide U-shaped. **LJ:** medium, wide U-shaped. **DF:** low, originates near or posterior to dorsal tail-body junction, tip rounded. **BS:** oval/depressed. **CP:** muted mottling. **TL:** 50–70/36. **NO:** postorbital and posteroventral integumentary glands, some species with keratinized spinules in the skin. **CI:** Inger (1985); Inger and Gritis (1983).
Aubria: **CO:** 2. **GR:** Cameroon–Zaire. **EMG:** exotroph: lentic: benthic. **LTRF:** 5(3–5)/3. **OA:** typical: anteroventral. **MP:** wide dorsal gap: multiserial laterally and uniserial ventrally. **SUP:** absent. **DEM:** absent. **NA:** nearer snout than eye. **VT:** dextral. **EP:** dorsal. **SP:** sinistral. **UJ:** medium, smooth arc. **LJ:** medium, V-shaped. **DF:** low, originates near dorsal tail-body junction, tip rounded. **BS:** oval/depressed. **CP:** totally black. **TL:** 40/36. **NO:** school in compact balls. **CI:** Schiøtz (1963).
Batrachylodes: **CO:** 8. **GR:** Solomon Islands. **EMG:** endotroph.
Ceratobatrachus: **CO:** 1. **GR:** Solomon Islands. **EMG:** endotroph.

Chaparana: **CO:** 6. **GR:** India–China. **EMG:** exotroph: lotic: clasping. **LTRF:** 8(3–8)/3. **OA:** typical: ventral. **MP:** wide dorsal gap: uniserial. **SUP:** few ventrolaterally and in a row distal to P-3. **DEM:** absent. **NA:** –. **VT:** dextral. **EP:** dorsal. **SP:** sinistral. **UJ:** medium, wide smooth arc. **LJ:** wide, V-shaped. **DF:** low, originates near dorsal tail-body junction. **BS:** oval/depressed **CP:** uniformly dark. **TL:** –. **CI:** C.-C. Liu and Hu (1961).
Conraua: **CO:** 6. **GR:** tropical sub-Saharan Africa. **EMG:** exotroph: lotic: clasping. **LTRF:** 7(5–7)/7(1), 8(5–8)/9, 14(7–14)/12(1–2). **OA:** typical: ventral. **MP:** narrow dorsal gap: biserial at least ventrally. **SUP:** abundant laterally and distal to posterior rows. **DEM:** lateral. **NA:** nearer eye than snout. **VT:** dextral. **EP:** dorsal. **SP:** sinistral. **UJ:** wide, medial edge straight, exceptionally long processes laterally. **LJ:** medium, V-shaped. **DF:** low, originates near dorsal tail-body junction, tip rounded. **BS:** oval/depressed. **CP:** body dorsum spotted, dorsum of tail muscle lightly banded. **TL:** 45–60/34. **CI:** Lamotte et al. (1959a); Lamotte and Zuber-Vogeli (1954a); Sabater-Pi (1985).
Discodeles: **CO:** 5. **GR:** Solomon Islands. **EMG:** endotroph.
Elachyglossa: **CO:** 1. **GR:** Thailand. Tadpole unknown.
Euphlyctis: **CO:** 4. **GR:** Arabia–Sri Lanka–Malaysia. **EMG:** exotroph: lentic: benthic. **LTRF:** 1/2[1]. **OA:** typical: anteroventral. **MP:** wide dorsal and medium ventral gaps: uniserial. **SUP:** few laterally. **DEM:** lateral. **NA:** nearer snout than eye. **VT:** dextral. **EP:** dorsal. **SP:** sinistral. **UJ:** medium, edge a smooth arc. **LJ:** medium, V-shaped. **DF:** low, originates near dorsal tail-body junction, tip pointed. **BS:** oval/depressed. **CP:** tail and body striped. **TL:** 35/36. **CI:** Annandale and Rao (1918); Kirtisinghe (1957).
Hildebrantia: **CO:** 3. **GR:** Ethiopia–Mozambique. **EMG:** exotroph: lentic: benthic. **LTRF:** 0/2. **OA:** typical: anteroventral. **MP:** wide dorsal gap: uniserial, widely spaced. **SUP:** absent. **DEM:** absent. **NA:** equidistant between snout and eye. **VT:** medial. **EP:** dorsal. **SP:** sinistral. **UJ:** wide, massive with medial convexity. **LJ:** wide, massive, open U-shaped. **DF:** low, originates near dorsal tail-body junction, tip pointed and attenuate. **BS:** oval/depressed. **CP:** uniformly dark. **TL:** 95/36. **CI:** Van Dijk (1966); Wager (1986).
Hoplobatrachus: **CO:** 5. **GR:** Angola–Ethiopia–southern Asia–China. **EMG:** exotroph: lentic: carnivore; lotic: carnivore. **LTRF:** 4(3–4)/4(1–2), 5(3–5)/5(1). **OA:** typical: anteroventral. **MP:** complete: uniserial. **SUP:** absent. **DEM:** absent. **NA:** nearer eye than snout. **VT:** dextral. **EP:** dorsal. **SP:** sinistral. **UJ:** wide, prominently medially incised. **LJ:** wide, prominently medially incised. **DF:** low, originates near dorsal tail-body junction. **BS:** oval/depressed. **CP:** uniformly brown. **TL:** 43/30. **NO:** M. A. Smith (1917) discussed variations and nonconcordance among sources that suggest other species are involved. **CI:** Annandale and Rao (1918); Lamotte and Zuber-Vogeli (1954a); M. A. Smith (1917).
Huia: **CO:** 4. **GR:** China–southeastern Asia–Greater Sundas. **EMG:** exotroph: lotic: gastromyzophorous. **LTRF:** several variations between 5/8 and 11/6. **OA:** typical: suctorial. **MP:** usually wide dorsal gap. **SUP:** absent. **DEM:** absent. **NA:** nearer eye than snout. **VT:** medial. **EP:** dorsal. **SP:** sinistral. **UJ:** wide, M-shaped. **LJ:** wide, V-shaped. **DF:** low, originates on tail posterior to dorsal tail-body junction, tip pointed. **BS:** oval/depressed **CP:** dark gray with pale spots, tail dark with pale spots at margins. **TL:** 68/39. **NO:** ventral fin originates on tail muscle, transverse keratinized patch on roof of sucker, clusters of glands on body, buccal papillae may be distinctive among *Amolops, Huia,* and *Meristogenys* (Inger 1985). **CI:** Inger (1966); Yang (1991)
Indirana: **CO:** 9. **GR:** India. **EMG:** exotroph: lentic: semiterrestrial. **LTRF:** 5(2–5)/3(1). **OA:** typical: ventral. **MP:** wide (entire upper labium) dorsal gap: uniserial. **SUP:** none. **DEM:** absent. **NA:** nearer eye than snout. **VT:** dextral. **EP:** dorsal. **SP:** sinistral. **UJ:** wide, strongly arched. **LJ:** strongly arched. **DF:** low, originates on tail muscle, tip rounded. **BS:** oval/depressed. **CP:** tail muscle banded. **TL:** 31/33. **CI:** Annandale (1918); Dubois (1985).
Ingerana: **CO:** 8. **GR:** western China–Thailand–Borneo–Philippines. **EMG:** endotroph.
Lanzarana: **CO:** 1. **GR:** Somalia. Tadpole unknown.
Limnonectes: **CO:** 62. **GR:** China–Southeast Asia–Philippines. **EMG:** exotroph: lotic: benthic; lentic: benthic. **LTRF:** 1/3(1), 1/4(1), 2(2)/3[1]. **OA:** typical: anteroventral. **MP:** wide dorsal gap: papillae often with elongate basal areas, narrow ventral gap some-

times present. **SUP:** few laterally. **DEM:** absent. **NA:** nearer snout than eye or reverse. **VT:** dextral. **EP:** dorsal. **SP:** sinistral. **UJ:** narrow to medium, smooth arc or slight medial convexity. **LJ:** narrow to medium, open U-shaped. **DF:** low, originates near dorsal tail-body junction, tip rounded to pointed. **BS:** oval/depressed. **CP:** variable, uniform to prominently spotted/mottled. **TL:** 25–40/36. **CI:** Inger (1954, 1966, 1985).

Meristogenys: **CO:** 9. **GR:** Borneo. **EMG:** exotroph: lotic: gastromyzophorous. **LTRF:** several variations between 6/6 and 8/7. **OA:** typical: suctorial. **MP:** complete: uniserial, reduced to crenulations dorsally, lateral papillae may be larger than elsewhere. **SUP:** few laterally. **DEM:** absent, but folds for expansion of the disc present. **NA:** nearer eye than snout. **VT:** medial. **EP:** dorsal. **SP:** sinistral. **UJ:** broadly divided, constructed of few cellular palisades (= ribbed). **LJ:** ribbed and sometimes broken medially. **DF:** low, originates near dorsal tail-body junction, tip pointed. **BS:** oval/depressed. **CP:** mottled, fins with little pigment. **TL:** 30–40/36. **NO:** ventral fin originates on tail muscle, skin glands present (e.g., behind eye, midlaterally, sometimes on tail); keratinized areas in roof of sucker. **CI:** Inger (1985); Yang (1991).

Micrixalus: **CO:** 6. **GR:** India. Tadpole unknown.

Nannophrys: **CO:** 3. **GR:** Sri Lanka. **EMG:** exotroph: lotic: semiterrestrial. **LTRF:** 2(2)/3(1). **OA:** typical: ventral. **MP:** wide dorsal gap: uniserial, very small ventrally. **SUP:** absent. **DEM:** absent. **NA:** nearer eye than snout. **VT:** dextral. **EP:** dorsal. **SP:** sinistral. **UJ:** wide, strongly arched, coarsely serrate. **LJ:** wide, strongly arched, coarsely serrate. **DF:** very low, originates well posterior to dorsal tail-body junction, tip rounded. **BS:** oval/depressed. **CP:** mottled. **TL:** 27/38. **NO:** ventral fin originates on tail muscle, eyes bulge noticeably, hind legs develop precociously relative to front legs and general development. **CI:** B. T. Clarke (1983); Kirtisinghe (1958).

Nannorana: **CO:** 3. **GR:** China. **EMG:** exotroph: lotic: benthic. **LTRF:** 3(2–3)/3(1). **OA:** typical: anteroventral. **MP:** wide dorsal gap: uniserial. **SUP:** throughout lateral and ventrolateral areas. **DEM:** lateral. **NA:** equidistant between snout and eye. **VT:** sinistral. **EP:** dorsal. **SP:** sinistral. **UJ:** wide, edge a smooth arc. **LJ:** wide, U-shaped. **DF:** low, originates near dorsal tail-body junction, tip rounded. **BS:** oval/depressed. **CP:** uniformly dark. **TL:** 36/36. **CI:** C.-C. Liu and Hu (1961).

Nyctibatrachus: **CO:** 10. **GR:** India. **EMG:** exotroph: lotic: benthic. **LTRF:** 0/0. **OA:** typical: anteroventral. **MP:** complete: uniserial. **SUP:** scattered throughout disc. **DEM:** dorsolateral and midventral. **NA:** nearer to eye than snout. **VT:** dextral. **EP:** dorsal. **SP:** sinistral. **UJ:** narrow, smooth arc. **LJ:** narrow, open U-shaped. **DF:** low, originates posterior of dorsal tail-body junction, tip bluntly pointed. **BS:** oval/depressed. **CP:** dark mottling, tail muscle banded. **TL:** 52/36. **NO:** see Rao (1923). **CI:** Pillai (1978).

Occidozyga: **CO:** 2. **GR:** Southeast Asia–Java. **EMG:** exotroph: lentic: carnivore. **LTRF:** 0/0. **OA:** oral disc and papillae reduced to fleshy rim, jaw sheaths recessed. **MP:** dorsal gap in fleshy, nonpapillate rim, small rounded flap middorsally. **SUP:** absent. **DEM:** absent. **NA:** equidistant between eyes and snout. **VT:** medial. **EP:** dorsal. **SP:** sinistral. **UJ:** –. **LJ:** distinctly U-shaped deeply semilunar in shape. **DF:** anterior portion prominently high, drops abruptly with remainder low, tip pointed. **BS:** oval/depressed. **CP:** black stripe through eye, tail variegated. **TL:** 30–50/36. **CI:** Inger (1985); M. A. Smith (1916b).

Paa: **CO:** 26. **GR:** India–Southeast Asia–China. **EMG:** exotroph: lotic: clasping. **LTRF:** 4(2–4)/3(1), 7(3–7)/3(1). **OA:** typical: anteroventral. **MP:** medium dorsal gap: uniserial. **SUP:** laterally and below P-3. **DEM:** absent. **NA:** nearer snout than eye. **VT:** dextral. **EP:** dorsal. **SP:** sinistral. **UJ:** wide, smooth arc. **LJ:** wide, smooth arc, open U-shaped. **DF:** low, originates near dorsal tail-body junction, tip round. **BS:** oval/depressed. **CP:** body uniformly dark, tail lighter with flecking. **TL:** 55/36. **CI:** Annandale (1912); C.-C. Liu and Hu (1961).

Palmatorappia: **CO:** 1. **GR:** Solomon Islands. **EMG:** endotroph.

Phrynoglossus: **CO:** 9. **GR:** Southeast Asia–East Indies. **EMG:** exotroph: lentic: carnivore. **LTRF:** 0/0. **OA:** oral disc and papillae reduced to fleshy rim, jaw sheaths recessed. **MP:** dorsal gap in fleshy, nonpapillate rim. **SUP:** absent. **DEM:** absent. **NA:** equidistant between eyes and snout. **VT:** medial. **EP:** dorsal. **SP:** sinistral. **UJ:** –. **LJ:** deeply semilunar in shape. **DF:** low, originates posterior to dorsal tail-body junction, tip pointed. **BS:** oval/depressed. **CP:** black marks form scalloped appearance along margins of fins.

TL: 51/36. **NO:** snout noticeably pointed, lower lip sometimes visible from above, otherwise similar to *Occidozyga*. **CI:** Inger (1985); M. A. Smith (1916b).
Platymantis: **CO:** 39. **GR:** Fiji–New Guinea–Philippines–China. **EMG:** endotroph.
Ptychadena: **CO:** 40. **GR:** sub-Saharan Africa. **EMG:** exotroph: lentic: benthic. **LTRF:** 3(2–3)/2, 3(2)/2, 2(2)/2, 1/2. **OA:** typical: anteroventral. **MP:** medium to wide dorsal gap: uni- and biserial. **SUP:** few laterally, row below distal most lower row. **DEM:** absent. **NA:** nearer snout than eye, or reverse. **VT:** medial. **EP:** dorsal. **SP:** sinistral. **UJ:** wide, medial convexity. **LJ:** wide, open U-shaped. **DF:** low with slight arc, originates near dorsal tail-body junction, tip pointed. **BS:** oval/depressed. **CP:** muted but variable. **TL:** 30–55/36. **CI:** Lamotte et al. (1958); Lamotte et al. (1959b); Lamotte and Perret (1961a); Van Dijk (1966); Wager (1986).

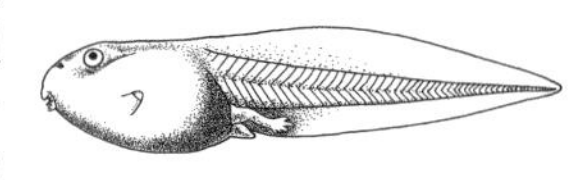
Ptychadena

Pyxicephalus: **CO:** 2. **GR:** sub-Saharan Africa. **EMG:** exotroph: lentic: benthic. **LTRF:** 5(3–5)/3. **OA:** typical: anteroventral. **MP:** wide dorsal gap: multiserial laterally, biserial ventrally. **SUP:** few laterally. **DEM:** lateral. **NA:** near eye than snout. **VT:** medial. **EP:** dorsal. **SP:** sinistral. **UJ:** wide, medial convexity. **LJ:** medium, open U-shaped. **DF:** medium with slight arc, originates near dorsal tail-body junction, tip bluntly pointed. **BS:** oval/depressed. **CP:** uniformly dark. **TL:** 70/36. **NO:** schools, sometimes in midwater and with other species. **CI:** Power (1927a); Van Dijk (1966).
Rana: **CO:** 224. **GR:** widespread. **EMG:** exotroph: lentic: benthic; lotic: benthic, clasping, gastromyzophorous; endotroph. **LTRF:** many variations between 1/2 and 9/8, 2/3 is common in North America, n/3 common elsewhere. **OA:** typical: anteroventral, ventral. **MP:** dorsal gap: usually uniserial, more rarely a narrow ventral gap, ventral papillae may be elongate. **SUP:** few laterally, absent. **DEM:** lateral, absent. **NA:** nearer eye than snout or reverse. **VT:** dextral. **EP:** dorsal. **SP:** sinistral. **UJ:** narrow to wide, smooth arc, medial part of edge straight. **LJ:** narrow to wide, narrow V-open U-shaped. **DF:** low, originates near dorsal tail-body junction, tip pointed to round. **BS:** oval/depressed. **CP:** variable. **TL:** 30–100/36. **NO:** a number of species groups have distinctive morphologies. **CI:** Boulenger (1892); Hillis (1982); Hillis and Frost (1985); Hillis and De Sá (1988); Lanza (1983); Scott and Jennings (1985); A. H. Wright and Wright (1949); Zweifel (1958).
Staurois: **CO:** 3. **GR:** Borneo–Philippines. **EMG:** exotroph: lotic: burrower. **LTRF:** 2(2)/9(1). **OA:** typical: ventral. **MP:** wide dorsal and narrow ventral gap: uniserial dorsolaterally, biserial ventrally. **SUP:** few dorsolaterally. **DEM:** lateral. **NA:** small, much nearer snout than eye. **VT:** medial. **EP:** dorsal. **SP:** sinistral. **UJ:** narrow, wide arc. **LJ:** medium, open V-shaped. **DF:** low, originates near dorsal tail-body junction, tip rounded. **BS:** oval/depressed. **CP:** very lightly pigmented, skin appears shiny. **TL:** 32/31. **NO:** convergent on the general centrolenid morphology, eyes particularly small. **CI:** Inger and Tan (1990); Inger and Wassersug (1990).
Taylorana: **CO:** 2. **GR:** India–Vietnam–Java. **EMG:** endotroph.
Tomopterna: **CO:** 13. **GR:** India–sub-Saharan Africa. **EMG:** exotroph: lentic: benthic. **LTRF:** 3(2–3)/3, 4(2–4)/3. **OA:** typical: anteroventral. **MP:** wide dorsal gap: biserial, except uniserial midventrally. **SUP:** few laterally. **DEM:** lateral. **NA:** nearer eye than snout or reverse. **VT:** dextral. **EP:** dorsal. **SP:** sinistral. **UJ:** wide, massive with medial convexity. **LJ:** wide, massive, open U-shaped. **DF:** low, originates at dorsal tail-body junction, tip pointed. **BS:** oval/depressed. **CP:** uniformly dark. **TL:** 36–44/36. **CI:** Grillitsch et al. (1988); Van Dijk (1966); Wager (1986).

RHACOPHORIDAE: CO: 12/292. **GR:** Africa–India–southeastern Asia. **EMG:** exotroph: lentic: benthic, arboreal, nektonic; lotic: benthic, clasping, suctorial; endotroph. **LTRF:** many variations between 1/0 and 8/3; n/3 most common. **OA:** typical: anteroventral or terminal. **MP:** wide dorsal gap, sometimes narrow ventral gap or complete: uni-, bi-, and multiserial. **SUP:** absent, laterally or below P-3, row(s) centripetal to marginal papillae. **DEM:** lateral, absent. **NA:** nearer snout than eye or reverse. **VT:** dextral, medial. **EP:** dorsal, lateral. **SP:** sinistral. **UJ:** variable including wide, medial margin straight or smooth arc coarsely serrate or not, absent. **LJ:** medium to wide, U- to V-shaped, coarsely serrate or not, sometimes small medial convexity. **DF:** low to moderately high, originates near dorsal tail-body junction, tip rounded to pointed. **BS:** round, oblong or oval/depressed. **CP:** variable. **TL:** 25–90/36.
Buergeriinae: CO: 1/4. **GR:** China–Japan. **EMG:** exotroph: lentic: benthic; lotic: benthic.

LTRF: 5(2–5)/3–4(1), 6(3–6)/4(1). **OA:** typical: anteroventral. **MP:** wide dorsal gap: uniserial. **SUP:** row(s) centripetal to most marginal papillae. **DEM:** lateral. **NA:** nearer snout than eye or reverse. **VT:** dextral. **EP:** dorsal. **SP:** sinistral. **UJ:** wide, medial margin straight, coarsely serrate. **LJ:** medium, V-shaped, coarsely serrate. **DF:** low, originates near dorsal tail-body junction, tip pointed. **BS:** oblong/depressed. **CP:** lightly pigmented with flecks and spots on tail. **TL:** 30–44/36.

Buergeria: **CO:** 4. **GR:** China–Japan. **CI:** Maeda and Matsui (1989); Okada (1931); Utsunomiya and Utsunomiya (1983).

Mantellinae: CO: 2/68. **GR:** Madagascar.

Mantella: **CO:** 10. **GR:** Madagascar. **EMG:** exotroph: lentic: benthic, arboreal. **LTRF:** 3(2–3)/3, 5(2–5)/3[1]. **OA:** typical: anteroventral. **MP:** wide dorsal gap: uniserial dorsolaterally, biserial ventrolaterally. **SUP:** absent. **DEM:** lateral. **NA:** nearer snout than eye. **VT:** dextral. **EP:** dorsal. **SP:** sinistral. **UJ:** medium, smooth arc. **LJ:** medium, open V-shaped. **DF:** low, tip rounded. **BS:** oval/depressed. **CP:** uniformly dark. **TL:** 20–30/36. **CI:** Arnoult (1965); Blommers-Schlösser and Blanc (1991); Glaw and Vences (1992).

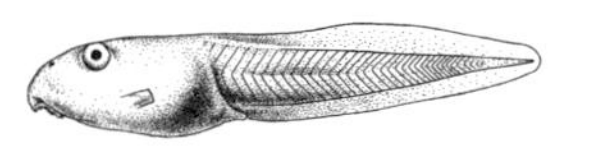

Mantidactylus; Mantidactylus sp.

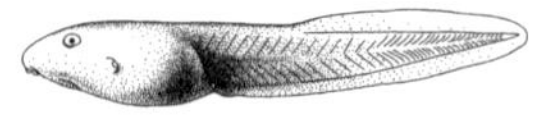

Mantidactylus; M. femoralis

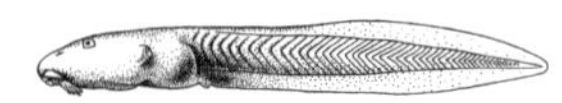

Mantidactylus; M. lugubris

Mantidactylus; M. opiparus

Mantidactylus: **CO:** 58. **GR:** Madagascar. **EMG:** exotroph: lentic: benthic, arboreal; lotic: benthic, psammonic, umbelliform; endotroph. **LTRF:** a number of variations between 0/0 and 6/3. **OA:** typical: anteroventral, terminal, ventral. **MP:** wide dorsal gap: uniserial, small ventral gap less common, complete. **SUP:** few laterally, row below P-3. **DEM:** lateral, absent. **NA:** close to snout then eye and reverse. **UJ:** absent, narrow to wide, smooth arc, sometimes with medial convexity, serrations fine to coarse, one species with no basal sheath and serrations elongated into series of spikelike projections. **LJ:** narrow to wide, V-shaped, spikelike projections. **DF:** tip pointed, sometimes proximal part low with a prominent increase in height more posteriorly. **BS:** oval/depressed. **CP:** quite variable, often uniformly dark. **TL:** 25–70/36. **NO:** *M. lugubris* without labial teeth but pigmented structures in rows on lower labium look like labial teeth. **CI:** Arnoult and Razarihelisoa (1967); Blommers-Schlösser and Blanc (1991); Glaw and Vences (1992).

Rhacophorinae: CO: 9/220. **GR:** Africa–Madagascar–Seychelle Islands–India–southeastern Asia.

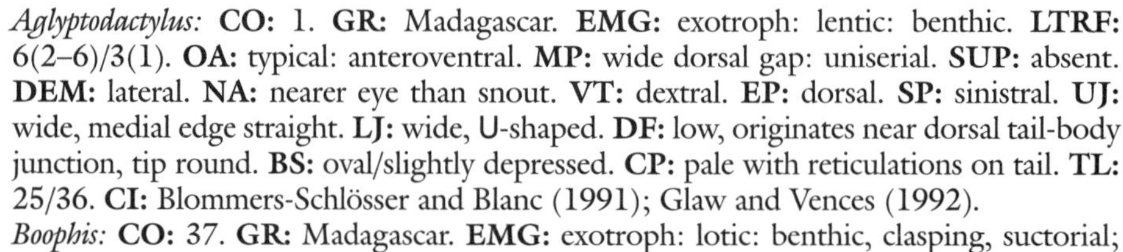

Aglyptodactylus: **CO:** 1. **GR:** Madagascar. **EMG:** exotroph: lentic: benthic. **LTRF:** 6(2–6)/3(1). **OA:** typical: anteroventral. **MP:** wide dorsal gap: uniserial. **SUP:** absent. **DEM:** lateral. **NA:** nearer eye than snout. **VT:** dextral. **EP:** dorsal. **SP:** sinistral. **UJ:** wide, medial edge straight. **LJ:** wide, U-shaped. **DF:** low, originates near dorsal tail-body junction, tip round. **BS:** oval/slightly depressed. **CP:** pale with reticulations on tail. **TL:** 25/36. **CI:** Blommers-Schlösser and Blanc (1991); Glaw and Vences (1992).

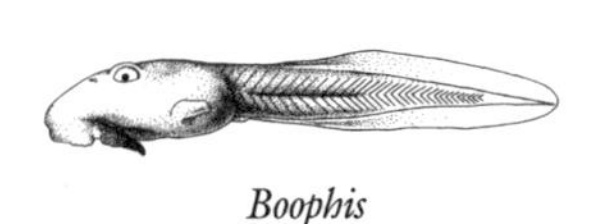

Boophis

Boophis: **CO:** 37. **GR:** Madagascar. **EMG:** exotroph: lotic: benthic, clasping, suctorial; lentic: benthic. **LTRF:** 3(2–3)/3, 5(2–5)/3(1), 6(2–6)/3[1], 6(3–6)/3(1), 7(2–7)/3(1), 75–7)/3, 8(3–8)/3, 8(5–8)/3. **OA:** typical: anteroventral, ventral. **MP:** wide dorsal gap, sometimes narrow ventral gap, uncommonly complete: uni- and biserial. **SUP:** common laterally and below P-3. **DEM:** lateral, absent. **NA:** various sites. **VT:** dextral. **EP:** dorsal. **SP:** sinistral. **UJ:** medium to wide and various shapes, absent. **LJ:** medium to wide, various shapes. **DF:** low, originates near dorsal tail-body junction, tip pointed to rounded. **BS:** oval/depressed. **CP:** variable mottled. **TL:** 30–90/36. **CI:** Blommers-Schlösser (1979b); Blommers-Schlösser and Blanc (1991); Glaw and Vences (1992).

Chirixalus: **CO:** 8. **GR:** Southeast Asia. **EMG:** exotroph: lentic: arboreal. **LTRF:** 5(2–5)/3, 4(2–4)/3(1), 2/2. **OA:** typical: terminal or anteroventral. **MP:** wide dorsal gap: multi- or uniserial midventrally, narrow ventral gap. **SUP:** absent. **DEM:** lateral, absent. **NA:** nearer snout than eye. **VT:** dextral. **EP:** dorsal. **SP:** sinistral. **UJ:** medium, massive construction, smooth arc. **LJ:** medium, V-shaped. **DF:** low, originates at dorsal tail-body junction, tip slightly rounded. **BS:** oval/depressed. **CP:** slightly speckled on tail, uniformly dark, distal part of tail dark. **TL:** 30/41. **NO:** *C. eiffingeri* oophagous. **CI:** Bourret (1942); Heyer (1971); Kam et al. (1996); Kuramoto and Wang (1987); Utsunomiya and Utsunomiya (1983).

Chiromantis: **CO:** 3. **GR:** sub-Saharan Africa. **EMG:** exotroph: lentic: benthic. **LTRF:** 3(2–3)/3, 5(2–5)/3. **OA:** typical: anteroventral. **MP:** wide dorsal gap, narrow ventral gap: uniserial dorsolaterally and biserial ventrally. **SUP:** absent. **DEM:** lateral. **NA:** small, nearer snout than eye. **VT:** medial. **EP:** dorsal. **SP:** sinistral. **UJ:** medium, smooth arc. **LJ:** narrow, V-shaped. **DF:** low, originates near dorsal tail-body junction, tip pointed.

BS: oval/depressed. **CP:** uniformly pale with blotches on tail. **TL:** 45–55/34. **CI:** Lamotte and Perret (1963a); Van Dijk (1966).

Nyctixalus: **CO:** 5. **GR:** Greater Sundas–Philippines. **EMG:** exotroph: lentic: arboreal. **LTRF:** 5(2–5)/3, 5(3–5)/3. **OA:** typical: anteroventral. **MP:** wide dorsal gap: biserial. **SUP:** absent. **DEM:** lateral. **NA:** equidistant between eye and snout, small medial projection. **VT:** medial. **EP:** dorsal. **SP:** sinistral. **UJ:** medial convexity, coarsely serrate. **LJ:** medium, V-shaped, coarsely serrate. **DF:** low, originates near dorsal tail-body junction, tip rounded. **BS:** oval/depressed. **TL:** 43/26. **CP:** uniformly purplish brown. **CI:** Alcala and Brown (1982); Inger (1966, 1985).

Philautus: **CO:** 87. **GR:** India–China–Philippines. **EMG:** exotroph: lentic: arboreal; endotroph. **LTRF:** 1/0. **OA:** typical: terminal. **MP:** complete: biserial. **SUP:** –. **DEM:** lateral. **NA:** tiny, nearer snout than eye. **VT:** medial. **EP:** dorsal. **SP:** sinistral. **UJ:** wide, without serrations. **LJ:** wide, coarsely serrate. **DF:** low, originates near dorsal tail-body junction, tip rounded. **BS:** oval/depressed, widest near plane of eyes, snout truncate. **CP:** uniformly pale brown. **TL:** 29/37. **CI:** Wassersug et al. (1981a).

Polypedates: **CO:** 19. **GR:** India–China–Japan–southeastern Asia–Philippines. **EMG:** exotroph: lentic: benthic. **LTRF:** 4(2–4)/3[1], 5(2–5)/3[1]. **OA:** typical: anteroventral. **MP:** wide dorsal gap: biserial, narrow ventral gap. **SUP:** few laterally, throughout extent of marginal papillae, absent. **DEM:** lateral. **NA:** nearer snout than eye. **VT:** dextral. **EP:** dorsal, lateral. **SP:** sinistral. **UJ:** medium, high smooth arc, medial convexity. **LJ:** medium, open V-shaped. **DF:** moderately high, originates near dorsal tail-body junction, tip pointed. **BS:** oval/depressed. **CP:** variable. **TL:** 25–45/36. **NO:** Inger (1966) shows tail sheath (see text chap. 3) in *P. macrotis* and *P. otilophus*. **CI:** Inger (1985).

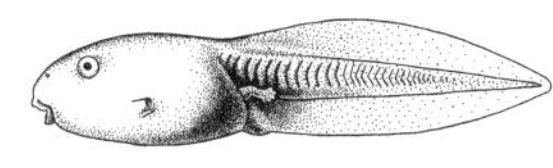

Polypedates

Rhacophorus: **CO:** 51. **GR:** India–China–Celebes Islands. **EMG:** exotroph: lentic: benthic, nektonic. **LTRF:** 4(2–3)/3(1), 5(2–5)/3[1]. **OA:** typical: anteroventral. **MP:** wide dorsal gap: biserial, sometimes small ventral gap. **SUP:** absent, few laterally. **DEM:** lateral. **NA:** nearer snout than eye. **VT:** dextral. **EP:** dorsal. **SP:** sinistral. **UJ:** medium to wide, smooth arc or small medial convexity. **LJ:** medium to wide, V-shaped. **DF:** low to medium, originates near dorsal tail-body junction, tip pointed. **BS:** oval/depressed. **CP:** usually unicolored, often pale. **TL:** 35–45/36. **CI:** Inger (1966, 1985).

Theloderma: **CO:** 9. **GR:** Burma–Malaysia. **EMG:** exotroph: lentic: arboreal. **LTRF:** 4(2–4)/4(1). **OA:** typical: nearly terminal. **MP:** wide dorsal gap: biserial. **SUP:** absent. **DEM:** lateral. **NA:** small, nearer snout than eye. **VT:** medial. **EP:** dorsal. **SP:** sinistral. **UJ:** medium, medial convexity, coarsely serrate. **LJ:** wide, more finely serrate, open U-shaped. **DF:** medium, erupts abruptly from near dorsal tail-body junction, tip broadly rounded. **BS:** round/depressed. **CP:** uniformly dark. **TL:** 28/28. **CI:** Wassersug et al. (1981a).

RHINODERMATIDAE: **CO:** 1/2. **GR:** southwestern South America.

Rhinoderma: **CO:** 2. **GR:** southwestern South America. **EMG:** endotroph.

RHINOPHRYNIDAE: **CO:** 1/1. **GR:** Texas–Costa Rica. **EMG:** exotroph: lentic: suspension feeder. **LTRF:** 0/0. **OA:** terminal, wide, slitlike mouth with multiple barbels. **MP:** not applicable. **SUP:** not applicable. **DEM:** not applicable. **NA:** small, nearer eye than snout. **VT:** medial. **EP:** lateral. **SP:** dual/lateral. **UJ:** not applicable. **LJ:** not applicable. **DF:** low, originates near dorsal tail-body junction, tip bluntly pointed. **BS:** oval/depressed. **CP:** uniformly pale or dark. **TL:** 50/36. **NO:** papilla-like structure at symphysis of infrarostral cartilages; looks grossly like a *Xenopus* tadpole.

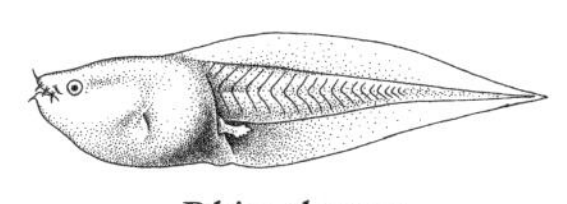

Rhinophrynus

Rhinophrynus: **CO:** 1. **GR:** Texas–Costa Rica. **CI:** Orton (1943); E. H. Taylor (1942).

SOOGLOSSIDAE: **CO:** 2/3. **GR:** Seychelle Islands.

Nesomantis: **CO:** 1. **GR:** Seychelle Islands. **EMG:** endotroph.

Sooglossus: **CO:** 2. **GR:** Seychelle Islands. **EMG:** endotroph.

Alphabetical List of Representative Morphotypes Presented in the Text: Genus, Guild, Species Drawn, and Source(s)

Acanthixalus (Arboreal Type 2); *A. spinosus;* after Lamotte et al. (1959a)
Acris (Benthic); *A. gryllus;* USNM 332799
Adelotus (Benthic); *A. brevis;* after Watson and Martin (1973)
Afrixalus (Nektonic); *Afrixalus* sp.; composite of USNM 308801-03
Ansonia (Clasping); *A. longidigita;* USNM 306204
Ascaphus (Suctorial); *A. truei;* USNM 297184
Astylosternus (Adherent); *Astylosternus* sp.; USNM 306967
Atelopus (Gastromyzophorous); *A. pachydermus;* USNM 286455
Bombina (Benthic); *B. bombina;* composite of cultured *B. orientalis* and after Angel (1946), Boulenger (1892), and Lanza (1983)
Boophis (Suctorial); *Boophis* sp.; USNM field number 107433A
Bufo (Benthic); *B. woodhousii;* USNM 320297
Cardioglossa (Benthic); *C. gracilis;* composite of USNM 306968 and *C.* sp. USNM 306969
Ceratophrys (Carnivorous Type 1); *C. cornuta;* USNM 299936
Chacophrys (Benthic); *C. pierottii;* after Faivovich and Carrizo (1992)
Cochranella (Fossorial); *C. granulosa;* composite of USNM 291045 and after *C. uranoscopa* of Heyer (1985)
Colostethus (Benthic); *C. trinitatis;* USNM 167515
Colostethus (Neustonic); *C. flotator;* composite of USNM 203704-07
Cyclorana (Benthic); *C. cultripes;* after Watson and Martin (1973)
Dendrobates (Arboreal Type 2); *D. quinquivittatus;* USNM 269075
Discoglossus (Benthic); *D. pictus;* after Boulenger (1892)
Duellmanohyla (Adherent); *D. salvavida;* USNM 304979
Hamptophryne (Suspension-feeder Type 1); *H. boliviana;* composite of USNM 298826, 299953, and 299957
Heleioporus (Benthic); *H. australiacus;* after Watson and Martin (1973)
Heleophryne (Suctorial); *H. regis;* USNM 300543
Hemisus (Benthic); *H. guttatum;* composite of USNM 308795 and 308798
Hyla (Suctorial); *H. armata;* USNM 346278
Hyla (Arboreal Type 1); *H. bromeliacia;* USNM 304997
Hyla (Arboreal Type 2); *H. picadoi;* USNM 331374
Hyla (Macrophagous); *H. sarayacuensis;* USNM 313525
Hymenochirus (Carnivorous Type 2); *H. curtipes;* USNM 140275
Insuetophrynus (Benthic); *I. acarpicus;* composite after Diaz and Valencia (1985) and Formas et al. (1980)
Kassina (Nektonic); *K. senegalensis;* composite of USNM 308809 and Wager (1986)
Lepidobatrachus (Carnivorous Type 2); *L. llanensis;* composite of USNM 307185 and Cei (1980)
Leptobrachella (Benthic); *L. gracilis;* after Inger (1966)
Leptobrachium (Benthic); *Leptobrachium* sp.; composite of *L. hasselti* and *L. nigrops* after Inger (1966) and Berry (1972)
Leptodactylodon (Neustonic); *Leptodactylodon* sp.; USNM 306971
Leptodactylus (Benthic); *L. melanonotus;* USNM 304941
Leptodactylus (Benthic, sometimes carnivorous); *L. pentadactylus;* USNM 203730
Leptopelis (Benthic); *L. natalensis;* composite of USNM 308812 and Wager (1986)
Leptophryne (Benthic); *L. borbonica;* after Berry (1972)
Mantidactylus (Benthic); *Mantidactylus* sp.; USNM field number 107426A
Mantidactylus (Fossorial); *M. femoralis;* USNM field number 107404
Mantidactylus (Fossorial); *M. lugubris;* USNM field number 107405
Mantidactylus (Neustonic); *M. opiparus;* USNM field number 11828
Megophrys (Neustonic); *Megophrys* sp.; USNM 292051
Microhyla (Neustonic); *M. heymonsi;* composite of USNM 206402-05
Nelsonophryne (Suspension-feeder Type 1); *N. aterrimus;* after Donnelly et al. (1990)
Otophryne (Psammonic); *O. pyburni;* composite of USNM 304280 and after Wassersug and Pyburn (1987)
Pelodytes (Benthic); *P. punctatus;* composite after E. N. Arnold and Burton (1978), Boulenger (1892), Lanza (1983), and Nöllert and Nöllert (1992)
Phasmahyla (Neustonic); *P. guttata;* USNM 241258
Phyllomedusa (Suspension-rasper); *P. vaillanti;* USNM 321149
Polypedates (Nektonic); *P. megacephala;* USNM 313968
Pseudis (Nektonic); *P. paradoxa;* USNM 302522
Ptychadena (Benthic); *Ptychadena* sp.; composite of USNM 308829 and Wager (1986)
Rhinophrynus (Suspension-feeder Type 2); *R. dorsalis;* composite of USNM 306925, 307148, and 307157
Scaphiophryne (Suspension-feeder but may rasp or feed as a carnivore); *Scaphiophryne* sp.; USNM field number 11691
Scaphiopus (Benthic); *S. couchii;* USNM 320302
Scarthyla (Nektonic); *S. ostinodactyla;* USNM 266112
Schismaderma (Benthic); *S. carens;* USNM 308793
Scinax (Nektonic); *S. sugillata;* composite of USNM 285280-88
Scinax (Nektonic); *S. elaeochroa;* USNM 330834
Scinax (Nektonic); *S. staufferi;* composite of USNM 306917 and 306921
Stephopaedes (Arboreal Type 3); *S. anotis;* after Channing (1978)
Thoropa (Semiterrestrial); *T. miliaris;* USNM 209372
Trichobatrachus (Suctorial); *T. robustus;* composite of USNM 306974 and 306976
Werneria (Suctorial); *W. preussi;* reconstructed after dorsal and ventral views in Mertens (1938)
Xenopus (Suspension-feeder Type 2); *X. laevis;* USNM 308822

Summary

Several interesting things came out of our attempt to characterize the morphological variation represented among all genera and families of anuran tadpoles. The use of a standard format and terminology gave us a much better appreciation of the distribution of larval traits among living frogs. This summation also uncovered the types of gaps in our knowledge. By our count, no information is available on the reproductive mode for 29 genera (~8%) of frogs, and only four of these are reasonable candidates for endotrophy. Most (86%) of these 43 species are in monotypic genera, and filling this gap has a high probability of uncovering different tadpole morphotypes. Although we currently do not know the total number of different tadpoles that have been described, we know that those of many species in genera from which some tadpoles are known remain to be discovered. Given past findings, we are confident that some of these undiscovered forms also will provide new insights into tadpole morphology. While this summation confirms that we know something about the reproductive mode of most frog species, the inclusion of data for a given taxon does not imply that comparable data are available for all species in that genus nor that the recorded data necessarily are representative of tadpoles of all included species. Less than a third of all tadpoles have been described reasonably well, and the number of entries that have no data, including endotrophs and exotrophs, shows that much work remains. We have a reasonably good foundation in certain aspects of tadpole morphological diversity, but there is much to learn.

Another spin-off from this compilation was a better appreciation of the geographical gaps in our knowledge. It seems clear to us that of the tropical regions of the world where frog diversity is high, the tadpole fauna of equatorial Africa stands out as one in need of much more study. Whether a broader sampling of larval morphology would shed new light on ranoid phylogeny is debatable, but the few glimpses we have had of morphological diversity and developmental patterns characteristic of some arthroleptid tadpoles cry out for attention. We are not aware of much current interest in tadpoles from tropical Africa and wonder if there are unworked larval collections in some European museums similar to the extensive holdings of reptiles that were amassed during periods of European colonial occupancy of tropical Africa. If not, efforts should be made to overcome the logistical and politico-economic problems that too often make field work in these areas so difficult and assemble new collections that are directed specifically at the tadpole fauna. To facilitate tadpole work there and in other regions of the world, we have summarized (table 12.4) the general distributions of anuran families as an aide to their identification and an adjunct to the comparative study of tadpole morphology. The following key also grew out of the summation and is intended not so much as an help to identification (which it may be) but more as an illustration of the distribution, including notable cases of convergence, of a few major larval traits among anuran families and subfamilies. All taxa are not accounted for in every case, and the key could have been compiled in several ways (i.e., the same or different characters in different sequences). Even so, the conveyed message in each case would have been the same—tadpoles in a few families can be keyed easily and those in another large group of families cannot be separated based on these few characters.

Table 12.4 World ranges of anuran families with exotrophic tadpoles in zoogeographical realms. Families are listed in approximate order of increasing number of species in each realm, and endemic families are in boldface

African: **Sooglossidae, Heleophrynidae, Hemisotidae,** Microhylidae, **Arthroleptidae,** Pipidae, Rhacophoridae, Bufonidae, Ranidae, and **Hyperoliidae**

Austropapuan: Ranidae, Hylidae, and **Myobatrachidae**

Nearctic: **Ascaphidae,** Rhinophrynidae, Leptodactylidae, Microhylidae, Pelobatidae, Bufonidae, Ranidae, and Hylidae

Neotropical: Rhinophrynidae, **Pseudidae,** Pipidae, Ranidae, Microhylidae, Bufonidae, **Centrolenidae, Dendrobatidae,** Leptodactylidae, and Hylidae

Oriental: Bombinatoridae, Hylidae, **Megophryidae,** Microhylidae, Bufonidae, Rhacophoridae, and Ranidae

Palearctic: **Pelodytidae,** Pelobatidae, Hylidae, Bombinatoridae, **Discoglossidae,** Bufonidae, and Ranidae

An Illustrative Key to Families of Anuran Tadpoles Based on a Few Major Characters

SECTION 1. SPIRACLE

A. Dual and lateral, centripetal wall absent	Section 2
B. Single, midventral on chest, abdomen, or near vent, centripetal wall absent except in cases where spiracle is near vent	Section 3
C. Single, low on left side, sometimes almost medial, centripetal wall often absent	Section 4
D. Single, sinistral, centripetal wall usually present, at least as a ridge	Section 5

SECTION 2. BARBELS AND ORAL APPARATUS

A. Single, long, motile, at corners of mouth; oral disc, jaw sheaths, and labial teeth absent	Pipidae: Pipinae and Xenopodinae
B. Multiple, short, nonmotile, around mouth margin; oral disc, jaw sheaths, and labial teeth absent	Rhinophrynidae
C. Barbels absent; oral disc and labial teeth absent; jaw sheaths present as few, spikelike projections	Leptodactylidae: Ceratophryinae (part)

SECTION 3. ORAL APPARATUS, NARES, AND SPIRACLE

A. Oral apparatus consists of large suctorial disc, large platelike upper jaw sheath, small lower sheath, and some rows of biserial labial teeth; nares present; spiracle on chest	Ascaphidae
B. Oral apparatus consists of semicircular oral flaps overhanging mouth and no jaw sheaths or labial teeth; nares absent; spiracle on abdomen or near vent	Microhylidae: Dyscophinae, Microhylinae (part), Phrynomerinae, and Scaphiophryninae (part)
C. Oral apparatus consists of large umbelliform oral disc derived mostly from the lower labium and no jaw sheaths or labial teeth; nares absent; spiracle on abdomen	Microhylidae: Microhylinae (also see section 5)

SECTION 4. MARGINAL PAPILLAE, TOOTH ROWS, A-1, AND LTRF

A. Marginal papillae complete or nearly so; tooth rows biserial; A-1 almost the same length as A-2; LTRF 2/3	Bombinatoridae and Discoglossidae
B. Marginal papillae complete or with dorsal gap; tooth rows uniserial; A-1 almost the same length as A-2; LTRF 2/3	Hylidae: Phyllomedusinae (part)
C. Marginal papillae complete or nearly so (i.e., narrow dorsal gap); tooth rows uniserial; A-1 much shorter than A-2; LTRF $> 2/3$	Megophryidae (part), Pelobatidae
D. Marginal papillae complete or with dorsal gap; tooth rows uniserial; A-1 almost the same length as A-2; LTRF $\leq 6/3$	Myobatrachidae: Myobatrachinae (part)

SECTION 5. ORAL DISC, JAW SHEATHS, LABIAL TEETH, AND BELLY

A. Oral disc reduced to rim around mouth, usually with dorsal gap; jaw sheaths present; labial teeth present or not; body wall unmodified	Hylidae Hemiphractinae (part), Hylinae (part) and Ranidae: Raninae (part)
B. Oral disc present, typical size or larger than normal but umbelliform (upturned); jaw sheaths present or not; labial teeth present (uniserial) or not; body wall not modified	Arthroleptidae: Astylosterninae (part), Dendrobatidae (part), Megophryidae (part), and Rhacophoridae: Mantellinae (part; also see section 3)
C. Oral disc normal; one or both jaw sheaths absent; labial teeth present; body wall unmodified	Section 6

D. Oral disc normal; both jaw sheaths present; labial teeth present; body wall unmodified	Section 7
E. Oral disc normal; both jaw sheaths present; labial teeth present; body wall variously modified	Section 8

SECTION 6. LTRF BALANCE VALUE

A. LTRF imbalanced negatively (3/14 = −11)	Heleophrynidae
B. LTRF imbalanced positively (9/3 = +6)	Rhacophoridae: Mantellinae (part)

SECTION 7. ORAL DISC AND LTRF

A. Oral disc enlarged, often with many marginal papillae small and closely spaced (i.e., rheophilous), may be complete or not; number of tooth rows 2/3–17/21; lotic habitats	Bufonidae (part), Hylidae: Hylinae and Pelodryadinae, Leptodactylidae: Telmatobiinae (part), Megophryidae (part), Myobatrachidae: Myobatrachinae, Ranidae: Raninae and Petropedetinae
B. Oral disc typical with larger, more widely spaced marginal papillae, usually with at least a dorsal gap (i.e., not rheophilous); number of tooth rows ≥ 1/3; leutic or lotic habitats	Bufonidae (part), Centrolenidae, Dendrobatidae (part), Hemisotidae, Hylidae: Hylinae, Pelodryadinae (part), and Phyllomedusinae (part), Hyperoliidae, Leptodactylidae: Telmatobiinae (part), Myobatrachidae: Limnodynastinae and Myobatrachinae, Pelodytidae, Pseudidae, and Ranidae: Raninae and Petropedetinae (part)

SECTION 8. BODY WALL MODIFICATION

A. Belly modified into actively operating sucker (i.e., center can be manipulated to form negative pressure between belly and substrate = gastromyzophorous)	Bufonidae (part) and Ranidae: Raninae (part)
B. Lateral and posterior body wall developed into free flap that cannot be manipulated to form negative pressure	Leptodactylidae: Telmatobiinae (part)
C. Lateral body wall with various forms and degrees of lateral sac development	Arthroleptidae: Astylosterninae (part)
D. Fleshy, hollow crown surrounding eyes and nares	Bufonidae (part)
E. Transverse, fleshy flap behind eyes	Bufonidae (part)
F. Fingerlike projection ventrolaterally at junction of opercular and abdominal cavities	Microhylidae: Melanobatrachinae

ACKNOWLEDGMENTS

Several persons helped with the preparation of this chapter in different ways. K. Spencer's sketches of many of the distinctive tadpole morphotypes enhanced the diversity presentation immeasurably. The sketch of *Leptobrachella* was redrawn from Inger (1966) and is reprinted by permission of Inger; the sketch of *Nelsonophryne* was redrawn from Donnelly et al. (1990a) and is reprinted courtesy of The American Museum of Natural History and De Sá. R. J. Richards supplied a number of pertinent papers on the Australian fauna at a moment's notice. Over the past thirty years the curators and staff at several museums have loaned specimens, permitted dissections and other special preparations, and collected or called to our attention material that has noticeably advanced our research. Without the cooperation of these and others our understanding and awareness of tadpole morphology would have been noticeably lessened.

GLOSSARY

The glossary includes terms that may have been defined or restricted by the terminology suggested in this book or are not commonly found in other sources. Refer to chapters 2 and 3 for further discussion of many morphological terms. Pertinent citations are often presented. Terms within a definition that are defined elsewhere in the glossary are in boldface. Synonymous terms that are accepted in this book are presented in parentheses, and unaccepted ones are presented in brackets. Unless noted otherwise, all staging notations pertain to Gosner (1960).

A-2 gap (Altig 1970) [medial interval or space]. Medial gap in second upper row in species with 2 upper rows; can be expanded to denote any row on either labium (e.g., A-4 gap, P-3 gap); see **A-2 gap ratio** and **labial tooth row formula.**

A-2 gap ratio (Altig 1970). Length of either section of the actual row of teeth in row A-2 divided by the length of the medial gap between the two parts of the row; number > 1 indicates a gap smaller than the length of a row section and vice versa; can be calculated for any row that is broken medially on either labium; see **A-2 gap** and **gap.**

abdominal length (Carr and Altig 1991). Medial, straight line distance from base of vent tube to the spiracular wall; this measurement is sometimes useful but has a large measurement error because neither landmark is consistently definable.

abdominal sucker (belly sucker) [belly or sucker disc]. See **belly sucker,** also **adhesive gland, belly flap,** and **gastromyzophorous.**

accessory labial tooth rows (R. G. Webb and Korky 1977). Short tooth rows oriented vertically or transversely and often positioned near the lateral ends of tooth rows; occur normally in some ranids and pelobatids, infrequently in other taxa; this modified LTRF shows four accessory rows between solidi: 3(2)/4/3(1).

acosmanura (Starrett 1973). Word synonym for Orton's (1953) Type 4 **tadpole;** ranoid of O. M. Sokol (1975).

adaptive peak. Combination of allele frequencies or morphologies that leads to the locally maximal fitness of a population.

additive genetic variance. That fraction of total phenotypic variation caused by alleles that are statistically additive in effect; can be acted upon by natural selection; see **nonadditive genetic variance.**

adherent. An **ecomorphological guild** that includes those **lotic tadpoles** that live in flowing water and have small, complete **marginal papillae, LTRF** commonly 2/3, and inhabit faster water than **tadpoles** in the **clasping** guild, position maintenance via **oral disc** common to continuously; see table 2.2, figures in chap. 12, **adherent, benthic, clasping, fossorial, gastromyzophorous, nektonic, neustonic, psammonic, rheophilous, semiterrestrial,** and **suctorial.**

adhesive gland [cement gland, oral or ventral sucker]. Transitory sticky gland of various shapes on the ventral surface of the head of an **embryo** or **hatchling** immediately posterior to the stomodeum or **mouth;** used to stabilize hatchlings before swimming abilities develop; connotations of "sucker" should not be used in this case.

afferent nerve or axon. Neuron that carries nerve impulses to the brain (= sensory) or into a brain **nucleus;** see **efferent nerve** or **axon.**

aggregation [school]. A general term that describes a group of tadpoles that are congregated in response to some other factor(s) than social interactions; qualities of a school (e.g., polarized orientation and coordinated movements) and the reason for the congregation are not implied; see **biosocial aggregation, school,** and **taxic aggregation.**

alar plate. Sensory portion of the brain stem, oriented vertically and located dorsal to the sulcus limitans along the lateral ventricular wall; see **basal plate.**

allelopathy. Production of chemicals into the environment by one individual that results in a reduction in the fitness of other organisms.

allochthonous. Originating outside the system; usually in reference to food materials.

amphigyrinid. Describes the condition of having dual, lateral **spiracles** as in pipids, rhinophrynids, and *Lepidobatrachus* (leptodactylid), although latter case not homologous with former two (Ruibal and Thomas 1988); see **laevogyrinid, mediogyrinid, paragyrinid,** and **sinistral.**

anamniote. Describes the condition of lacking an amnion; characteristic of fishes and amphibians.

anlage (n). An embryological precursor.

antagonistic pleiotropy. Negative genetic correlation between two traits caused by the fixation of pleiotropic genes that affect both traits in the adaptive direction, leaving those pleiotropic genes that affect one trait in the adaptive direction and the other in the maladaptive direction to contribute the genetic variation and covariation.

anus. See **vent** and **vent tube.**

aortic arches. Aortic vessels parallel with the posterior visceral arches and connecting the heart with the dorsal aorta; the major vessels carrying oxygenated blood.

AR-1 (Thibaudeau and Altig 1988). Expansion of the tooth row notation of Altig (1970) to designate the **tooth ridge** prior to tooth development in developmental studies; can be expanded to denote any tooth ridge.

arboreal. An **ecomorphological guild** that includes those **lentic tadpoles** that are modified in one of several morphological patterns for living and feeding in water-filled **phytotelmata** or similar arboreal sites; see table 2.2, figures in chap. 12, **benthic, carnivorous, macrophagous, nektonic, neustonic, suspension feeder,** and **suspension rasper.**

barbel [papilla]. Fleshy, filiform projection(s) at the corners of the mouth of pipid (long with internal support and musculature) and rhinophrynid (short, no internal support or musculature) **tadpoles,** although structures are not homologous between the two families (Cannatella and Trueb 1988a); barbel at the symphysis of the lower jaw of *Rhinophrynus* tadpoles is probably not homologous with either of the others.

basal plate. Motor portion of the brain stem oriented horizontally and located ventral to the sulcus limitans; see **alar plate.**

behavioral fever. Elevation of the preferred body temperature of an **ectotherm** in response to infection or inoculation with pathogenic bacteria, viruses, or products of these organisms.

behavioral thermoregulation. Regulation of body temperature (usually by an **ectotherm**) by movement or changes of posture or position within or between thermal regimes.

belly flap. Out-folding of lateral, posterior, or both parts of the body wall to form a flap; probably used in attachment by **semiterrestrial tadpoles** but only by increasing surface area for cohesion; see table 2.2, figures in chap. 12, **belly sucker,** and **gastromyzophorous.**

belly sucker (abdominal sucker) [belly or sucker disc]. The modified belly of a **gastromyzophorous tadpole,** includes a raised rim, musculature to raise the roof of the sucker to actually form a negative pressure, and papillate and sometimes keratinized areas on the roof; see table 2.2, figures in chap. 12, **belly flap, rheophilous,** and **suctorial.**

benthic [benthonic]. An **ecomorphological guild** that includes those **lentic** or **lotic tadpoles** that rasp food from submerged surfaces with keratinized **mouthparts;** mostly at or near the bottom, pools and backwaters in lotic sites, **body depressed,** eyes **dorsal, fins** low with rounded or slightly pointed tip, dorsal fin originates at or near the dorsal **tail-body junction;** see table 2.2, figures in chap. 12, **adherent, arboreal, carnivorous, clasping, fossorial, gastromyzophorous, macrophagous, nektonic, neustonic, psammonic, rheophilous, semiterrestrial, suctorial, suspension feeder,** and **suspension rasper.**

bicolored. Two colored, usually in reference to a lateral view of a tail muscle that is dark dorsally and pale ventrally.

biosocial aggregation. A general term that describes a group of tadpoles that are congregated because of some social interaction, although qualities of a school (i.e., polarized orientation and coordinated movements) are not implied; see **aggregation, school,** and **taxic aggregation.**

birth. (1) Release of some form of an immature individual from the reproductive tract of its mother, including **viviparous** and **ovoviviparous** amphibians. (2) By extension, used for the analogous release of a **froglet** from nonoviducal sites (e.g., *Assa, Gastrotheca, Rheobatrachus,* and *Rhinoderma*) in a parent's body; see **hatch.**

biserial. Describes structures occurring in two series at a site, as two rows of labial teeth per tooth ridge (e.g., some tooth rows in *Ascaphus* and all tooth rows in bombinatorids and discoglossids); see **multiserial** and **uniserial.**

blotch. Contrasting, usually dark, irregular marks of various sizes and shapes with indistinct margins; larger and more irregular than a **dot** or **spot;** see **mottled, punctate,** and **stellate.**

body. (1) [head-body] That part of a **tadpole** minus the **tail;** defined for measurement as the straight line distance from the tip of the snout posteriorly to the **body terminus;** see **body length.** (2) Poorly distinguishable part of a labial tooth (Gosner 1959) between the **head** (working surface) and **sheath** (embedded in **tooth ridge**).

body axis. Imaginary line in lateral view that extends from the tip of the snout to the **body terminus;** used as a reference for describing the relative locations of structures; see **tail axis.**

body length [head-body length]. Straight line, lateral measurement from the tip of snout to the **body terminus;** see fig. 3.1 and **tail length.**

body ratio (Altig 1970). **Total length** divided by **body length;** see **tail length** and **tail ratio.**

body terminus. Intersection of the posterior body wall and the axis of the tail myotomes; most accurately describes **body length** and **tail length** among diverse taxa in contrast to any measurement involving the **vent tube;** see chap. 3.

branchial arch(es). The three gill-supporting **visceral arches.**

branchial food trap (Wassersug 1976a, b). Glandular region between the ventral velum and filter plates of larval Types 3 and 4; see chap. 5.

buccal apparatus [oral apparatus]. All structures in the buccopharyngeal cavity taken as a unit; does not refer to any structures of the **mouthparts.**

buccal papilla(e). Collective term for all papillae in the buccal cavity; see **buccal apparatus.**

buccopharynx, anterior and posterior. Collective term for the buccal cavity (anterior) + pharynx (posterior).

carnivorous. An **ecomorphological guild** that includes those **lentic tadpoles** that feed on macroinvertebrates and conspecific and heterospecific **tadpoles,** either rasping the prey apart or engulfing them intact; keratinized **mouthparts** present and usually modified; does not include opportunistic scavenging of dead organisms; see tables 2.2 and 10.2, figures in chap. 12, **arboreal, benthic, macrophagous, nektonic, neustonic, suspension feeder,** and **suspension rasper.**

cauda equina. The spinal nerves that exit the spinal cord near the sacrum and then continue posteriorly as a series of fibers.

choana(e) (internal nares). Opening of the narial canal in the roof of the **buccopharynx.**

chromatophore. Pigment-containing cells derived from neuroectoderm; pigments contained in specific organelles; see **blotch, dot, melanin, melanophore, mottled, punctate, spot,** and **stellate.**

ciliary cushion (Wassersug 1976a, b; chap. 5). In larval Types 3 and 4, at the dorsolateral pharynx, structures hanging between the filter plates and consisting of ciliary cells and goblet cells originating from the esophagus; structurally like the **ciliary grooves** and in close contact with them; produces and transports mucus with entrapped food particles to the esophagus.

ciliary groove (Wassersug 1976a, b; chap. 5). A horizontal groove at the margin of the pharyngeal roof from the anterolateral corner of the ventral **velum** and eventually into the esophagus; positioned laterally in pipids and more posteriorly in larval Types 3 and 4; histology and function resembles the **ciliary cushions.**

clasping. An **ecomorphological guild** that includes those **lotic tadpoles** that live in flowing water and have **marginal papillae** with an anterior gap, LTRs commonly 5 but as numerous as 8/8 and usually with anterior rows more numerous than lower rows (e.g., 9/3); inhabit medium to slow currents, position maintenance via the **oral disc** minor; see table 2.2 and figures in chap. 12, **adherent, benthic, fossorial, gastromyzophorous, nektonic, neustonic, psammonic, rheophilous, semiterrestrial,** and **suctorial.**

clearance time. Period of time between ingestion of food and elimination of feces derived from that food.

cold hardening. Accommodation to a brief exposure of low temperature near the lethal limit such that the individual is more resistant to subsequent exposures to such temperatures; see **heat hardening.**

complex life cycle. A life cycle in amphibians characterized by two forms separated by an abrupt shift (= **metamorphosis**) in behavior, ecology, morphology, and physiology.

compressed [laterally compressed, depressed, or flattened]. Describes an imaginary cross-sectional view of a structure that is higher than wide; usually used in reference to body shape; see **cylindrical, depressed, fusiform,** and **globular.**

concentration refuge. That concentration or density of food below which a **tadpole** ceases to feed and thus the remaining prey survive and reproduce.

contralateral. Occurring on the opposite side of the body; see **ipsilateral.**

conversion efficiency. Ratio of mass of food ingested to body mass increase.

cranial nerve. One of about 14 nerves connected to the telencephalon (CNs 0–I), midbrain (CNs II–IV), and hindbrain (CNs V–XII); absolute number of cranial nerves is controversial; see **spinal nerve.**

crenulate. Describes an edge with a wavy margin as would be produced by a series of weakly differentiated papillae; see **emarginate** and **incised.**

critical oxygen tension (P_c). **PO_2** below which the rate of O_2 consumption of a metabolic O_2 regulator becomes dependent upon ambient **PO_2.**

critical thermal maximum (CT_{max}). Upper limit of the temperature tolerance range at which locomotor activity becomes disorganized and the animal loses its ability to escape from conditions that will promptly lead to its death; see **critical thermal minimum.**

critical thermal minimum (CT_{min}). Lower limit of the temperature tolerance range at which locomotor activity becomes disorganized and the animal loses its ability to escape from conditions that will promptly lead to its death; see **critical thermal maximum.**

cusp. Projection(s) of various size, shape, number, and orientation that occur on the working end (= **head**) of a **labial tooth;** does not refer to **serrations** on **jaw sheaths;** see **incised.**

cuspate. A term that has been used to describe jaw sheaths but should be abandoned (see **incised**) because of confusion with **cusps** of **labial teeth.**

cylindrical. Describes an imaginary cross-sectional view of a structure that approaches equidimensionality, therefore quasi-circular, neither depressed nor compressed; usually used in reference to body shape; see **compressed, depressed, fusiform,** and **globular.**

degree-day. Measure of cumulative potential biological activity, usually a sum of daily mean temperatures.

denticle. Unacceptable term that should be discarded in reference to the **labial teeth** of **tadpoles.**

depressed [dorsoventrally compressed or dorsoventrally flattened]. Describes an imaginary cross-sectional view of a structure that is wider than high; usually used in reference to body shape; see **compressed, cylindrical, fusiform,** and **globular.**

detritus. Dead organic matter.

developmental rate (DR). Change in form or stage through time.

dextral. To the right. Refers to a **vent tube** that opens to the right of the plane of the ventral fin or tail muscle; emergence from abdomen may be sagittal or not; see **medial** and **sinistral.**

diaphragm (peribranchial wall). The wall or septum separating the viscera from the **buccopharynx** of a **tadpole.**

differentiation rate. Generally, the inverse of time to develop between two given stages.

direct development. (1) Developmental mode of a frog that results in a **froglet** that developed without a **tadpole** morphotype from an egg not intimately associated with a parent's body. (2) An **ecomorphological guild** that includes species that have direct development; see table 2.2 and figures in chap. 12; see **endotrophy, exotrophy, exoviviparous, nidicolous, ovoviviparous, paraviviparous,** and **viviparous.**

dorsal. (1) Describes a category of eye position in which no part of the eye is included in a dorsal silhouette of the **tadpole** (i.e., the entire eye is positioned medial to the lateral surface of the **body**). (2) An anatomical descriptor meaning "toward top or dorsum" relative to a stated landmark; see **lateral.**

dorsal body terminus: See **body terminus** and **tail-body junction.**

dot. Small, rounded, dark marking with discrete edges; see **blotch, mottled,** and **spot.**

ecological species. A population of a species that is ecologically different from the rest of the population; occupies a separate niche.

ecomorphological guild. A categorization of a group of **tadpoles** from several taxa that share common morphological features that collectively suggest some sort of commonality in ecology; **tadpoles** with similar ecologies have similar morphologies regardless of their taxonomic relationships.

ectodermal-endodermal transition zone. Zone where the epiderm from the gills transition zone overlays the pharyngeal endoderm of the anlagen of filter plates.

ectotherm [cold-blooded, poikilotherm]. An organism that cannot retain and regulate metabolic heat, therefore, barring **behavioral thermoregulation,** the body temperature varies with the ambient temperature; see **behavioral fever** and **behavioral thermoregulation.**

efferent nerve or axon. Neuron that projects away from (= motor) the brain or a brain **nucleus;** see **afferent nerve** or **axon.**

elygium (Van Dijk 1966). A pigmented zone at the basal margin of the iris (= ocular elygium) or at the skin-cornea margin (= epidermal elygium); presumably shield the eye from excessive light; see **umbraculum.**

emarginate [indented, infolded, folded]. Describes a margin of the **oral disc** (most commonly lateral) with a real indentation(s), not one caused by folding or bending; distinction should be noted between emarginations, folds (= overlapping areas) and tucks (= marginal retractions caused by tensions of connective tissue and labial musculature).

embryo. An individual in any stage of development from fertilization until hatching; see **froglet, hatchling, immature, juvenile, larva, metamorph,** and **tadpole.**

endotroph, endotrophy (Altig and Johnston 1989). General developmental mode that results in an **embryo** or **larva** which derives its developmental energy either entirely from vitellogenic yolk or other parentally produced material, sometimes is a nonfeeding **tadpole;** see table 2.2, **direct development, exotrophy, exoviviparous, nidicolous, oviviparous, paraviviparous,** and **viviparous.**

exotroph, exotrophy (Altig and Johnston 1989). General developmental mode which results in a **larva** (**tadpole**) that feeds on various materials not derived from a parent or trophic eggs provided by subsequent ovulations of the parent, a feeding **tadpole;** see table 2.2, **direct development, endotrophy, exoviviparous, nidicolous, ovoviviparous, paraviviparous,** and **viviparous.**

exoviviparous (Altig and Johnston 1989). (1) Developmental mode of some frogs in which a hatched individual moves from where the eggs were oviposited to another site (often associated with male parent) from which it is eventually "birthed." (2) An **ecomorphological guild** that includes species that have some form of exoviviparous development; see table 2.2 and figures in chap. 12; see **direct development, endotrophy, exotrophy, nidicolous, ovoviviparous, paraviviparous,** and **viviparous.**

explosive breeding. All animals in a local population reproduce within a few days.

fictive swimming. Involuntary swimming induced in laboratory animals in order to study kinematic interactions (e.g., coordination of muscle groups or the cycle of firing of motor neurons).

filiform. Describes a long, narrow object (i.e., filament-like in form). (1) Often a projection that usually has uniform surfaces (e.g., shape of many **buccal papillae**). (2) Describes the shape of some **melanophores.**

fin [caudal or tail crest]. Unsupported flaps of epidermis-covered, loose connective tissue usually extending the length of the dorsal and ventral edge of the tail muscle.

final thermal preferendum (FTP). Temperature ultimately selected by an individual regardless of prior thermal experience; usually measured within 24–96 hr after placement in a thermal gradient.

flagellum (Altig and Johnston 1989) [filament]. Elongate posterior part of the tail that is often nonpigmented and may move independently of the remainder of the tail.

fluctuating selection. Adaptive value of a trait varies as a function of varying abiotic and biotic conditions.

froglet. A small, sexually immature frog produced by either metamorphosis of an **exotrophic tadpole** or by **direct development. Metamorph** is not applicable in the case of direct developing forms; see **embryo, hatchling, immature, juvenile, larva, metamorphic, metamorphic climax, premetamorphosis,** and **prometamorphosis.**

fossorial. An **ecomorphological guild** that includes those **lotic, fusiform tadpoles** that have **marginal papillae** with an anterior gap, LTRF 2/3 or less, and inhabit leaf mats in slow water areas; position maintenance via **oral disc** absent; see table 2.2, figures in chap. 12, **adherent, benthic, clasping, semiterrestrial, gastromyzophorous, nektonic, neustonic, psammonic, rheophilous,** and **suctorial.**

foliose. Foliage- or moss-like pattern or shape of individual or groups of **chromatophores.**

fracture plane. Suture-like area in the **body** of a **labial tooth** where the **head** can break off of the **sheath.** The next **replacement tooth** (see **tooth series**) protrudes through the sheath of the old tooth.

friction surface. Area that may be finely papillate, keratinized, or both on the roof of the **belly sucker** of some **gastromyzophorous tadpoles.**

functional constraint. Limit to evolution by natural selection caused by physical or mechanical laws.

funnel mouth. Has been used in reference to the upturned **oral disc** of some surface feeders (= **umbelliform**) **tadpoles** and the large, pendant (most easily visible in preserved specimens) **oral discs** of some lotic species; these usages should be abandoned to enhance clarification among the **mouthparts, oral disc,** and **mouth;** also see **suctorial.**

fusiform [vermiform]. Describes the **body** shape of **fossorial tadpoles,** extremely **depressed** with **dorsal** eyes, long **tail,** and low **fins;** see **compressed, cylindrical, depressed,** and **globular.**

ganglion. Enlargement in a nerve outside the central nervous system where pre- and postganglionic axons synapse or cell bodies of sensory nerves are located; see **nucleus.**

gap. Literally a break or discontinuity in a linear structure; used in reference to breaks, usually medial, in a tooth row; see **A-2 gap** and **A-2 gap ratio.**

gap ratio. See **A-2 gap ratio.**

gape limited. Describes a predator that cannot eat prey above a maximum size set by the size of its mouth; see **size refuge.**

gas exchange partitioning. Partitioning of O_2 uptake and CO_2 elimination among available external gas exchange surfaces (e.g., lungs, gills, skin, etc.) or between media (air and water).

gastromyzophorous. (1) Describes a **tadpole** that has the belly modified into an actual sucker (= forms negative pressure), like tadpoles of *Amolops* and *Atelopus* but not *Thoropa* and *Cycloramphus,* for position maintenance in fast and perhaps turbulent water. An **ecomorphological guild** that includes those **lotic** tadpoles that have such a modified belly; see table

2.2, figures in chap. 12, **adherent, belly flap, belly sucker, benthic, clasping, fossorial, nektonic, neustonic, psammonic, rheophilous, semiterrestrial,** and **suctorial.**

genetic correlation. Degree to which a gene(s) simultaneously influences two or more traits.

germ plasm. Hereditary material transmitted to the offspring through the germ cells (as opposed to the inheritance of acquired characteristics); in anuran germ line usually refers to a conspicuous juxtanuclear organelle aggregate consisting mostly of mitochondria.

glia. Supportive cells of the nervous system, some of which may provide nutrition to neurons or guide migrating neurons during development; may be 10–50 times more glial cells than neurites in the brain, astrocytes, oligodendrocytes (form myelin sheaths), and ependymal cells (line central canal of brain).

globular. The body shape common to many **tadpoles,** rounded and more or less spherical or elliptical; see table 2.2, figures in chap. 12, **compressed, cylindrical, depressed,** and **fusiform.**

glomus. A number of glomeruli grouped together.

growth rate (GR). Change in some measure of size over time.

hatch. In amphibians, to escape from the jelly membranes that surround an amphibian embryo; in most cases, a **hatchling** subsequently develops into a **tadpole,** but in some endotrophs, a **hatchling** actively moves to another site in or on a parent either temporarily or until released (e.g., *Assa;* see **birth**) after further development, or the parent actively relocates the **hatchling** (e.g., *Rhinoderma*). In **viviparous** and **ovoviviparous** forms, hatching occurs in the oviduct.

hatchling. Individual within a series of stages between hatching and a **tadpole** (stage 25); usually used in a generic sense to distinguish individuals in these ecologically unique developmental stages from an **embryo** or a **tadpole;** does not refer to a **metamorph;** see **juvenile, hatch,** and **larva.**

head. (1) Cranial part of **tadpole** body from the snout to the **peribranchial wall** or **diaphragm.** (2) The working surface of a **labial tooth** (Gosner 1959) that often bears **cusps;** see **body** and **sheath.**

heat hardening. Accommodation to a brief exposure of high temperature near the lethal limit such that the organism is more resistant to subsequent exposures to such temperatures; see **cold hardening.**

immature. A free-living individual prior to attainment of sexual maturity regardless of size or developmental stage; see **embryo, froglet, hatchling, juvenile, larva, metamorph,** and **tadpole.**

incised [cuspate]. A descriptor of the shape of the cutting edge of **jaw sheaths;** with one or more pronounced convexities in the upper sheath and sometimes matched by a concavity in the lower sheath, not forming a uniform arc; see **cusp, labial teeth,** and **serration.**

inclusive fitness. Genes contributed to the next generation by an individual indirectly by helping nondescendent kin, in effect creating relatives that would not have existed without the help of the individual.

infralabial prominence (Thibaudeau and Altig 1988). The U- or V-shaped medial protrusion of the lower jaw typical of many microhylid **tadpoles.**

internarial distance. Transverse distance between centers of the external narial openings; see chap. 3.

interorbital distance [interpupillary or interocular distance]. Transverse distance between the centers of the pupils of the eyes; see chap. 3.

ipsilateral. Occurring on the same side of the body; see **contralateral.**

jaws. Loose but inclusive term for both jaw sheaths and their supportive infrarostral and suprarostal cartilages; see chaps. 3 and 4.

jaw sheath (Altig 1970) [beak, maxilla (for upper jaw sheath), horny beak or jaws, labial mandible (for either lower or both jaw sheaths), mandible (for lower jaw sheath), suprarostradont (on suprarostral cartilage), infrarostradont (on infrarostral cartilage)]. Usually serrate, keratinized sheaths that overlie the infrarostral (lower; articulates posteriorly with Meckel's cartilage) and suprarostral (upper, articulates or attached to chondrocranium) cartilages that serve as cutting or abrasive feeding surfaces. Used as a separate term from "beak" because it is unlikely these structures are homologs of other structures termed beaks in birds and turtles; these structures overlie entirely different supportive elements (i.e., not dermal bones); see chap. 3.

juvenile. Postmetamorphic frog up to the time of attainment of sexual maturity; see **embryo, froglet, hatchling, immature, larva, metamorph,** and **tadpole.**

juxtaglomerular apparatus. A unit within the kidney composed of the macula densa (= part of the distal tubule segment), the lacis cells (= extraglomerular mesangium), and the epithelioid cells (= granular cells around the vas afferens).

Kugelzelle. "ball cell" or "unicellular gland" (chap. 5).

Kupffer cells (Sternzellen). Phagocytic cells of the **reticuloendothelial system** between the hepatocytes.

labia. see **labium.**

labial flap (Altig 1970; **oral flap**) [oral apron]. Fleshy flaps of various shapes and sizes that overhang the **mouth** in microhylid **tadpoles;** all keratinized structures absent; derivatives of upper labium (Thibaudeau and Altig 1988) of typical **tadpoles;** edges may be uniform or scalloped.

labial papilla (oral papilla). Collective term for reference to any **marginal** or **submarginal papillae** (but not **buccal papillae**) of the **oral disc.**

labial teeth (Altig 1970) [**denticle,** dermal, or labial denticle, keratodont, labial odontoid]. Keratinized tooth derived from cells produced in a mitotic zone in the base of the **tooth ridge;** composed of a **head** of various shapes and usually with **cusps, body,** and **sheath;** see chap. 3.

labial tooth row (**LTR**). Transverse linear array of **labial teeth** embedded in a **tooth ridge;** the erupted tooth usually has several **replacement teeth;** number, lengths, and distribution of rows vary on both **labia;** see **labial tooth row formula.**

labial tooth row formula (LTRF)(Altig 1970). Fractional notation for designating the number of tooth rows on the upper (numerator) and lower (denominator) labia and the disposition of rows with medial **gaps;** rows on the upper (anterior, superior) labium are numbered from the lip margin toward the **mouth** (distal-to-proximal) and rows on the lower (posterior, inferior) labium are numbered from the **mouth** posteriorly (proximal-to-distal). Numbers in parentheses indicate rows with medial **gaps,** and numbers in brackets indicate variation in the presence of a medial **gap.** A formula of 5(2–5)/3[1] denotes 5 anterior rows (rows 2 through 5 have medial gaps) and 3 posterior rows [row 1 may or may not have a medial gap]; see **accessory rows, labial tooth,** and **labial tooth row.**

labium [lip]. Fleshy, disc-shaped structures that more or less surround the **mouth;** usually with **labial teeth** arranged in transverse rows; margins mostly free and papillate; upper and lower labia together form the **oral disc;** see **labial teeth, labial tooth row, oral apparatus,** and **oral disc.**

laevogyrinid (sinistral). Describes a **tadpole** having a single **spiracle** on the left side; see **amphigyrinid, mediogyrinid,** and **paragyrinid.**

larva. Postembryonic, **hatched, immature** form of a normally metamorphosing (obligate process in most anurans) species; a larva of an anuran is specifically a **tadpole;** larviform adults (at least partial larviform morphology but sexually mature) occur in salamanders but not in anurans; see **embryo, hatchling, juvenile,** and **metamorph.**

larval transport. Parental transport over various periods of time that re-locates **exotrophic larvae** from the site of oviposition of terrestrial eggs to some aquatic site; larvae usually on the back of the parent; does not include **endotrophs.**

lateral. (1) Describes a category of eye position in which some part of the eye is included in a dorsal silhouette of the **tadpole** (i.e., the eyes protrude further laterally than the surrounding **body** surface). (2) An anatomical descriptor meaning "toward the side" relative to a stated landmark; see **dorsal.**

lateral line organ (neuromast). See **neuromast.**

lateral line system. All the pressure-sensitive **neuromasts** distributed over the body in specific patterns, usually in lines; see **lateral line organ** and **receptor unit.**

lateral process (Altig 1970). Poorly defined lateral portion of the upper jaw sheath beyond which serrations are small to absent; probably not a working surface for feeding.

lemmanura (Starrett 1973). Word synonym of Orton's (1953) Type 3 **tadpole;** discoglossoid of O. M. Sokol (1975).

lentic. Limnological description of any nonflowing water system; see **lotic.**

limbic system. Portion of forebrain, including the amygdala, habenula, hippocampus, and preoptic areas, that may form the highest correlative center in the brain.

limiting resource. Resource which is in critically short supply; an increase in the availability of a limiting resource should increase the size of a population.

lingual papilla (Wassersug 1980). Specifically those **buccal papillae** of various numbers, shapes, and sizes that rest on the lingual **anlage.**

lotic. Limnological description of any flowing water system; see **lentic.**

LTR. Abbreviation for **labial tooth row.**

LTRF. Abbreviation for **labial tooth row formula.**

lymphatic system. Lymph glands and vessels that function in immune responses.

macrophagous. An **ecomorphological guild** that includes those **lentic tadpoles** that presumably feed by taking larger bites (compared with smaller particles generated by rasping **tadpoles**) of attached materials on submerged substrates, sometimes at least facultatively oophagous; **oral disc** nearly terminal, jaw sheaths present, LTRF 0/0–0/1, **marginal papillae** greatly reduced to absent; see table 2.2, figures in chap. 12, **arboreal, benthic, carnivorous, nektonic, neustonic, suspension feeder,** and **suspension rasper.**

Magenhauptzelle. "stomach main cells" (chap. 5).

Malthner cells. Paired cells in the **basal plate** of **anamniotes** that mediate fast start and startle responses, especially in aquatic animals; largest cells in the brain stem and persist after metamorphosis in terrestrial forms of amphibians.

manicotto glandulare. Sleevelike glandular larval stomach of larval Types 3 and 4.

marginal papilla(e) (Altig 1970) [labial, oral, lower, or upper festoon or fringe; buccal, dermal, or peribuccal papillae; papillary border]. Describes any papilla(e) anywhere on the margin of the **oral disc;** commonly appear as a complete series [circumoral], around margin or with dorsal [rostral gap], ventral [mental gap], or dorsal and ventral **gaps.**

mature. Describes an individual that attained sexual maturity regardless of morphology; does not apply to a **tadpole** in latter stages of development; see **froglet, hatchling, larva,** and **metamorph.**

medial. (1) Refers to or toward the sagittal line relative to a reference. (2) Describes a **vent tube** that opens parallel with the plane of the ventral **fin** or tail muscle; see **dextral** and **sinistral;** G. F. Johnston and Altig (1986) discuss other variations of the **vent tube.**

median ridge (Wassersug 1980). Transverse flaplike ridge of various shapes suspended from the roof of the buccal cavity; see **buccal papillae.**

mediogyrinid (midventral spiracle). Having a single **spiracle** anywhere on the midventral, sagittal line (e.g., ascaphids and microhylids); may be on the chest, abdomen, or ventral to the vent; see **amphigyrinid, laevogyrinid, paragyrinid,** and **sinistral.**

melanin. Black or brown pigment that is a complex tyrosine-protein polymer; see **chromatophore, melanophore,** and **melanosome.**

melanophore. Dermal or epidermal **chromatophores** of various shapes that produce brown to black pigments that give colorations by the presence of **melanin;** see **blotch, dot, chromatophore, melanophore, melanosome, mottled, melanosome, spot, punctate,** and **stellate.**

melanosome (J. D. Taylor and Bagnara 1972). Pigment-containing organelle in a **melanophore;** see **melanin.**

mesangium. Tissue between the glomerular/glomular capillaries composed of mesangial cells and mesangial matrix.

metabolic depression. A normally large decrease in the **standard metabolic rate** that is usually triggered by some major change in physiological status, such as estivation or hibernation; caused by biochemical regulatory mechanisms that make this a controlled response.

metabolic oxygen conformer. Animal whose rate of O_2 consumption is independent of ambient $\mathbf{PO_2}$ over some specified range of ambient $\mathbf{PO_2}$; see **metabolic oxygen regulator.**

metabolic oxygen regulator. Animal whose rate of O_2 consumption is dependent on ambient $\mathbf{PO_2}$ over some specified range of ambient $\mathbf{PO_2}$; see **metabolic oxygen conformer.**

metabolic scope. The degree that a baseline metabolic rate, either the **standard metabolic rate** (SMR) or resting metabolic rate (RMR), can be increased by strenuous activity; may be expressed in relative (factorial increase) or absolute (incremental increase) terms.

metamorph. An immature frog that results from the **metamorphosis** of an **exotrophic tadpole;** see **embryo, hatchling, immature, juvenile, larva, metamorphic, metamorphic climax, premetamorphosis,** and **prometamorphosis.**

metamorphic. General term that describes either the process or an individual in the process of **metamorphosis;** see **larva, hatchling, metamorph, metamorphic climax, premetamorphosis, prometamorphosis,** and **tadpole.**

metamorphic climax. That period when **metamorphic** changes are the most profound and most rapid; Gosner 42–46; see **embryo, immature, juvenile, larva, metamorph, metamorphic, metamorphosis, premetamorphosis, prometamorphosis,** and **tadpole.**

metamorphosis. Process of a nidicolous **endotrophic** or **exotrophic tadpole** changing into a **froglet;** most changes occur during stages 42–46 of Gosner (1960); other definitions (chap. 2) are based on endocrinological data; often used in synonymy with **metamorphic climax;** see **metamorph, metamorphic, premetamorphosis,** and **prometamorphosis.**

minor metamorphosis. Metamorphic changes in morphology and ecology characterized by minimal variation between larval and postlarval stages (i.e., salamander metamorphosis as opposed to that of a frog).

monoculture. A population containing only a single species, usually as a cultured population.

mottled. A color pattern formed by haphazard arrays of usually dark streaks, lines, or blotches on a pale background or pale markings on a dark background; see **blotch, dot, punctate, spot,** and **stellate.**

mouth [**oral apparatus, oral disc,** and **mouthparts** (in reference to mouth]. Refers only to that opening formed by the rupture of the oropharyngeal membrane to connect the buccopharynx with the exterior; does not refer to any **mouthparts.**

mouthparts (oral apparatus). Collective term for all soft and keratinized feeding structures largely external to the **mouth** (e.g., **labia** and all associated structures, **tooth ridges, labial teeth,** and **jaw sheaths**); does not refer to any structures in the **buccopharynx.**

muciferous crypt [neuromast]. Refers to openings of unicellular glands but has also been used incorrectly in reference to **neuromasts;** should be abandoned because of these confusing usages.

multiserial. Describes structures occurring in three or more series at a site, as **tooth rows** on a **tooth ridge** (e.g., some rows of *Ascaphus*) or **marginal papillae;** see **biserial** and **uniserial.**

muscle height (Altig and Johnston 1989). Maximum height (vertical distance) measured from the junction of the body wall and the ventral margin of the tail musculature; see chap. 3.

muscle width (Altig and Johnston 1989). Maximum transverse width of the tail muscle measured in dorsal view at the same plane as **muscle height.**

naris, nares (nostril). Usually refers to the external openings of the nasal canal; internal opening in the roof of the **buccopharynx** is the **choana(e)** or internal nares.

nasolacrimal duct [lacrimal duct or gland, orbitonasal line]. A duct connecting the corner of the eye with the **naris;** appears as a contrastingly pigmented line during its formation in later stages of a **tadpole.**

nephrotome (nephromere, mesomere, intermediate plate). That part of the mesodermic somite that develops into excretory organs.

negative Bohr shift. Increase in the affinity of hemoglobin (decreased P_{50}) associated with a decrease in blood pH, as opposed to the normally observed decrease in affinity (increased P_{50}) that accompanies an increase in the acidity of the blood.

nektonic (pelagic). (1) An **ecomorphological guild** that includes those **lentic** or **lotic tadpoles** that rasp food from submerged surfaces with keratinized **mouthparts** and do so somewhere within the water column including quiet backwaters in lotic sites; **body compressed,** eyes **lateral, fins** high with pointed tip with or without a **flagellum;** dorsal fin often originates well anterior to dorsal **tail-body junction;** see table 2.2, figures in chap. 12, **adherent, arboreal, benthic, carnivorous, clasping, fossorial, gastromyzophorous, macrophagous, neustonic, psammonic, rheophilous, semiterrestrial, suctorial, suspension feeder,** and **suspension rasper.**

neuromast [minute glands, lateral line pore, muciferous crypt]. Complex (see chap. 6), pressure-sensitive organs arranged in various patterns on the **body** and **tail;** collectively form the **lateral line system.** Any definition using "pore" should not be used because the **lateral line system** of amphibians does not include pores or channels like in some fishes.

neustonic (surface feeder). An **ecomorphological guild** that includes those **lentic** or **lotic tadpoles** that filter particles in or near the surface film with upturned (= **umbelliform**) oral disc with or without keratinized **mouthparts; body depressed;** eyes usually **lateral;** tail long with low **fins;** see table 2.2, figures in chap. 12, **adherent, arboreal, benthic, carnivorous, clasping, fossorial, gastromyzophorous, macrophagous, nektonic, psammonic, rheophilous, semiterrestrial, suspension feeder,** and **suspension rasper.**

nidicolous. (1) Developmental mode of a frog whereby a **tadpole** morphotype of some form develops but the individual does not feed; morphology varies from a fully formed tadpole at one end of a continuum through many developmental variations that deviate progressively from the developmental pattern of a typical tadpole. (2) An **ecomorphological guild** that includes those species that have some form of nidicolous development; see table 2.2, figures in chap. 12, **endotrophy, exotrophy, direct development, exoviviparous, ovoviviparous, paraviviparous,** and **viviparous.**

nonadditive genetic variance. That fraction of the total phenotypic variation primarily caused by alleles that display dominance and epistasis; see **additive genetic variance.**

nucleus. (1) An organelle of a cell housing the genetic material and other structures. (2) Aggregation of cells within the brain that performs a similar function or operates toward a similar end; see **ganglion.**

opercular tube or canal. Single or paired structures that connect the branchial chamber to the external opening of the **spiracle** in **tadpoles** with midventral or ventrolateral spiracles.

opercular fold. Outgrowth from the hyoid arch in stages 20–24 that eventually forms a covering over the gills and associated structures and fuses with the body wall in patterns that produce several **spiracular** configurations; not a homolog of the piscine **operculum** or auditory elements; see **amphigyrinid, laevogyrinid, mediogyrinid, paragyrinid, peribranchial chamber,** and **sinistral.**

operculum. (1) The completed covering of the buccopharyngeal area of a **tadpole** provided by the development of the **opercular fold;** not homologous with the piscine gill cover of the same name; see **amphigyrinid, laevogyrinid, mediogyrinid, paragyrinid, sinistral,** and **spiracle.** (2) An auditory element in the fenestra ovalis; see chap. 4.

oral apparatus (mouthparts) [buccal apparatus, oral sucker]. Collective term applicable to all soft and keratinized structures mostly external to the **mouth;** see **barbel, labium, marginal papillae, mouth, oral disc,** and **submarginal papillae.**

oral disc (Altig 1970) [mouth disc, disk, or pad; oral sucker]. Composite structure composed of upper and lower **labia,** usually with transverse **tooth ridges** surmounted by rows of **labial teeth; marginal papillae** occur in various configurations at the edges and **submarginal papillae** occur in various patterns on the face of the disc; does not refer to the **mouth;** a major part of the **mouthparts;** see **oral apparatus.**

oral flap. See **labial flap.**

oral hitching (Altig and Brodie 1972). Describes the means of movement of **rheophilous** tadpoles by extension-retraction cycles of the **oral disc** whereby attachment to the substrate is maintained; largely independent from feeding activities; probably **gastromyzophorous** forms do similar actions.

oral papilla(e). See **labial papilla(e).**

oral sucker. A loose term for the **oral disc** of a **suctorial tadpole;** use for position maintenance in flowing water; see **belly flap, belly sucker, funnel mouth, gastromyzophorous, mouthparts, oral apparatus,** and **oral disc.**

oral tube (Lavilla 1990). Describes the shape of a protruding **oral apparatus** at a specific point in the feeding cycle of a **tadpole;** probably not a specific structure.

ovoviviparous. (1) Developmental mode of a frog whose **endotrophic larva** is retained within the female's reproductive tract and is born as a **metamorph** with no additional energy beyond vitellogenic yolk of the original egg being supplied by the mother. (2) An **ecomorphological guild** that includes those species with ovoviviparous development; see table 2.2, figures in chap. 12, **direct development, endotrophy, exotrophy, exoviviparous, nidicolous, paraviviparous,** and **viviparous.**

PR-1. See AR-1.

papilla(e). (1) A fleshy projection(s) on the margin or face of the **oral disc,** most often circular in cross section and with a blunt tip; see **barbel, marginal papillae,** and **submarginal papillae.** (2) A fleshy projection in general, like the various structures on the floor and roof of the **buccopharynx;** see chap. 5.

papillate, papilliferous. Describes structures or surface that have **papillae;** see **barbel, buccal papillae, marginal papillae,** and **submarginal papillae.**

paragyrinid (G. F. Johnston and Altig 1986). Describes a **tadpole** having a single **spiracle** on the left side but well below the **body axis,** sometimes nearly midventrally (e.g., phyllomedusine hylids); see **amphigyrinid, laevogyrinid, medigyrinid,** and **sinistral.**

paraviviparous (Altig and Johnston 1989). (1) Developmental mode of a frog whereby a **froglet** is hatched from a egg someplace other than the reproductive tract of a parent; female parent is usually involved and no additional energy is supplied beyond the vitellogenic yolk. (2) An **ecomorphological guild** that includes those species with some form of paraviviparous development; see table 2.2, figures in chap. 12, **direct development, endotrophy, exotrophy, exoviviparous, nidicolous, ovoviviparous,** and **viviparous.**

peribranchial chamber (opercular chamber). That space anterior to the **peribranchial wall** (= **diaphragm**) that separates the **buccopharynx** and abdominal cavities and delimited by the **operculum;** branchial arches, gills, filter plates, and other structures occur within this chamber; see **opercular fold** and **spiracle.**

peribranchial wall (diaphragm). See **peribranchial chamber.**

persistent epidermal gill (Viertel 1991). Until stage 24 external, from stage 25 internal gills persisting up to metamorphic climax; develop along the ventral parts of branchial arches I–IV; epidermis overlays the pharyngeal endoderm; see chap. 5 and **transient epidermal gills.**

phenology. Temporal sequence of events, often refers to seasonal changes in calling and breeding of adults and the presence of tadpoles.

phytotelm, phytotelmata. Water-holding cavity in some part of a plant (e.g., bromeliad cistern or tree hole) or plant product (e.g., a nut shell); often extended to refer to other small, isolated bodies of water (e.g., snail shells). A subset of **lentic** habitats.

plasticity. Variation in the phenotypic expression of a genotype over or influenced by a range of environmental conditions.

PO_2. Partial pressure of O_2 in the environment; see **$\dot{V}O_2$**

preferred body temperature (PBT). Mean (sometimes median or mode) body temperature selected by an animal in a laboratory thermal gradient or thermal shuttlebox.

premetamorphosis (Etkin 1968). Describes the developmental attainment of a tadpole in stages 25–35, the primary period of body growth; see **froglet, larva, metamorph, metamorphic, prometamorphic,** and **tadpole.**

prometamorphosis (Etkin 1968). Describes the developmental attainment of a tadpole in stages 36–41, the primary period of rapid limb growth; see **larva, metamorph, metamorphic, premetamorphic,** and **tadpole.**

psammonic. An **ecomorphological guild** that includes those **lotic tadpoles** that inhabit loose sand in slow-flowing streams; perhaps feed passively while buried; see table 2.2, chap. 3, figures in chap. 12, **adherent, benthic, clasping, fossorial, gastromyzophorous, nektonic, neustonic, rheophilous, semiterrestrial,** and **suctorial.**

punctate. (1) A pattern of small, often dark **spots** or **dots,** usually with distinct margins. (2) An individual **melanophore** that is more or less circular in outline; see **blotch, mottled,** and **stellate.**

Q_{10}. Rate of a process at temperature t (°C) divided by rate at t − 10; generally for any temperature interval Q_{10} = rate at T_1/rate at T_1 raised to the power of $(10/T_2 - T_1)$.

quantitative genetics. Science ". . . concerned with the inheritance of those differences between individuals that are of degree rather than of kind, quantitative rather than qualitative." (Falconer 1989).

rasp. A general feeding mode whereby a **tadpole** uses keratinized **mouthparts** to harvest particulate or small pieces of various materials from submerged surfaces or pick up materials from sedimented accumulations; see **suspension feeding** and **suspension rasper.**

receptor unit (Zakon 1984) [plaque, stitch]. Receptors of mechanical stimuli that form the **lateral line system;** located in tight clusters and developmentally derived from and aligned in the same direction as the single primary organ; see **neuromast.**

replacement tooth (Altig and Johnston 1989). One of several fully formed **labial teeth** interdigitated sequentially beneath the erupted tooth; see **labial teeth, tooth row,** and **tooth series.**

reticuloendothelial system. Connective and endothelial tissues related to the lymphatic system, produces immune responses, and functions in myelopoiesis.

rheophilous. Generic term that describes tadpoles of several **ecomorphological guilds** that are modified in various ways for inhabiting microhabitats in the flowing parts of **lotic** systems; see **adherent, benthic, clasping, fossorial, gastromyzophorous, nektonic, neustonic, psammonic, semiterrestrial,** and **suctorial.**

Riesenzellen. "Alarm cells" (Fox 1988); see chap. 5.

rim, rimmed. (1) A raised margin of an opening, usually the nares, with or without papillae. (2) With a rim.

Rohon-Beard cells. Cells that mediate mechanoreception from free nerve endings in the epidermis, gradually replaced by the dorsal root ganglion system; among the first sensory neurons to appear in the spinal cord of **anamniote** vertebrates and persist in amphibians until **metamorphosis.**

routine metabolic rate. Metabolic rate of an **ectotherm** that is acclimated to a specified temperature and exhibits normal activity.

Schleimköpfchenzellen. "Mucous-headed cell" (chap. 5).

school [aggregation]. A specific form of congregated tadpoles that involve some sort of social interaction and exhibit at least indications of polarized orientation and perhaps coordinated movements; see **aggregation, biosocial aggregation,** and **taxic aggregation.**

scoptanura (Starrett 1973). Word synonym of Orton's (1953) Type 2; microhyloid of O. M. Sokol (1975).

semiterrestrial [subaerial]. An **ecomorphological guild** that includes those **lotic tadpoles** that have **marginal papillae** with an anterior gap; LTRF usually 2/3, **jaw sheaths** usually with a high and narrow arch; inhabit rock faces, leaves, and the forest floor that provide damp or wet surfaces with little free water; see table 2.2, figures in chap. 12, **adherent, belly flap, belly sucker, benthic, clasping, fossorial, gastromyzophorous, nektonic, neustonic, psammonic, rheophilous,** and **suctorial.**

serration. Sawlike projections of various densities, orientations, shapes, sizes, and densities along the cutting edge of the **jaw sheaths;** should not be used in reference to **marginal papillae** or **labial teeth;** see **cusp.**

sheath. Hollow, basal part of a **labial tooth** (Gosner 1959) embedded in the tissue of the **tooth ridge;** see **head, body, tail sheath, tooth row,** and **tooth series.**

sinistral. To the left. (1) Describes a single **spiracle** on left side (**laevogyrinid**); see **amphigyrinid, laevogyrinid, mediogyrinid,** and **paragyrinid.** (2) Describes a **vent tube** with the aperture opening to the left of the plane of the ventral **fin;** see **dextral** and **medial.**

size refuge. Avoidance of predation by the attainment of a body size which reduces or eliminates predation by gape-limited predators; see **gape limited.**

spinal nerve. One of 20–29 nerves that arises from the spinal cord; typically formed by the confluence of a dorsal sensory root and a ventral motor root; the number reduces to about 11 after metamorphosis; see **cranial nerve.**

spiracle [atrial opening; spout]. One or two opening(s) of different shapes and positions for the exit of the water pumped into through the buccopharynx for respiration and feeding; not homologous with the elasmobranch spiracle; see **amphigyrinid, laevogyrinid, mediogyrinid, paragyrinid,** and **sinistral.**

spot. Dark, pigmented mark with relatively even margins but larger than a **dot** and more rounded and discrete than a **blotch;** see **mottled, punctate,** and **stellate.**

standard metabolic rate (SMR). Metabolic rate of an **ectotherm** under resting conditions and acclimated to a specified temperature; equivalent to basal metabolic rate of mammals.

stellate. (1) Describes a pigmented mark that is roughly star-shaped or at least has multiple projections from a central region. (2) An individual **chromatophore** that is similarly star-shaped; see **blotch, dot, spot, mottled,** and **punctate.**

Sternzellen. See **Kupffer cell** (chap. 5).

Stiftchenzellen. "pin-cells" (chap. 5).

submarginal papilla(e) (Altig 1970) [extramarginal, infralabial, inner or retromarginal papillae]. Fleshy projection(s) anywhere on the face of the **oral disc** except the margin and perhaps some other places in **tadpoles** that lack **jaw sheaths** (e.g., at the sites where jaw sheaths would be expected to occur); see **barbel, marginal papilla,** and **papilla.**

suctorial. (1) Describes the usually large, ventral-facing **oral apparatus** of tadpoles that maintain position and feed in fast water by adhering to rocks via the large **oral disc** (chap. 4); should not be used in reference to the **belly sucker** of **gastromyzophorous** tadpoles; see **belly flap.** (2) An **ecomorphological guild** that includes those **lotic** and **rheophilous tadpoles** with small, closely spaced **marginal papillae** in a complete series around the oral disc; LTRF > 2/3 to a maximum of 17/21; inhabits faster water than **adherent** and **clasping** tadpoles, position maintenance via oral disc continuous; see table 2.2, figures in chap. 12, **belly flap, benthic, fossorial, gastromyzophorous, nektonic, neustonic, psammonic,** and **semiterrestrial.**

suspension feeder, suspension feeding (= filter feeding). (1) Harvesting of naturally suspended particles by pumping water in through the **mouth,** over the buccopharyngeal filtering system and out the **spiracle**(s); tadpoles that specialize in this mode of feeding usually lack all keratinized and most soft **mouthparts.** (2) An **ecomorphological guild** that includes those **lentic tadpoles** that are behaviorally and morphologically (e.g., **lateral** eyes and **depressed** bodies) modified to feed in this manner; see table 2.2, figures in chap. 12, **arboreal, benthic, carnivorous, macrophagous, nektonic, neustonic,** and **suspension rasper.**

suspension rasper. An **ecomorphological guild** that includes those **lentic tadpoles** that seemingly feed by filtering suspended particles from within the water column and rasping submerged surfaces; jaw sheaths and labial teeth (usually 2/3) present, marginal papillae with anterior gap; eyes **lateral; flagellum** common **tail;** see table 2.2, figures in chap. 12, **arboreal, benthic, carnivorous, macrophagous, nektonic, neustonic,** and **suspension feeder.**

tadpole. Nonreproductive **exotrophic** or a **nidicolous endotrophic larva** of a frog between stages 25–41; does not refer to **immatures** of other amphibians; see **froglet, hatchling, juvenile, larva, mature,** and **metamorph.**

tail. That part of a **tadpole** minus the **body;** with a segmented musculature and dorsal and ventral **fins;** defined for measurement as the straight-line distance from the tail tip anteriorly to the **body terminus.**

tail axis (Van Dijk 1966). An imaginary line connecting the **body terminus** and the tip of the straightened tail; see **body axis.**

tail-body junction. Literally the junction of the **body** and **tail** of a **tadpole,** but for different purposes, this junction can be defined three ways. For most purposes of description and measurement, the site of the **body terminus** should be used. In some cases the ventral body terminus (where the ventral extent of the curvature of the posterior margin of the body contacts the plane of the ventral margin of the tail muscle) and the dorsal body terminus (where the dorsal extent of the curvature of the posterior margin of the body contacts the plane of the dorsal margin of the tail muscle) can be used.

tail flipping. A surprisingly efficient and accurate means of locomotion by **tadpoles** and particularly **hatchlings,** especially those that **hatch** from arboreal egg masses; move by frantic flips of the tail and not tail undulations.

tail height [tail depth]. Greatest vertical distance (i.e., height) of the tail muscle plus both fins; this site may not be the maximum dimension of both fins.

tail length. Straight line, longitudinal distance from the **body terminus** to the absolute tail tip, not just to the end of the musculature; see **body length.**

tail ratio. Total **length** divided by **tail length;** see **body ratio.**

tail sheath. Dense layer(s) of probable connective tissue in the proximal third of the tail of some tadpoles, sometimes visible in living specimens but more commonly seen in preserved specimens.

taxic aggregation (aggregation) [asocial aggregation]. A general term that describes a group of tadpoles that are congregated because of each individual responding to some sort (e.g., light or temperature) of taxis and not because of social interaction; qualities of a **school** (i.e., polarized orientation and coordinated movements) are not implied; see **biosocial aggregation** and **school.**

temperature acclimation. Accommodation by an organism in a laboratory setting by physiologically adjusting to the new temperatures; under natural conditions, it is called **temperature acclimatization.**

temperature acclimatization. See **temperature acclimation.**

tooth ridge [dental plate, labial ridge]. Serial, transverse fleshy ridges on the face of the oral disc surmounted by a row(s) of labial teeth generated in a mitotic zones in the base of the ridges; present in some **tadpoles** that never have **labial teeth** see **tooth row** and **tooth series.**

tooth row [fringed fold, tooth series, supralabial row on upper or anterior labium, infralabial row on lower or posterior labium]. Transverse, usually close-spaced row of **labial teeth** each embedded within a **tooth ridge.** Rows on either labium face the **mouth;** see **A-2 gap, gap, labial tooth row formula,** and **tooth series.**

tooth series. Collective term for the erupted **labial tooth** at a given site in a **tooth row** and each **replacement tooth** interdigitated sequentially below it; see **A-2 gap, gap, tooth ridge, tooth series, labial tooth row formula.**

total length [standard length]. Straight line distance measured from the tip of the snout to the tip of the tail; see **body length, body terminus,** and **tail length.**

transient epidermal gill (Viertel 1991). External gills of other authors, develop along the ventrolateral parts of branchial arches I–IV and atrophy by stage 25; see chap. 5 and **persistent epidermal gill.**

typhlosolis. Muscular invagination of the foregut.

umbelliform [funnel mouth]. Describes an upward-facing **oral disc,** a convergent trait in several families, used for harvesting particulate matter from the surface film regardless of actual **mouth** position or orientation; **oral disc** may be formed from parts of either **labium** and keratinized **mouthparts** often lacking see **neustonic** and **suctorial.**

umbraculum (Van Dijk 1966). Fleshy projection of the iris into or over part of the pupil in some ranids; assumed to protect eye from excessive light; see **elygium.**

uniserial. Describes structures occurring in one series or row per site, as one row of **labial teeth** per **tooth ridge** (e.g., most tadpoles); see **biserial** and **multiserial.**

Urwirbelfortsatz. Primordial vertebral process; see chap. 4.

velum (Wassersug 1976a, b; chap. 5). Glandular organ between the buccal cavity and the pharynx that regulates water flow into the pharynx.

vent [anal opening, **anus**]. Posterior intestinal opening of a **tadpole.** Because amphibians have a cloaca, this opening is not an anus, and all references as such should be omitted; see **vent flap** and **vent tube.**

vent flap [anal flap]. Fleshy flap of various shapes that extends from the body wall posteriorly and ventral to the **vent tube.** Sometimes encloses the hind limb buds and occurs most commonly in **rheophilous tadpoles.**

vent tube [anal or cloacal tube, anal tube piece, cloacal tail piece, proctodaeal tube]. A tube for voiding of feces with many variations in morphology; projects from the body wall distally to an opening that may lie parallel with the plane of the ventral **fin (medial)** or to the left **(sinistral)** or right **(dextral)** of that plane. In a few cases, the tube does not exit the body on the sagittal line; see **vent** and **vent flap.**

ventral body terminus. See **body terminus** and **tail-body junction.**

visceral arch. Cartilaginous or bony bars of the visceral skeleton, including the mandibular arch (= mandibulare and palatoquadratum), hyoid arch, and three branchial arches.

visceral pouch. Pharyngeal (endodermal) evaginations between the **visceral arches** from which the Eustachian tube is derived; pouches 2–4 open as gill slits; the filter plate anlagen and the branchiogene endocrines are derived from this endoderm.

viviparous (Altig and Johnston 1989). (1) Developmental mode of a frog which results in a larva (= fetus) that develops by consuming maternally derived, oviducal materials after the exhaustion of vitellogenic yolk. (2) An **ecomorphological guild** that includes species that have viviparous development; see table 2.2, figures in chap. 12, **direct development, endotrophy, exotrophy, exoviviparous, nidicolous, ovoviviparous,** and **paraviviparous.**

$\dot{V}O_2$. A measure of the volume of O_2 consumed per unit time; see **PO_2.**

whole-body O_2 conductance. Change in the rate at which O_2 can be consumed by a metabolic O_2 regulator per unit change in ambient **PO_2;** equivalent to the slope of a plot of **$\dot{V}O_2$** against ambient **PO_2** in the range of metabolic O_2 conformity.

xenoanura (Starrett 1973). Word synonym of Orton's (1953) Type 1 tadpole; pipoid of O. M. Sokol (1975).

LITERATURE CITED

Abe, A. S., and H. P. Godinho. 1991. Tolerance to high temperatures in tadpoles of *Leptodactylus fuscus* and *Hyla fuscovaria* in temporary ponds (Amphibia, Leptodactylidae, Hylidae). *Zool. Anz.* 226:280–284.

Abelson, A., T. Miloh, and Y. Loya. 1993. Flow patterns induced by substrata and body morphologies of benthic organisms, and their roles in determining availability of food particles. *Limnol. Oceanograph.* 38:1116–1124.

Abraham, M. H., and C. Rafols. 1995. Factors that influence tadpole narcosis. An LFER analysis. *J. Chem. Soc.* 10:1843–1851.

Abravaya, J. P., and J. F. Jackson. 1978. Reproduction in *Macrogenioglottus alipioi* Carvalho (Anura: Leptodactylidae). *Nat. Hist. Mus. Los Angeles Co. Contrib. Sci.* (298):1–9.

Accordi, F., and P. Cianfoni. 1981. Histology and ultrastructure of the adrenal gland of *Rhacophorus leucomystax* (Amphibia, Anura). *Boll. Zool.* 48:277–284.

Accordi, F., and E. Grassi-Milano. 1977. Catecholamine-secreting cells in the adrenal gland of *Bufo bufo* during metamorphosis and in the adult. *Gen. Comp. Endocrinol.* 33: 187–195.

———. 1990. Evolution and development of the adrenal gland in amphibians. pp. 257–268. In *Biology and physiology of amphibians. Progress in zoology 38. Proc. First Intl. Symp. Biol. Physiol. Amphibians,* Karlsruhe, Federal Republic of Germany, edited by W. Hanke. Stuttgart: Gustav Fischer-Verlag.

Adams, M. J., K. L. Richter, and W. P. Leonard. 1997. Surveying and monitoring amphibians using aquatic funnel traps. pp. 47–54. In *Sampling amphibians in lentic habitats,* edited by D. H. Olson, W. P. Leonard, and R. B. Bury. Northwest Fauna Number 4 Society for Northwestern Vertebrate Biology, Olympia, WA.

Adamson, L., R. G. Harrison, and I. Bayley. 1960. The development of the whistling frog *Eleutherodactylus martinicensis* of Barbados. *Proc. Zool. Soc. London* 133:453–469.

Adamson, M. L. 1981a. *Gyrinicola batrachiensis* (Walton, 1929) n. comb. (Oxyuroidea; Nematoda) from tadpoles in eastern and central Canada. *Canadian J. Zool.* 59:1344–1350.

———. 1981b. Seasonal changes in populations of *Gyrinicola batrachiensis* (Walton 1929) in wild tadpoles. *Canadian J. Zool.* 59:1377–1386.

Adler, K. 1970. The role of extraoptic photoreceptors in amphibian rhythms and orientation: A review. *J. Herpetol.* 4:99–112.

———. 1976. Extraocular photoreception in amphibians. *Photochem. Photobiol.* 23:275–298.

Adolph, E. F. 1931. The size of the body and the size of the environment in the growth of tadpoles. *Biol. Bull.* 61:350–375.

Affa'a, F. M. 1979. *Nyctositum ameita,* n. gen., n. sp., an endocommensal ciliate of tadpoles of *Acanthixalus spinosus. Protistologica* 15:333–336.

———. 1983. *Neonyctotherus* (Ciliophora, Heterotrichida), a new genus commensal in amphibian tadpoles (Anura) from Cameroons. *Protistologica* 19:141–147.

Affa'a, F. M., and J.-L. Amiet. 1994. Progrès récents dan la connaissance des Nyctothères (Protozoaires, Ciliés, Hétérotriches) associés aux Anoures. *Alytes* 12:75–92.

Agarwal, S. K., and I. A. Niazi. 1980. Development of the mouth-parts in the tadpoles of *Rana tigrina* (Daud.). *Proc. Indiana Acad. Sci.* 89:127–131.

Aggarwal, S. J., and A. Riggs. 1969. The hemoglobins of the bullfrog, *Rana catesbeiana* I. Purification, amino acid composition, and oxygen equilibria. *J. Biol. Chem.* 244:2372–2383.

Ahl, E. 1931. Anura III (Polypedatidae). *Das Tierreich* 55:1–475.
Ahlgren, M. O., and S. H. Bowen. 1991. Growth and survival of tadpoles (*Bufo americanus*) fed amorphous detritus derived from natural waters. *Hydrobiologia* 218:49–51.
Aho, J. M. 1990. Helminth communities of amphibians and reptiles: Comparative approaches to understanding patterns and processes. pp. 157–195. In *Parasite communities: Patterns and processes,* edited by G. W. Esch, A. O. Bush, and J. M. Aho. New York: Chapman and Hall.
Aichhorn, H., and A. Lametschwandtner. 1996. Vascular regression during amphibian metamorphosis—a scanning electron microscope study of vascular corrosion casts of the ventral velum in tadpoles of *Xenopus laevis* Daudin. *Scanning* 18:447–455.
Aichinger, M. 1987. Annual activity patterns of anurans in a seasonal neotropical environment. *Oecologia* 71:583–592.
Akin, G. C. 1966. Self-inhibition of growth in *Rana pipiens* tadpoles. *Physiol. Zool.* 39:341–356.
Alberch, P. 1982. The generative and regulatory roles of development in evolution. pp. 19–36. In *Environmental adaptation and evolution,* edited by D. Mossakowski and G. Roth. Stuttgart: Gustav Fischer-Verlag.
———. 1985. Problems with the interpretation of developmental sequences. *Syst. Zool.* 34:46–58.
———. 1987. Evolution of a developmental process—irreversibility and redundancy in amphibian metamorphosis. pp. 23–46. In *Development as an evolutionary process,* edited by R. A. Raff and E. C. Raff. New York: Alan R. Liss, Inc.
Alberch, P., and E. A. Gale. 1983. Size dependence during the development of the amphibian foot. Colchicine-induced digital loss and reduction. *J. Embryol. Exp. Morphol.* 76:177–197.
Alberch, P., S. J. Gould, G. Oster, and D. B. Wake. 1979. Size and shape in ontogeny and phylogeny. *Paleobiology* 5:296–317.
Alcala, A. C. 1962. Breeding behavior and early development of frogs of Negros, Philippine Islands. *Copeia* 1962:679–726.
Alcala, A. C., and W. C. Brown. 1957. Discovery of the frog *Cornufer guentheri* on Negros Island, Philippines, with observations on its life history. *Herpetologica* 13:182–184.
———. 1982. Reproductive biology of some species of *Philautus* (Rhacophoridae) and other Philippine anurans. *Philippine J. Biol.* 11:203–226.
Alcocer, I., X. Santacruz, H. Steinbeisser, K. H. Thierauch, and E. M. Del Pino. 1992. Ureotelism as the prevailing mode of nitrogen excretion in larvae of the marsupial frog *Gastrotheca riobambae* (Fowler)(Anura, Hylidae). *Comp. Biochem. Physiol.* 101A:229–231.
Alexander, R. D. 1974. The evolution of social behavior. *Ann. Rev. Ecol. Syst.* 5:325–383.
Alford, R. A. 1986a. Effects of parentage on competitive ability and vulnerability to predation in *Hyla chrysoscelis* tadpoles. *Oecologia* 68:199–204.
———. 1986b. Habitat use and positional behavior of anuran larvae in a northern Florida temporary pond. *Copeia* 1986:408–423.
———. 1989a. Effects of parentage and competitor phenology on the growth of larval *Hyla chrysoscelis. Oikos* 54:325–330.
———. 1989b. Competition between larval *Rana palustris* and *Bufo americanus* is not affected by variation in reproductive phenology. *Copeia* 1989:993–1000.
———. 1989c. Variation in predator phenology affects predator performance and prey community composition. *Ecology* 70:206–219.
———. 1994. Interference and exploitation competition in larval *Bufo marinus.* pp. 297–306. In *Advances in ecology and environmental sciences,* edited by P. C. Mishra, N. Behera, B. K. Senapati, and B. C. Guru. New Delhi: Ashish Publ. House, New Delhi.
Alford, R. A., and M. L. Crump. 1982. Habitat partitioning among size classes of larval southern leopard frogs, *Rana utricularia. Copeia* 1982:367–373.
Alford, R. A., and R. N. Harris. 1988. Effects of larval growth history on anuran metamorphosis. *American Nat.* 131:91–106.
Alford, R. A., and G. D. Jackson. 1993. Do cephalopods and the larvae of other taxa grow asymptotically? *American Nat.* 141:717–728.
Alford, R. A., and H. M. Wilbur. 1985. Priority effects in experimental pond communities: Competition between *Bufo* and *Rana. Ecology* 66:1097–1105.
Alley, K. E., and M. D. Barnes. 1983. Birth dates of trigeminal motoneurons and metamorphic reorganization of the jaw myoneuronal system in frogs. *J. Comp. Neurol.* 218:395–405.
Allison, J. D. 1992. Acoustic modulation of neural activity in the preoptic area and ventral hypothalamus of the green treefrog (*Hyla cinerea*). *J. Comp. Physiol.* 171A:387–395.
Al-Mukthar, K. A. K., and A. C. Webb. 1971. An ultrastructural study of primordial germ cells, oogonia, and early oocytes in *Xenopus laevis. J. Embryol. Exp. Morphol.* 26:195–217.
Altig, R. 1970. A key to the tadpoles of the continental United States and Canada. *Herpetologica* 26:180–207.
———. 1972. Notes on the larvae and premetamorphic tadpoles of four *Hyla* and three *Rana* with notes on tadpole color patterns. *J. Elisha Mitchell Sci. Soc.* 88:113–119.
———. 1973. Preliminary scanning electron observations of keratinized structures in amphibian larvae. *HISS News-J.* 1:129–131.
———. 1975. Freeze-drying anuran tadpoles for taxonomic examinations. *Herpetol. Rev.* 6:13.
———. 1987. Key to the anuran tadpoles of Mexico. *Southwest. Nat.* 32:75–84.
Altig, R., and R. A. Brandon. 1971. Generic key and synopsis for free-living larvae and tadpoles of Mexican amphibians. *Tulane Stud. Zool. Bot.* 17:10–15.
Altig, R., and E. D. Brodie, Jr. 1972. Laboratory behavior of *Ascaphus truei* tadpoles. *J. Herpetol.* 6:21–24.
Altig, R., and A. Channing. 1993. Hypothesis: Functional significance of colour and pattern of anuran tadpoles. *Herpetol. J.* 3:73–75.
Altig, R., and M. T. Christensen. 1981. Behavioral characteristics of the tadpoles of *Rana heckscheri. J. Herpetol.* 15:151–154.
Altig, R., and G. F. Johnston. 1986. Major characteristics of free-living anuran tadpoles. *Smithsonian Herpetol. Info. Serv.* (67):1–75.
———. 1989. Guilds of anuran larvae: Relationships among developmental modes, morphologies, and habitats. *Herpetol. Monogr.* (3):81–109.
Altig, R., and J. P. Kelly. 1974. Indices of feeding in anuran tadpoles as indicated by gut characteristics. *Herpetologica* 30:200–203.
Altig, R., and W. McDearman. 1975. Clearance times and per cent assimilation of five species of anuran tadpoles. *Herpetologica* 31:67–69.
Altig, R., and W. L. Pace. 1974. Scanning electron photomicrographs of tadpole labial teeth. *J. Herpetol.* 8:247–251.

Altman, J. S., and E. A. Dawes. 1983. A cobalt study of medullary sensory projections from lateral line nerves associated with cutaneous nerves, and the VIIIth nerve in adult *Xenopus*. *J. Comp. Neurol.* 213:310–326.

Alvarado, R. H., and T. C. Cox. 1985. Action of polyvalent cations on sodium transport across skin of larval and adult *Rana catesbeiana*. *J. Exp. Zool.* 236:127–136.

Alvarado, R. H., and S. R. Johnson. 1966. The effects of neurohypophysial hormones on water and sodium balance in larval and adult bullfrogs (*Rana catesbeiana*). *Comp. Biochem. Physiol.* 18A:549–561.

Alvarado, R. H., and A. Moody. 1970. Sodium and chloride transport in tadpoles of the bullfrog *Rana catesbeiana*. *American J. Physiol.* 218:1510–1516.

Amboroso, E. C. 1968. The evolution of viviparity. *Proc. Roy. Soc. Med.* 61:1188–1200.

Amer, F. I. 1968–1969. On the occurrence and nature of Bidder's organ in the Egyptian toad, *Bufo regularis* Reuss. *Zool. Soc. Egypt Bull.* 21:39–62.

———. 1972. Observations on the urinogenital ducts of the Egyptian toad, *Bufo regularis* Reuss. *Zool. Soc. Egypt Bull.* 24:15–19.

Amiet, J.-L. 1970. Morphologie et développement de la larve de *Leptodactylodon ventrimarmoratus* (Boulenger)(Amphibien Anoure). *Ann. Fac. Sci. Cameroun* 1970:53–71.

———. 1971. Le têtard d'*Astylosternus corrugatus* Boulenger (Amphibien Anoure). *Ann. Fac. Sci. Cameroun* 1971:85–98.

———. 1974a. La ponte de la larve d'*Opisthothylax immaculatus* Boulenger (Amphibien Anoure). *Ann. Fac. Sci. Cameroun* 1974:121–130.

———. 1974b. Le têtard d'*Hyperolius obstetricans* Ahl (Amphibien Anoure). *Bull. Inst. France Africa Noire* 36:973–981.

———. 1976. Les formes larvaires d'*Hyperolius krebsi* Mertens et *H. koehleri* Mertens (Amphibien Anoures). *Ann. Fac. Sci. Cameroun* 1976:159–169.

———. 1979. Description de l'adulte et de la larve d'*Hyperolius bopeleti* n. sp. (Amphibia Anura, Hyperoliidae). *Ann. Fac. Sci. Cameroun* 1979:113–124.

Anderson, J. D. 1972. Phototactic behavior of larvae and adults of two subspecies of *Ambystoma macrodactylum*. *Herpetologica* 28:222–226.

Andrén, C., L. Hendrickson, M. Olsson, and G. Nilson. 1988. Effects of pH and aluminum on embryonic and early larval stages of Swedish brown frogs *Rana arvalis, R. temporaria,* and *R. dalmatina*. *Holarctic Ecol.* 11:127–135.

Andrén, C., and G. Nilson. 1985. Breeding pool characteristics and reproduction in an island population of natterjack toads, *Bufo calamita* Laur., at the Swedish west coast. *Amphibia-Reptilia* 6:137–142.

Andres, G. 1963. Eine experimentelle analyse der Entwicklung der larvalen Pigmentmuster von fünf Anurenarten. *Zoologica* 40:1–112.

Andres, K. H. 1970. Anatomy and ultrastructure of the olfactory bulb in fish, Amphibia, reptiles, birds, and mammals. pp. 177–194. In *Taste and smell in vertebrates. CIBA Found. Symp.,* edited by G. E. W. Wolstenholme and J. Knight. London: J. and A. Churchill.

Andrews, R. M., and B. R. Rose. 1994. Evolution of viviparity: Constraints on egg retention. *Physiol. Zool.* 67:1006–1024.

Angel, F. 1946. *Faune de France. Reptiles et amphibiens.* Paris: Paul LeChevalier.

Anholt, B., S. Negovetic, and C. Som. 1998a. Methods for anaesthetizing and marking larval anurans. *Herpetol. Rev.* 29:153–154.

Anholt, B., S. Negovetic, C. Som, and R. Mulheim. 1998b. Marking tadpoles with VIE. <http://www.mp1pwrc.usgs/gov/marking/anholt.html>.

Anholt, B. R., D. K. Skelly, and E. E. Werner. 1996. Factors modifying antipredator behavior in larval toads. *Herpetologica* 52:301–313.

Anholt, B. R., and E. E. Werner. 1995. Interaction between food availability and predation mortality mediated by adaptive behavior. *Ecology* 76:2230–2234.

Annandale, N. 1912. Biological results of the Abor Expedition, 1911–12. *I. Batrachia. Rec. Indian Mus.* 8:7–36, plates II–IV.

———. 1917. The occurrence of *Rana pleskei* Günther, in Kashmir. *Rec. Indian Mus.* 13:417–418.

———. 1918. Some undescribed tadpoles from the hills of southern India. *Rec. Indian Mus.* 15:17–23, plate I.

———. 1919. The tadpoles of *Nyctibatrachus pygmaeus* and *Ixalus variabilis:* A correction. *Rec. Indian Mus.* 16:303.

Annandale, N., and S. L. Hora. 1922. Parallel evolution in the fish and tadpoles of mountains torrents. *Rec. Indian Mus.* 24:505–510.

Annandale, N., and C. R. N. Rao. 1918. The tadpoles of the families Ranidae and Bufonidae found in the plains of India. *Rec. Indian Mus.* 15:25–40, plate IV.

Antal, M., R. Kraftsik, G. Székely, and H. Van Der Loos. 1986. Distal dendrites of frog motor neurons: A computer aided electron microscopic study of cobalt-filled cells. *J. Neurocytol.* 15:303–310.

Antal, M., I. Tornai, and G. Székely. 1980. Longitudinal extent of dorsal root fibres in the spinal cord and brain stem of the frog. *Neuroscience* 5:1311–1322.

Archey, G. 1922. The habitat and life history of *Liopelma hochstetteri*. *Rec. Canterbury Mus.* 2:59–71.

Ardila-Robayo, M. C. 1979. Status sistematico del genero *Geobatrachus* Ruthven 1915 (Amphibia: Anura). *Caldasia* 12:383–495.

Armstrong, J. B., and G. M. Malacinski, eds. 1989. *Developmental biology of the axolotl.* New York: Oxford University Press.

Arnauld, E., and R. Cambar. 1970. Nouvelles observations sur les modalités d'induction de la morphogenèse du mésonéphros chez la larve de la grenouille agile (*Rana dalmatina*). *C. R. Hebd. Seances Acad. Sci. Paris* 270D:2563–2565.

Arnold, E. N. 1994. Do ecological analogues assemble their common features in the same order? An investigation of regularities in evolution, using sand-dwelling lizards as examples. *Philosoph. Trans. Roy. Soc. London* 344:277–290.

Arnold, E. N., and J. A. Burton. 1985. *A field guide to the reptiles and amphibians of Britain and Europe.* Glasgow: William Collins Sons & Co.

Arnold, S. J., and R. J. Wassersug. 1978. Differential predation on metamorphic anurans by garter snakes (*Thamnophis*): Social behavior as a possible defense. *Ecology* 59:1014–1022.

Arnoult, J. 1965. Contribution a l'étude des batraciens de Madagascar. Écologie et développement de *Mantella aurantiaca* Mocquard, 1900. *Bull. Mus. Natl. d'Hist. Nat.* 37:931–940.

Arnoult, J., and M. Lamotte. 1968. Les Pipidae de l'Ouest africain et du Cameroun. *Bull. Inst. France Afrique Noire* 30A: 270–306.

Arnoult, J., and M. Razarihelisoa. 1967. Contribution a l'étude des batraciens de Madagascar le genre *Mantidactylus* adultes et formes larvaires de *M. betsileanus* (Blgr.), *M. curtus* (Blgr.) et *M. alutus* (Peracca). *Bull. Mus. Natl. d'Hist. Nat.* 39:471–487.

Arrayago, M.-J., A. Bea, and B. Heulin. 1996. Hybridization

experiments between oviparous and viviparous strains of *Lacerta vivipara:* A new insight into the evolution of viviparity in reptiles. *Herpetologica* 52:33–342.

Arthur, W. 1982. A developmental approach to the problem of variation in evolutionary rates. *Biol. J. Linnean Soc.* 18:243–261.

Aschoff, L. 1924. Das Reticuloendotheliale System. *Ergebn. Innere Med. Kinderheilkde.* 26:1–118.

Ashby, K. R. 1969. The population ecology of a self-maintaining colony of the common frog (*Rana temporaria*). *J. Zool.* 158:453–474.

Ashley, H., P. Katti, and E. Frieden. 1968. Urea excretion in the bullfrog tadpole: Effect of temperature, metamorphosis, and thyroid hormones. *Develop. Biol.* 17:293–307.

Atkinson, B. G., and J. J. Just. 1975. Biochemical and histological changes in the respiratory system of *Rana catesbeiana* larvae during normal and induced metamorphosis. *Develop. Biol.* 45:151–165.

Atkinson, B. G., and G. H. Little. 1972. Growth and regression in tadpole pancreas during spontaneous and thyroid hormone-induced metamorphosis. 1. Rates of macromolecular synthesis and degradation. *Mech. Age Develop.* 1:299–312.

Atoda, K. 1950. Metamorphosis of the "nonaquatic frog" of the Palau Islands, Western Carolines. *Pacific Sci.* 4:202–207.

Atwell, W. J. 1918–1919. The development of the hypophysis of the Anura. *Anat. Rec.* 15:73–91.

Auburn, J. S., and D. H. Taylor. 1979. Polarized light perception and orientation in larval bullfrogs *Rana catesbeiana*. *Anim. Behav.* 27:658–668.

Audo, M. C., T. M. Mann, T. L. Polk, C. M. Loudenslager, W. J. Diehl, and R. Altig. 1995. Food deprivation during different periods of tadpole (*Hyla chrysoscelis*) ontogeny affects metamorphic performance differently. *Oecologia* 103:518–522.

Augert, D., and P. Joly. 1993. Plasticity of age at maturity between two neighbouring populations of the common frog (*Rana temporaria* L.). *Canadian J. Zool.* 71:26–33.

———. 1994. Dispersal of *Rana temporaria* tadpoles in large fishponds. *Alytes* 12:31–40.

Ax, P. 1984. *Das phylogenetische system: Systematisierung der lebenden Natur aufgrund ihrer Phylogenese.* Stuttgart: Gustav Fischer-Verlag.

Axelsson, E., P. Nustrom, and C. Brönmark. 1997. Crayfish predation on amphibian eggs and larvae. *Amphibia-Reptilia* 18:217–228.

Azevedo-Ramos, C., M. Van Sluys, J.-M. Hero, and W. E. Magnusson. 1992. Influence of tadpole movements on predation by odonate naiads. *J. Herpetol.* 26:335–338.

Babalian, A. L., and A. I. Shapovalov. 1984. Synaptic actions produced by individual ventrolateral tract fibres in frog lumbar motoneurons. *Exp. Brain Res.* 54:551–563.

Babbitt, K. J., and G. W. Tanner. 1997. Effects of cover and predator identity on predation of *Hyla squirella* tadpoles. *J. Herpetol.* 31:128–130.

Babbitt, K. J., and F. Jordan. 1996. Predation on *Bufo terrestris* tadpoles: Effects of cover and predator density. *Copeia* 1996:485–488.

Baccari, G. C., S. Minucci, and L. Dimatteo. 1990. Organogenesis of the Harderian gland in *Rana esculenta* and *Bufo viridis*. *Boll. Zool.* 57:221–224.

Bachmann, K. 1969. Temperature adaptations of amphibian embryos. *American Nat.* 103:115–130.

Baculi, B. S., and E. L. Cooper. 1968. Lymphomyeloid organs of Amphibia. IV. Normal histology in larval and adult *Rana catesbeiana*. *J. Morphol.* 126:463–476.

Bagenal, T. B. 1967. A method of marking fish eggs and larvae. *Nature* 214:113.

Bagnara, J. T. 1965. Pineal regulation of body blanching in amphibian larvae. *Progr. Brain Res.* 10:489–506.

———. 1972. Interrelationships of melanophores, iridophores, and xanthophores. pp. 171–180. In *Pigmentation: Its genesis and biologic control,* edited by V. Riley. New York: Appleton-Century-Crofts.

———. 1976. Color change. pp. 1–52. In *Physiology of the Amphibia.* Vol. 3, edited by B. Lofts. New York: Academic Press.

Bagnara, J. T., S. K. Frost, and J. Matsumoto. 1978. On the development of pigment patterns in amphibians. *American Zool.* 18:301–312.

Bagnara, J. T., J. Matsumoto, W. Ferris, S. K. Frost, W. A. Turner, Jr., T. T. Tchen, and J. D. Taylor. 1979. Common origin of pigment cells. *Science* 203:410–415.

Bailey, C. H., and P. Gouras. 1985. The retina and phototransduction. pp. 344–355. In *Principles of neural science.* 2d ed., edited by E. R. Kandel and J. H. Schwartz. New York: Elsevier.

Baker, C. L. 1949. The comparative anatomy of the aortic arches of the urodeles and their relation to respiration and degree of metamorphosis. *J. Tennessee Acad. Sci.* 24:12–40.

Baker, G. C., and T. J. C. Beebee. 1997. Microenvironmental effects on competition between *Rana* and *Bufo* larvae, and on the abundance of *Prototheca richardsi,* in small fish-ponds. *Herpetol. J.* 7:149–154.

Balinsky, B. I. 1972. The fine structure of the amphibian limb bud. *Acta Embryol. Exp. Suppl.* 1972:455–470.

———. 1981. *An introduction to embryology.* 5th ed. New York: W. B. Saunders.

Balinsky, J. B. 1970. Nitrogen metabolism in amphibians. pp. 519–624. In *Comparative biochemistry of nitrogen metabolism,* edited by J. W. Campbell. New York: Academic Press.

Balinsky. J. B., E. L. Choritz, C. G. L. Coe, and G. S. Van Der Schans. 1967. Urea cycle enzymes and urea excretion during the development and metamorphosis of *Xenopus laevis. Comp. Biochem. Physiol.* 22A:53–57.

Balinsky, J. B., T. L. Coetzer, and F. J. Mattheyse. 1972. The effect of thyroxine and hypertonic environment on the enzymes of the urea cycle in *Xenopus laevis. Comp. Biochem. Physiol.* 43B:83–95.

Balinsky, J. B., and R. J. Devis. 1963. Origin and differentiation of cytoplasmic structures in the oocytes of *Xenopus laevis. Acta Embryol. Morphol. Exp.* 6:55–108.

Banks, B., and T. J. C. Beebee. 1987. Spawn predation and larval growth inhibition as mechanisms for niche separation in anurans. *Oecologia* 72:569–573.

Bantle, J. A., J. N. Dumont, R. A. Finch, and G. Linder. 1990. Atlas of abnormalities: A guide for the performance of FETAX. *United States Army Med. Res. Develop. Command Publ.* DAMD17-88-c-8031.

Barch, S. H., J. R. Shaver, and G. B. Wilson. 1966. An electron microscopic study of the nephric unit in the frog. *Trans. American Micros. Soc.* 85:350–359.

Bargmann, W. 1937. Über sezernierende Zellelemente im Nephron von *Xenopus laevis. Zeit. Zellforsch. Mikros. Anat.* 25:764–768.

———. 1978. Niere und ableitende Harnwege. pp. 1–444. In *Handbuch der mikroskopischen Anatomie des Menschen.* Vol. 7, edited by W. Bargmann. New York: Springer-Verlag.

Bargmann. W., A. Knoop, and T. H. Schiebler. 1955. Histolog-

ische, cytochemische und elektronenmikroskopische Untersuchungen am Nephron (mit Berücksichtigung der Mitochondrien). *Zeit. Zellforsch. Mikros. Anat.* 42:386–422.
Barker, J., and G. C. Grigg. 1977. *A field guide to Australian frogs.* Adelaide, Australia: Rigby Ltd.
Bärlöcher, F., R. J. Mackay, and G. B. Wiggins. 1978. Detritus processing in a temporary vernal pool in southern Ontario. *Arch. Hydrobiol.* 81:269–295.
Barnes, M. D., and K. E. Alley. 1983. Maturation and recycling of trigeminal motoneurons in anuran larvae. *J. Comp. Neurol.* 218:406–414.
Barreto, L., and G. Moriera. 1996. Seasonal variation in age structure and spatial distribution of a savanna larval anuran assemblage in central Brazil. *J. Herpetol.* 30:87–92.
Barrington, E. J. W. 1942. Gastric digestion in the lower vertebrates. *Biol. Rev.* 17:1–27.
———. 1946. The delayed development of the stomach in the frog (*Rana temporaria*) and the toad (*Bufo bufo*). *Proc. Zool. Soc. London* 116:1–21.
———. 1957. The alimentary canal and digestion. In *The physiology of fishes.* Vol. 2, *Metabolism.* Edited by M. Brown. New York: Academic Press.
Barrio, A. 1953. Sistemática, morfologia y reproducción de *Physalaemus henselii* (Peters) y *Pseudopaludicola falcipes* (Hensel)(Anura, Leptodactylidae). *Physis* 20:379–389.
———. 1976. Observaciones etoecologicas sobre *Hylorina sylvatica* Bell (Anura, Leptodactylidae). *Physis* 27:153–157.
Barry, T. H. 1956. The ontogenesis of the sound-conducting apparatus of *Bufo angusticeps* Smith. *Morphol. Jahrb.* 97:477–544.
Barandun, J., and H.-U. Reyer. 1997. Reproductive ecology of *Bombina variegata:* Development of eggs and larvae. *J. Herpetol.* 31:107–110.
Bardsley, L., and T. J. C. Beebee. 1998. Interspecific competition between *Bufo* larvae under conditions of community transition. *Ecology* 79:1751.
Başoğlu, M., and N. Özeti. 1973. Türkiye amphibileri. *Ege Üniversitesi Fen Fakültesi Kitaplar Serisi* (50):1–155.
Bateson, P. 1983. Optimal outbreeding. pp. 257–277. In *Mate choice,* edited by P. Bateson. Cambridge: Cambridge University Press.
Battaglini, P., and P. Boni. 1967. Indigenous microbial flora and the large intestine in tadpoles. *Experientia* 23:950–951.
Batten, T. F. C., and P. M. Ingleton. 1987. The hypothalamus and pituitary gland. pp. 285–412. In *Fundamentals of comparative vertebrate endocrinology,* edited by I. Chester-Jones, P. M. Ingleton, and J. G. Phillips. New York: Plenum.
Bavay, A. 1873. Notes sur l'*Hylodes martinicensis* Tschudi et ses metamorphosis. *Rev. Sci. Nat.* 1:281–290.
Beach, D. H., and M. Jacobson. 1979a. Patterns of cell proliferation in the retina of the clawed frog during development. *J. Comp. Neurol.* 183:603–614.
———. 1979b. Influences of thyroxine on cell proliferation in the retina of the clawed frog at different ages. *J. Comp. Neurol.* 183:615–624.
Beachy, C. K. 1995. Effects of larval growth history on metamorphosis in a stream-dwelling salamander (*Desmognathus ochrophaeus*). *J. Herpetol.* 29:375–382.
Beams, H. W., and R. G. Kessel. 1974. The problem of germ cell determinants. *Intl. Rev. Cytol.* 39:413–479.
Beaumont, A. 1970. Apparition des cellules insulaires a dans le pancréas larvaire d'un Amphibien Anoure: *Discoglossus pictus* Otth. *C. R. Hebd. Seances Acad. Sci. Paris* 271D:1104–1106.
Beaumont, A., and J. Deunff. 1958. La kératinisation des dents et du bec larvaires chez *Alytes obstetricans* Laurenti. *Arch. Anat. Micros. Morphol. Exp.* 48:307–324.
Beck, C. W. 1997. Effect of changes in resource level on age and size at metamorphosis in *Hyla squirella. Oecologia* 112:187–192.
———. 1998. Mode of fertilization and parental care in anurans. *Anim. Behav.* 55:439–449.
Beckenbach, A. T. 1975. Influence of body size and temperature on the critical oxygen tension of some plethodontid salamanders. *Physiol. Zool.* 48:338–347.
Beddard, F. E. 1896. On the diaphragm and on the muscular anatomy of *Xenopus,* with remarks on its affinities. *Proc. Zool. Soc. London* 53:841–850.
Beebee, T. J. C. 1991. Purification of an agent causing growth inhibition in anuran larvae and its identification as a unicellular unpigmented alga. *Canadian J. Zool.* 69:2146–2153.
———. 1995. Tadpole growth: Is there an interference effect in nature? *Herpetol. J.* 5:204–205.
Beebee, T. J. C., and J. R. Griffin. 1977. A preliminary investigation into natterjack toad (*Bufo calamita*) breeding site characteristics in Britain. *J. Zool.* 181:341–350.
Beebee, T. J. C., and A. L.-C. Wong. 1992a. Leucine uptake by enterobacterial and algal members of larval anuran gut flora. *Comp. Biochem. Physiol.* 101B:527–530.
———. 1992b. *Prototheca*-mediated interference competition between anuran larvae operates by resource diversion. *Physiol. Zool.* 65:815–831.
Beigl, I. 1989. Untersuchung der Urogenitalverbindungen anurer Entwicklungsgeschichte des Amphibienherzens. *Morphol. Jahrb.* 51:355–412.
Beiswenger, R. E. 1972. Aggregative behavior of tadpoles of the American toad, *Bufo americanus,* in Michigan. Dissertation, University of Michigan.
———. 1975. Structure and function in aggregations of tadpoles of the American toad, *Bufo americanus. Herpetologica* 31:222–233.
———. 1977. Diel patterns of aggregative behavior in tadpoles of *Bufo americanus,* in relation to light and temperature. *Ecology* 58:98–108.
———. 1978. Responses of *Bufo* tadpoles (Amphibia, Anura, Bufonidae) to laboratory gradients of temperature. *J. Herpetol.* 12:499–504.
Beiswenger, R. E., and F. H. Test. 1966. Effects of environmental temperature on movements of tadpoles of the American toad, *Bufo terrestris americanus. Pap. Michigan Acad. Sci., Arts Letters* 51:127–141.
Bell, B. D. 1978. Observations on the ecology and reproduction of the New Zealand leiopelmid frogs. *Herpetologica* 34:340–354.
———. 1985. Development and parental-care in the endemic New Zealand frogs. pp. 269–278. In *Biology of Australian frogs and reptiles,* edited by G. Grigg, R. Shine, and H. Ehmann. Chipping Norton, New South Wales: Surrey Beatty.
Bello y Espinosa, D. 1871. Zoologiche Notizen aus Puerto Rico, nach dem Spanischen frei bearbeitel von Herrn E. von Martens in Berlin. *Zool. Garten* (Frankfurt a. M.) 12:348–351.
Belova, Z. A. 1964. On the feeding habits of the tadpoles of *Rana ridibunda* Pall. in the Volga Delta. *Zool. Zhur.* 453:1188–1192.
Benbassat, J. 1970. Erythroid cell development during natural amphibian metamorphosis. *Develop. Biol.* 21:557–583.
Bennett, A. F., and W. R. Dawson. 1976. Metabolism. pp. 127–223. In *Biology of the Reptilia.* Vol. 5. *Physiology A,* edited by C. Gans and W. R. Dawson. New York: Academic Press.

Bennett, A. F., and P. Licht. 1974. Anaerobic metabolism during activity in amphibians. *Comp. Biochem. Physiol.* 48A:319–327.

Benninghoff, A. 1921. Beiträge zur vergleichenden Anatomie und Entwicklungsgeschichte des Amphibienherzens. *Morphol. Jahrb.* 51:355–412.

Benson, W. W. 1971. Evidence for the evolution of unpalatability through kin selection in the Heliconiiae (Lepidoptera). *American Nat.* 105:213–226.

Bentley, P. J. 1971. *Endocrines and osmoregulation. A comparative account of the regulation of water and salt in vertebrates.* Vol. 1. *Zoophysiology and Ecology.* New York: Springer-Verlag.

Bentley, P. J., and G. P. Baldwin. 1980. Comparison of transcutaneous permeability in skins of larval and adult salamanders (*Ambystoma tigrinum*). *American J. Physiol.* 239R:5–9.

Bentley, P. J., and D. J. Greenwald. 1970. Neurohypophysial function in bullfrog (*Rana catesbeiana*) tadpoles. *Gen. Comp. Endocrinol.* 14:412–415.

Bernardini, G., P. Podini, R. Maci, and M. Camatini. 1990. Spermiogenesis in *Xenopus laevis:* From late spermatids to spermatozoa. *Mol. Reprod. Develop.* 26:347–355.

Bernasconi, A. F. 1951. Über den Ossifikationsmodus bei *Xenopus laevis* Daud. *Denkschr. Schweiz. Naturf. Ges.* 2:196–252.

Berninghausen, F. 1994. *Feld- und Bestimmungsschlüssel für Kaulquappen.* 2d ed. Naturschtzbund Deutschland.

———. 1997. *Welche Kaulquappe ist das?* Hanover: NABU.

Berns, M. W. 1965. Mortality caused by kidney stones in spinach-fed frogs (*Rana pipiens*). *BioScience* 15:297–298.

Berry, P. Y. 1972. Undescribed and little known tadpoles from west Malaysia. *Herpetologica* 28:338–346.

Berthold, G. 1980. Microtubules in the epidermal cells of *Carausius morosus* (Br.). Their pattern and relation to pigment migration. *J. Insect. Physiol.* 26:421–425.

Berton, J.-P. 1964. Anatomie vasculaire du rein de quelques amphibiens urodèles et anoures. *Ann. Sci. Nat. Zool.* 6:229–280.

Bertram, B. C. R. 1978. Living in groups: Predators and prey. pp. 64–96. In *Behavioural ecology: An evolutionary approach,* edited by J. R. Krebs and N. B. Davies. Oxford: Blackwell Sci. Publ.

Berven, K. A. 1982. The genetic basis of altitudinal variation in the wood frog *Rana sylvatica.* II. An experimental analysis of larval development. *Oecologia* 52:360–369.

———. 1987. The heritable basis of variation in larval developmental patterns within populations of the wood frog (*Rana sylvatica*). *Evolution* 41:1088–1097.

———. 1990. Factors affecting population fluctuations in larval and adult stages of the wood frog (*Rana sylvatica*). *Ecology* 71:1599–1608.

Berven, K. A., and B. G. Chadra. 1988. The relationship among egg size, density, and food level on larval development in the wood frog (*Rana sylvatica*). *Oecologia* 75:67–72.

Berven, K. A., and D. E. Gill. 1983. Interpreting geographic variation in life-history traits. *American Zool.* 23:85–97.

Berven, K. A., D. E. Gill, and S. J. Smith-Gill. 1979. Countergradient selection in the green frog, *Rana clamitans. Evolution* 33:609–623.

Bhaduri, J. L. 1953. A study of the urinogenital system of Salientia. *Proc. Zool. Soc. Bengal* 6:1–111.

Bhaduri, J. L., and S. L. Basu. 1957. A study of the urogenital system of Salientia, Part I. Ranidae and Hyperoliidae of Africa. *Ann. Mus. Roy. Congo Belge, Sci. Zool.* 55:1–56.

Bhaduri, J. L., and J. C. Daniel. 1956. The tadpoles of *Uperodon globulosum. J. Bombay Nat. Hist. Soc.* 53:713–716.

Bhalla, S. C., and R. R. Sokal. 1964. Competition among genotypes in the housefly at varying densities and proportions (the green strain). *Evolution* 18:312–330.

Binckley, C. A., B. Plesky, K. Werner, and S. Droege. 1998. Using the Visible Implant Flourescent Elastomer (VIE) tagging system to mark amphibians. <http://www.mp1pwrc.usgs.gov/marking/vie.html>.

Binkley, S., K. Mosher, F. Rubin, and B. White. 1988. *Xenopus* tadpole melanophores are controlled by dark and light melatonin without influence of time of day. *J. Pineal Res.* 5:87–98.

Birks, R., H. E. Huxley, and B. Katz. 1960. The fine structure of the neuromuscular junction of the frog. *J. Physiol.* 150:134–144.

Bishop, J. E. 1973. *Limnology of a small Malayan river Sungai Gombak.* The Hague: Dr. W. Junk.

Bixby, J. L., and N. C. Spitzer. 1982. The appearance and development of chemosensitivity in Rohon-Beard neurons of the *Xenopus* spinal cord. *J. Physiol.* 330:513–536.

Black, D. 1917. The motor nuclei of the cerebral nerves in phylogeny. A study of the phenomena of neurotaxis. *J. Comp. Neurol.* 28:379–427.

Black, J. H. 1971. The formation of tadpoles holes. *Herpetol. Rev.* 3:7.

———. 1974. Tadpoles nests in Oklahoma. *Oklahoma Geol. Notes* (3):105.

Blackburn, D. G. 1994. Discrepant usage of the term "ovoviviparity" in the herpetological literature. *Herpetol. J.* 4:65–72.

Blackler, A. W. 1970. The integrity of the reproductive cell line in the Amphibia. *Curr. Top. Develop. Biol.* 5:71–88.

Blair, W. F. 1961. Calling and spawning seasons in a mixed population of anurans. *Ecology* 42:99–110.

Blaustein, A. R. 1988. Ecological correlates and potential functions of kin recognition and kin association in anuran larvae. *Behav. Genet.* 18:449–464.

Blaustein, A. R., M. Bekoff, J. A. Byers, and T. J. Daniels. 1991. Kin recognition in vertebrates: What do we really know about adaptive value? *Anim. Behav.* 41:1079–1083.

Blaustein, A. R., M. Bekoff, and T. J. Daniels. 1987a. Kin recognition in vertebrates (excluding primates): Empirical evidence. pp. 287–331. In *Kin recognition in animals,* edited by D. J. C. Fletcher and C. D. Michener. New York: Wiley.

———. 1987b. Kin recognition in vertebrates (excluding primates): Mechanisms, functions, and future research. pp. 333–357. In *Kin recognition in animals,* edited by D. J. C. Fletcher and C. D. Michener. New York: Wiley.

Blaustein, A. R., K. S. Chang, H. G. Lefcort, and R. K. O'Hara. 1990. Toad tadpole kin recognition: Recognition of half siblings and the role of maternal cues. *Ethol. Ecol. Evol.* 2:215–226.

Blaustein, A. R., and R. K. O'Hara. 1981. Genetic control for sibling recognition. *Nature* 290:246–248.

———. 1982a. Kin recognition in *Rana cascadae* tadpoles: Maternal and paternal effects. *Anim. Behav.* 30:1151–1157.

———. 1982b. Kin recognition cues in *Rana cascadae* tadpoles. *Behav. Neural Biol.* 36:77–87.

———. 1983. Kin recognition in *Rana cascadae* tadpoles: Effects of rearing with nonsiblings and varying the strength of the stimulus cues. *Behav. Neural Biol.* 39:259–267.

———. 1986a. An investigation of kin recognition in red-legged frog (*Rana aurora*) tadpoles. *J. Zool.* 209:347–353.

———. 1986b. Kin recognition in tadpoles. *Sci. American* 254:108–116.

———. 1987. Aggregation behaviour in *Rana cascadae* tadpoles: Association preferences among wild aggregations and responses to non-kin. *Anim. Behav.* 35:1549–1555.

Blaustein, A. R., R. K. O'Hara, and D. H. Olson. 1984. Kin preference behaviour is present after metamorphosis in *Rana cascadae* frogs. *Anim. Behav.* 32:445–450.

Blaustein, A. R., and B. Waldman. 1992. Kin recognition in anuran amphibian larvae. *Anim. Behav.* 44:207–221.

Blaustein, A. R., and S. C. Walls. 1995. Aggregation behaviour and kin recognition. pp. 568–602. In *Amphibian biology,* Vol. 2. Chipping Norton, New South Wales: Surrey Beatty.

Blaustein, A. R., T. Yoshikawa, K. Asoh, and S. C. Walls. 1993. Ontogenetic shifts in tadpole kin recognition—loss of signal and perception. *Anim. Behav.* 46:525–538.

Blaustein, L., and B. P. Kotler. 1993. Oviposition habitat selection by the mosquito, *Culiseta longiareolata:* Effects of conspecifics, food, and green toad tadpoles. *Ecol. Entomol.* 18:104–108.

Blaustein, L., and J. Margalit. 1994. Mosquito larvae (*Culiseta longiareolata*) prey upon and compete with toad tadpoles (*Bufo viridis*). *J. Anim. Ecol.* 63:841–850.

———. 1995. Spatial distributions of *Culiseta longiareolata* (Culicidae: Diptera) and *Bufo viridis* (Amphibia: Bufonidae) among and within desert pools. *J. Arid Environ.* 29:199–211.

———. 1996. Priority effects in temporary pools: Nature and outcome of mosquito larva–toad tadpole interactions depend on order of entrance. *J. Anim. Ecol.* 65:77–84.

Bleakney, S. 1958. Cannibalism in *Rana sylvatica* tadpoles a well known phenomenon. *Herpetologica* 14:34.

Bles, E. J. 1905. The life history of *Xenopus laevis* Daud. *Trans. Roy. Soc. Edinburgh* 41:789–821.

Blight, A. R. 1976. The muscular control of vertebrate swimming movements. *Biol. Rev.* 52:181–218.

———. 1978. Golgi-staining of "primary" and "secondary" motoneurons in the developing spinal cord of an amphibian. *J. Comp. Neurol.* 180:679–690.

Blommers-Schlösser, R. M. A. 1975. Observations on the larval development of some Malagasy frogs, with notes on their ecology and biology (Anura: Dyscophinae, Scaphiophryninae and Cophylinae). *Beaufortia* 24:7–26.

———. 1979a. Biosystematics of the Malagasy frogs. I. Mantellinae (Ranidae). *Beaufortia* (352):1–76.

———. 1979b. Biosystematics of the Malagasy frogs II. The genus *Boophis* (Rhacophoridae). *Bijd. tot de Dierkunde* 49:261–312.

———. 1981. On endemic Malagasy frogs (Ranidae, Rhacophoridae and Hyperoliidae). *Monit. Zool. Italiano (N.S.) Suppl.* 15:217–224.

———. 1982. Observations on the Malagasy frog genus *Heterixalus* Laurent, 1944 (Hyperoliidae). *Beaufortia* 32:1–11.

Blommers-Schlösser, R. M. A., and C. P. Blanc. 1991. Amphibiens. *Faune de Madagascar* 75:1–379, 12 plates.

Blommers-Schlösser, R. M. A., and L. H. M. Blommers. 1984. The amphibians. pp. 89–104. In *Key environments Madagascar,* edited by A. Jolly, P. Oberle and R. Albinac. New York: Pergamon.

Blouin, M. S. 1992a. Comparing bivariate reaction norms among three species: Time and size at metamorphosis in three species of *Hyla* (Anura: Hylidae). *Oecologia* 90:288–293.

———. 1992b. Genetic correlations among morphometric traits and rates of growth and differentiation in the green tree frog, *Hyla cinerea. Evolution* 46:735–744.

Boas, J. E. V. 1882. Über den Conus arteriosus und die Arterienbögen der Amphibien. *Morphol. Jahrb.* 7:488–572.

Boatright-Horowitz, S. S., and A. M. Simmons. 1997. Transient "deafness" accompanies auditory development during metamorphosis form tadpole to frog. *Proc. Natl. Acad. Sci. United States of America* 94:14877–14882.

Bogart, J. P. 1981. How many times has terrestrial breeding evolved in anuran amphibians? *Monit. Zool. Italiano (N.S.) Suppl.* 15:29–40.

Bogert, C. M. 1969. The eggs and hatchlings of the Mexican leptodactylid frog *Eleutherodactylus decoratus* Taylor. *American Mus. Novitates* (2376):1–9.

Bokermann, W. C. A. 1963. Girinos de anfíbios brasileiros—2 (Amphibia, Salientia). *Rev. Brasil. Biol.* 23:349–353.

———. 1965. Notas sôbre as espécies de *Thoropa* Fitzinger (Amphibia, Leptodactylidae). *Anais Acad. Brasil. Cienc.* 37:525–537.

———. 1966. O gênero *Phyllodytes* Wagler, 1830 (Anura, Hylidae). *Anais Acad. Brasil. Cienc.* 38:335–344.

———. 1974. Observaçoes sobre desenvolvimento precoce em *Sphaenorhynchus bromelicola* Bok. 1966 (Anura, Hylidae). *Rev. Brasil. Biol.* 34:35–41.

Bolt, J. R. 1977. Dissorophoid relationships and ontogeny, and the origin of the Lissamphibia. *J. Paleontol.* 51:235–249.

———. 1979. *Amphibamus grandiceps* as a juvenile dissorophid: Evidence and implications. pp. 529–563. In *Mazon Creek fossils,* edited by M. H. Nitecki. New York: Academic Press.

Bond, A. N. 1960. An analysis of the response of salamander gills to changes in the oxygen concentration of the medium. *Develop. Biol.* 2:1–20.

Boonkoom, V., and R. H. Alvarado. 1971. Adenosinetriphosphatase activity in gills of larval *Rana catesbeiana. American J. Physiol.* 220:1820–1824.

Boord, R. L., and L. M. Eisworth. 1972. The central terminal fields of posterior lateral line and eighth nerves of *Xenopus. American Zool.* 12:727. (Abstr.)

Boord, R. L., L. B. Grochow, and L. S. Frishkopf. 1970. Organization of the posterior ramus and ganglion of the eighth cranial nerve of the bullfrog *Rana catesbeiana. American Zool.* 10:155. (Abstr.)

Booth, D. T. 1990. The significance of the boundary layer in cutaneous breathing, aquatic lungless salamanders. *Physiologist* 33:35 (Abstr.).

Boothby, K. M., and A. Roberts. 1995. Effects of site of tactile stimulation on the escape swimming responses of hatchling *Xenopus laevis* embryos. *J. Zool.* 235:113–125.

Bork, T., E. Schabtach, and P. Grant. 1987. Factors guiding optic fibers in developing *Xenopus* retina. *J. Comp. Neurol.* 264:147–158.

Borkhvardt, V. G., and Y. B. Malashichev. 1997. Position of the epicoracoids in arciferal pectoral girdles of the fire-bellies *Bombina* (Amphibia: Discoglossidae). *Russian J. Herpetol.* 4:28–30.

Borland, J. R. 1943. The production of experimental goiter in *Rana pipiens* tadpoles by cabbage feeding and methyl cyanide. *J. Exp. Zool.* 94:115–137, 3 plates.

Bossard, H. J. 1971. Experimentelle Analyse zur Entwicklung der chimären Vorniere bei *Triturus* und *Bombina. Wilhelm Roux' Arch. Entwicklungsmech. Org.* 168:282–303.

Boulenger, G. A. 1886. On the reptiles and batrachians of the Solomon Islands. *Trans Zool. Soc. London* 12:35–62.

———. 1892 (1891). A synopsis of the tadpoles of the European batrachians. *Proc. Zool. Soc. London* 1891:593–627, plates 45–47.

———. 1897–1898. The tailless batrachians of Europe. Parts I and II. London: Ray Society.

———. 1919. On *Aelurophyrne mammata* Gthr., an addition

to the batrachian fauna of Kashmir. *Rec. Indian Mus.* 16:469–470.

Bourges, M., and C. Bachelerie. 1974. Rôle des cellules épithéliales adjacentes aux colonnes dentaires dans la dynamique des odontoides cornés chez le têtard d'*Alytes obstetricans* Laurenti. *C. R. Soc. Biol. Paris* 168:1335–1339.

Bourne, G. R. 1997. Reproductive behavior of terrestrial breeding frogs *Eleutherodactylus johnstonei* in Guyana. *J. Herpetol.* 31:221–229.

———. 1998. Amphisexual parental behavior of a terrestrial breeding frog *Eleutherodactylus johnstonei* in Guyana. *Behav. Ecol.* 9:1–7.

Bournoure, L. 1937. Le sort de la lignée germinale chez la grenouille rousse après l'action des rayons ultra-violets sur le pôle inférieur de l'oeuf. *C. R. Hebd. Seances Acad. Sci. Paris* 104:1837–1839.

Bournoure, L., R. Aubry, and M.-L. Huck. 1954. Nouvelles recherches expérimentales sur les origines de la lignée reproductrice chez la grenouille rousse. *J. Embryol. Exp. Morphol.* 2:245–263.

Bourquin, O. 1985. A note on *Hemisus marmoratum,* compiled from the narrative of D. J. Bourquin. *South African J. Sci.* 81:210.

Bourret, R. 1942. Les batraciens de l'Indochine. *Mem. Inst. Oceanograph. Indochine* 6:1–547.

Bowen, S. H. 1980. Detrital nonprotein amino acids are the key to rapid growth in *Tilapia* in Lake Valencia, Venezuela. *Science* 207:1216–1218.

Bradfield, K. S. 1995. Do the ecological requirements of tadpoles always determine their distributions among habitats? Honours thesis, James Cook University, Townsville, Australia.

Bradford, D. F. 1983. Winterkill, oxygen relations, and energy metabolism of a submerged dormant amphibian, *Rana muscosa. Ecology* 64:1171–1183.

———. 1984a. Water and osmotic balance in overwintering tadpoles and frogs, *Rana muscosa. Physiol. Zool.* 57:474–480.

———. 1984b. Temperature modulation in a high-elevation amphibian, *Rana muscosa. Copeia* 1984:966–976.

———. 1989. Allotopic distribution of native frogs and introduced fishes in high Sierra Nevada lakes of California: Implications of the negative effect of fish introductions. *Copeia* 1989:775–778.

Bradford, D. F., and R. S. Seymour. 1985. Energy conservation during the delayed-hatching period in the frog *Pseudophryne bibroni. Physiol. Zool.* 58:491–496.

———. 1988a. Influence of water potential on growth and survival of the embryo, and gas conductance of the egg, in a terrestrial breeding frog, *Pseudophryne bibroni. Physiol. Zool.* 61:470–474.

———. 1988b. Influence of environmental PO_2 on embryonic oxygen consumption, rate of development, and hatching in the frog *Pseudophryne bibroni. Physiol. Zool.* 61:475–482.

Bradford, D. F., C. Swanson, and M. S. Gordon. 1994. Effects of low pH and aluminum on amphibians at high elevations in the Sierra Nevada, California. *Canadian J. Zool.* 72:1272–1279.

Bragg, A. N. 1949. Field preservation of amphibian eggs and larvae. *Turtox News* 27:262–263.

———. 1956. Dimorphism and cannibalism in tadpoles of *Scaphiopus bombifrons* (Amphibia, Salientia). *Southwest. Nat.* 1:105–108.

———. 1957. Aggregational feeding and metamorphic aggregations in tadpoles of *Scaphiopus hurteri* observed in 1954. *Wasmann J. Biol.* 15:61–68.

———. 1965. *Gnomes of the night: The spadefoot toads.* Philadelphia: University of Pennsylvania Press.

———. 1968. The formation of feeding schools in tadpoles of spadefoots. *Wasmann J. Biol.* 26:11–16.

Bragg, A. N., and W. N. Bragg. 1958. Variations in the mouth parts in tadpoles of *Scaphiopus* (*Spea*) *bombifrons* Cope (Amphibia: Salientia). *Southwest. Nat.* 3:55–69.

Bragg, A. N., and S. Hayes. 1963. A study of labial teeth rows in tadpoles of Couch's spadefoot. *Wasmann J. Biol.* 21:149–154.

Branch, L. C. 1983. Social behavior of the tadpoles of *Phyllomedusa vaillanti. Copeia* 1983:420–428.

Brattstrom, B. H. 1962. Thermal control of aggregation behavior in tadpoles. *Herpetologica* 18:38–46.

———. 1963. A preliminary review of the thermal requirements of amphibians. *Ecology* 44:238–255.

———. 1968. Thermal acclimation in anuran amphibians as a function of latitude and altitude. *Comp. Biochem. Physiol.* 24:93–111.

———. 1970. Amphibia. pp. 135–166. In *Comparative physiology of thermoregulation.* Vol. I. *Invertebrates and nonmammalian vertebrates,* edited by G. G. Whittow. New York: Academic Press.

———. 1979. Amphibian temperature regulation studies in the field and laboratory. *American Zool.* 19:345–356.

Brauer, A. 1898. Ein neuer Fall von Brutpflege bei Fröschen. *Zool. Jahrb. Abt. Syst., Geogr. Biol. Thiere* 12:89–94.

Braus, H. 1906. Vordere extremität und Operculum bie *Bombinator*-larven. Ein Beitrag zur Kenntnis morphogener correlation und regulation. *Sonderabd. Morphol. Jahrb.* 35:139–220.

Breden, F., and C. H. Kelly. 1982. The effect of conspecific interactions on metamorphosis in *Bufo americanus. Ecology* 63:1682–1689.

Breden, F., A. Lum, and R. Wassersug. 1982. Body size and orientation in aggregates of toad tadpoles *Bufo woodhousei. Copeia* 1982:672–680.

Bregman, B. S., and W. L. R. Cruce. 1980. Normal dendritic morphology of frog spinal motoneurons: A Golgi study. *J. Comp. Neurol.* 193:1035–1045.

Bresler, J. 1954. The development of labial teeth of salientian larvae in relation to temperature. *Copeia* 1954:207–211.

Bresler, J., and A. N. Bragg. 1954. Variations in the rows of labial teeth in tadpoles. *Copeia* 1954:255–257.

Bridges, C. M., and W. H. N. Gutzke. 1997. Effects of environmental history, sibship, and age on predator-avoidance responses of tadpoles. *Canadian J. Zool.* 75:87–93.

Briggs, R., and M. Davidson. 1942. Some effects of spinach feeding on *Rana pipiens* tadpoles. *J. Exp. Zool.* 90:401–411.

Brink, H. E. 1939. A histological and cytological investigation of the thyroids of *Arthroleptella bicolor villiersi* and *Bufo angusticeps* during normal and accelerated metamorphosis. *Proc. Linnean Soc. London* 151:120–125.

Britson, C. A., and R. E. Kissell, Jr. 1996. Effects of food type on developmental characteristics of an ephemeral pond-breeding anuran, *Pseudacris triseriata feriarum. Herpetologica* 52:374–382.

Brock, G. T. 1929. The formation and fate of the operculum and gill chambers in the tadpole of *Rana temporaria. Quart. J. Micros. Sci.* 73:335–343.

Brockelman, W. Y. 1969. An analysis of density effects and predation in *Bufo americanus* tadpoles. *Ecology* 50:632–644.

Brodie, E. D., Jr., and D. R. Formanowicz, Jr. 1983. Prey size preference of predators: Differential vulnerability of larval anurans. *Herpetologica* 39:67–75.

———. 1987. Antipredator mechanisms of larval anurans: Protection of palatable individuals. *Herpetologica* 43:369–373.

Brodie, E. D., Jr., and E. D. Brodie, III. 1978. The development of noxiousness of *Bufo americanus* tadpoles to aquatic insect predators. *Herpetologica* 34:302–306.

Brönmark, C., and P. Edenhamm. 1994. Does the presence of fish affect the distribution of treefrogs (*Hyla arborea*). *Conserv. Biol.* 8:841–845.

Brönmark, C., S. D. Rundle, and A. Erlandsson. 1991. Interactions between freshwater snails and tadpoles: Competition and facilitation. *Oecologia* 87:8–18.

Brook, A. J. 1980. The breeding seasons of frogs in Victoria and Tasmania. *Victorian Nat.* 97:6–11.

Browder, L. W., C. A. Erickson, and W. R. Jeffrey. 1991. *Developmental Biology.* 3d ed. New York: W. B. Saunders.

Brown, C. A., R. J. Wassersug, and T. Naitoh. 1992. Metamorphic changes in the vagal stimulation of the alimentary tract in the green frog, *Rana clamitans* (Ranidae). *Acta Anat.* 145:340–344.

Brown, D., A. Grosso, and R. C. De Sousa. 1981. The amphibian epidermis: Distribution of mitochondria-rich cells and the effect of oxytocin. *J. Cell Sci.* 52:197–213.

Brown, D. R., A. W. Everett, and M. R. Bennett. 1989. Compartmental and topographical distributions of axons in nerves to the amphibian (*Bufo marinus*) gluteus muscle. *J. Comp. Neurol.* 284:231–241.

Brown, H. A. 1967. Embryonic temperature adaptations and genetic compatibility in two allopatric populations of the spadefoot toad, *Scaphiopus hammondi. Evolution* 21:742–761.

———. 1969. The heat resistance of some anuran tadpoles (Hylidae and Pelobatidae). *Copeia* 1969:138–147.

———. 1990a. Temperature, thyroxine, and induced metamorphosis in tadpoles of the primitive frog, *Ascaphus. Gen. Comp. Endocrinol.* 79:136–146.

———. 1990b. Morphological variation and age-class determination in overwintering tadpoles of the tailed frog, *Ascaphus truei. J. Zool.* 220:171–184.

Brown, L. E., and R. R. Rosati. 1997. Effects of three different diets on survival and growth of larvae of the African clawed frog *Xenopus laevis. Progr. Fish-Cult.* 59:54–58.

Brown, R. M., and D. H. Taylor. 1995. Compensatory escape mode trade-offs between swimming performance and maneuvering behavior through larval ontogeny of the wood frog, *Rana sylvatica. Copeia* 1995:1–7.

Brown, S. C., E. A. Horgan, L. M. Savage, and P. S. Brown. 1986. Changes in body water and plasma constituents during bullfrog development: Effects of temperature and hormones. *J. Exp. Zool.* 237:25–33.

Brown, T. E., A. W. Morley, N. T. Sanderson, and R. D. Tait. 1983. Report of a large fishkill resulting from natural acid water conditions in Australia. *J. Fish Biol.* 22:335–350.

Brown, W. C., and A. C. Alcala. 1983. Modes of reproduction of Philippine anurans. pp. 416–428. In *Advances in herpetology and evolutionary biology,* edited by G. G. Whittow. Cambridge, MA: Mus. Comp. Zool.

Broyles, R. H. 1981. Changes in the blood during amphibian metamorphosis. pp. 461–490. In *Metamorphosis: A problem in developmental biology.* 2d ed., edited by G. G. Whittow. New York: Plenum.

Broyles, R. M., and M. J. Deutsch. 1975. Differentiation of red blood cells in vitro. *Science* 190:471–473.

Broyles, R. H., and E. Frieden. 1973. Sites of haemoglobin synthesis in amphibian tadpoles. *Nature New Biol.* 241:207–209.

Broyles, R. H., G. M. Johnson, P. B. Maples, and G. R. Kindell. 1981. Two erythropoietic microenvironments and two larval red cell lines in bullfrog tadpoles. *Develop. Biol.* 81:299–314.

Brust, D. G. 1993. Maternal brood care by *Dendrobates pumilio:* A frog that feeds its young. *J. Herpetol.* 27:96–98.

Buchan, A. M. J. 1985. Regulatory peptides in the amphibian pancreas. *Canadian J. Zool.* 63:2121–2124.

Buehr, M. L., and A. W. Blackler. 1970. Sterility and partial sterility in the South African clawed toad following the pricking of the eggs. *J. Embryol. Exp. Morphol.* 23:375–384.

Bullock, T. H. 1955. Compensation for temperature in the metabolism and activity of poikilotherms. *Biol. Rev.* 30:311–342.

Bullock, T. H., D. A. Bodznick, and R. G. Northcutt. 1983. The phylogenetic distribution of electroreception: Evidence for convergent evolution of a primitive sense modality. *Brain Res.* 6:25–46.

Burd, G. D. 1992. Development of the olfactory nerve in the clawed frog, *Xenopus laevis:* II. Effects of hypothyroidism. *J. Comp. Neurol.* 315:255–263.

Burggren, W. W. 1982. "Air-gulping" improves blood oxygen transport during aquatic hypoxia in the goldfish *Carassius auratus. Physiol. Zool.* 55:327–334.

———. 1984. Transition of respiratory processes during amphibian metamorphosis: From egg to adult. pp. 31–53. In *Respiration and metabolism of embryonic vertebrates,* edited by G. G. Whittow. Dordrecht: Dr. W. Junk.

———. 1985. Gas exchange, metabolism, and "ventilation" in gelatinous frog egg masses. *Physiol. Zool.* 58:503–514.

Burggren, W. W., and M. Doyle. 1986. Ontogeny of regulation of gill and lung ventilation in the bullfrog, *Rana catesbeiana. Respir. Physiol.* 66:279–291.

Burggren, W. W., and M. E. Feder. 1986. Effect of experimental ventilation of the skin on cutaneous gas exchange in the bullfrog. *J. Exp. Biol.* 121:445–449.

Burggren, W. W., M. E. Feder, and A. W. Pinder. 1983. Temperature and the balance between aerial and aquatic respiration in larvae of *Rana berlandieri* and *Rana catesbeiana.* Physiol. Zool. 56:263–273.

Burggren, W. W., and R. Fritsche. 1997. Amphibian cardiovascular development. pp. 166–182. In *Development of cardiovascular systems. Molecules to organisms,* edited by W. W. Burggren and B. B. Keller. Cambridge: Cambridge University Press.

Burggren, W. W., R. L. Infantino, Jr., and D. S. Townsend. 1990. Developmental changes in cardiac and metabolic physiology of the direct-developing tropical frog *Eleutherodactylus coqui. J. Exp. Biol.* 152:129–147.

Burggren, W. W., and J. J. Just. 1992. Developmental changes in physiological systems. pp. 467–530. In *Environmental physiology of the amphibians,* edited by M. E. Feder and W. W. Burggren. Chicago: University of Chicago Press.

Burggren, W., and R. Moalli. 1984. "Active" regulation of cutaneous gas exchange by capillary recruitment in amphibians: Experimental evidence and a revived model for skin respiration. *Respir. Physiol.* 55:379–392.

Burggren, W. W., and A. Mwalukoma. 1983. Respiration during chronic hypoxia and hyperoxia in larval and adult bullfrogs (*Rana catesbeiana*) I. Morphological responses of lungs, skin, and gills. *J. Exp. Biol.* 105:191–203.

Burggren, W. W., and N. H. West. 1982. Changing respiratory importance of gills, lungs, and skin during metamorphosis in the bullfrog *Rana catesbeiana*. *Resp. Physiol.* 47:151–164.

Burns, R. K. 1955. Urinogenital system. pp. 462–491. In *Analysis of development,* edited by B. H. Willier, P. A. Weiss, and V. Hamburger. Philadelphia: W. B. Saunders.

Burton, P. R. 1985. Ultrastructure of the olfactory neuron of the bullfrog: The dendrite and its microtubules. *J. Comp. Neurol.* 242:147–160.

Busk, M., E. H. Larsen, and F. B. Jensen. 1997. Acid-base regulation in tadpoles of *Rana catesbeiana* exposed to environmental hypercapnia. *J. Exp. Biol.* 200:2507–2512.

Byrd, C. A., and G. D. Burd. 1991. Development of the olfactory bulb in the clawed frog, *Xenopus laevis:* A morphological and quantitative analysis. *J. Comp. Neurol.* 314:79–90.

Bytinski-Salz, H., and H. Elias. 1938. Studî sui cromatofori dei Discolossidae. I. Melanofori paraepideremici di *Discoglossus pictus* (Amphibia, Anura). *Arch. Italiano Anat. Embriol.* 40:1–36.

Cadle, J. E., and R. Altig. 1991. Two lotic tadpoles from the Andes of southern Peru: *Hyla armata* and *Bufo veraguensis,* with notes on the call of *Hyla armata* (Amphibia: Anura: Hylidae and Bufonidae). *Neotrop. Fauna Environ.* 26:45–53.

Caldwell, J. P. 1974. A re-evaluation of the *Hyla bistincta* species group, with descriptions of three new species (Anura: Hylidae). *Occas. Pap. Mus. Nat. Hist., Univ. Kansas* (28):1–37.

———. 1982. Disruptive selection: A tail color polymorphism in *Acris* tadpoles in response to differential predation. *Canadian J. Zool.* 60:2818–2827.

———. 1986. A description of the tadpole of *Hyla smithii* with comments on tail coloration. *Copeia* 1986:1004–1006.

———. 1989. Structure and behavior of *Hyla geographica* tadpole schools with comments on classification of group behavior in tadpoles. *Copeia* 1989:938–948.

———. 1993. Brazil nut fruit capsules as phytotelmata: Interactions among anuran and insect larvae. *Canadian J. Zool.* 71:1193–1201.

———. 1994. Natural history and survival of eggs and early larval stages of *Agalychnis calcarifer* (Anura: Hylidae). *Herpetol. Nat. Hist.* 2:57–66.

Caldwell, J. P., and M. C. de Araújo. 1998. Cannibalistic interactions resulting from indiscriminate predatory behavior in tadpoles of poison frogs (Anura: Denbrodatidae). *Biotropica* 30:92–103.

Caldwell, J. P., and P. T. Lopez. 1989. Foam-generating behavior in tadpoles of *Leptodactylus mystaceus. Copeia* 1989:498–502.

Caldwell, J. P., J. H. Thorp, and T. O. Jervey. 1980. Predator-prey relationships among larval dragonflies, salamanders, and frogs. *Oecologia* 46:285–289.

Calef, G. W. 1973. Natural mortality of tadpoles in a population of *Rana aurora. Ecology* 54:741–758.

Callery, E. M., and R. P. Elinson. 1996. Developmental regulation of the urea-cycle enzyme arginase in the direct developing frog *Eleutherodactylus coqui. J. Exp. Zool.* 275:61–66.

Cambar, R. 1947a. Début du fonctionnement du pronéphros chez l'embryon de grenouille. *C. R. Seances Soc. Biol. Filosof.* 141:752–754.

———. 1947b. Destineé du canal de Wolff après ablation précoce du pronéphros chez les larves d'amphibiens anoures. *C. R. Hebd. Seances Acad. Sci. Paris* 225:644–646.

———. 1948. Recherches expérimentales sur les facteurs de la morphogénèse du mésonéphros chez les amphibiens anoures. *Bull. Biol. France-Belgique* 82:214–285.

———. 1952a. L'uretère primaire de la grenouille se développe, en direction postérieure, à partir d'un blastème voisin du pronéphros; démonstration expérimentale. *C. R. Seances Soc. Biol. Filosof.* 146:77–80.

———. 1952b. Essai expérimental de développement de l'uretère primaire des amphibiens anoures en milieu atypique. *C. R. Seances Soc. Biol. Filosof.* 146:453–455.

———. 1954. Influence de fonctionnement du pronéphros sur la morphogénèse de l'uretère primaire chez l'embryon de grenouille. *C. R. Hebd. Seances Acad. Sci. Paris* 238:2191–2192.

Cambar, R., and E. Arnauld. 1970. Les modalités d'acroissement du mésonéphros chez la larve de grenouille agile (*Rana dalmatina*). *C. R. Hebd. Seances Acad. Sci. Paris* 270D:1705–1707.

Cambar, R., and J.-D. Gipouloux. 1956a. Table cronologique du développement embryonaire et larvaire du crapaud commun, *Bufo bufo. Bull. Biol. France-Belgique* 90:198–217.

———. 1956b. Mesure expérimentale de la distance à laquelle l'uretère primaire peut induire la morphogénèse du mésonéphros (amphibiens anoures). *C. R. Hebd. Seances Acad. Sci. Paris* 242:2862–2865.

———. 1956c. Les blastèmes du mésonéphros possèdent une réactivité durable à l'action inductrice exercée par l'uretère primaire; démonstration expérimentale chez les amphibiens anoures. *C. R. Hebd. Seances Acad. Sci. Paris* 242:2992–2994.

———. 1970a. Les facteurs déterminant la migration des cellules de l'uretère primaire sont spécifiques; démonstration expérimentale chez les Amphibiens Anoures. *C. R. Hebd. Seances Acad. Sci. Paris* 270D:1359–1361.

———. 1970b. L'action inductrice exercée par l'uretère primaire sur la morphogénèse du mésonéphros est interspécifique; démonstration expérimentale chez les Amphibiens Anoures. *C. R. Hebd. Seances Acad. Sci. Paris* 270D:1607–1609.

———. 1973. Recherches expérimentales sur la localisation et la nature des facteurs déterminant la migration des cellules de l'uretère primaire chez la grenouille agile, *Rana dalmatina* Bon. (Amphibien Anoure). *Ann. Embryol. Morphol.* 6:309–316.

Cambar, R., and J. Garcin. 1962. Régenèration de l'uretère chez la grenouille agile (*Rana dalmatina* Bon.). *C. R. Hebd. Seances Acad. Sci. Paris* 254:3570–3572.

Cambar, R., and B. Marrot. 1954. Table chronologique du développement de la grenouille agile (*Rana dalmatina* Bon.). *Bull. Biol. France-Belgique* 88:168–177.

Cambar, R., and M. S. Martin. 1959. Table chronologique du développement embryonnaire et larvaire du crapaud accoucheur (*Alytes obstetricans* Laur.). *Actes Soc. Linneenne Bordeaux* 98:3–79.

Camerano, L. 1890. Richerche intorno allo sviluppo ed alle cause del polimorfismo dei girini degli Anfibi Anuri. *Atti Accad. Torino* 26:3–14.

———. 1892. Nuove richerche intorno allo sviluppo ed alle cause del polimorfismo dei girini Anfibi Anuri. II. Azione della luce. *Atti Accad. Torino* 28:2–17.

Campbell, F. R. 1970. Ultrastructure of the bone marrow of the frog. *American J. Anat.* 129:329–356.

Campbell, J. A., and B. T. Clarke. 1998. A review of frogs of the genus *Otophryne* (Microhylidae) with a description of a new species. *Herpetologica* 54:301–317.

Campbell, J. A. and T. M. Kubin. 1990. A key to the larvae of *Plectrohyla* (Hylidae), with a description of the tadpole presumed to be *Plectrohyla avia. Southwest. Nat.* 35:91–94.

Campbell, J. A., and E. N. Smith. 1992. A new frog of the genus *Ptychohyla* (Hylidae) from the Sierra de Santa Cruz, Guatemala, and description of a new genus of Middle American stream-breeding treefrogs. *Herpetologica* 48:153–167.

Campeny, R., and A. Casinos. 1989. Densities and buoyancy in tadpoles of midwife toad, *Alytes obstetricans. Zool. Anz.* 223:6–12.

Candelas, G. C., and M. Gomez. 1963. Nitrogen excretion in tadpoles of *Leptodactylus albilabris* and *Rana catesbeiana. American Zool.* 3:521–522. (Abstr.)

Cannatella, D.C. 1985. A phylogeny of primitive frogs (archaeobatrachians). Dissertation, University of Kansas.

Cannatella, D.C., and L. Trueb. 1988a. Evolution of pipoid frogs: Intergeneric relationships of the aquatic frog family Pipidae (Anura). *Zool. J. Linnean Soc.* 94:1–38.

———. 1988b. Evolution of pipoid frogs: Morphology and phylogenetic relationships of *Pseudhymenochirus. J. Herpetol.* 22:439–456.

Capranica, R. R. 1976. Morphology and physiology of the auditory system. pp. 551–575. In *Frog neurobiology, a handbook,* edited by R. Llinás and W. Precht. Berlin: Springer-Verlag.

Capranica, R. R., and A. J. M. Moffat. 1974. Frequency sensitivity of auditory fibers in the eighth nerve of the spadefoot toad, *Scaphiopus couchi. J. Acoust. Soc. America Suppl.* 55–85.

Caramaschi, U. 1979. O girino de *Odontophrynus carvalhoi* Savage & Cei, 1965 (Amphibia, Anura, Ceratophrydidae). *Rev. Brasil. Biol.* 39:169–171.

Caramaschi, U., J. Jim, and C. M. de Carvalho. 1980. Observações sobre *Aplastodiscus perviridis* A. Lutz (Amphibia, Anura, Hylidae). *Rev. Brasil. Biol.* 40:405–408.

Caramaschi, U., and G. Kisteumacher. 1989. O girino de *Crossodactylus trachystomus* (Reinhardt e Luetken, 1862)(Anura, Leptodactylidae). *Rev. Brasil. Biol.* 49:237–239.

Caramaschi, U., and I. Sazima. 1984. Uma nova espécies de *Thoropa* da Serra do Cipó, Minas Gerais, Brasil (Amphibia, Leptodactylidae). *Rev. Brasil. Zool.* 2:139–146.

———. 1985. Uma nova espécie de *Crossodactylus* da Serra do Cipó, Minas Gerais, Brasil (Amphibia, Leptodactylidae). *Rev. Brasil. Zool.* 3:43–49.

Cardoso, A. J., and C. F. B. Haddad. 1990. Redescripação e biologia de *Paratelmatobius gaigeae* (Anura, Leptodactylidae). *Pap. Avul. Zool.* 37:125–132.

Carlson, B. M. 1988. *Patten's foundations of embryology.* 5th ed. New York: McGraw-Hill.

Carmona-Osalde, C., M. A. Olvera-Novoa, A. Flores-Nava, and M. Rodriguez-Serna. 1996. Estimation of the protein requirement for bullfrog (*Rana catesbeiana*) tadpoles, and its effect on metamorphosis ratio. *Aquaculture* 141:223–231.

Carpenter, C. C. 1953. Aggregation behavior of tadpoles *Rana p. pretiosa. Herpetologica* 9:77–78.

Carpenter, K. L. 1978. Hemopoiesis in the pronephros of *Rana pipiens.* Thesis, Pennsylvania State University.

Carpenter, K. L., and J. B. Turpen. 1979. Experimental studies on hemopoiesis in the pronephros of *Rana pipiens. Differentiation* 14:167–174.

Carr, K. M., and R. Altig. 1991. Oral disc muscles of anuran tadpoles. *J. Morphol.* 208:271–277.

———. 1992. Configurations of the rectus abdominis of anuran tadpoles. *J. Morphol.* 214:351–356.

Carroll, E. J., Jr., and J. L. Hedrick. 1974. Hatching in the toad *Xenopus laevis:* Morphological events and evidence for a hatching enzyme. *Develop. Biol.* 38:1–13.

Carroll, E. J., Jr., A. M. Seneviratne, and R. Ruibal. 1991. Gastric pepsin in an anuran larvae. *Develop. Growth Differ.* 33:499–507.

Carver, V. H., and E. Frieden. 1977. Gut regression during spontaneous and triiodothyronine induced metamorphosis in *Rana catesbeiana* tadpoles. *Gen. Comp. Endocrinol.* 31:202–207.

Castanet, J., and M. H. Caetano. 1995. Influence of habitat condition on skeletal mass and bone structure of anuran amphibians. *Canadian J. Zool.* 73:234–242.

Castell, L. L. and R. Mann. 1994. Optimal staining of lipids in bivalve larvae with Nile Red. *Aquaculture* 119:89–100.

Casterlin, M. E., and W. W. Reynolds. 1977. Behavioral fever in anuran amphibian larvae. *Life Sci.* 20:593–596.

———. 1979. Behavioural thermoregulation in *Rana pipiens* tadpoles. *J. Therm. Biol.* 3:143–145.

Caston, J., and A. Bricout-Berthout. 1985. Influence of stimulation of the visual system on the activity of vestibular nuclear neurons in the frog. *Brain Behav. Evol.* 26:49–57.

Catton, W. T. 1976. Cutaneous mechanoreceptors. pp. 629–642. In *Frog neurobiology, a handbook,* edited by R. Llinás and W. Precht. Berlin: Springer-Verlag.

Cecil, S. C., and J. J. Just. 1979. Survival rate, population density, and development of a naturally occurring anuran larvae (*Rana catesbeiana*). *Copeia* 1979:447–453.

Cei, J. M. 1962. *Batracios de Chile.* Santiago: Edic. Univ. Chile.

———. 1964. The tadpole of *Batrachophrynus patagonicus* Gallardo. *Herpetologica* 20:242–245.

———. 1980. Amphibians of Argentina. *Monit. Zool. Italiano (N.S.) Monogr.* (2):1–609.

———. 1987. Additional notes to "Amphibians of Argentina": An update, 1980–1986. *Monit. Zool. Italiano (N.S.)* 21:209–272.

Cei, J. M., and L. Capurro. 1958. Biologia y desarrollo de *Eupsophus taeniatus* Girard. *Invest. Zool. Chilenos* (4):159–182.

Cei, J. M., and E. Crespo. 1982. Differences in larval morphology of allopatric isolated populations of the *Odontophrynus occidentalis* group from western Argentina. *Arq. Mus. Boçage* 1A:335–340.

Cei, J. M., and V. G. Roig. 1961. Batracios recolectados por la expedición biológica Erspamer en Corrientes y selva oriental de Misiones. *Notas Biol. Fac. Cien. Ex. Fis. Nat. Univ. Nac. Nordeste, Corrientes* 1:1–40.

Chacko, T. 1965. The development and metamorphosis of the hyobranchial skeleton in *Rana tigrina,* the Indian bull frog. *Acta Zool.* 46:311–328.

———. 1976. Development of the chondrocranium in *Rana tigrina* Daud. *J. Zool. Soc. India* 28:103–139.

Chang, L.-T., and C.-Y. Hsu. 1987. The relationship between the age and metamorphic progress and the development of tadpole ovaries. *Proc. Natl. Sci. Counc. Repub. China* 11B:211–217.

Channing, A. 1978. A new bufonid genus (Amphibia: Anura) from Rhodesia. *Herpetologica* 34:394–397.

———. 1984. Identification of ghost frog tadpoles. *Proc. Electr. Micros. Soc. South Africa* 14:89–90.

———. 1986. A new species of the genus *Strongylopus* Tschudi from Namaqualand, Cape Province, South Africa (Anura: Ranidae). *Ann. Cape Prov. Mus.* 16:127–135.

———. 1993. Observations on the natural history of *Stephopaedes anotis* (Bufonidae). *J. Herpetol.* 27:213–214.

Channing, A., and R. C. Boycott. 1989. A new frog genus and species from the mountains of the southwestern Cape, South Africa (Anura: Ranidae). *Copeia* 1989:467–471.

Channing, A., and M.-D. C. De Capona. 1987. The tadpole of *Hyperolius mitchelli* (Anura: Hyperoliidae). *South African J. Zool.* 22:235–237.

Channing, A., R. C. Boycott, and H. J. Hensbergen. 1988. Morphological variation of *Heleophryne* tadpoles from the Cape Province, South Africa (Anura: Heleophrynidae). *J. Zool.* 215:205–216.

Cherdantseva, E. M., and V. G. Cherdantsev. 1995. The organization of variability in serially homologous structures and its correlation with the growth rates (on examples of oral fields in the tadpoles of brown frogs). *Zool. Zhur.* 74:92–107.

Chesler, M., and C. Nicholson. 1985. Organization of the filum terminale in the frog. *J. Comp. Neurol.* 239:431–444.

Chester-Jones, I. 1987. Structure of the adrenal and interrenal glands. pp. 95–124. In *Fundamentals of comparative vertebrate endocrinology,* edited by I. Chester-Jones, P. M. Ingleton, and J. G. Phillips. New York: Plenum.

Chibon, P. 1960. Développement au laboratoire d'*Eleutherodactylus martinicensis* Tschudi, batracien anoure à développement direct. *Bull. Zool. France* 85:412–418.

———. 1962. Différenciation sexuelle de *Eleutherodactylus martinicensis* Tschudi batracien anoure a développement direct. *Bull Zool. France* 87:509–515.

Chieffi, G., M. Milone, and L. Iela. 1980. Amphibian reproduction: Reproductive physiology in the male *Rana esculenta* L. *Boll. Zool.* 47:63–70.

Chieffi, G., R. K. Rastogi, L. Iela, and M. Milone. 1975. The function of fat bodies in relation to the hypothalamo-hypophyseal-gonadal axis in the frog, *Rana esculenta. Cell Tiss. Res.* 161:157–165.

Chopra, D. P., and J. D. Simnett. 1969a. Changes in mitotic rate during compensatory renal growth in *Xenopus laevis* tadpoles after unilateral pronephrectomy. *J. Embryol. Exp. Morphol.* 21:539–548.

———. 1969b. Demonstration of an organ-specific mitotic inhibitor in amphibian kidney. The effects of adult *Xenopus* tissue extracts on the mitotic rate of embryonic tissue (in vitro). *Exp. Cell Res.* 58:319–322.

———. 1970a. Stimulation of mitosis in amphibian kidney by organ specific antiserum. *Nature* 225:657–658.

———. 1970b. Stimulation of cell division in pronephros of embryonic grafts following partial nephrectomy in the host (*Xenopus laevis*). *J. Embryol. Exp. Morphol.* 24:525–533.

———. 1971a. Stimulation of cell division in larval kidney (*Xenopus laevis*) by rat kidney antiserum. *Exp. Cell Res.* 64:396–402.

———. 1971b. Tissue-specific mitotic inhibition in the kidneys of embryonic grafts and partially nephrectomized host *Xenopus laevis. J. Embryol. Exp. Morphol.* 25:321–329.

Chou, W.-H., and J.-Y. Lin. 1997. Tadpoles of Taiwan. *Spec. Publ. Nat. Mus. Taiwan* (7):1–98.

Chovanec, A. 1992. The influence of tadpole swimming behaviour on predation by dragonfly nymphs. *Amphibia-Reptilia* 13:341–349.

Christman, S. P. 1974. Geographic variation for salt water tolerance in the frog *Rana sphenocephala. Copeia* 1974:773–778.

Ciantar, D. 1983. Models of morphogenesis. A study of the development of the nephric system in *Xenopus.* Dissertation, University of Southampton.

Cioni, C., F. de Palma, and A. Stefanelli. 1989. Morphology of afferent synapses in the Mauthner cell of larval *Xenopus laevis. J. Comp. Neurol.* 284:205–214.

Claas, B., B. Fritzsch, and H. Münz. 1981. Common efferents to lateral line and labyrinthine hair cells in aquatic vertebrates. *Neurosci. Lett.* 27:231–235.

Clarke, B. T. 1983. A morphological re-examination of the frog genus *Nannophrys* (Anura: Ranidae) with comments on its biology, distribution and relationships. *Zool. J. Linnean Soc.* 79:377–398.

Clarke, F. M., A. I. Ward, and J. R. Downie. 1995. Factors affecting the distribution and status of the golden tree frog, *Phyllodytes auratus,* in Trinidad. *British Herpetol. Soc. Bull.* (54):3–9.

Claussen, D. L. 1973. The thermal relations of the tailed-frog, *Ascaphus truei,* and the Pacific treefrog, *Hyla regilla. Comp. Biochem. Physiol.* 44A:137–153.

Clemens, D. T., and M. E. Feder. 1990. Effects of perfusion and medium-blood PO_2 difference on oxygen uptake across isolated perfused frog skin. *Physiologist* 33:70 (Abstr.).

Cline, M. J., and D. W. Golden. 1979. Cellular interactions in hematopoiesis. *Nature* 277:177–181.

Cocroft, R. B., and W. R. Heyer. 1988. Notes on the frog genus *Thoropa* (Amphibia: Leptodactylidae) with a description of a new species (*Thoropa saxatilis*). *Proc. Biol. Soc. Washington* 101:209–220.

Coggins, L. W. 1973. An ultrastructural and radioautographic study of early oogenesis in the toad *Xenopus laevis. J. Cell Sci.* 12:71–93.

Cohen, P. P. 1966. Biochemical aspects of metamorphosis: Transition from ammonotelism to ureotelism. *Harvey Lect.* 60:119–154.

———. 1970. Biochemical differentiation during amphibian metamorphosis. *Science* 168:533–543.

Cohen, P. P., R. F. Bruckner, and S. M. Morris. 1978. Cellular and molecular aspects of thyroid hormone action during amphibian metamorphosis. pp. 273–381. In *Hormonal proteins and peptides: Thyroid hormones.* Vol. 6. New York: Academic Press.

Collins, J. P. 1979. Intrapopulation variation in the body size at metamorphosis and timing of metamorphosis in the bullfrog, *Rana catesbeiana. Ecology* 60:738–749.

Collins, J. P., and M. A. Lewis. 1979. Overwintering tadpoles and breeding season variation in the *Rana pipiens* complex in Arizona. *Southwest. Nat.* 24:371–373.

Constantine-Paton, M. 1987. A neural pattern unfolding: Properties of retinotectal differentiation in frog tadpoles. *Neurol. Neurobiol.* 44:231–253.

Constantine-Paton, M., and R. R. Capranica. 1976. Axonal guidance of developing optic nerves in the frog. I. Anatomy of the projection from transplanted eye primordia. *J. Comp. Neurol.* 170:17–31.

Constantine-Paton, M., E. C. Pitts, and T. A. Reh. 1983. The relationship between retinal axon ingrowth, terminal morphology, and terminal patterning in the optic tectum of the frog. *J. Comp. Neurol.* 218:297–313.

Cooke, A. S. 1974. Differential predation by newts on anuran tadpoles. *British J. Herpetol.* 5:386–390.

Cooper, E. L. 1966. The lymphomyeloid organs of *Rana catesbeiana. Anat. Rec.* 154:456. (Abstr.)

———. 1967. Lymphomyeloid organs of Amphibia. I. Appearance during larval and adult stages. *J. Morphol.* 122:381–398.

———. 1976. Immunity mechanisms. pp. 163–272. In *Physiology of the Amphibia.* Vol. 3, edited by B. Lofts. New York: Academic Press.

Cooper, E. L., and R. K. Wright. 1976. The anuran amphibian spleen. An evolutionary model for terrestrial vertebrates. pp. 47–60. In *Immuno-aspects of the spleen,* edited by J. R. Battisto and J. W. Streilein. Amsterdam: Elsevier/North-Holland Biomedical Press.

Cooper, K. W. 1936. Demonstration of a hatching secretion in *Rana pipiens* Schreber. *Proc. Natl. Acad. Sci. United States of America* 22:433–434.

Cooper, R. S. 1943. An experimental study of the development of the larval olfactory organ of *Rana pipiens* Schreber. *J. Exp. Zool.* 93:415–452.

Cøpenhaver, W. M. 1955. Heart, blood vessels, blood, and entodermal derivatives. pp. 440–461. In *Analysis of development,* edited by B. H. Willier, P. A. Weiss, and V. Hamburger. Philadelphia: W. B. Saunders.

Corkran, C. C., and C. Thoms. 1996. *Amphibians of Oregon, Washington and British Columbia.* Edmonton: Lone Pine Press.

Corless, J. M., and R. D. Fetter. 1987. Structural features of the terminal loop region of frog retinal rod outer segment disk membranes: III. Implications of the terminal loop complex for disk morphogenesis, membrane fusion, and cell surface interactions. *J. Comp. Neurol.* 257:24–38.

Corless, J. M., R. D. Fetter, and M. J. Costello. 1987a. Structural features of the terminal loop region of frog retinal rod outer segment disk membranes: I. Organization of lipid components. *J. Comp. Neurol.* 257:1–8.

Corless, J. M., R. D. Fetter, O. B. Zampighi, M. J. Costello, and D. L. Wall-Buford. 1987b. Structural features of the terminal loop region of frog retinal rod outer segment disk membranes: II. Organization of the terminal loop complex. *J. Comp. Neurol.* 257:9–23.

Cornell, T. J., K. A. Berven, and G. J. Gamboa. 1989. Kin recognition by tadpoles and froglets of the wood frog *Rana sylvatica. Oecologia* 78:312–316.

Cortwright, S. A. 1988. Intraguild predation and competition: An analysis of net growth shifts in larval anuran prey. *Canadian J. Zool.* 66:1813–1821.

Cortwright, S. A., and C. E. Nelson. 1990. An examination of multiple factors affecting community structure in an aquatic amphibian community. *Oecologia* 83:123–131.

Corvaja, N., and P. d'Ascanio. 1981. Spinal projections from the mesencephalon in the toad. *Brain Behav. Evol.* 19:205–213.

Costa, H. H. 1967. Avoidance of anoxic water by tadpoles of *Rana temporaria. Hydrobiologia* 30:374–384.

Costa, H. H., and S. Balasubramanian. 1965. The food of the tadpoles of *Rhacophorus cruciger cruciger* (Blyth). *Ceylon J. Sci.* 5:105–109, 1 plate.

Cowden, R. R., B. M. Gebhardt, and E. P. Volpe. 1968. The histophysiology of antibody-forming sites in the marine toad. *Zeit. Zellforsch. Mikros. Anat.* 85:196–205.

Cowles, R. B., and C. M. Bogert. 1944. A preliminary study of the thermal requirements of desert reptiles. *Bull. American Mus. Nat. Hist.* 83:261–296.

Cox, T. C., and R. H. Alvarado. 1979. Electrical and transport characteristics of skin of larval *Rana catesbeiana. American J. Physiol.* 237:1274–1279.

———. 1983. Nystatin studies of skin of larval *Rana catesbeiana. American J. Physiol.* 244R:58–65.

Crawshaw, L. I., R. N. Rausch, L. P. Wollmuth, and E. J. Bauer. 1992. Seasonal rhythms of development and temperature selection in larval bullfrogs, *Rana catesbeiana* Shaw. *Physiol. Zool.* 65:346–359.

Cronin, J. T., and J. Travis. 1986. Size-limited predation on larval *Rana areolata* (Anura: Ranidae) by two species of backswimmer (Insecta: Hemiptera: Notonectidae). *Herpetologica* 42:171–174.

Crossland, M. R. 1997. Impact of eggs, hatchlings, and tadpoles of the introduced cane toad *Bufo marinus* (Anura: Bufonidae) on native aquatic fauna in northern Queensland, Australia. Dissertation, James Cook University, Townsville, Australia.

Crossland, M. R.,, and R. A. Alford. 1998. Evaluation of the toxicity of eggs, hatchlings, and tadpoles of the introduced toad *Bufo marinus* (Anura: Bufonidae) to native Australian aquatic predators. *Australian J. Ecol.* 23:129–137.

Crowder, W. C., M. Nie, and G. R. Ultsch. 1998. Oxygen uptake in bullfrog tadpoles (*Rana catesbeiana*). *J. Exp. Zool.* 280:121–134.

Crump, M. L. 1974. Reproductive strategies in a tropical anuran community. *Misc. Publ. Mus. Nat. Hist., Univ. Kansas* (61):1–68.

———. 1981. Energy accumulation and amphibian metamorphosis. *Oecologia* 49:167–169.

———. 1983. Opportunistic cannibalism by amphibian larvae in temporary aquatic environments. *American Nat.* 121: 281–287.

———. 1984. Ontogenetic changes in vulnerability to predation in tadpoles of *Hyla pseudopuma. Herpetologica* 40:265–271.

———. 1986. Cannibalism by younger tadpoles: Another hazard of metamorphosis. *Copeia* 1986:1007–1009.

———. 1989a. Effect of habitat drying on developmental time and size at metamorphosis in *Hyla pseudopuma. Copeia* 1989:794–797.

———. 1989b. Life-history consequences of feeding versus nonfeeding in a facultatively nonfeeding toad larva. *Oecologia* 78:486–489.

———. 1990. Possible enhancement of growth in tadpoles through cannibalism. *Copeia* 1990:560–564.

———. 1991. Choice of oviposition site and egg load assessment by a treefrog. *Herpetologica* 47:308–315.

———. 1992. Cannabilism in amphibians. pp. 256–276. In *Cannibalism. Ecology and evolution among diverse taxa,* edited by M. A. Elgar and B. J. Crespi. Oxford: Oxford University Press.

Crump, M. L., and M. Vaira. 1991. Vulnerability of *Pleurodema borelli* tadpoles to an avian predator: Effect of body size and density. *Herpetologica* 47:316–321.

Cullen, M. J., and H. D. Webster. 1979. Remodelling of optic nerve myelin sheaths an axons during metamorphosis in *Xenopus laevis. J. Comp. Neurol.* 184:353–362.

Culley, D. D., Jr. 1991. Bullfrog culture. pp. 185–205. In *Production in aquatic animals,* edited by C. E. Nash. Amsterdam: Elsevier Sci. Publ.

Culley, D. D., Jr., S. P. Meyers, and A. J. Doucette, Jr. 1977. A high density rearing system for larval anurans. *Lab Anim.* 6:34–41.

Culley, D. D., Jr., and P. K. Sotairidis. 1983. *Progress and problems associated with bullfrog tadpole diets and nutrition.* 3d Annual Dr. Scholl Conference on the Nutrition of Captive Wild Animals, Chicago.

Cummins, C. P. 1989. Interaction between the effects of pH and density on growth and development in *Rana temporaria* tadpoles. *Funct. Ecol.* 3:45–52.

Cupp, P. V., Jr. 1980. Thermal tolerance of five salientian amphibians during development and metamorphosis. *Herpetologica* 36:234–244.

Curtis, S. K., R. R. Cowden, and J. W. Nagel. 1979. Ultrastructure of the bone marrow of the salamander *Plethodon glutinosus* (Caudata: Plethodontidae). *J. Morphol.* 159:151–160.

Cusimano-Carollo, T. 1963. Investigation on the ability of the neural folds to induce a mouth in the *Discoglossus pictus* embryo. *Acta Embryol. Morphol. Exp.* 6:158–168.

———. 1969. Phenomena of induction by the transverse neural fold during the formation of the mouth in *Discoglossus pictus. Acta Embryol. Exp.* 1969:97–110.

———. 1972. The mechanism of the formation of the larval mouth in *Discoglossus pictus. Acta Embryol. Exp.* 1972:289–332.

Cusimano-Carollo, T., A. Fagone, and G. Reverbei. 1962. On the origin of the larval mouth in the anurans. *Acta Embryol. Morphol. Exp.* 5:82–103.

Czechura, G. V., G. J. Ingram, and D. S. Liem. 1987. The genus *Nyctimystes* (Anura: Hylidae) in Australia. *Rec. Australian Mus.* 39:333–338.

Czołowska, R. 1972. The fine structure of the "germinal cytoplasm" in the egg of *Xenopus laevis. Wilhelm Roux' Arch. Entwicklungsmech. Org.* 169:335–344.

Da Cruz, C. A. G. 1980. Descrição de uma nova espécie de Phyllomedusinae do Estado do Espírito Santo, Brasil (Amphibia, Anura, Hylidae). *Rev. Brasil. Biol.* 40:683–687.

———. 1982. Descrição de grupos espécies de Phyllomedusinae Brasileiras com base em caracteres larvários (Amphibia, Anura, Hylidae). *Arq. Univ. Fed. Rural Rio de Janeiro* 5:147–171.

Da Cruz, C. A. G., and O. L. Peixoto. 1978. Notas sobre o girino de *Dasypops schirchi* Miranda-Ribeiro (Amphibia, Anura, Microhylidae). *Rev. Brasil. Biol.* 38:297–299.

———. 1980a. Notas sobre o girino de *Sphaenorhynchus orophilus* (Lutz & Lutz, 1938)(Amphibia, Anura, Hylidae). *Rev. Brasil. Biol.* 40:383–386.

———. 1980b. Observações sobre a larva de *Proceratophrys appendiculata* (Günther 1873)(Amphibia, Anura, Leptodactylidae). *Rev. Brasil. Biol.* 40:491–493.

Dainton, B. H. 1988. Cold tolerance and thyroid activity in developing tadpoles of *Xenopus laevis. J. Herpetol.* 22:301–306.

———. 1991. Heat tolerance and thyroid activity in developing tadpoles and juvenile adults of *Xenopus laevis* (Daudin). *J. Therm. Biol.* 16:273–276.

Daniel, J. F., and A. B. Burch. 1933. A rotary disc for the observation of objects in profile. *Univ. California Publ. Zool.* 39:201–204.

Danos, S. O., J. S. Maki, and C. C. Remsen. 1983. Stratification of microorganisms and nutrients in the surface microlayer of small freshwater ponds. *Hydrobiologia* 98:193–202.

Darnell, R. M., Jr. 1949. The aortic arches and associated arteries of caudate Amphibia. *Copeia* 1949:18–31.

Das, I. 1994. The internal oral morphology of some anuran larvae from south India: A scanning electron microscopic study. *Amphibia-Reptilia* 15:249–256.

Dash, M. C., and A. K. Hota. 1980. Density effects on the survival, growth rate, and metamorphosis of *Rana tigrina* tadpoles. *Ecology* 61:1025–1028.

Dash, M. C., and B. K. Mahapatro. 1990. Effect of food quality and quantity on growth and metamorphosis of *Bufo stomaticus* larvae. *J. Ecobiol.* 2:67–76.

Dash, M. C., and P. K. Mishra. 1989. Energy cost of metamorphosis in a tropical tree frog *Polypedates maculatus* (Gray). *J. Ecobiol.* 1:131–136.

Dauça, M., R. Calvert, D. Ménard, J. S. Hugon, and J. Hourdry. 1982. Development of peroxisomes in amphibians. II. Cytochemical and biochemical studies on the liver, kidney and pancreas. *J. Exp. Zool.* 223:57–65.

———. 1983. Development of peroxisomes in amphibians. III. Study on liver, kidney, and intestine during thyroxine-induced metamorphosis. *J. Exp. Zool.* 227:413–422.

Dauça M., and J. Hourdry. 1983. Modifications du régime alimentaire chez les amphibiens anoures en métamorphose. *Soc. Zool. Paris* 108:409–415.

Dauça M., J. Hourdry, J. S. Hugon, and D. Ménard. 1980. Amphibian intestinal brush border enzymes during thyroxine-induced metamorphosis. A biochemical and cytochemical study. *Histochemistry* 70:33–42.

Davenport, C. B., and W. E. Castle. 1895. Studies on morphogenesis. III. On the acclimation of organisms to high temperature. *Arch. Entwicklungsmech. Mech. Org.* 2:227–249.

Davidson, E. H. 1986. *Gene activity in early development.* 3d ed. New York: Academic Press.

Davies, M. 1989. Ontogeny of bone and the role of heterochrony in the myobatrachine genera *Uperoleia, Crinia,* and *Pseudophryne* (Anura: Leptodactylidae: Myobatrachinae). *J. Morphol.* 200:269–300.

———. 1991. Descriptions of the tadpoles of some Australian limnodynastine leptodactylid frogs. *Trans. Roy. Soc. South Australia* 115:67–76.

Davies, M., and S. J. Richards. 1990. Developmental biology of the Australian hylid frog *Nyctimystes dayi* (Günther). *Trans. Roy. Soc. South Australia* 114:207–211.

Davis, M. R., and M. Constantine-Paton. 1983. Hyperplasia in the spinal sensory system of the frog. I. Plasticity in the most caudal dorsal root ganglion. *J. Comp. Neurol.* 221:444–452.

Davis, W. M., and R. F. Nunnemacher. 1974. Nerve fibers in the sympathetic trunk of *Rana pipiens:* A quantitative electron microscope study. *Anat. Rec.* 179:331–342.

Dawson, J. T. 1982. Kin recognition and schooling in the American toad (*Bufo americanus*). Dissertation, State University of New York, Albany.

De Albuja, C. M., M. Campos, and E. M. Del Pino. 1983. Role of progesterone in oocyte maturation in the egg-brooding hylid frog *Gastrotheca riobambae* (Fowler). *J. Exp. Zool.* 227:271–276.

De Bavay, J. M. 1993. The developmental stages of the sphagnum frog, *Kyarranus sphagnicolous* Moore (Anura, Myobatrachidae). *Australian J. Zool.* 41:151–201.

De Beer, G. R. 1926. Studies on the vertebrate head. II. The orbito-temporal region of the skull. *Quart. J. Micros. Sci.* 70:263–370.

———. 1951. *Embryos and ancestors.* Oxford: Clarendon Press.

———. 1985. *The development of the vertebrate skull.* Reprint, Chicago: University of Chicago Press.

DeBenedictis, P. A. 1974. Interspecific competition between tadpoles of *Rana pipiens* and *Rana sylvatica:* An experimental field study. *Ecol. Monogr.* 44:129–151.

Debski, E. A., and M. Constantine-Paton. 1990. Evoked pre- and postsynaptic activity in the optic tectum of the cannulated tadpole. *J. Comp. Physiol.* 167A:377–390.

Decker, R. S. 1976. Influence of thyroid hormones on neuronal death and differentiation in larval *Rana pipiens. Develop. Biol.* 49:101–118.

Degani, G. 1982. Amphibian tadpole interaction in a winter pond. *Hydrobiologia* 96:3–7.

———. 1986. Growth and behaviour of six species of amphibian larvae in a winter pond in Israel. *Hydrobiologia* 140:5–10.

Degani, G., and E. Nevo. 1986. Osmotic stress and osmoregulation of tadpoles and juveniles of *Pelobates syriacus*. *Comp. Biochem. Physiol.* 83A:365–370.
DeHaan, R. L. 1965. Morphogenesis of the vertebrate heart. pp. 377–400. In *Organogenesis*, edited by R. L. DeHaan and H. Ursprung. New York: Holt, Rinehart, and Winston.
De Jongh, H. J. 1967. Relative growth of the eye in larval and metamorphosing *Rana temporaria*. *Growth* 31:93–103.
———. 1968. Functional morphology of the jaw apparatus of larval and metamorphosing *Rana temporaria* L. *Netherlands J. Zool.* 18:1–103.
De la Riva, I. 1995. A new reproductive mode for the genus *Adenomera* (Amphibia: Anura: Leptodactylidae): Taxonomic implications for certain Bolivian and Paraguayan populations. *Stud. Neotrop. Fauna Environ.* 30:15–29.
Delbos, M. 1975. Observations sur l'ultrastructure de l'uretère primaire et du blastème mésonéphrétique avant sa differenciation chez *Rana dalmatina* (Amphibien Anoure). *C. R. Hebd. Seances Acad. Sci. Paris* 280D:893–895.
Delbos, M., J.-D. Gipouloux, and R. Cambar. 1971. Observations sur l'infrastructure des différents types cellulaires constituant la gonade larvaire chez la grenouille agile *Rana dalmatina* Bon. (Amphibien Anoure). *C. R. Hebd. Seances Acad. Sci. Paris* 272D:2372–2374.
Delbos, M., J.-D. Gipouloux, and S. Guennoun. 1981. Caractères ultrastructuraux et cytochimiques des cellules germinales primordiales des amphibiens anoures. *Arch. Anat. Micros. Morphol. Exp.* 70:117–128.
———. 1982. Intraendodermal and intramesenteric migration of anuran amphibian germ cells: Transmission and scanning electron microscopy. *J. Morphol.* 171:355–360.
Delbos, M., J. Lestage, and J.-D. Gipouloux. 1980. Localisation cytochimique d'une importante activité adénylate cyclasique dans le cordo-mésoderme embryonnaire des amphibiens anoures. *Arch. Anat. Micros. Morphol. Exp.* 69:47–56.
Delbos, M., K. R. Miller, and J.-D. Gipouloux. 1984. Freeze-fracture of *Rana pipiens* gonad anlage: Study of primordial germ cells and other cellular types. *Arch. Anat. Micros. Morphol. Exp.* 73:57–67.
Delfino, G. 1977. Il differenziamento delle ghiandole granulose cutanee in larve di *Bombina variegata pachypus* (Bonaparte)(Anfibio, Anuro, Discoglosside). Ricerca al microscopio ottico e al microscopio elettrinico. *Arch. Italiano Anat. Embriol.* 82:337–363.
Delfino, G., R. Brizzi, and C. Calloni. 1982. Development of cutaneous glands in *Salamandrina terdigitata* (Lacépède, 1788)(Amphibia: Urodela); findings by light and electron microscopy. *Zeit. Mikros. Anat. Forsch.* 96:948–971.
Delidow, B. C. 1989. Reevaluation of the effects of growth hormone and prolactin on anuran tadpole growth and development. *J. Exp. Zool.* 249:279–283.
Del Pino, E. M. 1975. Adaptaciones reproductivas para la vida terestre del sapo marsupial *Gastrotheca riobambae* (Fowler), Anura, Hylidae. *Rev. Univ. Catolica Ecuador* 3:119–140.
———. 1980a. Morphology of the pouch and incubatory integument in marsupial frogs (Hylidae). *Copeia* 1980:10–17.
———. 1980b. El mantenimiento y aspectos del comportaniento encautiverio del sapo marsupial *Gastrotheca riobambae* (Hylidae). *Rev. Univ. Catolica Ecuador* 8:41–49.
———. 1989a. Marsupial frogs. *Sci. American* May:110–118.
———. 1989b. Modifications of oogenesis and development in marsupial frogs. *Development* 107:169–187.
———. 1996. The expression of Brachyury (T) during gastrulation in the marsupial frog *Gastrotheca ribobambae*. *Develop. Biol.* 177:64–68.
Del Pino, E. M., I. Alcocer, and H. Grunz. 1994. Urea is necessary for the culture of embryos of the marsupial frog *Gastrotheca riobambae*, and is tolerated by embryos of the aquatic frog *Xenopus laevis*. *Develop. Growth Differ.* 36:73–80.
Del Pino, E. M., and R. P. Elinson. 1983. A novel development pattern for frogs: Gastrulation produces an embryonic disk. *Nature* 306:589–591.
Del Pino, E. M., and B. Escobar. 1981. Embryonic stages of *Gastrotheca riobambae* (Fowler) during maternal incubation and comparison of development with that of other egg-brooding hylid frogs. *J. Morphol.* 167:277–295.
Del Pino, E. M., M. L. Galarza, C. M. de Albuja, and A. A. Humphries, Jr. 1975. The maternal pouch and development in the marsupial frog *Gastrotheca riobambae* (Fowler). *Biol. Bull.* 149:480–491.
Del Pino, E. M., and A. A. Humphries, Jr. 1978. Multiple nuclei during early oogenesis in *Flectonotus pygmaeus* and other marsupial frogs. *Biol. Bull.* 154:198–212.
Del Pino, E. M., and S. Loor-Vela. 1990. The pattern of early cleavage of the marsupial frog *Gastrotheca riobambae*. *Development* 110:781–787.
Del Pino, E. M., C. Murphy, and P. H. Masson. 1992. 5S rRNA-encoding genes of the marsupial frog *Gastrotheca riobambae*. *Gene* 111:235–242.
Del Pino, E. M., H. Steinbeisser, A. Hofmann, C. Dreyer, M. Campos, and M. F. Trendelenburg. 1986. Oogenesis in the egg-brooding frog *Gastrotheca riobambae* produces large oocytes with fewer nucleoli and low RNA content in comparison to *Xenopus laevis*. *Differentiation* 32:24–33.
Delsol, M., and J. Flatin. 1972. *Anatomie du systéme vasculaire des têtards de batraciens*. Paris: Lib. Facult. Sci.
Dempster, W. T. 1930. The morphology of the amphibian endolymphatic organ. *J. Morphol. Physiol.* 50:71–126.
Denis, H. 1959. Sur l'induction des capsules nasales chez les Amphibiens. *Arch. Biol.* 70:851–873.
Denton, J., and T. J. C. Beebee. 1991. Palatability of anuran eggs and embryos. *Amphibia-Reptilia* 12:111–114.
Denver, R. J. 1997. Environmental stress as a developmental cue: Corticotropin-releasing hormone is a proximate mediator of adaptive phenotypic plasticity in amphibian metamorphosis. *Horm. Behav.* 31:169–179.
De Pérez, G. R., P. M. Ruiz-Carranza, and M. P. Ramirez-Pinilla. 1992a. Modificaciones tegumentarias de larvas y adultos durante el cuidado parental en *Minyobates virolinensis* (Amphibia: Anura: Dendrobatidae). *Caldasia* 17:75–86.
———. 1992b. Especializaciones del tegumento de incubacion de la hembra de *Cryptobatrachus boulengeri* (Amphibia: Anura: Hylidae). *Caldasia* 17:87–94.
De Quiroga, B. G., M. Lopez-Torres, and P. Gil. 1989. Hyperoxia decreases lung size of amphibian tadpoles without changing GSH-peroxidase or tissue peroxidation. *Comp. Biochem. Physiol.* 92A:581–588.
De Sá, R. O. 1983. Descripcion de la larva de *Argenteohyla siemersi* (Mertens 1937), (Anura: Hylidae). *Res. J. Cienc. Nat. Montevideo* 3:40–41.
———. 1988. Chondrocranium and ossification sequence of *Hyla lanciformis*. *J. Morphol.* 195:345–355.
De Sá, R. O., and S. Hill. 1998. Chondrocranial anatomy and skeletogenesis in *Dendrobates auratus*. *J. Herpetol.* 32:205–210.
De Sá, R. O., and E. O. Lavilla. 1997. The tadpoles of *Pseudis*

minuta (Anura: Pseudidae), an apparent case of heterochrony. *Amphibia-Reptilia* 18:229–240.

De Sá, R. O., and L. Trueb. 1991. Osteology, skeletal development, and chondrocranial structure of *Hamptophryne boliviana*. *J. Morphol.* 209:311–330.

De Saint-Aubain, M. L. 1981. Shunts in the gill filament in tadpoles of *Rana temporaria* and *Bufo bufo* (Amphibia, Anura). *J. Exp. Zool.* 217:143–145.

———. 1982. The morphology of amphibian skin vascularization before and after metamorphosis. *Zoomorphology* 100:55–63.

———. 1985. Blood flow patterns of the respiratory systems in larval and adult amphibians: Functional morphology and phylogenetic significance. *Zeit. Zool. Syst. Evol.-forsch.* 23:229–240.

Desser, S. S., J. Lom, and I. Dykova. 1986. Developmental stages of *Sphaerospora ohlmacheri* (Whinery 1893) n. comb. (Myxozoa: Myxosporea) in the renal tubules of bullfrog tadpoles, *Rana catesbeiana*, from Lake of Two Rivers, Algonquin Park, Ontario. *Canadian J. Zool.* 64:2231–2217.

Deuchar, E. M. 1975. *Xenopus: The South African clawed frog*. New York: Wiley.

Deunff, J., and A. Beaumont. 1959. Histogénèese des dents et du bec cornés chez les larves de *Discoglossus pictus* Otth. *C. R. Soc. Biol. Paris* 1959:1162–1164.

De Villiers, C. G. S. 1929a. Some observations on the breeding habits of the Anura of the Stellenbosch Flats, in particular of *Cacosternum capense* and *Bufo angusticeps*. *Ann. Transvaal Mus.* 13:123–141.

———. 1929b. Some features of the early development of *Breviceps*. *Ann. Transvaal Mus.* 13:142–151.

———. 1929c. The development of a species of *Arthroleptella* from Jonkershoek, Stellenbosch. *South African J. Sci.* 26:481–510.

De Vlaming, V. L., and R. B. Bury. 1970. Thermal selection in tadpoles of the tailed-frog, *Ascaphus truei*. *J. Herpetol.* 4:179–189.

De Waal, S. W. P. 1973. Die mesonefros van *Xenopus laevis* (Daudin) 'n histologiese oorsig. *Navors. Nas. Mus.* (Bloemfontein) 2:356–385.

De Witte, G. F. 1948. Faune de Belgique. Amphibiens et Reptiles. Bruxelles.

Deyrup, I. J. 1964. Water balance and kidney. pp. 251–328. In *Physiology of the Amphibia*. Vol. 1, edited by J. A. Moore. New York: Academic Press.

Diaz, N., M. Sallaberry, and H. Nuñez. 1983. The tadpole of *Telmatobufo venustus* (Anura: Leptodactylidae) with a consideration of generic relationships. *Herpetologica* 39:111–113.

Diaz, N. F., and J. Valencia. 1985. Larval morphology and phenetic relationships of the Chilean *Alsodes, Telmatobius, Caudiverbera* and *Insuetophrynus* (Anura: Leptodactylidae). *Copeia* 1985:175–181.

Diaz-Paniagua, C. 1985. Larval diets related to morphological characters of five anuran species in the biological reserve of Donaña (Huelva, Spain). *Amphibia-Reptilia* 6:307–322.

———. 1986. Reproductive period of amphibians in the biological reserve of Donaña (SW Spain). pp. 429–432. In *Studies in herpetology,* edited by Z. Roček. Prague: Charles University Press.

———. 1987. Tadpole distribution in relation to vegetal heterogeneity in temporary ponds. *Herpetol. J.* 1:167–169.

———. 1988. Temporal segregation in larval amphibian communities in temporary ponds at a locality in SW Spain. *Amphibia-Reptilia* 9:15–26.

———. 1989. Larval diets of two anuran species, *Pelodytes punctatus* and *Bufo bufo,* in SW Spain. *Amphibia-Reptilia* 10:71–75.

Di Bernardino, M. A. 1961. Investigations of the germ-plasm in relation to nuclear transplantation. *J. Embryol. Exp. Morphol.* 9:507–513.

DiBona, D. R., M. M. Civan, and A. Leaf. 1969. The anatomic site of the transepithelial permeability barriers of toad bladder. *J. Cell Biol.* 40:1–7.

Dickman, M. 1968. The effect of grazing by tadpoles on the structure of a periphyton community. *Ecology* 49:1188–1190.

Diener, E., and G. J. V. Nossal. 1966. Phylogenetic studies on the immune response. I. Localization of antigens and immune response in the toad, *Bufo marinus*. *Immunology* 10:535–542.

Diesel, R., G. Bäurle, and P. Vogel. 1995. Cave breeding and froglet transport: A novel pattern of anuran brood care in the Jamaican frog, *Eleutherodactylus cundalli*. *Copeia* 1995:354–360.

Dietz, T. H., and R. H. Alvarado. 1974. Na and Cl transport across gill chamber epithelium of *Rana catesbeiana* tadpoles. *American J. Physiol.* 226:764–770.

Di Grande, F. 1968. Sviluppo postmetamorfico delle gonadi e degli organi di Bidder in esemplari di *Bufo bufo* precocemente panirradiati con raggi X. *Rend. Accad. Naz. Lincei.* Ser. 8, 45:437–442.

———. 1987. Aspetti della morfogenesi delle gonadi e degli organi di Bidder in giovani esemplari di *Bufo bufo* sperimentalmente privati di gran parte delle loro cellule germinali. I. Stadi premetamorfici. *Arch. Italiano Anat. Embriol.* 92:159–177.

Di Grande, F., and O. Marescalchi. 1987. Aspetti della morfogenesi delle gonadi e degli organi di Bidder in giovani esemplari di *Bufo bufo* sperimentalmente privati di gran parte delle loro cellule germinali. II. Stadi premetamorfici postmetamorfici. *Arch. Italiano Anat. Embriol.* 92:179–208.

Dionne, J.-C. 1969. Tadpole holes: A true biogenic sedimentary structure. *J. Sediment. Petrol.* 39:358–360.

DiStefano, R. J., M. J. Roell, B. A. Wagner, and J. J. Descoske. 1994. Relative performances of four preservatives on fish and crayfish. *Trans. American Fish. Soc.* 123:817–823.

Ditrich, H., and A. Lametschwandtner. 1992. Glomerular development and growth of the renal blood vacular system in *Xenopus laevis* (Amphibia, Anura, Pipidae) during metamorphic climax. *J. Morphol.* 213:335–340.

Dixon, J. R., and W. R. Heyer. 1968. Anuran succession in a temporary pond in Colima, Mexico. *Bull. Southern California Acad. Sci.* 67:129–137.

Dixon, J. R., C. Mercolli, and A. A. Yanosky. 1995. Some aspects of the ecology of *Pseudis paradoxa* from northeastern Argentina. *Herpetol. Rev.* 26:183–184.

Dixon, J. R., and C. Rivero-Blanco. 1985. A new dendrobatid frog (*Colostethus*) from Venezuela with notes on its natural history and that of related species. *J. Herpetol.* 19:177–184.

Dixon, K. E. 1981. The origin of the primordial germ cells in the Amphibia. *Netherlands J. Zool.* 31:5–37.

Dodd, J. M. 1950. Ciliary feeding mechanisms in anuran larvae. *Nature* 165:283.

Dodd, M. H. I., and J. M. Dodd. 1976. The biology of metamorphosis. pp. 467–599. In *Physiology of the Amphibia*. Vol. 3, edited by B. Lofts. New York: Academic Press.

Donnelly, M. A., R. O. De Sá, and C. Guyer. 1990a. Description of the tadpoles of *Gastrophryne pictiventris* and *Nelson-*

ophryne aterrima (Anura, Microhylidae), with a review of morphological variation in free-swimming microhylid larvae. *American Mus. Novitates* (2976):1–19.

Donnelly, M. A., and C. Guyer. 1994. Patterns of reproduction and habitat use in an assemblage of neotropical hylid frogs. *Oecologia* 98:291–302.

Donnelly, M. A., C. Guyer, and R. O. De Sá. 1990b. The tadpole of a dart-poison frog *Phyllobates lugubris* (Anura: Dendrobatidae). *Proc. Biol. Soc. Washington* 103:427–431.

Donnelly, M. A., C. Guyer, D. M. Krempels, and H. E. Braker. 1987. The tadpole of *Agalychnis calcarifer* (Anura: Hylidae). *Copeia* 1987:247–250.

Donner, K. O., and T. Reuter. 1976. Visual pigments and photoreceptor function. pp. 251–277. In *Frog neurobiology, a handbook,* edited by R. Llinás and W. Precht. Berlin: Springer-Verlag.

Doughty, P., and R. Shine. 1998. Reproductive energy allocation and long term energy stores in a viviparous lizard (*Eulamprus tympanum*). *Ecology* 79:1073–1083.

Dowling, J. E. 1976. Physiology and morphology of the retina. pp. 278–296. In *Frog neurobiology, a handbook,* edited by R. Llinás and W. Precht. Berlin: Springer-Verlag.

Downes, H., and P. M. Courogen. 1996. Contrasting effects of anesthetics in tadpole bioassays. *J. Pharmacol. Exp.* 278:284–278.

Downie, J. R. 1990a. Temporal changes in the behavior of foam-making *Leptodactylus fuscus* tadpoles. *Herpetol. J.* 1:498–500.

———. 1990b. Functions of the foam in foam-nesting leptodactylids: Antipredator effects of *Physalaemus pustulosus* foam. *Herpetol. J.* 1:501–503.

———. 1994a. Developmental arrest in *Leptodactylus fuscus* tadpoles (Anura: Leptodactylidae). 1: Descriptive analysis. *Herpetol. J.* 4:29–38.

———. 1994b. Developmental arrest in *Leptodactylus fuscus* tadpoles (Anura: Leptodactylidae). 2: Does a foam-borne factor block development? *Herpetol. J.* 4:39–45.

———. 1996. A new example of female parental behaviour in *Leptodactylus validus,* a frog of the leptodactylid "*melanonotus*" species group. *Herpetol. J.* 6:32–34.

Downie, J. R., and A. Weir. 1997. Developmental arrest in *Leptodactylus fuscus* tadpoles (Anura: Leptodactylidae) III: Effect of length of arrest period on growth potential. *Herpetol. J.* 7:85–92.17.

Dressler, G. R., and P. Grusse. 1988. Do multigene families regulate vertebrate development? *Trends Genet.* 4:214–219.

Drewes, R. C. 1984. A phylogenetic analysis of the Hyperoliidae (Anura): Treefrogs of Africa, Madagascar, and the Seychelle Islands. *Occas. Pap. California Acad. Sci.* (139):1–70.

Drewes, R. C., R. Altig, and K. M. Howell. 1989. Tadpoles of three frog species endemic to the forest of the Eastern Arc Mountains, Tanzania. *Amphibia-Reptilia* 10:435–443.

Dreyer, T. F. 1915. The morphology of the tadpole of *Xenopus laevis. Trans. Roy. Soc. South Africa* 4:241–258.

Drysdale, T. A., and R. P. Elinson. 1993. Head ectodermal patterning and axial development in frogs. *American Zool.* 33:417–423.

Dubois, A. 1985. Diagnose préliminaire d'un nouveau genre de Ranoidea (Amphibien, Anoures) du sud de l'Inde. *Alytes* 4:113–118.

———. 1987. Miscellanea taxinomica batrachologia (I). *Alytes* 5:7–95.

———. 1992. Notes sur la clasification des Ranidae (Amphibiens Anoures). *Bull. Mensuel. Soc. Linneenne de Lyon* 61:305–352.

———. 1995. Keratodont formulae in anuran tadpoles: Proposals for a standardization. *J. Zool. Syst. Evol. Res.* 33:1–15.

Dubois, A., and A. Ohler. 1994. Formules dentaires des têtard d'Amphibiens Anoures: Une nouvelle méthode d'expression et de comparaison. *Ann. Sci. Nat. Zool., Paris* 15:163–166.

Dudley, R., V. A. King, and R. J. Wassersug. 1991. The implications of shape and metamorphosis for drag forces on a generalized pond tadpole (*Rana catesbeiana*). *Copeia* 1991: 252–257.

Duellman, W. E. 1970. The hylid frogs of Middle America. *Mus. Nat. Hist., Univ. Kansas Monogr.* (1):1–753, 72 plates.

———. 1972. A review of the neotropical frogs of the *Hyla bogotensis* group. *Occas. Pap. Mus. Nat. Hist., Univ. Kansas* (11):1–31.

———. 1978. The biology of an equatorial herpetofauna in Amazonian Ecuador. *Misc. Publ. Mus. Nat. Hist., Univ. Kansas* (65):1–352, 4 plates.

———. 1985. Reproductive modes in anuran amphibians: Phylogenetic significance of adaptive strategies. *South African J. Sci.* 81:174–178.

———. 1989. Alternative life-history styles in anuran amphibians: Evolutionary and ecological implications. pp. 101–126. In *Alternative life-history styles of animals,* edited by M. N. Bruton. Dordrecht: Kluwer Acad. Publ.

Duellmann, W. E., and R. Altig. 1978. New species of tree frogs (Hylidae) from the Andes of Colombia and Ecuador. *Herpetologica* 34:177–185.

Duellman, W. E., and J. A. Campbell. 1992. Hylid frogs of the genus *Plectrohyla:* Systematics and phylogenetic relationships. *Misc. Publ. Mus. Zool., Univ. Kansas* (181):1–32.

Duellman, W. E., and R. O. De Sá. 1988. A new genus and species of South American hylid frog with a highly modified tadpole. *Trop. Zool.* 1:117–136.

Duellman, W. E., and M. J. Fouquette, Jr. 1968. Middle American frogs of the *Hyla microcephala* group. *Univ. Kansas Publ. Mus. Nat. Hist.* 17:517–557.

Duellman, W. E., and P. Gray. 1983. Developmental biology and systematics of the egg-brooding hylid frogs, genera *Flectonotus* and *Fritziana. Herpetologica* 39:333–359.

Duellman, W. E., and D. M. Hillis. 1987. Marsupial frogs (Anura: Hylidae: *Gastrotheca*) of the Ecuadorian Andes: Resolution of taxonomic problems and phylogenetic relationships. *Herpetologica* 43:141–173.

Duellman, W. E., and M. S. Hoogmoed. 1984. The taxonomy and phylogenetic relationships of the hylid frog genus *Stefania. Misc. Publ. Mus. Nat. Hist., Univ. Kansas* (75): 1–39.

Duellman, W. E., and J. Lescure. 1973. Life history and ecology of the hylid frog *Osteocephalus taurinus,* with observations on larval behavior. *Occas. Pap. Mus. Nat. Hist., Univ. Kansas* (13):1–12.

Duellman, W. E., and J. D. Lynch. 1969. Descriptions of *Atelopus* tadpoles and their relevance to atelopodid classification. *Herpetologica* 25:231–240.

Duellman, W. E., and S. J. Maness. 1980. The reproductive behavior of some hylid marsupial frogs. *J. Herpetol.* 14:213–222.

Duellman, W. E., and A. Schwartz. 1958. Amphibians and reptiles of southern Florida. *Bull. Florida St. Mus.* 3:181–324.

Duellman, W. E., and L. Trueb. 1983. Frogs of the *Hyla columbiana* group: Taxonomy and phylogenetic relationships. pp. 33–51. In *Advances in herpetology and evolutionary biology,* ed-

ited by A. G. J. Rhodin and K. Miyata. Cambridge, MA: Mus. Comp. Zool.
———. 1986. *Biology of amphibians.* New York: McGraw-Hill.
———. 1989. Two new treefrogs of the *Hyla parviceps* group from the Amazon Basin in southern Peru. *Herpetologica* 45:1–10.
Dugès, A. 1834. Recherches sur l'ostéologie et la myologie des batraciens a leurs différens ages. *Mém. l'Acad. Roy. l'Inst. France, Sci. Math. Phys.* 6:1–216.
Duke, J. T., and G. R. Ultsch. 1990. Metabolic oxygen regulation and conformity during submergence in the salamanders *Siren lacertina, Amphiuma means,* and *Amphiuma tridactylum,* and a comparison with other giant salamanders. *Oecologia* 84:16–23.
Duke, K. L. 1978. Nonfollicular ovarian components. pp. 563–582. In *The vertebrate ovary comparative biology and evolution,* edited by R. E. Jones. New York: Plenum.
Dumas, P. C. 1964. Species-pair allopatry in the genera *Rana* and *Phrynosoma. Ecology* 45:178–181.
Dunlap, D. G. 1967. The development of the musculature of the hindlimb of the frog, *Rana pipiens. J. Morphol.* 119:241–258.
Dunlap, D. G., and C. K. Satterfield. 1985. Habitat selection in larval anurans: Early experience and substrate pattern selection in *Rana pipiens. Develop. Psychobiol.* 18:37–58.
Dunlop, S. A., and L. D. Beazley. 1981. Changing retinal ganglion cell distribution in the frog *Heleioporus eyrei. J. Comp. Neurol.* 202:221–236.
Dunn, E. R. 1924. Some Panamanian frogs. *Occas. Pap. Mus. Zool., Univ. Michigan* (151):1–16.
———. 1926. The frogs of Jamaica. *Proc. Boston Soc. Nat. Hist.* 38:111–130.
Dunn, R. F. 1978. Nerve fibers of the eighth nerve and their distribution to the sensory nerves of the inner ear in the bullfrog. *J. Comp. Neurol.* 182:621–636.
Dunson, W. A. 1977. Tolerance to high temperature and salinity by tadpoles of the Philippine frog, *Rana cancrivora. Copeia* 1977:375–378.
Dunson, W. A., and J. Connell. 1982. Specific inhibition of hatching in amphibian embryos by low pH. *J. Herpetol.* 16:314–316.
Dunson, W. A., and J. Travis. 1991. The role of abiotic factors in community organization. *American Nat.* 138:1067–1091.
Du Pasquier, L. 1968. Les protéins sériques et le complexe lymphomyéloide chez le têtard d'*Alytes obstetricans* normal and thymectomisé. *Ann. Inst. Pasteur* 114:490–502.
———. 1970. Ontogeny of the immune response in animals having less than one million lymphocytes: The larvae of the toad *Alytes obstetricans. Immunology* 19:353–362.
Dupré, R. K., J. J. Just, E. C. Crawford, Jr., and T. L. Powell. 1986. Temperature preference and responses of cutaneous temperature-sensitive neurons during bullfrog development. *Physiol. Zool.* 59:254–262.
Dupré, R. K., and J. W. Petranka. 1985. Ontogeny of temperature selection in larval amphibians. *Copeia* 1985:462–467.
Dupré, R. K., and S. C. Wood. 1988. Behavioral temperature regulation by aquatic ectotherms during hypoxia. *Canadian J. Zool.* 66:2649–2652.
Dutta, S. K., and P. Mohanty-Hejmadi. 1984. Ontogeny of teeth row structure in *Rana tigerina* tadpoles. *J. Bombay Nat. Hist. Soc.* 80:517–528.
Dziadek, M., and K. E. Dixon. 1975. Mitosis in presumptive primordial germ cells in post-blastula embryos of *Xenopus laevis. J. Exp. Zool.* 192:285–291.
———. 1977. An autoradiographic analysis of nucleic acid synthesis in the presumptive primordial germ cells of *Xenopus laevis. J. Embryol. Exp. Morphol.* 37:13–31.
Eakin, R. M. 1963. Ultrastructural differentiation of the oral sucker in the treefrog, *Hyla regilla. Develop. Biol.* 7:169–179.
Eakin, R. M., and F. E. Bush. 1957. Development of the amphibian pituitary with special reference to the neural lobe. *Anat. Rec.* 129:279–296.
Eaton, T. H. 1959. The ancestry of modern Amphibia: A review of the evidence. *Univ. Kansas Publ. Mus. Nat. Hist.* 12:157–180.
Ebbesson, S. O. E. 1970. On the organization of central visual pathways in vertebrates. *Brain Behav. Evol.* 3:178–194.
———. 1976. Morphology of the spinal cord. pp. 679–706. In *Frog neurobiology, a handbook,* edited by R. Llinás and W. Precht. Berlin: Springer-Verlag.
———. 1980. *Comparative neurology of the telencephalon.* New York: Plenum.
Ebenman, B. 1988. Competition between age classes and population dynamics. *J. Theor. Biol.* 131:389–400.
Eberhard, M. L., and F. H. Brandt. 1995. The role of tadpoles and frogs as paratenic hosts in the life cycle of *Dracunculus insignis* (Nematoda: Dracunculoidea). *J. Parasitol.* 81:792–793.
Echeverría, D. D. 1988. Oogenesis en las hembras juveniles de *Bufo arenarum* (Anura, Bufonidae). *Rev. Mus. Argentina Cienc. Nat. 'Bernardino Rivadavia' Inst. Nac. Invest. Cienc. Nat., Zool.* 15:57–75.
———. 1994 (1991). Metamorphosis en *Bufo arenarum* (Anura, (Bufonidae): Consideraciones acercade la salida de los miembros anteriores de las larvas. *Physis* 49B:1–4.
———. 1995. Microscopia electrónica de barrido del aparato bucal y de la cavidad oral de la larva de *Leptodactylus ocellatus* (Linnaeus 1758)(Anura, Leptodactylidae). *Alytes* 12:159–168.
———. 1996. Microscopía electrónica de barrido del aparato bucal y de la cavidad oral de la larva de *Hyla strigilata eringiophila* Gallardo, 1961 (Anura, Hylidae) con comentarios que facilitan su identifición durante la metamorphosis y postmetamorphosis. *Physis* 54B:1–6.
———. 1997. Microanatomy of the buccal apparatus and oral cavity of *Hyla minuta* Peters, 1872 larvae (Anura, Hylidae), with data on feeding habits. *Alytes* 15:26–36.
Echeverría, D. D., and A. M. Filipello. 1994 (1991). Consideraciones acerca de la formula dentaria y del aparato bucal de las larvas de *Odontophrynus occidentalis* (Berg, 1896)(Anura, Leptodactylidae). *Physis* 49B:59–63.
Echeverría, D. D., and L. E. Fiorito de Lopez. 1981. Estadios de la metamorfosis en *Bufo arenarum* (Anura). *Physis* 40B:15–23.
Echeverría, D. D., L. E. Fiorito de Lopez, and S. B. Montanelli. 1992. Esteromorfologia del aparato bucal y de la cavidad oral de las larvas de *Ololygon fuscovaria* (Lutz 1925)(Anura, Hylidae). *Rev. Mus. Argentina Cienc. Nat. "Bernardino Rivadavia" Inst. Nac. Invest. Cienc. Nat., Zool.* 16:2–13.
Echeverría, D. D., L. E. Fiorito de Lopez, S. B. Montanelli, O. B. Vaccaro, and A. M. Filipello. 1987. Consideraciones acerca de la formulas dentarias de las larvas de *Bufo arenarum* Hensel (Anura: Bufonidae). *Cuad. Herpetol.* 3:33–39.
Eddy, E. M. 1975. Germ plasm and the differentiation of the germ cell line. *Intl. Rev. Cytol.* 43:229–280.
Eddy, E. M., and S. Ito. 1971. Fine structural and radioautographic observations on dense perinuclear cytoplasmic material in tadpole oocytes. *J. Cell Biol.* 49:90–108.
Edgeworth, F. H. 1920. On the development of the hypo-

branchial and laryngeal muscles in Amphibia. *J. Anat.* 44:125–162.
———. 1930. On the masticatory and hyoid muscles of larvae of *Xenopus laevis*. *J. Anat.* 64:184–188.
———. 1935. *The cranial muscles of vertebrates.* Cambridge: Cambridge University Press.
Edwards, M. L. O., and E. B. Pivorun. 1991. The effects of photoperiod and different dosages of melatonin on metamorphic rate and weight gain in *Xenopus laevis* tadpoles. *Gen. Comp. Endocrinol.* 81:28–38.
Ehmann, H., and G. Swan. 1985. Reproduction and development in the marsupial frog, *Assa darlingtoni* (Leptodactylidae, Anura). pp. 279–285. In *The biology of Australasian frogs and reptiles,* edited by G. Grigg, R. Shine, and H. Ehmann. Chipping Norton, New South Wales: Surrey Beatty.
Eichler, V. B., and R. A. Porter. 1981. Rohon-Beard cells in frog development: A study of temporal and spatial changes in a transient cell population. *J. Comp. Neurol.* 203:121–130.
Eiselt, J. 1942. Der Musculus opercularis und die mittlere Ohrsphäre der anuren Amphibien. *Arch. Naturgesch. (n. ser.)* 10:179–219.
Ekblom, P. 1989. Developmentally regulated conversion of mesenchyme to epithelium. *FASEB J.* 3:2141–2150.
Ekman, G. 1913. Experimentelle Untersuchungen über die Entwicklung der Kiemenregion (Kiemenfäden und Kiemenspatten) einiger anuren Amphibien. *Morphol. Jahrb.* 47:419–592.
———. 1924. Experimentelle Beiträge zur Herzentwicklung der Amphibien. *Roux' Arch.* 106:320–352.
———. 1925. Neue experimentelle Beiträge zur frühesten Entwicklung des Amphibienherzens. *Acta Soc. Sci. Fennicae, Comm. Biol.* 1:1–36.
Eldred, W. D., T. E. Finger, and J. Nolte. 1980. Central projections of the frontal organ of *Rana pipiens,* as demonstrated by the anterograde transport of horseradish peroxidase. *Cell Tiss. Res.* 211:215–222.
Elepfandt, A. 1988a. Central organization of wave localization in the clawed frog, *Xenopus laevis.* I. Involvement and bilateral organization of the midbrain. *Brain Behav. Evol.* 31:349–357.
———. 1988b. Central organization of wave localization in the clawed frog, *Xenopus laevis.* II. Midbrain topology for wave directions. *Brain Behav. Evol.* 31:358–368.
Elias, H. 1937. Zur vergleichenden Histologie und Entwicklungsgeschichte der Haut der Anuren. *Zeit. Mikros. Anat. Forsch.* 41:359–416.
Elinson, R. P. 1987a. Change in development patterns: Embryos of amphibians with large eggs. pp. 1–21. In *Development as an evolutionary process. MBL Lect. Biol.* Vol. 8, edited by R. A. Raff and E. C. Raff. New York: Alan R. Liss.
———. 1987b. Fertilization and aqueous development of the Puerto Rican terrestrial-breeding frog, *Eleutherodactylus coqui*. *J. Morphol.* 193:217–224.
———. 1990. Direct development in frogs: Wiping the recapitulationist slate clean. *Seminars Develop. Biol.* 1:263–270.
———. 1994. Leg development in a frog without a tadpole (*Eleutherodactylus coqui*). *J. Exp. Zool.* 270:202–210.
Elinson, R. P., and E. M. Del Pino. 1985. Cleavage and gastrulation in the egg-brooding, marsupial frog, *Gastrotheca riobambae*. *J. Embryol. Exp. Morphol.* 90:223–232.
Elinson, R. P., E. M. Del Pino, D. S. Townsend, F. C. Cuesta, and P. Eichhorn. 1990. A practical guide to the developmental biology of terrestrial-breeding frogs. *Biol. Bull.* 179:163–177.
Emerson, S. B. 1976. A preliminary report on the superficial throat musculature of the Microhylidae and its possible role in tongue action. *Copeia* 1976:546–551.
———. 1986. Heterochrony and frogs: The relationship of a life history trait to morphological form. *American Nat.* 127:167–183.
Emerson, S. B., J. Travis, and M. Blouin. 1988. Evaluating a hypothesis about heterochrony: Larval life-history traits and juvenile hind-limb morphology in *Hyla crucifer. Evolution* 42:68–78.
Engel, W., and M. Schmid. 1981. H-Y antigen as a tool for the determination of the heterogametic sex in Amphibia. *Cytogenet. Cell Genet.* 30:130–136.
Engels, H. G. 1935. Über Umbildungsvorgänge im Kardinalvenensystem bei Bildung der Urniere. *Gegenbaurs Morphol. Jahrb.* 76:345–374.
Engermann, W.-E., J. Fritzsche, R. Günther, and F. J. Obst. 1993. *Lurche und Kriechtiere Europas.* Radebeul: Neuman Verlag.
Epple A., and T. L. Lewis. 1973. Comparative histophysiology of the pancreatic islets. *American Zool.* 13:567–590.
Erasmus, B. de W., B. J. Howell, and H. Rahn. 1970–1971. Ontogeny of acid-base balance in the bullfrog and chicken. *Respir. Physiol.* 11:46–53.
Erdmann, B. 1921. Über die Entwicklung der Atrioventrikularklappen bei den Anuren. *Morphol. Jahrb.* 51:339–354.
Escher, K. 1925. Das verhalten der Seitenorgane der Wirbeltiere und ihrer nerven beim Übergang zum Landleben. *Acta Zool.* 6:307–414.
Estes, R., Z. V. Špinar, and E. Nevo. 1978. Early Cretaceous pipid tadpoles from Israel (Amphibia: Anura). *Herpetologica* 34:374–393.
Estrada, A. R. 1987. Los nidos terrestres de dos especies de anfibios cubanos del género *Eleutherodactylus* (Anura: Leptodactylidae). *Poeyana* (352):1–9.
Etkin, W. 1934. The phenomena of anuran metamorphosis. II Oxygen consumption during normal metamorphosis. *Physiol. Zool.* 7:129148.
———. 1936. The phenomena of anuran metamorphosis. III. The development of the thyroid gland. *J. Morphol.* 59:69–89.
———. 1964. Metamorphosis. pp. 427–468. In *Physiology of the Amphibia.* Vol. 1, edited by J. A. Moore. London: Academic Press.
———. 1968. Hormonal control of amphibian metamorphosis. pp. 313–348. In *Metamorphosis,* edited by W. Etkin and L. I. Gilbert. New York: Meredith Corp.
Exbrayat, J. M., and S. Hraouibloquet. 1994. An example of heterochrony—the metamorphosis in Gymnophiona. *Bull. Soc. Zool. France* 199:117–126.
Fabrezi, M. 1985. Anatomic cranial de larvas de *Gastrotheca gracilis* (Anura: Hylidae). *Bol. Asoc. Herpetol. Argentina* 2:2.
Fabrezi, M., and G. García. 1994. Metamorfosis del aparato hiobranquial en *Pleurodema borellii* y *Ceratophrys cranwelli* (Anura: Leptodactylidae). *Acta Zool. Lilloana* 42:189–196.
Fabrezi, M., and E. O. Lavilla. 1992. Estructura del condrocráneo y esqueleto hiobranquial en larvas de algunos hílidos neotropicales (Anura: Hylidae). *Acta Zool. Lilloana* 41:155–164.
———. 1993. Anatomia del condrocraneo en larvas de tres especies de *Telmatobius* del grupo meridional (Anura: Leptodactylidae). *Physis* 48B:39–46.
Faier, J. M. 1972. Larval anuran teeth and some ecologic implications. Thesis, University of Nebraska.
Faivovich, J., and G. R. Carrizo. 1992. Descripción de la larva

de *Chacophrys pierotti* (Vellard, 1948) (Leptodactylidae, Ceratophryinae). *Alytes* 10:81–89.

Falconer, D. S. 1989. *Introduction to quantitative genetics.* 3d ed. London: Longwood.

Fales, D. E. 1935. Experiments on the development of the pronephros of *Amblystoma punctatum. J. Exp. Zool.* 72:147–173.

Fang, H., and E. P. Elinson. 1996. Patterns of distalless gene expression and inductive interaction in the head of the direct developing frog *Eleutherodactylus coqui. Develop. Biol.* 179:160–172.

Fanning, S. A. 1966. A synopsis and key to the tadpoles of Florida. Thesis, Florida State University.

Faragher, S. G., and R. G. Jaeger. 1998. Tadpole bullies: Examining mechanisms of competition in a community of larval anurans. *Canadian J. Zool.* 76:144–153.

Farel, P. B., and S. E. Bemelmans. 1985. Specificity of motoneuron projection patterns during development of the bullfrog tadpole (*Rana catesbeiana*). *J. Comp. Neurol.* 238:128–134.

———. 1986. Restoration of neuromuscular specificity following ventral rhizotomy in the bullfrog tadpole, *Rana catesbeiana. J. Comp. Neurol.* 254:125–132.

Farlowe, V. 1928. Algae of ponds as determined by an examination of intestinal contents of tadpoles. *Biol. Bull.* 55:443–448.

Farrell, A. P. 1997. Development of cardiovascular systems. pp. 101–113. In *Development of cardiovascular systems. Molecules to organisms,* edited by W. W. Burggren and B. B. Keller. Cambridge: Cambridge University Press.

Fauth, J. E. 1990. Interactive effects of predators and early larval dynamics of the treefrog *Hyla chrysoscelis. Ecology* 71:1609–1616.

Fauth, J. E., and W. J. Resetarits, Jr. 1991. Interactions between the salamander *Siren intermedia* and the keystone predator *Notophthalmus viridescens. Ecology* 72:827–838.

Fawcett, D. W., S. Ito, and D. B. Slautterback. 1959. The occurrence of intercellular bridges in groups of cells exhibiting synchronous differentiation. *J. Biophys. Biochem. Cytol.* 5:453–460.

Feder, M. E. 1981. Effect of body size, trophic state, time of day, and experimental stress on oxygen consumption of anuran larvae: An experimental assessment and evaluation of the literature. *Comp. Biochem. Physiol.* 70A:497–508.

———. 1982. Effect of developmental stage and body size on oxygen consumption of anuran larvae: A reappraisal. *J. Exp. Zool.* 220:33–42.

———. 1983a. Responses to acute aquatic hypoxia in larvae of the frog *Rana berlandieri. J. Exp. Biol.* 104:79–95.

———. 1983b. Effect of hypoxia and body size on the energy metabolism of lungless tadpoles, *Bufo woodhousei,* and air-breathing anuran larvae. *J. Exp. Zool.* 228:11–19.

———. 1983c. The relation of air breathing and locomotion to predation on tadpoles, *Rana berlandieri,* by turtles. *Physiol. Zool.* 56:522–531.

———. 1984. Consequences of aerial respiration for amphibian larvae. pp. 71–86. In *Respiration and metabolism of embryonic vertebrates,* edited by R. S. Seymour. Dordrecht: Dr. W. Junk.

———. 1985. Thermal acclimation of oxygen consumption and cardiorespiratory frequencies in frog larvae. *Physiol. Zool.* 58:303–311.

———. 1992. A perspective on environmental physiology of the amphibians. pp. 1–6. In *Environmental physiology of the amphibians,* edited by M. E. Feder and W. W. Burggren. Chicago: University of Chicago Press.

Feder, M. E., and W. W. Burggren. 1985. Cutaneous gas exchange in vertebrates: Design, patterns, control and implications. *Biol. Rev.* 60:1–45.

Feder, M. E., R. J. Full, and J. Piiper. 1988. Elimination kinetics of acetylene and Freon 22 in resting and active lungless salamanders. *Respir. Physiol.* 72:229–240.

Feder, M. E., A. G. Gibbs, G. A. Griffith, and J. Tsuji. 1984a. Thermal acclimation of metabolism in salamanders: Fact or artefact? *J. Therm. Biol.* 9:255–260.

Feder, M. E., and A. W. Pinder. 1988. Ventilation and its effect on "infinite pool" exchangers. *American Zool.* 28:973–983.

Feder, M. E., D. B. Seale, M. E. Boraas, R. J. Wassersug, and A. G. Gibbs. 1984b. Functional conflicts between feeding and gas exchange in suspension-feeding tadpoles, *Xenopus laevis. J. Exp. Biol.* 110:91–98.

Feder, M. E., and R. J. Wassersug. 1984. Aerial versus aquatic oxygen consumption in the larvae of the clawed frog, *Xenopus laevis. J. Exp. Biol.* 108:231–245.

Feinsinger, P., E. E. Spears, and R. W. Poole. 1981. A simple measure of niche breadth. *Ecology* 62:27–32.

Felsenstein, J. 1985. Phylogenies and the comparative method. *American Nat.* 125:1–15.

Feminella, J. W., and C. P. Hawkins. 1994. Tailed frog tadpoles differentially alter their feeding behavior in response to non-visual cues from four predators. *J. North American Benthol. Soc.* 13:310–320.

Feng, A. S., and W. Lin. 1991. Differential innervation patterns of three divisions of frog auditory midbrain (torus semicircularis). *J. Comp. Neurol.* 306:613–630.

Fermin, P. 1765. *Développement parfait du mystère de la génération du fameux crapaud de Surinam.* Maestricht.

Fernández, K. 1926. Sober la biología y reproducción de batracios argentinos (segunda parte). *Bol. Acad. Nac. Cienc. Córdoba* 29:271–320, 1 plate.

Fernández, K., and M. Fernández. 1921. Sobre la biología y reproducción de algunos batracios argentinos. I. Cystignathidae. *Ann. Soc. Cient. Argentina* 91:97–140, 3 plates.

Ferns, M. J., and A. H. Lamb. 1987. Regulation of cell numbers in the developing neuromuscular system in *Xenopus laevis. Neurosci. Lett. Suppl.* 27S:72.

Ferrari, L. 1998. Tolerance of high electrolytic and non-electrolytic osmolarities in *Bufo arenarum* premetamorphic tadpoles under organism density stress. *Alytes* 15:171–175.

Figge, F. H. J. 1936. The differential reaction of the blood vessels of a branchial arch of *Amblystoma tigrinum* (Colorado axolotl) I. The reaction to adrenalin, oxygen, and carbon dioxide. *Physiol. Zool.* 9:79–101.

Figiel, C. R., Jr., and R. D. Semlitsch. 1991. Effects of nonlethal injury and habitat complexity on predation in tadpole populations. *Canadian J. Zool.* 69:830–834.

Filipski, G. T., and M. V. H. Wilson. 1985. Staining nerves in whole cleared amphibians and reptiles using Sudan black B. *Copeia* 1985:500–502.

Fink, W. L. 1982. The conceptual relationship between ontogeny and phylogeny. *Paleobiology* 8:254–264.

Fiorito de Lopez, L. E., and D. D. Echeverría. 1984. Morfogenesis de los dientes larvales y pico corneo de *Bufo arenarum* (Anura, Bufonidae). *Zoología* 13:573–578.

———. 1989. Microanatomia e histogenesis del aparato bucal en las larvas de *Bufo arenarum* (Anura, Bufonidae). *Cuad. Herpetol.* 4:4–10.

Fisher, R. A. 1930. *The genetical theory of natural selection.* Oxford: Clarendon Press.

Fishwild, T. G., R. A. Schemidt, K. M. Jankens, K. A. Berven, G. J. Gamboa, and C. M. Richards. 1990. Sibling recognition by larval frogs (*Rana pipiens, R. sylvatica* and *Pseudacris crucifer*). *J. Herpetol.* 24:40–44.

Fite, K. V., ed. 1976. *The amphibian visual system: A multidisciplinary approach.* New York: Academic Press.

———. 1985. Pretectal and accessory-optic visual nuclei of fish, Amphibia, and reptiles: Theme and variations. *Brain Behav. Evol.* 26:71–90.

Fite, K. V., R. G. Carey, and D. Vicario. 1977. Visual neurons in frog anterior thalamus. *Brain Res.* 127:283–290.

Fletcher, K., and N. B. Myant. 1959. Oxygen consumption of tadpoles during metamorphosis. *J. Physiol.* 145:353–368.

Flock, Å. 1965. Electron microscopic and electrophysiologic studies on the lateral line canal organ. *Acta Otolaryngol. Suppl.* 199:1–90.

Flock, Å., and J. Wersäll. 1962. A study of the orientation of the sensory hairs of the receptor cells in the lateral line organ of fish, with special reference to the function of the receptors. *J. Cell. Biol.* 15:19–27.

Flores-Nava, A., M. A. Olvera-Novoa, and E. Gasca-Leyva. 1994. A comparison of the effects of three water-circulation regimes on the aquaculture of bullfrog (*Rana catesbeiana* Shaw) tadpoles. *Aquaculture* 128:105–114.

Flower, S. S. 1899. Notes on a second collection of reptiles and batrachians made in the Malay Peninsula and Siam, from November 1896 to September 1898, with a list of the species recorded from those countries. *Proc. Zool. Soc. London* 1899:885–916, plates 59–60.

Floyd, R. B. 1983. Ontogenetic change in the temperature tolerance of larval *Bufo marinus* (Anura: Bufonidae). *Comp. Biochem. Physiol.* 75A:267–271.

———. 1984. Variation in temperature preference with stage of development of *Bufo marinus* larvae. *J. Herpetol.* 18:153–158.

———. 1985. Effects of photoperiod and starvation on the temperature tolerance of larvae of the giant toad, *Bufo marinus. Copeia* 1985:625–631.

Ford, L. S., and D.C. Cannatella. 1993. The major clades of frogs. *Herpetol. Monogr.* (7):94–117.

Ford, T. D., and W. J. Breed. 1970. Tadpole holes formed during desciccation of overbank pool. *J. Sediment. Petrol.* 40:1044–1045.

Forehand, C. J., and P. B. Farel. 1982a. Spinal cord development in anuran larvae: I. Primary and secondary neurons. *J. Comp. Neurol.* 209:386–394.

———. 1982b. Spinal cord development in anuran larvae: II. Ascending and descending pathways. *J. Comp. Neurol.* 209:395–408.

Forge, P., and R. Barbault. 1977. Ecologie de la reproduction et due développement larvaire d'un amphibien déserticole, *Bufo pentoni* Anderson, 1893, au Senegal. *Terre Vie* 31:117–125.

Forman L. J., and J. J. Just. 1976. The life span of red blood cells in the amphibian larvae, *Rana catesbeiana. Develop. Biol.* 50:537–540.

Formanowicz, D. R., Jr. 1986. Anuran tadpole/aquatic insect predator-prey interactions: Tadpole size and predator capture success. *Herpetologica* 42:367–373.

Formanowicz, D. R., Jr., and M. S. Bobka. 1989. Predation risk and microhabitat preference: An experimental study of the behavioral responses of prey and predator. *American Midl. Nat.* 121:379–386.

Formanowicz, D. R., Jr., and E. D. Brodie, Jr. 1982. Relative palatabilities of members of a larval amphibian community. *Copeia* 1982:91–97.

Formas, J. R. 1972. A second species of Chilean frog genus *Telmatobufo* (Anura: Leptodactylidae). *J. Herpetol.* 6:1–3.

Formas, J. R., N. F. Diaz, and J. Valencia. 1980. The tadpole of the Chilean frog *Insuetophrynus acarpicus. Herpetologica* 36:316–318.

Formas, J. R., and E. Pugín. 1971. Reproduccion y desarrollo de: *Batrachyla antartandica* (Barrio)(Anura, Leptodactylidae). *Bol. Mus. Nac. Hist. Nat. Chile* 32:201–213.

———. 1978. Tadpoles of *Hylorina sylvatica, Eupsophus vittatus,* and *Bufo rubropunctatus* in southern Chile. *Herpetologica* 34:355–358.

Formas, J. R., and M. A. Vera. 1980. Reproductive patterns of *Eupsophus roseus* and *E. vittatus. J. Herpetol.* 14:11–14.

Fortman, J. R., and R. Altig. 1973. Characters of F_1 hybrid tadpoles between six species of *Hyla. Copeia* 1973:411–416.

Foster, M. S., and R. W. McDiarmid. 1982. Study of aggregative behavior of *Rhinophrynus dorsalis* tadpoles: Design and analysis. *Herpetologica* 38:395–404.

———. 1983. *Rhinophrynus dorsalis* (Alma de Vaca, Sapo Borracho, Mexican Burrowing Toad). pp. 419–421. In *Costa Rica Natural History,* edited by D. H. Janzen. Chicago: University of Chicago Press.

Foster, R. G., and A. Roberts. 1982. The pineal eye in *Xenopus laevis* embryos and larvae: A photoreceptor with a direct excitatory effect on behaviour. *J. Comp. Physiol.* 145A:413–419.

Fox, H. 1961. Quantitative analysis of normal growth of the pronephric system in larval Amphibia. *Proc. Zool. Soc. London* 136:301–315.

———. 1962a. Growth and degeneration of the pronephric system of *Rana temporaria. J. Embryol. Exp. Morphol.* 10:103–114.

———. 1962b. A study of the evolution of the amphibian and dipnoan pronephros by an analysis of its relationship with the anterior spinal nerves. *Proc. Zool. Soc. London* 138:225–256.

———. 1963. The amphibian pronephros. *Quart. Rev. Biol.* 38:1–25.

———. 1965. Early development of the head and pharynx of *Neoceratodus* with a consideration of its phylogeny. *J. Zool.* 146:470–554.

———. 1970. Tissue degeneration: An electron microscopic study of the pronephros of *Rana temporaria. J. Embryol. Exp. Morphol.* 24:139–157.

———. 1971. Cell death, thyroxine, and the development of *Rana temporaria* larvae with special reference to the pronephros. *Exp. Gerontol.* 6:173–177.

———. 1972. Tissue degeneration: An electron microscopic study of the tail skin of *Rana temporaria* during metamorphosis. *Arch. Biol. Liège* 83:373–394.

———. 1977. The anuran tadpole skin: Changes occurring in it during metamorphosis and some comparisons with that of the adult. pp. 269–289. In *Comparative anatomy of the skin. Proc. Symp. Zool. Soc. London (39),* edited by R. I. C. Spearman. London: Academic Press.

———. 1981. Cytological and morphological changes during amphibian metamorphosis. pp. 327–362. In *Metamorphosis: A problem in developmental biology.* 2d ed., edited by L. I. Gilbert and E. Frieden. New York: Plenum.

———. 1984. *Amphibian morphogenesis.* Clifton, NJ: Humana Press.

———. 1985. Changes in amphibian skin during larval development and metamorphosis. pp. 59–87. In *Metamorphosis,* edited by M. Balls and M. Bownes. Oxford: Clarendon Press.

———. 1986. Early development of caecilian skin with special reference to the epidermis. *J. Herpetol.* 20:154–167.

———. 1988. *Riesenzellen,* goblet cells, Leydig cells, and the large clear cells of *Xenopus,* in the amphibian larval epidermis: Fine structure and a consideration of their homology. *J. Submicros. Cytol. Pathol.* 20:437–451.

Fox, H., E. Bailey, and R. Mahoney. 1972. Aspects of the ultrastructure of the alimentary canal and respiratory ducts in *Xenopus laevis* larvae. *J. Morphol.* 138:387–406.

Fox, H., and L. Hamilton. 1964. Origin of the pronephric duct in *Xenopus laevis. Arch. Biol.* 75:245–251.

———. 1971. Ultrastructure of diploid and haploid cells of *Xenopus laevis* larvae. *J. Embryol. Exp. Morphol.* 26:81–98.

Fox, H., E. B. Lane, and M. Whitear. 1978. Sensory nerve endings and receptors in fish and amphibians. pp. 271–281. In *The skin of vertebrates,* edited by R. I. C. Spearman and P. A. Riley. London: Academic Press.

Fox, H., R. Mahoney, and E. Bailey. 1970. Aspects of the ultrastructure of the alimentary canal and associated glands of the *Xenopus laevis* larva. *Arch. Biol. Liège* 81:21–50.

Fox, H., and S. C. Turner. 1967. A study of the relationship between the thyroid and larval growth in *Rana temporaria* and *Xenopus laevis. Arch. Biol.* 78:61–90.

Fox, H., and M. Whitear. 1978. Observations on Merkel cells in amphibians. *Biol. Cell.* 32:223–232.

Foxon, G. E. H. 1964. Blood and respiration. pp. 151–209. In *Physiology of the Amphibia.* Vol. 1, edited by J. A. Moore. New York: Academic Press.

Franchi, L. L., A. M. Mandl, and S. Zuckermann. 1962. The development of the ovary and the process of oogenesis. pp. 1–88. In *The ovary.* Vol. 1, edited by S. Zuckermann, A. M. Mandl, and P. Eckstein. New York: Academic Press.

Frank, B. D., and J. G. Hollyfield. 1987a. Retinal ganglion cell morphology in the frog, *Rana pipiens. J. Comp. Neurol.* 266:413–434.

———. 1987b. Retina of the tadpole and frog: Delayed dendritic development in a subpopulation of ganglion cells coincident with metamorphosis. *J. Comp. Neurol.* 266:435–444.

Frank, G. 1988. Granulopoiesis in tadpoles of *Rana esculenta.* Survey of the organs involved. *J. Anat.* 160:59–66.

———. 1989a. Granulopoiesis in tadpoles of *Rana esculenta.* Ultrastructural observations on the developing granulocytes and on the development of eosinophil granules. *J. Anat.* 163:97–105.

———. 1989b. Granulopoiesis in tadpoles of *Rana esculenta.* Ultrastructural observations on the morphology and development of heterophil and basophil granules. *J. Anat.* 163:107–116.

Fraser, E. A. 1950. The development of the vertebrate excretory system. *Biol. Rev.* 25:159–187.

Freda, J. 1986. The influence of acidic pond water on amphibians: A review. *Water, Air, Soil Pollut.* 30:439–450.

Freda, J., and W. A. Dunson. 1984. Sodium balance of amphibian larvae exposed to low environmental pH. *Physiol. Zool.* 57:435–443.

———. 1985. Field and laboratory studies of ion balance and growth rates of ranid tadpoles chronically exposed to low pH. *Copeia* 1985:415–423.

———. 1986. The effect of prior exposure on sodium uptake in tadpoles exposed to low pH water. *J. Comp. Physiol.* 156B:649–654.

Freda, J., and D. G. McDonald. 1988. Physiological correlates of interspecific variation in acid tolerance in fish. *J. Exp. Biol.* 136:243–258.

Freeman, J. A. 1965. The cerebellum as a timing device: An experimental study in the frog. pp. 397–420. In *Neurobiology of cerebellar evolution and development,* edited by R. Llinás. Chicago: American Med. Assoc.

Friedman, R. T., N. S. LaPrade, R. M. Aiyawar, and E. G. Huf. 1967. Chemical basis for the [H^+] gradient across frog skin. *American J. Physiol.* 212:962–972.

Friet, S. C., and A. W. Pinder. 1990. Hypoxia during natural aquatic hibernation of the bullfrog. *American Zool.* 30:69A.

Fritzsch, B. 1981a. The pattern of lateral-line afferents in urodeles. A horseradish-peroxidase study. *Cell Tiss. Res.* 218:581–594.

———. 1981b. Electroreceptors and direction specific arrangement in the lateral line system of salamanders. *Zeit. Naturforsch.* 36C:493–495.

———. 1988. Phylogenetic and ontogenetic origins of the dorsolateral auditory nucleus of anurans. pp. 561–585. In *The evolution of the amphibian auditory system,* edited by B. Fritzsch, M. J. Ryan, W. Wilczynski, T. E. Hetherington, and W. Walkowiak. New York: Wiley.

Fritzsch, B., A. M. Nikundiwe, and U. Will. 1984. Projection patterns of lateral-line afferents in anurans: A comparative HRP study. *J. Comp. Neurol.* 229:451–469.

Fritzsch, B., M. J. Ryan, W. Wilczynski, T. E. Hetherington, and W. Walkowiak, eds. 1988a. *The evolution of the amphibian auditory system.* New York: Wiley.

Fritzsch, B., U. Wahnschaffe, and U. C. Bartsch. 1988b. Metamorphic changes in the octavolateralis system of amphibians. pp. 359–376. In *The evolution of the amphibian auditory system,* edited by B. Fritzsch, M. J. Ryan, W. Wilczynski, T. E. Hetherington, and W. Walkowiak. New York: Wiley.

Fröhlich, J., H. Aurin, and P. Kemnitz. 1977. Zur epidermalen Kugelzelle von Krallenfroschlarven. *Verh. Anat. Ges.* 71:1171–1175.

Frost, D. R. 1985. *Amphibian species of the world. A taxonomic and geographic reference.* Lawrence, KS: Assoc. Syst. Coll.

Frost, S. K., and S. J. Robinson. 1984. Pigment cell differentiation in the fire-bellied toad, *Bombina orientalis.* I. Structural, chemical, and physical aspects of the adult pigment pattern. *J. Morphol.* 179:229–242.

Frye, B. E. 1964. Metamorphic changes in the blood sugar and the pancreatic islets of the frog, *Rana clamitans. J. Exp. Zool.* 155:215–224.

Fujisawa, H. 1987. Mode of growth of retinal axons within the tectum of *Xenopus* tadpoles, and implications in the ordered neuronal connection between the retina and the tectum. *J. Comp. Neurol.* 260:127–139.

Fuller P. M., and S. O. E. Ebbesson. 1973. Central projections of the trigeminal nerve in the bullfrog (*Rana catesbeiana*). *J. Comp. Neurol.* 152:193–200.

Funkhouser, A. 1977. Plasma osmolarity of *Rana catesbeiana* and *Scaphiopus hammondi* tadpoles. *Herpetologica* 33:272–274.

Funkhouser, A., and S. A. Foster. 1970. Oxygen uptake and thyroid activity in *Hyla regilla* tadpoles. *Herpetologica* 26:366–371.

Funkhouser, A., and K. S. Mills. 1969. Oxygen consumption during spontaneous amphibian metamorphosis. *Physiol. Zool.* 42:15–21.

Gall, J. G. 1968. Differential synthesis of the genes for ribosomal RNA during amphibian oögenesis. *Proc. Natl. Acad. Sci. United States of America* 60:553–560.

Gallardo, J. M. 1961. Observaciones biológicas sobre *Hyla raddiana* Fitz., de la Provincia de Buenos Aires. *Cienc. Invest.* 17:63–69.

Gallien, L., and C. Houillon. 1951. Table cronologique de développement chez *Discoglossus pictus*. *Bull. Biol. France-Belgique* 85:373–375.

Gans, C., and G. De Gueldre. 1992. Striated muscle: Physiology and functional morphology. pp. 277–313. In *Environmental physiology of the amphibians,* edited by M. E. Feder and W. W. Burggren. Chicago: University of Chicago Press.

Gardener, L. W., and A. M. Peadon. 1955. The role of the thyroid gland in direct development in the anuran, *Eleutherodactylus martinicensis*. *Growth* 19:263–286.

Garstang, W. 1922. The theory of recapitulation: A critical restatement of the biogenetic law. *J. Linnean Soc. Zool.* 35:81–101.

Garth, S., ed. 1961. *Ovid's Metamorphoses, in fifteen books.* Translated by John Dryden et al. New York: Heritage Press.

Gartz, R. 1970. Adaptationsmorphologie der Melanophoren von Krallenfrosch-Larven. *Cytobiologie* 2:220–234.

Gascon, C. 1989a. The tadpole of *Atelopus pulcher* Boulenger (Anura, Bufonidae) from Manaus, Amazonas. *Rev. Brasil. Zool.* 6:235–239.

———. 1989b. Predator-prey size interaction in tropical ponds. *Rev. Brasil. Zool.* 6:701–706.

———. 1991a. Breeding of *Leptodactylus knudseni:* Responses to rainfall variation. *Copeia* 1991:248–252.

———. 1991b. Population- and community-level analyses of species occurrences of central Amazonian rainforest tadpoles. *Ecology* 72:1731–1746.

———. 1992a. Spatial distribution of *Osteocephalus taurinus* and *Pipa arrabali* in a central Amazonian forest. *Copeia* 1992:894–897.

———. 1992b. The effects of reproductive phenology on larval performance traits in a three-species assemblage of central Amazonian tadpoles. *Oikos* 65:307–313.

———. 1992c. Aquatic predators and tadpole prey in central Amazonia: Field data and experimental manipulations. *Ecology* 73:971–980.

———. 1995. Tropical larval anuran fitness in the absence of direct effects of predation and competition. *Ecology* 76:2222–2229.

Gascon, C., and J. R. Bider. 1985. The effect of pH on bullfrog, *Rana catesbeiana,* and green frog, *Rana clamitans melanota,* tadpoles. *Canadian Field Nat.* 99:259–261.

Gates, W. R. 1983. A comparison of the embryonic and larval development of the southern chorus frog, *Pseudacris nigrita nigrita* (LeConte), and the upland chorus frog, *Pseudacris triseriata feriarum* (Baird), in Alabama. Thesis, Auburn University.

Gatherer, D., and E. M. Del Pino. 1992. Somitogenesis in the marsupial frog *Gastrotheca riobambae*. *Intl. J. Develop. Biol.* 36:283–291.

Gatten, R. E., Jr., J. P. Caldwell, and M. E. Stockard. 1984. Anaerobic metabolism during intense swimming by anuran larvae. *Herpetologica* 40:164–169.

Gatten, R. E., Jr., K. Miller, and R. J. Full. 1992. Energetics of amphibians at rest and during locomotion. pp. 314–377. In *Environmental physiology of the amphibians,* edited by M. E. Feder and W. W. Burggren. Chicago: University of Chicago Press.

Gaupp, E. 1893. Beiträge zur Morphologie des Schädels. I: Primordial-Cranium und Kieferbogen von *Rana fusca*. *Morphol. Arb.* 2:275–481.

———. 1894. Beiträge zur Morphologie des Schädels. II. Das Hyobranchial-Skelet der Anuren und seine Umwandlung. *Morphol. Arb.* 3:399–437.

———. 1896. *A. Ecker's und R. Wiedersheim's Anatomie des Frosches.* Braunschweig: Vieweg und Sohn.

Gaze, R. M., and P. Grant. 1992. Spatio-temporal patterns of retinal ganglion cell death during *Xenopus* development. *J. Comp. Neurol.* 315:264–274.

Gee, J. H., and R. C. Waldick. 1995. Ontogenetic buoyancy changes and hydrostatic control in larval anurans. *Copeia* 1995:861–870.

Gegenbaur, C. 1898–1901. *Vergleichende Anatomie der Wirbeltiere, mit Berücksichtigung der Wirbellosen,* Vol. 2. Leipzig.

Gehrau, A., and R. De Sá. 1980. Comunicacion preliminar sobre larvas de *Limnomedusa macroglossa* (Amphibia: Leptodactylidae). *Res. J. Cienc. Nat. Montevideo* 1:85–86.

Gérard, P., and R. Cordier. 1933. Sur le méchanisme de transformation des néphrons ouverts en néphrons fermés chez la larve de *Discoglossus pictus*. *C. R. Acad. Roy. Belgique Cl. Sci.* 19:508–512.

———. 1934a. Esquisse d'une histophysiologie comparée du rein des vertébrés. *Biol. Rev.* 9:110–131.

———. 1934b. Recherches d'histophysiologie comparée sur le pro- et le mésonéphros larvaires des anoures. *Zeit. Zellforsch. Mikros. Anat.* 21:1–23.

Gerhardt, E. 1932. Die Kiemenentwicklung bei Anuren (*Pelobates fuscus, Hyla arborea*) and Urodelen (*Triton vulgaris*). *Zool. Jahrb. Anat.* 55:173–220.

Gerschenfeld, H. M., J. H. Tramezzani, and E. DeRobertis. 1960. Ultrastructure and function in neurohypophysis of the toad. *Endocrinology* 66:741–762.

Gesner, C. 1551–1604. *Historia Animalium.* Frankfurt: I. Wecheli.

Ghate, H. V., and A. D. Padhye. 1988. Predation of *Microhyla* tadpoles by *Gambusia*. *J. Bombay Nat. Hist. Soc.* 85:200–201.

Giarretta, A. A. 1996. Reproductive specializations of the bromeliad hylid frog *Phyllodytes luteolus*. *J. Herpetol.* 30:96–97.

Giarretta, A. A., W. C. A. Bokermann, and C. F. B. Haddad. 1993. A review of the genus *Megaelosia* (Anura: Leptodactylidae) with a description of a new species. *J. Herpetol.* 27:276–285.

Giaretta, A. A., and L. K. Castanho. 1990. Nova espécie de *Paratelmatobius* (Amphibia, Anura, Leptodactylidae) da Serra do Mar, Brasil. *Pap. Avul. Zool.* 37:133–139.

Gibley, C. W., Jr., and J. P. Chang. 1966. Fine structure of tubule cells in the functional pronephros of *Rana pipiens*. *American Zool.* 6:610. (Abstr.)

Giesel, J. T. 1976. Reproductive strategies as adaptations to life in temporally heterogeneous environments. *Ann. Rev. Ecol. Syst.* 7:57–79.

Giles, M. A., and E. M. Attas. 1993. Rare earth elements as internal batch marks for rainbow trout—retention, distribution, and effects on growth of injected Dysprosium, Europium, and Samarium. *Trans. American Fish. Soc.* 122: 289–297.

Gill, D. E. 1978. The metapopulation ecology of the red-spotted newt, *Notophthalmus viridescens* (Rafinesque). *Ecol. Monogr.* 48:145–166.

Gines, H. 1958. Representantes de la familia Pipidae (Amphibia, Salientia) en Venezuela. *Mem. Soc. Cienc. Nat. La Salle* 18:5–18.

Giorgi, P. P. 1974. Germ cell migration in toad (*Bufo bufo*): Effect of ventral grafting of embryonic dorsal regions. *J. Embryol. Exp. Morphol.* 31:75–87.

Gipouloux, J.-D. 1957. Le développement précoce du glomus est indépendant de la présence du pronéphros (Amphibiens, Anoures). *C. R. Hebd. Seances Acad. Sci. Paris* 244:1064–1065.

———. 1962. Mise en évidence du 'cytoplasme germinal' dans l'oeuf et l'embryon du Discoglosse: *Discoglossus pictus* Otth. *C. R. Hebd. Seances Acad. Sci. Paris* 254:2433–2435.

———. 1964. Une substance diffusible émanée des organes mésodermiques dorsaux attire les cellules germinales situées dans l'endoderme; démonstration expérimentale chez le Crapaud commun, *Bufo bufo* L. *C. R. Hebd. Seances Acad. Sci. Paris* 259:3844–3847.

———. 1970. Recherches expérimentales sur l'origine, la migration des cellules germinales, et l'édification des crêtes génitales chez les amphibiens anoures. *Bull. Biol. France-Belgique* 104:22–93.

———. 1975. "Cytoplasme germinal" et détermination germinale chez les amphibiens anoures. *Ann. Biol.* 14:475–487.

Gipouloux, J.-D., and R. Cambar. 1961. Recherches sur les facteurs déterminant la migration des cellules de l'uretère primaire (Amphibiens Anoures). *C. R. Hebd. Seances Acad. Sci. Paris* 252:3643–3645.

———. 1970a. La migration des cellules du blastème de l'uretère primaire est indépendante de la présence des somites, chez les embryons d'Amphibiens Anoures. *C. R. Hebd. Seances Acad. Sci. Paris* 270D:372–373.

———. 1970b. Nouvelles précisions sur la distance maximale à laquelle l'uretère primaire exerce son action inductrice sur le blastème mésonéphrogène chez les Amphibiens Anoures. *C. R. Hebd. Seances Acad. Sci. Paris* 270D:2329–2331.

Gipouloux, J.-D., and M. Delbos. 1977. Aspects ultrastructuraux de l'interaction uretère-mesenchyme, au cours de l'induction néphrogène, chez les amphibiens. *J. Embryol. Exp. Morphol.* 41:259–268.

Gipouloux, J.-D., and C. Girard 1986. Organogénèse de l'appareil urogénital et de l'organe interrénal. pp. 211–240. In *Traité de Zoologie, Anatomie, Systematique, Biologie,* Vol. 14, Fasc. 1B, edited by P.-P. Grassé and M. Delsol. Paris: Masson.

Gipouloux, J.-D., C. Girard, and M. Delbos. 1978. Effets d'un traitement par l'AMP cyclique sur la migration des cellules germinales primordiales dans l'embryon de *Xenopus laevis. C. R. Hebd. Seances Acad. Sci. Paris* 287D:1425–1427.

Gipouloux, J.-D., and J. Hakim. 1976. Etude comparée de la formation des ébauches mésonéphrogenes médullaire et interrénale chez les embryons du crapaud commun, *Bufo bufo* L., et du dactylètre du cap, *Xenopus laevis* Daudin. *Bull. Biol. France-Belgique* 110:283–298.

———. 1978. Le développement de l'appareil uro-génital après la greffe hétérotopique du mésoblaste latéral chez la neurula du crapaud commun *Bufo bufo* L. (Amphibien Anoure). I. Inversions séparées des axes antéro-postérieur et dorso-ventral de la lame latérale. *J. Embryol. Exp. Morphol.* 43:137–146.

Gitlin, D. 1944. The development of *Eleutherodactylus portoricensis. Copeia* 1944:91–98.

Glaw, R., and M. Vences. 1992. *A fieldguide to the amphibians and reptiles of Madagascar.* Leverkusen: Moos-Druck.

Gluecksohn-Waelsch, S. 1987. Regulatory genes in development. *Trends Genet.* 3:123–127.

Goater, C. P. 1994. Growth and survival of postmetamorphic toads: Interactions among larval history, density, and parasitism. *Ecology* 75:2264–2274.

Goater, C. P., and R. E. Vandenbos. 1997. Effects of larval history and lungworm infection on the growth and survival or juvenile wood frogs (*Rana sylvatica*). *Herpetologica* 53:331–338.

Goette, A. 1874. Atlas zur Entwicklung der Unke. *Arch. Mikros. Anat.* 9.

———. 1875. *Die Entwicklungsgeschichte der Unke* (Bombinator igneus). Leipzig: Verlag von Leopold Voss.

Goicoechea, O., O. Garrido, and B. Jorquera. 1986. Evidence for a trophic paternal-larval relationship in the frog *Rhinoderma darwinii. J. Herpetol.* 20:168–178.

Goin, C. J. 1947. Studies on the life history of *Eleutherodactylus ricordii planirostris* (Cope) in Florida: With special reference to the local distribution of an allelomorphic color pattern. *Univ. Florida Publ., Biol. Sci.* 4:1–66.

Goin, C. J., O. B. Goin, and G. R. Zug. 1978. *Introduction to herpetology.* San Francisco: W. H. Freeman.

Goin, O. B., C. J. Goin, and K. Bachmann. 1968. DNA and amphibian life history. *Copeia* 1968:532–540.

Goldacre, R. J. 1949. Surface films on natural bodies of water. *J. Anim. Ecol.* 18:36–39.

Gollman, B., and G. Gollman. 1991. Embryonic development of the myobatrachine frogs *Geocrinia laevis, Geocrinia victoriana,* and their natural hybrids. *Amphibia-Reptilia* 12:103–110.

———. 1993a. Hatching dynamics of larvae of Australian smooth froglets, *Geocrinia laevis* and *Geocrinia victoriana* (Amphibian, Anura, Myobatrachidae). *Zool. Anz.* 229:191–199.

———. 1993b. A laboratory experiment on interspecific competition between tadpoles of *Geocrinia victoriana* and *Pseudophryne semimarmorata* (Anura, Myobatrachidae). *Amphibia-Reptilia* 14:349–356.

———. 1994. Life history variation across a hybrid zone in *Geocrinia:* Embryonic development and larval growth (Amphibia, Anura, Myobatrachinae). *Acta Oecologica* 15:247–259.

———. 1995. Morphological variation in tadpoles of the *Geocrinia laevis* complex: Regional divergence and hybridization (Amphibia, Anura, Myobatrachidae). *J. Zool. Syst. Evol. Res.* 33:32–41.

———. 1996. Geographic variation of larval traits in the Australian frog *Geocrinia victoriana. Herpetologica* 52:181–187.

Gona, A. G., K. F. Hauser, and N.J. Uray. 1982. Ultrastructural studies on Purkinje cell maturation in the cerebellum of the frog tadpole during spontaneous and thyroxine-induced metamorphosis. *Brain Behav. Evol.* 20:156–171.

Goniakowska-Witalínska, L. 1984. Tubular myelin structures in the lungs of Amphibia. The mode of formation. *European J. Cell Biol.* 33:127–133.

———. 1986. Lung of the tree frog, *Hyla arborea* L. A scanning and transmission electron microcopic study. *Anat. Embryol.* 174:379–389.

Gonzalez, A., H. J. Ten Donkelaar, and R. De Boer-Van Huizen. 1984. Cerebellar connections in *Xenopus laevis.* An HRP study. *Anat. Embryol.* 169:167–176.

Goodyear, C. P., and R. Altig. 1971. Orientation of bullfrogs (*Rana catesbeiana*) during metamorphosis. *Copeia* 1971:362–264.

Gordon, F. J. 1934. Preservation of small Amphibia in gelatin. *Science* 80:457.

Gordon, J., and D. C. Hood. 1976. Anatomy and physiology of the frog retina. pp. 29–86. In *The amphibian visual system: A multidisciplinary approach,* edited by K. V. Fite. New York: Academic Press.

Gordon, M. S. 1962. Osmotic regulation in the green toad (*Bufo viridis*). *J. Exp. Biol.* 39:261–270.
Gordon, M. S., and V. A. Tucker. 1965. Osmotic regulation in the tadpoles of the crab-eating frog (*Rana cancrivora*). *J. Exp. Biol.* 42:437–445.
Gorham, E., S. J. Eisenreich, J. Ford, and M. V. Santelmann. 1985. The chemistry of bog waters. pp. 339–363. In *Chemical processes in lakes,* edited by W. Stumm. New York: Wiley.
Gorlick, D. L., and D. B. Kelley. 1987. Neurogenesis in the vocalization pathway of *Xenopus laevis. J. Comp. Neurol.* 257:614–627.
Görner, P. 1963. Untersuchungen zur Morphologie und Elektrophysiologie des Seitenlinieorgans vom Krallenfrosches (*Xenopus laevis* Daudin). *Zeit. Vergl. Physiol.* 47:316–338.
Gosner, K. L. 1959. Systematic variations in tadpole teeth with notes on food. *Herpetologica* 15:203–210.
———. 1960. A simplified table for staging anuran embryos and larvae with notes on identification. *Herpetologica* 16:183–190.
Gosner, K. L., and I. H. Black. 1955. The effects of temperature and moisture on the reproductive cycle of *Scaphiopus h. holbrooki. American Midl. Nat.* 54:192–203.
———. 1957a. Larval development in New Jersey Hylidae. *Copeia* 1957:31–36.
———. 1957b. The effects of acidity on the development and hatching of New Jersey frogs. *Ecology* 38:256–262.
Gotte, S. W., and R. P. Reynolds. 1998. Observations on the effects of alcohol vs. formalin storage of amphibian larvae. USGS Patuxent Wildlife Research Center homepage, <http://www.pwrc.nbs.gov/resshow/reynld1rs/amphlarv.html>.
Gould, S. J. 1977. *Ontogeny and phylogeny.* Cambridge: Harvard University Press.
Gradwell, N. 1968. The jaw and hyoidean mechanism of the bullfrog tadpole during aqueous ventilation. *Canadian J. Zool.* 46:1041–1052.
———. 1969a. The function of the internal nares of the bullfrog tadpole. *Herpetologica* 25:120–121.
———. 1969b. The respiratory importance of vascularization of the tadpole operculum in *Rana catesbeiana* Shaw. *Canadian J. Zool.* 47:1239–1243.
———. 1970. The function of the ventral velum during gill irrigation in *Rana catesbeiana. Canadian J. Zool.* 48:1179–1186.
———. 1971a. *Ascaphus* tadpole: Experiments on the suction and gill irrigation mechanism. *Canadian J. Zool.* 49:307–332.
———. 1971b. *Xenopus* tadpole: On the water pumping mechanism. *Herpetologica* 27:107–123.
———. 1972a. Comments on gill irrigation in *Rana fuscigula. Herpetologica* 28:122–125.
———. 1972b. Gill irrigation in *Rana catesbeiana.* Part I. On the anatomical basis. *Canadian J. Zool.* 50:481–499.
———. 1972c. Gill irrigation in *Rana catesbeiana.* Part II. On the musculoskeletal mechanism. *Canadian J. Zool.* 50:501–521.
———. 1973. On the functional morphology of suction and gill irrigation in the tadpole of *Ascaphus,* and notes on hibernation. *Herpetologica* 29:84–93.
———. 1974. Description of the tadpole of *Phrynomerus annectens,* and comments on its gill irrigation mechanism. *Herpetologica* 30:53–62.
———. 1975a. The clinging mechanism of *Pseudophryne bibroni* (Anura: Leptodactylidae) to an alga on glass. *Trans. Roy. Soc. South Australia* 99:31–34.
———. 1975b. Experiments on oral suction and gill breathing in five species of Australian tadpole (Anura: Hylidae and Leptodactylidae). *J. Zool.* 177:81–98.
———. 1975c. The bearing of filter feeding on the water pumping mechanism of *Xenopus* tadpoles (Anura: Pipidae). *Acta Zool.* 56:119–128.
Gradwell, N., and V. M. Pasztor. 1968. Hydrostatic pressures during normal ventilation in the bullfrog tadpole. *Canadian J. Zool.* 46:1169–1174.
Gradwell, N., and B. Walcott. 1971. Dual functional and structural properties of the interhyoideus muscle of the bullfrog tadpole (*Rana catesbeiana*). *J. Exp. Zool.* 176:193–218.
Grafen, A. 1990. Do animals really recognize kin? *Anim. Behav.* 39:42–54.
Grandison, A. G. C. 1978. The occurrence of *Nectophrynoides* (Anura: Bufonidae) in Ethiopia. A new concept of the genus with a description of a new species. *Monit. Zool. Italiano* 6:119–172.
———. 1980. Aspects of breeding morphology in *Mertensophryne microanotis* (Anura: Bufonidae): Secondary sexual characters, eggs, and tadpole. *Bull. British Mus. Nat. Hist. (Zool.)* 39:299–304.
———. 1981. Morphology and phylogenetic position of the West African *Didynamipus sjoestedti* Andersson, 1903 (Anura Bufonidae). *Monit. Zool. Italiano (N.S.) Suppl.* 15:187–215.
Grant, P., E. Rubin, and C. Cima. 1980. Ontogeny of the retina and optic nerve in *Xenopus laevis.* I. Stages in the early development of the retina. *J. Comp. Neurol.* 189:593–613.
Grant, S., and M. J. Keating. 1986. Ocular migration and the metamorphic and postmetamorphic maturation of the retinotectal system in *Xenopus laevis:* An autoradiographic and morphometric study. *J. Embryol. Exp. Morphol.* 92:43–69.
Grassé, P.-P. 1986. La spermatogénèse. pp. 1–20. In *Traité de Zoologie, Anatomie, Systematique, Biologie,* Vol. 14, Fasc. 1B, edited by P.-P. Grassé and M. Delsol. Paris: Masson.
Grassi Milano, E., and F. Accordi. 1983. Comparative morphology of the adrenal gland of anuran Amphibia. *J. Anat.* 136:165–174.
———. 1986. Evolutionary trends in the adrenal gland of anurans and urodeles. *J. Morphol.* 189:249–259.
Grassi Milano, E., F. Accordi, D. Antuzzi, and R. Rapisarda. 1979. Morphologic observations on the interrenal, Stilling and intermediary cells in the adrenal gland of *Rana esculenta*-complex. *Riv. Biol.* 72:46–69.
Gray, E. G. 1957. The spindle and extrafusal innervation of a frog muscle. *Proc. Roy. Soc. London* 146B:416–430.
Gray, J. 1936. Studies in animal locomotion IV. The neuromuscular mechanism of swimming in the eel. *J. Exp. Biol.* 13:170–180.
Gray, P. 1930. The development of the amphibian kidney. Part I. The development of the mesonephros of *Rana temporaria. Quart. J. Micros. Sci.* 73:507–546.
———. 1932. The development of the amphibian kidney. Part II. The development of the kidney of *Triton vulgaris* and a comparison of this form with *Rana temporaria. Quart. J. Micros. Sci.* 75:425–465.
———. 1936. The development of the amphibian kidney. Part III. The post-metamorphic development of the kidney, and the development of the vasa efferentia and seminal vesicles in *Rana temporaria. Quart. J. Micros. Sci.* 78:445–473.
Graybeal, A., and D.C. Cannatella. 1995. A new taxon of Bufonidae from Peru, with descriptions of two new species and a review of the phylogenetic status of supraspecific taxa. *Herpetologica* 51:105–131.
Green, N. B. 1938. The breeding habits of *Pseudacris*

brachyphona (Cope) with a description of the eggs and tadpole. *Copeia* 1938:79–82.

Greil, A. 1905. Ueber die Genese der Mundhöhlenschleimhaut der Urodelen. *Anat. Anz. Suppl. 27, Verh. Anat. Ges.* 19:25–37.

Greven, H. 1980. Ultrastructural investigations of the epidermis and the gill epithelium in the intrauterine larvae of *Salamandra salamandra* (L.)(Amphibia, Urodela). *Zeit. Mikros. Anat. Forsch.* 94:196–208.

Griffiths, I. 1961. The form and function of the fore-gut in anuran larvae (Amphibia, Salientia) with particular reference to the *manicotto glandulare. Proc. Zool. Soc. London* 137:249–283, 6 plates.

———. 1963. The phylogeny of the Salientia. *Biol. Rev.* 38:241–292.

Griffiths, I., and A. L. De Carvalho. 1965. On the validity of employing larval characters as major phyletic indices in Amphibia, Salientia. *Rev. Brasil. Biol.* 25:113–121.

Griffiths, R. A. 1986. Feeding niche overlap and food selection in smooth and palmate newts, *Triturus vulgaris* and *T. helveticus,* at a pond in mid-Wales. *J. Anim. Ecol.* 55:201–214.

———. 1991. Competition between common frog, *Rana temporaria,* and natterjack toad, *Bufo calamita,* tadpoles: The effect of competitor density and interaction level on tadpole development. *Oikos* 61:187–196.

———. 1995. Determining competition mechanisms in tadpole assemblages. *Herpetol. J.* 5:208–210.

Griffiths, R. A., and J. Denton. 1993. Interspecific associations in tadpoles. *Anim. Behav.* 44:1153–1157.

Griffiths, R. A., J. Denton, and A. L.-C. Wong. 1993. The effect of food level on competition in tadpoles: Interference mediated by protothecan algae? *J. Anim. Ecol.* 62:274–279.

Griffiths, R. A., P. W. Edgar, and A. L.-C. Wong. 1991. Interspecific competition in tadpoles: Growth inhibition and growth retrieval in natterjack toads, *Bufo calamita. J. Anim. Ecol.* 60:1065–1076.

Griffiths, R. A., L. Schley, P. E. Sharp, J. L. Dennis, and A. Román. 1998. Behavioural responses of Mallorcan midwife toad tadpoles to natural and unnatural snake predators. *Anim. Behav.* 55:207–214.

Grillitsch, B. 1992. Notes on the tadpole of *Phrynohyas resinifictrix* (Goeldi 1907). Buccopharyngeal and external morphology of a tree hole dwelling larva (Anura, Hylidae). *Herpetozoa* 5:51–66.

Grillitsch, B., and H. Grillitsch. 1989. Teratological and ontogenetic alterations to external oral structure in some anuran larvae (Amphibia: Anura: Bufonidae, Ranidae). *Fortschrit. Zool.* 35:276–282.

Grillitsch, B., H. Grillitsch, and H. Splechtna. 1988. Zur Morphologie der larve von *Tomopterna cryptotis* (Boulenger 1907)(Amphibia: Anura: Ranidae). *Herpetozoa* 1:31–46.

Grillitsch, B., H. Grillitsch, M. Häupl, and F. Tiedemann. 1983. *Lurche und Kriechtierre Neiderösterreichs.* Wien: Facultas-Verlag.

Grillner, S. 1974. On the generation of locomotion in the spinal dogfish. *Exp. Brain Res.* 20:459–470.

Grillner, S., and S. Kashin. 1976. On the generation and performance of swimming in fish. pp. 181–201. In *Neural control of locomotion,* edited by R. M. Herman, S. Grillner, P. S. G. Stein, and D. G. Stuart. New York: Plenum.

Grobstein, P., and C. Comer. 1983. The nucleus isthmi as an intertectal relay for the ipsilateral oculotectal projection in the frog, *Rana pipiens. J. Comp. Neurol.* 217:54–74.

Grobstein, P., C. Comer, and S. Kostyk. 1980. The potential binocular field and its tectal representation in *Rana pipiens. J. Comp. Neurol.* 190:175–185.

Gromko, M. H., F. S. Mason, and S. J. Smith-Gill. 1973. Analysis of the crowding effect in *Rana pipiens* tadpoles. *J. Exp. Zool.* 186:63–72.

Grover, B. G., and U. Grüsser-Cornehls. 1984. Cerebellar afferents in the frogs, *Rana esculenta* and *Rana temporaria. Cell Tiss. Res.* 237:259–267.

Grubb, J. C. 1972. Differential predation by *Gambusia affinis* on the eggs of seven species of anuran amphibians. *American Midl. Nat.* 88:102–108.

Gruberg, E. R., and S. B. Udin. 1978. Topographic projections between the nucleus isthmi and the tectum of the frog *Rana pipiens. J. Comp. Neurol.* 179:487–500.

Grüsser, O.-J., and U. Grüsser-Cornehls. 1976. Neurophysiology of the anuran visual system. pp. 297–385. In *Frog neurobiology, a handbook,* edited by R. Llinás and W. Precht. Berlin: Springer-Verlag.

Guibé, J., and M. Lamotte. 1958. Morphologie et reproduction par développement direct d'un Anoure du Mont Nimba, *Arthroleptis crusculum* Ang. *Bull. Mus. d'Hist. Nat. Paris* 30:125–133.

Guilford, T. 1988. The evolution of conspicuous coloration. *American Nat.* 131S:7–21.

Guillette, L. J., Jr. 1987. The evolution of viviparity in fishes, amphibians and reptiles: An endocrine approach. pp. 523–562. In *Hormones and reproduction in fishes, amphibians, and reptiles,* edited by D. O. Norris and R. E. Jones. New York: Plenum.

Gunasingh, A., S. Kasinathan, and S. Ramakrishnan. 1982. Endocrine control of the cycle of oviduct and urogenital papilla of the tropical frog *Rana hexadactyla* Lesson. *Indian J. Exp. Biol.* 20:285–289.

Guraya, S. S. 1972. Histochemical observations of the testis of frog, *Rana pipiens.* Short communication. *Acta Biol. Acad. Sci. Hungaricae* 23:327–329.

———. 1978. Maturation of the follicular wall of nonmammalian vertebrates. pp. 261–330. In *The vertebrate ovary. Comparative biology and evolution,* edited by R. E. Jones. New York: Plenum.

Gutzeit, E. 1889. Die Hornzähne der Batrachierlarven. Diss. Inaug. Zool. Univ. Albertina, Wilhelm Engelmann, Leipzig.

Guyer, M. F. 1907. A simple method for removing the gelatinous coats of eggs. *American Nat.* 41:400–401.

Guyetant, R. 1977a. Analysis of the group effect on *Alytes obstetricans* Laur. *Ann. Sci. Univ. Besancon* (14):11–21.

———. 1977b. Variations of the group-effect of *Alytes obstetricans* Laur. *Ann. Sci. Univ. Besancon* (14):23–38.

Haas, A. 1992. Anatomy and function of the buccal pump mechanism in *Phyllobates bicolor* (Anura: Dendrobatidae). *Proc. Second Intern. Symp. Sonderforschungsbereich* 230:87–92.

———. 1993. Das Cranium der Larve von *Phyllobates bicolor* Bibron, 1841 (Anura: Dendrobatidae): Ein Beitrag zum phylogenetischen System der Anuran basierend auf Larvalmerkmalen. Diplomarbeit, University of Tübingen.

———. 1995. Cranial features of dendrobatid larvae (Amphibia: Anura: Dendrobatidae). *J. Morphol.* 224:241–264.

———. 1996a. Nonfeeding and feeding tadpoles of hemiphractine frogs: Larval head morphology, heterochrony, and systematics of *Flectonotus goeldii* (Amphibia: Anura: Hylidae). *J. Zool. Syst. Evol. Res.* 34:163–171.

———. 1996b. Das larvale Cranium von *Gastrotheca riobambae* und seine Metamorphose (Amphibia, Anura, Hylidae). *Verh. Naturwiss. Ver. Hamburg* 36:33–162.

———. 1997a. The larval hyobranchial apparatus of discoglossoid frogs: Its structure and bearing on the systematics of the Anura (Amphibia: Anura). *J. Zool. Syst. Evol. Res.* 35:179–197.

———. 1997b. Der larvale Hyobranchial-Apparat von *Ascaphus truei* (Anura: Ascaphidae) und das Grundmuster der Anura. *Verhandlungen der Deutschen Zoologischen Gesellschaft, Mainz* 90:172.

Haertel, J. D., and R. M. Storm. 1970. Experimental hybridization between *Rana pretiosa* and *Rana cascadae*. *Herpetologica* 26:436–446.

Hah, J. H. 1974. Alkaline phosphatase activity in the developing pronephros and mesonephros of the frog *Bombina orientalis*. *Korean J. Zool.* 17:177–184.

Hailman, J. P. 1976. Oil droplets in the eyes of adult anuran amphibians: A comparative study. *J. Morphol.* 148:453–468.

Hairston, N. G. 1989. Hard choices in ecological experimentation. *Herpetologica* 45:119–122

Hakim, J., and J.-D. Gipouloux. 1978. Le développement de l'appareil urogénital après la greffe hétérotopique du mésoblaste latéral chez la neurula du crapaud commun *Bufo bufo* L. (Amphibien Anoure). II. Inversion simultanée des axes antéro-postérieur et dorso-ventral de la lame latérale. *J. Embryol. Exp. Morphol.* 44:113–119.

Hall, J. A., J. H. Larsen, Jr., and R. E. Fitzner. 1997. Postembryonic ontogeny of the spadefoot toad, *Scaphiopus intermontanus* (Anura: Pelobatidae): External morphology. *Herpetol. Monogr.* (11):124–178.

Hall, J. A., J. H. Larsen, Jr., D. E. Miller, and R. E. Fitzner. 1995. Discrimination of kin- and diet-based cues by larval spadefoot toads, *Scaphiopus intermontanus* (Anura: Pelobatidae), under laboratory conditions. *J. Herpetol.* 29:233–243.

Hall, J. C., and A. S. Feng. 1987. Evidence for parallel processing in the frog's auditory thalamus. *J. Comp. Neurol.* 258:407–419.

Hall, R. W. 1904. Development of the mesonephros and the Mullerian ducts of Amphibia. *Bull. Mus. Comp. Zool.* 45:29–125.

Hamburger, V. 1988. *The heritage of experimental embryology: Hans Spemann and the organizer.* New York: Oxford University Press.

Hamilton, W. D. 1964a. The genetical evolution of social behavior I, II. *J. Theor. Biol.* 7:1–52.

———. 1964b. The genetical evolution of social behaviour. I. *J. Theor. Biol.* 7:1–16.

———. 1964c. The genetical evolution of social behaviour. I. *J. Theor. Biol.* 7:17–52.

———. 1971. Geometry for the selfish herd. *J. Theor. Biol.* 31:295–311.

Hammerman, D. L. 1965. Development of the tongue of *Rana clamitans*. *American Zool.* 5:250–251. (Abstr.)

———. 1969a. The frog tongue: I. General development and histogenesis of filiform papillae and mucous glands in *Rana catesbeiana*. *Acta Zool.* 50:11–23.

———. 1969b. The frog tongue: II. Histogenesis of fungiform papillae in *Rana catesbeiana*. *Acta Zool.* 50:25–33.

———. 1969c. The frog tongue: III. Histogenesis and regeneration following complete and partial extirpation of anlagen. *Acta Zool.* 50:215–232.

Hammerschlag, R., S. Maines, and M. Ando. 1989. Sensory ganglia from tadpoles but not adult bullfrogs synthesize heat shock-like proteins in vitro at non-heat shock temperatures. *J. Neurosci. Res.* 23:416–424.

Hampton, S. H., and E. P. Volpe. 1963. Development and interpopulation variability of the mouthparts of *Scaphiopus holbrooki*. *American Midl. Nat.* 70:319–328.

Hanke, W. 1976. Neuroendocrinology. pp. 975–1022. In *Frog neuroendocrinology,* edited by R. Llinás and W. Precht. Berlin: Springer-Verlag.

Hanken, J. 1983. Miniaturization and its effects on cranial morphology in plethodontid salamanders, genus *Thorius* (Amphibia, Plethodontidae): II. The fate of the brain and sense organs and their role in skull morphogenesis and evolution. *J. Morphol.* 177:255–268.

———. 1984. Miniaturization and its effects on cranial morphology in plethodontid salamanders, genus *Thorius* (Amphibia: Plethodontidae). I. Osteological variation. *Biol. J. Linnean Soc.* 23:55–75.

———. 1986. Developmental evidence for amphibians origins. pp. 389–417. In *Evolutionary biology,* Vol. 20, edited by M. K. Hecht, B. Wallace, and G. T. Prance. New York: Plenum.

Hanken, J., and B. K. Hall. 1984. Variation and timing of the cranial ossification sequence of the Oriental fire-bellied toad, *Bombina orientalis* (Amphibia, Discoglossidae). *J. Morphol.* 182:245–255.

———. 1988. Skull development during anuran metamorphosis: I. Early development of the first three bones to form—the exoccipital, the parasphenoid, and the frontoparietal. *J. Morphol.* 195:247–256.

Hanken, J., D. H. Jennings, and L. Olsson. 1997. Mechanistic basis of life-history evolution in anuran amphibians: Direct development. *American Zool.* 37:160–171.

Hanken, J., M. W. Klymkowsky, C. H. Summers, D. W. Seufert, and N. Ingebrigtsen. 1992. Cranial ontogeny in the direct-developing frog, *Eleutherodactylus coqui* (Anura: Leptodactylidae), analyzed using whole-mount immunohistochemistry. *J. Morphol.* 211:95–118.

Hanken, J., and C. H. Summers. 1988. Developmental basis of evolutionary success: Cranial ontogeny in a direct-developing anuran. *American Zool.* 28:12. (Abstr.)

Hanken, J., and D. B. Wake. 1993. Miniaturization of body size: Organismal consequences and evolutionary significance. *Ann. Rev. Ecol. Syst.* 24:501–519.

Hardy, J. D., Jr. 1984. Frogs, egg teeth, and evolution: Preliminary comments on egg teeth in the genus *Eleutherodactylus*. *Bull. Maryland Herpetol. Soc.* 20:1–11.

Harkey, G. A., and R. D. Semlitsch. 1988. Effects of temperature on growth, development, and color polymorphism in the ornate chorus frog *Pseudacris ornata*. *Copeia* 1988:1001–1007.

Harpur, R. P. 1968. Osmoregulation et metabolisme de l'uree: Comparison entre *Bufo viridis* et *Rana temporaria*. *Canadian J. Zool.* 46:295–301.

Harris, R. N. 1987. Density-dependent paedomorphosis in the salamander *Notophthalmus viridescens dorsalis*. *Ecology* 68:705–712.

———. 1989. Ontogenetic changes in size and shape of the facultatively paedomorphic salamander *Notophthalmus viridescens dorsalis*. *Copeia* 1989:35–42.

Harris, R. N., R. A. Alford, and H. M. Wilbur. 1988. Density and phenology of *Notophthalmus viridescens dorsalis* in a natural pond. *Herpetologica* 44:234–242.

Harris, R. N., R. D. Semlitsch, H. M. Wilbur, and J. E. Fauth. 1990. Local variation in the genetic basis of paedomorphosis in the salamander *Ambystoma talpoideum*. *Evolution* 44:1588–1603.

Harrison, J. D. 1987. Food and feeding relations of common

frog and common toad tadpoles (*Rana temporaria* and *Bufo bufo*) at a pond in mid-Wales. *Herpetol. J.* 1:141–143.

Harrison, P. H. 1990. Induction of locomotion in spinal tadpoles by excitatory amino acids and their agonists. *J. Exp. Zool.* 254:13–17.

Harrison, R. G. 1904. Experimentelle Untersuchungen über die Entwicklung der Sinnesorgane der Seitenlinie bei den Amphibien. *Arch. Mikros. Anat. Entwickl.* 63:35–149.

Hart, M. W. 1996. Evolutionary loss of larval feeding: Development, form and function in a facultatively feeding larva, *Brisaster latifrons. Evolution* 50:174–187.

Hartl, D. L., and A. G. Clark. 1989. *Principles of population genetics.* Sunderland, MA: Sinauer.

Harvey, P. H., and P. J. Greenwood. 1978. Anti-predator defense strategies: Some evolutionary problems. pp. 129–151. In *Behavioral ecology: An evolutionary approach,* edited by J. R. Krebs and N. B. Davies. Oxford: Blackwell Sci. Publ.

Harvey, P. H., and and M. D. Pagel. 1991. *The comparative method in evolutionary biology.* Oxford: Oxford University Press.

Hassinger, D. D. 1972. Early life history and ecology of three congeneric species of *Rana* in New Jersey. Dissertation, Rutgers University.

Hastings, D., and W. Burggren. 1995. Developmental changes in oxygen consumption regulation in larvae of the South African clawed frog *Xenopus laevis. J. Exp. Biol.* 198:2465–2475.

Hathaway, E. S. 1927. Quantitative study of the changes produced by acclimatization in the tolerance of high temperatures by fishes and amphibians. *Bull. United States Bur. Fish.* 43:169–192.

Hauser, K. F., N. J. Uray, and A. G. Gona. 1986a. Stellate cell development in the frog cerebellum during spontaneous and thyroxine-induced metamorphosis. *J. Comp. Neurol.* 244:229–244.

———. 1986b. Granule cell development in the frog cerebellum during spontaneous and thyroxine-induced metamorphosis. *J. Comp. Neurol.* 253:185–196.

Havenhand, J. N. 1993. Egg to juvenile period, generation time, and the evolution of larval type in marine invertebrates. *Mar. Ecol. Progr. Ser.* 97:247–260.

Hawkins, C. P., L. J., Gottschalk, and S. S. Brown. 1988. Densities and habitat of tailed frog tadpoles in small streams near Mt. St. Helens following the 1980 eruption. *J. North American Benthol. Soc.* 7:246–252.

Hay, J. M., I. Ruvinsky, S. B. Hedges, and L. R. Maxson. 1995. Phylogenetic relationships of amphibian families inferred from DNA sequences of mitochondrial 12S and 16S ribosomal RNA genes. *Mol. Biol. Evol.* 12:928–937.

Hayes, T. B. 1995. Interdependence of corticosterone and thyroid hormones in larval toads (*Bufo boreas*). 1. Thyroid hormone-dependent and independent effects of corticosterone on growth and development. *J. Exp. Zool.* 271:95–102.

Hayes, T. B., R. Chan, and P. Licht. 1993. Interactions of temperature and steroids on larval growth, development, and metamorphosis in a toad (*Bufo boreas*). *J. Exp. Zool.* 266:206–215.

Hayes, T. B., and P. Licht. 1995. Factors influencing testosterone metabolism in anuran larvae. *J. Exp. Zool.* 271:112–119

Hayes, T. B., and T. H. Wu. 1995. Interdependence of corticosterone and thyroid hormone in toad larvae (*Bufo boreas*). 2. Regulation of corticosterone and thyroid hormones. *J. Exp. Zool.* 271:103–111.

Hayes-Odum, L. A. 1990. Observations on reproduction and embryonic development in *Syrrhophus cystignathoides campi* (Anura: Leptodactylidae). *Southwest. Nat.* 35:358–361.

Hazard, E. S., III, and V. H. Hutchison. 1978. Ontogenetic changes in erythrocytic organic phosphates in the bullfrog, *Rana catesbeiana. J. Exp. Zool.* 206:109–118.

Hearnden, M. N. 1992. The reproductive and larval ecology of *Bufo marinus* (Anura: Bufonidae). Dissertation, James Cook University, Townsville, Queensland, Australia.

Heasman, J., R. O. Hynes, A. P. Swan, V. Thomas, and C. C. Wylie. 1981. Primordial germ cells of *Xenopus embryos:* The role of fibronectin in their adhesion during migration. *Cell* 27:437–447.

Heasman, J., T. Mohum, and C. C. Wylie. 1977. Studies on the locomotion of primordial germ cells from *Xenopus laevis in vitro. J. Embryol. Exp. Morphol.* 42:149–161.

Heasman, J., and C. C. Wylie. 1978. Electron microscopic studies on the structure of motile primordial germ cells of *Xenopus laevis in vitro. J. Embryol. Exp. Morphol.* 46:119–133.

Heasman, J., C. C. Wylie, and S. Holwill. 1985. The importance of basement membranes in the migration of primordial germ cells in *Xenopus laevis. Develop. Biol.* 112:18–29.

Heathcote, R. D., and A. Chen. 1991. Morphogenesis of adrenergic cells in a frog parasympathetic ganglion. *J. Comp. Neurol.* 308:139–148.

Heatwole, H. 1962. Contributions to the natural history of *Eleutherodactylus cornutus maussi. Stahlia* (2):1–11.

Heatwole, H., S. B. De Austin, and R. Herrero. 1968. Heat tolerances of tadpoles of two species of tropical anurans. *Comp. Biochem. Physiol.* 27:807–815.

Hegner, R. W. 1922. The effects of changes in diet on the incidence, distribution and numbers of certain intestinal Protozoa of frog and toad tadpoles. *J. Parasitol.* 9:51–67.

Heisler, N. 1986. Acid-base regulation in fishes. pp. 309–356. In *Acid-base regulation in animals,* edited by N. Heisler. Amsterdam: Elsevier.

Heisler, N., G. Forcht, G. R. Ultsch, and J. F. Anderson. 1982. Acid-base regulation in response to environmental hypercapnia in two aquatic salamanders, *Siren lacertina* and *Amphiuma means. Respir. Physiol.* 49:141–158.

Helff, O. M. 1932. Studies on amphibian metamorphosis. X. Hydrogen ion concentration of the blood of anuran larvae during involution. *Biol. Bull.* 63:405–418.

Helff, O. M., and M. C. Mellicker. 1941. Studies on amphibian metamorphosis. XIX. Development of the tongue in *Rana sylvatica,* including histogenesis of "premetamorphic" and filiform papillae and the mucous glands. *American J. Anat.* 68:339–369.

Helff, O. M., and K. I. Stubblefield. 1931. The influence of oxygen tension on the oxygen consumption of *Rana pipiens* larvae. *Physiol. Zool.* 4:271–286.

Hendricks, F. S. 1973. Intestinal contents of *Rana pipiens* Schreber (Ranidae) larvae. *Southwest. Nat.* 18:99–101.

Hendrickson, A. E., and D. E. Kelly. 1971. Development of the amphibian pineal organ; fine structure during maturation. *Anat. Rec.* 170:129–142.

Henrickson, B.-I. 1990. Predation on amphibian eggs and tadpoles by common predators in acidified lakes. *Holarctic Ecol.* 13:201–206.

Hensley, F. R. 1993. Ontogenetic loss of phenotypic plasticity of age at metamorphosis in tadpoles. *Ecology* 74:2405–2412.

Hero, J.-M. 1990. An illustrated key to tadpoles occurring in the central Amazona rainforest, Manaus, Amazonas, Brasil. *Amazoniana* 11:201–262.

———. 1991. Predation, palatability, and the distribution of tadpoles in the Amazon rainforest. Dissertation, Griffith University, Brisbane, Queensland, Australia.

Hero, J.-M., and A. Mijares-Urrutia. 1995. The tadpole of *Scinax rostrata* (Anura: Hylidae). *J. Herpetol.* 29:307–311.

Héron-Royer, and C. Van Bambeke. 1881. Sur les caractères fournis par la bouche des têtards des batraciens anoures d'Europe. *Bull. Soc. Zool. France* 6:75–81.

———. 1889. Le vestibule de la bouche chez les têtards des batraciens anoures d'Europe. *Arch. Biol.* 9:185–309, plates 12–24.

Herreid, C. F., II, and S. Kinney. 1966. Survival of Alaskan woodfrog (*Rana sylvatica*) larvae. *Ecology* 47:1039–1041.

———. 1967. Temperature and development of the woodfrog, *Rana sylvatica,* in Alaska. *Ecology* 48:579–590.

Herrick, C. J. 1909. The nervus terminalis (nerve of Pinkus) in the frog. *J. Comp. Neurol.* 19:175–190.

———. 1948. *The brain of the tiger salamander.* Chicago: University of Chicago Press.

Hetherington, T. E. 1987. Timing of development of the middle ear of Anura (Amphibia). *Zoomorphology* 106:289–300.

———. 1988. Metamorphic changes in the middle ear. pp. 339–357. In *The evolution of the amphibian auditory system,* edited by B. Fritzsch, M. J. Ryan, W. Wilczynski, T. E. Hetherington, and W. Walkowiak. New York: Wiley.

Heusser, H. 1970a. Spawn eating by tadpoles as possible cause of specific biotope preferences and short breeding times in European anurans (Amphibia Anura). *Oecologia* 4:83–88.

———. 1970b. Laich-Fressen durch Kaulquappen als mögliche Ursache spezifischer Biotoppräferenzen und kurzer Laichzeiten bei europäischen Froschlurchen (Amphibia, Anura). *Oecologia* 4:83–88.

———. 1971. Laich-Raubern und Kannibalismus bei sympatrisher Anuran-Kaulquappen. *Experientia* 27:474–475.

Hewitt, J. 1927. Further descriptions of reptiles and batrachians from South Africa. *Rec. Albany Mus.* 3:371–415.

Hews, D. K. 1988. Alarm response in larval western toads, *Bufo boreas:* Release of larval chemicals by a natural predator and its effect on predator capture efficiency. *Anim. Behav.* 36:125–133.

———. 1995. Overall predator feeding rates and relative susceptibility of large and small tadpoles to fish predation depend on microhabitat: A laboratory study. *J. Herpetol.* 29:142–145.

Hews, D. K., and A. R. Blaustein. 1985. An investigation of the alarm response in *Bufo boreas* and *Rana cascadae* tadpoles. *Behav. Neural Biol.* 43:47–57.

Heyer, W. R. 1969. The adaptive ecology of the species groups of the genus *Leptodactylus* (Amphibia, Leptodactylidae). *Evolution* 23:421–428.

———. 1971. Descriptions of some tadpoles from Thailand. *Fieldiana Zool.* 58:83–91.

———. 1973a. Systematics of the *marmoratus* group of the frog genus *Leptodactylus* (Amphibia, Leptodactylidae). *Nat. Hist. Mus. Los Angeles Co. Contrib. Sci.* (251):1–50.

———. 1973b. Ecological interactions of frog larvae at a seasonal tropical location in Thailand. *J. Herpetol.* 7:337–361.

———. 1974. Niche measurements of frog larvae from a seasonal tropical location in Thailand. *Ecology* 55:651–656.

———. 1976a. Studies in larval amphibian habitat partitioning. *Smithsonian Contrib. Zool.* (242):1–27.

———. 1976b. The presumed tadpole of *Paratelmatobius lutzi* (Amphibia, Leptodactylidae). *Pap. Avul. Zool.* 30:133–135.

———. 1979. Annual variation in larval amphibian populations within a temperate pond. *J. Washington Acad. Sci.* 69:65–74.

———. 1983a. Variation and systematics of frogs of the genus *Cycloramphus* (Amphibia, Leptodactylidae). *Arq. Zool. Mus. Univ. São Paulo* 30:235–239.

———. 1983b. Notes on the frog genus *Cycloramphus* (Amphibia: Leptodactylidae), with descriptions on two new species. *Proc. Biol. Soc. Washington* 96:548–559.

———. 1985. Taxonomic and natural history notes on frogs of the genus *Centrolenella* (Amphibia: Centrolenidae) from southeastern Brasil and adjacent Argentina. *Pap. Avul. Zool.* 36:1–21.

Heyer, W. R., M. A. Donnelly, R. W. McDiarmid, L.-A. C. Hayek, and M. S. Foster, eds. 1994. *Measuring and monitoring biological diversity. Standard methods for amphibians.* Biological Diversity Handbook Series. Washington, D.C.: Smithsonian Institution Press.

Heyer, W. R., R. W. McDiarmid, and D. L. Weigmann. 1975. Tadpoles, predation, and pond habitats in the tropics. *Biotropica* 7:100–111.

Heyer, W. R., A. S. Rand, C. A. G. Da Cruz, O. L. Peixoto, and C. E. Nelson. 1990. *Frogs of Boracéia. Arq. Zool., Mus. Zool., Univ. São Paulo* 31:231–410.

Heyer, W. R., and P. A. Silverstone. 1969. The larva of the frog *Leptodactylus hylaedactylus* (Leptodactylidae). *Fieldiana Zool.* 51:141–145.

Hibiya, T. 1982. *An atlas of fish histology. Normal and pathological features.* Stuttgart: Gustav Fischer-Verlag.

Higgins, G. M. 1920. The nasal organ in Amphibia. *Illinois Biol. Monogr.* 6:3–90, 6 plates.

Hillis, D. M. 1982. Morphological differentiation and adaptation of the larvae of *Rana berlandieri* and *Rana sphenocephala* (*Rana pipiens* complex) in sympatry. *Copeia* 1982:168–174.

Hillis, D. M., and R. O. De Sá. 1988. Phylogeny and taxonomy of the *Rana palmipes* group (Salientia: Ranidae). *Herpetol. Monogr.* (2):1–26.

Hillis, D. M., and J. S. Frost. 1985. Three new species of leopard frogs (*Rana pipiens* complex) from the Mexican Plateau. *Occas. Pap. Mus. Nat. Hist., Univ. Kansas* (117):1–14.

Hinckley, M. H. 1880. Notes on the eggs and tadpoles of *Hyla versicolor. Proc. Boston Soc. Nat. Hist.* 21:104–107, plate 3.

———. 1881. On some differences in the mouth structure of the anourous batrachians found in Milton, Mass. *Proc. Boston Soc. Nat. Hist.* 21:307–315, plate 5.

———. 1882. Notes on the development of *Rana sylvatica* Leconte. *Proc. Boston Soc. Nat. Hist.* 22:104–107.

———. 1884. Notes on the peeping frog, *Hyla pickeringii,* Leconte. *Mem. Boston Soc. Nat. Hist.* 3:311–318, plate 28.

Hirakow, R. 1989. Origin and differentiation of the chordate heart. pp. 261–263. In *Trends in vertebrate morphology. Progr. Zool. 35,* edited by H. Splechtna and H. Hilgers. Stuttgart: Gustav Fischer-Verlag.

Hirakow, R., S. Komazaki, and T. Hiruma. 1987. Early cardiogenesis in the newt embryo. *Scan. Micros.* 1:1367–1376.

Hirakow, R., and Y. Sugi. 1990. Intercellular junction and cytoskeletal organization in embryonic chick myocardial cells. pp. 95–113. In *Developmental cardiology: Morphogenesis and function,* edited by E. B. Clark and A. Takao. Mount Kisco, NY: Futura.

Hird, D. W., S. L. Diesch, R. G. McKinnell, E. Gorham, F. B. Martin, S. W. Kurtz, and C. Dubrovolny. 1981. *Aeromonas*

hydrophila in wild-caught frogs and tadpoles (*Rana pipiens*) in Minnesota. *Lab. Anim. Sci.* 31:166–169.
Hirsch, G. C. 1938. Die Nieren der Amphibien. pp. 775–805. In *Handbuch der vergleichenden Anatomie der Wirbeltiere,* Vol. 5, edited by E. B. Clark and A. Takao. Berlin: Urban and Schwarzenberg.
Hisaoka, K. K., and J. C. List. 1957. The spontaneous occurrence of scoliosis in larvae of *Rana sylvatica. Trans. American Micros. Soc.* 76:381–387.
Hochachka, P. W. 1988. Metabolic suppression and oxygen availability. *Canadian J. Zool.* 66:152–158.
Hochachka, P. W., and G. N. Somero. 1984. *Biochemical adaptation.* Princeton: Princeton University Press.
Hoegh-Guldberg, O., and R. B. Emlet. 1997. Energy use during the development of a lecithotrophic and a planktotrophic echinoid. *Biol. Bull.* 192:27–30.
Hoff, K. 1986a. Undulatory swimming in anuran larvae without waves of muscular contraction. *J. Biomech.* 19:478–479.
———. 1986b. Morphological correlates of fast-start performance in anuran larvae. *American Zool.* 26:132. (Abstr.)
———. 1988a. Evasive maneuvers of frogs and tadpoles in complex environments. *American Zool.* 28:155 (Abstr.).
———. 1988b. Morphological determinants of fast-start performance in anuran tadpoles. Dissertation, Dalhousie University.
———. 1989. Changes in evasive maneuvers during metamorphosis in bullfrogs. *American Zool.* 29:43 (Abstr.).
Hoff, K., and D. Ingle. 1988. Frogs have short term memory of obstacles. *Soc. Neurosci. Abstr.* 14:692.
Hoff, K. vS., V. A. King, N. Huq, and R. J. Wassersug. 1989. Kinematics of larval salamander swimming (Ambystomatidae). *Canadian J. Zool.* 67:2756–2761.
Hoff, K., and R. Wassersug. 1985. Do tadpoles swim like fishes? pp. 31–34. In *Functional morphology in vertebrates. Fortschrit. Zool. 30,* edited by H.-R. Duncker and G. Fleischer. Stuttgart: Gustav Fischer-Verlag.
———. 1986. The kinematics of swimming in larvae of the clawed frog, *Xenopus laevis. J. Exp. Biol.* 122:1–12.
Hofmann, M. H., and D. L. Meyer. 1989. Central projections of the nervus terminalis in four species of amphibians. *Brain Behav. Evol.* 34:301–307.
Hokit, D. G., and A. R. Blaustein. 1994. The effects of kinship on growth and development in tadpoles of *Rana cascadae. Evolution* 48:1383–1388.
———. 1995. Predator avoidance and alarm-response behaviour in kin-discriminating tadpoles (*Rana cascadae*). *Ethology* 101:280–291.
———. 1997. The effects of kinship on interaction between tadpoles of *Rana cascadae. Ecology* 78:1722–1735.
Holder, N.J., I. D. W. Clarke, and D. Tonge. 1987. Pathfinding by dorsal column axons in the spinal cord of the frog tadpole. *Development* 99:577–587.
Hollyfield, J. G. 1966. The origin of erythroblasts in *Rana pipiens* tadpoles. *Develop. Biol.* 14:461–480.
Holomuzki, J. R. 1998. Grazing effects by green frog tadpoles (*Rana clamitans*) in a woodland pond. *J. Freshw. Ecol.* 13:1–8.
Honda, E., T. Hirakawa, and S. Nakamura. 1992. Tactile and taste sensory maturation in the frog larvae tongue. *Brain Res.* 595:74–78.
Honegger, R. E., C. Schneider, and E. Zimmerman. 1985. Notizen zur aufzucht von Schmuckhornfroschen *Ceratophrys ornata* (Bell 1843). *Salamandra* 21:70–83.
Hoogmoed, M. S., and J. E. Cadle. 1991. Natural history and distribution of *Agalychnis craspedopus* (Funkhouser 1957) (Amphibia: Anura: Hylidae). *Zool. Med.* 65:129–142.
Hoppe, D. M. 1978. Thermal tolerance in tadpoles of the chorus frog *Pseudacris triseriata. Herpetologica* 34:318–321.
Hora, S. L. 1923. Further observations on the oral apparatus of the tadpoles of the genus *Megophrys. Rec. Indian Mus.* 30:139–145.
Horat, P., and R. D. Semlitsch. 1994. Effects of predation risk and hunger on the behaviour of two species of tadpoles. *Behav. Ecol. Sociobiol.* 34:393–401.
Horiuchi, S., and Y. Koshida. 1989. Effects of foodstuffs on intestinal length in larvae of *Rhacophorus arboreus* (Anura: Rhacophoridae). *Zool. Sci.* 6:321–328.
Horseman, N. D., A. H. Meier, and D. D. Culley, Jr. 1976. Daily variations in the effects of disturbance on growth, fattening and metamorphosis in the bullfrog (*Rana catesbeiana*) tadpole. *J. Exp. Zool.* 198:353–357.
Horton, J. D. 1971a. Ontogeny of the immune system in amphibians. *American Zool.* 11:219–228.
———. 1971b. Histogenesis of the lymphomyeloid complex in the larval leopard frog, *Rana pipiens. J. Morphol.* 134:1–20.
Horton, P. 1983. Reproductive system. pp. 84–92. In *The gastric brooding frog,* edited by M. J. Tyler. Beckenham: Croom Helm, Ltd.
Hosoi, M., S. Niida, Y. Yoshiko, S. Suemune, and N. Maeda. 1995. Scanning electron microscopy of horny teeth in the anuran tadpole Rhacophoridae, *Rhacophorus arboreus* and *Rhacophorus schlegelii. J. Electron Micros.* 44:351–357.
Hota, A. K., and M. C. Dash. 1981. Growth and metamorphosis of *Rana tigrina* larvae: Effects of food level and larval density. *Oikos* 37:349–352.
———. 1983. Evidence of interspecific predation among larval anurans: Predation of *Rana tigrina* larvae on *Bufo melanostictus* larvae. *Biol. Bull. India* 5:54–55.
———. 1986. Growth and metamorphosis of anuran larvae: Effect of diet and temperature. *Alytes* 5:165–172.
Hourdry, J. 1974. Étude des branchies "internes" puis de leur régression au moment de la métamorphose, chez la larve de *Discoglossus pictus* (Otth), amphibien anoure. *J. Micros.* 20:165–182.
———. 1993. Passage to the terrestrial life in amphibians. 2. Endocrine determinism. *Zool. Sci.* 10:715–731.
Hourdry, J., and R. Guyetant. 1979. Group breeding effect on the iodine concentration in the colloid of thyroid follicles of *Alytes obstetricans* tadpole, and evaluation of the iodine content of the thyroid colloid. *C.R. Hebd. Seances Acad. Sci. Paris* 289D:323–326.
Hourdry, J., L. Hermite, and R. Ferrand. 1996. Changes in the digestive tract and feeding behavior of anuran amphibians during metamorphosis. *Physiol. Zool.* 69:219–251.
Hower, R. O. 1967. The freeze-dry preservation of biological specimens. *Proc. United States Natl. Mus.* 119:1–24.
Hoyer, M. H. 1905a. Untersuchungen über das Lymphgefässystem der Froschlarven. I. Teil. *Bull. Acad. Cracov.* 1905:417–430.
———. 1905b. Untersuchungen über das Lymphgefässystem der Froschlarven. II. Teil. *Bull. Acad. Cracov.* 1905:451–464.
Hoyer, W. J., J. N. Shafer, J. E. Mauldin, and H. T. Corbett. 1971. Discriminated avoidance and escape conditioning with the tadpole (*Rana pipiens*). *Psychon. Sci.* 24:247–248.
Hrbâçek, J. 1950. On the flight reaction of tadpoles of the common toad caused by chemical substances. *Experientia* 6:100–101.

Huey, R. B. 1980. Sprint velocity of tadpoles (*Bufo boreas*) through metamorphosis. *Copeia* 1980:537–540.
Huey, R. B., and R. D. Stevenson. 1979. Integrating thermal physiology and ecology of ectotherms: A discussion of approaches. *American Zool.* 19:357–366.
Huey, R. B., and T. P. Webster. 1976. Thermal biology of *Anolis* lizards in a complex fauna: The *cristatellus* group on Puerto Rico. *Ecology* 57:985–994.
Hughes, A. 1957. The development of the primary sensory system in *Xenopus laevis* (Daudin). *J. Anat.* 91:323–338.
———. 1959. Studies in embryonic and larval development in Amphibia. I. The embryology of *Eleutherodactylus ricordii* with special reference to the spinal cord. *J. Embryol. Exp. Morphol.* 7:22–38.
———. 1961. Cell degeneration in the larval ventral horn of *Xenopus laevis* (Daudin). *J. Embryol. Exp. Morphol.* 9:269–284.
———. 1962. An experimental study of the relationships between limb and spinal cord of the embryo of *Eleutherodactylus martinicensis. J. Embryol. Exp. Morphol.* 10:575–601.
———. 1966. The thyroid and the development of the nervous system in *Eleutherodactylus martinicensis:* An experimental study. *J. Embryol. Exp. Morphol.* 16:401–430.
———. 1976. Metamorphic changes in the brain and spinal cord. pp. 856–863. In *Frog neurobiology, a handbook,* edited by R. Llinás and W. Precht. Berlin: Springer-Verlag.
Hughes, A., and M. C. Prestige. 1967. Development of behaviour in the hindlimb of *Xenopus laevis. J. Zool.* 152:347–359.
Hughes, A., and P. Reier. 1972. A preliminary study of the effects of bovine prolactin on embryos of *Eleutherodactylus ricordii. Gen. Comp. Endocrinol.* 19:304–312.
Hulsebus, J., and E. S. Farrar. 1985. Insulin-like immunoreactivity in serum and pancreas of metamorphosing tadpoles. *Gen. Comp. Endocrinol.* 58:114–119.
Hurley, M. B. 1958. The role of the thyroid in kidney development in the tadpole. *Growth* 22:125–166.
Huschke, E. 1826. Über die Umbildung des Darmkanals und der Kiemen der Froschkaulquappen. *Isis von Oken:* 613–627.
Hutchinson, T. 1796. *The natural history of the frog fish of Surinam.* 8 p., 4 plates. York: G. Peacock.
Hutchison, V. H., and R. K. Dupré. 1992. Thermoregulation. pp. 206–249. In *Environmental physiology of the amphibians,* edited by M. E. Feder and W. W. Burggren. Chicago: University of Chicago Press.
Hutchison, V. H., and C. G. Hill. 1978. Thermal selection of bullfrog tadpoles (*Rana catesbeiana*) at different stages of development and acclimation temperatures. *J. Therm. Biol.* 3:57–60.
Hutchison, V. H., and J. D. Maness. 1979. The role of behavior in temperature acclimation and tolerance in ectotherms. *American Zool.* 19:367–384.
Huxley, J. 1942. *Evolution: The modern synthesis.* New York: Harper and Brothers.
Huxley, T. H. 1858. On the theory of the vertebrate skull. *Proc. Roy. Soc. London* 9:381–457.
Icardo, J. M. 1997. Morphogenesis of vertebrate hearts. pp. 114–126. In *Development of cardiovascular systems. Molecules to organisms,* edited by M. E. Feder and W. W. Burggren. Cambridge: Cambridge University Press.
Ichizuya-Oka, A., and A. Shimozawa. 1987. Development of the connective tissue in the digestive tract of the larval and metamorphosing *Xenopus laevis. Anat. Anz.* 164:81–93.
Ideker, J. 1976. Tadpole thermoregulatory behavior facilitates grackle predation. *Texas J. Sci.* 27:244–245.
Ijiri, K. L., and N. Egami. 1975. Mitotic activity of germ cells during normal development of *Xenopus laevis* tadpoles. *J. Embryol. Exp. Morphol.* 34:687–694.
Ikenishi, K. 1980. Cortical granules persisting in germ cells of *Xenopus laevis* after the cleavage stages. *Develop. Growth Differ.* 22:669–678.
———. 1982. A possibility of an in vitro differentiation of primordial germ cells (PGCs) from blastomeres containing "germinal plasm" of early cleavage stage in *Xenopus laevis. Develop. Growth Differ.* 24:205–215.
Ikenishi, K., and M. Kotani. 1975. Ultrastructure of the "germinal plasm" in *Xenopus* embryos after cleavage. *Develop. Growth Differ.* 17:101–110.
Ikenishi, K., M. Kotani, and K. Tanabe. 1974. Ultrastructural changes associated with UV irradiation in the "germinal plasm" of *Xenopus laevis. Develop. Biol.* 36:155–168.
Infantino, R. L., Jr. 1989. Ontogeny of gill and lung ventilatory responses to oxygen and carbon dioxide in the bullfrog, *Rana catesbeiana. American Zool.* 29:57. (Abstr.)
Inger, R. F. 1954. Systematics and zoogeography of Philippine Amphibia. *Fieldiana Zool.* 34:389–424.
———. 1956. Some amphibians from the lowlands of North Borneo. *Fieldiana Zool.* 34:389–424.
———. 1966. The systematics and zoogeography of the Amphibia of Borneo. *Fieldiana Zool.* 52:1–402.
———. 1967. The development of a phylogeny of frogs. *Evolution* 21:369–384.
———. 1969. Organization of communities of frogs along small rainforest streams in Sarawak. *J. Anim. Ecol.* 38:123–148.
———. 1983. Larvae of southeast Asian species of *Leptobrachium* and *Leptobrachella* (Anura: Pelobatidae). pp. 13–32. In *Advances in herpetology and evolutionary biology,* edited by A. Rhodin and K. Miyata. Cambridge, MA: Mus. Comp. Zool.
———. 1985. Tadpoles of the forested regions of Borneo. *Fieldiana Zool. (N.S.)* (26):1–89.
———. 1986. Diets of tadpoles living in a Bornean rain forest. *Alytes* 5:153–164.
———. 1992a. Variation of apomorphic characters in stream-dwelling tadpoles of the bufonid genus *Ansonia* (Amphibia, Anura). *Zool. J. Linnean Soc.* 105:225–237.
———. 1992b. A bimodal feeding system in a stream-dwelling larva of *Rhacophorus* from Borneo. *Copeia* 1992:887–890.
———. 1996. Commentary on a proposed classification of the family Ranidae. *Herpetologica* 52:241–246.
Inger, R. F., and K. J. Frogner. 1979. New species of narrow-mouth frogs (genus *Microhyla*) from Borneo. *Sarawak Mus. J.* 27:311–322.
Inger, R. F., and P. A. Gritis. 1983. Variation in Bornean frogs of the *Amolops jerboa* species group, with descriptions of two new species. *Fieldiana Zool. (N.S.)* (19):1–13.
Inger, R. F., and F. L. Tan. 1990. Recently discovered and newly assigned frog larvae (Ranidae and Rhacophoridae) from Borneo. *Raffles Bull. Zool.* 38:3–9.
Inger, R. F., and H. K. Voris. 1988. Taxonomic status and reproductive biology of Bornean tadpole-carrying frogs. *Copeia* 1988:1060–1061.
Inger, R. F., H. K. Voris, and K. J. Frogner. 1986a. Organization of a community of tadpoles in rain forest streams in Borneo. *J. Trop. Ecol.* 2:193–205.
Inger, R. F., H. K. Voris, and P. Walker. 1986b. Larval transport in a Bornean ranid frog. *Copeia* 1986:523–525.

Inger, R. F., and R. J. Wassersug. 1990. A centrolenid-like anuran larvae from southeast Asia. *Zool. Sci.* 7:557–561.

Ingerman, R. L. 1992. Maternal-fetal oxygen transfer in lower vertebrates. *American Zool.* 32:322–330.

Ingle, D. 1976. Behavioral correlates of central visual function in anurans. pp. 435–451. In *Frog neurobiology, a handbook,* edited by R. Llinás and W. Precht. Berlin: Springer-Verlag.

Ingram, G. J., M. Anstis, and C. J. Corben. 1975. Observations on the Australian leptodactylid frog, *Assa darlingtoni. Herpetologica* 31:425–429.

Ingram, V. M. 1972. Embryonic red blood cell formation. *Nature* 335–339.

Inokuchi, T., K. I. Kobayashi, and S. Horiuchi. 1991. Acid proteinases of the fore-gut in metamorphosing tadpoles of *Rana catesbeiana. Comp. Biochem. Physiol.* 99B:653–662.

Ishizuya-Oka, A., and A. Shimozawa. 1990. Changes in lectin-binding pattern in the digestive tract of *Xenopus laevis* during metamorphosis. II. Small intestine. *J. Morphol.* 205:9–15.

Ison, R. E. 1968. The development of the heart of *Rana temporaria* L. *J. Nat. Hist.* 2:449–457.

Istock, C. A. 1967. The evolution of complex life cycle phenomena: An ecological perspective. *Evolution* 21:592–605.

———. 1984. Boundaries to life history variation and evolution. pp. 143–168. In *A new ecology: Novel approaches to interactive systems,* edited by P. W. Price, C. N. Slobodchikoff, and W. S. Gaud. New York: Wiley.

Iwasawa, H., and M. Kobayashi. 1976. Development of the testis in the frog *Rana nigromaculata,* with special reference to germ cell maturation. *Copeia* 1976:461–467.

Iwasawa, H., and K. Yamaguchi. 1984. Ultrastructural study of gonadal development in *Xenopus laevis. Zool. Sci.* 1:591–600.

Izecksohn, E. 1997. Novo enero de Hylidae Brasileiro (Amphibia, Anura). *Rev. Univ. Rural, Sér. Ciênc. Vida* 18:47–52.

Izecksohn, E., and C. A. G. Da Cruz. 1972. Notas sobre os girinos de *Dendrophryniscus leucomystax* Izecksohn e *D. brevipollicatus* Espada (Amphibian, Anura, Bufonidae). *Arq. Univ. Fed. Rural Rio de Janeiro* 2:63–69.

———. 1976. Nova espécies de Phyllomedusinae do Estado do Espírito Santo, Brasil (Amphibia, Anura, Hylidae). *Rev. Brasil. Biol.* 36:257–261.

Izecksohn, E., C. A. G. Da Cruz, and O. L. Peixoto. 1979. Notas sobre o girino de *Proceratophrys boiei* (Wied)(Amphibia, Anura, Leptodactylidae). *Rev. Brasil. Biol.* 39:233–236.

Izecksohn, E., and S. T. De Albuquerque. 1972. Algumas observações sobre o desenvolvimento de *Eleutherodactylus venancioi* B. Lutz (Amphibia, Anura). *Arq. Univ. Fed. Rural Rio de Janeiro* 2:13–15.

Izecksohn, E., J. Jim, S. T. De Albuquerque, and W. F. De Mendonça. 1971. Observações sôbre o desenvolvimento e os hábitos de *Myersiella subnigra* (Miranda-Ribeiro). *Arq. Mus. Nac. Rio de Janeiro* 54:69–73.

Jacobson, A. G., and J. T. Duncan. 1968. Heart induction in salamanders. *J. Exp. Zool.* 167:79–103.

Jacobson, C. M. 1968. The development of the chondrocranium in two species of the Australian anuran genus *Pseudophryne* Fitzinger. *Australian J. Zool.* 16:1–15.

Jacoby, J., and K. Rubinson. 1983. The acoustic and lateral line nuclei are distinct in the premetamorphic frog, *Rana catesbeiana. J. Comp. Neurol.* 216:152–161.

Jaeger, C. B., and D. E. Hillman. 1976. Morphology of gustatory organs. pp. 588–606. In *Frog neurobiology, a handbook,* edited by R. Llinás and W. Precht. Berlin: Springer-Verlag.

Jaeger, R. G., and J. P. Hailman. 1976. Ontogenetic shift of spectral phototactic preferences in anuran tadpoles. *J. Comp. Physiol. Psychol.* 90:930–945.

Jaeger, R. G., and S. C. Walls. 1989. On salamander guilds and ecological methodology. *Herpetologica* 45:111–119.

Jaffee, O. C. 1954a. Morphogenesis of the pronephros of the leopard frog (*Rana pipiens*). *J. Morphol.* 95:109–123.

———. 1954b. Phenol red transport in the pronephros and mesonephros of the developing frog (*Rana pipiens*). *J. Cell. Comp. Physiol.* 44:347–364.

———. 1963. Cellular differentiation in the anuran pronephros. *Anat. Rec.* 145:179–181.

Jägersten, G. 1972. *Evolution of the metazoan life cycle: A comprehensive theory.* London: Academic Press.

Jande, S. S. 1966. Fine structure of lateral-line organs of frog tadpoles. *J. Ultrastruct. Res.* 15:496–509.

Janes, R. G. 1934. Studies on the amphibian digestive system. I. Histological changes in the alimentary tract of anuran larvae during involution. *J. Exp. Zool.* 67:73–91.

Jasieński, M. 1988. Kinship ecology of competition: Size hierarchies in kin and nonkin laboratory cohorts of tadpoles. *Oecologia* 77:407–413.

Jaslow, A. P., T. E. Hetherington, and R. E. Lombard. 1988. Structure and function of the amphibian middle ear. pp. 69–91. In *The evolution of the amphibian auditory system,* edited by B. Fritzsch, M. J. Ryan, W. Wilczynski, T. E. Hetherington, and W. Walkowiak. New York: Wiley.

Jayatilaka, A. D. P. 1978. An ultrastructural study of the thyroid gland in the pre-metamorphic *Xenopus laevis* (Daudin) tadpole. *J. Anat.* 125:579–591.

Jennings, D. H. 1994. Thyroid hormone mediation of embryonic development in a non-metamorphosing frog, *Eleutherodactylus coqui. J. Morphol.* 220:250.

Jennings, D. H., and J. Hanken. 1998. Mechanistic basis of life history evolution in anuran amphibians: Thyroid gland development in the direct-developing frog, *Eleutherodactylus coqui. Gen. Comp. Endocrinol.* 111:225–32.

Jennings, R. D., and N.J. Scott, Jr. 1993. Ecologically correlated morphological variation in tadpoles of the leopard frog, *Rana chiricahuensis. J. Herpetol.* 27:285–293.

Jenssen, T. A. 1967. Food habits of the green frog, *Rana clamitans,* before and during metamorphosis. *Copeia* 1967:214–218.

Jia, X. X., and W. W. Burggren. 1997a. Developmental changes in chemoreceptive control of gill ventilation in larval bullfrogs (*Rana catesbeiana*). I. Reflex ventilatory responses to ambient hyperoxia, hypoxia, and NaCn. *J. Exp. Biol.* 200:2229–2236.

———. 1997b. Developmental changes in chemoreceptive control of gill ventilation in larval bullfrogs (*Rana catesbeiana*). II. Site of O_2-sensitive chemoreceptors. *J. Exp. Biol.* 200:2237–2248.

Jiang, T., and A. Holley. 1992. Morphological variations among output neurons of the olfactory bulb in the frog (*Rana ridibunda*). *J. Comp. Neurol.* 320:86–96.

Johansen, K., and C. Lenfant. 1972. A comparative approach to the adaptability of O_2-HB affinity. pp. 750–783. In *Oxygen affinity of hemoglobin and red cell acid base status. Alfred Benzon Symp.,* edited by M. Rorth and P. Astrup. New York: Academic Press.

John, K. R., and D. Fenster. 1975. The effects of partitions on the growth rates of crowded *Rana pipiens* larvae. *American Midl. Nat.* 93:123–130.

John, K. R., and J. M. Fusaro. 1981. Growth and metamorphosis of solitary *Rana pipiens* tadpoles in confined space. *Copeia* 1981:737–741.

John-Alder, H. B., and P. J. Morin. 1990. Effects of larval density on jumping ability and stamina in newly metamorphosed *Bufo woodhousii fowleri*. *Copeia* 1990:856–860.

Johnson, D. S. 1967. On the chemistry of freshwaters in southern Malaya and Singapore. *Arch. Hydrobiol.* 63:477–496.

Johnson, L. M. 1991. Growth and development of larval northern cricket frogs (*Acris crepitans*) in relation to phytoplankton abundance. *Freshw. Biol.* 25:51–59.

Johnston, G. F. 1982. Functions of the keratinized oral features in anuran tadpoles with an analysis of ecomorphological tadpole types. Thesis, Mississippi State University.

———. 1990. Feeding function and anatomy of the oral apparatus of anuran larvae. Dissertation, Southern Illinois University.

Johnston, G. F., and R. Altig. 1986. Identification characteristics of anuran tadpoles. *Herpetol. Rev.* 17:36–37.

Johnston, L. M. 1991. Growth and development of larval northern cricket frogs (*Acris crepitans*) in relation to phytoplankton abundance. *Freshw. Biol.* 25:51–59.

Jolivet-Jaudet, G., and J. Leloup-Hatey. 1984. Variations in aldosterone and corticosterone plasma levels during metamorphosis in *Xenopus laevis* tadpoles. *Gen. Comp. Endocrinol.* 56:59–65.

Jollie, M. 1982. Ventral branchial musculature and synapomorphies questioned. *Zool. J. Linnean Soc.* 74:35–47.

Jonas, L., and P. Röhlich. 1970. Elektronenmikroskopischer Nachweis saurer Mucopolysaccharide in den Flaschenzellen der *Xenopus*-Niere. *Zeit. Zellforsch. Mikros. Anat.* 104:56–68.

Jonas, L., and L. Spannhof. 1971. Elektronenmikroskopischer Nachweis von Mucoproteiden in den Flaschenzellen der Urniere von *Xenopus laevis* Daudin mittels Phosphorwolframsäurefärbung. *Acta Histochem.* 41:185–192.

Jones, R. E., A. M. Gerrard, and J. J. Roth. 1973. Estrogen and brood pouch formation in the marsupial frog *Gastrotheca riobambae*. *J. Exp. Zool.* 184:177–184.

Jones, R. M. 1980. Nitrogen excretion by *Scaphiopus* tadpoles in ephemeral ponds. *Physiol. Zool.* 53:26–31.

Jordan, H. E. 1933. The evolution of blood-forming tissues. *Quart. Rev. Biol.* 8:58–76.

Jørgensen, C. B. 1973. Mechanisms regulating ovarian function in amphibians (toads). pp. 133–151. In *The development and maturation of the ovary and its functions,* edited by H. Peters. Amsterdam: Excerpta Medica.

———. 1986. Effects of fat body excision in female *Bufo bufo* on the ipsilateral ovary, with a discussion of fat body-gonad relationships. *Acta Zool.* 67:5–10.

Jørgensen, C. B., L. O. Larsen, and B. Lofts. 1979. Annual cycles of fat bodies and gonads in the toad *Bufo bufo* (L.), compared with cycles in other temperate zone anurans. *Biol. Skr. (Copenhagen)* 22:1–27.

Jorquera, B., O. Garrido, and E. Pugín. 1982. Comparative studies of the digestive tract development between *Rhinoderma darwinii* and *R. rufum*. *J. Herpetol.* 16:204–214.

Jorquera, B., and L. Izquierdo. 1964. Tabla de desarrollo normal de *Calyptocephalella gayi* (Rana chilena). *Biologica* 36:43–53.

Jorquera, B., E. Pugín, and O. Y. Goicoechea. 1972. Tabla de desarrollo normal de *Rhinoderma darwinii*. *Arch. Med. Vet.* 4:5–19.

Joseph, B. S., and D. G. Whitlock. 1968. Central projections of selected spinal dorsal roots in anuran amphibians. *Anat. Rec.* 160:279–288.

Juncá, F. A., R. Altig, and C. Gascon. 1994. Breeding biology of *Colostethus stepheni,* a dendrobatid frog with a nontransported nidicolous tadpole. *Copeia* 1994:747–750.

Jungfer, K.-H. 1985. Beitrag zur Kenntnis von *Dendrobates speciosus* O. Schmidt, 1857 (Salientia: Dendrobatidae). *Salamandra* 21:263–280.

———. 1996. Reproduction and parental care of the coronated treefrog, *Anotheca coronata* (Steindachner 1864)(Anura: Hylidae). *Herpetologica* 52:25–32.

Jungfer, K.-H., and L. C. Schiesari. 1995. Description of a central Amazonian and Guianan tree frog, genus *Osteocephalus* (Anura, Hylidae), with oophagous tadpoles. *Alytes* 13:1–13.

Junqueira, L. C. U., P. P. Joazeiro, S. M. O. Vieira, A. B. Costa Silva, A. Cais, and G. S. Montes. 1984. Specific attachment of morphologically-distinct intermediate filaments to desmosomes and hemidesmosomes in the epidermis of the tadpole of the anuran, *Pseudis paradoxus*. *J. Submicros. Cytol.* 16:643–648.

Jurgens, J. D. 1971. The morphology of the nasal region of Amphibia and its bearing on the phylogeny of the group. *Ann. Univ. Stellenbosch* 46A:1–146.

Just, J. J., and B. G. Atkinson. 1972. Hemoglobin transitions in the bullfrog, *Rana catesbeiana,* during spontaneous and induced metamorphosis. *J. Exp. Zool.* 182:271–280.

Just, J. J., R. N. Gatz, and E. C. Crawford. 1973. Changes in respiratory functions during metamorphosis of the bullfrog, *Rana catesbeiana*. *Respir. Physiol.* 17:276–282.

Just, J. J., R. Sperka, and S. Strange. 1977. A quantitative analysis of plasma osmotic pressure during metamorphosis of the bullfrog, *Rana catesbeiana*. *Experientia* 33:1503–1504.

Justis, C. S., and D. H. Taylor. 1976. Extraocular photoreception and compass orientation in larval bullfrogs, *Rana catesbeiana*. *Copeia* 1976:98–105.

Kaiser, H., and R. Altig. 1994. The atypical tadpole of the dendrobatid frog, *Colostethus chalcopis,* from Martinique, French Antilles. *J. Herpetol.* 28:374–378.

Kalt, M. R. 1973a. Ultrastructural observations on the germ line of *Xenopus laevis*. *Zeit. Zellforsch. Mikros. Anat.* 138:41–62.

———. 1973b. Morphology and kinetics of spermatogenesis in *Xenopus laevis*. *J. Exp. Zool.* 195:393–408.

Kalt, M. R., and J. G. Gall. 1974. Observations on early germ cell development and premeiotic ribosomal DNA amplification in *Xenopus laevis*. *J. Cell Biol.* 62:460–472.

Kam, Y.-C., Z.-C. Chuang, and C.-F. Yen. 1996. Reproduction, oviposition-site selection, and tadpole ecology of an arboreal nester, *Chirixalus eiffingeri* (Rhacophoridae), from Taiwan. *J. Herpetol.* 30:52–59.

Kamat, N. D. 1962. On the intestinal content of tadpoles and algae of small ponds. *Curr. Sci.* 321:300–310.

Kamimura, M., K. Ikenishi, M. Kotani, and T. Matsuno. 1976. Observations on the migration and proliferation of gonocytes in *Xenopus laevis*. *J. Embryol. Exp. Morphol.* 36:197–207.

Kamimura, M., M. Kotani, and K. Yamagata. 1980. The migration of presumptive primordial germ cells through the endodermal cell mass in *Xenopus laevis*. A light and electron microscopic study. *J. Embryol. Exp. Morphol.* 59:1–17.

Kampmeier, O. F. 1915. On the origin of lymphatics in *Bufo*. *American J. Anat.* 17:161–209.

———. 1922. The development of the anterior lymphatics and lymph hearts in anuran embryos. *American J. Anat.* 30:61–131.

———. 1958. On the lymphatics system of *Ascaphus:* Its evolutionary significance. *Anat. Rec.* 132:343–364.

———. 1969. *Evolution and comparative morphology of the lymphatic system.* Springfield, IL: Charles C. Thomas.
Kanamadi, R. D., and S. K. Saidapur. 1988. Effect of fat body ablation on spermatogenesis in the Indian skipper frog *Rana cyanophlyctis* Schn. *Zool. Anz.* 220:79–81.
Kaplan, M. 1994. A new species of frog of the genus *Hyla* from the Cordillera Oriental in northern Colombia with comments on the taxonomy of *Hyla minuta. J. Herpetol.* 28:79–87.
Kaplan, R. H. 1992. Greater maternal investment can decrease offspring survival in the frog *Bombina orientalis. Ecology* 73:280–288.
Karlson, P. 1972. *Kurzes Lehrbuch der Biochemie.* Stuttgart: Georg Thieme-Verlag.
Karlstrom, E. L. 1962. The toad genus *Bufo* in the Sierra Nevada of California: Ecological and systematic relationships. *Univ. California Publ. Zool.* 62:1–104.
Karnovsky, M. J. 1965. A formaldehyde-glutaraldehyde fixative of high osmolality for use in electron microscopy. *J. Cell. Biol.* 27A:137–138.
Kasinathan, S., A. Gunashing, and S. L. Basu. 1978. Fat bodies and spermatogenesis in South Indian green frog *Rana hexadactyla* Lesson. *Boll. Zool.* 45:15–22.
Kats, L. B., J. W. Petranka, and A. Sih. 1988. Antipredator defenses and the persistence of amphibian larvae with fishes. *Ecology* 69:1865–1870.
Katz, L. C., M. J. Potel, and R. J. Wassersug. 1981. Structure and mechanisms of schooling in tadpoles of the clawed frog, *Xenopus laevis. Anim. Behav.* 29:20–33.
Katz, M. J. 1980. Allometry formula: A cellular model. *Growth* 44:89–96.
Kaung, H. C. 1975. Development of the horny jaws of *Rana pipiens* larvae. *Develop. Biol.* 11:25–49.
———. 1981. Immunocytochemical localization of pancreatic endocrine cells in frog embryos and young larvae. *Gen. Comp. Endocrinol.* 45:204–211.
———. 1983. Changes of pancreatic beta cell population during larval development in *Rana pipiens. Gen. Comp. Endocrinol.* 49:50–56.
Kaung, H. C., and J. J. Kollros. 1976. Cell turnover in the beak of *Rana pipiens. Anat. Rec.* 188:361–370.
Kawada, J., R. E. Taylor, Jr., and S. B. Barker. 1969. Measurement of Na-K-ATPase in the separated epidermis of *Rana catesbeiana* frogs and tadpoles. *Comp. Biochem. Physiol.* 30:965–975.
Kawai, A., J. Ibeya, and K. Yoshizato. 1994. A three-step mechanism of action of thyroid hormone and mesenchyme in metamorphic changes in anuran larval skin. *Develop. Biol.* 166:477–488.
Kawamura, T., and M. Kobayashi. 1960. Studies on hybridization in amphibians. VII. Hybrids between Japanese and European brown frogs. *J. Sci. Hiroshima Univ.* 18B:221–238.
Keating, M. J., and R. M. Gaze. 1970. The ipsilateral retinotectal pathway in the frog. *Quart. J. Exp. Physiol.* 55:284–292.
Kehr, A. I. 1989. Factores dependientes de la densidad y influencia sobre el crecimiento individual de los estados larvales de *Hyla pulchella pulchella* (Amphibia Anura). *Limnobios* 2:757–761.
———. 1997. Stage-frequency and habitat selection of a cohort of *Pseudacris ocularis* tadpoles (Hylidae: Anura) in a Florida temporary pond. *Herpetol. J.* 7:103–109.
Kehr, A. I., and E. O. Adema. 1990. Crecimento corporal y analises estadistico de la frequencia por clases de edades de los estadios larvales de *Bufo arenarum* en condiciones naturales. *Neotrópica* 36:67–81.
Kehr, A. I., and N. G. Basso. 1990. Description of the tadpole of *Lysapsus limellus* (Anura, Pseudidae) and some considerations on its biology. *Copeia* 1990:573–575.
Kehr, A. I., and J. A. Schnack. 1991. Predator-prey relationship between giant water bugs (*Belostoma oxyurum*) and larval anurans (*Bufo arenarum*). *Alytes* 9:61–69.
Kehr, A. I., and J. D. Williams. 1990. Larvas de anuros de la Republica Argentina. *Cuad. Herpetol. Monogr.* (2):1–44.
Keiffer, H. 1888. Recherches sur la structure et le développement des dents et du bec cornés chez *Alytes obstetricans. Arch. Biol.* 9:55–81.
Keller, R. 1991. Early embryonic development of *Xenopus laevis.* pp. 62–113. In *Xenopus laevis: Practical uses in cell and molecular biology. Meth. Cell Biol.* Vol. 36, edited by B. K. Kay and H. B. Peng. New York: Academic Press.
Kemali, M. 1974. Ultrastructural asymmetry of the habenular nuclei of the frog. J. Hirnforsch. 15:419–426.
Kemali, M., and V. Braitenburg. 1969. *Atlas of the frog's brain.* Berlin: Springer-Verlag.
Kemali, M., and V. Guglielmotti. 1977. An electron microscope observation of the right and the two left portions of the habenular nuclei of the frog. *J. Comp. Neurol.* 176:133–148.
———. 1987. A horseradish peroxidase study of the olfactory system of the frog, *Rana esculenta. J. Comp. Neurol.* 263:400–417.
Kemp, A. 1977. The pattern of tooth plate formation in the Australian lungfish, *Neoceratodus forsteri* (Krefft). *Zool. J. Linnean Soc. London* 60:223–258.
Kemp, N. E. 1949. Development of intestinal coiling in the anuran tadpole. *Anat. Rec.* 105:88. (Abstr.)
———. 1951. Development of intestinal coiling in anuran larvae. *J. Exp. Zool.* 116:259–287.
Kemp, N. E., and J. A. Hoyt. 1969. Sequence of ossification in the skeleton of growing and metamorphosing tadpoles of *Rana pipiens. J. Morphol.* 129:415–444.
Kenny, J. S. 1969a. The Amphibia of Trinidad. *Stud. Fauna Curaçao Caribbean Isl.* 29:1–78.
———. 1969b. Pharyngeal mucous secreting epithelia of anuran larvae. *Acta Zool.* 50:143–153.
———. 1969c. Feeding mechanisms in anuran larvae. *J. Zool.* 157:225–246.
Kenward, R. E. 1978. Hawks and doves: Factors affecting success and selection in goshawk attacks on woodpigeons. *J. Anim. Ecol.* 47:449–460.
Kerr, J. B., and K. E. Dixon. 1974. An ultrastructural study of germ plasm in spermatogenesis of *Xenopus laevis. J. Embryol. Exp. Morphol.* 32:573–592.
Kerr, J. G. 1905. The embryology of certain of the lower fishes, and its bearing upon vertebrate morphology. *Proc. Roy. Soc. Edinburgh* 1905–1906, Vol. 16.
———. 1919. *Textbook of embryology.* London: MacMillan Publ. Co.
Kessel, R. G., H. W. Beam, and C. Y. Shih. 1974. The origin, distribution and disappearance of surface cilia during embryonic development of *Rana pipiens* as revealed by scanning electron microscopy. *American J. Anat.* 141:341–360.
Kesteven, H. L. 1944. The evolution of the skull and the cephalic muscles: A comparative study of the development and adult morphology. Part II. The Amphibia. *Mem. Australian Mus. Sydney* 8:133–236.
Kett, N. A., and E. D. Pollack. 1985. Retention of lateral motor

column neurons during the phase of rapid cell loss after limb amputation in *Rana pipiens* tadpoles. *J. Exp. Zool.* 236:59–66.

Khan, M. S. 1982a. Key to the identification of amphibian tadpoles from the plains of Pakistan. *Pakistan J. Zool.* 14:133–145.

———. 1982b. Collection, preservation, and identification of amphibian eggs from the plains of Pakistan. *Pakistan J. Zool.* 14:241–243.

Khan, M. S., and S. A. Malik. 1987. Buccopharyngeal morphology of the tadpole of *Rana hazarensis* Dubois and Khan 1978, and its torrenticole adaptations. *Biologia (Lahore)* 33:45–60.

Khan, M. S., and S. A. Mufti. 1994. Oral disc morphology of amphibian tadpole and its functional correlates. *Pakistan J. Zool.* 26:25–30.

Kicliter, E. 1979. Some telencephalic connections in the frog, *Rana pipiens. J. Comp. Neurol.* 185:75–86.

Kicliter, E., and S. O. E. Ebbesson. 1976. Organization of the "non-olfactory" telecephalon. pp. 946–972. In *Frog neurobiology, a handbook,* edited by R. Llinás and W. Precht. Berlin: Springer-Verlag.

Kicliter, E., and R. G. Northcutt. 1975. Ascending afferents to the telencephalon of ranid frogs: An anterograde degeneration study. *J. Comp. Neurol.* 161:239–254.

Kiesecker, J. 1996. pH-mediated predator-prey interactions between *Ambystoma tigrinum* and *Pseudacris triseriata. Ecol. Appl.* 6:1325–1331.

Kilham, P. 1982. The biogeochemistry of bog ecosystems and the chemical ecology of *Sphagnum. Michigan Bot.* 21:159–168.

Kindahl, M. 1938. Zur Entwicklung der Exkretionsorgane von Dipnoërn und Amphibien mit Anmerkungen bezüglich Ganoiden und Teleostier. *Acta Zool.* 19:1–190.

Kindred, J. E. 1927. Cell division and ciliogenesis in the ciliated epithelium of the pharynx and esophagus of the tadpole of the green frog, *Rana clamitans. J. Morphol.* 43:267–297.

King, M. 1981. A cytotaxonomic analysis of Australian hylid frogs of the genus *Litoria. Proc. Herpetol. Symp., Melbourne* 169–175.

Kingsbury, B. F. 1895. The lateral line system of sense organs in some American Amphibia, and comparison with the dipnoans. *Trans. American Micros. Soc.* 17:115–154.

Kinney, E. C. 1964. Extent of acid mine pollution in the United States affecting fish and wildlife. *United States Fish Wildl. Circ.* 191. 27 p.

Kirtisinghe, P. 1957. *Amphibia of Ceylon.* Published privately, Colombo.

———. 1958. Some hitherto undescribed anuran tadpoles. *Ceylon J. Sci.* 1:171–176

Kiseleva, E. I. 1984. Responses of behavioral reactions in the *Rana temporaria* tadpoles to chemical stimuli under the conditions of pretreatment. *Zool. Zhur.* 63:1046–1054.

———. 1992. Amino acid sensitivity in tadpoles of the spadefoot toad *Pelobates fuscus. J. Evol. Biochem. Physiol.* 28:527–530.

———. 1997. Aspects of the chemical ecology and chemosensory guided behavioral reactions in anuran tadpoles. *Adv. Amphib. Res. Form. Soviet Union* 2:95–101.

Kiseleva, E. I., and Y. B. Manteifel. 1982. Behavioral reactions of *Bufo bufo* and *Rana temporaria* tadpoles to chemical stimuli. *Zool. Zhur.* 61:1669–1681.

Klein, S. L., and P. P. C. Graziadei. 1983. The differentiation of the olfactory placode in *Xenopus laevis:* A light and electron microscope study. *J. Comp. Neurol.* 217:17–30.

Kluge, A. G. 1988. The characterization of ontogeny. pp. 57–81. In *Ontogeny and systematics,* edited by C. J. Humphries. New York: Columbia University Press.

———. 1989. A concern for evidence and a phylogenetic hypothesis of relationships among *Epicrates* (Boidae, Serpentes). *Syst. Zool.* 38:7–25.

Kluge, A. G., and J. S. Farris. 1969. Quantitative phyletics and the evolution of anurans. *Syst. Zool.* 18:1–32.

Kluge, A. G., and R. E. Strauss. 1985. Ontogeny and systematics. *Ann. Rev. Ecol. Syst.* 16:247–268.

Kluger, M. J. 1978. The evolution and adaptive value of fever. *American Sci.* 66:38–43.

Klymkowsky, M. W., and J. Hanken. 1991. Whole-mount staining of *Xenopus* and other vertebrates. *Meth. Cell Biol.* 36:419–441.

Kobayashi, M. 1975. Sexual and bilateral differences in germ cell numbers in the course of the gonadal development of *Rana nigromaculata. Zool. Mag. (Tokyo)* 84:109–114.

Kok, D., L. H. Du Preez, and A. Channing. 1989. Channel construction by the African bullfrog: Another anuran parental care strategy. *J. Herpetol.* 23:435–437.

Kollros, J. J. 1961. Mechanisms of amphibian metamorphosis: Hormones. *American Zool.* 1:107–114.

———. 1981. Transitions in the nervous system during amphibian metamorphosis. pp. 445–459. In *Metamorphosis: A problem in developmental biology.* 2d ed., edited by L. I. Gilbert and E. Frieden. New York: Plenum.

———. 1984. Growth and death of cells of the mesencephalic fifth nucleus in *Rana pipiens* larvae. *J. Comp. Neurol.* 224:386–394.

Kollros, J. J., and V. M. McMurray. 1956. The mesencephalic V nucleus in anurans. II. The influence of thyroid hormone on cell size and cell number. *J. Exp. Zool.* 131:1–26.

Kollros, J. J., and M. L. Thiesse. 1985. Growth and death of cells of the mesencephalic fifth nucleus in *Xenopus laevis* larvae. *J. Comp. Neurol.* 233:481–489.

Kopsch, F. 1952. *Die Entwicklung des braunen Grasfrosches* Rana fusca *Roesel (Dargestellt in der Art der Normentafeln zur Entwicklungsgeschichte der Wirbeltiere).* Stuttgart: Georg Thieme-Verlag.

Korky, J. K., and R. G. Webb. 1973. The larva of the Mexican toad *Bufo cavifrons* Wiegmann. *J. Herpetol.* 7:47–49.

———. 1991. Geographic variation in larvae of mountain treefrogs of the *Hyla eximia* group (Anura: Hylidae). *Bull. New Jersey Acad. Sci.* 36:7–12.

———. 1994. A scanning electron microscope study of the larval labial teeth of *Rana montezumae. Bull. Maryland Herpetol. Soc.* 30:126–131.

———. 1996. Morphological variation in larvae of *Rana temporaria* L. (Anura: Ranidae) in the Republic of Ireland. *Irish Biogeograph. Soc. Bull.* (19):159–181.

Koskela, P. 1973. Duration of the larval stage, growth, and migration in *Rana temporaria* L. in two ponds in northern Finland in relation to environmental factors. *Ann. Zool. Fennici* 10:414–418.

Kothbauer, H., and H. Schenkel-Brunner. 1981. Immunochemical investigations on fish eggs and toad spawn: On the appearance of glycosidases during the hatching process. *Zool. Anz.* 206:354–360.

Kotthaus, A. 1933. Die Entwicklung des Primordial-cranium von *Xenopus laevis* bis zue Metamorphose. *Zeit. Wiss. Zool.* 144:510–572.

Kraemer, M. 1974. La morphogenèse du chondrocrâne de *Dis-*

coglossus pictus Otth. (Amphibien, Anoure). *Bull. Biol.* 108:211–228.

Kramer, D. L. 1987. Dissolved oxygen and fish behavior. *Environ. Biol. Fish.* 18:81–92.

Kramer, D. L., and J. B. Graham. 1976. Synchronous air breathing, a social component of respiration in fishes. *Copeia* 1976:689–697.

Kramer, H., and G. M. Windrum. 1955. The metachromatic staining reaction. *J. Histochem. Cytochem.* 3:227–237.

Kratochwill, K. 1933. Zur Morphologie und Physiologie der Nahrungsaufnahme der Froschlarven. *Zeit. Wiss. Zool.* 144:421–468.

Kress, A., and U. M. Spornitz. 1972. Ultrastructural studies of oogenesis in some European amphibians. I. *Rana esculenta* and *Rana temporaria*. *Zeit. Zellforsch. Mikros. Anat.* 128:438–456.

Krügel, P. 1993. *Biologie und Ökologie der Bromelienfauna von* Guzmania weberbaureri *im amazonischen Peru ergänzt durch eine emfassende. Bibliographie der Bromelien-Phytotelmata*. Wien: Österreichische Akad. Wissenshaft.

Krügel, P., and S. Richter. 1995. *Syncope antenori*—a bromeliad breeding frog with free-swimming, non-feeding tadpoles (Anura, Microhylidae). *Copeia* 1995:955–963.

Krujitzer, E. M. 1931. De ontwikkeling van het chondrocranium en enkele kopzenuwen van *Megalophrys montana*. Dissertation, Leiden.

Kruse, K. C., and M. G. Francis. 1977. A predation deterrent in larvae of the bullfrog, *Rana catesbeiana*. *Trans. American Fish. Soc.* 106:248–252.

Kunst, J. 1936. Vergleichende Untersuchungen an Anurennieren insbesondere an der Niere von *Xenopus laevis* Daud. *Zool. Jahrb. Abt. Anat. Ontog. Tiere* 61:51–76.

Kunte, K. 1998. *Ahaetulla nasuta* feeding on tadpoles. *Hamadryad* 22:124–125.

Kuntz, A. 1924. Anatomical and physiological changes in the digestive system during metamorphosis in *Rana pipiens* and *Amblystoma tigrinum*. *J. Morphol.* 38:581–598.

Künzle, H. 1983. Spinocerebellar projections in the turtle. Observations on their origin and terminal organization. *Exp. Brain Res.* 53:129–141.

Kupferberg, S. J. 1997a. The role of larval diet in anuran metamorphosis. *American Zool.* 37:146–159.

———. 1997b. Facilitation of periphyton production by tadpole grazing: Functional differences between species. *Freshw. Biol.* 37:427–439.

———. 1997c. Bullfrog (*Rana catesbeiana*) invasion of a California river: The role of larval competition. *Ecology* 78:1736–1751.

———. 1998. Predator mediated patch use by tadpoles (*Hyla regilla*): Risk balancing or consequences of motionlessness? *J. Herpetol.* 32:84–92.

Kupferberg, S. J., J. C. Marks, and M. E. Power. 1994. Effects of variation in natural algal and detrital diets on larval anuran (*Hyla regilla*) life-history traits. *Copeia* 1994:446–457.

Kuramoto, M., and C.-S. Wang. 1987. A new rhacophorid treefrog from Taiwan, with comparisons to *Chirixalus eiffingeri* (Anura, Rhacophoridae). *Copeia* 1987:931–942.

Kuzmin, S. L. 1995. The problem of food competition in amphibians. *Herpetol. J.* 5:252–256.

Laird, A. K., S. A. Tyler, and A. D. Barton. 1965. Dynamics of normal growth. *Growth* 29:233–248.

Lajmanovich, R. C. 1994. Contribución al conocimiento de la alimentación de larvas de la rana criolla *Leptodactylus ocellatus* (Amphibia, Leptodactylidae) en el Paraná medio, Argentina. *Stud. Neotrop. Fauna Environ.* 29:55–61.

Lajmanovich, R. C., and J. Faivovich. 1998. Dieta larval de *Phyllomedusa tetraploidea* Pombal and Haddad, 1992 en la provincia de Misiones (Argentina). *Alytes* 15:137–144.

Lamar, W. W., and E. R. Wild. 1995. Comments on the natural history of *Lithodytes lineatus* (Anura: Leptodactylidae), with a description of the tadpole. *Herpetol. Nat. Hist.* 3:135–142.

La Marca, E. 1985. Systematics and ecological observations on the neotropical frogs *Hyla jahni* and *Hyla platydactyla*. *J. Herpetol.* 19:227–237.

La Marca, E., and A. Mijares-U. 1988. Description of the tadpole of *Colostethus mayorgai* (Anura: Dendrobatidae) with preliminary data on the reproductive biology of the species. *Bull. Maryland Herpetol. Soc.* 24:47–57.

Lamb, A. H. 1981. Target dependency of developing motoneurons in *Xenopus laevis*. *J. Comp. Neurol.* 203:157–171.

Lamb, A. H., M. J. Ferns, and K. Klose. 1989. Peripheral competition in the control of sensory neuron numbers in *Xenopus* frogs reared with a single bilaterally innervated hindlimb. *Develop. Brain Res.* 45:149–154.

Lamberti, G. A., S. V. Gregory, C. P. Hawkins, R. C. Wildman, L. R. Ashkenas, and D. M. Denicola. 1992. Plant-herbivore interactions in streams near Mt. St. Helens. *Freshw. Biol.* 27:237–247.

Lambertini, G. 1929. Il manicotto glandulare di *Rana esculenta*. *Ric. Morfol. Roma* 9:71.

Lambiris, A. J. L. 1988a. A review of the amphibians of Natal. *Lammergeyer* (39):1–210.

———. 1988b. *Frogs and toads of the Natal Drakensberg*. Ukhahlamba Series no. 3. Pietermaritzburg: University of Natal Press.

———. 1989. The frogs of Zimbabwe. *Mus. Region. Sci. Nat. Torino Monogr.* (10):1–246.

Lamborghini, J. E. 1980. Rohon-Beard cells and other large neurons in *Xenopus* embryos originate during gastrulation. *J. Comp. Neurol.* 189:323–333.

———. 1987. Disappearance of Rohon-Beard neurons from the spinal cord of larval *Xenopus laevis*. *J. Comp. Neurol.* 264:47–55.

Lametschwandtner, A., U. Albrecht, and H. Adam. 1978. The vascularization of the kidneys in *Bufo bufo* (L.), *Bombina variegata* (L.), *Rana ridibunda* (L.) and *Xenopus laevis* (D.) (Amphibia, Anura) as revealed by scanning electron microscopy of vascular corrosion casts. *Acta Zool.* 59:11–23.

Lametschwandtner, A., A. Laminger, and W. Slattenschek. 1982. Eine experimentelle Anordunung zur Erfassung des Crowd-Effektes bie Kaulquappen des Suedafrikanischen Krallenfrosches *Xenopue laevis* (Daudin). *Zool. Anz.* 209: 145–155.

Lamotte, M. 1961. Contribution à l'étude des Batraciens de l'Ouest africain. XII. Les formes larvaires de *Cardioglossa leucomystax* Blgr. *Bull. Inst. France Afrique Noire* 23:211–216.

———. 1963. Contribution à l'étude des Batraciens de l'Ouest africain. XVII. Le développement larvaire de *Hymenochirus* (*Pseudhymenochirus*) *merlini* Chabanaud. *Bull. Inst. France Afrique Noire* 25:944–953.

———. 1964. Contribution à l'étude des Batraciens de l'Ouest africain. XVIII. Le développement larvaire de *Phrynomantis microps* Peters. *Bull. Inst. France Afrique Noire* 26:228–237.

Lamotte, M., and S. Dzieduszycka. 1958. Contribution à l'étude des Batraciens de l'Ouest africain. VII. Le développement larvaires de *Phrynobatrachus francisci*. *Bull. Inst. France Afrique Noire* 20:1071–1086.

Lamotte, M., S. Dzieduszycka, and G. Lauwarier. 1958. Contribution a l'etude des Batraciens de l'Ouest africain. VIII. Les formes larvaires de *Ptychadena submascareniensis, Pt. tournieri* et *Pt. trinodis*. *Bull. Inst. France Afrique Noire* 20:1464–1482.

Lamotte, M., R. Gaçon, and F. Xavier. 1973. Recherches sur le développement embryonnaire de *Nectophrynoides occidentalis* Angel amphibien anoure vivipare. 2. Le développement des gonades. *Ann. Embryol. Morphogen.* 6:271–296.

Lamotte, M., and J. Lescure. 1977. Tendances adaptives a l'affranchissement due milieu aquatique chez les amphibiens anoures. *Terre Vie* 31:225–312.

———. 1989. Les têtards rhéophiles et hygropétriques de l'Ancien et due Nouveau Monde. *Ann. Sci. Nat., Zool., Paris* 10:111–122, 125–144.

Lamotte, M., and J.-L. Perret. 1961a. Contribution à l'étude des Batraciens de l'Ouest africain. XI. Les formes larvaires de trois espêcies de *Ptychadena: Pt. maccarthyensis* And., *Pt. perreti* G. et L., et *Pt. mascareniensis* D. et B. *Bull. Inst. France Afrique Noire* 23A:192–210.

———. 1961b. Contribution à l'étude de Batraciens de l'Ouest africain. XIII. Les formes larvaires de quelques especes de *Leptopelis: L. aubryi, L. viridis, L. anchietae, L. ocellatus* et *L. calcaratus*. *Bull. Inst. France Afrique Noire* 23:855–885.

———. 1963a. Contribution à l'étude de Batraciens de l'Ouest africain. XIV. Les développement larvaire de *Chiromantis rufescens* Günther. *Bull. Inst. France Afrique Noire* 25:266–276.

———. 1963b. Contribution à l'étude de Batraciens des l'Ouest africain. XV. Le développement direct de l'espèce *Arthroleptis poecilonotus* Peters. *Bull. Inst. France Afrique Noire* 25:277–284.

———. 1963c. Contribution à l'étude de Batraciens de l'Ouest africain. XVI. Les formes larvaires de cínq espêcies d'*Hyperolius* de Cameroun: *H. hieroglyphycus, H. steindachneri pardalis, H. acutirostris, H. viridiflavus, H. tuberculatus*. *Bull. Inst. France Afrique Noire* 25:545–558.

Lamotte, M., J.-L. Perret, and S. Dzieduszycka. 1959a. Contributions à l'étude des Batraciens de l'Ouest africain. IX. Les formes larvaires de *Petropedetes palmipes, Conraua goliath* et *Acanthixalus spinosus*. *Bull. Inst. France Afrique Noire* 21:762–776.

———. 1959b. Contributions à l'étude des Batraciens de l'Ouest africain. X. Les formes larvaires de *Cryptothylax greshoffi, Leptopelis notatus* et *Ptychadena taeniocelis*. *Bull. Inst. France Afrique Noire* 21:1336–1350.

Lamotte, M., and F. Xavier. 1972. Recherches sur le développement embryonnaire de *Nectophrynoides occidentalis* Angel, amphibien anoure vivipare. I. Les principaux traits morphologiques et biometriques de développement. *Ann. Embryol. Morphol.* 5:315–340.

———. 1973. Le développement des gonades et la différentiation sexuelle chez l'embryon de *Nectophrynoides occidentalis* Angel. *Symbiosis* 4:259–261.

Lamotte, M., and M. Zuber-Vogeli. 1954a. Contribution à l'étude des Batraciens de l'Ouest africain. II. Le développement larvaire de *Bufo regularis* Reuss, de *Rana occidentalis* Günther et de *Rana crassipes* (Buch. et Peters). *Bull. Inst. France Afrique Noire* 16:940–954.

———. 1954b. Contribution à l'étude des Batraciens de l'Ouest-africain. III. Les développement larvaire de deux espêcies rhéophiles, *Astylosternus diadematus* et *Petropedetes natator*. *Bull. Inst. France Afrique Noire* 16:1222–1233.

———. 1956. Les formes larvaire de *Chiromantis rufescens, Afrixalus leptosomus fulvovittatus,* and *Kassina senegalensis*. *Bull. Inst. France Afrique Noire* 18:863–876.

Lancaster, J., and A. G. Hildrew. 1993. Characterizing instream flow refugia. *Canadian J. Fish. Aquat. Sci.* 50:1663–1675.

Lannoo, M. J. 1985. Neuromast topography in *Ambystoma* larvae. *Copeia* 1985:535–539.

———. 1987. Neuromast topography in anuran amphibians. *J. Morphol.* 191:115–129.

———. 1988. The evolution of the amphibian lateral line system and its bearing on amphibian phylogeny. *Zeit. Zool. Syst. Evol.-Forsch.* 26:128–134.

Lannoo, M. J., and S. C. Smith. 1989. The lateral line system. pp. 176–186. In *Developmental biology of the axolotl,* edited by J. B. Armstrong and G. M. Malacinski. Oxford: Oxford University Press.

Lannoo, M. J., D. S. Townsend, and R. J. Wassersug. 1987. Larval life in the leaves: Arboreal tadpole types, with special attention to the morphology, ecology, and behavior of the oophagous *Osteopilus brunneus* (Hylidae) larva. *Fieldiana Zool. (N.S.)* (38):1–31.

Lanza, B. 1983. *Guide per il riconoscimento delle specie animali delle acque interne Italiane 27.* Anfibi, Rettili. Consiglio Nazionale delle Ricerche.

Largen, M. J. 1975. The status of the genus *Kassina* (Amphibia Anura Hyperoliidae) in Ethiopia. *Monit. Zool. Italiano Suppl.* 6:1–28.

Larsell, O. 1925. The development of the cerebellum in the frog (*Hyla regilla*) in relation to the vestibular and lateral-line systems. *J. Comp. Neurol.* 39:249–289.

———. 1934. The differentiation of the peripheral and central acoustic apparatus in the frog. *J. Comp. Neurol.* 60:473–527.

———. 1967. *The comparative anatomy and histology of the cerebellum from myxinoids through birds.* Edited by J. Jansen. Minneapolis: University of Minnesota Press.

Larson, P., and R. O. De Sá. 1998. Chondrocranial morphology of *Leptodactylus* larvae (Leptodactylidae: Leptodactylinae): Its utility in phylogenetic reconstruction. *J. Morphol.* 238:287–305.

Larson-Prior, L. J., and W. L. R. Cruce. 1992. The red nucleus and mesencephalic tegmentum in a ranid amphibian: A cytotechtonic and HRP connectional study. *Brain Behav. Evol.* 40:273–286.

Lataste, F. 1879. Étude sur le Discoglosse. *Actes Soc. Linneenne Bordeaux* 33:275–341, plates 3–5.

Latsis, R. V., and N. Y. Saraeva. 1980. Development of hemopoietic tissue in *Rana temporaria*. *Arkh. Anat. Gistol. Embriol.* 78:100–105.

Lauder, G. V. 1981. Form and function: Structural analysis in evolutionary morphology. *Paleobiology* 7:430–442.

Laurens, H. 1914. The reactions of normal and eye-less amphibian larvae to light. *J. Exp. Zool.* 16:195–210.

Laurila, A., and T. Aho. 1997. Do female common frogs choose their breeding habitat to avoid predation on tadpoles? *Oikos* 78:585–591.

Laurila, A., J. Kujasalo, and E. Ranta. 1997. Different antipredator behavior in two anuran tadpoles: Effects of predator diet. *Behav. Ecol. Sociobiol.* 40:329–336.

Lavilla, E. O. 1984a. Larvas de *Telmatobius* (Anura: Leptodac-

tylidae) de la Provincia de Tucuman (Argentina). *Acta Zool. Lilloana* 38:69–79.
———. 1984b. Redescripcion de larvas de *Hyla pulchella andina* (Anura: Hylidae) con un analisis de la variabilidad interpoblacional. *Neotrópica* 30:19–30.
———. 1985. Diagnosis generica y agrupacion de las especies de *Telmatobius* (Anura: Leptodactylidae) en base a caracteres larvales. *Physis* 43:63–67.
———. 1987. La larva de *Rhinoderma darwinii* D. & B. (Anura: Rhinodermatidae). *Acta Zool. Lilloana* 39:81–88.
———. 1988. Lower Telmatobiinae (Anura: Leptodactylidae): Generic diagnoses based on larval characters. *Occas. Pap. Mus. Nat. Hist., Univ. Kansas* (124):1–19.
———. 1990. The tadpole of *Hyla nana* (Anura: Hylidae). *J. Herpetol.* 24:207–209.
———. 1991. Condrocáneo y esqueleto visceral en larvas de *Cycloramphus stejnegeri* (Leptodactylidae). *Amphibia-Reptilia* 12:33–38.
———. 1992a. Estructura del condrocráneo y esqueleto visceral de larvas de *Alsodes barrioi* (Anura: Leptodactylidae). *Acta Zool. Lilloana* 42:13–17.
———. 1992b. The tadpole of *Dermatonotus muelleri* (Anura: Microhylidae). *Boll. Mus. Region. Sci. Nat.* 10:63–71.
Lavilla, E. O., and I. De la Riva. 1993. La larva de *Telmatobius bolivianus* (Anura, Leptodactylidae). *Alytes* 11:37–46.
Lavilla, E. O., and M. Fabrezi. 1987. Anatomie de larvas de *Hyla pulchella andina* (Anura: Hylidae). *Physis* 45B:77–82.
———. 1992. Anatomia craneal de larvas de *Lepidobatrachus llanensis* y *Ceratophrys cranwelli* (Anura: Leptodactylidae). *Acta Zool. Lilloana* 42:5–11.
Lavilla, E. O., and J. A. Langone. 1991. Ontogenetic changes in spiracular and proctodeal tube orientation in *Elachistocleis bicolor* (Anura: Microhylidae). *J. Herpetol.* 25:119–121.
Lawler, K. L., and J.-M. Hero. 1997. Palatability of *Bufo marinus* tadpoles to a predatory fish decreases with development. *Wildl. Res.* 24:327–334.
Lawler, S. P. 1989. Behavioral responses to predators and predation risk in four species of larval anurans. *Anim. Behav.* 38:1039–1047.
Lawler, S. P., and P. J. Morin. 1993. Temporal overlap, competition, and priority effects in larval anurans. *Ecology* 74:174–182.
Lawson, D. P. 1993. The reptiles and amphibians of Korup National Park Project, Cameroon. *Herpetol. Nat. Hist.* 1:27–93.
Lázár, G. Y., Z. S. Liposits, P. Tóth, S. L. Trasti, J. L. Maderdrut, and I. Merchenthaler. 1991. Distribution of galanin-like immunoreactivity in the brain of *Rana esculenta* and *Xenopus laevis*. *J. Comp. Neurol.* 310:45–67.
Lázár, G. Y., P. Tóth, G. Y. Csank, and E. Kicliter. 1983. Morphology and location of tectal projection neurons in frogs: A study with HRP and cobalt-filling. *J. Comp. Neurol.* 215:108–120.
Leaf, A. 1965. Transepithelial transport and its hormonal control in toad bladder. *Physiol. Biol. Chem. Exp. Pharmacal.* 56:216–263.
Lebedinskaya, I. I., T. L. Padzyukevich, and G. A. Lasledov. 1989. Morphofunctional characteristics of myotomal muscle fibers from the tails of *Rana temporaria* tadpoles. *J. Evol. Biochem. Physiol.* 25:223–228.
Lee, A. K. 1967. Studies in Australian Amphibia II. Taxonomy, ecology, and evolution of the genus *Heleioporus* Gray (Anura: Leptodactylidae). *Australian J. Zool.* 15:367–439.
Lee, J. C. 1982. Accuracy and precision in anuran morphometrics: artifacts of preservation. *Syst. Zool.* 31:266–281.
———. 1996. *The amphibians and reptiles of the Yucatán Peninsula.* Ithaca, NY: Comstock Publ. Assoc.
Lee, R. K. K., and R. C. Eaton. 1989. Segmental template for reticulospinal organization. *Soc. Neurosci. Abstr.* p. 32.
Lefcort, H. 1996. Adaptive, chemically mediated flight response in tadpoles of the southern leopard frog, *Rana utricularia*. *Copeia* 1996:455–459.
———. 1998. Chemically mediated fright response in southern toad (*Bufo terrestris*) tadpoles. *Copeia* 1998:445–450.
Lefcort, H., and A. R. Blaustein. 1995. Disease, predator avoidance, and vulnerability to predation in tadpoles. *Oikos* 74:469–474.
Lefcort, H., and S. M. Eiger. 1993. Antipredatory behaviour of feverish tadpoles—implication for pathogen transmission. *Behaviour* 126:13–27.
Leibovitz, H. E., D. D. Culley, Jr., and J. P. Geaghan. 1982. Effects of vitamin C and sodium benzoate on survival, growth, and skeletal deformities of intensively cultured bullfrog larvae *Rana catesbeiana* reared at two pH levels. *J. World Maricult. Soc.* 13:322–328.
Leips, J., and J. Travis. 1994. Metamorphic responses to changing food levels in two species of hylid frogs. *Ecology* 75:1345–1356.
Leloup-Hatey, J., M. Buscaglia, G. Jolivet-Jaudet, and J. Leloup. 1990. Interrenal function during the metamorphosis in anuran Amphibia. pp. 139–154. In *Biology and physiology of amphibians. Progress in Zoology 38. Proc. First Intl. Symp. Biol. Physiol. Amphibians* (Karlsruhe, Federal Republic of Germany), edited by W. Hanke. Stuttgart: Gustav Fischer-Verlag.
Lemckert, F. 1996. An observation of aggregation behaviour by tadpoles of the great barred river frog (*Mixophyes fasciolatus*). *Herptofauna* 26:43–44.
Leon-Ochoa, J., and R. Donoso-Barros. 1970. Desarrollo embrionario y metamorfosis de *Pleurodema brachyops* (Cope) (Salientia-Leptodactylidae). *Bol. Soc. Biol. Concepcion* 42: 355–379.
Le Quang Trong, Y. 1973. Structure et développement de la peau et des glandes cutanées de *Bufo regularis* Reuss. *Bull. Soc. Zool. France* 98:449–485.
———. 1974. Étude de la peau et des glandes cutanées de *Xenopus tropicalis* Gray. *Bull. Inst. France Afrique Noire* 26:407–426.
Lescure, J. 1968. Le comportement social des batraciens. *Rev. Comport. Anim.* 2:1–33.
———. 1976. Etude de deux têtards de *Phyllobates* (Dendrobatidae): *P. femoralis* (Boulenger) et *P. pictus* (Bibron). *Bull. Soc. Zool. France* 101:299–304.
———. 1981. Contribution à l'étude des amphibiens de Guyane française. IX. Le têtard gastromyzophore d'*Atelopus flavescens* Duméril et Bibron (Anura, Bufonidae). *Amphibia-Reptilia* 2:209–215.
———. 1984. Las larvas de dendrobatidae [sic]. *[Proc.] II. Reunión Iberoamer. Cons. Zool. Vert.* 37–45.
Leucht, T. 1987. Effects of weak magnetic fields on background adaptation in *Xenopus laevis*. *Naturwissenschaften* 74:192–194.
Levine, R. L. 1980. An autoradiographic study of the retinal projection in *Xenopus laevis* with comparisons to *Rana*. *J. Comp. Neurol.* 189:1–29.
Lewinson, D., M. Rosenberg, and M. R. Warburg. 1987. Ultrastructural and ultracytochemical studies of the gill epithelium in the larvae of *Salamandra salamandra* (Amphibia, Urodela). *Zoomorphology* 107:17–25.
Lewis, E. R. 1976. Surface morphology of the bullfrog amphibian papilla. *Brain Behav. Evol.* 13:196–215.

Lewis, E. R., and R. E. Lombard. 1988. The amphibian inner ear. pp. 93–123. In *The evolution of the amphibian auditory system,* edited by B. Fritzsch, M. J. Ryan, W. Wilczynski, T. E. Hetherington, and W. Walkowiak. New York: Wiley.

Lewis, S., and C. Straznicky. 1979. The time of origin of the mesencephalic trigeminal neurons in *Xenopus. J. Comp. Neurol.* 183:633–646.

Lewis, W. M., Jr. 1970. Morphological adaptations of cyprindontoids for inhabiting oxygen deficient waters. *Copeia* 1970:319–326.

Lewontin, R. C. 1955. The effects of population density and composition on viability in *Drosophila melanogaster. Evolution* 9:27–41.

Leydig, F. 1853. *Anatomisch-histologische Untersuchungen über Fische und Reptilien.* Berlin.

———. 1857. *Lehrbuch der Histologie des Menschen und der Thiere.* Frankfurt: Meidinger Sohn & Co.

———. 1876. Ueber die Allgemeinen Bedeckungen der Amphibien. *Arch. Mikros. Anat.* 12:119–242.

Li, J. C., and C. S. Lin. 1935. Studies of the "rain frog" *Kaloula borealis* II. The food and feeding of the embryos and adults. *Peking Nat. Hist. Bull.* 10:45–53.

Licht, L. E. 1967. Growth inhibition in crowded tadpoles: Intraspecific and interspecific effects. *Ecology* 48:736–745.

———. 1968. Unpalatability and toxicity of toad eggs. *Herpetologica* 24:93–98.

———. 1969. Palatability of *Rana* and *Hyla* eggs. *American Midl. Nat.* 88:296–299.

———. 1971. Breeding habits and embryonic thermal requirements of the frogs, *Rana aurora aurora* and *Rana pretiosa pretiosa,* in the Pacific Northwest. *Ecology* 52:116–124.

———. 1974. Survival of embryos, tadpoles and adults of *Rana aurora* and *Rana pretiosa* sympatric in southwestern British Columbia. *Canadian J. Zool.* 52:613–627.

Lieberkund, I. 1937. *Vergleichende Studien über die Morphologie und Histogenese der Larvalen Haftorgane bei den Amphibien.* Copenhagen: C. A. Reitzels Forlag.

Liem, K. F. 1961. On the taxonomic status and the granular patches of the Javanese frog *Rana chalconota* Schlegel. *Herpetologica* 17:69–71.

Liem, K. F., and D. B. Wake. 1985. Morphology: Current approaches and concepts. pp. 366–377. In *Functional vertebrate morphology,* edited by M. Hildebrand, D. M. Bramble, K. F. Liem, and D. B. Wake. Cambridge: Belknap Press.

Ligname, A. H. 1964. Fixation method for amphibian eggs. *J. Roy. Microscop. Soc.* 83:481–482.

Lillie, R. D. 1965. *Histopathologic technic and practical histochemistry.* 3d ed. New York: McGraw-Hill.

Lillywhite, H. B. 1970. Behavioral temperature regulation in the bullfrog, *Rana catesbeiana. Copeia* 1970:158–168.

Lillywhite, H. B., P. Licht, and P. Chelgren. 1973. The role of behavioral thermoregulation in the growth energetics of the toad, *Bufo boreas. Ecology* 54:375–383.

Limbaugh, B. A., and E. P. Volpe. 1957. Early development of the Gulf Coast toad, *Bufo valliceps* Wiegmann. *American Mus. Novitates* (1842):1–32.

Lindquist, E. D., and T. E. Hetherington. 1998. Tadpoles and juveniles of the Panamanian golden frog, *Atelopus zeteki* (Bufonidae), with information on development of coloration and patterning. *Herpetologica* 54:370–376.

Ling, R. W., J. P. Van Amberg, and J. K. Werner. 1986. Pond acidity and its relationship to larval development of *Ambystoma maculatum* and *Rana sylvatica* in upper Michigan. *J. Herpetol.* 20:230–236.

Linss, W., and G. Geyer. 1964. Über die elektronenmikroskopische Struktur der Nierentubul von *Rana esculenta. Anat. Anz.* 115:282–296.

Lips, K. R., and J. M. Savage. 1996. Key to the known tadpoles (Amphibia: Anura) of Costa Rica. *Stud. Neotrop. Fauna Environ.* 31:17–26.

Littlejohn, M. J. 1963. The breeding biology of the Baw Baw frog *Philoria frosti* Spencer. *Proc. Linnean Soc. New South Wales* 88:273–276.

Littlejohn, M. J., and A. A. Martin. 1964. The *Crinia laevis* complex (Anura: Leptodactylidae) in south-eastern Australia. *Australian J. Zool.* 12:70–83.

———. 1965. A new species of *Crinia* (Anura: Leptodactylidae) from South Australia. *Copeia* 1965:319–324.

Liu, C.-C. 1940. Tadpoles of West China Salientia. *J. W. China Border Res. Soc.* 12:7–62, 4 plates.

———. 1950. Amphibians of western China. *Fieldiana Zool. Mem.* 2:1–400, 10 plates.

Liu, C.-C., and S. Q. Hu. 1961. *The Anura of China.* Peking: Science Press.

Liu, H., R. Wassersug, and K. Kawachi. 1996. A computational fluid dynamics study of tadpole swimming. *J. Exp. Zool.* 199:1245 – 1260.

———. 1997. The three-dimensional hydrodynamics of tadpole locomotion. *J. Exp. Biol.* 200:2807–2819.

Lofts, B. 1974. Reproduction. pp. 107–218. In *Physiology of the Amphibia,* Vol. 2, edited by B. Lofts. New York: Academic Press.

Lopez, K. 1989. Sex differentiation and early gonadal development in *Bombina orientalis* (Anura, Discoglossidae). *J. Morphol.* 199:299–311.

Löschenkohl, A. 1986. Niche partitioning and competition in tadpoles. pp. 399–402. In *Studies in herpetology,* edited by Z. Roček. Prague: Charles University Press.

Løvtrup, S., and A. Pigón. 1968. Observations on the gill development in frogs. *Ann. Embryol. Morphol.* 2:3–13.

Løvtrup, S., and B. Werdinius. 1957. Metabolic phases during amphibian embryogenesis. *J. Exp. Zool.* 135:203–220.

Lowe, D. A. 1986. Organization of lateral line and auditory areas in the midbrain of *Xenopus laevis. J. Comp. Neurol.* 245:498–513.

———. 1987. Single-unit study of lateral line cells in the optic tectum of *Xenopus laevis:* Evidence for bimodal lateral line/optic units. *J. Comp. Neurol.* 257:396–404.

Lowe, D. A., and I. J. Russell. 1982. The central projections of the lateral line and cutaneous sensory fibres (VII and X) in *Xenopus laevis. Proc. Roy. Soc. London* 216B:279–297.

Lucas, E. A., and W. A. Reynolds. 1967. Temperature selection by amphibian larvae. *Physiol. Zool.* 40:159–171.

Luckenbill, L. M. 1964. Fine structure and development of horny jaws and teeth of the tadpole (*Rana pipiens*). Dissertation, Brown University

———. 1965. Morphogenesis of the horny jaws of *Rana pipiens* larvae. *Develop. Biol.* 11:25–49.

Ludwig, D., and L. Rowe. 1990. Life history strategies for energy gain and predator avoidance under time constraints. *American Nat.* 135:686–707.

Lum, A. M., R. J. Wassersug, M. J. Potel, and S. A. Lerner. 1982. Schooling behavior of tadpoles: A potential indicator of ototoxicity. *Pharmacol. Biochem. Behav.* 17:363–366.

Luna, L. G., ed. 1968. *Manual of histologic staining methods of the Armed Forces Institute of Pathology.* New York: McGraw-Hill Book.

Luther, A. 1914. Über die vom N. trigeminus versorgte Musku-

latur der Amphibien mit einem vergleichenden Ausblick über den Adductor mandibulae der Gnathostomen, und einem Beitrag zum Versändnis der Organisation der Anurenlarven. *Acta Soc. Sci. Fennicae* 44:1–151, 1 plate.
Lutz, B. 1944a. Observations on frogs without aquatic larvae: The hatching of *Eleutherodactylus parvus* Girard. *Bol. Mus. Nac. Rio de Janeiro* (15):5–7.
———. 1944b. Biologia e taxonomia de *Zachaenus parvulus*. *Bol. Mus. Nac. Rio de Janeiro* (17):1–66.
———. 1948. Ontogenetic evolution in frogs. *Evolution* 2:29–39.
Lyapkov, S. M. 1996. The influence of predators on brown frogs during and after metamorphosis. *Adv. Amphib. Res. Form. Soviet Union* 1:140–159.
Lynch, J. D. 1971. Evolutionary relationships, osteology, and zoogeography of leptodactyloid frogs. *Misc. Publ. Mus. Nat. Hist., Univ. Kansas* (53):1–238.
———. 1973. The transition from archaic to advanced frogs. pp. 133–182. In *Evolutionary biology of the anurans. Contemporary research on major problems,* edited by J. L. Vial. Columbia: University of Missouri Press.
———. 1986. Notes on the reproductive biology of *Atelopus subornatus*. *J. Herpetol.* 20:126–129.
Lynch, J. D., P. M. Ruiz, and J. V. Rueda. 1983. Notes on the distribution and reproductive biology of *Centrolene geckoideum* Jimenez De la Espada in Colombia and Ecuador (Amphibia: Centrolenidae). *Stud. Neotrop. Fauna Environ.* 18:239–244.
Lynch, K. 1984. Growth and metamorphosis of the rectus abdominis muscle in *Rana pipiens*. *J. Morphol.* 182:317–337.
Lynn, W. G. 1936. A study of the thyroid in embryos of *Eleutherodactylus nubicola*. *Anat. Rec.* 64:525–539.
———. 1940a. The development of the skull in the nonaquatic larva of the tree-toad *Eleutherodactylus nubicola*. *Biol. Bull.* 79:375. (Abstr.)
———. 1940b. The embryonic origin and development of the pharyngeal derivatives in *Eleutherodactylus nubicola*. *Biol. Bull.* 79:376. (Abstr.)
———. 1942. The embryology of *Eleutherodactylus nubicola,* an anuran which has no tadpole stage. *Carnegie Inst. Washington Contrib. Embryol.* 30:27–62.
———. 1948. The effects of thiourea and phenylthiourea upon the development of *Eleutherodactylus ricordii*. *Biol. Bull.* 94:1–15.
———. 1961. Types of amphibian metamorphosis. *American Zool.* 1:151–161.
Lynn, W. G., and A. Edelman. 1936. Crowding and metamorphosis in the tadpole. *Ecology* 17:104–109.
Lynn, W. G., and B. Lutz. 1946a. The development of *Eleutherodactylus guentheri* Stdnr. 1864 (Salientia). *Bol. Mus. Nac. Rio de Janeiro* (71):1–46.
———. 1946b. The development of *Eleutherodactylus nasutus* Lutz. *Bol. Mus. Nac. Rio de Janeiro* (79):1–30.
Lynn, W. G., and A. M. Peadon. 1955. The role of the thyroid gland in direct development in the anuran, *Eleutherodactylus martinicensis*. *Growth* 19:263–286.
Lynn, W. G., and Sister Alfred De Marie. 1946. The effect of thiouracil upon pigmentation in the tadpole. *Science* 104:31.
Macgregor, H. C., and E. M. Del Pino. 1982. Ribosomal gene amplification in multinucleate oocytes of the egg brooding hylid frog *Flectonotus pygmaeus*. *Chromosoma* 85:475–488.
Mackay, W. C., and B. Schmidt-Nielsen. 1969. Osmotic and diffusional water permeability in tadpoles and frogs, *Rana clamitans*. *Bull. Mount. Desert Is. Biol. Lab.* 9:26–27.
Maeda, N., and M. Matsui. 1989. *Frogs and toads of Japan*. Tokyo: Bun-Ichi Sogo Shuppan Co., Ltd.
Maetz, J. 1974. Aspects of adaptation to hypo-osmotic and hyper-osmotic environments. pp. 1–167. In *Biochemical and biophysical perspectives in marine biology,* Vol. 1, edited by D.C. Malins and J. R. Sargent. New York: Academic Press.
Magnin, E. 1959. Anatomie du têtard d'*Alytes obstetricans*. *Actes Soc. Linneenne Bordeaux* 98:33–79.
Magnusson, W. E., and J.-M. Hero. 1991. Predation and the evolution of complex oviposition behaviour in Amazon rainforest frogs. *Oecologia* 86:310–318.
Mahapatro, B. K., and M. C. Dash. 1987. Density effects on growth and metamorphosis of *Bufo stomaticus* larvae. *Alytes* 6:88–98.
Maher, S. W. 1962. Primary structures produced by tadpoles. *J. Sediment. Petrol.* 32:138–139.
Main, A. R. 1957. Studies on Australian Amphibia. 1. The genus *Crinia* Tschudi in south-western Australia, and some species from south-eastern Australia. *Australian J. Zool.* 5:30–55.
———. 1968. Ecology, systematics and evolution of Australian frogs. pp. 37–86. In *Advances in ecological research,* Vol. 5. Edited by J. B. Cragg. New York: Academic Press.
Majecki, J., and K. Majecka. 1996. Predation by *Oligotricha striata* caddis larvae on amphibian eggs: Effects of a high quality food on growth rate. *Netherlands J. Aquat. Ecol.* 30:21–25.
Malone, B. S. 1985. Mortality during the early life history stages of the Baw Baw frog, *Philoria frosti* (Anura: Myobatrachidae). pp. 1–5. In *The biology of Australasian frogs and reptiles,* edited by G. Grigg, R. Shine, and H. Ehmann. Chipping Norton, New South Wales: Surrey Beatty.
Malvin, G. M. 1985a. Adrenoreceptor types in the respiratory vasculature of the salamander gill. *J. Comp. Physiol.* 155B:591–596.
———. 1985b. Vascular resistance and vasoactivity of gills and pulmonary artery of the salamander, *Ambystoma tigrinum*. *J. Comp. Physiol.* 155B:241–249.
———. 1988. Gill structure and function amphibian larvae. pp. 121–151. In *Comparative pulmonary physiology: Current concepts,* edited by S. C. Wood. New York: Marcel Dekker.
Malvin, G. M., and R. G. Boutilier. 1985. Ventilation-perfusion relationships in Amphibia. pp. 114–124. In *Circulation, respiration, and metabolism current comparative approaches,* edited by R. Gilles. Berlin: Springer-Verlag.
Malvin, G. M., and M. P. Hlastala. 1986a. Effects of lung volume and O_2 and CO_2 content on cutaneous gas exchange in frogs. *American J. Physiol.* 20R:941–946.
———. 1986b. Regulation of cutaneous gas exchange by environmental O_2 and CO_2 in the frog. *Respir. Physiol.* 65:99–111.
———. 1988. Effects of environmental O_2 on blood flow and diffusing capacity in amphibian skin. *Respir. Physiol.* 76:229–242.
Malvin, G. M., and C. Riedel. 1990. Autonomic regulation of cutaneous vascular resistance in the bullfrog *Rana catesbeiana*. *J. Exp. Biol.* 152:425–439.
Manahan, D. T. 1990. Adaptations by invertebrate larvae for nutrient acquisition from seawater. *American Zool.* 30:147–160.
Mangia, F., G. Procicchiani, and H. Manelli. 1970. On the development of the blood island in *Xenopus laevis* embryos. *Acta Embryol. Exp.* 14:163–184.
Mangold, O. 1936. Experimente zur Analyse der Zusammenarbeit der Keimblätter. *Naturwissenschaften* 24:753–760.

Maniatis, G. M., and V. M. Ingram. 1971a. Erythropoiesis during amphibian metamorphosis. I. Site of maturation of erythrocytes in *Rana catesbeiana*. *J. Cell Biol.* 49:372–379.

———. 1971b. Erythropoiesis during amphibian metamorphosis. II. Immunochemical study of larval and adult hemoglobins in *Rana catesbeiana*. *J. Cell Biol.* 49:380–389.

———. 1971c. Erythropoiesis during amphibian metamorphosis. III. Immunochemical detection of tadpole and frog hemoglobins (*Rana catesbeiana*) in single erythrocytes. *J. Cell Biol.* 49:390–404.

Manis, M. L., and D. L. Claussen. 1986. Environmental and genetic influences on the thermal physiology of *Rana sylvatica*. *J. Therm. Biol.* 11:31–36.

Manning, M. J., and G. M. Al Johari. 1985. Immunological memory and metamorphosis. pp. 420–433. In *Metamorphosis,* edited by M. Balls and M. Bownes. Oxford: Clarendon Press.

Manning, M. J., and J. D. Horton. 1969. Histogenesis of lymphoid organs in larvae of the South African clawed toad, *Xenopus laevis* (Daudin). *J. Embryol. Exp. Morphol.* 22:265–277.

———. 1982. RES structure and function of the Amphibia. pp. 423–460. In *The reticuloendothelial system. A comprehensive treatise,* Vol. 3. *Phylogeny and ontogeny,* edited by N. Cohen and M. M. Sigel. New York: Plenum.

Manteifel, Y. 1995. Chemically mediated avoidance of predators by *Rana temporaria* tadpoles. *J. Herpetol.* 29:461–463.

Manteifel, Y., and A. V. Zhushev. 1996. Avoidance of predator chemical cues by tadpoles of four East European anuran species (*Bufo bufo, Rana arvalis, R. lessonae,* and *R. temporaria*). *Adv. Amphib. Res. Form. Soviet Union* 1:161–180.

Manteuffel, G., O. Wess, and W. Himstedt. 1977. Messungen am dioptrischen Apparat von Amphibienaugen und Berechnung der Sehrschärfe im Wasser und Luft. *Zool. Jahrb.* 81:395–406.

Marcus, E. 1930. Zur Entwicklungsgeschichte des Vorderdarmes der Amphibien. *Zool Jahrb. Anat.* 52:405–486.

Marcus, H. 1908. Beiträge zur Kenntnis der Gymnophionen. I. Über das Schlundspaltengebiet. *Arch. Mikros. Anat. Entwickl.-Gesch.* 77:695–774.

Marian, M. P., K. Sampath, A. R. C. Nirmala, and T. J. Pandian. 1980. Behavioural response of *Rana cyanophylictis* tadpole exposed to changes in dissolved oxygen concentration. *Physiol. Behav.* 25:35–38.

Marinelli, M., S. Tei, D. Vagnetti, and M. P. Sensi. 1985. SEM observation on the development and the regression of the oral disc of *Bufo vulgaris* larva. *Acta Embryol. Morphol. Exp.* 6:31–39.

Marinelli, M., and D. Vagnetti. 1988. Morphology of the oral disc of *Bufo bufo* (Salientia: Bufonidae) tadpoles. *J. Morphol.* 195:71–81.

Maritz, M. F., and R. M. Douglas. 1994. Shape quantization and the estimation of volume and surface area of reptile eggs. *J. Herpetol.* 28:281–291.

Marlier, G. 1973. Limnology of the Congo and Amazon rivers. pp. 223–238. In *Tropical forest ecosystems in Africa and South America: A comparative review,* edited by N. Cohen and M. M. Sigel. Washington, D. C.: Smithsonian Institution Press.

Marshall, E., and G. C. Grigg. 1980a. Acclimation of CTM, LD_{50}, and rapid loss of acclimation of thermal preferendum in tadpoles of *Limnodynastes peronii* (Anura: Myobatrachidae). *Australian Zool.* 20:447–455.

———. 1980b. Lack of metabolic acclimation to different thermal histories by tadpoles of *Limnodynastes peroni* (Anura: Leptodactylidae). *Physiol. Zool.* 53:1–7.

Marshall, G. A., R. L. Amborski, and D. D. Culley, Jr. 1980. Calcium and pH requirements in the culture of bullfrog (*Rana catesbeiana*) larvae. *Proc. World Maricult. Soc.* 11:445–453.

Martin, A. A, and A. K. Cooper. 1972. The ecology of terrestrial anuran eggs, genus *Crinia* (Leptodactylidae). *Copeia* 1972:163–168.

Martin, A. A., and M. J. Littlejohn. 1966. The breeding biology and larval development of *Hyla jervisiensis* (Anura: Hylidae). *Proc. Linnean Soc. New South Wales* 91:47–57, 1 plate.

Martin, W. 1991. Calling phenology of savannah woodland frogs in northern Queensland, Australia. Honours Thesis, James Cook University, Townsville, Australia.

Martinez, I., R. Alvarez, and I. Herraez. 1992. Skeletal malformations in hatchery reared *Rana perezi* tadpoles. *Anat. Rec.* 233:314–320.

Martinez, I., R. Alvarez, J. M. Maiega, and M. P. Herraez. 1990. Estudio preliminar de los factores que afectan al desarrollo embrionarioi de *Rana perezi* en cautividad. *Actas III Congreso Nac. Acuicult.* 621–629.

Martinez, I., M. P. Herraez, and R. Alvarez. 1994. Response of hatchery-reared *Rana perezi* larvae fed different diets. *Aquaculture* 128:235–244.

Matesz, C. 1979. Central projection of the VIIIth cranial nerve in the frog. *Neuroscience* 4:2061–2071.

Matesz, C., and G. Székely. 1978. The motor column and sensory projections of the branchial cranial nerves in the frog. *J. Comp. Neurol.* 178:157–176.

Mathies, T., and R. M. Andrews. 1996. Extended egg retention and its influence on embryonic development and egg water balance: Implications for the evolution of viviparity. *Physiol. Zool.* 69:1021–1035.

Mathis, U., F. Schaeffel, and H. C. Howland. 1988. Visual optics in toad (*Bufo americanus*). *J. Comp. Physiol.* 163A:201–213.

Matsurka, Y. 1935. Über die Entwicklung des Vornierenkanälchensystems der japanischen Kröte (*Bufo formosus*). *Folia Anat. Japonica* 13:417–448.

Matthews, L. H. 1957. Viviparity in *Gastrotheca* (Amphibia: Anura) and some considerations on the evolution of viviparity. *Bull. Soc. Zool. France* 82:317–320.

Maturana, H. R., J. Y. Lettvin, W. S. McCulloch, and W. H. Pitts. 1960. Anatomy and physiology of vision in the frog (*Rana pipiens*). *J. Gen. Physiol. Suppl.* 2:129–175.

Matz, G. 1975. Les grenouilles de Genre *Mantella* (Ranidae). *Aquarana* 9:20–84.

Maurer, F. 1888. Die Kiemen und ihre Gefässe bei anuren und urodelen Amphibien und die Umbildungen der beiden ersten Arterienbogen bei Teleostiern. *Gegenbaurs Morphol. Jahrb.* 14:175–222.

May, R. M. 1974. *Stability and complexity in model ecosystems.* 2d ed. Princeton: Princeton University Press.

Mayhew, W. W. 1965. Adaptations of the amphibian, *Scaphiopus couchi,* to desert conditions. *American Midl. Nat.* 74:95–109.

Maynard Smith, J. 1964. Group selection and kin selection. *Nature* 201:1145–1147.

McCarrison, R. 1921. Observations on the effects of fat excess on the growth and metamorphosis of tadpoles. *Proc. Roy. Soc. London* 92B:295–303.

McClanahan, L. L. 1975. Nitrogen excretion in arid-adapted amphibians. pp. 106–116. In *Environmental physiology of de-*

sert organisms, edited by N. F. Hadley. Stroudsburg, PA: Dowden, Hutchinson and Ross.

McClendon, J. F. 1910. The development of isolated blastomeres of the frog's egg. *American J. Anat.* 10:425–430.

McCollum, S. A. 1993. Ecological consequences of predator-induced polyphenism in larval hylid frogs. Dissertation, Duke University.

McCollum, S. A., and J. D. Leimberger. 1997. Predator-induced morphological changes in an amphibian: Predation by dragonflies affects tadpole color, shape, and growth rate. *Oecologia* 109:615–621.

McCollum, S. A., and J. Van Buskirk. 1996. Costs and benefits of a predator-induced polyphenism in the gray treefrog *Hyla chrysoscelis. Evolution* 50:583–593.

McCormick, C. A. 1982. The organization of the octavolateralis area in actinopterygian fishes: A new interpretation. *J. Morphol.* 171:159–181.

———. 1988. Evolution of auditory pathways in the Amphibia. pp. 587–612. In *The evolution of the amphibian auditory system,* edited by B. Fritzsch, M. J. Ryan, W. Wilczynski, T. E. Hetherington, and W. Walkowiak. New York: Wiley.

McCranie, J. R., L. D. Wilson, and K. L. Williams. 1989. A new genus and species of toad (Anura: Bufonidae) with an extraordinary stream-adapted tadpole from northern Honduras. *Occas. Pap. Mus. Nat. Hist., Univ. Kansas* (129):1–18.

McCutcheon, F. H. 1936. Hemoglobin function during the life history of the bullfrog. *J. Cell. Comp. Physiol.* 8:63–81.

McDiarmid, R. W. 1978. Evolution of parental care in frogs. pp. 127–147. In *The development of behavior: Comparative and evolutionary aspects,* edited by R. M. Burkhardt and M. Bekoff. New York: STMP Press.

McDiarmid, R. W., and R. Altig. 1990. Description of a bufonid and two hylid tadpoles from western Ecuador. *Alytes* 8:51–60.

McDiarmid, R. W., and S. Gorzula. 1989. Aspects of the reproductive ecology and behavior of the tepui toads, genus *Oreophrynella* (Anura, Bufonidae). *Copeia* 1989:445–451.

McDonald, D. G., J. L. Ozog, and B. P. Simons. 1984. The influence of low pH environments on ion regulation in the larval stages of the anuran amphibian, *Rana clamitans. Canadian J. Zool.* 62:2171–2177.

McEdward, L. R., and D. A. Janies. 1993. Life cycle evolution in asteroids: What is a larva? *Biol. Bull.* 184:255–268.

McIndoe, R., and D. G. Smith. 1984a. Functional anatomy of the internal gills of the tadpole of *Litoria ewingii* (Anura, Hylidae). *Zoomorphology* 104:280–291.

———. 1984b. Functional morphology of gills in larval amphibians. pp. 55–69. In *Respiration and metabolism of embryonic vertebrates,* edited by R. S. Seymour. Dordrecht: Dr. W. Junk.

McLachlan, A. 1981a. Interactions between insect larvae and tadpoles in tropical rain pools. *Ecol. Entomol.* 6:175–182.

———. 1981b. Food sources and foraging tactics in tropical rain pools. *Zool. J. Linnean Soc.* 71:265–277.

McLennan, I. S. 1988. Quantitative relationships between motoneuron and muscle development in *Xenopus laevis:* Implications for motoneuron cell death and motor unit formation. *J. Comp. Neurol.* 271:19–29.

Medewar, P. B. 1953. Some immunological and endocrinological problems raised by the evolution of viviparity in vertebrates. *Symp. Soc. Exp. Biol.* 7:328–338.

Méhëly, E. G. 1901. Beiträge zur Kenntniss der Engystomatiden von Neu-Guinea. *Természetr. Füz.* 24:169–271.

Mendelson, J. R., III. 1997. A new species of toad (Anura: Bufonidae) from the Pacific highlands of Guatemala and southern Mexico, with comments on the status of *Bufo valliceps macrocristatus. Herpetologica* 53:14–30.

Menke, M. E., and D. L. Claussen. 1982. Thermal acclimation and hardening in tadpoles of the bullfrog, *Rana catesbeiana. J. Therm. Biol.* 7:215–219.

Menzies, J. I. 1967. An ecological note on the frog *Pseudhymenochirus merlini* Chabanaud in Sierra Leone. *J. West African Sci. Assoc.* 12:23–28.

Menzies, J. I., and R. G. Zweifel. 1974. Systematics of *Litoria arfakiana* of New Guinea and sibling species (Salientia, Hylidae). *American Mus. Novitates* (2558):1–16.

Merchant-Larios, H. 1978. Ovarian differentiation. pp. 47–81. In *The vertebrate ovary comparative biology and evolution,* edited by R. E. Jones. New York: Plenum.

Merchant-Larios, H., and I. Villalpando. 1981. Ultrastructural events during early gonadal development in *Rana pipiens* and *Xenopus laevis. Anat. Rec.* 199:349–360.

Mertens, R. 1938. Herpetologische Ergebnisse einer Reise nach Kamerun. *Abh. Senckenberg. Naturf. Ges.* 443:1–52.

———. 1960. Die Larven der Amphibien und ihre evolutive Bedeutung. *Zool. Anz.* 164:337–358.

Meseguer, J., A. García-Ayala, and A. López. 1996. Structure of the amphibian mesonephric tubule during ontogenesis in *Rana ridibunda* L. tadpoles: Early ontogenetic stages, renal corpuscle formation, neck segment, and peritoneal funnels. *Anat. Embryol.* 193:397–406.

Meseguer, J., M. T. Lozano, and B. Aguilleiro. 1985. Ultrastructure of the renal granulopoietic tissue of the *Rana ridibunda* tadpole. *J. Submicros. Cytol.* 17:391–401.

Metter, D. E. 1964. A morphological and ecological comparison of two populations of the tailed frog, *Ascaphus truei* Stejneger. *Copeia* 1964:181–195.

———. 1967. Variation in the ribbed frog, *Ascaphus truei* Stejneger. *Copeia* 1967:634–649.

Meyer, M., J. Fröhlich, and I. Nagel. 1975a. Zur epidermalen Kugelzelle der Larve von *Ascaphus truei* Stejneger (Amphibia, Salientia, Ascaphidae): Ein biometrischer Beitrag. *Zeit. Mikros. Anat. Forsch.* 89:1086–1098.

Meyer, M., R. Guderjahn, and J. Fröhlich. 1975b. Über Kugelzellen und die Struktur der larvalen Epidermis von Froschlurchen. I. *Ascaphus truei* Stejneger. *Zeit. Mikros. Anat. Forsch. Leipzig* 89:79–107.

Michael, M. I., and M. A. Al Adhami. 1974. The development of the thyroid glands in anuran amphibians of Iraq. *J. Zool.* 174:315–323.

Michael, M. I., and A. Y. Yacob. 1974. The development, growth and degeneration of the pronephric system in anuran amphibians of Iraq. *J. Zool.* 174:407–417.

Michael, S. F. 1997. Captive breeding of *Eleutherodactylus antillensis* (Anura: Leptodactylidae) from Puerto Rico with notes on behavior in captivity. *Herpetol. Rev.* 28:141–143.

Michalowski, J. 1966. Studies on the relationship of *Bombina bombina* (Linnaeus) and *Bombina variegata* (Linnaeus) II. Some taxonomic characters of tadpoles of both species and of tadpoles obtained from cross under laboratory conditions. *Acta Zool. Cracov.* 11:67–94.

Mijares-Urrutia, A. 1990. The tadpole of *Centrolenella andina* (Anura: Centrolenidae). *J. Herpetol.* 24:410–412.

Millamena, O. M., E. J. Aujero, and I. G. Borlongan. 1990. Techniques for algae harvesting and preservation for use in culture and as larval food. *Aquacult. Engineer.* 9:295–304.

Millard, N. 1942. Abnormalities and variations in the vascular

system of *Xenopus laevis* (Daudin). *Trans. Roy. Soc. South Africa* 29:9–28.
———. 1945. The development of the arterial system of *Xenopus laevis,* including experiments on the destruction of the larval aortic arches. *Trans. Roy. Soc. South Africa* 30:217–234.
———. 1949. The development of the venous system of *Xenopus laevis. Trans. Roy. Soc. South Africa* 32:55–96.
Miranda, L. A., D. A. Paz, and A. Pisanó. 1991. Efecto de la actividad tiroidea en la tolerancia al frio de larvas de *Bufo arenarum. Rev. España Herpetol.* 6:55–60.
Miranda, L. A., and A. Pisanó. 1993. Efecto de la densidad poblacional en larvas de *Bufo arenarum* producido a traves de señales visuales. *Alytes* 11:64–76.
Mishra, P. K., and M. C. Dash. 1984. Larval growth and development of a tree frog, *Rhacophorus maculatus* Gray. *Trop. Ecol.* 25:203–207.
Mitchell, N., and R. Swain. 1996. Terrestrial development in the Tasmanian frog, *Bryobatrachus nimbus* (Anura: Myobatrachidae): Larval development and a field staging table. *Pap. Proc. Roy. Soc. Tasmania* 130:75–80.
Mitchell, S. L. 1983. Spatial distribution of eight species of anuran tadpoles. Thesis, Mississippi State University.
Mitchell, S. L., and R. Altig. 1983. The feeding ecology of *Sooglossus gardineri. J. Herpetol.* 17:281–282.
Miyamoto, M. M. 1985. Consensus cladograms and general classifications. *Cladistics* 1:186–189.
Mobbs, I. G., V. A. King, and R. J. Wassersug. 1988. Prostaglandin E_2 does not inhibit metamorphosis of tadpole tails in tissue culture. *Exp. Biol.* 47:151–154.
Mocquard, F. 1890. Recherches sur la faune herpetologique des Iles de Bornéo et de Palawan. *Nouv. Arch. Mus. Natl. Hist. Nat.* (3):115–168, plates 7–11.
Mohanty, S. N., and M. C. Dash. 1986. Effects of diet and aeration on the growth and metamorphosis of *Rana tigrina* tadpoles. *Aquaculture* 51:89–96.
Mohanty-Hejmadi, P., S. K. Dutta, and S. C. Mallick. 1979. Life history of Indian frogs II The marbled balloon frog *Uperodon systoma* Schneider. *J. Zool. Soc. India* 31:65–72.
Mohawald, A. P., and S. Hennen. 1971. Ultrastructure of the germ plasm in eggs and in embryos of *Rana pipiens. Develop. Biol.* 24:37–53.
Monayong Ako'o, M. 1981. L'evolution du tube digestif chez la larve de *Phrynodon sandersoni* Parker (Amphibia, Anura, Phrynobatrachinae). *Ann. Fac. Sci. Cameroun* 28:105–115.
Montgomery, N. M. 1988. Projections of the vestibular and cerebellar nuclei in *Rana pipiens. Brain Behav. Evol.* 31:82–95.
Montgomery, N. M., and K. V. Fite. 1989. Retinotopic organization of central optic projections in *Rana pipiens. J. Comp. Neurol.* 283:526–540.
Montgomery, N. M., K. V. Fite, and A. M. Grigonis. 1985. The pretectal nucleus *lentiformis mesencephali* of *Rana pipiens. J. Comp. Neurol.* 234:264–275.
Montgomery, N., K. V. Fite, M. Taylor, and L. Bengston. 1982. Neural correlates of optokinetic nystagmus in the mesencephalon of *Rana pipiens:* A functional analysis. *Brain Behav. Evol.* 21:137–150.
Mookerjee, H. K. 1931. On the development of the vertebral column of Anura. *Philosoph. Trans. Roy. Soc. London* 219B:165–196.
Mookerjee, H. K., and S. K. Das. 1939. Further investigation on the development of the vertebral column in Salientia (Anura). *J. Morphol.* 64:167–209.
Moore, J. A. 1949a. Patterns of evolution in the genus *Rana.* pp. 315–338. In *Genetics, paleontology, and evolution,* edited by G. L. Jepsen, E. Mayr, and G. G. Simpson. Princeton: Princeton University Press.
———. 1949b. Geographic variation of adaptive characters in *Rana pipiens* Schreber. *Evolution* 3:1–24.
———. 1961. Frogs of eastern New South Wales. *Bull. American Mus. Nat. Hist.* 121:149–386.
———. 1972. *Heredity and development.* 2d ed. New York: Oxford University Press.
Moore, M. K., and S. G. Faragher. 1998. Tadpole jello: A technique for monitoring food consumption and manipulating the chemical composition of larval foodstuffs. *Herpetol. Rev.* 29:18–19.
Moore, M. K., and V. R. Townsend, Jr. 1998. The interaction of temperature, dissolved oxgyen and predation pressure in an aquatic predator-prey system. *Oikos* 81:329–336.
Moran, N. A. 1994. Adaptation and constraint in the complex life cycles of animals. *Ann. Rev. Ecol. Syst.* 25:573–600.
Moreira, G., and A. P. Lima. 1991. Seasonal patterns of juvenile recruitment and reproduction in four species of leaf litter frogs in central Amazonia. *Herpetologica* 47:295–300.
Morgan, B. E., N. I. Passmore, and B. C. Fabian. 1989. Metamorphosis in the frog *Arthroleptella lightfooti* (Anura: Ranidae) with emphasis on neuro-endocrine mechanisms. pp. 347–370. In *Alternative life-history styles of animals,* edited by M. N. Burton. Dordrecht: Kluwer Acad. Publ.
Morgan, M. J., J. M. O'Donnell, and R. F. Oliver. 1973. Development of the left-right asymmetry in the habenular nuclei of *Rana temporaria. J. Comp. Neurol.* 149:203–214.
Morin, P. J. 1981. Predatory salamanders reverse the outcome of competition among three species of anuran tadpoles. *Science* 212:1284–1286.
———. 1983. Predation, competition, and the composition of larval anuran guilds. *Ecol. Monogr.* 53:119–138.
———. 1986. Interactions between intraspecific competition and predation in an amphibian predator-prey system. *Ecology* 67:713–720.
———. 1987. Predation, breeding asynchrony, and the outcome of competition among treefrog tadpoles. *Ecology* 68:675–683.
———. 1989. New directions in amphibian community ecology. *Herpetologica* 45:124–128.
Morin, P. J., and E. A. Johnson. 1988. Experimental studies of asymmetric competition among anurans. *Oikos* 53:398–407.
Morin, P. J., S. P. Lawler, and E. A. Johnson. 1988. Competition between aquatic insects and vertebrates: Interaction strength and higher-order interactions. *Ecology* 69:1401–1409.
———. 1990. Ecology and breeding phenology of larval *Hyla andersonii:* The disadvantages of breeding late. *Ecology* 71:1590–1598.
Morris, J. L., and G. Campbell. 1978. Renal vascular anatomy of the toad (*Bufo marinus*). *Cell Tiss. Res.* 189:501–514.
Moss, B., and V. M. Ingram. 1968a. Hemoglobin synthesis during amphibian metamorphosis. I. Chemical studies on the hemoglobins from the larval and adult stages of *Rana catesbeiana. J. Mol. Biol.* 32:481–492.
———. 1968b. Hemoglobin synthesis during amphibian metamorphosis. II. Synthesis of adult hemoglobin following thyroxine administration. *J. Mol. Biol.* 32:493–504.
Moulton, J. M., A. Jurand, and H. Fox. 1968. A cytological study of Mauthner's cell in *Xenopus laevis* and *Rana temporaria* during metamorphosis. *J. Embryol. Exp. Morphol.* 19:415–431.

Moury, J. D., and J. Hanken. 1995. Early cranial neural crest migration in the direct-developing frog, *Eleutherodactylus coqui*. *Acta Anat.* 153:243–253.
Mudry, K. M., and R. R. Capranica. 1980. Evoked auditory activity within the telencephalon of the bullfrog (*Rana catesbeiana*). *Brain Res.* 182:303–311.
———. 1987. Correlation between auditory evoked responses in the thalamus and species-specific call characteristics. I. *Rana catesbeiana* (Anura: Ranidae). *J. Comp. Physiol.* 160A:477–489.
Mudry, K. M., M. Constantine-Paton, and R. R. Capranica. 1977. Auditory sensitivity of the diencephalon of the leopard frog *Rana pipiens*. *J. Comp. Physiol.* 114A:1–13.
Mullally, D. P. 1953. Observations on the ecology of the toad *Bufo canorus*. *Copeia* 1953:182–183.
Müller, F., and P. Sprumont. 1972. Entwicklung des Aditus laryngis bei *Rana ridibunda*. Licht-und elektronenoptische Befunde. *Rev. Siusse Zool.* 79:1114–1120.
Munn, N. L. 1940a. Learning experiments with larval frogs. A preliminary report. *J. Comp. Physiol. Psychol.* 29:97–108.
———. 1940b. Learning experiments with larval frogs. *J. Comp. Physiol. Psychol.* 53:443–445.
Munro, A. F. 1953. The ammonia and urea excretion of different species of Amphibia during their development and metamorphosis. *Biochem. J.* 54:29–36.
Muntz, L. 1975. Myogenesis in the trunk and leg during development of the tadpole of *Xenopus laevis* (Daudin 1802). *J. Embryol. Exp. Morphol.* 33:757–774.
Muntz, L., J. E. Hornby, and M. R. A. Dalooi. 1989. A comparison of the distribution of muscle type in the tadpole tails of *Xenopus laevis* and *Rana temporaria:* An histological and ultrastructural study. *Tiss. Cell* 21:773–781.
Muntz, W. R. A. 1962a. Microelectrode recordings from the diencephalon of the frog (*Rana pipiens*) and a blue sensitive system. *J. Neurophysiol.* 25:699–711.
———. 1962b. Effectiveness of different colors of light in releasing positive phototactic behavior of frogs, and a possible function of the retinal projection to the diencephalon. *J. Neurophysiol.* 25:712–720.
Muntz, W. R. A., and T. Reuter. 1966. Visual pigments and spectral sensitivity in *Rana temporaria* and other European tadpoles. *Vision Res.* 6:601–618.
Murray, D. L. 1990. The effects of food and density on growth and metamorphosis in wood frogs (*Rana sylvatica*) from central Labrador. *Canadian J. Zool.* 68:1221–1226.
Muto, Y. 1971. Reduction in size or closure of the spiracle and defective digital malformations in the toad larva reared in aerated water kept at a high temperature. *Bull. Aichi Univ. Edu.* 20:139–149.
Myers, C. W., A. Paolillo O., and J. W. Daly. 1991. Discovery of a defensively malodorous and nocturnal frog in the family Dendrobatidae: Phylogenetic significance of a new genus and species from the Venezuelan Andes. *American Mus. Novitates* (3002):1–33.
Nachtigall, W. 1974. *Biological mechanisms of attachment: The comparative morphology and bioengineering of organs for linkage, suction, and adhesion.* New York: Springer-Verlag.
———. 1982. Biophysik der Fortbewegung. pp. 608–622. In *Biophysik,* edited by W. Hoppe, H. M. Lohmann, and H. Ziegler. Berlin: Springer-Verlag.
Nagai, Y., S. Nagai, and T. Nishikawa. 1971. The nutritional efficiency of cannibalism and an artificial feed for the growth of tadpoles of Japanese toad (*Bufo vulgaris* sp.). *Agr. Biol. Chem.* 35:697–703.
Naito, I. 1984. The development of glomerular capillary tufts of the bullfrog kidney from a straight interstitial vessel to an anastomosed capillary network. A scanning electron microscopic study of vascular casts. *Arch. Histol. Japonica* 47:441–456.
Naitoh, T., A. Miura, and H. Akiyoshi. 1990. Movements of the large intestine in the anuran larvae, *Xenopus laevis. Comp. Biochem. Physiol.* 97C:201–208.
Naitoh, T., R. J. Wassersug, and R. A. Leslie. 1989. The physiology, morphology, and ontogeny of emetic behavior in anuran amphibians. *Physiol. Zool.* 62:819–843.
Nakata, K., M. Sokabe, and R. Suzuki. 1982. A model for the crowding effect in the growth of tadpoles. *Biol. Cybern.* 42:169–176.
Nanba, H. 1972. Ultrastructural and cytochemical studies on the thyroid gland of normal metamorphosing frogs (*Rana japonica* Guenther) I. Fine structural aspects. *Arch. Histol. Japonica* 34:277–291.
Nathan, J. M., and V. G. James. 1972. The role of protozoan in the nutrition of tadpoles. *Copeia* 1972:669–679.
Nation, J. L. 1983. A new method using hexamethyldisilazane for preparation of soft insect tissues for scanning electron microscopy. *Stain Tech.* 58:347–351.
Naue, H. 1890. Ueber Bau und Entwicklung der Kiemen der Froschlarven. *Zeit. Naturwiss.* 63:129–176.
Neary, T. J. 1988. Forebrain auditory pathways in ranid frogs. pp. 233–252. In *The evolution of the amphibian auditory system,* edited by B. Fritzsch, M. J. Ryan, W. Wilczynski, T. E. Hetherington, and W. Walkowiak. New York: Wiley.
Neary, T. J., and R. G. Northcutt. 1983. Nuclear organization of the bullfrog diencephalon. *J. Comp. Neurol.* 213:262–278.
Neary, T. J., and W. Wilczynski. 1977. Autoradiographic demonstration of hypothalamic efferents in the bullfrog, *Rana catesbeiana. Anat. Rec.* 187:665.
———. 1986. Auditory pathways to the hypothalamus in ranid frogs. *Neurosci. Lett.* 71:142–146.
Nelson, C. E., and R. Altig. 1972. Tadpoles of the microhylids *Gastrophryne elegans* and *G. usta. Herpetologica* 28:381–383.
Nelson, C. E., and H. S. Cuellar. 1968. Anatomical comparison of the tadpoles of the genera *Hypopachus* and *Gastrophryne. Copeia* 1968:423–424.
Nelson, G. 1978. Ontogeny, phylogeny, paleontology, and the biogenetic law. *Syst. Zool.* 27:324–345.
Newman, R. A. 1987. Effects of density and predation on *Scaphiopus couchii* tadpoles in desert ponds. *Oecologia* 71:301–307.
———. 1988a. Adaptive plasticity in development of *Scaphiopus couchii* tadpoles in desert ponds. *Evolution* 42:774–783.
———. 1988b. Genetic variation for larval anuran (*Scaphiopus couchii*) development time in an uncertain environment. *Evolution* 42:763–783.
———. 1989. Developmental plasticity of *Scaphiopus couchii* tadpoles in an unpredictable environment. *Ecology* 70:1775–1787.
———. 1994. Effects of changing density and food level on metamorphosis of a desert amphibian, *Scaphiopus couchii. Ecology* 75:1085–1096.
———. 1995. Genetic variation for phenotypic plasticity in the larval life history of spadefoot toads (*Scaphiopus couchii*). *Evolution* 48:1773–1785.
Newman, R. A., and A. E. Dunham. 1994. Size at metamorphosis and water loss in a desert anuran (*Scaphiopus couchii*). *Copeia* 1994:372–381.
Ńguenga, D., G. G. Teugels, M. Legendre, and F. Ollevier.

1997. First data on the experimental evaluation of predation on *Heterobranchus longifilis* larvae (Siluroidei; Clariidae) by toad tadpoles (*Bufo regularis*). *Aquacult. Res.* 28:335–339.

Nichols, H. W. 1975. Growth media—freshwater. pp. 7–24. In *Handbook of phycological methods. Culture methods and growth measurements,* edited by J. R. Stein. Cambridge: Cambridge University Press.

Nichols, R. J. 1937. Taxonomic studies on the mouth parts of larval Anura. *Illinois Biol. Monogr.* 15(4):1–73.

Nie, M., J. D. Crim, and G. R. Ultsch. 1999. Dissolved oxygen, temperature, and habitat selection by bullfrog (*Rana catesbeiana*) tadpoles. *Copeia* 1999:153–162.

Nieuwkoop, P. D., and J. Faber. 1956. *Normal tables of* Xenopus laevis *(Daudin).* Amsterdam: North-Holland.

———. 1967. *Normal table of* Xenopus laevis *(Daudin): A systematical and chronological survey of the development from the fertilized egg till the end of metamorphosis.* 2d ed. Amsterdam: North-Holland.

Nieuwkoop, P. D., and L. A. Sutasurya. 1976. Embryological evidence for a possible polyphyletic origin of the recent amphibians. *J. Embryol. Exp. Morphol.* 35:159–167.

Nieuwenhuys, R., and P. Opdam. 1976. Structure of the brain stem. p. 811–855. In *Frog neurobiology, a handbook,* edited by R. Llinás and W. Precht. Berlin: Springer-Verlag.

Nikeryasova, E. N., and V. A. Golichenkov. 1988. Dynamics of pigment granule distribution in dermal melanophores of larval anurans. 2. Aggregation. *Soviet J. Develop. Biol.* 19:406–412.

Nikundiwe, A. M., and R. Nieuwenhuys. 1983. The cell masses in the brainstem of the South African clawed frog *Xenopus laevis:* A topographical and topological analysis. *J. Comp. Neurol.* 213:199–219.

Nina, H. L., and E. M. Del Pino V. 1977. Estructura histologica del ovario del sapo *Eleutherodactylus unistrigatus* y observaciones sobre el desarrollo embrionario. *Rev. Univ. Catolica Ecuador* 5:31–41.

Nishikawa, A., M. Kaiho, and K. Yoshizato. 1989. Cell death in the anuran tadpole tail: Thyroid hormone induces keratinization and tail-specific growth inhibition of epidermal cells. *Develop. Biol.* 131:337–344.

Nishikawa, K., and R. Wassersug. 1988. Morphology of the caudal spinal cord in *Rana* (Ranidae) and *Xenopus* (Pipidae) tadpoles. *J. Comp. Neurol.* 269:193–202.

———. 1989. Evolution of spinal nerve number in anuran larvae. *Brain Behav. Evol.* 33:15–24.

Noble, G. K. 1917. The systematic status of some batrachians from South America. *Bull. American Mus. Nat. Hist.* 37:793–814.

———. 1925a. An outline of the relation of ontogeny to phylogeny within the Amphibia, I, II. *American Mus. Novitates* (165):1–17, (166):1–10.

———. 1925b. The integumentary, pulmonary and cardiac modifications correlated with increased cutaneous respiration in the Amphibia: A solution to the "hairy frog" problem. *J. Morphol.* 40:341–416.

———. 1926a. The hatching process of *Alytes, Eleutherodactylus* and other amphibians. *American Mus. Novitates* (229):1–7.

———. 1926b. The importance of larval characteristics in the classification of South African Salientia. *American Mus. Novitates* (237):1–10.

———. 1927. The value of life history data in the study of the evolution of the Amphibia. *Ann. New York Acad. Sci.* 30:31–128, plate 9.

———. 1929. The adaptive modifications of the arboreal tadpoles of *Hoplophryne* and the torrent tadpoles of *Staurois. Bull. American Mus. Nat. Hist.* 58:291–334.

———. 1931. *The biology of the Amphibia.* New York: McGraw-Hill.

Noden, D. M., and A. De Lahunta. 1985. *The embryology of domestic animals. Developmental mechanisms and malformations.* Baltimore: Williams and Wilkins.

Nodzenski, E., and R. F. Inger. 1990. Uncoupling of related structural changes in metamorphosing torrent-dwelling tadpoles. *Copeia* 1990:1047–1054.

Nodzenski, E., R. J. Wassersug, and R. F. Inger. 1989. Developmental differences in visceral morphology of megophryine pelobatid tadpoles in relation to their body form and mode of life. *Biol. J. Linnean Soc.* 38:369–388.

Noland, R., and G. R. Ultsch. 1981. The roles of temperature and dissolved oxygen in microhabitat selection by the tadpoles of a frog (*Rana pipiens*) and a toad (*Bufo terrestris*). *Copeia* 1981:645–652.

Nöllert, A., and C. Nöllert. 1992. *Die Amphibien Europas. Besstimmung–Gefährdung–Schutz.* Stuttgart: Franckh-Kosmos.

Nomura, S., Y. Shiba, Y. Muneoka, and Y. Kanno. 1979a. A scanning and transmission electron microscope study of the premetamorphic papillae: Possible chemoreceptive organs in the oral cavity of an anuran tadpole (*Rana japonica*). *Arch. Histol. Japonica* 42:507–516.

———. 1979b. Developmental changes of premetamorphic and fungiform papillae of the frog (*Rana japonica*) during metamorphosis: A scanning electron microscopy. *Hiroshima J. Med. Sci.* 28:79–86.

Nordlander, R. H. 1984. Developing descending neurons of the early *Xenopus* tail spinal cord in the caudal spinal cord of early *Xenopus. J. Comp. Neurol.* 228:117–128.

———. 1986. Motoneurons of the tail of young *Xenopus* tadpoles. *J. Comp. Neurol.* 253:403–413.

———. 1987. Axonal growth cones in the developing amphibian spinal cord. *J. Comp. Neurol.* 263:485–496.

Nordlander, R. H., S. T. Baden, and T. Ryba. 1985. Development of early brainstem projections to the tail spinal cord of *Xenopus. J. Comp. Neurol.* 231:519–529.

Nordlander, R. H., J. W. Gazzerro, and H. Cook. 1991. Growth cones and axon trajectories of a sensory pathway in the amphibian spinal cord. *J. Comp. Neurol.* 307:301–250.

Nordlander, R. H., J. F. Singer, and M. Singer. 1981. An ultrastructural examination of early ventral root formation in Amphibia. *J. Comp. Neurol.* 199:535–551.

Nordlander, R. H., and M. Singer. 1982a. Spaces precede axons in *Xenopus* embryonic spinal cord. *Exp. Neurol.* 75:221–228.

———. 1982b. Morphology and position of growth cones in the developing *Xenopus* spinal cord. *Develop. Brain. Res.* 4:181–193.

Northcutt, R. G. 1974. Some histochemical observations on the telencephalon of the bullfrog, *Rana catesbeiana* Shaw. *J. Comp. Neurol.* 157:379–390.

———. 1980. Central auditory pathways in anamniotic vertebrates. pp. 79–118. In *Proceedings in life sciences. Comparative studies of hearing in vertebrates,* edited by A. N. Popper and R. R. Fay. New York: Springer-Verlag.

———. 1981. Evolution of the telencephalon in nonmammals. *Ann. Rev. Neurosci.* 4:301–350.

———. 1986. Lungfish neural characters and their bearing on sarcopterygian phylogeny. *J. Morphol. Suppl.* 1:277–297.

———. 1989. The phylogenetic distribution and innervation of craniate mechanoreceptive lateral lines. pp. 17–78. In *The mechanosensory lateral line. Neurobiology and evolution,* edited by S. Coombs, P. Görner, and H. Münz. New York: Springer-Verlag.

Northcutt, R. G., and G. J. Royce. 1975. Olfactory bulb projections in the bullfrog *Rana catesbeiana. J. Morphol.* 145:251–268.

Nowogrodzka-Zagórska, M. 1974. The organization of extraocular muscles in Anura. *Acta Anat.* 87:22–44.

Nyman, S. 1986. Mass mortality in larval *Rana sylvatica* attributable to the bacterium, *Aeromonas hydrophila. J. Herpetol.* 20:196–201.

Odendaal, F. J., and C. M. Bull. 1983. Water movements, tadpole competition, and limits to the distribution of the frogs *Ranidella riparia* and *R. signifera. Oecologia* 57:361–367.

Odendaal, F. J., C. M. Bull, and R. C. Nias. 1982. Habitat selection in tadpoles of *Ranidella signifera* and *R. riparia* (Anura: Leptodactylidae). *Oecologia* 52:411–414.

Ogielska, M., and E. Wagner. 1990. Oogenesis and development of the ovary in European green frog, *Rana ridibunda* (Pallas). I. Tadpole stages until metamorphosis. *Zool. Jahrb. Abt. Anat. Ontog. Tiere* 120:211–221.

O'Hara, R. K. 1974. Effects of developmental stage and prior experience on habitat selection in three species of anuran larvae. Thesis, Michigan State University.

———. 1981. Habitat selection behavior in three species of anuran larvae: Environmental cues, ontogeny, and adaptive significance. Dissertation, Oregon State University.

O'Hara, R. K., and A. R. Blaustein. 1981. An investigation of sibling recognition in *Rana cascadae* tadpoles. *Anim. Behav.* 29:1121–1126.

———. 1982. Kin preference behavior in *Bufo boreas* tadpoles. *Behav. Ecol. Sociobiol.* 11:43–49.

———. 1985. *Rana cascadae* tadpoles aggregate with siblings: An experimental field study. *Oecologia* 67:44–51.

———. 1988. *Hyla regilla* and *Rana pretiosa* tadpoles fail to display kin recognition behaviour. *Anim. Behav.* 36:946–948.

Ohtani, O., and I. Naito. 1980. Renal microcirculation of the bullfrog, *Rana catesbeiana.* A scanning electron microscope study of vascular casts. *Arch. Histol. Japonica* 43:319–330.

Ohzu, E., N. Abe, and I. Nakatani. 1987. Scanning electron microscopic observations of the development of hatching gland cells in amphibian embryos. *Bull. Yamagata Univ., Nat. Sci.* 11:431–438.

Oka, Y., R. Ohtani, M. Satou, and K. Ueda. 1989. Location of forelimb motoneurons in the Japanese toad (*Bufo japonicus*): A horseradish peroxidase study. *J. Comp. Neurol.* 286:376–383.

Okada, Y. 1931. The tailless batrachians of the Japanese Empire. *Imp. Agr. Exp. Sta.* 1–207, 29 plates.

Oku, U., K. Katakura, J. Nagatsuka, and M. Kayima. 1979. Possible role of the tadpole of *Rana chensinensis* as an intermediate host of *Angiostrongylus cantonensis* parasitic to rodents. *Japanese J. Vet. Res.* 27:1–4.

Okutomi, K. 1937. Die Entwicklung des Chondrocraniums von *Polypedates buergeri schlegelii. Zeit. Anat. Entwickl.* 107:28–64.

Oldham, R. S. 1977. Terrestrial locomotion in two species of amphibian larvae. *J. Zool.* 181:285–295.

Olson, D. H. 1988. The ecological and behavioral dynamics of breeding in three sympatric anuran amphibians. Dissertation, Oregon State University.

Olson, D. H., W. P. Leonard, and R. B. Bury, eds. 1997. Sampling amphibians in lentic habitats: Methods and approaches for the Pacific Northwest. *Northwest Fauna (4), Soc. Northw. Vert. Biol.* Olympia, WA.

Omerza, F. F., and K. E. Alley. 1992. Redeployment of trigeminal motor neurons during metamorphosis. *J. Comp. Neurol.* 325:124–134.

Opatrny, E. 1973. Clay "nests" of toad tadpoles. *British J. Herpetol.* 4:338–339.

Opdam, P., M. Kemali, and R. Nieuwenhuys. 1976. Topological analysis of the brain stem of the frogs *Rana esculenta* and *Rana catesbeiana. J. Comp. Neurol.* 165:307–332.

Orlando, K., and A. W. Pinder. 1995. Larval cardiorespiratory ontogeny and allometry in *Xenopus laevis. Physiol. Zool.* 68:63–75.

Orr, P. R. 1955. Heat death. II. Differential response of entire animal (*Rana pipiens*) and several organ systems. *Physiol. Zool.* 28:294–302.

Orton, G. L. 1943. The tadpole of *Rhinophrynus dorsalis. Occas. Pap. Mus. Zool., Univ. Michigan* (472):1–7, 1 plate.

———. 1944. Studies on the systematic and phylogenetic significance of certain larval characters in the Amphibia Salientia. Dissertation, University of Michigan.

———. 1946. The unknown tadpole. *Turtox News* 24:131–132.

———. 1949. Larval development of *Nectophrynoides tornieri* (Roux), with comments on direct development in frogs. *Ann. Carnegie Mus.* 31:257–276.

———. 1952. Key to the genera of tadpoles in the United States and Canada. *American Midl. Nat.* 47:382–395.

———. 1953. The systematics of vertebrate larvae. *Syst. Zool.* 2:63–75.

———. 1954. Dimorphism in larval mouthparts in spadefoot of the *Scaphiopus hammondi* group. *Copeia* 1954:97–100.

———. 1957. The bearing of larval evolution on some problems in frog classification. *Syst. Zool.* 6:79–86.

Osborne, P. L., and A. J. McLachlan. 1985. The effect of tadpoles on algal growth in temporary, rain-filled rock pools. *Freshw. Biol.* 15:77–87.

Ott, M. E., N. Heisler, and G. R. Ultsch. 1980. A reevaluation of the relationship between temperature and the critical oxygen tension in freshwater fishes. *Comp. Biochem. Physiol.* 67A:337–340.

Ottoson, D. 1976. Morphology and physiology of muscle spindles. pp. 643–675. In *Frog neurobiology, a handbook,* edited by R. Llinás and W. Precht. Berlin: Springer-Verlag.

Ovalle, W. K. 1979. Neurite complexes with Merkel cells in larval tentacles of *Xenopus laevis. Cell Tiss. Res.* 204:233–341.

Ovaska, K. 1991. Reproductive phenology, population structure, and habitat use of the frog *Eleutherodactylus johnstonei* in Barbados, West Indies. *J. Herpetol.* 25:424–430.

Overton, J. 1959. Studies on the mode of outgrowth of the amphibian pronephric duct. *J. Embryol. Exp. Morphol.* 7:86–93.

Oxner, W. M., J. Quinn, and M. E. DeMont. 1993. A mathematical model of body kinematics in swimming tadpoles. *Canadian J. Zool.* 71:407–413.

Padhye, A. D., and H. V. Ghate. 1988. Effect of altered pH on embryos and tadpoles of the frog *Microhyla ornata. Herpetol. J.* 1:276–279.

Paicheler, J.-C., F. De Broin, J. Gaudant, C. Mourer-Chauvire, J.-C. Rage, and C. Vergnaud-Grazzini. 1978. Le bassin lacustre Miocène de Bes-Konak (Anatolie-Turquie): Géologie et introduction à la paléontologie des vertébrés. *Géobios* (11):43–65.

Pancharatna, M., and S. K. Saidapur. 1985. Ovarian cycle in the frog *Rana cyanophlyctis:* A quantitative study of follicular kinetics in relation to body mass, oviduct, and fat body cycles. *J. Morphol.* 186:135–147.

Pandian, T. J., and M. P. Marian. 1985a. Predicting anuran metamorphosis and energetics. *Physiol. Zool.* 58:538–552.

———. 1985b. Time and energy costs of metamorphosis in the Indian bullfrog *Rana tigrina*. *Copeia* 1985:653–662.

———. 1985c. Physiological correlates of surfacing behaviour—effect of aquarium depth on surfacing, growth, and metamorphosis in *Rana tigrina*. *Physiol. Behav.* 35:867–872.

Papernal, I., and R. Lainson. 1995. *Alloglugea bufonis* nov. gen., nov. sp. (Microsporea: Glugeidae), a microsporidian of *Bufo marinus* tadpoles and metamorphosing toads (Amphibia: Anura) from Amazonian Brasil. *Dis. Aquat. Org.* 23:7–11.

Parichy, D. M., and R. H. Kaplan. 1992a. Developmental consequences of tail injury on larvae of the Oriental fire-bellied toad, *Bombina orientalis*. *Copeia* 1992:129–137.

———. 1992b. Maternal effects on offspring growth and development depend on environmental quality in the frog *Bombina orientalis*. *Oecologia* 91:579–586.

———. 1995. Maternal investment and developmental plasticity: Functional consequences for locomotor performance of hatchling frog larvae. *Funct. Ecol.* 9:606–611.

Parker, G. E. 1967. The influence of temperature and thyroxine on oxygen consumption in *Rana pipiens* tadpoles. *Copeia* 1967:610–616.

Parker, H. W. 1934. *A monograph of the Microhylidae*. London: Jarrold and Sons, London (67 figs).

———. 1936. A collection of reptiles and amphibians from the mountains of British New Guinea. *Ann. Mag. Nat. Hist.* Ser. 10. 17:66–93.

Parker, W. K. 1871. On the structure and development of the skull of the common frog (*Rana temporaria* L.). *Philosoph. Trans. Roy. Soc. London* 161B:137–211.

———. 1876. On the structure and development of the skull in the Batrachia. Part II. *Proc. Zool. Soc. London* 1876:601–669.

———. 1881. On the structure and development of the skull in the Batrachia. Part III. *Philosoph. Trans. Roy. Soc. London* 172:1–266.

Partridge, B. L. 1982. The structure and function of fish schools. *Sci. American* 246:114–123.

Partridge, B. L., and T. J. Pitcher. 1980. The sensory basis of fish schools: Relative roles of lateral line and vision. *J. Comp. Physiol.* 135A:315–325.

Paterson, N. F. 1939a. The head of *Xenopus laevis*. *Quart. J. Micros. Sci.* 81:161–234.

———. 1939b. The olfactory organ and tentacles of *Xenopus laevis*. *South African J. Sci.* 36:390–404.

Patterson, J. W., and A. J. McLachlan. 1989. Larval habitat duration and size at metamorphosis in frogs. *Hydrobiologia* 171:121–126.

Pattist, M. J., ed. 1956. *Ovid's Metamorphoses*. 2d ed. Groningen: J. B. Wolters.

Pavignano, I. 1989. Method employed to study the diet of anuran amphibians larvae. *Amphibia-Reptilia* 10:453–456.

———. 1990. Niche overlap in tadpole populations of *Pelobates fuscus insubricus* and *Hyla arborea* at a pond in northwestern Italy. *Boll. Zool.* 57:83–87.

Peacor, S. D., and E. E. Werner. 1997. Trait-mediated indirect interactions in a simple aquatic food web. *Ecology* 78:1146–1156.

Pearman, P. B. 1995. Effects of pond size and consequent predatory density on two species of tadpoles. *Oecologia* 102:1–8.

Pearman, P. B., and H. M. Wilbur. 1990. Changes in population dynamics resulting from oviposition in a subdivided habitat. *American Nat.* 135:708–723.

Pearse, A. G. 1968. *Histochemistry: Theoretical and applied*. Vols. I and II. Boston: Little, Brown and Co.

Pehek, E. L. 1995. Competition, pH, and the ecology of larval *Hyla andersoni*. *Ecology* 76:1786–1793.

Peixoto, O. L. 1981. Notas sobre o girino de *Crossodactylodes pintoi* Cochran (Amphibia, Anura, Leptodactylidae). *Rev. Brasil. Biol.* 41:339–341.

———. 1982. Observações sobre a larva de *Pleurodema diplolistris* (Peters 1870)(Amphibia, Anura, Leptodactylidae). *Rev. Brasil. Biol.* 42:631–633.

Peixoto, O. L., and C. A. G. Da Cruz. 1983. Girinos de espécies de *Hyla* do grupo "*Albomarginata*" do Sueste Brasileiro (Amphibia, Anura, Hylidae). *Arq. Univ. Fed. Rural Rio de Janeiro* 6:155–163.

Pelaz, M. P. 1987. On the identity of the so-called "algae like cells" in tadpole cultures of European green frogs (*Rana ridibunda*). *Alytes* 6:23–26.

Pelaz, M. P., and C. Rougier. 1990. A comparative study of buccal pump volumes and the cartilagenous branchial basket skeleton of *Rana ridibunda* and *Rana dalamatina* tadpoles. *Copeia* 1990:658–665.

Pérez-Rivera, R. A., and E. Nadal. 1993. Dos nidos de coquíes (*Eleutherodactylus*) en nidos de Aves. *Caribbean J. Sci.* 29:128.

Perret, J.-L. 1966. Les amphibiens du Cameroun. *Zool. Jahrb. Syst.* 93:289–464.

Persson, L. 1988. Asymmetries in competitive and predatory interactions in fish populations. pp. 203–218. In *Size-structured populations: Ecology and evolution*, edited by B. Ebenmann and L. Persson. Berlin: Springer-Verlag.

Petersen, H. 1922. Zur Konstruktionanalyse des Kieferapparates der Froschlarven [V. Beitrag zur Mechanik des Tierkörpers.]. *Beitr. Zeit. Ges. Anat.* 64:22–28.

Peterson, J. A., and A. R. Blaustein. 1991. Unpalatability in anuran larvae as a defense against natural salamander predators. *Ethol. Ecol. Evol.* 3:63–72.

Petranka, J. W. 1985. Does age-specific mortality decrease with age in amphibian larvae? *Copeia* 1985:1080–1083.

———. 1989. Chemical interference competition in tadpoles: Does it occur outside laboratory aquaria? *Copeia* 1989:921–930.

———. 1995. Interference competition in tadpoles: Are multiple agents involved? *Herpetol. J.* 5:206–207.

Petranka, J. W., M. E. Hopey, B. T. Jennings, S. D. Baird, and S. J. Boone. 1994. Breeding habitat segregation of wood frogs and American toads: The role of interspecific tadpole predation and adult choice. *Copeia* 1994:691–697.

Petranka, J. W., and L. Hayes. 1998. Chemically mediated avoidance of a predatory odonate (*Anax junius*) by American toad (*Bufo americanus*) and wood frog (*Rana sylvatica*) tadpoles. *Behav. Ecol. Sociobiol.* 42:263–271.

Petranka, J. W., J. J. Just, and E. C. Crawford. 1983. Hatching of amphibian embryos; the physiological trigger. *Science* 217:257–259.

Petranka, J. W., J. J. Just, E. C. Crawford, L. B. Kats, and A. Sih. 1987. Predator-prey interactions among fish and larval amphibians: Use of chemical cues to detect predatory fish. *Anim. Behav.* 35:420–425.

Petranka, J. W., A. W. Rushlow, and M. E. Hopey. 1998. Predation by tadpoles of *Rana sylvatica* on embryos of *Ambystoma maculatum*: Implications of ecological role reversals by *Rana* (predator) and *Ambystoma* (prey). *Herpetologica* 54:1–13.

Petranka, J. W., and D. A. G. Thomas. 1995. Explosive breeding reduces egg and tadpole cannibalism in the wood frog, *Rana sylvatica. Anim. Behav.* 50:731–739.

Petrini, S., and F. Zaccanti. 1998. The effects of aromatase and 5 alpha-reductase inhibitors, antiandrogen, and sex steroids on Bidder's organ development and gonadal differentiation in *Bufo bufo* tadpoles. *J. Exp. Zool.* 280:245–259.

Pettigrew, A. G. 1981. Brainstem afferents to the torus semicircularis of the Queensland cane toad (*Bufo marinus*). *J. Comp. Neurol.* 202:59–68.

Pfeiffer, W. 1966. Die Verbreitung der Schreckreaktion bei Kaulquappen und die Herkunft des Schreckstoffes. *Zeit. Vergl. Physiol.* 52:79–98.

———. 1974. Pheromones in fishs and Amphibia. pp. 269–296. In *Pheromone: Frontiers of biology.* Vol 32, edited by M. C. Birch. New York: North Holland.

Pfennig, D. W. 1990a. The adaptive significance of an environmentally cued developmental switch in an anuran tadpoles. *Oecologia* 85:101–107.

———. 1990b. "Kin selection" among spadefoot toad tadpoles: A side-effect of habitat selection? *Evolution* 44:785–798.

———. 1992a. Polyphenism in spadefoot toad tadpoles as a locally adjusted evolutionarily stable strategy. *Evolution* 46:1408–1420.

———. 1992b. Proximate and functional causes of polyphenism in an anuran tadpole. *Funct. Ecol.* 6:167–174.

Pfenning, D. W., and W. A. Frankino. 1997. Kin-mediated morphogenesis in facultatively cannibalistic tadpoles. *Evolution* 51:1993–1999.

Pfenning, D. W., W. A. Frankino, A. Mabry, and D. Orange. 1991. Environmental causes of correlations between age and size at metamorphosis in *Scaphiopus multiplicatus. Ecology* 72:2240–2248.

Pfennig, D. W., H. K. Reeve, and P. K. Sherman. 1993. Kin recognition and cannibalism in spadefoot toad tadpoles. *Anim. Behav.* 46:87–94.

Pflugfelder, O., and G. Schubert. 1965. Elektronenmikroskopische Untersuchungen an der Haut von Larven-und Metamorphosestadien von *Xenopus laevis* nach Kaliumperahlanatbehandlung. *Zeit. Zellforsch. Mikros. Anat.* 67:96–112.

Piatka, B. A., and C. W. Gibley, Jr. 1967. Alkaline phosphatase activity in the developing pronephros of the frog (*Rana pipiens*). *Texas Rept. Biol. Med.* 25:83–89.

Piavaux, A. 1972. β-polysaccharidases du tube digestif de *Rana temporaria* L. pendant la croissance et la métamorphose. *Arch. Intl. Physiol. Biochim.* 80:153–160.

Pick, J. 1970. *The autonomic nervous system.* Philadelphia: J. B. Lippincott.

Picker, M. D. 1985. Hybridization and habitat selection in *Xenopus gilli* and *Xenopus laevis* in south-western Cape Province. *Copeia* 1985:574–580.

Picker, M. D., C. J. McKenzie, and P. Fielding. 1993. Embryonic tolerance of *Xenopus* (Anura) to acidic blackwater. *Copeia* 1993:1072–1081.

Pickersgill, M. 1984. Three new *Afrixalus* (Anura: Hyperoliidae) from south-eastern Africa. *Durban Mus. Novitates* 13:207–220.

Pierce, B. A. 1985. Acid tolerance in amphibians. *BioScience* 35:239–243.

Pierce, B. A., and J. M. Harvey. 1987. Geographic variation in acid tolerance of Connecticut wood frogs. *Copeia* 1987:94–103.

Pierce, B. A., J. B. Hoskins, and E. Epstein. 1984. Acid tolerance in Connecticut wood frogs (*Rana sylvatica*). *J. Herpetol.* 18:159–167.

Piiper, J. 1982. Respiratory gas exchange at lungs, gill, and tissues: Mechanisms and adjustments. *J. Exp. Biol.* 100:5–22.

Piiper, J., and P. Scheid. 1977. Comparative physiology of respiration: Functional analysis of gas exchange organs in vertebrates. pp. 219–253. In *Intl. Rev. Physiol., Respir. Physiol.* II, Vol. 14, edited by J. G. Widdicombe. Baltimore: University Park Press.

Pillai, R. S. 1978. On *Nyctibatrachus major* Boul. (Ranidae) with a description of its tadpole. *Bull. Zool. Surv. India* 1:135–140.

Pillai, R. S., and G. M. Yazdani. 1973. *Bufoides,* a new genus for the rock-toad, *Ansonia meghalayana* Yazdani and Chanda, with notes on its ecology and breeding habits. *J. Zool. Soc. India* 25:65–70.

Pinder, A. W., and W. W. Burggren. 1983. Respiration during chronic hypoxia and hyperoxia in larval and adult bullfrogs (*Rana catesbeiana*). *J. Exp. Biol.* 105:205–213.

Pinder, A. W., and M. E. Feder. 1990. Effect of boundary layers on cutaneous gas exchange. *J. Exp. Biol.* 143:67–80.

Pinder, A. W., B. Pelster, and R. G. Boutilier. 1990. Regulation of blood flow in isolated amphibian skin. *American Zool.* 30:136 (Abstr.).

Pinder, A. W., K. B. Storey, and G. R. Ultsch. 1992. Estivation and hibernation. pp. 250–274. In *Environmental physiology of the amphibians,* edited by M. E. Feder and W. W. Burggren. Chicago: University of Chicago Press.

Pintar, T. 1987. Zur Kenntnis des Tomatenfrosches *Dyscophus antongili* (Grandidier 1877)(Anura: Microhylidae). *Salamandra* 23:106–121.

Pitcher, T. J. 1986. *The behavior of teleost fishes.* Baltimore: John Hopkins University Press.

Plasota, K. 1974. The development of the chondrocranium (neurocranium and the mandibular and hyoid arches) in *Rana temporaria* L. and *Pelobates fuscus* (Laur.). *Zool. Poloniae* 24:99–168.

Plassman, W. 1980. Central neuronal pathways in the lateral line system of *Xenopus laevis. J. Comp. Physiol.* 136:203–213.

Płytycz, B., and J. Bigaj. 1983. Amphibian lymphoid organs: A review. *Folia Biol.* (Kraków) 31:225–240.

Płytycz, B., J. Dulak, and A. Pecio. 1984. Genetic control of length of larval period in *Rana temporaria. Folia Biol. (Kraków)* 32:155–166.

Płytycz, B., and M. Klimek. 1992. Effects of diet on transplantation immunity of tadpoles of the European common frog, *Rana temporaria. J. Nutr. Immunol.* 1:19–30.

Poinar, G. O., Jr. 1988. Infection of frog tadpoles (Amphibia) with insect parasitic nematodes. *Experientia* 44:528–531.

Polis, G. A. 1984. Age structure component of niche width and intraspecific resource partitioning: Can age groups function as ecological species? *American Nat.* 123:541–564.

———. 1988. Exploitation competition and the evolution of interference, cannibalism, and intraguild predation in age/size-structured populations. pp. 185–202. In *Size-structured populations: Ecology and evolution,* edited by B. Ebenmann and L. Persson. Berlin: Springer-Verlag.

Polis, G. A., and C. A. Myers. 1985. A survey of intraspecific predation among reptiles and amphibians. *J. Herpetol.* 19:99–107.

Poole, T. J., and M. S. Steinberg. 1977. SEM-aided analysis of morphogenetic movements: Development of the amphibian pronephric duct. pp. 43–52. In *Scanning electron microscopy/1977.* Vol. 2, edited by O. Johari. Chicago: ITT Res. Inst.

Pope, C. H. 1931. Notes on the amphibians from Fukein, Hai-

nan, and other parts of China. *Bull. American Mus. Nat. Hist.* 61:397–611.
———. 1940. *China's animal frontier.* New York: Viking.
Porges, R., L. A. Van Den Berg, and D. G. Ballinger. 1986. Reassessing an old problem—acid mine drainage. *Proc. American Soc. Civil Engr.* (San. Engr. Div.) 92(SA-1):69–83.
Poska-Teiss, L. 1930. Über die larvale Amphibienepidermis. *Zeit. Zellforsch. Mikros. Anat.* 11:445–483.
Postek, M. T., K. Howard, A. H. Johnson, and K. L. McMichael. 1980. *Scanning electron microscopy: A student's handbook.* Burlington, VT: Ladd Res. Ind., Inc.
Potel, M. J., and R. J. Wassersug. 1981. Computer tools for the analysis of schooling. *Environ. Biol. Fish.* 6:15–19.
Potter, H. D. 1965. Mesencephalic auditory region of the bullfrog. *J. Neurophysiol.* 28:1132–1154.
Potthoff, T. L., and J. D. Lynch. 1986. Interpopulation variability in mouthparts in *Scaphiopus bombifrons* in Nebraska (Amphibia: Pelobatidae). *Prairie Nat.* 18:15–22.
Pough, F. H., and S. Kamel. 1984. Post-metamorphic change in activity metabolism of anurans in relation to life history. *Oecologia* 65:138–144.
Pouyet, J. C., and A. Beaumont. 1975. Ultrastructure du pancréas larvaire d'un Amphibien Anoure, *Alytes obstetricans* L. en culture organotypique. *C. R. Soc. Biol. Paris* 169:846–850.
Powell, G. V. N. 1974. Experimental analysis of the social value of flocking by starlings (*Sturnus vulgaris*) in relation to predation and foraging. *Anim. Behav.* 22:501–]505.
Powell, T. L., and J. J. Just. 1985. Development of the urinary bladder in *Rana catesbiana* [sic] tadpoles. *American Zool.* 25:99. (Abstr.)
———. 1987. Ultrastructure and function of the developing urinary bladder in *Rana catesbeiana. American Zool.* 27:3. (Abstr.)
Power, J. H. 1927a. Some tadpoles from Griqualand West. *Trans. Roy. Soc. South Africa* 14:249–251, 14 plates.
———. 1927b. On the herpetological fauna of the Lobatsi-Linokana area. *Trans Roy. Soc. South Africa* 14:405–422.
Power, J. H., and W. Rose. 1929. Notes on the habits and life-histories of some Cape Peninsula Anura. *Trans. Roy. Soc. South Africa* 17:109–115.
Prakash, R. 1954. The heart and its conducting system in the tadpoles of the frog, *Rana tigrina* Daudin. *Proc. Zool. Soc. Calcutta* 7:28–36.
Prasadmurthy, Y. S., and S. K. Saidapur. 1987. Role of fat bodies in oocyte growth and recruitment in the frog *Rana cyanophlyctis* Schn. *J. Exp. Zool.* 243:153–162.
Precht, W. 1976. Physiology of the peripheral and central vestibular systems. pp. 481–512. In *Frog neurobiology, a handbook,* edited by R. Llinás and W. Precht. Berlin: Springer-Verlag.
Prestige, M. C. 1967. The control of cell number in the lumbar ventral horns during the development of *Xenopus laevis* tadpoles. *J. Embryol. Exp. Morphol.* 18:359–387.
———. 1973. Gradients in time of origin of tadpole motoneurons. *Brain Res.* 59:400–404.
Preston, F. W. 1968. The shape of bird eggs: Mathematical aspects. *Auk* 85:454–463.
Pretty, R., T. Naitoh, and R. J. Wassersug. 1995. Metamorphic shortening of the alimentary tract in anuran larvae (*Rana catesbeiana*). *Anat. Rec.* 242:417–423.
Price, G. C. 1897. Development of the excretory organs of a myxinoid *Bdellostoma stouti* Lockington. *Zool. Jahrb. Abh. Anat. Ontog. Tiere.* 10:205–226.
Prigioni, C. M., and D. Arrieta. 1992. Descripcion de la larva de *Melanophryniscus sanmartini* Klappenbach, 1968 (Amphibia: Anura: Bufonidae). *Bol. Soc. Zool. Uruguay* 7:57–58.
Prigioni, C. M., and J. A. Langone. 1990. Descripcion de la larva de *Melanophryniscus orejasmirandai* Prigioni & Langone, 1986 (Amphibia, Anura, Bufonidae). *Comun. Zool. Mus. Hist. Nat. Montevideo* 12:1–9.
Pronych, S., and R. Wassersug. 1994. Lung use and development in *Xenopus laevis* tadpoles. *Canadian J. Zool.* 72:738–743.
Prosser, C. L. 1973. *Comparative animal physiology.* Philadelphia: W. B. Saunders.
Prosser, C. L., and F. A. Brown, Jr. 1961. *Comparative animal physiology.* 2d ed. Philadelphia: W. B. Saunders.
Púgener, L. A., and A. M. Maglia. 1997. Osteology and skeletal development of *Discoglossus sardus* (Anura: Discoglossidae). *J. Morphol.* 233:267–286.
Pulliam, H. R. 1973. On the advantages of flocking. *J. Theor. Biol.* 38:419–422.
Pulliam, H. R., and T. Caraco. 1984. Living in groups: Is there an optimal group size? pp. 122–147. In *Behavioural ecology: An evolutionary approach,* edited by J. R. Krebs and N. B. Davis. Sunderland, MA: Sinauer Assoc.
Punzo, F. 1976. The effects of early experience on habitat selection in tadpoles of the Malayan painted frog, *Kaloula pulchra* (Anura: Microhylidae). *J. Bombay Nat. Hist. Soc.* 73:270–277.
———. 1991. Group learning in tadpoles of *Rana heckscheri* (Anura: Ranidae). *J. Herpetol.* 25:214–217.
Pusey, H. K. 1938. Structural changes in the anuran mandibular arch during metamorphosis, with reference to *Rana temporaria. Quart. J. Micros. Sci.* 80:479–552.
———. 1943. On the head of the liopelmid frog, *Ascaphus truei.* I. The chondrocranium, jaws, arches, and muscles of a partly-grown larva. *Quart. J. Micros. Sci.* 84:105–185.
Pyburn, W. F. 1967. Breeding and larval development of the hylid frog *Phrynohyas spilomma* in southern Veracruz, Mexico. *Herpetologica* 23:184–194.
———. 1975. A new species of microhylid frog of the genus *Synapturanus* from southeastern Colombia. *Herpetologica* 31:439–443.
———. 1980. An unusual anuran larvae from the Vaupés region of southeastern Colombia. *Pap. Avul. Zool.* 33:231–238.
Pyles, R. A. 1987. Morphology and mechanics of the jaws of anuran amphibians. Dissertation, University of Kansas.
Qualls, C. P., and R. Shine. 1995. Maternal body-volume as a constraint on reproductive output in lizards: Evidence from the evolution of viviparity. *Oecologia* 103:73–78.
———. 1996. Reconstructing ancestral reaction norms: An example using the evolution of reptilian viviparity. *Funct. Ecol.* 10:688–692.
Quinn, D., and W. Burggren. 1983. Lactate production, tissue distribution, and elimination following exhaustive exercise in larval and adult bullfrogs *Rana catesbeiana. Physiol. Zool.* 56:597–613.
Rabb, G. B., and M. S. Rabb. 1960. On the mating and egg-laying behavior of the Surinam toad, *Pipa pipa. Copeia* 1960:271–276.
Race, J., Jr., C. Robinson, and R. J. Terry. 1966. The influence of thyroxine on the normal development of the pancreas in *Rana pipiens* larvae. *J. Exp. Zool.* 162:181–192.
Rada de Martinez, D. 1990. Contribucion al conocimiento de la larvas de anfibios de Venezuela. *Mem. Soc. Cienc. Nat. La Salle* 49/50:391–403.

Radakov, D. V. 1973. *Schooling in the ecology of fish. Israel Programme for Scientific Translations.* New York: Wiley.
Raff, R. A. 1987. Constraint, flexibility, and phylogenetic history in the evolution of direct development in sea urchins. *Develop. Biol.* 119:6–19.
Raff, R. A., J. A. Amstrom, J. E. Chin, K. G. Field, M. T. Ghiselin, D. J. Lane, G. J. Olsen, N. R. Pace, A. L. Parks, and E. C. Raff. 1987. Molecular and developmental correlates of macroevolution. pp. 109–138. In *Development as an evolutionary process. MBL Lect. Biol. 8.* New York: Alan R. Liss.
Raff, R. A., and T. C. Kaufmann. 1983. *Embryos, genes, and evolution. The developmental-genetic basis of evolutionary change.* New York: Macmillan.
Raines, D. E., S. E. Korten, and A. G. Hill. 1993. Anesthetic cutoff in cycloalkanemethanols: A test of current theories. *Anesthesiology* 78:918–927.
Ramaswami, L. S. 1938. Connections of the pterygoquadrate in the tadpoles of *Philautus variabilis* (Anura). *Nature* 192:577.
———. 1940. Some aspects of the chondrocranium of the South Indian frogs. *J. Mysore Univ.* 1:15–41.
———. 1941. Some aspects of the head of *Xenopus laevis. Proc. Indian Sci. Congr.* 28:183–184.
———. 1943. An account of the chondrocranium of *Rana afghana* and *Megophrys,* with a description of the masticatory musculature of some tadpoles. *Proc. Natl. Inst. Sci. India* 9:43–58.
———. 1944. The chondrocranium of two torrent-dwelling anuran tadpoles. *J. Morphol.* 74:347–374.
Rao, C. R. N. 1923. Notes on Batrachia. *Rec. Indian Mus.* 28:439–447.
Rappaport, R., Jr. 1955. The initiation of pronephric function in *Rana pipiens. J. Exp. Zool.* 128:481–487.
Rastogi, R. K., L. Iela, M. Di Meglio, L. Di Matteo, S. Minucci, and I. Izzo-Vitiello. 1983. Initiation and kinetic profiles of spermatogenesis in the frog, *Rana esculenta* (Amphibia). *J. Zool.* 201:515–525.
Rathke, H. 1832. *Anatomisch-philosophische Untersuchungen über den Kiemenapparat und das Zungenbein der Wirbeltiere.* Riga Dorpat, E. Frantzens Buchhaundlung.
Ratiba, R., C. Girard, and J.-D. Gipouloux. 1981. Mise en evidence de catécholamines dans la corde dorsale de embryons d'amphibiens anoures. Relations possibles aves l'activité adénylate-cyclasique de cet organe et la migration de cellules germinales primordiales. *Arch. Anat. Micros. Morphol. Exp.* 70:149–160.
Rau, R. E. 1978. The development of *Xenopus gilli* Rose & Hewitt. *Ann. South African Mus.* 76:247–263.
Rauthner, M. 1967. Pharynx and epitheliale Organe der Pharynxwand. 1. Kiemen der Anamnier-Kiemendarmderivate der Cyclostomen und Fische. pp. 211–278. In *Handbuch der vergleichenden Anatomie der Wirbeltiere.* Band 3, edited by L. Bolk, E. Göppert, E. Kallius, and W. Lubosch. Amsterdam: Asher.
Rautio, S. A., E. A. Bura, K. A. Berven, and G. J. Gamboa. 1991. Kin recognition in wood frog tadpoles (*Rana sylvatica*)—factors affecting spatial proximity to siblings. *Canadian J. Zool.* 69:2569–2571.
Razarihelisoa, M. 1969. Observations sur les Batraciens aquatiques du centre des hauts plateaux de Madagascar. *Verh. Intl. Verein. Limnol.* 17:949–955.
———. 1973a. Contribution a l'étude des Batraciens de Madagascar. Ecologie et développement larvaire de *Gephyromantis methueni,* Angel, Batracien à biotope végétal sur les *Pandanus. Bull. Acad. Malgache* 51:113–128.
———. 1973b. Contribution a l'étude des Batraciens de Madagascar. Ecologie et développement larvaire de *Mantidactylus brevipalmatus* Ahl., Batracien de eaux courantes. *Bull. Acad. Malgache* 51:113–128.
———. 1979. Contribution à l'étude biologique de quelgues Batraciens de Madagascar. Dissertation, University of Paris.
Read, A. W. 1994. Treatment of redleg in tadpoles of *Pleurodeles. British Herpetol. Soc. Bull.* (49):7.
Reading, C. J. 1990. Palmate newt predation on common frog, *Rana temporaria,* and common toad, *Bufo bufo,* tadpoles. *Herpetol. J.* 1:462–465.
Redshaw, M. R. 1972. The hormonal control of the amphibian ovary. *American Zool.* 12:289–306.
Redshaw, M. R., and T. J. Nicholls. 1971. Oestrogen biosynthesis by ovarian tissue of the South African clawed toad, *Xenopus laevis* Daudin. *Gen. Comp. Endocrinol.* 16:85–96.
Reed, S. C., and H. P. Stanley. 1972. Fine structure of spermatogenesis in the South African clawed toad *Xenopus laevis* Daudin. *J. Ultrastruct. Res.* 41:277–295.
Regös, J., and A. Schlüter. 1984. Erste Ergebnisse zur Fortpflanzungsbiologie von *Lithodytes lineatus* (Schneider 1799). *Salamandra* 20:252–261.
Reh, T. A., E. C. Pitts, and M. Constantine-Paton. 1983. The organization of the fibers in the optic nerve of normal and tectum-less *Rana pipiens. J. Comp. Neurol.* 218:282–296.
Reilly, S. M., and G. V. Lauder. 1990. Metamorphosis of cranial design in tiger salamanders (*Ambystoma tigrinum*): A morphometric analysis of ontogenetic change. *J. Morphol.* 204:121–137.
Reinbach, W. 1939. Untersuchungen über die Entwicklung des Kopfskeletts von *Calyptocephalus gayi* (mit einem Anhand über das Os supratemporale der anuren Amphibien). *Zeit. Naturwiss. (Jena)* 72:211–362.
Reiss, J. O. 1997. Early development of chondrocranium in the tailed frog *Ascaphus truei* (Amphibia: Anura): Implications for anuran palatoquadrate homologies. *J. Morphol.* 231:63–100.
Rengel, D., and A. Pisanó. 1981. Il differenziamento sessuale della gonade di *Phyllomedusa sauvagii. Rend. Accad. Sci. Fis. Mat.* 48:329–335.
Reques, R., and M. Tejedo. 1995. Negative correlation between length of larval period and metamorphic size in natural populations of natterjack toads (*Bufo calamita*). *J. Herpetol.* 29:311–314.
Resetarits, W. J., Jr. 1998. Differential vulnerability of *Hyla chrysoscelis* eggs and hatchlings to larval insect predation. *J. Herpetol.* 32:440–443.
Resetarits, W. J., Jr., and H. M. Wilbur. 1989. Choice of oviposition site by *Hyla chrysoscelis:* Role of predators and competitors. *Ecology* 70:220–228.
Retallick, R. W., and J.-M. Hero. 1998. The tadpoles of *Taudactylus eungellensis* and *T. liemi* and a key to the stream-dwelling tdpoles of the Eungella Rainforest in east-central Queensland, Australia. *J. Herpetol.* 32:304–309.
Reynolds, W. W., and M. E. Casterlin. 1979. Behavioral thermoregulation and the "final preferendum" paradigm. *American Zool.* 19:211–224.
Rhodin, J. A. G., and A. Lametschwandtner. 1993. Circulatory pattern and structure in the tail and tail fins of *Xenopus laevis* tadpoles. *J. Submicros. Cytol. Pathol.* 25:297–318.
Rice, T. M., S. E. Walker, B. J. Blackstone, and D. H. Taylor.

1998. A new method for marking individual anuran larvae. *Herpetol. Rev.* 29:92–93.
Richards, C. M. 1958. The inhibition of growth in crowded *Rana pipiens* tadpoles. *Physiol. Zool.* 31:138–151.
Richards, S. J. 1992. The tadpole of the Australian frog *Litoria nyakalensis* (Anura, Hylidae), and a key to the torrent tadpoles of northern Queensland. *Alytes* 10:99–103.
Richards, S. J., and C. M. Bull. 1990a. Nonvisual detection of anuran tadpoles by odonate larvae. *J. Herpetol.* 24:311–313.
———. 1990b. Size-limited predation on tadpoles of three Australian frogs. *Copeia* 1990:1041–1046.
Richmond, J. E. 1968. Changes in serum proteins during the development of bullfrogs (*Rana catesbeiana*) from tadpoles. *Comp. Biochem. Physiol.* 24:991–996.
Richmond, N. D. 1947. Life history of *Scaphiopus holbrookii holbrookii* (Harlan). Part I: Larval development and behavior. *Ecology* 28:53–67.
Richter, S. 1989. The amphibian pronephros in interaction with the arterial and venous system during ontogenesis. pp. 88–92. In *Trends in vertebrate morphology. Progress in Zoology.* Vol. 35, edited by H. Splechtna and H. Hilgers. Stuttgart: Gustav Fischer-Verlag.
———. 1992. Ontogenese des Exkretionssystems bei *Rana esculenta* (Ranidae, Anura). Histo- und morphogenetische Analyse der Anurenniere. Dissertation, University of Wien.
———. 1995. The opisthonephros of *Rana esculenta* (Anura). I. Nephron development. *J. Morphol.* 226:173–187.
Richter, S., and H. Splechtna. 1990. Morphogenesis of the Malphighian corpuscle in the anuran opisthonephros (*Rana esculenta*). p. 469. In *Verhandlungen der Deutschen Zoologischen Gesellschaft, 83. Jahresversammlung 1990, Frankfurt am Main,* edited by H. D. Pfannenstiel. Stuttgart: Gustav Fischer-Verlag.
Ridewood, W. G. 1897a. On the structure and development of the hyobranchial skeleton and larynx in *Xenopus* and *Pipa;* with remarks on the affinities of the Aglossa. *J. Linnean Soc. London* 26:53–128.
———. 1897b. On the structure and development of the hyobranchial skeleton of the parsley-frog (*Pelodytes punctatus*). *Proc. Zool. Soc. London* 1897:577–595.
———. 1898a. On the development of the hyobranchial skeleton of the midwife-toad (*Alytes obstetricans*). *Proc. Zool. Soc. London* 1898:4–12.
———. 1898b. On the larval hyobranchial skeleton of the anurous batrachians with special reference to the axial parts. *J. Linnean Soc. London* 26:474–487.
Rieger, R. M. 1994. The biphasic life cycle—a central theme of metazoan evolution. *American Zool.* 34:484–491.
Riggs, A. 1951. The metamorphosis of hemoglobin in the bullfrog. *J. Gen. Physiol.* 35:23–40.
Riis, N. 1991. A field study of survival, growth, biomass, and temperature dependence of *Rana dalmatina* and *Rana temporaria* larvae. *Amphibia-Reptilia* 12:229–243.
Rist, L., R. D. Semlitsch, H. Hotz, and H. U. Reyer. 1997. Feeding behaviour, food consumption, and growth efficiency of hemiclonal and parental tadpoles of the *Rana esculenta* complex. *Funct. Ecol.* 11:735–742.
Ritke, M. E., and R. L. Mumme. 1993. Choice of calling sites and oviposition sites by gray treefrogs (*Hyla chrysoscelis*)—a comment. *Ecology* 74:623–626.
Rivero, J. A. 1976 (1978). Notas sobre los anfibios de Venezuela. II. Sobre los *Colostethus* de Los Andes venezolanos. *Mem. Soc. Cien. Nat. La Salle* 35:327–344.
Roberts, A. 1976. Neuronal growth cones in an amphibian embryo. *Brain Res.* 118:526–530.
Roberts, A., and S. T. Alford. 1986. Descending projections and excitation during fictive swimming in *Xenopus* embryos: Neuroanatomy and lesion experiments. *J. Comp. Neurol.* 250:253–261.
Roberts, A., and J. D. W. Clarke. 1982. The neuroanatomy of an amphibian embryo spinal cord. *Philosoph. Trans. Roy. Soc. London* 296B:195–212.
Roberts, A., and B. P. Hayes. 1977. The anatomy and function of the free nerve endings in an amphibian skin sensory system. *Proc. Roy. Soc. London* 196B:415–429.
Roberts, A., and D. T. Patton. 1985. Growth cones and the formation of central and peripheral neurites by sensory neurones in amphibian embryos. *J. Neurosci. Res.* 13:23–28.
Roberts, A., S. R. Soffe, J. D. W. Clarke, and N. Dale. 1984. Initiation and control of swimming in amphibian embryos. pp. 261–284. In *Neural origin of rhythmic movements. SEB Symp. (37).* New York: Cambridge University Press.
Roberts, A., and J. S. H. Taylor. 1983. A study of the growth cones of developing embryonic sensory neurites. *J. Embryol. Exp. Morphol.* 75:31–47.
Roberts, J. D. 1981. Terrestrial breeding in the Australian leptodactylid frog *Myobatrachus gouldii* (Gray). *Australian Wildl. Res.* 8:451–462.
———. 1984. Terrestrial egg deposition and direct development in *Arenophryne rotunda* Tyler: A myobatrachid frog from coastal sand dunes at Shark Bay, Western Australia. *Australia Wildl. Res.* 11:191–200.
Roberts, J. D., P. Horwitz, G. Wardell-Johnson, L. R. Maxson, and M. J. Mahony. 1997. Taxonomy, relationships, and conservation of a new genus and species of myobatrachid frog from the high rainfall region of southwestern Australia. *Copeia* 1997:373–381.
Robinson, D.C. 1961. The identify of the tadpole of *Anotheca coronata. Copeia* 1961:495.
Roček, Z. 1981 (1980). Cranial anatomy of frogs of the family Pelobatidae Stannius, 1856, with outlines of their phylogeny and systematics. *Acta Univ. Carolinae Biol.* 1980:1–164.
———. 1989. Developmental patterns of the ethmoidal region of the anuran skull. pp. 412–420. In *Trends in vertebrate morphology,* edited by H. Splechtna and H. Helgers. Stuttgart: Gustav Fischer-Verlag.
———. 1990. Ethmoidal endocranial structures in primitive tetrapods: Their bearing on the search for anuran ancestry. *Zool. J. Linnean Soc.* 99:389–407.
Roček, Z., and M. Veselý. 1989. Development of the ethmoidal structures of the endocranium in the anuran *Pipa pipa. J. Morphol.* 200:300–319.
Rödel, M.-O. 1996. *Amphibien der westafrikanischen Savanne.* Frankfurt: Edition Chimaira.
———. 1998. A reproductive mode so far unknown in African ranids: *Phrynobatrachus guineensis* Guibé & Lamotte, 1961 breeds in tree holes (Anura: Ranidae). *Herpetozoa* 11:19–26.
Rödel, M.-O., and K. E. Linsenmair. 1997. Predator-induced swarms in the tadpoles of an African savanna frog, *Phrynomantis microps. Ethology* 103:902–914.
Root, R. B. 1967. The niche exploitation pattern of the Blue-gray Gnatcatcher. *Ecol. Monogr.* 37:317–350.
Rösel von Rosenhof, A. J. 1753–1758. *Historia naturalis Ranarum nostratium.* Nürenberg.
Rose, M. J. 1982. Antagonistic pleiotropy, dominance, and genetic variation. *Heredity* 48:63–78.

Rose, S. M. 1959. Failure of survival of slowly growing members of a population. *Science* 129:1026.
———. 1960. A feedback mechanism of growth control in tadpoles. *Ecology* 41:188–199.
Rosen, D. E., P. L. Forey, B. G. Gardiner, and C. Patterson. 1981. Lungfishes, tetrapods, paleontology, and plesiomorphology. *Bull. American Mus. Nat. Hist.* 167:163–275.
Rosenberg, E. A., and B. A. Pierce. 1995. Effect of initial mass on growth and mortality at low pH in tadpoles of *Pseudacris clarkii* and *Bufo valliceps. J. Herpetol.* 29:181–185.
Rosenthal, B. M., and W. L. R. Cruce. 1985. The dendritic extent of motoneurons in frog brachial spinal cord: A computer reconstruction of HRP-filled cells. *Brain Behav. Evol.* 27:106–114.
Rossi, A. 1959. Tavole cronologiche dello sviluppo embrionale e larval del *Bufo bufo* (L.). *Monit. Zool. Italiano* 66:1–11.
Rostand, J. 1934. *Toads and toad life.* Translated by J. Fletcher. London: Methuen.
Roth, A. H., and J. F. Jackson. 1987. The effect of pool size on recruitment of predatory insects and on mortality in a larval anuran. *Herpetologica* 43:224–232.
Roth, G. 1987. *Visual behavior in salamanders.* Berlin: Springer-Verlag.
Rouf, M. A. 1969. Hematology of the leopard frog, *Rana pipiens. Copeia* 1969:682–687.
Roux, W. 1888. Contributions to the developmental mechanisms of the embryo. On the artificial production of half-embryos by destruction of one of the first two blastomeres, and the later development (postgeneration of the missing half of the body). *Virchows Arch. Path. Anat. Physiol. Klin. Med.* 114:113–153. [English translation in B. H. Willier and J. M. Oppenheimer. 1974. *Foundations of experimental embryology.* 2d ed. New York: Hafner Press.]
Rovira, J., A. C. Villaro, and M. E. Bodegas. 1993. Structural study of the frog *Rana temporaria* larval stomach. *Tiss. Cell* 25:695–707.
Rowe, C. L., and W. A. Dunson. 1995. Impacts of hydroperiod on growth and survival of larval amphibians in temporary ponds of central Pennsylvania. *Oecologia* 102:397–403.
Rowe, C. L., O. M. Kinney, A. P. Fiori, and J. D. Congdon. 1996. Oral deformities in tadpoles (*Rana catesbeiana*) associated with coal ash deposition: Effects of grazing ability and growth. *Freshw. Biol.* 36:723–730.
———. 1998a. Oral deformities in tadpoles of the bullfrog (*Rana catesbeiana*) caused by conditions in a polluted habitat. *Copeia* 1998:244–246.
Rowe, C. L., O. M. Kinney, R. D. Nagle, and J. D. Congdon. 1998b. Elevated maintenance costs in an anuran (*Rana catesbeiana*) exposed to a mixture of trace elements during the embryonic and early larval periods. *Physiol. Zool.* 71:27–35.
Rowe, C. L., W. J. Sadinski, and W. A. Dunson. 1994. Predation on larval and embryonic amphibians by acid-tolerant caddisfly larvae (*Ptilostomis postica*). *J. Herpetol.* 28:357–364.
Rowe, L., and D. Ludwig. 1991. Size and timing of metamorphosis in complex life cycles: Time constraints and variation. *Ecology* 72:413–427.
Rowedder, W. 1937. Die Entwicklung des Geruchsorgans bei *Alytes obstetricans* und *Bufo vulgaris. Zeit. Anat. Entwickl.* 107:91–123.
Rubin, D. I., and L. M. Mendell. 1980. Location of motoneurons supplying individual muscles in normal and grafted supernumerary limbs of *Xenopus laevis. J. Comp. Neurol.* 192:703–715.
Rubinson, K., and B. Friedman. 1977. Vagal afferent projections in *Rana pipiens, Rana catesbeiana,* and *Xenopus mulleri* with a note on lateral line and VIII nerve projection zones. *Brain Behav. Evol.* 14:368–380.
Rubinson, K., and M. P. Skiles. 1975. Efferent projections of the superior olivary nucleus in the frog, *Rana catesbeiana. Brain Behav. Evol.* 12:151–160.
Ruby, J. R., R. F. Dyer, R. G. Skalko, and E. P. Volpe. 1970. Intercellular bridges between germ cells in the developing ovary of the tadpole, *Rana pipiens. Anat. Rec.* 167:1–10.
Rugh, R. 1934. The space factor in the growth rate of tadpoles. *Ecology* 15:407–411.
———. 1951. *The frog. Its reproduction and development.* Philadelphia: Blakiston Co.
Ruibal, R. 1959. The ecology of a brackish water population of *Rana pipiens. Copeia* 1959:315–322.
———. 1962. Osmoregulation in amphibians from heterosaline habitats. *Physiol. Zool.* 35:133–147.
Ruibal, R., and E. Thomas. 1988. The obligate carnivorous larvae of the frog, *Lepidobatrachus laevis* (Leptodactylidae). *Copeia* 1988:591–604.
Ruiz Garcia, F. N. 1980. La larva de *Bufo longinasus* (Stejneger) (Amphibia: Bufonidae) en el laboratorio. *Poeyana* (207):1–8.
Rusconi, M. 1826. *Développment de la grenouille commune.* Paris.
Russell, A. P., and A. M. Bauer. 1993. *The amphibians and reptiles of Alberta.* Calgary, Alberta: University of Calgary Press and University of Alberta Press.
Russell, I. J. 1976. Amphibian lateral line receptors. pp. 513–550. In *Frog neurobiology, a handbook,* edited by R. Llinás and W. Precht. Berlin: Springer-Verlag.
Ruthven, A. G. 1915. Observations on the habits, eggs, and young of *Hyla fuhrmanni* Peracca. *Occas. Pap. Mus. Zool., Univ. Michigan* (14):1–4.
Ruthven, A. G.,, and H. T. Gaige. 1915. The breeding habits of *Prostherapis subpunctatus* Cope. *Occas. Pap. Mus. Zool., Univ. Michigan* (10):1–17.
Ryan, M. J. 1986. Synchronized calling in a treefrog (*Smilisca sila*). *Brain Behav. Evol.* 29:196–206.
———. 1988. Constraints and patterns in the evolution of anuran acoustic communications. pp. 637–677. In *The evolution of the amphibian auditory system,* edited by B. Fritzsch, M. J. Ryan, W. Wilczynski, T. E. Hetherington, and W. Walkowiak. New York: Wiley.
Ryke, P. A. J. 1953. The ontogenetic development of the somatic musculature of the trunk of the aglossal anuran *Xenopus laevis* (Daudin). *Acta Zool.* 34:1–70.
Sabater-Pi, J. 1985. Contribution to the biology of the giant frog (*Conraua goliath,* Boulenger). *Amphibia-Reptilia* 6:143–153.
Saber, P. A., and W. A. Dunson. 1978. Toxicity of bog water to embryonic and larval anuran amphibians. *J. Exp. Biol.* 204:33–42.
Sabnis, J. H., and K. S. M. Kuthe. 1980. Observations on food and growth of *Bufo melanostictus* tadpole. *J. Bombay Nat. Hist. Soc.* 77:21–25.
Sahu, A. K., and M. K. Khare. 1988. Food and feeding habits of *Rana alticola* Boulenger (Anura: Ranidae) during different stages of metamorphosis. *J. Adv. Zool.* 9:97–104.
Saidapur, S. K. 1983. Patterns of testicular activity in Indian amphibians. *Indian Rev. Life Sci.* 3:157–184.
———. 1986. Patterns of ovarian activity in Indian amphibians. *Indian Rev. Life Sci.* 6:231–256.
Saidapur, S. K., and V. B. Nadkarni. 1974. Steroid-synthesizing cellular sites in amphibian ovary. A histochemical study. *Gen. Comp. Endocrinol.* 22:459–462.

Saint-Ange, M. 1831. Recherches anatomiques et physiologiques sur les organes transitoires et la metamorphose des batraciens. *Ann. Sci. Nat.* 34.

Salthe, S. N., and W. E. Duellman. 1973. Quantitative constraints associated with reproductive modes in anurans. pp. 229–249. In *Evolutionary biology of the anurans. Contemporary research on major problems,* edited by J. L. Vial. Columbia: University of Missouri Press.

Salthe, S. N., and J. S. Mecham. 1974. Reproductive and courtship patterns. pp. 309–351. In *Physiology of the Amphibia.* Vol. 2, edited by B. Lofts. New York: Academic Press.

Salvador, A. 1985. *Guia de campo de los anfibios y reptiles de la Peninsula Iberica, Isla Baleares y Canarais.* Madrid: Unigraf.

Salvatorelli, G. 1970. Osservazioni citometriche sulgi eritrociti di *Bufo bufo* durante il periodo larvale e adulto. *Ann. Univ. Ferrara (Ser. Anat. Comp., Sezione 13)* 3:17–27.

Sampson, L. V. 1900. Unusual modes of breeding and development among Anura. *American Nat.* 34:687–715.

———. 1904. A contribution to the embryology of *Hylodes martinicensis. American J. Anat.* 3:473–504.

Sarikas, S. N. 1977. Histochemistry of egg membranes and thickness of egg membranes related to sperm dimensions in two species of salamander of the genus *Ambystoma.* Thesis, Southern Illinois University.

Sarnat, H. B., and M. G. Netsky. 1981. *Evolution of the nervous system.* 2d ed. Oxford: Oxford University Press.

Saruwatari, T., J. A. López, and T. W. Pietsch. 1997. Cyanine Blue: A versatile and harmless stain for specimen observation. *Copeia* 1997:840–841.

Sasaki, F., and T. Kinoshita. 1994. Adult precursor cells in the tail epidermis of *Xenopus* tadpoles. *Histochemistry* 101:391–396.

Sasaki, H., and H. Mannen. 1981. Morphological analysis of astrocytes in the bullfrog (*Rana catesbeiana*) spinal cord with special reference to the site of attachment of their processes. *J. Comp. Neurol.* 198:13–35.

Satel, S. L., and R. J. Wassersug. 1981. On the relative sizes of buccal floor depressor and elevator musculature in tadpoles. *Copeia* 1981:129–137.

Sater, A. K., and A. G. Jacobson. 1989. The specification of heart mesoderm occurs during gastrulation in *Xenopus laevis. Development* 105:821–830.

———. 1990. The role of the dorsal lip in the induction of heart mesoderm in *Xenopus laevis. Development* 108:461–470.

Sato, K. 1924. Über die Metamorphose von *Bufo vulgaris. Zeit. Anat. Entwickl.* 71:41–184.

Savage, J. M. 1960. Geographic variation in the tadpole of the toad *Bufo marinus. Copeia* 1960:233–235.

———. 1968. The dendrobatid frogs of Central America. *Copeia* 1968:745–776.

———. 1980a. A synopsis of the larvae of Costa Rican frogs and toads. *Bull. Southern California Acad. Sci.* 79:45–54.

———. 1980b. The tadpole of the Costa Rican fringe-limbed tree-frog, *Hyla fimbrimembra. Proc. Biol. Soc. Washington* 93:1177–1183.

Savage, R. M. 1935. The ecology of young tadpoles, with specific reference to some adaptations to the habit of mass spawning in *Rana temporaria temporaria* L. *Proc. Zool. Soc. London* 1935:605–610.

———. 1952. Ecological, physiological, and anatomical observations of some species of anuran tadpoles. *Proc. Zool. Soc. London* 122:467–514.

———. 1955. The ingestive, digestive and respiratory systems of the microhylid tadpole, *Hypopachus aquae. Copeia* 1955:120–127.

———. 1962. *The ecology and life history of the common frog.* New York: Hafner Publ. Co.

Saxén, L. 1987. *Organogenesis of the kidney.* Cambridge: Cambridge University Press.

Sazima, I., and W. C. A. Bokermann. 1978. Cinco novas espécies de *Leptodactylus* do centro e sudeste Brasileiro (Amphibia, Anura, Leptodactylidae). *Rev. Brasil. Biol.* 38:899–912.

———. 1982. Anfíbios da Serra do Cipó, Minas Gerais, Brasil 5: *Hylodes otavioi* sp. n. (Anura, Leptodactylidae). *Rev. Brasil. Biol.* 42:767–771.

Scadding, S. R. 1993. Victoria Blue B staining for cartilage. p. 20. In *Handbook of methods. International workshop on the molecular biology of axolotls and other urodeles.* Indianapolis, IN.

Scalia, F. 1976. Structure of the olfactory and accessory olfactory systems. pp. 213–233. In *Frog neurobiology, a handbook,* edited by R. Llinás and W. Precht. Berlin: Springer-Verlag.

Scalia, F.,, G. Gallousis, and S. Roca. 1991a. A note on the organization of the amphibian olfactory bulb. *J. Comp. Neurol.* 305:435–442.

———. 1991b. Differential projections of the main and accessory olfactory bulb in the frog. *J. Comp. Neurol.* 305:443–461.

Scalia, F. M. Halpern, H. Knapp, and W. Riss. 1968. The efferent connexions of the olfactory bulb in the frog: A study of degenerating unmyelinated fibers. *J. Anat.* 103:245–262.

Schechtman, A. M. 1934. Unipolar ingression in *Triturus torosus,* a hitherto undescribed movement in the pregastrular stages of a urodele. *Univ. California Publ. Zool.* 39:303–310.

Scheel, J. J. 1970. Notes on the biology of the African tree-toad, *Nectophryne afra* Buchholz & Peters, 1875 (Bufonidae, Anura) from Fernando Poo. *Rev. Zool. Bot. Africaines* 81:225–236.

Schijfsma, K. 1932. Notes on some tadpoles, toads, and frogs from Java. *Truebia* 14:43–72.

Schiøtz, A. 1963. The amphibians of Nigeria. *Vidensk. Medd. Dansk Naturh. Foren.* 125:1–125, 4 plates.

———. 1967. *The treefrogs (Rhacophoridae) of west Africa.* Copenhagen: Bianco Lunos Bogtrykkeri.

———. 1975. *The treefrogs of eastern Africa.* Copenhagen: Steenstrupia.

Schlaepfer, M. A. 1998. Use of a flourescent marking technique on small terrestrial anurans. *Herpetol. Rev.* 29:25–26.

Schley, L., R. A. Griffiths, and A. Román. 1998. Activity patterns and microhabitat selection of Mallorcan midwife toad (*Alytes muletensis*) tadpoles in natural torrent pools. *Amphibia-Reptilia* 19:143–151.

Schlosser, G., and G. Roth. 1995. Distribution of cranial and rostral spinal nerves in tadpoles of the frog *Discoglossus pictus* (Discoglossidae). *J. Morphol.* 226:189–212.

———. 1997a. Evolution of nerve development in frogs. I. The development of the peripheral nervous system in *Discoglossus pictus* (Discoglossidae). *Brain Behav. Evol.* 50:61–93.

———. 1997b. Evolution of nerve development in frogs. II. Modified development of the peripheral nervous system in the direct-developing frog *Eleutherodactylus coqui* (Leptodactylidae). *Brain Behav. Evol.* 50:94–128.

Schluga, A. 1974. Beitrag zur Entwicklung des Exkretionssystems bei Amphibien. Dissertation, University of Wien.

Schlüter, A. 1990. Reproduction and tadpole of *Edalorhina perezi* (Amphibia, Leptodactylidae). *Stud. Neotrop. Fauna Environ.* 25:49–56.

Schlüter, A., and A. W. Salas. 1991. Reproduction, tadpoles, and ecological aspects of three syntopic microhylid species from Peru (Amphibia: Microhylidae). *Stuttgarter Beitr. Naturk.* (458):1–17.

Schmalhausen, I. I. 1968. *The origin of terrestrial vertebrates.* New York: Academic Press.

Schmidt, B. R. 1993. Are hybridogenetic frogs cyclical parthogens? *Trends Ecol. Evol.* 8:271–272.

Schmidt-Nielsen, B., and W. C. Mackay. 1970. Osmotic and diffusional water permeability in metamorphosing *Rana clamitans* tadpoles. *Bull. Mount Desert Is. Biol. Lab.* 10:71–75.

Schmidt-Nielsen, K. 1983. *Animal physiology: Adaptation and environment.* London: Cambridge University Press.

Schmuck, R., W. Geise, and K. E. Linsenmair. 1994. Life cycle strategies and physiological adjustments of reedfrog tadpoles (Amphibia, Anura, Hyperoliidae) in relation to environmental conditions. *Copeia* 1994:996–1007.

Schmutzer, W. 1977. Histopathologische Veranderungen durch *Nephrotrema truncatum* (Leuckart 1942)(Digenea: Troglotrematidae) in der Schwanzregion von Froshlarven (*Rana temporaria* L.). *Zool. Anz.* 198:355–368.

Schneider, L. 1957. Beinflussung des Wachstums durch im Thymus enthaltene Stoffe. I. Herstellung fraktionierter Thymusextrakte und ihre Prüfung im Kaulquappenversuch. Wilhelm Roux' Arch. *Entwicklungsmech. Org.* 149:644–683.

Schoonbee, H. J., J. F. Prinsloo, and J. G. Nxiweni. 1992. Observations on the feeding habits of larvae, juvenile, and adult stages of the African clawed frog, *Xenopus laevis,* in impoundments in Transkei. *Water South Africa* 18:227–236.

Schorr, M. S., R. Altig, and W. J. Diehl. 1991. Populational changes of the enteric protozoans *Opalina* spp. and *Nyctotherus cordiformis* during the ontogeny of anuran tadpoles. *J. Protozool.* 37:479–481.

Schulze, F. E. 1888. Über die inneren kiemen der Batrachierlarven. I. Mittheilung. Über das epithel der Lippen, der Mund-, Rachen- und Kiemenhöhle erwachsener larven von *Pelobates fuscus. Abh. Preuss. Akad. Wiss. Berlin* 1888:1–59, 4 plates.

———. 1892. Über die inneren Kiemen der Batrachierlarven. II. Mitteilung. Skelet, Musculatur, Blutgefässe, Filterapparat, respiratorische Anhänge und Athmungsbewegungen erwachsener Larven von *Pelobates fuscus. Abh. Preuss. Akad. Wiss.* 1892:1–66.

Schütte, F., and A. Ehrl. 1987. Zur Haltung und Zucht der großen südamerikanischen Waßenkröte *Pipa pipa* (Linnaeus 1758). *Salamandra* 23:256–268.

Schütte, M., and S. G. Hoskins. 1993. Ipsilaterally projecting retinal ganglion cells in *Xenopus laevis:* An HRP study. *J. Comp. Neurol.* 331:482–494.

Schwartz, A., and R. W. Henderson. 1991. *Amphibians and reptiles of the West Indies. Descriptions, distributions, and natural history.* Gainesville: University of Florida Press.

Schwartz, J. J., and K. D. Wells. 1983. An experimental study of acoustic interference between two species of neotropical treefrogs. *Anim. Behav.* 31:181–190.

Scott, N.J., Jr., and R. D. Jennings. 1985. The tadpoles of five species of New Mexican leopard frogs. *Occas. Pap. Mus. Southw. Biol.* (3):1–21.

Scott-Birabén, M. T., and K. Fernández-Marcinowski. 1921. Variaciones locales de caracteres específicos en larvas de anfibios. *Ann. Soc. Cient. Argentina* 92:129–144.

Seale, D. B. 1980. Influence of amphibian larvae on primary production, nutrient flux, and competition in a pond ecosystem. *Ecology* 61:1531–1552.

———. 1982. Obligate and facultative suspension feeding in anuran larvae: Feeding regulation in *Xenopus* and *Rana. Biol. Bull.* 162:214–231.

———. 1987. Amphibia. pp. 467–552. In *Animal energetics.* Vol. 2, edited by T. J. Pandian and E. J. Vernberg. New York: Academic Press.

Seale, D. B., and N. Beckvar. 1980. The comparative ability of anuran larvae (genera: *Hyla, Bufo,* and *Rana*) to ingest suspended blue-green algae. *Copeia* 1980:495–503.

Seale, D. B., K. Hoff, and R. Wassersug. 1982. *Xenopus laevis* larvae (Amphibia, Anura) as model suspension feeders. *Hydrobiologia* 87:161–169.

Seale, D. B., and R. J. Wassersug. 1979. Suspension feeding dynamics of anuran larvae related to their functional morphology. *Oecologia* 39:259–272.

Sedra, S. N. 1950. The metamorphosis of the jaws and their muscles in the toad, *Bufo regularis* Reuss, correlated with changes in the animal's feeding habits. *Proc. Zool. Soc. London* 120:405–449.

———. 1951. On the morphology of the suprarostral system and the mandibular arch of *Bufo lentiginosus. Proc. Egyptian Acad. Sci.* 7:128.

Sedra, S. N., and M. I. Michael. 1956a. The development of the skeleton and muscles of the head from stage 55 till the end of metamorphosis. pp. 55–68. In *Normal table of* Xenopus laevis *Daudin,* edited by P. D. Nieuwkoop and J. Faber. Amsterdam: North-Holland.

———. 1956b. The structure of the hyobranchial apparatus of the fully developed tadpole larva of *Bufo regularis* Reuss. *Proc. Egyptian Acad. Sci.* 12:38–46.

———. 1957. The development of the skull, viscera arches, larynx, and visceral muscles in the South African clawed toad *Xenopus laevis* (Daudin) during the process of metamorphosis (from stage 55 to stage 66). *Verh. Konink. Nederland Akad. Wetensch. Natuurk.* 51:1–80.

———. 1958. The metamorphosis and growth of the hyobranchial apparatus of the Egyptian toad, *Bufo regularis* Reuss. *J. Morphol.* 103:1–30.

———. 1959. The ontogenesis of the sound conducting apparatus of the Egyptian toad, *Bufo regularis* Reuss, with a review of this apparatus in Salientia. *J. Morphol.* 104:359–375.

———. 1961. Normal table of the Egyptian toad, *Bufo regularis* Reuss, with an addendum on the standardization of the stages considered in previous publications. *Ceskoslovenska Morfol.* 9:333–351.

Sedra, S. N., and A. A. Moursi. 1958. The ontogenesis of the vertebral column of *Bufo regularis* Reuss. *Morfologie* 6:7–32.

Seigel, R. A. 1983. Natural survival of eggs and tadpoles of the wood frog, *Rana sylvatica. Copeia* 1983:1096–1098.

Sekar, A. G. 1992. A study of the food habits of six anuran tadpoles. *J. Bombay Nat. Hist. Soc.* 89:9–16.

Selenka, E. 1882. Der embryonale excretionsapparat des kiemlosen *Hylodes martinicensis. Sitzungsber. Kön. Preuss. Akad. Wiss. Berlin* 117–124.

Semlitsch, R. D. 1990. Effects of body size, sibship, and tail injury on the susceptibility of tadpoles to dragonfly predation. *Canadian J. Zool.* 68:1027–1030.

———. 1993. Asymmetric competition in mixed populations of tadpoles of the hybridogenetic *Rana esculenta* complex. *Evolution* 47:510–519.

Semlitsch, R. D., and J. P. Caldwell. 1982. Effects of density on growth, metamorphosis, and survivorship in tadpoles of *Scaphiopus holbrooki. Ecology* 63:905–911.

Semlitsch, R. D. and S. Gavasso. 1992. Behavioural responses of *Bufo bufo* and *Bufo calamita* tadpoles to chemical cues of vertebrate and invertebrate predators. *Ethol. Ecol. Evol.* 4:165–173.

Semlitsch, R. D., and J. W. Gibbons. 1988. Fish predation in size-structured populations of treefrog tadpoles. *Oecologia* 75:321–326.

Semlitsch, R. D. and H.-U. Reyer. 1992a. Modification of anti-predator behaviour in tadpoles by environmental conditioning. *J. Anim. Ecol.* 61:353–360.

———. 1992b. Performance of tadpoles from the hybridogenetic *Rana esculenta* complex: Interactions with pond drying and interspecific competition. *Evolution* 46:665–676.

Semlitsch, R. D., and S. Schmiedehausen. 1994. Parental contribution to variation in hatchling size and its relationship to growth and metamorphosis in tadpoles of *Rana lessonae* and *Rana esculenta*. *Copeia* 1994:406–412.

Semlitsch, R. D., D. E. Scott, and J. H. K. Pechmann. 1988. Time and size at metamorphosis related to adult fitness in *Ambystoma talpoideum*. *Ecology* 69:184–192.

Semlitsch, R. D., D. E. Scott, J. H. K. Pechmann, and J. W. Gibbons. 1993. Phenotypic variation in the arrival time of breeding salamanders: Individual repeatability and environmental influences. *J. Anim. Ecol.* 62:334–340.

———. 1996. Structure and dynamics of an amphibian community. Evidence from a 16-year study of a natural pond. pp. 217–248 In *Long-term studies of vertebrate communities*. New York: Academic Press.

Semlitsch, R. D., and H. M. Wilbur. 1989. Artificial selection for paedomorphosis in the salamander *Ambystoma talpoideum*. *Evolution* 43:105–112.

Senn, D. G. 1972. Development of tegmental and rhombencephalic structures in a frog (*Rana temporaria* L.). *Acta Anat.* 82:525–548.

Serra, J. A., and R. M. Albuquerque. 1963. Anfibios de Portugal. *Rev. Portuguesa Zool. Biol. Geral* 4:75–227.

Sesama, P., J. Rovira, A. C. Villaro, M. E. Bodegas, and E. Valverde. 1995. Metamorphic changes in the stomach of the frog *Rana temporaria* tadpoles. *Tiss. Cell* 27:13–22.

Sever, D. M. 1994. Comparative anatomy and phylogeny of the cloacae of salamanders (Amphibia: Caudata). VII. Plethodontidae. *Herpetol. Monogr.* (8):276–337.

Severtsov, A. S. 1969a. Food-seizing mechanism in anuran larvae. *Dokl. Akad. Nauk SSSR* 187:211–214.

———. 1969b. Origin of the basal elements of the hyobranchial skeleton of larvae of amphibians. *Dokl. Akad. Nauk SSSR* 187:677–680.

Sexton, O. J., and C. Phillips. 1986. A qualitative study of fish-amphibian interactions in 3 Missouri ponds. *Trans. Missouri Acad. Sci.* 20:25–35.

Seymour, R. S. 1973. Gas exchange in spadefoot toads beneath the ground. *Copeia* 1973:452–460.

Seymour, R. S., and D. F. Bradford. 1995. Respiration of amphibian eggs. *Physiol. Zool.* 68:1–25.

Seymour, R. S., F. Geiser, and D. F. Bradford. 1991a. Gas conductance of the jelly capsules of terrestrial frog eggs correlates with embryonic stage, not metabolic demand or ambient PO_2. *Physiol. Zool.* 64:673–687.

———. 1991b. Metabolic costs of development in terrestrial frog eggs (*Pseudophryne bibronii*). *Physiol. Zool.* 64:688–690.

Seymour, R. S., M. J. Mahony, and R. Knowles. 1995. Respiration of embryos and larvae of the terrestrially breeding frog *Kyarranus loveridgei*. *Herpetologica* 51:369–379.

Seymour, R. S., and J. D. Roberts. 1991. Embryonic respiration and oxygen distribution in foamy and nonfoamy egg masses of the frog *Limnodynastes tasmaniensis*. *Physiol. Zool.* 64:1322–1340.

Shaffer, H. B. 1984. Evolution in a paedomorphic lineage. II. Allometry and form in the Mexican ambystomatid salamander. *Evolution* 38:1207–1218.

Shaffer, H. B., R. A. Alford, B. D. Woodward, S. J. Richards, R. G. Altig, and C. Gascon. 1994. Quantitative sampling of amphibian larvae. pp. 130–141. In *Measuring and monitoring biological diversity. Standard methods for amphibians*. Biological Diversity Handbook Series, edited by W. R. Heyer, M. A. Donnelly, R. W. McDiarmid, L.-A. C. Hayek, and M. S. Foster. Washington, D.C.: Smithsonian Institution Press.

Shaffer, L. R., and D. R. Formanowicz, Jr. 1966. A cost of viviparity and parental care in scorpions: Reduced sprint speed and behavioural compensation. *Anim. Behav.* 51:1017–1024.

Shaw, E. 1978. Schooling fishes. *American Sci.* 66:166–175.

Shaw, J. P. 1979. The time scale of tooth development and replacement in *Xenopus laevis* (Daudin). *J. Anat.* 129:323–342.

Shelton, P. M. J. 1970. The lateral line system at metamorphosis in *Xenopus laevis* (Daudin). *J. Embryol. Exp. Morphol.* 24:511–524.

———. 1971. The structure and function of the lateral line system in larval *Xenopus laevis*. *J. Exp. Zool.* 178:211–232.

Sheratt, T. N., and I. F. Harvey. 1989. Predation by larvae of *Pantala flavescens* (Odonata) on tadpoles of *Phyllomedusa trinitatis* and *Physalaemus pustulosus:* The influence of absolute and relative density of prey on predator choice. *Oikos* 56:170–176.

Sherman, E. 1980. Ontogenetic change in thermal tolerance of the toad *Bufo woodhousii fowleri*. *Comp. Biochem. Physiol.* 65A:227–230.

Shield, J. W., and P. J. Bentley. 1973. Respiration of some urodele and anuran Amphibia. I. In water, role of the skin and gills. *Comp. Biochem. Physiol.* 46A:17–28.

Shine, R. 1995. A new hypothesis for the evolution of viviparity in reptiles. *American Nat.* 145:809–823.

Shirane, T. 1970. On the formation of germ cells of the frog's larvae from eggs irradiated with ultraviolet rays. *Mem. Fac. Gen. Educ. Hiroshima Univ. III.* 4:24–37.

———. 1972. On the formation of primordial cells of the Japanese pond frog, *Rana brevipoda* Ito. *Mem. Fac. Gen. Educ. Hiroshima Univ. III.* 5:17–27.

———. 1982. Gonadal differentiation in frogs, *Rana japonica* and *Rana brevipoda,* raised from UV irradiated eggs. *J. Exp. Zool.* 223:165–173.

———. 1986. A new, early, morphological indication of sex differentiation in Anura, *Rana japonica,* and *Rana nigromaculata*. *J. Exp. Zool.* 240:113–118.

Shiriaev, B. I., and O. V. Shupliakov. 1986. Synaptic organization of dorsal root projections to lumbar motoneurons in the clawed toad (*Xenopus laevis*). *Exp. Brain Res.* 63:135–142.

Shoemaker, V. H., and L. L. McClanahan. 1973. Nitrogen excretion in the larvae of a land-nesting frog (*Leptodactylus bufonius*). *Comp. Biochem. Physiol.* 44A:1149–1156.

———. 1982. Enzymatic correlates and ontogeny of uricotelism in tree frogs of the genus *Phyllomedusa*. *J. Exp. Zool.* 220:163–169.

Shofner, W. P., and A. S. Feng. 1984. Quantitative light and scanning electron microscopic study of the developing auditory organs in the bullfrog: Implications on their functional characteristics. *J. Comp. Neurol.* 224:141–154.

Shumway, W. 1940. Stages in the normal development of *Rana pipiens*. I. *External form*. *Anat. Rec.* 78:139–147.
Shvarts, S. S., and O. A. Pyastolova. 1970. Regulators of growth and development of amphibian larvae. I. Specificity of effects. *Ekologiya* 1:58–62.
Siegfried, W. R., and L. G. Underhill. 1975. Flocking as an antipredator strategy in doves. *Anim. Behav.* 23:504–508.
Siekmann, J. M. 1949. A survey of the tadpoles of Louisiana. Dissertation, Tulane University.
Sih, A., P. Crowley, M. McPeek, J. Petranka, and K. Strohmeier. 1985. Predation, competition, and prey communities: A review of field experiments. *Ann. Rev. Ecol. Syst.* 16:269–311.
Silver, M. L. 1942. The motoneurons of the spinal cord of the frog. *J. Comp. Neurol.* 77:1–40.
Silverstone, P. A. 1975. A revision of the poison-arrow frogs of the genus *Dendrobates* Wagler. *Sci. Bull., Nat. Hist. Mus. Los Angeles Co.* 21:1–55.
Simnett, J. D., and D. P. Chopra. 1969. Organ specific inhibitor of mitosis in the amphibian kidney. *Nature* 222:1189–1190.
Simon, M. P. 1983. The ecology of parental care in a terrestrial breeding frog from New Guinea. *Behav. Ecol. Sociobiol.* 14:61–67.
Simpson, H. B., M. L. Tobias, and D. B. Kelley. 1986. Origin and identification of fibers in the cranial nerve IX-X complex of *Xenopus laevis:* Lucifer yellow backfills in vitro. *J. Comp. Neurol.* 244:430–444.
Sin, G., and L. Gavrila. 1977. Study of the diet of *Rana ridibunda* tadpoles. *Stud. Cercet. Biol.* 29:93–98.
Sinsch, U. 1990. Froschlurche (Anura) der zentral-peruanischen Anden: Artdiagnose, Taxonomie, Habitate, Verhaltensökologie. *Salamandra* 26:177–214.
———. 1997. Effects of larval history and microtags on growth and survival of natterjack (*Bufo calamita*) metamorphs. *Herpetol. J.* 7:163–168.
Sive, H., and L. Bradley. 1996. A sticky problem: The *Xenopus* cement gland as a paradigm for anteroposterior patterning. *Develop. Biol.* 205:265–280.
Sivula, J. C., M. C. Mix, and D. S. McKenzie. 1972. Oxygen consumption of *Bufo boreas boreas* tadpoles during various developmental stages of metamorphosis. *Herpetologica* 28:309–313.
Skelly, D. K. 1990. Behavioral and life-historical responses of larval American toads to an odonate predator. *Ecology* 71:2313–2322.
———. 1991. Field evidence for a cost of behavioral antipredator response in a larval amphibian. *Ecology* 73:704–708.
———. 1995a. A behavioral trade-off and its consequences for the distribution of *Pseudacris* treefrog larvae. *Ecology* 76:150–164.
———. 1995b. Competition and the distribution of spring peeper larvae. *Oecologia* 103:203–207.
———. 1996. Pond drying, predators, and the distribution of *Pseudacris* tadpoles. *Copeia* 1996:599–605.
Skelly, D. K., and E. E. Werner. 1990. Behavioral and life-historical responses of larval American toads to an odonate predator. *Ecology* 71:2313–2322.
Slade, N. A., and R. J. Wassersug. 1975. On the evolution of complex life cycles. *Evolution* 29:568–571.
Slater, D. W., and E. J. Dornfeld. 1939. A triple stain for amphibian embryos. *Stain Tech.* 14:103–104.
Slater, P., and A. R. Main. 1963. Notes on the biology of *Notaden nichollsi* Parker (Anura, Leptodactylidae). *Western Australian Nat.* 8:163–166.
Smirnov, S. V. 1990. Evidence of neoteny: A paedomorphic morphology and retarded development in *Bombina orientalis* (Anura, Discoglossidae). *Zool. Anz.* 225:324–332.
———. 1992. The influence of variation in larval period on adult cranial diversity in *Pelobates fuscus* (Anura: Pelobatidae). *J. Zool.* 226:601–612.
———. 1994. Postmaturation skull development in *Xenopus laevis* (Anura, Pipidae): Late-appearing bones and their bearing on the pipid ancestral morphology. *Russian J. Herpetol.* 1:21–29.
———. 1995. Extra bones in the *Pelobates* skull as evidence of the paedomorpic origin of the anurans. *Zhur. Obs. Biol.* 56:317–328.
———. 1997. Additional dermal ossifications in the anuran skull: Morphological novelties or archaic elements? *Russian J. Herpetol.* 4:17–27.
Smith, C. L., and E. Frank. 1988a. Peripheral specification of sensory connections in the spinal cord. *Brain Behav. Evol.* 31:227–242.
———. 1988b. Specificity of sensory projections to the spinal cord during development in bullfrogs. *J. Comp. Neurol.* 269:96–108.
Smith, D.C. 1983. Factors controlling tadpole populations of the chorus frog (*Pseudacris triseriata*) on Isle Royale, Michigan. *Ecology* 64:501–510.
———. 1986. Enhanced growth in full sib populations of chorus frog tadpoles. *American Zool.* 26:8. (Abstr.)
———. 1987. Adult recruitment in chorus frogs: Effects of size and date at metamorphosis. *Ecology* 68:344–350.
———. 1990. Population structure and competition among kin in the chorus frog (*Pseudacris triseriata*). *Evolution* 44:1529–1541.
Smith, D.C., and J. Van Buskirk. 1995. Phenotypic design, plasticity, and ecological performance in two tadpole species. *American Nat.* 145:211–233.
Smith, L. D. 1965. Transplantation of the nuclei of primordial germ cells into enucleated eggs of *Rana pipiens. Proc. Natl. Acad. Sci. United States of America* 54:101–107.
———. 1966. The role of a germinal plasm in the formation of primordial germ cells in *Rana pipiens. Develop. Biol.* 14:330–347.
Smith, L. D., and M. Williams. 1979. Germinal plasm and germ cell determinants in anuran amphibians. pp. 167–197. In *Proc. 4th Symp. British Soc. Develop. Biol.*, edited by D. R. Newth and M. Balls. Cambridge: Cambridge University Press.
Smith, M. A. 1916a. Descriptions of five tadpoles from Siam. *J. Nat. Hist. Soc. Siam* 2:37–43.
———. 1916b. On the frogs of the genus *Oxyglossis. J. Nat. Hist. Soc. Siam* 2:172–176.
———. 1917. On tadpoles from Siam. *J. Nat. Hist. Soc. Siam.* 2:261–275.
———. 1930. The Reptilia and Amphibia of the Malay Peninsula. *Bull. Raffles Mus.* (3):1–149.
Smith, R. J. F. 1977. Chemical communication as an adaptation: Alarm substances of fish. pp. 303–320. In *Chemical signals in vertebrates,* edited by D. Muller-Schwarze and M. M. Mozell. New York: Plenum.
Smith, S. C., M. J. Lannoo, and J. B. Armstrong. 1988. Lateral-line neuromast development in *Ambystoma mexicanum* and a comparison with *Rana pipiens. J. Morphol.* 198:367–379.
———. 1990. Development of the mechanoreceptive lateral-line system in the axolotl: Placode specification, guidance of migration, and the origin of neuromast polarity. *Anat. Embryol.* 182:171–180.

Smithberg, M. 1954. The origin and development of the tail in the frog, *Rana pipiens*. *J. Exp. Zool.* 127:397–425.
Smith-Gill, S. J., and K. A. Berven. 1979. Predicting amphibian metamorphosis. *American Nat.* 113:563–585.
Smith-Gill, S. J., and D. E. Gill. 1978. Curvilinearities in the competition equations: An experiment with ranid tadpoles. *American Nat.* 112:557–570.
Snetkova, E., N. Chelnaya, L. Serova, S. Saveliev, E. Cherdanzova, S. Pronych, and R. Wassersug. 1995. The effects of space flight on *Xenopus laevis* larval development. *J. Exp. Zool.* 273:21–32.
Sobotka, J. M., and R. G. Rahwan. 1994. Lethal effect of latex gloves on *Xenopus laevis* tadpoles. *J. Pharmacol. Toxicol. Meth.* 32:59.
Sokal, R. R., and I. Huber. 1963. Competition among genotypes in *Tribolium castaneum* at varying densities and gene frequencies (the *sooty* locus). *American Nat.* 97:169–184.
Sokal, R. R., and I. Karten. 1964. Competition among genotypes in *Tribolium castaneum* at varying densities and gene frequencies (the *black* locus). *Genetics* 49:195–211.
Sokol, A. 1984. Plasticity in the fine timing of metamorphosis in tadpoles of the hylid frog, *Litoria ewingi*. *Copeia* 1984:868–873.
Sokol, O. M. 1959. Studien an pipiden Fröschen. I. Die Kaulquappe von *Hymenochirus curtipes* Noble. *Anat. Anz.* 162:154–160.
———. 1962. The tadpole of *Hymenochirus boettgeri*. *Copeia* 1962:272–284.
———. 1975. The phylogeny of anuran larvae: A new look. *Copeia* 1975:1—23.
———. 1977a. The free swimming *Pipa* larvae, with a review of pipid larvae and pipid phylogeny (Anura: Pipidae). *J. Morphol.* 154:357–426.
———. 1977b. A subordinal classification of frogs (Amphibia: Anura). *J. Zool.* 182:505–508.
———. 1981a. The filter apparatus of larval *Pelodytes punctatus* (Amphibia: Anura). *Amphibia-Reptilia* 2:195–208.
———. 1981b. The larval chondrocranium of *Pelodytes punctatus,* with a review of tadpole chondrocrania. *J. Morphol.* 169:161–183.
Song, J., and L. R. Parenti. 1995. Clearing and staining whole fish specimens for simultaneous demonstration of bone, cartilage and nerves. *Copeia* 1995:114–118.
Sotelo, C. 1976. Morphology of the cerebellar cortex. pp. 864–892. In *Frog neurobiology, a handbook,* edited by R. Llinás and W. Precht. Berlin: Springer-Verlag.
Spallanzani, L. 1785. *Expériences pour servir à l'histoire de la génération des animaux et des plantes.* Genève: Chez Barthelemi Chirol, Libraire.
Spannhof, I., and L. Spannhof. 1971. Beobachtungen zur Brutbiologie und Larvenentwicklung von *Gastrotheca marsupiata*. *Wiss. Zeit. Univ. Rostock Math. Naturwiss. Reihe* 20:97–104.
Spannhof, L. 1956. Zur Morphologie und Histologie muzinhaltiger Zellen in den Nierentubuli des Krallenfrosches. *Zool. Anz. Suppl.* 19:291–296.
Spannhof, L., and S. Dittrich. 1967. Histophysiologische Untersuchungen an den Flaschenzellen der Urniere von *Xenopus laevis* Daudin unter experimentellen Bedingungen. *Zeit. Zellforsch. Mikros. Anat.* 81:407–415.
Spannhof, L., and L. Jonas. 1969. Elektronenmikroskopische Untersuchungen zur Genese und Sekretbildung in den Flaschenzellen der Urniere vom Krallenfrosch. *Zeit. Zellforsch. Mikros. Anat.* 95:134–142.
Spemann, H. 1898. Über die Entwicklung der tuba Eustachi und des Kopfskeletts von *Rana temporaria*. *Zool. Jahrb.* 11:389–416.
Sperry, D. C. 1987. Relationship between natural variations in motoneuron number and body size in *Xenopus laevis:* A test for size matching. *J. Comp. Neurol.* 264:250–267.
Sperry, D. C., and P. Grobstein. 1983. Postmetamorphic changes in the lumbar lateral motor column in relation to muscle growth in the toad, *Bufo americanus*. *J. Comp. Neurol.* 216:104–114.
———. 1985. Regulation of neuron numbers in *Xenopus laevis:* Effects of hormonal manipulation altering size at metamorphosis. *J. Comp. Neurol.* 232:287–298.
Sperry, R. W. 1963. Chemoaffinity in the orderly growth of nerve fiber patterns and connections. *Proc. Natl. Acad. Sci. United States of America* 50:703–710.
Špinar, Z. V. 1972. *Tertiary frogs from central Europe.* Prague: Academia.
———. 1980. Fossile Raniden aus dem oberen Pliozan vonwillershausen (Niedersachsen). *Stuttgarter Beitr. Naturk.* 53B:1–53.
Spitzer, J. L., and N. C. Spitzer. 1975. Time of origin of Rohon-Beard neurons in the spinal cord of *Xenopus laevis*. *American Zool.* 15:781. (Abstr.)
Spray, D.C. 1976. Pain and temperature receptors of anurans. pp. 607–628. In *Frog neurobiology, a handbook,* edited by R. Llinás and W. Precht. New York: Springer-Verlag.
Starr, R. C., and J. A. Zeikus. 1993. UTEX–The culture collections of algae at the University of Texas at Austin. *J. Phycol. Suppl.* 29:1–106.
Starrett, P. H. 1960. Descriptions of tadpoles of Middle American frogs. *Misc. Publ. Mus. Zool., Univ. Michigan* (110):1–38, 1 plate.
———. 1967. Observations on the life history of frogs of the family Atelopodidae. *Herpetologica* 23:195–204.
———. 1968. The phylogenetic significance of the jaw musculature in anuran amphibians. Dissertation, University of Michigan.
———. 1973. Evolutionary patterns in larval morphology. pp. 251–271. In *Evolutionary biology of the anurans. Contemporary research on major problems,* edited by J. L. Vial. Columbia: University of Missouri Press.
Stauffer, H. P., and R. D. Semlitsch. 1993. Effects of visual, chemical and tactile cues of fish on the behavioural responses of tadpoles. *Anim. Behav.* 46:355–364.
Stebbins, R. C. 1985. *A field guide to western reptiles and amphibians. Field marks of all species in western North America, including Baja California.* Boston: Houghton Mifflin.
Stebbins, R. C., and J. R. Hendrickson. 1959. Field studies of amphibians in Colombia, South America. *Univ. California Publ. Zool.* 56:497–540.
Steele, C. W., S. Strickler-Shaw, and D. H. Taylor. 1991. Failure of *Bufo americanus* tadpoles to avoid lead-enriched water. *J. Herpetol.* 25:241–243.
Stefanelli, A. 1951. The Mauthnerian apparatus in the Ichthyopsida: Its nature and function and correlated problems of neurohistogenesis. *Quart. Rev. Biol.* 26:17–34.
Stehouwer, D. J. 1987. Effect of tectotomy and decerebration on spontaneous and elicited behavior of tadpoles and juvenile frogs. *Behav. Neurosci.* 101:378–384.
———. 1988. Metamorphosis of behavior in the bullfrog (*Rana catesbeiana*). *Develop. Psychobiol.* 21:383–395.
Stehouwer, D. J., and P. B. Farel. 1980. Central and peripheral controls of swimming in anuran larvae. *Brain. Res.* 195:323–335.

———. 1981. Sensory interactions with a central motor program in anuran larvae. *Brain Res.* 218:131–140.

Stein, J. R., ed. 1975. *Handbook of phycological methods. Culture methods and growth measurements.* Cambridge: Cambridge University Press.

Steinke, J. H., and D. G. Benson, Jr. 1970. The structure and polysaccharide chemistry of the jelly envelopes of the egg of the frog, *Rana pipiens. J. Morphol.* 130:57–66.

Steinman, R. M. 1968. An electron microscopic study of ciliogenesis in developing epidermis and trachea in the embryo of *Xenopus laevis. American J. Anat.* 122:19–56.

Steinwascher, K. 1978a. The effect of coprophagy on the growth of *Rana catesbeiana* tadpoles. *Copeia* 1978:130–134.

———. 1978b. Interference and exploitation competition among tadpoles of *Rana utricularia. Ecology* 59:1039–1046.

———. 1979a. Competitive interactions among tadpoles: Responses to resource level. *Ecology* 60:1172–1183.

———. 1979b. Host-parasite interaction as a potential population-regulating mechanism. *Ecology* 60:884–890.

———. 1981. Competition for two resources. *Oecologia* 49:415–418.

Steinwascher, K., and J. Travis. 1983. Influence of food quality and quantity on early larval growth of two anurans. *Copeia* 1983:238–242.

Steneck, R. S., and L. Watling. 1982. Feeding capabilities and limitation of herbivorous mollusc: A functional group approach. *Mar. Biol.* 68:299–319.

Stensaas, L. J., and S. S. Stensaas. 1971. Light and electron microscopy of motoneurons and neuropile in the amphibian spinal cord. *Brain Res.* 31:67–84.

Stephenson, E. M. T. 1951. The anatomy of the head of the New Zealand frog, *Leiopelma. Trans. Zool. Soc. London* 27:255–305.

Stephenson, N. G. 1951a. Observations on the development of the amphicoelous frogs, *Leiopelma* and *Ascaphus. J. Linnean Soc. (Zool.)* 42:18–28.

———. 1951b. On the development of the chondrocranium and visceral arches of *Leiopelma archeyi. Trans. Zool. Soc. London* 27:203–252.

———. 1955. On the development of the frog, *Leiopelma hochstetteri* Fitzinger. *Proc. Zool. Soc. London* 124:785–795.

Sterba, G. 1950. Über die morphologischen und histogenetischen Thymusprobleme bei *Xenopus laevis* Daudin nebst einigen Bemerkungen über die Morphologie der Kaulquappen. *Abh. Sächs. Akad. Wiss. Leipzig* 44:1–55.

Stevenson, H. M. 1976. *Vertebrates of Florida. Identification and distribution.* Gainesville: University Presses of Florida.

Stiffler, D. F. 1996. Exchanges of calcium with the environment and between body compartments in Amphibia. *Physiol. Zool.* 69:418–434.

Stilling, H. 1898. Zur Anatomie der Nebenniere. *Arch. Mikros. Anat. Entwicklungsmech.* 52:176–195.

Stirling, R. V., and E. G. Merrill. 1987. Functional morphology of frog retinal ganglion cells and their central projections: The dimming detectors. *J. Comp. Neurol.* 258:477–495.

Stöhr, P. 1882. Zur Entwicklungsgeschichte des Anurenschädels. *Zeit. Wiss. Zool.* 36:68–103.

Stokely, P. S., and J. C. List. 1954. The progress of ossification in the skull of the cricketfrog *Pseudacris nigrita triseriata. Copeia* 1954:211–217.

———. 1955. Observations on the development of the anuran urostyle. *Trans. American Micros. Soc.* 74:112–115.

Stone, L. S. 1922. Experiments on the development of the cranial ganglia and the lateral line sense organs in *Amblystoma punctatum. J. Exp. Zool.* 35:421–496.

———. 1933. The development of the lateral-line sense organs in amphibians observed in living and vital-stained preparations. *J. Comp. Neurol.* 57:507–540.

Storm, R. M. 1960. Notes on the breeding biology of the red-legged frog (*Rana aurora aurora*). *Herpetologica* 16:251–259.

Straka, H. and N. Dieringer. 1991. Internuclear neurons in the ocular motor system of frogs. *J. Comp. Neurol.* 312:537–548.

Strathmann, R. R. 1978. The evolution and loss of feeding larval stages of marine invertebrates. *Evolution* 32:894–906.

———. 1993. Hypotheses on the origins of marine larvae. *Ann. Rev. Ecol. Syst.* 24:89–117.

Strathmann, R. R., L. Fenaux, and M. F. Strathman. 1992. Heterochronic developmental plasticity in larval sea urchins and its implication for evolution of nonfeeding larvae. *Evolution* 46:972–986.

Strauss, R. E., and R. Altig. 1992. Ontogenetic body form changes in three ecological morphotypes of anuran tadpoles. *Growth Develop. Aging* 56:3–16.

Straw, R. M. 1958. Experimental notes on the Deep Springs toad *Bufo exsul. Ecology* 39:552–553.

Strawinski, S. 1956. Vascularization of respiratory surfaces in ontogeny of the edible frog, *Rana esculenta* L. *Zool. Poloniae* 7:327–365.

Straznicky, C., and D. Tay. 1982. Retinotectal map formation in dually innervated tecta: A regeneration study in *Xenopus* with one compound eye following bilateral optic nerve section. *J. Comp. Neurol.* 206:119–130.

Streb, M. 1967. Experimentelle Untersuchungen uber die Beziehung zwischen Schilddruse und Hypophyse wahrend der Larvalentwicklung und Metamorphose von *Xenopus laevis* Daudin. *Zeit. Zellforsch.* 82, 407–433.

Strickler-Shaw, S., and D. H. Taylor. 1990. Sublethal exposure to lead inhibits acquisition and retention of discriminate avoidance learning in green frog (*Rana clamitans*) tadpoles. *Environ. Toxicol. Chem.* 9:47–52.

———. 1991. Lead inhibits acquisition and retention learning in bullfrog tadpoles. *Neurotoxicol. Teratol.* 13:167–173.

Strijbosch, H. 1979. Habitat selection of amphibians during their aquatic phase. *Oikos* 33:363–372.

Strong, D. R., Jr. 1983. Natural variability and the manifold mechanisms of ecological communities. *American Nat.* 122:636–660.

Stuart, L. C. 1941. Another new *Hypopachus* from Guatemala. *Proc. Biol. Soc. Washington* 54:125–128.

———. 1948. The amphibians and reptiles of Alta Verapaz, Guatemala. *Misc. Publ. Mus. Zool., Univ. Michigan* (69):1–109.

———. 1961. Some observations on the natural history of tadpoles of *Rhinophrynus dorsalis* Dumeril and Bibron. *Herpetologica* 17:73–79.

Stuesse, S. L., W. L. R. Cruce, and K. S. Powell. 1983. Afferent and efferent components of the hypoglossal nerve in the grass frog, *Rana pipiens. J. Comp. Neurol.* 217:432–439.

———. 1984. Organization within the cranial IX-X complex in ranid frogs: A horseradish peroxidase transport study. *J. Comp. Neurol.* 222:358–365.

Sullivan, B. 1974. Reptilian hemoglobins. pp. 77–122. In *Chemical zoology.* Vol. 9. *Amphibia and Reptilia.* Edited by M. Florkin and B. Scheer. New York: Academic Press.

Summers, K., and W. Amos. 1997. Behavioral, ecological, and molecular genetic analyses of reproductive strategies in the

Amazonian dart-poison frog, *Dendrobates ventrimaculatus. Behav. Ecol.* 8:260–267.
Summey, M. R., and A. Mathis. 1998. Alarm responses to chemical stimuli from damanged conspecifics by larval anurans: Tests of three neotropical species. *Herpetologica* 54:402–408.
Süssbier, W. 1936. Anlage und Metamorphose der Drüsen der Darmwandung bei den Anuren. *Zeit. Wiss. Zool.* 148:309–349.
Sutherland, R. M., and R. F. Nunnemacher. 1981. Fibers in spinal nerves of treefrogs: *Eleutherodactylus* and *Hyla. J. Comp. Neurol.* 202:415–420.
Swammerdam, J. 1737–1738. *Biblia Naturae.* Vol. II. Leyden: I. Severinus.
Swanepoel, J. H. 1970. The ontogenesis of the chondrocranium and of the nasal sac of the microhylid frog *Breviceps adspersus pentheri* Werner. *Ann. Univ. Stellenbosch* 45A:1–119.
Swart, C. C., and R. O. De Sá. 1999. The chondrocranium of the Mexican burrowing toad, *Rhinophrynus doralis* (Anura: Rhinophrynidae). *J. Herpetol.* 33:23–28.
Sweet, S. S. 1980. Allometric inference in morphology. *American Zool.* 20:643–652.
Swofford, D. L., and G. J. Olsen. 1990. Phylogeny reconstruction. pp. 411–501. In *Molecular Systematics,* edited by D. M. Hillis and C. Moritz. Sunderland, MA: Sinauer Associates.
Syuzyamova, L. M., and N. L. Iranova. 1988. Morphological characteristics and energy metabolism in the tadpoles of *Rana arvalis* bread [sic] at different diets. *Ekologiya* 1988:18–24.
Szabo, T., J.-P. Denzot, S. Blähser, M. Veron-Ravaille, and D. Rouilly. 1990. The olfactoretinalis system does not correspond to the "terminal nerve" as ascribed to F. Pinkus. *Soc. Neurosci. Abstr.,* p. 129.
Szarski, H. 1957. The origin of the larva and metamorphosis in Amphibia. *American Nat.* 91:283–301.
Székely, G. 1976. The morphology of motoneurons and dorsal root fibers in the frog's spinal cord. *Brain Res.* 103:275–290.
Székely, G., N. Antal, and T. Görcs. 1980. Direct dorsal root projection onto the cerebellum in the frog. *Neurosci. Lett.* 19:161–165.
Székely, G., and G. Czéh. 1976. Organization of locomotion. pp. 765–792. In *Frog neurobiology, a handbook,* edited by R. Llinás and W. Precht. New York: Springer-Verlag.
Tachibana, T. 1978. The Merkel cell in the labial ridge epidermis of anuran tadpole. I. Fine structure, distribution, and cytochemical studies. *Anat. Rec.* 191:487–502.
Tachihama, H., F. Sasaki, T. Tachibana, H. Iseki, T. Horiguchi, T. Kinoshita, and K. Watanabe. 1987. Fine structure and degeneration of the horny teeth and the epidermis of the labial ridge of the anuran tadpole. *Tsurumi Univ. Dental J.* 13:49–63.
Tahin, Q. S., N. C. Meirelles, and S. B. Gargoggini. 1979. Ammoneotelism and ureotelism of terrestrial and aquatic amphibians. *Cienc. Cult.* (São Paulo) 31:771–774.
Taigen, T. L., and F. H. Pough. 1985. Metabolic correlates of anuran behavior. *American Zool.* 25:987–997.
Taigen, T. L., F. H. Pough, and M. M. Stewart. 1984. Water balance of terrestrial anuran *Eleutherodactylus coqui* eggs: Importance of parental care. *Ecology* 65:248–255.
Takahashi, H. 1971. Sex differentiation and development of the gonad in *Rana rugosa. Zool. Mag. (Tokyo)* 80:15–21.
Takasu, T., and H. Iwasawa. 1983. Ultrastructural studies on the testicular differentiation in the frog *Rhacophorus arboreus. Zool. Mag. (Tokyo)* 92:501. (Abstr.)
Takisawa, A., Y. Ōhara, and K. Kanō. 1952a. Die Kaumuskulatur der Anuren (*Bufo vulgaris japonicus*) während der Metamorphose. *Okajimas Folia Anat. Japonica* 24:1–28.
Takisawa, A., Y. Ōhara, and Y. Sunaga. 1952b. Über die Umgestaltung der Mm. intermandibulares der Anuren während der Metamorphose. *Okajimas Folia Anat. Japonica* 24:217–241.
Takisawa, A., and Y. Sunaga. 1951. Über die Entwicklung des M. depressor mandibulae bei Anuren im Laufe der Metamorphose. *Okajimas Folia Anat. Japonica* 23:273–293.
Tanaka, S., and M. Nishinara. 1987. Foam nest as a potential food source for anuran larvae: A preliminary experiment. *J. Ethol.* 5:86–88.
Tang, Y.-Y., and C. M. Rovainen. 1996. Cardiac output in *Xenopus laevis* tadpoles during development and in response to an adenosine agonist. *American J. Physiol.* 270R:997–1004.
Tanimura, A., and H. Iwasawa. 1986. Development of gonad and Bidder's organ in *Bufo japonicus formosus:* Histological observation. *Sci. Rept. Niigata Univ.* (23)D:11–21.
———. 1987. Germ cell kinetics in gonadal development in the toad *Bufo japonicus formosus. Zool. Sci.* 4:657–664.
———. 1988. Ultrastructural observations on the origin and differentiation of somatic cells during gonadal development in the frog *Rana nigromaculata. Develop. Growth Differ.* 30:681–691.
———. 1989. Origin of somatic cells and histogenesis in the primordial gonad of the Japanese tree frog *Rhacophorus arboreus. Anat. Embryol.* 180:165–173.
———. 1991. Proliferative activity of somatic cells during gonadal development in the Japanese pond frog, *Rana nigromaculata. J. Exp. Zool.* 259:365–370.
Tatum, J. B. 1975. Egg volume. *Auk* 92:576–580.
Taugner, R., A. Schiller, and S. Ntokalou-Knittel. 1982. Cells and intercellular contacts in glomeruli and tubules of the frog kidney: A freeze-fracture and thin section study. *Cell Tiss. Res.* 226:589–608.
Taxi, J. 1976. Morphology of the autonomic nervous system. pp. 93–150. In *Frog neurobiology, a handbook,* edited R. Llinás and W. Precht. New York: Springer-Verlag.
Taylor, A. C., and J. J. Kollros. 1946. Stages in the normal development of *Rana pipiens* larvae. *Anat. Rec.* 94:7–24.
Taylor, C. L., and R. Altig. 1995. Oral disc kinematics of four rheophilous anuran tadpoles. *Herpetol. Nat. Hist.* 3:101–106.
Taylor, C. L., R. Altig, and C. R. Boyle. 1995. Can anuran tadpoles choose among foods that vary in quality? *Alytes* 13:81–86.
———. 1996. Oral disc kinematics of four lentic anuran tadpoles. *Herpetol. Nat. Hist.* 4:49–56.
Taylor, D. H., and D. E. Ferguson. 1970. Extraoptic celestial orientation in the southern cricket frog *Acris gryllus. Science* 168:390–392.
Taylor, D. H., C. W. Steele, and S. Strickler-Shaw. 1990. Responses of green frog (*Rana clamitans*) tadpoles to lead-polluted water. *Environ. Toxicol. Chem.* 9:87–93.
Taylor, E. H. 1942. Tadpoles of Mexican Anura. *Univ. Kansas Sci. Bull.* 28:37–55.
———. 1954. Frog-egg-eating tadpoles of *Anotheca coronata* (Stejneger)(Salientia, Hylidae). *Univ. Kansas Sci. Bull.* 36:589–596.
Taylor, J. D., and J. T. Bagnara. 1972. Dermal chromatophores. *American Zool.* 12:43–62.
Taylor, J. S. H., and A. Roberts. 1983. The early development of the primary sensory neurones in an amphibian embryo: A scanning electron microscope study. *J. Embryol. Exp. Morphol.* 5:49–66.

Taylor, R. E., Jr., and S. B. Barker. 1965. Transepidermal potential difference: Development in anuran larvae. *Science* 148:1612–1613.

Taylor, R. J. 1976. Value of clumping to prey and the evolutionary response of ambush predators. *American Nat.* 110: 13–29.

———. 1979. The value of clumping to prey when detectability increases with group size. *American Nat.* 113:299–301.

Taylor, W. R. 1977. Observations on specimen fixation. *Proc. Biol. Soc. Washington* 90:753–763.

Tejedo, M. 1991. Effects of predation by two species of sympatric tadpoles on embryo survival in natterjack toads (*Bufo calamita*). *Herpetologica* 47:322–327.

———. 1992. Absence of the trade-off between the size and number of offspring in the natterjack toad (*Bufo calamita*). *Oecologia* 90:294–296.

Tejedo, M., and R. Reques. 1992. Effects of egg size and density on metamorphic traits in tadpoles of the natterjack toad (*Bufo calamita*). *J. Herpetol.* 26:146–152.

———. 1994a. Does larval growth history determine timing of metamorphosis in anurans? A field experiment. *Herpetologica* 50:113–118.

———. 1994b. Plasticity in metamorphic traits of natterjack tadpoles: The interactive effects of density and pond duration. *Oikos* 71:295–304.

Ten Donkelaar, H. J., and R. De Boer-Van Huizen. 1982. Observations on the development of descending pathways from the brain stem to the spinal cord in the clawed toad *Xenopus laevis*. *Anat. Embryol.* 163:461–473.

Teran, H. R., and A. M. De Cerasuolo. 1988. Estudio histomorfológico del tracto digestivo larval de *Gastrotheca gracilis* Laurent (Anura Hylidae). *Neotrópica* 34:115–123.

Terent'ev, P. V., and S. A. Chernov. 1965. *Key to the amphibians and reptiles.* 3d enlarged ed. [English Translation]. Israel Program for Scientific Translation, Jerusalem.

Terhivuo, J. 1988. Phenology of spawning for the common frog (*Rana temporaria* L.) in Finland from 1846 to 1986. *Ann. Zool. Fennici* 25:165–175.

Test, F. H., and R. G. McCann. 1976. Foraging behavior of *Bufo americanus* tadpoles in response to high densities of micro-organisms. *Copeia* 1976:576–578.

Tevis, L., Jr. 1966. Unsuccessful breeding by desert toads (*Bufo punctatus*) at the limit of their ecological tolerance. *Ecology* 47:766–775.

Theil, E. C. 1970. Red blood cell replacement during the transition from embryonic (tadpole) hemoglobin to adult (frog) hemoglobin in *Rana catesbeiana*. *Comp. Biochem. Physiol.* 33:717–720.

Thibaudeau, D. G., and R. Altig. 1988. Sequence of ontogenetic development and atrophy of the oral apparatus of six anuran tadpoles. *J. Morphol.* 197:63–69.

Thibier-Fouchet, C., O. Mulner, and R. Ozon. 1976. Progesterone biosynthesis and metabolism by ovarian follicles and isolated oocytes of *Xenopus laevis*. *Biol. Reprod.* 14:317–326.

Thiele, J. 1888. Der Haftapparat der Batrachierlarven. *Zeit. Wiss. Zool.* 46:67–79.

Thomas, E. I. 1972. Structure of the heart in the tadpole larvae of *Rana tigrina* Daud. *Ann. Zool., Agra* 8:41–50.

Thompson, R. L. 1996. Larval habitat, ecology, and parental investment of *Osteopilus brunneus* (Hylidae). pp. 259–269. In *Contributions to West Indian herpetology: A tribute to Albert Schwartz,* edited by R. Powell and R. W. Henderson. Contrib. Herpetol., vol. 12. Ithaca, NY: Society for the Study of Amphibians and Reptiles.

Thrall, J. H. 1972. Food, feeding, and digestive physiology of the larval bullfrog, *Rana catesbeiana* Shaw. Dissertation, Illinois State University.

Tihen, J. A. 1965. Evolutionary trends in frogs. *American Zool.* 5:309–318.

Tinbergen, N. 1968. *Curious naturalists.* Garden City, NY: Natural history library, Doubleday & Co.

Toal, K. R., III, and J. R. Mendelson, III. 1995. A new species of *Hyla* (Anura: Hylidae) from cloud forest in Oaxaca, Mexico, with comments on the status of the *Hyla bistincta* group. *Occas. Pap. Mus. Nat. Hist., Univ. Kansas* (174):1–20.

Toews, D. P., and R. G. Boutilier. 1986. Acid-base regulation in the Amphibia. pp. 265–308. In *Acid-base regulation in animals,* edited by N. Heisler. Amsterdam: Elsevier.

Toloza, E. M., and J. M. Diamond. 1990a. Ontogenetic development of nutrient transporters in bullfrog intestine. *American J. Physiol.* 258G:760–769.

———. 1990b. Ontogenetic development of transporter regulation in bullfrog intestine. *American J. Physiol.* 258G:770–773.

Toner, P. G., and K. E. Carr. 1968. *Cell structure. An introduction to biological electron microscopy.* Edinburgh: E. and S. Livingstone Ltd.

Tosney, K. W. 1978. The early migration of neural crest cells in the trunk region of the avian embryo: An electron microscopic study. *Develop. Biol.* 62:317–333.

Townsend, D. S. 1986. The costs of male parental care and its evolution in a neotropical frog. *Behav. Ecol. Sociobiol.* 19:187–195.

———. 1996. Patterns of parental care in frogs of the genus *Eleutherodactylus*. pp. 229–239. In *Contributions to West Indian herpetology: A tribute to Albert Schwartz,* edited by R. Powell and R. W. Henderson. Contrib. Herpetol., vol. 12. Ithaca, NY: Society for the Study of Amphibians and Reptiles.

Townsend, D. S., and W. H. Moger. 1987. Plasma androgen levels during male parental care in a tropical frog (*Eleutherodactylus*). *Horm. Behav.* 21:93–99.

Townsend, D. S., and M. M. Stewart. 1985. Direct development in *Eleutherodactylus coqui* (Anura: Leptodactylidae): A staging table. *Copeia* 1985:423–436.

———. 1986. The effect of temperature on direct development in a terrestrial-breeding, neotropical frog. *Copeia* 1986: 520–523.

Townsend, D. S., M. M. Stewart, and F. H. Pough. 1984. Male parental care and its adaptive significance in a neotropical frog. *Anim. Behav.* 32:421–431.

Townsend, D. S., M. M. Stewart, F. H. Pough, and P. F. Brussard. 1981. Internal fertilization in an oviparous frog. *Science* 212:469–471.

Tracy, C. R. 1976. A model of the dynamic exchanges of water and energy between a terrestrial amphibian and its environment. *Ecol. Monogr.* 46:293–326.

Trainor, F. R., S. Wallett, and J. Grochowski. 1991. A useful algal growth medium. *J. Phycol.* 27:460–461.

Travis, J. 1980a. Genetic variation for larval specific growth rate in the frog *Hyla gratiosa*. *Growth* 44:167–181.

———. 1980b. Phenotypic variation and the outcome of interspecific competition in hylid tadpoles. *Evolution* 34:40–50.

———. 1981a. A key to the tadpoles of North Carolina. *Brimleyana* 6:119–127.

———. 1981b. Control of larval growth variation in a population of *Pseudacris triseriata* (Anura: Hylidae). *Evolution* 5:423–432.

———. 1983a. Variation in development patterns of larval anurans in temporary ponds. I. Persistent variation within a *Hyla gratiosa* population. *Evolution* 37:496–512.

———. 1983b. Variation in growth and survival of *Hyla gratiosa* larvae in experimental enclosures. *Copeia* 1983:232–237.

———. 1984. Anuran size at metamorphosis: Experimental test of a model based on intraspecific competition. *Ecology* 65:1155–1160.

Travis, J., S. B. Emerson, and M. Blouin. 1987. A quantitative-genetic analysis of larval life-history traits in *Hyla crucifer. Evolution* 41:145–156.

Travis, J., W. H. Keen, and J. Juilianna. 1985a. The effects of multiple factors on viability selection in *Hyla gratiosa* tadpoles. *Evolution* 39:1087–1099.

———. 1985b. The role of relative body size in a predator-prey relationship between dragonfly naiads and larval anurans. *Oikos* 45:59–65.

Travis, J., and J. C. Trexler. 1986. Interactions among factors affecting growth, development, and survival in experimental populations of *Bufo terrestris* (Anura: Bufonidae). *Oecologia* 69:110–116.

Treisman, M. 1975. Predation and the evolution of gregariousness. I. Models for concealment and evasion. *Anim. Behav.* 23:779–800.

Trenerry, M. P. 1988. The ecology of tadpoles in a tropical rainforest stream. Honours thesis, James Cook University, Townsville, Queensland, Australia.

Trewavas, E. 1933. The hyoid and larynx of the Anura. *Trans. Roy. Philosoph. Soc. London* 222:401–527.

Trueb, L. 1973. Bones, frogs, and evolution. pp. 65–132. In *Evolutionary biology of the anurans. Contemporary research on major problems,* edited by J. L. Vial. Columbia: University of Missouri Press.

———. 1979. Leptodactylid frogs of the genus *Telmatobius* in Ecuador with the description of a new species. *Copeia* 1979:714–733.

———. 1985. A summary of osteocranial development in anurans with notes on the sequence of cranial ossification in *Rhinophrynus dorsalis* (Anura: Pipoidea: Rhinophrynidae). *South African J. Sci.* 81:181–185.

———. 1993. Patterns of cranial diversity among the Lissamphibia. pp. 255–343. In *The vertebrate skull,* edited by J. Hanken and B. Hall. Chicago: University of Chicago Press.

Trueb, L., and D.C. Cannatella. 1982. The cranial osteology and hyolaryngeal apparatus of *Rhinophrynus dorsalis* (Anura: Rhinophrynidae) with comparisons to Recent pipid frogs. *J. Morphol.* 171:11–40.

———. 1986. Systematics, morphology, and phylogeny of genus *Pipa* (Anura: Pipidae). *Herpetologica* 42:412–449.

Trueb, L., and W. E. Duellman. 1970. The systematic status and life history of *Hyla verrucigera* Werner. *Copeia* 1970:601–610.

Trueb, L., and J. Hanken. 1993. Skeletal development in *Xenopus laevis* (Anura: Pipidae). *J. Morphol.* 214:1–41.

Tschugunova, T. J. 1981. Interfrontalia in *Bombina orientalis* (Blgr.) and *Bombina bombina* (L.). pp. 117–121. In *Herpetological investigations in Siberia and the Far East,* edited by L. J. Borkin. Leningrad: Akad. Nauk SSSR.

Tsuneki, K., H. Kobayashi, R. Gallardo, and P. K. T. Pang. 1984. Electron microscopic study of the innervation of the renal tubules and urinary bladder epithelium in *Rana catesbeiana* and *Necturus maculosus. J. Morphol.* 181:143–154.

Tubbs, L. O. E., R. Stevens, M. Wells, and R. Altig. 1993. Ontogeny of the oral apparatus of the tadpole of *Bufo americanus. Amphibia-Reptilia* 14:333–340.

Tung, T. C. 1935. On the time of determination of the dorsoventral axis of the pronephros in *Discoglossus. Peking Nat. Hist. Bull.* 10:115.

Tung, T. C., and S. H. Ku. 1944. Experimental studies on the development of the pronephric duct in anuran embryos. *J. Anat.* 78:52–57.

Turner, R. J. 1969. The functional development of the reticulo-endothelial system in the toad, *Xenopus laevis* (Daudin). *J. Exp. Zool.* 170:467–480.

Turner, R. J., and M. J. Manning. 1972. Response of the toad, *Xenopus laevis,* to circulating antigens. I. Cellular changes in the spleen. *J. Exp. Zool.* 183:21–34.

———. 1974. Thymic dependence of amphibian antibody responses. *European J. Immunol.* 4:343–346.

Turpen, J. B., and C. M. Knudson. 1982. Ontogeny of hematopoietic cells in *Rana pipiens:* Precursor cell migration during embryogenesis. *Develop. Biol.* 89:138–151.

Turpen, J. B., C. M. Knudson, and P. S. Hoefen. 1981a. Embryonic origin of hematopoietic precursor cells from the region of the presumptive mesonephros in *Rana pipiens. Anat. Rec.* 199:261. (Abstr.)

———. 1981b. The early ontogeny of hematopoietic cells studied by grafting cytogenetically labeled tissue anlagen: Localization of a prospective stem cell compartment. *Develop. Biol.* 85:99–112.

Turpen, J. B., C. J. Turpen, and M. Flajnik. 1979. Experimental analysis of hematopoietic cell development in the liver of larval *Rana pipiens. Develop. Biol.* 69:466–479.

Turnipseed, G., and R. Altig. 1975. Population density and age structure of three species of hylid tadpoles. *J. Herpetol.* 9:287–291.

Twitty, V. C. 1955. Eye. pp. 402–414. In *Analysis of development,* edited by B. H. Willier, P. A. Weiss, and V. Hamburger. Philadelphia: W. B. Saunders.

———. 1966. *Of scientists and salamanders.* San Francisco: W. H. Freeman.

Twombly, S. 1996. Timing of metamorphosis in a freshwater crustacean: Comparison with anuran models. *Ecology* 77:1855–1866.

Tyler, M. J. 1962. On the preservation of anuran tadpoles. *Australian J. Sci.* 25:222.

———. 1963. A taxonomic study of amphibians and reptiles of the Central Highlands of New Guinea, with notes on their ecology and biology. 2. Anura: Ranidae and Hylidae. *Trans. Roy. Soc. South Australia* 86:105–130.

———. 1968. Papuan hylid frogs of the genus *Hyla. Zool. Verh.* (96):1–203, 4 plates.

———. 1971. The phylogenetic significance of vocal sac structure in hylid frogs. *Univ. Kansas Publ. Mus. Nat. Hist.* 19:319–360.

———. 1985. Reproductive modes in Australian Amphibia. pp. 265–267. In *Biology of Australasian frogs and reptiles,* edited by G. Grigg, R. Shine, and H. Ehmann. Chipping Norton, New South Wales: Surrey Beatty.

———. 1989. *Australian frogs.* Penguin Books Australia, Ringwood, Victoria.

Tyler, M. J., and M. Anstis. 1975. Taxonomy and biology of frogs of the *Litoria citropa* complex (Anura: Hylidae). *Rec. South Australia Mus.* 17:41–50.

Tyler, M. J., M. Anstis, G. A. Crook, and M. Davies. 1983. Reproductive biology of the frogs of the Magela Creek system, Northern Territory. *Rec. South Australia Mus.* 18:415–440.

Tyler, M. J., and M. Davies. 1978. Species-groups within the Australopapuan hylid frog genus *Litoria* Tschudi. *Australian J. Zool. Suppl.* (63):1–17.

———. 1983. Larval development. pp. 44–57. In *The gastric brooding frog,* edited by M. J. Tyler. London: Croom Helm.

Tyler, M. J., A. A. Martin, and M. Davies. 1979. Biology and systematics of a new limnodynastine genus (Anura: Leptodactylidae) from north-western Australia. *Australian J. Zool.* 27:135–150.

Uchiyama, H., T. A. Reh, and W. K. Stell. 1988. Immunocytochemical and morphological evidence for a retinopectal projection in anuran amphibians. *J. Comp. Neurol.* 274:48–59.

Uchiyama, M., S.-T. Iwasaki, and T. Murakami. 1991. Surface and subsurface structures of neuromasts in tadpoles of the crab-eating frog, *Rana cancrivora. J. Morphol.* 207:157–164.

Uchiyama, M., T. Murakami, and H. Yoshizawa. 1990. Notes on the development of the crab-eating frog, *Rana cancrivora. Zool. Sci.* 7:73–78.

Uchiyama, M., and H. Yoshizawa. 1992. Salinity tolerance and structure of external and internal gills in tadpoles of the crab-eating frog, *Rana cancrivora. Cell Tiss. Res.* 267:35–44.

Udin, S. B., and M. D. Fisher. 1985. The development of the nucleus isthmi in *Xenopus laevis.* I. Cell genesis and the formation of connections with the tectum. *J. Comp. Neurol.* 232:25–35.

Udin, S. B., M. D. Fisher, and J. J. Norden. 1992. Isthmotectal axons make ectopic synapses in the monocular regions of the tectum in developing *Xenopus laevis* frogs. *J. Comp. Neurol.* 322:461–470.

Ueck, M. 1967. Der manicotto glandulare ("Drüsenmagen") der Anurenlarve in Bau, Funktion und Beziehung zur Gesamtlänge des Darmes. Eine mikroskopisch-anatomische, histochemische, und electronenoptische Studie an der omnivoren und mikrophagen larve von *Xenopus laevis* und der carnivoren und makrophagen larve von *Hymenochirus boettgeri* (Anura, Pipidae). *Zeit. Wiss. Zool.* 176:173–270.

Ueda, H. 1986. Reproduction of *Chirixalus eiffingeri* (Boettger). *Sci. Rept. Lab. Amphib. Biol., Hiroshima Univ.* 8:109–116.

Uhlenhuth, E. 1919. Relation between thyroid gland, metamorphosis, and growth. *J. Exp. Zool.* 1:473–482.

———. 1921. The internal secretions in growth and development of amphibians. *American Nat.* 55:193–221.

Ultsch, G. R. 1973. The effects of water hyacinths (*Eichhornia crassipes*) on the microenvironment of aquatic communities. *Arch. Hydrobiol.* 72:460–473.

———. 1976a. Ecophysiological studies of some metabolic and respiratory adaptations of sirenid salamanders. pp. 287–312. In *Respiration of amphibious vertebrates,* edited by G. M. Hughes. New York: Academic Press.

———. 1976b. Gas exchange and metabolism in the Sirenidae (Amphibia: Caudata). I. Oxygen consumption of submerged sirenids as a function of body size and respiratory surface area. *Comp. Biochem. Physiol.* 47A:485–498.

———. 1987. The potential role of hypercarbia in the transition from water-breathing to air-breathing in vertebrates. *Evolution* 41:442–445.

———. 1989. Ecology and physiology of hibernation and overwintering among freshwater fishes, turtles, and snakes. *Biol. Rev.* 64:435–516.

———. 1996. Gas exchange, hypercarbia and acid-base balance, paleocology, and the evolutionary transition from water-breathing to air-breathing among vertebrates. *Palaeogeograph. Palaeoclimatol. Palaeoecol.* 123:1–27.

Ultsch, G. R., and J. T. Duke. 1990. Gas exchange and habitat selection in the aquatic salamanders *Necturus maculosus* and *Cryptobranchus alleganiensis. Oecologia* 83:250–258.

Ultsch, G. R., R. W. Hanley, and T. R. Bauman. 1985. Responses to anoxia during simulated hibernation in northern and southern painted turtles. *Ecology* 66:388–395.

Ultsch, G. R., and D.C. Jackson. 1996. pH and temperature in ectothermic vertebrates. *Bull. Alabama Mus. Nat. Hist.* 18:1–41.

Ultsch, G. R., and D.C. Jackson, and R. Moalli. 1981. Metabolic oxygen conformity among lower vertebrates: The toadfish revisited. *J. Comp. Physiol.* 142:439–443.

Ultsch, G. R., M. E. Ott, and N. Heisler. 1980. Standard metabolic rate, critical oxygen tension, and aerobic scope for spontaneous activity of trout (*Salmo gairdneri*) and carp (*Cyprinus carpio*) in acidified water. *Comp. Biochem. Physiol.* 67A:329–335.

Underhill, D. K. 1966. An incidence of spontaneous caudal scoliosis in tadpoles of *Rana pipiens* Schreber. *Copeia* 1966:582–583.

———. 1967. Spontaneous silvery color and pituitary abnormality in tadpoles from a laboratory cross of *Rana pipiens. Copeia* 1967:673–674.

Uray, N.J. 1985. Early stages in the formation of the cerebellum in the frog. *J. Comp. Neurol.* 232:129–142.

Uray, N.J., and A. G. Gona. 1979. Golgi studies on Purkinje cell development in the frog during spontaneous metamorphosis. II. Details of dendritic development. *J. Comp. Neurol.* 185:237–252.

———. 1982. Golgi studies on Purkinje cell development in the frog during spontaneous metamorphosis. III. Axonal development. *J. Comp. Neurol.* 212:202–207.

Uray, N.J., A. G. Gona, and K. F. Hauser. 1987. Autoradiographic studies of cerebellar histogenesis in the premetamorphic bullfrog tadpole I. Generation of the external granular layer. *J. Comp. Neurol.* 266:232–246.

———. 1988. Autoradiographic studies of cerebellar histogenesis in the premetamorphic bullfrog tadpole. II. Formation of the interauricular granular band. *J. Comp. Neurol.* 269:118129.

Utsunomiya, Y., and T. Utsunomiya. 1983. On the eggs and larvae of the frog species occurring in the Ryukyu Archipelago. *J. Fac. Appl. Biol. Sci., Hiroshima Univ.* 22:255–270.

Utsunomiya, Y., T. Utsunomiya, S. Katsuren, and M. Toyama. 1983. Habitat segregation observed in the breeding of five frog species dwelling in a mountain stream of Okinawa Island. *Annot. Zool. Japonenses* 56:149–153.

Valentine, J. W., and D. Jablonski. 1983. Larval adaptations and patterns of brachiopod diversity in space and time. *Evolution* 37:1052–1061.

Valett, B. B., and D. L. Jameson. 1961. The embryology of *Eleutherodactylus augusti latrans. Copeia* 1961:103–109.

Van Bambeke, C. 1863. Recherches sur la structure de la bouche chez les têtards des batraciens anoures. *Bull. Acad. Roy. Belgique* Ser. 2 16:339–354.

Van Bergeijk, W. A. 1959. Hydrostatic balancing mechanism of *Xenopus* larvae. *J. Acoustic. Soc. America* 31:1340–1347.

Van Beurden, E. K. 1980. Energy metabolism of dormant Australian water-holding frogs (*Cyclorana platycephalus*). *Copeia* 1980:787–799.

Van Buskirk, J. 1988. Interactive effects of dragonfly predation in experimental pond communities. *Ecology* 69:857–867.

Van Buskirk, J., S. A. McCollum, and E. E. Werner. 1997. Natural selection for environmentally induced phenotypes in tadpoles. *Evolution* 51:1983–1992.

Van Buskirk, J., and K. L. Yurewicz. 1998. Effects of predators on prey growth rate: Relative contributions of thinning and reduced activity. *Oikos* 82:20–28.
Vance, W. H., G. L. Clifton, M. L. Applebaum, W. D. Willis, Jr., and R. E. Coggeshall. 1975. Unmyelinated preganglionic fibers in frog ventral roots. *J. Comp. Neurol.* 164:117–126.
Van de Kamer, J. C. 1965. Histological structure and cytology of the pineal complex in fishes, amphibians and reptiles. *Progr. Brain Res.* 10:30–48.
Van Den Broek, A. J. P., G. J. Van Oordt, and G. C. Hirsch. 1938. Das Urogenitalsystem. pp. 683–854. In *Handbuch der vergleischenden Anatomie dere Wirbeltiere,* edited by L. Bolk, E. Goeppert, E. Kallius, and W. Lubosch. Berlin: Urban & Schwarzenberg.
Van Der Horst, C. J. 1934. The lateral line nerves of *Xenopus. Psychiat. Neurol. Bl. Amsterdam* 3:426–435.
Van Der Linden, J. A. M., and H. J. Ten Donkelaar. 1987. Observations on the development of cerebellar afferents in *Xenopus laevis. Anat. Embryol.* 176:431–439.
Van Der Linden, J. A. M., H. J. Ten Donkelaar, and R. De Boer-van Huizen. 1988. Development of spinocerebellar afferents in the clawed toad, *Xenopus laevis. J. Comp. Neurol.* 277:41–52.
Van Der Westhuizen, C. M. 1961. The development of the chondrocranium of *Heleophryne purcelli* Sclater, with special reference to the palatoquadrate and the sound-conducting apparatus. *Acta Zool.* 42:1–72.
Van Dijk, E. D. 1959. On the cloacal region of Anura in particular of larval *Ascaphus. Ann. Univ. Stellenbosch* 35:169–249.
———. 1966. Systematic and field keys to the families, genera, and described species of southern African anuran tadpoles. *Ann. Natal Mus.* 18:231–286.
———. 1972. The behavior of southern African anuran tadpoles with particular reference to their ecology and related external morphological features. *Zool. Africana* 7:49–55.
———. 1981. Material data other than preserved specimens. *Monit. Zool. Italiano (N.S.) Suppl.* 15:393–400.
———. 1985. *Hemisus marmoratum* adults reported to carry tadpoles. *South African J. Sci.* 81:209–210.
Van Eeden, J. A. 1951. The development of the chondrocranium of *Ascaphus truei* Stejneger with special reference to the relations of the palatoquadrate to the neurocranium. *Acta Zool.* 32:41–176.
Van Gansen, P. 1986. Ovogènese des amphibiens. pp. 21–55. In *Traité de Zoologie, Anatomie, Systematique, Biologie.* Vol. 14, Fasc. 1B, edited by P. P. Grassé and M. Delsol. Paris: Masson.
Van Geertruyden, J. 1946. Recherches expérimentales sur la formation du mésonéphros chez les amphibiens anoures. *Arch. Biol.* 57:145–181.
———. 1948. Des premiers stades de développement du mésonéphros chez les amphibiens anoures. *Acta Morphol. Neerlando Scandinavica* 6:1–17.
Van Kampen, P. N. 1923. *The Amphibia of the Indo-Australian archipelago.* Leiden: E. J. Brill, Ltd. (29 figs).
Van Mier, P., and H. J. Ten Donkelaar. 1988. The development of primary afferents to the lumbar spinal cord in *Xenopus laevis. Neurosci. Lett.* 84:35–40.
Van Mier, P., R. Van Rheden, and H. J. Ten Donkelaar. 1985. The development of the dendritic organization of primary and motoneurons in the spinal cord of *Xenopus laevis.* An HRP study. *Anat. Embryol.* 172:311–324.
Vannini, E., and P. P. Giorgi. 1969. Organogenesi del l'apparato urogenitale degli anfibi: Agenesi ed interruzione del dotto di Wolff in embrioni di *Bufo bufo. Arch. Italiano Anat. Embriol.* 74:111–143.
Vannini, E., and A. Sabbadin. 1954. The relation of the interrenal blastema to the origin of the somatic tissue of the gonad in frog tadpoles. *J. Embryol. Exp. Morphol.* 2:275–289.
Van Oordt, P. G. W. J. 1974. Cytology of the adenohypophysis. pp. 53–106. In *Physiology of the Amphibia.* Vol. 2, edited by B. Lofts. New York: Academic Press.
Van Pletzen, R. 1953. Ontogenesis and morphogenesis of the breast-shoulder apparatus of *Xenopus laevis. Ann. Univ. Stellenbosch* 29A:137–184.
Van Seters, W. H. 1922. Le développement du chondrocrâne d'*Alytes obstetricans* avant la métamorphose. *Arch. Biol.* 32:373–491.
Vaz-Ferreira, R., and A. Gehrau. 1975. Comportamiento epimeletico de la rana comun, *Leptodactylus ocellatus* (L.)(Amphibia, Leptodactylidae). I. Atencion de la cria y actividades alimentaria y agresivas relacionadas. *Physis* 34:1–14.
Veltkamp, C. J., J. C. Chubb, S. P. Birch, and J. W. Eaton. 1994. A simple freeze dehydration method for studying epiphytic and epizooic communities using the scanning electron microscope. *Hydrobiologia* 288:33–40.
Verma, N., and B. A. Pierce. 1994. Body mass, developmental stage, and interspecific differences in acid tolerance of larval anurans. *Texas J. Sci.* 46:319–327.
Vermeij, G. J. 1976. Interoceanic differences in vulnerability of shelled prey to crab predation. *Nature* 260:135–136.
Vesselkin, N. P., A. L. Agayan, and L. M. Nomokonova. 1971. A study of the thalamo-telencephalic afferent systems in frogs. *Brain Behav. Evol.* 4:295–306.
Via, S., and R. Lande. 1985. Genotype-environment interaction and the evolution of phenotypic plasticity. *Evolution* 39:505–522.
Viertel, B. 1982. The oral cavities of central European anuran larvae (Amphibia) morphology, ontogenesis, and generic diagnosis. *Amphibia-Reptilia* 4:327–360.
———. 1984a. Filtration, eine Strategie der Nahrungsaufnahme der Larven von *Xenopus laevis, Rana temporaria* und *Bufo calamita* (Amphibia, Anura). *Verh. Ges. Ökologie (Berlin)* 12:563575.
———. 1984b. Habit, melanin pigmentation, oral disc, oral cavity, and filter apparatus of the larvae of *Baleaphryne muletensis.* pp. 21–43. In *Life history of the Mallorcan midwife toad,* edited by H. Hemmer and J. A. Alcover. Ciutat de Mallorca.
———. 1984c. Suspension feeding of the larvae of *Baleaphryne muletensis.* pp. 153–161. In *Life history of the Mallorcan midwife toad,* edited by H. Hemmer and J. A. Alcover. Ciutat de Mallorca.
———. 1985. The filter apparatus of *Rana temporaria* and *Bufo bufo* larvae (Amphibia, Anura). *Zoomorphology* 105:345–355.
———. 1987. The filter apparatus of *Xenopus laevis, Bombina variegata,* and *Bufo calamita* (Amphibia, Anura): A comparison of different larval types. *Zool. Jahrb. Anat.* 115:425–452.
———. 1990. Suspension feeding of anuran larvae at low concentrations of *Chlorella* algae (Amphibia, Anura). *Oecologia* 85:167–177.
———. 1991. The ontogeny of the filter apparatus of anuran larvae (Amphibia, Anura). *Zoomorphology* 110:239–266.
———. 1992. Functional response of suspension feeding anuran larvae to different particle sizes at low concentrations. *Hydrobiologia* 234:151–173.
———. 1996. Der Buccopharynx und der Filterapparat der Anurenlarven—Morphologie, Ultrastructur, Ontogenese,

Funktion, Leistung bei der Nahrungsaufnahme und ökologische Bedeutung. *Mainzer Naturwiss. Archiv.* 34:187–263.
Vietti, M., A. Cianca Perone, and A. Guardabassi. 1973. Pronephric degeneration in normal and prolactin treated larvae from *Rana temporaria,* in correlation with metamorphosis. *Boll. Zool.* 40:401–404.
Vigny, C. 1979. Morphologie larvaire de 12 espèces et sous-espèces due genre *Xenopus. Rev. Suisse Zool.* 86:877–891.
Villa, J., R. W. McDiarmid, and J. M. Gallardo. 1982. Arthropod predators of leptodactylid frog foam nests. *Brenesia* 19/20:577–589.
Villa, J., and C. E. Valerio. 1982. Red, white, and brown. Preliminary observations on the color of the centrolenid tadpole (Amphibia: Anura: Centrolenidae). *Brenesia* 19/20:1–16.
Villy, F. 1890. The development of the ear and accessory organs in the common frog. *Quart. J. Micros. Soc.* 30:523–550.
Vilter, V., and A. Lugand. 1959. Recherches sur le déterminisme interne et externe du corps jaune gestatif chez le crapaud vivipare du Mont Nimba, le *Nectophrynoides occidentalis* Ang. de la Haute Guinée. *C. R. Soc. Biol. Paris* 153:294–297.
Vine, I. 1973. Detection of prey flocks by predators. *J. Theor. Biol.* 40:207–210.
Viparina, S., and J. J. Just. 1975. The life period, growth, and differentiation of *Rana catesbeiana* larvae occurring in nature. *Copeia* 1975:103–109.
Visser, J. 1985. The fanglike teeth of the early larvae of some *Heleophryne. South African J. Sci.* 81:200–202.
Vitalis, T. Z. 1990. Pulmonary and cutaneous oxygen uptake and oxygen consumption of isolated skin in the frog, *Rana pipiens. Respir. Physiol.* 81:391–400.
Vizotto, L. D. 1967. *Desenvolvimento de anuros da região norte-ocidental do Estado de São Paulo.* Fac. Filosof. Ciênc. Letr. Zool. São Jose Rio Prêto No. Especial. 161 p.
Vogt, C. C. 1842. *Untersuchungen über die Entwickelungsgeschichte der Geburtshelferkröte.* Solothurn.
Volkov, V. I. 1982. Morphological characteristics of the frog (*Rana ridibunda*) myocytes in the process of cardiogenesis. *Arch. Anat. Gistol. Embryol.* 82:73–80.
Voris, H. K., and J. P. Bacon, Jr. 1966. Differential predation on tadpoles. *Copeia* 1966:594–598.
Wachtel, S. S., G. C. Koo, and E. A. Boyse. 1975. Evolutionary conservation of H-Y ("male") antigen. *Nature* 254:270–272.
Wager, V. A. 1929. The breeding habits and life-histories of some of the Transvaal Amphibia.—II. *Trans. Roy. Soc. South Africa* 17:125–135.
———. 1965. *The frogs of South Africa.* Cape Town and Johannesburg: Purnell & Sons Pty., Ltd.
———. 1986. *Frogs of South Africa. Their fascinating life stories.* Craighall: Delta Books.
Wagner, E., and M. Ogielska. 1990. Oogenesis and development of the ovary in European green frog, *Rana ridibunda* (Pallas). II. Juvenile stages until adults. *Zool. Jahrb. Abt. Anat. Ontog. Tiere* 120:223–231.
Wakahara, M. 1978. Induction of supernumerary primordial germ cells by injecting vegetal pole cytoplasm into *Xenopus* eggs. *J. Exp. Zool.* 203:159–164.
———. 1982. Chronological changes in the accumulation of Poly(A)+RNA in developing cells of *Xenopus laevis,* with special reference to PGCs. *Develop. Growth Differ.* 24:311–318.
Wake, D. B. 1970. Aspects of vertebral evolution in the modern Amphibia. *Forma et Functio* 3:33–60.
Wake, D. B., and J. Hanken. 1996. Direct development in the lungless salamanders: What are the consequences for developmental biology, evolution and phylogenesis? *Int. J. Develop. Biol.* 40:859–869.
Wake, D. B., K. C. Nishikawa, U. Dicke, and G. Roth. 1988. Organization of the motor nuclei in the cervical spinal cord of salamanders. *J. Comp. Neurol.* 278:195–208.
Wake, M. H. 1977. Fetal maintenance and its evolutionary significance in the Amphibia: Gymnophiona. *J. Herpetol.* 11:379–386.
———. 1978. The reproductive biology of *Eleutherodactylus jasperi* (Amphibia, Anura. Leptodactylidae), with comments on the evolution of live-bearing systems. *J. Herpetol.* 12:121–133.
———. 1979. The comparative anatomy of the urogenital system. pp. 555–614. In *Hyman's comparative vertebrate anatomy,* edited by M. H. Wake. 3d ed. Chicago: University of Chicago Press.
———. 1980. The reproductive biology of *Nectophrynoides malcolmi* (Amphibia: Bufonidae), with comments on the evolution of reproductive modes in the genus *Nectophrynoides. Copeia* 1980:193–209.
———. 1982. Diversity within a framework of constraints. Amphibian reproductive modes. pp. 87–106. In *Environmental adaptation and evolution,* edited by D. Mossakowski and G. Roth. New York: Gustav Fischer-Verlag.
———. 1989. Phylogenesis of direct development and viviparity in vertebrates. pp. 235–250. In *Complex organismal functions: Integration and evolution in vertebrates,* edited by D. B. Wake and G. Roth. New York: Wiley.
———. 1993a. The skull as a locomotor organ. pp. 197–240. In *The skull.* Vol. 3. *Functional and evolutionary mechanisms.* Edited by J. Hanken and B. K. Hall. Chicago: University of Chicago Press.
———. 1993b. Evolutionary scenarios, homology and convergence of structural specializations for vertebrate viviparity. *American Zool.* 32:256–263.
———. 1993c. Evolution of oviductal gestation in amphibians. *J. Exp. Zool.* 266:394–413.
Wake, M. H., and J. Hanken. 1982. Development of the skull of *Dermophis mexicanus* (Amphibia: Gymnophiona), with comments on skull kinesis and amphibian relationships. *J. Morphol.* 173:203–223.
Wakeman, J. M., and G. R. Ultsch. 1976. The effects of dissolved O_2 and CO_2 on metabolism and gas-exchange partitioning in aquatic salamanders. *Physiol. Zool.* 48:348–359.
Waldman, B. 1981. Sibling recognition in toad tadpoles: The role of experience. *Zeit. Tierpsychol.* 56:341–358.
———. 1982a. Sibling association among schooling toad tadpoles: Field evidence and implications. *Anim. Behav.* 30:700–713.
———. 1982b. Adaptive significance of communal oviposition in wood frogs (*Rana sylvatica*). *Behav. Ecol. Sociobiol.* 10:169–174.
———. 1984. Kin recognition and sibling association among wood frog (*Rana sylvatica*) tadpoles. *Behav. Ecol. Sociobiol.* 14:171–180.
———. 1985a. Olfactory basis of kin recognition in toad tadpoles. *J. Comp. Physiol.* 156:565–577.
———. 1985b. Sibling recognition in toad tadpoles: Are kinship labels transferred among individuals? *Zeit. Tierpsychol.* 68:41–57.
———. 1986a. Preference for unfamiliar siblings over familiar nonsiblings in American toad (*Bufo americanus*) tadpoles. *Anim. Behav.* 34:48–53.
———. 1986b. Chemical ecology of kin recognition in anuran

amphibians. pp. 225–242. In *Chemical signals in vertebrates.* Vol. 4, edited by D. Duval, D. Müller-Schwarze, and R. M. Silverstein. New York: Plenum.

———. 1989. Do anuran larvae retain kin recognition abilities following metamorphosis? *Anim. Behav.* 37:1055–1058.

———. 1991. Kin recognition in amphibians. pp. 162–219. In *Kin recognition,* edited by P. G. Hepper. Cambridge: Cambridge University Press.

Waldman, B., and K. Adler. 1979. Toad tadpoles associate preferentially with siblings. *Nature* 282:611–613.

Wallace, R. A. 1978. Oocyte growth in nonmammalian vertebrates. pp. 469–502. In *The vertebrate ovary. Comparative biology and evolution,* edited by R. E. Jones. New York: Plenum.

Wallace, R. L., and L. V. Diller. 1998. Length of the larval cycle of *Ascaphus truei* in coastal streams of the Redwood Region, northern California. *J. Herpetol.* 32:404–409.

Walls, S. C., and A. R. Blaustein. 1995. Larval marbled salamanders, *Ambystoma opacum,* eat their kin. *Anim. Behav.* 50:537–545.

Walters, B. 1975. Studies of interspecific predation within an amphibian community. *J. Herpetol.* 9:267–279.

Wangersky, P. J. 1976. The surface film as a physical environment. *Ann. Rev. Ecol. Syst.* 7:161–176.

Waringer-Löschenkohl, A. 1988. An experimental study of microhabitat selection and microhabitat shifts in European tadpoles. *Amphibia-Reptilia* 9:219–236.

Warkentin, K. M. 1992a. Effects of temperature and illumination on feeding rates of green frog tadpoles (*Rana clamitans*). *Copeia* 1992:725–730.

———. 1992b. Microhabitat use and feeding rate variation in green frog tadpoles (*Rana clamitans*). *Copeia* 1992:731–740.

———. 1995. Adaptive plasticity in hatching age: A response to predation risk trade-offs. *Proc. Nat. Acad. Sci., United States of America* 92:3507–3510.

Warner, S. C., W. A. Dunson, and J. Travis. 1991. Interaction of pH, density, and priority effects on the survivorship and growth of two species of hylid tadpoles. *Oecologia* 88:331–339.

Warner, S. C., J. Travis and W. A. Dunson. 1993. Effect of pH variation on interspecific competition between two species of hylid tadpoles. *Ecology* 74:183–194.

Warren, E. 1922. Observations on the development of the non-aquatic tadpole of *Anhydrophryne rattrayi* Hewitt. *South African J. Sci.* 19:254–262, plates II and III.

Wasserman, W. J., and L. D. Smith. 1978. Oocyte maturation in nonmammalian vertebrates. pp. 443–468. In *The vertebrate ovary comparative biology and evolution,* edited by R. E. Jones. New York: Plenum.

Wassersug, R. 1971. On the comparative palatability of some dry-season tadpoles from Costa Rica. *American Midl. Nat.* 86:101–109.

———. 1972. The mechanism of ultraplanktonic entrapment in anuran larvae. *J. Morphol.* 137:279–288.

———. 1973. Aspects of social behavior in anuran larvae. pp. 273–297. In *Evolutionary biology of the anurans. Contemporary research on major problems,* edited by J. L. Vial. Columbia: University of Missouri Press.

———. 1974. Evolution of anuran life cycles. *Science* 185:377–378.

———. 1975. The adaptive significance of the tadpole stage with comments on the maintenance of complex life cycles in anurans. *American Zool.* 15:405–417.

———. 1976a. Oral morphology of anuran larvae: Terminology and general description. *Occas. Pap. Mus. Nat. Hist., Univ. Kansas* (48):1–23.

———. 1976b. Internal oral features in *Hyla regilla* (Anura: Hylidae) larvae: An ontogenetic study. *Occas. Pap. Mus. Nat. Hist., Univ. Kansas* (49):1–24.

———. 1976c. The identification of leopard frog tadpoles. *Copeia* 1976:413–414.

———. 1980. Internal oral features of larvae from eight anuran families: Functional, systematic, evolutionary, and ecological considerations. *Misc. Publ. Mus. Nat. Hist., Univ. Kansas* (68):1–146.

———. 1984. The *Pseudohemisus* tadpole: A morphological link between microhylid (Orton Type 2) and ranoid (Orton Type 4) larvae. *Herpetologica* 40:138–149.

———. 1986. How does a tadpole know when to metamorphose? A theory linking environmental and hormonal cues. *J. Theor. Biol.* 118:171–181.

———. 1989a. Locomotion in amphibian larvae (or "Why aren't tadpoles built like fishes?"). *American Zool.* 29:65–84.

———. 1989b. What, if anything is a microhylid (Orton type II) tadpole? *Forsch. Zool.* 35:534–538.

———. 1992. The basic mechanisms of ascent and descent by anuran larvae (*Xenopus laevis*). *Copeia* 1992:890–894.

———. 1996. The biology of *Xenopus* tadpoles. pp. 195–211. In *The biology of* Xenopus, edited by R. C. Tinsley and H. R. Kobel. Oxford: Oxford University Press.

———. 1997a. Where the tadpole meets the world—observations and speculations on biomechanical and biochemical factors that influence metamorphosis in anurans. *American Zool.* 37:124–136.

———. 1997b. Assessing and controlling amphibian populations from the larval perspective. pp. 271–281. In *Amphibians in decline: Canadian studies of a global problem,* edited by D. M. Green. *Herpetol. Conserv.* (1). Victoria, British Columbia

Wassersug, R., and W. E. Duellman. 1984. Oral structures and their development in egg-brooding hylid frog embryos and larvae: Evolutionary and ecological implications. *J. Morphol.* 182:1–37.

Wassersug, R. J., and M. E. Feder. 1983. The effects of aquatic oxygen concentration, body size and respiratory behavior on the stamina of obligate aquatic (*Bufo americanus*) and facultative air-breathing (*Xenopus laevis* and *Rana berlandieri*) anuran larvae. *J. Exp. Biol.* 105:173–190.

Wassersug, R. J., K. J. Frogner, and R. F. Inger. 1981a. Adaptations for life in tree holes by rhacophorid tadpoles from Thailand. *J. Herpetol.* 15:41–52.

Wassersug, R. J., and C. M. Hessler. 1971. Tadpole behaviour: Aggregation in larval *Xenopus laevis. Anim. Behav.* 19:386–389.

Wassersug, R. J., and W. R. Heyer. 1983. Morphological correlates of subaerial existence in leptodactylid tadpoles associated with flowing water. *Canadian J. Zool.* 61:761–769.

———. 1988. A survey of internal oral features of leptodactyloid larvae (Amphibia: Anura). *Smithsonian Contrib. Zool.* (457):1–99.

Wassersug, R. J., and K. Hoff. 1979. A comparative study of the buccal pumping mechanism of tadpoles. *Biol. J. Linnean Soc.* 12:225–259.

———. 1982. Developmental changes in the orientation of the anuran jaw suspension. A preliminary exploration into the evolution of anuran metamorphosis. pp. 223–246. In *Evolutionary Biology,* Vol. 15, edited by M. Hecht, B. Wallace, and G. Prace. New York: Plenum.

———. 1985. The kinematics of swimming in anuran larvae. *J. Exp. Biol.* 119:1–30.
Wassersug, R. J., A. M. Lum, and M. J. Potel. 1981b. An analysis of school structure for tadpoles (Anura: Amphibia). *Behav. Ecol. Sociobiol.* 9:15–22.
Wassersug, R. J., and A. M. Murphy. 1987. Aerial respiration facilitates growth in suspension-feeding anuran larvae (*Xenopus laevis*). *Exp. Biol.* 46:141–147.
Wassersug, R. J., R. Paul, and M. E. Feder. 1981c. Cardio-respiratory synchrony in anuran larvae (*Xenopus laevis, Pachymedusa dacnicolor,* and *Rana berlandieri*). *Comp. Biochem. Physiol.* 70A:329–334.
Wassersug, R. J., and W. F. Pyburn. 1987. The biology of the Pe-ret' toad, *Otophryne robusta* (Microhylidae), with special consideration of its fossorial larva and systematic relationships. *Zool. J. Linnean Soc.* 91:137–169.
Wassersug, R., and K. Rosenberg. 1979. Surface anatomy of branchial food traps of tadpoles: A comparative study. *J. Morphol.* 159:393–426.
Wassersug, R. J., and E. A. Seibert. 1975. Behavioral responses of amphibian larvae to variation in dissolved oxygen. *Copeia* 1975:86–103.
Wassersug, R. J., and K. A. Souza. 1990. The bronchial diverticula of *Xenopus* larvae: Are they essential for hydrostatic assessment? *Naturwissenschaften* 77:443–445.
Wassersug, R. J., and D. G. Sperry. 1977. The relationship of locomotion to differential predation on *Pseudacris triseriata* (Anura: Hylidae). *Ecology* 58:830–839.
Wassersug, R. J., and B. Viertel. 1993. Do cilia drive water through the buccopharyngeal and opercular cavities in the fossorial *Otophryne robusta* tadpole? *Amphibia-Reptilia* 14:83–86.
Wassersug, R. J., and D. B. Wake. 1995. Fossil tadpoles from the Miocene of Turkey. *Alytes* 12:145–157.
Watanabe, K., F. Sasaki, and H. Takahama. 1984. The ultrastructure of oral (buccopharyngeal) membrane formation and rupture in the anuran embryo. *Anat. Rec.* 210:513–524.
Watkins, T. B. 1997. The effect of metamorphosis on the repeatability of maximal locomotor performance in the Pacific tree frog, *Hyla regilla*. *J. Exp. Biol.* 200:2663–2668.
Watson, G. F., and A. A. Martin. 1973. Life history, larval morphology, and relationships of Australian leptodactylid frogs. *Trans. Roy. Soc. South Australia* 97:33–45.
Watt, K. W. K., and A. Riggs. 1975. Hemoglobins of the tadpole of the bullfrog, *Rana catesbeiana*. *J. Biol. Chem.* 250:5934–5944.
Watt, P. J., S. F. Nottingham, and S. Young. 1997. Toad tadpole aggregation behaviour: Evidence for a predator avoidance function. *Anim. Behav.* 54:865–872.
Webb, J. F., and D. M. Noden. 1993. Ectodermal placodes: Contributions to the development of the vertebrate head. *American Zool.* 33:343–447.
Webb, P. W. 1975. Hydrodynamics and energetics of fish propulsion. *Bull. Fish. Res. Bd. Canada* 190:1–158.
———. 1978. Fast-start performance and body form in seven species of teleost fish. *J. Exp. Biol.* 74:211–226.
———. 1984a. Form and function in fish swimming. *Sci. America* July:71–82.
———. 1984b. Body form, locomotion, and foraging in aquatic vertebrates. *American Zool.* 24:107–120.
Webb, R. G. 1963. The larva of the casque-headed frog, *Pternohyla fodiens* Boulenger. *Texas J. Sci.* 15:89–97.
Webb, R. G., and J. K. Korky. 1977. Variation in tadpoles of frogs of the *Rana tarahumarae* group in western Mexico (Anura: Ranidae). *Herpetologica* 33:73–82.
Weber, M. 1898. Ueber auffallende ecaudaten-larven von Tjibodas (Java). *Ann. Jard. Bot. Buitenzorg Suppl.* 2:5–10.
Weigmann, D. L. 1982. Relationships between production and abundance of ranid (*Rana pipiens, R. catesbeiana,* and *R. clamitans*) tadpoles and the effects of tadpole grazing on algal productivity and diversity and on phosphorus, nitrogen, and organic carbon. Dissertation, Michigan State University.
Weisz, P. B. 1945a. The development and morphology of the larva of the South African clawed toad, *Xenopus laevis* I. The third-form tadpole. *J. Morphol.* 77:163–191.
———. 1945b. The development and morphology of the larva of the South African clawed toad, *Xenopus laevis* II. The hatching and the first- and second-form tadpoles. *J. Morphol.* 77:193–217.
Wells, K. D. 1978. Courtship and parental behavior in a Panamanian poison-arrow frog (*Dendrobates auratus*). *Herpetologica* 34:148–155.
———. 1980. Evidence for growth of tadpoles during parental transport in *Colostethus inguinalis*. *J. Herpetol.* 14:426–428.
Wells, K. D., and K. M. Bard. 1988. Parental behavior of an aquatic-breeding tropical frog, *Leptodactylus bolivianus*. *J. Herpetol.* 22:361–364.
Wenig, J. 1913. Der Albinismus bei den Anuren, nebst Bemerkungen über den Bau des Amphibien integuments. *Anat. Anz.* 43:113–135.
Werner, E. E. 1986. Amphibian metamorphosis: Growth rate, predation rate, and the optimal size at transformation. *American Nat.* 128:319–341.
———. 1988. Size, scaling, and the evolution of complex life cycles. pp. 60–81. In *Size-structured populations: Ecology and evolution,* edited by B. Ebenman and L. Persson. Berlin: Springer-Verlag.
———. 1991. Nonlethal effects of a predator on competitive interactions between two anuran larvae. *Ecology* 72:1709–1720.
———. 1994. Ontogenetic scaling of competitive relations: Size-dependent effects and responses in two anuran larvae. *Ecology* 75:197–213.
Werner, E. E., and B. R. Anholt. 1996. Predator-induced behavioral indirect effects: Consequences to competitive interactions in anuran larvae. *Ecology* 77:157–169.
Werner, E. E., and J. F. Gilliam. 1984. The ontogenetic niche and species interactions in size-structured populations. *Ann. Rev. Ecol. Syst.* 15:393–425.
Werner, E. E., and M. A. McPeek. 1994. Direct and indirect effects of predators on two anuran species along an environmental gradient. *Ecology* 75:1368–1382.
Werschkul, D. F., and M. T. Christensen. 1977. Differential predation by *Lepomis macrochirus* on the eggs and tadpoles of *Rana*. *Herpetologica* 33:237–241.
Wessels, N. K. 1968. Problems in the analysis of determination, mitosis and differentiation. pp. 132–151. In *Epithelia-mesenchymal interactions. 18th Hahnemann Symp.,* edited by R. Fleischmajer and R. E. Billingham. Baltimore: Williams and Wilkins.
Wessenberg, H. S. 1978. Opalinata. pp. 552–581. In *Parasitic protozoa*. Vol. II. *Intestinal flagellates, histomonads, trichomonads, amoeba, opalinids, and ciliates,* edited by J. P. Kreier. New York: Academic Press.
West, N. H., and W. W. Burggren. 1982. Gill and lung ventilatory responses to steady-state aquatic hypoxia and hyperoxia in the bullfrog tadpole. *Respir. Physiol.* 47:165–176.
———. 1983. Reflex interactions between aerial and aquatic gas exchange organs in larval bullfrogs. *American J. Physiol.* 244R:770–777.

Westerfield, M. 1994. *The zebrafish book. A guide for the laboratory use of zebrafish* (Brachydanio rerio). Eugene, OR: Institute of Neuroscience, University of Oregon.

Wever, E. G. 1973. The ear and hearing in the frog, *Rana pipiens. J. Morphol.* 141:461–477.

———. 1985. *The amphibian ear.* Princeton, NJ: Princeton University Press.

Weygoldt, P. 1980. Complex brood care and reproductive behavior in captive poison-arrow frogs, *Dendrobates pumilio* O. Schmidt. *Behav. Ecol. Sociobiol.* 7:329–332.

———. 1989. Feeding behavior of the larvae of *Fritziana goeldi* (Anura, Hylidae). *Amphibia-Reptilia* 10:419–422.

Weygoldt, P., and S. P. De Carvalho e Silva. 1991. Observations on mating, oviposition, egg sac formation and development in the egg-brooding frog, *Fritziana goeldi. Amphibia-Reptilia* 12:67–80.

White, B. A., and C. S. Nicoll. 1981. Hormonal control of amphibian metamorphosis. pp. 363–396. In *Metamorphosis a problem in developmental biology.* 2d ed., edited by L. I. Gilbert and E. Frieden. New York: Plenum.

Whitear, M. 1976. Identification of the epidermal "Stiftchenzellen" of frog tadpoles by electron microscopy. *Cell Tiss. Res.* 175:391–402.

———. 1977. A functional comparison between the epidermis of fish and of amphibians. *Symp. Zool. Soc. London* 39:291–313.

———. 1983. The question of free nerve endings in the epidermis of lower vertebrates. *Acta Biol. Hungaricae* 34:303–319.

Whitford, W. G., and K. H. Meltzer. 1976. Changes in O_2 consumption, body water, and lipid in burrowed desert juvenile anurans. *Herpetologica* 32:23–25.

Whitington, P. M., and K. E. Dixon. 1975. Quantitative studies of germ plasm and germ cells during early embryogenesis of *Xenopus laevis. J. Embryol. Exp. Morphol.* 33:57–74.

Wiens, J. A. 1970. Effects of early experience on substrate pattern selection in *Rana aurora* tadpoles. *Copeia* 1970:543–548.

———. 1972. Anuran habitat selection: Early experience and substrate selection in *Rana cascadae* tadpoles. *Anim. Behav.* 20:218–220.

Wiens, J. J. 1989. Ontogeny of the skeleton of *Spea bombifrons* (Anura: Pelobatidae). *J. Morphol.* 202:29–51.

Wilbur, H. M. 1972. Competition, predation, and the structure of the *Ambystoma-Rana sylvatica* community. *Ecology* 53:3–21.

———. 1976. Density-dependent aspects of metamorphosis in *Ambystoma* and *Rana sylvatica. Ecology* 57:1289–1296.

———. 1977a. Density-dependent aspects of growth and metamorphosis in *Bufo americanus. Ecology* 58:196–200.

———. 1977b. Interactions of food level and population density in *Rana sylvatica. Ecology* 58:206–209.

———. 1980. Complex life cycles. *Ann. Rev. Ecol. Syst.* 11:67–93.

———. 1984. Complex life cycles and community organization in amphibians. pp. 195–224. In *A new ecology: Novel approaches to interactive systems,* edited by P. W. Price, C. N. Slobodchikoff, and W. S. Gaud. New York: Wiley.

———. 1987. Regulation of structure in complex systems: Experimental temporary pond communities. *Ecology* 68:1437–1452.

———. 1988. Interactions between growing predators and growing prey. pp. 157–172. In *Size-structured populations; ecology and evolution,* edited by B. Ebenman and L. Persson. New York: Springer-Verlag.

———. 1989. In defense of tanks. *Herpetologica* 45:122–123.

Wilbur, H. M., and R. A. Alford. 1985. Priority effects in experimental pond communities: Responses of *Hyla* to *Bufo* and *Rana. Ecology* 66:1106–1114.

Wilbur, H. M., and J. P. Collins. 1973. Ecological aspects of amphibian metamorphosis. *Science* 182:1305–1314.

Wilbur, H. M., and J. E. Fauth. 1990. Experimental aquatic food webs: Interactions between two predators and two prey. *American Nat.* 135:176–204.

Wilbur, H. M., P. J. Morin, and R. N. Harris. 1983. Salamander predation and the structure of experimental communities: Anuran responses. *Ecology* 64:1423–1429.

Wilbur, H. M., and R. D. Semlitsch. 1990. Ecological consequences of tail injury in *Rana* tadpoles. *Copeia* 1990:18–24.

Wilczynska, B. 1981. The structure of the alimentary canal and the dimensions of the mucosa in ontogenetic development of some Anura. *Acta Biol. Cracov.* 23:13–46.

Wilczynski, W. 1981. Afferents to the midbrain auditory center in the bullfrog, *Rana catesbeiana. J. Comp. Neurol.* 198:421–433.

———. 1988. Brainstem auditory pathways in anuran amphibians. pp. 209–231. In *The evolution of the amphibian auditory system,* edited by B. Fritzsch, M. J. Ryan, W. Wilczynski, T. E. Hetherington, and W. Walkowiak. New York: Wiley.

———. 1992. The nervous system. pp. 9–39. *Environmental physiology of the amphibians,* edited by In M. E. Feder and W. W. Burggren. Chicago: University of Chicago Press.

Wilczynski, W., and R. G. Northcutt. 1977. Afferents to the optic tectum of the leopard frog: An HRP study. *J. Comp. Neurol.* 173:219–230.

———. 1983a. Connections of the bullfrog striatum: Afferent organization. *J. Comp. Neurol.* 214:321–332.

———. 1983b. Connections of the bullfrog striatum: Efferent projections. *J. Comp. Neurol.* 214:333–343.

Wild, E. R. 1995. New genus and species of Amazonian microhylid frog with a phylogenetic analysis of New World genera. *Copeia* 1995:837–849.

———. 1996. Natural history and resource use of four Amazonian tadpole assemblages. *Occas. Pap. Nat. Hist., Univ. Kansas* (176):1–59.

———. 1997. Description of the adult skeleton and developmental osteology of the hyperossified horned frog, *Ceratophrys cornuta* (Anura: Leptodactylidae). *J. Morphol.* 232:169–206.

Wiley, E. O. 1979. Ventral gill arch muscles and the interrelationships of gnathostomes, with a new classification of the vertebrata. *Zool. J. Linnean Soc.* 67:149–179.

———. 1981. *Phylogenetics: The theory and practice of phylogenetic systematics.* New York: Wiley.

Wilhelm, G. B., and R. E. Coggeshall. 1981. An electron microscope analysis of the dorsal root in the frog. *J. Comp. Neurol.* 196:421–429.

Will, U. 1982. Efferent neurons of the lateral-line system and the VIIIth cranial nerve in the brainstem of anurans. *Cell Tiss. Res.* 225:673–685.

———. 1986. Mauthner neurons survive metamorphosis in anurans: A comparative HRP study on the cytoarchitecture of Mauthner neurons in amphibians. *J. Comp. Neurol.* 244:111–120.

———. 1988. Organization and projections of the area octavolateralis in amphibians. pp. 185–208. In *The evolution of the amphibian auditory system,* edited by B. Fritzsch, M. J. Ryan, W. Wilczynski, T. E. Hetherington, and W. Walkowiak. New York: Wiley.

Will, U., and B. Fritzsch. 1988. The eighth nerve of amphibians: Peripheral and central distribution. pp. 159–183. In *The evolution of the amphibian auditory system,* edited by B. Fritzsch, M. J. Ryan, W. Wilczynski, T. E. Hetherington, and W. Walkowiak. New York: Wiley.

Will, U., B. Fritzsch, G. Luhede, and P. Görner. 1985. The area octavo-lateralis in *Xenopus laevis.* Part II. Second order projections and cytoarchitecture. *Cell Tiss. Res.* 239:163–175.

Willey, A. 1893. Note on the position of the cloacal aperture in certain batrachian tadpoles. *Trans. New York Acad. Sci.* 12:242–245.

Willhite, C., and P. V. Cupp, Jr. 1982. Daily rhythms of thermal tolerance in *Rana clamitans* (Anura: Ranidae) tadpoles. *Comp. Biochem. Physiol.* 72A:255–257.

Williams, G. C. 1964. Measurement of consociation among fishes and comments on the evolution of schooling. *Publ. Mus. Michigan St. Univ.* 2:351–383.

Williams, J. D., and E. Gudynas. 1987. Descripcion de la larva de *Elachistocleis bicolor* (Valenciennes 1838)(Anura: Microhylidae). *Amphibia-Reptilia* 8:225–229.

Williams, M. A., and L. D. Smith. 1971. Ultrastructure of the "germinal plasm" during maturation and early cleavage in *Rana pipiens. Develop. Biol.* 25:568–580.

Williamson, I., and C. M. Bull. 1989. Life history variation in a population of the frog *Ranidella signifera:* Egg size and early development. *Copeia* 1989:349–356.

Wills, I. A. 1936. The respiratory rate of developing Amphibia with special reference to sex differentiation. *J. Exp. Zool.* 73:481–510.

Wilson, D. J., and H. Lefcort. 1993. The effect of predator diet on the alarm response of red-legged frog, *Rana aurora,* tadpoles. *Anim. Behav.* 46:1017–1019.

Wilson, L. D., and J. R. McCranie. 1993. Preliminary key to the known tadpoles of anurans from Honduras. *Life Sci. Occas. Pap. Roy. Ontario Mus.* (40):1–12.

Wiltshire, D. J., and C. M. Bull. 1977. Potential competitive interactions between larvae of *Pseudophryne bibroni* and *P. semimarmorata* (Anura: Leptodactylidae). *Australian J. Zool.* 25:449–454.

Wimberger, P. H. 1992. Plasticity of fish body shape. The effects of diet, development, family, and age in two species of *Geophagus* (Pisces: Cichlidae). *Biol. J. Linnean Soc.* 45:197–218.

———. 1993. Effects of vitamin C deficiency on body shape and skull osteology in *Geophagus brasiliensis:* Implications for interpretations of morphological plasticity. *Copeia* 1993:343–351.

Winklbauer, R., and P. Hausen. 1983a. Development of the lateral line system in *Xenopus laevis.* I. Normal development and cell movement in the supraorbital system. *J. Embryol. Exp. Morphol.* 76:265–281.

———. 1983b. Development of the lateral line system in *Xenopus laevis.* II. Cell multiplication and organ formation in the supraorbital system. *J. Embryol Exp. Morphol.* 76:283–296.

———. 1985a. Development of the lateral line system in *Xenopus laevis.* III. Development of the supraorbital system in triploid embryos and larvae. *J. Embryol. Exp. Morphol.* 88:183–192.

———. 1985b. Development of the lateral line system in *Xenopus laevis.* IV. Pattern formation in the supraorbital system. *J. Embryol. Exp. Morphol.* 88:193–207.

Wischnitzer, S. 1976. The lampbrush chromosomes: Their morphology and physiological importance. *Endeavour* 35:27–31.

Wise, R. W. 1970. An immunological comparison of tadpole and frog hemoglobins. *Comp. Biochem. Physiol.* 32:89–95.

Witschi, E. 1949. The larval ear of the frog and its transformation during metamorphosis. *Zeit. Naturforsch.* 4B:230–242.

———. 1950. The bronchial diverticula of *Xenopus laevis* Daudin. *Anat. Rec.* 108:590.

———. 1955. The bronchial columella of the ear of larval Ranidae. *J. Morphol.* 96:497–511.

———. 1956. *Development of vertebrates.* Philadelphia: W. B. Saunders.

———. 1962. Equivalent numerical designations for staging systems: Amphibians and fishes. pp. 272–273. In *Growth, including reproduction and morphological development,* edited by P. L. Altman and D. S. Dittmar. Washington, D. C.: Fed. American Soc. Exp. Biol.

Wollmuth, L. P., and L. I. Crawshaw. 1988. The effect of development and season on temperature selection in bullfrog tadpoles. *Physiol. Zool.* 61:461–469.

Wollmuth, L. P., L. I. Crawshaw, R. B. Forbes, and D. A. Grahn. 1987. Temperature selection during development in a montane anuran species, *Rana cascadae.* Physiol. Zool. 60:472–480.

Wollweber, L., R. Stracke, and U. Gothe. 1981. The use of a simple method to avoid cell shrinkage during SEM preparation. *J. Micros.* 121:185–189.

Wong, A. L.-C., and T. J. C. Beebee. 1994. Identification of a unicellular, nonpigmented alga that mediates growth inhibition in anuran tadpoles: A new member of the genus *Prototheca* (Chlorophyceae: Chlorococcales). *Hydrobiologia* 277:85–96.

Wong, S., and D. T. Booth. 1994. Hypoxia induces surfacing behaviour in brown-striped frog (*Limnodynastes peronii*) larvae. *Comp. Biochem. Physiol.* 109A:437–445.

Woodward, B. D. 1982. Tadpole competition in a desert anuran community. *Oecologia* 54:96–100.

———. 1983. Predator-prey interactions and breeding-pond use of temporary-pond species in a desert anuran community. *Ecology* 64:1549–1555.

———. 1986. Paternal effects on juvenile growth in *Scaphiopus multiplicatus* (the New Mexico spadefoot toad). *American Nat.* 128:58–65.

———. 1987a. Clutch parameters and pond use in some Chihuahuan desert anurans. *Southwest. Nat.* 32:13–19.

———. 1987b. Intra- and interspecific variation in spadefoot toad (*Scaphiopus*) clutch parameters. *Southwest. Nat.* 32:127–156.

———. 1987c. Interactions between Woodhouse's toad tadpoles (*Bufo woodhousii*) of mixed sizes. *Copeia* 1987:380–386.

———. 1987d. Paternal effects on offspring traits in *Scaphiopus couchi* (Anura: Pelobatidae). *Oecologia* 73:626–629.

Woodward, D. B., and J. Travis. 1991. Paternal effects on juvenile growth and survival in spring peepers (*Hyla crucifer*). *Evol. Ecol.* 5:40–51.

Woodward, D. B., J. Travis, and S. Mitchell. 1988. The effects of the mating system on progeny performance in *Hyla crucifer* (Anura: Hylidae). *Evolution* 42:784–794.

Workman, G., and K. C. Fisher. 1941. Temperature selection and the effect of temperature on movement in frog tadpoles. *American J. Physiol.* 133P:499–500.

Wratten, S. D. 1994. *Video techniques in animal ecology and behaviour.* London: Chapman and Hall.

Wray, G. A. 1996. Parallel evolution of nonfeeding larvae in echinoids. *Syst. Biol.* 45:308–322.

Wright, A. A., and A. H. Wright. 1933. *Handbook of frogs and toads—the frogs and toads of the United States and Canada. Handbooks of American Natural History.* Vol. 1. Ithaca, NY: Comstock Publ.

———. 1942. *Handbook of frogs and toads—the frogs and toads of the United States and Canada. Handbooks of American Natural History.* Vol. 1. 2d ed. Ithaca, NY: Comstock Publ.

Wright, A. H. 1914. North American Anura—life-histories of the Anura of Ithaca, New York. *Carnegie Inst. Washington Publ.* (197):1–98

———. 1929. Synopsis and description of North American tadpoles. *Proc. United States Natl. Mus.* 74, 11 (2756):1–70, plates 1–9.

———. 1932. *Life-histories of the frogs of Okefinokee Swamp, Georgia.* North American Salientia (Anura) no. 2. New York: Macmillan Publ.

Wright, A. H., and A. A. Wright. 1949. *Handbook of frogs and toads of the United States and Canada.* 3d ed. Ithaca, NY: Comstock Publ.

———. 1995. *Handbook of frogs and toads of the United States and Canada.* Ithaca, NY: Comstock Classic Handbooks, Cornell Paperbacks, Comstock Publ. Assoc.

Wright, M. L., L. S. Blanchard, S. T. Jorey, C. A. Basso, Y. M. Myers, and C. M. Paquette. 1990. Metamorphic rate as a function of the light/dark cycle in *Rana pipiens* larvae. *Comp. Biochem. Physiol.* 96A:215–220.

Wright, M. L., L. J. Cykowski, L. Lundrigan, K. L. Hemond, D. M. Kochan, E. E. Faszewski, and C. M. Anuszewski. 1994. Anterior pituitary and adrenal cortical hormones accelerate or inhibit tadpole hindlimb growth and development depending on stage of spontaneous development or thyroxine concentration in induced metamorphosis. *J. Exp. Zool.* 270:175–188.

Wright, P. M., and P. A. Wright. 1996. Nitrogen metabolism and excretion in bullfrog (*Rana catesbeiana*) tadpoles and adults exposed to elevated environmental ammonia levels. *Physiol. Zool.* 69:1057–1078.

Wright, S. 1931. Evolution in Mendelian populations. *Genetics* 16:97–159.

Wylie, C. C., and J. Heasman. 1976a. The formation of the gonadal ridge in *Xenopus laevis.* I. A light and transmission electron microscope study. *J. Embryol. Exp. Morphol.* 35:125–138.

———. 1976b. The formation of the gonadal ridge in *Xenopus laevis.* II. A scanning electron microscope study. *J. Embryol. Exp. Morphol.* 35:139–148.

Wylie, C. C., J. Heasman , A. Snape, M. O'Driscoll, and S. Holwill. 1985. Primordial germ cells of *Xenopus laevis* are not irreversibly determined early in development. *Develop. Biol.* 112:66–72.

Wylie, C. C., S. Heasman, A. P. Swan, and B. H. Anderton. 1979. Evidence for substrate guidance of primordial germ cells. *Exp. Cell Res.* 121:315–324.

Wylie, C. C., and T. B. Roos. 1976. The formation of the gonadal ridge in *Xenopus laevis.* III. The behaviour of isolated primordial germ cells *in vitro. J. Embryol. Exp. Morphol.* 35:149–157.

Xavier, F. 1977. An exceptional reproductive strategy in Anura: *Nectophrynoides occidentalis* Angel (Bufonidae), an example of adaptation to terrestrial life by viviparity. pp. 545–553. In *Major patterns in vertebrate evolution,* edited by M. K. Hecht, P. C. Goody, and B. M. Hecht. New York: Plenum.

Yang, D.-T. 1991. Phylogenetic systematics of the *Amolops* groups of ranid frogs of southeastern Asia and the Greater Sunda Islands. *Field. Zool. (N.S.)* (1423):1–42.

Ye, C., and L. Fei. 1995. Taxonomic studies on the small type *Megophrys* in China including descriptions of the new species (subspecies) (Pelobatidae: genus *Megophrys*). *Acta Herpetol. Sinica* 6:72–81.

Yong, H. S., C. K. Ng, and R. Ismail. 1988. Conquest of the land: Direct development in a Malaysian *Philautus* tree frog. *Nat. Malaysiana* 13:4–7.

Yoshizaki, N. 1973. Ultrastructure of the hatching gland cells in the South African clawed toad, *Xenopus laevis. J. Fac. Sci. Hokkaido Univ., Ser. VI, Zool.* 18:469–480.

———. 1974. Ultrastructural cytochemistry of hatching gland cells in *Xenopus* embryos in relation to the hatching process. *J. Fac. Sci. Hokkaido Univ., Ser. VI, Zool.* 19:309–314.

———. 1975. Ultrastructure of the frog hatching gland cell in relation to its role in the hatching process. *Zool. Mag. (Tokyo)* 84:39–47.

———. 1991. Changes in surface ultrastructure and proteolytic activity of hatching gland cells during development of *Xenopus* embryo. *Zool. Sci.* 8:295–302.

Yoshizaki, N., and C. Katagiri. 1975. Cellular basis for the production and secretion of the hatching enzyme by frog embryos. *J. Exp. Zool.* 192:203–212.

Yoshizaki, N., and M. Yamamoto. 1979. A stereoscan study of the development of hatching gland cells in the embryonic epidermis of *Rana japonica. Acta Embryol. Exp.* 979:339–348.

Yoshizaki, N., and H. Yamasaki. 1991. Morphological and biochemical changes in the fertilization coat of *Xenopus laevis* during the hatching process. *Zool. Sci.* 8:303–308.

Yoshizato, K. 1986. How do tadpoles lose their tails during metamorphosis? *Zool. Sci.* 3:219–226.

———. 1992. Death and transformation of larval cells during metamorphosis of Anura. *Develop. Growth Differ.* 34:607–611.

Youn, B. W., R. E. Keller, and G. M. Malacinski. 1980. An atlas of notochord and somite morphogenesis in several anuran and urodelean amphibians. *J. Embryol. Exp. Morphol.* 59:223–247.

Youngstrom, K. A. 1938. Studies on the developing behavior of Anura. *J. Comp. Neurol.* 68:351–380.

Yung, E. 1904. De l'influence du regime alimentaire sur la longeur de l'intestin chez les larves de *Rana esculenta. C. R. Hebd. Seances Acad. Sci. Paris* 134:749–751.

Zaccanti, F., G. Tognato, and S. Tommasini. 1977. Proposta di un metodo di valutazione del grado di differenziamento sessuale dei corpi genitale in corso di sviluppo di anfibi anuri. *Arch. Italiano Anat. Embriol.* 82:233–252.

Zaccanti, F., and C. Vigenti. 1980. Quantitative observations on the development of *Bufo bufo* genital bodies. *Arch. Italiano Anat. Embriol.* 85:353–362.

Zakon, H. H. 1984. Postembryonic changes in the peripheral electrosensory system of a weakly electric fish: Addition of receptor organs with age. *J. Comp. Neurol.* 228:557–570.

Zamachowski, W. 1985. Changes in the body content of water in the common frog, *Rana temporaria* (L.), during its ontogenesis. *Acta Biol. Cracov.* 27:11–18.

Zamorano, B., A. Cortés, and A. Salibián. 1988. Ammonia and urea excretion in urine of larval *Caudiverbera caudiverbera* (L.), (Anura, Amphibia). *Comp. Biochem. Physiol.* 91A:153–155.

Zamorano, B., and A. Salibián. 1994. Ionic net fluxes through the in situ epithelia of larval *Caudiverbera caudiverbera* (Anura, Leptodactylidae). *Alytes* 12:135–144.

Zettergren, L. D., H. Kubugawa, and M. D. Cooper. 1980. Development of B-cells in *Rana pipiens*. pp. 177–185. In *Phylogeny of immunology memory*, edited by M. J. Manning. Amsterdam: Elsevier/North-Holland.

Zimmerman, H., and E. Zimmerman. 1981. Socialverhalten, Fortpflanzungsverhalten und Zucht der Färberfrösche *Dendrobates histrionicus* und *D. lehmanni* sowie einiger anderer Dendrobatiden. *Zeit. Köhlner Zoo* 24:83–99.

Züst, B., and K. E. Dixon. 1975. The effect of UV irradiation of the vegetal pole of *Xenopus laevis* eggs on the presumptive primordial germ cells. *J. Embryol. Exp. Morphol.* 34:209–220.

———. 1977. Events in the germ cell lineage after entry of the primordial germ cells into the genital ridges in normal and u.v.-irradiated *Xenopus laevis*. *J. Embryol. Exp. Morphol.* 41:33–46.

Zweifel, R. G. 1958. Ecology, distribution, and systematics of frogs of the *Rana boylei* group. *Univ. California Publ. Zool.* 54:207–276.

———. 1964. Life history of *Phrynohyas venulosa* (Salientia: Hylidae) in Panama. *Copeia* 1964:201–208.

———. 1970. Descriptive notes on larvae of toads of the *debilis* group, genus *Bufo*. *American Mus. Novitates* (2407):1–13.

———. 1977. Upper thermal tolerances of anuran embryos in relation to stage of development and breeding habits. *American Mus. Novitates* (2617):1–21.

Zwilling, E. 1940. An experimental analysis of the development of the anuran olfactory organ. *J. Exp. Zool.* 84:291–323.

———. 1941. The determination of the otic vesicle in *Rana pipiens*. *J. Exp. Zool.* 86:333–342.

AUTHOR INDEX

SUBJECT INDEX

TAXONOMIC INDEX

Names of families and subfamilies are indexed only for the taxonomic accounts in chapter 12.